T0190310

CAMBRIDGE LIBRARY COLLECTION

Books of enduring scholarly value

Technology

The focus of this series is engineering, broadly construed. It covers technological innovation from a range of periods and cultures, but centres on the technological achievements of the industrial era in the West, particularly in the nineteenth century, as understood by their contemporaries. Infrastructure is one major focus, covering the building of railways and canals, bridges and tunnels, land drainage, the laying of submarine cables, and the construction of docks and lighthouses. Other key topics include developments in industrial and manufacturing fields such as mining technology, the production of iron and steel, the use of steam power, and chemical processes such as photography and textile dyes.

Electric Illumination

Two years after Thomas Edison patented his electric light bulb, the 1881 International Exposition of Electricity in Paris, featuring many spectacular lighting displays, showcased the potential of this technology for commercial and domestic use. The accompanying International Congress of Electricians also agreed on international standards for units of electrical resistance, potential and current. In its wake, James Dredge (1840–1906), editor of the British periodical *Engineering*, compiled this illustrated overview of electrical technology and its application to lighting. First published in two volumes between 1882 and 1885, and using material that had previously appeared in *Engineering*, as well as new articles by various contributors, this substantial work reflects the complexities and possibilities of a propitious technological development. Among other topics, Volume 2 covers electrical measurement, standard textbooks, photometry, and recent developments in lamps and dynamos. The appendices give abstracts of British electrical patents from 1873 to 1882.

Cambridge University Press has long been a pioneer in the reissuing of out-of-print titles from its own backlist, producing digital reprints of books that are still sought after by scholars and students but could not be reprinted economically using traditional technology. The Cambridge Library Collection extends this activity to a wider range of books which are still of importance to researchers and professionals, either for the source material they contain, or as landmarks in the history of their academic discipline.

Drawing from the world-renowned collections in the Cambridge University Library and other partner libraries, and guided by the advice of experts in each subject area, Cambridge University Press is using state-of-the-art scanning machines in its own Printing House to capture the content of each book selected for inclusion. The files are processed to give a consistently clear, crisp image, and the books finished to the high quality standard for which the Press is recognised around the world. The latest print-on-demand technology ensures that the books will remain available indefinitely, and that orders for single or multiple copies can quickly be supplied.

The Cambridge Library Collection brings back to life books of enduring scholarly value (including out-of-copyright works originally issued by other publishers) across a wide range of disciplines in the humanities and social sciences and in science and technology.

Electric Illumination

VOLUME 2

EDITED BY JAMES DREDGE

CAMBRIDGE
UNIVERSITY PRESS

CAMBRIDGE
UNIVERSITY PRESS

University Printing House, Cambridge, CB2 8BS, United Kingdom

Cambridge University Press is part of the University of Cambridge.

It furthers the University's mission by disseminating knowledge in the pursuit of
education, learning and research at the highest international levels of excellence.

www.cambridge.org
Information on this title: www.cambridge.org/9781108070645

© in this compilation Cambridge University Press 2015

This edition first published 1885
This digitally printed version 2015

ISBN 978-1-108-07064-5 Paperback

This book reproduces the text of the original edition. The content and language reflect
the beliefs, practices and terminology of their time, and have not been updated.

Cambridge University Press wishes to make clear that the book, unless originally published
by Cambridge, is not being republished by, in association or collaboration with,
or with the endorsement or approval of, the original publisher or its successors in title.

["ENGINEERING" SERIES.]

ELECTRIC ILLUMINATION.

BY

JAMES DREDGE.   DR. M. F. O'REILLY.

AND

H. VIVAREZ.

EDITED BY

JAMES DREDGE.

(Partly compiled from "ENGINEERING.")

―――――――

With Abstracts of Specifications having reference to Electric Lighting.

PREPARED BY

W. LLOYD WISE,

Member of Council of the Institute of Patent Agents.

VOL. II.

LONDON:

OFFICES OF "ENGINEERING," 35 & 36, BEDFORD STREET, STRAND, W.C.

NEW YORK: JOHN WILEY AND SONS.

1885.

PREFACE.

—:-:—

IN the preface to the first volume of "Electrical Illumination," a reference was made to the headlong impetuosity with which the too credulous public—encouraged by speculators and sanguine or astute inventors—threw enormous sums of money into electric lighting enterprises, and it was predicted that the time was fast approaching when a disastrous reaction must set in, which would have the inevitable effect, not only of overthrowing all those numerous companies that had been established on an insecure basis, but also of shaking to their foundations the more solid and responsible associations. The course of events has fully justified the views expressed, as well as the consequences indicated, and public confidence which had been so misplaced, was succeeded by doubt and mistrust that have checked materially the commercial progress of electric lighting. By effecting large reductions in capital, and by vigorous retrenchment in expenses, some of the more powerful electric lighting companies have weathered the storm, and are to day doing a considerable and a more or less profitable business; for although none of the extensive projects for lighting on a large scale, which occupied the attention of promoters two years ago, have been realised, yet installations on a modest scale have become more numerous, and isolated plants are now largely in use. The use of electricity for lighting ocean steamers may now be said to be general, and will soon be universal, and the details involved by this special application, have been worked out in a manner that leaves little to be desired.

The time having past when the inventors of a dynamo or an incandescence or arc lamp, could hope to obtain large sums by the sale of their patents, it has followed as a necessary consequence, and by the law of the survival of the fittest, that the number of types of dynamos in use has been reduced, and that invention has followed more closely the narrower path towards improvement in detail and increase of efficiency. With several well-known systems the latter object has been achieved, and practically

nothing further can be hoped towards the attainment of a higher percentage of useful work.

It would of course be absurd to suppose that science has no more to do in effecting further economy, or that inventors must confine themselves to the perfection of details, but a stage has now been reached in the progress of electric illumination, when cost can be counted and comparisons be made with the older modes of lighting. And the result of such comparison may be briefly stated as follows : For large commercial and industrial establishments, electricity, employed for arc lighting, is cheaper than gas, and even when applied by means of incandescence, the extra trouble and expense incurred are often more than repaid by the purer and increased light obtained, and by the total absence of deleterious effects to employés and material. The same remark applies with equal force to the lighting of theatres by incandescence lamps ; for although experience obtained in this application tends to prove that gas is by far the cheaper illuminant, the advantages of electric illumination are so great that its adoption for this purpose will be soon general. Its use in private houses supplied with isolated plants and secondary batteries, is in all respects a luxury, and has to be paid for as such, and it does not appear at all probable that any scheme of general district lighting can, with our present knowledge, be made to compete with gas, the price of which would be largely reduced, if competition rendered such a step necessary.

Whilst electric lighting must thus, for private uses, be regarded as a luxury, it has for many purposes become an absolute necessity, and at the same time the practicability of employing it as a steady and reliable means of illuminating, has been demonstrated at the two great Exhibitions held at South Kensington. At the Health Exhibition this year more than four millions of people have witnessed the solution of a problem which would have been considered impracticable four or five years ago—the illumination night after night of a series of vast halls and public gardens without any trouble or interruption. The actual cost of this demonstration, it must be admitted, is probably impossible to ascertain. Since their inauguration at the Palais de l'Industrie in 1881, exhibitions having the same object, have been held at Vienna, Münich, London, Philadelphia, and elsewhere, but all these differ from the Fisheries and the Health Exhibitions, in this respect, that the electric lighting of the two last named was an incidental feature, and a commercial undertaking.

The foregoing remarks certainly do not apply to the United States of

North America, where the high price of gas, and the facility with which new enterprises take root, and make rapid growth, have combined to render electric lighting rather the rule than the exception. The system of distribution of electrical energy from fixed centres, has already been largely carried into practice, and is supported by thousands of subscribers as well as by municipal authorities. But though it must be admitted that America is far in advance of Europe in this application of electrical science, it should be remembered that she possesses a richer and more easily cultivated field for the development of the industry than can be hoped for either in England or on the Continent.

While it has to be conceded that, upon the whole, the progress made in the science of electric lighting during the last two years, in England at any rate, has not been so rapid as its previous growth had suggested, it may be confidently asserted that its future prospects are improved rather than injured by delay.

In the present pages an effort has been made to bring the subjects treated of in the first volume down to the present time, and to present in a convenient form some of the subjects inseparably connected with the conversion of mechanical into electrical energy.

The first section, contributed by Dr. M. F. O'Reilly, treats of the principal measurements which occur in electrical engineering. It opens with a brief *exposé* of the fundamental principles necessary for a good comprehension of the practical methods. In deducing and explaining these, as well as the theories of the standard instruments, a knowledge of elementary mathematics is implied such as every one should possess who aims at working intelligently at applied electricity. In two cases only has the calculus been used, and that on account of the rapidity of the method and importance of the results.

The section on Photometry is chiefly the work of M. H. Vivarez, and that on Dynamometers, which has appeared substantially in the pages of *La Lumière Electrique*, is by M. Gustave Richard. The pages devoted to the descriptions and illustrations of modern dynamos and lamps, contain, it is hoped, all, or nearly all, systems and arrangements introduced to the public since the publication of the first volume, and which are worthy of notice.

It is almost superfluous to speak of the Appendix containing the abstracts of specifications relating to electrical matters. This arduous work has been done carefully and conscientiously by Mr. W. Lloyd Wise, and will, we have no hesitation in saying, be found of very high value. It was

impossible within the limits of the present volume to bring these abstracts down to the end of the year 1883, but it is hoped that this may be done on a subsequent occasion.

Some errors have, it is feared, escaped notice. On page 15, the multiplying power of the shunt should be $n+1$, which will make S, its resistance, $\dfrac{G}{n}$; on page 29, for the phrase "A being an axis," read "A being on axis;" on page 76, for ($\frac{1}{10}$th ampère) read (10 ampères); on page 88, the formula given should read $w_2\ (t_2 - t_1)$ J ; on page 158, in the heading of the table, read "wick variable and chimney constant," for "wick constant and chimney variable."

The writer desires to acknowledge the assistance rendered by Mr. Conrad W. Cooke, in contributing the notice of the Hochhausen system, by Mr. Thomas Wilkins, and by Mr. B. A. Raworth, Dr. A. Borns, and Mr. J. Munro.

JAMES DREDGE.

Offices of Engineering, 35 and 36, Bedford Street, Strand,
November, 1884.

CONTENTS.

————:-:————

SECTION I.

PAGE

ELECTRICAL MEASUREMENT.—General Principles; Ohm's Law; Resistance; Unit Resistance; Conductivity; Specific Conductivity; Specific Resistance; Resistance in Series and Multiple Arc; Internal Resistance; Electromotive Force; The Volt: Standard Cells; Electromotive Force of Batteries; Current; Battery Currents; Internal Resistance of Batteries; Galvanometer Shunts; Capacity; Condensers; Thomson's Quadrant Electrometer; Thomson's Replenisher; The Tangent Galvanometer; Horizontal Component; Ratio of Torsion; Time of Vibration; Deflexion Experiment; Working Formulæ; Reflecting Galvanometer; Electro-Galvanometer; Bifilar Suspension; Siemens's Electro-Dynamometer; Potential Galvanometer; Graduation of Galvanometer; Verification of Magnet Strength; Current Galvanometer; Siemens's Torsion Galvanometer; Constant of Torsion Galvanometer; The Wheatstone Bridge; Kirchoff's Laws; Slide Wire or Metre Bridge; Resistances; Calibrating Resistances; Resistance Coils; Wheatstone's Rheostat; Poggendorff's Rheocord; Measurement of Resistance; Specific Resistance; Mance's Method of Measuring Specific Resistance; Lodge's Method of Measuring Specific Resistance; Munro and Muirhead's Method of Measuring Specific Resistance; High Resistances; Insulation Resistance; Time for Half Charge; Discharge Deflections from Shunted Galvanometers; Discharge Deflections; Electromotive Force; Condenser Method; Poggendorff's Method; Clark's Compensation Method; Superior Limit of Electromotive Force; Clark's Potentiometer; Capacity of Condensers; Absolute Capacity; Capacity from Galvanometer Throw; Capacity of Submarine Cables; Capacity from Rate of Fall of Potential; Air Condenser; Quantity; Work; Electric Work; Electric Activity; The Watt; The Watt Meter; Heat Energy; The Unit of Heat; The Joule; Joule's Calorimetric Method; Development of Energy and its Distribution in a Circuit; Horse-Power Absorbed and Wasted; Efficiency of Dynamos; The Electric Arc; Electric Storage of Energy; Series and Shunt Dynamos; The Characteristic of Dynamos; Rapidity and Economy of Working; Electric Motors and High Potentials; The Electromotive Force of Clark's Standard Cells ... ... 3 to 95

STANDARD WORKS ON ELECTRICITY AND MAGNETISM.—Professor S. P. Thompson's "Elementary Lessons in Electricity and Magnetism;" Dr. Ferguson's "Electricity," Revised and Enlarged by Professor J. Blyth; Deschanel's "Natural Philosophy" (Everett); Professor Fleeming Jenkin's "Electricity and Magnetism;" Professor Tyndall's "Notes on Electrical Phenomena and Theories;" Gordon's "Physical Treatise on Electricity and Magnetism;" Cumming's "Introduction to the Theory of Electricity;" Professor G. Chrystal's Electricity and Magnetism ("Encyclopædia Britannica);" Clerk-Maxwell's "Treatise on Electricity and Magnetism;" Mascart and Joubert's "Treatise on Electricity and Magnetism" (Atkinson's Translation); Maxwell's "Elementary Treatise on Electricity;" Kohlrausch's "Introduction to Physical Measurements;" Gray's "Absolute Measurements in Electricity and Magnetism;" Kempe's "Handbook of Electrical Testing;" Day's "Examples in Electrical and Magnetic Measurement;" Day's "Electric Lighting Arithmetic;" Everett's "Units and Physical Constants;" List of Publishers ... ... 96 to 98

SECTION II.

PAGE

PHOTOMETRY.—General Introduction; Fundamental Laws; Law of the Squares of Distances; Illumination of Oblique Surfaces; General Formula; Bouguer's Method; Bouger's Photometer; Ritchie's Photometer; Foucault's Photometer; Degrand's Photometer; Wolff's Photometer; Cornu's Photometer; Masson's Photometer; Hammerl's Photometer; Rumford's Photometer; Bunsen's Photometer; Wheatstone's Photometer; Arago's Photometer; Duboscq's Photometer; Leslie's Photometer; Leslie's Photometer Modified by Ritchie; Balfour Stewart's Modification of Leslie's Photometer; Chemical Photometers; Dr. Draper's Experiments; Bunsen and Roscoe's Experiments; Bunsen and Roscoe's Chemical Photometer; Eden's Photometer; Selenium Photometer; Properties of Selenium; Siemens's Artificial Eye; Siemens's Selenium Photometer; Coulon's Photometer; Stevenson's Photometer; Spectro-Photometers; Professor Crova on Spectro-Photometry; Crova's Spectro-Photometer; Ayrton and Perry's Photometer; Luminous Standards; Electric Lighting and Luminous Standards; Candles; Bougies de l'Etoile; Girout's Comparison of Luminous Standards; Gas Jets; The Composition of Standard Candles; The Carcel Lamp; Modifications of the Carcel Lamp; Andouin and Bérard's Experiments with the Carcel Lamp; Consumption of Oil in the Carcel Lamp with Wick Varying and Chimney Fixed; Consumption of Oil in the Carcel Lamp with Wick Fixed and Chimney Variable; Dumas and Regnault's Recommendations in Using the Carcel Lamp; Wick and Oil for the Carcel Lamp; Dumas and Regnault's Photometer; Gas Jet Standards; Sautter and Lemonnier's Unit; Crova on Gas Jet Units; Seven-Carcel Gas Standard; Other Luminous Standards; Draper's Incandescent Platinum Standard; Zöllner's Platinum Standard; Schwendler's Platinum Standard; Crova on Platinum Standards; Violle and Cornu's Platinum and Silver Standards; the Electrical Congress on Photometric Standards; Photometric Measurements; The Positions of Carbons for the Voltaic Arc; Results of Photometric Experiments by Sautter and Lemonnier ... ... 101 to 173

THE EYE AS A PHOTOMETRIC INSTRUMENT.—The Distribution of Light; The Structure of the Eye; Refraction; Indices of Refraction of Different Substances; The Action of Lenses in Causing Refraction; The Lenses of the Eye; Chromatic Aberration; Ratios of Dispersive Power; Achromatism; Power of Accommodation of the Eye; Adjustment for Distance; The Rods and Cones of the Eye; The Blind Spot of the Eye; Intrinsic Brightness; Light, Heat, and Chemical Rays; The Waves of Light; Lengths of Wave Undulations; Oscillations of Light Rays per Second ... 174 to 190

SECTION III.

DYNAMOMETERS.—Introductory Remarks; The Prony Brake; Kretz's Brake; Easton and Anderson's Dynamometer; Emery; Brauer; Amos; Imray; Deprez; Bramwell; Carpentier; Raffard's Dynamometric Balance; Thiabaud; Weyher and Richmond; Félu and Deliége; Raffard; Emerson; Froude's Inertia Brake; Brown; Raffard; German Dynamometer; King; Silver and Gay; Smith; Bourdon; Hirn's Torsion Dynamometer; Belt Dynamometers; Froude; Tatham; Farcot; Parsons; Hefner Alteneck; King; The Royal Agricultural Society's Dynamometer; Morin; Bourry; Megy; Ruddick; Valet; Taurines; Neer; Ayrton and Perry; Latchinoff; Darwin; Matter; Emerson; Smith's Ergometer; Marcel Deprez's Dynamo - Electric Brake .. .. 193 to 241

A TESTING INSTALLATION.—Testing the Heinrichs Dynamo; General Arrangement of Testing Room; The Morin Dynamometer; The Prony Brake; Modifications of the Prony Brake; Testing the Prony Brake; Kempe and Ferguson's Electrical Revolution Counter; Harding's Speed Indicator; Young's Speed Indicator; Method of Testing; Tables of Tests of Heinrichs' Dynamo, Series Wound (A), Values Measured; Table of Tests (B), Values Calculated; Table of Tests (C), External Circuit; Table of Tests

(D), Percentage of Efficiency ; Mr. Preece's Experiment ; Friction of Dynamo Calcu-lated from Indication of the Morin Dynamometer and the Prony Brake ; Efficiency of the Heinrichs Dynamo   ...   .   242 to 258

SECTION IV.

RECENT DYNAMO MACHINES AND LAMPS.—THE WESTON SYSTEM OF ELECTRIC LIGHTING : Recent Types of Weston Dynamo ; The Weston Armature ; The Weston Commutator ; Methods of Coiling the Weston Armature ; Connection of Weston Armature and Commutator ; The Weston Field Magnets ; Weston Dynamo for Incandescence Lamps ; The Weston Rheostat ; The Weston Duplex Arc Lamp ; Feeding Mechanism of the Weston Arc Lamp ; Brake Wheel Feeding Mechanism ; Weston's Automatic Cut-Out ; Weston's Fusible Plug Cut-Out ; Weston's Central Station Indicator ; Maxim's Automatic Cut-Out for Incandescence Lamps ; Weston's Safety Devices ; Weston's Brackets for Incandescence Lamps. THE HOCHHAUSEN SYSTEM : Hochhausen's Dynamo for Arc Lighting ; The Hochhausen Commutator ; Core of the Hochhausen Armature ; Winding the Hochhausen Armature ; Mode of Coupling the Armature Coils ; Hochhausen's Automatic Regulator ; Motor for Auto-matic Regulator ; Hochhausen Arc Lamps ; Single Carbon Lamp ; Double Carbon Lamp ; Device for Starting Second Carbon into Action. THE FERRANTI-THOMSON DYNAMO : The Ferranti-Thomson Armature ; Efficiency of the Ferranti-Thomson Dynamo ; The Ferranti 5000-Lamp Machine ; Armature of the Ferranti 5000-Lamp Machine ; Collectors and Field Magnets ; Ferranti Continuous Current Dynamo ; Commutator. BRUSH INCANDESCENCE SYSTEM : Allen's Automatic Regulator ; Brush Incandescence Lamp and Fittings ; Wright and Mackie's Glass-Blowing Machine ; Fusible Cut-Outs ; Brush Regulator for Theatres ; Brush Electric Meter. THE EDISON SYSTEM : Dynamo for Shipboard ; Dynamos K, L, Z ; Hopkinson-Edison Dynamo ; Edison Regulator ; Brackets and Fittings ; 1200-Light Dynamo. GAULARD AND GIBBS' SYSTEM OF ELECTRIC DISTRIBUTION ; Installation on the Metropolitan Railway. Gramme's Multipolar Dynamo. Gordon's Dynamo ; Frame and Magnets ; Coils ; Armature ; Circuits ; Regulating Apparatus. Ganz's Dynamo. THE ZIPERNOWSKY SYSTEM : The National Theatre at Budapest ; The Life of Incandescence Lamps ; The Zipernowsky Dynamo. The Chertemps-Dandeu Dynamo ; Coils and Magnets ; Sabine's Experi-ments ; Self-Regulating Arrangements. Ball's Unipolar Dynamo ; Sabine's Tests ; Constructive Details ; Alabaster, Gatehouse, and Co.'s Tests. Matthews' Multipolar Dynamo ; Arrangement of Armature. The Hopkinson-Muirhead Dynamo. Schwerdt's Dynamo. Crompton's Step-Wound Armature. ELECTRIC ARC LAMPS : André ; Hawkes ; Crompton ; Bürgin and Crompton ; Abdank ; Solignac ; Clark and Bowman ; Lever ; Breguet ; Werdermann ; Varley ; Egger-Kremenetsky ; Zipernowsky ; Piette and Krizik. INCANDESCENCE LAMPS : Woodhouse and Rawson ; Müller ; Brush ; Beeman ; Bernstein ; Crookes ; Defries ; Gatehouse and Alabaster ; Knowles ; Edison ; Fergus-son ; Cherrill ; Soward ; Guest ; Harrison ; Swan ...   .   261 to 439

INDEX   ...   .   441 to 455

APPENDICES.

Abstracts of Electrical Patents from January 1, 1873, to June 30, 1882 ...   .   i to ccclxv
INDEX ...   .   ccclxvii to cccxciv
A.—Electrical and Photometric Tests made at the Munich Electrical Exhibition...   ...   ...   i to xiii

LIST OF ILLUSTRATIONS.

——:o:——

SECTION I.—ELECTRICAL MEASUREMENT.

			PAGE
Fig. 1	...	Diagram of Joint Resistance of Conductors	7
Fig. 2	...	De la Rue's Standard Cell	9
Fig. 3	...	Daniell's Modified Standard Cell	9
Fig. 4	...	Diagram of Division of Current in Two Conductors	12
Fig. 5	...	Galvanometer and Shunt...	13
Fig. 6, 7	...	Capacity of Two Concentric Spheres	16, 17
Fig. 8	..	Condensers in Cascade or Series	19
Fig. 9	...	Condensers of ⅓ m.f. Capacity	19
Fig. 10	...	Sir William Thomson's Quadrant Electrometer (Elevation)	20
Fig. 11	...	,, ,, ,, (Section)	21
Fig. 12	...	,, ,, ,, (Connection of Quadrants) ...	22
Fig. 13	...	,, ,, ,, (Diagram showing Action of Needle)	23
Fig. 14	...	Guard Plate of Sir William Thomson's Quadrant Electrometer	25
Fig. 15	...	Sir William Thomson's Replenisher	25
Fig. 16	...	,, ,, Inductor...	25
Fig. 17	...	Plan of Sir William Thomson's Inductor	26
Fig. 18	...	Sir William Thomson's Replenisher	27
Fig. 19	...	Helmholtz's Tangent Galvanometer	28
Fig. 20	...	Diagram of Needle Deflection in Tangent Galvanometer	28
Fig. 21	...	Diagram of Strength of Currents...	29
Fig. 22	...	,, Tangent Galvanometer	31
Figs. 23, 24	...	Diagrams of Deflection Experiment	36
Fig. 25	...	Diagram of Bifilar Suspension	40
Fig. 26	...	Siemens's Electro Dynamometer	41
Fig. 27 to 29	...	Thomson's Potential Galvanometer	42, 43
Fig. 30	...	Diagram of Poggendorff's Method for Finding Resistance of Batteries	45
Figs. 31, 32	...	The Current Galvanometer (Contact Maker)	46
Fig. 33	...	Siemens's Torsion Galvanometer	47
Fig. 34	...	Diagram showing Method of Ascertaining Battery Constants	48
Fig. 35	...	,, of Wheatstone Bridge	50
Figs. 36 to 39	...	Diagrams Illustrating Kirchoff's Laws	51, 52
Figs. 40 to 42	...	,, of Slide Wire or Metre Bridge	53
Fig. 43	...	Calibrating Resistances	56
Figs. 44, 45	...	Winding Resistance Coils	57
Fig. 46	...	Resistance Box	58
Fig. 47	...	Plan of Resistance Box	58

PAGE

Fig. 48 ... Wheatstone's Rheostat 59
Fig. 49 ... Measurement of Resistance 60
Fig. 50 ... Measuring Resistance of Galvanometer Coils 64
Fig. 51 ... Mance's Method for Measuring the Internal Resistance of a Battery ... 65
Fig. 52 ... Lodge's Method for Measuring Battery Resistance 66
Fig. 53 ... Munro and Muirhead's Method for Measuring Battery Resistance 66
Fig. 54 ... Diagram of Measurement of High Resistance 68
Fig. 55 ... ,, ,, Incandescence Lamps when Hot 68
Fig. 56 ... The Measurement of Discharge Deflections 71
Fig. 57 ... Poggendorff Compensation Method 73
Fig. 58 ... The Comparison of Electromotive Force 75
Fig. 59 ... Determining the Capacity of a Condenser 77
Fig. 60 ... The Capacity of Submarine Cables 80
Fig. 61 ... Comparing the Capacities of Condensers... 80
Fig. 62 ... Characteristic Dynamo Curve (Hopkinson) 92
Fig. 63 ... ,, ,, (Adams) 93
Fig. 64 ... Dynamo Efficiency Diagram (Thompson) 94

SECTION II.—PHOTOMETRY.

Fig. 65 ... Diagram Illustrating the Law of the Squares of Distances 103
Fig. 66 ... ,, ,, ,, Quantity 104
Fig. 67 ... ,, ,, Intensity of Illumination 105
Fig. 68 ... ,, ,, Fundamental Principle of Photometer... 105
Fig. 69 ... Bouguer's Photometer 106
Fig. 70 ... Ritchie's Photometer 106
Fig. 71 ... Foucault's Photometer 106
Figs. 72, 73 ... Degrand's Photometer 108
Figs. 74, 75 ... Details of Degrand's Photometer 109
Fig. 76 ... Wolff's Photometer 111
Figs. 77 to 79... Cornu's Photometer... ... ... 111, 113, 114
Figs. 80, 81 ... Masson's Photometer 114, 115
Fig. 82 ... Rumford's Photometer 116
Figs. 83 to 85... Bunsen's Photometer 118, 119
Fig. 86 ... Wheatstone's Photometer 120
Figs. 87 to 90... Arago's Photometer... ... ... 121, 123
Figs. 91 to 93... Duboscq's Photometer 125, 126
Figs. 94, 95 ... Leslie's Photometer 127
Fig. 96 ... Ritchie's Photometer (Leslie's Modified)... 127
Figs. 97, 98 ... Bunsen and Roscoe's Chemical Photometer 132, 136
Fig. 99 ... Siemens's Artificial Eye 139
Figs. 100 to 103 Siemens's Selenium Photometer 139, 140, 141
Fig. 104 ... Nacks's Photometer... 142
Figs. 105, 106... Coulon's Photometer 143
Figs. 107, 108... Stevenson's Absorption Photometer 145
Figs. 109, 110... Ayrton and Perry's Photometer... 150
Figs. 111, 112... Girout's Comparison of Luminous Standards ; Gas Jets 153
Figs. 113, 114... ,, ,, ,, Candles 154
Figs. 115, 116... Dumas and Regnault's Photometer 161
Figs. 117, 118... Seven Carcel Gas Jet Photometric Standard 164
Figs. 119 to 139 Sautter and Lemonnier's Photometric Experiments 169, 170
Fig. 140 ... Diagram of Negative and Positive Carbons 171
Figs. 141, 142... Positions of Negative and Positive Carbons 171

THE EYE AS A PHOTOMETRIC INSTRUMENT.

Fig. 143 ... Distribution of Light over a Hollow Sphere 174
Fig. 144 ... Vertical Section of the Eye 175

PAGE

Fig. 145	...	Horizontal Section of the Eye	176
Figs. 146, 147...		Diagram Illustrating Refraction of Light	178, 179
Figs. 148, 149...		„ Action of Lenses in Causing Refraction	179
Fig. 150	...	The Lenses of the Eye	180
Fig. 151	...	Optical Adjustment for Distance	183
Fig. 152	...	Focussing of Rays	183
Fig. 153	...	The Rods and Cones of the Eye	184
Fig. 154	...	Diagram of Heat, Light, and Chemical Rays	187

SECTION III.—DYNAMOMETERS.

Fig. 155	...	Kretz's Dynamometer	194
Figs. 156, 157...		Easton and Anderson's Dynamometer	195
Fig. 158	...	Emery's Dynamometer	196
Fig. 159	...	Brauer's Dynamometer	197
Fig. 160	...	Amos's Dynamometer	197
Fig. 161	...	Imray's Dynamometer	198
Fig. 162	...	Deprez's Dynamometer	198
Fig. 163	...	Bramwell's Dynamometer	198
Figs. 164, 165...		Carpentier's Dynamometer	199
Fig. 166	...	Raffard's Dynamometric Balance	200
Fig. 167	...	Thiabaud's Dynamometer	201
Fig. 168	...	Weyher and Richmond's Dynamometer	201
Fig. 169	...	Felu and Deliège's Dynamometer	201
Fig. 170	...	Raffard's Dynamometer	202
Fig. 171	...	Emerson's Dynamometer	202
Figs. 172 to 175		Froude's Inertia Brake	203
Fig. 176	...	Brown's Dynamometer	205
Fig. 177	...	Raffard's Dynamometer	206
Fig. 178	...	German Dynamometer	206
Fig. 179	...	King's Dynamometer	207
Fig. 180	...	Silver and Gay's Dynamometer	207
Fig. 181	...	Smith's Dynamometer	209
Figs. 182, 183...		Bourdon's Dynamometer	210
Figs. 184 to 186		Hirn's Torsion Dynamometer	211
Figs. 187, 188...		Froude's Belt Dynamometer	213
Figs. 189 to 191		Tatham's Dynamometer	215, 217
Fig. 192, 193...		Farcot's Dynamometer	218, 220
Fig. 194	...	Parsons' Dynamometer	222
Figs. 195 to 197		Hefner Alteneck's Dynamometer	223, 224
Figs. 198 to 200		Application of the Hefner Alteneck Dynamometer	225, 226
Fig. 201	...	King's Dynamometer	227
Figs. 202 to 204		The Royal Agricultural Society's Dynamometer	227
Figs. 205, 206...		Registering Apparatus of the Royal Agricultural Society's Dynamometer	228
Fig. 207	...	Morin's Dynamometer	229
Figs. 208 to 210		Bourry's Dynamometer	230
Fig. 211	...	Megy's Dynamometer	231
Figs. 212, 213...		Ruddick's Dynamometer	233
Figs. 214 to 217		Valet's Dynamometer	233, 234
Figs. 218 to 220		Taurines' Dynamometer	235
Fig. 221	...	Neer's Dynamometer	235
Figs. 222 to 225		Ayrton and Perry's Dynamometer	236
Figs. 226, 227...		Darwin's Dynamometer	237
Figs. 228 to 232		Matter's Dynamometer	238
Figs. 233, 234...		Emerson's Dynamometer	239
Figs. 235 to 237		Smith's Ergometer	240
Fig. 238	...	Marcel Deprez's Dynamo Electric Brake	241

A TESTING INSTALLATION.

PAGE
Fig. 239 ... Heinrichs' Dynamo Testing Installation... ... ... To adjoin page 242
Figs. 240 to 242 Modifications of the Prony Brake 245
Figs. 243 to 245 Kempe and Ferguson's Electrical Counter 249
Figs. 246 to 248 Harding's Speed Indicator... 251
Figs. 249 to 251 Young's Speed Indicator 252

SECTION IV.—RECENT DYNAMOS AND LAMPS.

Fig. 252 ... Weston's Dynamo Machine 262
Figs. 253 to 255 The Weston Armature 263, 264
Figs. 256 to 259 Coiling the Weston Armature 264, 265
Fig. 260 ... Connecting the Weston Armature and Commutator 266
Fig. 261 ... The Weston Rheostat 269
Fig. 262 ... Weston Double-Carbon Lamp 270
Figs. 263 to 265 Feeding Mechanism of Weston's Arc Lamp 271, 272
Fig. 266 ... Weston's Single Arc Lamp 273
Figs. 267, 268... Brake Wheel Mechanism for Feeding Weston's Arc Lamp 274, 275
Fig. 269 ... Weston's Automatic Cut-out 277
Fig. 270 ... ,, Fusible Plug Cut-out 278
Figs. 271, 272... ,, Central Station Indicator 279
Fig. 273 ... Maxim's Automatic Cut-out 280
Figs. 274, 275... Weston's Safety Devices 281
Fig. 276 ... ,, Bracket for Incandescence Lamps 281
Fig. 277 ... The Hochhausen Dynamo for Arc Lighting 283
Fig. 278 ... Core of the Hochhausen Armature 285
Fig. 279 ... Construction of the Hochhausen Armature 285
Figs. 280, 281... Method of Winding the Hochhausen Armature 285
Fig. 282 ... Longitudinal Section of the Hochhausen Armature 286
Figs. 283, 284... Motor for Working the Hochhausen Automatic Regulator 288
Figs. 285 to 288 Hochhausen Automatic Regulator 289, 290
Figs. 289, 290... ,, Single-Carbon Lamp 295
Fig. 291 ... ,, Double-Carbon Lamp 297
Fig. 292 ... Hochhausen's Device for Setting Carbons into Action 299
Fig. 293 ... The Ferranti 5000-Light Dynamo 303
Fig. 294 ... ,, Armature 304
Fig. 295 ... Field Magnets of the Ferranti Dynamo 304
Figs. 296 to 301 The Ferranti Continuous Dynamo 307, 308, 309
Figs. 302 to 313 Details of the Ferranti Continuous Dynamo 310
Figs. 314, 315... The Brush Company's Dynamo for Incandescence Lighting... ... ... 313, 314
Fig. 316 ... ,, ,, with Allen's Automatic Regulator... 315
Fig. 317 ... ,, ,, ,, Three Armatures 316
Figs. 318, 319... Allen's Automatic Regulator 317
Fig. 320 ... The Brush Company's Incandescence Lamp 319
Figs. 321 to 323 Attachments for the Brush Company's Incandescence Lamps 319
Figs. 324 to 326 Methods of Mounting the Brush Company's Incandescence Lamps... 320
Fig. 327 ... Wright and Mackie's Glass Blowing Machine 321
Figs. 328 to 330 Details of Wright and Mackie's Glass Blowing Machines 322, 323
Fig. 331 ... Brush Company's Fusible Cut-out for Six Circuits 324
Figs. 332, 333... ,, ,, ,, Large Installation 325
Fig. 334 ... The Brush Company's Regulator for Theatres 326
Fig. 335 ... ,, ,, Electric Meter 327
Fig. 336 ... Edison Dynamo and Brotherhood Engine 330
Fig. 337 ... ,, K Dynamo 331
Fig. 338 ... ,, L ,, ... 332
Fig. 339 ... ,, Z ,, ... 333
Fig. 340 ... ,, Electric Regulator 334

PAGE

Fig. 341 ... Edison Combined Gas Bracket and Fittings 334
Fig. 342 ... ,, Jointed Bracket 335
Fig. 343 .. Epergne Fitted with Edison Lamps 336
Fig. 344 ... Group of Ten Edison Lamps enclosed in Globe 337
Fig. 345 ... Edison's Branch and Main Line Junction 338
Fig. 346 ... Combined Gas and Edison Lamp Bracket 338
Figs. 347, 348... Edison Lamp Fittings 339
Fig. 349 ... The Edison 1200-Light Dynamo ... ... To adjoin page 340
Fig. 350 ... Gaulard and Gibbs's Secondary Generator 345
Fig. 351 ... Section of Gaulard and Gibbs's Secondary Generator 345
Fig. 352 ... Diagram of Circuit, Gaulard and Gibbs's System 345
Figs. 353 to 358 Gramme Multipolar Dynamo 351, 352, 353
Fig. 359 ... Gordon's Dynamo 355
Figs. 360, 361... Ganz's Dynamo and Engine ... ... To adjoin page 361
Fig. 362 ... Armature of Ganz's Dynamo 361
Fig. 363 ... Ganz's Method of Driving Dynamos 364
Fig. 364 ... Zipernowsky's Arc Lamp 364
Fig. 365 ... Block Plan of Budapest Theatre 364
Figs. 366 to 368 Budapest Installation (Zipernowsky System) 365
Fig. 369 ... Incandescence Lamp Fed by Continuous Current 365
Figs. 370, 371... Budapest Theatre, Mode of Hanging Lamps 367
Fig. 372 ... The Zipernowsky Dynamo... ... ... To adjoin page 368
Figs. 373 to 375 The Chertemps-Dandeu Dynamo 371
Figs. 376 to 378 Ball's Multipolar Dynamo 378
Fig. 379 ... Matthews's Multipolar Dynamos... 382
Figs. 380 to 394 Details of Armature for Ball's Multipolar Dynamo 383
Figs. 395 to 401 Hopkinson-Muirhead's Dynamo 385, 386
Figs. 402, 403... Armature of Hopkinson-Muirhead's Dynamo 387
Figs. 404, 405... Schwerdt's Dynamo 388
Figs. 406 to 414 Crompton's Step-Wound Armature 390

ARC LAMPS.

Figs. 415 to 424 André's Arc Lamp 392
Figs. 425 to 429 Hawkes's Arc Lamp 396
Figs. 430 to 434 Crompton's Arc Lamp 400
Figs. 435 to 437 ,, ,, (Duplex) 401
Figs. 438, 439... Crompton and Bürgin's Arc Lamp 403
Figs. 440 to 443 Abdank's Arc Lamp 405
Figs. 444, 445... Cut-Off for Abdank's Arc Lamp 406
Figs. 446, 447... Solignac's Arc Lamp 409
Figs. 448 to 450 Clark and Bowman's Arc Lamp 413
Figs. 451 to 455 Lever's Arc Lamp 415
Fig. 456 ... Breguet's Arc Lamp 417
Figs. 457, 458... Werdermann's Arc Lamp 419
Figs. 459 to 462 Varley's Arc Lamp... 421
Fig. 463 ... Egger-Kremenetsky's Arc Lamp 422
Figs. 464 to 467 Zipernowsky's Arc Lamp 423, 424
Fig. 468 ... Piette and Krizik's Arc Lamp 425

INCANDESCENCE LAMPS.

Fig. 469 ... Woodhouse and Rawson's Incandescence Lamp 428
Fig. 470 ... Miller's Incandescence Lamp 429
Figs. 471, 472... Brush's Incandescence Lamp 429
Figs. 473, 474... Beeman's Incandescence Lamp 429
Figs. 475 to 477 Bernstein's Incandescence Lamp 430, 431
Figs. 478, 479... Crookes's Incandescence Lamp 431

PAGE

Figs. 480, 481... Defries's Incandescence Lamp 432

Figs. 482, 483... Gatehouse and Alabaster's Incandescence Lamp 432

Fig. 484 ... Knowles's Incandescence Lamp 433

Figs. 485, 486... Edison's Incandescence Lamp 434

Figs. 487, 488... Fergusson's Incandescence Lamp 434

Fig. 489 ... Cherrill's Incandescence Lamp 435

Fig. 490 ... Soward's Incandescence Lamp 435

Fig. 491 ... Guest's Incandescence Lamp 436

Fig. 492 ... Harrison's Incandescence Lamp 437

Figs. 493 to 500 Swan's Incandescence Lamp 437, 438, 439

SECTION I.

——————•——————

ELECTRICAL MEASUREMENT.

1. *General Principles.*

2. *Principal Instruments.*

3. *Practical Methods.*

B

ELECTRIC ILLUMINATION.

I.

ELECTRICAL MEASUREMENT.

GENERAL PRINCIPLES.--The adoption of a system of units (which cannot yet be regarded as perfect) has developed the study of electricity from a mere observation of phenomena into a science of exact measurement. The electrical quantities of Resistance, Electromotive Force, Strength of Current, and Capacity can be accurately determined; and--thanks to the labours of successive Committees of the British Association and to the Congress of Electricians--expressed in practical units.

The electrician can now give the capacity of a submarine cable in Microfarads, or its insulation resistance in Megohms; and he can determine the electrical efficiency of a dynamo, or the mechanical energy expended in an arc, or in an incandescence, lamp with ease and precision. Such measurements involve a knowledge of certain general principles, which it will be useful to briefly consider before describing the more important instruments and practical methods commonly used in this branch of physical science.

The fundamental law of this science was first enunciated by Dr. G. S. Ohm, in a work on the mathematical theory of the electric current, published in 1827.

It establishes the relation connecting the Electromotive Force, the Resistance, and the Current, viz., $C = \dfrac{E}{R}$.

This law shows:

(a) That when a given E.M.F. is impressed at the ends of different conductors, the current through these varies inversely as their respective resistances; and

(b) That for a given current passing through the conductors, the difference of the potentials at their ends varies directly as their resistance.

This great electrical law cannot be deduced from purely theoretical considerations. It was experimentally, though roughly, verified by Ohm himself, and subsequently by several eminent investigators, including Pouillet, Becquerel, Fechner, and Kohlrausch. More recently it has occupied the attention of a Committee of the British Association, in consequence of its accuracy being suspected by Weber, Lorenz, and Schuster.

The experimental verification was entrusted to Mr. G. Chrystal, of the Cavendish Laboratory, now Professor of Mathematics in the University of Edinburgh; and most carefully conducted experiments have shown it to be not merely a close approximation, but a strict physical law.*

Kohlrausch and Nippoldt, in a classical series of experiments, have shown it to hold for electrolytes throughout a great range of electromotive force.

Besides direct verifications, we have evidence for the accuracy of the law in the many discoveries to which it has led, as well as in a great variety of measurements which have been made with a degree of precision rarely approached in other physical determinations.

1. *Resistance.*—The idea implied in the term "*electrical resistance*" is that of an actual force opposing the E.M.F. which maintains the current. It is analogous to friction in mechanics, for it tends to diminish the available energy of the current.

Ohm's law furnishes a definition of resistance, and consequently also of unit resistance. It defines the resistance of a conductor to be the ratio of the numerical value of the E.M.F. to that of the current which it produces; and hence a conductor of unit resistance is one in which unit current is produced by unit electromotive force.

The resistance of a conductor may also be defined as the work done in it by the passage of unit current for unit time; for, as will be seen further on,

$$W = C\,T\,E = C\,T.C\,R = C^2\,R\,T.$$

Hence, if C be unit current and T unit time, we have

$$W = R,$$

that is, the work done is numerically equal to the resistance.

The resistance of metallic conductors depends in no way upon the current which is passing, or the E.M.F. which is impressed. It is a physical

* B.A. Report for 1876.

quantity, which is perfectly constant so long as the molecular conditions of the conductor remain unaltered.

The resistance of a conductor is affected by heat, strain, tempering, and even by magnetisation.

Resistance, in the electromagnetic system, has the dimensions of, and is therefore expressible as, a velocity.* The standard resistance, constructed in 1863 by the Committee of the British Association, is called the *ohm*, and was intended to represent the velocity of a body which, in one second of time, would move over a quadrant of a terrestrial meridian; that is to say, a velocity of 10^9 centimetres per second.

Some doubt having arisen as to the accuracy of this standard, re-determinations of its value have been made by Kohlrausch, Lorenz, H. Weber, Rowland, Glazebrook, Carey Foster, and Lord Rayleigh. The numerical results obtained by different methods are not perfectly concordant. That given by Lord Rayleigh from his latest determination (1883) makes the B. A. unit $= .9868 \; \dfrac{\text{earth quadrant}}{\text{second}}$ or .9868 of the theoretic ohm.

A *megohm* is a million ohms; a *microhm* is the millionth part of an ohm.

Conductivity is plainly the reciprocal of resistance. No special unit has yet been introduced.

Let A B be a conductor of length l and sectional area s, then the flow of electricity (the current) along A B will vary

(a) directly as the difference of potentials $V_a - V_b$;

(b) directly as s;

(c) directly as a constant c depending upon the material of the conductor; and

(d) inversely as l, *i.e.* :

$$C = \frac{V_a - V_b}{l} \, s \, c$$

$$= \frac{V_a - V_b}{\dfrac{l}{s\,c}} = \frac{V_a - V_b}{R}$$

---

* A coil of wire, in revolving about a vertical axis, cuts the horizontal component H of the earth's magnetic force. If C be the induced current, r the mean radius of the coil, n the number of windings, R the resistance, and ω the angular velocity of the coil, we have $C = \dfrac{H \cdot r^2 \pi \, n \cdot \omega}{R}$. Also, from the deflection of the magnet suspended at the centre of the spinning coil, we have $C = \dfrac{r \, H}{2 \pi n} \tan \phi$. By equating these values, we get $R = 2 \, \pi^2 \, n^2 \, r \, \omega \cot \phi$; that is, R is expressed as a velocity.

where

$$R = \frac{l}{s\,c} \qquad . \qquad . \qquad . \qquad . \qquad . \qquad . \qquad . \quad (1)$$

From (1) we conclude that the resistance of a conductor varies directly as its length and inversely as its cross section. If l = unit length (1 cm.) and s = unit area (1 sq. cm.), then R, the resistance of 1 cm. cube, is called the *specific resistance* of the conductor; and therefore, also, $c = \frac{1}{R}$ will be the *specific conductivity*.

The resistance of a cylindrical conductor may be defined with reference to its weight. Thus:

The weight W of a body is numerically the product of its mass M by the acceleration due to gravity g, *i.e.*, $W = M\,g$.

M is the product of the volume V by the density ρ, *i.e.*, $M = V\,\rho$, whence

$$W = V\,\rho\,g = s\,l\,\rho\,g \qquad . \qquad . \qquad . \qquad . \qquad . \qquad . \quad (2)$$

The specific gravity of a body is the ratio of its weight to that of an equal volume of some standard substance. For solid bodies the standard of reference is water at the temperature of maximum density, 4° C.

In the centimetre-gramme-second (the c.g.s.) system of units, the gramme is the unit of mass; and as 1 cub. cm. of water at 4° C. weighs 1 gramme, the specific gravity of any body is numerically the weight of one cub. cm. of that body.

Putting $l\,s = 1$, equation (2) becomes $W = \rho\,g$, *i.e.*, $\rho\,g$ is the specific gravity of the body. Denoting it by σ, we have

$$W = l\,s\,\sigma \quad \therefore \quad s = \frac{W}{l\,\sigma}$$

And since R varies as $\dfrac{l}{s}$, R will also vary as $\dfrac{l^2\,\sigma}{W}$.

Ex. Let it be required to compare the resistances R_1 and R_2 of two copper wires, one of 2 metres long and weighing $\frac{1}{4}$ gm., and the other 5 metres long and weighing $\frac{2}{3}$ of a gm. Assuming that the wires have the same specific gravity σ,

$$\begin{aligned}
R_1 \quad : \quad R_2 &= \frac{(2)^2\,\sigma}{\frac{1}{4}} \quad : \quad \frac{(5)^2\,\sigma}{\frac{2}{3}} \\
&= 16\,\sigma \quad : \quad 37.5\,\sigma \\
&= 32 \quad : \quad 75
\end{aligned}$$

Ex. It is required to find the specific resistance of a wire 437 mm. long, which has a resistance of .1257 ohm, and which weighs .411 gm. in air and .365 gm. in water.

Here $.411 - .365 = .046 =$ weight of an equal volume of water $\therefore$ specific gravity $= \dfrac{.411}{.046} = \dfrac{411}{46}$

But $W = V \sigma \therefore .411 = V \times \dfrac{411}{46} \therefore V = .046$ cub. cm.; and $V = r^2 \pi l$

$= r^2 \pi \times 43.7 \therefore r^2 \pi = \dfrac{.046}{43.7} = .001053$

The resistance of 43.7 cm. of the wire of area $.001053 = .1257$ ohm $\therefore$ the resistance of conductor 1 cm. long and 1 cm. sectional area,

$$= \dfrac{.1257}{43.7} \times .001053 \text{ ohm} = .000003029 \text{ ohm}$$

$$= 3.029 \text{ microhms}$$

$$= \text{ specific resistance required.}$$

The *joint resistance* of any number of conductors arranged *in series* is plainly the sum of their separate resistances, that is, $R = r_1 + r_2 + r_3 \ldots$

When arranged in *multiple arc*, the joint resistance is found as follows:

Let V_a and V_b be the potentials at A and B respectively.

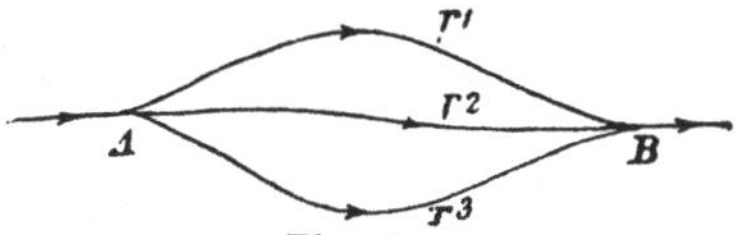

Then $V_a - V_b = E =$ the E.M.F. producing the current C.

Fig. I.

But $C = \dfrac{E}{R} \therefore C R = E.$

Let r_1, Fig. 1, be the resistance of one conductor and c_1 the current through it, with similar notation for the other branches. Then

$$V_a - V_b = c_1 r_1 = c_2 r_2 = c_3 r_3$$

$$= \dfrac{c_1}{\dfrac{1}{r_1}} = \dfrac{c_2}{\dfrac{1}{.r_2}} = \dfrac{c_1}{\dfrac{1}{r_3}}$$

$$= \dfrac{c_1 + c_2 + c_3}{\dfrac{1}{r_1} + \dfrac{1}{r_2} + \dfrac{1}{r_3}}$$

But $c_1 + c_2 + c_3 = C.$

Whence $\dfrac{1}{r_1} + \dfrac{1}{r_2} + \dfrac{1}{r_3} = \dfrac{1}{R}.$ This gives the resistance R of the multiple arc.

Ex. Find the joint resistance of three conductors of 10, 12, 18 ohms arranged in multiple arc. Here

$$\dfrac{1}{R} = \dfrac{1}{10} + \dfrac{1}{12} + \dfrac{1}{18} = \dfrac{43}{180}$$

$$\therefore R = \dfrac{180}{43} = 4.18 \text{ ohms.}$$

The resistance of metallic conductors is invariably increased by heat. Dr. Matthiessen[*] and Sir William Siemens[†] have given formulæ for finding the resistance of a conductor throughout an extensive range of temperature.

It is important to notice that the resistance of carbon falls with a rise of temperature. Thus, the resistance of a filament when cold was 71 ohms, and when heated to incandescence it was found to be only 47 ohms. The decrease when white hot is generally about $\frac{1}{3}$ the resistance when cold.

The electrical properties of selenium and tellurium, in the crystalline state, are modified by exposure to light, their resistance being considerably and instantaneously diminished. Professor W. G. Adams has shown that, under certain circumstances, light can even set up an E.M.F. and start a current in a bar of selenium.[‡]

The *internal resistance* of a voltaic cell practically depends upon the nature of the liquids, the size of the plates, and their distance apart. There appears to be some reason for thinking that the resistance is also a function of the current itself.[§]

The internal resistance is diminished

(*a*) by increasing the active area of the plates, and

(*b*) by bringing them closer together.

Heat affects the resistance of liquids, a rise of temperature diminishing the resistance of electrolytes generally.

Mr. W. H. Preece has recently shown that of the batteries in general use, the Daniell is most affected by variations of temperature. It becomes then necessary, where accurate measurement is required, to keep the temperature of the battery constant, or else to make frequent determinations of resistance and allow for the change.

2. *Electromotive Force.*—This is the name given to that force which tends to displace electricity from one point of a conductor to another.

Electromotive, must be distinguished from ordinary mechanical, force: the latter tends to set a certain *mass* in motion; the former to cause the transfer of *electricity* constituting the phenomenon of the electric current.

The nature of this force[‖] and its causes are entirely unknown, whilst even its seat is yet a debatable subject among physicists.

---

[*] B. A. Report, 1864.   [†] Bakerian Lecture, 1871.   [‡] Phil. Trans. 1877.
[§] Prof. G. Chrystal, in Encycl. Brit., art. Electricity, p. 50.
[‖] According to Newton's definition of force, the term electromotive force is a misnomer.

Unit Electromotive Force exists between two points when it requires the expenditure of unit work to bring a unit of positive electricity from one of the points to the other.

This unit is exceedingly small, and accordingly a " practical" unit called the *volt* has been adopted, which is intended to represent 10^8 such absolute units. The E.M.F. of a Daniell's cell is a little more than a volt.

There is yet no absolutely constant standard of E.M.F.; a Daniell, a Clark, or a De la Rue cell is generally used when a definite E.M.F. is required.

Latimer Clark's standard cell consists of pure mercury, the surface of which is covered over with a mercurous paste; on this rests a plate of pure zinc. The positive pole is the platinum wire which is in contact with the mercury. The E.M.F. of this cell is given as 1.457 volts.

Dr. De la Rue's cell is shown in Fig. 2, where A is a rod of chemically pure zinc, B a cylinder of silver chloride into which is let a silver electrode C; B is usually enclosed in a small bag of parchment paper. The active liquid is a solution of pure sal-ammoniac in water. The E.M.F. is said to be 1.068 volts.

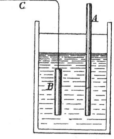

Fig. 2.

At the Southampton Meeting of the British Association, 1882, Sir William Thomson proposed a modified form of Daniell's cell to serve as a standard of E.M.F.

The zinc plate, Fig. 3, rests on the bottom of the vessel, and is immersed in a saturated solution of zinc sulphate. On the surface of this liquid is poured a quantity of copper sulphate, only half saturated, so that it may not rapidly flow down and intermix with the zinc sulphate. The copper plate is suspended horizontally in the upper stratum.

Fig. 3.

The vessel is furnished with a bent glass tube, drawn out at one extremity to a fine bore, and connected at the other with a piece of india-rubber tubing attached to a funnel.

By raising the funnel, the copper sulphate solution which it contains flows gently into the vessel and forms a clear horizontal surface of separation; by lowering it, the liquid is withdrawn.

This must be done as soon as a measurement is finished, and the cell no longer required.

The E.M.F. of this cell has been carefully determined, and found to be, at ordinary temperatures, 1.07 volts.

The E.M.F. of a battery, unlike its resistance, is but very little affected by heat. Latimer Clark found that the E.M.F. of his cell decreased slightly with rise of temperature. Mr. Preece has recently* shown that the E.M.F. of the batteries in common use—the Daniell, Bichromate, Leclanché—is practically unaffected by the ordinary variations of temperature.

The experiments of Beetz and others show, however, that the counter E.M.F. set up in secondary batteries during the process of charging, and consequently also the charge itself, is sensibly diminished by heat; it becomes, then, very important to keep such cells cool while charging.

The E.M.F. of a cell depends upon the nature of the materials used, but not upon the size of the plates. Thus, the E.M.F. of a Bunsen pint cell is the same as that of a quart cell. But when two similar cells are connected up in series, the resulting E.M.F. is twice that of either of them.

Let E be the E.M.F. of one cell, then that of n cells arranged in parallel circuit is simply E; if arranged in series, it is n E.

The E.M.F. of a battery also depends upon the resistance through which it is worked, being greatest when this is infinite, that is, when the poles are free. The E.M.F. of a battery is defined as the difference of potentials of its poles when these are free or insulated. When worked through a low resistance, polarisation sets up an appreciable inverse E.M.F., which diminishes that of the battery. Hence standard cells should never be used over a low resistance. They are especially useful for determining the value of electrometer readings, charging condensers, and for standards of comparison as in the "compensation methods."

3. *Current.*—The strength of the current is given by Ohm's law $C = \dfrac{E}{R}$, E being measured in volts and R in ohms, C is given in ampères.

Let there be n similar cells, each of E.M.F. E and resistance R, connected up in parallel circuit and maintaining a current through an external resistance r. The E.M.F. of the battery is E, and its internal resistance $\dfrac{R}{n}$, then $C = \dfrac{E}{\dfrac{R}{n} + r}$. If the poles of the battery be connected by a short thick

---

* Proc. Roy. Soc., Feb. 1883.

wire, its resistance r may be neglected, and $C = \dfrac{E}{\dfrac{R}{n}} = \dfrac{n\,E}{R}$, *i.e.*, the current

from the battery is n times that given by a single cell.

Let the n cells be arranged in series, then the E.M.F. of the battery is $n\,E$, and its internal resistance $n\,R$.

$$\therefore C = \frac{n\,E}{n\,R + r} = \frac{E}{R + \dfrac{r}{n}}.$$

If r be small, $\dfrac{r}{n}$ may be neglected, then $C = \dfrac{E}{R}$, *i.e.*, the current from the battery is no greater than that furnished by a single cell whose poles are connected up by a short thick wire. But if R be small in comparison with r, then $C = \dfrac{E}{\dfrac{r}{n}} = \dfrac{n\,E}{r}$, *i.e.*, the current from the battery is approximately pro-

portional to the number of cells.

It is often required to find the best arrangement of a battery of n similar cells for a known external resistance r.

Let there be q rows, each containing p cells. The E.M.F. of the p cells is $p\,E$, and their resistance $p\,R$, the current from one row is then

$$= \frac{p\,E}{p\,R + r}$$

The q rows have the same E.M.F. ($p\,E$) as any one of them, and a resistance $= \dfrac{p\,R}{q}$.

$$\therefore C = \frac{p\,E}{\dfrac{p\,R}{q} + r} = \frac{E}{\dfrac{R}{q} + \dfrac{r}{p}}$$

But

$$p\,q = n \therefore q = \frac{n}{p}$$

$$\therefore \text{ by substitution, } C = \frac{E}{\dfrac{p\,R}{n} + \dfrac{r}{p}}$$

Now C is greatest when the denominator of above fraction is least. Let

$$u = \frac{p\,R}{n} + \frac{r}{p}$$

$$\therefore \frac{d u}{d p} = \frac{R}{n} - \frac{r}{p^2} = 0 \text{ (for a minimum)}$$

Whence

$$r = \frac{p^2\,R}{n} = \frac{p\,R}{q}$$

i.e., the best arrangement is that which will make the resistance of the battery $\dfrac{p\,\mathrm{R}}{q}$ equal to the given external resistance r.

Ex. Let it be required to find the most advantageous combination of 12 cells of 6 ohms resistance each, the external resistance being 8 ohms.

Here

$$n = 12 = p\,q \qquad r = 8 \quad \text{and } \mathrm{R} = 6$$

But

$$r = \frac{p^2\,\mathrm{R}}{n} \;\therefore\; 8 = \frac{6\,p^2}{12} \;\therefore\; p^2 = 16 \text{ and } p = 4$$

And since

$$p\,q = 12 \;\therefore\; q = 3$$

$\therefore$ the best arrangement is 3 rows of 4 cells each.

The internal resistance of the battery is greatest when the cells are in series. It is then $n\,\mathrm{R}$. If the external resistance r equal $n\,\mathrm{R}$, or be greater than $n\,\mathrm{R}$, then the arrangement in series is the best. If r be less than $n\,\mathrm{R}$, p and q must be chosen so as to make the internal resistance as nearly equal to the external as possible.

From $\mathrm{C} = \dfrac{\mathrm{E}}{\mathrm{R}}$ we see that the current through a conductor is inversely as its resistance. This principle has many important applications.

Thus, let it be required to find how a current C divides in meeting two conductors joined up in multiple arc, Fig. 4.

Let c_1 be the current through the conductor, whose resistance is r_1, with similar notation for the other branches.

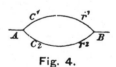

Fig. 4.

Then

$$\mathrm{C} = c_1 + c_2 \;\therefore\; c_2 = \mathrm{C} - c_1$$

Also

$$c_1\,r_1 = c_2\,r_2 = \text{difference of potentials between A and B} \;\therefore\; c_1\,r_1 = r_2\,(\mathrm{C} - c_1)$$

Whence

$$c_1 = \frac{r_2}{r_1 + r_2}\,\mathrm{C}$$

Similarly,

$$c_2 = \frac{r_1}{r_1 + r_2}\,\mathrm{C}$$

Let the current be 7 ampères and the resistances be 30 and 40 ohms respectively.

Then

$$c_1 = \frac{7 \times 40}{40 + 30} = \frac{280}{70} = 4 \text{ ampères}$$

And

$$c_2 = \frac{7 \times 30}{40 + 30} = \frac{210}{70} = 3 \text{ ampères}$$

If three wires were joined up in multiple arc, we should have, as above,

$$C = c_1 + c_2 + c_3$$

And

$$c_1\, r_1 = c_2\, r_2 = c_3\, r_3$$

Whence

$$c_2 = \frac{c_1\, r_1}{r_2} \quad \text{and} \quad c_3 = \frac{c_1\, r_1}{r_3}$$

$$\therefore C = c_1 + \frac{c_1\, r_1}{r_2} + \frac{c_1\, r_1}{r_3}$$

$$\therefore c_1 = \frac{r_2\, r_3}{r_1\, r_3 + r_1\, r_2 + r_2\, r_3}\, C$$

From the symmetry of this expression, we may at once write the values of c_2 and c_3.

When a delicate galvanometer is used, even a weak current may send the spot of light off the scale. To get a readable deflection, it is necessary to reduce the sensitiveness of the instrument. This is done by introducing a *shunt*, *i.e.*, a coil of wire connected up in multiple arc with the galvanometer, as shown in Fig. 5.

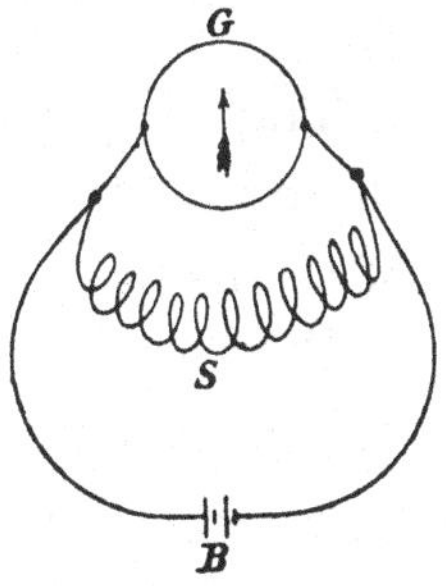

Let c_1 and c_2 be the parts of C which pass through the galvanometer, resistance G, and the shunt, resistance S, respectively :

Fig. 5.

Then

$$C = c_1 + c_2$$

And

$$c_1 : c_2 = \frac{1}{G} : \frac{1}{S}$$

Whence

$$c_1 = \frac{S}{G + S}\, C,$$

And

$$c_2 = \frac{G}{G + S}\, C$$

From these expressions, we may deduce the current C if we know c_1, for

$$C = c_1\, \frac{G + S}{S}.$$

The fraction $\dfrac{G + S}{S}$ by which we multiply c_1 in order to get C is called the *multiplying power* of the shunt.

The shunts in common use are the $\frac{1}{9}$th, $\frac{1}{99}$th, and $\frac{1}{999}$th. The $\frac{1}{9}$th shunt has a resistance equal to $\frac{1}{9}$th that of the galvanometer, so that nine parts of the current flow through the shunt, and only one through the galvanometer. If the deflection with a $\frac{1}{9}$th shunt be d divisions of the scale, then the deflection without the shunt would be 10 d. The introduction of the shunt, however, reduces the resistance of the circuit, and thereby increases the current. The resistance of the circuit may, and in some cases must, be maintained constant by the addition of a suitable resistance, called the *compensating* resistance.

Suppose we have in circuit a galvanometer of 3000 ohms, a battery whose resistance is negligible, and a box with 7000 ohms unplugged. The total resistance is thus 10,000 ohms. Now let a $\frac{1}{99}$th shunt be introduced. The joint conductivity of galvanometer and shunt is $\dfrac{1}{G} + \dfrac{1}{S} = \dfrac{G+S}{GS}$

$$\therefore \text{ their joint resistance} = \frac{GS}{G+S} = \frac{G \times \dfrac{G}{99}}{G + \dfrac{G}{99}} = \frac{G}{100} = \frac{3000}{100} = 30 \text{ ohms.}$$

The compensating resistance will thus be $3000 - 30 = 2970$ ohms; and, to maintain the same current with as without the shunt, we must unplug 9970 ohms in the box.

When using a delicate galvanometer, it is advisable, unless working with very weak currents, to begin with the $\frac{1}{999}$th shunt, and use successively the $\frac{1}{99}$th and the $\frac{1}{9}$th shunt. If required, the sensitiveness of the galvanometer may be still further increased by throwing out the shunt altogether.

It is of primary importance that the shunts be accurately adjusted. Even when this condition is fulfilled, some electricians object to their use, on the ground that variations of temperature may affect them and the galvanometer unequally, and thus alter their multiplying power. When currents are to be *measured*, shunts can hardly be relied upon.

It is often required to prepare a shunt of a certain multiplying power. Thus :

Let the resistance of a galvanometer be 3000 ohms, it is required to find the resistance of the shunt which will reduce its sensitiveness a hundredfold.

As the sensitiveness is to be reduced 100 times, only the $\frac{1}{100}$th part of the current must pass through the galvanometer, the remaining $\frac{99}{100}$ths passing through the shunt. Since the currents are inversely as the resis-

tances, the resistance of the shunt must be $\frac{1}{99}$th of that of the galvanometer.

$$i.e.,\ \text{S} = \frac{3000}{99} = 30\frac{10}{33}\ \text{ohms.}$$

This could also be got from $\frac{\text{G}+\text{S}}{\text{S}}$. Since we know that the multiplying power is 100,

$$\therefore\ \frac{\text{G}+\text{S}}{\text{S}} = 100\ \therefore\ \text{S} = \frac{\text{G}}{99}$$

In general, let the shunt required be the $\frac{1}{n}$th. Then the multiplying power is n,

$$\therefore\ \frac{\text{G}+\text{S}}{\text{S}} = n$$

Whence S, the resistance of the shunt, $= \dfrac{\text{G}}{n-1}$.

4. *Capacity.*—The capacity of a condenser, in electrostatic measure, is defined as the quantity of electricity necessary to raise the potential of one coating from zero to unity, the other being put to earth. If a charge Q be required to raise the potential from V_1 to V_2, then $C = \dfrac{Q}{V_2 - V_1}$.

Kohlrausch experimentally established the very important fact that the ratio of the instantaneous discharge Q of a condenser to its potential V is constant; and it is this ratio $\dfrac{Q}{V}$, a constant measurable quantity, that is termed the capacity of the condenser.

From a series of experiments, Boltzmann concludes that condensers of ebonite, paraffin, and sulphur take their full charge instantaneously, whilst those in which glass and gutta-percha are used require time for their maximum charge.

In the condensers commonly used, paraffined paper is the dielectric; the capacities of such condensers may be compared by charging and discharging them through a suitable (ballistic) galvanometer as rapidly as convenient.

The same holds for condensers in which plates of mica are used to insulate the conducting sheets of tinfoil.

The capacity in electrostatic measure may in certain cases be calculated from the dimensions of the condenser or conductor.

a. To find the capacity of an air-condenser consisting of two equal parallel plates at potentials V_1 and V_2, the distance between them being t.

In the mathematical theory of electricity, it is shown that the force just outside an electrified conductor is $4\pi\rho$, where ρ is the surface density on the conductor. This is also the rate of change of potential along t.

$$\therefore \frac{V_1 - V_2}{t} = 4\pi\rho \quad \text{and} \quad \rho = \frac{V_1 - V_2}{4\pi t}$$

If S denote the area of either plate, the charge Q is given by

$$\rho S = Q \quad \therefore Q = \frac{V_1 - V_2}{4\pi t} S$$

But

$$\text{Capacity} = \frac{Q}{V} \quad \therefore C = \frac{S}{4\pi t}$$

For condenser with a dielectric whose specific inductive capacity is k,

$$C = \frac{k S}{4\pi t} = \frac{S}{4\pi \frac{t}{k}}$$

The fraction $\frac{t}{k}$ is called *the reduced thickness*, i.e., the thickness of the dielectric reduced to its equivalent of air.

b. A sphere. The capacity of a sphere is numerically equal to its radius R. For the potential is constant throughout the sphere, and, therefore, the same as at the centre. The potential being $\Sigma\left(\frac{q}{r}\right)$, its value at the centre is $\frac{Q}{R}$.

$$\therefore \text{Capacity } C = \frac{Q}{\frac{Q}{R}} = R.$$

In the case of an electrified *shell*, the potential is also single-valued throughout the mass of the shell and the enclosed space. It is, therefore, the same as at the centre, viz., $\frac{Q}{R}$ where R is the radius of the outer surface.

Therefore, also,

$$C = \frac{Q}{\frac{Q}{R}} = R.$$

c. Two concentric spheres. Let r_1, Fig. 6, be the radius and Q the charge of the inner sphere. An equal charge of opposite sign, $-Q$, will be induced on the inside, and a charge $+Q$ on the outside of the spherical shell, whose internal radius is r_2 and external r_3. The potential at the centre will then be

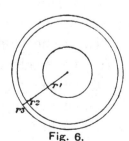

Fig. 6.

$$V = \frac{Q}{r_1} - \frac{Q}{r_2} + \frac{Q}{r_3} = Q\left(\frac{1}{r_1} - \frac{1}{r_2} + \frac{1}{r_3}\right)$$

If we put the outside to earth, then the charge $+Q$ is removed, and

$$V = Q\left(\frac{1}{r_1} - \frac{1}{r_2}\right)$$

$$\therefore C = \frac{Q}{Q\left(\frac{1}{r_1} - \frac{1}{r_2}\right)} = \frac{1}{\frac{1}{r_1} - \frac{1}{r_2}} = \frac{r_1 r_2}{r_2 - r_1}$$

And for any other dielectric than air

$$C = k\,\frac{r_1 r_2}{r_2 - r_1}.$$

Kohlrausch used an isolated sphere as a standard of comparison in his determination of the ratio of the electrostatic to the electromagnetic units. Gibson and Barclay used a condenser consisting of two concentric spheres in their researches on the specific inductive capacity of paraffin.

d. An important case is that of a condenser consisting of two circular concentric cylinders, one of which is put to earth. The quantity of electricity on an element of a narrow longitudinal strip, Fig. 7, is $\rho\, dy\, ds$, and the force F exerted on unit electricity at A, situated on the axis of the cylinder, is $-\frac{\rho\, ds\, dy}{r^2}$. The component along AO is

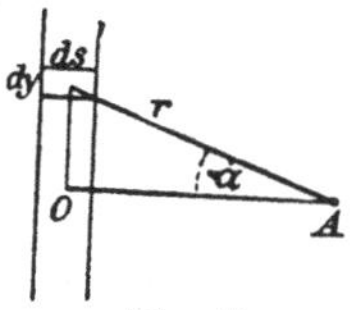

Fig. 7.

$$-\frac{\rho\, dy\, ds}{r^2}\cos a. \quad \text{Now } r\cos a = \text{radius of cylinder} = R$$

And

$$\frac{y}{R} = \tan a \quad \therefore \frac{dy}{da} = R\sec^2 a$$

$$\therefore F = -\frac{\rho\, ds.\, R\sec^2 a\, da}{R^2 \sec^2 a}\cos a$$

$$= -\frac{\rho\, ds}{R}\cos a\, da$$

For whole strip of indefinite length the limits of integration will be $+\frac{\pi}{2}$ and $-\frac{\pi}{2}$. Hence

$$F = -\frac{\rho\, ds}{R}\int_{-\frac{\pi}{2}}^{+\frac{\pi}{2}}\cos a \,.\, da = \frac{2\,\rho\, ds}{R}.$$

This will also be the expression for the force F when the length L of the cylinder is so great with respect to its radius that we may neglect the action of distant parts.

c

But this force, with its sign changed, is the differential of the potential

$$\therefore \frac{d\,V}{d\,R} = -\frac{2\,\rho\,ds}{R}$$

$$\therefore V = -2\,\rho\,ds\int\frac{d\,R}{R}$$

As there is an equal negative charge on the outer cylinder, radius R_2, then R_1 denoting the radius of the inner, the integral gives

$$V = 2\,\rho\,ds\left(\log._\epsilon\frac{1}{R_1} - \log._\epsilon\frac{1}{R_2}\right)$$

$$= 2\,\rho\,ds\,\log._\epsilon\frac{R_2}{R_1}$$

But the charge q on the strip is $L\,.\,ds\,.\,\rho$.
Whence

$$V = \frac{2\,q}{L}\,\log._\epsilon\frac{R_2}{R_1}$$

And for the whole charge Q

$$V = \frac{2\,Q}{L}\,\log._\epsilon\frac{R_2}{R_1}$$

$$\therefore C = \frac{Q}{V} = \frac{Q}{\dfrac{2\,Q}{L}\,\log._\epsilon\dfrac{R_2}{R_1}} = \frac{L}{2\,\log._\epsilon\dfrac{R_2}{R_1}}$$

which is the usual expression for the capacity of a cylindrical air condenser whose length is great compared with its radius.

For any other dielectric

$$C = \frac{k\,L}{2\,\log._\epsilon\dfrac{R_2}{R_1}}$$

This formula is used in telegraphy for calculating the capacity and charge of submarine cables.

The formula may be further simplified. Let $t =$ thickness of dielectric.
Then

$$R_2 = R_1 + t$$

And

$$\log._\epsilon\frac{R_2}{R_1} = \log._\epsilon\frac{R_1 + t}{R_1} = \log._\epsilon\left(1 + \frac{t}{R_1}\right) = \frac{t}{R_1} - \frac{1}{2}\left(\frac{t}{R_1}\right)^2 = \frac{t}{R_1}\left(1 - \frac{1}{2}\frac{t}{R_1}\right)$$

$$\therefore C = \frac{L}{2\,\dfrac{t}{R_1}\left(1 - \frac{1}{2}\dfrac{t}{R_1}\right)} = \frac{L\,R_1}{2\,t\left(1 - \frac{1}{2}\dfrac{t}{R_1}\right)}$$

Multiplying by $1 + \frac{1}{2}\dfrac{t}{R_1}$ we get

$$C = \frac{L\,R_1}{2\,t}\left(1 + \frac{1}{2}\frac{t}{R_1}\right)$$

But surface $S = 2\,\pi\,R_1\,L$

$$\therefore \; C = \frac{S}{4\,\pi\,t}\left(1 + \frac{t}{2\,R_1}\right) = \frac{S}{4\,\pi\,t} \text{ when } t \text{ is small compared with } R_1.$$

A Leyden jar fulfils approximately the above conditions. Hence its capacity may be calculated from $C = \dfrac{S\,k}{4\,\pi\,t}$, where k is the specific inductive capacity of the glass used.

When several condensers are simultaneously charged, their similar coatings are usually connected together. Their joint capacity is then plainly the sum of their separate capacities, *i.e.*, $C = c_1 + c_2 + c_3$.

Sometimes the *cascade* or *series* arrangement is adopted. Let there be three jars connected up in cascade, Fig. 8. Then, the last being to earth E,

$$Q = c_1\,(V_1 - V_2) \; \therefore \; \frac{Q}{c_1} = V_1 - V_2$$

$$= c_2\,(V_2 - V_3) \; \therefore \; \frac{Q}{c_2} = V_2 - V_3$$

$$= c_3\,(V_3 - 0) \; \therefore \; \frac{Q}{c_3} = V_3 - 0$$

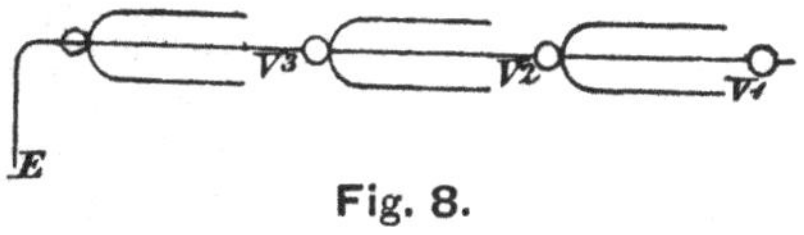

Fig. 8.

$$\therefore \text{ By adding } Q\left(\frac{1}{c_1} + \frac{1}{c_2} + \frac{1}{c_3}\right) = V_1$$

Now Capacity $(C) = \dfrac{\text{Charge } (Q)}{\text{Potential } (V)} \; \therefore \; C = \dfrac{Q}{Q\left(\dfrac{1}{c_1} + \dfrac{1}{c_2} + \dfrac{1}{c_3}\right)} = \dfrac{1}{\dfrac{1}{c_1} + \dfrac{1}{c_2} + \dfrac{1}{c_3}}$

Whence

$$\frac{1}{c} = \frac{1}{c_1} + \frac{1}{c_2} + \frac{1}{c_3}$$

This gives C.

From the formula $C = \dfrac{Q}{V}$ we may deduce the value in electromagnetic measure of the practical unit of capacity. Thus

$$\text{Unit Capacity} = \frac{\text{Unit of Quantity}}{\text{Unit of Potential}} = \frac{\text{Coulomb}}{\text{Volt}} = \frac{10^{-1}}{10^{8}} = 10^{-9} \text{ C.G.S. units.}$$

To this unit the name Farad has been given, in honour of the great electrician Michael Faraday.

The farad is much too large for ordinary purposes. Its millionth part, the Microfarad (m.f.) is the practical unit. It equals $\dfrac{10^{-9}}{1{,}000{,}000} = 10^{-15}$ C.G.S. units of capacity.

Condensers of $\frac{1}{3}$ m.f. capacity are frequently used: Fig. 9 shows such a condenser. The brass terminals are connected respectively with the two coatings of the condenser. When the plug is inserted the condenser is discharged.

Subdivided condensers are now made by Elliott

Fig. 9.

Brothers, the parts being arranged like the coils of a resistance box, so that

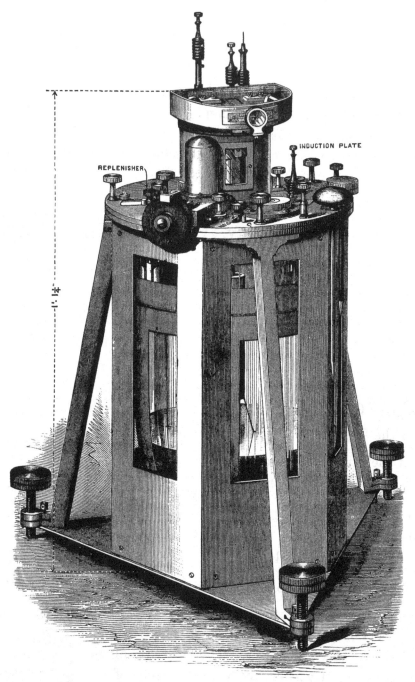

Fig. 10.   Sir William Thomson's Quadrant Electrometer.

by inserting plugs capacities may be had varying from a thousandth to a whole microfarad.

2. PRINCIPAL INSTRUMENTS.—The quadrant electrometer, Figs. 10 and 11,*

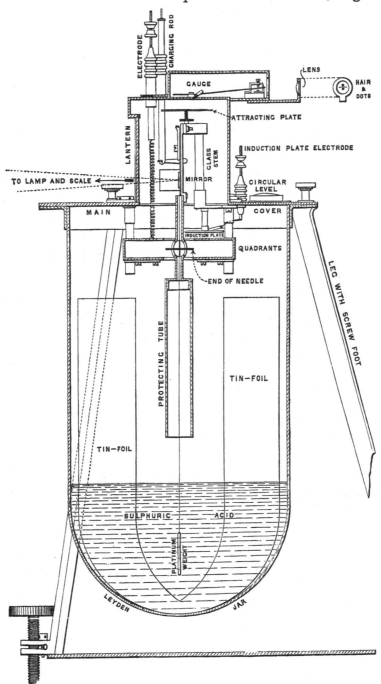

Fig. II. Sir William Thomson's Quadrant Electrometer.

of Sir William Thomson, is a most useful and delicate instrument for measuring and comparing differences of potential. Its essential parts are :

* Reproduced by permission from Gordon's "Electricity and Magnetism."

(1) A Leyden jar.
(2) Two pairs of quadrants.
(3) A light movable needle.
(4) An idiostatic gauge.
(5) A replenisher.

The Leyden jar is of peculiar construction, the outer coating consisting of a number of broad strips of tinfoil with intervening spaces. The inner coating is strong sulphuric acid, which, by its hygrometric property, keeps the air within the jar dry and thereby the interior well insulated.

The four brass quadrants are arranged around a common centre, inclosing a cylindrical box-like space, within which the needle is suspended. By this means the needle is kept in a truly constant electrical field, and is protected against external influence. The angular deviation of the needle will then be constantly proportional to the twisting couple due to the electric forces, as it is also directly proportional to the couple arising from the bifilar suspension of the needle. It is necessary that the quadrants be placed symmetrically with respect to the needle. The distance between them may be varied, and accurately adjusted by means of radial slots and a micrometer screw fixed on the cover of the instrument, Fig. 10.

The quadrants diametrically opposite each other are connected by a fine wire; thus A and C, Fig. 12, are in electrical communication by the wire W, B and D by W[1]. The quadrants are connected respectively to the two electrodes, Fig. 10.

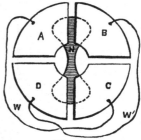

Fig. 12.

The flat dumb-bell shaped needle of aluminium foil is supported by a bifilar suspension of unspun silk. Connected rigidly with it is a small slightly concave mirror of silvered glass, the angular displacement of the movable system being indicated by the movement of a spot of light over a divided scale. The needle is kept at a constant high potential by means of a platinum wire with which it is connected, and which dips into the sulphuric acid forming the inner coating of the Leyden jar.

All the movable parts are guarded against external influences; the mirror by a sort of hood, and the platinum wire by a metal "guard tube."

When the electrodes are connected with conductors at the same potential, the bifilar suspension is so adjusted that the median line of the needle lies evenly between the adjacent pairs of quadrants.

When there is a difference of potential the needle will be deflected through an angle proportional to this difference.

The mathematical expression for the force on the needle may be found as follows :

Let B, Fig. 13, be a plate of area S symmetrically situated with respect to the fixed conductors A and C, and movable about a vertical axis ; then ρ being the electrical density on B, the force exerted by A on an element of charge of B is $2 \pi \rho$, and the force F_1, on the whole charge ρ S, is $2 \pi \rho \cdot \rho S = 2 \pi \rho^2 S$.

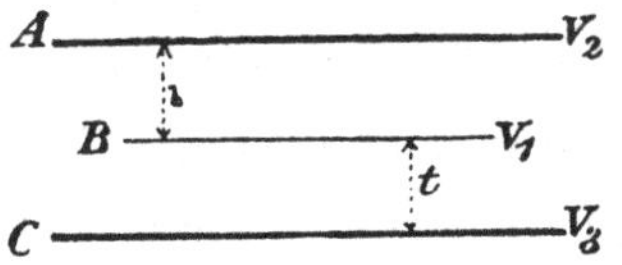

Fig. 13.

But the force anywhere in the space between A and B is $4 \pi \rho$, and as electrical force is the rate of change of potential we have

$$4 \pi \rho = \frac{V_1 - V_2}{t}, \text{ and } \rho = \frac{V_1 - V_2}{4 \pi t}$$

$$\therefore F_1 = 2 \pi S \frac{(V_1 - V_2)^2}{16 \pi^2 t^2} = \frac{S (V_1 - V_2)^2}{8 \pi t^2}$$

Similarly the force F_2 exerted by C on the movable plate B is

$$\frac{S (V_1 - V_3)^2}{8 \pi t^2}$$

$$\therefore \text{ the resultant force } F = F_1 - F_2 = \frac{S}{8 \pi t^2} \left\{ (V_1 - V_2)^2 - (V_1 - V_3)^2 \right\}$$

Observing that the expression within brackets is the difference of two squares, we have, after reducing,

$$F = \frac{S}{4 \pi t^2} \left(V_1 - \frac{V_2 + V_3}{2} \right) (V_3 - V_2)$$

Which may be written

$$F = a \left(V_1 - \frac{V_2 + V_3}{2} \right) (V_3 - V_2)$$

This expression is independent of the form of the symmetrical conductors A and C, and is therefore applicable to the quadrant electrometer.

If the potential V_1 of the needle be very high with respect to that of the quadrants V_2 and V_3, as is usually the case, we have

$$F = a V_1 (V_3 - V_2)$$

which shows that the deflecting couple varies directly as the potential of the needle. This suggests a means of increasing the sensitiveness of the instrument, viz., by raising the potential of the needle. If this be kept constant we may write

$$F = \beta (V_3 - V_2)$$

But the deflecting force F is proportional to the angular deviation θ. Hence we ultimately have

$$\theta = \beta \, (V_3 - V_2)$$

The constant β is the factor by which the readings must be multiplied in order to convert them into absolute measure. It may be determined, once for all, by comparison with an absolute electrometer.

Another method for reducing the electrometer readings to absolute or to practical units, is to compare the deflection with that given by a standard cell. Thus, it being required to find the E.M.F. of a certain electro-chemical apparatus which gave a deflection of 114 scale divisions, a Latimer Clark's cell was connected with the electrodes of the electrometer, and gave a deflection of 93 divisions. Assuming the E.M.F. of the cell to be 1.457 volts, we have the simple proportion $93 : 114 = 1.457 : x$.

Whence

$$x = 1.785 \text{ volts} = \text{the E.M.F. of the apparatus.}$$

Projecting from the main cover of the instrument, to the right of the lantern, Fig. 11, is an electrode, which is in connection with a small movable plate called the *induction plate*. By means of this appendage we may reduce the sensitiveness of the electrometer. This is especially useful when working with high potentials : for instance, suppose we are measuring the insulation resistance of a condenser, and that when we put the terminals of the charged condenser in connection with the electrodes, we get a deflection which sends the spot of light off the scale. To obtain a readable deflection, we must reduce the sensitiveness of the instrument ; and we may do this by connecting one terminal of the condenser with an electrode of the electrometer, and the other terminal with the induction plate. In this case, the charge of the underlying quadrant will be entirely induced and necessarily less than before and the deflection will be reduced proportionally. There are other ways of using the induction plate for the same purpose.

An important accessory to the quadrant electrometer is the *idiostatic gauge*, Fig. 11, the object of which is to keep the jar at a constant potential, an indispensable condition for a series of measurements to be comparable *inter se*.

The gauge is simply an attracted-disc or "trap-door" electrometer. It consists of two circular metal plates, the lower of which, the "suspension plate," is electrically connected with the needle and inside coating of the Leyden jar. Parallel to this is the "guard plate," containing a central rectangular aperture ; the square end of a spade-like appendage of sheet

aluminium, Fig. 14, fits into this aperture, and
is freely movable about a horizontal axis; the
other end of this lever terminates in two
prongs, across which is stretched a fine hair.
The horizontal axis consists of a tightly
stretched platinum wire A B, which, by its
torsion, keeps the "trap" (in the uncharged
condition of the electrometer) above the surface of the guard plate and, consequently, the
sighting hair below the dots. These are made
on a piece of white enamel, and are viewed
through a plano-convex lens. By keeping the
line of sight normal to the centre of the lens,

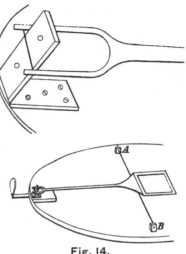

Fig. 14.

errors due to parallax are avoided. When the instrument is charged,
the "trap" is acted upon inductively by the underlying plate, and is
attracted downwards. The force of attraction (F) is given by $F =$
$\dfrac{S\,(V_1 - V_2)^2}{8\,\pi\,t^2}$, which, since the guard plate is put to earth, becomes $\dfrac{S\,V_1^2}{8\,\pi\,t^2}$,
where S is the area of the "trap" and V_1 the potential of the suspension
plate. It is, however, unnecessary to estimate this force.

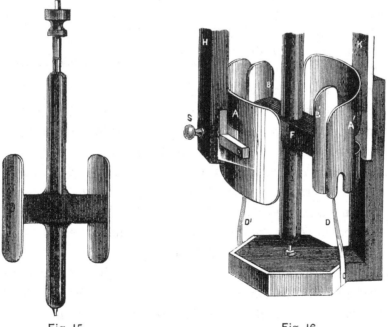

Fig. 15.   Fig. 16.

Owing to this depression of one end of the lever, the index rises;

the potential of the jar is gradually increased by means of the replenisher until the hair lies exactly between the two dots. The aluminium square is then accurately in the plane of the guard plate, and the normal and easily reproduced charge is attained.

The *Replenisher* is virtually a small electrical induction machine. It consists of a vulcanite stem, Fig. 15, which can be spun round a vertical axis by twirling a small milled head. Attached to this stem is a vulcanite cross piece, furnished at its extremities with two brass plates called "carriers."

The inductor A, Fig. 16, is fixed to an insulating support H, and is connected with the acid of the jar, while A^1 is in electrical connection with the cover of the electrometer, and is therefore to earth. By means of the charging electrode, a slight initial positive charge is given to the acid.

At one part of their revolution the carriers just graze two springs C and C^1, Fig. 17, which are attached to, and are in metallic connection with, the inductors ; at another part of their revolution they touch two other springs, D and D^1, which are connected together, but otherwise insulated. The forward edges of the carriers touch the springs D and D^1 while yet they are under cover of the inductors, and, therefore, well within their influence. The positive electricity of A induces a nega-

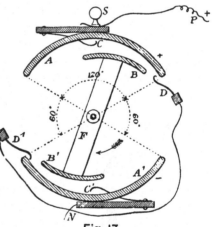

Fig. 17.

tive charge on B, and repels positive electricity to B^1 through the connecting springs D and D^1. The next instant they are insulated from each other, B retaining its negative and B^1 its positive charge.

Continuing their revolution, B^1 meets the spring C giving up its positive charge to the inductor, whilst B delivers its negative charge to A^1 through the spring C^1. Passing on to D, the carrier B^1 comes under the inductive influence of A, which is now at a higher potential than before, and the same cycle of mutual actions is repeated.

In this way the carrier, emerging from the inductor connected with the inner coating of the Leyden jar, carries a negative charge round to the receiving spring connected with the outside coating; while the other carrier, emerging from the inductor connected with the outside coating, carries a positive charge round to the receiving spring in connection with the inside

coating. As the charge induced in the carriers at every half-revolution is a constant fraction of the charge then on the inductors (which is itself increased at every half-turn by that same amount), it follows that a very small original electrification may be developed, by this sort of compound-interest reaction, into a very strong charge.

The initial charge may be given by a small electrophorus or a Leyden jar; and if the glass of the electrometer be carefully chosen, this initial charge may be retained for an indefinite period, the slightest difference of potential between the two inductors sufficing for gradually building up a high electrostatic charge.

It may be further observed that the replenisher is perfectly reversible, *i.e.*, if the motion of the spindle be reversed, the potential of the jar will be lowered at each half-revolution, at exactly the same rate at which it was increased by the opposite motion.

The increments or decrements being very small, it is easy to regulate

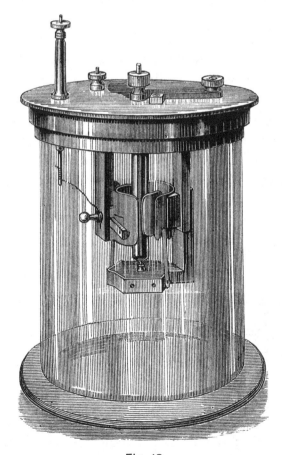

Fig. 18.

the charge of the needle so that the idiostatic gauge shall indicate precisely the normal potential of the electrometer.

The replenisher is sometimes made as a separate instrument, Fig. 18.

In damp weather, or when a high state of insulation is required, some strong sulphuric acid should be placed at the bottom of the jar.

The Tangent Galvanometer.—The tangent galvanometer is used for comparing currents; when its "constant" is known, its readings can be converted into absolute units.

This instrument is sometimes made as a single copper ring, at the centre of which the needle is suspended. The objection to this form is that, except at and near the centre, the field of force is not uniform, and that those lines of force to which the deflection is mainly due are not normal to the plane of the ring—an assumption which, however, is made in deducing the equation of equilibrium of the needle. For a relatively *short* needle, however, the lines of force are sensibly normal to the ring.

The French physicist Gaugain sought to improve this form by suspending the magnet excentrically on the axis of the coil, at a distance from its centre equal to half the radius. Besides the want of uniformity in the field near the point of suspension, a grave objection to this instrument was the difficulty of making the linear measurements required for determining the point of suspension. Helmholtz removed these difficulties, and converted the arrangement into a practical instrument by adding a second coil, similar to the first and symmetrically situated with respect to the magnet, Fig. 19; the distance between the mean planes of the coils is, therefore, equal to their radius. The lines of force due to two such circular coils are,

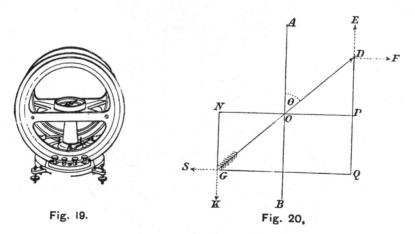

Fig. 19. Fig. 20.

in the neighbourhood of the magnet, parallel and equidistant straight lines,

so that the needle will be kept throughout the deflection in a truly uniform field—a condition essential to perfect accuracy.*

Let the needle (length l) of a single ring galvanometer be deflected through an angle θ, Fig. 20, by a current C. Then, if m denote the strength of the magnet-pole, the intensity of the earth's directive force will be H m, H being the horizontal component of the earth's magnetic force), and will act along D E and G K parallel to the magnetic meridian.

The magnetic action of the current will, if r be the radius of the ring, be $\dfrac{2\,r\,\pi\,\mathrm{C}\,m}{r^2}$, and will act along D F and G S, perpendicular to the meridian.

The arm of the couple due to the earth's force is G Q $= l \sin \theta$.

∴ the moment of this couple is H m. $l \sin \theta$.

The arm of the couple due to the current is D Q $= l \cos \theta$.

∴ the moment of this couple is $\dfrac{2\,\pi\,r\,\mathrm{C}\,m\,l\cos\theta}{r^2}$

The needle being in equilibrium under the action of these two couples, we have

$$\frac{2\,r\,\pi\,\mathrm{C}\,m}{r^2}\,l\cos\theta \;=\; \mathrm{H}\,m\,l\sin\theta$$

Whence

$$\mathrm{C} \;=\; \frac{r\,\mathrm{H}}{2\,\pi}\,\tan\theta$$

The strength of the current in absolute units may be found in a similar manner in terms of measurable quantities, in the case of the double coil galvanometer. As this is often required, we shall give several methods for determining the constant of the instrument.

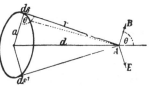

Fig. 21.

Let ϕ be the angle between the tangent to the current at an element ds and the line joining the element to the magnet pole. The effect of each element is shown experimentally to vary as sin ϕ.

∴ the action of an element $= \dfrac{ds \,.\, \mathrm{C} \,.\, m}{r^2} . \sin\phi = \dfrac{ds \,.\, \mathrm{C} \,.\, m}{r^2}$ for A being an axis, $\phi = 90°$

This force acts along A B, that is, at right angles to the plane containing the pole and element ds.

The symmetrical element ds_1 would give an equal force acting along A E.

The components of these forces along the axis are equal and similar; those perpendicular to the axis are equal and opposite. Hence the force effectual in deflecting the needle is $2\,\dfrac{ds \,.\, \mathrm{c} \,.\, m}{r^2}\cos\theta$. For one complete

---

* These lines are given in Maxwell, Vol. II., Plate XIX.

turn of the wire, we must replace $2\,ds$ by $2\,a\,\pi$, and if there be n turns in all on both coils, the expression for the deflecting force becomes

$$\mathbf{F} = \frac{\mathrm{C}\,.\,m}{r^2}\,.\,2\,a\,\pi\,n\,.\,\cos\theta\ =\frac{\mathrm{C}\,m}{r^2}\,.\,2\,a\,\pi\,n\,.\,\frac{a}{r},\ \left(\text{for}\ \frac{a}{r}\ =\ \cos\theta\right)$$

∴ the moment of the couple due to the current producing a deflection δ.

$$= \frac{2\,a^2\,\pi\,.\,\mathrm{C}\,.\,n}{r^3}\,l\,m\ \cos\delta$$

The moment of the couple due to the earth's force is $\mathrm{H}\,.\,m\,l\,.\,\sin\delta$

$$\therefore\ \frac{2\,a^2\,\pi\,\mathrm{C}\,n}{r^3}\,l\,m\ \cos\delta\ =\ \mathrm{H}\,.\,m\,l\,.\,\sin\delta$$

$$\therefore\ \mathrm{C}\ =\ \frac{\mathrm{H}\,r^3}{2\,a^2\,\pi\,n}\ \tan\delta$$

Now $r = \sqrt{a^2 + d^2}$

$$\therefore\ \mathrm{C} = \frac{\mathrm{H}\,(a^2 + d^2)^{\frac{3}{2}}}{2\,a^2\,\pi\,n}\ \tan\delta\ =\ \frac{\mathrm{H}\ \tan\delta}{\dfrac{2\,a^2\,\pi\,n}{(a^2 + d^2)^{\frac{3}{2}}}}$$

The horizontal component H of the earth's force is variable with respect to place and time. For very accurate experiments it would be necessary to find its value for the room in which the tangent galvanometer is used. The method will be given further on.

The principal constant*, usually called *the constant*, of the galvanometer is $\dfrac{2\,a^2\,\pi\,n}{(a^2 + d^2)^{\frac{3}{2}}}$, and this may be calculated if we can measure with precision a and d (in centimetres), and ascertain the number (n) of windings.

It should be noticed here that the quantity which is called the constant of the galvanometer, viz., $\dfrac{2\,a^2\,\pi\,n}{(a^2 + d^2)^{\frac{3}{2}}}$, and which is usually denoted by G or by the Greek letter Γ, is also the strength of the magnetic field due to unit current at the point (A) on the common axis which is equally distant from the plane of the coils, and at which the needle is suspended.

Where practicable, it is preferable to find the constant by comparison with a standard pair of coils, large and carefully constructed so that all the linear measurements may be made with great accuracy.

* The electromagnetic constants of a coil are

    (1) The magnetic force at the centre due to unit current,

    (2) The magnetic moment of the coil due to unit current,

    (3) The co-efficient of self-induction,

        And, in case of a pair of coils,

    (4) The co-efficient of mutual induction.

Fig. 22 represents the tangent galvanometer placed concentrically with the standard pair of coils. The planes of the coils must be placed in the magnetic meridian, the instruments carefully levelled, and the connections made, so that the currents through the standard coils and through the galvanometer may oppose each other in their effects upon the needle. By means of the resistances R and r, these currents

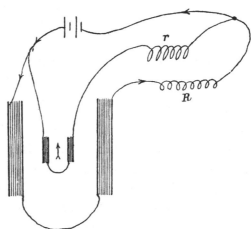

Fig. 22.

are varied until the one exactly neutralises the other. The needle then remains steadily at zero when the circuit is closed.

Let C_1 be the current flowing through the standard coils, whose constant is G_1. Then

$$C_1 = \frac{H \cdot \tan \delta}{G_1} \quad \therefore \ C_1 \, G_1 = H \tan \delta$$

As the galvanometer tends to produce an equal and opposite deflection, we have, C denoting the current through the galvanometer and G its constant,

$$C \, G = H \tan \delta \quad \therefore \ C \, G = C_1 \, G_1 \text{ and } G = G_1 \frac{C_1}{C}$$

Also, by Ohm's law,

$$\frac{C_1}{C} = \frac{\rho + r}{D + R},$$ where ρ is the resistance of the galvanometer, and D that of the coils.

$$\therefore \ G = \frac{\rho + r}{D + R} \, G_1,$$ which gives the required constant.

There is yet a third method for converting the readings of a tangent galvanometer into absolute measure. It gives good results and is easy of application. It is derived from the laws of electrolysis.

The electro-chemical equivalent ϵ of a substance is the mass in grammes liberated at the negative electrode by the passage of unit current for unit time.

The electro-chemical equivalent of hydrogen is .000105

,,	,,	copper	.003307
,,	,,	zinc	.003412

As an ampère is $\frac{1}{10}$th the absolute unit of current, the mass in grammes deposited, per second, by an ampère would be :

For hydrogen	.	.	.0000105
„ copper	.	.	.0003307
„ zinc	.	.	.0003412

A current C in time T would liberate a mass W, given by

$$W = \epsilon \, C \, T$$

The tangent galvanometer, the vessel containing the electrolyte (say, copper sulphate), and a key are connected up in series with the battery.

The preliminary operation consists in levelling the galvanometer, placing the plane of its coils in the meridian, and weighing with a delicate balance either both electrodes or only the negative. The galvanometer should be read every minute or half-minute for about 20 minutes. If proper means have been taken, the readings will not vary much.

Then if W_1 be the weight of the negative electrode at the beginning, and W_2 at the end, the mass deposited by C ampères in T seconds is $W_2 - W_1$.

$$\therefore W_2 - W_1 = \epsilon \, C \, T$$

This will give C in ampères. Now $C = k \tan \phi$, whence the reduction-factor k is known.

Ex. Let W_1 be 6.2813 gm., W_2 be 6.9175 gm., time T 14 minutes, and the mean reading of the galvanometer 35° 50′. The ϵ of copper being 0003307, we have :

$$6.9175 - 6.2813 = .6362 = .0003307 \times C \times 14 \times 60$$

Whence

$$C = 2.29 \text{ ampères}$$

But

$$C = k \tan \phi = k \tan 35° 50′ \quad \therefore 2.29 = k \times .722 \quad \therefore k = 3.1717$$

If we want the current in absolute units, we must multiply the tangent of the deflection by $\dfrac{3.1717}{10}$, *i.e.*, by .31717.

It is plain that the value of k thus found is constant only so far as the numerical value of the horizontal component H of the earth's magnetic force remains constant, an assumption that must not be made when accurate results are required. It would be necessary in such cases to redetermine k at intervals, and allow for any changes that may take place.

In using a tangent galvanometer, it is advisable, where possible, to arrange the battery and adjust the resistances so as to get a deflection as near 45° as possible—that being the *angle of maximum sensitiveness.*

For

$$C = \frac{H}{G} \tan \phi \quad \therefore \tan \phi = C \frac{G}{H} \qquad . \qquad . \qquad . \qquad . \qquad . \quad (1)$$

Since the sensitiveness is measured by the increase of the deflection due to an increment of current, we must differentiate the above equation with respect to ϕ and C.

$$\therefore \; \sec^2 \phi \; \frac{d\phi}{dC} = \frac{G}{H} \qquad . \qquad . \qquad . \qquad . \qquad . \qquad . \quad (2)$$

By inverting (1) we have $\cot \phi = \frac{1}{C} \frac{H}{G} \qquad . \qquad . \qquad . \qquad . \qquad . \quad (3)$

$$\therefore \text{ by multiplying (2) and (3) } \; \sec^2 \phi \; . \; \cot \phi \frac{d\phi}{dC} = \frac{1}{C}$$

Whence

$$\frac{d\phi}{dC} = \frac{1}{C \sec^2 \phi \cot \phi} = \frac{\cos^2 \phi \; . \; \sin \phi}{C \; . \; \cos \phi} = \frac{2 \cos \phi \sin \phi}{2 \, C} = \frac{\sin 2\phi}{2 \, C}$$

This is plainly a maximum when $\phi = 45°$.

To find the *horizontal component* H we determine—

(1) The time of vibration T of a magnet swinging under the action of the earth's magnetic force ; and

(2) The angle through which this magnet, placed at a known distance, deflects a freely suspended magnet or needle from the meridian.

A swinging magnet is like a pendulum, and obeys similar laws. Its time of vibration is given by

$$T = 2\pi \sqrt{\frac{K}{M H (1 + \theta)}}$$

Where K is the moment of inertia of the magnet, M its magnetic moment, and θ the ratio of torsion of the suspending fibre.

The magnet is usually a cylinder or a rectangular bar. If l be the length of the *cylinder* in centimetres, r its radius, and W its weight in grammes, then its moment of inertia about an axis perpendicular to the axis of the cylinder at its middle point is $K = W \left(\frac{l^2}{12} + \frac{r^2}{4} \right)$. If l be the length and b the breadth of the *bar*, its moment of inertia about an axis passing through its centre of gravity and perpendicular to the plane containing l and b, is $K = W \left(\frac{l^2 + b^2}{12} \right)$.

The first operation therefore consists in weighing the magnet and measuring its dimensions with great care.

The magnet is next suspended from a torsion-head by one or more fibres of unspun silk, and is protected from currents of air by a glass case. A lamp and scale are adjusted so that a sharp image of a fine wire stretched over the aperture, is reflected from the magnet-mirror and is visible on the scale. The torsion circle is then turned through an angle a alternately to

D

the right and the left, the deflections of the magnet being noted and their mean taken.　But the angular displacement of the spot of light is twice that of the mirror; and if n be the number of scale divisions over which the light has moved, and r the distance from the scale to the mirror, then the circular measure $\left(\dfrac{\text{arc}}{\text{radius}}\right)$ of the angle through which the magnet has turned is $\dfrac{n}{2\,r}$, neglecting a small correction for the rectangular character of the scale.　But $2\,\pi$ is the circular measure of $360°$; therefore $57\frac{1}{3}°$, that is $\dfrac{360°}{2\,\pi}$, is the unit of angular measurement, and $\dfrac{n}{2\,r} \times 57\frac{1}{3}$ will be the angular measure of the displacement of the magnet.　Call this ϕ.　If τ denote the force of torsion, H the horizontal component of the earth's force, and m the strength of a pole of the magnet, then the ratio of the torsional force to the directive force on the magnet is $\dfrac{\tau}{\text{H}\,m}$, and this is equal to $\dfrac{\phi}{a - \phi}$, which is, for shortness, called the *ratio of torsion.*

To find T, note exactly the time when the image of the wire crosses the middle of the scale; count (say) 5 complete vibrations, and note the time of the last passage.　Repeat this 5, 7, or 10 times.　After a short interval, during which the required reductions may be made, a second series may be taken, and then a third.

The reductions will be best understood from an example.　The first transit noted occurred at 3h. 48′ 20″, the fifth vibration was completed at 3h. 49′ 29″, the tenth at 3h. 50′ 38″, and the rest as in the following table :

TIME OF PASSAGE.

I.			II.			III.		
h.	m.	s.	h.	m.	s.	h.	m.	s.
3	48	20	4	6	14	4	15	10
3	49	29	4	7	22	4	16	18
3	50	38	4	8	32	4	17	27
3	51	47	4	9	40	4	18	35
3	52	55	4	10	49	4	19	45

An approximate value for T is found by dividing the time between any two transits (the best being the first and the last) by the number of vibrations.　Thus

$$\underset{\text{3 52 55}}{\overset{\text{h. m. s.}}{}} - \underset{\text{3 48 20}}{\overset{\text{h. m. s.}}{}} = 275 \text{ seconds, and } T = \frac{275}{20} = 13.75 \text{ seconds.}$$

$$\text{The second series gives T} = 13.75 \quad \text{,,}$$
$$\text{The third series gives　T} = 13.75 \quad \text{,,}$$
$$\therefore \text{ the mean value obtained in this way is } 13.75 \quad \text{,,}$$

The time at which the *middle passage* occurred is found by taking the mean of transits equally distant from the middle.

Thus

	h.	m.	s.		h.	m.	s.		h.	m.	s.
From the two extremes we get ½ (3	48	20	+	3	52	55)	=	3	50	37.5
From the second and fourth						.	.	3	50	38
And from direct observation					.	.	.	3	50	38
∴ the true time of middle vibration from I. is			.	.	.	.	3	50	37.8 (1)
Similarly II. gives			.	.	.	.	.	4	8	31.5 (2)
And III. gives		.	.	.	.	.	.	4	17	27.0 (3)

The interval between (1) and (2) is 1073.7 seconds, and this divided by 13.75 should give a *whole* number of vibrations. Performing the operation we get $\frac{1073.7}{13.75} = 78.06$. Therefore the number ($n$) of vibrations was 78. The exact time of vibration will then be $\frac{1073.7}{78} = 13.756$ seconds.

In like manner (1) and (3) give $n = 117$ and $T = 13.755$; and (2) and (3) give $n = 39$ and $T = 13.743$.

The mean of these is 13.751 seconds, which is the true time of vibration.

In this determination we may dispense with lamp and scale, using instead a small telescope provided with cross hairs. The telescope of a theodolite, a level, or a cathetometer answers very well. In this case, it is necessary to have a mark on the magnet (an arrow on a piece of paper will do), or else to reckon the arc of vibration from one turning point to the other. It is necessary that the arc of vibration be small, otherwise the periods would not be isochronous, the time increasing slightly with the amplitude.

The corrected time T is obtained from the observed time t by the formula

$$T = t - t\left(\frac{1}{4}\sin^2\frac{a}{4} + \frac{5}{64}\sin^4\frac{a}{4}\right)$$

where a is the number of degrees in the arc of vibration of the magnet.

The value of M H is now deducible from

$$T = 2\pi\sqrt{\frac{K}{M H (1 + \theta)}}$$

The object of the second experiment is to find $\frac{M}{H}$. For this purpose, the magnet whose T was previously determined is placed at a certain distance (r_1) from the centre of a second magnet or needle. The deflection of the spot of light on the scale is noted, the magnet is then reversed and the second deflection observed. By taking the mean of these two, we allow for any want of symmetry in the magnetisation of the magnet. The

deflecting magnet is now placed at an equal distance on the opposite side of the needle and two more readings are taken, in order to eliminate the effects of any unsymmetrical distribution in the needle itself. The mean of these four readings is taken as the true deflection.

The same four observations are repeated at a second distance (r_2), in order to correct for any unknown distribution in the magnet or the needle.

Let n be the mean reading, r being the distance from the mirror to the scale. Then the tangent of the angle of deflection of the magnet is approximately $= \dfrac{n}{2\,r}$.

A closer approximation is given by the first two terms of the expansion for tan θ, viz.,

$$\tan \theta = \frac{n}{2\,r}\left[1 - \left(\frac{n}{2\,r}\right)^2\right]$$

The ratio of torsion (ϑ) for the fibre from which the needle is suspended must be determined as antecedently.

Then the ratio $\dfrac{\mathrm{M}}{\mathrm{H}}$ will be found from

$$\frac{\mathrm{M}}{\mathrm{H}} = \tfrac{1}{2}\,(1 + \vartheta)\,r^3 \tan \theta$$

for let N S, Fig. 23, be the deflecting magnet, $n\,s$ the needle, m and μ the strength of the poles, $2\,l$ the length of the magnet, and r the distance from A to n.

The attraction of S and n

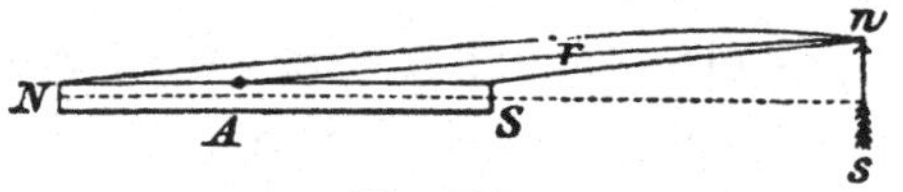

$$= \frac{m\,\mu}{(\mathrm{S}\,n)^2} = \frac{m\,\mu}{(r-l)^2} \text{ for S } n \text{ is very nearly } = r - l.$$

Fig. 23.

The repulsion between N and n is $\dfrac{m\,\mu}{(r+l)^2}$

$\therefore$ the resultant force $\mathrm{F} = \dfrac{m\,\mu}{(r-l)^2} - \dfrac{m\,\mu}{(r+l)^2} = \dfrac{4\,l\,m\,\mu}{r^3} \quad \dfrac{2\,\mu}{r^3}\,2\,l\,m = \dfrac{2\,\mu\,\mathrm{M}}{r^3}$

where M is the moment ($2\,m\,l$) of the magnet.

If θ be the angle through which the needle is deflected, we see from Fig. 24 that the arm of the couple $\dfrac{2\,\mu\,\mathrm{M}}{r^3}$ is (calling λ the length of the needle) $\lambda \cos \theta$ and its moment $2\dfrac{\mu\,\mathrm{M}}{r^3}\,\lambda \cos \theta$, whilst the arm of the couple $\mathrm{H}\mu$ is $\lambda \sin \theta$ and its moment is $\mathrm{H}\,\mu\,.\,\lambda \sin \theta$.

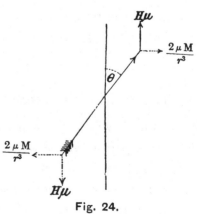

Fig. 24.

$$\therefore \frac{2\,\mu\,\mathrm{M}}{r^3}\,\lambda\cos\theta = \mathrm{H}\,\mu\,\lambda\sin\theta$$

Whence

$$\frac{\mathrm{M}}{\mathrm{H}} = \tfrac{1}{2}\,r^3\tan\theta$$

Allowing for the torsion of the fibre, we have

$$\frac{\mathrm{M}}{\mathrm{H}} = \tfrac{1}{2}\,(1+\vartheta)\,r^3\tan\theta.$$

A more accurate formula is

$$\tfrac{1}{2}\,(1+\vartheta)\,r^3\tan\theta = \mathrm{A} + \frac{\mathrm{B}}{r^2}$$

where A is the ratio $\dfrac{\mathrm{M}}{\mathrm{H}}$ and B is a constant depending upon the distribution of magnetism in the deflecting magnet.

Writing this formula for the two distances r_1 and r_2, and eliminating B, we have

$$\mathrm{A} = \tfrac{1}{2}\,\frac{r_2^5\tan\theta_2 - r_1^5\tan\theta_1}{r_2^2 - r_1^2}\,(1+\vartheta)$$

which is a good working formula.

As an illustration of the above, let it be required to find H and also M from the following measurements :

Length of bar magnet		10 cm.	
Breadth	,,	.6 cm.	
Weight	,,	28.308 gm.	

$$\therefore \mathrm{K} = 28.308\,\frac{(10)^2 + (.6)^2}{12} = 236.75$$

T was found previously to be 13.751 seconds.

To determine θ, the torsion circle was turned through 360°, to the right and the left, the mean deflection of the magnet, converted into angular measure, being 2°.4,

$$\therefore \theta = \frac{2°.4}{360° - 2°.4} = .0066$$

$$\therefore 13.751 = 2\,\pi\,\sqrt{\frac{236.75}{\mathrm{M}\,\mathrm{H}\,(1.0066)}}$$

whence M H = 49.144.

In the deflection experiment, the scale was at a distance of 52.7 cm. from the mirror, r_1 was 30 cm., and the mean of the four deflections was 11.97.

$$\therefore \tan\theta_1 = \frac{11.97}{2 \times 52.7} = .113$$

By using $\tan\theta_1 = \dfrac{n}{2\,r}\left[1 - \left(\dfrac{n}{2\,r}\right)^2\right]$ the correction which is subtractive is .00146, and may, in the present case, be neglected.

At distance $r_2 = 40$ cm., the mean of the deflections was 4.97.

$$\therefore \tan \theta_2 = \frac{4.97}{2 \times 52.7} = .047$$

The ratio of torsion (ϑ) was found to be .0035,

$$\therefore \frac{M}{H} = \tfrac{1}{2} \times \frac{(40)^5 \times .047 - (30)^5 \times .113}{(40)^2 - (30)^2} \times 1.0035$$
$$= 1481.525$$

And as M H = 49.144
$$\therefore \text{M} = 269.828$$
and H = .18213

The distances (r_1 and r_2) chosen should be in the ratio of 3 to 4, as errors of observation will then have the least influence on the results.

The Reflecting Galvanometer.—In the common form of this instrument, there are two circular coils within which is suspended the astatic pair of needles. It is provided with a curved magnet, for the purpose of obtaining various degrees of sensitiveness. When this magnet is turned so as to oppose the action of the earth, it may be gradually slid along its supporting rod until it exactly neutralises the earth's field; there will then be no directive force on the needles, and they will remain in any position in indifferent equilibrium. Displacing the regulating magnet slightly, there will be just sufficient force to keep the needles N and S; the deflection produced by a given current will then be the greatest possible, and the instrument is adjusted to its *maximum sensitiveness.*

Reflecting galvanometers are frequently used as galvanoscopes merely to detect the passage of a current, or, as in the *null* methods, to show that a balance has been obtained.

When used for comparing currents, it should be remembered that these are proportional to the tangent of the deflections of the magnet; that is,

$$C_1 : C_2 = \tan \tfrac{1}{2} \theta_1 : \tan \tfrac{1}{2} \theta_2$$

where θ denotes the angular displacement of the spot of light. We know $\tan \theta$, for it equals $\frac{n}{d}$, n denoting the number of scale divisions in the deflection, and d the distance from the mirror to the scale. From this we may deduce $\tan \tfrac{1}{2} \theta$.

For

$$\tan \theta = \frac{2 \tan \tfrac{1}{2} \theta}{1 - \tan^2 \tfrac{1}{2} \theta}$$
$$\therefore \tan \theta - \tan \theta \tan^2 \tfrac{1}{2} \theta = 2 \tan \tfrac{1}{2} \theta$$
$$\text{or } \tan \theta \tan^2 \tfrac{1}{2} \theta + 2 \tan \tfrac{1}{2} \theta = \tan \theta$$
or, by writing a for $\tan \theta$ and x for $\tan \tfrac{1}{2} \theta$, $a x^2 + 2 x = a$

$$\therefore x = \frac{\sqrt{1 + a^2} - 1}{a}$$

$$i.e. \quad \tan \tfrac{1}{2}\, \theta = \frac{\sqrt{1 + \tan^2 \theta} - 1}{\tan \theta}$$

Now

$$\tan \theta = \frac{n}{d}$$

$$\therefore \tan \tfrac{1}{2}\, \theta = \frac{\sqrt{1 + \frac{n^2}{d^2}} - 1}{\frac{n}{d}} = \frac{\sqrt{n^2 + d^2} - d}{n}$$

Hence

$$C_1 : C_2 = \frac{\sqrt{n_1^2 + d^2} - d}{n_1} \; : \; \frac{\sqrt{n_2^2 + d^2} - d}{n_2} \qquad . \qquad . \qquad . \quad (1)$$

Expanding and taking only the first two terms, we get

$$C_1 : C_2 = n_1 : n_2 \qquad . \qquad . \qquad . \qquad . \qquad . \quad (2)$$

This proportion gives rise to no appreciable error when the deflections do not differ largely ; but when this is not the case, (1) should be used.

The Electro-Dynamometer.—Currents may also be measured in practical or in absolute units by means of an electro-dynamometer. This instrument was invented by Weber, and used by him in verifying Ampère's electro-dynamical theory. Weber's electro-dynamometer consisted of a large fixed coil and a small movable one with bifilar suspension, and furnished with a mirror, by means of which the deflections of the coil could be read off in the usual subjective way with telescope and scale.

When an electro-dynamometer is used, it is adjusted (1) so that the planes of the coils are at right angles to each other, that of the movable coil being (2) also perpendicular to the magnetic meridian.

When a current passes through the instrument, the suspended coil tends to turn round and place itself parallel to the fixed coil, and by (1) the mutual action of the coils is made as great as possible, whilst by (2) the action of terrestrial magnetism is reduced to a minimum.

The couples acting on the inner coil when deflected are,

    (*a*) that due to the bifilar suspension, and

    (*b*) that due to the electro-dynamic action of the coils.

It is important to remember that (*a*) is proportional to the sine of the angular deflection of the coil,

Let D and E, Fig. 25, be the points of suspension of 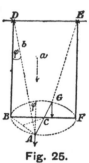 the wires, B and F the points at which they are attached to the coil. Then the weight (W) of the coil acting vertically downwards at C is equivalent to $\frac{W}{2}$ at B and F respectively. Let the coil be turned through an angle B C A (θ), the corresponding displacement of the wire being ϕ. Then, considering only one wire, we see that the weight $\frac{W}{2}$ at A

Fig. 25.

may be resolved into $\frac{W}{2} \sec \phi$ along the wire D A, and $\frac{W}{2} \tan \phi$ in the direction A B. As ϕ is very small, we may replace $\tan \phi$ by $\sin \phi$.

The perpendicular from C on A B is $a \cos \frac{\theta}{2}$, where A C = a

$\therefore$ the moment (M) of this force about C is $\frac{W}{2} \sin \phi . a \cos \frac{\theta}{2}$.

If l denote the length of the wire D B, then

$$l \sin \phi = \text{A B} = 2 a \sin \frac{\theta}{2} \therefore \sin \phi = \frac{2a}{l} \sin \frac{\theta}{2}$$

$$\therefore \text{M} = \frac{W}{2} . \frac{2a}{l} \sin \frac{\theta}{2} . a \cos \frac{\theta}{2} = \frac{Wa^2}{2l} \sin \theta$$

And finally for both wires $\text{M} = \frac{W a^2}{l} \sin \theta = \beta \sin \theta$ (say).

The electro-dynamical action of two circuits traversed by currents γ_1 and γ_2 is proportional to the product $\gamma_1 \gamma_2$, and, therefore, to γ^2 in the case which we are considering.

The couple due to the current may be written $a \gamma^2$ where a depends upon the geometrical data of the coils. The moment of this couple, when the small coil is deflected through an angle θ, is $a \gamma^2 \cos \theta$. Therefore for equilibrium we have

$$a \gamma^2 \cos \theta = \beta \sin \theta$$

And

$$\gamma^2 = \frac{\beta}{a} \tan \theta = \lambda \tan \theta.$$

If we take account of the torsion of the wire, its effect for each wire will be $\mu \theta$, μ being the couple of torsion for 1°. The above equation would then become

$$a \gamma^2 \cos \theta = \beta \sin \theta + 2 \mu \theta.$$

To completely eliminate the effect of the earth's magnetism, we must take four observations for $\tan \theta$—viz., (1) on making, and (2) on reversing the current, (3) and (4) on sending the currents round the coils in opposite directions.

The electro-dynamometer may also be used as a galvanoscope; it was thus employed by Kohlrausch and Nippoldt in their researches on the resistance of electrolytes.

A convenient form of electro-dynamometer has been devised by Sir W. Siemens, and is extensively used in electric lighting measurements.

The instrument is shown in Fig. 26. There are two fixed coils, one of thin and the other of thick insulated copper wire. The latter is used when continuous and the former when alternating currents are to be measured. The movable coil is reduced to a single loop of very stout wire, in order to reduce as much as possible the effect of terrestrial magnetism; this loop is suspended by a thread and spiral spring, and in its zero position its plane is at right angles to that of the fixed coils. Its ends dip into separate mercury cups, by means of which the rotations are facilitated and circuit completed.

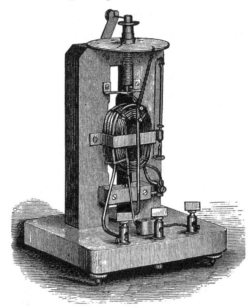

Fig. 26.

The deflections of the loop are indicated by a bent index-arm, which is brought back to the zero of graduation by turning the upper end of the spring, the number of degrees of torsion being indicated by a pointer.

The couple tending to bring back the index-arm to zero is proportional simply to the angle of torsion a. Therefore

$$\gamma^2 = \lambda\, a \text{ where } \lambda \text{ is a constant.}$$

The value of λ is determined for each instrument and marked in the table accompanying it.

It is useful to verify λ from time to time, as the resiliency of the spring may slightly change with use or accidental rough handling. This verification is readily made by comparison with a tangent galvanometer whose reduction factor has been carefully determined.

From the readings of the tangent galvanometer we have

$$C = \frac{H}{G} \tan \theta, \text{ or } C = k \tan \theta$$

From the electro-dynamometer,

$$C^2 = \lambda\, a$$

Whence

$$\lambda = \frac{k^2 \tan^2 \theta}{a}$$

In an actual measurement, θ was 47° 30′, a was 12.75°, and K 3.171 ; *i.e.*, this was the number by which the tangent of the deflection should be multiplied in order to give the current in ampères. If C were required in absolute units, we should simply have to divide the number of ampères by 10.

$$\therefore \lambda = \frac{(3.171 \times 1.0913)^2}{12.75} = .938$$

The number given in the table was .923.

Of course the value of λ should not be taken from any one measurement, but deduced from a series of careful observations.

If the currents which we are measuring be suddenly reversed, the currents through the loop and the coil will be simultaneously reversed and the deflection will not be altered. Hence an electro-dynamometer may be used in measuring rapidly alternating induction currents.

The Potential Galvanometer.—Sir William Thomson has recently devised two instruments of great value to the practical electrician, viz., the potential and the current galvanometer. The former might have been described in an earlier part of this chapter, but it was thought better to place it after the tangent galvanometer by comparison with which it is graduated.

The potential galvanometer, Figs. 27 and 28, consists of a coil A of

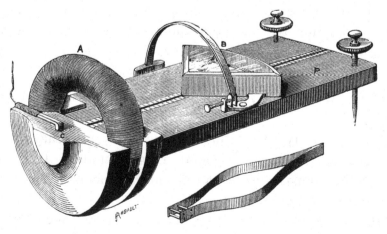

Fig. 27.   Fig. 28.

very high resistance (6000 or 7000 ohms) and a movable magnetometer B,

Four small magnets form the "needle" and are rigidly connected with a long light aluminium pointer, Fig. 28. An image of this pointer is formed in an underlying plane mirror, and errors in reading due to parallax are avoided by placing one's self so that the image and the needle are seen coincident in the same vertical plane.

The deflections are kept within convenient limits by increasing the intensity of the magnetic field in which the needle moves. This is effected by a semicircular steel magnet whose field is carefully determined and marked on the magnet.

As a preliminary, it may be necessary to show that when a coil of high resistance (R) is connected with two points A and B, Fig. 29, of a conductor, through which a current (C) is passing, the difference of potentials (V) between the two points is practically unaltered.

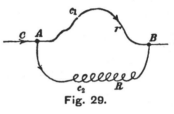
Fig. 29.

Calling c_1 and c_2 the currents in the branches of the multiple arc, we have

$$C = c_1 + c_2$$

And

$$c_1 r = c_2 R = V$$

$$\therefore c_2 R = r (C - c_2) \text{ and } c_2 = \frac{r}{R + r} C$$

But by Ohm's law,

$$C = \frac{V}{r} \text{ (before R was inserted).}$$

$$\therefore c_2 = \frac{r}{R + r} \times \frac{V}{r} = \frac{V}{R + r}$$

$$= \frac{V}{R \left(1 + \frac{r}{R} \right)} = \frac{V}{R}, \text{ when } \frac{r}{R} \text{ is negligible.}$$

and the difference of potentials (V) deduced from $c_2 = \frac{V}{R}$ will differ from the true value by a quantity which may be made vanishing.

To graduate the instrument P G, we may connect it up with a tangent galvanometer T and a known resistance (r). A sufficiently strong battery must be used in order to give a deflection of nearly 45° on T, the magnetometer (its magnet being removed) must be slid along its platform until nearly the same deflection (45 divisions) is obtained.

From T we have

$$C = k \tan$$

whence C is known (say) in ampères. Multiplying by r, expressed in ohms, we get in volts the difference of potential (V) at the terminals A and B of the potential galvanometer.

The rest of the operation will be best understood from an example. Let the deflection of P G be 42 divisions, that of T, 43° 30 ; its k, 3.171, and the resistance (r) over which the fall is required 10 ohms. Then

$$C = 3.171 \times .885 = 2.81 \text{ ampères,}$$

And

$$2.81 \times 10 = 28.1 \text{ volts.}$$

Therefore,

$$1 \text{ volt would give } \frac{42}{28} \text{ divisions in the earth's field,}$$

And

$$1 \text{ volt in unit field would give } \frac{42}{28} \times .18 = .27 \text{ divisions.}$$

Hence in that position of the magnetometer, the number to be marked on the platform of the instrument is .27.

The position corresponding to .25 or $\frac{1}{4}$ may be deduced as above, viz.,

$$\frac{x}{28} \times .18 = \tfrac{1}{4}, \text{ whence } x = 38.9.$$

This is the deflection which the magnetometer should give at the number $\frac{1}{4}$; it must, therefore, be moved back until this reading is obtained, and then the fraction $\frac{1}{4}$ is marked on the platform.

To ascertain the position for $\frac{1}{2}$, 1, 2, 4, &c., we must diminish the battery power (otherwise the deflection would be excessive) and proceed exactly as before. For $\frac{1}{8}$, $\frac{1}{16}$, &c., we must, on the contrary, increase the battery in order to keep the deflections as near 45 as possible, that being the angle of maximum sensitiveness.

The above formula may now be generalised. Let d be the reading of the magnetometer, V the required difference of potential, H the strength of the earth's magnetic field, and n the number marked on the platform. Then from above

$$\frac{d\,H}{V} = n, \text{ whence } V = \frac{d\,H}{n}.$$

When the semicircular magnet is used the resultant field is the sum of that due to the earth and the magnet. Calling this f, we have

$$V = \frac{d\,f}{n}.$$

Ex. Required the difference of potential at the ends of a resistance of 50 ohms through which a current from a battery of 15 Grove's cells is

passing, the reading of the potential galvanometer being 9.5 at the number 4 on the platform and the strength of the magnet 11.12 c.g.s. units of intensity.

Here

$$V = \frac{9.5\,(11.2 + .18)}{4} = 26.88 \text{ volts.}$$

In magnetic measurements we cannot assume that the moment of a carefully manipulated magnet remains constant for, say, two consecutive hours, still less may we assume this constancy in the case of the semi-circular magnet of the magnetometer which, in spite of all precautions, will be sometimes handled with but little delicacy, left lying about in the neighbourhood of other magnets or masses of iron, and in all positions with respect to the magnetic meridian.

It is, therefore, imperative to re-determine, from time to time, the strength of the field produced by the magnet or by the magnet and earth combined, *i.e.* the f of the formula.

This verification may be easily made by a method suggested by Poggendorff's method of compensation for finding the resistance of batteries.

Fig. 30 shows the various connections. B is a battery of 12 or 15 Grove's cells; P G the potential galvanometer whose resistance is g (7000 ohms); D is a standard cell, say, one of Latimer Clark's whose E.M.F. (E_2) is 1.457 volts; F is a zero galvanometer, and r an adjustable resistance. This r is varied until F shows no deflection, then

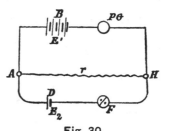

Fig. 30.

we know that the fall of potential between A and H, due to the battery B, is equal to that due to the cell D.

The current (C) produced by the battery is $\dfrac{E_1}{g + r}$ neglecting the resistance of the battery, which is small in comparison with the resistances in circuit. Therefore, the difference of potential between A and H is $\dfrac{r\,E_1}{g + r}$

$$\therefore \frac{r\,E_1}{g + r} = E_2, \text{ whence } E_1 = \frac{g + r}{r}\,E_2$$

But C × g = difference of potential at terminals of galvanometer

$$= \frac{\text{deflection} \times \text{field}}{\text{number on platform}} = \frac{d\,f}{n}$$

That is

$$\frac{E_1}{g + r} \times g = \frac{d\,f}{n}$$

For the unknown E.M.F. (E_1) substitute the value found above and we get

$$\frac{g}{g+r} \times \frac{(g+r)\,E_2}{r} = \frac{g\,E_2}{r} = \frac{d\,f}{n}$$

Whence

$$f = \frac{n\,g}{d\,r}\,E_2$$

which gives f in terms of known quantities.

By subtracting the earth's horizontal force (H) from f, we obtain the intensity of the field due to the magnet, *i.e.*, the number which should be marked upon it.

The Current Galvanometer.—The current galvanometer differs from the potential galvanometer only in its coil, which is one of low resistance. For graduation, it is connected up in series with a tangent galvanometer and a resistance-box. The resistances in circuit must be varied so as to keep the deflection, in all positions of the magnetometer, as near 45 as possible.

The current in ampères is given by the tangent galvanometer, and the graduation may be effected in a manner quite analogous to that used for the potential galvanometer.

Before an observation, the potential and current galvanometers should be levelled, the magnet removed, and needle set free. The platform must then be gradually turned until the pointer stands at zero. Replacing the magnet, the instrument is ready for use.

The methods of making contact are ingenious. The leading wires or "leads" are terminated at one extremity by two thick copper strips, Fig. 31, kept apart by an insulating material. When inserted at C, Fig. 27, contact is made with another similar pair of plates connected respectively with the two ends of the coil. The other extremity of each "lead" is furnished with a pincer-like appendage, Fig. 32.

Fig. 31.   Fig. 32.

The upper part consists of a tongue of boxwood, and is kept, by means of a tightly fitting ring, pressing against a copper plate which is soldered to the lead. When a wire is clipped by this apparatus, it is held in electrical contact with the lead.

The Torsion Galvanometer.—We may here describe Siemens's torsion

galvanometer, an instrument which is extensively used in electric lighting to measure differences of potentials.

It consists, Fig. 33, of one of Dr. Werner Siemens's bell-shaped magnets suspended by a thread and spiral spring between two vertical coils of fine

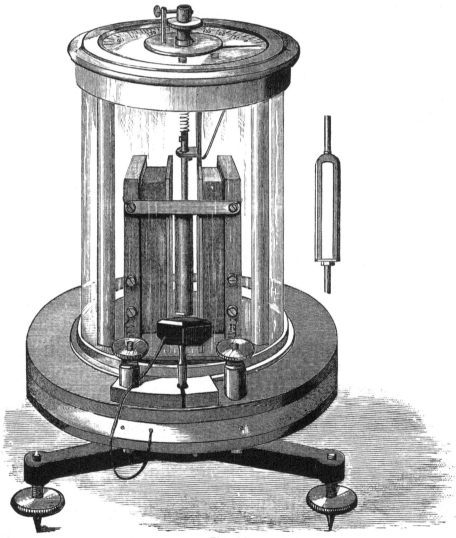

Fig. 33.

wire and high resistance. The lower end of the spring is attached to the magnet, and the upper to the axis of a pointer that moves over a graduated circle.

When a current C passes through the coils, the magnet is deflected, and is brought back to its zero position by twisting the spring. The

number of degrees of torsion is indicated by the pointer, and is proportional to the force (more strictly, the couple) of torsion.

In the electro-dynamometer, an increase of current in the fixed coil is attended by a corresponding increase in the movable one, and hence the mutual action of the two parts of the same circuit varies as C^2.

In the torsion galvanometer, on the other hand, an increase of current in the vertical coils is not attended by a similar change in the strength of the magnet. The mutual action, then, between the current and the magnet varies as the first power of C. The theory of this instrument may be thus deduced :

By Ohm's law,
$$C = \frac{E}{R}.$$

The laws of elasticity give
$$C = \lambda\,\theta$$

where λ is the couple of torsion for $1°$ and θ is the number of degrees of torsion.
$$\therefore\ \frac{E}{R} = \lambda\,\theta \text{ and } E = R\,\lambda\,\theta = \mu\,\theta \text{ (say),}$$

i.e., the difference of potentials, as measured by the instrument, varies directly as the angle of torsion.

The constant μ may be determined as antecedently by comparison with a standard cell or with a tangent galvanometer.

We shall here describe another method which, in some circumstances, may be found more convenient than the others. It is, moreover, susceptible of numerous applications.

It requires a condenser, a standard cell, and a reflecting galvanometer adapted for measuring transient currents, that is one in which the needle has a large moment of inertia, and consequently a long period of vibration. By this means, we insure that the whole discharge from the condenser will pass through the galvanometer coils before the needle has appreciably moved away from its position of equilibrium. Such a modification constitutes what is called a "ballistic"* galvanometer.

The connections for this test are shown in Fig. 34, where T G is the torsion galvanometer, with its terminals joined up to the binding screws of a resistance-box R.

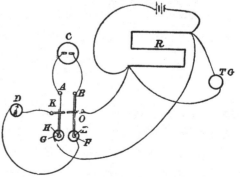

Fig. 34.

* See Gordon's Electricity, Vol. I., p. 255 (Second Edition).

C is a $\frac{1}{3}$ m.f., and D a galvanometer, both connected with a discharge key. A and B are binding screws fixed on two brass strips A H and B E, which are kept, in their normal condition, pressing against the metallic pieces K and O by their own elasticity; when H and E are depressed, they make contact with specially arranged binding screws G and F.

The diagram shows one coating of the condenser connected to B and, through O, to one end of the resistance R. Depressing H, the other coating is connected through A and the brass strip with the other end of the resistance. A momentary depression of H suffices to charge the condenser up to the given E.M.F., viz., the difference of potentials at the ends of the resistance R.

Releasing H and putting down E, the condenser is discharged through the galvanometer. Such is the play of the key.

To obtain the "constant" of a torsion galvanometer, we may proceed as follows :

Take a battery of (say) 5 Grove's cells, and unplug some resistance (say 50 ohms) in the box. On closing the battery circuit, the needle of T G will be deflected ; let it require 25° of torsion to bring it back to zero ; charge and discharge the condenser by successively depressing and releasing H and E, using a shunt if necessary. With a $\frac{1}{9}$th shunt let the deflection be 59 scale divisions ; therefore, without the shunt the deflection would have been $59 \times 10 = 590$ divisions.

The given E.M.F. may be evaluated by charging the condenser from a Latimer Clark's cell and noting the discharge deflection.

Let this be 48.9 divisions, or 489 without the $\frac{1}{9}$th shunt. We now have the proportion

$$489 : 590 = 1.457 : x$$

Whence

$$x = 1.75 \text{ volts} = \text{the charging E.M.F.}$$

This must also be the number deduced from the reading of the torsion galvanometer.

Hence

$$1.75 = 25 \, \mu$$
$$\therefore \, \mu = 0.70$$

i.e., the number .070 is the constant factor by which the observed angle of torsion must be multiplied in order to obtain the difference of potentials in volts.

From a series of similar determinations the value of μ may be accurately found.

E

In the usual construction of the instrument, a torsion of about 15° corresponds to an E.M.F. of 1 volt. By inserting a plug, Fig. 33, a resistance may be thrown out of circuit, and then a torsion of 15° corresponds to 0.1 volt.

The Wheatstone Bridge.—The " bridge " method of measuring electrical resistance was introduced in 1833, by Mr. Hunter Christie, of the Royal Military Academy, Woolwich. It remained unnoticed until 1843, when it was taken up and greatly developed by Sir Charles Wheatstone.

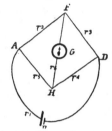

Fig. 35.

Let four conductors, A F, F D, D H, H A, be joined up in circuit with battery B and galvanometer G, Fig. 35.

Then the fall of potential between A and D along A F D is the same as along A H D. By adjusting one of the resistances (say r_2), the deflection of the galvanometer may be reduced and finally eliminated. As no current then passes through G, the potential at F must be the same as at H. The fall over A F must consequently bear to that over F D the same ratio that the fall over A H bears to that over H D. But, by Ohm's law $\left(C = \dfrac{E}{R} \therefore C R = E\right)$, the fall of potential is directly proportional to the resistance over which it takes place ; hence we have, from the previous considerations,

$$r_2 : r_3 = r_5 : r_4 \quad \text{and} \therefore r_2 r_4 = r_3 r_5$$

which is the equation of the bridge.

When this condition is fulfilled, the battery circuit may be opened or closed without affecting G, and reciprocally the galvanometer circuit may be opened or closed without affecting B. The two branches F H and A B D are then said to be *conjugate* to each other.

The above important relation may be deduced in another way. Let us suppose the bridge balanced and the galvanometer circuit open.

The resistance of A F and F D $= r_2 + r_3$
That of   A H and H D $= r_5 + r_4$

Therefore their joint resistance in multiple arc is $\dfrac{(r_2 + r_3)\ (r_4 + r_5)}{(r_2 + r_3) + (r_4 + r_5)}$

When the galvanometer circuit is closed, its resistance (r_6) has no effect on the current since the bridge is balanced.

The whole resistance in the bridge may also be considered as made up of the two multiple arcs A F, A H and D F, D H arranged in series.

The resistance of A F, A H is $\dfrac{r_2 r_5}{r_2 + r_5}$

That of D F, D H is $\dfrac{r_3\,r_4}{r_3+r_4}$

And their joint resistance in series is $\dfrac{r_2\,r_5}{r_2+r_5} + \dfrac{r_3\,r}{r_3+r_4}$

Hence the equation

$$\frac{(r_2+r_3)\;(r_4+r_5)}{(r_2+r_3)+(r_4+r_5)} = \frac{r_2\,r_5}{r_2+r_5} + \frac{r_3\,r_4}{r_3+r_4}$$

which reduces to the complete quadratic

$$(r_2\,r_4)^2 + (r_3\,r_5)^2 - 2\,r_2\,r_3\,r_4\,r_5 = 0$$

Taking the square root

$$r_2\,r_4 - r_3\,r_5 = 0$$

Whence

$$r_2\,r_4 = r_3\,r_5$$

This relation is often deduced from Kirchoff's laws of current distribution in a network of linear conductors. These laws, moreover, enable us to find the current-strength in any branch of the bridge, and thence the best conditions for any required measurement of resistance.

Kirchoff's laws are two in number :

(I.) The algebraical sum of all the currents meeting at any node of the network is zero.

This follows, at once, from the principle of continuity ; for as there is no accumulation of electricity at the node, the sum of all the currents flowing *to* it must equal that of all the currents flowing *from* it ; *e.g.*, in Fig. 36,

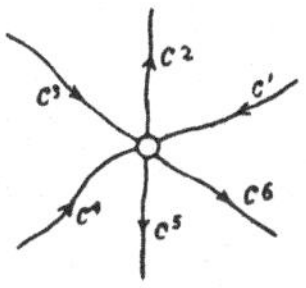

Fig. 36.

$$c_1 + c_3 + c_4 = c_2 + c_5 + c_6$$

Reckoning the former direction positive, the latter becomes negative, and therefore

$$c_1 + c_3 + c_4 = - (c_2 + c_5 + c_6)$$

or,

$$c_1 + c_3 + c_4 + (c_2 + c_5 + c_6) = 0$$
$$i.e.,\quad \Sigma c = 0$$

(II.) If we go round any closed circuit of the network, then, reckoning currents and electromotive forces in one direction positive, and those in the opposite direction negative, the whole electromotive force in the circuit is equal to the sum of the products of each current by the resistance through which it passes.

Let Fig. 37 represent a closed circuit including the electromotive forces E_1, E_2, and E_3. Let c_1 be the current through the conductor whose resistance is r_1, a and g denoting the potentials at its two extremities ; and let there be similar notation for the other parts of the circuit.

E 2

Then, by Ohm's law, $\quad c_1 r_1 = a - g$

Similarly, $\qquad\qquad\quad c_2 r_2 = d - b$

And $\qquad\qquad\qquad c_3 r_3 = f - e$

$\therefore c_1 r_1 + c_2 r_2 + c_3 r_3 = a - g + d - b + f - e$

$\qquad\qquad\qquad\quad = (a-b) + (d-e) + (f-g)$

$\qquad\qquad\qquad\quad = E_1 \quad + \quad E_2 \quad + \quad E_3$

i.e. $\Sigma\, c\, r = \Sigma\, E$, which is the law.

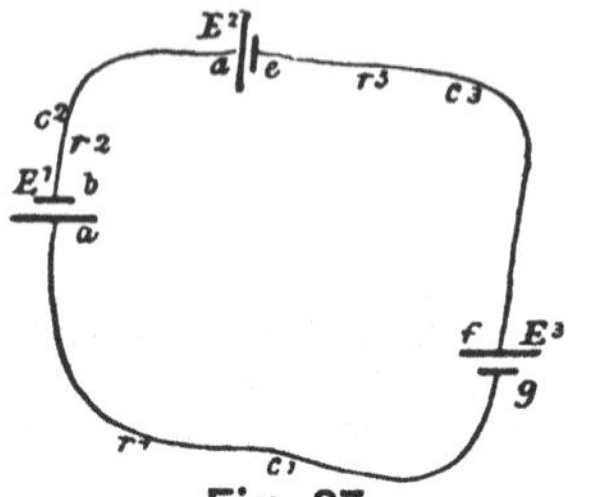

Fig. 37.

If we consider a circuit such as shown in Fig. 38, in which there is no impressed E.M.F., we have a, b, d denoting the potentials at A, B, D respectively,

$$c_1 r_1 = a - b$$
$$c_2 r_2 = b - d$$
$$c_3 r_3 = d - a$$
$$\therefore c_1 r_1 + c_2 r_2 + c_3 r_3 = 0$$

Fig. 38.

$\therefore \Sigma\, c\, r = 0$, which is in accordance with the law.

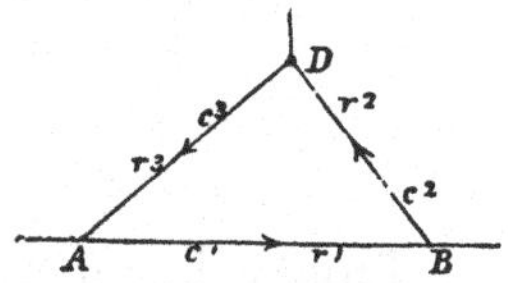

Now, in Fig. 39, let r_1, r_2, &c., be the resistances of the different parts of the bridge, c_1, c_2, &c., the currents through them, and E the electromotive force of the battery B. Then applying Kirchoff's laws to the circuit B A F D B, we get

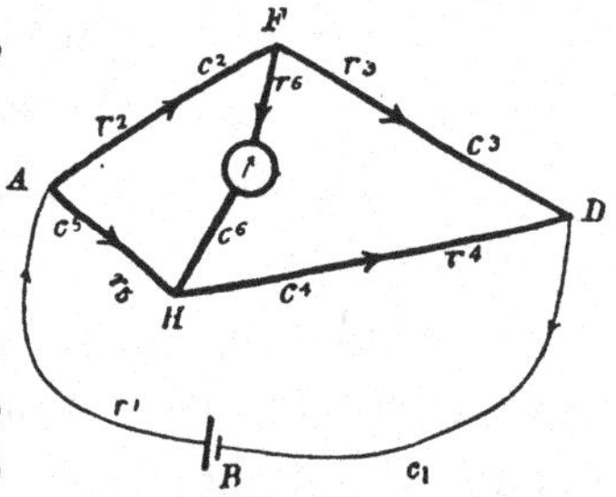

Fig. 39.

$$c_1 r_1 + c_2 r_2 + c_3 r_3 = E \quad . \quad (1)$$

From A F H, $\quad c_2 r_2 + c_6 r_6 - c_5 r_5 = 0 \quad . \quad (2)$

From F D H, $\quad c_6 r_6 + c_4 r_4 - c_3 r_3 = 0 \quad . \quad (3)$

At A, $\qquad\quad c_1 = c_2 + c_5 . \quad . \quad . \quad (4)$

At F, $\qquad\quad c_2 = c_3 + c_6 . \quad . \quad . \quad (5)$

At D, $\qquad\quad c_1 = c_3 + c_4 . \quad . \quad . \quad (6)$

It is necessary to find the current through the galvanometer (c_6) in terms of the battery current c_1 and the resistances in the various branches.

From (6) $c_4 = c_1 - c_3$

From (4) $c_5 = c_1 - c_2 = c_1 - (c_3 + c_6)$ from (5)

Substituting in (3) and also in (2) we get

$$c_6 r_6 + r_4 (c_1 - c_3) - c_3 r_3 = 0$$
$$r_2 (c_3 + c_6) + r_6 c_6 - r_5 (c_1 - c_3 - c_6) = 0$$

These last two equations may be written

$$c_6 r_6 + r_4 c_1 - c_3 (r_3 + r_4) = 0 \quad . \quad . \quad . \quad . \quad (7)$$
$$c_6 (r_2 + r_5 + r_6) + c_3 (r_2 + r_5) - r_5 c_1 = 0 \quad . \quad . \quad (8)$$

Between (7) and (8) eliminating c_3, we get

$$c_6 = \frac{(r_3 r_5 - r_2 r_4)\, c_1}{r_2 r_4 + r_2 r_3 + r_2 r_6 + r_3 r_5 + r_3 r_6 + r_4 r_5 + r_4 r_6 + r_5 r_6}$$

$$\quad = \frac{(r_3 r_5 - r_2 r_4)\, c_1}{(r_2 + r_5)(r_3 + r_4) + r_6 (r_2 + r_3 + r_4 + r_5)}$$

The current through the galvanometer (c_6) will therefore vanish when

$$r_3 r_5 - r_2 r_4 = 0 \qquad \text{or} \quad r_3 r_5 = r_2 r_4$$

The relative *sensitiveness* of the bridge in any particular measurement is practically determined by finding the greatest change in the adjustable resistance that corresponds to the least observable galvanometer-deflection.

It may be shown* that, for accurate determinations, all the resistances on the bridge should be as nearly as possible equal to the resistance to be measured, an approximate value for the latter being obtained from a preliminary test.

The mathematical theory of the bridge also shows it to be most sensitive when, of the battery and galvanometer, the one which has the higher resistance connects the junction of the two greater with that of the two smaller resistances.

The practical connections corresponding to Fig. 39 are shown in Fig. 40, where r_2 and r_3 are the "arms" of the bridge, and r_4 the resistance that is varied until the balance is obtained.

The *Slide-Wire*, or *Metre* Bridge, Fig. 41, is a modification of the bridge due to Kirchoff, and is especially useful for the measurement of low resis-

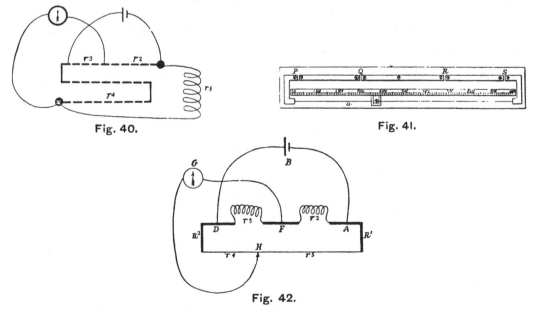

Fig. 40.

Fig. 41.

Fig. 42.

tances. A galvanometer of correspondingly low resistance should be used The common astatic galvanometer of about half an ohm answers well for a variety of purposes. With a little care, very accurate results may be obtained.

* Maxwell, § 348; or Kempe's Handbook of Electrical Testing, p. 120.

The practical connections corresponding to Fig. 39 are shown in Fig. 42. The galvanometer must always be put between F and H, and the battery between A and D, irrespective of the theoretical conditions of sensitiveness, otherwise the surface of the wire would be gradually injured by making and breaking contact. This wire is made of an alloy of platinum and iridium, which is hard and tough, not easily scratched or otherwise altered with respect to its surface conditions. Its resistance is also but little affected by changes of temperature.

When a telephone is used instead of the galvanometer G, it may not be always possible to obtain complete silence ; in this case, that position of the movable contact H, Fig. 42, which gives a minimum of sound, corresponds to the balance.

When the bridge is balanced, we have

$$r_3 : r_4 = r_2 : r_5 \quad \text{or} \quad r_2 r_4 = r_3 r_5$$

This equation is true only in so far as the resistance R_1, R_2 of the connecting pieces is negligible. When measuring high resistances, R_1 and R_2 may be neglected. In the case of low resistances this neglect would cause an error which would increase with the smallness of the resistance. R_1 and R_2 may be calculated as follows :

For the sake of clearness, in the balanced bridge, let a, b, c, and d denote respectively r_2, r_3, r_4, and r_5.

Interchange r_2 and r_3 and balance again, and let the new value of r_4 be e, and that of r_5 be f.

Then we have

$$\frac{a}{d + R_1} = \frac{b}{c + R_2}$$

And

$$\frac{b}{f + R_1} = \frac{a}{e + R_2}$$

Eliminating R_2 between these equations,

$$R_1 = \frac{a b (e - c) + b^2 d - a^2 f}{a^2 - b^2}$$

But

$$c + d = e + f \quad \therefore \quad e - c = d - f$$

$$\therefore R_1 = \frac{a b (d - f) + b^2 d - a f}{a^2 - b^2}$$

Working out and factoring

$$R_1 = \frac{(a + b)(b d - a f)}{a^2 - b^2} = \frac{b d - a f}{a - b}$$

Similarly,

$$R_2 = \frac{b\,e - a\,c}{a - b}$$

Professor G. Carey Foster has shown how the difference of two nearly equal resistances may be accurately found independently of those of the connections.

Let Q and R, Fig. 41, be unchangeable, and P and S the interchangeable resistances. It is required to find $P - S$ in terms of the slide-wire, its length being l, and μ the resistance of unit length.

When the bridge is balanced, we have

$$\frac{Q}{P + r_1 + \mu\,a_1} = \frac{R}{S + r_2 + \mu\,(l - a_1)}$$

or,

$$\frac{Q}{R} = \frac{P + r_1 + \mu\,a_1}{S + r_2 + \mu\,l - \mu\,a_1}$$

Therefore, also,

$$\frac{Q}{Q + R} = \frac{P + r_1 + \mu\,a_1}{P + S + r_1 + r_2 + \mu\,l}$$

Similarly, when P and S are interchanged,

$$\frac{Q}{Q + R} = \frac{S + r_1 + \mu\,a_2}{P + S + r_1 + r_2 + \mu\,l}$$

Whence

$$P + r_1 + \mu\,a_1 = S + r_1 + \mu\,a_2$$

And

$$P - S = \mu\,(a_2 - a_1)$$

If $S = 0$, that is if one of the gaps is closed by a conductor of insensible resistance, we have

$$P = \mu\,(a_2 - a_1)$$

i.e., the difference of the two scale readings gives the length of the wire whose resistance is equal to that of the conductor to be measured.

It is evident the resistance of this latter must be small, and expressible in parts of the standard wire.

The value of the coefficient μ may be readily found by the preceding method. Thus a wire whose resistance R is a little less than that of the standard slide-wire is taken, and its resistance carefully determined as above. It is next joined up in multiple arc with a standard unit of resistance S, and the measurement repeated.

If c and d be the difference of the scale readings in the two cases respectively, we have

$$\mu\,c = R$$

And

$$\mu\,d = \frac{R\,S}{R + S}$$

Whence
$$\mu = \frac{c-d}{c\,d}\,\text{S.}$$

If the slide-wire should not be of uniform resistance throughout, the values for the different parts, and also a mean value for the wire as a whole, can be found by giving small varying values to R.

If the instrument has been subjected to rough usage, or if it be required for very delicate work, it will be found necessary or advantageous to *calibrate* it, that is, to divide it into parts of equal resistance.

This may be done by an elegant method due to Professor G. Carey Foster.* Let E F, Fig. 43, be the wire to be calibrated. The battery is connected to the binding-screws a and b, the terminals of the galvano-

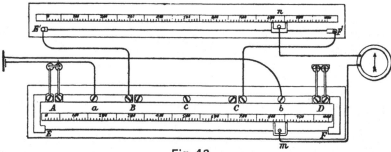

Fig. 43.

meter to two movable contact-makers m and n, the former on the principal and the latter on a compensating wire $E^1 F^1$ The ends of the compensator are connected with binding-screws near the gaps B and C. The end gap A is closed by a short thick copper conductor, that at D by a length of German silver wire whose resistance is equal to that of the part of the slide-wire which it is desired to take at a time. This serves as a sort of unit, or gauge or standard of comparison.

The gauge being inserted at D, the contact m is placed as near F as possible, and the slider n is moved along the compensator until the galvanometer is balanced. A and D are then interchanged, and the balance is restored by displacing m.

It follows from above that the resistance of the length of F E over which m has been moved, is equal to that of the gauge, or, more correctly, to the difference between the resistances of the gauge and the thick copper connector.

These are now restored to their original position, and n is shifted until a balance is again obtained. The gauge and the copper connector are then interchanged, and the balance restored by displacing m. This gives a second

* Jour. Soc. Tel. Engineers, Oct., 1872.

length of the wire equal in resistance to the first. The operation is continued in this way until the slider *m* is near E, when the calibration of the wire is completed.

Resistance Coils.—The usual material for resistance coils is German silver wire, this alloy changing but little with time and the ordinary variations of temperature. For coils of low resistance, comparatively thick wire is preferred, whilst for high resistance, for the sake of compactness, thin wire is used.

The wire is covered with silk, and then steeped in melted paraffin wax. This keeps the silk dry, and at the same time secures good insulation ; but it has the drawback—paraffin wax having only a low conducting power for heat—of preventing the coils from readily assuming the temperature of the surrounding medium.

The silk-covered wire is wound double, Fig. 44, round a hollow reel of insulating material—generally ebonite—and then enclosed within a brass cylindrical case. When immersed in a vessel of water, such a coil gradually attains the temperature of the liquid.

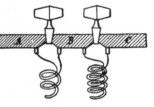

Fig. 44.

Several modifications of the above method of constructing resistance coils have been suggested. Dr. Fleming proposes to wind the wire, uncovered with silk, on an insulating cross-shaped frame, and to then enclose it within a flat nickel case. When placed in a vessel of water, such a coil soon takes the temperature of the liquid, which temperature may be easily determined and kept constant.

Where practicable, it is advantageous to use thick wire for coils, as it will be less heated by the passage of the current.

It will also be found useful for many measurements to have at one's disposal an ohm of stout wire, without any insulating covering, and wound in a long spiral. By its size and exposure, it will readily take the temperature of the room in which the measurements are made.

The object of winding the wire of resistance coils double, is to eliminate the effects of self-induction. Thus, if a current flowing through a conductor A B, Fig. 45, from A to B is suddenly interrupted, an induced current will

Fig. 45.

be started in the wire in the same direction A B. Similarly, the secondary due to the cessation of a current in the conductor D C flowing from D to C, will be in the direction D C. If the two conductors A B and C D be joined up in one circuit, the same induction phenomena will still take place

so that the " extra" current on breaking set up in the doubled wire will produce, in adjacent windings as also on independent neighbouring conductors, equal and opposite electromagnetic and inductive effects.

The same is true of the extra current produced on closing the circuit.

The effects on a coil due to the making and breaking of circuits, to the movement of electrified conductors and of magnets in its vicinity, will also be eliminated by such a method of double winding.

The connections of the successive coils in a resistance-box with the metal blocks, Figs. 44 and 46, should be made so as to have a very small and constant resistance.

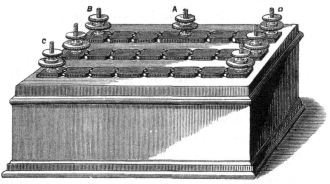

Fig. 46.

To insure good contact, the brass plugs should occasionally be cleaned with emery paper, care being taken to rub off any particles of grit that may be left adherent to the plug. For the same reason it is well, before using a box, to take out all the plugs and dust the brass and ebonite portions carefully.

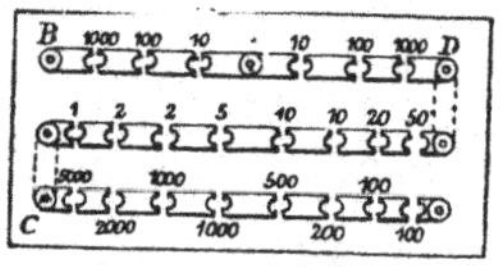

The usual numerical arrangement of the coils in a resistance-box is shown in Fig. 47.

Sir Charles Wheatstone devised a convenient little apparatus, the *rheostat,* for altering the strength of a current without interrupting it. It consists, Fig. 48, of two equal parallel cylinders, one of which, B, is made of brass, and the other, L, of non-conducting material (well dried wood). The latter is cut with a spiral groove throughout its length, so that the successive convolutions of the wire shall be well insulated from one another. When all the wire is wound on B, it is plain there is practically no rheostat resistance in circuit. By turning clockwise the handle M, supposed attached to the axis of L, the wire will be gradually and regularly coiled on the grooved cylinder, and just so much

Fig. 47.

additional resistance will be thrown into circuit as will be wound on L. This is read off on a graduated bar placed between the two cylinders, any

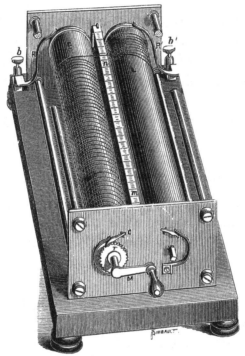

Fig. 48.

fraction of a revolution being indicated on a dial-plate not shown in the figure. To decrease the resistances, the handle is tranferred to T and turned counter-clockwise.

The current enters at the binding-screw *b*, passes up the spring R across the brass cylinder and the stretched wire to L, where it passes round the various insulated windings to a second spring and binding-screw, whence it returns through the rest of the circuit to the battery.

It is not easy to determine, with any degree of precision, the length of the wire thus introduced, and hence the rheostat is unsuited for delicate work.

Somewhat similar to Wheatstone's rheostat is Poggendorff's *rheocord*. It consists of two stretched parallel platinum wires connected with a pair of binding-screws, to facilitate introduction into circuit. A wooden box containing a small quantity of mercury is so arranged that it slides easily along the wires, the longitudinal displacement being read off on an adjacent scale. When thrown into circuit, the current passes through one wire over the mercury bridge to the other, whence it re-enters the main circuit. The

rheocord resistance is obviously that of the wire traversed by the current. By sliding the bridge backward or forward, a very nice adjustment may be obtained.

The use of powerful currents for electric lighting has led to the construction of special resistances. These are usually made of iron or carbon. In the former case, helices of iron about $\frac{1}{4}$-inch in diameter, or zigzags of hoop iron, are employed; and in the latter, carbon rods similar to those used in arc lamps.

Resistance-boxes usually go up to 10,000 ohms. Messrs. Warden and Muirhead make some of 100,000 ohms, and even of 1,000,000 ohms, or megohm.

Platinum-iridium wire is now often used in the construction of megohms. Mr. Willoughby Smith has used selenium for the same purpose; but the varying resistance of this substance under the influence of light and with the passage of currents, as well as the structural changes superinduced by time, render it unreliable and unfit it for use in the construction of electrical standards.

A useful means of making very high resistances has been suggested by Mr. S. A. Phillips, viz., by drawing a thick lead-pencil line on a non-conducting surface such as ebonite, varnished glass, or even ordinary paper, taking care to make good connections with binding-screws. A line, a few inches long, may have a resistance comparable with a megohm. Special paper is now manufactured, in which plumbago is mixed with the pulp. A strip of this *carbon paper*, 20 inches by $\frac{1}{2}$-inch, may give a resistance of nearly 50,000 ohms.

In some continental works, resistances are given in *Siemens's unit*. This unit is the resistance of a column of pure mercury 1 metre long and 1 square millimetre in section. Its value was found (1878) by H. F. Weber to be $.955 \times 10^9$ absolute units, or .955 of the theoretical ohm.

Measurement of Resistance.—The practical connections for measuring a resistance x are shown in Fig. 49.

When the bridge is balanced we have

$$r_2\, x\ =\ r_1\, r_3$$

Whence

$$x\ =\ \frac{r_1\, r_3}{r_2}$$

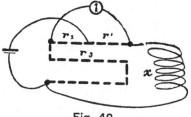

Fig. 49.

For a preliminary test, we may take $r_1 = r_2$, then x will be equal to r_3. Knowing the approximate value, we may alter

the ratio of the arms, taking care to attend to the condition of maximum sensitiveness.

If we make $r_1 = 10$, and $r_2 = 100$, then we shall be able to read to a tenth of an ohm ; for,

$$100\, x = 10\, r_3, \quad \text{whence } x = \frac{r_3}{10}$$

Similarly, if we make $r_1 = 10$ and $r_2 = 1000$, we may, supposing such arms to be consistent with sensitiveness, read to the 100th of an ohm.

It not unfrequently happens that even when every means available with the bridge has been taken, we cannot get a perfect balance. In such a case, it would be useful to introduce a rheocord or a calibrated wire. It is usual, however, to determine the resistance by a sort of method of interpolation.

For instance, suppose that with equal arms in the bridge and 11 ohms in r_3, we get a permanent deflection of 34 scale divisions to the right, and that with 12 ohms unplugged, we get a deflection of 16 divisions to the left, then the true resistance ($x = r_3$) would be $11 + \dfrac{34}{34 + 16} = 11.68$ ohms.

Small resistances are sometimes compared by a method which involves a knowledge of what is called the *logarithmic decrement* of a galvanometer needle.

When a magnetic needle oscillates, it sets up in neighbouring conductors induced currents whose electromagnetic action tends to gradually diminish the amplitude of the swing and ultimately to bring the needle to rest. There is also a further dissipation of the energy of the swinging needle arising from the resistance of the air and, to a small extent, from the viscosity of the suspending fibre.

The oscillations in such cases are said to be *damped*. Theory requires and experiment shows that the arcs form a decreasing geometrical progression ; the ratio of the amplitude of any two successive vibrations is called the *ratio of damping*, and its logarithm the *logarithmic decrement*.

In the mathematical theory of damping, it is shown that the expression for the angular elongation of the magnet contains a term involving the Napierian logarithm of the decrement. This latter quantity is denoted by λ and is, for the sake of brevity, called *the* logarithmic decrement.

It is, in most measurements, desirable that the galvanometer needle should, when a current is sent through the coils, quickly indicate the permanent deflection. In some instruments, this is imperfectly attained by

winding the wire on a copper frame ; in some others, great damping is effected by surrounding the magnet with a mass of copper, whilst in another class of galvanometers very considerable damping is attained by greatly multiplying the windings of the wire. In these galvanometers, the magnetic field due to the currents induced by the motion of the needle, is correspondingly increased, as is also the reaction of the field on the swing of the needle. With long coils in which the resistance may rise to 30,000 or 40,000 ohms, the free vibratory motion of the needle may be completely checked.

The same object is also attained by enclosing the magnet-mirror in a small brass tube, glazed in front and behind, and just large enough to allow the mirror to swing freely. The resistance of the closely confined air effectually checks any violent motion of the mirror, which, accordingly, turns to its position of equilibrium and stops there dead. Hence the name *Dead-Beat* by which all such galvanometers are known.

If a_m and a_n be the amplitudes, of the mth and nth oscillations, then the ratio of damping is $\left(\dfrac{a_m}{a_n}\right)^{\frac{1}{n-m}}$, and the log. decrement is $\dfrac{1}{n-m}\log_\epsilon\left(\dfrac{a_m}{a_n}\right)$

Errors of observation will have the least effect on the result when $\dfrac{a_m}{a_n}$ is equal to the base ϵ of the Napierian system of logarithms, *i.e.*, to 2.718, or 3 nearly.

It should be observed, that as part of the damping is due to the viscosity of the air and of the suspending fibre, the damping due to the galvanometer coils is the difference of that observed (1) when the galvanometer circuit is open and (2) when it is closed by a wire of no perceptible resistance.

To find the log. decrement of the needle of a small astatic galvanometer, the following observations of the turning-points were made :

With open Circuit.		With closed Circuit.	
50.6°	48°	46°	40°
44	42	38	33
40	38	32	28
35	33	26	23
31	29	22	19
27	26	18	16
24	23	15	14
22	20	13	11
19	18		

From the above we get the arcs of vibration to be

With open Circuit.	With closed Circuit.
98.6°	86°
86	71
78	60
68	49
60	41
53	34
47	29
42	24
37	

Therefore,

$$\lambda_0 \text{ (with open circuit) is } \tfrac{1}{8} \log_\epsilon \frac{98.6}{37} = .12$$

And

$$\lambda_1 \text{ (with closed circuit) is } \tfrac{1}{7} \log_\epsilon \frac{86}{24} = .18$$

$$\therefore \text{ the damping due to the coils is } .18 - .12 = .06$$

The damping depends upon the strength of the needle and on the form of the coil and the number of windings. The mathematical theory shows also that the logarithmic decrement varies inversely as the resistance in circuit. This suggests a method of measuring, or comparing, *small* resistances. In the case of high resistances, the damping becomes feeble and difficult of accurate measurement.

Let it be required, for example, to find the resistance G_1 of a galvanometer. If

λ_0 be the log. decrement with open circuit,

λ_1 ,, ,, ,, closed circuit,

And

λ_2 ,, ,, ,, known resistance R_2 in circuit.

Then we have

$$\lambda_1 - \lambda_0 : \lambda_2 - \lambda_0 = \frac{1}{G_1} : \frac{1}{G_1 + R_2}$$

Which gives G_1 in terms of R_2.

Ex. To determine the resistance of the galvanometer whose log. decrement was found above to be .06, an ohm was inserted between its terminals, and the log. decrement was again observed and found to be .02.

Then

$$.06 : .02 = \frac{1}{G_1} : \frac{1}{G_1 + 1}$$

Whence

$$G_1 = \tfrac{1}{2} \text{ ohm.}$$

If λ_3 be the log. decrement with an external resistance R_3 in circuit, we have

$$\lambda_1 - \lambda_0 : \lambda_2 - \lambda_0 : \lambda_3 - \lambda_0 = \frac{1}{G_1} : \frac{1}{G_1 + R_2} : \frac{1}{G_1 + R_3}$$

which gives R_3 in terms of R_2 and known quantities.

If a table of Napierian logarithms be not ready to hand, we may use common logarithms, multiplying them by 2.3;

For let

$$a^x = b^y = m \text{ (any number)}.$$

Then

$$a^{\frac{x}{y}} = b \quad \therefore \quad \frac{x}{y} = \log_a b \text{ and } x = y \log_a b \text{ i.e., } \log_a m = \log_b m \times \log_a b$$

$$\therefore \log_\epsilon m = \log_{10} m \times \log_\epsilon 10 \quad . \quad . \quad . \quad . \quad (1)$$

$$\text{Let } \epsilon^x = 10, \text{ then } x = \log_\epsilon 10.$$

Also taking the logarithm of both sides of the exponential, we have

$$x \log_{10} \epsilon = \log_{10} 10 = 1$$

Whence

$$x = \frac{1}{\log_{10} \epsilon} = \frac{1}{.434} = 2.3$$

Therefore, by (1)

$$\log_\epsilon m = \log_{10} m \times 2.3$$

which proves the above rule.

To determine the *specific resistance* of a wire or cylindrical conductor, we must find the resistance R of a length l and radius r. Then the specific resistance ρ being that of one centimetre cube of the material, that is, of a cylinder 1 cm. long and 1 sq. cm. in sectional area, we have

$$\rho = \frac{R}{l} \pi r^2$$

The diameter of the wire is measured by a fine-adjustment micrometer screw.

The resistance of a *galvanometer coil* may be found by the Wheatstone bridge, just as that of an ordinary conductor, provided we have a second instrument to use as a galvanoscope. If we have not, the resistance of the galvanometer may be determined by a method suggested to Sir William Thomson by Mance's method for finding the resistance of batteries.

The galvanometer is placed, Fig. 50, in one branch C D of the bridge, and a key in D F. On closing the battery circuit, the needle of the galvanometer is permanently deflected. When the bridge is balanced, D F and A B C are conjugate, and hence making or breaking contact at K will not affect the deflection.

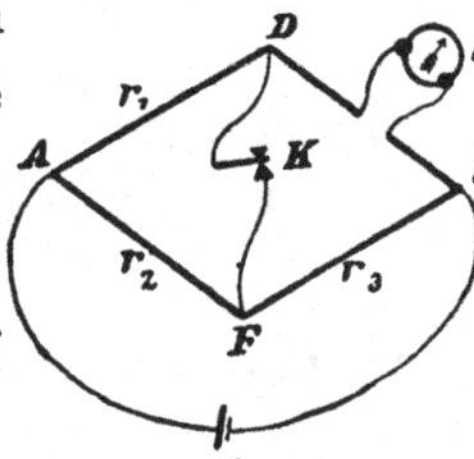

Fig. 50.

If G denote the resistance of the galvanometer, we have

$$G r_2 = r_1 r_3 \quad \text{whence } G = \frac{r_1 r_3}{r_2}$$

Besides dispensing with the use of an auxiliary galvanoscope, this method has the advantage of being independent of the resistance of the battery.

Mance's method for finding the *internal resistance* of a battery is shown in the theoretical diagram, Fig. 51. On closing the battery circuit, the galvanometer is permanently deflected, and the resistances are adjusted until the deflection is (if possible) unaffected by opening and closing K. Then B denoting the battery resistance, we have

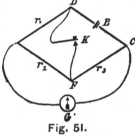

Fig. 51.

$$B\, r_2 = r_1\, r_3 \qquad \text{whence } B = \frac{r_1\, r_3}{r_2}$$

As the galvanometer is most sensitive when its deflection is small, a compensating magnet is usually placed near it, in order to bring the needle back to zero. It should be observed, however, that the needle is then in an unfavourable condition, being constrained by the strong field due to the magnet.

In practice it is found extremely difficult in most cases to so adjust the resistances that the galvanometer shall be absolutely unaffected by making and breaking contact at K. This difficulty arises from the change in the electromotive force of the battery occasioned by the alteration in the total resistance which opposes it. This variation, however, does not take place instantaneously. It requires time before becoming perceptible, and it is gradual in its development.

If the bridge be very nearly balanced, then on depressing K the needle will start off with a sudden jerk which will be followed by a slow increasing swing. The first effect is due to the deviation of the bridge from balance, and the second to the fall in the electromotive force.

To find the battery resistance as accurately as possible by this method, it is necessary to alter the resistances so that the sudden jerk and the gradual swing shall be in opposite directions. Should the latter be considerable, it may be reduced to a convenient limit by inserting a resistance in the battery circuit. The former is then worked down by varying the adjustable resistance until the needle is observed to start off gently on its swing.

Then $B = \dfrac{r_1\, r_3}{r_2}$, in which B denotes the whole resistance in the branch D C. By subtracting the additional resistance that may have been introduced, we obtain that of the battery.

Dr. O. J. Lodge has devised* a means of making this test which is practically independent of the variation of the electromotive force. The

---

* Phil. Mag., June, 1877.

theoretical arrangements are shown in Fig. 52, where L is a condenser inserted in the galvanometer circuit. When K is depressed the condenser is charged; there is a corresponding *throw* of the galva-nometer needle, after which it returns to its position of rest. Should any variation of charge occur, it will manifest itself on the galvanometer by a "kick" pro-portional to the extent of the variation. The ampli-tude of the initial throw is observed, and then worked down by altering the adjustable resistance until it is finally eliminated. Then the galvanometer will be absolutely unaffected by making and breaking contact at K.

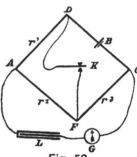

Fig. 52.

This test requires two keys, one in the battery and the other in the galvanometer circuit, and they should be constructed so that the latter may be broken the instant after the former is short-circuited. In this way the test may be completed in an exceedingly short period of time, and before any appreciable change in the electromotive force can take place.

With this modification, Mance's becomes a stricly null method.

The amplitude of the throw of the needle depends upon the galvanometer constant, which is itself a function of the number of windings. Hence a gal-vanometer with a great many turns of wire is the most sensitive for this test.

Another excellent method has been lately introduced by Mr. J. Munro and Dr. A. Muirhead. It is a modification of Kempe's bat-tery resistance test. The battery, condenser, and galvano-meter are connected up as shown in Fig. 53.

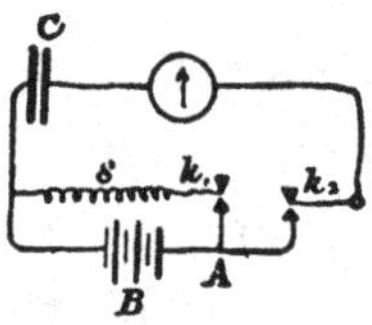

On striking down k_2, there is a rush of electricity which is equivalent to an instantaneous current, and which charges up the condenser C through the galvanometer, giving a throw d_2.

Fig. 53.

If R be the insulation resistance of the condenser, G the resistance of the galvanometer, and r that of the battery, then

$$d_2 = \frac{E}{G+R+r} \quad \text{whence } E = d_2\,(G+R+r)$$

Still keeping k_2 closed, k_1, which introduces the shunt S placed between the poles of the battery, is depressed and the resulting throw d_1 read off. This current, which is opposite in direction to the former, is due to the fall of potential caused by the introduction of the shunt.

The reduced charge of the condenser would, therefore, give a discharge deflection equivalent to $d_2 - d_1$.

By inserting the shunt, the interpolar resistance consisting of the

multiple arc formed by the condenser R and galvanometer G as one branch and the shunt S as the other, is

$$\frac{(G+R)\,S}{G+R+S}$$

And the battery current is

$$\frac{E}{r + \dfrac{(G+R)\,S}{G+R+S}}$$

The current which charges the condenser and would give the throw $d_2 - d_1$ is, therefore,

$$\frac{E}{r + \dfrac{(G+R)\,S}{G+R+S}} \times \frac{S}{G+R+S}$$

$$\therefore \frac{d_2}{d_2 - d_1} = \frac{r\,(G+R+S) + (G+R)\,S}{S\,(r+R+G)}$$

or dividing down by R

$$= \frac{r\left(\dfrac{G}{R} + 1 + \dfrac{S}{R}\right) + \left(\dfrac{G}{R} + 1\right)S}{S\left(\dfrac{r}{R} + 1 + \dfrac{G}{R}\right)}$$

Since R is very great in comparison with the other quantities, we may put

$$\frac{G}{R} = 0 = \frac{S}{R} = \frac{r}{R}$$

$$\therefore \frac{d_2}{d_2 - d_1} = \frac{r+S}{S}$$

Whence

$$r = \frac{d_1}{d_2 - d_1}\,S$$

The determination of d_1 can be made within a small fraction of a second, and consequently before any perceptible polarisation can set in.

The values d_1 and $d_2 - d_1$ may be directly observed by manipulating as follows :

First close k_1 and immediately afterwards k_2. The deflection obtained corresponds to $d_2 - d_1$. Then open k_2 and immediately after that open k_1. On again closing k_2, the new deflection will give d_1.

This method of operating has the practical advantage of giving the deflections in one and the same direction.

Ex. With three small Leclanché cells, a condenser of $\frac{1}{3}$ m.f. capacity and a shunt of 40 ohms, the initial throw d_2 was 220 divisions of the scale ; on shunting the battery, the throw d_1 was 80.

The resistance of the battery was therefore

$$\frac{80}{140} \times 40 = 22.8 \text{ ohms.}$$

Very high resistances may be determined (1) from the fall of potential over a known resistance with which the unknown is connected ; and

(2) From the rate of fall of potential of a condenser of known capacity when discharging through the unknown resistance.

Let R, Fig. 54, be the known and x the unknown resistance. One pole of the battery is connected to A, the other pole as well as the end C being put to earth, or connected up together.

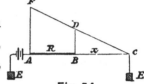

Fig. 54.

The fall of potential from A to C is gradual. By connecting the electrometer electrodes to A and C, the deflection is proportional to the fall over A C. Let this be denoted by the perpendicular A F. Shifting one electrode from A to B, the new deflection will be proportional to the fall over B C, viz., B D. As the fall from A to C is gradual, F D C will be a straight line, and by similar triangles we have

$$A\,F : B\,D = R + x : x$$

whence x is known.

For the electrometer we may substitute a condenser. Connecting the terminals to A and C, the discharge deflection will be proportional to the difference of potentials between A and C, viz., A F. By charging up over B C, the discharge deflection will give B D. whence x may be found as above.

The insulation resistance of a submarine cable may be determined by this method.

If required to find the resistance of an incandescence lamp when hot, we might connect up as in Fig. 55, where B C denotes the lamp, T G a torsion galvanometer, and R a known resistance made of thick uncovered wire.

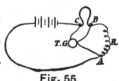

Fig. 55.

By connecting T G to end of the resistance A and the lamp terminal C, we get the fall of potential over the known and unknown resistances. Then disconnecting at A and joining up with B, we get the fall over the lamp alone.

As an example, let R be 2.7 ohms, and the fall over R and lamp give a discharge deflection of 38 divisions, whilst that over lamp alone gives 34.

We then have

$$38 : 34 = 2.7 + x : x$$

whence $x = 23$ ohms = resistance of lamp when hot.

The torsion galvanometer may be replaced in this measurement by a voltmeter or a potential galvanometer.

The *insulation resistance* of condensers as well as of cables is usually found from the rate of fall of potential. Thus let a condenser whose capacity in microfarads is C be charged up to a potential V_1, and after T seconds let this have fallen to V_2. Then the resistance R of the condenser, expressed in megohms, is given by

$$R = \frac{T}{C \log_\epsilon \frac{V_1}{V_2}}$$

Ex. A condenser whose capacity was $\frac{1}{57}$th of a microfarad when charged by a Daniell's cell, gave an instantaneous discharge V_1 of 49 scale divisions, after one second of 41 divisions, and after four seconds of 25 divisions.

Here

$$R = \frac{1}{\frac{1}{57} \log_\epsilon \frac{49}{41}} = \frac{1}{\frac{1}{57}\left(\log_{10}\frac{49}{41}\right) \times 2.3} = \frac{1 \times .43}{\frac{1}{57} \log_{10}\frac{49}{41}} = 318 \text{ megohms.}$$

Ex. To find the insulation resistance of a cable, one end of the core was insulated, the other being connected with the positive pole of a battery. The negative pole was connected with a metal strip placed in the water in which the cable was immersed. The cable ($\frac{1}{90}$ m.f.) when charged up to the full potential of the battery, gave a discharge deflection V_1 of 61 scale divisions. On being recharged and then insulated for 30 seconds, the discharge deflection V_2 was 36. The resistance of the cable was therefore

$$\frac{30 \times .43}{\frac{1}{90} \log_{10}\frac{61}{36}} = 504 \text{ megohms.}$$

This is the mean resistance of the cable during the 30 seconds; for, owing to what is called *electrification*, the insulation appears to improve with time. Thus if a battery, cable, and galvanometer be joined up in circuit, it will be observed on closing the circuit that the deflection of the galvanometer decreases very rapidly at first, then very gradually, becoming practically stationary after a certain time.

This phenomenon is but imperfectly understood. Whether due to polarisation or not, its electrical effect is equivalent to an increase in the resistance of the insulating material of the cable.

The insulation resistance of *conductors* covered with gutta-percha for electric lighting or subterranean telegraph lines is determined, in practice, by the same method which is used for measuring the insulation of ordinary telegraph lines.

It consists in insulating one end of the conductor which is then charged

through a galvanometer. The leakage of electricity from the charged conductor to the earth or water with which the insulating surface may be in contact, will manifest itself on the galvanometer by a steady current. By comparing this current with that given when the charging battery is closed over a known resistance, we may deduce the insulation of the conductor.

Thus suppose that with 10,000 ohms in circuit, we get a deflection of 15 divisions of the scale, then the resistance required for a deflection of one division would be $10,000 \times 15 = 150,000$ ohms. This gives what is called the *constant*.

If the insulating material of a conductor charged from the same battery, gave a deflection of five divisions, we should infer that its resistance is $\frac{150,000}{5} = 30,000$ ohms. If its length were 100 yards, and we assume that the leakage is uniform over the line, then the resistance per yard would be $30,000 \times 100 = 3,000,000$ ohms or 3 megohms. This is obvious, for the greater the length, the greater also will be the leakage and the less the insulation. Therefore the total insulation resistance varies inversely as the length of the conductor.

The preceding formula enables us to find the time required for the charge of a condenser, or a cable, to fall to any fraction of its initial value. In cable testing, the time of falling to half charge is often taken as a unit of comparison.

This might be found directly from experiment; but as such a determination would entail much time and trouble, it is usual to calculate it from two observations of potential and the known interval of time between them.

Thus from

$$R = \frac{.43\,T}{C \log_{10} \frac{V_1}{V_2}}$$

we get

$$T_1 = \frac{R\,C}{.43} \log \frac{V_1}{V_2}. = \lambda \log \frac{V_1}{V_2}$$

where λ denotes the constant factor $\dfrac{R\,C}{.43}$.

The time T_2 required to fall from V_1 to V_3 is

$$= \lambda \log \frac{V_1}{V_3}$$

Whence

$$\frac{T_2}{T_1} = \frac{\log \frac{V_1}{V_3}}{\log \frac{V_1}{V_2}}$$

For half-charge, $V_3 = \frac{1}{2} V_1$

$$\therefore \frac{T_2}{T_1} = \frac{\log 2}{\log \frac{V_1}{V_2}}$$

Whence

$$T_2 = T_1 \frac{\log 2}{\log \frac{V_1}{V_2}}$$

which gives the required time T_2 in terms of known quantities.

Applying this to the first example on page 69, we get

$$T_2 = 1 \times \frac{.3010300}{\log \frac{49}{41}} = \frac{.3010300}{.0774122} = 3.89 \text{ seconds.}$$

From the experimental data, we see that in four seconds the potential fell from 49 to 25.

In measuring discharge deflections, it is often necessary, in order to get a scale reading, to shunt the galvanometer, the true deflection being taken as equal to the reading multiplied by the multiplying power of the shunt.

It will, however, be found that the deflection without the shunt, when on the scale, is slightly greater than the deflection thus calculated.

This discrepancy was first pointed out by Mr. Latimer Clark[*], and traced by him to the induced currents set up in the galvanometer coils by the swing of the needle.

This effect should be estimated and allowed for in all discharge deflections with a shunted galvanometer.

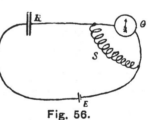

Fig. 56.

If C_1 be the current through the galvanometer whose resistance is denoted by G, Fig. 56, β that through the shunt, resistance S, and a that through the condenser, resistance R, we have

$$a = \beta + C_1 \quad . \quad . \quad . \quad . \quad . \quad . \quad . \quad (1)$$

Also from the circuit formed by the condenser, galvanometer, and battery, we have

$$a R + C_1 G = E - e_1 \quad . \quad . \quad . \quad . \quad . \quad (2)$$

where e_1 is the opposing E.M.F. produced by the swing of the needle.

From the circuit formed by the condenser, shunt, and battery, we have

$$a R + \beta S = E \quad . \quad . \quad . \quad . \quad . \quad (3)$$

Taking the value of β from (1) and substituting it in (3), we get

$$a R + (a - C_1) S = E$$

---

[*] Jour. Soc. Tel. Engineers, January, 1873.

or,
$$a(R+S) - S C_1 = E \qquad . \qquad . \qquad . \qquad . \qquad . \qquad . \qquad (4)$$

Eliminating a between (2) and (4), we get
$$R(C_1 G + e_1) + S(C_1 G + e_1) + S(R C_1 - E) = 0$$

or,
$$(R+S)(C_1 G + e_1) + S(R C_1 - E) = 0 \qquad . \qquad . \qquad (5)$$

On removing the shunt, the discharge gives
$$C_2 = \frac{E - e_2}{R + G}$$

where e_2 is the new opposing E.M.F.

Solving for E and substituting in (5), we get
$$(R+S)(C_1 G + e_1) + R S C_1 - S(C_2 R + C_2 G - e_2) = 0$$

As the opposing E.M.F. is proportional to the discharge, we may put
$$e_1 = k\, C_1 \text{ and } e_2 = k\, C_2.$$
$$\therefore (R+S)(C_1 G + k\, C_1) + R S C_1 - S(C_2 R + C_2 G - k\, C_2) = 0$$

Dividing down by R, this becomes
$$\left(1 + \frac{S}{R}\right)(G+k)\, C_1 + S\, C_1 - S\, C_2 \left(1 + \frac{G}{R} - \frac{k}{R}\right) = 0$$

Observing that R is comparatively very great, we may put
$$\frac{S}{R} = \frac{G}{R} = \frac{k}{R} = 0$$

Therefore finally
$$(G+k)\, C_1 + S\, C_1 - S\, C_2 = 0$$

Whence
$$C_2 = \frac{G + k + S}{S}\, C_1$$

To determine k, we charge a condenser and, using a shunt equal in resistance to that of the galvanometer, note the discharge deflection d_1.

Taking a second condenser of half the capacity of the former, we charge it up from the same battery, and observe the deflection d_2 obtained without a shunt.

It is then obvious that $2\, d_2$ is the deflection which we should have got from the first condenser had no shunt been used, and this must equal d_1 multiplied by the multiplying power of the shunt. That is
$$2\, d_2 = d_1 \left(\frac{G + k + S}{S}\right) = d_1 \left(\frac{2\, G + k}{G}\right)$$

This gives the value of k.

Ex. Using a condenser of a microfarad and a shunt equal in resistance to the galvanometer (3000 ohms), d_1 was 156 scale divisions. With $\frac{1}{2}$ m.f. and no shunt, d_2 was 160 divisions.

Therefore,
$$320 = 156 \left(\frac{6000 + k}{3000}\right)$$

Whence

$$k = 154 \text{ ohms.}$$

When, therefore, this particular galvanometer is used to measure discharge deflections, the multiplying power of any shunt should be taken as

$$\frac{G + 154 + S}{S}$$

The *electromotive force* of a battery and the difference of potentials between two points of a conductor conveying an electric current, may be found in absolute measure by means of an absolute electrometer or by an ordinary electrometer whose constant is known.

It is seldom, however, that an absolute electrometer is available, and hence the comparative methods which are used.

The E.M.F. of a battery is often found by comparison with a standard cell. Thus,

Connecting an ordinary quadrant electrometer with a Latimer Clark's element, a deflection of 60 scale divisions was obtained, whilst a certain cell whose electromotive force E was required, gave a deflection of 54.

Therefore we have

$$60 : 54 = 1.457 : E$$

Whence

$$E = 1.31 \text{ volts.}$$

In all such measurements, we may replace the electrometer by a condenser and suitable galvanometer. Thus a $\frac{1}{3}$ m.f. charged from a Daniell's cell, gave a discharge deflection of 180 scale divisions, whilst the deflection when the condenser was charged from a Planté's secondary cell was 390.

Here

$$180 : 390 = 1.08 : E$$

Whence

$$E = 2.34 \text{ volts.}$$

In the above, it is assumed that the electromotive force of a Clark's cell is 1.457 volts and that of the standard Daniell 1.08 volts.

A method of *compensation* due to Poggendorff is often used for directly comparing the electromotive forces of two batteries.

The principle of this test will be understood from Fig. 57 where E_1 and E_2 are the electromotive forces which we have to compare, r_1 and r_5 the resistances of the batteries, whilst r_2 and r_3 are adjustable resistances. In the circuit of E_2 there is a zero galvanometer of resistance r_4. The connections are made so that E_1 and E_2 tend to send a current in the same direction.

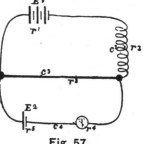

Fig. 57.

From Kirchoff's laws we have

$$c_2 + c_4 = c_3 \qquad . \qquad . \qquad . \qquad . \qquad . \qquad . \quad (1)$$

$$c_2 (r_1 + r_2) + c_3 r_3 = E_1 \qquad . \qquad . \qquad . \qquad . \quad (2)$$

$$c_4 (r_4 + r_5) + c_3 r_3 = E_2 \qquad . \qquad . \qquad . \qquad . \quad (3)$$

Subtracting (2) from (3) we get

$$c_4 (r_4 + r_5) - c_2 (r_1 + r_2) = E_2 - E_1 \qquad . \qquad . \qquad . \quad (4)$$

Substituting in (2) the value of c_3 from (1), we get

$$c_2 (r_1 + r_2 + r_3) + c_4 r_3 = E_1 \qquad . \qquad . \qquad . \quad (5)$$

Eliminating c_2 between (4) and (5),

$$c_4 = \frac{(E_2 - E_1)(r_1 + r_2 + r_3) + E_1 (r_1 + r_2)}{r_3 (r_1 + r_2) + (r_4 + r_5)(r_1 + r_2 + r_3)}$$

When the resistances are adjusted so that the galvanometer indicates no current, we must have $c_4 = 0$. Therefore,

$$(E_2 - E_1)(r_1 + r_2 + r_3) + E_1 (r_1 + r_2) = 0$$

Whence

$$\frac{E_1}{E_2} = \frac{r_1 + r_2 + r_3}{r_3} \qquad . \qquad . \qquad . \qquad . \qquad . \quad (6)$$

This expression contains the battery resistance r_1. We may eliminate this quantity by a second measurement, in which r_2 is changed and r_3 altered until there is again no galvanometer deflection.

Let the new values of r_2 and r_3 be R_2 and R_3 respectively.

Then

$$\frac{E_1}{E_2} = \frac{r_1 + R_2 + R_3}{R_3} \qquad . \qquad . \qquad . \qquad . \quad (7)$$

From (6),

$$E_2 (r_1 + r_2 + r_3) = E_1 r_3$$

From (7),

$$E_2 (r_1 + R_2 + R_3) = E_1 R_3$$

By subtracting,

$$E_2 (R_2 + R_3 - r_2 - r_3) = E_1 (R_3 - r_3)$$

Whence

$$\frac{E_1}{E_2} = \frac{(R_2 + R_3) - (r_2 + r_3)}{R_3 - r_3}$$

which gives the required ratio in terms of known quantities, and without involving the battery resistance.

This method would be perfect if we had an absolutely constant standard (E_2) of E.M.F., and if the electromotive force of batteries were independent of the resistance in circuit. We know, however, that E_1 does depend upon the interpolar resistance r_2 and r_3, and hence the ratio $\frac{E_2}{E_1}$ will be slightly different for different values of these variables.

This source of inaccuracy may be eliminated by a valuable modification,

due to Mr. Latimer Clark, of the former method, in which the comparison is made when neither the trial battery nor the standard cell is sending a current, and consequently also errors arising from polarisation are entirely avoided.

For the sake of brevity, equation (6) may be written

$$\frac{E_1}{E_2} = \frac{R}{r_3}$$

where R is the total resistance in the circuit of the standard cell E_1.

In Fig. 58, E_2 and E_3 are the electromotive forces to be compared, F is an auxiliary battery whose electromotive force is E_1, and the total resistance in its circuit is denoted by R.

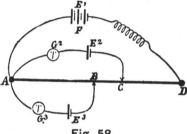

Fig. 58.

When the point C on the slide resistance A D has been found such that the galvanometer G_2 is undeflected, we have

$$\frac{E_1}{E_2} = \frac{R}{A\,C}$$

Similarly when the point B has been taken such that the needle of G_3 stands at zero, we have

$$\frac{E_1}{E_3} = \frac{R}{A\,B}$$

Whence

$$\frac{E_2}{E_3} = \frac{A\,C}{A\,B}$$

and this expression is independent of the variations of E_1 and of R.

This relation may also be deduced in the following simple manner : The current C given by the battery F is $\frac{E_1}{R}$.

Also the potential-difference E_3 between A and B.

$$= C \times A\,B = \frac{E_1}{R} \times A\,B$$

Similarly the potential-difference E_2 between A and C,

$$= C \times A\,C = \frac{E_1}{R} \times A\,C$$

Therefore,

$$\frac{E_2}{E_3} = \frac{A\,C}{A\,B}.$$

If E_3 were absolutely constant, the electromotive force of any battery could be very accurately given in terms of such a standard.

For many purposes Latimer Clark's cell is used as a standard, its electromotive force being taken as 1.457 volts.

For the ready application of this method, Mr. Latimer Clark has devised his *Potentiometer*. The slide resistance consists of a platinum-iridium wire of 40 ohms, wound helically round an ebonite cylinder. The edge of this cylinder is accurately divided into 1000 parts; and, as the wire makes twenty complete turns, it is practically divided into 20,000 equal parts, each having a resistance of $\frac{1}{500}$th $\left(\frac{40}{20,000}\right)$ of an ohm. By means of a very delicate galvanometer, it is found possible with this instrument to attain an accuracy of the $\frac{1}{1,000,000}$th part of the electromotive force of a Daniell's cell.

The *superior limit* of the electromotive force of a battery may be determined from the energies of combination of the metals and the electrolyte. This was first pointed out by Sir William Thomson in his two remarkable papers on the dynamical theory of electrolysis, published in the *Philosophical Magazine* for 1851.

Thus in a Daniell's cell we may consider the chemical work done to be, on the one hand, the combination of zinc with sulphuric acid, forming zinc sulphate; and, on the other, the precipitation of copper in the metallic state from the sulphate. The former action is attended by the evolution and the latter by the absorption of a definite quantity of heat.

According to Julius Thomsen, 1670 thermal units, or water-gramme-degrees Centigrade, are given out when one gramme of zinc, and 881 when one gramme of copper dissolves in sulphuric acid.

If unit current ($\frac{1}{10}$th ampère) pass through the cell for unit time, one electro-chemical equivalent of zinc (.003412 gm.) will dissolve, thus giving out 003412×1670 thermal units, while an electro-chemical equivalent of copper (.003307 gm.) will be deposited, thereby absorbing $.003307 \times 881$ thermal units.

If we assume that the difference between these two quantities is entirely transformed into the energy of the electric current, we shall have for the superior limit of the resulting electromotive force E of a Daniell's cell

$$(.003412 \times 1670 - .003307 \times 881) \times 42 \times 10^6 \text{ absolute units,}$$

where 42×10^6 is Joule's mechanical equivalent for one water-gramme-degree Centigrade.

Therefore,

$$E = 1.165 \times 10^8 = 1.165 \text{ volts.}$$

The *capacity* of a condenser is usually determined in practice by comparison with a standard condenser of known capacity.

It may, however, be found independently in absolute measure.

It may be shown* that the quantity Q of electricity in a transient current such as that of a condenser, is given by

$$Q = \frac{H}{G} \frac{\tau}{\pi} 2 \sin \tfrac{1}{2} \theta \qquad . \qquad . \qquad . \qquad . \qquad . \qquad . \quad (1)$$

Where H denotes the earth's horizontal force.

G   ,,   constant of the galvanometer.

τ ,,   time of a semi-vibration of the galvanometer needle.

θ ,,   first elongation or throw of the needle.

Then if E be the electromotive force and R the resistance in the circuit of the battery, we have the current strength

$$C = \frac{E}{R} = \frac{H}{G} \tan \phi,$$

where ϕ denotes the angular deflection of the galvanometer.

Therefore,

$$\frac{H}{G} = \frac{E}{R \tan \phi}.$$

Also the charge $Q = F E$, where F denotes the capacity of the condenser.

Therefore from (1)

$$F E = \frac{E \tau}{\pi R \tan \phi} 2 \sin \tfrac{1}{2} \theta$$

And

$$F = \frac{\tau}{\pi R \tan \phi} 2 \sin \tfrac{1}{2} \theta \qquad . \qquad . \qquad . \qquad . \qquad . \quad (2)$$

This assumes that the needle swings in an unresisting medium. There is, however, always some damping arising from the reaction of the induced currents, caused by the swinging magnet as well as from the viscosity of the air and of the suspending fibre. If the logarithmic decrement λ be not great, these disturbing effects will be practically eliminated by writing in (2)

$$(1 + \tfrac{1}{2} \lambda) \theta \text{ for } 2 \sin \tfrac{1}{2} \theta.$$

We therefore get for the working formula

$$F = \frac{\tau (1 + \tfrac{1}{2} \lambda)}{\pi R \tan \phi} \theta.$$

The connections for this determination may be made as in Fig. 59. Unplugging a very high resistance, say 10,000 ohms, and inserting the galvanometer in the battery circuit, we find $\tan \phi$. The galvanometer is then connected with the binding screws D and K of a suitable discharge key.

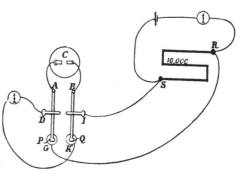

Fig. 59.

* Maxwell, Art. 748.

The figure shows one terminal of the condenser C connected with S through the binding screw B and the cross-piece I; the other coating of the condenser is connected with P through A, and the other end of the resistance R with the binding-screw G.

Consequently by depressing P the condenser is charged, the charging electromotive force being the potential-difference between the points S and R.

Releasing P and snapping down K, we discharge the condenser through the galvanometer and obtain the elongation θ. While the needle is swinging, we note down its successive turning points and thence deduce the log decrement

Ex. To find in absolute measure the capacity of a certain condenser, a Daniell's cell, a galvanometer of 3000 ohms, and a resistance box were connected up in simple circuit. To keep the spot of light on the scale a $\frac{1}{99}$th shunt was used. The resistance of the shunted galvanometer was thus

$$\frac{1}{\frac{1}{G}+\frac{99}{G}} = \frac{G}{100} = 30 \text{ ohms.}$$

Hence to maintain a circuit resistance of 10,000 ohms, a resistance of 9970 ohms was unplugged in the box.

The deflection of the galvanometer was 9.5 millimetre divisions, the scale being 730 mm. from the mirror.

Therefore,

$$\tan \phi = \frac{9.5}{730 \times 2}$$

neglecting a small correction.* Discharging the condenser through the galvanometer the throw was 5 scale-divisions.

Hence

$$\theta = \frac{5}{730 \times 2}$$

The number of divisions traversed by the spot of light was successively

$$93 \qquad 78 \qquad 65 \qquad 54 \qquad 46 \qquad 39 \qquad 33$$

from which a mean value for λ was found to be .171.

The time τ was determined from a number of observations of the time of 20 semi-vibrations, the mean value being 1.8 seconds.

The capacity F of the given condenser, expressed in farads, is therefore

$$\frac{1.8 \times 1.085 \times \dfrac{5}{730 \times 2}}{\dfrac{22}{7} \times 10,000 \times \dfrac{9.5}{730 \times 2} \times 100.}$$

* *Ante* page 36.

The last factor (100) in the denominator is inserted to get the deflection which would have been obtained, had it been possible to dispense with the shunt.

Therefore,

$$\text{F, in farads,} = \frac{327}{10^9} = .327 \text{ microfarads.}$$

Having a condenser of know capacity, that of any other condenser may be found from the throw of the galvanometer; for

$$F_1 = \frac{\tau (1 + \frac{1}{2} \lambda)}{\pi \, R \, \tan \phi} \, 2 \sin \tfrac{1}{2} \, \theta_1$$

And

$$F_2 = \frac{\tau (1 + \frac{1}{2} \lambda)}{\pi \, R \, \tan \phi} \, 2 \sin \tfrac{1}{2} \, \theta_2$$

Therefore,

$$\frac{F_1}{F_2} = \frac{\sin \frac{1}{2} \, \theta_1}{\sin \frac{1}{2} \, \theta_2}.$$

If n be the number of scale divisions in the deflection, and r be the distance from the scale to the mirror, we know that

$$\sin \tfrac{1}{2} \, \theta = \frac{n}{4 \, r} \left[1 - \frac{11}{2} \left(\frac{n}{4 \, r} \right)^2 \right]$$

As the angular displacement of the magnet is always small, we may neglect the second term in the above expression and write

$$\sin \tfrac{1}{2} \, \theta = \frac{n}{4 \, r}$$

Whence

$$\frac{F_1}{F_2} = \frac{n_1}{n_2}.$$

Ex. A condenser of $\frac{1}{3}$ m.f. when charged over a certain resistance through which a current was passing, gave a throw of 75 divisions of the scale, whilst a second condenser when charged over the same resistance gave a deflection of 50 scale divisions.

Therefore,

$$75 : 50 = \tfrac{1}{3} : F.$$

Whence

$$F = \tfrac{2}{9} \text{ m.f.}$$

The capacity F_2 of a condenser is often found by sharing its charge with a condenser of known capacity F_1.

The charge

$$Q = F_1 \, V_1$$

When this is shared, the potential falls to V_2 and we have

$$Q = (F_1 + F_2) \, V_2$$

Therefore,

$$F_1 \, V_1 = (F_1 + F_2) \, V_2$$

Whence
$$F_2 = \frac{(V_1 - V_2)}{V_2} F_1.$$

Ex. A $\frac{1}{3}$ m.f. charged from a Daniell's cell gave a discharge deflection V_1 of 62 divisions; when the charge was shared with a small mica condenser the deflection V_2 fell to 38.

Therefore,
$$F_2 = \tfrac{1}{3} \times \tfrac{24}{38} = .211 \text{ m.f.}$$

The capacity of a submarine cable may be found in the same way. One end L, Fig. 60, of the cable is insulated; a metal strip S being immersed in the water serves to make contact with the outer coating (the water) of the cable. To avoid leakage over the surface,

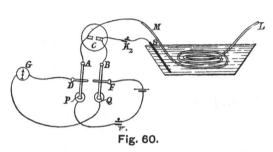

Fig. 60.

the gutta-percha near the insulated end L is pared, and then steeped in melted paraffin wax.

Closing K_1 and then successively depressing and relieving P and Q, we charge and discharge the condenser, thus getting a deflection corresponding to V_1. Inserting the plug in the condenser, we insure that there is no residual accumulation. Putting down P, we again charge C; relieving P and closing K_2, we share the charge with the cable. Opening K_2 and snapping down Q we get the discharge deflection corresponding to V_2. This completes the data required for determining the capacity of the cable.

Ex. A $\frac{1}{3}$ m.f. when fully charged gave a discharge deflection of 62 divisions. After recharging and connecting with a piece of cable, the second discharge was 60 divisions.

The capacity of the cable was, therefore, $\frac{1}{3} \times \frac{2}{60} = \frac{1}{90}$ microfarad.

The capacity of condensers may also be compared by charging them up over known resistances. These should be constructed so as to eliminate the effects of self-induction. Ordinary resistance coils may be used.

Let the distribution of potentials at any instant be as marked in Fig. 61. The condition that there shall be no current through the galvanometer is that

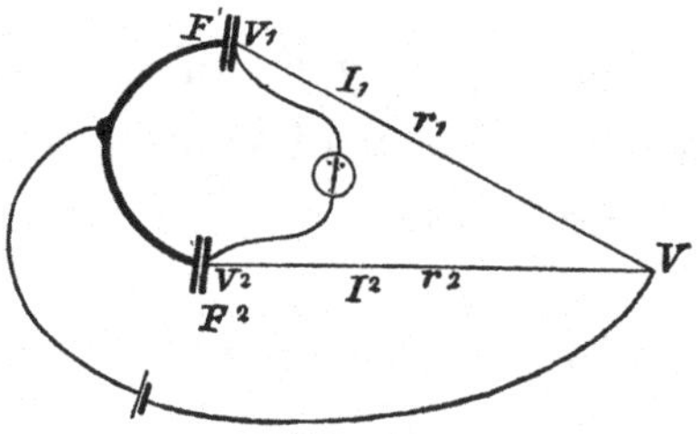

$$V_1 = V_2$$

Therefore,

$$V - V_1 = V - V_2 \qquad . \quad . \quad . \quad . \quad . \quad . \quad (1)$$

Fig. 61.

We also have the current

$$I_1 = \frac{V - V_1}{r_1} \qquad . \qquad . \qquad . \qquad . \qquad . \qquad .$$

And

$$I_2 = \frac{V - V_2}{r_2} \qquad . \qquad . \qquad . \qquad . \qquad . \qquad .$$

$\left.\vphantom{\begin{array}{c} \\ \\ \\ \\ \end{array}}\right\}(2)$

But in a small time dt the quantity of electricity that flows through r_1 is $I_1\, dt$, and this gives the condenser a charge $F_1\, d\, V_1$

$$\therefore\ F_1\, d\, V_1 = I_1\, dt$$

And similarly,

$$F_2\, d\, V_2 = I_2\, dt.$$

If the increments of potential are equal, as they must be for no galvanometer deflection, we have by dividing these last equations,

$$\frac{F_1}{F_2} = \frac{I_1}{I_2} = \frac{V - V_1}{r_1} \times \frac{r_2}{V - V_2} = \frac{r_2}{r} \ \text{(by 2)}$$

Whence

$$F_1\, r_1 = F_2\, r_2.$$

When this relation holds, we know that $V_1 = V_2$ throughout the period of charging, and, consequently, there can be no deflection in the galvanometer.

Ex. Using a Leclanché cell and unplugging 2200 ohms (r_1) in one branch of a resistance box which was connected with a condenser of capacity F_1, the galvanometer deflection was gradually worked down by varying the resistance (r_2) in the other branch which was connected with a condenser of capacity F_2, the needle standing at zero when r_2 was 1465 ohms.

Therefore,

$$\frac{F_1}{F_2} = \frac{1465}{2200} = \tfrac{2}{3} \ \text{very nearly.}$$

The condenser whose capacity is denoted by F_2 was $\tfrac{1}{2}$ m.f.

Therefore,

$$F_1 = \tfrac{1}{2} \times \tfrac{2}{3} = \tfrac{1}{3} \ \text{m.f.}$$

The capacity of a condenser may also be found from the rate at which it discharges through a known resistance.

We have already seen* that a very high resistance may be determined from

$$R = \frac{.434\ T}{S \log_{10} \dfrac{V_1}{V_2}}$$

where S is the capacity of a condenser and T the time required for the

* See page 69.

potential to fall from V_1 to V_2, the condenser discharging itself through a resistance R.

If we know R, then S can be found from

$$S = \frac{.434 \, T}{R \log_{10} \frac{V_1}{V_2}}$$

As it is the ratio of the potentials that appears in the above expression, these may be reckoned in any, the same, arbitrary unit.

This ratio should be taken nearly equal to 3 or to ϵ (2.718) the base of the Napierian logarithms, as errors of observation will then produce the least effect.

If R be estimated in ohms, S will be given in farads; if in megohms, the resulting capacity will be in microfarads.

Ex. The initial discharge V_1 of a condenser was 150 scale-divisions; on recharging and allowing the condenser to discharge itself for 90 seconds through a resistance of 180 megohms, the deflection V_2 obtained was 60.

$$\text{Here the capacity } S = \frac{.434 \times 90}{180 \times \log \frac{150}{60}} = .623 \text{ microfarads.}$$

If this method be applied to cables, the capacity deduced can only be approximate, because the insulation resistance varies with the time from charging; it is its mean value that enters the above formula.

It was pointed out by Clerk-Maxwell that, owing to electric absorption, the only safe dielectric for a standard condenser is air. Accordingly, a Committee of the British Association has been appointed to prepare a number of such standards.

At the Southport meeting, 1883, it was stated that Dr. A. Muirhead had already completed three such air-condensers.

It is seldom that the *Quantity* Q has to be determined as a separate element. It may, however, be deduced from observations on a tangent galvanometer whose constant is known.

The quantity Q of electricity that passes in a time T through a given section of a conductor traversed by a constant current C, is C T. If C be measured in ampères and T in seconds, then Q will be given in coulombs. The coulomb being the quantity of electricity conveyed by an ampère in a second is therefore $\frac{1}{10}$th of the absolute unit of quantity.

When a current is very variable we may not assume that the mean of many readings of a tangent galvanometer corresponds to the equivalent constant current, and thence deduce the quantity of electricity that has

passed; but we must evaluate $\int \gamma \, dt$ over small periods of time during each of which the current remained sensibly constant. The sum of these separate integrals will give Q.

Q may also be determined from the decomposition of an electrolyte whose electro-chemical equivalent ϵ is known.* Thus in the case of copper sulphate, we know that an ampère in a second, *i.e.*, a coulomb, throws down on the negative electrode ·0003307 grammes. Therefore a deposit of W grammes would require $\dfrac{W}{.0003307}$ coulombs.

It is by means of the copper voltameter that it has been proposed to measure the quantity of electricity absorbed in domestic electric lighting. The negative electrode would be carefully weighed at the beginning and at the end (say) of a month; then if the difference in grammes be W, we have $\dfrac{W}{\frac{\epsilon}{10}} = \dfrac{10\,W}{\epsilon}$ the number of coulombs absorbed.

The quantity contained in a transient current, such as induction currents and also discharge currents from condensers, may be calculated from

$$Q = \frac{H}{G} \frac{\tau}{\pi} \, 2 \sin \tfrac{1}{2} \, \theta,$$

which formula has already been discussed.

The strength C of induction currents which rapidly change their sign and are not rectified by a commutator such as those of an alternate-current machine may be measured, as previously shown, by an electro-dynamometer; and consequently the quantity of available electricity developed by such a machine in a time T may be found when necessary by multiplying the current expressed in ampères by the time in seconds.

WORK.—When a force acts upon a body so that its point of application is displaced, in the direction of the force, then the force, or more properly the agent applying it, is said to do work. This work may manifest itself as a motion of translation or of rotation; it may appear as a deformation of the body, or it may assume more subtle forms, such as heat and electricity.

Work is estimated quantitatively as the product of the force and the effective distance through which it moves. The effective distance is always measured in the line of action of the force.

In lifting a stone, we do work against gravity; in starting a train, the expansive force of steam does work against inertia and friction; in bending a steel blade, work is done against molecular forces.

* Lord Rayleigh, in a communication (March 8, 1884) to the London Physical Society on the electro-chemical equivalent of silver, gave as the mean result of careful determinations made in the Cavendish Laboratory, ϵ=.01118. This would make the ϵ of copper .003261; and, therefore, a coulomb would deposit .0003261 grammes,

Whenever work is done against opposing forces, energy is conferred. The raised stone, the moving train, the bent blade, are capable of doing work, and are, therefore, said to possess energy.

The principle of work is applicable to electrical as it is to mechanical forces.

To bring up a particle charged with a unit of positive electricity from one point of an electric field due to a quantity of positive electricity to any other point, work is spent in overcoming electric force, viz., that due to the repulsion of similar electrifications. In like manner, to remove a particle charged with a unit of negative electricity from one point of the above field to another, work is done in overcoming the electric force arising from the attraction of opposite electrifications.

The *potential* at any point of an electric field is defined as the work which must be done in order to bring a particle charged with a unit of positive electricity from the infinitely distant boundary of the field up to that point. The work done in bringing up a small body charged with Q units would, therefore, be Q times the potential at that point, *i.e.*, Q V. So also the work done in bringing Q units from a point at which the potential is V_1 to a point at which it has the higher value V_2, is

$$Q (V_2 - V_1) \text{ or simply } Q E,$$

where E denotes the electromotive force due to the potential-difference between the points.

Since work is force multiplied by distance, unit work is unit force multiplied by unit distance,

$$= \text{dyne} \times \text{centimetre}$$
$$= \text{the } erg.$$

The force with which a body is drawn to the earth, that is, the force with which it gravitates, and which is measured by its weight W, depends upon the mass M of the body, and also upon the acceleration g due to gravity.

Therefore we may write

$$W = M g.$$

Since unit force (the dyne) is that force which, acting on one gramme of matter for one second, imparts to it an acceleration of one centimetre per second, and as g may be taken to be 981 centimetres per second*, it follows

---

* The value of g at Paris is 980.94
    ,,   ,,   Greenwich is 981.17
    ,,   ,,   New York is 980.23.
    ,,   ,,   Baltimore 980.05

that the force which is measured statically by the weight of one gramme is equal to 1 × 981 units of force,

$$= 981 \text{ dynes};$$

and the work done in lifting one gramme vertically through one centimetre,

$$= 981 \times 1 \text{ units of work},$$
$$= 981 \text{ ergs}.$$

As g varies with geographical position, so also do the work done in lifting one gramme through one centimetre and the force which is represented by the weight of one gramme. For accuracy, therefore, the numerical value of g at the place of observation, should be given.

A horse-power is a larger unit for estimating the rate (*i.e.* work per second) at which work is done by machines, steam-engines, &c. It is taken as equivalent to the work required to raise 550 pounds through one foot, or, which is the same thing, 76 kilogrammes through one metre.

Therefore,

$$1 \text{ H.P.} = 76 \text{ kilogramme-metres},$$
$$= 76 \times 1000 \text{ gramme-metres},$$
$$= 76 \times 1000 \times 100 \text{ gramme-centimetres},$$
$$= 76 \times 10^5.$$

But 1 gramme-centimetre is equivalent to 981 ergs,

$$\therefore 1 \text{ H.P.} = 76 \times 10^5 \times 981$$
$$= 746 \times 10^7 \text{ ergs}.$$

Electric work is measured, as shown above, by $Q\,V$ or by $C\,T\,V$, where C denotes the current strength and T the time.

Hence the rate of doing work, or as Sir William Thomson proposes to call it *the activity*, is $C\,V$. Now the ampère is 10^{-1} of the absolute unit of current and the volt is 10^8 absolute units of potential. Therefore, electric work reckoned in absolute units or ergs is

$$C \times 10^{-1} \times V \times 10^8,$$

And the H.P.

$$= \frac{C\,V \times 10^{-1} \times 10^8}{746 \times 10^7} = \frac{C\,V}{746} = \frac{C^2\,R}{746}$$

When using this formula, it must be remembered that C is estimated in ampères, V in volts, and R in ohms.

The late Sir William Siemens suggested* to take as unit of power, the power conveyed by a current of an ampère through the difference of potential of a volt, and to call this new unit a *Watt*. A watt, therefore, expresses the rate of an ampère multiplied by a volt and

$$= 10^{-1} \times 10^8 = 10^7 \text{ ergs}.$$

* Presidential Address, British Association, 1882.

Hence electric work in watts

$$= \frac{C \times 10^{-1} \times V \times 10^8}{10^7} = C\ V.$$

Sir William Siemens also introduced a modified form of his electro-dynamometer, in order to measure directly the energy absorbed in any part of a circuit. For this purpose the connections are arranged so that the single loop of thick wire is inserted in the main circuit whilst the coil of thin wire is used as a shunt to that portion, say an incandescence lamp, in which the energy to be measured is absorbed.

The current which passes through the coil is proportional to the potential-difference $V_1 - V_2$ at the terminals of the lamp, that which passes through the movable loop is the main current C; hence the force tending to deflect the loop varies as $C(V_1 - V_2)$; and, as the torsion necessary to bring the index-arm back to zero also varies as the same quantity, the number of degrees of torsion will be proportional to the electric energy absorbed in the lamp.

The constant of the instrument given in the table which accompanies it, may be verified at any time by comparing the readings with the values for the work found by using (say) a tangent galvanometer to determine the current and a potential galvanometer to measure the difference of potentials.

The instrument is sometimes called a power-meter, and as its readings when multiplied by its constant give watts, it is frequently called a watt-meter.

The *heat* developed in a conductor by the passage of a current may also be estimated quantitatively.

When a current flows through a conductor without doing any external work such as that of driving an electro-magnetic engine, deflecting magnets, inducing currents in neighbouring conductors, all the energy of the current is converted into heat which appears in the various parts, liquid as well as metallic, of the circuit.

If C be the current, V_1 and V_2 the potentials at any two points of a conductor, then in T seconds $C\ T$ units of electricity are conveyed from the higher potential V_1 to the potential V_2, and the corresponding work is

$$C\ T\ (V_1 - V_2)\ \text{or}\ C\ T\ E.$$

If R denote the resistance of the conductor included between the two points, the potential difference or electromotive force $E = C\ R$, whence the expression for the work becomes $C^2\ R\ T$.

The work per second or activity, is therefore $C^2 R$; or, in absolute units

$$C^2 \times 10^{-2} \times R \times 10^9 = C^2 R \times 10^7.$$

If we take for heat-unit the quantity of heat necessary to raise one gramme of cold water one degree Centigrade, its value in absolute units, as determined by Dr. Joule, is 4.2×10^7; hence the heat-work, expressed in water-gramme-degrees Centigrade, will be

$$\frac{C^2 R \times 10^7}{4.2 \times 10^7} = \frac{C^2 R}{4.2} = .0238\, C^2 R.$$

As this heat-unit is not very definite, Sir William Siemens suggested a new unit based upon the electro-magnetic system and which he proposed to call the *Joule*. The joule would be defined as the heat generated in one second by a current of an ampère flowing through the resistance of an ohm. Its value would be $(10^{-1})^2 \times 10^9 = 10^7$ C. G. S. units.

Therefore the heat per second, expressed in joules, is

$$\frac{C^2 R \times 10^7}{10^7} = C^2 R = C\, V.$$

The increase in the temperature of a conducting wire may also be expressed numerically.

Let w be the weight of the wire in grammes,

s „   specific heat of the material of the wire,

and τ „   elevation of temperature;

then the energy, in absolute units, obviously is

$$w\, s\, \tau \times 4.2 \times 10^7.$$

And this must equal

$$C^2 \times 10^{-2} \times R \times 10^9 \times T$$

Whence

$$\tau = \frac{C^2 R T}{4.2\, w\, s}.$$

The electrical resistance of a conductor may also be deduced from thermal data. The method is commonly known as *Joule's calorimetric method.*

A copper cylindrical vessel, or calorimeter, is nearly filled with water, and in it is placed the wire wound for convenience in the form of a helix and also previously steeped in melted paraffin wax for the sake of better insulation. The ends are soldered to two thick copper rods which pass through the wooden cover of the calorimeter and serve as terminals for connection with the battery circuit. The cover is further perforated to admit a thermometer and a stirrer, the latter for the purpose of equalising the temperature throughout the liquid. A battery of about five Grove's

cells and a plug-key are connected up in circuit with a tangent galvano-meter and the calorimeter.

If t_1 be the initial temperature of the water,

t_2 ,,   temperature after T seconds,

w_2 ,,   mass of the water and water-equivalent of the calorimeter combined,

and s ,,   specific heat of copper,

then the energy absorbed by the water and calorimeter is

$$w\,(t_2 - t_1)\,s\,J$$

where J denotes Joule's dynamical equivalent, viz., 4.2×10^7.

This must equal the energy, in absolute units, developed by the current, that is

$$C^2\,R\,T \times 10^{-2} \times 10^9.$$

Therefore,

$$R = \frac{w\,(t_2 - t_1)\,s \times 4.2 \times 10^7}{C^2\,T \times 10^7} = \frac{w\,(t_2 - t_1)\,s \times 4.2}{C^2\,T}.$$

The loss of heat by radiation must be determined and, if appreciable, allowed for. Account must also be taken of the capacity for heat, or water-equivalent, of the wire itself, the stirrer and the immersed thermometer.

Ex. To find the resistance of a length of German silver wire, the following weighings and observations were made :

Weight of calorimeter . .	128 grammes
,, calorimeter and water	782 ,,
Initial temperature of water .	24°
Final temperature . . .	41°.7
Time	17 minutes
Galvanometer-deflection . .	61°
Galvanometer-constant for ampères	3.171

The heat, in absolute units, absorbed by the water and calorimeter is, therefore,

$$(654 \times 17.7 + 128 \times 17.7 \times .095)\, 4.2 \times 10^7$$

The heat developed by the passage of the current is

$$\left(\frac{1.827 \times 3.171}{10}\right)^2 \times R \times 10^9 \times 17 \times 60$$

Equating these two expressions and reducing, we get

$$R = 1.446 \text{ ohms.}$$

This is the resistance of the wire at $41.7°$ C. ; a small correction is necessary to reduce it to its value at $0°$ C.

If we suppose the resistance R to be known, then the current strength can be found. This method was used by Dr. J. Hopkinson in his investiga-

tions of the relation between the current and the corresponding electromotive force developed by a dynamo running at a constant speed and with varying resistances in the external circuit.

The following examples will serve to illustrate the development of energy and its distribution in the various parts of a circuit.

Ex. A current of 1.3 ampère was sent for three minutes through a wire of 2.8 ohms resistance; the heat-energy developed was, therefore,

$$\frac{1.3 \times 1.3 \times 2.8 \times 3 \times 60}{4.2} = 322 \text{ gramme-degrees C.}$$

$$= 322 \times 4.2 = 1352 \text{ joules.}$$

Ex. The H.P. absorbed in an incandescence lamp through which a current of 3.5 ampères was passing, the difference of potentials at the terminals of the lamp being 1.8 volt, is

$$\frac{3.5 \times 1.8}{746} = .0084$$

$$= 6.3 \text{ watts.}$$

Ex. Ten arc-lamps, each having a resistance of 1.87 ohm, are placed on a circuit 400 yards in length, the resistance of the circuit being .004 ohm per yard; it is required to find the H.P. which is used and that which is wasted respectively.

$$\text{The H.P. used} = \frac{C^2}{746} \times 18.7;$$

$$\text{that wasted in the circuit} = \frac{C^2}{746} \times 1.6$$

Also the ratio of the wasted to the total energy spent in the circuit

$$= \frac{1.6}{18.7 + 1.6} = \frac{1.6}{20.3}$$

$$= 8 \text{ per cent. nearly.}$$

Ex. To determine the efficiency of an electric installation, the following measurements were made:*

The work spent in driving the dynamo, measured by a dynamometer, was 6.66 H.P.; the current, measured by a current galvanometer, was 2.523 ampères; whilst the potential-difference at the terminals of the machine, measured by a potential galvanometer with the aid of a high supplementary resistance, was 1447 volts.

$$\text{The electric energy in H.P. at the terminals of the dynamo} = \frac{2.523 \times 1447}{746} = 4.89$$

---

* ENGINEERING, March 23, 1883.

The efficiency of a machine being the ratio of the useful work to that expended, the electric efficiency of the above dynamo was

$$\frac{4.89}{6.66} = .73$$

i.e., 73 per cent. of the whole mechanical work spent on the machine is converted into the energy of the electric current.

The same dynamo being used to drive an electric motor, it was found that the difference of potentials at the terminals of the motor was 1037 volts, whilst the mechanical work measured by a Prony brake was 2.32 H.P.

Here the electric energy at the terminals of the motor is

$$\frac{2.523 \times 1037}{746} = 3.50 \text{ H.P.}$$

The efficiency (mechanical) of the motor* is, therefore,

$$\frac{2.32}{3.50} = .67 = 67 \text{ per cent. };$$

the efficiency (electric) of the circuit is

$$\frac{3.50}{4.89} = 71 \text{ per cent. };$$

and that of the whole installation is

$$\frac{2.32}{6.66} = 35 \text{ per cent.}$$

The electric conditions of the *voltaic arc* are determined when we know the current strength C and the difference V of potentials at the ends of the carbons.

The ratio of these two quantities $\dfrac{V}{C}$ is usually called " the resistance of the arc," by which is meant the ohmic resistance that would replace the arc without altering the current. Such a use of the term resistance, however, is not strictly scientific, for Ohm's law, on which is based the definition of resistance, does not hold for the electric arc, the difference of potentials on the two sides being, for given material of the electrodes, distance between them and atmospheric pressure, approximately constant.†

The energy in watts developed in the arc is C V and the ratio of this to the energy expended may be taken to represent the efficiency of the arrangement.

---

* Of the power C V supplied to the motor, part is frittered away in heat, the remainder being converted into mechanical energy of motion. In like manner, a portion of the energy of the running motor is wasted in overcoming mechanical friction at the bearings and collecting brushes, as also in air friction and what Professor Ayrton (Jour. Soc. Tel. Engineers, May, 1883) calls "magnetic friction," *i.e.*, the resistance arising from the electromagnetic reactions of the movable and stationary parts of the machine. If W denote the useful work as measured by any suitable ergometer, then the efficiency of the motor is given by $\dfrac{W}{C\,V}$.

† Proc. Inst. Mechanical Engineers, April, 1880, page 272.

The energy stored up in *secondary batteries,* or "accumulators," is also expressible in practical units. For this purpose we measure the charging current C in ampères and the difference of potentials V between the terminals of the battery in volts ; then the

$$\text{H.P.} = \frac{\text{C V}}{746}$$
$$= \text{C V watts.}$$

The energy given out may be measured in the same way, and the ratio of these two will give the efficiency of the battery.

If the charging or the discharge current be not sufficiently constant, we must make frequent determinations of the current and potential-difference, and evaluate the energy for each. The sum of these evaluations will give the integral energy.

The term *electric accumulator* is open to objection because it suggests that electricity as such may be stored up and that an indefinite quantity of energy may be accumulated in the cell, whereas we know this latter to be limited by the amount of chemical work which may be done on the formed leaden plates; nor is the term *secondary* battery a very happy one, since the electrical action of such a battery is quite analogous to that of any voltaic element or electro-chemical apparatus.

When a current is sent through such a " battery," work is done (1) in overcoming the counter-electromotive force *e* called out by the resolution of the electrolyte into its constituents, and (2) in heating the various parts of the circuit. The product C *e* measures the activity or work done per second in the battery ; consequently the higher the value of the counter-electromotive force the greater will be the energy stored up.

Experiments made by Beetz and others show that heat diminishes the counter-electromotive force set up by the passage of a current through an electrolytic cell. Hence the importance of keeping a secondary battery cool during the period of charging.

The energy wasted in the battery as local heat is C (E−*e*), from which expression it is manifest that the nearer the charging electromotive force E is to the counter-electromotive force *e* the less will be this dissipation of energy and therefore the greater the efficiency.

Repeated measurements have shown that for a fully charged Faure accumulator, *e* is very nearly 2.2 volts. Hence to charge ten such accumulators we should require little more than an E.M.F. of 22 volts.

This rule cannot be strictly followed when a dynamo is used to furnish the charging current ; for it is not economical to drive the machine at a

very low speed, the various losses arising from frictional resistances and the development of heat in the conductors would more than equal the advantage gained by the low electromotive force generated; and in addition to this there is the possibility, in the case of any accidental slackening of speed, of the battery discharging itself through the coils of the dynamo and thus, more or less, seriously injuring their insulation.

The power which a dynamo is capable of developing, when working in given conditions of circuit, may be deduced from its *characteristic*. This curve is constructed by varying the external resistance, and taking the current for abscissa and the difference of potentials between the terminals of the machine for ordinate.

Fig. 62 shows such a curve constructed from Dr. J. Hopkinson's
measurements with a Siemens's
medium-size *series* dynamo driven at
a speed of 720 revolutions per minute.* We may remark, in passing,
that a series dynamo is one in which

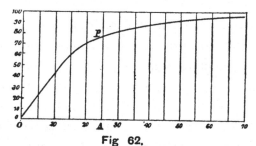

Fig 62.

the whole current is sent through the external circuit and also through the coils of the field-magnets. In another class of machines, the *ordinary shunt* dynamos, the current divides on leaving the armature, part going through the main circuit and part round the magnets. In a third class, the *compound shunt* or *self-regulating* dynamo, there are two distinct circuits, one of fine wire and high resistance going round the exciting magnets and the other of thick wire forming the extrapolar circuit and going likewise round the magnets. It will be seen from Fig. 62

(1) that the early part of the curve differs but little from a straight
line; and

(2) that the electromotive force does not increase without limit, the
curve evidently approaching a horizontal asymptote.

The name *critical current* is given by Dr. Hopkinson to that value of the current for which the curve ceases to be rectilinear. At that particular point the electromotive force attains about two-thirds of its maximum value.

The characteristics of the same machine drawn for various speeds differ only in the scale of ordinates which is greater the higher the speed. Hence for a given dynamo, the critical current is independent of the velocity of

* Proc. Inst. Mechanical Engineers, April, 1879.

rotation. In the figure, it is 15 ampères; the corresponding electromotive force is 60 volts, the greatest being about 96 volts.

It will also be noticed that the curve cuts the axis of ordinates a little above the origin. This displacement is attributed to the residual magnetism of the field-magnets, which is necessary for starting the machine.

If we take any point P on the curve, the ordinate gives the electromotive force; the abscissa, the current; and the slope of P O $\left(\dfrac{E}{C}\right)$, the resistance of the circuit. The energy in watts is denoted by the area of the rectangle having P A and A O for adjacent sides.

Another method of drawing the characteristic of a dynamo has been suggested by Professor W. G. Adams.* The resistance is measured along the axis of abscissæ, and the electromotive force along the axis of ordinates. The curve corresponding to Fig. 62, when plotted by this method, is shown in Fig. 63.

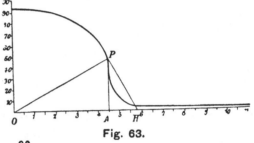

Fig. 63.

For any point P on the curve, the current $\left(\dfrac{E}{R}\right)$ is given by the slope of P O, viz., tan A O P. Drawing H P at right angles to P O, the heat developed in the circuit is represented by A H; for

$$\frac{A\ H}{A\ P} = \tan A\ P\ H = \tan A\ O\ P = \text{current C.}$$

Therefore,

$$A\ H = A\ P \times C = E\ C = C^2\ R = \text{heat per second in joules.}$$

In the case of an electric motor driven by a constant current such as that from a battery or from a steadily working magneto machine, the whole work per second or activity is C E. If R be the total resistance of the circuit and e the counter-electromotive force set up by the rotation of the armature of the motor, then

$$C = \frac{E - e}{R};$$

This back electromotive force e is a variable which, for a constant driving current, increases regularly with the speed of the motor. The power absorbed per second by the motor and converted into mechanical energy of rotation is C e.

This may be written

$$\frac{e\ (E - e)}{R}.$$

* Proc. Inst. Mechanical Engineers, April, 1879,

For a maximum we must equate to zero the differential coefficient of this expression taken with respect to the variable e.

Therefore,

$$E - 2e = 0$$

Whence

$$e = \tfrac{1}{2} E$$

And

$$\frac{e}{E} = \tfrac{1}{2}$$

i.e., the rate of utilizing work in the motor is greatest when the battery current is reduced by the back electromotive force to one-half its original value. This, which is Jacobi's law, is the condition for rapidity of work.

For economy, the ratio $\dfrac{e}{E}$ should be nearly equal to unity; and to attain this, the motor must rotate at such a speed that the inverse E.M.F. be almost equal to that of the driving current.

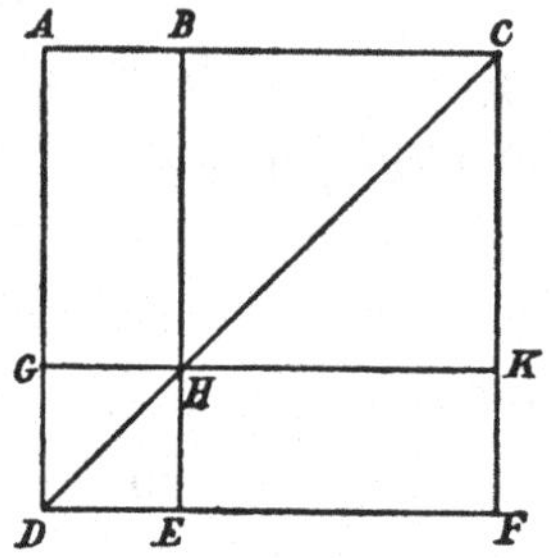

These results may be graphically represented by a simple method due to Professor S. P. Thompson.*

Take A C, Fig. 64, to represent E the constant external electromotive force; describe the square A F, mark off D G equal to e, and complete the figure.

Fig. 64.

The electric energy C E expended per second is $\dfrac{E(E-e)}{R}$, and that absorbed by the motor $\dfrac{e(E-e)}{R}$. Omitting the constant R, these quantities are obviously represented by A K and H F respectively. The ratio $\dfrac{HF}{AK}$ would denote the efficiency if the motor were a perfect electric engine; and as F H = H A, the square B K represents the energy which is wasted. This waste (B K) plainly diminishes as e rises. Now it may be readily shown that H F is a maximum when H is the middle point of the diagonal D C, in which case $\dfrac{HF}{AK} = \tfrac{1}{2}$.

If a dynamo be used to generate the current which feeds the motor, the electromotive force will vary

(1)   as the speed of the dynamo n,

(2)   as a constant a depending upon the form and arrangement of the field-magnets, and

(3)   as some function of the current $\phi(C)$.

Therefore for the generator, we may write

$$E = n_1\, a_1\, \phi(C)$$

* Phil. Mag., February, 1883.

And for the motor,

$$e = n_2\, a_2\, \phi\, (\text{C}).$$

The efficiency of the system $\dfrac{e}{\text{E}}$ is consequently

$$\frac{n_2\, a_2}{n_1\, a_1}$$

If the motor be similar in construction to the generator, a_1 will be equal to a_2; and the efficiency will be denoted by $\dfrac{n_2}{n_1}$. Consequently, for economical working, the two machines must rotate at nearly the same speed.

It is also obvious that the greater the speed of the motor, the greater also will be the efficiency; for then the more will the ratio $\dfrac{n_2}{n_1}$ tend to unity.

If $\text{E} - e$ have a constant value, the efficiency will increase with the electromotive force E; for the energy which is wasted as heat is $\text{C}^2\,\text{R}$, and its ratio to the whole electrical energy is

$$\frac{\text{C}^2\,\text{R}}{\text{C}\,\text{E}} = \frac{\text{C}\,\text{R}}{\text{E}} = \frac{\text{R}}{\text{E}} \times \frac{\text{E} - e}{\text{R}} = \frac{\text{E} - e}{\text{E}} = \frac{\text{constant}}{\text{E}}$$

This waste manifestly diminishes as E increases, whence it follows that the efficiency is greater with high than it is with low potentials.

This may also be seen from Fig. 64. Let the driving electromotive force be 100 volts, and the constant potential difference $\text{E} - e$ be 30 volts. Then the area $\text{A K} = 100 \times 30 = 3000$; also $\text{H F} = 30 \times 70 = 2100$.

$$\therefore \text{ the theoretic efficiency } = \frac{2100}{3000} = \frac{7}{10} = \frac{42}{60}$$

Let the driving electromotive force be now raised to 200 volts. Then $\text{A K} = 200 \times 30 = 6000$; and $\text{H F} = 30 \times 170 = 5100$.

$$\therefore \text{ the efficiency } = \frac{5100}{6000} = \frac{51}{60}.$$

Therefore by increasing the E.M.F., the theoretic efficiency of the motor has also been increased.

NOTE.—Some doubt having arisen as to the constancy of Latimer Clark's standard cells, Lord Rayleigh recently subjected those which are in the Cavendish Laboratory to careful tests, and he obtained, as he stated before the London Physical Society, March 8, 1884, very satisfactory results. The cells were always worked through a resistance comparable with a megohm. The value of the E.M.F. given by Latimer Clark is 1.457 volts, that found by Lord Rayleigh is 1.453 volts. This supposes the B. A. unit to be absolutely correct. According to Lord Rayleigh, however, it is only .9867 of the theoretic ohm. Assuming this determination to be accurate, the E.M.F. of a Clark's cell is 1.453 × .9867 = 1.434 volts.

WORKS ON ELECTRICITY AND ELECTRICAL MEASUREMENT.

A GOOD elementary knowledge of the general phenomena and fundamental laws of electricity and magnetism, may be obtained by a careful study of any one of the three following books : Professor S. P. Thompson's " Elementary Lessons in Electricity and Magnetism ;"[1] Dr. Ferguson's " Electricity," revised and enlarged by Professor J. Blyth ;[2] and Part III. of Deschanel's " Natural Philosophy," translated and edited by Professor J. D. Everett.[3] Professor Fleeming Jenkin's " Electricity and Magnetism"[4] will be found useful, as also Professor Tyndall's " Notes on Electrical Phenomena and Theories."[5]

A comprehensive account of what has been done in the several parts of the vast domain of electrical science is given in J. E. H. Gordon's " Physical Treatise on Electricity and Magnetism."[6] This extensive work reviews, and generally at no inconsiderable length, the more important recent researches, explaining the methods followed and stating the results obtained. It is specially remarkable for the excellence of the views and explanatory diagrams of standard electrical and magnetic instruments.

The student who seeks to acquire a sound knowledge of theory, mathematical and physical, will do well to familiarise himself with L. Cumming's " Introduction to the Theory of Electricity ;"[7] after this he may read, with advantage, Professor G. Chrystal's two admirable articles on Electricity and Magnetism in the new edition of the " Encyclopædia Britannica." The parts of the encyclopædia in which these articles occur, may be had separately from the publishers.[8]

This course of reading will prepare for the severe pages of Clerk-Maxwell's " Treatise on Electricity and Magnetism."[9] Some sections of

this great work require an acquaintance with only elementary mathematics ; but no one can master the advanced parts who is not well skilled in the highest forms of analysis.

As an auxiliary to Clerk-Maxwell, one might use Mascart and Joubert's " Treatise on Electricity and Magnetism," recently translated by Dr. E. Atkinson. This work is announced to be in two volumes ; the first is the only one yet (March, 1884) published. It is devoted exclusively to theory, and follows closely the methods and views of its English prototype.

Aware of the difficulties with which his treatise bristles, Professor Maxwell, in the interest of students generally, began to prepare his " Elementary Treatise on Electricity ;" [10] but was prevented by premature death from finishing it. The manuscript, in its fragmentary form, was entrusted to Professor W. Garnett, who has completed it by the addition of a number of articles selected from the first volume of the larger work.

For practical purposes we have, as yet, no one book to guide a student through the various departments of electrical and magnetic measurements. It is hoped that the forthcoming volume by Professors Mascart and Joubert will go far towards supplying this want. Meanwhile Kohlrausch's " Introduction to Physical Measurements"[11] will be found serviceable.

Andrew Gray, chief assistant to the Professor of Natural Philosophy in the University of Glasgow, has recently published an excellent little manual entitled " Absolute Measurements in Electricity and Magnetism." [12] It is not, however, a complete treatise, being designed to give, by means of some typical determinations, a clear account of the absolute system of units. It contains much valuable information on such subjects as the theory of continuous and alternating-current machines, including the effects of self-induction, the efficiency of electric motors, and the measurement of intense magnetic fields.

The specific requirements of telegraphy are fully treated in H. R Kempe's " Handbook of Electrical Testing." [13]

This book not only describes the principal methods in use, but also gives the theory and thence deduces the best conditions for making the various tests.

Nothing is better adapted to give a clear perception and firm grasp of the laws of any branch of science than to work out a series of numerical exercises on them. In this connection, two little books by R. E. Day will be found very helpful and suggestive, viz., " Examples in Electrical and Magnetic Measurement" [14] and " Electric Lighting Arithmetic." [15]

H

Another useful book, one almost indispensable, is Professor J. D. Everett's " Units and Physical Constants."[16] It gives clear and ample information about the fundamental and derived units whether mechanical, electrical, or magnetic, and contains tables of the most important constants in the several branches of physical science.

[1] Macmillan and Co., London.

[2] W. and R. Chambers, London and Edinburgh.

[3] Blackie and Son, London, Dublin, and Edinburgh.

[4] Longmans and Co., London.

[5] Longmans and Co., London.

[6] Sampson Low, Marston, Searle, and Rivington, London.

[7] Macmillan and Co., London.

[8] A. and C. Black, Edinburgh.

[9] The Clarendon Press, Oxford.

[10] The Clarendon Press, Oxford.

[11] J. and A. Churchill, New Burlington Street, London.

[12] Macmillan and Co., London.

[13] E. and F. N. Spon, Charing Cross, London.

[14] Longmans and Co., London.

[15] Macmillan and Co., London.

[16] Macmillan and Co., London.

SECTION II.

PHOTOMETRY.

PHOTOMETRY.

AS its name, derived from the Greek (φῶς, φωτός, light, and μέτρον, measure), indicates, photometry is that branch of optics which treats of the various methods used for comparing the intensities of different sources of light. Amongst the numerous problems connected with the measurements of energy, there is not one that presents greater material difficulties. Although based on a geometrical law, no practical method has been found, up to the present time, by which mathematical accuracy can be attained. In addition to this, of all the units of measurement employed in various countries, or proposed by physicists, there is not one that possesses the qualities of absolute identity, permanence, convenience, and simplicity, which are indispensable in industrial investigations which have to do with money expenditure. Thus in spite of the dissertations which occupied several *séances* of the Electrical Congress that met in Paris in 1881, no agreement could be arrived at as to the choice of a photometer, and of an international standard. The problem remains therefore unsolved, and should be studied afresh by a committee of electricians representing all countries interested ; it was proposed by the Congress that the French Government should invite such a committee, but no steps have yet (March, 1884) been taken in the matter.

We propose to give a general *exposé* of the problems to be solved, a rapid description of various methods in use or proposed, and to notice their advantages and inconveniences. The subject is one of especial interest and importance in connection with the measurement of the electric light ; but it will be necessary to consider it in its general aspect, which is the evaluation of all luminous intensities, paying at the same time particular attention to its special development in relation to electric lighting. This is all the more necessary, since, in consequence of recent invention, electric lighting comprises at the present time a large series, ranging from the

powerful lights devised for lighthouses, down to the small incandescence lamps of Swan or Edison.

In dealing as far as possible with the whole subject, within its widest limits, we find the following natural divisions :

1. An examination of the various photometric methods.

2. The study of luminous standards, those in use, as well as those which are proposed.

3. A discussion of the results obtained with various electric lamps, and their application to the more general problem of electric lighting.

Apart from various special methods, which have but little practical value, the processes actually employed by all engaged in artificial illumination, consist in replacing a direct estimation of the luminous source, by an indirect comparison with a given unit.   Two contiguous surfaces are illuminated to the same degree—one by light from the luminous source, the other by light from the unit source ; the two sources being placed at unequal distances from the surfaces illuminated, when the illumination by one does not exactly equal that by the other ; this method is easier than that of direct comparison, but is still liable to error.   The difference in the value of the two sources of light, thus becomes a simple function of the ratio of the distances which gives the proportion of the intensities.   Speaking generally, this function will be the same when the luminous sources produce an identical physical effect by neutralising their lighting properties, on a screen placed at a fixed distance from each of them. This will become clearer as we proceed.   Confining ourselves to our first proposition it will be seen that the process consists in substituting for a difference of luminous emission at the same distances, an equality of lighting at different distances.   The eye can perceive with sufficient accuracy for *general* purposes whether two similar surfaces reflect equal quantities of light, and the point where this approximate equality of illumination is produced, is easy of detection.   The eye, however, is not an absolute organ, its sensitiveness varies with each observer, and changes with different conditions of health or fatigue, in the same observer.   It results from this that the most careful experiments are liable to involuntary errors of no inconsiderable range.

The differences in colour which are generally observed in lights that are being compared, introduces another difficulty.   In the particular case of the electric light, the difference in luminous value—often very large—between the unit and the intensity to be measured, still further complicates the experi-

ments, and increases the uncertainty of the results. Nevertheless, it is this method, imperfect, complicated, and full of chances of error, which is generally employed, for want of a better, in a great variety of ways. We will proceed to explain this system with the various improvements that have been introduced.

Law of the Squares of Distances.—The principle of the method was first enunciated by Kepler the astronomer, and. is known as the law of the squares of the distances. It may be stated conveniently as follows : *When a luminous source is placed successively at various distances from a screen which it illuminates, the quantities of light received on this surface vary in an inverse ratio with the squares of the successive distances of the luminous source from the screen.* Thus if the distance is doubled, the intensity of the light will be four times less. It will be nine times feebler if the distance is trebled, &c. In a word, if Q represents the quantity of light received by the screen at unit distance, the quantity Q^1 received at the distance D will be given by the formula

$$Q^1 = \frac{Q}{D^2}$$

This law may be demonstrated as follows : Suppose that L, Fig. 65, is a luminous point emitting light in all directions in the interior of two spheres, having radii R and R^1. It is clear that the total quantity of light received by the interior surfaces of the two spheres will be the same, since each intercepts the same number of luminous rays. If we take on each sphere segment of surface S, the quantity of light intercepted by this surface on each sphere will be in the ratio of S to the total surfaces of the sphere. We shall thus have

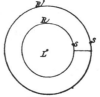

Fig. 65.

$$\frac{\text{Light received by S on the sphere R}}{\text{Light received by the sphere R}} = \frac{S}{4\,\pi\,R^2}$$

In the same way

$$\frac{\text{The light received by S on the sphere } R_1}{\text{Light received on the sphere } R_1} = \frac{S}{4\,\pi\,R_1{}^2}$$

Whence

$$\frac{\text{The light received by S on the sphere R}}{\text{Light received by S on the sphere } R^1} = \left(\frac{R_1}{R}\right)^2$$

Another geometrical law allows the evaluation of the quantities of light to be made, which emanate from a luminous point, and are received on a screen at various angles. *The quantity of light received by an element of surface, varies as the cosine of the angle between the normal to this element of surface and the ray of light.*

This law can be demonstrated as easily as the preceding. Let L, Fig. 66, be a luminous point, A B A₁ the section of an element of surface lighted normally by L. Let us suppose a second position C B C₁ of the same element making an angle ω with the first.

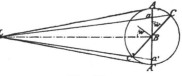

Fig. 66.

The element, in this new position, will not receive more than the rays comprised in the figure a L a_1 represented by the ratio

$$\frac{a\,a_1}{A\,A_1}$$

or,

$$\frac{a\,B}{A\,B}$$

if the element A A₁ is sufficiently small in proportion to the distance L B. Under these conditions it may be admitted that the angle C a B is a right angle, and consequently in the right-angled triangle C a B we have

$$\frac{a\,B}{C\,B} = \cos \omega = \frac{a\,B}{A\,B} = \frac{a\,a_1}{A\,A_1}$$

which proves the theorem.*

These laws serve as a basis to the problem of photometry. They may be all written in one formula. Calling S an element of surface, I the intensity of the rays emitted normally, ω the angle with which the pencil of rays is emitted in the plane, ω₁ the angle of incidence of the rays on the plane, D the distance between the luminous point and the illuminated surface; then the illumination E of this surface will be expressed by the formula

$$E = \frac{S.I.\,\cos\omega\,\cos\omega_1}{D^2}$$

* This law may also be expressed thus : *The intensity of illumination of a surface lighted obliquely is proportional to the section of the rays falling upon it, taken at a plane perpendicular to their direction.* Let A be the area of an oblique surface, the whole of which is lit by rays, the area of a section of which, taken in a plane perpendicular to their direction, is B; then I the intensity of illumination of the oblique surface equals $\frac{B}{A}$.

For if the oblique surface $a\,b$ be illuminated by rays falling upon it in the direction of the lines 1, 2 . . . 6, to which $b\,c$ is perpendicular, then only the same area of light that would be intercepted at $b\,c$ is intercepted at $a\,b$. Hence the intensity of illumination per unit of area of the surface $a\,b$ is less than that which would obtain per unit of surface $b\,c$; the ratio being $\frac{b\,c}{a\,b}$. Now $a\,c\,b$ being a right angle, $b\,c$ equals cosine of angle $a\,b\,c$. Again, $b\,d$ is the normal or perpendicular to the oblique surface $a\,b$; and it may be proved geometrically that the angle $d\,b\,c$ equals the angle $a\,b\,c$. Let the angle $d\,b\,c$ be called ω: then the intensity of illumination of the surface $a\,b$ per unit of area is proportional to the section of the rays at the plane $b\,c$; or what is the same thing, is proportional to the cosine of angle ω which the direction of the rays makes with the normal to the inclined surface.

This will perhaps be more readily understood by a simple example:
Let *a b*, Fig. 67, represent the ground-glass side of a box in which a light is enclosed, and let *c d* represent a ground-glass screen. Rays of light would pass

Fig. 67.

from *a b* to *c d* in the direction *e f*; *e g* and *f h* being the respective normals to the surfaces, and ω ω_1 the respective angles with those normals. Suppose that both ω and ω_1 are angles of 30°, of which the cosine is $\frac{1}{2}$; that the intensity of illumination of *a b* per unit surface is 20; and also that the distance between *a b* and *c d* is 5: then $\dfrac{1 \times 20 \times \frac{1}{2} \times \frac{1}{2}}{5^2} = \frac{1}{5} =$ relative intensity of illumination of surface *c d*.

We have already said that the eye appreciates with difficulty the relative values in the intensities of two different luminous sources; but it can, on the other hand, appreciate with more accuracy when the two surfaces, brought near to each other, are lighted equally by two sources placed at different distances, and emitting their rays through the same angles. In this case, if we call E the illumination of the two equal screens by two sources I and I_1, placed at distances D and D_1, we shall have

$$E = \frac{I}{D^2} = \frac{I_1}{D_1^2}$$

Whence

$$\frac{I}{I_1} = \left(\frac{D}{D_1}\right)^2$$

This formula contains the fundamental principle of the photometric method very generally adopted on account of its simplicity. It consists in illuminating two adjacent screens C B and $C_1 B_1$, Fig. 68, by two lights *l* and L directed against the screens, the lines *m l*, m_1 L making with the plane *x y*, normal to the two screens, equal and sufficiently small angles. The plane *x y* separates the two sources of light in such a way that each screen is illuminated only by one of them. The standard light *l* is displaced progressively along the line *m l* until the observer placed at *y* sees that the two screens are equally illuminated.

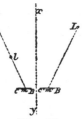

Fig. 68.

Then the distances *m l* and m_1 L are measured, and the proportion $\left(\dfrac{m_1\,\mathrm{L}}{m\,l}\right)^2$ represents the ratio of the luminous intensity of L to the unit *l*.

This method which we have briefly explained is due to Bouguer, a French physicist who lived at the commencement of the eighteenth century, and who has fairly a claim to be considered as the inventor of the photometer.

It was described by him for the first time in a pamphlet published in 1729, under the title of " Essai Optique sur la Gradation de la Lumière."

Bouguer's Photometer.—The photometer devised by Bouguer is an application of the simplest conditions of the law of the square of the distance. It consists of a wooden frame containing two equal apertures covered with semi-transparent paper, and an opaque partition separating the lights and at right angles to the screen. Fig. 69 shows the general arrangement of the apparatus. The light l is shifted until the observer on the remote side of the screen sees the two apertures equally lighted, the distances d, D of the lamps

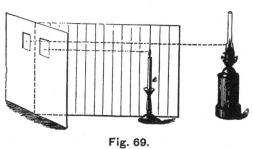

Fig. 69.

from the screen are then measured, and the ratio $\left(\dfrac{D}{d}\right)^2$ gives that of the intensities.

Ritchie's Photometer.—This is a simple modification of the preceding apparatus. The two lights to be compared, are placed on opposite sides of a rectangular tube A B blackened on the inside, and a part of one surface of this tube is replaced by a sheet of oiled paper or a screen of ground glass $a\,b$, Fig. 70, on which two plane mirrors M M₁, inclined at an angle of 45°, reflect the rays from the two lights.

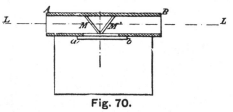

Fig. 70.

The observer examines the screen through a second tube also blackened on the inside, which limits the field of view and cuts off all extraneous light. The mirrors may be replaced by sheets of white paper.

Foucault.—Foucault's photometer, which is largely in use, is based on the same principles as the preceding. It consists of a vertical frame of blackened wood, in which is formed a small opening at the height of the eye; Fig. 71 shows a front and a side elevation of the important parts of the apparatus. This frame A B C D is generally furnished with

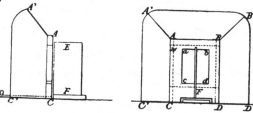

Fig. 71.

inclined wings A₁ B₁ C₁ D₁, and receives between two lateral grooves a plate of translucent glass, one part of which appears behind the opening $a\,b\,c\,d$. A screen E F is placed perpendicularly to the glass, dividing it into

two equal parts. It is furnished with a graduated scale, which slides in a horizontal guide, and traverses its front face, in such a manner that, in moving it, the operator can move the screen, and adjust its distance from the glass plate, so as to reduce to a single line the shadow separating the two luminous surfaces. To make the tints uniform, the glass plate is often covered with a second piece of transparent coloured glass. It is not altogether an easy matter to select suitable glass for the comparison of the luminous surfaces. It should be sufficiently translucent to allow the rays to pass, without the lights themselves becoming visible, and it should appear uniformly illuminated over its whole surface. In this respect ground glass is scarcely suitable, because the grain on the surface is too large. Opal glass possesses another objection pointed out by M. Crova in a recent work. The relative opacity of glass of this nature is obtained by pulverulent matters held in suspension, more or less equally distributed, which diffuse in all directions blue and violet rays, and is freely traversed by the red rays, in such a manner that by diffusion, such glass appears of a blue tint, while by transmission, it is red. What is far more satisfactory is a clear glass made opalescent by a very uniform deposit of finely powdered starch reduced by levigation.

A condition essential for accuracy in results, consists in removing the photometer and the eye from all other lights but those which are to be compared, and in eliminating diffused and reflected light. This is done, first, by examining the luminous surfaces through a cone or cylinder of blackened cardboard which has the incidental advantage of keeping the eye at a fixed distance from the plate; next, in fixing the apparatus, and in placing the operator in the greatest obscurity possible. For this purpose the simplest mode is to fix the photometer at the end of a dark room, entirely closed on one side behind the lamp, and on the other behind the operator. Such a room can be very easily made with a light wood framing, over which a black opaque fabric is stretched, so as not to admit any external light. In this way the photometer is illuminated only by the lights that are to be compared. This process requires usually two operators, one observing the intensities, and the other the displacement of the standard lamp, according to the indications of the former. The light, of which the value is to be ascertained, is fixed at a determined distance, measured accurately on the line, passing through its centre, and that of the photometer. The standard lamp is movable on a graduated scale, and a line drawn from its centre to that of the photometer, makes the same angle as the light to be measured. It is

advisable to decide upon the equality in tint of the glass only after two operations; first, in placing the standard lamp at too great a distance, and then moving it up gradually to the right point; the other, in placing it at too small a distance, and removing it gradually.

The Degrand Photometer.—By some special modifications of the Foucault photometer the services of an assistant may be dispensed with, as well as all the preliminary adjustments, by fixing, correctly and finally, the position of the standard. Other modifications permit measurements to be taken under different angles. We can best indicate how these special conditions

Fig. 72.

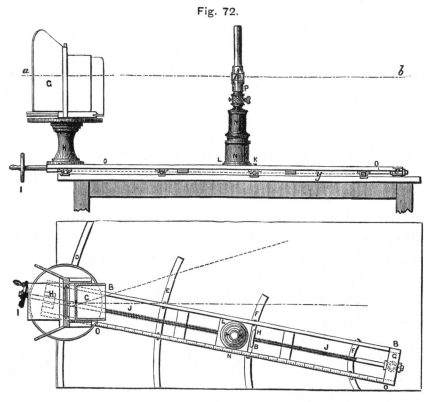

Fig. 73.

are supplied in the Degrand photometer, by referring to a notice published by MM. Sautter, Lemonnier, and Co., relative to photo-electric apparatus employed by the French Navy. The details are shown in Figs. 72, 73, 74, and 75. The standard lamp M is placed on a small carriage L, fitted with an index N, corresponding to the axis of the lamp, which is displaced along the graduated scale O O, whilst the carriage is shifted between the grooves of the frame B B, O O. This movement is produced by means of the long screw J, which the operator moves by means of the small handwheel I, and

a nut fixed under the carriage.   The frame B B, O O is free to move laterally
on the curved guides D E F, in such a way as will vary the angle of incidence of the rays on the plate of the photometer.   This latter G is only a
Foucault photometer mounted on a foot H, about which it can be revolved.
The source to be measured is placed on a special bracket, Figs. 74 and 75,
allowing rays from it at all angles comprised between 70° above the horizontal and 70° below, to be directed towards the photometer.   The rays, how-

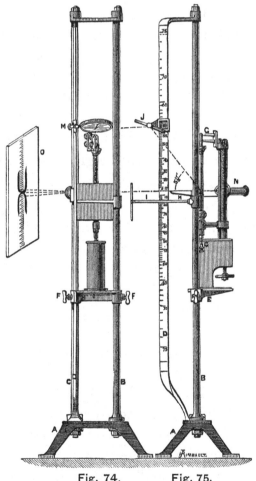

Fig. 74.   Fig. 75.

ever, especially those from the light at but a small angle from the horizontal,
fall upon the photometer screen almost perpendicularly.   The mirror simply
deflects a ray of any required angle, from the source of light to the photometer ; the chief use of this photometer being the comparison of the illuminating power of rays at various angles from the horizontal line.   The bracket is
placed upon a line traced on the ground at an angle corresponding to the angle

of the frame B O. Figs. 74 and 75 show the different portions of this support. It consists of a foot A carrying two guides B C, between which can be shifted the carriage E, on which the electric lamp is placed, and it is fixed in position by means of the locking screw F; a small index H determines the height of the luminous point. This slides in front of the graduated scale D, the divisions of which correspond to one millimetre, the distance apart of the carbons. The experiment is controlled as follows: The electric lamp is raised above the screen I to make a first observation on a horizontal plane; afterwards the position of the mirror frame is altered, so as to bring it successively to divisions marked in advance on the graduated scale, and corresponding to angles of emission rising by increments of 10°. For each measurement, the observer displaces the standard lamp by means of the handwheel I, until the luminous bands produced on the screen appear to him to have the same intensity. He then reads on the scale O the distance d; D being the distance between the electric lamp and the photometer; the intensity will be given by the ratio $\left(\dfrac{D}{d}\right)^2$. It is necessary to correct the results obtained by multiplying them by a coefficient which depends upon a the angle of incidence. Taking $a = 1$ for a direct observation on a horizontal plane, then with a silvered mirror the coefficients for the undermentioned angles will be:

$$
\begin{aligned}
\text{For } 10° \; a &= 1.470 \\
\text{„ } 20 \; a &= 1.350 \\
\text{„ } 30 \; a &= 1.240 \\
\text{„ } 40 \; a &= 1.176 \\
\text{„ } 50 \; a &= 1.173 \\
\text{„ } 60 \; a &= 1.167 \\
\text{„ } 70 \; a &= 1.162*
\end{aligned}
$$

Wolff's Photometer.—Mr. C. Wolff described in the "Journal de Physique" (Vol. I., 1872, p. 81) a photometric apparatus derived from those which we have described, and which was especially arranged for astronomical observations. The following is a brief description: On a table two collimators a and b, Fig. 76, are fixed horizontally, and a telescope c, the object-glasses having each a focal length of about forty centimetres. In the direction

---

* It will be noticed from the illustration that the angles of incidence and reflection at the mirror are much larger when nearly horizontal rays are taken, than when those are taken which more nearly approach a vertical direction. It should, however, be remembered that the reflecting power of a mirror depends not only upon the degree of polish, and nature of its surface, but increases with the obliqueness of the rays incident upon it; for the loss of intensity of illumination due to obliqueness of incidence is accompanied by increased intensity of reflection.

of the axis of the collimator *b* is placed on a second table, a small moderator
lamp *h*, mounted on a carriage that
slides on two rails, and which the
operator can move to and fro by means
of the cords *i* and *j*. A second lamp
g, which is to be measured, is placed
on a movable bracket.   These two
lamps have the bases of their glass
chimneys surrounded with a cardboard
sleeve containing a circular opening
at the height of the centre of the
flame.   The rays passing from the
lamp *g* traverse the collimator *a*, and

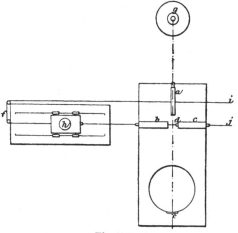

Fig. 76.

fall on a small silvered mirror *e*, movable around a graduated circle, and
which can be shifted in such a way as to reflect the luminous ray on a totally
reflecting prism *d*, which covers one-half the object-glass of the telescope *c*.
The rays proceeding from the lamp *h* are received direct.   The plates closing
the tubes of the collimators contain two semicircular openings arranged in
such a way that the observer can see by means of the telescope two openings,
which appear to the observer side by side, with their vertical diameters
adjacent.   The comparison of intensities is thus made without difficulty.

Cornu's Photometer.—Another French physicist, M. A. Cornu, has
published in *La Lumière Electrique* (November, 1881), a series of photometric
researches which should be referred to here, because the apparatus that he
employed is based upon observations made with two adjacent luminous
surfaces, as in the photometers of Bouguer, Foucault, and those derived
from them.   Nevertheless it differs notably from these, in the fundamental
principle, of the method, which utilises the discovery made by Bouguer,
referring to a property of lenses, which may be thus defined.   The focal
image of an illuminating body produced by a lens is, as to form, independent
of the size and shape of the aperture ; and, as to intensity, is proportional
to the area of this aperture.   The photometer constructed by M. Cornu to
carry out this principle, is formed of two identical achromatic object-glasses,
the optical axes of which intersect at about double the common focal distance.
Each of these objectives throws upon a white screen the image of two
diaphragms, and behind these are the luminous sources to be compared.   On
each object-glass is a double metal plate carrying a small, peculiarly constructed diaphragm.   This is composed of two plates superposed, each

having a square opening. These plates can be displaced relatively to each other by means of a small pinion, arranged as shown in Fig. 77, which causes the two square openings to move in opposite directions. When the two plates cover each other exactly, the two openings coincide, and give passage to the maximum number of luminous rays. By turning the pinion the openings are displaced, and the size of the resultant opening is reduced, the square form and the same central point being always maintained until one of the holes is completely covered. A small graduated scale indicates at each moment the length of side of the corresponding square. The lights to be compared being placed near the diaphragms,

Fig. 77

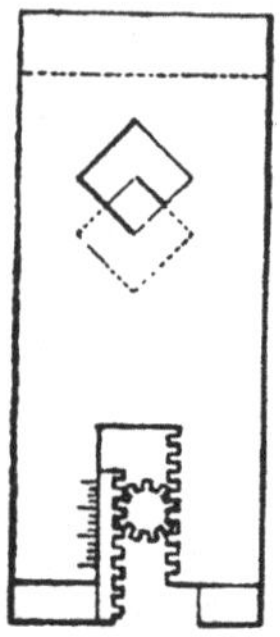

the apparatus is adjusted so as to have very distinct images, the edges of which must coincide. Equal intensity is then sought by opening the diaphragm full in front of the feeblest light, and by adjusting the pinion of the other diaphragm. In this way the ratio of the intensities is obtained by taking the inverse ratio of the surfaces of the corresponding squares. The experiment can be repeated by testing it on a smaller opening of the first diaphragm. In this way several results can be obtained, the mean of which may be taken. Any differences that proceed from any variation in the two object-glasses can also be corrected, by exchanging them and taking the average of reversed observations. If the luminous sources are of different colours, they can be equalised with tinted glass, and the divergence of hue may be reduced by weakening the intensities of the images on the screen, since, as M. Cornu says, "the eye, in experiencing a gradual reduction in its impressions, appears to lose the perception of colour before losing that of intensity." Either by the use of a coloured glass, or by reducing the intensities, two ratios are obtained, the mean of which may be taken. This method may be modified so as to apply, not to images thrown on a screen, but to aërial images themselves, which are brighter and better defined. In order that the eye may observe them simultaneously by one observation, it is necessary that the optical axes of the two object-glasses should be made to coincide. M. Cornu effects this by means of a glass placed at an angle of 45°, which transmits the rays of one of the two object-glasses, and diverts the others by reflection, so as to obtain two contiguous images. The equality in the intensity of these two images does not give direct, the ratio of the illuminating power of the sources being compared, because the reflection and the refraction of the glass do not reduce the luminous rays in the same proportion. It is necessary in this case to use an auxiliary source, and to employ

the following indirect method. Let **K** be the ratio of intensities transmitted by reflection and refraction; $I_1 I_2$ the intensities of the two lights. In making the two experiments we shall have

$$K \frac{I_1}{I_2} = m. \quad \text{And } K \frac{I_2}{I_1} = n$$

Whence

$$\left(\frac{I_1}{I_2}\right)^2 = \frac{m}{n}$$

It should also be remarked that the glass polarises the rays, and if the two incident rays be previously polarised, the method would cease to present the same simplicity. Moreover, the two faces of the plate give each of them an independent image, but one of these may be suppressed by using a plate of sufficient thickness. Fig. 78 shows the arrangement adopted by M. Cornu. A B is a piece of black glass placed at an angle of 45° to the direction of the optical axes of the object-glasses L L₁—placed themselves at right angles to each other— and in such a way that the edge B is at the intersection of these axes, perpendicularly to their plane. The two pencils of rays from L and L₁ will thus produce two images on one side and the other of the axis B C, one to the right, visible by transmission, the other to the left, visible by reflection, and separated by the image of the edge B. By placing a microscope at *c*, provided with a

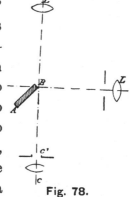

Fig. 78.

circular diaphragm c^1, the field of vision will be two semicircles separated by a vertical diameter. One-half will represent the fixed intensity of the auxiliary source, the other will vary according to the opening of the photometric screen. M. Cornu points out that according to the principle of the apparatus, in order that the method should be correctly applicable, it is necessary for the eye to receive all the light emanating from this source, or emanating from the opening of the diaphragm; it is necessary, therefore, to be sure that the minimum square opening of the photometer L is wholly visible in the circular ring of the microscope, and that the opening in the objective L, or the image of the source, is also wholly visible and concentric with the image of the square opening.

In a further note M. Cornu has given an application of the preceding method for investigating the intensity of the electric light, by replacing the direct observation by an indirect one, that suppresses, or at least reduces the oscillations inseparable from the voltaic arc. It consists in setting the

luminous source in the middle of a spherical chamber painted white inside in such a way as to make it more uniform by reducing the direct lighting.   Let A B, Fig. 79, be the position of the spherical surface placed in a dark chamber X Y ; O its centre, and E the electric lamp. Placed symmetrically with E is a tube, blackened inside, and provided with an opening G G₁, widened out so as to receive the rays of a part of the sphere on a screen H H of ground glass.   Behind this tube the photometer above described, is placed with its glass M N admitting through the diaphragm $e\,e_1$, rays from the standard light K.   By placing successively at E, the various luminous sources to be compared, the ratios of their intensities can be obtained.

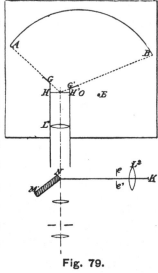

Fig. 79.

Masson's Photometer.—The following is the description of a photometer of a totally different character devised by M. Masson, and described by him in a long article published in Vol. XV. of the " Annales de Physique et de Chimie."   This photometer is based on the following principle : If we take a disc of white cardboard on which alternate black and white sectors are drawn, as shown in Fig 80, and if we impart to this disc a rapid movement of rotation by any convenient mechanism, it will appear to be of a uniform tint.   If this revolving disc be illuminated suddenly by a light, and for a sufficiently short interval, the disc will have the appearance of being motionless, and the sectors will be seen separate and distinct.   If the effect of this instantaneous illumination

Fig. 80.

be gradually reduced by removing the disc, then at a certain distance the excess of illumination which the light throws on the disc will become too feeble for the sectors to be distinguished, and the disc will again become of a uniform tint.   The distance at which the sectors cease to be rendered-visible is variable with each observer, but, according to M. Masson, is always sensibly constant for the same eye, if the conditions of vision remain the same.   The special object with which this photometer was designed was to measure flashing, and especially electric lights.   It will be found that by making the flashes to be compared, play successively on the disc at the fixed distances when the sectors cease to be visible, the ratio of the squares of these distances will give the ratios of the intensities.   The measure of the

absolute intensity of the flashing in relation to the permanent light, depends on the number of sectors and the speed of rotation. The greater or less accuracy of the measurements will depend on the capacity of the eye to appreciate the precise point where the sectors cease to be visible. Experiments prove that this limit is very clearly marked, and that the point can be seized without difficulty.

To convert the apparatus into an ordinary photometer capable of comparing permanent lights, it is sufficient to direct two similar flashes on the disc, illuminated by the two continuous lights, and in each case to adjust them gradually until the sectors cease to be visible during the continuance of the flash. The experimenter will note in every case the distance of the source from the screen, and the ratio of the squares of the distance will give the ratio of intensities. The photometer of M. Masson consists of a disc, as shown in Fig. 81, about 8 centimetres in diameter, on which are drawn sixty black and white sectors. This disc is secured to a copper support fixed to an axis connected to clockwork in the box a. By means of a detent the disc may be made to revolve at a speed of from 200 to 250 revolutions

per second. At B, Fig. 81, is a box in the form of a lantern opposite the disc, and pierced on one side with an opening to allow the rays from a Carcel lamp placed inside, to pass. This box can be moved along a graduated scale $c\,d$, which gives the distance of the flame from the centre of the disc. The scale is set at an angle of 45° to the plane of the disc. In another direction, at

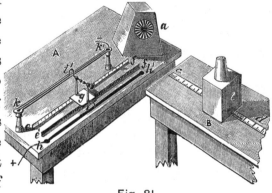

Fig. 81.

an angle of 45°, are placed on the table, two scales ef, e^1f^1, in the form of rails, on which slides a small carriage g, carrying two cylindrical glass rods fitted with two balls, between which the sparks play, and which can be set closer together or farther apart by a micrometer screw. One of these is connected by the copper hook i with an insulated mercury trough $h\,h^1$; the other is in electrical connection with a second raised channel $k\,k^1$, in which dips a metal strip i^1 connected to the second ball. The terminals are connected at h and k in such a manner that the circuit remains constant, no matter what may be the position of the carriage g. The spark is derived from a Leyden jar placed in an adjoining room with the electrical machine.

By substituting for the Carcel, the electric or other lamp which it is desired to compare, all the elements of the experiment are obtained. M. Tchikoleff, the Russian physicist, who is a member of the Electrical Congress, has drawn attention to this somewhat forgotten photometer, the use of which he recommended.

Hammerl's Photometer.—Dr. Hammerl makes use of revolving sectors in photometry in quite a different way, namely, to reduce the light coming from any intense source, as from an electric arc, before it is allowed to fall on the screen. For this purpose he uses a revolving disc with three sector-like openings cut in it. This disc is placed in the path of the beam, and interrupts a certain proportion of the rays, corresponding to the area of the solid portion, while the remainder pass onward to the screen. By employing two discs, placed in contact and adjustable with relation to one another, the size of the openings can be varied at will by causing the solid parts of one disc to overlap the openings of the other to a greater or less extent, and thus the brilliancy of the light under examination can be greatly varied.

Instead of employing two discs, one may be used having openings whose angular apertures gradually diminish from the centre to the circumference. By this arrangement the circle of light on the screen gradually diminishes in brilliancy from the centre outwards. It is projected on to a grease line, instead of on to a spot, and at one particular point on that line balances the standard light, consequently from a computation of the area of the opening in the disc at the radius corresponding to that point, the intensity of the light can be determined.

Rumford's Photometer.—The photometer of Count Rumford is based on the indications given by Lambert in a work published in 1760, called "Photometria sive de Mensura et Gradibus Luminis, Colorum et Umbræ." It is prior to that of Foucault, and in a chronological point of view ought to find a place by the side of that of Bouguer. Fig. 82 explains the arrangement and principle clearly. A B is a screen before which is placed a vertical rod C. On two straight lines forming equal angles with the plane of the rod C, perpendicular to the screen, the lights L L¹ are placed, which are to be compared. These two lights

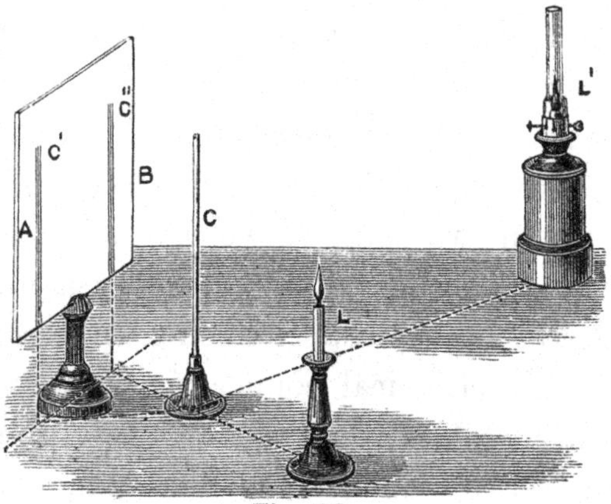

Fig. 82.

project on the screen two shadows C′ C″ of the rod ; one of the lamps is shifted until the shadows appear of equal intensity. Then it is clear that the shadow C cuts off from L as much light as C¹ cuts off from L¹. The ratios of the intensities will then be given by the ratios of the squares of the distances of each light from its shadow. This process is very convenient for rapid measurements, which do not require any great degree of accuracy.*

Bunsen's Photometer.—This photometer is one of those most in use. It is based on the fact that if a paper screen on which is a semi-transparent grease spot, is placed between two lights, the spot will disappear when equally illuminated on both sides. The photometer constructed on this principle consists essentially (Figs. 83 and 84) of a graduated scale from 3 metres to 3.50 metres long, on which slides a small carriage carrying a frame, and an index which records its position. The frame is fitted with a paper diaphragm treated with a solution of spermaceti in benzine, with the exception of a circular spot in the centre. Sometimes the opposite arrangement is employed, a translucent spot being made in the middle of the opaque diaphragm. In this photometer a candle is generally employed as a standard. In one of the arrangements shown in Fig. 84, the candle is placed on the arm of a balance which regulates the hourly consumption of spermaceti. In this case, as in the preceding, the screen is enclosed in a thin metallic cylinder open at both ends, and having two longitudinal slots, by which the two sides of the paper disc can be observed. It is more convenient, as shown in Fig. 85, to place on each side of the diaphragm, two mirrors inclined at an angle of 45°, which allow the experimenter to observe the reflections on both sides of the screen at the same time. Fig. 85 shows how the apparatus is adapted for using the Carcel lamp as a standard. The lamp and the system of screen and mirrors are rigidly connected and move together. In all these arrangements the frame carrying the diaphragm is shifted on the scale until the spot ceases to be visible, and the division on the graduated scale as marked by the index, is then noted. The distance of the screen from the lights is thus recorded, and the ratio of the squares of those distances will give that of the intensities. The three illustrations

---

* A full description of Rumford's method of using his apparatus, with plates showing the nature of the screen employed, will be found in "The Complete Works of Count Rumford," published by the American Academy of Arts and Sciences, Boston, 1875, Vol. IV. Count Rumford was an American by birth and education, his real name being Benjamin Thompson. He was created Count by the King of Bavaria, to whom he rendered many important services. It may be of interest to add that he was the founder of the Royal Institution of Great Britain.

of Bunsen's photometer show instruments adapted for measuring the
intensity of gas jets, but the arrangement would be identical for measur-

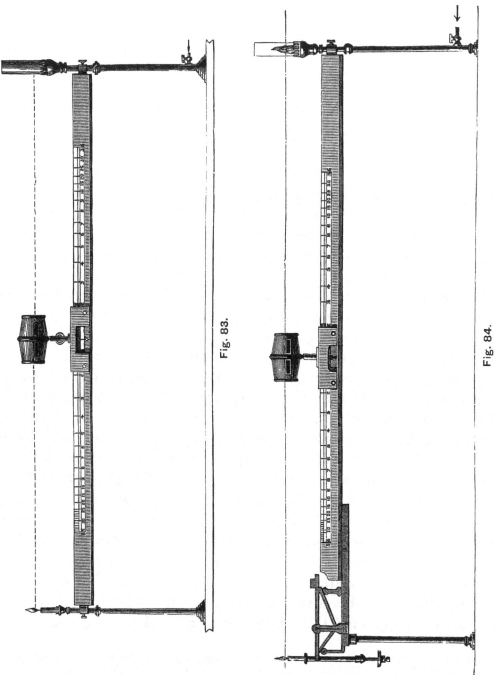

Fig. 83.

Fig. 84.

ing that of an electric light. The illustrations are borrowed from the
portfolio of the Continental Company for the Manufacture of Gas Counters
and other apparatus.

The German physicist Carstaedt has described in Poggendorff's "Annalen" (Vol. CL., page 551) an experimental verification of the law of the squares of the distances, founded on the use of the Bunsen photometer. If Q and Q_1 be the quantities of incident light received from the two sources; R, R_1, the fractions reflected by the opaque portion of the diaphragm; and V, V_1, the fractions transmitted by the oiled portions, at the moment when the spot disappears, we have evidently

$$R Q + V Q = R_1 Q_1 + V_1 Q_1.$$

That is

$$Q (R + V) = Q_1 (R_1 + V_1)$$

Whence

$$\frac{Q}{Q_1} = \frac{R_1 + V_1}{R + V}$$

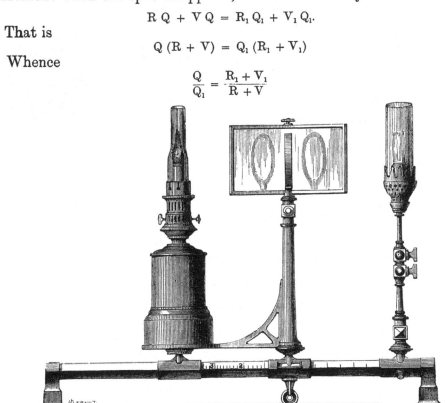

Fig. 85.

The coefficients R, R_1, V and V_1 depending wholly on the photometer, it follows that the ratio $\frac{Q}{Q_1}$ is constant. Now, if we admit as correct the law of the squares of the distances, we shall have, employing the ordinary notation,

$$Q = \frac{I}{d^2} ; \; Q_1 = \frac{I}{d_1^2}$$

Whence

$$\frac{Q}{Q_1} = \frac{I}{I_1} \times \left(\frac{d_1}{d}\right)^2$$

The ratio $\frac{I}{I_1}$ remaining the same during a series of experiments, it

follows then that $\left(\dfrac{d_1}{d}\right)^2$ must be equally constant.　It is this that M. Car-staedt has verified and illustrated by numerous tabulated experiments.*

Wheatstone's Photometer.—The following is the description of a very rarely used instrument, founded on the permanence of luminous impressions on the retina, for periods of about one-tenth of a second.　A bright steel bead A (Fig. 86) is fixed on a blackened disc B, carried on a small moving pinion C, which can travel round the inner circumference of a toothed ring D.　If the pinion set in movement by the handle E makes a complete turn in less than one-tenth of a second, and if the bead is lighted by any luminous source, the eye will perceive permanently the luminous epicycloid traced out by the revolving bead.　If the bead is illuminated simultaneously by two lights, two

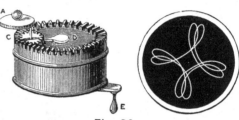

Fig. 86.

* The foregoing mathematical explanation supposes the diaphragm of the Bunsen photometer to be quite opaque in one portion, and perfectly transparent in the greased part.　It is, however, semi-opaque and semi-transparent; for if the diaphragm is illuminated on one side only, to an observer on the same side as the source of light the greased portion appears darker than the surrounding paper, because more light is transmitted through the semi-transparent greased part than through the remaining semi-opaque part.　To an observer on the other side of the diaphragm the greased part appears the brighter.　Hence there is both reflected and transmitted light, on all parts of both sides of a diaphragm illuminated by a standard light, and one to be compared with it.　When the grease spot disappears to an observer, it is evident that the reflected and transmitted light received by him from each unit of surface area of the semi-opaque part, are together equal to the transmitted, plus the reflected light received by him from the semi-transparent part.　Suppose a standard source of light S; we have from the side of diaphragm next that light,

R = light reflected per unit of surface of semi-opaque part,
r = ,,　　　,,　　　　　　,,　　　semi-transparent part;

and on the opposite side of diaphragm,

V = light transmitted　　　　,,　　　semi-transparent part,
v = ,,　　　,,　　　　　　,,　　　semi-opaque part.

And if Q = the quantity of light received on the diaphragm from S, $\dfrac{R}{Q}$, $\dfrac{r}{Q}$, $\dfrac{V}{Q}$, and $\dfrac{v}{Q}$ will be the respective fractions of Q transmitted or reflected, as the case may be, per unit area.　Similarly from a source S^1 to be compared with the standard, a similar series of fractions $\dfrac{R_1}{Q_1}$, $\dfrac{r_1}{Q_1}$, $\dfrac{V_1}{Q_1}$, $\dfrac{v_1}{Q_1}$ will be obtained.

It will be obvious that when the greased portion appears to an observer, on the side next the standard light, to be similarly illuminated to the semi-opaque portion, it is because $\dfrac{R}{Q} + \dfrac{v_1}{Q_1} = \dfrac{V_1}{Q_1} + \dfrac{r}{Q}$; whence $\dfrac{R-r}{Q} = \dfrac{V_1 - v_1}{Q_1}$; giving $\dfrac{Q}{Q_1} = \dfrac{R-r}{V_1 - v_1}$

Similarly, to an observer on the other side, $\dfrac{R_1}{Q_1} + \dfrac{v}{Q} = \dfrac{V}{Q} + \dfrac{r_1}{Q}$; giving $\dfrac{Q}{Q_1} = \dfrac{V-v}{R_1 - r_1}$

When the spot, as observed from both sides of the diaphragm, appears of the same character, it is evidently because $Q = Q_1$.　If then the quantity of light received is the same on both sides of the diaphragm, it must be because I, the intensity of illumination of the standard source, is in the same ratio to I_1, the intensity of the source to be measured, as are the squares of the distances d of the standard source and d_1 of the other source from the diaphragm.　For, by the law of the squares of distances $Q = \dfrac{I}{d^2}$; and $Q_1 = \dfrac{I_1}{d_1^2}$

Whence when $Q = Q_1$, $\dfrac{I}{d^2} = \dfrac{I_1}{d_1^2}$. 　$\therefore$ 　$\dfrac{I}{I_1} = \left(\dfrac{d}{d_1}\right)^2$

brilliant spots will be seen, and two epicycloids of different luminosity will be formed, the brightness of which will be equalised by placing the apparatus at a suitable distance from both lights. When this is effected, the ratio of the squares of the distances will give the ratio of intensities of the two sources.

Arago's Photometer.—The celebrated physicist **Arago** published in Vol. X. of his scientific memoirs, several notes on photometry, which have resulted in two methods, one of which is based on the phenomenon of Newton's rings, and the other on polarisation. In describing these instruments, we may recall the experiments on which they rest, and give a brief explanation of them.

1. Coloured rings. Let us suppose that on a glass plate A B (Fig. 87) we place a very flat lens C D ; we shall then have an optical system that will produce by reflection and refraction, a series of concentric coloured rings.

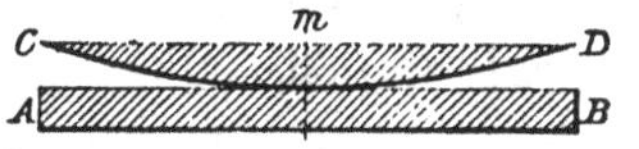

Fig. 87.

When the eye, placed above *m*, receives the rays falling on the apparatus, it perceives at the point of contact of the lens and the plate, a dark spot, surrounded by a series of regularly recurring iris-coloured rings, produced entirely by interference. These rings become oval if the eye deviates from its normal position. If, on the other hand, the eye is placed so as to receive the rays by transmission through the system of plate and lens, the bright rings seen will appear less brilliant, the central space will be bright, and the rings will succeed each other as in the previous experiment. If monochromatic light be employed, the rings will appear alternately bright and dark, the former, seen by reflection, corresponding in position to the latter when seen by transmission, and *vice versâ.* If we arrange the lens as shown in Fig. 88, that is to say in such a manner that the eye placed at O observes, on a paper screen A B, the rings seen by transmitted and by reflected light simultaneously, the rings disappear, proving that the same rings of the two systems have complementary colours of equal intensities. This experiment is due to Young and to Arago. If we intercept the rays coming from one of the sides, only one of the systems of rings will disappear.

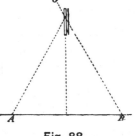

Fig. 88.

Arago deduced from this experiment the following photometric method. The two lights to be compared are placed one on each side of the lens and plate, and the brighter light is moved until all appearance

of the rings has disappeared. The illumination is then the same on each side, and it only remains to apply the law of the square of the distances to find the ratio of intensities.

2. It is also to Arago that we owe the polarising photometer, the principles of which, viz., those of double refraction and polarisation, it may be useful to explain. Double refraction is a phenomenon discovered by Bartholinus in 1670. It is exhibited in a high degree by Iceland spar. which is crystallised calcic carbonate. These crystals possess the remarkable property of giving two refracted rays and two images. The laws of this phenomenon were investigated by Huyghens. Crystals of Iceland spar, suitably cut, take the form of an oblique parallelopipedon, of which the limiting sides are parallelograms with equal angles, arranged so that two opposite edges of the parallelopipedon are the vertices of regular trihedral angles. The axis of these trihedral angles, *i.e.*, the line equally inclined to the three edges, is called the axis of the crystal, and along this axis and all lines parallel to it, optical phenomena are distributed symmetrically. Any plane containing the axis or parallel to it, is called a *principal plane.* When a ray falls on such a crystal it is split up into two elementary rays, one of which follows the law of sines, and is consequently called the *ordinary* ray, whilst the other which does not follow that law, is called the *extraordinary* ray. When the incident ray is perpendicular to the face of the crystal, the ordinary ray is an extension of it, and the extraordinary ray is refracted laterally. The two corresponding images appear of equal intensity, so long as the incident ray coincides with the normal.

If we allow the two emergent rays to fall on a second crystal, they will give rise to four images of different intensities, according to the relative positions of the two crystals. It should be remarked that two of these images always disappear simultaneously.

It will be seen that the ordinary and extraordinary rays possess this property, that if they are received on a second crystal, then in two rectangular positions of the principal plane of this second crystal, the emergent rays are successively extinguished two by two. The light is polarised, and the plane of polarisation is that plane to which the principal plane of the crystal which receives the polarised ray is parallel, when the emergent extraordinary ray is extinguished. It is hardly necessary to add that Iceland spar is not the only substance which possesses this property. Double refraction is a phenomenon common in a greater or less degree to all crystals.

We may now proceed to examine how Arago used the phenomena of polarisation for photometric measurements. His first object was to demonstrate the accuracy of the method; and it may not be out of place to

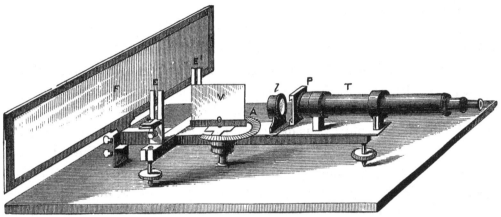

Fig. 89.

explain how he arrived at this result, by analysing rapidly one of his papers on photometry, published in the first volume of his scientific memoirs. Figs. 89 and 90 illustrate the apparatus he constructed: F is a large sheet of white paper stretched on a frame and illuminated uniformly from behind. A plate of glass is shown at V, and T is a tube fitted with a diaphragm and a vertical slit. This tube is movable in a horizontal plane around an axis situated about the middle of the plate. It is carried by a light frame, and can be displaced angularly around the point O, its angular displacement being read on the graduated circle A; *l* is a ring movable within a second ring, very minute motions being susceptible of accurate measurement. It can also be displaced longitudinally to varying distances from the opening

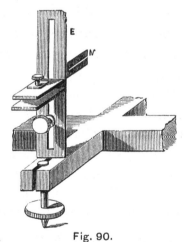

Fig. 90.

of the tube, by sliding upon the frame. At E E[1], on each side of the plate, are two small supports, which hold between their clips two screens N, formed of two blackened strips and slit longitudinally to admit the light transmitted by the large screen F[1]. If the tube T is placed somewhat obliquely to the plate, the eye, at its extremity, will see on the glass the reflected image of the slit E[1], and the transmitted image of the slit E. The tube may be shifted until the two images have the same intensity. Then the transmitted, will be equal to the reflected, light. Instead of illuminated

slits, **Arago** also used blackened wires, which intercept, the one at E a portion of the light received by reflection, and the other that transmitted to it by the reflected light. The two lights will be of equal intensity when the two shadows are equally dark.

With the apparatus this equality is found when the axis of the telescope is inclined at an angle of 11° 8′ to the glass plate.

To find the obliquity when one of the images is half or a quarter of the other, a prism of rock-crystal is placed at P. It is turned round until the ordinary and extraordinary images are of equal intensity. The tube T is then displaced until the transmitted, is just equal to either of the reflected, images, and its displacement is noted. By means of a second prism, the intensity of either image may be again halved, and the transmitted image may be reduced to one-fourth the intensity of the primitive image. A series of angles may be determined in this way, and the corresponding intensities tabulated.

The preceding theory supposes that no radiations are quenched or absorbed in the various reflections and refractions, *i.e.*, that the incident ray is precisely equal in intensity to the sum of the reflected and refracted rays. Arago verified this by inclining T at an angle of 11° 8′, and placing a second plate parallel to the first, but at some distance, in order to send back the reflected light. The fixed plate thus receives in place of direct rays, the integral of the rays reflected and transmitted by the second plate, and it is always found that the two images are equal when the inclination of T is exactly 11° 8. This being done, in order to check the accuracy of the law of polarisation of Malus, Arago placed one of the small screens above the plate V in such a way as to throw into the tube only the rays coming directly from the illuminated paper. The other screen placed lower down allowed the rays to pass, which are transmitted from the plate. The rays were again divided into two groups of equal intensity by a bi-refracting prism. The ordinary ray was received on a second prism parallel to the first, and was divided into two, and it was necessary to ascertain if the ordinary ray varies as the square of the cosine of the angle between the principal planes of the two prisms. To do this, the tube T was displaced until the intensity of the reflected light was equal to that of the light coming from the illuminated screen across the two prisms. The tables give the measure of the first intensity, and the second is then known. The principal planes of the prisms are displaced to different angles, and by proceeding in the same way the corresponding intensities are obtained. In this way it

may be shown that they agree with the results obtained by calculation, and the correctness of the law is thereby established.

Dubosc's Photometer.—The formulæ and deductions of Malus have served as the basis for a series of instruments devised by various physicists for different photometric observations. We need not attempt to describe all, because the greater number have for their object the measurement of the intensity of solar, lunar, or stellar light. We may, however, refer to the arrangement devised by M. Dubosc, and based on the same principles. Let us consider two bi-refracting prisms B B^1, Fig. 91, upon which impinge rays from two sources of intensities I and I^1. The ordinary and extraordinary images given by these prisms, seen through an analyser, will have for their intensities,

$$I_0 = I \cos^2 a \qquad I_0^1 = I^1 \cos^2 a$$
$$I_e = I \sin^2 a \qquad I_e^1 = I^1 \sin^2 a$$

If the two prisms are turned round, B 45° to the left, and B^1 45° to the right, the intensities will become,

$$I_0 = I \cos^2 (45 - a) \qquad I_0^1 = I^1 \cos^2 (45 + a)$$
$$I_e = I \sin^2 (45 - a) \qquad I_n^1 = I \sin^2 (45 + a).$$

If we suppose that by any means we can arrive at suppressing the extraordinary image, we shall see in the analyser only two images, the intensities of which will be

$$I_0 = I \cos^2 (45 - a) \qquad I_0^1 = I^1 \cos^2 (45 + a).$$

By turning the analyser to a suitable angle we can make $I_0 = I_e$, from which we shall have

$$I \cos^2 (45 - a) = I^1 \cos^2 (45 + a),$$

or,

$$I \sin^2 (45 + a) = I^1 \cos^2 (45 + a),$$

and

$$\frac{I^1}{I} = \frac{\sin^2 (45 + a)}{\cos^2 (45 + a)} = \tan^2 (45 + a).$$

The ratio of intensities is thus reduced to a simple reading of angles.

The photometer itself consists of a copper tube 50 millimetres in diameter and 35 centimetres long, Fig. 92. It carries at the end, which is turned towards the light to be compared, the two prisms B and B^1. At the other end is an eye-piece containing the analyser. This latter is enclosed in a tube which turns around its axis by means of a rack gearing into the circumference of a large circular plate that acts as a screen. Within the tube everything is symmetrical around a central vertical plane. In the diagram Fig. 92, n n is a screen that divides into two equal parts the plate

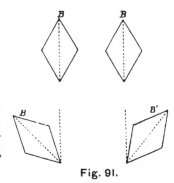

Fig. 91.

of ground glass $m\,m^1$, and separates the field of each of the lights to be compared.   A A is a double prism of glass intended to make the images

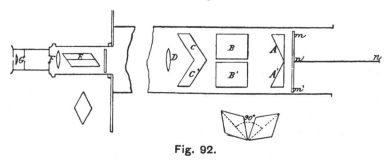

Fig. 92.

diverge, and to correct any imperfect achromatism in the prisms B B^1; the part $c\,c^1$ consists of two plates of glass.   The extraordinary ray is extinguished by the glass when it falls within a convenient angle given to the glass plate, and this curve (Fig. 93) rises very rapidly beyond 25° or 30° in such a way that a slight variation of only the ordinary ray is allowed to pass.   D is a bi-convex lens, and E the analysing prism ; F G is the eye-piece forming with the lens D, a microscope focussed on the screen $m\,m^1$.   If we make $\tan^2(45+a)=v$ and

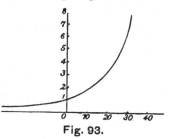

Fig. 93.

construct the curve of which the abscissæ would be a, and the ordinates v, we shall see a small change in the angle a gives a great variation in the ratio $\dfrac{I}{I_1}$.

It is necessary, therefore, to arrange the light in such a way as to produce almost the same illumination on the screen, correcting, however, the result obtained by the ratio of the squares of the distances.   The process has this advantage, that when the lights are once arranged, they need not be again displaced.   It is sufficient to turn the analyser by which the maximum, minimum, and mean variations of the light, can be easily measured.

The various photometric methods that we have described are those most generally employed.   In addition to them there are many others based on different assumptions more or less correct, and to which it will be interesting to refer in general terms.

Leslie's Photometer.—Leslie has attempted to transform his differential thermometer into a photometer.   The thermometer is composed, as is well known, of thin glass bulbs connected by a small tube in which is placed a suitable quantity of a coloured liquid which rises or falls according to the difference in temperature of the bulbs.   In the photometer (Figs. 94 and 95)

one of these is gilt, and the apparatus is placed so that both are at the same distance from the light to be measured. The calorific effect is to displace the level of the liquid, and this would also represent the luminous value if the two phenomena were strictly proportional. This, however, is not the case, and the instrument can only be usefully employed to measure different intensities of the same light, and even with such observations, only approximate results are obtained.

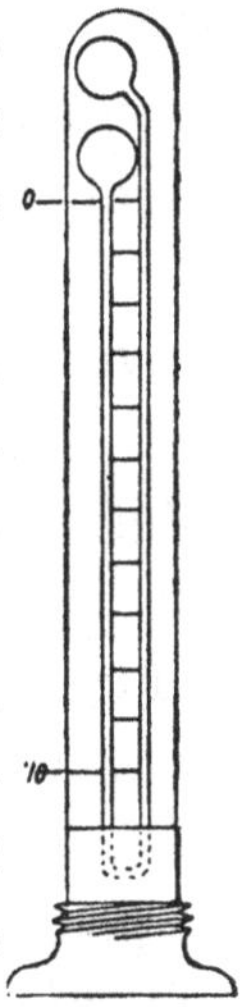

In his " Inquiry into the Nature and Propagation of Heat," London, 1804, Leslie describes this instrument as being best constructed with one bulb of black enamel glass, the other of plain glass. He gives two illustrations, and prefers that shown in Fig. 94 for laboratory use. The other is a pocket instrument.

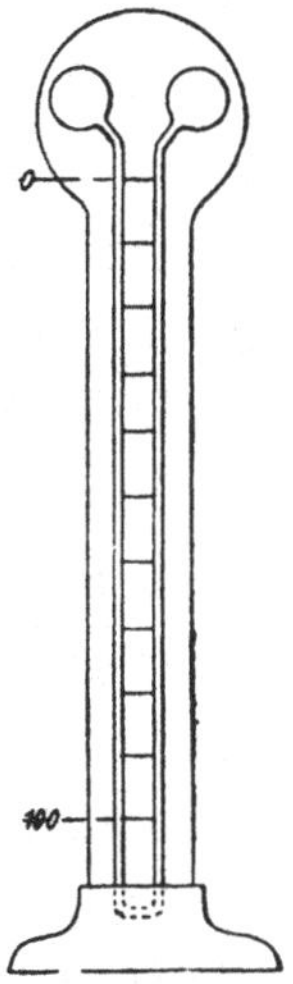

Fig. 94.   **Fig. 95.**

The following is his description of its action. " The incident rays freely traverse the clear ball without exciting any effect. They are, however, detained and absorbed at the surface of the black ball; and there, assuming a latent form, act as heat; hence the temperature of the black ball continues to rise till the increasing dispersion of heat caused by refrigeration, becomes equal to the regular supply derived from the incessant influx of light."

Ritchie's Photometer.—Ritchie, in 1825, suggested a modification of Leslie's instrument. He supposes that as heat-rays when they acquire sufficient periodic frequency are able to give the sensation of light, light-rays are converted into heat when a substance arrests them which can decrease the rapidity of their vibration, and so render them invisible. His instrument is based on the opinion that equal volumes of air would be equally expanded by an equal absorption of light rays by black surfaces, and on the principle that the intensity of light varies inversely as the square of the distance from the luminous body.

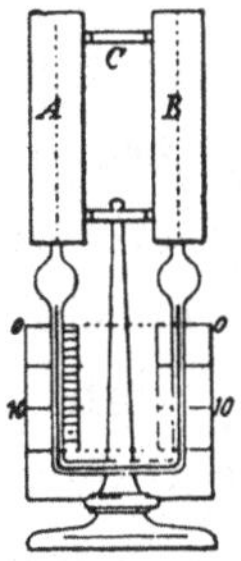

A B (Fig. 96) are two shallow cylinders of tin-plate, from 2 in. to 12 in. in diameter, and from $\frac{1}{4}$ in. to 1 in. deep. One end of each is closed by tin-plate soldered air-tight. The two outer faces are closed by the thickest and finest plate-glass, also fitted air-tight. More effectually to cut off heat-rays, double plates of glass, parallel with each other and with water between them, were sometimes used. The connecting pieces C are glass

Fig. 96.

rods. Half way between the inner surfaces of the glass and the metal ends is stretched in both cylinders a sheet of blackened bibulous paper. The ∪ tube contains sulphuric acid tinged with carmine. The scale is divided into equal parts. Accuracy depends on the perfect equality of the two ends, to ascertain which, the instrument is placed between two steady flames, and moved nearer one or the other till the liquid remains at zero. The instrument is then turned half round upon a suitable turn-table, so that it is not shifted as regards the lights. If the liquid remains at zero the instrument is correct. Ritchie mentions that he had constructed instruments which were affected by a single candle at a distance of 30 ft., while they were not sensibly affected by a mass of hot iron affording considerably more heat. An instrument found to be correct by the above method, placed between two groups of unequal numbers of candles, each giving light of equal intensity to each of the rest, was found to stand at zero when distant from the groups as the square roots of the number of lights in those groups. Thus nine lights at 3 ft. distance balanced four lights at 2 ft. distance. Leslie's instrument depended for its action upon a difference of temperature being maintained in the two bulbs, by one of the sources of lights to be measured—separate measurements being taken to compare one source with another. In Ritchie's instrument the two sources are directly balanced one against the other by inequality of distance. There is no reason why means should not be adopted to eliminate the heat rays and the chemical rays by suitable screens or prisms. The results of experiments with the instrument so arranged would give the requisite data for ascertaining whether the instrument could be depended upon for ascertaining the values of the luminous rays. Ritchie says he was able with this instrument to compare the light of the sun directly with a single candle. Certainly, if this be the case, and the results are such as can be depended upon, this almost forgotten instrument is worthy of some further trials.

Professor Balfour Stewart, of Owen's College, Manchester, recently invented a modification of Leslie's photometer. He blackened a hemisphere of the globe at one extremity of a ∪ tube, and whitened a hemisphere of the globe at the other. The ∪ tube contained oil. The difference in the reaction upon the contents of the globes, caused respectively by light being absorbed at the black hemisphere, but reflected at the white, was shown by a difference in the densities of the contents of the globes, which was in turn shown by a disturbance of the level of the oil in the ∪

tube. This apparatus is described in the Proceedings of the Royal Society, London, for March 23, 1876. It may be mentioned that Leslie had tried oil and various other liquids, and with various gases in the bulbs instead of air. He however preferred strong sulphuric acid tinged with carmine, the bulbs being filled with air.

Chemical Photometers.—A system of photometrical measurement is based on the chemical properties of luminous rays. This method, simple as it is, contains fundamental errors and objections that render it useless for accurate measurements. Chemical and optical phenomena are not comparable, and two pencils of rays may have such characteristics that the one possessing the highest chemical energy, develops the lowest lighting power, and *vice versâ*. This consideration is sufficient to show the inaccuracy of such a method. The first physicists who employed it, MM. Fizeau and Foucault, were well aware of the defect, and freely admitted its gravity and extent. Nevertheless, it can be made to give useful results for comparing different intensities of the same kind of light. It has also a particular interest which justifies a somewhat detailed description. The chemical method was the first employed to measure the intensity of carbon rendered incandescent by a battery current, or in other words, of the voltaic arc. The "Annales de Chimie et de Physique," 1844, Vol. XI., page 370, contain a detailed description of the manner in which the distinguished physicists above named, reduced their method to practice. If the focal image of a luminous surface, obtained by means of a lens, is received on a sensitive plate, the latter will undergo changes varying with the intensity of the image and the time of exposure. From the data thus obtained, comparisons can be made between two lights, by equalising these changes, provided the relations connecting the intensity of the focal image with that of the luminous surface are known, Now, if we call the latter I, i denoting the focal intensity, r the radius of the aperture, d the focal distance of the lens, and $2a$ the angle under which the aperture of the lens is seen from the luminous surface, these various elements are connected by the relation

$$i = \frac{I\,r^2}{d^2} = I \tan^2 a.$$

If two lights of intensities I and I^1 give at the lens the angles $2a$ and $2a^1$, the focal images producing at the same time the same degree of chemical change on two identical resistent films, we shall have

$$i = i^1, \text{ whence } I \tan^2 a = I^1 \tan^2 . a^1$$

Whence

$$\frac{I}{I^1} = \frac{\tan^2 a^1}{\tan^2 a}$$

K

But it is very difficult to select lenses, the elements of which are such as to bring the intensities I and I^1 to exactly equal values at the focal image, whilst it is more convenient to extend the time of exposure until the changes are equal. It would be sufficient then to assume that the periods of exposure should be inversely in proportion to the intensities. MM. Fizeau and Foucault endeavoured to ascertain up to what point this law, which appears plausible at first sight, has a practical application. They assumed that it would hold for the times t and t^1, when t is less than ten times t^1. In this case it may be admitted that $\dfrac{i}{i^1} = \dfrac{t^1}{t}$. If then we obtain an equal degree of change during the periods t and t^1, we shall have

$$\frac{i}{i^1} = \frac{t^1}{t}, \text{ or } \frac{I \tan^2 a}{I^1 \tan^2 a^1} = \frac{t^1}{t}$$

or,

$$\frac{I}{I^1} = \frac{t^1 \tan^2 a^1}{t \tan^2 a}$$

It now only remains to explain how MM. Fizeau and Foucault estimated the equality of change in the sensitive plate at points affected by the two images. This is precisely the delicate part of the operation. The sensitive solution employed was the silver iodide used by Daguerre, and the degree of alteration, that at which the mercury vapours begin to condense and develop the photographic image. The sensitised plate was placed in the focus of the lens in a dark chamber, and the image of the light was projected upon it. By producing an image sufficiently small, it was possible to shift the axis of the instrument without throwing it out of the focal plane, and by raising the screen during a time measured by a chronometer, it was possible to obtain on the plate five or six successive images, corresponding to different periods of exposure. Afterwards, by treating the plate with vapour of mercury, several images of different intensity were produced, and it was easy to choose from among them the image corresponding to the time of a certain exposure. The same process was carried out for the luminous source taken as the standard, and in this way all the necessary elements were obtained for comparing the intensities.

Bunsen and Roscoe have also carried out a number of photometrical experiments based on the chemical action of the luminous rays on the mixture of chlorine and hydrogen. These gentleman, in the account of their experiment published in Phil. Trans., 1857, mention that the only previous attempt to refer the chemical action of light to a standard measure, was by Dr. Draper. He collected hydrogen evolved by electrolysis

over hydrochloric acid saturated with chlorine. This, by diffusion of the dissolved chlorine and by added chlorine, gave approximately equal quantities of hydrogen and chlorine mixed, which gases almost entirely disappeared under the action of light. The alteration of volume of the gases was read off on a scale, and being proportional to the time of the exposure, within certain limits, served as a measure of the chemical rays.

But the gases thus prepared were not of constant composition, and there were no means of ascertaining the relative proportions of the gases employed. From these causes an exact agreement between the various indications taken with Draper's instrument was impossible. Another source of error was due to variations in the density of the gases.

Draper had asserted that it was impossible to obtain hydrogen and chlorine in equal volumes by electrolysis. This Messrs. Bunsen and Roscoe disproved. A current of three or four zinc carbon elements was passed through hydrochloric acid of 1.148 sp. gr., carbon poles being used. After a time evolution of chlorine was visible, and the amount gradually increased. The colour of the liquid became deeper, and at length a point was obtained, more or less quickly according to the strength of the current and the quantity of hydrochloric acid, when the free gases were in a condition of statical equilibrium with those absorbed by the liquid. From that point the composition of the free and absorbed gases remained constant, provided temperature and pressure were not varied, and also provided the hydrochloric acid in the liquid electrolyte did not diminish below a certain amount.

The composition of the gas absorbed by the electrolyte was found to vary with every change of temperature; but that of the free gas remained the same, provided the equilibrium alluded to was not disturbed. When once equilibrium was established, the gas was invariably a mixture of one volume of hydrogen with one volume of chlorine. Numerous experiments and analyses were made to determine that fact, and also to ascertain that the water holding the hydrochloric acid in solution did not undergo decomposition, and that no other chlorine compounds were formed.

The apparatus they used, both for preparing the mixture of gases and ascertaining the effect of light upon that mixture, is illustrated in Fig. 97, and was adopted as being the best method of excluding all disturbing causes, and of referring the chemical action of light not merely to comparative, but even to an absolute measurement.

The glass tube a contained two carbon electrodes with terminal

platinums fused into the glass. This electrolytic cell was connected by wires *b b* with the four-cell zinc-carbon battery C. Another electrolytic cell *d* of considerable resistance could be introduced into the circuit by interrupting the shorter circuit D; and by this means the evolution of gases in *a* could be readily reduced from a maximum to a minimum. The tube *f* which conveyed the gases from *a* was ground into the neck of *a*, which was formed so as to make a water lute. at *g*. The mouth of the vessel *a* was kept forced down upon mercury in the jar A, by means of its own weight and that of the apparatus to which it was connected. From *f* the gases could pass two ways: Through a safety manometer *m*, and by the pipe *p* into F, which contained water which served as a pressure regulator, by shifting the position of *p* in the sliding collar *t*; the waste gases passing into G, where they were condensed by alternate layers of slaked lime and

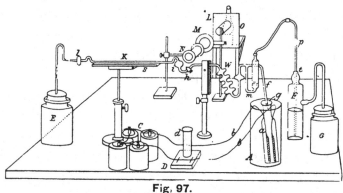

Fig. 97.

charcoal. Or, through the washing bulbs W; past the glass stop-cock *h*; into *i*, where they were subject to the action of light when required; through K and *l* into E, which was a similar condenser to G. The vessel *i* contained 2 or 3 cu. cm. of water, and was attached to the tubes by ground-glass joints, perfectly air-tight. When the stop-cock *h* was open, the gas passed by W K into E, the pressure at F being regulated to cause this. After passing for a time the hydrochloric acid of 1.148 sp. gr. in *a* and the liquids in W, *i* and *l* became saturated. During this time the composition of the free gas was undergoing a continual change, which diminished as the liquid approached the saturation point. When this point was reached an equilibrium of absorption was established, and then the composition of the free gas remained perfectly constant, being a mixture of one volume of hydrogen with one volume of chlorine.

It will be seen that the gas in different parts of the apparatus is subject to different pressures. At the electrodes, it is under the pressure of the

liquids in a, W, i, and l; in the tube f it is under the pressure of the liquids in W, i, and l. As any alteration of pressure alters the composition of the free gas and necessitates a fresh saturation, it was necessary to stop the passage of the gases through i without altering the pressure there. This is effected by closing the stop-cock h, which cuts off the current of gas without alteration of pressure on the side towards i. The gas being evolved then passes through m into F and G; but the evolution at the electrodes is immediately checked by diverting the current at D as before explained. A sufficient evolution, however, continues to keep the liquids in a and W ready saturated for a fresh experiment. The light which was to act upon the gas in i was placed upon a support at o, and several arrangements of screens were adopted to shade i from the heat of the light, and from that of the body of the experimentalist. One of the arrangements for shading i from the light is shown. There is a screen L, with a tube through which the rays pass to the convex lens M, and thence through the cylinder N which contained water, the ends being closed by glass plates. Of course corrections had to be made for the absorption of light by the various media through which it was passed.

It was found necessary to evolve 10 or 12 litres of gas to saturate the 10 or 12 grammes of water in the apparatus.

As the platinum wires were destroyed by the chlorine in a few days, carbon plates were cut off the carbon of a zinc-carbon cell, which had been long in use, and these were boiled in aqua regia, and then made white-hot in a current of dry chlorine till all sublimation of volatile chlorides had ceased. Platinum wires were then fixed in holes bored in the carbon with a needle; a glass capillary tube was passed over the wire, and its melted end pressed round the wire into the carbon; the rest of the tube was fused on to the wire. The upper ends of the carbons were then saturated with white wax. The platinum wire, thus protected with a glass enamel, was then fused into the upper part of a. By these means the platinum was protected from contact with the chlorine.

Another important point to be observed was that the electrodes should never be allowed to come into contact with the gases above the liquid in a; for, acting in a similar manner to spongy platinum, they would bring about an explosion.

The combination of the mixed gases brought about in i by the action of the light, of course diminished the quantity; consequently some of the liquid in l was driven by atmospheric pressure into K. The amount of gas

thus acted upon was ascertained by means of the scale S, upon which K is placed.

Professors Bunsen and Roscoe remark in the paper from which these particulars are taken, that this apparatus fulfils the following conditions necessary to secure exact photo-chemical measurement :

(1) The gas which is exposed to the action of the light is composed of equal volumes of hydrogen and chlorine.

(2) It contains no foreign impurities.

(3) It nowhere comes in contact with caoutchouc or other organic matter that would alter its composition.

(4) The change of pressure during a series of observations is imperceptible, owing to the volume of water in l being much larger than would fill the tube K.

(5) Statical equilibrium between the free and absorbed gases can be perfectly established.

(6) The surface of water in i does not alter its position during the experiment, so that the hydrochloric acid produced is always absorbed under precisely the same conditions.

(7) The vessel i is covered with black varnish outside to the level of the contained water, so that no alteration of free gas dissolved in water occurs.

(8) The volume of gas entering i from the tube K is very small compared with the total volume of the gas exposed to light. In most of the experiments it·was $\frac{1}{5000}$ for every division of the scale.

(9) The disturbing action of radiant heat is fully eliminated.

Professors Bunsen and Roscoe found that the first action of light upon the mixture of chloride and hydrogen, is accompanied by a phenomenon which they term photo-chemical induction. Chemical action does not at once commence at a maximum rate. It is at first very slight, almost, if not quite, imperceptible; but it gradually increases till a permanent maximum effect is reached. This maximum is reached in about seven to ten minutes after the vessel i has been exposed to the action of the light.

An enlargement in the combining power is also seen when the light is suddenly increased.

When the vessel i is exposed to the action of the light, the absorption should be observed by noting the readings on the scale S at regular intervals of one minute. When the maximum rate of combination is reached, the

mean of every three or four readings is taken, and the mean of three or four such means gives the result with great accuracy, eliminating the small irregularities due to the capillary resistance of the tube.

The apparatus is easily filled, and is operated as follows :

In the morning, about 55 cu. cm. of hydrochloric acid of 1.148 sp. gr. are placed in the electrolytic cell a, and gas is made to evolve at such a rate as to cause about two bubbles per second to pass through the washing bulbs and the bulb i. In the afternoon the acid is replaced by a fresh supply, and the operation is repeated with this important exception: the gas evolved during the first hour is not passed through the bulb W, but the stop-cock h is kept closed, and the gas passes through the manometer into the condenser G. All this time the apparatus is kept protected from light. After passing the gas for about three hours more, the stop-cock h is closed, and the apparatus is left for the night. In the morning the operations of the previous afternoon are repeated, with fresh acid, allowing the impure gas which passes off during the first hour to pass into condenser G. Readings may then be taken with a standard lamp as the source of light. It will be from three to six, or even nine days, according to circumstances, before the maximum effect of the light of the lamp is noted; but this once attained, the instrument can be used for comparative measures for several months. All that is necessary is to put a fresh supply of acid into a, pass the gas evolved during the first hour into G, and then take the readings with the standard flame, after the gas has passed for fifteen minutes through i. The stop-cock h is of course shut during an observation; and when it is closed, the operator should not forget to at once diminish the current, through or by means of the resistance introduced at D, without which precaution, the equilibrium of the gases dissolved in the acid is soon destroyed, owing to combination of H and a at the surfaces of the carbon electrodes.

Access of air to the washing bulbs and to i must be always carefully prevented; for if the atmosphere of i be exposed to the air but for a few minutes, from two to six days' saturation will be necessary to regain the maximum effects. The amount of air or other gases which, mixed with the hydrogen and chlorine, would render the readings incorrect, is said to be less than a billionth part of the total gaseous contents of i.

Care must be taken also to reduce the intensity of the light to be compared with the standard, to such a degree as will not explode the mixed gases. A thin glass bulb about the size of a pigeon's egg, filled with these gases, exploded instantly on exposure to the diffused daylight of an over-

clouded sky at an open window.　Parts of the apparatus have also exploded when exposed to evening light after the sun had sunk below the horizon.

Numerous experiments were undertaken to ascertain the loss of intensity of light due to absorption and reflection at the various screens and lenses employed, with a view to ascertaining the exact photo-chemical effect of the standard lamp employed.

Coal gas was used as the source of standard light.　It was at first burnt in a Scott's zinc burner, but afterwards a platinum burner p (Fig. 98) in the top of a small-pressure box r was used.　Gas was taken to this box through the tube S, which had a metre scale upon it.　A platinum wire was arranged to project from the holder on a graduated rod t, to measure the height of the flame, and a small water manometer was attached to the box r, a pressure of but 1 mm. of gas being maintained in that box.　The lamp was enclosed in a larger box u, which was fitted with a cover that could be adjusted so as to leave an opening above the flame, whatever the position of the burner.　Air was admitted through the bottom of the box.　The box was blackened

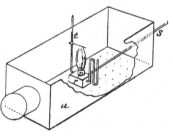

Fig. 98.

inside, and at one end was a water cylinder, the ends of which were closed by parallel plates of glass.　The flame was thus somewhat like that of the well-known Bunsen burner when deprived of air; indeed, the Bunsen burner was designed during the progress of these experiments, and was used to heat small carbon cylinders which had been immersed in solutions of various salts, so that the effect of various coloured flames might be ascertained.　It may here be mentioned that the red, violet, yellow, and green flames produced by the chlorides of lithium, strontium, potassium, sodium, and barium gave no more photo-chemical action than the colourless flame of the Bunsen burner; whereas the green flame of chloride of copper and the pale flame of chloride of antimony gave considerably more effect than the standard luminous coal gas flame—in some cases nearly double as much— the flames of course being the same size.

With respect to the effects of variations in the composition of the coal gas, it was found that experiments made at intervals of three or four months with the standard lamp, varied but slightly more than one per cent.

The bulb i was carefully protected by screens, and enclosed in a small cylindrical box attached to one of the screens, having at one end, glass through which the light rays entered, while the cover had but a very small

hole, to enable the operator to see the flame through the bulb i, so as to note any variations in its height. Hence when an air thermometer with a bulb of the same substance and capacity as i was put in its place, no expansion was noted when light was suffered to fall upon its bulb ; and it was ascertained that variations of temperature in the dark room from 64° to 80° Fahr. made no sensible difference in the readings of the apparatus, neither were they affected by variations of upwards of half an inch in the barometer.

M. Eden has recently published in the "Comptes Rendus" of the Academy of Sciences of Vienna, a photometric process for measuring the intensity of the ultra violet rays of sunlight. It consists in exposing to light a liquid composed of 40 grammes of oxalate of ammonia dissolved in 1 litre of water, and 50 grammes of corrosive sublimate dissolved in the same quantity of water. Two volumes of the former are mixed with one volume of the latter, and under the action of light, a black precipitate is obtained, the weight of which depends on the intensity of the light.

Selenium Photometers.—The curious property of selenium, by which its electrical resistance changes under the action of light, has led to a new method of photometric measurement. As it is independent of the eye, it is free from the " personal error" of the experimenter.

It was on February 12, 1873, that Mr. Willoughby Smith announced, at a meeting of the Society of Telegraph Engineers, the observation made by Mr. May, a telegraph clerk at Valentia, that exposure to light diminishes the resistance of a plate of selenium. This observation was corroborated by the Earl of Rosse and others. Professor W. G. Adams made an exhaustive investigation of the electrical properties of selenium, and embodied his remarks in a paper communicated to the Royal Society in 1877. He then showed that the most effective rays are the greenish yellow, and also that the change of resistance is proportional to the square root of the illumination.

Mr. Shellford Bidwell discussed before the Physical Society, in November, 1882, a series of experiments made by him to ascertain whether the phenomenon was due to the heat rays or the light rays, or to both. His experiments have finally disposed of the theory held by some that the fall of resistance in a selenium cell is due to a microphonic action at the junctions between the selenium and the metals forming the electrodes. According to this theory the heat rays increase the volume of the substance, thus causing it to make better contact and at the same time diminishing proportionally the resistance of the circuit. Mr. Shellford Bidwell has shown that the heat

rays actually *increase*, while the light rays *decrease*, the resistance. Hence, when a ray of white light falls upon a selenium cell, the change observed is the resultant of the luminous and thermal effects.

This curious substance, selenium, was discovered by Berzelius in 1817 as a by-product in the distillation of iron pyrites. It is fusible and combustible; it belongs to the sulphur group, enjoying properties similar to those of sulphur, phosphorus, and tellurium. When melted and rapidly cooled, it becomes an amorphous mass with conchoidal fracture, and is practically a non-conductor of electricity. If, however, it is heated to 100° Cent. (it melts at 217° Cent.), and is kept at that temperature for a certain time it becomes crystalline, and in that condition it slightly conducts electricity, its conductivity increasing, as Professor Adams has shown, with the battery power.

It is in this crystalline condition that selenium is sensitive to light.

In February, 1876, a series of interesting experiments was made in the late Sir William Siemens's laboratory at Woolwich by Dr. Obach with the following results:

Selenium in	Relative Conductivities.		Resistance in Ohms.
	Deflections.	Ratio.	
1. Dark	32	1	10,070,000
2. Diffused daylight .	110	3.4	2,930,000
3. Lamplight . . .	180	5.6	1,790,000
4. Sunlight	470	14.7	680,000

In testing the selenium plate in different parts of the spectrum, it was shown that the actinic rays exercised no appreciable effect, and that the effect gradually increased towards the red, or least refrangible rays.

It should be here remarked that the resistance of a bar of selenium is not constant, but varies by the action of the currents sent through it, and also with the structural changes brought about by time. Thus a plate of selenium whose resistance was one day 613,000 ohms was 670,000 the next day. The resistance of a certain specimen was 14,900 ohms one year and 19,000 the next, whilst that of another specimen fell from 7,600,000 to 745 in the same period of time.

Sir William Siemens exhibited the photo-electrical properties of selenium in a lecture given at the Royal Institution in March, 1876, by means of an artificial "eye." This interesting apparatus (Fig. 99) consists of a hollow ball

with two circular apertures opposite each other, the one being furnished with a lens and the other with an adjustable stopper carrying the sensitive plate which is connected in circuit with a Daniell's cell and galvanometer. The lens is covered by two slides, representing the eyelids, the ball itself being the body of the eye, and the sensitive plate corresponding to the retina.

Having placed a white illuminated screen in front of the "eye," a strong deflection was observed as soon as the eyelid was opened. A black screen gave an exceedingly small deflection, a blue screen a greater, and a red a still greater one. All of these, how-

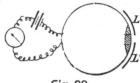

Fig. 99.

ever, fell short of that produced by the reflected light of the illuminated screen.

Dr. Werner Siemens, of Berlin, has successfully applied the properties of selenium to photometry.

His photometer is formed of a copper tube A B (Fig. 100) 3 centimetres in diameter and 15 long, the inner surface of which is blackened.

One of the ends is furnished with a diaphragm, the other carries a plate of selenium which may be exposed to, or withdrawn from the action of light by means of a screen, movable from outside. This tube, fixed on a vertical spindle, is attached to a circular support by three locking points, one of which is furnished with a screw, by which it can be shifted at will. To this support is connected a graduated scale L M, on which can be moved a shifting candle-holder N carrying the standard light, the flame of which ought to be in the axis of tube A B. The selenium plate is connected to two terminals H, insulated from the tube, to which are connected the conducting wires G, which are coupled by means of the ebonite plate F and a movable piece situated in the axis to the terminals D E, which are in the circuit of the battery and of the galvanometer. This arrangement, which the sketch Fig. 100 explains, insures contact whatever may be the

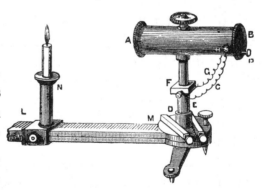

Fig. 100.

position of the tube A B. The galvanometer employed is a Thomson reflecting instrument mounted on a small wooden stand, which also carries a little petroleum lamp enclosed in a lantern. This lantern contains a lens, which concentrates the luminous rays, and directs them on a totally reflect-

ing prism carried by the galvanometer. This prism can be moved in such a way as to cause the mirror to reflect the spot of light on a graduated scale.

The manipulation of the selenium photometer just described, involves two operations. First, the installation of the reflecting galvanometer. For this, the scale should be placed away from the influence of the light, and in such a position as to be at least half its length from the centre, and perpendicular to the axis of the galvanometer, the centre of the scale, which is the zero point, being on the prolongation of the axis. The copper tube containing the mirror is introduced within the coil of the galvanometer in such a way that the small wires fixed to the mirror and serving as an axis of rotation, may be almost vertical. The plane of the flame of the petroleum lamp is placed in a line joining the wire to the centre of the lens. To obtain this result the prism fixed to the galvanometer coil is raised in such a way that the light falls on a piece of white paper placed at a distance of from 1 to 2 metres; by shifting the lens an image can be obtained of the wire on the paper. The lamp is then moved until the image of the wire is clearly defined and without coloured fringes. The prism is then put back in its place, and is turned until the reflected ray falls on the mirror; this can be adjusted by looking into the mirror directly opposite it. The direction of the ray reflected by the mirror is then followed by means of a slip of paper, on which the ray is allowed to fall, and the prism is fixed in such a way that the light reflected by it falls on the centre of the mirror, and that reflected by the mirror returns directly above the prism (see

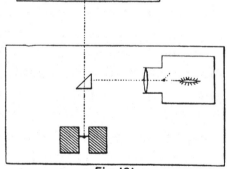

Fig. 101.

Fig. 101). By turning the screw fixed to the scale, the ray is brought to the height of the scale, and by turning the directing magnet above the galvanometer, it is brought to the division of the scale exactly desired. Finally, the lens is moved until the image of the wire reflected on the scale is sharply defined, and the copper tube containing the magnetic mirror is turned until the image moves horizontally along the scale. By raising or lowering the directing magnet, the sensitiveness of the galvanometer is increased or diminished.

Fig. 102.

This latter can be placed in any convenient plane. In place of the stretched wire in the lantern, a diaphragm in which a slit is formed can be employed.

Sometimes the lantern is placed behind the scale, the image of the wire being directly projected upon the mirror without the intervention of the reflecting prism, the centre of the lamp being slightly below the horizontal plane of the centre of the galvanometer mirror, and the scale being slightly above it. By this means, if the mirror be vertical, the angles of incidence and reflection a_1 and a being equal, the image of the wire is reflected upon the scale (see Fig. 103).

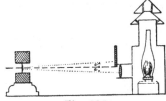

Fig. 103.

The second operation is the use of the instrument. The selenium photometer is disposed in such a manner that the movable tube when it is perpendicular to the scale L M, Fig. 100, may be directed towards the luminous source, the intensity of which is to be measured. To do this accurately, the horn obturator, which contains the slip of selenium, is removed and replaced by a cover having a cross marked on a transparent ground. The front face is furnished with a cover which contains an appropriate opening. The photometer is then turned until the image of this opening is projected on the middle of the cross. Then the two covers are removed, and the rear one is replaced by the selenium. A battery of from 12 to 24 cells filled with acidulated water is included in the circuit with the selenium and galvanometer, but the circuit is not closed until the observation is made. As soon as the circuit is closed, a deviation of the galvanometer is produced, and the image is often thrown off the scale, but by adjusting the height of the directing magnet, or by a shunt or a resistance box included in the battery circuit, it is brought within convenient limits. Under the influence of the light the resistance of the selenium diminishes, and the deviation produced by the current on the galvanometer increases. The plate of selenium is exposed to the light by opening the shutter in the selenium box, when the tube is turned towards the luminous source to be measured. Note is taken of the corresponding deviation of the magnetic mirror, and the selenium is directed towards the candle, which is then moved along the scale, until the same deviation is obtained. The luminous intensities are in the inverse proportions of the squares of the distances. The selenium photometer is not absolutely correct, and it only gives results comparable for lights of the same colour.

Dr. T. Nacks has somewhat improved this method in the following manner. He placed in the circuit of a battery a galvanometer, a selenium cell, and a variable resistance (see Fig. 104). The whole system being in the dark, he regulated the resistance in such a way as to obtain a very slight

deviation of the needle of the galvanometer.   Then the selenium being illuminated by rays from the luminous body placed at unit distance, he noted the
corresponding deviation a_1.   He then noted the deviations a_2, a_3, a_4, corresponding to distances 2, 3, and 4, and consequently to the intensities $\frac{1}{4}, \frac{1}{9}$, and
$\frac{1}{16}$. From these data he deduced a regular graduation.   The luminous
unit proposed by Dr. Nacks was thus defined.   It is
that which increases by one electro-magnetic unit, the
strength of a current of which the electromotive force is
equal to 10, and the resistance of which is equal to a
Siemens's unit.   Let l equal this quantity, I the strength
of the current when the apparatus is in darkness, and I^1

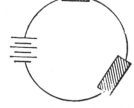

Fig. 104.

when it is lighted.   Then $\dfrac{I^1 - I}{l}$ will be the luminous intensity of the source.
If E be the electromotive force of the battery, and R the total resistance,
we shall have $I = \dfrac{E}{R}$.   When the apparatus has been exposed to the light,
the resistance of the sensitive surface, assumed to be equal to unity, will
give r, and we shall have $I^1 = \dfrac{E}{R - r}$.   Hence the value of $\dfrac{I^1 - I}{l}$ may be
obtained by substituting the above value of I^1 and I for $I^1 - I =$
$\dfrac{E}{R - r} - \dfrac{E}{R} = \dfrac{E\,r}{R\,(R - r)}$; whence $\dfrac{I^1 - I}{l} = \dfrac{E\,r}{l\,R\,(R - r)}$.

The values E and R are known, r is determined by means of the
resistance.   The latter is used to bring the total resistance back to that
of the apparatus in the dark.   This resistance introduced, converted into
Siemens's units, gives the value of r.

Photometer of Raimond Coulon.—The methods just described are of
high interest, but they lack as yet the test of practice ; and the same remark
applies to the photometer of M. Raimond Coulon, which was described by
him in " La Lumière Electrique."*

This photometer is based on the use of Crookes's radiometer.   As is
well known, the radiometer starts into motion as soon as rays of light are
allowed to impinge upon its vanes.   If then the two opposite faces of a
radiometer are lighted by two sources and maintained in exactly similar
thermal conditions, the apparatus will revolve in virtue of the difference of
the illumination on its two faces.   If it remain motionless, it may be concluded that the two faces receive equal quantities of light, and the law of
the squares of distances will then give the ratio of the intensities.   The

* See " La Lumière Electrique," Vol. IV., p. 344.

apparatus devised by M. Coulon consists of a glass globe, carefully exhausted, mounted on a stand and containing a disc free to revolve about a vertical axis. One half of this disc is painted white, and the other half is lamp-blacked. This radiometer is placed in a cubical metallic chamber filled with water, which is kept at a uniform temperature by four small lamps placed at the corners of the chamber. Small screens placed between the lamps and the radiometer protect the latter from extraneous radiation, and serve to equalise the temperature of the water. In the middle of each face of the chamber, is an opening fitted with glasses, two of which, opposite each other, allow the radiometer to be examined, while the two opposite ones permit the rays to enter and fall on the vane. Under the differential action of the latter, the radiometer will revolve. The motion is gradually stopped by removing the stronger light to a greater distance. When this is done, the radiometer will be at rest, and the disc will be seen edgeways by the observer. After a first experiment it is advisable to change the relative

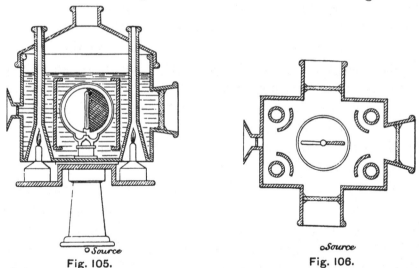

Fig. 105.   Fig. 106.

positions of the two sources, to correct the error that may be set up by any inequality in the two faces of the radiometer. A mean of the two results may then be taken as the true one. Figs. 105 and 106 show the construction of this photometer very clearly. M. Coulon does not, however, confine himself to this particular form; he has constructed a second instrument, to avoid the difficulties resulting from the displacement of the standard lamp, and the source to be measured. He divides the photometric measurements into two series; the first consisting of those made, once for all, in the laboratory, whilst the others, conducted in the photometer room, have for their object to vary the action of the luminous source of unknown

intensity until the results obtained are equal to those obtained with the standard lamp.

The apparatus consists of a very light aluminium needle, carrying at each end two small plates of soft iron weighing 3 decigrammes. This needle is supplied with two disc of mica, the faces of which are painted black and white alternately. This system is delicately suspended in the centre of an exhausted globe, placed, as in the former case, in an enclosure of constant temperature; this globe is terminated above and below by two tubes. In the upper one is placed, so that it can be moved from the outside by a magnet, a small soft iron ring acting as a carrier; within this ring there is a cork, through which passes a glass rod, carrying the silk fibre from which the needle is suspended. This ring is held by four light springs. By means of the magnet, the ring can be lowered until the needle reaches a rest fitted in the lower tube, which thus relieves the thread from the weight of the needle. Finally, a magnetised bar, mounted on a cork-lined ring, and placed within the upper tube near the top, determines the normal direction of the needle. This latter becomes displaced under the action of the light, and will assume an intermediate position depending on the two opposing forces. Besides being used as a photometer, this apparatus can be employed both as a thermometer and an electrometer. The arrangement is certainly original, but it appears to be too delicate to be capable of giving practical results, and it would seem more difficult of application than the form previously described. At present it must be considered as a purely experimental apparatus.

In concluding his description,* M. Coulon intimates that the system can be used to measure not merely the intensity of a light, but also the general lighting of an apartment illuminated by many lamps. It should be observed that this general problem involves the use of a special unit, because it is impossible to compare a body receiving the light on all sides with one receiving it on one surface only. M. Coulon proposes the following theoretical unit: Every point in space receiving the light produced from four standards, spaced one metre apart, and placed at the extremities of two rectangular axes, will be considered as receiving a unit of diffused light. In this case the photometer contains two crossed needles. Such a method of comparative measurement is certainly ingenious, and deserves to be placed on record.

Stevenson's Photometers.—The action of the photometers devised by

---

* See "La Lumière Electrique," Vol. V., p. 67.

Mr. Thomas Stevenson, F.R.S.E., are of the latter kind, and depend on the interposition between the light and the eye of a liquid absorbing medium, the thickness of which can be increased or diminished. The simplest form of this instrument is shown by Figs. 107 and 108 in section and end elevation, and consists of one tube sliding telescopically within a second. The inner tube is closed with a disc of plain glass at the further end G, and carries a screen or guard for the eye at the end H. The outer tube is also closed with glass at its extremity A, and at the other end is provided with a stuffing-box E packed with sponge, through which the smaller tube moves water-tight. The liquid absorbing medium between the two tubes, is a mixture of writing ink and water ; it rises and falls in the vacant space J as the inner tube is pushed in or out. To use the instrument it is turned towards the source of light, and the inner tube is drawn backwards until the layer of interposing fluid is so thick that the image just fades away. The distance between the two plain glasses is then a measure of the intensity of the light falling upon the outer one.

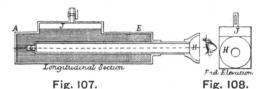

Fig. 107. Fig. 108.

Mr. Stevenson has worked out the same idea in another way, by fitting the two plain glasses in the opposite sides of a collapsible india-rubber bag. One glass is secured to the framework of the instrument, and the other to a sliding tube, so that the two are maintained in parallel planes and are capable of being made to approach towards or recede from each other. The upper part of the bag communicates with a reservoir into which the excess of liquid is delivered as the volume of the bag is reduced by the motion of the eye tube.

In another arrangement one of the plain glasses, the object-glass, is replaced by a non-transparent cover, and the light is admitted to the tube through two prisms by which it is twice totally reflected. These prisms are mounted in a sliding frame which travels along a slot in the upper surface of a rectangular tube. The prism which first receives the light is in the air above the surface of the tube, and the other is plunged in the liquid. The adjustment is effected by sliding the prism frame along its slot until the image on the eye-glass fades away. By a modification, this instrument may be made to indicate by means of simultaneous comparison. Two slots are made in the upper surface of the tube, both converging upon the eye-piece. A prism frame is mounted upon each, and the images of the

two sources of light are projected side by side upon the eye-glass. By moving the prisms along the grooves, the lengths of fluid through which the rays proceeding from the two objects will pass, will be varied, either until there is an equalisation in the intensity of the images, or until extinction has been effected, when the relative lengths of fluid are noted. In order to adapt the instrument for measuring lights in different azimuths from the observer, and at different elevations from each other, the upper prisms are made capable of rotation round their vertical and horizontal axes, so as to present their first refracting surfaces in any required direction.

Another modification, approaching still more closely to the usual methods of comparison, comprises two independent tubes, each throwing an image on to a screen. These screens are placed at an angle of 45° to each other, and the prism frames are moved backwards and forwards until they are equally illuminated.

Mr. Stevenson has successfully used several of these forms of instrument when experimenting on the powers of different lighthouses, but he has found that, like other instruments of the same kind, a good deal of practice is necessary before the eye becomes capable of estimating correctly the equality of any two lights. He has also devised a photometer entirely different from the above, and designed to supply a meteorological rather than an optical want, though it also admits of adaptation to other photometric purposes. It was made for the purpose of measuring varying amounts of daylight, with a view of determining whether there is any connection between dark gloomy weather and heavy gales.

In this photometer a minute portion of daylight is allowed to pass through a small hole pierced in a diaphragm, behind which the light, spreading into the dark chamber in concentric spherical shells, diverges over nearly 180°. The intensity of this diverging light is ascertained by moving a transparent diaphragm near enough the aperture to allow the eye to decipher any characters that may be inscribed on it; the distances of the diaphragm from the aperture necessary for producing distinct vision represent, in the inverse duplicate ratio, the intensties of the rays. By arranging numbers in any order, unknown to the observer, the possibility of mistake, arising from his fancying that he has distinct vision when he has not, may be prevented. In order to adapt this very simple instrument to the measurement of artificial light, or of direct sunlight, a small piece of tube, having a lens of short focal length fixed in it, is placed before the fine orifice. Rays incident upon the lens will convey a focus in the minute

orifice, after which they will pass in a diverging cone, and thus be made susceptible of having their intensities measured.

Spectro-Photometers.—An examination of the various photometric processes, shows that a most important element in the problem, and one that cannot be efficiently dealt with, is the difference of colours in the lights which are being compared. No solution of this difficulty has yet been found, and the best that has been done is to diminish the difficulties, as far as possible, with more or less success in the results obtained. Many efforts have been made to replace the direct measurement of the ratio of two lights. of different tints, by that of the ratio of two simple wave-lengths : this method is known by the name of spectro-photometry. It has been particularly studied and advocated by the French physicist, M. Crova, professor of physics at Montpellier, and set forth by him in several papers, especially those communicated to the Academy of Sciences on the 26th of September and 5th of December, 1881, in his memoir published in the "Annales de Chimie et de Physique," Fifth Series, Vol. XIX., 1880, and in various articles in the "Revue Scientifique."

The errors to which the experimenter is liable when endeavouring to compare two lights of different colours, are illustrated by the following experiment. If we illuminate a Foucault photometer by two sources, one of which gives yellow and the other a blue light, and as a preliminary trial adjust the two to approximate equality, then moving forward each light to half its original distance, the equality ought not to be changed; it will, however, be seen that the yellow side is more strongly lighted than the blue. This shows that the impression on the retina is a function of the luminous intensity, which varies with the quality of the light. This observation is due to Purkinje. In completing the experiment it will be seen that the further the sources are removed from the screen, the more the physiological effects tend to be proportional to the luminous intensities. From this it follows that in measuring intensities of different colours, it is necessary to work under conditions of low illumination, that is to say, with the two sources placed as far as possible from the screen. Let us suppose now that we have to compare a white light with a yellow one, for instance, that of an electric arc with a Carcel lamp. We decompose the two lights into their simple elements, so as to produce two contiguous spectra, and regulate the distances of the two lamps so that the lights on the screen may be of equal intensity. The spectrum of the electric beam will be very intense towards the violet, but less intense towards the red end, while the reverse effect will take

place in the spectrum of the Carcel lamp. If, then, we lay down the total intensity of each of the lights by a curve, the abscissæ representing the wave lengths, and the ordinates, the corresponding luminous intensities, these curves ought to have equal areas, since the total lighting is the same. The known ordinate corresponds to a wave length, similar in both spectra, and the ratio of intensities of these rays of equal wave length in the two spectra gives the ratio of total intensities. In practice, it is necessary to ascertain first the simple ratio that represents proportionally the intensity of the source. To find this, M. Crova recommends taking as an index of the illuminating power, the greater or less distinctness with which fine lines drawn on a white surface can be distinguished, reducing the light gradually until the eye, placed at a constant distance, fails to distinguish them. The slit of a spectroscope is covered with a series of very fine parallel lines that cease to be visible when the light is reduced by varying the angles of the principal planes of the two Nicol prisms placed in the spectroscope. The curve of elementary intensities for each of the lights to be compared, may then be constructed, and the ratio of ordinates corresponding to the same abscissæ, which is equal to the ratio of the areas, may be found. This having been done, the simple ratio for each of the spectra must be deduced. M. Crova proposes two methods : " It is possible," he says, " to produce by means of a spectro-photometer the contiguous spectra of two sources to be compared, their rays being projected on the halves of the opening, by two reflecting prisms, and separated afterwards by means of a slit placed before the eye-piece of the spectroscope ; by means of two Nicol prisms, the brighter light can be reduced in such a way as to render it equal to the feebler one. This method is very exact, but rather delicate in application ; and a simpler one is desirable for general use. The following is a description of such a method : If we observe a spectrum projected on a white screen, placing before the eye two Nicol prisms, the right sections of which are rectangular, and having between them a plate of quartz 9 millimetres thick, we shall see two wide black bands, due to the crystalline plate darkening the extremities of the spectrum (the red and the violet being extinguished) ; while the yellow, orange, and blue are very considerably weakened ; in such a way that there remains only a narrow band, in which the green, due to a certain wave length, is preserved, while the colours situated on both sides are, at first greatly reduced, and then extinguished. By slightly turning one of the prisms, these bands are displaced along the spectrum, and we can arrange their position in such a way as to preserve the maximum intensity in that

part of the spectrum where the ratio of intensities is equal to that of the lighting. This done, we observe the screen of the Foucault photometer, the halves of which are illuminated by rays of very different tints, and for which the curve referred to above has been drawn ; it will appear to us of a greenish tint, identical in the two adjacent fields, if the two Nicol prisms have been arranged as we have just described. The angle varies with the nature of the two sources of light to be compared, but if one of them is constant (standard Carcel), it will be sufficient to trace on the mounting of the movable prism, a line which will be level with the divisions on the fixed tube, and each of which corresponds to a light of a known character—the sun, different electric lamps, &c. Lastly, if the two lights are of the same tint, their spectral composition is identical, and the photometric composition gives the same results, whatever may be the position of the Nicol prism ; in this case the arrangement is still useful, because it will throw on the photometric screen, the colour least fatiguing to the eye. In the apparatus constructed by M. Duboscq, the arrangement of prism, and quartz plate, is fixed in an eye-piece adjusted to the end of the conical tube of the photometer, and through which the operator examines the screen." It will be seen from the above that the problem of photometry can be solved even in the most general cases, in which comparisons of luminous sources of different tints have to be made.

Ayrton and Perry's Photometer.—This is a dispersion photometer, in which a concave lens (instead of increased distance) is used for the purpose of weakening the light from an electric lamp, and thus bringing the measurements within suitable distances. A second feature of the instrument is the introduction of a plain mirror to reflect the rays of the electric beam on to the diverging lens.

The instrument is shown in Figs. 109 and 110, where E is a tripod stand supporting a small table, susceptible of exact levelling. F is a pin, directly under the centre of the mirror H, around which the photometer may be turned without altering the distance between the mirror and the electric lamp. This rotation of the photometer becomes necessary when the inclination of the incident beam is changed, a small horizontal displacement being sufficient for this adjustment.

The horizontal axis of the mirror is placed at 45° to its reflecting surface, in consequence of which rays coming at any angle from the lamp can be measured without introducing errors arising from varying absorption for different angles of reflection. The mirror carries a pointer which indicates on the graduated arc G, the inclination of the beam to the horizontal.

A is a black rod, which is used as in early photometers, its shadow being received on a screen of blotting-paper B.

Lighting the standard candle in the holder D, two shadows of the rod will be cast side by side on the screen, one by the candle and the other by the electric light. The latter has been reflected from the mirror, and afterwards dispersed by the lens C. The lens is then displaced until the shadows are of equal intensity, (1) when viewed through a sheet of red glass, and, (2) when viewed through a sheet of green glass. We then read on the fixed graduated scale, the distance d of the lens from the screen, and also the distance c of the candle. Then let f denote the focal length of the lens, D the distance of the electric lamp from the screen, *i.e.*, the distance of the lamp from the mirror, together with the distance of the mirror from the screen, and let L be the intensity of the light in standard candles, then

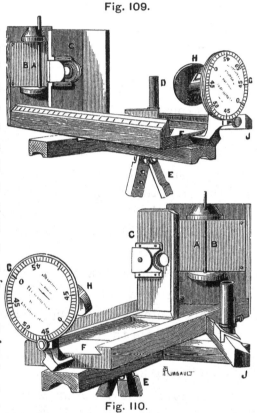

Fig. 109.

Fig. 110.

$$L = \frac{1}{c^2} \left\{ D + \frac{d\,(D-d)}{f} \right\}^2$$

Professors Ayrton and Perry state that from 30 to 40 per cent. of the incident light at 45° is absorbed, whether the light is of ruby-red or signal-green colour and practically none by the lens, so that from the intensity found, the normal intensity may be readily calculated.

The horizontal beam being measured, the electric lamp is fixed in any new position. The mirror is then turned until its centre ray passes exactly through the centre of the lens. The distance a from the screen to the mirror being invariable, then if λ be the distance from the centre of the mirror to the vertical from the lamp, and θ the angle of elevation, we have

$$D = a + \lambda \sec \theta.$$

Putting this value of D in the above formula and making the percentage correction for absorption, the true intensity of the light, in the terms of the standard, is known.

We shall conclude this portion of our subject by referring to the arrangement adopted by the committee appointed by the Franklin Institute (U.S.A.) in their photometric measurements made in 1877-78. To make these measurements accurately, it was found necessary to cut off all reflected or diffused light. For this purpose the electric lamp was enclosed in a box, open at the back for convenience of access, but closed with an opaque screen during experiments. In front of the box was placed a wooden tube with its inner surface blackened to prevent reflection ; and a similar wooden tube was placed at a proper distance from the first and holding in its further end the standard candle. This last tube contained the photometer paper (Bunsen's) mounted on a slide so as to be easily adjusted between the two sources of light. A small slit in the side of the tube allowed the observer to see the paper disc.

The difficulties arising from the difference of colour were at first thought to be considerable, but the committee found that practice and experience enabled the observer to overcome these to such an extent that the error arising from them was inconsiderable, being far less than that due to the fluctuations and unsteadiness in the electric arc itself.

LUMINOUS STANDARDS.

The selection of a suitable luminous unit has for many years occupied the attention of those physicists and engineers occupied with problems of lighting. On account of its importance, this question was naturally brought before the Congress of Electricians for consideration and discussion ; unfortunately, however, no satisfactory solution was arrived at, and for want of something better it was necessary to stop at the least objectionable of the standards generally used—the Carcel lamp—and to recommend it provisionally, with the object of obtaining results sufficiently uniform to render them comparable. As we have already said, an International Commission ought to be formed to investigate the problem and determine the rules which should be definitely adopted in photometric investigations. The determination of a luminous standard will certainly be one of the most delicate parts of its programme, because the qualities that should be combined in such a standard are so numerous, that it would appear very difficult to realise them simultaneously.

First of all let us consider the particular case of electric lighting, from which intensities of the most varied character are obtained—from incandescence lamps, each of a fraction of a candle, to the great Brush arc light

of many thousand candles, such as was exhibited at the Crystal Palace in December, 1881. Can so wide a range of intensities be referred for measurement to one standard unit, or will it be possible to find several units, absolutely multiples of each other, and applicable to the various electric lamps? Such a standard by no means exists at present, and we are now obliged to resort to several units more or less variable, of very uncertain ratios, and the permanent nature of which can be acquired only under numerous conditions. Moreover, the least objectionable of them all is too feeble to measure high intensities. Practically, if an error δ is made in the distance d of the standard lamp of the photometer, the ratio $\left(\dfrac{D}{d}\right)^2$ which gives the measure of intensity to be valued, will become $\dfrac{D^2}{(d+\delta)^2}$, and the ratio of the two evaluations, correct and erroneous, will be :

$$\frac{\left(\dfrac{D^2}{d^2}\right)}{\dfrac{D^2}{(d+\delta)^2}} = \left(1 + \frac{\delta}{d}\right)^2$$

which will more nearly approximate to the unit, as the ratio $\dfrac{\delta}{d}$ is smaller.

We have seen that Messrs. Ayrton and Perry have attempted a correction of this error by placing a diverging lens before the powerful luminous source, which spreads and reduces the light per unit surface, in a known ratio. This is one of the numerous propositions on which the International Commission will have to decide. The inability of the Congress of 1881 to deal with this subject is evidence of the difficulties that the problem presents. They will be made more evident from the analysis which we propose to make of the various units employed.

The chief of these units are " candles " of various types, oil lamps, and gas burners.

Candles.—These form the first units of light employed in photometry. In France they have been superseded by the Carcel lamp, and are only used exceptionally under the name of *Bougies de l'Etoile* for measurements of very low intensities. In England and Germany the candle is, on the other hand, very largely used. In this country they are made of spermaceti, while in Germany, paraffin is employed. This difference in types naturally produces great confusion in making comparisons, because the three candles are not of equal intensity. Moreover, in each type the candles have not always the same power, and during combustion, variations occur which often

affect the luminous values to a considerable extent. A French engineer, M. Girout, who recommended as a standard, a gas jet with a pressure regulator, of which we shall speak presently, has published in the *Paris Journal of Gas Lighting*, a *résumé* of very careful experiments, which he carried out, on the variations in intensity of photometric candles. These experiments are summarised very clearly in a series of graphic tables, some of which we reproduce in Figs. 111 to 114; in each of them the vertical lines represent the successive positions occupied by the movable screen of a

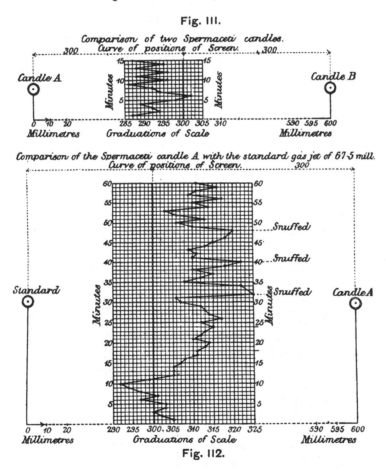

Fig. III.

Fig. 112.

Bunsen photometer placed between the two lights to be compared, at the moment when the spot disappeared. The ordinates give the series of observations made from minute to minute, the time being reckoned by a seconds' registering chronometer. The candles were weighed to within 1.5 grains, and moulded just before each experiment, in such a way as to leave a length of one centimetre at the carbonised part of the wick. Fig. 111 gives a comparison between two spermaceti candles, weighing 73.35 grammes,

and 73.40 grammes, and absolutely identical in appearance.   It would be supposed, *à priori*, that they would yield an equal light, but from the diagram it will be seen that candle B burnt with more intensity than candle A.   The mean ratio of the intensities was B = 1.113 A, with variations

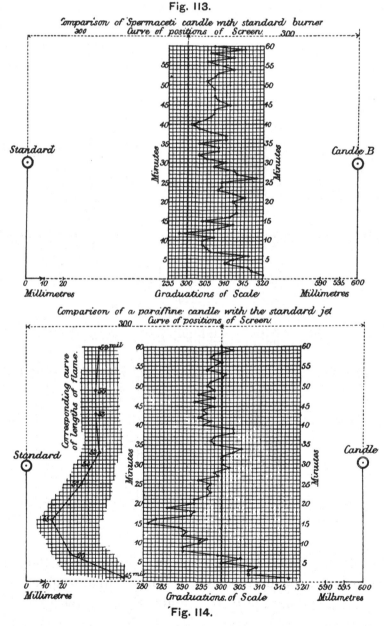

ranging from B = 0.973 A and B = 1.3 A.   In the course of the experiment, the candle B lost 7.824 grammes and candle A lost 7.534 grammes. The experiments, Figs. 112 and 113, show the candles A and B compared

with a standard gas jet, burnt from a single hole burner nominally equal in intensity to .1 of the normal carcel, and having a flame 2.65 in. in height. In the first case, the mean intensity was .0875 carcel, with a maximum variation of 55 per cent. from the lowest intensity. In the second it was .087 carcel, with a variation of 34 per cent. from the minimum value. The experiment corresponding to the curve, in Fig. 114, results from a comparison of the Girout standard, with a paraffin candle of which the successive lengths of flame were measured. The intensities are practically subject to the same variations as the lengths of the flame, ranging to 62 per cent. above the lowest value, with an average corresponding to .1041, or about one-tenth of a carcel.

The variations shown by these experiments are easily explained if we consider the numerous conditions which an invariable intensity requires—homogeneousness of the fatty body forming the candle, absolutely equal distribution around the wick, uniformity in structure of wicks, a constant rate of combustion, the absence of all currents of air. If we consider the mode of making each type of candle, it will be seen that the materials of which they are composed are never absolutely homogeneous, either on account of differences in chemical composition, or because of irregular mechanical distribution. Practically, stearine candles contain very variable proportions of wax and paraffin. Spermaceti candles are mixed with about 3 per cent. of wax and paraffin, and the latter material in its turn has a very variable composition; its melting point ranges from 113° C. to 140°, according to its mode of manufacture, and the nature of the raw material from which it was extracted. Stearine and wax are also sometimes mixed with the paraffin. Each of the bodies composing the three types of candles is thus a compound of very variable composition; besides this, the process of moulding presents great difficulties. In cooling, fatty bodies, especially paraffin, contract and form cavities around the wick, or flake and develop a tendency to crystallisation, which renders the mass quite heterogeneous. It is not necessary to insist further on the causes of variations in intensity and duration arising from differences in the material of which the candles are composed. The wicks of the candles are generally formed of from 60 to 80 woven strands of cotton, and treated with substances to facilitate combustion, such as nitrate of potash, and with materials such as boric acid, borax, phosphate of ammonia, &c., which form heavy cinders coming away from the end of the wick. These materials are dissolved in a great excess of water, which prevents the wick from burning down to the level of the body of the candle.

It will be seen at once that the structure of the wick, its length, the form it assumes at different periods of combustion, and its more or less eccentric position, exercise a great influence on the intensity of the light. The slightest current of air that causes the flame to flicker, will still further affect the variations due to unequal combustion and to changes in length of the wick.

For all these reasons, then, it is impossible to trust absolutely to the results obtained by the comparison of luminous sources with candles. We have seen that according to the experiments of M. Girout, the variations sometimes exceed 50 per cent. of the minimum value, so that an electric light measured by means of a candle standard, might vary in nominal value between 1000 and 1500 candles. Under these conditions photometric measurement offers no guarantee of accuracy.

To summarise this part of the question, the mean values adopted for the intensities of various types of candles are as follows : The French stearine candle (called *L'Etoile*) weighs one-fifth of a pound, and burns at the rate of 7.60 grammes per hour. According to Péclet this standard has a value of about one-seventh of a normal carcel, and according to Becquerel from one-eighth to one-ninth. These differences may be taken as due to variations in the length of the wick, which if not trimmed may lose 12 per cent. of its lighting power. It is recommended that the wicks should be kept at a constant length of one centimetre. We have seen above that the luminous value of the spermaceti candle is .087 carcel, and that of the paraffin candle .1 carcel. It is almost unnecessary to point out that these values are always uncertain.

The Carcel Lamp.—The Carcel lamp has this advantage over candles, in that it provides a more constant unit and one of a higher absolute value. The principle of its construction is a modification of that devised by Aimé Argand in 1789; previous to that date little or nothing had been done in improving the arrangement of lamps. Argand substituted for the flat wicks then in use, annular wicks, with a central current of air, burning the oil more completely on account of the increase in surface of the wick in contact with the air. By this means he suppressed the smoke and consequent waste produced from other wicks and arising from the insufficiency of air in the centre of the flame. The central draught that Argand at first tried to produce by a small metal chimney placed at a certain height above the flame, was soon replaced by a cylindrical glass chimney, by means of which the height of the flame, and therefore its intensity, could be increased, while

absolute steadiness was secured. The form of the chimney was modified a short time after by Lange, who conceived the idea of contracting the glass at the point nearest the flame in such a way as to break up the current of air, and to direct it on to the flame in order to produce more complete combustion. The lamp devised by Carcel in 1800 added to the arrangement of Argand, a regular feeding of the wick. In this lamp the oil reservoir is placed in the base, and the oil is raised up to the wick by means of a small pump, actuated by clockwork. The improvements afterwards made by Perrot brought these lamps into a practical form, excellent in principle, but somewhat complicated. Perrot's modification consisted in replacing the pump by a simple leather membrane. This device incorporated in the so-called moderator lamps, effected a large economy over the complicated mechanism of Carcel. The moderator was devised by Franchot, who substituted for the clockwork, a spring, the gradual unwinding of which, forced up the oil by means of a piston placed above it; the fall of the piston caused the oil to rise by a tube up to the level of the wick. The piston is formed of two thin discs of iron, with a leather washer fastened between them, that allows the oil poured into the lamp to descend into the lower reservoir, and afterwards forces it up a tube that the piston carries with it in its movement. This tube, which is very small, fits into a larger one attached to the burner, and which carries at its centre a rod that enters the tube bringing up the oil, the penetration of the rod varying with the position of the piston. It is this part of the apparatus that gives the name of moderator to the lamp. The force of the spring is so adjusted that the amount of oil supplied to the burner remains almost constant, whatever the position of the piston.

This brief description is necessary as an introduction to a study of the value of oil lamps as photometric standards. First employed for this purpose by Arago and Fresnel, they were afterwards adapted by MM. Dumas and Regnault for the photometric experiments carried out to regulate the illuminating power of the gas supplied to the city of Paris.

The intensity of the Carcel lamp depends on various elements, the principal of which are the composition of the oil, the nature and the length of the wick, the position of the contraction in the chimney above the flame, &c. These elements have been studied by MM. Paul Andouin and Paul Bérard, in a paper published in the " Annales de Physique et de Chimie" (1862, Vol. LXV., Third Series, page 423). Their mode of experimenting consisted in successively varying one of the elements, keeping the others

constant, in order to ascertain to what degree these variations influenced the intensity and expenditure of oil in the Carcel lamp. The latter being placed on a balance at a distance of one metre from the photometer, a ring gas jet with twenty holes was placed at an equal distance. The consumption of gas could be regulated so as to make the intensity of the flame exactly equal to that of the lamp. A gas counter recorded the gas used. The lighting power of the Carcel was reduced by a simple proportion to an expenditure of 42 grammes per hour. The consumption of gas increased with an increase in the intensity of the lamp. When this had reached a maximum, the gas jet, regulated in such a way as to give an equal light, consumed a maximum amount of gas. A first series of experiments had for their object to investigate how the consumption varies with the height of the wick, the height of the bend in the chimney remaining constant, and 7 millimetres above the top of the wick. The first of the subjoined Tables gives the results obtained.

OIL CONSUMED IN CARCEL LAMP, WITH WICK CONSTANT AND CHIMNEY VARIABLE.

Length of Wick.	Small Wick.		Medium Wick.		Large Wick.	
	Consumption of Oil per Hour.	Consumption of Gas calculated to equal the Carcel of 42 Grammes.	Consumption of Oil per Hour.	Consumption of Gas calculated to equal the Carcel of 42 Grammes.	Consumption of Oil per Hour.	Consumption of Gas calculated to equal the Carcel of 42 Grammes.
mill.	grammes.	litres.	grammes.	litres.	grammes.	litres.
4	27	96	30	155	32	99
6	33	175	36	193	36	159
8	38	196	42	185	42	192
10	40	190	42	200	45	194
12	35	170	40	193	48	212
14	38	177	40	...	51	216
16	36	180	45	186	48	189
18	31	153	...	...	42	192

The second series of experiments relates to variations in consumption according to the position of the bend in the chimney, the wick being kept at a fixed height of 7 millimetres.

From the accompanying Tables the following deductions may be made. 1. By increasing the height of the wick, the consumption of oil and the intensity of light increases up to 10 millimetres, and then diminishes. 2. The more the bend in the chimney is raised, the more the consumption of oil is increased, but at a certain moment the consumption continuing to increase, the intensity diminishes. There is therefore a height of the neck of the chimney corresponding a maximum intensity. 3. The average wick

corresponds, with equal consumption, to the maximum intensity; it is therefore the most advantageous. 4. Except when the consumption was nearly 42 grammes of oil, where the numbers representing the intensities show but little differences, there is no constant ratio between the expenditure of oil and of gas. This ratio increases up to a maximum which it attains towards the consumption of 42 grammes. And when photometric measurements are taken with the Carcel as a standard, if the consumption is not exactly 42 grammes, it is impossible to reduce the results to the consumption of 42 grammes by any simple calculation, until the actual consumption lies

CONSUMPTION OF OIL IN CARCEL LAMP, WITH WICK FIXED AND CHIMNEY VARIABLE.

Height of Bend in Chimney above Level of Wick.	Small Wick.		Medium Wick.		Large Wick.	
	Consumption of Oil per Hour.	Consumption of Gas calculated to equal the Carcel of 42 Grammes.	Consumption of Oil per Hour.	Consumption of Gas calculated to equal the Carcel of 42 Grammes.	Consumption of Oil per Hour.	Consumption of Gas calculated to equal the Carcel of 42 Grammes.
mill.	grammes.	litres.	grammes.	litres.	grammes.	litres.
− 2	18	14	18	11	15	23
+ 3	25	63	21	57	27	
7	36	187	39	161	48	175
12	39	199	42	200	50	186
19	42	151	45	175	51	164
24	46	315	45	161	54	140
29	...	...	51	133		

between the limits of 40 and 44 grammes. Outside of those limits the results given are inexact. It is impossible to fix absolutely the condition in which a Carcel lamp will have a constant expenditure. The temperature, the agitation of the air, the duration of lighting, the quantity of oil in the reservoir, all of these causes combine to vary the consumption, and therefore the intensity. Before commencing a series of photometric measurements it is expedient, if great precision is desired, to subject the lamp to preliminary tests, to ascertain under what condition it gives the best results.

The use of the standard Carcel has been especially studied in reference to its application in verifying the lighting power of gas made by the Compagnie Parisienne, and the recommendations of MM. Dumas and Regnault, in connection with the subject, have been much followed in conducting official tests in France of the luminous value of the electric light. We may therefore recall the practical instructions prepared by these two celebrated physicists, and explain at the same time the ordinary mode of measuring

the values of gas jets. According to these instructions the elements of the Carcel lamp should be as follows:

		Millimetres.
Exterior diameter of burner		23.5
Interior „ inner air current		17
„ „ outer „		45.5
Total height of chimney		290
Distance from elbow to base of glass		61
Exterior diameter at level of bend		47
Interior „ of glass at top of chimney		34
Mean thickness of glass		2

Wick.—A medium wick should be employed, of the type known as lighthouse wick. It should be woven with 75 strands, and should weigh 3.6 grammes per decimetre of length. The wicks should be kept in a dry place, or should be preserved in a box with a double bottom containing quick lime that ought to be renewed from time to time.

Oil.—Purified colza should be used.

The photometer of MM. Dumas and Regnault is illustrated by Figs. 115 and 116; it is essentially the same as that of Foucault. The parts comprising the complete apparatus are placed by a table attached to a frame by means of four adjusting screws. The Carcel lamp is put on one scale of a balance, and the photometric screen is set behind the glass O. The luminous field can be varied by means of the small screw placed above the observing glass O, which displaces the transverse partition, and allows the two illuminated fields to be brought in contact. In this apparatus the gas is consumed as in a Bengal burner, fitted with a gauge to indicate the pressure at which it flows; G is a very accurate counter divided into twenty-eight equal parts, each one of which corresponds to the consumption of one litre of gas. Each division is sub-divided into ten equal parts, and one-fourth of each sub-division can be estimated by the eye, which thus gives an approximation to one-fortieth of a litre. The counter carries two indices, one of which is set in motion as soon as the gas begins to flow, and the other, loose on the axis, can be made fast by a lever controlling a seconds' counter. At the moment of making an experiment, the loose pointer of the counter, and that of the seconds' counter, are set at zero, and the two needles are started. The system V R is a clepsydra, which serves to control the indications of the counter with great exactness. The lighting of the lamp comprises the two following operations: Putting in a new wick and cutting it to the level of the holder; filling the lamp with oil, exactly to the level of the gallery,

and placing the lamp in position. While lighting, the wick is kept at a height of 5 or 6 millimetres, and the chimney is then put on. In order to regulate the consumption of oil the wick is raised to a height of 10 millimetres, and

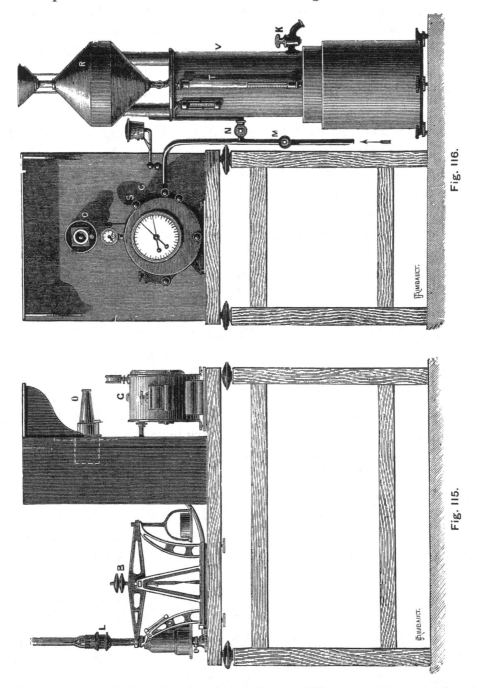

Fig. 116.

Fig. 115.

the glass is so fixed that the bend may be 7 millimetres above the level of the wick. To secure these conditions a little gauge adapted to the wick-

M

holder, is employed, the lower point of which is placed on the wick itself, and the upper point is adjusted to a line cut on the neck of the chimney. The lamp now ought to consume 42 grammes of oil an hour, and it is necessary that this rate of consumption should be maintained. When the quantity burnt falls below 38, or rises above 46 grammes, the experiment is stopped.

The scale carrying the lamp is provided with a sector, balanced at the centre. At the moment when equilibrium is produced, the arrangement starts a small hammer which strikes a bell, and notifies the operator that the lamp has consumed the quantity of 42 grammes, counterbalanced carefully beforehand, by means of a weight. The lamp, as well as the gas jet, is lighted half an hour before the operation. As regards the gas jet, it may be assumed that when the burner has not been warmed, it gives five per cent. less light than when it is heated. Then the lamp is carefully weighed by placing it in the cylinder fixed on one of the scale pans of the balance, the equilibrium being adjusted by small shot. In the scale carrying the lamp a small supplementary weight of ten grammes is added. Communication with the arm of the balance and the bell is then established, and a careful observation is taken as to the exact adjustment of the lamp and gas flames at the same level, and at the same distance from the screen. The movable needle on the axis of the counter is then turned to zero, as well as that of the seconds' counter. The experiments comprise the following details. The operator first places himself behind the observing glass, and to obtain equal illumination of both halves of the screen, the gas consumption is varied by means of a screw tap placed on the counter. In order to determine more accurately the relative intensities of the two lights, it is advisable to use two small plates adjustable by means of a screw ; these serve to diminish the field of the instrument. When the hammer strikes the bell, the counter needle is released by means of the lever which puts both needles in movement. The 10 gramme weight is then put in the scale pan, and the bell connection is re-established. During the whole duration of the trial, it is necessary to observe carefully through the object-glass if the equality of both lights is maintained ; if necessary, it is adjusted by regulating the gas supply with the screw cock before mentioned. At the moment when the hammer again strikes on the bell, the lever is pressed to stop the two needles. The expenditure of gas is then read on the dial of the counter, and the pressure on the gauge fitted to the service pipe, and which ought to mark 2 or 3 millimetres of water.

To give an example, let it be assumed that the counter indicates a con-

sumption of 24.5 litres. The supplementary weight being 10 grammes, the consumption of gas for 42 grammes of oil will be $2.45 \times 42 = 102.9$ litres. This test should be repeated three times at intervals of half an hour. The lamp and gas jet lighted at the commencement of the experiment will serve under the same conditions during the other tests. The average of the three results will then be taken. The normal consumption of the lamp being 42 grammes of oil an hour, 10 grammes will be burnt in 14 minutes 17 seconds. Thus the seconds' counter allows the determination, in each experiment of the consumption of oil per hour in the lamp, and shows whether it is within the limits mentioned above. For example, let the seconds' counter mark 15 minutes 30 seconds, then

$$\frac{10}{15.5} = \frac{x}{60} \text{ whence } x = 38.7 \text{ grammes per hour.}$$

Gas Jets.—From the foregoing it will be seen what minute precautions are necessary in using the Carcel lamp as a luminous standard. It was, therefore, natural to propose, as a substitute, a gas jet burning under a constant pressure. We have already spoken of the gas standard devised by M. Girout, in referring to the use made of it by MM. Sautter and Lemonnier as a unit in measuring electric light intensities, and we may now analyse the observations made by these gentlemen in their paper already quoted, upon photo-electric apparatus. The "candle jet" has an opening one millimetre in diameter, and the flame of 67.5 millimetres in height corresponds normally to one-tenth of a carcel. Slight variations in the diameter of the opening make no sensible differences in the height of the flame, and only exercise an effect on its intensity in the proportion of 1.5 per cent. for each increase of one-twentieth of a millimetre. The variations in the quantity of carbon, and in the purity of the gas, cause the height of the flame to vary, and one millimetre of increase in the height of the flame, corresponds to .022 carcel increase in lighting power. It appears, however, somewhat unlikely that these figures should be absolutely correct, and the formula deduced from them can scarcely be fully accepted. M. Crova, whose work in photometry is well known, maintains that gas gives out very uncertain intensities, and even when burnt with equal and constant pressures with identical burners, light of very different values is obtained. Lighting gas is not a definite and constant mixture, but a combination of a number of elements produced in changing proportions at different stages of the distillation process, from the commencement, when the hydro-carbons are produced, up to the end when pure hydrogen without luminous energy is given off. However, if pending

a complete discussion and fuller experiments conducted by the International Committee, we accept the data given above as being accurate, it follows that in employing a gas burner equal to seven carcels as a standard for the electric light, we may correct the lighting intensity by a simple calculation of proportion to the mean of the flame of the "candle burner" fed by the same gas. Thus, if we measure the differences in height of the flame of the burner to about one millimetre, the error for the burner of seven carcels will be .022 c. × 7 = .154 carcels. The standard of seven carcels may thus be used for measuring electric lights of high power. On the same service pipe as the standard of seven carcels the "candle burner" is fixed, to serve as a means of correction. Figs. 117 and 118 show sections of these two apparatus.

The gas is supplied into the receiver C, and passes thence by an opening to the burner. The flow is sustained under the influence of a constant pressure fixed by the weight of the bell C and of the tube A. The chimney of the "candle burner" carries a circular screen, movable along a scale graduated in millimetres, the centre of which corresponds to the end of the burner. In raising it so that its lower edge corresponds to the extremity of the flame, an exact measurement of its height is obtained. The subjoined Table gives the value in carcels of the normal seven-carcel burner for heights corresponding to that of the "candle burner."

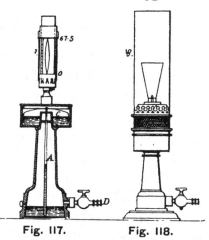

Fig. 117.　　　　Fig. 118.

TABLE SHOWING LUMINOUS VALUES OF THE SEVEN-CARCEL GAS STANDARD.

Heights in Millimetres of the "Candle Burner."	Actual Value of the Seven Carcel Burner. Carcels.
72	7.693
71.5	7.616
71	7.539
70.5	7.462
70	7.38
69.5	7.308
69	7.231
68.5	7.154
68	7.007
67.5	7
67	6.923
66.5	6.846
66	6.769
65.5	6.692
65	6.615
64.5	6.538

Heights in Millimetres of the "Candle Burner."	Actual Value of the Seven Carcel Burner. Carcels.
64	6.461
63.5	6.384
63	6.307
62.5	6.230
62	6.153
61.5	6.076
61	5.999
60.5	5.922
60	5.845

The preceding information will be completed by giving the equivalent in carcels, and "star" candles of the principal types of gas burners adopted by the Paris Municipality. The comparison will be found useful whenever, in arranging an electric light installation, a certain number of gas jets have to be replaced by electric lamps.

The Municipality of Paris employs three principal types, to which has now to be added the so-called "intense" burners.

	Carcels.	"Star" Candles.
The 1st series of burners consumes 100 litres of gas per hour and gives a light of	77 or	5.39
The 2nd series of burners consumes 140 litres of gas per hour and gives a light of	1.10 ,,	.7.70
The 3rd series of burners consumes 200 litres of gas per hour and gives a light of	1.72 ,,	12.04

In reducing the lighting power to 100 litres of gas it will be seen that the first series gives

$$\frac{100 \times 5.39}{100} = 5.39 \text{ candles per 100 litres of gas burnt.}$$

The second series gives

$$\frac{100 \times 7.70}{140} = 5.50 \text{ candles per 100 litres of gas burnt.}$$

The third series gives

$$\frac{100 \times 12.04}{200} = 6.00 \text{ candles per 100 litres of gas burnt.}$$

This shows the advantage gained in employing burners consuming larger quantities of gas, since the utilisation is better, which is also proved by the fact that large burners of the latest type consume 1400 litres of gas per hour, and give 13 carcels, or 91 candles, which is equivalent to a duty per 100 litres of

$$\frac{91 \times 100}{1400} = 6.5 \text{ candles.}$$

Other Luminous Standards.—The uncertainty attending the employment of the various standards in use, the difficulty of placing them under permanently identical conditions, and of establishing for them constant

ratios, long since attracted the attention of physicists, and determined them to seek for a new and more perfect unit. The development in artificial lighting, through recent discoveries, renders the research more necessary and more urgent than ever. Solutions to the problem have been attempted, but none have attained any high degree of success, as we shall see in reviewing the various standards proposed.

First we may refer in passing to petroleum as a light-giving substance. Hydro-carbons have a composition far too variable to be used for any purpose where constantly uniform light is concerned. Petroleum lamps combine the faults of gas jets and those of colza oil lamps, with special drawbacks of their own.

Mr. Draper in 1844, and Mr. Zöllner in 1859, recommended as a unit of light the energy developed by a wire or strip of platinum of determined dimensions, when rendered incandescent by the passage of a constant current of known intensity. Mr. Schwendler more recently adopted this, and advised the comparison of luminous intensities with a plate of platinum 2 millimetres wide and 36.28 millimetres long curved in the form of a ∪, and traversed by a current of 6 ampères, maintained constant by means of a mercury rheostat. Such a light would be equal to nearly two-thirds of a spermaceti candle, forming a standard of very low value, which of itself is a grave objection to its use. M. Crova has examined this standard very carefully, and he criticises it as follows, in a paper published by him in the *Revue Scientifique*: "It is necessary to maintain the current absolutely constant, which requires a constant supervision : bearing in mind that the quantity of heat given off by a conductor increases as the square of the intensity of the current, we shall see that a small variation in the current will produce a relatively considerable change in the temperature ; moreover, the temperature varying within very narrow limits, the light emitted will be subjected to changes in intensity as much greater, as the temperature is higher (Draper and E. Becquerel); it follows from this, that almost imperceptible variations in the current will produce relatively great changes in the intensity of the light emitted.

"The surface of the incandescent plate of platinum has of itself an emissive power, feeble at first when it is new and polished; but by use, this surface disintegrates and becomes dull; its emissive properties then increase, so that the same current traversing the plate will give a luminous value varying with time. I have tried to render this power constant by depositing electrically on the strip, a film of platinum black.

So treated, the emission was at first increased, then it diminished, and the strip became bright on the surface in consequence of the slow volatilisation of the film. It is probable that the change in the structure of the platinum and its slow volatilisation at a high temperature, especially in a vacuum, are the principal reasons which have led to the disuse of incandescence lamps with platinum or iridium wires in a vacuum, as suggested by Mr. Edison. Lastly, this light has a much redder tinge than that from a moderator lamp. I have shown *("* Annales de Chimie et de Physique," 5ème Série, Vol. XIX., p. 538) practically, that the temperature of these lamps is about 2000 deg. Cent. The temperature of the Schwendler standard being necessarily lower than that of the fusion of platinum which, according to M. Violle is 1775 deg. Cent., is still lower than that of the flame of the Carcel standard, and gives consequently a redder light. It will, therefore, be seen that this standard is still less suitable than the Carcel burner, to compare with the relatively white electric light, such as that from electric candles or arc lamps."

MM. Violle and Cornu have proposed as standards : first, the quantity of light emitted by a square centimetre of surface of platinum brought to a state of fusion; and second, that emitted by the same surface of molten silver, a metal less costly and more fusible. These standards are more theoretical than practical, and were rejected *à priori* by the Congress, on account of the difficulty attending their application. They are also open to the objection pointed out by M. Crova in the extract from his paper given above, that they give a light redder than that from the flame of an oil lamp, a condition especially unfavourable for the measure of electric arc lamps.

We may conclude this part of our investigation by indicating the conclusions arrived at on the general question of photometry by the Congress of 1881. To show the value of the resolutions, it is sufficient to mention the names of those who took part in the discussions: Messrs. Warren De la Rue. Helmholtz, J. D. Dumas, Becquerel, William and Werner Siemens, Gladstone, Cornu, Crova, Allard, Leblanc, Violle, Tchikoleff, Ayrton, &c. The discussion occupied the meetings of September 16th, 18th, 22nd, and 23rd of a Commission specially nominated for the purpose; and the full séance of September 21st, Mr. Spottiswoode acting as President.

The first question submitted to the examination of the Commission was as follows : In the absence of an absolute unit for the measurement of luminous intensities, does there exist any system that can be recommended

as an international standard ? And is it possible to establish simple rules for photometric measurements ?

These questions resulted in three resolutions presented to the full meeting on September 21st.

1. The third section moved the resolution, that until a photometric unit of absolute value could be determined upon, it is advisable to recommend the use of the Carcel lamps of the so-called lighthouse type in preference to candles.

2. The Congress decided that all photometric determinations of electric lamps ought to comprise as an essential element, the formula of the lamp, that is to say, the relation existing between the luminous intensity and the direction of the rays. This resolution was presented by M. Rousseau.

3. An International Commission should be nominated to test the different photometric methods and to propose the adoption of that which appears the most practical. This resolution was presented by M. Crova, professor of physics at Montpellier.

The discussion of these resolutions having resulted in some observations, the Congress referred the discussion to the third section for their more careful consideration.

The sittings of the 22nd and 23rd of September were occupied in this investigation under the presidency of Professor Hughes. It resulted in a general proposition emanating from Dumas, Warren De la Rue, and Sir W. Siemens, and was as follows : " The Congress recommends to the jury the employment of the Carcel lamp, in photometric comparisons made between the various electric light apparatus exhibited. The Congress begs the French Government to place itself in relation with foreign Governments to effect the formation of an International Commission, to be charged with the determination of the definite luminous standard, and the arrangements to be observed in the execution of experiments of comparison. These resolutions were definitely adopted on the 24th of September, 1881.

Photometric Measurements.—When the operator has decided which is the most efficient photometer at his disposal, and has settled on the standard of light best adapted for the comparison of the intensity to be measured, he should devote himself to a series of experiments in order to determine the power of the rays emitted in all directions, and to deduce formulæ for estimating the value of the luminous source. It is also necessary that he should study the effect produced by one or several lights upon the ground,

in the working plane, and in the variable directions which may change in each particular case. In a word, not only the light emitted must be considered, but also the light received, the absolute intensity of the source employed, the useful distribution of the rays, the suppression of

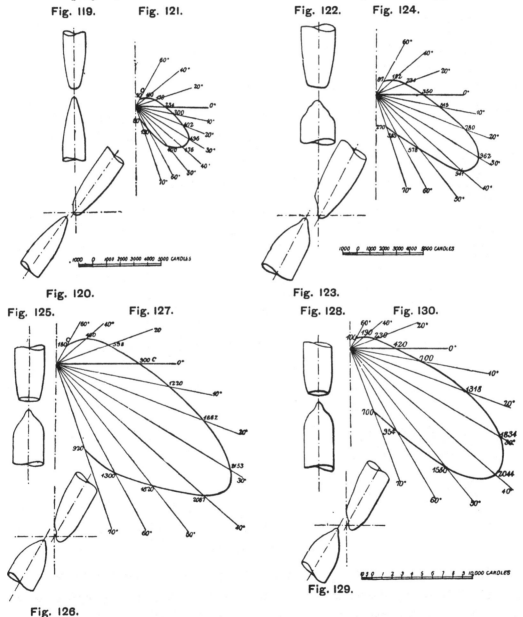

Fig. 119. Fig. 121. Fig. 122. Fig. 124.

Fig. 120. Fig. 123.

Fig. 125. Fig. 127. Fig. 128. Fig. 130.

Fig. 126. Fig. 129.

shadows, and the uniformity of the light on all points illuminated. The investigator may then pass from a theoretical study—that of light—to a practical question—that of lighting. We shall not consider here the question of the cost of illumination obtained with different varieties of

electric burners, but shall confine ourselves to certain brief indications on the luminous efficiency of some of these sources of illumination. In doing this we shall be able to show the application of different photometric methods, and the results to which they have led.

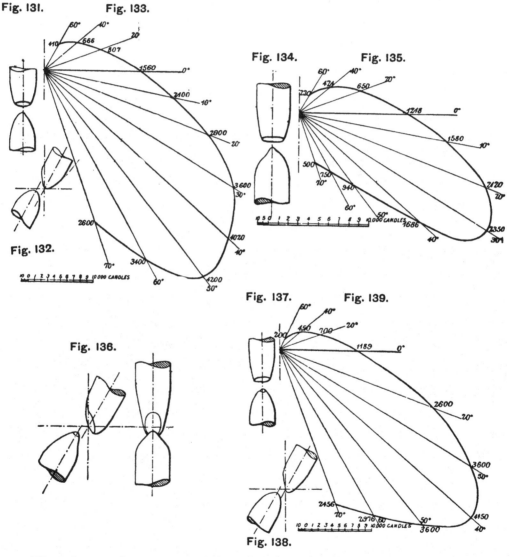

Fig. 131. Fig. 133.

Fig. 134. Fig. 135.

Fig. 132.

Fig. 136.

Fig. 137. Fig. 139.

Fig. 138.

Electric arc lamps may be divided into two classes, those fed by direct and those by alternating, currents. In observing the projection of a voltaic arc upon a screen it will be seen that the two carbons are not equally efficient in illuminating power. The negative carbon becomes pointed, the positive on the contrary is hollowed out into a crater (Fig. 140), and of the total quantity of light emitted, about 5 per cent. only is attri-

butable to the arc itself, 10 per cent. to the negative carbon, and 85 per cent. to the positive carbon. It becomes necessary therefore, if it be desired to place such a light under the ordinary condition of service —that is to say throw the light downwards—to make use of regulating lamps in which the positive carbon is above the negative, so that the crater may act as a reflector to throw the rays upon the ground in every direction, except vertically under the negative carbon. The amount of light distributed in any given direction will be increased, as the extent of the crater that can be seen is greater. If the axis of the two carbons be identical and the points be separated by a space of a few millimetres, it is evident that the boundary of maximum intensity will be a cone, the opening of which is greater or smaller according to the conditions of the the experiment. Outside and within this cone of maximum brilliancy, the intensity will decrease gradually under symmetrical conditions, in such a way that the illuminated zones on a plane perpendicular to the axis of the lamp and placed below it, will be concentric on each side of a circle of maximum light.

Fig. 140

When instead of uniform distribution, it is desired to obtain a maximum beam in a given direction, such as is necessary for lighthouses and other special installations, it becomes necessary to incline the lamp in such a way that the direction in which it is desired to obtain the greatest intensity, corresponds to the angle of the greatest emission of light. It is easy to obtain a higher efficiency by the simple device of shifting the axes of the carbons, as shown in Figs. 141 and 142, so that the positive carbon may present an oblique crater, almost wholly visible in the desired direction. Again, by inclining the carbons in such a way that the crater may be vertical, a maximum intensity in a horizontal plane is obtained, the opposite side presenting the minimum of brightness. MM. Sautter, Lemonnier, and Co., Paris, the well-known engineers, have constructed several regulating lamps on this principle for lighthouses, and for naval and military purposes. We borrow from the publication already referred to,* a certain number of practical results obtained by means of the Degrand photometer, and a gas jet of seven carcels, regulated by the Giroud rheometer. Figs. 119 to 139 show the curves of distribution of intensities, and the luminous values with the carbons arranged vertically, and also inclined in such a way as to throw the

Fig. 141.

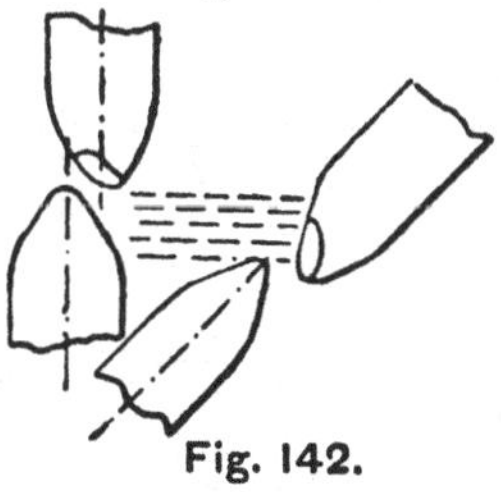

Fig. 142.

* " Appareils Photo-Electriques Employés par les Marines Militaires,"

RESULTS OF PHOTOMETRIC EXPERIMENTS BY MM. SAUTTER, LEMONNIER, AND CO.

(See Figs. 119 to 139.)

Particulars.	Units.	Figs. 119 to 121.	Figs. 122 to 124.	Figs. 125 to 127.	Figs. 128 to 130.	Figs. 131 to 133.	Figs. 134 to 136.	Figs. 137 to 189.
Speed of armature	Revolutions per min.	1600	820	880	675	670	1380	475
Diameter of carbons	mill. and inches	9=.35	13=.51	13=.51	18=.71	18=.71	18=.71	20=.79
Coppered or non-coppered		coppered	+ coppered − non-coppered	+ coppered − non-coppered	coppered	non-coppered	coppered	coppered
Mean length of arc	mill. and inches	3=.12	4=.16	4=.16	4=.16	5=.2	4.5=.177	6=.24
Resistance of armature	ohms	1.374	.658	...	.240	.120	.069	.178
,, electro-magnets	,,	3.716	.458	...	.585	.288	.184	.246
,, total of generator	,,	5.090	1.116	.568	.825	.408	.253	.424
,, conductor	,,	.127	.350	.240	.245	.240	.280	.173
,, carbons and arc	,,	1.449	1.867	1.213	.725	.629	.791	.774
Total resistance	,,	6.666	3.428	2.021	1.795	1.277	1.324	1.371
Strength of current	ampères	13.5	24.5	42	48	70	65	70
Electromotive force	volts	90	80	85	86.160	89	86	96
Mean luminous value	carcels	226 (Fig. 119)	490 (Fig. 122)	1185 (Fig. 125)	1015 (Fig. 128)	2382 (Fig. 131)	1241 (Fig. 134)	2198 (Fig. 137)
,, from projector with carbons inclined	,,	625 (Fig. 119)	1200 (Fig. 123)	2600 (Fig. 126)	2500 (Fig. 129)	4600 (Fig. 132)	3300 (Fig. 135)	6000 (Fig. 138)

PARTICULARS OF GRAMME GENERATORS EMPLOYED AND LIGHT OBTAINED.

Type of Generator, Power in Carcels.	Wire in Armature. Diameter.	Wire in Armature. Length.	Wire in Field Magnets. Diameter.	Wire in Field Magnets. Length.	No. of Revolutions per Minute.	Resistance in Ohms of Armatures.	Resistance in Ohms of Field Magnets.	Current in Ampères.	Electromotive Force.	Resistance of Conductor.	Resistance of Arc.	Mean Quantity of Light.	Maximum Quantity of Light.	Power Absorbed, Horse-Power.
	mm.	ft.	mm.	ft.		ohms.	ohms.		volts.	ohms.	ohms.	carcels.	carcels.	
M 200	1.2	...	1.8	...	1600	1.374	3.716	13.5	90	.127	1.44	226	625	1.25
A C 600	1.8	...	3.4	...	820	.658	.458	24.5	80	.350	1.867	490	1200	2.75
C T 1600	2.8	...	3.4	...	675	.240	.585	48	86.160	.245	.725	1015	2500	5.25
C Q 2500	3.65	...	3.4	...	1380	.069	.184	65	86	.280	.791	1241	3300	8
D Q 4000	4.3	...	3.8	...	475	.178	.246	70	96	.173	.774	2198	6000	12
Two generators A G coupled in quantity	1.8	...	3.4	...	880	.330	.238	42	85	.240	1.213	1185	2600	5.50
Two generators C T coupled in quantity	2.8	...	3.4	...	670	.120	.288	70	89	.240	.629	2382	4600	10.5

maximum beam in a horizontal plane. We give on page 172 a Table showing the principal particulars of the Gramme generators used in the French marine, and with which the experiments were made.

The last two lines of the Table compared with the second and third, and Figs. 127 and 133 compared with Figs. 124 and 135, show a gain in light by coupling two generators in quantity.

In the foregoing Table the mean quantities of light are the averages of quantities observed following the different angles. The maximum quantities are attained in the projectors by employing inclined lamps, and placing the face of the positive carbon opposite the observer. The annexed Table contains further information upon the same experiments, and completes the information given in the diagrams Figs. 119 to 139.

An examination of the preceding figures will show that the system of lighting with superposed carbons presents this inconvenience, that beneath the apparatus is a dark zone, where it is often of importance to have light. Manufacturers have reduced the extent of this zone by placing above the arc, the mechanism feeding the carbons, which was at one time generally fixed underneath, but the negative carbon and its holder, however small they may be, always throw a shadow. To avoid this arose the idea of placing the carbons horizontally, which also allows the employment of rods of greater length, and a consequent increased duration of lighting.

THE EYE AS A PHOTOMETRIC INSTRUMENT.

THE important part fulfilled by the observer's eye in almost every method of photometrical measurement, has probably, not escaped the notice of our readers. Even when photography is brought into use, as in MM. Fizeau and Foucalt's method, the eye has ultimately to determine the value of the different effects produced. It is only in such purely chemical methods as that of Bunsen and Roscoe that the eye has not to measure results. It may, therefore, well be asked: How does light effect the eye; and within what limits can that organ judge differences of illumination?

Light is given off from every visible surface; being either reflected therefrom, transmitted through the substance of which it is a surface, or actually produced by molecular vibrations of the substance, as in the case of luminous points on the surface of molten iron. A luminous point gives off rays of light in every direction, and those rays travel in straight lines. This is obvious; for from whatever position a luminous point is viewed, it is visible so long as no opaque object is interposed; but the rays do not travel by a curved line, and so pass round an obstacle.

It will be obvious that a luminous point at the centre of a hollow sphere will illuminate every equal portion of the internal surface of the sphere to an equal extent. Consequently the concave surface of the segment A B, Fig. 143, of a sphere A B C, will receive from the luminous point G the same quantity of light as the concave surface D E of the sphere D E F; and that quantity will, if the outer sphere be twice the size of the inner, be spread over an area four times as great; for the surface areas of spheres vary as the squares

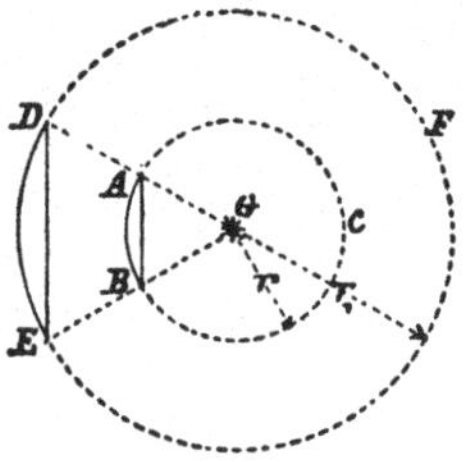

Fig. 143.

of the radii, the formula for ascertaining the area being $4 \pi r^2$; so if r equals 1, and r_1 equals 2, the surfaces would be as 1 to 4. Thus it is proved that the intensity of illumination of a surface is inversely as its distance from the source of light.

Considering this law, it is, at first thought, strange, that objects should appear of almost exactly similar brightness at a considerable distance as when near at hand; but that they do appear similar is due to the fact that the areas of the images of such objects upon the retina, likewise vary inversely as the squares of their distances from the observer; hence the physiological effect upon the portion of retina affected is similar, whatever the distance at which the objects may be from the observer. Any difference in the apparent brightness to an observer of two lights of similar size and intensity, but at different distances from him, is due to the absorption of light by the medium through which the rays of light pass from their sources to the observer.

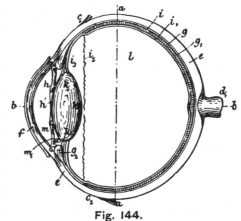

Fig. 144.

In order the better to explain the construction of the human eye, a vertical section, Fig. 144, and a horizontal section, Fig. 145, are given. It will be seen that the eye is a nearly spherical ball. The transverse diameter is said to be slightly greater than either the vertical or the longitudinal diameters, which latter are approximately equal. The eye lies, embedded laterally and behind in fat, in a bony socket called the orbit, in which it is turned by means of six muscles. The attachments of four of these muscles to the ball are marked, respectively, c_1, c_2, c_3, c_4, and they direct the eye upon the object at which it is desired to look. Their other ends are attached to the edges of the hole in the back of the orbit, through which hole the optic nerve d passes; its sheath d_1 being a continuation of the outer coat of the eye. The other two muscles are attached rather to the rear of the point c_3, above and below it respectively. One passes above, and the other below the eye, and they pull it forward towards the inner front part of the orbit.

The eye consists of a thick horny outer shell, the opaque portion of which, e, forms what is known as the white of the eye; it is called the sclerotica. The transparent portion, f, similar in form to a watch-glass, is the cornea. It forms a segment of a smaller sphere than the eye, and is let into, and continuous with the sclerotica.

Immediately inside the sclerotica is the choroid coat g, which consists of a close network of capillaries, joining numerous arteries and veins which lie outside the network, the whole being united by connective tissue layers of stellate cells containing black pigment. The inner surface of the choroid coat is a single layer g_1 of flattened hexagonal pigment cells, and, strictly speaking, forms part of the retina. The choroid coat is principally useful to absorb, by means of the black pigment, the rays of light which pass through the

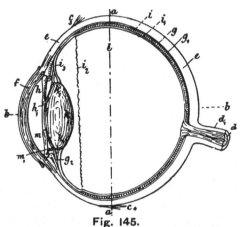

Fig. 145.

transparent retina. It ends, internally, in front, in a number of radial folds or ridges called cilliary processes, g_2. Externally, it is continuous with the iris h, which is an annular opaque curtain, the inner diameter of which is variable, thus altering the size of the pupil h_1, by which means the quantity of rays of light entering the eye is regulated.

The retina, a very delicate membrane, varying from the $\frac{1}{160}$th to the $\frac{1}{80}$th part of an inch in thickness, lies next inside the choroid coat. It is an expansion of the optic nerve d, and is the only part of the eye which, under the influence of light, conveys the sensation of vision to the brain.

The outer portion of the retina, next the pigment cells g_1, consists of a layer i of rods and cones, which are arranged radially, and by means of which the vibrations called light stimulate the fibres of the optic nerve. Inside the layer of rods and cones, which is barely $\frac{1}{6}$th of the whole thickness of the retina, is the membrane i_1, in which lie the nerve fibres from the optic nerve, and the blood vessels of the retina. The nervous elements of the retina end at the *ora serata*, i_2, a sort of wavy border line, beyond which only a delicate membrane of connective tissue, i_3, passes on to the crystalline lens k.

From the choroid coat proceeds the elastic suspensory ligament k_1, which supports the lens k. The lens, too, is very elastic, especially in youth, but it hardens with age ; and when the eye is at rest, it is kept in tension by the ligament k_1. The front surface of the lens is then less convex than when the eye is adjusted for viewing near objects. The lens and its ligament divide the interior of the eye into two chambers, the larger of which, l, is filled with a delicate jelly called the vitreous humour. The smaller

chamber is itself sub-divided by the iris into the anterior and posterior aqueous chambers m and m_1 respectively; so called because they are filled with watery fluid. These contents keep the eye in proper shape.

If, when the eye is fixed upon some object in front of the observer, a slight motion of any object takes place, even so far as 90° or 95° from the line of direct vision, the eye notices it, and an impulse is given to turn the eye and head in the direction of the moving object, to view it more directly. It will thus be seen that the eyes are conscious of movements occurring in fully—if not more than—a semicircle about the observer. That a ray of light coming from the moving object on one side should be bent from its straight course, and pass through the pupil on to the retina, is due to the fact that, although the ray moves in a straight line, so long as it is passing through the same medium, part of the light is reflected from the surface at which the rest enters another medium, while that which enters the second medium, moves through it at an angle to its original course in the first medium. This bending of the course of the ray is termed refraction. Whatever may be the angle at which a ray strikes a surface, the sine of the angle which it makes with the perpendicular to that surface, and the sine of the angle which it makes with the same perpendicular while passing through the second medium, are always in the same ratio. Taking a vacuum as the standard of comparison, we have

$$\frac{\text{Sine of angle of incidence in vacuum}}{\text{Sine of angle of refraction in a medium}} =$$

A constant quantity, termed the index of refraction, for that medium.

The bending that takes place when a ray passes from a vacuum into the air or into gases is slight; on entering water, or the media of the eye, it is much more, and on entering glass still more. The following are some of the indices of refraction:

Hydrogen	1.000138
Air	1.000294
Chlorine	1.000772
Pure water	1.336
Alcohol	1.372
The aqueous humour of the eye	1.3366
„ vitreous „ „	1.339
„ cornea	1.377
„ outer coat of crystalline lens	1.337
„ inner „ „	1.379
„ central portion „	1.399
Crown glass; average „	1.547
Flint „ „	1.609

The same law of sines holds good between any two media; and the

relative index of refraction for those media is the ratio of their indices with respect to a vacuum. Thus from air to water the relative index $= \frac{1.336}{1.000294} = 1.335607$. Hence the bending is almost as great from air as from a vacuum.

On leaving the medium in which refraction has occurred, and returning into the medium through which it was previously passing, the ray resumes a course parallel to its original course, if the surfaces at which it enters and leaves the intermediate medium are parallel. If those surfaces are not parallel there is a permanent bending of the ray. The differences of the refractive indices of the various media of the eye are very small; consequently, the bending, after the ray has once entered the cornea, is slight. A ray, bending from its path through the air on entering the cornea, would on leaving the cornea return slightly towards its original course; but in passing through the lens it would be bent further from that course, again returning slightly to it on entering the vitreous humour. Although the refractive index of glass is greater than those of the media through which light passes in the eye, a ray being bent more on entering glass than on entering the eye, yet much may be learnt concerning the eye from experiments with glass prisms and lenses, for it is to refraction that the phenomena produced by them are due.

If $a\,b$, Fig. 146, be a ray falling upon the surface A B of the prism A B C, and in the plane of a principal section of the prism, and if i be the angle of incidence, and r the refracted angle, μ being the relative index of refraction; then $\sin i = \mu \sin r$. Similarly on leaving the surface B C, the ray takes a course $e\,g$, and here $\sin i_1 = \mu \sin$ r_1. The deviation produced at b is $i-r$, and at e it is i_1-r_1. The total deviation is the angle h; consequently

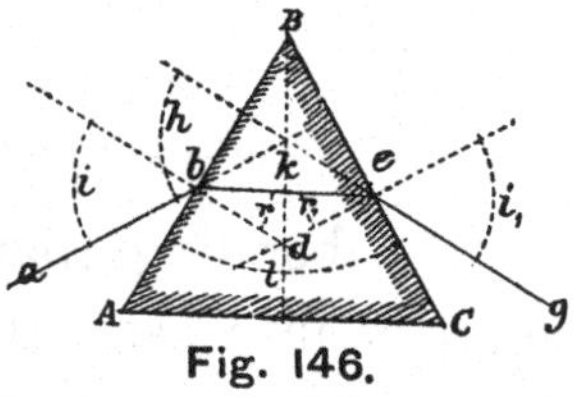

Fig. 146.

$$h = i_1 - r + i - r_1 \quad . \qquad . \qquad , \qquad (1)$$

A perpendicular let fall from B upon the point of intersection of the rays at k divides the angle l into two parts, of which A B $k = r$ and C B $k = r_1$; for B b is perpendicular to $b\,d$, and B k is also perpendicular to $b\,e$; and the angle contained by the perpendiculars to two lines is equal to the angle contained by the lines themselves. Consequently the angle $l = r + r_1$, which value being substituted in equation (1) gives

$$h = i + i_1 - (r + r_1) = i + i_1 - l \quad . \qquad . \qquad . \qquad (2)$$

When the ray $b\,e$ through the prism makes the angles at the two surfaces b and e equal, the angles r and r_1, with the perpendiculars to those

surfaces, must be equal; and consequently $i = i_1$; for μ is of the same value in both cases. Thus equation (2) becomes $h = i + i_1 - l = 2i - l$,

Whence

$$i = \frac{h + l}{2} \qquad . \qquad . \qquad . \qquad . \qquad (3)$$

Moreover,

$$\mu = \frac{\sin i}{\sin r} = \frac{\sin \dfrac{h + l}{2}}{\sin \dfrac{h}{2}} \qquad . \qquad . \qquad . \qquad (4)$$

Thus the index of refraction is obtainable by calculation from measurement of the angle l of the prism, and the angle h of total deviation.

Again, if $a\,b\,c$, Fig. 147, represent the section of a prism, and if the perpendiculars $d\,e, f\,g$ to the faces $a\,b, a\,c$ be continued to cut a line continued from the base $b\,c$ at h and i, and if arcs $k\,e\,l, m\,g\,l$ be struck from h and i, then the path of a ray, $n\,e\,g\,o$, would not be altered. Hence it will be obvious that with curved surfaces a tangent must be drawn at the point of incidence of the ray, and that the angles of incidence and refraction must be taken from perpendiculars to the tangents drawn from the points of incidence and exit respectively. Thus, if the ray $A\,B$, Fig. 148, is incident at B upon the curved surface $C\,D$, and if $E\,F$ is the tangent at B, and $G\,H$ the perpendicular thereto, then $\dfrac{\sin\,\alpha}{\mu} = \sin\,\beta$, which gives the direction of $B\,I$ the refracted ray.

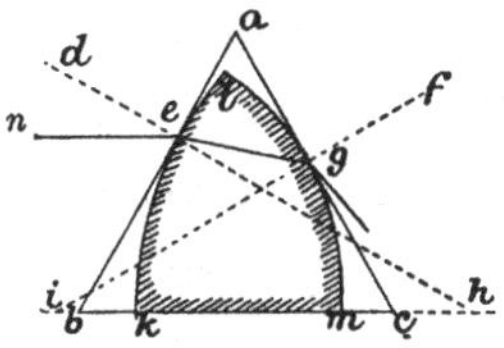

Fig. 147.

These remarks will make plain the action of lenses in causing refraction, for a lens may be regarded as the revolution of a figure similar to $k\,e\,l\,g\,m$, Fig. 147, about an axis $k\,m$. Such a lens is called a double convex lens, and the crystalline lens of the eye belongs to that class. But in the eye

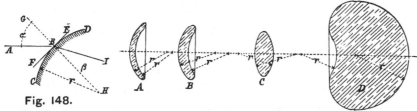

Fig. 148.

Fig. 149.

there are lenses of other forms. First, the cornea A, Fig. 149, which is of the class called a converging meniscus; then the aqueous humour B, which is also a lens of similar class but of different refractive power; next there is the crystalline lens C; and finally the vitreous humour D, which is practically a very thick diverging meniscus. These four lenses, when built

up into one compound lens, have a similar effect to that a globular lens would have, if made of materials of sufficiently high refractive index, to focus parallel rays at twice the radius of curvature—viz., at the opposite surface to that at which the rays enter. The mean refractive index of the various media of the eye is, however, but little more than that of water, and this has to be compensated for, in order that the rays may be focussed upon the retina, which fits closely on the outer surface of the lens D, Fig. 149. This compensation is effected by the projection of the cornea beyond the surface of the larger sphere, and by its surface having a smaller radius of curvature.

The effect of such a lens as the eye may be studied by means of a small globular flask filled with water. The flask may have an imitation of the cornea secured upon its surface with cement, a small air-hole being left at the point of junction, by means of which the space a, Fig. 150, B, can be filled by exhausting an air-pump receiver, in which the flask is placed, submerged in water; or a spot about the size of a shilling may be heated near the neck of a flask, and upon blowing into the flask, the surface of the spot can be forced outward to the de-

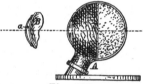

Fig. 150.

sired form. The opposite hemisphere of the flask should then be ground, and the rest of its surface, between the ground part and the artificial cornea, painted black. The flask when filled with water will be found to form upon the ground surface a picture of the objects towards which the imitation cornea is directed, provided of course that the surface representing the cornea has the proper radius struck from a centre at suitable distance from the centre of the flask. A, Fig. 150, represents such a lens, secured to a stand.

It may then be asked, What is the use of four variously shaped lenses? The answer to this is two-fold. First, to correct the chromatic aberration; next, to provide means for keeping the focal length constant, whether looking at near or at distant objects, in order that in both cases the rays refracted at the surface of the cornea may be focussed on the retina. It has been calculated by Olbers that without this accommodation, the difference in focal length for objects as near as 4 in. from the eye, and so far away that the rays might be considered parallel, would be only .143 of an inch, and as the whole retina is only from 00625 to 0125 thick, while the length of the rods and cones, which are influenced by light, is only one-sixth of that, or at most .0021 of an inch, it will be seen that this ability to accommodate

the eye to both near and distant objects is very important. It is very wonderful that the eye itself is so formed that the limit of variation in focal length should be so small as .143 of an inch, and that at the same time it should contain within itself this power of accommodation.

Means of correcting the chromatic aberration are needed, because without them the rays of various colours would not be coincident upon the retina, and the images of objects there formed, would have a blurred outline. Light rays of different colours are bent or refracted to a different extent. When a ray of white light, which is a combination of the various coloured rays of the spectrum, passes into a medium of different density to that through which it has been passing, the rays of different colours being refracted to different extents, take different paths; and the ray of white light is decomposed into its variously coloured constituents, which become visible at different parts of the screen on which the ray falls. This phenomenon is termed chromatic aberration. Now the material of the eye itself is not only of low refractive power as compared with glass or the diamond, and consequently does not disperse the coloured rays to so great an extent, but the dispersion, or as it is termed deviation that does occur, is corrected by the combination of lenses formed of materials having different refractive powers. The rays do not deviate far through one medium; a different medium with surfaces of different curvatures collects them again; and thus the eye is rendered practically achromatic.

The angular difference of deviation between the brightest red ray and the brightest violet ray transmitted by a prism, is called the dispersion of the prism. If μ represent the index of refraction for the brightest part of the spectrum, when a prism is in the position of minimum deviation—which is the position in which the rays are less refracted from the original path than any other—and if $\mu_{\prime}$ and $\mu_{\prime\prime}$ represent the indices of the two coloured rays under consideration; then for such a prism as that represented in Fig. 146 (see page 178 *ante*), the deviations would be $(\mu-1)\, l$, $(\mu_{\prime}-1)\, l$, and $(\mu_{\prime\prime}-1)\, l$. The difference of deviation between the red and violet rays would be $(\mu_{\prime\prime}-1)\, l-(\mu_1-1)\, l=(\mu_{\prime\prime}-\mu_{\prime})\, l$; and the ratio between that difference and the deviation for the brightest part of spectrum would be

$$\frac{(\mu_{\prime\prime}-\mu_{\prime})\, l}{(\mu-1)\, l}=\frac{\mu_{\prime\prime}-\mu_{\prime}}{\mu-1}$$

This is called the dispersive power of the substance of which the prism is made. The value of this ratio is .033 for crown glass, and .052 for flint glass.

Newton supposed this ratio the same for all substances, and that achromatism was impossible. It was, however, shown by Hall, and afterwards by Dolland, that this opinion was erroneous; and advantage is taken of this difference of dispersive power, to construct lenses for optical instruments. Achromatic lenses usually consist of a diverging meniscus of flint glass, and a double convex lens of crown glass, fitted one into the other. The so-called achromatic eye-pieces of microscopes and telescopes, however, are made of two lenses of crown glass, having different curvatures, and so arranged that the one corrects the spherical and achromatic aberration of the other, not by making the red and violet images coincide, but by causing one to cover the other at the point where the eye perceives the image. The compound lens of the eye has, however, already been shown to be of more complex structure.

To provide an optical instrument with means for accommodation to different distances, the optician would either vary the distance between the surface at which the rays were refracted, and the screen on which they were focussed, or he would provide lenses of varying convexity, or of materials having different refractory indexes. Helmholtz, who studied this subject very fully, constructed an ophthalmeter with which he could measure alterations in the relative positions of the images reflected from the surfaces of the various lenticular media of the eye. These measurements gave data for calculations, and it is ascertained that there is a change in the shape of the crystalline lens. During accommodation to near objects, the curvature of the cornea and of the posterior surface of the crystalline lens, remains unaltered, but the anterior surface of the lens becomes more convex, and approaches the cornea. There is but little doubt that the power by which this change is effected, is supplied by the ciliary muscle, which arises at the inner surface of the sclerotic, close outside the margin of the cornea, the muscular fibres passing to the ciliary processes in which they are inserted. There are also some circular fibres. When excited the ciliary muscle draws forward the choroid, and thus the tension of the suspensory ligament is lessened, and the anterior surface of the lens, in virtue of the elasticity inherent to the lens, assumes a more convex form, it being kept in tension when the eye is at rest. The ligament, it will be seen in Figs. 144 and 145, is attached to the margin of the anterior surface, and, when in tension, pulls it to a flatter form without altering the posterior surface.

In every eye there is a limit to the power of accommodation. If two pinholes be made in a card, $\frac{1}{16}$th of an inch apart, and a small needle be

viewed through the holes, one eye being closed, and the card held close to
the other, the needle can be
clearly focussed up to a moderate distance. Nearer than 6 in.
or 8 in. two images are seen, or
the image is ill defined. The

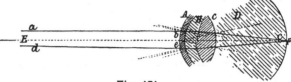

Fig. 151.

point at which this occurs is called the near point. Fig. 151 shows the alteration that takes place in the form and position of the front face of the crystalline lens on adjustment for long and short distances. The various lenses
are lettered as in Fig. 149. Above the line E F the lens is shown adjusted for
distant objects, and the ray $a\,b$ is focussed at c. A ray $d\,e$ from the centre of
an object at about 4 in. distance, is shown below line E F, and is also focussed
at c. Rays coming from a point lying at an angle with the line of vision,
taking that line to be a prolongation of the line E F, are focussed on the
retina at a point in a plane containing also the line E F, and the point
from which the ray comes. The rays are focussed on the retina on the
opposite side of the line E F to that on which lies the point from which the

rays proceed. In Fig. 152, the eye is
represented looking at the edge of a
rule held horizontally before it at a
distance of 4 in. ; two rays are shown
coming from A ; and, passing through
the pupil at its opposite sides, they
are focussed at a. Similarly two rays
from B are focussed at b. In this
way an image of the edge of the rule
is formed along the retina. Two rays

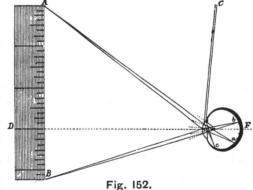

Fig. 152.

are also shown coming in the direction C from a point on a line at an angle
of 95° with the axis D E of the eye, and they are focussed at c near the
margin of the retina.

If the rule were at a greater distance the image on the retina would be
correspondingly shorter. When, therefore, it is remembered that a black line
$\frac{1}{500}$th of an inch wide can be distinctly seen upon a white ground at a
distance of fully three feet, it will be obvious that the width of the image
of that line upon the retina cannot be more than $\frac{1}{35,000}$th of an inch wide.
It is said that brilliantly illuminated specks of gold dust of half that diameter can be distinctly seen. How delicate then must be the minute organs
which are stimulated to send intelligence to the brain of the presence

of such tiny images. The eye cannot, however, distinguish between two such minute objects, if their images on the retina lie closer than $\frac{1}{12,000}$th of an inch, and that is found, by microscopical measurement, to be the diameter of the cones in the central part of the retina. Hence, although there is a consciousness of an image which covers but $\frac{1}{30}$th the area of a cone, unless two such images lie upon different cones, the eye does not distinguish between them. The rods and cones are shown at H, Fig. 153, which is a diagrammatical representation of a section of the retina. A is a very delicate membrane next the vitreous humour; B a layer of nerve fibres radiating from the optic nerve, with an outer layer of ganglion cells, lying in the spaces of a connective tissue framework; C is a layer of connective tissue, traversed by very fine fibres from the ganglion cells; D a layer of small round nerve cells embedded in connective tissue; E a layer of connective tissue traversed by nerve fibrils; F several strata of small nerve cells connected by fibrils with the rods and cones; G is a membrane which defines the internal limit of the rods and cones, and H the rods and cones; I the pigment cells of the choroid coat in which the ends of the rods and cones are embedded.

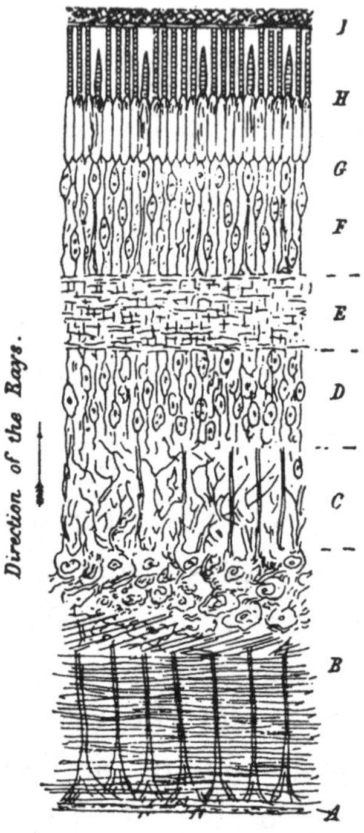

Fig. 153.

At the central portion of the retina, all the layers are much reduced in thickness, except that of the rods and cones which is much thicker than elsewhere. The depression thus formed is of a yellow hue, and is called the yellow spot. The cones here are very close together and smaller in diameter as well as much longer, than in other parts of the retina. This, too, is the most sensitive portion of the retina. Beyond the margin of the yellow spot, there are spaces between the cones, and these spaces are filled up with rods surrounding each cone. In the outer part of the retina the cones are much fewer in number than near the yellow spot. In the rod and cone layer of the retinas of birds there are usually many more cones than rods. In man, the reverse is the case. In nocturnal birds, however, as the owl, and in many nocturnal and burrowing animals, as the mole, mouse, and bat, only rods are present. Fibres from the rods and cones pass through the connective tissue to the ganglion cells, and these in turn are connected with the optic nerve.

It is not however at the rods and cones that the individual is conscious of light, but at the brain; for where the eye has been destroyed by accident or disease, the sensation of light may be produced by pinching or otherwise exciting the optic nerve. Nevertheless, that it is the rods and cones which convert the light vibrations into a stimulus to be conveyed to the brain by the nerves, is proved by the fact that if in a dark room a candle be moved to and fro at the side of and close to the eye, while the eyes look steadily forward into darkness, a branching figure of dark lines on a reddish ground is seen. As the candle is moved, the figure moves in an opposite direction. This branching figure is the shadow of the retinal vessels, which have already been shown to be inside the rods and cones, cast upon the retina by the candlelight; and it is obvious that the elements of the retina which send the knowledge of the presence of this shadow to the brain must be outside the vessels.

Moreover, just at the point where the optic nerve enters the eye is a circular space where there are no rods or cones, and from this portion of the retina there is no sensation of light. This may be proved by marking upon a card a small cross 3 in. distant from a circular spot $\frac{1}{10}$th of an inch in diameter. Hold the card about 15 in. from the eye and look steadily at the cross, the spot being towards the outer side. The spot will be plainly seen, but when the card is moved to 11 in. or 12 in. from the eye the spot is invisible, but is again seen when the card is brought to 7 in. or 8 in. from the eye. This clearly proves that while the image of the spot was traversing the point of entrance of the optic nerve, the brain was not rendered conscious of the presence of that image. This portion of the retina is called the blind spot.

The eye, therefore, may be compared to a camera, the iris acting as a diaphragm to prevent spherical aberration by obstructing the passage of all marginal rays from an object in front of the eye which it is desired to examine closely. Distinctness of vision is best obtained when abundant light falls upon the object viewed, when the object is held as near the eye as accommodation will allow, and when the marginal rays are cut off by contraction of the pupil. It is further secured by the interior of the eye being coated with black pigment cells, which absorb the rays and prevent internal reflection. Achromatism is secured by a system of compound lenses varying in refractive power; while accommodation to near and distant objects is provided for by the crystalline lens being capable of alteration in form and position.

In estimating the brightness of a luminous surface which is not large

enough to fill the retina, it is necessary to distinguish between effective brightness and intrinsic brightness. The latter exceeds the former in the same ratio as the area of the pupil exceeds the area of the beam passing through it. When a beam of light fills the area of the pupil the effective brightness equals the intrinsic brightness.

Intrinsic brightness may be thus measured: If A be the area of the pupil, and a that of a small surface directly opposite the pupil, and perpendicular to the line of vision; and if d be the distance between A and a, then the quantity of light Q passing in unit time from a to A, varies jointly as the area A, the apparent area of a (which, as it will vary as the square of the distance, may be expressed $\dfrac{a}{d^2}$), and the intrinsic brightness I. Thus

$$Q = A \; I \; \frac{a}{d^2} = a I \frac{A}{d^2}.$$

If ω represent the solid angle of light subtended by the pupil from the central point of the surface a, ω will vary as $\dfrac{A}{d^2}$, then $Q = a\,I\,\omega$, therefore $I = \dfrac{Q}{a\,\omega}$. This is the reason why, if no light be stopped by the air or other medium intervening between the surfaces and the eye, surfaces of equal brightness appear equally bright at all distances; for the area of the image formed on the retina varies with the square of the distance, as also does the quantity of light from each point of the surface, as already shown. Therefore, as the distance increases, a larger portion of luminous surface is focussed on the same area of the retina, in exactly the same proportion as the quantity of light from each luminous point of the surface is decreased. Hence in photometrical measurements the rays from lights to be compared are generally received upon a screen or screens; for then the eye sees simply the effect of the lights upon the screens. The images of the lights are not formed on the retina, but images of the screens, upon which by various arrangements the lights are made to give effects of apparently equal value. It is very difficult to estimate fractional differences of illumination. To discern the difference of illumination between one candle and two candles, at equal distances from a screen, is easy, but as lesser differences occur, the eye is a less accurate measure, and it is hard to distinguish much difference between the illumination afforded by nine candles and by ten, at equal distances from a screen. It is, however, said that a difference in intensity of light or shadow of only one-sixtieth is perceived, if the fields of light and shade are each uniform, and if they are very near together. It

must be remembered that the question is not one of affecting more rods and cones, but of discerning the differences of the physiological effects produced upon the same number of rods and cones by nine candles and by ten. The point is analogous to judging the difference in weights. One can readily perceive the difference between the heaviness of two articles weighing respectively two pounds and four pounds, but it would be difficult, if not impossible, to tell, by balancing in the hand, whether an article weighed 3.6 lb. or 4 lb. For this reason one cannot help coming to the conclusion that attempts to directly compare lights of very great difference of intensity, are likely to lead to error. Even with the steelyard no attempt is made to weigh grains and tons with the same lever and weight. The most perfect photometric measurements will be made, when lights of from nearly equal intensities to those whose intensities are about as 1 : 4 are directly compared. Hence as lights are of very great difference of intensity it follows that there should be a series of light measures, multiples of some universally recognised standard unit, just as there are standard measures of weight, length, and capacity, suitable for measuring tons or grains, chains or hundredths of an inch, gallons or minims.

Light rays are accompanied by heat rays and by chemical rays, and it is found that the heat rays are less refrangible than the light rays, while the chemical rays are more refrangible. Fig. 154 gives the intensity curves

Fig. 154.

of light and heat. The chemical rays have their maximum effect in that part of the spectrum in which the violet rays are situated, and extend beyond the visible spectrum to a distance of about two-thirds its length; all the chemical rays being more refrangible than the rays of the brightest part of the visible spectrum. The extent of their blackening power varies with different substances. The line showing the blackening effect upon bromide of silver must not be taken as an intensity curve, it simply shows the length and position of the blackened part as compared with the visible spectrum, the point of maximum effect, and the somewhat rapid shading off at both ends. Some other substances are blackened more towards the orange and more beyond the violet. The heat curve shows the extent to which the heat rays are refracted, as compared with the same visible spectrum.

It has already been noticed that coloured rays being differently re-fracted, refraction is largely prevented in the eye. Still, the eye is not equally affected by light of all colours; and it has been ascertained that the area of the retina capable of being impressed by any particular colour, differs slightly for the various colours. One observer may also think an object to be of one shade of colour, while another will pronounce it of a different shade, although the eyes of both are equally sensitive to colour.

It is generally agreed that there are three distinct physiological actions developed in the retina, which, variously combined, give various colour sensa-tions; and that each of the three is excited to its maximum by a light of a particular wave length, and affected only to a slight degree by light of other wave lengths. One of these sensations is considered to be best excited by the green, another by the deepest red, and the third by the violet. Helmholtz suggests that different nerve fibrils go to different strata in the length of the rods and cones, and that the coloured rays, being variously refracted, are focussed at these different depths in the retina.

It will thus be seen that it is difficult for the eye to accurately compare the intensity of lights of different colours. Spectro-photometry is probably the best method of comparing such lights.

In conclusion, it may be observed that considering these circumstances, and that heat rays are of one series of wave lengths, the various visible rays of another series, and the chemical rays of another, the retina can hardly be sensible to many of the chemical rays or heat rays; for, owing to the difference of refrangibility, most of the chemical rays would be focussed in front of the retina, while such of the heat rays as were not absorbed by the media of the eye, would be focussed beyond it. For these reasons, in com-paring the illuminating effect of lights by chemical photometry, the chemical effects produced would not be a measure of the illuminating effects, unless it could be shown that both varied in the same ratios. It is not known what is the nature of the action upon the retina which gives rise to the sensations by which the mind becomes conscious of objects through the sense of vision. It is but a theory that variations in the temperature of any body are accompanied by variations in the rate of a supposed motion of the molecules of bodies, and that radiation is the imparting of the various motions of those molecules to a supposed impalpable æther pervading all things and all space, giving rise to a series of undulations in that æther, the undulations becoming mixed with those from other bodies of different tem-

perature, tending to produce a uniform series of undulations and uniform temperature. This wave theory is adopted for heat, for light, and for certain chemical effects accompanying radiation. Various phenomena support this theory; but refraction plainly shows that the undulations which give the sensation of heat, of light, and of chemical action, differ from each other in a somewhat similar way to that in which the undulations which, acting upon the retina, cause the sensation of the presence of a red light, differ from those which indicate violet light. It has been mentioned that light rays are capable of different degrees of refraction, the red being least refracted, and the violet most so. Fig. 154 also shows that the rays of greatest intensity of light are refracted to a position in the spectrum intermediate between the points to which the heat rays and the chemical rays are respectively refracted. A system of measurement of what is termed the wave lengths of the various coloured rays, has been based upon the phenomenon styled diffraction, which is a modification that light undergoes when it passes through a small opening, or by the edge of a body. If a very small pencil of light be allowed to enter a dark room through red glass and be received upon a condensing lens of short focal length, a luminous cone of rays passes from the lens, and may be caused to illuminate a screen. If, then, the sharp edge of an opaque screen, as, for instance, a knife-blade, be caused to intercept one-half of the cone of rays, that part of the screen which it might be supposed would be left uniformly illuminated, will exhibit a series of alternate light and dark bands, gradually fading one into the other, from the respective points of maximum and minimum intensity. All the colours of the spectrum give rise to similar bands; but, using the same apparatus, and simply changing the coloured glass, it will be found that the fringes are broader in proportion as the less refrangible rays are experimented with. Taking the same number of bands respectively in the fringes of red, green, and blue light, the widths of the fringes are about 27 : 20 : 17. The wave length is determined by a calculation based upon half the distance between the centres of the first pair of dark bands, the distance of the screen from the edges causing diffraction, a narrow slit, or a series of slits termed a grating, being used in preference to a single knife edge, and upon the width of the said slit or slits. The following wave lengths are given by Ganot :

				Length of Undulations in Decimals of an Inch.
Dark line B of spectrum	...	Bright red light	... ...	.0000271
„ C „	...	Orange „	... ...	.0000258
„ D „	...	Yellow „	... ...	.0000244

				Length of Undulations in Decimals of an Inch.
Dark line E of spectrum	Green ,,	...	...	.0000207
,, F ,, ...	Dark green ,,	...	...	.0000191
,, G ,, ...	Blue ,,	...	...	.0000169
,, H ,, ...	Violet ,,	...	...	.0000155

It has also been calculated, by dividing the velocity of light by the above wave lengths, that the number of oscillations per second corresponding to the bright red light of line B, is 434,420,000,000,000 per second, and to the violet 758,840,000,000,000. It will, of course, be understood that the undulations of light are supposed to be a regular series of states of condensation and rarefaction of the æther through which a ray passes, just as sound waves are such states of the air. They are not oscillations to and from the direct central line of the ray. Supposing the analogy to hold good for heat rays and for chemical rays, the following result is obtained : With undulations of a certain length the greatest intensity of radiant heat is produced ; the presence of undulations of a greater length is unaccompanied by a corresponding increase of heat, but rapidly increases the intensity of light up to undulations of a certain length, beyond which the presence of undulations of greater length does not produce a corresponding increase of either heat or light, but the effect of such radiant undulations is to produce certain chemical results. Thus, when heat is radiated there may be undulations between certain limits of length. When light is radiated, the undulations are between certain other limits of length. And when chemical power is radiated the undulations are between certain other limits of length. Not only are the undulations, which give rise to the sensation, of a peculiar colour, separated from the rest of the light rays, but it is quite possible for undulations, which cause the sensation of light, to occur without the presence of those giving heat or chemical actions. Even when all three classes occur together, they can be separated by various processes ; and the eye is not affected differently as regards the visual rays, whether they be accompanied by chemical rays or heat rays, or both. The white light from snow is painfully brilliant to the eye, while a piece of black-hot iron sends hardly any light rays to the retina, and is visible chiefly by comparison with the surface on which it lies, just as a hole is seen. Still, the black-hot iron would send many more heat rays to the eye, than the snow. Consequently as intensity of heat is no measure of intensity of light, it may be doubted whether intensity of chemical effect can be.

SECTION III.

———•———

DYNAMOMETERS.

DYNAMOMETERS.*

IN an electric light installation on a practical scale, energy is exhibited under three principal and very different forms—electricity, light, and mechanical power. The measurement of the former two has furnished the subjects of the preceding chapters, and now we come to consider how the work given off by a motor, or absorbed by a machine, is to be estimated, in order that the loss involved in the change from one mode of motion to another may be determined. In the following pages will be found an account, which we believe to be fairly complete, of all the dynamometers which have had a useful existence, and of some which have fallen short of this. The number and variety of the designs testify to the difficulty of the subject to which they apply, and to the real need which they were intended to fulfil, while the scarcity of the instruments themselves in the workshops of engineers and manufacturers, is significant evidence of the difficulty of using them, or of their untrustworthiness. Since the commencement of the present electric lighting era, greatly increased attention has been paid to the subject of power measurement, and we now seem to be within a measurable distance of obtaining a dynamometer which shall be easy to manipulate and be worthy of credence.

Dynamometers are of two main classes, first, those which absorb, in friction, the whole power of a motor, of which the Prony brake is the characteristic example, and, second, those which receive power from a motor and transmit it to a machine, measuring it as it passes through them; and it is in this order that we will consider them.

Prony.—As long ago as 1821, Piobert and Fardy applied a brake as a dynamometer to determine the power of water-wheels; but M. Prony was the first to employ a brake to estimate the power transmitted by steam. It had been demonstrated that friction had a uniform resisting power that might be intensified by pressure. Prony contrived a brake to apply this retarding power to revolving shafts, to obtain a

---

* The greater portion of this section is due to M. Gustave Richard, who published a series of admirable articles on the subject in *La Lumière Electrique.*

required uniform velocity, so that the power applied by, or conserved in, a machine, might be measured.   Prony's brake, in its simplest form, consisted of a rotating horizontal shaft grasped between a pair of wooden jaws or blocks, pressed together by bolts, whose tension could be easily adjusted. One of these blocks was lengthened to form a lever, and carried at its further end a scale pan for the reception of weights.   The method of using this brake was as follows : Supposing the shaft to be revolving uniformly, and driving certain machinery, its speed was noted, and then the machinery was disconnected, and the brake fixed on the shaft.   The lever was held between fixed stops, which allowed it a certain amount of play, and the bolts were tightened until the speed of the shaft was the same as when driving the machinery.   At this time the lever would be pressing hard against one of the stops, the scale pan being empty.   Weights were then added until the lever descended—leaving the upper stop—and assumed a position of equilibrium between the two stops, when the weights in the pan, together with the effective weight of the lever, represented the load.   The latter quantity was found, once for all, by substituting a knife edge for the shaft, and suspending the other end of the lever from a spring balance.

The jaws were not always applied directly to the shaft, but a drum of cast-iron was fixed upon it by screws or keys, and acted as a brake pulley. In any case the power in foot-pounds per minute was equal to

$$2 \pi R L V,$$

where R is the length of the lever, measured from the centre of the shaft, L the load, and V the revolutions per minute.

Kretz.—One of the first improvements introduced into the Prony brake

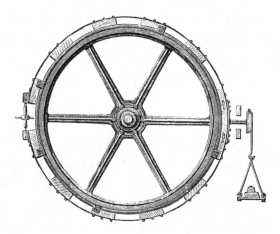

Fig. 155.

was that by M. Kretz (Fig. 155); it consisted in largely increasing the dia-

meter of the pulley, thereby reducing the pressure of the band per unit of surface in contact, and rendering the brake more accurate and more easily fixed, and in arranging it so that the centre of gravity of the brake proper fell within the axis of the drum, and consequently the sensitiveness of the brake was independent of its weight.

Easton and Anderson.—Figs. 156 and 157 illustrate the brake con-

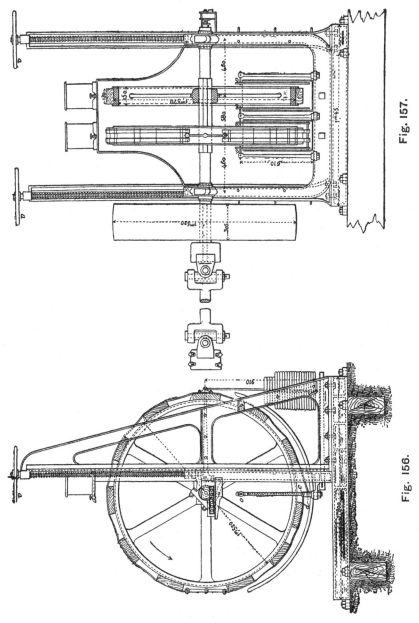

Fig. 157.

Fig. 156.

structed by Messrs. Easton and Anderson, especially for engine tests at the Vienna Exhibition of 1873, but which, we believe, was never used for that

purpose ; many modifications of this arrangement have been introduced in more recent dynamometers. As will be seen, the shaft on which the brake drums are mounted can be raised or lowered, by means of the screw and hand-wheels *v*, to accommodate the apparatus to the various heights of the machines which it is desired to test. The dynamometer is also fitted with Appoldt's automatic regulator, which consists in articulating the ends of the brake straps at the points *c* and *d* to the ends of a lever, free to turn around the axis *o*. The effect of these so-called differential levers is to relieve the brakes if the weight is raised by any increase in the coefficient of friction, and, on the other hand, to draw the bands tighter if they commence to slip. By this arrangement, the brake is made self-regulating for a certain range in the coefficient of friction, to a constant resistance, and the weighted lever is kept in a horizontal position.

Emery.—In the arrangement proposed by Mr. Emery, of New York (see Fig. 158), the ends of the band of the brake are articulated at *b* and *c* to the lever L, connected either to the balance S, or to the system *a* M W for transmitting movement. In the former case the weight W is balanced by the friction F of the brake, and the traction *a* of the spring S. If the friction F increases, L is drawn towards S, the tension *a* of which decreases ; the reverse of this takes place if the friction diminishes, in such a way that there is always a balance between the weight W and the action $F + a$. In the second arrangement

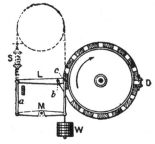

Fig. 158.

the force *a* acts automatically on W by the lever M. For example, if we call *a b c* the forces acting at their point, and F the friction necessary to balance *w*, we have at each instant the relation, $F = w + c - b$, $b = a + c$; whence $F = w - a$.

Brauer.—In this arrangement (see Fig. 159) the oscillations of the index H, are limited by the action of the cord F, attached to one of the guides G of the brake band. The friction of this band, once regulated by the screw D, is maintained constant automatically by means of the crank C turning in a loop of the cord L, in such a way as to tend to tighten or slacken A, according to whether the weight P rises or falls.

Amos.—The arrangement of compensating levers in the Amos brake (see Fig. 160), is similar to the Appoldt regulator ; the rim of the brake wheel is channelled ⌐__⌐, and the cold water introduced to absorb the heat arising from friction is fed in at the bottom of the rim, in which it is

retained, by centrifugal action, until it escapes through a discharge pipe, whose mouth is curved to dip into the channel.

Imray.—Here the axis B of the brake (see Fig. 161) carries a balanced lever C, one of the ends of which terminates by a sector D of the same diameter as the pulley A. The part F W of the brake band bears on this

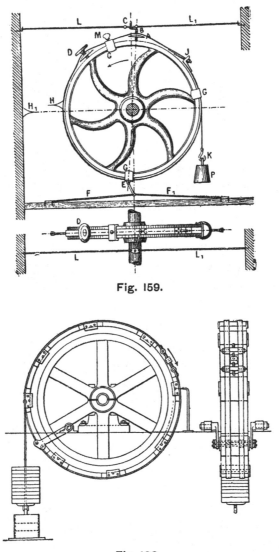

Fig. 159.

Fig. 160.

sector. When the friction increases the sector D rises, and the surface of the brake band in contact diminishes in proportion, the length E P increasing ; the reverse takes place if the friction diminishes, when W falls ; in a word the angle a of the brake band varies every instant, in an inverse direction to the coefficient of friction f, in such a way as to maintain the local

friction of the band always equal to the difference between the weights W and P.

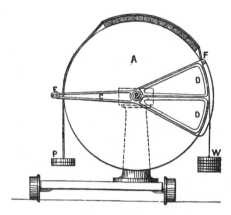

Fig. 161.

Marcel Deprez.—In this machine (see Fig. 162) the shaft A, the work of which is to be measured, carries a pulley B, embraced by the jaws C, of the lever D F E, turning upon centres fixed to the sleeve H G, and com-

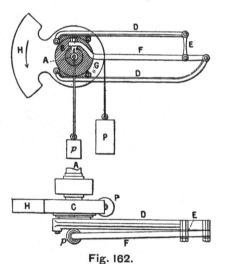

Fig. 162.

pressed by the weight p, the cord of which, attached to F, passes through the centre of A. The counterweight H balances the system of levers, in such a way that only the action of the weight P tends to turn the piece G around the pulley B, drawing with it the system of levers, in spite of the friction of the jaws C. The mean balance once established, if the friction increases, the lever F inclines in the direction of the weight p, the action of which diminishes, as well as the grip of the jaws C, in such a way as to establish automatically the equilibrium between their friction, and the action of the dynamometric weight P. It is sufficient to insure the working of the apparatus, to give to p such a value that its action may be at first somewhat superior to that of the weight P.

Bramwell.—We may notice here (see Fig. 163) a very convenient arrangement, suggested by Sir F. K. Bramwell, for measuring small powers. The band or cord of the brake wheel a is connected to a spring c, the tension of which increases or diminishes as the weight b rises or falls. It results from this

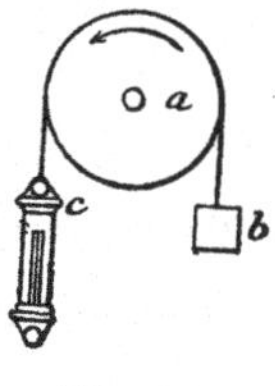

Fig. 163.

that the variations of the differences $b-c$ make the friction of the brake constant, whatever the variation of the coefficient of friction may be, in the same way as the variation of the angle, or band surface in contact, in some of the preceding apparatus.

Carpentier.—The so-called funicular brake of M. Carpentier is illustrated in Figs. 164 and 165. On an extension of the shaft of the motor that is to be tested are two pulleys of the same diameter; one of these, A, is fixed, and the other, B, is loose. From a hook on the latter is a cord which is rolled round the fixed pulley, and carries, at its free end, a weight p. A second cord, attached to the same hook, passes round the loose pulley, and terminates in a heavier weight P. The fixed pulley, turning with the shaft in the direction of the arrow, sets up, between itself and the cord to which the weight p is hung, a certain friction, which tends to turn the loose pulley and raise the weight P. If the friction increases, the weight P continues to be raised, but at the same time the length of circumference in contact with the cord is reduced; if the friction is diminished, the weight P falls, but at the same time the length of circumference in contact with the cord p is increased. In this way a varying compensation is established, in such a way that

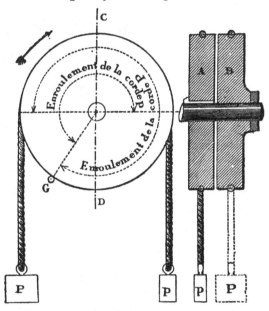

Fig. 164.   Fig. 165.

the power exerted is always equal to $P-p$; and to ascertain the work T all that is necessary is to measure the number of revolutions of the shaft during a unit of time. The formula for the work per second is

$$T = \frac{\pi\,d\,n}{60}(P-p)\,;$$

n representing the number of revolutions per minute.

This brake is especially adapted for very small powers, and a modification of the arrangement, rendering it applicable on a larger scale, has been worked out by M. Raffard, under the name of the Dynamometric Balance. This modification is shown in Fig. 166. The apparatus is not on the shaft of the machine to be tested, but is mounted on a separate frame, the spindle carrying the pulleys having another pulley for connection by a strap to the

machine.   There are three pulleys altogether, the two outer ones, D D′, being loose, and the centre one, D″, fast on the shaft.   A bar B, bent twice at right angles, and forked at both ends, is supported on the spindle by these

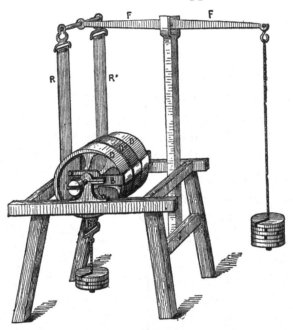

Fig. 166.

forks.   The bar is balanced by a weight placed at the end, as shown, and to it are attached the steel ribbons R R′ R″.   The middle one of these, R′, passes over the fixed pulley, and carries at its lower end the weight p.   The two other ribbons pass under the loose pulleys, and are attached to one end of a scale beam carried by a vertical support forming a part of the frame.   At the other end of this beam is an iron rod loaded with a weight P, heavier than the weight p.   When the apparatus, arranged as shown in the figure, revolves in a direction opposite to the hands of a watch, the length of circumference in contact with the ribbons R R¹ varies as the friction of the ribbon R″ increases or decreases, and compensation is thus established exactly as in the Carpentier brake.   The work to be measured can be ascertained by the foregoing formula, to simplify which, the circumference of the pulleys is made exactly one metre, so that $T = \dfrac{n}{60}(P - p)$.

It is sufficient, therefore, to measure the number of revolutions by a counter and to multiply the speed by the difference between P and p.   In order to maintain constant the friction between the pulleys and the ribbons, both are immersed in a zinc trough filled with water.   This apparatus appears to be useful for measurements up to three horse-power.

Thiabaud.—The Prony brake has received several improvements de-signed to effect the cooling of the pulley by a constant circulation of water. In the brake of M. Thiabaud (Fig. 167) water enters at A and, after traversing the passages *a*, B and *e*, escapes at C; the brake pulley is fixed upon the driving shaft by the apparatus M M, and the brake D is held upon the pulley by the screws G. M. Thiabaud is able to mea-sure from 20 to 250 horse-power by his brake.

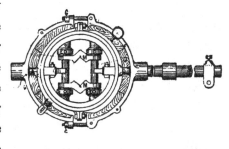

Fig. 167.

Weyher and Richmond.—MM. Weyher and Richmond enclose the brake pulley between two plates of iron, and in the drum thus formed they pro-duce a circulation of water following the course *t t′ t″*, Fig. 168. The weights

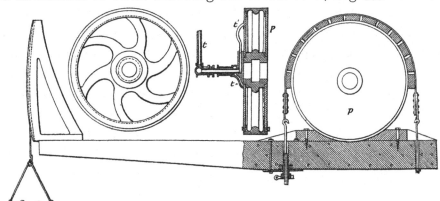

Fig. 168.

are suspended from a sector struck from the centre of the brake pulley, so as to render the length of the lever arm sensibly constant in spite of its oscillation.

Felu and Deliège.—The compensating mechanism for the dynamometer

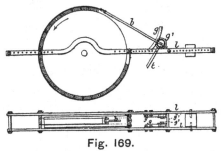

Fig. 169.

brake employed by the firm of Felu and Deliège is one of the simplest. The

rollers g^1 (Fig. 169), whose axes are connected to the ground by the rods t, roll upon the guides g, fixed on the brake lever, and are formed of such a curvature that the link b extends upon an increase of friction, and contracts if the friction should decrease.

Raffard.—M. Raffard has proposed to neutralise the variations of the coefficient of friction of the strap of the brake by fixing upon it the armature of an electro-magnet, the current of which varies, in an inverse sense, to the variation of the coefficient.   In Fig. 170 E represents the electro-magnet; it is traversed by a current from a battery s, which also passes through the liquid interposed between the plate p, and the disc on the end of the rod H.   When the friction decreases the brake falls, the disc

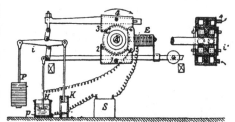

Fig. 170.

H approaches the plate P, the resistance H P diminishes, and the armature attaches itself more strongly to the pulley.   The inverse takes place when the friction increases.   The brake is also provided with an automatic compensating tightening apparatus P i.   When the brake falls suddenly the resistance of the dash-pot K tightens the brake slightly, and when the lever rises it slackens it.   The circulation of water takes place in the passages 1, 2, 3, 4.

Emerson.—Emerson's dynamometer, which is specially designed for the trial of turbines, is composed of a Prony brake (Fig. 171), with a bronze band,

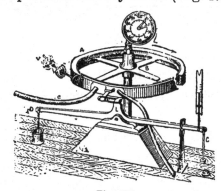

Fig. 171.

furnished with arrangements for the circulation of water, and acting almost directly upon the dynamometer lever. This lever carries, at one of its extremities, a weight rod C, and is connected at its other end with a dash-pot D, whose piston is provided with screw holes that can be more or less opened according to the degree of sensibility required. The pulley B is

mounted directly upon the turbine shaft, and carries a counter. The brake band is tightened by a handwheel V, with a universal joint, so arranged that it can slide upon the axis in such wise, that, when it is being acted upon, no resistance is offered to the drawing up of the brake.

In addition to the dynamometers we have already noticed in this class, there remains one nearly related to them which presents features of originality that are peculiar to itself. This is Froude's inertia dynamometer.

Froude's Inertia Dynamometer. — The apparatus consists of a cast-iron casing, containing two gun-metal dishes forming two circular channels marked A A A A in Fig. 173. Identical gun-metal dishes are fixed to an internal cast-iron disc keyed to the shaft B B, which goes right through the casing. Blades, radial as to their edges, but inclined as to their planes, are

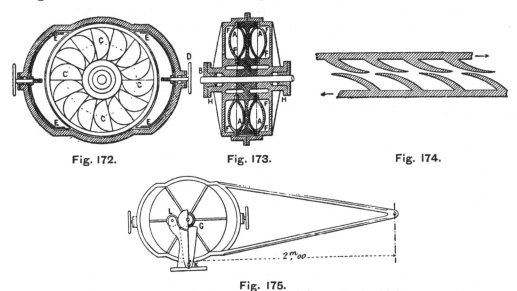

Fig. 172.   Fig. 173.   Fig. 174.

Fig. 175.

fixed in the channels. For convenience in our description, the internal disc will be called the fan.

The channels with their blades, if developed in the line of the circular axis of the channel (see dotted line C C C C, in Fig. 172), and taken in longitudinal section, would present the form shown by the development, Fig. 174, continued indefinitely in each direction. The channels are taken out to such a section as will make the line of intersection of the blades with the channel a semicircle (see the fourteen semicircles in Fig. 172). The casing is filled with water. If the casing be held fast, and the fan made to revolve within it so as to make the relative motions of the blades be in the directions shown by the arrows in the longitudinal development, a great

internal resistance is set up, and the casing becomes a brake to the revolving fan.

At D D are two wheels for regulating the brake action of the apparatus in the following manner : E E E E are plates, there being two pairs, but one of each pair only appearing in Fig. 172. Each pair are coupled together by the curved backs, which appear in section in Fig. 173. The spindles to which the regulating wheels D D are attached are screwed, and go through screwed bosses in the curved back above mentioned ; thus, by turning the wheels D D the four plates are pushed forward between the fan and the casing. The communication between the water in the fixed, and in the moving channel and blades, is thus interrupted, and the brake action is diminished.

The space F F F F is a water chamber ; water enters through two cocks from outside, and is brought through holes through the plane of the blades to the centre of the channels. A water exit is also provided at the top of the casing, and by this means an interchange of water is effected.

When an engine is to be tested by the dynamometer, the shaft of the engine is coupled to the dynamometer shaft B B, and the effort of the engine is balanced by weights attached to the end of the lever (Fig. 175). The casing rests on two rockers on points H H (Fig. 173), the necks of the stuffing-boxes ; G, in Fig. 175, shows the edge of one of the rockers. Both rockers are on one shaft, and the end of this shaft appears at K, in Fig. 175. Four small rollers L touch the circular lugs of the stuffing-boxes to prevent the casing from moving laterally. The rocker gives a very "tender" suspension to the casing, and insures that the weight at the end of the lever shall represent the whole of the effort of the engine.

The interchange of the water is necessary to the carrying away of the heat which results from the absorption of the work of the engine under trial. A complete analysis of the principles by which the resistance is set up in the apparatus will be found in a paper read by the late Mr. W. Froude, F.R.S., before the Institution of Mechanical Engineers at Bristol, in July, 1877.

The maximum power-absorbing capability of a moderate sized apparatus is as follows :

						Horse-power.
At 100 revolutions ...	...	...	...	...	...	20
„ 200 „	...	...	...	...	...	160
„ 300 „	...	...	...	...	...	540

It will be seen from this Table that the power absorbed varies as the cube of the number of revolutions per minute. At the same time it must be remembered that the regulating plates will reduce the resistance as low as may be required. The length of the lever, namely 6.56 ft., is chosen, so that an eight-pound weight, with the brake running at 100 revolutions, represents one horse-power.

The next class refers to apparatus that measure the power transmitted through them, without absorbing it, otherwise than by the friction of their moving parts. The earliest forms of the apparatus were those in which the measurement was effected by noting the reaction between the teeth of wheels.

Brown.—This dynamometer (see Fig. 176) is driven by gearing ; in it the work is transmitted from the motor pulley A to the driven pulley B, of the same diameter, by the equal sized wheels C and E, and the intermediate pinion D hung to the beam of the balance *f* F. It is evident that the wheel D exerts on this beam a pressure (neglecting friction and resistances of the machine) double the motive effort measured on the circumference of the wheel C, and it is easy so to graduate F as to read direct the motive

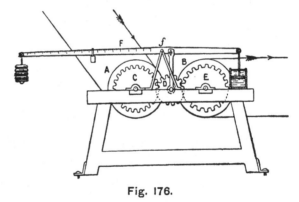

Fig. 176.

effort corresponding to the value and position of the weight required to keep it horizontal. The oil cylinder shown is employed to deaden the vibrations of the beam.

Raffard.—This apparatus (see Fig. 177) exhibits in a modified form the same principle as Brown's. The driving wheel A actuates the internally-geared wheel C, connected to the machine by a clutch or by a belt, through the intermediate pinion D of equal diameter to C; the shaft of D is connected to the beam F by the lever B *f*, movable around the axis *b* in the same plane as the axis of C, and consequently tangential to the circumference of D. With this arrangement, the reaction of the wheels A and D exerts no turning action on the lever B, which only tends to move by the reaction of D on C, the friction of which is so slight that the effort exerted at the point *f* may be considered as practically the same thing as that at the circumference of C.

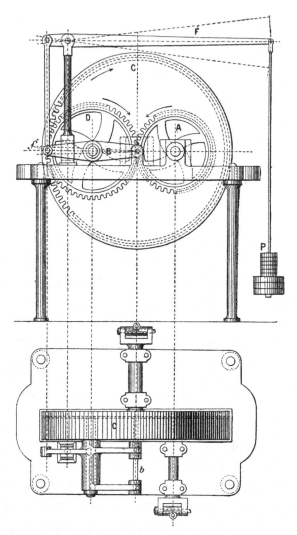

Fig. 177.

German Dynamometer.—In this apparatus (Fig. 178), which does not bear the name of the designer, the internal toothed wheel D, transmits the motive power of O to the shaft of G by the two equal pinions K. A clutch on the frame upon which the pinions are mounted forms a connection with the boss of the pulley L, which is coupled to the dynamo-meter spring by the band Z;

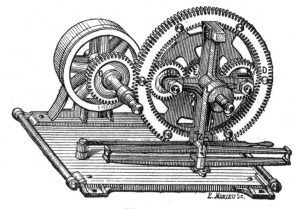

Fig. 178.

this arrangement, which is very compact, possesses the drawback of having very considerable resistance.

King.—The dynamometer of Mr. King is distinguished from the preceding by the use of bevel wheels. In it the driving and driven pulleys A and B (Fig. 179), loose upon their shaft, are connected by the pinions a, d, b. The reactions of a and b upon d, each sensibly equal to the motive force exerted at the pitch circle of the pinions, are balanced by the weight p, in such a way that the turning power t of the pulley A is given

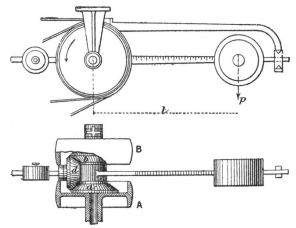

Fig. 179.

approximately, that is neglecting friction, by the expression $t = \pi \, l \, p$.

Silver and Gay.—The dynamometer of Messrs. Silver and Gay also

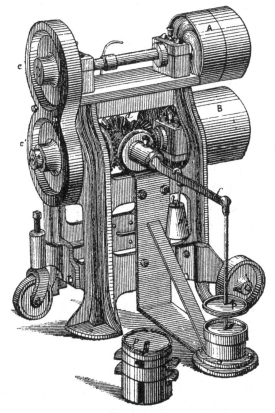

Fig. 180.

belongs to the class in which the power is transmitted through bevel wheels.

A (Fig. 180) is the driving pulley, B the pulley from which the strap is taken to the machine the power of which it is desired to measure, d the train of bevel pinions, f the steelyard, P the scale pan, p^1 the movable weight. The rod which carries the scale pan terminates in a piston, moving in the air cylinder c, whose office is to lessen the vibrations and regulate the oscillations of the steelyard. The pulley A drives the pulley B through the intermediary of the toothed wheel $c\,e^1$, and the pinions d; the amount of weight placed in the scale pan, and the position of the runner p^1 necessary to keep the beam horizontal, is a measure of the power transmitted. From this the necessary deduction must be made for the friction of the apparatus.

Smith.—The dynamometer of Mr. Smith, of Taunton, is based on the same principle as the King dynamometer. It comprises two pulleys which, in transmitting the movement, act on a spring in place of a weight, and extend it more or less according to the effort exerted. Fig. 181 shows the apparatus. The two pulleys P and P' are mounted on the same shaft; they are loose on this shaft, but are always connected by bevel gearing. Between the two pulleys, in the middle of the shaft, and loose on it, is an open sector M. Within this is a bevel pinion on a vertical spindle, gearing into the two bevel wheels; and, as the sector M is loose, it acts as a planet wheel. At the bottom of the sector is a band kept stretched by a spring in the box A. The sector is extended above the shaft by a wooden sector E: a counterweight placed behind the sector balances this latter with M. If the pulley P is driven in the reverse direction of the hands of a watch, the system of bevel gearing will transmit to the pulley P' a movement in an opposite direction, and this pulley P' puts in movement the machine to be tested. But if this machine exerts a certain resistance to the motion, the intermediate toothed wheel, besides transmitting the movement, will be displaced towards the right, and draw on the spring in the box A, and this displacement will increase as the effort exerted becomes more considerable. By measuring the tension of the spring and the number of revolutions of the spindle, the work developed will be ascertained. The number of revolutions is given by an ordinary counter shown in the figure. Two toothed wheels of equal diameter between the pulley P and the front bearing of the shaft, transmit the movement to this counter. The different degrees of tension of the spring are recorded as follows : On the circumference of the sector E is placed horizontally a wooden rod N' H, which at H passes freely through an opening in the vertical support. Two threads N' and O', attached to the sector and

the rod, guide this latter in such a way as to follow all the movements of the sector E, and, therefore, of the sector M. On the rod is a style T, formed of a glass tube drawn down to a point, and filled with aniline ink; this bears on a sheet of paper rolled round a wooden cylinder. On the

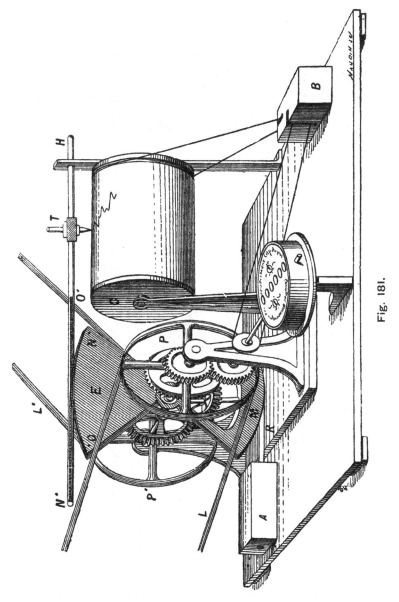

Fig. 181.

counter spindle is placed a small pulley, a cord from which drives an endless screw in the box B, and this, gearing into a wormwheel, actuates a second pulley, from which the cylinder is set in motion. The various parts are so proportioned that the cylinder has a speed bearing a known relation to that

P

of the pulleys. When the spring A is extended, the sector M is drawn towards the right, while the rod and style T are, of course, moved toward the left; and there is traced upon the paper mounted on the revolving cylinder, a curve representing the effort exerted, and the time during which

Fig. 182.

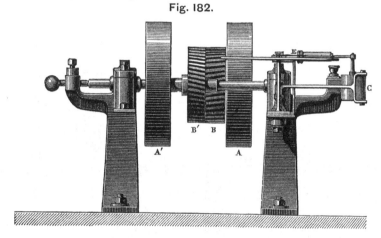

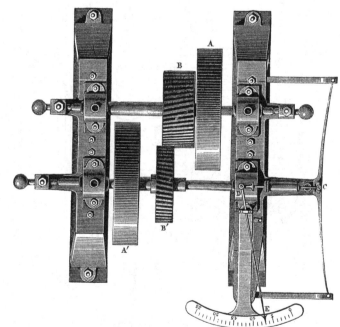

Fig. 183.

it is continued; the work done can be ascertained by computing the area of the diagram thus produced, which is preferably deduced by the design of this apparatus, by cutting out and weighing the paper enclosed by the curve. The number of seconds that any given experiment lasts is recorded by a chronographic attachment also marking on the cylinder.

Bourdon.—It is possible to greatly diminish the friction of dynamometers in which toothed wheels are employed by the use of helicoidal teeth. The Bourdon dynamometer, Figs. 182 and 183, is constructed on this plan : The motion is transmitted from the driving pulley A to the driven pulley A¹ through the helicoidal toothed wheels B B¹. If the incli-

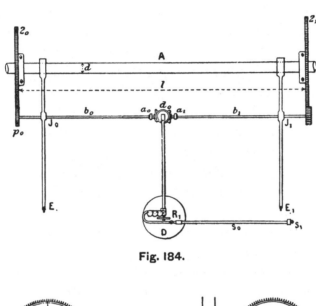

Fig. 184.

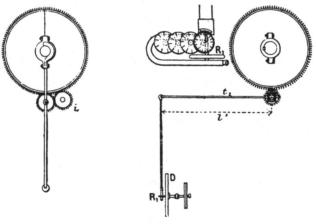

Fig. 185.   Fig. 186.

nation of the teeth with respect to the axis be called i, and the tangential force at their pitch circles P, the component of the force, parallel to the axis of B¹, equal to P cot. i, will act directly on the arched spring C and its index finger D E.

Hirn's Torsion Dynamometer.—Hirn's torsion dynamometer measures

the power transmitted by a shaft by noting the angle of torsion through which a portion, or the whole of it, is deflected. The shaft A, Figs. 184, 185, and 186 through which the power is passing, carries two equal toothed wheels 2_0 2_1, gearing the one directly through the wheel p_0, and the other through the intermediate wheel i, with two shafts b_0 b_1, which are thus rotated in opposite directions. The shafts are coupled together by the differential train a_0 a_1, d_0 d_1, of which the latter two wheels are carried on the rod t_1, which controls the position of the roller R_1 upon the totaliser disc D. This roller, guided by the long bar S_0, movable about S_1, is moved away from the centre of the disc as the torsion increases, that is, as the power transmitted is augmented, and at the same time the disc rotates at a speed which is proportional to that of the shaft; from this it results that the number registered by the counter, which is driven by the roller, is proportional to the work transmitted by the shaft A.

The bearings J_0 J_1 of the shaft b_0 b_1 are fixed upon levers E_0 E_1 whose extremities rest upon rollers free to run in a horizontal plane. By this arrangement the vibrations transmitted to the wheels 2_0 2_1 by the play of the shaft A in its bearings are communicated to the wheels p_0 p_1, and do not affect the indications of the roller.

This simple and ingenious instrument has the distinguishing characteristic that it may often be applied without modification of the existing method of transmission. Since its introduction in 1867 it has always been found to give accurate results.

Belt Dynamometers.—The principle of belt dynamometers consists in replacing the transmission, the work on which it is desired to measure, by a temporary transmission, the belt of which is so arranged that the observer can register, during the whole trial, the variations in the differences in tension of the sides of the belt and its speed at each moment. In this way are obtained the elements of the work transmitted to the belt of the dynamometer; to deduce the useful work transmitted by this belt to the machine being driven, it is necessary to deduct the work absorbed by the resistance of the dynamometer, and by twisting, slipping, and stretching of the belt. The work absorbed by the resistance of the dynamometer can be ascertained with sufficient practical accuracy by measuring it when idle, that is to say, by making it drive a loose pulley, or better still a pulley of known resistance, at a normal speed; the other resistances which affect the accuracy of the apparatus are practically indeterminable, but they can be reduced to a minimum by a careful selection of a belt, and by arranging it to

embrace so large a proportion of the circumference of the pulley as to render dipping almost *nil*.

Froude.—One of the oldest and best studied apparatus of this type is the Froude dynamometer, in which the belt passes from the driving pulley A (Fig. 187), to the driven pulley B, being taken over two guide wheels C D on its way. These guide wheels are carried by a balanced arm E, movable around the axis F. Between this arm and the pulley B, are introduced guide pulleys G and H. One only of these pulleys comes into operation, the selection depending upon the position and size of the driven pulley, which may be as shown at B, or may be at a higher level, or of any diameter up to three times as great as A. The object

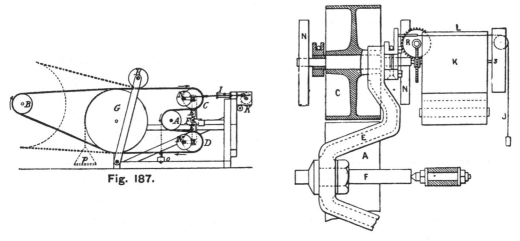

Fig. 187.

Fig. 188.

of these guide pulleys is to preserve the parts of the belt between G and the dynamometer parallel. From this arrangement it follows that, when once the apparatus is in working order, and supposing its friction to be insignificant, the axis of the pulley D will be subjected to a strain equal to the sum of the tension of the two sides of the belt around it, or to double the tension of the following side, and that the axis of the guide wheel C will receive a force equal to double the tension of the driving side; the dynamometric spring I will thus have to sustain an effort equal to double the difference of tension in the sides of the belt employed. The style *s* connected to the balanced lever E by a cord (Fig. 188) moves along the face of the cylinder K, actuated by the gearing R, with a rotating motion proportional to that of the driving pulley; this style traces a curve of force, the area of which is proportional to the work done by the belt, and the contour of which indicates all the variations of this work. An

accumulating counter can, of course, be added to this instrument, to indicate constantly the work done from the commencement of the trial, and its value in horse-power. Referring to the illustration, it will be seen that the axes of the wheels C and D rest upon anti-friction rollers N N; that the balanced lever E is in the form of **Z** to compensate, by a slight bending of the arm, any displacement of the axes, and to preserve them parallel to F. The spring is connected to a rod on which is a piston working in an oil cylinder, the opposite ends of which are connected by a pipe controlled by a tap, so that the resistance opposed to the vibration of the spring can be regulated without affecting the precision of its general indication. The apparatus can be adjusted by replacing the belt by a graduated belt fixed to the frame at o and loading it at p with a known weight, the effect produced being indicated by a pointer.

Tatham.—This apparatus is founded on that of Froude; its general forms are shown in Figs. 189 and 190. In the former, the balanced lever of Froude is replaced by two semicircles B B' free to turn on the knife edges C C', and connected by links bearing on the knife edges D D'. The driving pulley is shown at A, and the sides of the belt $a\ a'$ pass on one side of the knife edges C C' over the guide wheels E E', and thence over F F' to the driven pulley. By this arrangement the sides $a\ a'$ exert no turning effort on the pieces B B', which only tend to turn by the difference in tension of the sides $b\ b'$, so that the effect of friction is almost removed. The second arrangement, Fig. 190, has for its special object the partial avoidance of the disturbances due in the wearing of the spindles of the rollers, the effect of such wear being to increase the distance apart of the tangent points on the right of the edges C C'. This is avoided by the arrangement shown in the diagram, where A is the driving pulley, and M the driven pulley; the lever B, the arms of which are equal to the effective diameter of the wheels E E', is pulled at H by an amount due to the difference in the tension of the sides $b\ b'$ of the belt. In this type of dynamometer any variation in the thickness of the belt, the planes of the sides $a\ a'$ of which ought to pass through the axis C, does not appreciably affect the accuracy of the results.

Tatham.—Another form of the Tatham dynamometer, which has been constructed for the use of the Franklin Institute, is represented in Fig. 191. It consists of double wooden A frames, braced together and carrying four pulleys between them. Of these, the lower is driven from the engine or prime mover, and transmits motion to the one

immediately above it, which is connected, by an elongation of its shaft not

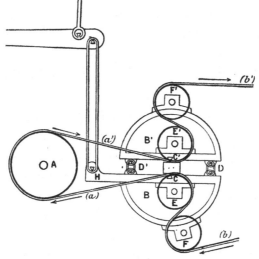

Fig. 189.

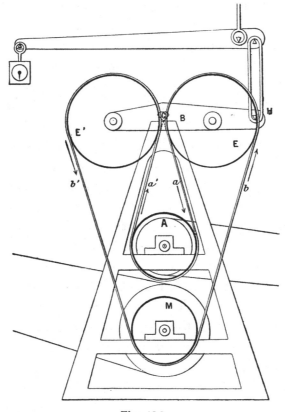

Fig. 190.

shown in the view, to the machine whose consumption of power is to be

determined. The remaining two pulleys are idle, and are carried in separate lever frames. Each frame is suspended, at its outer end, upon knife edges carried upon a crossbar, and is connected at its inner end by a link to a scale beam above it. These two links are attached to the scale beam on opposite sides of its pivot (1.9 in. therefrom) and consequently their action upon it may be balanced by a weight that corresponds to their difference. The beam is capable of weighing up to 300 lb., and is graduated to 25 lb., by pounds and tenths; it may be balanced by dead weights, or by a spring balance, the latter being used when an indicator diagram is being produced. From the centre of the beam to the extreme right-hand is $32\frac{1}{4}$ in., from which it may be deduced that the scale of the illustration is about $\frac{1}{18}$th.

The idle pulleys are each 7 in. broad, and the largest and smallest circumference of each is $27\frac{3}{32}$ in. and 27 in., whence it follows that the average radius is 4.30375 in. The axis of each pulley is situated 4.39 in. from a line joining the knife edges on which its lever frame turns, the effective radius of the pulley having been found to be 4.387256 in. The distance of the axis from the suspension link is 8.78 in. The middle pulley, which drives the counter and indicator card, is 7 in. broad, and has a maximum circumference of $38\frac{3}{4}$ in., and a minimum circumference of $38\frac{7}{16}$ in., or 38.59375 in. on the average. Careful measurements show that the delivery of belt per revolution is 39.595 in. The driving pulley is 30 in. in diameter and 7 in. broad, and runs in journals that can be adjusted vertically to tighten the belt. The latter is 16 ft. long, 6 in. wide, and $\frac{2}{9}$ in. thick, of oak-tanned leather. The splices are scarfed, glued, and rivetted, so as to preserve a uniform thickness. The flesh side of the belt is next the driving pulley and the two idle pulleys, and the hair side is next the upper middle pulley. The belt runs in the direction of the arrows on the outside, down on the left and up on the right. But in describing its operation it is better to follow the tension of the belt in a direction contrary to the motion of the belt itself. The tension, originating at the lower driving wheel, acts vertically upon the left-hand idle pulley at the extremity of its effective radius, and cuts a line joining the two knife edges, which constitute the fulcrum of the lever frame. The effect of this part of the belt upon the scale beam is, therefore, *nil*.

Losing enough force to overcome the friction of the idle pulley, the remaining tension acts vertically, first by reaction upon the lever frame carrying the idle pulley at a point corresponding to the extremity of the

inside effective radius of the pulley, and thence through the link upon the positive side of the scale beam; and, second, upon the middle pulley representing the machine on trial. These forces are equal and opposite. The tension acting on the middle pulley then performs the work that is to be

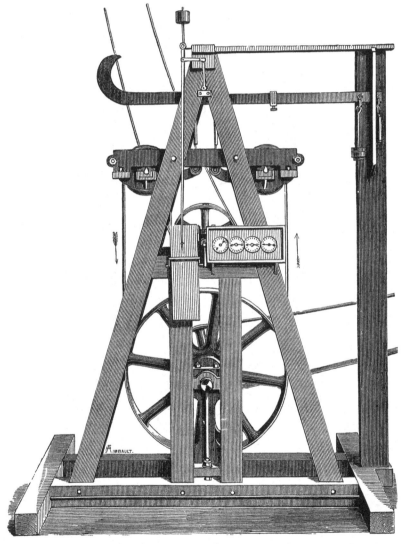

Fig. 191.

measured and is reduced thereby. The remainder acts, first, by reaction on the middle pulley; and, second, directly upon the lever frame carrying the right-hand idle pulley, and thence through the link to the negative side of the scale beam. These two forces are also equal and opposite. The tension then passes over the idle pulley through the fulcrum as before to the

place of beginning. It is evident, therefore, that the only forces bearing upon the scale beam are the tension of the tight belt on the positive side of the beam, and the tension of the slack belt on the negative side. The beam weighs the difference between the two, and if its indication in pounds be multiplied by the number of revolutions of the middle pulley per minute,

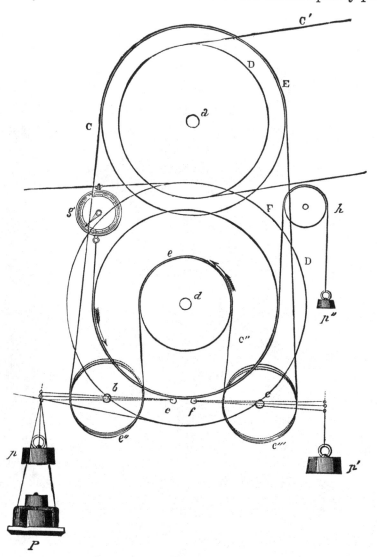

Fig. 192.

and the product be divided by 10,000, the result will be horse-power and decimals.

The principal centre of the scale beam is lengthened towards the observer, and at its nearest extremity carries a lever arm attached to a horizontal link, connecting it with a long vertical index lever, which carries

a pencil at its lower end. This pencil moves horizontally as the end of the lever vibrates vertically, and marks upon a ribbon of paper, caused to move vertically between two revolving rollers, which are driven by the worm upon the prolongation of the shaft of the middle pulley. If the scale beam be attached to the spring balance the ordinates of the curve traced by the pencil will represent the force employed, while the abscissæ will represent the motion.

Farcot.—M. Farcot's transmission dynamometer is also designed for the continuous measurement of the work transmitted to any machine driven by a belt, so as to record the work absorbed, its action depending on the measurement of the tension of one side of the belt going from the source of power, to the driven machine. Fig. 192 illustrates the general arrangement of the apparatus. The two pulleys D′ E are fixed on the shaft *d*; the former is driven from left to right by the belt C′ from the motor, on the latter is the transmission belt C, which afterwards passes round the pulley *e′* and the tightening wheels *e″ e‴*. On the shaft *d′*, which carries the pulley *e′*, is also a pulley D, which drives the machine to be tested by a belt. The tightening wheels are loose on their spindles, which run in frames oscillating round the points *e* and *f*. When the machine is at rest the frames are loaded with two equal weights *p* and *p′* which produce the tension on the belt. If, without connecting the apparatus with the machine to be tested, the pulley D is put in rotation, the side C of the belt will be placed in tension, and lifting the tightening wheel *b* will destroy the equilibrium. To restore this it is necessary to add to the scale pan, hung at P, a weight *g* that represents the effort produced by the apparatus itself. If, now, by means of the belt over the pulley D, the machine to be tested is set in motion, the resistance offered by this machine will exert a further tension of the belt at C, the tightening pulley *b* will be lifted and equilibrium again be disturbed, requiring another weight in the scale pan to restore it. When this has been done the tension of the belt at *e′ e″* will be equal to the effort exerted on the circumference of the pulley to put the machine in movement. The work absorbed can then be obtained by multiplying this tension by the speed of a point on this circumference, that is to say, by $2 \pi R n$, calling *n* the number of revolutions per second, and R the radius of the pulley *e′*. The tightening pulley *b* being in the middle of its frame, the vertical reaction applied at its centre to balance the weight P will equal 2 P, and, as this force is divided between the sides C and *e′ e″* of the belt, if these are assumed to be vertical, which is nearly the case, the

tension of each of them will be equal to P. If from this latter be deducted the tension g equal to the effort necessary to set the pulleys in movement, the work absorbed will be equal to $(P-g) \times 2\pi R n$. In this formula P and g are given by direct observation, R is a fixed and known quantity, and n is ascertained by an ordinary counter. We have assumed that the belts C

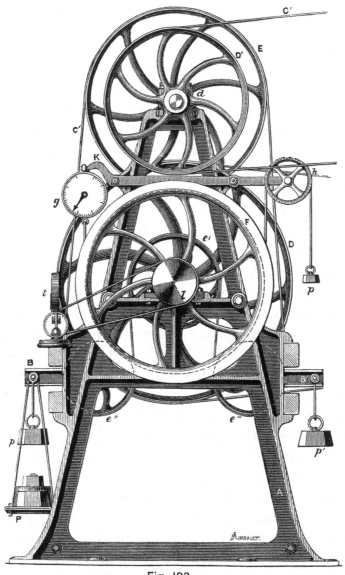

Fig. 193.

and c' c'' are vertical and parallel. Actually this is not the case, and the belt makes a certain angle with the vertical, which must be allowed for to obtain a correct result. If we call $2 T'$ the vertical force applied at b to balance the weight P, we have $2 P = 2 T'$; again, calling T the tension of the

belt $e'\ e''$, then $T = T' \cos a$, whence $T' = \dfrac{T}{\cos a}$; as $P = T'$ it follows that $P = \dfrac{T}{\cos a,}$ whence $T = P \cos a$. Evidently the smaller the angle a the more nearly will T approximate to P, and in many cases the difference may be neglected, but it can be always easily computed for accurate results. The effect of the machine itself already referred to varies with the speed, and a test of these variations due to different speeds is convenient. In a standard type of the machine the tares are as follows :

Tare for 100 revolutions		...	...	...	2200 lb.
,,	120	,,	...	...	...	2640 ,,
,,	140	,,	...	...	...	3080 ,,
,,	160	,,	...	...	...	3300 ,,
,,	180	,,	...	...	...	3410 ,,
,,	200	,,	...	...	...	3520 ,,
,,	220	,,	...	...	...	3740 ,,
,,	240	,,	...	...	...	3960 ,,
,,	260	,,	...	...	...	4180 ,,
,,	280	,,	...	...	...	4400 ,,
,,	300	,,	...	...	...	4840 ,,

In ascertaining the above amounts, the weights were tested by means of a very simple brake shown in the figures. It consists of a cord passing over a grooved pulley F, fixed on the spindle d; this cord is secured to a spring balance g, the other end passing over the pulley h, and being held by the weight p''. Fig. 193 gives a general view of the apparatus, the same letters applying to similar parts in the diagram. In this machine the dimensions of the principal pulleys are $e' = 23.62$ in.; E, 2.76 in.; D, 43.3 in.; and the tightening pulleys are each 15.75 in.; the total height of the apparatus is 10 ft. 2 in. In the arrangement shown a Prony brake can be connected with a pulley on the shaft d, its object being to check the accuracy of the instrument from time to time. To do this, the machine to be tested is first measured by the dynamometer, as above described, and then, when the weight P has been ascertained, this weight is left in the scale, and connection with the machine to be tested, is discontinued; the Prony brake is then adjusted and loaded until equilibrium has been restored. The work indicated by the brake ought to be equal to that previously indicated by the dynamometer.

Parsons.—This dynamometer, which is extremely simple, is similar in its action to that of Froude. The driving pulley, Fig. 194, is A, and C is the driven pulley. The apparatus being set in motion, the weights at Q

are varied in such a manner as to maintain, as far as possible at one level, the axes of the suspended pulleys B and D; there then exists constantly

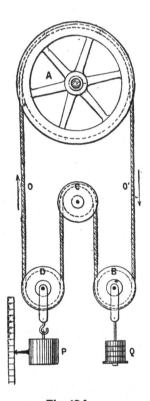

Fig. 194.

between the tensions T_1 and T_2 of the sides o and o^1 of the belt, the relation

$$T_2 - T_1 = \frac{P - Q}{2}.$$

Hefner Alteneck.—In its simplest form, that in which it was first brought out in 1872, the Von Hefner Alteneck dynamometer consisted of two rollers connected rigidly to one another, and embracing between them the two parts of the belt extending between two pulleys, one of which was fixed on the motor and the other on the machine to be tested, the two sides of the belt being brought near to each other by the rollers. The dynamometer was fixed to a support in such a way that the two rollers, and the portions of the belt in contact with them, occupied a position perfectly symmetrical to the line joining the centres of the two pulleys. The frame carrying the rollers could be shifted slightly perpendicular to the centre line, the motion being limited by stops, and the central position exactly fixed by marks. When not in motion the two parts of the belt,

being equally stretched, the two rollers maintained their central position, but as soon as the belt began to transmit power, the driving side became more tense and displaced the rollers. The force necessary to bring the system into its normal position was, therefore, proportional to the difference of tension between the two sides, that is to say, to the power trans-

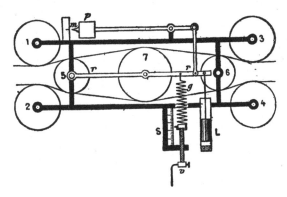

Fig. 195.

mitted at the moment. This force was measured by a spring, which was extended by a screw adjustment, until the index was brought to its normal position. The extension of this spring measured on a graduated scale, the power sought for. The force indicated by the spring, divided by the sum of the sines of the two angles, formed by the centre line, and the two

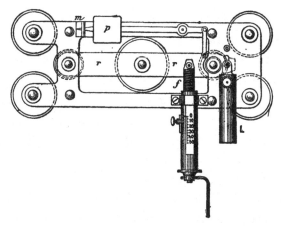

Fig. 196.

portions of the belt on each side of the same roller, gave the difference in tension between the two sides of the belt, that is to say, the power transmitted, P.

Although the dimensions required for the equation by which the power was worked out were easily measured, the accuracy of the deductions

depended on the precision with which they are taken, so that considerable care was necessary in using the apparatus, which, moreover, required a new measurement for each installation. To avoid these drawbacks, Mr. Hefner Alteneck has greatly simplified the arrangement. With the new apparatus direct readings can be taken of the difference of tension, and the mode of mounting on the belt does not affect the accuracy of the measurement. Fig. 195 illustrates the principle and Figs. 196 and 197 the instrument itself. The two parts of the belt rest, as shown, upon seven rollers, of which six (1 to 6) are fixed in the frame of the apparatus, the seventh roller, that in the centre, is carried in a frame $r\,r$ movable about the spindle of the roller 5 in such a way that it can travel slightly on each side of its central position. The apparatus is fixed on the driving belt between two pulleys in such a way that the spring g may be on the stretched side of the belt. The two small rollers 5 and 6 are employed to maintain the two angles of the belt always equal. The weight of the roller 7 and of the movable frame $r\,r$ is balanced by

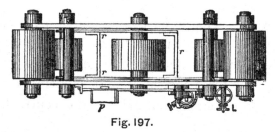

Fig. 197.

the counterweight p fixed to the end of a lever, and connected by a rigid rod to the frame $r\,r$. By this arrangement the equilibrium is maintained in every position of the apparatus. This lever is also used as an indicator; when the point m is opposite to the mark on the piece projecting from the frame, the roller o is in the centre of the apparatus. This modified form operates in the same way as the original apparatus, the spring g tending to bring the roller 7 to a central position, but as the angle of the belts remains constant, the coefficient of the apparatus can be determined once for all, and the spring graduated beforehand. As a rule, a millimetre on this scale corresponds to a kilogramme in difference of tension. The dynamometer can be fixed as may be most convenient, but it should be placed in the same plane as the two pulleys, so as to avoid any slipping of the belt. To prevent any sudden and violent action of the movable part, the frame is connected to the piston rod of a small pump filled with water which absorbs shocks. In order to attach the apparatus without unfastening the belt, one side of the instrument can be removed and replaced easily.

When, at any given moment, the difference of tension between the sides of the belt has been read on the scale, the work transmitted at that moment is ascertained by multiplying the number found by the circumferential speed of the pulley. This can be measured by any of the ordinary methods, either on the transmitting or receiving shaft, according to whether it is desired to allow for the loss caused by the friction of the belt. To check the accuracy of the instrument, care is first taken that all the rollers are running quite free, and that at any angle, the index of the spring being

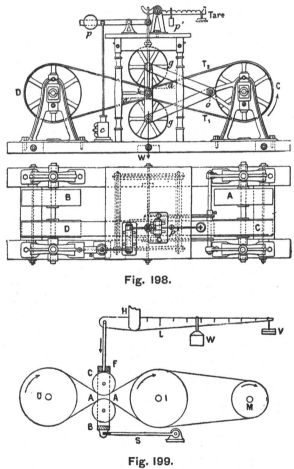

Fig. 198.

Fig. 199.

at zero, the other index shall coincide with the mark m. This adjustment can be secured by shifting the spring index, and, if necessary, the counter-weight p. The instrument is then placed almost vertical, and a small belt is brought in contact with the roller. The sides of this belt are loaded with weights, and additional weights are applied to the side next the spring. The apparatus is then balanced by means of the spring g, and the index of

this latter should indicate a force exactly corresponding to the difference of the weights employed.

Figs. 198 to 200 illustrate different modes of applying the Hefner Alteneck dynamometer.   In Fig. 198, the driving pulley A transmits motion to the pulley B by the two auxiliary pulleys C D, the belt passing

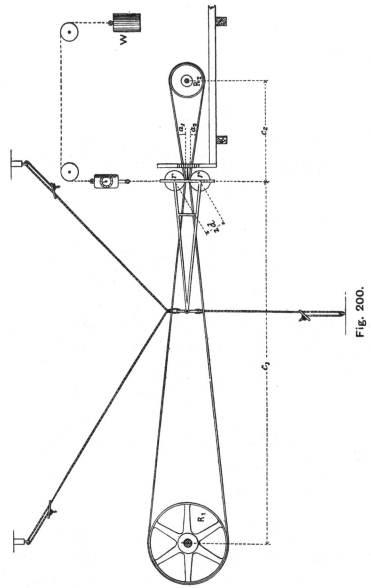

Fig. 200.

over which is kept stretched by the guide wheels *g*, kept always at the same distance apart in the frame on which they are mounted, which turns around *o*; this movable portion is balanced by the weight *p*, on the horizontal lever, to which a rod is connected, having a piston on its

end working in an oil cylinder, to deaden the vibration of the system. When the apparatus is at rest, the movable system is balanced by the weight p, and the equal tension of the belt ; as soon as the apparatus is set in motion, equilibrium is destroyed by the difference $T_1 - T_2$ in the tension of the driving and driven sides of the belt, and the value of these differences can be ascertained by the index i, which moves with the turning frame against a graduated scale. Figs. 199 and 200 show two modifications, the latter adopted by Dr. J. Hopkinson in a series of trials with dynamo-electric generators, and the former proposed by Mr. Elihu Thomson in the Journal of Franklin Institute.

King.—King's dynamometer is applied directly to the ends $c\ c^1$ (Fig. 201) of the driving strap, and acts as a coupling between them, transmitting

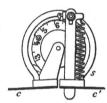

the strain from one to the other. As the dynamometer is alternately carried from the slack to the tight side of the belt the horseshoe shaped spring s opens and closes, and registers its movements by aid of a pawl on a graduated scale. Thus the work of the strap in a given time is totalised, but the results are not to be relied upon too closely.

Fig. 201.

Perhaps the best known of the transmission dynamometers are those in which the turning force is measured by the flexure of springs of various kinds by means of adjustable weights. Of these, the most familiar example in this country, is that of the Royal Agricultural Society.

Royal Agricultural Society's Dynamometer.—The arrangement of this apparatus is indicated by Figs. 202 to 206. The driving pulley A, put in motion by a belt, or by the clutch E, actuates the loose pulley, connected to the

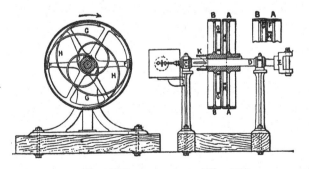
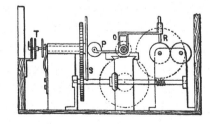

Fig. 202.   Fig. 203.   Fig. 204.

machine to be measured, by means of the springs G and H, which are so arranged that the angular displacement of A in relation to B, may be proportional to the power transmitted by B. The springs H and G are curved

in opposite directions, in such a way as to neutralise the effect of centrifugal force upon their blades. The registering apparatus, Figs. 205 and 206, consists of a cylinder R, on which the force curves are traced, of a totaliser *s*, and of a counter T; all these parts are controlled by the pinion V, driven by the shaft D. The pencil working on the cylinder R, and the wheel P of the totaliser, are fixed on the carriage O, and are actuated by the spindle N, which passes freely through the shaft C D, and bears at L on the spiral

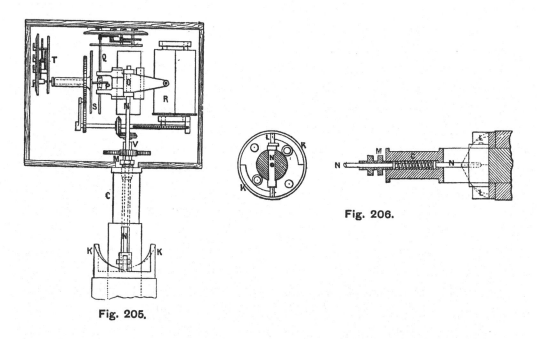

Fig. 205.

Fig. 206.

path K, fast on the boss of the pulley B; this arrangement insures that the movement of the pencil along the cylinder, and of the wheel P, are proportional to those of the pulley B on its shaft, or, in other words, to the variations in the power transmitted; the counter Q, controlled by the wheel P, indicates the amount of work done.

Morin.—The Morin dynamometer is remarkable for its extreme simplicity; the pulley D (Fig. 207), provided with a disc E, actuates the driven pulley B, by a wheel G, fixed to the disc; the arm *b'* is connected to the pulley B and the steel band *c*; this band compresses the coiled springs F, guided by rods fast on the disc E, to an extent proportional to the driving effort. The pallet *p* actuates a radial rack moved by the disc, and the movement of which is transmitted by a pinion, also carried by the disc, to a second cylindrical rack, movable on the axis of the shaft A. The movement of this rack is transmitted, either to a simple indicator, to the

pencil of a work recorder L, or to the wheel of a totaliser.   The friction of

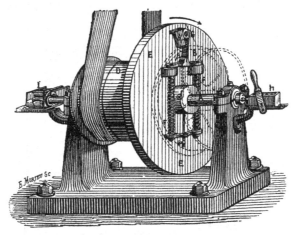

Fig. 207.

the instrument may be neglected, because the driving effort is transmitted almost direct from the driving pulley to the spring.

Bourry.—Figs. 208 and 209 show the Bourry dynamometer. The driving pulley A commands B, by the bent levers *e e'*, which move on the centres *o* and *o''*, and are jointed to the rods *b*, of the disc P.   This disc, free to slide on the shaft of B, compresses the springs R, to an extent which varies with the effort transmitted ; in sliding it draws with it the plate *t* of the totaliser *t'*, the rack of the dynamometer frame *d*, and the style *s* which traces the curve of work ; at the same time it can, through the levers *f g*, act on the steam distribution, and so regulate the speed of the engine.

A slightly modified form of this dynamometer, applicable as a coupling between two shafts, instead of for a belt transmission, is shown at Fig. 210. The shaft A drives the shaft B by means of a cross-piece H, and links on coupling rods *b* upon the elbow levers *l l'* ; these are pivotted on the axes *o* and *o'*, and articulated at *a a'* to the links *b b'* connected to the disc B. This disc, free to slide upon the shaft B, is connected by the springs R to the collar *m*, which is keyed upon B, and carries the axes *o, o'*.   The greater the resistance the more the springs are deflected in moving the disc P and its registering apparatus, which consists of a totalising disc *t*, a scale *t'*, a counter *c*, a dynamometer scale *d*, and a style *s* which describes the curve of effort.

Megy.—This dynamometer, manufactured by MM. Sautter, Lemonnier, and Co., of Paris, is designed to record the number of foot-pounds or kilo-grammetres transmitted by the steam engine to the machine to be tested

by a simple multiplication of a coefficient into the number registered by the counter.　As in the Morin dynamometer, there are two pulleys on the shaft

Fig. 208.

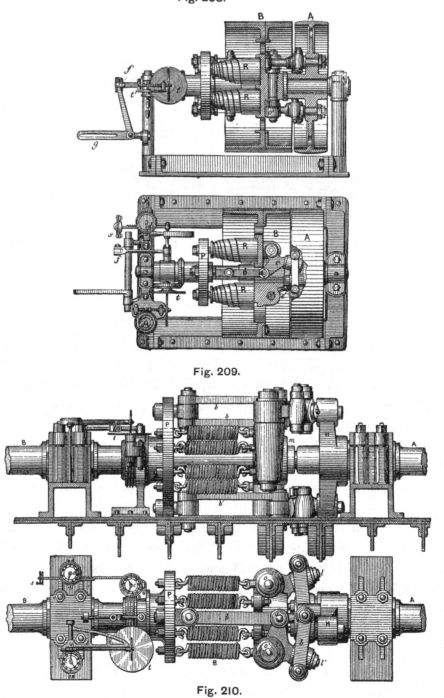

Fig. 209.

Fig. 210.

B; one of these, A, Fig. 211, is fast, and the other C loose on the shaft. Around

the latter are four projecting pins D D that can be brought into contact with four springs E fixed to a sleeve that is fast upon the shaft B. When the fast pulley A is driven by the motor, the shaft B, the sleeve F, and the springs E are also caused to revolve, and these springs, pressing against the pins placed around the wheel C, make this latter to turn, but the effort causes the springs to bend, and it is this bending that is made use of to record the work done. To effect this the following mechanism is introduced : around the sleeve F is placed a brass cylinder G, cut with a quick threaded screw, and receiving a brass cylinder, similarly threaded inside.

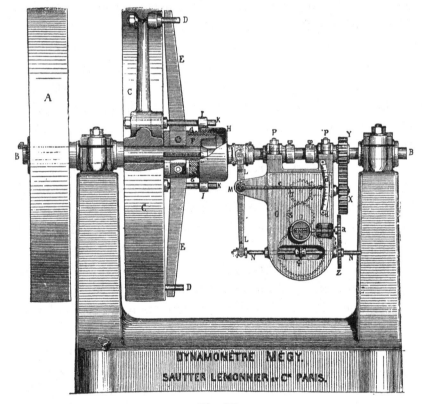

Fig. 211.

On this cylinder are the projections I I, through which pass the rods K K, fast on the boss of the loose pulley. The cylinder G being fixed to the sleeve carrying the lower ends of the springs, and the cylinder H to the loose pulley, it follows that any deformation of the springs will cause the cylinder H to travel on the screw thread G to an extent proportional to the movement of the springs, and, therefore, to the force transmitted. As will be seen from the engraving, the cylindrical nut H terminates in an extension which is passed over the shaft B, and is grooved outside to

receive the end of a forked lever L, centred at M on a pin in the frame of the registering mechanism, which is hung free upon the shaft B. The lower end of the lever L is articulated to the rod N, on the middle of which the disc Q is fast, the periphery resting in contact with the disc R as shown. The toothed wheel Y, fast on the shaft B, and the gearing X S T, transmit motion to the disc R, the speed of which, of course, depends on that of the shaft, which is recorded by a counter through the wheels *a z*. As soon as a traversing motion of the cylindrical nut H takes place, the lever L is set in motion, to an extent recorded by the index and graduated scale *c*, fast on the turning pin of the lever. At the same time the movement of the lever L shifts the disc Q from its normal position to a point more or less removed from the centre of the disc R, according to the force applied to the fast pulley A. If, now, the disc R is rotating it will, by friction, move the disc Q, and consequently the lever N, with a speed proportional to the force. But, through the agency of the pinion Y fast on the shaft B, and the gearing S T X, the disc R is revolved at a speed proportional to that of the shaft, and from this it follows that the speed of the disc Q, which varies according to its distance from the centre, which is due to the movement of the lever L, is proportional to the effort transmitted and the distance traversed, that is to say, to the work. The number of revolutions of the disc Q are recorded on the counter *b* by the gearing *a z*. The reading of this counter multiplied by a coefficient gives the work transmitted. In the machines made by MM. Sautter, Lemonnier, and Co., this coefficient is marked on the springs, as well as another coefficient, which, multiplied by the number indicated at any time by the deflection of the index *c*, gives the effort exerted at that moment in kilogrammes. This dynamometer, which appears to work well and accurately for high speeds, has done good service in recording the power absorbed by dynamo-electric generators driven at a speed of 1500 or 1600 revolutions per minute.

Hamilton Ruddick.—Ruddick's dynamometer is one of those which are fixed directly on the shaft, taking the place of the pulley that would otherwise be employed to transmit the power. It consists of a pulley C (Figs. 212 and 213), a disc capable of turning about its boss, and of two arms A and B, fixed, the one to the dynamometer pulley and the other to the shaft H, and connected to the springs D, which are deflected proportionally to the power transmitted. The extremity of the arm B, which is the one keyed to the shaft, is connected at G to an index, coupled by a drag link to the dynamometer pulley. The opposite extremity of the index carries a pencil,

which approaches more nearly to the centre of the disc H, as the power exerted is greater. This disc has a motion of rotation imparted to it relatively to the rest of the apparatus, by ratchet mechanism, put in operation by

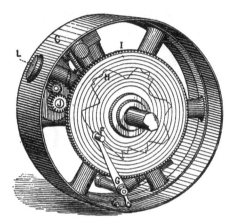

Fig. 212.

Fig. 213.

a button L, which is forced inwards by the shaft at each revolution of the pulley. The circles traced on the disc are each marked with two numbers, one indicating the turning force, and the other the corresponding horse-power.

Valet.—In the dynamometer, illustrated in Figs. 214 to 217, which is the invention of M. Valet, the disc D, keyed upon the shaft E, drives, by means of the springs F and snugs *a*, the pulley A, the work of which is to be measured. The flexure of the springs is transmitted by I *i* J K, to the roller L of a

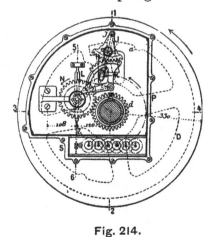

Fig. 214.

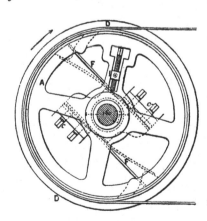

Fig. 215.

totaliser, whose disc N, held against the roller by the spring *f*, gears with the wheel *d*, mounted loosely upon the shaft, and held from rotating by the support G. The roller L is connected to the spindle of the counter S, and the whole of the recording apparatus is enclosed within a box P to protect

it from dust. The position of the axis I can be adjusted by a screw; the

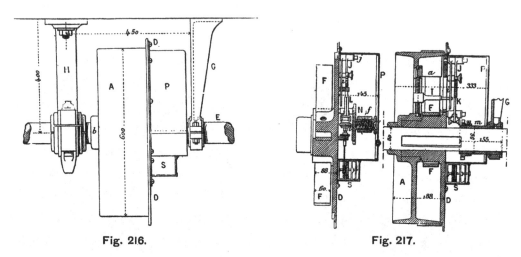

Fig. 216.　　　　　　　　　　　　　　　Fig. 217.

further it is from the centre the greater the motion imparted to the roller L by the flexure of the springs.

Taurines.—The dynamometer of M. Taurines, which has been adopted in the French Navy, is remarkable for its originality of action, and for the ingenuity of its construction. The driving shaft A, Figs. 218 to 220, is coupled to the driven shaft A′, by means of a double arm B, keyed on the shaft A, an intermediate cross-piece C′, loose upon A′, and the dynamometrical springs D, connecting C′ to the double arm B′, keyed upon A′. The double arm B drives the cross-piece C′ by the studs *b b*, the links C and the studs *c*, into which are fitted the springs D; these springs act upon the studs *b′*, which are free in the holes *b″* of the cross-piece, but are fixed without play in B′. The movements of the springs D are increased by the auxiliary springs F, and the indicating springs H, connected to the rod *h* of the registering apparatus. The bearing *c′* serves to isolate the dynamometer by locking the studs *b′* in the holes *b″* of the cross-piece C′, in such a way that C′ is directly commanded by B′.

It will be noticed that, in this dynamometer, the springs operate in a longitudinal direction, and their movements are angular deflections, or variations of the angle formed by the tangents drawn to their extremities. The deformation of springs submitted to this kind of strain are exactly proportional to the force they transmit, which force may be very considerable without injury to their elasticity. This method of employing springs and amplifying their movements was first adopted by M. Taurines, and reflects great credit upon the inventor.

Fig. 218.

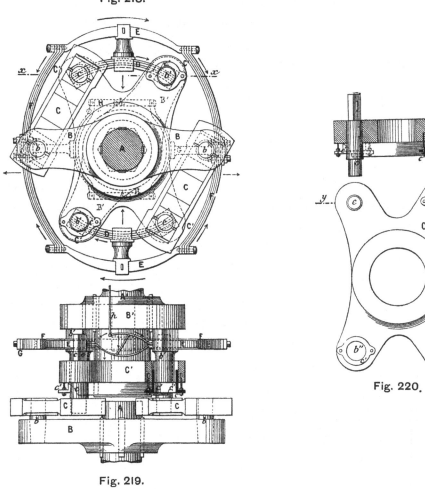

Fig. 220.

Fig. 219.

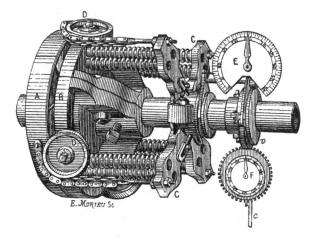

Fig. 221.

Neer.—Neer's dynamometer, like that of Taurines', is fixed on the shaft, the work of which is to be estimated, in place of a coupling. The motive power is transmitted from the disc A (Fig. 221) to the disc B, by chains which pass round the pulleys D, and act upon the helical springs attached to the cross C; these springs are compressed according to the amount of force that is being exerted, and their movements are exhibited upon the indicating dial E. The two dials E and F are mounted upon a collar loose upon a shaft, and are prevented from revolving by the finger *c*. The endless screw *v* operates the index of the counter F.

Ayrton and Perry.—The dynamometer designed by Professors Ayrton and Perry for use in the Guilds Laboratory is shown on Figs. 222 to 225 B B' is a loose pulley used to drive by means of a strap, any machine, such as a dynamo-electric machine. F F" is a boss keyed on the shaft driven by the motor. This boss has four stout arms, Fig. 223, which are attached by four spiral springs to the rim of the loose pulley. When power is being transmitted through the springs they elongate, and their motion is indicated by the following means : H C is an arm turning about a pivot C on the rim

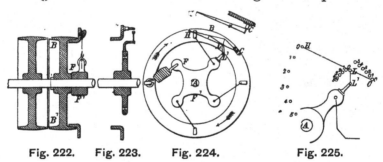

Fig. 222.   Fig. 223.   Fig. 224.   Fig. 225.

of the loose pulley and moved by a link L L', attached at L by a pivot to the arm at L" and at the other end to one spoke of the keyed boss F F'. If the loose pulley and the keyed boss receive an angular twist, the end H of the arm, to which a bright bead is attached, moves almost radially towards the shaft. The exact path of the bead H is shown in Fig. 225 for different positions of C. To measure the power transmitted by the engine to the machine, all that is necessary is to observe on a scale placed in a suitable position the radius of the circle traced out by the bright bead H, and the number of rotations per minute.

Latchinoff.—In the dynamometer of M. Latchinoff, as in that of Messrs. Perry and Ayrton, the driving pulley A (Fig. 226) actuates the pulley B by the tension of the springs C, and the effort transmitted is measured by the amount that the line *n*, upon the pulley B, diverges from zero of the scale

marked upon A. This divergence can be measured, owing to the phenomenon of persistence of vision, by regarding the line n through the slit h, that is, provided the speed of the pulleys is sufficiently rapid.

Darwin.—In this apparatus the springs are replaced by the action of a weight W (Fig. 227) articulated to the parallel motion B G O O′, of which O O′ are the fixed points. The driving pulley N controls the loose pulley R by a chain passing from M over the pulley K, the spindle of which L, carried by N, has a second pulley J connected by a pitched chain, which passes

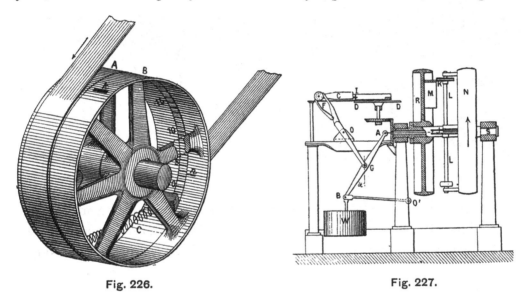

Fig. 226.　　　　　　　　　　　　　**Fig. 227.**

through the shaft S to the end A of the lower lever B G. If B C = G A = G O, the effort p exerted at A, is given by the formula

$$p = W \tan a.$$

As the displacements of the point F, controlling the totaliser C I D are also proportionate to tan a, it follows that the number of revolutions of the wheel I, shown by the counter, is proportional to the work transmitted to the pulley R.

Matter.—The apparatus of M. Matter, Figs. 228 to 232, is another example of dynamometers of transmission. The power is transmitted from the pulley C, keyed upon A, to the pulley D, loose upon A, by a series of springs E, which bend around guides F, so designed that the angular displacements of the pulley D may be proportional to the power which it receives from the strap whose work it is desired to calculate. The displacement is transmitted by I I¹ H G to the platform K, the pencil of which L traces upon the drum M the corresponding curve of power ; the drum being driven by means clearly shown on the figure, at a speed

proportional to that of the shaft A. The same means causes a paddle wheel T to rotate in a movable vessel U, which is full of water, and tends to move the vessel in spite of the antagonism of the spring A^1, as the speed

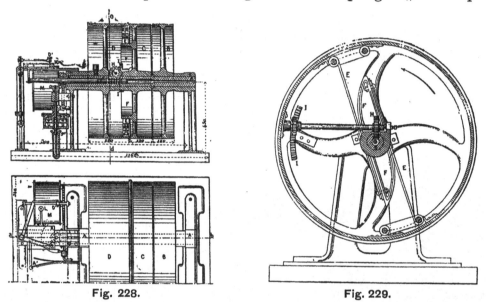

Fig. 228.   Fig. 229.

of the dynamometer increases. The flexure of the spring, suitably magnified by the styles D^1 and C^1, traces upon the cylinder M the curve of speeds, and upon the diagram K a series of arcs, of which the radii, measured from B, vary

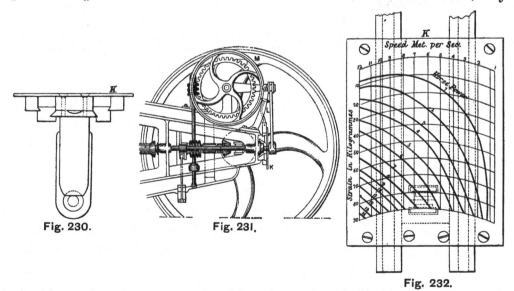

Fig. 230.   Fig. 231.   Fig. 232.

inversely as the power transmitted by the dynamometer, and turn through angles increasing with the speed. Fig. 232 represents the graduation of the diagram, the arcs represent curves of equal effort, the ordinates are

proportional to the speed, the curves in thick lines are isodynamical, or curves of equal powers, that is to say, the loci of points for which the products of the effort multiplied by the speed are constant. The outline of the curve traced by C^1 upon this system allows of all the variations of the work being easily followed. This dynamometer of M. Matter has given excellent results in the establishments of MM. Dolfus Mieg, of Mulhouse.

Emerson.—The apparatus of Mr. Emerson, Figs. 233 and 234, is much used in the United States, and is founded upon the same principle as his portable dynamometer described above. The pulley A, whose work is to be measured, is driven from the shaft by means of the levers O, pivotted at N to its rim, at M to the arms J, and at their other extremities to a bell-crank which is connected to the motor shaft by the two-armed coupling J; this coupling lever is pivotted at *z*. One of the arms of this bell-crank, by means of the rods G, commands a collar C, which is coupled to the fork of the dynamometer lever G G¹, whose other end is connected to the pulley D of a quadrant arm suspended by the rod V. It results from the play of the

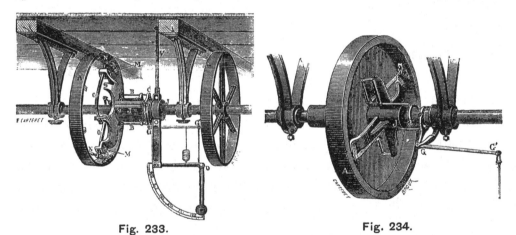

Fig. 233.   Fig. 234.

levers, which can be followed more clearly on Fig. 234, that the position of the index indicates at each instant the intensity of the power transmitted to the pulley A. The dash-pot G G moderates the oscillations of the pendulum, and the rods N allow the dynamometer to be connected to pulleys of different diameter without any change in the arms of the lever.

Smith.—In the apparatus (Fig. 235) to which the inventor, Mr. F. J. Smith, has given the name of Ergometer,* the motor pulley drives through the intermediary of pinions M N K, the dynamometer pulley G, keyed upon the shaft B, and connected to the machine whose resistance is to be estimated.

* Phil. Mag., February, 1883.

The bosses of the pinions M and N are connected to straps E, and the cross-head P to the dynamometer spring A F, situated within the spindle B, in such a way as not to be influenced by the centrifugal force. The tensions are indicated by the index Q, and the revolutions by the counter R. Mr. Smith has also proposed to measure the power expended by means of the water passed through a pump D (Fig. 236), worked by a slide oscillating about the point B, the latter being driven by a connecting rod C and an eccentric upon the dynamometer. The block is slidden towards the point A by an amount proportional to the stress on the dynamometer spring, to which it is connected by the rod F.

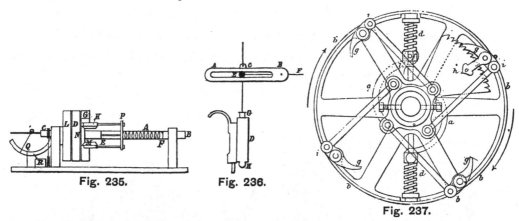

Fig. 235.   Fig. 236.

Fig. 237.

In the dynamometer represented in Fig. 237, the pulley *b* drives the shaft, whose work it is desired to measure, through the double cam *a* keyed on the shaft, and impelled by the pressure of rollers pressed upon it by springs *d*. The cam carries a quadrant *k*, with pawls *g*, which prevent the pulley from recoiling after a stoppage, and a graduated arc upon which a finger *o* indicates the work. To disengage the pawls *g*, it is only necessary to slacken slightly the strap upon their interior ends *i*.

Marcel Deprez's Dynamo-Electric Brake.— This dynamometer is especially adapted for measuring the work absorbed by dynamo-electric machines, particularly those of the Gramme type. The principle of the arrangement will be understood from Fig. 238: It consists in suspending the Gramme machine A by means of two strong brackets mounted on the knife edge F F, the lower edges of which are in the axis of the ring; the bearings I of the shaft of the ring are removed, and replaced by the bearings D independent of the frame, from which the brushes are also separated, so that (once balanced by a weight placed at the end of the lever K) it has a tendency to turn upon the knife edges, only by the reaction

of the magnetic field developed by the inductors in the ring; the field acts on the ring like the jaws of an invisible brake. If we call P the weights added to the lever k to maintain the long indicating needle l at zero, and d the distance of the point of suspension of these weights from the axis of the ring, the work T developed per revolution of the ring is given by the

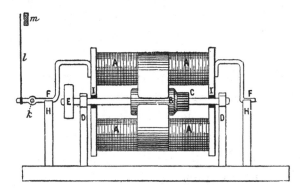

Fig. 238.

expression $T = 2 \pi P d$. By means of this arrangement M. Deprez has tested the accuracy of the law expressed by the formula

$$T \frac{E I}{\omega g}$$

in which E equals the electromotive force of the current produced by the machine, I the intensity of the current, ω the angular velocity of the ring, and g the effort of gravity.

A TESTING INSTALLATION.

WE have in the three foregoing sections endeavoured to investigate with sufficient completeness for practical purposes, the various methods by which energy, in the three forms with which this volume has to do, can be measured, and to explain the principles on which the methods we have noticed are based. Before concluding, we propose to illustrate our subject by describing a very complete installation for taking electrical and mechanical measurements, recently completed by Mr. C. F. Heinrichs. The plant was laid down for the purpose of an exhaustive inquiry into the efficiency of the Heinrichs dynamo machine,* and the experiments were by Messrs. Alabaster, Gatehouse, and Co.

The investigation extended over a period of more than five weeks, during the first three of which, numerous tests were taken for the purpose of verifying the first results obtained, and for calibrating the instruments. Several apparatus, ordinarily reputed efficient, had to be superseded by others better adapted to the circumstances and conditions under which the tests were made.

We believe that the following condensed description of the tests, and of Mr. Heinrichs' experiments, will be found valuable as trustworthy examples of the methods in which trials of dynamo-electric machines should be conducted.

In the illustration, Fig. 239, we give a general perspective view of the entire plant employed in these investigations. In the large room there were fixed the engine, dynamo machine D, Morin dynamometer, lamps, &c., while in the smaller room was the electrical testing apparatus, consisting of the following instruments, arranged on the table marked T 1 : a Siemens electro-dynamometer C_1 for current measurements, an Ayrton and Perry voltmeter E_1 for measuring the difference of potential at the terminals of the machine, while, for the measurement of the resistance of the circuit a Wheatstone bridge, placed on the same table, could be connected up at any moment.

---

* For a description of this machine, see " Electric Illumination," Vol. I., p. 234.

Fig. 239. Heinrichs' Electrical Testing Plant.

To adjoin page 243.

On the table marked T 2 in the large room, several coils of thick insulated copper wire R were connected up to a large compound switch S by which they could in succession be thrown into the circuit as required. Two lamp circuits were so arranged that they could be used either independently or together, with or without additional resistance. A second Siemens electro-dynamometer C_2 was placed upon this table, and included in the circuit to act as a check upon the other, or in the event of the first breaking down. Upon the table T 3, were placed an Ayrton and Perry's voltmeter, a key K_2, and a small switch S; these instruments were connected with one of the eight straight carbon lamps, and with one of Mr. Heinrichs' circular carbon lamps, so that the E.M.F. of each lamp could readily be noted during the experiments. The whole test plant and the circuits were permanently arranged in such a manner as to allow all readings to be taken simultaneously. We shall refer later on to these apparatus, and to the methods by which the readings were verified and compared with each other.

A Morin transmission dynamometer was employed for measuring the mechanical power absorbed by the machine, and, in order to prevent unnecessary and possibly erroneous calculation of the friction of intermediate shafts, the dynamometer M was placed next to the dynamo machine.

The Morin Dynamometer.—This apparatus is composed of two distinct parts, the one for receiving and transmitting the force to be measured, and the other for indicating its value. The first part consists of a cast-iron plate E (Fig. 207, page 229 *ante*), which is keyed on the dynamometer shaft, and has mounted on one of its faces a pair of spiral springs F F, placed symmetrically at either side of the shaft and parallel to each other. These springs are placed upon two rods, connected by two crossbars. The upper crossbar is fixed to a steel band p which passes over a pulley g, and is attached to the arm k (partly shown in the drawing). The arm k is fixed to the pulley B (shown in dotted lines). The upper crossbar is further fixed to a rack which is, through a pinion, in gear with a rod moving freely in the centre of the shaft, and extending at one end from the shaft. On power being applied to the pulley D, the force exerted will be transmitted through the steel band p to the pulley B, at the same time compressing the springs F F to an extent corresponding to the tractive force applied through the driving belt and the resistance opposed by the machine to be driven. The compression of the springs represents the value of the force exerted; and since the upper crossbar is geared to the rod in the centre of the shaft, the

amount of compression of the springs will be indicated by the rod being drawn into the shaft as the springs become compressed. The rod terminates in a flat head, bevelled to an acute angle at its circumference, which serves as an index to a graduated scale fixed above it, and upon which the variation in the pressure of the spiral springs can be read off.

To the shaft of this dynamometer, and revolving with it, an electrical contact breaker was fixed, by which one of Mr. Kempe's electrical revolution counters was actuated; this instrument is shown on the table near the electro-dynamometer. Repeated tests were taken of the dynamo, and the horse-power absorbed was calculated from the indications of the Morin apparatus. During these tests it appeared that the dynamo gave varying percentages of efficiency, as regards the power converted into current. This was against Mr. Heinrichs' expectation and calculation, since, whatever the commercial efficiency might prove to be, he had deduced from his own theory of the action of a ring armature, that his machine must give at all numbers of revolutions (within usual limits) *approximately the same percentage of efficiency in converting mechanical into electrical energy*, whilst machines of the Gramme principle could only show the highest efficiency in that respect at the highest speed when the armature acquires but little magnetism.

The Morin dynamometer was then subjected to careful examinations and tests, during which it appeared that the centrifugal force of its moving parts increased considerably with an increase of speed. It was then decided to take dynamic tests with a Prony brake, and to compare the readings with each other. Great difficulties had to be overcome before the Prony brake could be relied upon, but it was eventually proved that the Morin dynamometer gave, with high speeds, figures below the truth, and at low speeds, figures above the truth. This led to the abandonment of all direct calculations of horse-power from the indications of the Morin apparatus. These indications were observed and recorded during the electrical tests, and afterwards a Prony brake was put on the dynamo, and the readings previously obtained on the dynamometer, when the power of the machine was converted into electrical energy, were reproduced when the power was being opposed by the friction of the Prony brake, which could be exactly measured.

In this arrangement it matters little what kind of transmission dynamometer is employed, since, whatever disturbing influence the centrifugal force of the moving parts of the dynamometer has upon the indication during the electrical tests, the same will take place whenever the Prony brake absorbs the same power, at a given speed, as the electrical work did.

The Prony Brake.—Notwithstanding the great simplicity of the Prony brake, it was found that considerable difficulties attend its use in inexperienced hands, and that little assistance is to be gained from books, as few data as to the proper size and conditions of the brake for varying values have been published. Fig. 240 is an illustration of the ordinary pattern of this well-known brake. From experiments made by M. Poncelet in France it was found that under some circumstances the long lever may itself tighten the grip of the jaws. In order to remove this inconvenience, which sometimes produces false results, M. Poncelet made the two levers of equal length, and placed one screw only near the point A, Fig. 241. The flexibility of the wood allows gradual tightening of the jaws, which con-

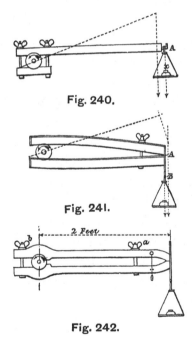

Fig. 240.

Fig. 241.

Fig. 242.

tributes greatly to the maintenance of the uniform friction, without which no measurement can be made. In order to maintain a stable equilibrium of the brake lever, M. Poncelet did not place the weight at the point A, as in Fig. 240, but at the end of a vertical bar B, fixed to the one jaw. Thus whenever the lever of the construction shown in Fig. 240 rises, the effective weight becomes diminished, but in Fig. 241 the arm through which the weight acts, increases slightly, as can be seen from the dotted lines. During the application of the brake it happens sometimes that the lever rises through some trifling irregularity, and if the weight be diminished, the difficulty increases; but when, as in Fig. 241, the weight is increased, it

will overcome the momentary irregularity, and bring the lever to its position of equilibrium. For the purpose of this investigation Mr. Heinrichs decided to employ a similar brake, and adopted the form we show in Fig. 242.

The second bolt is here retained to prevent the levers bending outwards. A small plummet is arranged near the end of the brake lever to indicate when it is in the level position, both during the experiment and during the subsequent process of weighing the lever and scale.

The friction of the brake during the test must be perfectly uniform, so that the equilibrium of the levers can be maintained for some considerable time, to allow the weights to be properly adjusted in the scale. Although the construction of the lever is of great importance, the proportion of the brake, as will be shown, is still more important.

At first a brake was employed 2 feet long and $1\frac{1}{2}$ inch wide, the brake pulley being 4 inches in diameter. It was used during a test when the pulley made 1615 revolutions per minute. The total weight on the scale was 4.11 pounds, and the power measured was 2.5 horse-power. During this test the friction of the brake was uniform, and a perfect balance of the lever was obtained, the lever being sensitive to $\frac{1}{2}$ an ounce. But in a subsequent experiment, in using the same brake for a measurement of 6 horse-power, the apparatus failed, the friction becoming so irregular that the belts were thrown off. At first it was thought that the small tank w_2 did not contain a sufficient quantity of soap and water, and that the irregularity in the friction was due to the soap and water becoming warm. A second and larger tank was then connected with the small one, but without affecting the result. A larger brake was then made, being 2 feet long by $2\frac{1}{2}$ inches wide, with a brake pulley of 6 inches diameter, and this brake was used during the following tests :

Number of Test.	Number of Revolutions of Brake Pulley.	Weight on the Scale in Pounds.	Horse Power Measured.
1	960	15.75	5.69
5	1680	12.1	7.74
6	1680	10.4	6.66

During the tests 5 and 6 the brake acted admirably, the lever responding freely when one ounce was added or withdrawn from the scale.

During the test No. 1 the friction appeared slightly irregular, showing that the limit of the brake had been reached; still a reliable test was obtained.

On repeating the first-mentioned test of 2.5 horse-power with the large brake it was found that the brake acted indifferently, and it was not sensitive at all, half a pound of large shot could be added or withdrawn from the scale, but the lever did not respond, and it became clear that the frictional surface of the brake was too large for so small a measurement.

It follows, evidently, from what has been stated, that a certain size of brake can only be used within a limited range.

The following dimensions of brakes are deduced from the experiments :

To Measure	Number of Revolutions of Brake Pulley.	Size of Brake.	Diameter of Brake Pulley.
From 2 to 5 horse-power	1200 to 1800	2 ft. long, 1½ in. wide	4 inches
	700 to 1200	2 ft. long, 2½ in. wide	6 inches
From 5 to 8 horse-power	1200 to 1600	2 ft. long, 2½ in. wide	6 inches
	800 to 1200	2 ft. long, 4 in. wide	6 inches

The following remarks will probably be found useful in making measurements with this apparatus, for they explain the conditions to be observed in order to obtain reliable results :

If the frictional surface of the brake be too small, the friction will be found very irregular, and sudden and powerful shocks will be noticed, which are produced by the wooden jaws seizing the brake pulley for want of proper lubrication.

But if, on the other hand, the friction surface of the brake be too large, the brake will act indifferently, a considerable weight may be added to, or withdrawn from, the scale without the position of the lever being affected. The brake was found to be most sensitive whenever the pressure per unit of surface was such that a firm grip was obtained of the pulley, and yet the rubbing surfaces were kept lubricated. If this pressure were exceeded, the lubrication was stopped, the brake pulley became hot, and expanded, and the wood swelled, consequently the grip became irregular, and sometimes rigid, stopping the rotation instantly. If the pressure be so small as to allow jarring or vibration to be established

between the frictional surfaces, then the instrument loses its sensitiveness, and there is great liability to get the most untrue results. In making a series of tests the most convenient plan will be to take a very wide brake pulley, and several brake levers of different widths. At the commencement, rather a narrow brake should be used, and if the friction be found irregular, then a broader one, until a very substantial friction grip be obtained, and yet without any symptoms of seizing. The most favourable pressure is characterised by extreme sensitiveness, so that if one ounce be added to, or withdrawn from the scale, the lever will respond freely. A very great quantity of cold soap water must be on hand, and must continually be applied to the brake pulley, which must be kept perfectly cool.

The horse-power absorbed by the brake is found by the following well-known formula : n being the number of revolutions of the brake pulley per minute, l the length of the lever in feet, and w the effective weight in pounds :

$$\text{HP.} = \frac{\pi \times 2\,l \times w \times n}{33,000} = \frac{l\,w\,n}{5252}$$

omitting decimals.

We shall now describe the apparatus employed for counting the exact number of revolutions of the dynamo. As before stated, Mr. Kempe's revolution counter was connected up with a revolving contact breaker, which was fixed on the shaft of the Morin dynamometer. The counter was fixed upon table T 2 and a key interposed in the circuit. On depressing the key for one minute the exact number of revolutions of the Morin dynamometer was read off the counter, and since it had appeared from tests made with mechanical counters, that for one revolution of the Morin dynamometer the dynamo made exactly two, thus from one reading of the counter the number of revolutions of the machine and those of the dynamometer were known at the same time. Mr. Kempe's electrical counter is no doubt a most convenient instrument, for it can be placed far away from the machinery and the most exact reading be taken. The very ingenious construction allows all the pointers to be put at zero after each test, and so obviates unnecessary calculations.

Kempe and Ferguson's Electrical Revolution Counter.—The metal frame F F F F (Figs. 243 to 245), in which is mounted the whole mechanism, is arranged within a wooden casing. In the frame F F are fitted seven pinions a, b, c, d, e, f, g ; to each of the pinions b, c, d, e, f, g is fitted a small cam o ; and to each of the pinions c, d, e, f, g is fitted a multiplying wheel having ten times as many teeth as the pinions. These multiplying wheels do not gear

into the pinions as in other counters of similar construction, but are free of each other. Upon the frame F F is fixed, free to move, a second frame A A, which carries five wheels h, i, k, l, m ; these five wheels are of the same size as the multiplying wheels, and are so arranged as. to connect the five

Fig. 243.

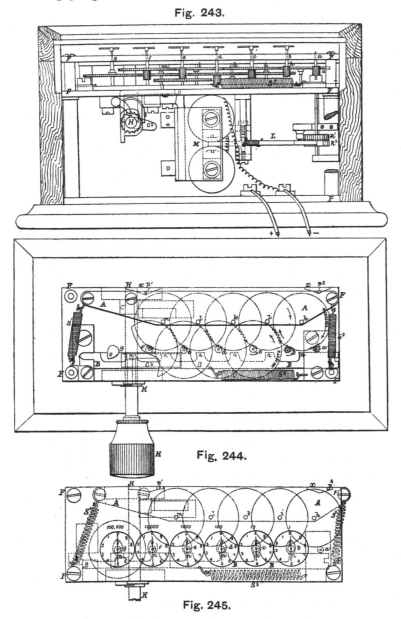

Fig. 244.

Fig. 245.

multiplying wheels and the seven pinions with each other, and whenever the instrument is in operation, and the magnet M attracts and releases its armature, the escapement lever L actuates the ratchet wheel R, and so sets the pinions a in motion ; the counter will thus act as any ordinary electrical

counter. We shall now describe the very ingenious mechanism by which the pinions b, c, d, e, f, g, with their pointers, are placed at zero.

On turning the shaft H by the handle the çams C^1 and C^2 fixed to the shaft, are brought into action. The cam C^1 pushes the frame A A to the right, causing it to recede from the main set of pinions, since the two pins p' and p^2 are fitted into the frame A A, and slide in the V formed recess x x in the frame F F. Thus the main set of pinions and multiplying wheels are disengaged and become free. On further turning the handle, the second cam C^2 becomes engaged in the bar B B. This bar has seven recesses in which, when at rest, the small cams o, o, o, o, o, o can turn unhindered; but, whenever this bar is pushed by the cam C^2 to the left, the small cams o, o will be forced to turn, so that their flat parts slide upon the bar B B, and thus place the pinions and the pointer at zero, as shown in Fig. 245. On further turning the handle H the cam C^1 disengages itself from the frame A A, and the bar B B is pulled back by the spiral springs S^1 S^2; the five wheels then engage again into the pinions and five multiplying wheels, which are still held by the bar B B. But as soon as the whole wheels are again in gear, the cam C^2 releases the bar B B, which is then pulled back by the spiral spring S^3, thus leaving the small cams o, o, o, o, o, o free, and the counter is ready for a new experiment.

This very interesting and original instrument has been of great service during the tests before us, and whenever a perfectly steady working engine is used during an experiment, no further and more reliable counter is required.

The engine employed by Mr. Heinrichs in these tests was a portable eight-horse power engine by Messrs. Marshall, of Gainsborough. The speed could not be always maintained so steady as was desired, particularly during the measurement with the Prony brake, when the weight on the brake, the number of revolutions, and the indications on the Morin dynamometer, had to be observed at the same instant of time. Too much time was wasted in re-counting the number of revolutions, and it became evident that an instantaneous speed indicator was required. A Harding's speed indicator was then selected, fixed upon table T 2, and connected to the machine by a belt as shown in Fig. 239.

Harding's Speed Indicator.—This instrument consists mainly of two fans arranged independently of each other in a casing. One of the fans is rotated by the machine, of which the speed is to be indicated. The rotation of this fan produces a current of air, which tends to rotate the

second fan, that is attached to the second pointer spindle, but its rotation is counteracted by the action of a spring.

Fig. 247 of the accompanying drawings is a vertical section of the apparatus, in which the shaft 1 driven through the pulley 2 carries the first fan 3. By the side of this, but not touching it, the second similar fan 4 is carried by a small steel shaft, sensitively mounted, one end of which projects through the front of the case, where it carries the needle or index 6.

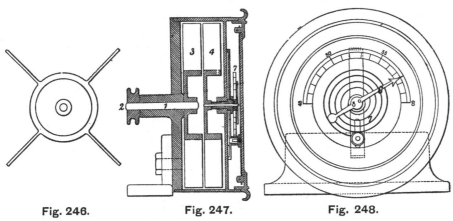

Fig. 246.   Fig. 247.   Fig. 248.

As this index is kept back to the zero point on the dial by the action of the spring 7 (best seen in the front view, Fig. 248) the greater or less speed imparted to the first fan 3 overcomes more or less the resistance of the spring 7 which retains the second fan 4, and the position of the needle 6 on the dial shows the increase of the speed.

This instrument seems to answer very well the usual demands, but it is somewhat sluggish in its backward indications.

A Young's speed indicator was then tried, of which favourable reports had been given. The apparatus is used as an ordinary mechanical speed counter, though it shows instantaneously the number of revolutions of the shaft.

Young's Speed Indicator.—It consists of a small high-speed centrifugal governor, so constructed as to prevent a shock to the apparatus by it being suddenly brought into contact with a shaft rotating at a high speed. This centrifugal governor acts upon a pointer on the top of the apparatus, and thus indicates the number of revolutions.

In Figs. 249, 250, and 251, A, A¹, are the governor balls which are provided with arms, and are connected by means of screws to the links *b b* ; these are also coupled in a similar manner to the sleeve or sliding bush B, the arms being coupled to the flywheel P in the manner shown. The flywheel P

is fitted upon the spindle D in such a manner that when the said spindle is rotated, the flywheel will be carried round with it by friction as soon as the inertia of the flywheel is overcome. The spindle D passes through the casing to receive the carrier L^1. The pinion l^2 is fixed upon the spindle D, and gears with the spurwheel m fixed on the spindle M, which passes through the casing to receive the said carrier.

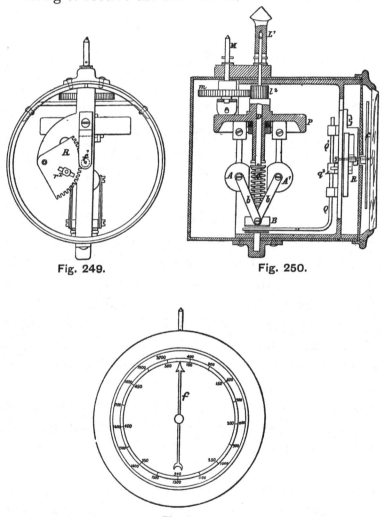

Fig. 249.   Fig. 250.

Fig. 251.

The sleeve or sliding bush B is constructed with a circular groove, into which fit two arms of the rod Q, the said rod being carried in lugs on the frame Q^1, and being free to slide therein. R is a toothed segment pivotted upon the frame Q^1, and provided with a slot through which passes the pin r^2, the said pin being capable of adjustment in the said slot, and being secured by means of a nut upon one end of it. The pin r^2 at its other end fits into a

slotted arm q^2 fixed on the rod Q. By means of this arrangement, motion is imparted by the sleeve or sliding bush B to the segment R, which actuates the index or pointer f by means of the pinion f^1; this gears with the toothed segment R, and is fixed on the spindle of the index.

This instrument operates as follows : The carrier being placed upon either of the spindles D or M is thrust against the centre of one end of the shaft whose speed is to be indicated, and the governor is thereby caused to rotate owing to the friction between the spindle D and flywheel P ; the balls A, A¹, flying outwards by reason of the centrifugal force, draw the sleeve or sliding bush B towards the flywheel P, thereby compressing the spiral spring O, and at the same time acting upon the rod Q to cause it to move in the bearings ; by this movement the slotted arm q^2 is caused to act upon the pin r^2, and thus turn the toothed segment R upon its pivot ; the toothed segment acts upon the pinion f^1, and thereby moves the index or pointer f, so that it indicates upon the outer figures of the dial the number of revolutions per minute of the governor, and (if the carrier is upon the spindle D, as shown) of the shaft against which such carrier is thrust. If the carrier is placed upon the spindle M, the speed of the shaft against which it is thrust will be indicated upon the inner figures of the dial.

This instrument has been of the greatest value during the tests with the Prony brake. It showed most correctly the right number of revolutions, as was proved during repeated check tests of the Young's speed indicator, Kemp's electrical counter, and a mechanical counter.

Having now explained all the important apparatus which were used during these extremely interesting tests, we shall now describe the experiments themselves, and manner in which the tests were made.

The Tests.—Preliminary tests were taken on each occasion, and the results compared with each other. Calculations of the electromotive force from the current, multiplied by the resistance ; calculations of the resistance from the electromotive force divided by the current, and of the current from the electromotive force divided by the resistance, were made, and when all calculations closely corresponded with the results, read off the instruments, then the final tests were taken off. The voltmeters were several times compared by Messrs. Alabaster, Gatehouse, and Co. with a standard of electromotive force of known value. The results of the final tests were also compared with each other, and only those readings were accepted of which the cross calculations showed them to be correct, and of such tests only were power measurements taken. All

readings of the several instruments were taken simultaneously by Mr. Heinrichs' staff, and by the staff of Messrs. Alabaster, Gatehouse, and Co. conjointly, thus insuring the correctness of the results. Our large illustration (Fig. 239) shows the whole staff at work, taking power measurements. The observer on the farther side of the dynamo, takes with the Young's speed indicator the number of revolutions of the generator, whilst the observer at the Morin dynamometer reads its indications and calls them out, enabling the operator at the Prony brake to adjust the weight on the scale.

Whenever the correct number of revolutions of the dynamo and the indication of the Morin dynamometer were reproduced from the previous electrical test, and the lever of the Prony brake swung in the centre between the two stops, *i.e.*, in the level position, then the same power was absorbed by the Prony brake as was absorbed in the production of the current in the previous experiments. Having adjusted the weights in the scale, the speed of the engine was first reduced, and then again increased very gradually until the required speed was attained. At this point the reading of the Morin

TABLES OF TESTS OF HEINRICHS' DYNAMO-ELECTRIC MACHINE.
(SERIES WOUND).
A.—VALUES MEASURED.

Number of Test.	E.M.F. at Terminals of Dynamo.	Current in Circuit in Ampères.	RESISTANCE IN OHMS. (WARM).			PRONY BRAKE.			MORIN DYNAMOMETER.		
			Dynamo.	External Circuit.	Total.	Number of Revolutions of Brake Pulley.	Weight on Scale in Pounds.	Length of Lever in Feet.	Indication when Loaded.	Indication when Driving Dynamo Empty.	Indication produced by the Centrifugal Force.
No.	e.	C.	$R - r$.	r.	R.	n.	x.	l.	K^t.	K^t.	K^o.
1	94	29	1.68	3.23	4.91	960	15.57	2	8	1.5	1
2	120.5	17.4	1.83	6.82	8.25	1100	8.5	2	6	2	1.5
3	171	17.3	1.64	9.9	11.54	1400	8.843	2	8	3.5	3
4	168.7	17	1.73	—	—	1500	8.3	2	8.5	4.5	4
5	202	23	1.84	—	—	1680	12.1	2	13.75	5.5	5
6	197.6	21.01	1.76	9.38	11.14	1680	10.422	2	12	5.5	5
7	212	12.63	1.73	—	—	1850	5.718	2	12	8.40	7.75
*	187	25.9	—	—	—	1680	—	—	—	—	—
*	188	21	1.88	9.11	10.99	1660	10.375	2	12	5.5	5

B.—VALUES CALCULATED.

Number of Tests.	Total E.M.F.	Resistance in Ohms. (Warm). External Circuit.	Total.	Circumferential Speed of Armature per Minute in Feet.	Total Number of Watts.	Number of Watts of External Circuit.	Value of Total Current in HP.	Value of Current Appearing in External Circuit in HP.	HP. Measured by the Prony Brake.	HP. Absorbed by the Friction of the Dynamo.	Total HP. Absorbed by the Dynamo.
No.	E.	r.	R.	s.	W.	w.	E. HP.	E. HP.	HP.	*HP.	HP.
1	142	—	—	2100	.4129	2716.4	5.53	3.64	5.69	.43	6.12
2	150.2	—	—	2400	2619	2065	3.51	2.767	3.56	.44	4.0
3	200	—	—	3100	3454	2966	4.63	3.97	4.7	.5	5.2
4	198	8.17	9.9	3300	3361	2867	4.51	3.84	4.59	.5	5.09
5	234	8.78	10.52	3700	5619	4646	7.53	6.23	7.74	.51	8.24
6	234	—	—	3700	4917	4141	6.59	5.55	6.66	.51	7.17
7	234	16.7	18.43	4000	2953	2677.6	3.95	3.58	4.03	.7	4.73
*	—	—	—	—	—	—	—	—	—	—	—
*	230·8	—	—	3700	4817	4017.5	6.45	5.4	6.55	.5	7.05

C.—EXTERNAL CIRCUIT. **D.—PERCENTAGE OF EFFICIENCY.**

Number of Test.	Description.	E.M.F. of Lamp.	Absorbed Power Converted into Current.	Total Power Converted into Current.	Ratio of Total Current appearing in External Circuit.	Ratio of Absorbed Power appearing in External Circuit.	Ratio of Total Power appearing in External Circuit.
No.	r.	e′	I.	II.	III.	IV.	V.
1	Wire resistance . . .	—	97	90	65.8	64	60
2	,, ,, . .	—	98.4	88	78.5	77.4	69
3	,, ,, . .	—	98.5	89	85.7	84.5	76.4
4	Circular carbon lamp and wire resistance }	43	98	88.6	85.4	83.8	75.5
5	8 lamps in 2 parallel circuits, 4 lamps in each circuit . . }	45	97.2	92.3	82.6	80.4	75.6
6	Wire resistance . . .	—	98.9	91	84.2	83.2	77.5
7	4 Lamps in series . .	48.2	98	83.7	90	88.5	75.8
*	8 Lamps in 2 parallel circuits, 4 lamps in each circuit . . }	40.5	—	—	—	—	—
*	Wire resistance . . .	—	98.4	91.8	83	82.4	76.5

dynamometer was again compared with the weight on the scale, the latter adjusted, and, if correct, the lever was maintained for several minutes in its position of equilibrium. This process was repeated several times, and the sensitiveness of the lever tested by putting very small weights into the scale, or withdrawing them. If the lever freely responded, the experiment was considered satisfactory, and the brake was then taken to the weighing arrangement Table T 4 (Fig. 239).

This consisted of a pulley, of the same size as the brake pulley, moving freely upon a very small pin. A pair of scales were so arranged that, when the brake was fixed upon the pulley, the brake could be attached to one of the scales, and the weight of the brake scale be accurately determined; this arrangement was found to be necessary, as it was impossible to keep small quantities of soap water out of the scale during the tests, and these had to be accounted for, in such accurate experiments, when the weight on the scale was ascertained to $\frac{1}{4}$ of an ounce. The plummet on the brake here also indicated the level positions of the brake during the weighing.

From the above description it will be seen that throughout this investigation, the greatest precautions were taken in order to obtain accurate results.

The last two tests, marked with an asterisk, were taken by Mr. Heinrichs' staff, in the presence of Mr. W. H. Preece, F.R.S., on March 24th, 1884. Mr. Preece supervised the operation, and expressed his satisfaction with the manner in which the tests were conducted.

We add here calculations of one of the above tests.

Values Calculated of Test No. 3 (Table B).

$$\mathrm{E} = \mathrm{C} \times \mathrm{R} = 17.3 \times 11.54 = 200 \text{ volts.}$$

$$\mathrm{W} = \mathrm{C}^2 \times \mathrm{R} = 17.3^2 \times 11.54 = 3453.8 \text{ watts.}$$

$$\mathrm{w} = \mathrm{C}^2 \times r = 17.3^2 \times 9.91 = 2966 \text{ watts.}$$

$$\mathrm{E.HP.} = \frac{\mathrm{W}}{746} = \frac{3453.8}{746} = 4.63 \text{ electrical horse-power.}$$

$$\mathrm{E.HP.} = \frac{\mathrm{w}}{746} = \frac{2966}{746} = 3.97 \text{ electrical horse-power.}$$

Absorbed Horse-Power Calculated from the Prony Brake Readings.

$$\mathrm{HP.} = \frac{3.1416 \times 2\, l \times x \times n}{33,000} = \frac{n\, l\, x}{5252} = \frac{1400 \times 2 \times 8.343}{5252} = 4.7 \text{ horse-power.}$$

Friction of the Dynamo Calculated from the Indication on the Morin Dynamometer and Prony Brake.

$$\overset{*}{\mathrm{HP.}} = (\mathrm{K}^t - \mathrm{K}\, f) : \mathrm{HP.} : (\mathrm{K}\, f - \mathrm{K}\, o) : x = 4.5 : 4.7 : : .5 : .5 \text{ horse-power.}$$

$$\mathrm{HP.} = \mathrm{HP.} + \overset{*}{\mathrm{HP.}} = 4.7 + .5 = 5.2 \text{ total horse power.}$$

Percentage of Efficiency.

1. $\dfrac{\text{E.HP.}}{\text{HP.}} = \dfrac{4.63}{4.7} = 98.5$ per cent.

2. $\dfrac{\text{E.HP.}}{\text{HP.}} = \dfrac{4.63}{5.2} = 89$ per cent.

3. $\dfrac{\text{w}}{\text{W}} = \dfrac{2966}{3453.8} = 85.7$ per cent.

4. $\dfrac{\text{E.HP.}}{\text{HP.}} = \dfrac{3.97}{4.7} = 84.5$ per cent.

5. $\dfrac{\text{E.HP.}}{\text{HP.}} = \dfrac{3.97}{5.2} = 76.4$ per cent.

The Tables on pages 254 and 255 show eight complete tests taken of the same machine, and we can thus gather the conditions under which the machine will give the best results. In most other tests, only one measurement of horse-power appears to have been taken, and it must always be open to doubt whether the result may not have been obtained through accident. Repetition is the only sure means of obtaining accuracy in measurements, and there can be no doubt that this precaution has been well attended to in the tests of the Heinrichs machine.

The efficiency of the machine has been shown in five divisions in Table D, viz. :

I. Conversion of absorbed power into electric current. The tests have proved that the machine shows at the different speeds approximately the same percentage (an average of 98 per cent). This fact was predicted by Mr. Heinrichs as a conclusion from his theory of the action of a ring armature. It also appears that in this machine hardly any Foucault currents are generated ; nor is there any opposing electromotive force set up ; thus the highest electrical efficiency is obtained.

II. The conversion of the total horse-power given to the machine by the prime mover (*i.e.*, the Morin dynamometer in this case) into current. We find the percentage almost as high as most machines of the same type show in the first division.

III. The ratio of total electrical energy appearing in the external circuit is 65.8 per cent. with a current of 29 ampères (evidently not the most suitable condition of the machine), and rises up to 90 per cent., showing the limit on the other side. This machine sets up a very high electromotive force with very little copper wire of low resistance, and it ought to give the highest results when working in conjunction with incandescence lamps.

IV. The ratio of the absorbed power appearing in the external circuit. The percentage obtained in this division has wrongly, in most tests, been

called the commercial efficiency; but, in fact, the power absorbed by the friction of the machine is eliminated from the calculations. Those who wish to compare the commercial efficiency of other machines with the Heinrichs must refer to this column.

V. The commercial efficiency, *i.e.*, the ratio of the total power given to the machine by the prime mover (in this case the Morin dynamometer) to that appearing in the external circuit, can be taken as 75 per cent., which is indeed the highest percentage shown by any machine of the same size and type as the Heinrichs, considering that the figure stands supported by repeated tests.

In regard to economy, we gather the following : The machine gives to the external circuit a current of 48 watts per pound of copper wire (field magnets and armature), and a current of 280 watts for every pound of copper wire on the armature. Further, it supplies one light requiring a current of 12 ampères and 40 volts of electromotive force for every two pounds of copper wire on the armature, and for every twelve pounds of wire on the machine. We are informed that it is a very inexpensive machine, and can be sold at the rate of 10*l.* per light of 2000 candle-power.

SECTION IV.

RECENT DYNAMO MACHINES AND LAMPS.

THE WESTON SYSTEM OF ELECTRIC LIGHTING.*

AS is well known, the United States Illuminating Company has adopted the Weston arc system, and is using for its incandescence system, the Maxim incandescence lamp and the Weston machine. Both systems appear to be working in the most satisfactory manner, and are going very largely into general use.

The dynamo embodying Mr. Weston's recent improvements (see Fig. 252) has been very carefully studied, not only in its general electrical and mechanical design, but in all of its details. The general construction and arrangement of the Weston machine are already well known,† and are retained in the more modern type, as will be seen by the illustration. Very important modifications and improvements have been made, however, which add greatly to its efficiency and simplicity. The mechanical design has been perfected by casting the supports for the armature bearings integral with the pole-pieces, thus securing strength and rigidity for the frame. At the commutator end, the armature bearing is supported only from the lower pole-piece, as the strain upon the bearing at that end is comparatively slight, and the commutator is thus left open, and readily accessible. The bearing next to the driving pulley is, however, supported by projections from both pole-pieces, arranged in the form of an arch. But as this would produce a closed magnetic circuit between the pole-pieces, to bring the projecting supports together at the shaft, and thus divert the lines of force from the armature, a heavy brass bushing, about the shaft, is interposed between the supports, and firmly secured to them by bolts. The

---

* The following description of the Weston System of Electric Lighting first appeared in the New York *Electrician.*

† See "Electric Illumination," Vol. I., p. 231.

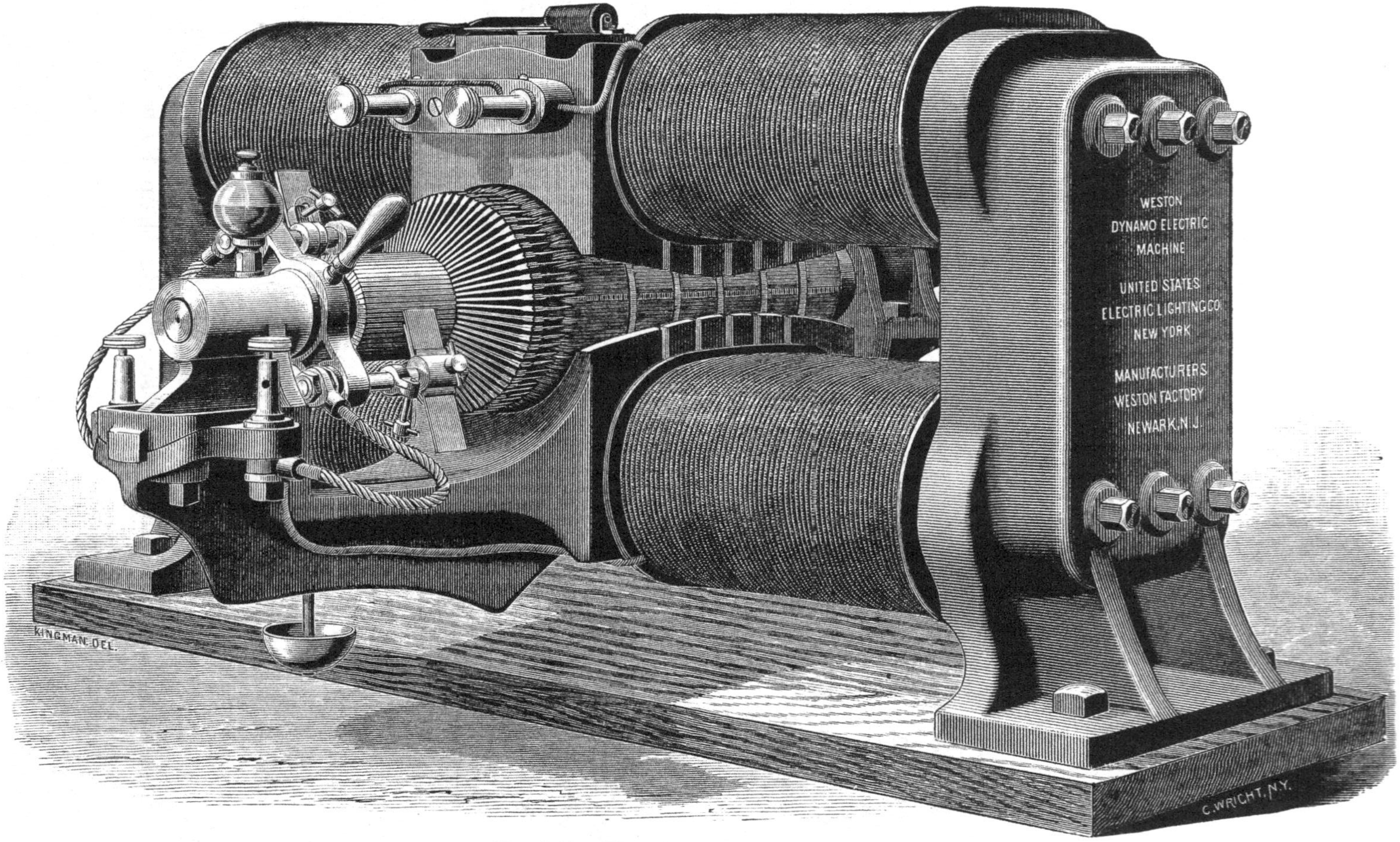
WESTON
DYNAMO ELECTRIC
MACHINE
UNITED STATES
ELECTRIC LIGHTING CO.
NEW YORK
MANUFACTURERS
WESTON FACTORY
NEWARK, N.J.
KINGMAN DEL.
C. WRIGHT, N.Y.

Fig. 252. The Weston Dynamo Machine.

general arrangement of the machine is extremely symmetrical, and the design leaves nothing to be desired mechanically or electrically.

Probably the most striking distinctive feature of the Weston machine is the sectional armature, which is still retained in the present machine. As will be seen, however, the grooves for holding the wire are made somewhat shallower than before, in order to bring the coils up nearer to the pole-pieces of the field, and the end plates have been changed in shape. The

Fig. 253.

Fig. 254

armature core is, as shown in Figs. 253 and 254, built up of iron discs, of the form shown in the former figure.

These are secured together upon the armature shaft, but separated somewhat from each other so as to leave spaces between them. These spaces serve to break up the continuity of the core, and thus prevent the formation of induced currents; they also form ventilating spaces. By a very ingenious arrangement, the armature is made to act as a centrifugal blower, to maintain a circulation of air through the core and about the coils, and thus whatever heat may be generated in them is dissipated. The coils are spread apart, where they pass across the heads of the armature, by flanged plates (shown somewhat removed from the head of the armature in Fig. 254) so as to leave an opening about the shaft for the admission of air, which is taken into the interior of the armature and thrown out between the coils by centrifugal force. With a sectional armature, and this system of ventilation, no trouble whatever is experienced from heating of the core or coils, although, as is well known, such heating was so great with machines of this type, having solid armature cores, as to be an almost insuperable obstacle to their use. The armature complete, with the coils connected to the commutator, is shown in Fig. 255.

As will be seen, the number of sections in the commutator has been very greatly increased, in order to reduce the spark and prevent any

tendency of the current to discharge across from one brush to the other. The number of sections formerly used was eight; in the present machines, from forty-eight to one hundred and forty are used.

As the amount of spark at the commutator depends largely upon the method in which the armature coils are wound and connected to the com-

Fig. 255.

mutator, Mr. Weston has bestowed much attention upon different systems of winding, as applied to cylindrical armatures. The continuous winding of the coils in a single closed circuit, with loops taken out to the commutator sections, which was introduced by Gramme, was undoubtedly a very great improvement upon any system which had before been used; but this winding cannot be applied to a cylindrical armature without considerable modification, since in winding coils upon a cylinder, each coil fills up two diametrically opposite spaces, and the entire surface of the cylinder is covered in winding half way round, while only half as many loops are taken off at the junctions

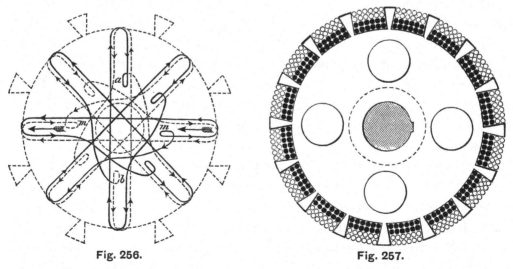

Fig. 256.   Fig. 257.

of the coils, as there should be sections in the commutator. To overcome this difficulty Mr. Weston devised the system of winding, of which a diagram is given in Fig. 256.

For the sake of simplicity, only eight coils are shown, but it is obvious

that the same principle may be applied to any other number. Supposing, for instance, that the winding commences at *a*, the coils follow the course indicated by the full lines, and at every place where the wire passes from one coil to another, a loop is taken off for connection with a corresponding segment of the commutator. After winding four coils, it will be observed that all of the spaces on the armature have been occupied, while there are only four loops for attachment to the commutator, and there should be eight; to supply the remaining loops a second set of coils is wound, as shown by the broken lines, and loops are taken off in the same way until the entire surface of the armature has again been covered, when the last terminal of the second set of coils is connected to the entering terminal of the first set at *a*. Supposing now that the brushes are in contact with the commutator

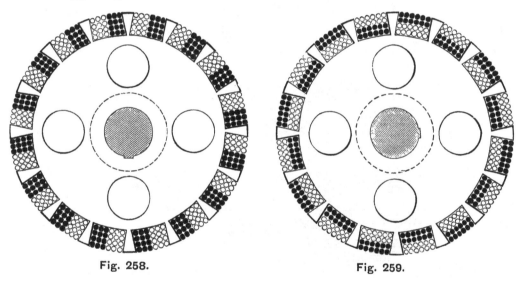

Fig. 258.   Fig. 259.

plates connected with *m, m,* the course of the current through the coils may be traced by following the direction indicated by the arrows, and it will be seen that the current divides and passes through the coils on each side of the dividing line from *m* at the right to *m* on the other side. If any other diametrically opposite loops be taken, the current will be found to follow a similar course. By covering the entire surface of the armature with the first set of coils, and then winding the second, the second coils are superposed upon the first in the manner shown in the diagram Fig. 257, which represents a cross section of the armature.

The white circles indicate the wires of the second set of coils, and the black ones those of the first. This was found to be objectionable, as the coils of the second set were not only longer than those of the first, but were

brought up nearer to the poles of the field, and moved faster, being further from the shaft. In consequence, the two sets were not electrically balanced, and an objectionable spark was produced at the commutator. This difficulty was overcome by winding the coils of the two sets, side by side, as shown in the diagram Fig. 258.

With this winding, all of the coils occupy the same relative position with respect to the armature core and the field ; and a perfect electrical balance is obtained, by which the spark at the commutator is reduced to a minimum. A modification of this, which is shown in diagram Fig. 259, has been found simpler to wind, and in some respects more efficient.

In this, the coils of the two sets are alternately superposed, so that both sets are, as a whole, under precisely the same conditions, and are electrically balanced. With machines of the improved type, wound in this way, the spark on the commutator is hardly perceptible.

In Fig. 260 is shown an ingenious system of connecting the coils with

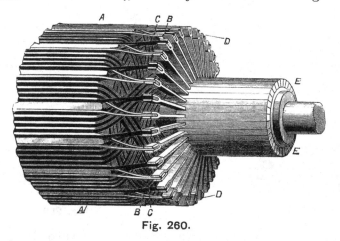

Fig. 260.

the commutator, for use in machines in which the electromotive force is high. Coils are sometimes burned out in machines having commutators of this general type, by accidental short-circuiting of two adjacent commutator sections. In the armature, of which Fig. 260 is a diagram, there are two distinct sets of coils, represented by the white and the black lines, which alternate about the circumference ; and they are connected to alternate plates of the commutator as shown. It is obviously impossible to short-circuit any coil by connecting adjacent sections of the commutator, and the chances of accidental connection between three consecutive sections are very slight.

The great rigidity of the frame, and the accuracy with which the

mechanical work on these machines is done, admits of running the armature with an extremely small clearance between its surface and the pole-pieces of the field.

The cores and pole-pieces of the field magnets are made very heavy, so as to maintain an extremely intense field, with comparatively little expenditure of current energy; and the pole projections, on the armature, bring a part of the magnetic material of the core almost directly into contact with the pole-pieces, so as to concentrate the lines of force of the field directly upon the armature. All of these features of construction contribute to produce the requisite electromotive force with very low internal resistance, and low speed.

The ratio between the resistance of the armature and the normal resistance of the working circuit, in these machines, is about one to forty, indicating the remarkably small loss of only $2\frac{1}{2}$ per cent. of the current energy in the armature.

Another important modification, shown in the Weston machine of the present type, is the placing of the field coils in derived circuit, instead of in the main circuit. This arrangement of the field was suggested quite early in the history of the dynamo machine, but its advantages have not been properly appreciated until quite recently. The field coils are made of very high resistance, so as to divert into the field circuit only a very small part of the entire current (varying from $2\frac{1}{2}$ to 5 per cent. in the different machines). From the construction of the machine there can be but little air friction, and the friction at the journals is quite small. The twenty-light machine runs at 900 turns per minute, and gives a current of about 18 ampères, with twenty lamps in circuit, having a resistance of $1\frac{1}{2}$ ohms each, or in all, 30 ohms. The power required is about 14 horse-power. This indicates an extremely high efficiency.

The Weston machine for running incandescence lamps is of the same general type, and is equally efficient. It has been very carefully studied in all its details, and in one respect at least calls for special notice.

Incandescence lamps being run in multiple arc, instead of in series, it is necessary that the electromotive force at the machine should be kept constant, and that the current should vary directly as the number of lights in circuit.

It is obvious that with a generator of constant electromotive force the system would be entirely self-regulating; for, since $Q = \dfrac{E}{R}$, if E be constant

Q will vary inversely as R; but in the multiple arc system of distribution the resistance varies inversely as the number of lamps in circuit; and accordingly with E constant, Q varies directly as the number of lamps, assuming, of course, that the lamps are uniform.

Mr. Weston has succeeded in constructing a machine of which the electromotive force is constant, and practically independent of the quantity of the current. The most violent fluctuations in quantity seem to exert no appreciable influence upon the electromotive force. As, for instance, with the hundred-light machine, ninety-nine of the lamps may be switched out instantly without endangering the remaining lamp, or perceptibly affecting its brilliancy. This is a most remarkable result, and its importance for the incandescence system can hardly be over-estimated.

It is frequently very desirable to have some device for varying the electromotive force of the machine within certain limits. For example, in putting in an incandescence plant it is seldom that the speed of the armature will be exactly that required to give the precise electromotive force to bring the lamps up to the standard illuminating power. It is also nearly impossible to make the incandescence lamps of perfectly uniform resistance, and this necessitates a change in the electromotive force to suit the different grades of lamps. Sometimes there are places in which the illuminating power of the lamps must be varied as a whole. This is frequently the case in theatres and large buildings, where the full light is not required at all times.

In the arc system it is absolutely necessary that the electromotive force of the machine shall admit of wide variation, since this is the only really satisfactory method of adjusting the machine to a varying number of lights. A special device has been constructed to meet these requirements.

It consists of the rheostat, which is shown in Fig. 261, interposed in the field circuit. The contact plates shown in front are connected with the terminals of resistance coils of German silver wire enclosed in the box. The resistance of the field circuit, and consequently the amount of current passing through it, may be varied by turning the handle of the rheostat, so as to adjust the machine for running any number of lights, from one up to its full complement. This system of regulation does not at all disturb the proper operation of the machine, as is the case where the commutator brushes are turned, or the field partially shunted out of the circuit. The twenty-light or thirty-light machine works as well with one lamp as with its full complement. The same form of rheostat is used with machines for running incandescence lights for the purposes we have stated above; but

unless it is desired to vary the illuminating power of all the lights, it is not used after the first adjustment is made.

A striking peculiarity of the Weston arc system is the shortness of the arc used. The Weston arc is somewhat less than $\frac{1}{32}$nd of an inch in length. The arc in the Brush and most other systems, is nearly $\frac{1}{8}$th of an inch. In the Weston system a current of about 18 ampères is used, while in the Brush and most other systems the current does not exceed 11. The only disadvantage of the short arc system is the increased size of the line conductors necessary for conveying the increased quantity of current without too great loss; but this disadvantage is more than counterbalanced by very decided advantages. More light is obtained for a given expenditure of current energy, and the light produced is of much better colour. When a long arc

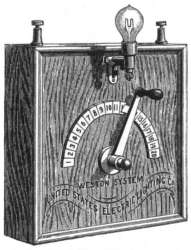

Fig. 261.

is used there is a marked preponderance of the rays at the violet end of the spectrum, giving the light a bluish tinge, which is extremely objectionable. With a short arc, the light approaches more nearly a pure white, and is much more soft and agreeable to the eye. It is also more diffusive—that is, it produces more perfect illumination at a distance from the lamp. The arc is also much steadier. Slight impurities in the carbon, which produce serious disturbances with the long arc, do not appear to affect the short arc.

But by far the greatest advantage of the short arc system is the largely decreased liability of injury to persons coming in contact with the conductors. We have no doubt that the dangers to be apprehended from electric lighting apparatus, as at present constructed, have been largely over-estimated; and such dangers as really exist can be overcome by proper care in the use of

the apparatus.   Experience has demonstrated, however, that fatal accidents are possible with many of the systems now in use.   Where currents of very high electromotive force are used, the most extreme care is necessary for

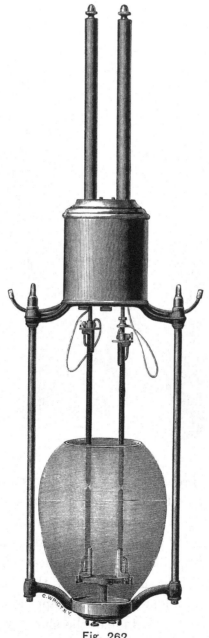

Fig. 262.

avoiding accidents; and it is desirable, on this account, that the electromotive force should be kept as low as possible.   It seems, on the other hand, to be essential for economy in distribution to run a very large number

of lights in the same circuit. The resistance of the Weston arc lamp, being only about one ohm and a half, as compared with a resistance of from four to six ohms in most of the other systems now in use, it is obvious that the electromotive force required is correspondingly low, and the danger of accident is greatly decreased.

Different forms of Weston arc lamps are used. Fig. 262 is a duplex or double carbon lamp.

The feeding mechanism of this lamp is shown in Figs. 263, 264, and

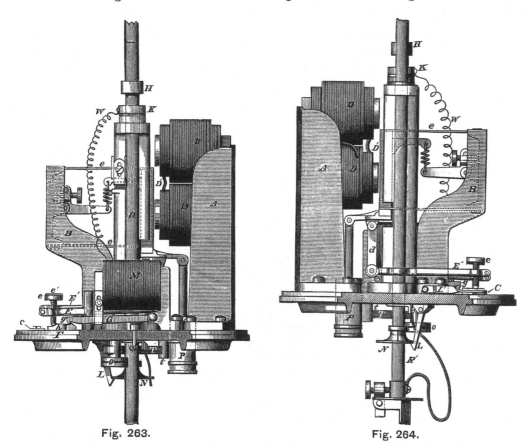

Fig. 263.   Fig. 264.

265. In this there is but one electro-magnet D D for controlling the feeding of both sets of carbons. It is differentially wound with two sets of coils, one of coarse wire, being included in the arc circuit, and the other of fine wire in a derived circuit of high resistance, so as to adapt the lamp for use in series.

The lower terminal of the coarse wire helix is electrically connected with both upper carbon carriers, and the current and the feeding mechanism are shifted simultaneously, at the proper time, to the second set of carbons

by the shifting magnet M, which is included in a derived circuit of high resistance. The shifting lever C carries wedge-shaped slides *h h'*, which are inserted under the end of one clutch or the other, so as to trip it, and prevent it from engaging with its rod.

While the first set of carbons is burning, the circuit of M is open ; the upper carbon R', of the second set, is held up by the hook L, and the shifting lever is locked in the proper position to leave the first clutch free and trip the second. When the first set of carbons is consumed, the circuit of M is completed by a stop H on the upper rod R coming into contact with K, and the shifting magnet drawing up its armature G, lifts the detent from C, allowing it to swing over, and at the same time reverse the position

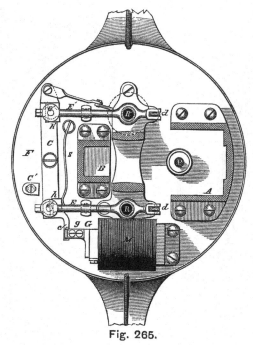

Fig. 265.

of the slides under the clutches, and release the upper carbon of the second set. As the upper carbon R of the first set is supported, out of contact with its lower carbon, by the stop, the current is diverted to the second set of carbons as soon as they come into contact, and the feeding magnet now works the second clutch instead of the first. This is done so quickly as to cause hardly a flicker in the light.

The feeding mechanism of the single lamp is the same as that of the duplex lamp, omitting of course the duplicate parts and the shifting mechanism. Its construction and operation are too well known to require detailed description. The extraordinary sensitiveness and certainty of

operation of this feeding mechanism are probably due, principally, to the peculiar arrangement of the armature of the electro-magnet D, and to the construction of the clutch. It will be observed that the armature D, which controls the movements of the clutch and upper carbon carrier, is suspended (in front of the poles of the electro-magnet), but somewhat out of a sym-

Fig. 266.

metrical position, with respect to them, by the flexible strips *e e*, which admit of only a vertical movement of the armature. When the magnet is excited the armature tends to come to a symmetrical position with respect to the poles, and rises into a more intense part of the field of the magnet. At the same time it recedes somewhat laterally from the poles. These two

T

movements are so related to each other that the increase of the power of the magnet due to one, is very nearly compensated by its decrease of power due to the other movement of the armature, and consequently the lifting power of the magnet is approximately uniform through a very wide range of movement of the armature. This is an especially desirable feature in a lamp to be used in series.

The clutch, as will be observed, is provided with a long tail-piece, and

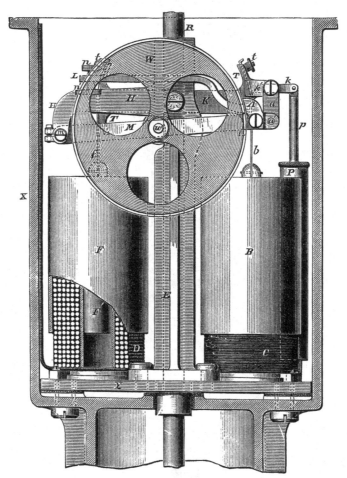

Fig. 267.

the point of contact of the clutch with the floor of the lamp, is at the end of the tail-piece. This gives the detaching point of the clutch great leverage over the lifting point, and consequently it requires only a very small amount of power to detach the clutch from the rod and allow it to feed. Of course the sensitiveness of the feeding mechanism in the lamp depends largely upon the amount of force which is necessary to lock or unlock the

feeding mechanism, as this force is derived solely from variations in the strength of the current, due to fluctuations in the length of the arc.

In the form of lamp frame shown in Fig. 266, the electrical connections are placed inside of the tubular support, by which the globe is suspended from the casing above, and there is no exposed or accessible part of the lamp, in electrical connection with the circuit. This form of frame

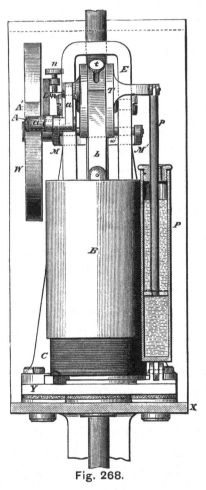

Fig. 268.

may be used for duplex or single lamps, the same as the form shown in Fig. 262, and the same feeding mechanism may be used.

In Figs. 267 and 268, is shown a new form of feeding mechanism, which Mr. Weston has devised. The feeding of the carbons is controlled by a brake-wheel, to the shaft of which the upper carbon carrier is attached by a cord and pulley, or rack and pinion movement, so that the rod cannot be moved without turning the wheel. The shaft of the wheel is mounted upon a swinging lever M, which is pivotted to a fixed support at m, and at

its other end is linked to an oscillating lever K.   Two solenoids, F and B, of peculiar form, are used for controlling the position of the oscillating lever.   One of these solenoids is included in the main circuit, and the other in a derived circuit about the arc; and they are attached to the ends of the lever by flexible metallic straps f and b.   The wheel W is controlled by a brake A, which is pivotted to the lever M, just outside of the periphery of the wheel, and has a short arm at the right, to which the lifting link a is attached at a'.   It also has a long tail-piece L, extending over and resting upon the frame of the lamp at n'.   The construction of the solenoids used is shown at the left, in Fig. 267, where the shell and part of the coil are cut away to show the construction.   It will be seen that the solenoid has the ordinary core, and in addition, an iron shell surrounding the coil, and made in one piece with the core.   Mr. Weston has found, that with a solenoid constructed in this way, an extremely long range of movement is obtained with very uniform power; and that the solenoid is about as powerful as an ordinary bi-branched electro-magnet, having double the amount of wire upon it.   The great disadvantage of ordinary solenoids is their comparative feebleness in proportion to their size.

Supposing the lamp to be at rest, with the carbons in contact, upon first applying the current the solenoid F will preponderate in force, and draw down its end of the oscillating lever; this movement carries the brake upward, and locks it upon the wheel, as soon as its tail-piece L is raised sufficiently to lift it from the stop n', and the further movement of the oscillating lever carries the lever M and the wheel up bodily, so as to separate the carbons and form the arc.   As the arc increases in length, F weakens, and B strengthens, and a condition of equilibrium is soon reached, in which the end of the lever L rests upon the stop with just sufficient force to raise the brake and allow the wheel to revolve.   The slightest decrease in the length of the arc will immediately set the brake, and the slightest increase release it.   It will be observed that the brake is so arranged with reference to the wheel, that the weight of the carbon and its carrier, tends to lock the brake when its detaching lever is lifted from the stop, so that its grip on the wheel is very positive; but the wheel being large, and the detaching point of the brake lever having great leverage over the brake, the device works with the utmost delicacy.   The feeding is, under ordinary conditions, so nearly continuous, that it is impossible to detect the movement at any one time.

In Fig. 269 is shown the automatic cut-out ordinarily used with these

lamps. It is a modification of the well-known Varley cut-out, and consists of an electro-magnet in the same circuit with the arc, and an armature, so arranged as to close a short circuit about the lamp when the armature is released by the magnet. This has been put into very compact and convenient form, and is usually attached to the support above the lamp, but may be included in the casing with the feeding mechanism. The diagram indicates clearly the electrical connections. About the coil of the magnet are wound a few convolutions of insulated German silver wire, which are included in the shunt circuit. The current in this supplementary coil flows in the same direction as in the main coil, and the object of the supplementary coil is to cause the electro-magnet to act more quickly to open the shunt circuit when the lamp is lighted. The

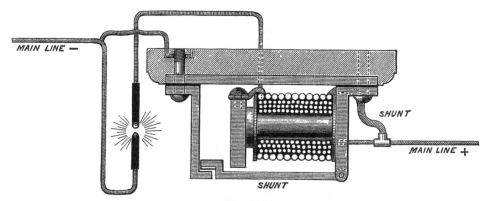

Fig. 269.

resistance of the shunt coil is very slight (about the same as that of the lamp with the carbons in contact), and, with the shunt closed, and the lamp circuit open, it is obvious that the entire current must pass through the resistance, but there being few convolutions, and these at a distance from the core of the magnet, too little magnetism is developed to lift the armature. Whenever the lamp circuit is closed, however, the current at first divides between the two coils on the magnet, and both co-operating lift the armature very quickly. As soon as the armature is lifted the shunt is broken, and the entire current passes through the lamp. This form of cut-out has been found very efficient under ordinary circumstances, but where a very large number of lamps are used in the same circuit, the form of cut-out shown in Fig. 270 is used, for the sake of additional precaution to guard against any dangerous elongation of the arc.

In the diagram, P and N are insulated conductors, leading from the binding-posts of the lamp, and terminating in contact surfaces p and n, of

a spring switch, located above the arc, as shown in the illustration. The switch, when left free, is closed by a spring, but is normally held open by a plug a, of easily fusible alloy, arranged between the jaws of the switch, as shown in the plan view, in the lower part of the figure. The switch is supported near the upper carbon, but at a sufficient distance above the arc to prevent melting of the fusible plug while the arc is of normal length. As is well known, any marked increase in the arc produces a very great increase in heat. When this occurs the fusible plug is melted, and allows the switch to

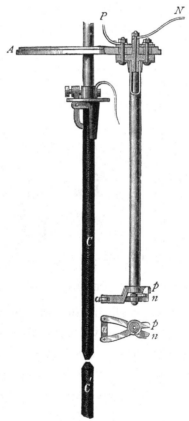

Fig. 270.

close, and cut the lamp out before the arc becomes dangerously long. This is very positive in its action, and has been found extremely reliable. It may be easily readjusted by inserting another fusible plug. It is, however, rarely brought into action in practice, as accidents to the feeding mechanism seldom occur.

In Figs. 271 and 272 is shown an indicator, for use in central stations, and other places where the lamps are at a considerable distance from the machine. The coils of the magnet, shown in Fig. 272, are included in the

circuit with the lamps, and the retractable spring of the Z-shaped armature is so adjusted as to bring the needle, which is attached to the shaft, to zero on the scale, when the current is of normal strength. Should the current exceed or fall below the normal strength, however, either the magnet or the spring would preponderate, and the needle would swing in one direction or the other, indicating the direction of the change; and in case of any considerable variation, one of the contact springs on the lever would close the battery circuit, by making contact with a stop-screw in the upper part of the box, and ring a bell at the top to call the attention of the attendant.

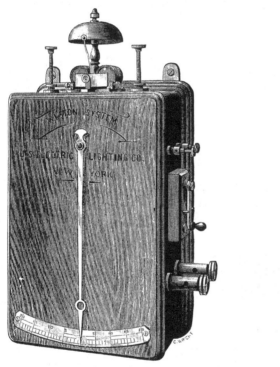

Fig. 271.

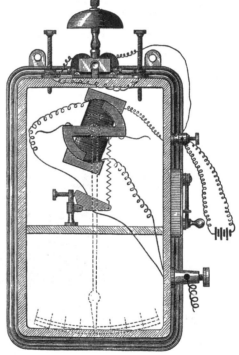

Fig. 272.

By means of this indicator, the attendant at the central station can see, upon starting the machine, that the normal current is sent to the lamps, and is informed at once when any accident occurs to the circuit, or when lamps are put into circuit or cut out, and can immediately adjust the machine to compensate for the change.

We may conclude with a few words upon some devices connected with the incandescence lighting as considered by the United States Lighting Company.

Fig. 273 shows a device, designed by Mr. Maxim, for automatically cutting the carbon of an incandescence lamp out of circuit at the proper

time when it is being built up in a hydro-carbon atmosphere. An electro-magnet K is included in the same circuit with the carbon under treatment, which is represented at C. The surrounding vessel B contains a rarefied atmosphere of hydro-carbon gas. The wires A A lead to the generator, which should be worked under the same conditions during the process. It is obvious that as the resistance of the carbon C decreases, owing to the deposition of carbon upon it, more current flows through the coils of the magnet; and the retractile spring M is so adjusted, that the magnet will draw down its armature when the resistance of the carbon has fallen to the standard fixed. When this occurs, the hammer G is released, and, falling upon F, breaks the circuit suddenly at T, and cuts out the carbon; the spring F is prevented from rising again by a detent R. In a modification of this device, which is somewhat more sensitive, instead of the retractile

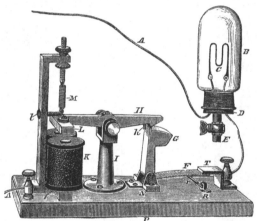

Fig. 273.

spring M, another electro-magnet, included in a derived circuit about the carbon in course of treatment, is made to pull down the other end of the armature lever A. By this apparatus the carbons are made of very uniform resistance. As is well known, the Maxim lamp is held in a socket or holder, which may be attached to an ordinary gas fixture, or other suitable support. The socket contains the terminals of the circuit leading to the lamp, and these are so arranged that the ends of the leading-in wires of the lamp are brought in contact with them, when the lamp is inserted in the socket; merely placing the lamp in position makes the necessary electrical connections.

In Figs. 274 and 275 are shown different forms of safety devices which Mr. Weston has invented. With the multiple arc system, it is obvious that the distribution of the current among the branches depends upon their relative

resistance. If from any accident one of the branches should happen to be short-circuited, an abnormal amount of current would be diverted to it from the main circuit, and this might be sufficient to dangerously heat, or even fuse the branch wire. A safety device, of the form shown in the illustration, is introduced in each branch, to guard against this danger. It consists of a strip of easily fusible alloy introduced into the circuit, and arranged in convenient form for renewal. The alloy used is a compound of tin, lead, cadmium, and bismuth, and fuses at about 155 deg. Fahr. The strip is made of sufficient size

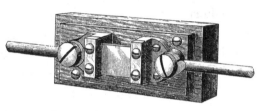 

Fig. 274.   Fig. 275.

to convey the normal current without heating it to its fusing point; but in case of any abnormal flow of current, the alloy section of the branch would, of course, melt before the copper wire became dangerously heated, if the section of alloy were made of equal, or even somewhat greater conductivity than the rest of the circuit. As the fusing point of the alloy is below the point of ignition of any combustible, the drops of melted metal can do no harm in case the safety device is called into operation. These devices are made in various forms, some of which are shown in the cuts, and are located at any convenient place in the circuit, so as to be readily accessible for the renewal of the alloy strip when necessary.

Fig. 276 shows the wall plate belonging to a bracket, devised by Mr. Weston, for incandescence lights. This plate is permanently attached to the wall, and the circuit wires connected at *c* and *e″*. When the bracket is placed on the plate, its electrical terminals form contact with the springs *c* and *d*, and the wires pass from them through the fixture to the lamp. At *e*, in the wall plate, is a fusible cut-off. This bracket can be swung from side to side without affecting the circuit, as the connections are made through the pivots upon which it swings. Mr. Weston has also devised forms of double swing brackets, drop-lights, chandeliers, and various other fixtures, for both arc and incandescence lights.

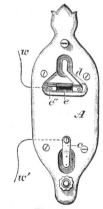

Fig. 276.

THE HOCHHAUSEN SYSTEM.

AMONG the many systems of electric lighting with which the buildings of the International Fisheries Exhibition, held at South Kensington in 1883, were illuminated, the system of Mr. Hochhausen was that which attracted most scientific attention on account of its novelty and the many points of special interest involved in the construction of its apparatus, and in its method of working. The system is the invention of Mr. William Hochhausen, of New York, and has been introduced into this country by Mr. Henry Edmunds, whose name is well known in connection with the extraordinary progress which illumination by electricity has made during the last few years, he having introduced into this country both the Wallace-Farmer and Brush systems, and having taken an important part in the development of the Swan mode of lighting by incandescence, and its successful introduction both here and abroad.

The Hochhausen systems are for both arc and incandescence lighting, the currents utilised for the two methods of illumination being produced by machines which are somewhat different from one another. The arc system is complete in itself, that is to say, it has its own machine, its own lamp, and has many special features of the conductors and general installation peculiar to itself.

The dynamo-electric arc-light machine of Mr. Hochhausen, a general view of which is given in Fig. 277, belongs to that great class of generators in which an armature of annular or tubular form is employed, a group, the parentage of which undoubtedly belongs to Dr. Antonio Pacinotti, of Florence,[*] and in a certain degree to a still earlier inventor, Herr Elias, of Haarlem.[†] It has, moreover, been embodied more or less in the machines of Worms de Romilly,[‡] Gramme,[§] Schuckert,[||] Brush,[¶] Maxim,[**] Gülcher,[††] Jürgensen,[‡‡] De Meritens,[§§] Heinrichs,[||||] Bürgin,[¶¶] and several others.

* See "Electrical Illumination," Vol. I., p. 126. † Ibid., p. 102. ‡ Ibid., p. 672. § Ibid., p. 147.
|| Ibid., p. 299. ¶ Ibid., p. 204. ** Ibid., p. 224. †† Ibid., p. 297. ‡‡ Ibid., p. 301. §§ Ibid., p. 187.
|| || Ibid., p. 234. ¶¶ Ibid. p. 218.

The armature rotates on a horizontal shaft within an intense magnetic field, produced between the poles of two vertical electro-magnets, whose common axis passes through that of the armature and perpendicularly to it, while their magnetic intensity is to a certain extent further reinforced

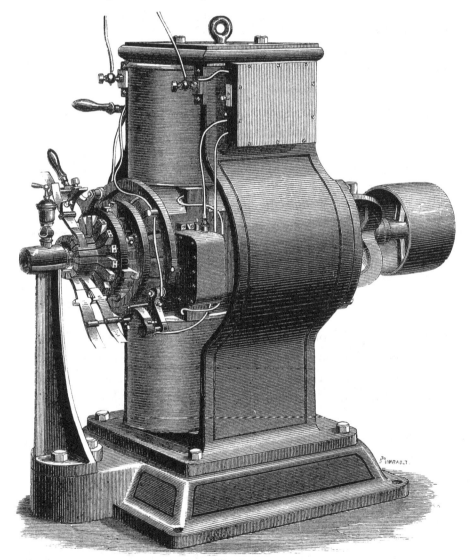

Fig. 277.

inductively by the action of two massive soft iron bars connecting the free poles of the two magnets.

The coils of the armature are connected to a cylindrical commutator of very ingenious construction, which, in principle, is identical with the commutators of several of the direct current machines of other inventors, such

as Siemens, Gramme, Edison, and others, and the current is carried into the external circuit by two pairs of elastic metallic springs, which press against the sectors of the commutating cylinder in the usual way.

A special feature of the Hochhausen machine, and one to which a great proportion of its undoubtedly steady working and economy is due, is the very beautiful contrivance by which its conversion of dynamic, into electric energy is regulated automatically, to the work it has to do, or in other words, by which the load on the driving engine is adjusted to the number of lamps in actual operation.

The core of the Hochhausen armature (a perspective view of which, with two coils removed, is shown in Fig. 278), consists of a cylindrical cage

Fig. 278.

built up of flat wrought-iron quadrants; without its coils, it bears some resemblance to the cylindrical sieve of a rotary screening machine. The quadrants are stamped out of sheet iron, about one-sixteenth of an inch in thickness, and have dovetailed ends. These are shown in detail in Fig. 279, in which A A A are the quadrants, and B and B' are bars having dovetail grooves along their outer sides, into which the ends of the quadrants are fitted with distance-pieces of the same thickness between them. There are four pairs of these bars distributed at equal angular distances around the armature, as is well shown in Fig. 280. The bars of each pair are held together by a piece of the form shown at C, Fig. 279, having a dovetail groove cut along its length, and which slides over the dovetail formed at the top of the bars. They are further held together, and at the same time are rigidly attached to the outer framework of the armature, by two projecting half-studs D and D¹, which, when in position, form a single stud, and these studs, passing through corresponding holes in the end frames, are secured by nuts on the outer sides.

From the above description it will be clear that the armature can be taken into four pieces at any moment with great facility, and the coils are

so constructed that they can be taken off and on by simply sliding them over the quadrants when the armature is so dissected.

The method by which the coils are wound in the Hochhausen machine is a special feature of novelty and interest, because by it is solved a difficulty which has hitherto been inseparable from the coiling of annular armatures, whose helices are either contiguous, as in the Gramme, or very close together, as in several other types of machines. This difficulty is a mechanical and manipulative one, arising from the geometrical fact, that the internal

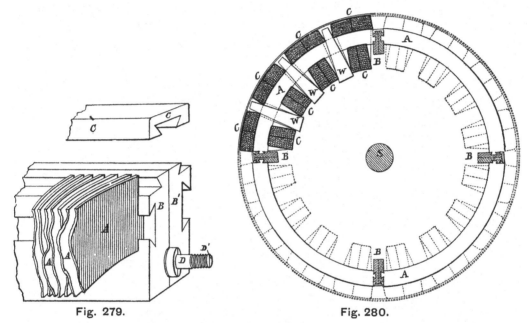

Fig. 279.   Fig. 280.

circumference of a solid ring is smaller than its external circumference, and that therefore the spaces occupied by the coils must be more or less wedge-shaped, and thus while the strands passing over the outer circumference had hitherto to be more and more spread out in order to be uniformly distributed, they had for the same reason to be inconveniently crowded together over the inner circumference. To overcome this difficulty various devices have been suggested, but none of them can, in our opinion, surpass in simplicity and efficiency the method adopted by Mr. Hochhausen, which is indicated in Fig. 280, but which will be understood more clearly by referring to the diagram Fig. 281, which represents a section of one of the coils, made by a plane cutting through the armature perpendicularly to its axis. In the coil represented in the diagram, there are forty-two convolutions; but while the lower part of the

Fig. 281.

helix is laid in six layers of seven strands each, the upper portion (representing the outer circumference) is laid in three layers of fourteen strands. In this way the cross section of the bundle of wires of the helix, both at the outer and at the inner circumference of the armature, is always maintained a parallelogram, and at the same time the width of the outer bundle is wider than that towards the axis. In Fig. 280 this is well shown, as well as the general construction of the armature.

Referring to this figure, and to Fig. 282, A A A A are the quadrants of which the compound annular core is built up, and B B B B the dovetail bars by which the four sets of quadrants are held together and attached to the end frames H H, which are keyed on to the driving shaft S. The helices

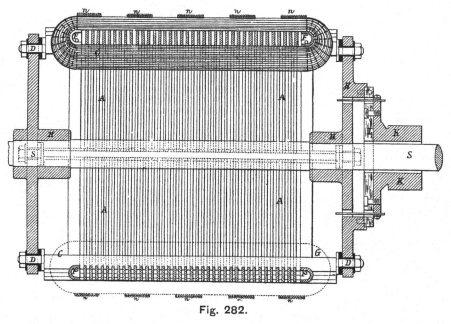

Fig. 282.

are shown at C C C C, which, from the method by which the core is constructed, can either be wound independently upon a separate mandrel, or upon hollow bobbins, and slid over the free end of the quadrant, or they can be wound direct upon the quadrants, each quarter of the core being wound separately, and the four quarters put together afterwards—a form of construction which will at once commend itself to every practical engineer, for, by it, the principal difficulty in coiling annular armatures is completely eliminated, namely, the having to pass the spool containing the wire, to be wound through the ring for every convolution laid on. In order to avoid sharp bends in the wire at the two ends of the armature, and to keep the coiling as compact as possible, the core is finished off by quadrants of

⊔-shaped cross section indicated at F F F F, Fig. 282, and also shown in Fig. 278.

In the armature illustrated, there are sixteen coils, that is to say, each of the four quadrants carries four helices, which are kept symmetrically in their places by wooden wedges W W W, driven between them from the inside, and in order to prevent any displacement of the convolutions, or distortion of the armature by centrifugal force, the whole is bound together by coiled hoops of phosphor-bronze wire *n n n n n*, Fig. 282, shown in perspective in Fig. 278.

The armature coils of the Hochhausen machine are coupled up in series, and their junctions are also connected to the sectors of a cylindrical commutator, which is constructed in a very simple and practical manner, consisting simply of a crown of sixteen ∟-shaped gun-metal or copper blocks K, Fig. 282, screwed by their basis to a circular disc of slate L, their form and arrangement being as shown in Fig. 278, as well as in the general view of the machine, Fig. 277. The only insulation between the blocks is obtained by air spaces, and upon their cylindrical surfaces, the collecting springs shown in Fig. 277 are pressed. It is by the position of these springs with respect to the magnetic field that the electromotive force of the machine is automatically controlled and adjusted to the number of lamps in operation.

We may now proceed to refer to some of the arrangements and apparatus belonging to this system, for regulating the current of the machine to the work which it may at any time be called upon to do. The object in view is to control the electromotive force of the machine, without altering its velocity of rotation, by varying, either automatically or not, the intensity of the magnetic field within which the armature coils are being rotated. This object has been arrived at by various devices, very different in both principle and construction from one another; in some, the commutator brushes are automatically shifted around the commutator, by means of a small supplementary electric motor rotating under the influence of the magnetic field; in others, a similar motor is employed to throw in or cut out certain lengths of the exciting helices of the field magnets; again, in others, the main shaft of the machine is caused to give motion to the brush-holders, their direction of angular displacement, as well as the period when it takes place, being determined by the action of a supplementary electro-magnet, which is controlled by a relay.

Although these regulating devices are included in Mr. Hochhausen's

patents, they are by no means essential to his system, and as a matter of fact a great number of the machines are not supplied with any automatic system of current regulation; but, as these contrivances undoubtedly render the Hochhausen system a very complete one, and are very interesting and ingenious, we shall treat them as if they formed an integral part of the system we are describing.

The machine by which the principal promenade at the late Fisheries Exhibition was illuminated, and which is illustrated in Fig. 277, was fitted with the automatic regulator, the principle and construction of which is shown in Figs. 283 to 286. By this apparatus the electromotive force is, as in the Maxim machine, controlled by a shifting of the collecting springs or brushes around the commutator to positions of greater or less efficiency, not, however, by a "step-by-step" motion, as in the Maxim machine, the effect

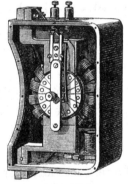

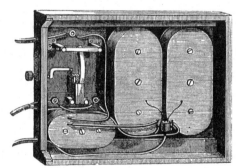

Fig. 283.   Fig. 284.

of which comes too late to be of practical use, but directly, by means of a small electric motor working under the influence of the magnetic field, and having its direction of rotation determined by a small magnetic switch or relay, the position of which is controlled by the strength of the magnetic field, and indirectly by that of the current in the external or lamp circuit of the machine.

The motor, which is shown in Fig. 283, consists of a small annular armature of the Pacinotti type, capable of rotation upon a horizontal axis within a magnetic field, and enclosed in a metal box, which is shown in the front of Fig. 277. Fig. 285 is a diagram in which the arrangement of the parts, as well as the connections, are clearly indicated. In this figure B is the annular armature of the motor, the spindle of which carries a pinion G^2, which either gears directly into a circular rack attached to the ring R, which carries the commutator brushes (see Fig. 277), or moves the ring R through the intervention of a train of wheels. The magnetic field for the motor is

obtained by two pole-pieces A and A², which partly surround the armature (see Fig. 285), and which, by being attached to the pole-pieces of the machine by iron brackets W W, become sufficiently polarised by induction to produce a very good field of magnetic force between them ; the general disposition of the armature, pole-pieces, and polar extensions are shown in the diagram, Fig. 287. The direction of rotation of the motor, and therefore of that of the brush-holders, is determined by a magnetic relay, combined with suitable resistance coils contained in a small rectangular box (shown in

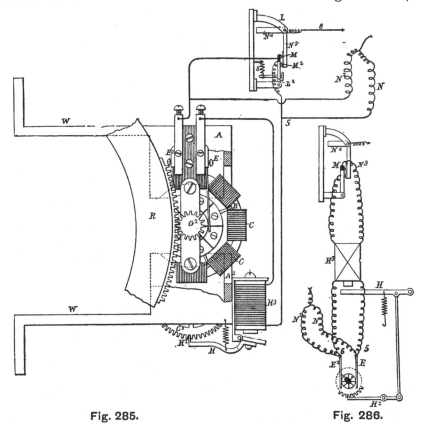

Fig. 285.   Fig. 286.

Fig. 277), attached just below the top plate of the upper field magnet, and illustrated in Fig. 284, as well as in the upper portions of Figs. 285, 286, and 288. This relay consists of a small soft iron armature bar N⁴, to which is attached, perpendicularly to it, an arm N³, thus constituting an ∟-shaped lever or bell-crank pivotted at its right angle. The lower end of the arm N³, which is electrically connected to the wire of the magnet helix, makes contact with one or both of two contact pieces connected respectively with the two brushes by which the current is transmitted into the armature of the motor. These two contact pieces, although insulated from one another,

U

form a lever pivotted at its lower extremity, and pressing against the lower end of the bar N^3, with a pressure which can be adjusted to any required amount by varying the tension of the adjustable spring S, which is also seen in Fig. 284. The armature bar being well within the region of magnetic influence of the upper extremity of the field magnets, would place itself in a position as nearly along the line of magnetic force which passes through its pivot as its suspension would permit, but it is opposed partly by its own weight and partly by the pressure against its lower bar, by the spring lever to which we have referred. It therefore takes up a position determined by the varying intensity of the magnetic field and by the

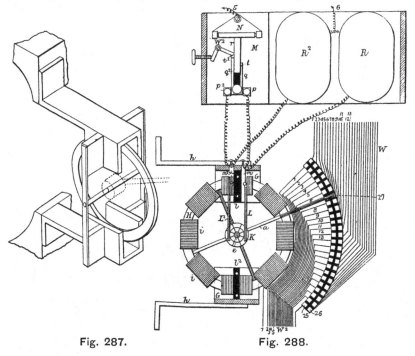

Fig. 287.   Fig. 288.

constant tension of the adjustable spring S; and this spring is so adjusted that, when the machine is generating its normal current, and the commutator brushes are in their corresponding position with respect to the neutral axis, the bar N^3 is in contact with both contact pieces of the lower lever; and the connections and circuits are so arranged that the current divides itself between two branches of equal resistance; in this position the brushes of the motor are connected to points in the circuit of equal potential, and therefore no current is transmitted through its armature and no rotation takes place; the commutator brushes, therefore, remain fixed, and the action of the apparatus is in every respect normal. If, however, resistance be added

to the external circuit by the cutting out of lamps or by any other cause, the current decreases, and there will be a corresponding diminution in the intensity of the magnetic field; the tension of the spring S will, therefore, overcome the directive influence of the field upon the armature bar, which will drop, and from the mutual position and action of the bar N^3 and spring contact bar, the circuit will be preserved between the upper contact stud and the bar N^3, while the lower contact stud will be lifted away from N^3. A portion of the current will, therefore, be transmitted through the armature of the motor, rotation will take place, and will continue until the commutator brushes have been shifted to such a position as to restore the magnetic field to its normal intensity; the relay armature will then assume its normal position, and both sides becoming again in contact with N^3, the motor will be brought to rest and the brushes will be held in their last position until the conditions are again changed. If, on the other hand, the resistance of the main circuit should be diminished, whereby its current is increased, the armature of the relay will be drawn up by reason of the greater intensity of the magnetic field, and the lever N^3, rocking on the compound contact bar, will break the circuit at the upper stud, while that at the lower stud, and therefore between the wires 5 and 6 (Fig. 288), will be maintained. The current will, therefore, have two circuits open to it, one through the resistance R, and the other through the armature of the motor and the resistance R^2. In this case the current will traverse the motor in the opposite direction to that in which it was transmitted before, and the commutator brushes are moved round in the contrary direction through positions of less efficiency, until a normal state of things is again brought about; the relay armature again becomes horizontal, contact with both studs is restored, and the motor and brush-holders are brought once more to rest.

It would be obviously outside the objects of this notice to enter into all the details of this interesting and highly ingenious method of automatically starting and reversing the motor, but there are several points connected with these devices which constitute special features of interest. One of these is the very elegant and ingenious way in which Mr. Hochhausen makes use of artificial resistances for the threefold object of protecting the motor from injury by the transmission of too strong a current through its coils, for using as little of the current generated by the machine as will effectively drive its regulating apparatus, and for the complete elimination of sparking, and consequent destruction of the contacts in the connections of the motor. Upon referring to Fig. 284, it will be

seen that there are three resistance coils or shunts, the two upright coils to the right of the figure correspond to N and N^2, in Fig. 285, and to R and R^2, in Fig. 288, and they form two branches of equal resistance for the current to divide itself over, their opposite ends being connected respectively to the two brushes of the motor, and to the two contact studs of the compound bar. When these are connected by the contact piece of the relay, currents of equal strength flow through the two branches of the circuit, and the motor brushes being of the same potential, no current passes between them, and no rotation takes place ; but directly one or the other of the contact studs is separated from the relay bar, the balance of current in the two branches is disturbed, and the excess flows through the coils of the motor in the direction of from the higher to the lower potential.

The coil to the left-hand lower corner of Fig. 284 is a shunt placed between the relay bar and the outer ends of the other two resistances, and its resistance is equal to that of the other two together in parallel circuit, that is to say, it is of half the resistance of one of them, and is so proportioned to that of the motor, as to allow of just sufficient current to be diverted through its coils as will drive the apparatus, and no more. If we have described the arrangement and connections with sufficient clearness, it will be observed that the current is always closed, and that whenever the circuit through any branch is interrupted, there is always another path open to it, and by this means Mr. Hochhausen has completely eliminated sparking in the apparatus, and has at the same time rendered the regulation more instantaneous and effective.

Referring to Fig. 284, it will be observed that the current in traversing the shunt coil is not utilised in any way ; and in some forms of his apparatus, shown in the diagrams Figs. 285 and 286, Mr. Hochhausen has replaced the shunt coil by an electro-magnet H_3 of equal resistance, and has employed the magnetism so produced to attract or release an armature lever H, and thus, by putting a detent H_2 out of, or into gear, with the toothed wheel to effect the release or locking of the motor, and so prevent the regulating apparatus moving too far through its own inertia, or under any other disturbing influence.

Fig. 288 is a diagram illustrating a current regulator devised by Mr. Hochhausen upon a different principle. In this case there is, it is true, an electric motor and a magnetic relay by which the direction of rotation is controlled, but the action of the motor is not to alter the position of the commutator brushes with respect to the neutral axis, but to throw in or cut

out of circuit, certain portions of the helices by which the field magnets are excited, and by that means to control the intensity of the magnetic field, and with it the electromotive force of the machine.

Referring to the figure, H is the armature of the motor capable of moving around its axis, and between the supplementary pole-pieces G G and brackets *h h.* Attached to the axis of the motor is an arm which carries at its outer extremity a contact piece 27, which traverses during its angular displacement, an arc built up of a number of contact blocks in two concentric rows marked respectively 25 and 26. The contact piece 27 makes connection between a block of the outer row, and its corresponding block of the inner row, and thus permits a current of electricity to be transmitted from one to the other. The outer blocks are by means of as many wires W, 1, 2, 3, 4, &c., connected respectively to different portions of the upper magnet helix, and the inner series of blocks are, by the wires W_2, similarly connected to portions of the lower helix, and the connections are so arranged that when the circuit is completed through the pair of blocks at 1, all the convolutions of both magnet coils are traversed by the current; if the connection be made at 2 a smaller number of convolutions will be in circuit, fewer still at 3, and so on, more and more convolutions being cut out as the lever attached to the motor causes the contact piece 27 to traverse the blocks from the top to the bottom of the arc. As the direction of rotation of the motor is controlled by apparatus identical with that which we have already described, we need not again refer to it here, beyond pointing out that it, as well as its connections with the regulating switch, are illustrated in Fig. 288.

The action of the apparatus is very simple. If, by the increase of the strength of the current caused by the diminution of resistance in the external circuit, the magnetic field becomes more intense, the magnetic relay transmits a current through the motor, setting it into rotation and thereby causing its arm to traverse the set of blocks, cutting out of circuit more and more of the exciting helix, and this goes on until the magnetic field is so diminished in intensity that the relay taking up its normal position brings the motor to rest, and it will remain in that position until another change of current causes these conditions to be disturbed, and a readjustment takes place. Should, on the other hand, the current decrease, a similar operation is repeated, but this time in the reverse direction.

In addition to these arrangements, Mr. Hochhausen has devised regulating apparatus in which no electric motor is employed at all, and in which the current is controlled by a shifting of the commutator brushes around

the commutator by ordinary mechanical means. In one of these a small friction wheel, mounted in a swinging frame, is brought by means of a controlling magnet in contact with one or the other, or neither, of two disc wheels which are geared directly on to the revolving axle of the machine ; and the mechanical arrangements are so made that when the friction wheel is in contact with one disc, the brushes are moved in one direction, and when against the other disc, the brushes are rotated in the opposite direction, there being a position of the magnet in which the disc is held free of both discs, and the brushes in that case remain in a fixed position.

The arc lamps employed under the Hochhausen system are constructed in two forms : the first, called the single lamp, holding carbons of sufficient length to burn for eight hours ; and the second, or double-carbon lamp, employing two pairs of carbons so arranged that upon the consumption of the first pair, the second is automatically thrown into circuit, and in this way the period during which the lamp is capable of burning without a fresh supply of carbon rods, is extended to sixteen hours. Mr. Hochhausen has, moreover, a focus-keeping lamp for employment in lighthouses, or for other purposes, in which it is important for the centre of illumination to remain fixed in one spot.

The motive power in all these lamps is obtained by the weight of the upper carbon-holder acting by gravity upon a train of wheels, by which its descent is controlled, and the electrical regulating apparatus depends for its working upon the differential action of two pairs of solenoids, one of which is included in the main circuit of the arc, and the other, which is of much higher resistance, in a shunt circuit thereto. In each of these is a pair of iron cores, which are drawn more or less within the coils, according to the strength of the current passing through them. The core of the solenoid which is included in the main circuit, is weighted so as to drop the moment that the current is interrupted, and so allow the upper carbon to descend on to the lower, but as long as the lamp is working in its normal condition this weighted core is held up within its coil, and the length of the arc is controlled by the movements of the core of the shunt coil, and this, in its turn, is affected by the variations of current passing through its coil, which are inverse to those in the arc.

Referring to the illustrations, Fig. 289 is a general view of the single carbon lamp, the doors of the case being open in order to show the internal mechanism, which is shown more in detail in the diagram Fig. 290, all the superfluous parts and fittings being omitted for the sake of greater clearness.

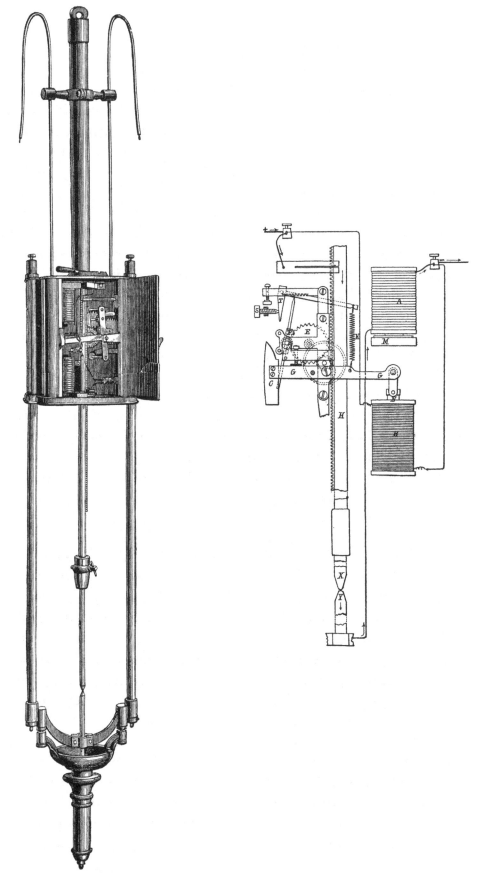

Figs. 289 and 290. Hochhausen's Single Carbon Lamp.

In this figure, X and Y are the upper and lower carbons respectively. The terminals of the lamp at which the current enters and leaves it, are marked + and −, the arrows indicating the directions of the currents in traversing the apparatus. H is the upper carbon-holder to which is attached a rack (shown in the figure), and which gears into a pinion forming part of the feeding and controlling mechanism. A and B are the regulating solenoids, the upper of which, wound with thick wire, is, it will be seen, included in the circuit of the arc, and when in action, holds up within it, its core, to which the weight M is attached. To the core of the shunt solenoid B— which is wound with fine wire of a high resistance—is attached, by means of a pair of links, the lever G G, which tends to draw the core out of the solenoid, by the joint action of the dead weight C and the spring K, the tension of which is adjustable by the screw S and bell-crank lever T. The feeding and releasing mechanism, the framework of which is attached to the lever G G, consists of a train of wheels driven by the rack on the upper carbon-holder, and controlled by the vibrating pallet P and escape-ment wheel E, with which it works, and this pallet and wheel can be fixed or released by a locking bar (shown above the lever T) which engages with the point of a vibrating fly attached to the pallet.

The action of the apparatus is briefly as follows : The current, in traversing the lamp, has two routes open to it, one through the carbons across the arc and through the solenoid A, the other as a shunt to the arc circuit, through the high-resistance helices B. If the arc become too long, a larger proportion of the current will be diverted through the shunt coils B, and their cores will be attracted, drawing down with them the lever G, and therefore the train of wheels controlling the feed. The vibrating arm of the pallet will thereby be released, and in its oscillations will allow the escapement wheel to rotate tooth by tooth, and so will feed the upper carbon downwards, until, through the diminution of the resistance of the arc, a smaller proportion of the current is transmitted through the shunt coil B, which is no longer capable of attracting its core against the opposing influence of the counterweight C and tension spring K ; the core therefore rises, and with it the controlling train, the pallet arm coming into contact with the locking bar is held in its place, the train is stopped, and the upper carbon is held in a fixed position until another change in the resistance of the arc takes place, when the process is repeated.

It will be observed that as long as there is a current passing from one carbon to the other it must traverse the coils of the solenoids A,

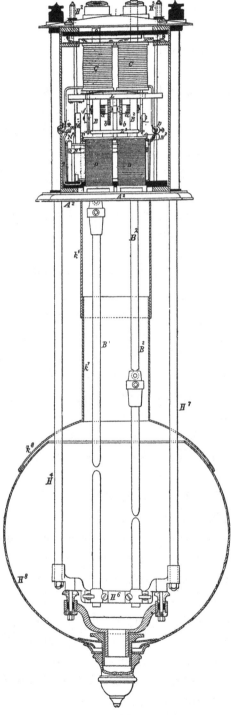

Fig. 291. Hochhausen's Double Carbon Lamp.

which are included in the same circuit, and the weight M is held up in the position shown in the diagram. The moment, however, that the current is interrupted, either through the breaking of a carbon rod or from any other cause, the weight M drops on to the end of the lever G, performing suddenly that which, under ordinary circumstances, the solenoids B do gradually; the upper carbon is released entirely from the regulating train, and dropping on to the lower carbon, restores the circuit through the carbons and the coils A; the weight M is once more lifted off G, and the spring K is then able to separate the carbons and establish the arc.

In order to guard against the very remote possibility of a series of lamps becoming extinguished through the feed of one of them being stopped by any derangement or obstruction in the regulating mechanism, each lamp is provided with an automatic switch (not shown in the drawing), whereby the offending lamp is instantly cut out of circuit without affecting the performance of other lamps operated by the same current; and, by another simple device, this same switch or "cut-out" is thrown into action when the carbons, in their consumption, become so short that there is danger of the fittings of the lamp being injured by the heat of the arc through either radiation or conduction.

Fig. 291 is a sectional elevation of the double carbon or sixteen-hour lamp, together with its globe and globe fittings. The principle of action, as well as the details of construction, of the feeding and controlling mechanism of this lamp, are identical with those of the single lamp just described, but it is, of course, larger, and there are in it certain devices and points of interest which are not possessed by the smaller instrument. B and B^2 are the two upper carbon-holders, while H^6 is the holder common to both the lower carbons. C C are the main circuit solenoids, and L the weight attached to their cores. The two shunt helices are shown at D D, while a b G and G^4 represent some of the parts of the controlling apparatus, which is repeated for each pair of carbons, the two trains being quite separate and distinct from one another. The lower part of the figure illustrates the method of fixing the lower carbons, as well as the fittings of the opal glass globe H^8. The whole arrangement is very compact and ingenious, and gives great facilities for fixing or removing the globe or replacing the carbons when consumed. Above the globe is a hood K^8, surmounted by a tube or chimney K^7, constructed of thin sheet metal, which slides after the manner of a telescope into a similar tube K^6, which is attached to the base-plate A^2 of the regulating portion of the lamp. By this arrangement the carbon-

holders are protected from dust and injury, and the arc is shielded from disturbing currents of air.

It is obvious that, as the object of employing two pairs of carbons in one apparatus is to double the period during which a lamp will keep in operation without attention, a lamp would be useless for that purpose, unless it were provided with an automatic arrangement, whereby the one pair of carbons held in reserve should start into operation upon the consumption of the second pair; and it will be remembered* that in the double-carbon lamp devised by Mr. Brush, this was effected by a couple of jamming washers of peculiar construction encircling the carbon-holders, and acted on by a lever and solenoid. In the arrangement adopted by Mr. Hochhausen the upper carbon of the second pair is held up by a detent until the first pair has descended to a certain point, when it is released

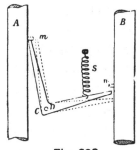

Fig. 292.

mechanically, and is immediately started into action. This arrangement is shown in Fig. 292. A and B are the two upper carbon-holders, of which A is that belonging to the reserve carbon, and B is the holder of that which has nearly burnt out. Each is provided with a projecting stud or stop marked m and n respectively, and the rod A is kept from descending by its stop m, resting against the point of the bell-crank detent D, which is pivoted at C, and held up against m, by the tension spring S. The vertical distance between m and n, when both carbons are new, is equal to the length of carbon to be consumed, and when the carbon attached to B is approaching its finish, the projecting stud n, pressing on the detent lever D, and overcoming the tension of the spring S, depresses the right-hand arm of D, and thereby releases m, allowing A to descend and establish the arc between the carbons of its corresponding pair.

In all the Hochhausen lamps the framework of the apparatus is carefully insulated from every part through which the current is transmitted, and the suspension rods H⁴ and H⁷ (Fig. 291) are thickly glazed with porcelain, so that accidents from shocks through handling the lamp are rendered impossible.

* See " Electric Illumination," Vol. I., page 429.

THE FERRANTI-THOMSON DYNAMO.

THE Ferranti-Thomson machine—which is the joint invention, or to speak more correctly, a combination of the inventions of Mr. S. Ziani de Ferranti and of Sir William Thomson—is an alternating-current machine of an exceedingly interesting and original type, the special characteristic feature being the armature, which contains neither wire nor iron, and is of such small dimensions and weight that it can be rotated at a very high velocity between the magnetic poles (which in small machines are only separated by a distance of three-quarters of an inch), and therefore moves within a magnetic field of very great intensity.

In general external appearance, the Ferranti machine has some resemblance to the Siemens alternating, and Wilde machines, but the resemblance is only superficial, for not only are the armatures, and construction and arrangement of the field magnets, quite distinct from either of those machines, but the principle of action is altogether different. The form and construction of the Ferranti armature, which will be described in detail further on, may be understood by imagining an eight-toothed pinion, 18 in. in diameter, and $\frac{1}{2}$ in. thick, around which are loosely wound twelve turns of copper ribbon $\frac{1}{2}$ in. wide, having an insulating band of the same width wound with it, so as to insulate contiguous convolutions from one another; if now, by any mechanical contrivance (such, for example, as causing the pinion so coiled to gear into another similar wheel) the copper ribbon be forced into the spaces between the teeth, an undulating outline, which will be parallel to that of the circumferential surface of the pinion on which it is wound, will be given to it, and if the pinion be then removed, the copper strip will be left as a hollow pinion-shaped star, which, when mounted upon a spindle passing through its axis would, to all intents and purposes, be a Ferranti armature.

The armature is composed of a continuous strip of thin copper ribbon,

coiled upon itself with an insulating tape between its convolutions, and having the undulating or pinion-shaped outline which we have already described. There is no iron core to this armature, the interior space being quite open, thus permitting of free circulation of air in and around the copper coil while the machine is running. The weight of this portion of the apparatus is insignificant, and this is perhaps the most interesting feature of the machine.

The small size of the armature, and especially its thinness (in small ones only half an inch), enables it to be run within a very narrow magnetic field, and therefore one of exceptionally high intensity. The field magnets are arranged in two crowns around a pair of discs of cast iron which form the outer framework of the machine, their free ends which constitute the poles of one set of magnets, being directed towards the free ends of the opposite set, and towards the armature which rotates between them. Each crown is built up of sixteen electro-magnets, which are, like the armature coils of the Wallace-Farmer machine,* sector-shaped in cross section, and the exciting coils with which they are wound are also of similar shape; in fact, if a picture of the Wallace-Farmer armature were altered so as to show a crown of sixteen coils instead of twenty-three, it would convey a very fair idea of the arrangement of field magnets in the Ferranti machine. As, however, the current generated by this machine is alternating in character, the field magnets have to be excited by a separate machine, and for this purpose a small Siemens dynamo may be employed. The magnets are each wound with insulated copper wire of about No. 10 B.W.G., and are all connected together in series in such a manner as to produce alternate polarity around the crown, so that any point in the armature passes alternately a north pole and a south pole in its rotation around its axis.

At one of the first installations of this machine, it was driven by one of Messrs. Fowler's semi-portable 16-horse engines, developing at the flywheel about 26 horse-power, and by a countershaft, at a speed of 1900 revolutions per minute, giving a mean velocity to the armature of about 7600 ft. per minute, a velocity which, by reason of the extreme lightness of the moving parts, was in no way inconvenient, and gave no trouble, and as there were sixteen pairs of magnets of alternating polarity, it follows that there were no less than 500 reversals of the current every second, or 30,000 per minute.

There is no commutator to the machine, the armature being connected

* See " Electric Illumination," Vol. I., page 183.

to the outside circuit in a manner precisely similar to that adopted in some of the early Saxton machines—that is to say, one end of the armature conductor is permanently connected to one metallic collar fixed to, and forming part of, the spindle upon which it rotates, and the other end is attached to another similar collar fixed to the spindle, but insulated from it. Upon the circumference of each of these collars, is pressed a piece of metal like a brake block, attached to the end of a short lever, so as to maintain a good rubbing contact, and these two levers are connected, the one to a terminal screw of the machine, and the other to a terminal screwed into the framework of the apparatus.

On the occasion of the first exhibition of this dynamo, 320 twenty-candle Swan lamps were arranged in 107 groups, that is to say, 106 groups of three lamps, each in series, and one group having two lamps in series, and the constancy of illumination at a constant speed of the engine was all that could be desired. With 300 lights in use, the electrical energy being thrown into the circuit was about $22\frac{1}{2}$ horse-power.

The following are some of the figures connected with this very interesting machine and installation :

Resistance of armature	.0265 ohms.
,, ,, lamp circuit	.7735 ,,
Total resistance	.8 ,,
E.M.F. in main circuit	125 volts.
Current ,, ,, ...	156 ampères.
Resistance of lamps, each	31.5 ohms.
Current absorbed by each lamp	1.3 ampères.
E.M.F. taken ,, ,, ...	41 volts.

The weight of machine and exciter was about 1500 lb., giving 13 volt-ampères per pound weight of apparatus.

The three special features of the Ferranti dynamo are :
1. Small weight of moving conductor.
2. High velocity of ditto.
3. Intense magnetic field.

The first is obtained by the peculiar shell-like character of the armature which has no iron cores to add to the weight, or to absorb energy in the production of local currents, and therefore of heat, and these conditions control the second element, enabling the machine to be run with advantage at a high velocity. The third element of efficiency is rendered possible to an exceptional degree by the form and dimensions of the armature, for, as that part of the apparatus is so very thin, the poles of

the magnets can be approached to within a small fraction of an inch of that distance, and the convolutions of the armature are thus under the best possible conditions to be influenced by electro-magnetic induction ; this fact is alone sufficient to account for a large proportion of the efficiency of the Ferranti machine.

Fig. 293 shows the arrangement of a Ferranti dynamo made to supply 5000 incandescence lamps, each requiring a current having an electro-

Fig. 293.

motive force of 200 volts, and a current of .33 ampères. It comprises three principal parts, each of which is self-contained; the armature (Fig. 294) and the two sets of field magnets (Fig. 295). The former consists of two discs or bosses insulated from each other and from the shaft, and carrying between them an insulated brass ring. To the periphery of this ring there are fastened at equal intervals eight copper strips $1\frac{1}{4}$ in. wide and 1.75 mm. thick, and these strips are then bent into the undulating form shown in

Fig. 294, each strip making the circuit of the sinuous periphery of the armature twice, and being secured at its outer end to a rivet in electrical

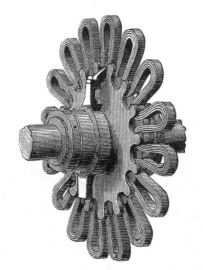

Fig. 294.

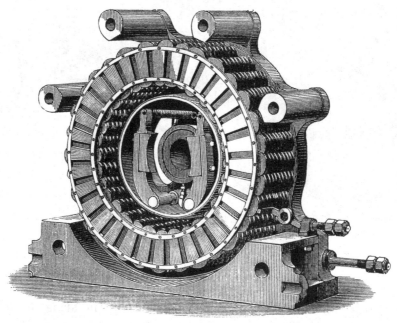

Fig. 295.

contact with one of the discs. Thus, as the strips are divided from each other by insulating material, there are eight parallel circuits around the armature, each starting from the internal brass ring and ending, after two

complete revolutions, at one of the bosses. At either side of the armature is a collector ring, one of them being joined by a heavy copper connection (Fig. 294) to the internal brass ring, and the other to the opposite disc, so that the two rings form the terminals of the armature circuit. The current is not taken off by brushes in the usual way, but by solid rubbing pieces (Fig. 295) which embrace both sides of the collector rings, and are held up to their places by springs. The current is led from these collectors down heavy copper bars, enclosed in a casing on the outside of the framing (Fig. 293) to two terminal studs, by which it is transmitted to the main leads. The outside diameter of the armature is 36 in., and its speed rather less than 1000 revolutions per minute, at which rate it will furnish about 2000 ampères of current with an electromotive force of 200 volts. As the machine is separately excited from an independent generator, and as its internal resistance is exceedingly small in comparison with that of the outside circuit, to which it will be applied, it follows that the difference of potential at the terminals will be sensibly constant for a given speed, whether there be few or many lamps in action, and that these will burn with the same brilliancy whether a part or the whole of them be lighted. Further, as the amount of current in circulation varies exactly with the number of lamps in action, the power absorbed by the machine rises and falls with the amount of light produced, and there is no waste of fuel at the time when the demands for current are but slight.

Although the armature shows considerable improvement in details from the form in which it originally appeared, and as above described, yet it is in the field magnet that the most striking alterations are to be seen. There are thirty-two magnets on each side, alternately north and south, each north pole being confronted by a south pole, and likewise having a south pole on either side of it. The cores of the magnets are cast with the framing, which, as shown in Fig. 295, is made in halves, and thus by the removal of six bolts the machine may be divided in the centre, and the whole of the internal parts laid open for inspection, while at the same time there is no danger of accidents from parts jarring loose and getting into the path of the armature. The magnet cores are excited by a current circulating among them through a system of copper bars. These are of rectangular section $\frac{3}{4}$ in. by $\frac{7}{8}$ in., and are bent in a special machine to a shape somewhat like that of the conductor in the armature. Supposing the cores to be numbered 1, 2, 3, &c., the first bar will pass over 1, under 2, over 3, under 4, and so on ; the second bar will pass under 1, over 2, under 3, over 4, and so on for

x

the whole nine bars and thirty-two cores, and thus each core will be completely surrounded both at its sides and ends by copper conductors. The exciting current will flow in the front or outermost bar down between 1 and 2, up between 2 and 3, down between 3 and 4, and so on; in the next bar it will follow the same course, and so through the whole nine, the connections being made in such a way as to secure that the current in all the conductors between each pair of cores shall flow in the same direction. In each undulation of the bars there are four plugs of insulating material inserted, and these act as distance pieces to keep the metal out of contact. Likewise a groove is cut along each corner of the cores, and a key of insulating material inserted in it to keep the copper and the iron apart. The current from the exciter enters at one of the upper studs (Fig. 293), traverses one set of copper bars, is then conveyed to the other set by a connecting bar omitted in the figure, and finally emerges at the second terminal.

The various details of the generator have been designed with great care. Each bearing has an oil cup of the usual form, and in addition to this, three oil pipes in connection with a pump driven by the countershaft, and through these latter lubricants can be forced if the ordinary means fail to keep the bearings cool; all the parts are made interchangeable, and the whole machine can be entirely taken to pieces in a short time.

Figs. 296 to 313 illustrate the Ferranti continuous current dynamo, which is used to generate the current circulating in the field magnet conductors of the large machine. From these drawings it will be seen that very great care and thought have been expended over the mechanical construction, and that everything has been done that was possible in the way of providing ample bearing surfaces and efficient means of lubrication, to insure that the original conception shall not suffer from being carried out under unsatisfactory mechanical conditions.

The general design is that of a Ferranti machine having five branches or sinuosities in the armature, and ten field magnets on either side of it. The magnet cores are cast with the frames and are wrapped with copper strips $1\frac{1}{2}$ in. wide by $\frac{3}{8}$ in. thick, bent edgeways so that they pass in a zigzag fashion backwards and forwards between the cores, each strip passing not round and round one particular core as is usual in the conductors of electromagnets, but under one, over the next, under the succeeding one, and so on in the manner already described with reference to the larger machine. The alternate strips are applied to break joint, as it were, and thus each core is completely surrounded with conductors, not only at the sides, but also at

the top and bottom. The magnet poles are alternately of north and south polarity, the current flowing in one direction between any two cores, producing a north pole in the right-hand core, and a south pole in the left-hand core, the connections between the strips being so arranged that the currents

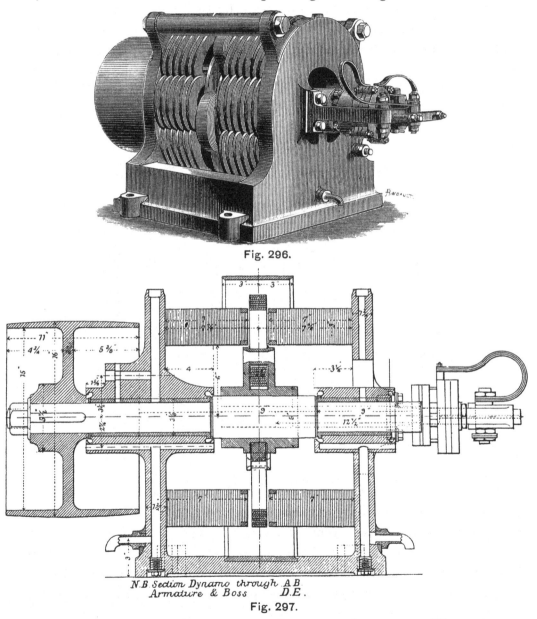

Fig. 296.

Fig. 297.

in them all tend to produce the same effect on a given core. The strips are insulated from one another and from the iron cores by pieces of vulcanised fibre, and after they have been laid in their places, are secured by two brass hoops, one encircling the outer periphery of all the cores and one lying

x 2

within them. The exciting current enters at one terminal (Fig. 299), traverses the strips on one set of magnets, is then led by a copper connecting piece (Fig. 300) to the other strips, and thence to the second terminal.

The armature consists of a boss having five double arms (Figs. 297 and 298) carrying four copper strips 1½ in. wide by two millimetres thick, bent into a sinuous form somewhat resembling the outline of a wheel with five blunt-pointed teeth. Each strip makes two complete courses round the armature, and is connected at either end to an insulated conductor passing through the main spindle (Fig. 301). The whole mass of the strips is fixed to the boss by thimbles and rivets (Fig. 297).

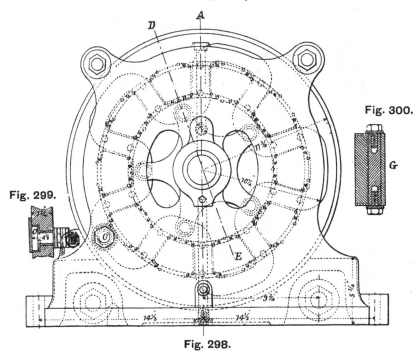

Fig. 299.

Fig. 300.

Fig. 298.

Coming now to the commutator, it will be seen that it differs in construction entirely from any previous device for the same purpose. The rubbing surfaces are not set around the circumference of the commutator as is usual, but at its end, and they move in contact with solid collecting pieces in place of brushes. There are ten fixed collecting pieces arranged in a circle, and connected alternately to the positive and negative leads, and there are six moving contact pieces sliding over them, three jointed to one terminal of the armature, and three to the others. At a given moment when a current is being generated in the armature in one direction, three moving contact pieces connected to one end of it, stand opposite three of

the fixed collecting pieces which are in communication with (say) the positive lead, and the contact pieces in connection with the other end are similarly in communication with the negative lead. When the armature has rotated 36 deg. the current in its conductor becomes opposite to that just imagined, and at the same time the three contact pieces which were touching the positive collecting pieces, become transferred to the negative collecting pieces, and *vice versâ*. Between the two instants when the

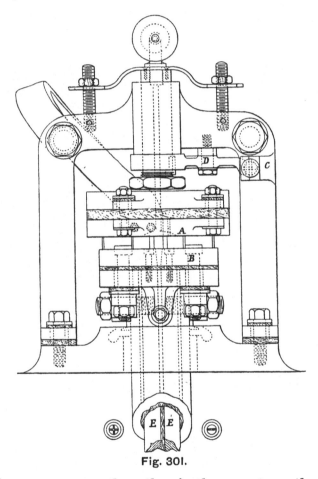

Fig. 301.

current is flowing one way or the other in the armature, there is a moment of rest or neutrality, and at that time the contact pieces are passing over insulated blocks interposed between the positive and negative collecting pieces.

The general appearance of the commutator is to be seen in the perspective view Fig. 296, while Fig. 297 is a plan of it. The main spindle is hollow for a portion of its length, and within it there are placed two insulated conductors E E¹ (Figs. 301 and 311), one connected to each end of the armature. Two insulated terminals are screwed into the conductors (Figs. 309 and 311),

and convey the current to two insulated half discs B B (Fig. 313) fixed to the end of the spindle. Upon these half discs there are six raised contact pieces, 1 to 6, the odd numbers being connected to one pole and the even numbers to the other. Opposite to the half discs there is mounted a stationary plate A (Figs. 301, 304, and 307), in whose face there are 15 sectors marked *a, b, c,* of which five are connected to the positive terminal of the machine, five to the negative terminal, and five are insulated. Those

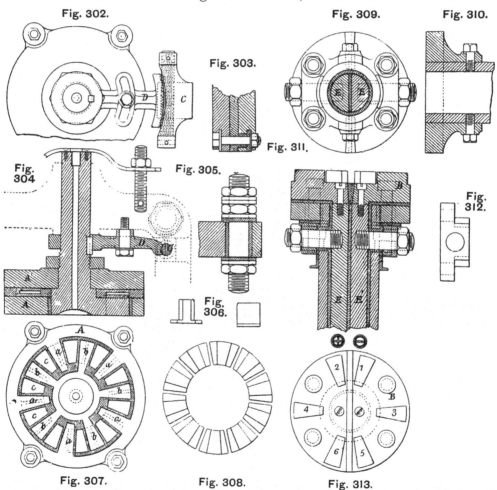

Fig. 302.   Fig. 303.   Fig. 309.   Fig. 310.

Fig. 304.   Fig. 305.   Fig. 311.   Fig. 312.

Fig. 306.

Fig. 307.   Fig. 308.   Fig. 313.

marked *a* form part of the front plate A, which is a ring with five fingers pointing inwardly towards the centre; those marked *b* constitute the star-like end of the spindle which carries the plate A, while those marked *c* (Figs. 306, 307, and 308), are separate blocks fixed by their flanges to the insulating material between the plates A and A¹ (Fig. 304). A bundle of copper strips connects the plate A to one terminal (Figs. 301 and 305), while a second bundle joins the plate A¹ to the second terminal (Fig. 301). There

is a worm G and a worm segment D (Fig. 302), by which the collectors can be set to the place of least sparking.

A moment's reflection will show that if the disc B be applied to A in such a position that the contact piece 1 (Fig. 313) coincides with a sector *a* (Fig. 307), then the pieces 3 and 5 will also coincide with two other sectors *a*, and the pieces 2, 4, and 6 will coincide with three of the sectors *b*. At such a time the ten radial portions of the armature will be passing in front of the poles of the magnetic field and the current will be at its maximum. From this point, a rotation of 18 deg. will carry the armature bars into the neutral spaces between the poles, and during this motion the current will have fallen from its maximum to zero, and at the same time all the pieces 1 to 6 will have been transferred from the sectors *a* and *b* to the insulated sector *c*, and thus all communication between the armature and the external circuit will be cut off. Another 18 deg. of rotation exactly reverses the first conditions, and as the current is now flowing in the opposite direction in the armature, it follows that it maintains its old course in the external circuit, while the impulses follow each other with such rapidity that they form practically a continuous current. The machine is intended to run from 300 to 400 revolutions per minute, and at that speed gives a current of 800 ampères with an electromotive force of 10 volts.

THE BRUSH INCANDESCENCE SYSTEM.

THE Brush Company was formed at a time when incandescence lighting was almost unheard of, and at once achieved an amount of success that kept them fully employed, leaving them but little time for the production of novelties. As the march of events, however, quickly showed that a firm which was not in a position to supply lamps for domestic purposes, would be left behind in the active competition for public favour, the company acquired the Lane-Fox patents, and was thus presumably in a position to execute contracts for any description of electric lighting. But for some reason or other, Brush incandescence installations were few and far between, and the public, while freely buying their generators and arc lamps, conceived the opinion that they did not excel in the newer department, and this feeling was signally confirmed when the Metropolitan Brush Company was found to have erected a number of Swan lamps fed by a Siemens generator. The passing of the Electric Lighting Act, and the demands of the sub-companies, who quickly found that the business in incandescence lighting far exceeded that upon which the Brush Company had made its reputation, brought the matter prominently before the Board, and consequently their attention, and that of their staff under the direction of Mr. Percy B. Allen, became devoted to the elaboration of a complete system including generators, regulators, lamps, holders, switches, and all the various details connected with the generation and control of the current.

As is well known, the Brush Company own several dynamo patents in addition to the one by which they are best known, and, therefore, when it became a question of manufacturing a machine expressly for incandescence lighting, they did not need to produce an entirely new design, but selected the Shuckert generator from their repertory, modifying it to suit the altered conditions under which it was required to work. It will be remembered

that the Shuckert machine* has an armature consisting of a disc wound Gramme-fashion, and revolving between two pole-pieces that have deep extensions embracing the sides of the disc. The field magnets, commutator, &c., are similar to those of the Gramme machine, indeed the only difference is the substitution of a disc for the ring in the armature. Originally intended to feed a number of arc lamps arranged in series, the machine furnished a current of very moderate intensity and of high potential, conditions which were totally inapplicable to incandescence lighting. The long coils of fine wire had therefore to be replaced by a much coarser conductor, and at the same time the diameter of the disc was increased, and the number of field poles was raised from two to four. This, of course, produced four neutral points in the field, and necessitated the use of

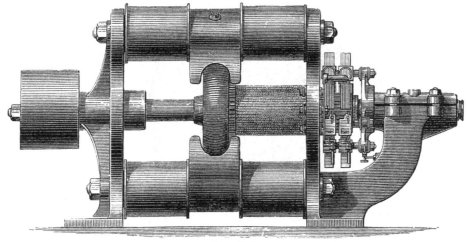

Fig. 314.

four brushes on the commutator to collect the currents, which could be employed to feed two independent circuits, or be coupled parallel into one circuit. Figs. 314, 315, and 316, show the modified Shuckert generator as now constructed, while Fig. 317 illustrates a further modification, in which three armature discs are mounted upon one spindle, each revolving within corresponding polar extensions, which are hidden from view in the figure. A considerable number of the former variety of these machines has been made, and very good results have been realised. They give a good return in current from the power expended, and maintain the lights with greater uniformity.

When the field magnets of the generators are "shunt wound," their production will vary to a certain extent with the demands of the circuit,

* See "Electric Illumination," Vol. I., p. 299.

but yet not sufficiently so, as to allow more than a moderate proportion of the lamps to be turned out without danger of injury to the remainder. To remedy this defect, and to render it possible that any number of the lamps on a circuit might be extinguished without regard to the rest, Mr. Allen devised an automatic regulator which acts upon the field magnets of the generator in such a way that the difference of potential at the poles is kept constant, whether the current be strong or weak. Fig. 316 shows the regulator and the dynamo as they appear when at work. Fig. 318 is an elevation partly in section of one part of the regulator, and Fig. 319 is a diagrammatic view, explanatory of its method of action. Referring to this latter figure it will be seen that the magnet coils of the generator form a shunt to the main circuit, and have connected with them two piles of carbon plates through

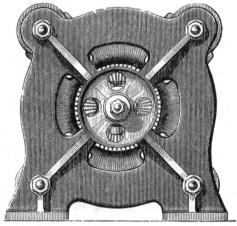

Fig. 315.

which the magnetising current must flow. These plates are contained within a vertical cylindrical case (Figs. 316 and 318), and have an electrical resistance varying according to the pressure brought to bear upon them. For every change in the number of lamps burning on the circuit a change is made in the pressure upon these plates; when the lamps are increased the plates are forced tighter together, and when the lamps are diminished the pressure is relaxed. This is effected in the first place by the solenoid *d* (Fig. 319), which is shown upon the shelf in Fig. 316. This solenoid is placed in derivation from the main circuit, and continually tends to draw down its core in opposition to the pull of an adjustable spring. The core is connected to a balanced lever *c*, which plays between two contact pieces *a* and *b*, but which, when the difference of potential between the mains is normal, touches neither of them. Should, however, this difference increase, the core would

be drawn down, and the lever would rest upon the lower contact, thus establishing a circuit from the main conductor, round the right-hand magnet *f*, and back again. The armature *g* would then attract and put the clutch into gear with one of the bevel pinions, which are constantly kept running

Fig. 316.

by a band from the axis of the dynamo (Fig. 316). The pinion immediately begins to rotate the wheel with which it is in gear, and through a worm and wormwheel, and a screwed spindle, to lower the platform upon which the carbon plates are mounted. This increases the resistance of the pile, and

consequently diminishes the current flowing in the field circuit of the machine, and the total amount of current generated. On the other hand, should the difference of potential between the main leads diminish, the contrary effect takes place, and the plates are forced together, until at length, if the demand of the circuit increase up to the full capacity of the generator, the upper plate is brought into contact with a fixed stop, and the carbon pile short-circuited. It is surprising with what readiness this regulator works ; if a couple of lamps be added to, or subtracted from, the circuit, the wheels go into gear, and, after a turn or two, come out again,

Fig. 317.

remaining stationary until a fresh change is made, and, if the lights be turned out to the last one or two, but very slight difference is to be detected in their brilliancy. By adjusting the tension of the spring which supports the core of the solenoid, the condition of the whole circuit, as regards potential, may be raised or lowered as desired.

The most important requisite in incandescence lighting is a good lamp, and at the same time it is the most difficult to attain. This is evidenced by the fact that, while there are numerous reliable generators in the market, yet only two lamps (the Swan and Edison) have come into comparatively extensive use in this country. Hence one of the points claiming the atten-

tion of the Brush Company was the reorganisation of the lamp works, and the elaboration of a system that would produce lamps of regular and uniform quality, of great endurance, and endued with a capacity for sustaining a high degree of incandescence. Abandoning the bass-broom fibre, advocated by Mr. Lane-Fox, they adopted prepared cotton thread as the material of their filaments, following in other respects the general lines of Mr. Fox's patent.

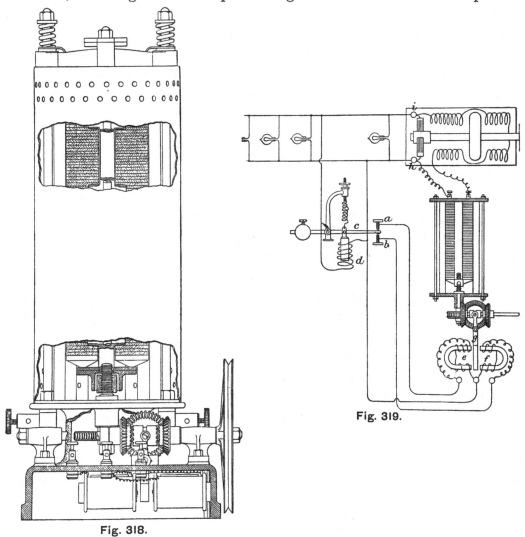

Fig. 318.

Fig. 319.

But it is well known that the efficiency of an incandescence lamp depends far less upon the patent under which it is made, than the skill and care expended upon its manufacture, and the perfection with which the air and occluded gases are removed from the globe and conductors. The general features of the Lane-Fox process of making incandescence lamps, as now carried out, is well understood, and need only brief mention. Cotton thread

is wound upon a block of carbon of pear-shaped section, and, after being packed with powdered graphite in a crucible, is exposed to a red heat for a considerable time. After it has become cool, the thread is broken into lengths at the point where it was bent over the sharp edge of the block, and each piece, after it has been cut to the desired dimensions, serves as the filament of a lamp. It is necessary first, however, to reject all those whose resistance exceeds certain limits of divergence from a given standard, and this is determined by sending a current through each, while in circuit with a galvanometer. The mounting of the filaments upon the platinum conducting wires and the whole of the preparatory work, is done by girls, and it is only in the sealing, exhausting, and flashing processes, that men are employed. The adjustment of the lamps to a standard resistance, so that all may give the same light under similar conditions, is done at two operations, the first bringing them nearly to the required point, and the second making the final reduction. Each filament is placed in a bottle through which a stream of hydro-carbon gas is passing, and is connected to two leads maintained at a constant potential, while momentary currents are sent through it by means of a key, flashing it into incandescence and depositing the carbon of the gas upon it, particularly on the thinner parts which attain the highest temperature. When the resistance is brought nearly to the desired point, the filament is mounted in its globe, the conductors sealed into their places, and then the same process of adjustment is repeated with a more delicate measuring apparatus, until the exact required resistance is attained.

The exhaustion is effected by the Lane-Fox air pump, which is very simple and gives a high vacuum, but at the same time entails a good deal of manual labour, as the mercurial reservoir has to be raised shoulder high at each stroke, and the escape valve opened and closed by hand. After the vacuum has been formed, a current is sent through the lamp to drive out the occluded gases, and the pumping is renewed from time to time over a considerable period, until it is seen that there is no discharge whatever. The lamp is then sealed up at the exhaustion tube and is complete. It remains, however, to provide terminals for the conducting wires, and for this purpose a metal ferrule, stopped at its lower end by a wooden plug, is slipped over the lower neck of the globe, and secured by plaster-of-paris. The ends of the conducting wires are brought through holes in the plug, and twisted under cheese-headed screws, which thus are made to form contact-pieces to which the positive and negative leads must be connected.

As in the Swan and Edison systems, so in the Brush system, the act of fixing the lamp in its holder or socket makes all the necessary electrical connections. The socket is convolvulus shaped, and is designed to fit the screw of an ordinary gas bracket at its lower end, while at its upper end (Fig. 320) its points or petals exert an elastic pressure upon the globe,

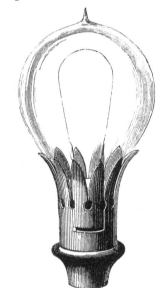

Fig. 320.

Fig. 321.　　　Fig. 322.　　　Fig. 323.

preventing any vibration. The two leading wires are led up the centre of the holder, and are secured to two segmental curved springs fixed to a non-conducting plug in the body of the socket. When the lamp is dropped into position and rotated to lock the bayonet joint (Fig. 320), the heads of its two terminal screws slide over the above-mentioned curved springs, forcing

them down towards the plug, to which they are attached at one end, and making a good contact, which is rubbed clean every time the lamp is removed or inserted. Figs. 321, 322, and 323 show the same socket constructed in a more substantial manner for attachment to a wall or a piece of furniture. In Fig. 322 the lamp is replaced by a plug provided with flexible

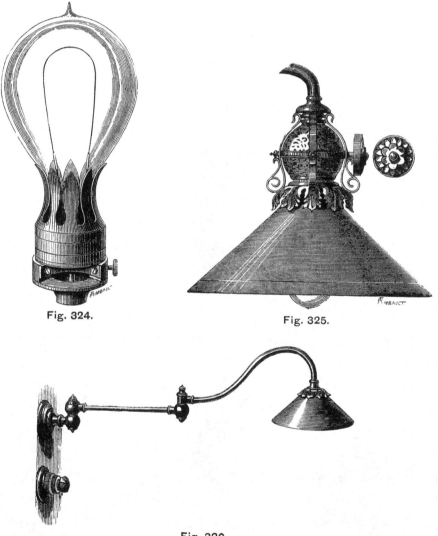

Fig. 324.   Fig. 325.

Fig. 326.

wires for connection to a portable bracket or motor, and fitted with terminals similar to those upon a lamp neck. Fig. 324 illustrates Grindle's patent socket, which has been used by some of the Brush subsidiary companies. This has no bayonet joint, but is provided with a set screw to fix the lamp, while the contact springs are set in a vertical plane instead of a horizontal

one, the terminals of the lamp being correspondingly modified in accordance. By rotating the lamp 90 deg., the electrical connections are broken.

In Fig. 325 there is shown a Brush incandescence lamp mounted in an

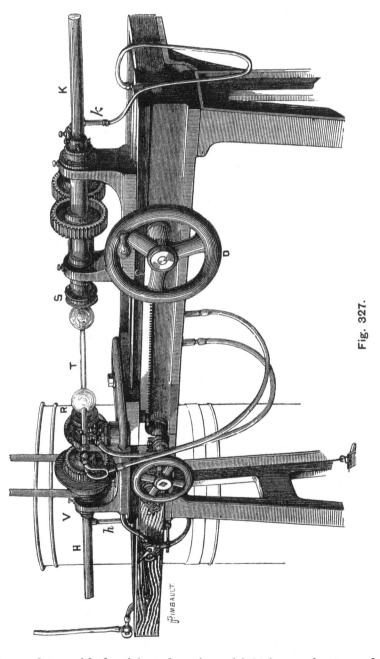

Fig. 327.

opal shade, and provided with a key by which it can be turned completely in and out, or have its intensity regulated from full incandescence to a feeble glow. The bulb above the shade contains a number of carbon discs, which,

Y

when the handle is full over in one direction, are tightly compressed, and at the same time short-circuited by a metallic bye-pass. As the handle is turned the metallic circuit is first broken, so that the lamp current has to flow through the discs, and the pressure on the latter is gradually relaxed,

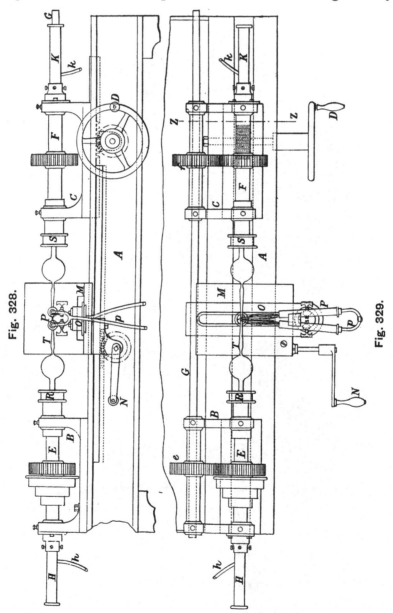

so that they offer a steadily increasing resistance to the current, and thus diminish its intensity and that of the light. In the final position the circuit is broken and the light extinguished   Fig. 326 shows a jointed bracket with the regulator or turn-down switch fixed below it.

In connection with the manufacture of incandescence lamps, we may here refer to an extremely ingenious apparatus for mechanically producing the glass bulbs. The apparatus is the joint invention of Mr. Frank Wright and Mr. M. W. W. Mackie, electrical engineers, London, and its action will be understood from the accompanying illustrations, Fig. 327 being a general view, Fig. 328 a side elevation, Fig. 329 a plan, and Fig. 330 a transverse section along Z Z. On a metal bed, like that of an ordinary lathe, are fitted two headstocks B and C, one of which, B, is fixed, while the other, C, is capable of being slid to and fro by a rack and pinion worked by a handle D.

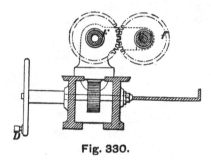

Fig. 330.

Each headstock is fitted with tubular mandrels E and F; the mandrel E being revolved by a belt from a motor, and F being also revolved at the same speed by gearing from E.

The gearing for this purpose consists of a horizontal spindle G, carrying a pinion *e* driven by a toothed wheel on E, and another pinion *f* driving a toothed wheel on F. The pinion *f* is fitted with a key or feather on the spindle G, so that it must revolve with it, but may slide freely along it with the headstock C.

At the ends of the headstocks are fixed tubes H and K, closed at their outer ends, but communicating freely with the respective tubular mandrels, and to each of the tubes H and K there is a communication by a flexible tube *h* and *k* from a reservoir of compressed air, each communication being provided with a cock or valve, enabling the operator to open or close it at pleasure.

Between the two headstocks B and C is fitted a slide M, which can be moved along the bed by a rack and pinion worked by a handle N. On the slide is fitted a transverse slide O carrying a blowpipe P, which may be double, as shown, and supplied with gas and air by flexible pipes *p* provided with valves and stop-cocks to regulate the blast. The blowpipe turns on a vertical axis on the slide O, enabling the flames to be directed on the glass at any angle.

Each mandrel E F is fitted with a chuck R S at its inner end to grasp the glass tubes inserted into it. The chucks are lined internally with soft elastic packing, such as felt or caoutchouc, and are arranged to clamp a tube without straining it unequally, while at the same time they prevent the escape of air from the hollow mandrel. The body of the chuck is of iron; it dissipates its heat rapidly, and shuts up square on the glass tube inserted within it.

Fig. 331.

To work the machine, a glass tube or rod T is inserted into and clamped in the two clutches R S, and caused to revolve by starting the mandrels. The blowpipe flame is then directed on any part of it, and the heated part is drawn out thinner, or pressed in thicker, by moving the headstock C away from, or towards, B. By admitting air under pressure into either or both of the pipes H K, the glass tube can be blown at the heated part into a bulb, which can be readily lengthened or flattened by moving the headstock C. The air is supplied by a bellows feeding a reservoir, which keeps an equal pressure, and the supply can be regulated, as we have said, at will.

In the same way two glass rods or tubes can be joined to each other by clamping the pieces in the two chucks, and bringing the free ends together while they are heated by the blowpipe. This operation is a tedious and expensive one when done by hand; and it requires great skill on the part of a blower to make even a clumsy joint between two long tubes. But with the new machine a neat joint can be made by an unskilled person in a very short time, a fact which will materially reduce the cost of many mechanical and philosophical glass instruments.

The question of safety fuses or safety switches, which constitute an essential portion of every incandescence installation, has been worked out

by the Brush Company in an exceedingly neat form. Fig. 331 shows a fuse box containing six fuses, and suitable for the control of several circuits proceeding to the various rooms of a house or building. Each consists of a loop of fusible wire stretched round a block of vulcanised fibre, and held between a fixed and a spring stop. Should the fuse be destroyed, it is the work of a moment to remove it by the handle at the top, and replace it by another, the elasticity of the bow spring sufficing to maintain the con-

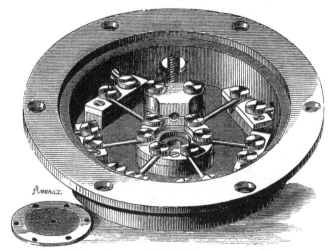

Fig. 332.

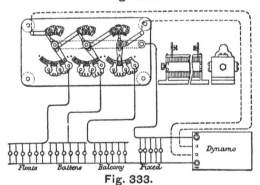

Fig. 333.

nections without the aid of screws. Fig. 332 shows a larger fuse box intended to be inserted at the connections of a main with a system of submains.

Under no circumstances is more skill and care required in the design of an incandescence installation than when it is to be erected in a theatre. The lamps in each part of the house have to be under perfect and prompt control, while the sudden variation or extinction of one set must have no effect upon the brilliancy of the remainder. At the Savoy this result is attained by the use of independent machines for each department of the lighting, and the insertion and withdrawal of resistances to and from the

field magnet circuits of each. But such an arrangement is not always convenient, and it will often be more advantageous to feed all the circuits from one generator. It is evident that a regulator such as described above in connection with the modified Schuckert generator would not be sufficiently rapid in its action to prevent fluctuations when 30 or 50 per cent. of the lamps are turned down at once, and that, therefore, the regulation of the generator must be effected by the same act as the adjustment of any particular circuit. To effect this result the Brush Company has devised a keyboard, Figs. 333 and 334, containing as many switches as there are circuits to be controlled, each switch operating upon two sets of resistances, one in the circuit to which it corresponds, and one in the field magnet circuit of the generator. Thus in the same proportion as the flow of current along any one circuit is checked, in the same proportion is the generation of current checked by the diminu-

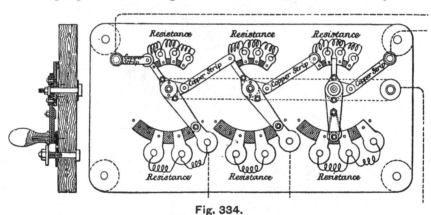

Fig. 334.

tion of the intensity of the magnetic field. Referring to the illustrations, Fig. 333 is a diagrammatic view of the installation, and Fig. 334 a view of the keyboard. There are four lamp circuits, named respectively "floats," "battens," "balcony," and "fixed," the latter being in the passages, dressing-rooms, refreshment-rooms, &c., and requiring no variation. One lead from the dynamo is connected directly to all the circuits, while the other is joined to a copper strip on the underside of the keyboard. Pivotted to this strip are three levers, the two arms of each of which have no electrical connection with each other, and through the long arms the current passes to the three variable lamp circuits, either directly or through one or both of the resistances, indicated by twirls, and also shown in section to the right of Fig. 334, according to the positions of the levers. The field magnet circuit of the generator is shown by dotted lines; it is led to a binding post at the left side of the keyboard, thence by a copper strip to the first set of

resistances, and along the short arm of a lever and a fixed copper strip to the next set of resistances, and so on, back to the generator. Thus it will be seen that if the left-hand switch, for instance, were moved to its extreme position, two sets of resistances would be inserted in the "float circuit," and three in the magnet circuit, and if these be properly proportioned, no effect should be produced upon the luminous intensity of the lamps in any of the remaining circuits.

The commercial measurement of a supply of electricity appears to be a problem that is not easy of satisfactory solution, and although many inventors have worked at it, nothing has yet been produced that promises to approach the simplicity and accuracy of the gas meter. The Brush Company, in its search after a measuring apparatus, appears to have thought

Fig. 335.

it preferable to produce a meter that might be relied upon to do its work fairly well, if it received a reasonable amount of attention, than one which, while theoretically self-acting, could not command the confidence of the public. Therefore, discarding the use of electric motors, which have figured so often in previous meters, they have boldly adopted clockwork as the going power by which the record is effected. We fear it would be trusting too much to the honesty of the public to place the control of the clocks in their hands, but a monthly visit from an inspector would not be any great nuisance in a house, the more so as, unlike a gas meter, the instrument could be placed in the hall, and so fulfil a useful purpose in showing the time. The clockwork drives a number of counter dials (Fig. 335) by means of a lever and a silent ratchet feed, the amount of feed at each stroke being regulated by the position of a core, which is drawn into a solenoid to a

depth proportional to the intensity of the current flowing round it. Hence, when many lamps are at work the rotation of the dials will be rapid, and when they are all extinguished it will cease. At the same time a pencil connected to the solenoid core may be made to produce a graphic record upon a moving paper, and thus to check the reading of the dials. This instrument is, of course, merely an ampèremeter and not an energy meter, but the regulations of the Board of Trade, as regards variation of electro-motive force, are so strict that the measurement of the current will probably suffice for all practical purposes.

THE EDISON SYSTEM.

SINCE the publication of the first volume of "Electric Illumination" no radical changes have been made in the Edison dynamo, although several modifications have been introduced into these machines to adapt them to special requirements, for which the demand had not arisen in the early days of their manufacture. In this country, especially, the Edison Company have made many installations on shipboard, in which special conditions have to be provided for. Fig. 336 illustrates what may be considered as a standard arrangement for this class of installation. It was fitted on the steamships Tarawera and Waihora, belonging to the Union Steamship Company of New Zealand.

The dynamo and engine were placed in a small room, 12 ft. by 11 ft. and 7 ft. high, opening into the engine-room, the floor being on a level with the gallery running round the top of the cylinders of the main engines. The wiring of the vessel was done in the most careful manner, nothing but the most highly insulated wire, protected by an india-rubber covering, and over that a waterproof covering of waxed thread, being used. The wires were all laid under the woodwork of the vessel in wood beading, made for the purpose, and at proper places safety catches, with fusible lead wires, were placed to obviate any danger of over-heating. The fixtures to carry the lamps were nickel-plated and of an elaborate character; in the saloon, arrangements were made by which the incandescence lamps were suspended inside the shade of the ordinary oil lamps, but could be removed at pleasure at a moment's notice; the state-rooms were lit by a single lamp, controlled by a switch placed within hand reach of the occupant of the berth; the lamps, which were enclosed in opalescent glass shades, were all the ordinary Edison 16-candle type, requiring an electromotive force of 96 volts to make each yield its normal light. In the gangways and the ladies' drawing-

room the lamps were suspended from the ceiling enclosed in opalescent glass globes. The lamp in each state-room was controlled by its own tap on the socket; the lamps in the saloon were all controlled by a single switch placed in the engine-room, so that the light can be shut out at eleven o'clock, leaving the lamps burning in the state-rooms.

The generator was one of the Edison 150-light dynamos, which had been slightly modified in order to reduce the normal speed; it was capable of being driven practically and conveniently without belting. The following data supply the necessary information as to the dynamo.

Fig. 336.

Resistance of armature, 0.1 ohm; resistance of each leg of field magnets, 20 ohms; speed, 475 revolutions per minute; electromotive force, 96 volts; current, 120 ampères; length of armature, 5 ft. 5½ in.; diameter of armature, 10 in.; 150 lamps of 16 measured candle power were maintained, each lamp having a resistance of 125 ohms hot, and taking 0.8 ampères of current.

The dynamo was driven by one of Mr. Peter Brotherhood's well-known three-cylinder engines, 7 in. diameter of cylinders, 4½ in. stroke, working up to 20 horse-power. Both generator and engine were fixed on the

same base-plate, which was bolted down to a teak bedding 3 in. thick, through the deck beams, the dynamo being further stayed by cross-stays holding it in position. The extreme size of the bed-plate was 9 ft. by 3 ft., the height of the machine over all being 6 ft. 6 in.

Figs. 337, 338, and 339 show three different Edison dynamos designed for isolated plants, which have been largely adopted in the United States. They are designated by the makers K, L, and Z machines, and are intended to feed 250, 150, and 60 lamps respectively. The general characteristics of

Fig. 337.

the three machines are very similar, the most noticeable difference, apart from their size, lying in the number of field magnets employed, which varies in the examples before us from two to six. In every case a switch apparatus is mounted upon the crossbar of the magnets to enable the circuit to be broken or otherwise manipulated. In the two larger examples the brushes are not laid tangentially to the commutators as is usual, but are fixed horizontally opposite the line of highest potential, sufficient area of contact

being obtained by making the brushes very thick. The whole of the
arrangement carrying the brushes can be moved round the commutator to
find the region of least sparking. In the Z or smaller machine the brushes
are laid tangentially in the usual way.

The generators are the type employed by the Edison Company for
Isolated Electric Lighting in America, and differ to a marked degree in their
capacity, although not in efficiency, from those now being manufactured in
this country by Messrs. Mather and Platt, of Salford. In the hands of

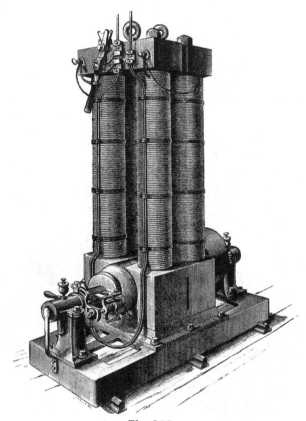

Fig. 338.

Dr. Hopkinson the Z machine has been so modified that it will now feed
200 lights in place of 60, with a proportionate increased expenditure of
power. In January, 1882, Mr. John W. Howell, of the Stevens Institute
of Technology, published a report of tests made on one of these machines in
which, as the mean of three different sets of experiments, the electrical
efficiency appeared as 95.8 per cent., and the commercial efficiency as 89.2
per cent. Lately Mr. F. S. Sprague has carried out a series of tests upon
the Hopkinson-Edison dynamo, and has found the mean efficiencies to be
94.8 and 86 per cent. respectively, results which practically correspond with

those of **Mr.** Howell. But the amount of power absorbed, and the value of the current generated, was nearly three times as great in the latter case as in the former, the cause of the difference lying in the fact that the intensity of the magnetic field had been immensely increased by the new proportions and disposition of the field magnets.

The Edison generators are shunt wound, that is, the coils of the field magnets are excited from a derived circuit branching from the poles of the

Fig. 339.

armature, and hence they are to a certain extent self-regulating. But the regulation is far from perfect, and when any considerable proportion of the lamps are turned out, say 20 per cent., there is a marked increase in the brilliancy of the remainder, and the current passing through them attains an intensity which, if allowed to continue, would shorten the life of the filament. To prevent this a regulator is supplied with each installation, and this requires to be adjusted from time to time when the variations of a

pilot light, or the deflections of a galvanometer, indicate that the difference of potential between the poles differs from the normal. This regulator (see Fig. 340) consists of a box of resistance coils joined up in series, and so arranged, that by the movement of the handle, any number of them can be

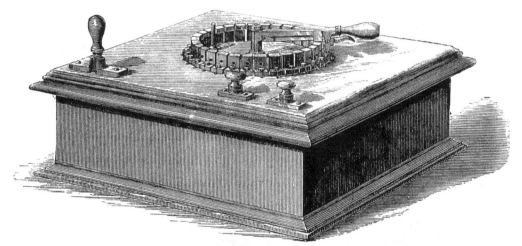

Fig. 340.

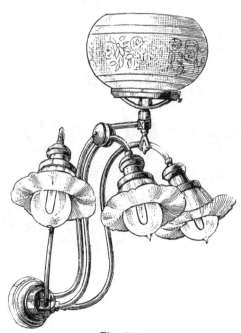

Fig. 341.

included in the field magnet circuit, and so add to its electrical resistance, causing a corresponding decrease in the current flowing through it, and consequently reducing the intensity of the magnetic field. From the junction between each adjacent pair of coils, a conductor is led to a contact piece on

the top of the box, just as in a Gramme ring a wire connects the coupling piece of each pair of bobbins to the commutator. The commencement of the first coil is connected to the first contact piece, and the end of the last coil to the last contact piece. The positive conductor from the pole of the armature is coupled also to the first contact piece, and the conductor leading to the coils is joined to the spindle upon which the handle turns. Thus, when the handle is at zero, the current passes through it from the first contact piece without traversing any of the resistance coils; when it is moved, say, 90 deg., one-fourth of the coils are included in the circuit, and

Fig. 342.

the current must circulate through them before it can reach the handle and proceed to the magnets. In this country the use of self-regulating generators is gradually displacing devices of this kind, but in theatres and places of amusement where many graduations of illumination are required, they will probably always be used.

Fig. 341 illustrates a combined electric and gas bracket, as erected in the Bijou Theatre, Boston, U.S.A., by the Edison Company for Isolated Lighting. The state of nervous apprehension which has been developed in large audiences by the fearful disasters which have occurred in theatres during the last few years, has rendered them peculiarly liable to panic, and

there are but few cases where electric light installations are so organised that total extension is an impossibility. Therefore, as managers naturally decline to place entire dependence upon electricity, the Edison Company has boldly faced the situation and has brought out a fitting which admits of the use of both illuminants, either of which can be turned up or down with equal facility.

Figs. 342, 343, and 344 are illustrations of three other varieties of electric light fittings used by the Edison Company for Isolated Lighting. Fig. 342

Fig. 343.

shows an inexpensive form of jointed bracket with shade. The joints are mechanical, not electrical, the current being carried by a flexible cord, and they are provided with thumb-screws by which they can be set in any position, a very convenient provision, as it allows of the use of long branches. In designing such fittings there is much greater scope than in gas brackets, for, although it is both unhealthy and dangerous to bring a gas flame down to the level of the table, an incandescence lamp may be placed in the most convenient position which can be found, provided the direct rays are screened

from the eye. Fig. 343 shows an epergne with four lamps, the design representing four luminous flowers, set in a bouquet of metal-work. Fig. 344 is a fitting of 10 lamps, the whole enclosed in large globe to protect them from dust and injury.

Figs. 345 to 348 illustrate some of the fittings used by the Edison Company for Isolated Lighting in America. They do not present any great novelty, but they are interesting as showing the latest practice of one of the very few companies which has secured a commercial success. Fig. 345 is a view of the junction of a main line and branch line within a building. Each

Fig. 344.

is laid under a wooden moulding, provided with two grooves on its under side, and the junction is made at a block situated at the intersection of the two lines. Each of the branch circuits is provided with a safety catch consisting of a short piece of lead, which will melt at a temperature between 450 deg. Fahr. and 500 deg. Fahr., or 1700 deg. lower than the point at which the copper conductors will give way. This is enclosed in a brass case similar in shape to the brass band fixed to the neck of an Edison lamp, and connects the two terminals with which that case is provided, standing in the same relation to them, that the carbon filament does to the terminals of a

z

lamp.   When the wire is fused by the passage of an abnormal current, it can be instantly renewed by withdrawing the case and inserting a new one.   The branch moulding at the left-hand side of the figure is shown

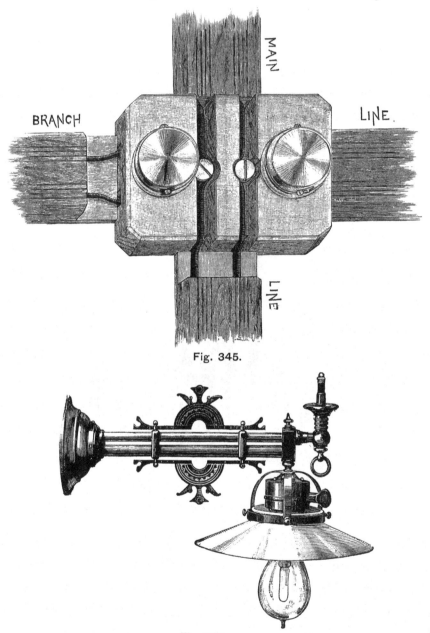

Fig. 345.

Fig. 346.

drawn back a short distance to display the relative positions of the two conductors beneath it.

Fig. 346 is a combined electric and gas bracket without any swivelling

arrangement. The lamp has an opal shade over it, and is provided with the old form of key socket. A more recent arrangement is shown in Fig. 347. Here the two conductors are made fast to two screws, one of which is connected to a plate on which the bottom terminal of the lamp socket rests, and the other to a spring bracket which touches the hollow screw that receives

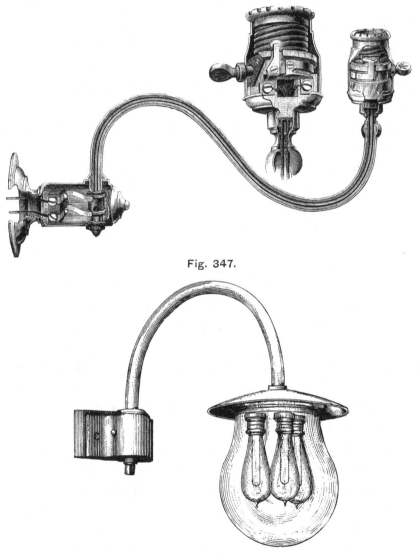

Fig. 347.

Fig. 348.

the screwed neck of the lamp, except when it is moved out of contact with it by a cam on the end of the key. Under these conditions the circuit is broken and the lamp extinguished. The bracket shown in the figure is a swivelling one, the connection between the fixed conductors and the movable ones being made by springs and two cylindrical rubbing surfaces. Fig. 348

shows three lamps enclosed in a strong globe or lantern. This arrangement
is suitable for use in exposed positions, and in mines, in which latter case
the globe can be filled with water to prevent the ignition of firedamp by
sparks and minute arcs.

Fig. 349 illustrates the great 1200-light Edison dynamo in use at
the Milan central distributing station. Milan is peculiarly well adapted
for electric lighting in its present state. The principal shops, theatres,
cafés, and clubs are all in or near the magnificent "Galleria" of Victor
Emmanuel. The lighting station is in a three-story brick building erected
specially for the purpose on the site of the Theatre of Saint Radegonde.
The lower floor is about 11 ft. below the ground level. This floor is occupied
by the dynamos and regulators, and by a test board for 1500 lamps. The
boilers are located on the floor above, and are supported directly on cast-
iron columns resting on a solid foundation. The upper floor is used as store-
room, and a small portion is divided off for a laboratory. In a vaulted room
under the courtyard is located an engine for driving a centrifugal blower,
and a pump to be used as an adjunct to the Körting injectors feeding the
boilers.

Four Edison dynamos, each capable of supplying 1200 Edison lamps
and of the type shown in Fig. 349, are already in place, and provision is
made for adding several more, so that the design when carried out will
embrace a greater number of lamps than even the station at New York.
The armatures of the machines are 27.3 in. in diameter, and 61 in. long.
The shaft is of steel $7\frac{3}{4}$ in. in diameter, and of a total length of 10 ft. 3 in. ;
the journals are $6\frac{1}{2}$ in. in diameter, and 15 in. long, running in Babbitt
metal bearings supplied with a continuous water circulation. Provision is
made for an air blast to be applied to the armature to keep it cool. The
commutator is so nicely adjusted to the requirements that, notwithstanding the strength of the current, no sparks of appreciable size are perceptible.
The steam is furnished by Babcock and Wilcox boilers aggregating 1000
horse-power, and worked under a pressure of 120 lb. All of the engines
are connected directly to the shafts of the armatures. Two of the engines
are of the Porter-Allen type with cylinders $11\frac{3}{16}$ in. in diameter, and 16 in.
stroke ; they are run at 350 revolutions per minute, and have a piston speed
of 933 ft. per minute. The other two engines already in use are by Arming-
ton and Sims. It is supposed that the latter will give better results when
two dynamos are run together on the same circuit. They have cylinders
13 in. diameter and 13 in. stroke.

Fig. 349. Edison Twelve-Hundred Light Dynamo Machine.

To adjoin page 840.

The weight of each dynamo with its engine is as follows :

							lb.
Base-plate	...	...	...	...	...	...	10,300
Dynamo	...	...	...	...	...	...	44,800
Engine	...	...	...	...	...	...	6,450
							61,550

The regulation is effected by changing the intensity of the current circulating about the field magnets. The dynamos are all regulated together by means of a variable resistance introduced into the field. Each switch is driven by a bevelled gear working into a corresponding gear keyed to a shaft.

This gearing insures an even regulation. The indicator of the strength of the current passing into the lamps, consists of a high resistance magnet placed in derivation and balanced against a tensile spring with variable adjustment. The magnet is supplied with "high" and "low" points working separate lamp circuits through a small relay. Two coloured lamps are in these circuits ; one or the other burns as the current is too strong or too weak. When either of the lamps is burning a vibrating bell sounds an alarm. A change in current sufficient to sound an alarm is, however, scarcely sufficient to be noticeable to the eye. The conductors are copper rods with connections specially designed for the purpose, so that no mistake can easily be made in laying them even by inexperienced workmen.

GAULARD AND GIBBS' SYSTEM OF ELECTRICAL DISTRIBUTION.

THE Electric Lighting Act expressly provides that the "undertakers" shall not prescribe to householders and others, to whom they supply current, the use of any particular kind or form of lamp. Nominally, every one is free to select his own fittings, and to use either Swan, Edison, Brush, or any other kind of lamp he pleases. But, although the undertakers may not insist upon the use of their own lamps, they may determine, within limits, the potential at which the mains shall be maintained, and thus they have it in their power to favour one class of lamp at the expense of another. In the experiments of the Committee at the Paris Exhibition of 1881, the difference of potential at the terminals of incandescence lamps varied from 48 to 91 volts, according to the resistances of the filaments, there being for each maker's lamp a particular point at which its incandescence is thoroughly luminous, and yet not dangerously intense. A divergence of ten per cent. above or below this standard is injurious to the life of the lamp, or greatly reduces its lighting power ; and we may be sure that it will not be the object of an undertaker who favours the sale of high resistance lamps to provide a current of low potential, or *vice versâ*, and that if a system of town lighting were now established, the householder's freedom of selection would be very limited.

But there are other points to be considered in a system of distribution, of greater importance than the prospects of rival filaments. The possibility of practically effecting arc and incandescence lighting from one source, and of combining with these, electric heating and motive power, becomes greatly complicated when there is a possibility of a different electric potential ruling in the different districts of a town. As matters now stand there is a prospect that a lamp which burnt well in the City would be dull red if transferred to Hampstead, and quite black in Hanover-square, and that the power to be

obtained from a motor would vary 50 per cent. on opposite sides of Temple Bar. Under whatever conditions the early installations are made, it is certain that there will be a steady movement in favour of supplying currents of high potential, and that lamp makers will be stimulated to furnish an article that will permit of this. The first cost, and consequently also the annual charges, on generators, leads, and troughs are in an inverse ratio to the potential at which the current is delivered to the consumer, and consequently, as interest and depreciation figure so heavily in all electric lighting estimates, it is of importance that means should be devised whereby such an important factor of economy may be utilised.

In arc lighting and the transmission of power, the advantages of high tension currents were quickly perceived. The Brush machines upon their introduction to this country, came rapidly to the front, because they would maintain many lamps upon one widely extended circuit, while other apparatus required a separate conductor for each light.

The employment of high tension currents for arc lamps is an easy matter, but with a system of incandescence lighting, in which the lamps are arranged in parallel arc, such currents have hitherto been inadmissible. Various plans were propounded before the Committee that sat to consider the Electric Lighting Bill, for the transformation of currents from a high to a low potential, by the aid of secondary batteries, but these batteries have not as yet attained a commercial existence in this country. In New York, however, we hear Mr. Brush has elaborated a scheme of this kind, and that he proposes to employ the generators that at night are used in the illumination of the streets, during the daytime to charge batteries placed in private houses.

One of the latest inventions for the distribution of electric energy is that of Messrs. Gaulard and Gibbs. It has a two-fold object; firstly, it aims at rendering it practicable for undertakers to supply current at the most economical potential permitted by the terms of their provisional orders; and, secondly, it is intended to make the user independent of the producer, and to enable him to apply the current he receives to any purpose he may please, such as arc lighting, incandescence lighting, the generation of power, or of heat. According to this invention the current in the main conductors, which proceed from the central station, is employed solely for the generation of other or secondary currents at each spot at which they may be required. Thus, if there were five hundred consumers in connection with one station, each would have his own separate generator, actuated or energised by the

main current, and he would have it in his power to generate currents at a potential of 45, 60, 91, or other number of volts as he chose. The main current would act merely as a carrier of power, taking the energy of the steam engine, and distributing it in small quantities all over the town, each subscriber absorbing any amount he required, up to the maximum capacity of his generator.

The secondary generator of Messrs. Gaulard and Gibbs is based upon the same principle as the Rhumkorff coil. In magneto machines producing alternating currents, a coil of insulated copper wire, with or without an iron core, is rapidly carried through a number of magnetic fields, in such a way that each end is exposed to quick alternations of north and south polarity; in a Rhumkorff coil there is, of course, no mechanical motion, but, in place of it, a magnetic or polar motion is induced within the coil by the alternate making and breaking of the primary current which flows round an iron core, each end of which becomes alternately polar and neutral. If, in place of a broken or pulsating current, an alternating current be employed in the primary circuit, the analogy to a magneto-electric generator becomes perfect, and magnetic fields of alternate polarity are induced within the coil in rapid succession. This magnetic or polar motion induces, in the convolutions of the coil, currents whose electromotive force depends upon the number of turns and on the intensity of the magnetic field, and the quantity of which varies with the resistances of the circuit of which they form a part. If the secondary coil be made in several parts, each with independent terminals, these parts may be variously combined either in parallel arc, or compound parallel arc, or series, according to the conditions under which the second, or locally generated current, is to be employed, and thus it is possible to generate by means of one secondary generator, currents of high or low potential. This in itself is not entirely new, for in 1877 M. Jablochkoff patented the combined use of a generator, induction coils, and lamps. The coils were placed in the main circuit, the current of which was interrupted by a make and break, and the lamps were situated in series in each of the secondary circuits. What measure of success he attained we do not know, but we believe that the system was actually tried in this country. M. de Meritens and also Mr. Bright have provisionally protected similar schemes, but neither seemed to think them worth prosecuting.

The illustrations explain this system of distribution. Fig. 350 is a perspective view of the apparatus as it was first shown at the Westminster Aquarium, Fig. 351 is a diagrammatic longitudinal section of one of the

coils, and Fig. 352 a diagram of an installation consisting of one central primary generator and eleven secondary generators. Referring to Fig. 350,

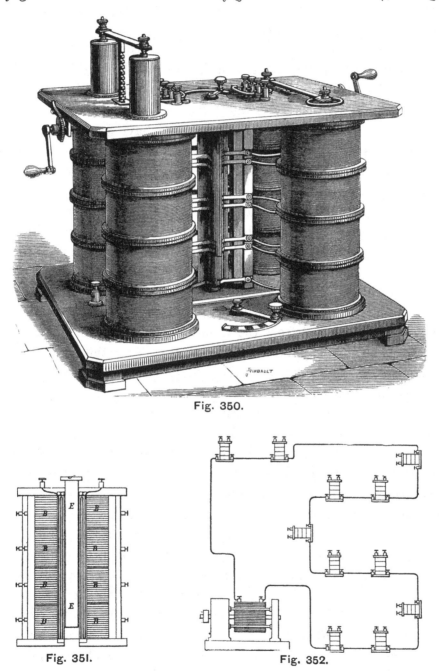

Fig. 350.

Fig. 351.   Fig. 352.

it will be seen that the apparatus consists of four vertical compound bobbins, and as these are identical in their construction the description of one will serve for all. A core of soft iron fits easily within a wooden or paper tube,

upon which is wound an insulated copper conductor in three layers of spirals, as in an ordinary solenoid. The diameter of this conductor is three millimetres. Over the cylinder thus formed, there are placed four bobbins, each wound with an insulated strand composed of six insulated copper wires, each half a millimetre in diameter. These bobbins constitute the secondary circuit of the coil, and have each two spring terminals, Fig. 350, which rest upon a vertical commutator. This consists of a polygonal piece of non-conducting material, capable of being turned about its vertical axis by a handle shown at the top of the apparatus, and having conducting metal strips let into its surface. In one angular position the positive end of one bobbin is connected to the negative end of the next, and so on ; in another position all the positive ends are connected together and all the negative ends are likewise connected, while in a third position the bobbins are grouped in two series of two each. The upper and lower ends of the secondary conductor of each vertical set of four bobbins, are connected to the terminals on the upper board, and by means of short pieces of wire these terminals can be coupled together so as to group the columns in series, or in parallel arc, or in two sets of two each. The drawing shows that the cores can be raised from the interior of the primary coils by racks and pinions, and in this way another means of varying the secondary current is obtained, as the polar field may be partly transferred to a region where it has little or no inductive effect upon the coils. The main or actuating current enters the apparatus at the left-hand terminal on the bottom board, and leaves by a similar terminal on the right side, after traversing the primary circuits of the four columns. By means of the central handle the primary coils can be successively cut out of circuit, and the current sent along a by-pass conductor.

The value of this invention will, of course, depend upon its economic efficiency. Its inventors claim that it gives very high results, but they do not appear to have the records of any trials made under exact conditions. At the Aquarium they laboured under considerable disadvantages, as they derived their current from one circuit on a Siemens W_3 alternate current machine, the other circuits of which were feeding lamps in the building. Added to this the current was too great for the capacity of their instrument, and an arc lamp had to be placed in the main circuit to modify its intensity. Now it is well known that each of the four circuits of such a machine will, in ordinary practice, feed fifty-five Swan lamps, therefore, subtracting eight lamps as the equivalent of the arc, it should in this case feed forty-

nine. It did actually feed twenty-six, but it is evident that it would have maintained a larger number—how much larger we do not know; but of course it is impossible but that there should be a loss in the secondary generation. This loss must represent an increased expenditure of coal, and it would not be difficult to determine its minimum value for a given hypothetical installation. Against this must be placed the saving of first cost and annual charges to the undertakers, that is, if they do not provide the secondary generators. We do not know at what price it is proposed to supply these, but it is certain that their aggregate value in an installation of 20,000 lamps, distributed among 1000 houses, would form a formidable item to be placed against the saving at the central station. But Messrs. Gaulard and Gibbs are far from basing the claims of their invention to the appreciation of the public upon a comparison of its capabilities with those of the small ventures in electric lighting already attempted ; they put it forward as a solution of the problem " of the further industrial development of absolute distribution, that is to say, a system of distribution limited neither by the distance of the central factory from the point of consumption, nor by the number of consumers to be supplied."

The National Company for the Distribution of Electricity by Secondary Generators have, with considerable success, completed the lighting of a portion of the Metropolitan Railway, seven miles in length, from a central station, by the system we have just described. The chief apparatus employed in the installation, although identical in principle with the former, differs greatly from it in appearance, and has developed from a scientific instrument into an industrial plant, capable of application in connection with all forms of electric lighting. The distinctive feature of this system is that the electric current generated at the central station does not pass through any of the lamps which owe their light to it, but circulates in a closed circuit, which, starting from the primary generator, traverses the whole district to be lighted, carrying the energy developed by the engine and absorbed by the generator, into every part of it. · The visible manifestation of this energy in the form of light, is due to the secondary generators, which give the name to the company. The first current is an alternating one, and, in the secondary generators, induces at each reversal of its direction, a momentary current in wires lying beside the one it is traversing. The intensity and electromotive force of the secondary current depend upon the relation of the two wires, and consequently the apparatus will, if suitably designed, give currents of any desired quality, either for arc or incandescence light-

ing, or for other purposes. The qualities of the primary current are not determined by the conditions under which the lamps will burn, because it is only a carrier of energy, and consequently it is in the power of the electrician to select those conditions which are the most economical both as regards the first cost of the conductor and the loss due to its resistance. The size of the conductor and the loss, depend entirely upon the intensity of the current, and consequently in this installation a moderate current of 10 ampères, with a high electromotive force of 2000 volts, is employed.

The central station is at Edgware-road, where the primary current is generated by a Siemens W_o alternating machine. Although the tension of the current is, as we have said, 2000 volts, yet any part of the circuit may be touched without inconvenience so long as the insulation in the other parts is maintained perfect.

The total length of the primary circuit is fifteen miles. It consists of a copper strand of No. 8 B. W. G., carried on insulators fixed to the brickwork of the tunnel, and is visible from the carriage windows on the right-hand side going westwards. There are at present four stations lighted, Aldgate, Gower-street, Edgware-road, and Notting Hill Gate, the total number of lamps being about 100 Swans and 6 arcs. At each station there is a secondary generator, placed in a cellar or vault or in any odd corner, as it requires no mechanical power or attention. It occupies a floor space of 4 ft. square, stands about 5 ft. high, and consists of 16 vertical columns. Each column is composed of an iron core, wound with a cable formed of a central wire of No. 8 B. W. G., surrounded by 48 fine wires, each separately insulated, and all connected together at their ends. The terminals of both primary and secondary wires are led to ingeniously and well-arranged switch boards, by which the columns can be grouped together in any desired order to obtain the required conditions. For instance, one group of four columns can be arranged in series to feed a number of Jablochkoff candles, also arranged in series. Another group may be arranged parallel to drive a powerful arc lamp, while the remaining eight columns may be put in compound parallel groups of two each to supply high resistance incandescence lamps. Minute variations are obtained by raising and lowering a copper shield, which covers a portion of the iron core of each column. With a primary current of 10 ampères, each secondary generator will absorb 8 electrical horse-power, the loss being stated to be only 10 per cent. The power required to overcome the resistance of the primary conductor will be one-quarter of a horse-power per mile, with a wire of 4 millimetres diameter, and

a current of 10 ampères, and as long as these conditions are maintained the loss is constant for any power that may be transmitted.

Thus in the present installation four electrical horse-power are consumed in driving the current through the primary circuit, but if another hundred lamps were added in the stations, or if other stations were lighted, there would be no increase in the loss, as additional circuits would be brought into play in the primary generator, or its speed would be increased until the difference of potential at its terminals was sufficient to maintain a current of 10 ampères in the line as before.

It is impossible not to be struck by the novelty and singular adaptability of Messrs. Gaulard and Gibbs' system. The main already laid is sufficient to stretch round the whole Metropolitan Railway circuit, and if the electromotive force of the current were adequately increased, it would carry energy to light every station without greater actual loss on the way than occurs already. To what extent the potential can be raised is a moot question, but appearances tend to prove that it can be multiplied to several times its present figure if the insulation is carefully attended to. The cost of a conductor to carry a direct current for lighting by incandescence, on a circuit of fifteen miles, would be perfectly prohibitory to any scheme.

MULTIPOLAR GRAMME DYNAMO.

THE multipolar Gramme continuous current generator has two series of electro-magnets of twelve each, mounted in circles upon two parallel frames, and enclosing between them a thin Gramme armature. The poles of the magnets are alternately of north and south polarity on each frame, but a north pole on one frame faces a north pole on the other, and a south pole faces a south pole, opposite poles being always of like name. In this respect the arrangement differs from the usual multipolar machines, which, as a rule, are of the alternate current type.

The armature is mounted upon wooden discs connected to the axis by cast-iron flanges, and its position between the magnetic poles is regulated by collars in the bearings, so that the clearance space can be divided exactly in halves.

All the positive brushes are held in fixed cases in a metallic disc, and all the negative brushes in cases attached to a second metallic disc, Fig. 357. These two discs, electrically insulated, are bolted together. They can turn freely upon a central circular bearing cast in one with the framing. They are further held against the framing by two slides (Fig. 354), which allow them to be displaced angularly by a certain amount, when it is desired to vary the position of the brushes in relation to the neutral magnetic axis. The regulation, fixing, and disconnection of the brushes is effected by means of pins mounted in bearings, which latter are fixed to the periphery of two discs, likewise electrically insulated, and bolted together like the first pair (Fig. 358). The two discs turn freely upon a central bearing cast upon the support of the main shaft. The pins penetrate into a series of notches cut in the brush holders, one of the regulating discs receiving the pins designed to operate the positive brushes, and the other those for the negative brushes (Figs. 353 and 354), and thus each set is coupled in quantity.

Ordinarily the four discs, the brush holders, and the brushes are rigidly connected; and to bring each brush successively to the upper part of the

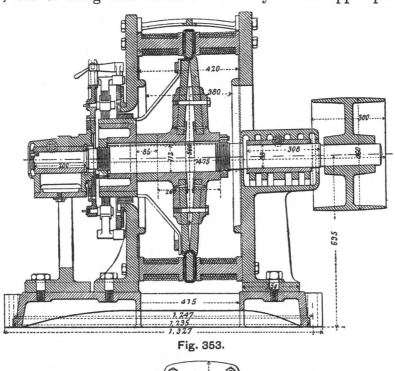

Fig. 353.

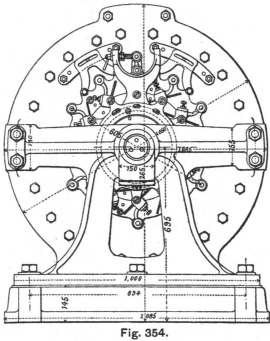

Fig. 354.

collector, it is only necessary to withdraw the bolts which fix the first discs to the slide (Fig. 357) of the framing. Thus the lengths of the brushes can

be exactly regulated before they are fixed in their respective cases.   This
completed, the bolts are returned to the slides, and the topmost brush is
placed in the best position with regard to the neutral line.   All the brushes

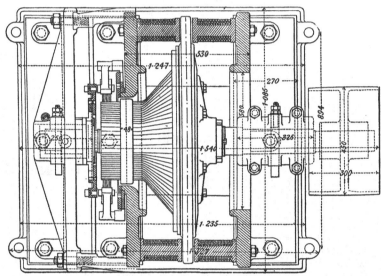

Fig. 355.

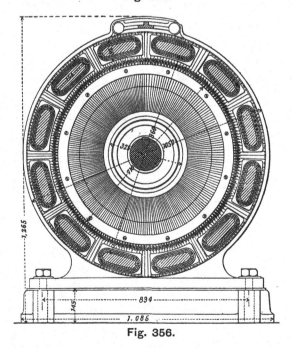

Fig. 356.

are thus placed in the best position with relation to the polar armatures,
but as yet they do not touch the armature.   To put the brushes in action
it is necessary slightly to rotate the discs carrying the regulating pins, by

means of the screw shown to the left of the centre line in Fig. 354. It results that the pins cause all the brush holders to rotate simultaneously, and to give a uniform and graduated pressure to the brushes. With a little practice all the twelve brushes can be adjusted as easily as two ordinarily are.

Two machines of this type have been completed, and are still in the experimental stage; consequently it is impossible to give exact figures as

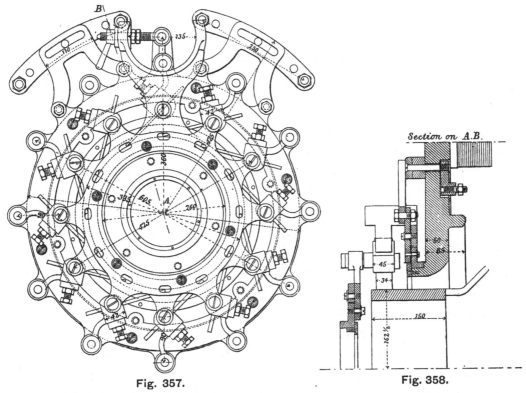

Fig. 357.   Fig. 358.

to their performance. In the early trials they were arranged as generator and motor, the former running at 834 revolutions, and the latter at 645. The work of the motor, as measured by a brake, was 40.5 horse-power. The return was more than 60 per cent., the power being transmitted to a distance of a kilometre.

GORDON'S DYNAMO-ELECTRIC MACHINE.

MR. GORDON'S dynamo-electric machine at the works of the Telegraph Construction and Maintenance Company, Greenwich, is without doubt the largest that has yet been constructed. The object of the inventor was to produce a plant which would take, as it were, the place of large gas works for supplying light from a central station, and with this object the dynamo illustrated by Fig. 359 was constructed to supply from 5000 to 7000 Swan incandescence lights of 20-candle power each.

In this machine the inducing magnets are revolved, and the induced coils are fixed, a principle which has been followed before by many other inventors. Such a plan allows the revolving wheel in the Gordon machine to be built up of massive wrought-iron plates. A disadvantage of machines in which the armature itself moves, is the difficulty of constructing a suitable rubbing contact for currents of from 1000 to 10,000 ampères or more in strength; but when the magnets themselves are caused to revolve this difficulty does not appear. A further difference in the pattern for large and small dynamos is to be found in the fact that about the same electromotive force is required, whether the machine be large or small. The armature of a small dynamo is wound with a great length of fine wire; while the conductors on the armature of a large machine must be much shorter and of much greater sectional area. In the machine about to be described the conductor is about 130 ft. long and has a total cross section of 3.44 square inches.

These short, thick conductors may be constructed in two ways: either they may be made of copper rod or ribbon, or they may consist of a great number of separate coils of wire of moderate diameter, all connected in quantity. The latter plan has been adopted in the Gordon machine, where there are 128 coils, each wound with about 130 ft. of No. 7 copper wire, and all so connected. The diameter of each wire is 0.185 in., and its

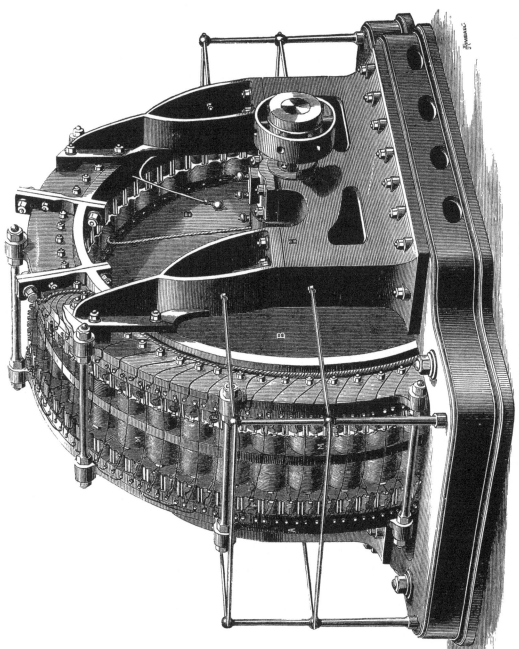

Fig. 359. Gordon's Dynamo-Electric Machine.

cross-section is .0269 square inch, giving 3.44 square inches for the cross section of the conductor on the total number of coils.

As will be seen from the figure, the dynamo consists of a strong revolving wheel W, built up of wrought-iron boiler plates, and consisting of two central discs A and two hollow cones B, whose bases fit upon the central discs. The main shaft C passes through the apices of these cones. The discs are kept apart at the rim by a wrought-iron ring, and at the centre by a cast-iron distance piece. The cones and discs are separated at the centre by massive cast-iron bosses, and the whole frame is properly strengthened by angle pieces.

This ring carries a series of thirty-two electro-magnets M M round its periphery, the bobbins forming the poles projecting from its sides. The poles run alternately north and south on each side. These magnets are formed of cylindrical wrought-iron cores, fitted with brass bobbins containing the coils. Pole-pieces of boiler plate, made in the form of a V with sides following radii of the wheel, are fitted to the magnets. The shaft runs in bearings of phosphor-bronze carried by the side frames H. There is a large opening in the sole-plate of the machine, to allow the wheel to turn in a pit below the floor level, so as to bring its centre of gravity low, and increase the stability of the structure

The end thrust is taken by two loose iron collars placed on the shaft and pressed gently against the inside ends of the journals by set screws projecting from the ends of the cast-iron bosses. The collars carry contact rings for conveying the exciting current to the magnets. These rings are of phosphor-bronze, and are insulated from the iron collars by split rings of vulcanite. The usual copper brushes press on the contact rings, and convey the exciting current to them. The electro-magnets are all connected in series, so that the current from the exciting machines flows right through them. At Greenwich these machines are the ordinary Bürgin type, supplied by Mr. R. E. Crompton.

The revolving wheel thus constructed turns between two parallel sets of fixed coils F F carried by the cast-iron frame, which is made in three segments, one smaller than the rest. This segment can be removed to allow of a defective magnet being got at by revolving the central wheel until the faulty magnet comes in sight. There are 128 fixed coils in all, or sixty-four on each side, and these are acted on by the thirty-two poles of the magnet wheel facing them. There is thus twice the number of fixed coils to moving magnets. If the machine had an equal number of coils and magnets the

induction of one coil on the next would be such as to reduce the efficiency of the machine, so that if a certain number of lamps were being fed by one coil, the closing of the circuit in the one next to it on one side, would reduce the light of the lamps on the first coil by about 25 per cent., and closing the circuit of the next on the other side, would still further reduce it by a like amount. This is due to the fact that as the currents in adjacent coils circulate in opposite directions, they are in the same direction in those parts of the coils nearest each other; and as they increase together they retard each other mutually.

The coils are painted red and blue alternately, and the red coils form a separate circuit from the blue ones. The magnets act alternately on the red and blue sets. For instance, when the poles of sixteen magnets are having their maximum actions on the red coils 1, 3, 5 up to 31, the other blue coils, 2, 4, 6 up to 32, are practically idle, and act as a metallic screen to the injurious induction of the red coils on each other. Coils 1 and 5 still tend to reduce the current in 3; but being separated from it by the thickness of coils 2 and 4, their action is so small as to be unnoticeable. Coils 1 and 3 induce electromotive forces in the two sides of 2, but these are equal and in opposite directions and so annul each other, leaving coil 2 undisturbed.

In coupling several coils together, either in quantity or series, the red coils are only connected with red coils, and the blue with blue ones. Each coil can be used to feed a separate lamp circuit, or groups of coils can be joined together at will; but in general practice the coils would be connected in two separate circuits, the red in one and the blue in the other.

To economise space these coils are made wedge-shaped, with their sides radial to the centre of the shaft. The cores are of wrought iron, as Mr. Gordon finds it capable of higher magnetisation than cast iron. The flanges F F of these fixed coils, next to the revolving wheel, are of German silver, a material which has great rigidity and a high resistance to the circulation of induced currents in it. To further prevent the circulation of these currents the flanges are slotted, in the manner shown, at right angles to the direction such currents would take.

These coils are fixed to the cast-iron frames by prolonging the cores outwards, and securing them to a fixed iron ring-shaped plate. To prevent the waste of power by induced currents in this plate, it is set back from the coils to a calculated distance, which reduces the induction currents in the plate to nearly *nil* for the whole of the coils, and thus the frame does not develop heat. Blocks of wood are inserted between the plate and coils

to keep the frame rigid. The clearance between the revolving wheel and fixed coils is about ⅛ in.

All the bobbins, both of the magnets and fixed coils, are wound with No. 7 copper wire, having a double covering of cotton. The bobbins, when wound, are completely soaked in shellac varnish, and baked at a high temperature. They are then painted over with asbestos paint, the parts liable to heat receiving several coats. This paint is peculiarly serviceable in such a case, as it does not tend to blister.

The scale on which the machine is constructed, may be gathered from the fact that the diameter of the revolving wheel is 8 ft. 9 in., and its weight 7 tons. The sole-plate is 13 ft. 4 in. by 7 ft., and the total weight is about 18 tons.

The current is an alternating one, and therefore is not adapted to charge secondary batteries; but the alternations are so rapid, and the machine so manageable, that accumulators are not strictly necessary to it, since it can be kept going by day, supplying a few lights with a proportionate absorption of energy.

The machine serves to light the entire factory and grounds of the Telegraph Construction and Maintenance Company at East Greenwich. The installation comprises 1300 Swan lamps of 20-candle power each, distributed through the premises, among the various fitting shops, cable machines, testing rooms, and offices, as well as the avenues of approach, where the lamps are enclosed in sets of four within ship's lanterns, to serve as street lamps.

In feeding 1300 lamps, the 128 excited coils are joined up four in series, and thirty-two in quantity. The current in the magnets, from the two Bürgin machines, is about 17 ampères, and the electromotive force is about 88 volts. The current in each armature wire is $24\frac{1}{4}$ ampères. For a larger number of lamps all the coils would be connected in quantity, and the speed would be 200 revolutions per minute, the current in the magnets being about 48 ampères, with the same electromotive force as before. Then with 5200 lamps in circuit the current in each armature wire would be $24\frac{1}{4}$ ampères as before, but this wire will easily carry 40 ampères, which Mr. Gordon claims would feed 8600 lamps.

It will be seen that the speed of revolution is a slow one compared with that of other dynamos, and hence the machine is less likely to deteriorate. When more current is required the power of the exciting magnets is augmented by increasing the exciting current. The operations by which

this is effected are conducted in a dark room, in which a photometer is kept, and through which pass the steam pipes of the two engines driving the dynamo and the exciting machines. The regulating apparatus consists of a shadow photometer, a pressure gauge, a Hearson strophometer, a Perry and Ayrton ammeter, and stop valves applied to the steam pipes, with their wheels so placed that a man can control them while reading the photometer. The strophometer indicates the speed in turns per minute of the large dynamo, and the ammeter shows the strength of current in the exciting magnets. One wheel works the stop valve of the engine driving the generator, and the other works that of the engine driving the exciting dynamos. The control is effected in the following manner: Two Swan lamps are fitted up on the photometer table, one in the red circuit and one in the blue. By means of suitable switches, the current is turned on to one of these lamps, and the intensity of the shadow cast by it on the screen of the photometer is compared with that thrown by a standard candle. If it is too dim, showing that the lamp is too feeble, the operator turns the valve wheel of the engine driving the exciter, and admits more steam, thus quickening the engine and the exciting machines. The result is that the exciting currents in the magnets are increased in strength, as the ammeter in circuit shows. The intensification of the magnetic field of the generator which ensues, of course causes the generator to go more slowly, but this slackening of speed is corrected by admitting more steam to its driving engine by working the other valve wheel. The strophometer indicates when the old speed has been regained, and the photometer shows that the lamp has recovered its normal brightness. If a lamp should become too bright, the same process, in the contrary direction, is followed out, and one circuit after the other is tested thus by means of the switches. The speed of the generator is kept practically constant, the regulation being effected by increasing or diminishing the strength of current in the exciting magnets. This mode of regulation allows of a great number of lights being put out at a time, and enables the machine to feed few or many. To avoid any sudden change of light in regulating, a slow motion is given to the valve wheels by means of tangent screws turned by hand.

GANZ'S ELECTRIC GENERATOR.

THE enormous electric generator shown by Messrs. Ganz and Co., of Budapest, at the Vienna Electrical Exhibition of 1883, has now been permanently erected at the Central Station at the Hungarian States Railway at Budapest. It is employed to supply current for 70 arc and 600 incandescence lamps, and will form part of one of the most important and interesting installations in Austria. The machine itself and the inverted compound engine by which it is driven, are illustrated in Figs. 360 and 361, the two being combined on one bed-plate, and the moving part of the generator forming the flywheel of the engine.

The currents are alternating, and therefore the field magnets need to

Fig. 362.

be separately excited. Consequently the machine consists of two separate parts—viz., an alternate current generator and an exciting dynamo, both mounted concentrically on the same shaft. The armature of the main generator is composed of a wire drum, on the inside of which there are mounted thirty-six induction coils, in the manner adopted by Messrs. Ganz and Co., and as shown in Fig. 362. These are wound with wire of 3.8 mm. (.150 in.) diameter. The thirty-six field magnets rotating inside the armature coils form a flywheel 8 ft. $2\frac{1}{2}$ in. in diameter, and 18 in. wide. The fixed magnets are wound with wire .138 in. in. diameter.

The exciting machine (Fig. 360) is a six-poled Gramme ring of

To adjoin page 361.   Fig. 360. Engine and Dynamo by Messrs. Ganz and Co., Budapest.

The material originally positioned here is too large for reproduction in this reissue. A PDF can be downloaded from the web address given on page iv of this book, by clicking on 'Resources Available'.

To adjoin page 361.

Fig. 361. Engine and Dynamo by Messrs. Ganz and Co., Budapest.

The material originally positioned here is too large for reproduction in this reissue. A PDF can be downloaded from the web address given on page iv of this book, by clicking on 'Resources Available'.

3 ft. 11 in. mean diameter, revolving within twelve pairs of field magnets. The ring is wound with wire .98 in. in diameter, and the magnets with wire .138 in. in diameter. The commutator is 11.8 in. in diameter, and consists of 180 sectors.

A noticeable feature of this machine is that the whole of the fixed parts, that is, the induction coils of the alternate current machine and the field magnets of the dynamo, are removable sideways to such an extent that a man may enter for the purpose of making repairs, though it is scarcely likely that this provision will be needed. To this end the fixed parts are mounted upon two stout horizontal bars, along which they can be drawn by two screws, geared together and to a handwheel. By the aid of this arrangement one man can perform the work of removing the drum in a few minutes.

The following are the electrical measurements in connection with this machine :

Resistance of 36 induction coils in parallel circuit	0.0039 ohms.
„ 1200 lamps and leads	0.038 „
„ field magnets of alternate current machine	0.44 „
„ Gramme ring	0.165 „
„ field magnets of dynamo	0.24 „
Number of revolutions	180 per min.
Difference of potential at terminals of alternate current machine ...	57.6 volts.
Difference of potential at terminals of exciter	36.4
Current in alternate current machine	1516 ampères.
„ dynamo	88.8 „
Watts of alternate current	96,297
„ exciting current	6,663
„ in lamp circuit	87,330
Electrical efficiency	85 per cent.

The weight of the machine is about 15 tons. It was designed and constructed in the short space of three months, and during the time of the Exhibition was employed without any interruption to feed 1200 Swan lamps of 20-candle power each, of which 900 were placed in the theatre.

THE ZIPERNOWSKY SYSTEM.

IN 1878, Messrs. Ganz and Co. devoted a part of their extensive engineering works to the manufacture of electric light machinery, and adopted the system patented by Mr. Zipernowsky, to whom they also entrusted the management of this department. The Zipernowsky system embraces both arc and incandescence lamps and the dynamo machines necessary to produce the current for the lights, as well as all accessory appliances required for a complete installation. The first machine constructed on this system in 1879, was designed especially with a view to feed several lamps, arranged either in parallel circuit or in series, and to admit of lights being worked at very great distances from the dynamo machines.

Considerable progress had already been made by Messrs. Ganz and Co. in 1880, especially in agricultural districts, where electric light was used for various kinds of farming work during the night. The first experiments having been highly satisfactory, several installations were fitted up on the Government farm at Mezöhegyes, and these have now been in regular use for four years, and have met with general approval. The advantages of working thrashing machines at night, particularly during harvest time, are too well known to require special mention here; suffice it to say, that the director of this Government farm, Mr. Gluzek, found as the result of three years' experience, that twelve thrashing plants fitted with electric light, would produce more work than sixteen without it. The mode of driving the dynamos direct from the portable engine and without the use of an additional loose pulley is shown in Fig. 363, where the large circle A represents the driving pulley or flywheel of the portable, B that of the dynamo, and C on the right that of the thrashing machine. The two driving belts work one over the other on the engine flywheel, and the

dynamo is fixed on a special frame at a convenient distance between the engine and the thrashing machine.

When, in May, 1881, the Crown Prince of Austria was received at Budapest, Messrs. Ganz and Co. embraced the opportunity to make a grand show of electric lights, and erected thirty arc lamps of 600-candle power each, which attracted universal attention, and proved to the

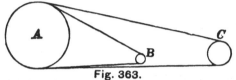

Fig. 363.

Austro-Hungarians that electric illumination was considerably beyond the experimental stage. Orders for a large number of installations soon rewarded the enterprise of Messrs. Ganz and Co. The arc lamp used in the Zipernowsky system is shown in diagram in Fig. 364, and is manufactured in three different sizes, for burning five, eight, and sixteen hours. The latter are double lamps, with two sets of carbons, and are fitted with an automatic shunt, which puts into circuit the second pair of carbons after the first are burned down. The first permanent electric light installation

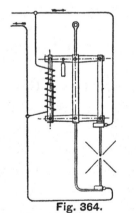

Fig. 364.

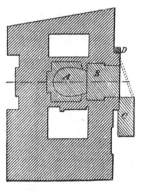

Fig. 365.

carried out by Messrs. Ganz and Co., after keen competition with foreign companies, was that at the skating rink in Budapest, with ten arc lamps. This was followed by the harbour of Fiume with eight, the shops of the Hungarian State Railway with fourteen lamps, and the Szlatina mines with twenty arc lamps. All these installations, although the first in execution, are still in constant use, and give thorough satisfaction, while the lighting of Fiume Harbour may be quoted as an example of rapid work, the lamps being in working condition within one week of the order being received at the works.

Since 1882 the firm have successfully carried out a number of installations with incandescence lamps, one of the largest being that of the National

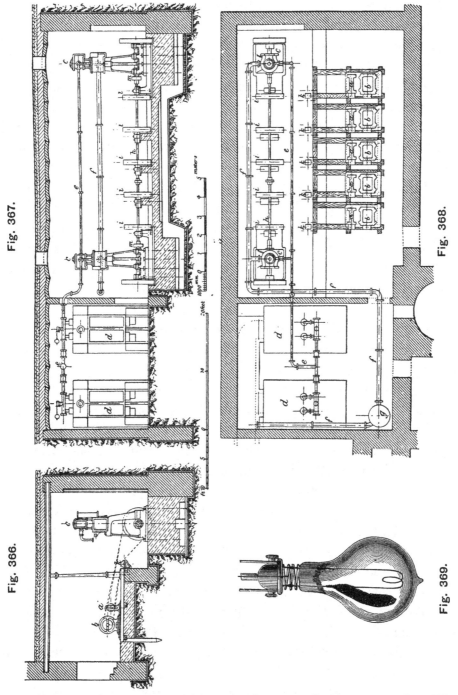

Fig. 367.

Fig. 368.

Fig. 366.

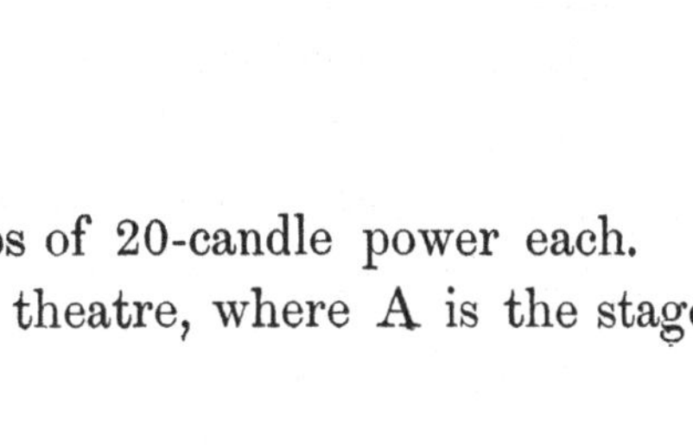

Fig. 369.

Theatre in Budapest, with 1000 lamps of 20-candle power each. Fig. 365 represents a small block plan of this theatre, where A is the stage, B the

auditorium, C the engine and machine-house, and D the chimney. The engine-house is in a building adjoining the theatre, where a spacious basement was available; this basement is divided by a wall into two parts, the smaller of which contains two water-tube boilers *d d*, of Büttner's type, one only being required for regular work, the second forming a stand-by. The steam supplies two vertical compound engines by Messrs. J. and H. Gwynne, which are placed one at each end of the engine-room, the main shafting being arranged between them, with a friction coupling at each extremity, by means of which either one or the other, or both engines, may be used to drive the shafting. To the latter are keyed five carefully balanced driving pulleys *i i*, running at 180 revolutions per minute, and these carry the belts for five alternating current machines, specially designed for lighting theatres with incandescence lamps. Each machine is capable of supplying current for 250 twenty-candle power incandescence lamps, requiring a difference of potential of 56 volts, and a current of 1.4 ampères each. Four machines are kept at work, the fifth serving as a reserve, and being for this reason electrically connected in such a manner that it can replace any one of the others. Each dynamo machine *b* is fixed with its exciting machine *a* on a framework, by means of which the continuous driving belts can be stretched while at work, an arrangement which was considered expedient for the safety of the installation.

The dynamo machines have each twelve induction coils; twelve circuits are taken off these, and are arranged parallel to supply the different lamps. The wires are led to a switch board fixed on the stage, from whence a larger or smaller number of coils can be thrown into circuit, and the intensity of the lamps thereby varied. The arrangement in the auditorium, where 128 lamps are fixed, admits of nine different grades of light, while for the stage, where the lights are divided, are seven flies with sixty lamps each, see Figs. 370 and 371, the intensity of which can be varied in twenty-one different grades. In addition, each fly, as a whole, can be varied in three degrees of light intensity, and each one of the flies is fitted, as shown in Figs. 370 and 371, with three rows of coloured lamps, sixty in each row, so as to enable the effects of morning or evening, or bright daylight, to be produced easily.

The dynamos used in this installation are alternate current machines of Mr. Zipernowsky's design; these have been very generally adopted by Messrs. Ganz and Co. for incandescence lighting, since they maintain that the lamps last much longer if supplied by alternate current machines than if supplied by continuous current machines. Experiments showed

that with a continuous current, the carbon filaments of incandescence lamps showed signs of wear at the point where the positive current enters after about 500 to 600 hours, producing a black spot on the globe, see Fig. 369. On the other hand, when working with alternate current machines, the lamps burn without deterioration for 1200 hours and upwards, after which time they begin to show signs of wear throughout the carbon, a slight deposit is formed on the globes, the carbon becomes thinner, the resistance greater, and the light less intense.

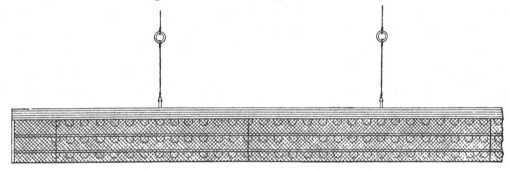

Fig. 370.

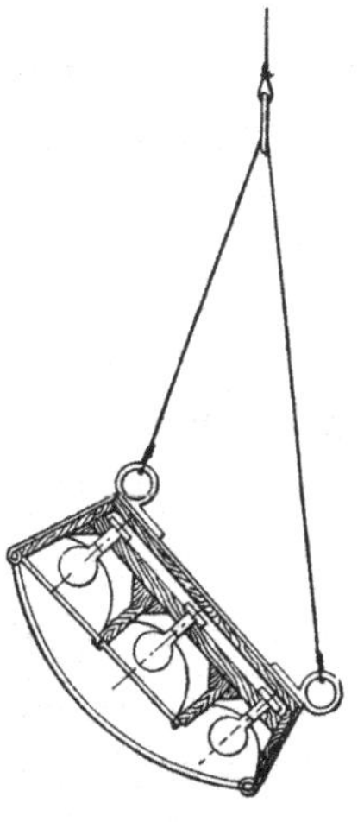

Fig. 371.

These experiences have been partly gained at the works, partly at an installation consisting of 220 incandescence lamps at the Gisella steam mills at Budapest, and also at another smaller installation of sixty-four lamps at the flour mills of Mr. Stefan Schwarz, in Erlau. At the first-named place, after ninety-five days' working, the 220 lamps had been burning for 980 hours each, in which time fifteen had failed, while six more had been accidentally broken. At the latter establishment the lamps burned from January to June, 1883, almost all of them having been alight for over

1400 hours, during which time, according to the manager's statement, not one lamp had been destroyed. These results are very strong evidence in favour of alternate current machines for incandescence lighting, since the cost for light is reduced almost to that of coal and lubricants.

The Zipernowsky dynamo, although constructed as a self-exciting machine, is generally fitted with a separate exciter, as in the installation of the Budapest Theatre, see Figs. 366 and 368, where *a* is the exciting machine, and *b* the dynamo. In case it is to be used as a self-exciting machine, the dynamo is fitted with a switch, designed by Mr. Déri, of the same firm, by means of which the potential difference at the terminals of this machine can be made to remain constant and independent of the number of lamps in the circuit; when, however, a separate exciting machine is used, a rheostat designed by Mr. Zipernowsky is employed to gain the same end.

The same dynamo machine is equally suitable for arc lamps. At the Electric Exhibition of Trieste, Messrs. Ganz and Co. illuminated the Csárda with eight lamps so successfully that the Exhibition Commissioners gave an order for lighting the whole of the grounds with thirty-two arc lamps of 600-candle power each. The dynamos supplying this installation were placed at Lloyd's Arsenal, a distance of over three miles from the Exhibition grounds, which distance did not, however, interfere with the successful results of the installation.

The most brilliant effects in connection with this Exhibition were, however, produced on board the Lloyd steamer Berenice, which, on the occasion of a visit by the crowned heads of Austria, was most magnificently illuminated by Messrs. Ganz and Co., with sixty-two incandescence and four arc lamps. The ship's deck was transformed into a ball saloon, beautifully adorned and artistically and most successfully lighted; this installation brought decorations to both the firm and Mr. Zipernowsky at the Exhibition in Trieste. The arrangement of engine and dynamo in this installation is somewhat interesting, and we give in Fig. 372 a general illustration of the same. A vertical Gwynne engine is coupled direct on one side to the dynamo machine, and on the other side to the exciting machine, which latter is of a larger pattern than would have otherwise been necessary; but since the speed, 750 revolutions, could not well be increased in this arrangement, a larger machine was used. The current supplies forty incandescence lamps of 20-candle power, and four arc lamps of 600-candle power each. The exciting machine was also used to supply a signal light; but during a series of experiments carried out with this installation, the current,

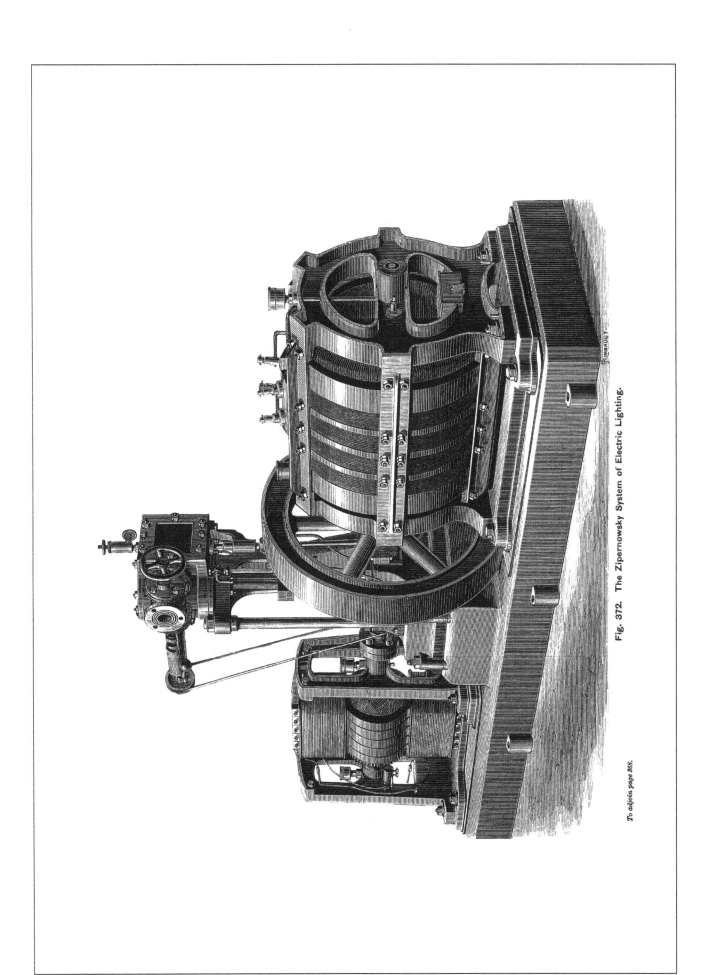

Fig. 372. The Zipernowsky System of Electric Lighting.

To adjoin page 368.

The material originally positioned here is too large for reproduction in this

otherwise employed for the incandescence lamps, was used to supply a large signal arc lamp of 8000-candle power.

So successful were these experiments that Messrs. Ganz and Co. immediately received orders for several installations for Lloyd's steamers, the principal one of which was that on board the Elektra. Several other successful installations carried out by this firm deserve mention here, one on board the troopship Custazza with 130 incandescence lamps, which has given great satisfaction, and one for street illumination in Szegedin. The main street, nearly a mile long, leading to the railway station, is lighted by 20-candle incandescence lamps placed at distances of 130 ft., while the place in front of the station is lighted by four arc lamps. This was the first installation for street lighting in Austria in which the two systems of lighting were adopted side by side, and the result is said to be highly satisfactory.

THE CHERTEMPS-DANDEU DYNAMO.

IN any system intended for private or domestic electric lighting, some means of regulating the generative power of the machines in accordance with the requirements of the users, has to be considered. Without efficient regulation in a circuit of arc lamps, the extinction of one or two is usually followed by an increased brilliancy in the remainder, while if a large proportion be cut out; there is great danger of burning the insulation of the generator. With incandescence lamps, arranged in parallel arc, the same effect is produced, except that it is usually the lamps that suffer from the increased current. The leading manufacturers of the electric light apparatus have each had their own method of dealing with this difficulty. Mr. Brush long ago devised a plan of regulating the current, by means of a variable carbon resistance, arranged as a shunt across the magnet coils. This was subject to a pressure obtained by passing the current that was to be regulated, round an electro-magnet, and allowing a portion of the exciting current, dependent on the current in the main circuit, to run to waste. Mr. Edison varies the productive power of his generators by modifying the resistance and consequently the current in the circuit of the field magnets. This is effected by an attendant, who, using a lamp, and a photometer as a measure of the difference of electric potential between the two main leads, switches resistance coils in and out of the magnetising circuit, as required. Messrs. Siemens' method of accomplishing this object consisted at first in making the magnetising coils a shunt to the main leads, and in so proportioning the diameter of the wire and the number of convolutions, that the electromotive force of the current was sensibly constant, but now they, and indeed practically all other manufacturers, have adopted compound winding, that is the

use of two sets of magnetising coils, one set in the main circuit, as in a " series " machine, and the other set arranged as a shunt to the main circuit. The Swan Company have often used the Siemens alternate current machine in their larger installations of incandescence lights, as have also Messrs. Siemens Brothers themselves. This is a separately excited machine, and consequently if its internal resistance be small, as compared with the combined resistance of all the lamps and leads, the difference of potential at its

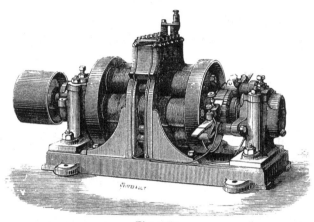

Fig. 373.

terminals is pretty constant, and it will permit of a considerable proportion of the lamps being extinguished without unduly increasing the current supplied to the others.

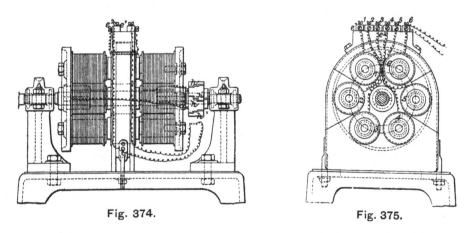

Fig. 374. Fig. 375.

The dynamo of MM. Chertemps and Dandeu, is designed with the intention of its being employed to feed arc lamps arranged in series, or incandescence lamps in parallel arc, the number of such lamps being capable of variation from the full number, corresponding to the maximum power of the

B B 2

machine, down to one, without any appreciable change in the illuminative power of each lamp. This is effected without any mechanical operation, the intensity or the electromotive force of the current automatically changing at each alteration in the circuit without any attention, and without any variation in the speed of rotation.

Referring to the illustrations, Figs. 373, 374, and 375, which show the machine in side and end elevation and also in perspective, it will be seen that it belongs to that class in which the armature bobbins are stationary, while the field magnets revolve. The former are fixed in a wooden frame erected on the middle of the base-plate. Lengthways of the machine, and passing through the centre of the armature frame, is a shaft having mounted upon it two metal discs, to each of which a set of field magnets, equal in number to the armature bobbins, is affixed. These magnets, as is usual in alternate current machines, are alternately north and south at their free ends, and each pole of one set is opposite to a contrary pole of the other set. The terminal wires of each of the armature bobbins are carried up to a grouping table, and attached to separate binding screws. This permits of them being joined up in series or parallel, or divided into independent circuits as may be desired, even without stopping the machine. One of the armature bobbins is appropriated to the work of exciting the field magnets, its current being led first to a commutator, by which it is transformed into a continuous current, and then round all the magnet coils, the whole forming a closed circuit. The commutator is so constructed that one section does not leave the brush before the next section is in contact with it. It may be described as composed of two hollow cylinders, each with a closed end, their peripheries being cut away so as to permit of them interlocking to the full length of the cylinders. There are as many interlocking parts as there are magnets in each field, and each part is cut into steps, as shown, so that at the junction, the collecting brush can bear upon both segments, and thus short-circuit the exciting bobbin at the moment when the field magnets are passing the spaces between the armature bobbins. It will thus be seen that there are usually two circuits on the machine, one consisting of all the armature coils (except one) and the external or lamp circuit, and the other of the single remaining armature coil and the field magnets. Two of these machines, a small and medium-sized one, have been tested by Mr. Robert Sabine. As regards the small one, he reports that it was employed to feed eight Jablochkoff candles, which were added to the circuit two by two, and that with two, four, six, and eight lamps respectively, the power expended

was 1.28, 2.27, 3.60, and 4.25 horse-power, the speed of rotation remaining constant. The medium machine was tested both with incandescence and arc lights. In the latter experiment ten Jablochkoff candles were employed, which were added to the circuit two at a time. Annexed, in a tabular form, are the results obtained by Mr. Sabine in the latter trial; they are very curious, and afford a clue to the action of the machine. It will be noticed that the current slightly increased as candles were added to the circuit, being 5 ampères with two candles and 5.5 with ten. At the same time the electromotive force of the current went up by leaps, the difference of

CHERTEMPS' (MEDIUM SIZED) DYNAMO MACHINE TESTED WITH JABLOCHKOFF
CANDLES CONNECTED IN SERIES.

Number of lamps in circuit $=n=$	10	8	6	4	2
I. *Main Circuit.*					
Current in ampères $=c=$	5.5	5.5	5.3	5.0ʹ	5.0
Internal resistance in ohms $=r=$	7.3	7.3	7.3	7.3	7.3
Horse-power accounted for internally $=\dfrac{c^2 r}{746}=w=$	0.30	0.30	0.27	0.24	0.24
External potential, difference of volts, volts $=$ E $=$	430	344	258	168	88
Horse-power accounted for in lamps $=\dfrac{\text{E } c}{746}=w_1=$	3.16	2.53	1.93	1.15	0.60
II. *Magnetising (Field) Circuit.*					
Current (ampères) $=c_1$	5.4	5.0	4.6	4.1	3.7
Resistance of wire (ohms) $=r_1$	7.4	7.4	7.4	7.4	7.4
Horse-power accounted for in field circuit $\dfrac{c_1^2\, r_1}{746}=w_2=$	0.29	0.25	0.21	0.17	0.14
Speed of dynamo	1242	1230	1236	1260	1218
Pull on dynamometer in kilogs. $=k$	42	36	32	28	24
Speed of dynamometer, revolutions per minute $=s$	665	666	678	676	678
Indicated horse-power $=3.21 \times 10^{-4}\, s\, (k-19)=w$	4.9	3.7	3.0	2.0	1.2
Horse-power accounted for electrically $=w+w_1+w_2=$	3.75	3.08	2.41	1.56	0.98
Proportion of horse-power electrically accounted for $=$	0.77	0.83	0.81	0.78	0.81
Proportion accounted for in lamps $=$	0.64	0.68	0.64	0.57	0.50
Horizontal front candle power of one lamp*	289	334	321	308	279

* The carbons used in these measurements were apparently of inferior quality.

potential at the machine terminals increasing, on an average, 85 volts as each pair of candles was added, and the pull on the dynamometer, situated between the engine and the generator, experiencing a corresponding change. The exciting or field magnet current, which we have seen circulates in an independent circuit, responded to the introduction of each additional pair of lamps, increasing .4 of an ampère each time, and representing, with ten candles, rather more than twice the power that it did when there were only two.

The figures show that the general performance of the machine was good, as it maintained ten Jablochkoff candles, having an average lighting power each of 310 candles, with an expenditure of 4.9 actual horse-power. The machine itself is small and compact, and is one that presents no difficulties in construction.

The theory of the self-regulating action is thus explained by the inventors :

When the field magnets are in motion, each magnet, as it approaches one of the bobbins of the armature, exercises an inductive action on the core of the latter, inducing in its corresponding extremity a polarity opposite to its own, and generating a current within its convolutions. The direction of this current is such that it reduces the polarity of the core, and thus the magnetic intensity of the interior soft iron is the resultant of the two opposing actions of the armature coil and of the field magnet. At the same time the polarity of the core has an influence upon the polarity of the field magnets, and in this way affects the exciting current in a marked degree. To make this plain, let us trace the action of one of the field magnets, indued with, say, north polarity at its free end, upon one of the armature coils, when the external circuit is open, that is, when it has an infinite resistance. As the magnet approaches the end of the bobbin core, it induces a south pole in it, and the two are mutually attracted. As it leaves the core the attraction is still continued, but is exerted in the opposite direction, as regards the driving power. Thus when the field magnet passes the core there is no action upon it except such as would be produced by the presentation and withdrawal of a piece of soft iron from its pole. But, now suppose the external circuit be completed by the introduction of ten arc lamps in series. As the magnet approaches the core not only does the latter become magnetised, but a current is generated in its surrounding coils that opposes the increase of its polarity, consequently it no longer behaves like a piece of soft iron, but rather as a weak permanent magnet that must have its natural polarity reversed by the induction of the field magnet before it will fulfil its office. But this reversal distorts the lines of force around which the magnet coils are grouped, and in so doing sets up an opposing electromotive force, within the convolutions, that reduces the exciting current, and consequently the intensity of the magnetic field. This reacts upon the exciting bobbin, and the magnetising current is further reduced, resulting in a decrease of current or of electromotive force in the external circuit. If the resistance of the circuit be again decreased by short-circuiting one of the

lamps there will be a momentary increase of current, but this will further oppose the magnetisation of the cores, and will again affect the diminution of the intensity of the magnetic field. The process may be continued until the machine is short-circuited, when, if it be properly proportioned, the influences of the armature and field magnets will be almost equal, and the exciting current will nearly disappear.

BALL'S UNIPOLAR DYNAMO.

THIS machine is the invention of Mr. C. E. Ball, of Philadelphia, and is illustrated by Figs. 376 and 377; the large perspective view (Fig. 378) shows a later form of generator, designed to burn fifteen lights, but the difference between the two is merely constructive, both acting in precisely the same way. Referring to the larger view, it will be seen that the machine resembles two dynamos of the Pacinotti type placed end to end, each armature being partly surrounded by one pole-piece only, instead of two, as is usual. The magnets are rectangular bars carrying bobbins wound with insulated wire in such a way as to produce consequent poles, at a distance of about one-third their length from one end. Thus the pole-piece on the upper magnets is, say, of north polarity, and that on the lower magnets of south polarity, and each influences one armature only. In the smaller machine the two parallel bars $C\,C^1$ constitute the cores of the field magnets, and for constructive reasons are left naked opposite the armatures, whether intended to receive a pole-piece or not. The direction of the current is shown on the drawing; leaving the commutator F it passes in succession through the left-hand coil and the middle coil of the bar c, through the coil c^2, through the external circuit, through the coils c^3, c^4, c^5, to the brush F^1, and armature D^3, and by the brush F through the armature D^2. This arrangement may be modified in various ways; for instance, both the pole-pieces may be of the same polarity, and the armatures may rotate in the same or opposite directions.

When the dynamo was brought here from Philadelphia it was placed in the hands of Mr. Robert Sabine, who made a number of dynamometrical, photometrical, and electrical tests, of which the following are the main results:

Current in circuit ... ... 15 ampères.
Difference of potential between terminals ... ... 195 volts.
Resistance of dynamo ... ... 4.5 ohms.
Speed of dynamo ... ... 1650 to 1715 rev. per min.
    ,,   dynamometer... ... ... 332 rev. per min.
Indicated pull on dynamometer, less pull when circuit
    was open ... ... 52 kilos.

From these values it follows that the useful work imparted to the dynamo, above that absorbed by the friction of the spindle, was 5.68 horse-power; and the work electrically accounted for in the outside circuit was 3.92 horse-power, and in the inner circuit 1.35 horse-power. The proportion accounted for electrically was, therefore,

$$\frac{5.27}{5.68} = 0.92.$$

of which

$$\frac{3.92}{5.68} = 0.69.$$

was found in the lamp circuit. Photometric measurements were made with one of the lamps, placed apart from the others in a dark recess, the light of the five remaining lamps not being measured. The following mean results were obtained during the observations:

Current in lamps ... ... 13.9 ampères.
Potential difference ... ... 40.1 volts.
Horse-power accounted for in lamp ... ... 0.75 HP.

The illuminating effects were as follows:

Position of Arc with regard to Photometer.	Illuminating Power.	
	Standard Candles.	Candles per Horse-Power accounted for in Lamp.
22 deg. below	602	802
11 ,,	626	835
Horizontal	887	1183
11 deg. above	1213	1617
22 ,,	1831	2441
31 ,,	1859	2479
35 ,,	1690	2253

The dynamo machine upon which the test was made had two armatures, each 8 in. in diameter, rotating in opposite directions, and each wound with about 19 lb. of No. 14 (American gauge) copper wire. The field magnet

coils consisted of 130 lb. of No. 8 wire, the total weight of copper on the machine being 168 lb., which, divided into the number of volt-ampères, gives 17.4 watts per pound of copper. In later machines much better results have been obtained, the most recent form, designed for five lights, having but 88 lb. of copper in it, and weighing only 350 lb. The fifteen-light machine has two armatures 12 in. in diameter, wound with 3000 ft. of No. 14 gauge copper wire, weighing 80 lb. The six magnet bars, each 2 in. by 2 in. in cross section, are wound with 4444 ft. of No. 8 wire, weighing 200.5 lb.

Fig. 376.

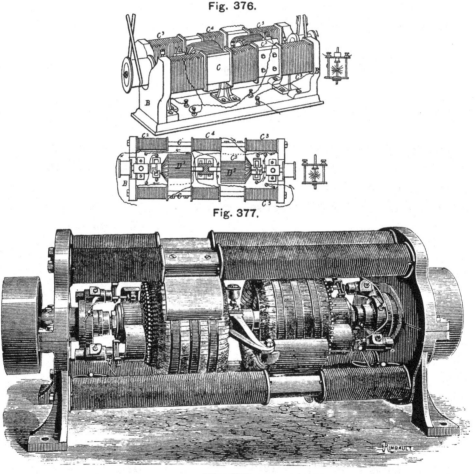

Fig. 377.

Fig. 378.

The total weight of the machine is 997.5 lb., or 66 lb. per lamp, and its internal resistance about 7 ohms. It is difficult to see what advantage can be gained by the use of one pole-piece in place of two, as the magnetic field must necessarily be weaker, but the experiments showed that it is not attended by any loss, and that when accompanied by a high speed of rotation it does not necessitate the use of specially large machines.

Since the experiments above referred to were carried out, a new series of tests have been conducted by Messrs. Alabaster, Gatehouse, and Co., which confirm, and, indeed, sensibly exceed the results first obtained, as will be seen from the following extracts from the later report.

"The machine tested was nominally adapted for supplying ten arc lamps. It was, however, capable of actuating a larger number. The potential difference of the machine between terminals was measured by the condenser method, a Clark's standard cell being employed for comparisons of electromotive forces. Two Siemens' electro-dynamometers, one of these being supplied by us, indicated the current strength, each instrument giving identically the same readings. The external resistances, and that of the machine, were measured immediately after each test. Horse-power measurements were taken by means of a Morin transmission dynamometer, and a Prony brake. Countershafting was placed between the dynamometer and the dynamo, the former being so arranged as to run at about the same number of revolutions per minute for varying speeds of the dynamo, this being accomplished by using pulleys of different diameter on the latter. The Prony brake, which was balanced, gave admirable readings, there being but a difference of two ounces in the weight applied at the end of the lever in three distinct measurements.

"The results obtained by us are tabulated as follows:

TABLE OF TESTS. DOUBLE ARMATURE, BALL "UNIPOLAR" DYNAMO.
Date, April 29, 1884.

| Date. | Number of Test. | Current in Ampères. | RESISTANCE IN OHMS. | | | ELECTRICAL WORK. | | | | MECHANICAL WORK. | | | Revolutions of Dynamometer. | MORIN DYNAMOMETER. | | | PERCENTAGE OF EFFICIENCY. | | | | Revolutions of Dynamo. |
			External R−r.	Dynamo (warm) r.	Total R.	Watts. C² R.	Horse-Power, External.	Horse-Power, Internal.	Horse-Power, Total.	Prony Bk. Weight in lbs.	Length of Lever, feet.	Horse-Power absorbed.		Pull, Machine loaded, kilog.	Pull, Machine empty, kilog.	Horse-Power, calculated	Conversion.	Appearing in External Circuit.	Commercial calculated from absorbed HP.	Commercial.	
1884.																					
Ap.25	A	8.07	56.0	17.15	73.15	4764	4.888	1.497	6.385	...	...	...	378	69	14	6.83	93.5	76.5	71.5	63.3	1290
Ap.25	B	8.07	68.8	16.93	85.73	5583	6.00	1.479	7.479	...	...	...	365	89	20	8.28	90.3	80.3	72.4	...	1620
Ap.26	A*	8.07	59.33	17.12	76.45	4980	5.179	1.494	6.674	13.94	2.0	6.9	384	75	18	7.16	93.2	77.6	72.3	67.3	1300
										Prony Brake							96.7	...	75.0	70.0	...

"*Observations.*—Test A* is nearly a reproduction of test A, and to it

was applied the Prony brake; in three separate measurements at varying intervals the weights applied at the end of the lever were 13 lb. 15 oz., 13 lb. 14 oz., and 14 lb. respectively.   The formula is

$$\frac{n \times L \times x}{5252}$$

where n is the number of revolutions per minute, L the length of lever, and x the weight.   The formula for the Morin transmission dynamometer is

$$\frac{1.5 \times k \times s}{60 \times 76.04}$$

where s is the speed, and k the indicated pull in kilos. on the dynamometer less the pull when the circuit is open.   When the Prony brake was employed to measure the absorbed horse-power, the Morin instrument was used as an indicator.   Thus when electrical work was being produced with the machine running at 1300 revolutions, the Morin indicated 75 K, this pull was then reproduced in the Prony brake test with the machine revolving at the same speed.

"The commercial efficiency of the machine is calculated both from the absorbed horse-power above, and with the addition of the horse-power which the machine requires to run at 1300 revolutions unloaded = 0.54 HP. Test B, at a higher speed, was taken at the suggestion of Mr. Ball.

"In test A* the Prony brake was resting by its own weight upon one of the pulleys of the dynamo when doing electrical work, and also when running empty.   The external circuit was in each case made up of a wire resistance.

"It will be seen that the data obtained by us agree very nearly with the March figures taken by Mr. Robert Sabine on one of your machines as long ago as 1883; an improvement, however, being observable in the latter form of machine.   It would appear to be more economical to run this machine at the higher speed from an electrical point of view; for, although the percentage of conversion is somewhat lower, a larger percentage of energy is accounted for in the external circuit.   From what we can gather concerning its mechanical construction it appears to be a remarkably cheap dynamo to build up, at the same time its constituent parts are very unlikely to get out of order.   The double commutator may be considered, from the contractors' point of view, as a disadvantage, but electrically it is decidedly advantageous.   By spreading a given amount of wire over two armatures, as in your machine, double the cooling surface is provided, and the pole-pieces of the field magnets thereby brought much nearer to the armature core."

MATTHEWS' MULTIPOLAR DYNAMO.

THE dynamo machines invented and manufactured by Mr. Matthews are of the multipolar type, in which a disc armature revolves between two sets of electro-magnets, arranged each in a circle, as will be seen on reference to the perspective view Fig. 379, which represents an electric generator and a Matthews three-cylinder engine mounted on one bed-plate. The general design of the machine follows that of the Ferranti or the Siemens alternate current generator, the features of novelty being confined to the armature. In its simplest form this armature, as designed for small alternate current machines, consists of a number of radial bars attached to a disc, and connected in pairs alternately at the top and bottom. Thus each set of bars represents a zig-zag bent into a circle. Two zig-zags are combined in such a way that their radial bars and their outer connecting bars lie in the same plane, as shown in Fig. 380, where B C B D B is one zig-zag, and $B^1 C^1 B^1 D^1$ the other. The alternate sinuosities of the ziz-zags are of different sizes, so that when the two are placed together they will lie alternately within each other. As the bars D^1 cross the bars B B, it is evident that they cannot lie in the same plane, and consequently the bars B^1 are bent, near their inner ends, to cause them to clear. Two arrangements for this purpose are shown in Fig. 381, while in Fig. 383 the bars are thinned down so as to half-lap. Several pairs of ziz-zags may be mounted upon the same disc (Fig. 381). If a high electromotive force be required, the end of one zig-zag is coupled to the next in which the current circulates in a corresponding direction; the course of the current is shown by the arrows in Fig. 380.

In small machines the conductors are stamped complete out of sheet copper, and in large machines they are made of square or rectangular copper bars (Figs. 388 and 390), the inner connecting parts $D D^1$ being bent into the forms shown in Figs. 387 and 389. The current is led to the commutator or contact rings P P by wires *o o*. For a 1000-light dynamo the armature

is made of copper bars of square section $\frac{1}{2}$ in. by $\frac{1}{2}$ in., after the design of Fig. 390.

Instead of the zig-zags, independent frames may be used, corresponding to the bobbins of wire usually employed. These are shown at Fig. 392; the current flows up the thick central member, divides right and left along the circumferential portion, and descends by the two extreme bars. These frames are coupled together by connections, shown in the figure by thin

Fig. 379.

lines, so as to gain the requisite electromotive force, the thick bars being joined successively to the thin ones.

Another variety of the Matthews machine has the radial bars or strips, not in the same, but in parallel, planes (Figs. 384 and 385), the bars being wound or built upon a core I, so as to constitute a kind of irregular coarsely pitched thread, the number of internal enclosures in one convolution being equal to the number of like magnetic fields. After one coil has been wound on, as shown to the right of Fig. 385, another and another is added until the core is covered, as shown to the left of the figure. The ends J J are coupled together, according to the nature of the current required. When a large electromotive force is desired other layers of bars may be laid over the first. In some cases for alternating currents the coils are wound in such a manner

round the core that the radial parts of two contiguous coils are opposite to each other, one on each side of the core, the circumferential parts being necessarily alternating, and lying in a diagonal direction.

For continuous currrent machines with armatures of the kind described with reference to Fig. 380, where the number of part coils is equal to the number of alternating fields, or equal to twice the number of the same polar fields, the discs are placed in advance of each other as in Fig. 393, so that the pairs of part coils or zig-zags are not at their maximum induction at the same time. For continuous currents with armatures having but one set of bars, the number of spaces are arranged to be less or greater by any even number

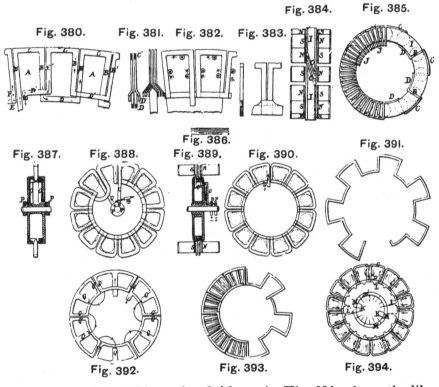

Fig. 380. Fig. 381. Fig. 382. Fig. 383. Fig. 384. Fig. 385.

Fig. 386.

Fig. 387. Fig. 388. Fig. 389. Fig. 390. Fig. 391.

Fig. 392. Fig. 393. Fig. 394.

than twice the number of like polar fields, as in Fig. 394, where the like fields, represented by the dotted circles, are nine in number, and the spaces and part coils are sixteen. The inner parts of the coils $D^1 D^1$ are coupled up and connected to a commutator k which is made up of 144 bars, each of the terminals L^1 being coupled to nine bars of the commutator spaced at equal intervals, two such connections being shown by the rings M M[1], which have nine branches k leading to the bars of the commutator, so that when all the connections are made there will be 144 bars. By this arrangement only one pair of brushes are required, instead of one pair to each pair of magnets, as is usual in multipolar direct current machines.

THE HOPKINSON - MUIRHEAD DYNAMO.

THE continuous current and the alternating current generators invented by Dr. J. Hopkinson and Dr. Alexander Muirhead, are peculiarly interesting as being probably the first in which the bobbins of the armature were wound with copper ribbon and arranged on a disc armature, much in the same way as was afterwards done by Sir William Thomson and by Mr. Ferranti. In the Muirhead-Hopkinson machine the armature coils are attached to a soft iron ring, whereas in the Ferranti the iron core is dispensed with, and a gain of lightness in the armature or rotating part is thus effected. The general form of this generator is clearly shown by the side and end elevations, Figs. 395 and 396.

The armature is made by taking a pulley and encircling it with a rim of sheet-iron bands, each insulated from the other by asbestos paper. On one or both sides of the rim thus formed, radial slots are cut to admit radial coils of insulated copper wire or ribbon, so that they lie in planes parallel to the plane of the pulley. In the continuous current machine coils are placed on both sides of the iron rim and arranged alternately, that on the one side always covering the gap between two on the other side. In this way, when a coil on one side of the rim is at its "dead point" and yields its minimum of current, the corresponding coil on the other side is giving out its maximum.

The field magnets are made in a similar manner to the armature, and run in circles parallel to the rim of the latter. The cores may be built of wrought iron, as the rim of the armature is ; but it is found cheaper to make them of solid wrought or cast iron. To stop the local induced currents in the core, however, grooves are cut in the faces of the iron cores, and filled up with sheet-iron strips insulated from each other, similar to the sheet-iron ring of the armature.

The coils, both in the armature and electro-magnets, are packed as closely as possible to each other, and have thus a compressed or quadrilateral shape. The arrangement is shown in Figs. 397 and 398, which represent, in side view and plan, the armature pulley with the soft iron rim and coils attached. In them *a* is the pulley, which is keyed to the shaft of the machine, and is encircled with bands of sheet iron *b*, insulated from each other by ribbons of asbestos paper laid between every two bands. When the rim has been built up in this way, radial holes are drilled through it from the outer edge inwards, and the whole rim is bound together by bolts *d* inserted in the holes and secured by cotters *e*. Radial slots are then cut on each side of the rim all round, and the coils of wire mounted in them.

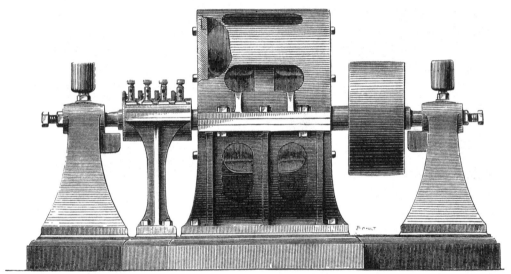

Fig. 395.

Figs. 397 and 398 show the armature of the continuous current dynamo, with the coils on one side of the rim, half way between the coils on the other side, so as to give a more continuous current. In the alternating current machine the slots on the opposite faces are face to face.

Figs. 399 and 402 illustrate the complete continuous current machine, the latter showing the internal arrangement of the field magnets, and Fig. 399 the external frame of cast iron supporting them. In these figures *a* is the armature already described, *b b* are the cores of the electro-magnets with a strong cast-iron backing *c c*; *d d* are the exciting coils or field magnets, so connected that the poles presented to the armature are alternately north and south, thus bringing a south pole on one side of the armature opposite a north pole on the other side.

The commutator e is arranged to prevent sparking when the brushes leave a contact piece. This is done by splitting up the brushes into several parts, and inserting resistances between the part which leaves the contact piece last and the rest of the circuit. This resistance checks the current ere the final rupture of contact takes place.

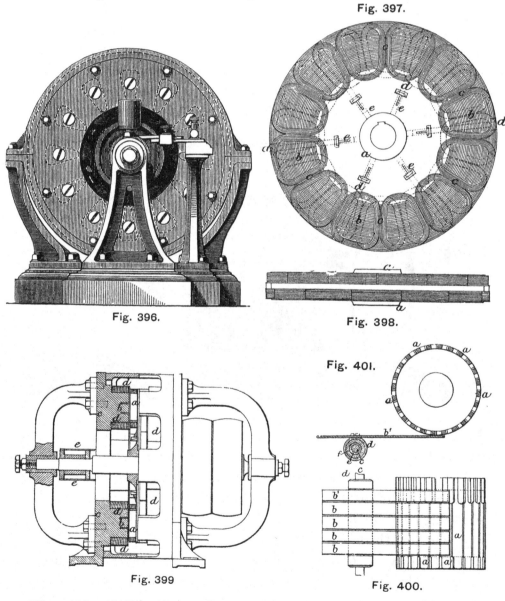

Fig. 397.

Fig. 396.

Fig. 398.

Fig. 401.

Fig. 399.

Fig. 400.

Figs. 400 and 401 will explain the structure of the commutator. Here $a\,a\,a$ are the segment or contact pieces insulated from each other, and $b^1\,b\,b$ are the collecting brushes carried on a spindle $c\,c$. One of these brushes b^1 is connected to the spindle c through an electrical resistance of plumbago,

arranged as shown in Fig. 401, where $d\ e$ are metal cylinders, d being in contact with the brush b^1, while e is in contact with the spindle c. The space f between these two cylinders $d\ e$ is filled with a mixture of plumbago and lampblack of suitable resistance, confined at the ends by ivory discs. The brush b^1 is adjusted, by bending, till it remains in contact with any segment of the commutator for a short time after the other brushes have left contact with that segment, and thus instead of sudden break of circuit and consequent sparking, a resistance is introduced, and contact is not broken until the current has been considerably reduced.

The contact segments are supported at both ends by solid insulating discs; but they are insulated from each other in the central portions by the air spaces between them.

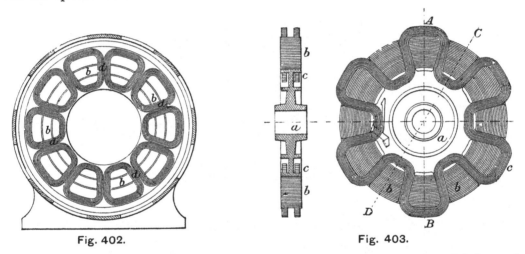

Fig. 402.   Fig. 403.

The alternating current dynamo of Drs. Hopkinson and Muirhead differs little in general construction from that we have described; except that the commutator is very much simplified, and the armature bobbins are placed opposite each other on both sides of the rim. Instead of forming the coils into complete bobbins, Dr. Muirhead prefers to wind them in a zig-zag form round the grooved iron rim, after the manner shown in Fig. 403, which represents a plan and section of the alternating current armature. This arrangement is simpler in construction than the bobbin winding, and is less liable to generate self-induction currents in the armature. In this figure a is the pulley fixed to the spindle of the machine, $b\ b$ is the iron rim, and $c\ c$ are the zig-zag coils of copper ribbon. The field magnets are also wound in a similar manner.

SCHWERDT'S DYNAMO.

THE Schwerdt system, comprising a dynamo machine and arc lamp, presents nothing new, as regards principle. The generator which we illustrate by a longitudinal section (Fig. 404), and a transverse view (Fig. 405), consists of a ring R of the Gramme type, carried by the four arms *l* of a star of red copper, which revolves in the box-like extensions M¹ M² of the field magnets. These boxes enclose the outer portion of the ring, and a part of its inner surface. The inventor hopes in this way to increase

Fig. 404.

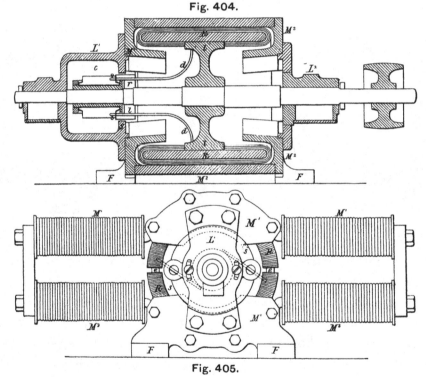

Fig. 405.

the extent of the magnetic field, and to utilise to a maximum the effect of the electro-magnets. The pole-pieces are of cast-iron; they are connected on each side by a bronze armature which receives the bearings carrying the axis on which the ring revolves. The current is transmitted to the collector by the wires *d* which traverse a wooden disc *r* placed on the shaft. From this short description it will be seen that the machine possesses originality only in arrangement.

CROMPTON'S STEP-WOUND ARMATURE.

IN all ring armatures the number of turns of wire that can be applied is determined by the inner circumference of the ring, and there is consequently, vacant space at the periphery equal to the difference between the inner and outer circumferences. In machines of the Gramme type this does not amount to much, but with armatures whose radial width is considerable in relation to their width measured parallel to the axis, the difficulty becomes serious, and has hitherto been a bar to their use. Mr. Crompton has lately devised a method of winding, which he calls step-winding, whereby this waste of space is avoided, and the whole of both faces of the ring or disc are covered with wire, as well as the inner and outer circumferences. The method is as follows. The disc is divided into segments equal in number to the intended separate coils, and these segments are wound with as many equal and parallel turns of wire as the length of the inner circumferential arc will admit of; the winding is then continued through a series of holes pierced through the disc, or steps cut in the segment, in such manner and order that each successive turn of wire, or groups of turns of wire, is rather shorter than the one preceding it. In this manner the otherwise unoccupied triangles are filled up with winding in a series of turns, or groups of turns, arranged step-wise, so that the whole of the wedge-shaped segments are completely covered.

Fig. 406 illustrates one method of carrying out the invention. In this the winding of one of the segments is commenced at the oblong hole a, and is continued, as above described, until the hole a is filled with wire. After this the winding is continued through the holes $c_1 c_2 c_3$, until the full length of the outer bounding arc of the segment is filled. The next segment is then commenced at the hole b, and the winding continued all the way round in a similar manner, each section being formed of as many layers as may be convenient. Figs. 407 and 408 show another method of winding applicable to a disc core built up of segments. Each section is of malleable

cast iron or other magnetisable material, and is formed stepwise along one edge, as indicated at *a*. The segments are secured together in pairs in such manner that they form a rigid compound segment, capable of withstanding the strain caused by the rotation of the disc. By this arrangement the respective segments can be wound separately and built up into the disc, and, if necessary, be subsequently removed and replaced without disturbing the other segments. The winding is similar to that shown in

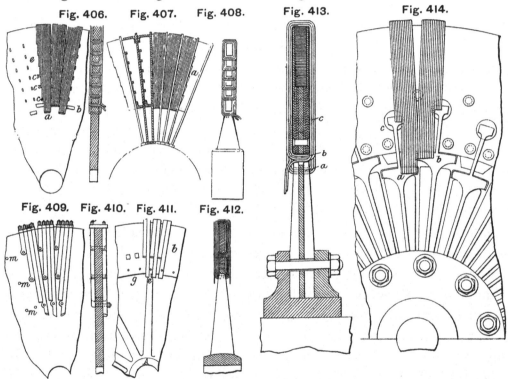

Fig. 406.   Fig. 407.   Fig. 408.   Fig. 413.   Fig. 414.

Fig. 409.   Fig. 410.   Fig. 411.   Fig. 412.

Fig. 406, and each segment can be coiled before it is bolted to the next. Figs. 409, 410, 411, and 412 show two disc armatures in which the conductors are thick copper bars *m m*, each of which consists of a stirrup of rectangular section. The connections between adjacent turns are made by copper bolts passing through holes in the disc, which may be made either of several thicknesses of iron plate rivetted together, or of a rim of non-magnetisable metal covered with iron plates as in Fig. 413. Figs. 413 and 414 show a modification in which the holes *a* and *b*, corresponding to the holes *a* and *b* in Fig. 406, are connected to radial slots extending to the inner circumference of the plates. This affords facility in winding, as by this device the wire can be wound on a bobbin and passed through the central hole of the disc.

ELECTRIC ARC LAMPS.

SINCE the first volume of " Electric Illumination" was published, a large number of patents have been granted for various forms of arc lamps. Of these, few have been brought into anything like extensive use, and the selection given in the following pages is in consequence a very limited one. It is believed, however, that it comprises all, or nearly all, that have attained any degree of practical success. The pages of the Appendix attached to the present volume, and containing abstracts of all lamps which have been patented in this country, up to the middle of 1882, may be consulted for the remainder.

ANDRÉ.

IN André's arc lamp the regulation is effected by a special current, set in operation by sensitive mechanism operated by a shunt current from the arc. The shunt current traverses an electro-magnet and attracts a swinging armature, which is balanced by a counterweight, so that its movement always corresponds to a definite length or resistance of arc. As the armature swings round it switches the feed current, which for a moment every half-minute flows along a special conductor from the dynamo, into the feed electro-magnets, where it works the regulator and adjusts the carbons. The feed circuit is a direct branch from the generator, and as the current does not exceed 0.5 ampères in intensity, a very small wire is sufficient for its transmission. The negative cable forms a common return conductor for both currents.

The general form of the lamp is shown in Figs. 415 and 416. On the base-plate A, which is insulated from the lantern by the ring

B, a metal standard C is mounted, for the purpose of supporting the upper carbon.　On the base-plate is also fixed a double-armed electro-magnet D, each bobbin consisting of four layers of No. 12 insulated copper wire wound on a wrought-iron core $\frac{5}{8}$ in. in diameter, and $2\frac{1}{2}$ in. long ; the resistance of the whole magnet being 0.06 ohms.　An armature E is suspended over the poles of this magnet from a metal stem, which, at its upper end, acts on

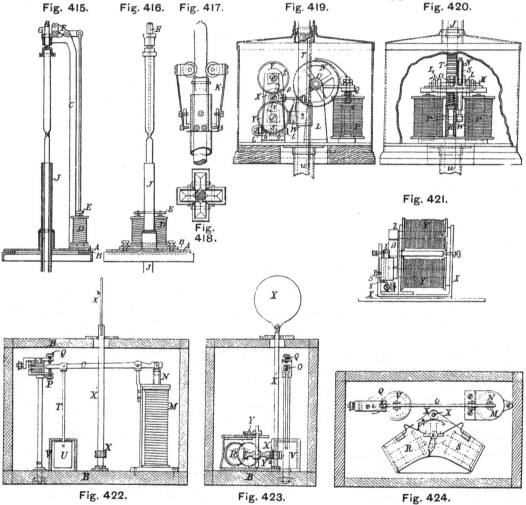

Fig. 415.　Fig. 416.　Fig. 417.　Fig. 419.　Fig. 420.

Fig. 418.

Fig. 421.

Fig. 422.　　　　Fig. 423.　　　　Fig. 424.

a lever F attached by two straps to a tilting ring G, which is thus tilted whenever the armature E is attracted to the poles of the magnet D. Through this ring slides a rod H, which, at its lower part, is fitted with a socket for holding the butt-end of the upper or negative carbon.　From the centre of the base-plate A, but insulated from it, rises a long brass tube J, which contains the positive carbon, held in a central position by means of the roller contacts K, mounted on springs as shown in Fig. 417.

As soon as the main current is turned on, it traverses both carbons and passes through the electro-magnet D by the metal standard C. The armature E is then attracted and the ring G tilted by means of the lever F. The negative carbon is thus raised and the arc formed. On the other hand, when the current is turned off, the armature is released, the carbon-holder H is allowed to slip through the ring G, and the carbons come together again.

The brass tube J, containing the lower carbon, is continued downwards, as will be seen from Fig. 419, until it reaches a box containing the feed apparatus and regulator. The feed apparatus is shown in detail in Figs. 419 and 420, the former being a side and the latter a front elevation. It consists of two **A**-shaped frames supporting a shaft M, to which are attached a ratchet wheel N and pinion O. In front of the two frames is a pair of electro-magnets P, joined below by a breech-piece. Each coil consists of sixteen layers of No. 24 insulated copper wire, wound on a wrought-iron core $\frac{1}{2}$ in. in diameter and 2 in. long, the resistance of the coils in series being about 4 ohms. At the ends of the shaft M and outside the frames L, are two arms working loosely on the shaft, and carrying an armature, with their free ends adjusted to the proper position by the spring on the pillar R. This armature carries on its upper surface the ratchet S, gearing into the ratchet wheel N, and thus capable of driving the shaft M. The pinion O drives the long rack P, which carries at its upper end the lower carbon, whilst the lower end is protected by a brass tube u fixed to the base-plate of the whole arrangement. A pawl W prevents the ratchet wheel from running backwards.

The ends of the feed magnet coils are connected to the regulator, which, as shown in Fig. 421, consists of a brass frame X, between the arms of which are fixed two coils Y, of insulated copper wire, 6 millimetres in diameter and wound on a wrought-iron core $\frac{5}{16}$ in. in diameter, and $2\frac{1}{8}$ in. long, joined at the back by a breech-piece. The other end of each core passes through the front end of the frame X and terminates in a pole-piece Z, the inner surface being cylindrical and described about the central axis of the pair of magnets. An armature β, rigidly connected to a lever γ, is pivotted about the same central axis and is free to swing near the top pole-piece Z, which it can never touch owing to its cylindrical shape. The lower end of the lever γ carries two platinum pins $\delta \delta^1$ which play between the ends of two screws ϵ and ϵ^1 fixed to the frame X, but insulated from it, one of which is pointed with platinum, the other with an insulating

material. The lever γ also carries a rod and an adjustable weight ʃ, by means of which any desired counterbalance can be given to the attraction between the magnets and the armature.

The coils Y are a shunt to the main circuit, and when the arc increases to a certain resistance, they attract the armature β in opposition to the counterbalance ʃ, and the lever γ is inclined towards the right, thus bringing the pin δ¹ into contact with the screw ε¹, which has an insulated point. The feed current enters at the platinum pointed screw ε, and thence traverses the feed magnets, which have a resistance of 6 ohms. When, however, the weight ʃ overpowers the attraction of the magnets on the armature, the platinum pin δ rests against the platinum pointed screw ε, thereby opening a passage of almost no resistance through the frame X to the next lamp.

The positive carbon employed is one meter long and 12 millimetres in diameter, the negative carbon is only 8 in. in length, but is 18 millimetres in diameter. A metre carbon, copper-plated, keeps the lamp burning for thirty-six hours continuously, or for a week of five hours each night. To avoid the resistance of this whole length of positive carbon, the current enters by roller contacts provided near its points as shown at K, Fig. 417. The resistance of the lamp is thus kept as low as 0.1 ohm, when unlit, and 2 ohms when lit ; the resistance of the arc being about 1.9 ohms. A current of 12 ampères is found to give a good light of about a 1000 candles in these lamps, and at least ten of them can be worked in series from a Gramme machine of H type.

The intensity of the light is regulated by throwing in or taking out resistance from the circuit of the field magnets, to the extent of a fraction of an ohm. This resistance is contained in frames of stout copper wire. Mr. André also employs a current indicator of ingenious device to show the engineer when the current rises above or falls below 12 ampères. This is illustrated in Figs. 422, 423, and 424 ; it consists of a solenoid M through which the current passes, acting by magnetic suction on a soft iron core N suspended from one end of the pivotted lever O. The other arm of this lever carries an adjustable stem T fitted with a counterweight U, which can be increased by the weight of the outer case V when the stem is raised. The end of this arm of the lever plays between two contact pins P Q, in circuit with a local battery, an electric bell, and the electro-magnets R S, Fig. 424. These magnets are capable of attracting a curved iron armature Y, and thereby rotating the vertical shaft X¹ by means of the toothed segment

Y² and the pinion X. According as one or other magnet is excited, the armature is pulled to one side or the other, and the rotated shaft displays the front or back of the signal disc X², Fig. 423, to the engineer. When the current is under 12 ampères, the core N rises in the solenoid M, and the lever O falls on the lower pin P, thereby sending the local current to ring a bell and actuate the magnet R, so as to make the shaft X present the green face of the signal disc to the engineer. On the other hand, if the current is over 12 ampères, the upper contact is made, the bell rings, and the magnet S revolves the shaft so as to bring the red face of the disc into view. When, however, the current keeps its proper strength of 12 ampères, the lever rests between the pins, and the signal disc presents its edge as shown in Fig. 422.

HAWKES.

THE mechanism of the Hawkes lamp is illustrated in Figs. 425 and 426, which represent it in action. It is a clutch lamp, and is adapted to burn long carbons if required, so that it can be used for the lighting of streets as well as halls, factories, engineering works, and so on.

Figs. 425 and 426 are two vertical sections of the lamp in planes crossing each other at right angles. Fig. 427 is a sectional plan of the lamp along the line 1—2 in Figs. 425 and 426, and Fig. 428 is a sectional plan taken in the line 3—4 of the same figures. Fig. 429 shows a special device for cutting the lamp out of circuit.

The frame of the lamp consists of three metal base-plates A, A¹, and A², rigidly connected by vertical stems A³, A⁴, and A⁵. The stem A³, which is in connection by the terminal below, with the positive wire from the generator, is insulated from these metal plates, and the stem A⁴, which is in connection with the negative wire through the plates A, A¹, is insulated from the upper plate A² only. This plate supports the mechanism for striking the arc and regulating the advance of the upper carbon C and holder C¹, while the plate A supports the mechanism for regulating the advance of the lower carbon D and its holder D¹. This carbon traverses

a long tube B, pendent from the bottom of the lamp, and prolonged at B¹ by a liquid reservoir into which a piston descends.

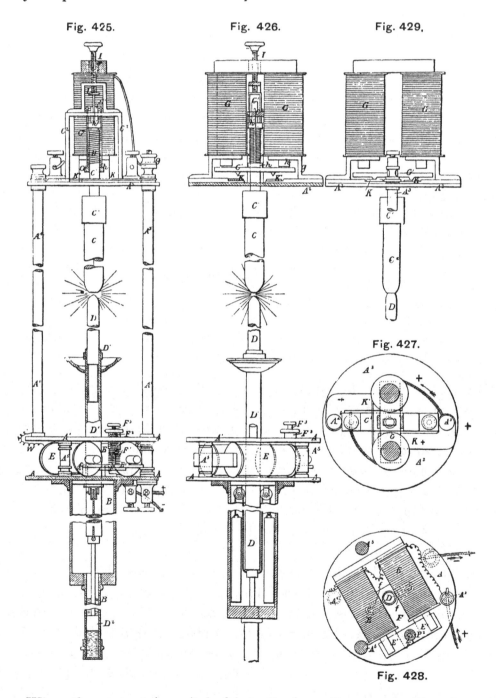

Fig. 425.   Fig. 426.   Fig. 429.

Fig. 427.

Fig. 428.

When the current is switched into the lamp it passes by the stem A^3 and terminal g to the electro-magnet G, and thence by the terminal g^1 to

the plate A², which is in metallic contact with the carbon-holder C¹. It thus traverses the carbons and completes its circuit through the lower holder D¹ and the plates A¹ A. In traversing the double-coiled electro-magnet G, it attracts the armature G¹, and thereby separates the upper carbon from the lower and forms the arc. Mr. Hawkes, however, has not contented himself with the mere striking of the arc on the introduction of this electro-magnet in the main circuit. He has also arranged that it shall regularly adjust the position of the carbon C. To this end he employs the following device : H is a coiled spring surrounding the upper part of the carbon-holder C¹ and retained between two collars h h¹, which also surround the carbon-holder. The lower collar h is suspended from the guide frame C² by means of pendent rods which pass freely through the head of the guide frame, and are each provided with a head to take a bearing on the guide frames. The upper collar h¹ is attached to a pair of guide rods which, after passing up through the guide frame C², are connected together by a crosshead, upon which bears a regulating screw I, which regulates the amount of compression put upon the spring. The spring is intended to press back the armature of the electro-magnet when it is attracted by the current, and consequently to determine the position at which the upper carbon will come to rest. On the strength of the main current decreasing through the lengthening of the arc, the spring comes into action, and forces the armature and carbon-holder downwards, until its further expansion is checked by the pendent collar h. After this the armature simply falls by its own weight, and the upper carbon drops into contact with the lower one, thereby closing the main circuit. The electro-magnet again comes into full play and re-establishes the arc as before. Thanks, however, to the spring device just described, this sudden restoration of the arc need seldom occur.

The lower carbon is advanced to maintain the arc by the. pull of two counterweights shown in Fig. 426, the cords from the weights passing over two pulleys and being attached to D¹ This advance is controlled by a clutch device of ingenious construction. The carbon-holder D¹ passes through a hole in the plate A¹ A, and between the poles of an electro-magnet E, which is in a branch or derived circuit, as shown by the wires W W. Attached to the armature E¹ of this electro-magnet is a gripping piece E, which is carried by pivot pins *f*, and has an opening through it to allow of the free passage upwards of the carbon-holder when the armature E¹ is attracted by its magnet, and the gripping piece is thereby raised as shown in Fig. 425.

Should, however, the strength of the current in the electro-magnet become enfeebled, owing to a too rapid rise of the carbon, the reacting force of the spring F^1 presses down the armature and the gripping piece arrests the upward motion of the carbon-holder, until the arc attains its proper length, and the derived current in the electro-magnet E its normal strength. The lower carbon is then unlocked by the superior attraction of the electro-magnet over that of the spring, and the carbon advances as before.

The action of the spring E^1 is regulated so as to limit the upward play of the armature E^1, and to prevent its being drawn to the poles of the magnet. To serve th s double purpose a hollow regulating screw F^2 is provided. On passing through the plate A^1 this screw F^2 bears on the coiled spring F^1, which is made fast to the armature E^1. The hollow screw is tapped to receive an elongated screw F^3, which projects down through the coiled spring so as to form a stop to the rising armature. By the adjustment of the screw F^2 the force of the spring is regulated so as to counteract the power of the electro-magnet to any desired extent, and the upward range of the armature is properly limited. The piston D^2 moving in a well (B^1) of viscous liquid, such as glycerine and water, gives additional smoothness to the working of the lamp.

The device employed by Mr. Hawkes to cut out a lamp from a circuit when some accident has befallen the carbons, is a very simple and ingenious one. Insulated from the top plate A^2, but resting on it, is a strip of metal K, which is in metallic connection with the stem A^3, and underlies one end of the armature G^1. A similar strip of insulated metal K^1 is connected to the other stem A^4. The armature G^1 at the parts which overlie the strips $K K^1$, is fitted with platinum points. While the lamp is in action, the armature being held up by the attraction of its electro-magnet, is clear of these insulated metal pieces, but as soon as an undue resistance is offered to the main current, the power of the electro-magnet is overcome by the spring H, and the armature is dropped down and short-circuits the current by the strips $K K^1$, as shown in Fig. 429. The lamp is thus cut out of the general circuit.

Mr. Hawkes also constructs lamps of the same type having double carbons, and these are fitted with an automatic device for diverting the current from one pair of carbons to the other.

The upper or positive carbon, is ordinarily twenty millimetres in diameter, and copper coated; while the lower or negative carbon, is thirteen millimetres in diameter. The lower burns at the rate of an inch per hour,

and is long enough to burn for forty or fifty hours. The lamp is stated to yield a light of 6000-candle power when fed by an A Gramme machine consuming $2\frac{1}{2}$ horse-power.

CROMPTON.

FIGS. 430 to 434 show one of the later developments of Mr. Crompton's arc lamp. The salient features of the design are that, instead of the carbons being provided with holders or rods, they are guided near the focus by earthenware bushes, and at their remote ends by sliding blocks. The feed is controlled by a train of wheels governed by a brake block, held in contact with the corresponding wheel by a spring. This gearing is mounted on the end of a vibrating lever, the position of which is determined by the core of a differential solenoid. As the resistance of the arc increases the core rises until the brake lever meets a fixed stop and the gearing is free to revolve. The arc is also established by the motion of the same frame, by means of a cord passing round a sheave or pulley on the first spindle.

Referring to the illustrations it will be seen that the upper carbon passes through a porcelain guide near the arc, and, at its upper end, is fixed in a sliding clip c that runs between the two stationary guide rods E E^1. The lower carbon is similarly held, and moves within the tube D; it is connected to the upper electrode by a cord, or other flexible connection, passing round the six pulleys h, h^1, h^2, h^3, h^4, h^5, and consequently the two pencils make simultaneous motions, and thus keep the focus constant in position. The pulley h^4 is mounted on the first spindle of a train of three wheels, the last spindle of which carries a brake wheel, normally subject to the action of a brake lever g. The train is mounted on a frame fixed on the forked lever k, which is pivotted at K (Figs. 431 and 432). The position of this lever is controlled by the core m of the differential solenoid M N. When the circuit is completed through the lamp the coarse coil of the solenoid draws down the core, and with it the gearing frame and the pulley h^4, around which the bight of the cord $i\,i$ passes. This separates the carbons and establishes the arc. As the resistance of the arc increases the core m rises, until the brake lever g meets a fixed stop on the frame. The wheels are then free to revolve, and the upper carbon, by its superior

weight, then descends, raising the lower one by means of the connecting cord, at half the speed at which it falls itself. The fumes emitted from the arc are prevented from entering the lamp by the guard case C.

Fig. 433 shows a modification, in which two coils M and N have inde-

Fig. 430.　　Fig. 432.　　Fig. 431.

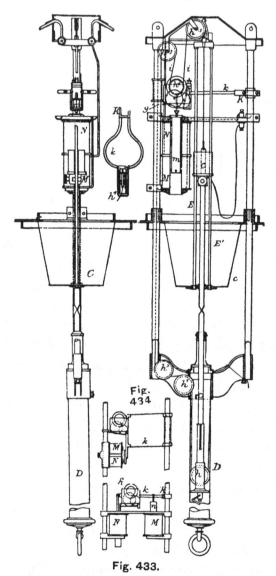

Fig.
434

Fig. 433.

pendent cores, and Fig. 434 illustrates a lamp in which magnets are used in place of solenoids.

Messrs. Crompton and Co. have recently introduced a very simple form of arc lamp which has proved remarkably successful in practice. It is known as their D D pattern, and its mechanism consists solely of a rack,

a pinion, a brake wheel, a brake lever, and a solenoid. In Fig. 435 it is shown arranged as a duplex lamp, and consequently many of the parts appear twice, taking away somewhat from the apparent simplicity of the arrangement. Each upper carbon-holder, B or B¹, slides vertically through bearings in the framing, and is provided at one side with rack teeth. Upon the holder there slides a gun-metal sleeve E carrying a short horizontal spindle, upon which is a pinion gearing into the rack, and a brake wheel. The sleeve is prevented from sliding down the holder by means of a brake lever L, pivotted at one end to the side rod of the lamp, and at the other

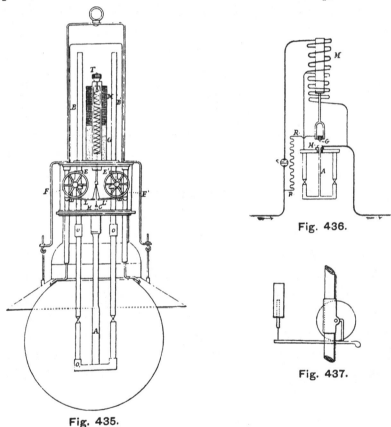

Fig. 435.

Fig. 436.

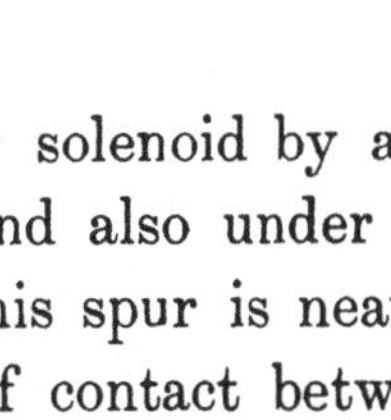

Fig. 437.

end connected to the core of the regulating solenoid by a chain. This lever stands directly under the brake wheel, and also under a spur or peg F¹ projecting downwards from the sleeve. This spur is nearer the centre of oscillation of the lever than is the point of contact between the wheel and the lever, consequently when the lever is inclined downwards the weight of the sleeve is borne on the spur, and when the lever is inclined upwards the weight is borne by the wheel. The practical effect of this arrangement is that in the former position of the lever, the wheel is free to

D D

revolve, and therefore the holder can slide down through the sleeve; in the latter position the wheel is held against rotation, and there can be no relative motion of the sleeve and the holder. Nevertheless they are free to move together, and advantage is taken of this to establish the arc. As soon as the current circulates in the solenoid, the core is drawn strongly upwards. The lever is raised first to the horizontal position, and then above it, carrying with it the sleeve, the carbon-holder, and the carbon, and thus forming the arc. As the carbons consume, the lever gradually falls to the horizontal position, when the smallest additional depression releases the brake wheel, and allows the carbon to feed downwards. The details of the feeding mechanism are shown to a larger scale in Fig. 437.

In a duplex lamp it is necessary that one pair of carbons should remain out of action until the other are consumed. This is attained in this case by making one spur rather longer than the other, so that one carbon feeds, when the lever is just above the horizontal line. When this carbon is nearly exhausted, the further descent of the holder is prevented by a stop, and the solenoid core descends until the second holder begins to feed.

Fig. 436 illustrates the electrical connections for a lamp intended to be used in series, the main circuit being shown in thick lines, and the shunt circuit in fine lines. S is a switch by which the lamp can be cut out and replaced by a suitable resistance R R. When the circuit is broken, either purposely or by any accidental cause, such as the sticking of a rack, the main portion of the solenoid M loses its power and the core falls, closing the circuit across the contacts G and H, which thus form an automatic cut-out. For single lamp, or for lamps arranged parallel, the shunt coil G of the solenoid is omitted or temporarily disconnected. The lamp is regulated by the screw T, which varies the tension of the spring by which the core is suspended.

BÜRGIN AND CROMPTON.

FIGS. 438 and 439 show a form of the Bürgin and Crompton lamp. The two carbons are shown at A and B; the lower one is fixed in the frame of the lamp with which it is in electrical communication. The upper carbon slides in a tube insulated from the outer tube c, and is hung from the

cord E. This traverses the armature F, passes over the pulley G, and is wound round the drum H. A second cord K is wound round the same drum. This carries at its end a copper ring L by which the upper carbon can be lifted. The armature F is surrounded by fine wires connected to the terminal P and to the body of the lamp; it is free to move before the electro-magnet M N, wound with coarse wire. This forms the vertical and movable side of an articulated parallelogram, the opposite side of which is fixed to the frame of the lamp in such a way that any attraction between the electro-magnet M N and the armature F causes a vertical movement of the upper carbon. In the diagram the terminals are shown at P P¹; Q

Fig. 438.   Fig. 439.

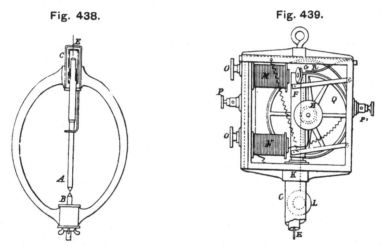

is a fly with a spring brake R. The current on entering the lamp divides and passes round the coiled armature and the electro-magnet to an extent determined by the resistance of each of them. The latter attracts the armature and raises the upper carbon. During this movement, the fly rests against the brake R, and the cord not being free to unroll, the carbon follows the motion of the armature until the resistance of the spring R balances the attractive power of the electro-magnet. In proportion as the carbons consume, the resistance increases in the main circuit, which includes the electro-magnet. The strength of the current decreases, while the shunt increases in the armature, which opposes the attraction of the magnet, and descending, releases the spring brake, allowing the drum to turn and the carbon to fall by its weight until equilibrium is restored.

ABDANK.*

M. ABDANK separates his regulator from his lamp. The regulator may be fixed anywhere, within easy inspection and manipulation, and away from disturbing influences of the lamp. The lamp can be fixed in any desired place.

The bottom or negative carbon is fixed, but the top or positive carbon is movable in a vertical line. It is screwed at the point C to a brass rod T (Fig. 441), which moves freely inside the tubular core of an electro-magnet K. This rod is clutched and lifted by the soft iron armature A B, when a current passes through the coil M M. The mass of the iron in the armature is distributed so that the greater portion is at one end B much nearer the pole than the other end, hence this portion is attracted first; the armature assumes an inclined position maintained by a brass button t, which prevents any adhesion between the armature and the core of the electro-magnet. The electric connection between the carbon and the coil of the electro-magnet is maintained by the flexible wire S.

The electro-magnet A (Fig. 440) is fixed to a long and heavy rack C, which falls by its own weight and by the weight of the electro-magnet and carbon fixed to it. The length of the rack is equal to the length of the two carbons. The fall of the rack is controlled by a friction brake B (Fig. 442) which acts upon the last of a train of three wheels put in motion by the above weight. The brake B is fixed at one end of a lever B A, the other end carrying the soft iron armature F, easily adjusted by three screws. This armature is attracted by the electro-magnet E E (whose resistance is 1200 ohms), whenever a current circulates through it. The length of play is regulated by the screw V. The spring L applies tension to the brake.

The regulator consists of a balance and a cut-off. The balance (Figs. 443 and 444) is made with two solenoids S S¹, whose relative distance is adjustable. S conveys the main current, and is wound with thick wire, having practically no resistance, and S¹ is traversed by a shunt current, and is wound with fine wire, having a resistance of 600 ohms. In the axes of these two coils, a small and light iron tube (2 mm. diameter and 60 mm. length) freely moves in a vertical line between two guides. When magnetised it has one pole in the middle, and the other at each end. Its

---

* This description is taken from a paper read by Mr. W. H. Preece before Section G of the Southampton meeting of the British Association, 1882.

upward motion is controlled by the spring N T. This spring rests upon the screw **H**, with which it makes contact by platinum electrodes. This contact is broken whenever the little iron rod strikes the spring N T.

The positive lead from the dynamo is attached to the terminal **B**, then passes through the coil S to the terminal B¹, whence it proceeds to the

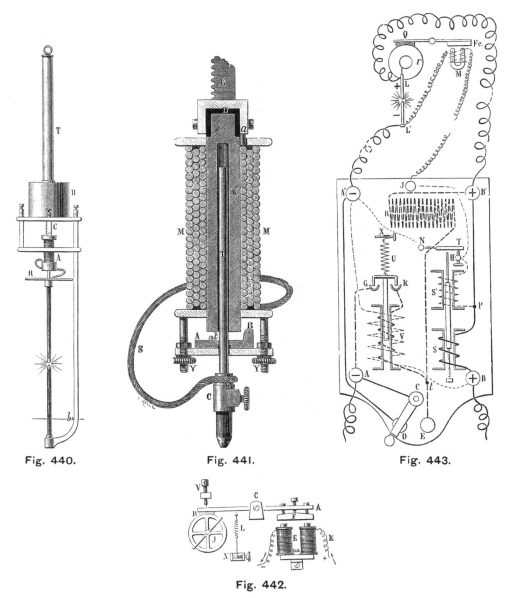

Fig. 440.   Fig. 441.   Fig. 443.

Fig. 442.

lamp. The negative lead is attached to terminal **A**, passing directly to the other terminal A¹, and thence to the lamp.

The shunt which passes through the fine coil S¹ commences at the point P ; the other end is fixed to the screw H, whence it has two paths,

the one offering no resistance through the spring T N to the upper negative terminal A¹, the other through the terminal J to the electro-magnet of the brake M, and thence to the negative terminal of the lamp L¹.

The last part of the apparatus to be described is the " cut-off," which is used when there are several lamps in series.   It is brought into play by the

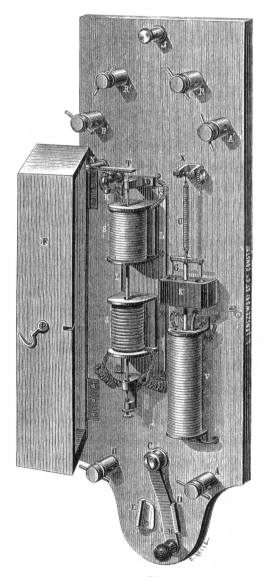

Fig. 444.

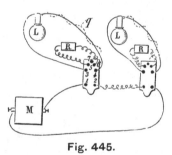

Fig. 445.

switch C D, which can be placed at E or D, Fig. 444.   When it is at E, the negative terminal A is in communication with the positive terminal B through a resistance R, equal to the resistance of the lamp, which is there-fore out of circuit.   When it is at D, the cut-off acts automatically to do

the same thing when required. This is done by a solenoid V which has two coils, the one of thick wire offering no resistance, and the other of 2000 ohms resistance. The fine wire connects the terminals A^1 and B. The solenoid has a movable soft iron core suspended by the spring U. It has a cross-piece of iron which can dip into two mercury cups G and K when the core is sucked into the solenoid. When this is the case, which happens when any accident occurs to the lamp, the terminal A is placed in connection with the terminal B through the thick wire of V and the resistance R in the same way as it was done by the switch C D.

When the current enters the balance it passes through the coil S, magnetising the iron core, and drawing it downwards (Fig. 443). It then passes to the lamp L, through the carbons, then returns to the balance and proceeds back to the negative terminal of the machine; a small portion of the current is shunted off at the point P, passing through the coil S^1, through the contact spring T N to the terminal A^1, and drawing the iron core in opposition to S. The carbons are then in contact, but in passing through the lamp the current magnetises the electro-magnet M (Fig. 443), which attracts the armature A B, that lifts up the rod with the upper carbon a definite and fixed distance that is easily regulated by adjusting screws. The arc then is formed, and will continue to burn steadily as long as the current remains constant. But the instant the current falls, due to the increased resistance of the arc, a greater proportion passes through the shunt S^1 (Fig. 443), increasing its magnetic moment on the iron core, while that of S is diminishing. The result is that a moment arises when equilibrium is destroyed, the iron rod strikes smartly and sharply upon the spring N T, contact between T and H is broken, and the current passes through the electro-magnet of the brake in the lamp. The brake is released for an instant, and the carbons approach each other. But the same rupture of contact introduces in the shunt a new resistance of considerable magnitude (viz., 1200 ohms), that of the electro-magnet of the brake. Then the strength of the shunt current diminishes considerably and the solenoid G recovers briskly its drawing power upon the rod, and contact is restored. The carbons approach during these periods only about .01 mm. to .02 mm.; if this is not sufficient to restore equilibrium, it is repeated continually till equilibrium is obtained, the result being that the carbon-holder is continually falling by a motion invisible to the eye, but sufficient to provide for the consumption of the carbons.

The contact between N T and H is never completely broken; the

sparks are very feeble and the contacts do not oxidise. The resistances inserted are so considerable that heating cannot occur, while the portion of the current abstracted for the control is so small that it may be neglected.

The balance acts like the key of a Morse machine and the brake precisely like the sounder receiver so well known in telegraphy. It emits the same kind of sound and acts automatically like a skilled and faithful telegraphist.

This regulation, by very small short and successive steps, offers several advantages. 1. It is imperceptible to the eye. 2. It does not affect the main current. 3. Any sudden instantaneous variation of the main current does not allow a too near approach of the carbon point.

Should an accident occur—for instance, a carbon be broken—at once the automatic cut-off acts, the current passes through the resistance R instead of passing through the lamp, the current through the fine coil is suddenly increased, the rod is drawn in, contact is made at G and K, and the current is sent through the resistance R. As soon as contact is again made by the carbons, the current in the coil S is increased, that in V diminished, and the antagonistic spring U breaks the contact at G and K.

SOLIGNAC.

IN the first volume of " Electric Illumination" (page 508) will be found the description of a very ingenious lamp devised by M. Solignac, in which the distance apart of the carbon points is regulated by the gradual softening of two small glass rods fixed below the carbons and abutting one against the other. To this lamp M. Solignac added an automatic relighting device, which we should have described if it had not been, despite its originality, abandoned and replaced by a new lamp in some respects still more simple, and much more convenient and economical in application. In pursuing his investigations with this lamp, M. Solignac was led to study the application of glass rods to a vertical carbon lamp. It occurred to him to use one rod only, employing one carbon much larger than the other, and fixed in a refractory casing, which descends freely, is consumed very slowly, and the under surface of which is constantly maintained in a fixed position

with reference to the lower carbon. The sketch, Fig. 447, shows the operation of this apparatus. Subsequently M. Solignac suppressed the glass rod altogether, and his new lamp has nothing in common with that

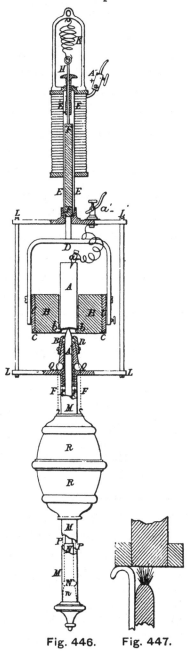

Fig. 446.   Fig. 447.

from which it was evolved. The construction of these lamps will be clearly understood with the aid of the engraving, Fig. 446, which is one-fifth full size. This lamp produces a light of about 600 carcels when fed by the

smallest size direct-current Gramme machine, absorbing one horse-power. In the illustration, A is the negative carbon, formed of a cylinder $2\frac{1}{2}$ centimetres diameter, through which the current is led by a small terminal fixed in the carbon as shown. The terminal is connected by a wire to the terminal A^1 insulated from the frame on which it is fixed, and thence to the source of electricity. This large carbon fits freely in a hollow cylinder of refractory material, the axial hole being from 1 to 1.5 millimetres larger than the carbon, which can thus slide without friction. It is held at the bottom by a small projecting lip in the refractory cylinder. This latter is contained in a copper cylinder C also made with a projection. The cylinder is hung between the arms of a bracket D, which is fixed to the end of an iron rod F, moving freely inside the brass tube E E. Around this tube is a solenoid formed with 30 metres of copper wire 1.5 millimetres diameter, and insulated with gutta-percha. The rod F moves in the axis of the solenoid, and abuts against a screwed rod through which passes the spindle H, hung from the top of the stirrup by the spring K. It will be seen that by moving the screw G in the nut belonging to it, the time when the rod F abuts against it can be hastened or retarded, and the movement of the negative carbon can thus be limited. By this arrangement, if the solenoid has an action in excess of that actually required, the movement of the rod may be restricted to what is necessary for regulating the arc. The solenoid receives the current from the terminal A''; the current passes from the tube E to the bracket D, which is insulated from the large carbon, and to the frame carrying the positive carbon. This latter A slides in a long copper sheath. The lower part is contained in a socket M provided with two springs $n\,n$ that press against the inside of the tube, and collect the current. This socket is carried by two small chains F F passed over the rollers Q and fixed to the counterweight R, which slides over the tube M. The upper end of the carbon A' traverses a nickel tube R, similar to those used in carriage lamps. The opening at the end of the tube is a little smaller than the diameter of the carbon, so that only the coned end projects. As this carbon is consumed, the counterweights raise it towards the upper carbon.

The action of the lamp is as follows : At the moment that the apparatus is put in circuit, that is when it is in the position shown in Fig. 446, the solenoid begins to act, lifting the iron rod and the negative carbon, which it maintains at a distance from the positive carbon fixed in advance by the screw G, according to the conditions of lighting, and the number of lamps

that are to be placed in the circuit. The arc passes from the positive carbon to the negative, which it rapidly heats, until the whole of its lower surface is luminous. To obtain steadiness of arc the positive electrode is formed of a so-called wick carbon, that is a tubular carbon filled with carbon. With this type of pencil the arc always passes axially with the cylindrical opening. Contrary to what may be imagined on first seeing this lamp, the refractory cylinder surrounding the large carbon has no function in the Solignac lamp, similar to that of the Soleil. On the contrary, it serves to cool the electrode, and thus to prevent the unequal wear of the carbon, which, without this device, would speedily become pointed, and fall out of the opening in the surrounding cylinder. Under the actual conditions, the absorption of heat insures an even wear of the carbon, which falls gradually, always maintaining its lower surface practically flat. The heat arrives at its maximum a few seconds after lighting, and preserves this temperature as long as a balance is maintained between the heat produced by the arc and the loss resulting from the radiation from the large surface of the refractory ring. This is constantly bathed by the hot air which surrounds the arc and spreads over the base of the refractory cylinder. It will thus be seen that the large carbon is raised to a very high temperature, and that while the wear is extremely slow, it preserves on account of its steady motion, due to its weight, a constant distance from the positive carbon. The refractory cylinder also acts as a reflector, but as it rapidly blackens by the deposit and absorption of dust, M. Solignac proposes to substitute a cylinder of enamelled porcelain.

From the foregoing description it will be seen that this lamp is simple in its arrangement and working. As regards the light produced, it should be pointed out, that owing to the construction of the lower part, and to the presence of the counterweight, a marked reduction of light is noticed beneath the lamp, though this does not amount to an actual shadow on account of the large size of the upper carbon. This disadvantage will probably be removed by modifying the details of the lamp, in which attention has been chiefly devoted to the production of a fixed light. The counterweight may be easily shifted to the top of the lamp, and the interception of the light will be thus avoided.

IN a very successful electric light installation at Brixton, the arc lamps used are the invention of Messrs. Clark and Bowman; the mechanism is illustrated by Figs. 448 to 450, where Fig. 448 shows the parts as they appear when the regulator is out of action, or the arc is of the normal length; Fig. 449 shows them in the act of feeding the carbon; and Fig. 450 is a perspective view of the clutch mechanism. The electro-magnet j is situated in a shunt circuit round the arc. The amount of current traversing this circuit varies, as is well understood, with the resistance of the arc, the two increasing and decreasing together. Over the magnet is hung a pendulum armature k, which is drawn back by the light flat spring k^1 against the attraction of the magnet. It is consequently very sensitive to any change in the current circulating in the coils, and responds to its smallest variation. On the front of the armature is a contact spring l, which in its extreme position meets with an adjustable screw f, and in so doing completes a second circuit through the magnet d, which operates the feeding mechanism and puts it into action. Thus the office of the magnet j and its armature is to act as a relay, completing a circuit when the resistance of the arc exceeds a certain determined amount, and breaking it again when this resistance is reduced, and it is to this arrangement that the lamp owes its great sensitiveness. The regulating device does nothing but regulate; none of the heavier work of actually feeding the carbon is thrown on to it, and consequently it acts with great promptitude and certainty. In practice the lamp works with regular feeds of about $\frac{1}{100}$ in. at periodic intervals, without giving rise to any jumping or winking.

When the circuit is closed between the contact pieces l and f, a current is sent through the wire $f^1 f^1$ to the contact spring x, and thence through the adjustable screw f^2 to the coils of the magnet d. This latter immediately attracts its armature b, drawing it down, and at the same time breaking the circuit at x. The distance the armature will descend before the magnet loses its power, depends upon the flexibility of the spring x. As soon, however, as the contact is broken, the armature ascends and again completes the circuit, when, if the contact at $l f$ is still maintained, the feeding mechanism makes another stroke, and so on until the arc is reduced to the proper length. As a matter of practice one stroke is usually sufficient, and when the circuit breaks at x it also breaks at k^1, and the

parts come to rest until another infinitesimal portion of the carbon is consumed.

The feeding mechanism is shown in Figs. 448 and 449 as a clip washer and brake spring, as it is sometimes made, but in Fig. 450 it is illustrated as it is actually constructed in the Brixton installation. The armature lever b has an eye b^1 which surrounds the carbon-holder y, and grips it when in an inclined position. It is pivotted between two lugs on the plate C

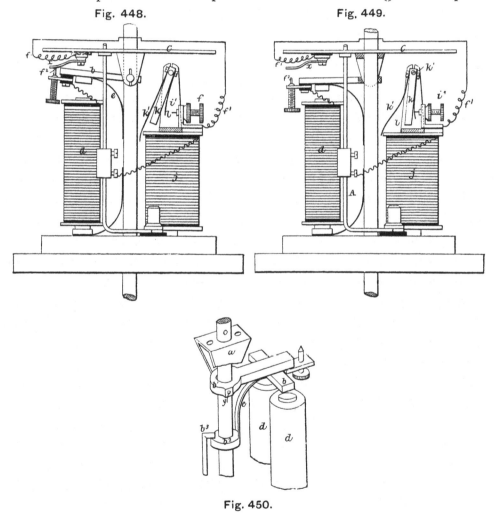

Fig. 448.   Fig. 449.

Fig. 450.

(Fig. 448). From the under side of the lever there extends a spring to a second washer ring b^2, so that when the lever is drawn down and the upper washer loses its grip, the lower one takes hold of the electrode, forcing it towards the arc positively, and not allowing it to slide, as is usual. On the return stroke the washer b^1 again takes hold, and b^2 loses its grasp as it comes against the fixed hook b^3. By this arrangement all necessity for

flies, pendulums, or glycerine cylinders is avoided, and all chance of the carbons sliding together is prevented, while it is no longer indispensable for the lamp to be hung vertically, as in those regulators which depend upon gravity to effect the feeding. The arc is established by an electro-magnet connected to the lower carbon-holder.

LEVER.

STATED in general terms, the action in this lamp is similar to that of Brush with the main regulating magnet replaced by a spring. The general arrangement is clearly shown in Fig. 451, while the action may be more clearly seen in the diagram Fig. 452. F F are the two branches of an electro-magnet, whose coils form a shunt circuit of high resistance round the arc. The attraction of these magnets upon their armature A is opposed by the adjustable spring D. Resting loosely on a projection of the lever L, to which the armature is attached, is a clip B, which is a washer encircling the upper carbon-holder C. When the washer is horizontal the carbon-holder is free to slide through it under the action of gravity, but when it is tilted by the spring, it grips the holder and retains it. The action of the lamp is as follows : When no current is passing the washer holds the rod, and the two carbons remain apart until the completion of the circuit, when the magnet is excited, and the armature drawn down. This releases the upper carbon, and the two come in contact, whereupon, a new path of low resistance being opened for the current, the magnet loses its power, and releases its hold on the armature. The latter is then immediately raised by the spring, carrying with it the washer and upper carbon, and thus the arc is established. The lamp then burns until the increasing resistance of the arc diverts a greater proportion of the current through the magnet coils, thus augmenting their power to attract the armature and diminishing the bite of the washer. This action continues until the carbon-holder is free, and falls through a slight space, usually so small as to be almost imperceptible, when it is again gripped, until an increase of the arc produces a fresh adjustment, and so on. Fig. 453 shows a modification of the invention, in which two spring levers are substituted for the washer. These are connected directly to the armature of the magnet, and are so arranged that when in a

horizontal attitude, or thereabouts, as shown in the illustration, they nip the carbon-holder between their ends, but when drawn down by the attraction of the magnet on the armature, they release it. It will be noticed that when the lamps are first put in action the carbons are separated, and consequently the only path for the current is through the whole of the shunt coils in succession. These have of necessity a high resistance, and conse-

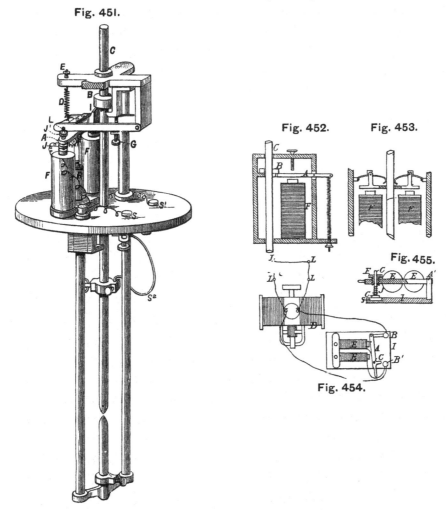

Fig. 451.

Fig. 452.   Fig. 453.

Fig. 455.

Fig. 454.

quently the dynamo machine, unless it be shunt wound, cannot generate a current of sufficient intensity to excite its field magnets, or those in the lamp. To obviate this difficulty Mr. Lever has devised an automatic apparatus, which has the effect of short-circuiting the generator through a shunt of moderate resistance, until the current has attained a determined intensity. The shunt circuit is then broken, and the full force of the current is thrown into the lamps, with the effect of instantly exciting their magnets

and establishing the arcs. Figs. 454 and 455 illustrate this apparatus. From the terminals of the generator there run two circuits, one comprising four lamps L L, indicated by circles, and the other the automatic shunt. This includes an electro-magnet, whose coils E E have a resistance about equal to that of the lamps, a pivotted armature A, and a sliding contact C, shown to a larger scale in Fig. 455. At first the course of the current is through the magnet coils to the terminal B^1 and the fixed portion of the contact, and then across the armature to the terminal B and back to the generator. As soon, however, as the armature is drawn up to the magnet, one part C of the contact piece slides off the other S, and the circuit is broken, whereupon the whole current flows through the lamps.

BREGUET.

FIG. 456 illustrates the Breguet arc lamp. The regulation of the arc is effected by an electro-magnet placed in derivation to the current which traverses the apparatus, and the feed of the carbons is produced by the weight of the carbon-holder. At its upper part is placed gearing, worked by a carbon-holder C, and commanded by an electro-magnet E^1 of 700 ohms resistance. The armature A^1 of this magnet carries at one part a finger D, which arrests the last wheel by engaging a star-wheel I, and at another part a rod T, of which we shall speak further on. The lower part comprises a platform P^1, to which there is fixed a second electro-magnet of very feeble resistance, whose armature A carries the lower carbon. When the lamp is not in action this armature is maintained at a certain distance from the electro-magnet by a spring R, whose tension can be regulated by the screw e; it is also connected with a lever L, which operates a rod T contained in the side rod M. This rod carries at its upper end a nut c^1, which acts upon the rod D, when it is at the top of its stroke. The side rod M, which carries the negative terminal B^1, is insulated by ebonite rings from the two platforms, which it connects. The opposite side rod which carries the positive terminal, is in metallic contact with the lower platform and is insulated from the upper platform.

The action of the lamp is as follows : The terminal B is connected to the positive pole of a generator, and the terminal B^1 with the negative pole.

The current enters at B, passes to P and divides itself into two parts, as indicated by the arrows on the figure. The first part passes through the carbons, circulates around the electro-magnet E, remounts by the side rod, and leaves by the terminal B¹. The second part traverses the electro-

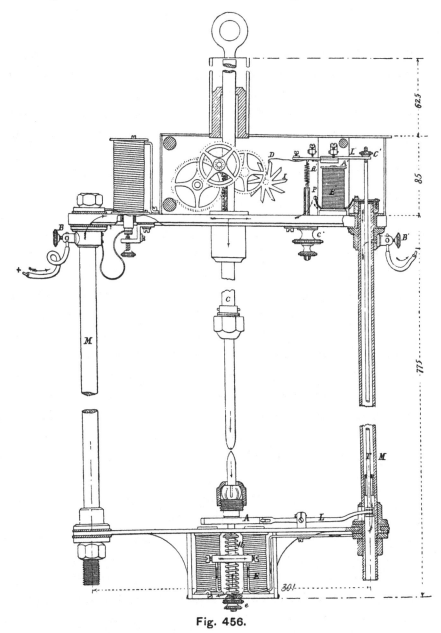

Fig. 456.

magnet E¹, and also leaves by the terminal B¹, the magnet being in a derived circuit around the arc. In circulating in the electro-magnet E, the current magnetises its soft iron cores, the armature A is attracted, and the voltaic

arc is established between the two carbons.   The armature A in descending raises, by the intermediary of the lever L and the rod T, the screw c^1, which thus ceases to act any longer on the lever L^1.   It will be seen that this part of the mechanism is inactive so long as the current circulates in the apparatus.   Further on we shall explain the part played by this device upon the extinction of the lamp.

The fraction of the current which traverses the electro-magnet E^1 magnetises also its cores, but the tension of the spring balances the attraction which they exercise on the armature, and the upper carbon remains fixed.   When the arc increases by the wasting of the carbons, the intensity of the current which traverses them diminishes, while at the same time, the current in the electro-magnet E^1 augments.   At last there arrives a moment when the influence exercised by the electro-magnet upon the armature is greater than the tension of the spring R^1.   The finger D is then raised, and the carbon-holder, becoming free, descends until the arc recovers its primitive resistance ; the tension of the spring then draws the finger D into contact with one of the branches of the starwheel I, and the carbon-holder becomes stationary.   The method of action we have just described is the one which obtains if care be taken to place the carbons in contact by hand at the moment of lighting.   The use of the rod T and the lever L is to dispense with this trouble, by bringing the carbons automatically into contact as soon as the current ceases to circulate in the lamp.   At that moment the armature A is suddenly raised by the springs R, the nut c^1 then presses against the lever L^1, the finger D is raised, and the upper carbon descends. As soon as it touches the lower carbon, the weight of the carbon-holder lowers slightly the armature A, the nut c^1 is raised, and the upper carbon arrested.   The gearing being locked, the two carbons cannot possibly cross each other.

To regulate the lamp it is necessary in the first place to fix the stroke of the armature A, and the position of the nut c^1.   By acting upon the nut e, the distance of the electro-magnet from its armature is adjusted ; this should be about three millimetres.   We have already seen that when the armature A is at the top of its stroke the upper carbon can descend.   It is therefore necessary, when the armature A has been adjusted, to operate c^1 until the gearing begins to move.   At the slightest pressure exercised upon the armature A, the nut c^1 ought to rise, and the gearing to move.   In order to adjust the nut conveniently, it is necessary to draw it into a position such that on a very small displacement upwards from below, the apparatus will

be put in action. That part of the regulation ought to be effected before the current circulates in the lamp. As has been already mentioned, when the current traverses the apparatus the armature is subject to two forces, the tension of the spring and the attraction of the electro-magnet. It is therefore evident that the more the spring R^1 is tightened, the greater must be the attraction of the electro-magnet to unlock the gearing; further, the greater the tension of the spring the longer will be the arc.

In starting the lamp a current is made to traverse it, and the distance between the carbons is observed. If this is too great the spring is relaxed, and if on the contrary it is too small, the spring is extended. Each of the adjusting screws carries a counter screw, which it is very important to tighten energetically to prevent the variations to which the regulator may be exposed.

WERDERMANN.

THE details of the Werdermann lamp are shown in Figs. 457 and 458. The upper carbon is held between three rollers a^1, a^2, a^3, of which a^3 is carried on a jockey frame, and is kept up to the pencil by the pressure of

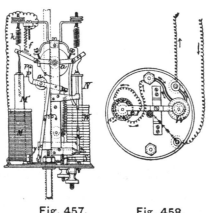

Fig. 457.   Fig. 458.

a spring. a^1 has a disc attached to it by which it can be rotated in either direction by means of two gripping levers, one operated by the coarse coil

W to establish the arc, and the other by the shunt coil M to give the feed. These levers are jointed to independent arms pivotted to the axis of the pulley a^1. The solenoid M is in a shunt circuit around the arc, and is wound with two lengths of wire, so arranged that the current may traverse either one, or both of them. The outer end of the inner length and the inner end of the outer length are connected to the plate f, which lies in the path of a terminal on the feeding lever. When the circuit is completed the core N of the solenoid W is drawn down to the armature S and there retained, the pulley a^1 being at the same time rotated and the arc established. Just before the core N reaches the bottom of its stroke, a lever, fixed on the axis of the gripping lever, comes in contact with the frame, and, rotating the axis, releases the pulley a^1, which is at the same time nipped by the opposite lever, operated by the core M^1. As the arc lengthens the power of the solenoid M increases, and its core descends until the contact at f is broken, when the additional resistance of the second coil reduces the shunt current, and allows the core to rise again ready for another stroke. Should the carbons break or the arc fail, the gripping lever meets the stud p, and, drawing the jockey frame backwards, allows the carbon to drop. As it is the carbon itself, and not a holder, that passes through the rollers, very long pencils can be used in this lamp without any undue elongation of the framing or increase of resistance.

VARLEY.

THE lamp of Messrs. C. F. and F. H. Varley is illustrated in the accompanying figures. There is no upper carbon-holder, but in place of it, the pencil is held between four rollers D D, and is forced downwards by the spring C, under the modifying influence of a dash piston, which is formed of discs of paper and felt, and is provided with an adjustable air valve. The cores of the differential solenoids L^1 L are connected together by a lever at the top, and simultaneously operate two arms below the rollers D. When the circuit is completed these arms rise together, and, lifting the rollers and carbon, establish the arc. As the arc lengthens

the arms fall clear of the rollers, allowing them to revolve and the pencil to descend. The lower carbon is carried by the arm E¹, of which Fig. 460 is a

Fig. 459.

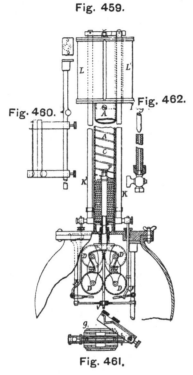

Fig. 460.

Fig. 462.

Fig. 461.

separate view. The holder is made to swivel (Fig. 461) in order that it may be turned out of the way, when the upper electrode is inserted.

EGGER-KREMENETSKY.

IN the Egger-Kremenetsky arc lamp, which is illustrated by Fig. 463, A and B are the two solenoids, through the coils of the first of which the principal current passes and forms the arc; the second receives the derived current and regulates the arc. The cores a and b of the solenoid are connected by an articulated system which includes two symmetrical fingers that bear against the upper carbon. When the core a is raised, the fingers turn around their axes, clip the carbon-holder and lift it. In falling they come in contact with two forks which open them, and, liberating the carbon-holder, allow it to slide. In the rod of the upper

carbon-holder are two lateral grooves to receive cords, which pass round the pulleys R R', and hold up the lower carbon-holder ; *s s* are two wheels which roll against the bars F F', and preserve the axial position of the carbon. A cylinder, filled with glycerine, acts as a slide for the lower carbon-

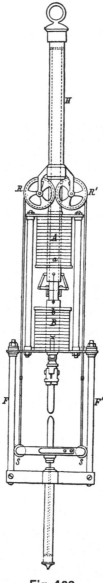

Fig. 463.

holder, which is formed of a brass rod with a piston at the end. The upper carbon-holder slides in the tube H, which serves as a support of the lamp. The current flows to the lamp through the solenoid A, traverses the coils, and attracts the core *a*. The fingers then grasp the upper carbon-

holder, lift it, and form the arc. As this lengthens, through the wear of the carbons, a larger part of the current flows through the derivation B; the core b is attracted, the fingers open and allow the upper carbon to slide until the arc becomes normal. When lamps are used in series, each is, in addition, provided with an electro-magnet, partly traversed by the principal current and partly by the derived current, and which, when one lamp becomes extinguished from any cause, introduces a corresponding resistance between the terminals.

ZIPERNOWSKY.

WE have already referred to the electric lighting system of Messrs. Ganz and Co., of Budapest, but we did not describe the Zipernowsky arc lamp, which they employ largely. Fig. 464 is a diagram illustrating the manner in which this lamp works. As will be seen, it comprises an articulated parallelogram suspended from a rod. The two principal sides of this parallelogram are the negative carbon-holder, and the positive carbon-holder.

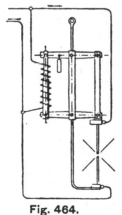

Fig. 464.

This latter passes through a fine wire solenoid in derivation on the main circuit. The action of this solenoid tends to raise the lower carbon, and at the same time, by the action of the parallelogram, to depress the upper carbon to an equal extent. This effect is counterbalanced by that of a small weight, which can be regulated so as to impart the desired length to the arc. The combustion of the carbons increases the length of the arc, and

consequently the strength of the current passing through the solenoid, which then acts with greater energy, and restores the arc to the position of equilibrium. Figs. 465, 466, and 467 illustrate the mechanical details of the lamps, as carried out. Without entering into a minute description, we may explain the solenoid S, in which there moves the hollow iron coil carrying the piece which holds the lower carbon. In this hollow iron core there is a small piston by which the movement is controlled. The spiral spring *f*, by which the tension is regulated, by aid of the screw *m*, placed under the upper plate

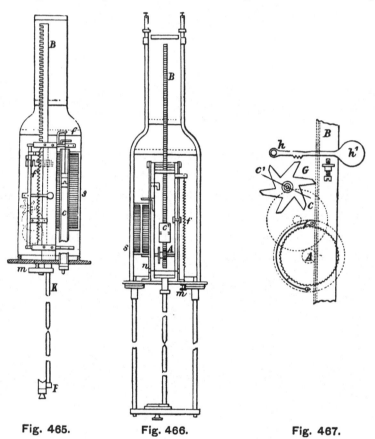

Fig. 465.   Fig. 466.   Fig. 467.

of the lamp, acts in the direction of the attraction of the solenoid S, and in an opposite direction to the weight of the core C. The upper carbon-holder is connected to a rack gearing with a pinion A, Fig. 467, the motion of which is regulated by the starwheel G and the small fly *c c*[1]. When, by reason of the wear of the carbons, the rack is at the bottom of its stroke, it allows the piece *h h*[1], movable around *h*, to fall; this arrests the motion of the star-wheel of the gearing, and of the whole system. The small coarse wire coil S is thrown into the circuit by means of the copper spring *n*, at the moment

when the carbons are completely consumed, and the lamp only resumes its functions when new ones are introduced. M. Zipernowsky makes lamps with two pairs of carbons which burn for sixteen hours.

PIETTE AND KRIZIK.

IN the first volume of "Electrical Illumination" we described the Piette and Krizik, or Pilsen, lamp; since then certain modifications have been introduced which are shown in Fig. 468, where P P are the terminals, $E^1 E^2$ the iron solenoid cores, H the coarse wire sole-

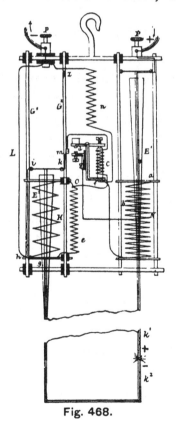

Fig. 468.

noid in the main circuit, N the solenoid in the derived circuit, with some turns of coarse wire and a number of fine wire coils, c an electro-magnet, n a resistance of German silver, and $G^1 G^2$ guides insulated from the core E^2.

Assuming that the carbons are separated and that it is desired to light the lamp, the current enters by the terminal P and goes by the frame of the lamp to the point a ; it then traverses the coil of the solenoid N, passes from b by the platinum contact d and thence to the point f; finally it leaves the lamp, after having traversed the resistance n and the insulated terminal P. The core E^1 is attracted by the solenoid $a\,b$ N, and the carbon k^1 falls into contact with the carbon k^2. At this moment the current has two courses open to it; that described above and a second as follows : The lamp, the carbons, the insulated guides $G^1\ G^2$, thence by the rollers i and k at m, the coils of c as far as O, the coil of the solenoid H, the wire L, and the negative terminal P. The chief part of the current will pass by this route, which offers a slight resistance. The solenoid H will raise the core E^2 and separate the carbons thus forming the arc. The contact magnet c will attract its armature and break the contact at d. That portion of the current passing by the first circuit being very slight, will take the following course :

From P across the lamp at a, then through the coarse and fine wire coils of the solenoid N, through the coils of the contact magnet c, and thence to the main circuit at m. When the arc lengthens by the combustion of the carbons, the resistance augments ; the derived current then increases in strength, the core E is attracted in the solenoid N, and the carbons will be brought nearer, which will restore the normal condition of working. When the carbons are burnt out, one of the conducting rollers of the core E^2 comes in contact with a small ivory plate x fixed to the guide. Then the main current passes from P across the lamp, through the resistance e as far as o, the solenoid H and the wire L to the terminal P. The derived current goes from a to c and d as far as f, and across the German silver resistance n to P. It then traverses the contact $c\,d$, which is restored by the magnet contact c.

INCANDESCENCE LAMPS.

DURING the past two years, electric lighting by incandescence lamps has found more numerous and extensive applications than that of the voltaic arc, and the manufacture and use of the former has to-day attained very considerable dimensions. This class of electrical lighting has indeed become a practical as well as a commercial success, as it is peculiarly adapted to those special purposes for which the demand is large and increasing—for theatres, ships, hotels, workshops, and public exhibitions, such as that of the Fisheries at South Kensington in 1883, and the Health Exhibition of the present year, where, combined with arc lighting, many varieties of luminous effects have been produced. But although lighting by incandescence is thus rapidly growing in use, the lamps employed are made only by a few makers, and the systems which first proved themselves successful, are still most largely, almost exclusively used. These systems are those of Edison, Swan, Lane-Fox, modified by the Brush Company, and Woodhouse and Rawson. Nevertheless inventors have been busy during the past two years, either in scheming new forms of lamps, or in devising modifications of existing types. In the following pages particulars will be given of these inventions, some of which have, or will doubtless have, a practical value, while all necessarily possess a certain interest.

Woodhouse and Rawson.—As already mentioned, Messrs. Woodhouse and Rawson manufacture incandescence lamps on a large scale. As will be seen from Fig. 469, their globe is supported by three light fingers or springs attached to the holder, while the contact with the platinum terminals is made by two spiral springs coupled to the binding posts at either side. By this arrangement an even, steady pressure is maintained upon the platinum loops, in spite of any vibration of the lamp, an important point in some installations, especially on board ship, where the rolling of the vessel is apt to break the contacts, and give rise to minute arcs, if special precautions be not taken.

In the lamp itself the chief novelty lies in the method adopted for connecting the filament to the metallic conductors. These are bent to a right angle at a little distance from their extremities, and their ends are flattened into discs or plates, which are then rolled up to form tubes into which the legs of the filament are inserted. Good contact is insured at the junction by means of a carbonaceous cement, which has the further advantage of increasing the cross section of the conductor, and so reducing its temperature in the immediate vicinity of the platinum wire. The outer ends of the metallic conductors are turned into loops and sealed into the glass by the method known as "pinching in," as is now usual in incan-

Fig. 469.

descence lamps. The lamp shown in Fig. 469 is of twenty-candle power, and requires a difference of potential of sixty volts to drive it to its full safe capacity. Messrs. Woodhouse and Rawson, in addition to their standard lamps, which vary from three-candle power upwards, and are adapted to work at potentials from three to sixty volts, manufacture lamps for inventors and others who desire to have them made to their own specification or according to their own processes. One form of which they make a specialty, is a coloured incandescence lamp for theatrical and ornamental purposes. The tinting of the globe is effected after the other parts of the manufacture are complete, and consequently the process can be applied to the lamps of other makers or to ordinary globes.

Müller.—The carbons are shaped so as to form two cylindrical or conical spirals crossing one another ; these are jointed at the one end, and the free ends connected to the conducting wires. Fig. 470 represents the form of the carbon filament. The carbons are manufactured as follows : Fine strips of sugar cane are placed in a receptacle, which is hermetically closed and exhausted and then filled with a solution of hydrate of carbon in water, whereby the pores of the cane are impregnated, and the fibre is made more flexible. The fibre is then shaped to the required form and carbonised as usual, after which it is impregnated with a solution of cellulose or similar material in the same manner as before.

Fig. 470.

Brush.—Incandescence electric lamps are constructed according to this arrangement, with a suitable screen, preferably of glass or mica, between the limbs of the bent filament, whereby the " radiant matter " projected from any part of the filament is prevented from exerting its destructive action on other portions. Fig. 471 shows the screen E (preferably of glass) in the form of a rod or tube lying in the plane of the filament loop *d d.* Fig. 472 shows the screen E, which may be split at one side, arranged

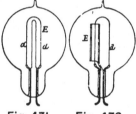

Fig. 471.   Fig. 472.

as a tube enclosing one limb of the filament *d.* In order to produce the same effect, the filament *d* may have such a form in cross section that " radiant matter " from each limb of the filament being projected in various directions normal to the surface, will not strike the other limb.

Beeman.—Mica, talc, or any other non-conducting semi-transparent matter of convenient thickness and size, is used to act as a support for the carbon filament. This support may, by acting as a non-conducting diaphragm, tend to arrest the deposit of carbon particles on the various parts of the filament, and may be used to act as a reflective surface, either alone or in combination with silvered glass, &c. Figs. 473 and 474 show a method of fixing the shield of mica or talc C, to serve as a support to the carbon filament B, and

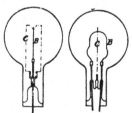

Fig. 473. Fig. 474.

to tend to arrest the deposit of particles of carbon from one portion of the filament on to the other.

Bernstein.—This lamp is illustrated in Fig. 475. The wirework frame

$r^2\ r^3$ supports the block of carbon k. A carbon rod t^2 is pressed against this block by means of the spring o^2 and movable carbon block k^2. The current enters by a platinum coil cast in the neck of the lamp, and passes to the blocks $k\ k^2$. The electrode is made of porous non-conducting carbon, which is coated with a dense layer of the same material, and treated at a high temperature in an atmosphere of hydro-carbon vapour.

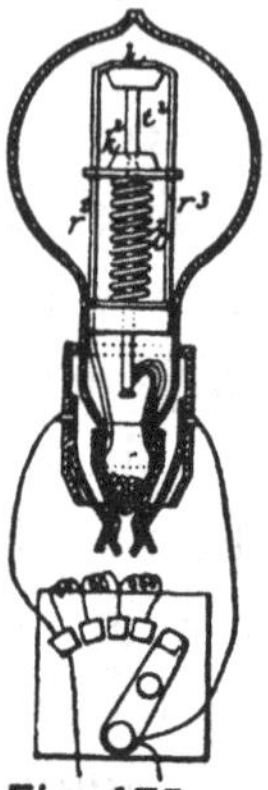

Fig. 475.

In another arrangement proposed by Mr. Bernstein a hollow carbon cylinder is employed as the light-emitting part of the lamp. Referring to the illustration, the light-giving carbon B is made of a hollow cylinder, and supported at its ends by socket pieces or blocks of solid carbon C and D. The conducting wires E and F enter the globe either at one end, through the neck, or at opposite ends, and one, E, is made rigid, whilst the other, F, is made flexible, being composed of a large number of thin wires. Or several hollow carbons of the same length are supported by small blocks at one end connected to the flexible conductor, and by a single large block at the other end connected to the rigid conductor. The cylinders of different lengths may be placed in series, and connected at the adjacent ends by carbon blocks, the flexible conductor being connected to a block at one end and a rigid conductor to each of the other blocks, so that, by means of a switch, the current can be passed through one or any number of the cylinders in series. A large number of these lamps may be connected

Fig. 476.

in a combined series and multiple arc arrangement, a switch being provided with each lamp, and so arranged that the force of the current can be concentrated on one lamp or divided among a number.

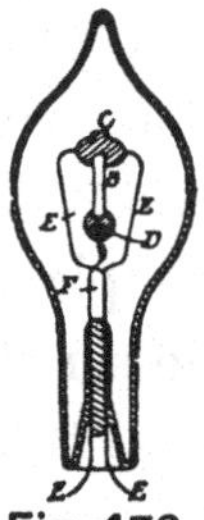

A third proposal of Mr. Bernstein relates to the manufacture of hollow carbon cylinders, and to the manner of uniting these carbons to the conducting wires. Suitable organic substance, such as paper, is cut into sheets, rolled into the desired cylindrical form, and carbonised. Preferably the fabric or other organic substance having the form of sheets, is cut into strips ; one side of the strip is pasted over with a carbonaceous cement such as gum or paste, and the strips are rolled upon a core or mandrel of suitable size, so as to form two or three superposed layers, and as soon as the cement is dry, the cores are withdrawn, and the cylinders placed in iron boxes filled with powdered plumbago, or charcoal, and carbonised. As shown in Fig. 477, the carbon cylinder is held in a vertical position by large carbon sockets D and E, which connect it to the conductors B and C. If a curved hollow

carbon is desired, knitted or braided tubes made from organic fibres are put upon cores, covered with a carbonaceous cement, and as soon as the cement is dry, the cores are withdrawn and the tubes are carbonised as before. The sockets D and E are provided with holes at each end, one large to receive the filament, and the other smaller to receive the conductor, the two holes being united by a countersunk tapering head adapted to fit this countersunk part. One conducting wire is connected to a metallic socket, whilst the other passes through a screw in the centre of the socket and terminates in a flat head.

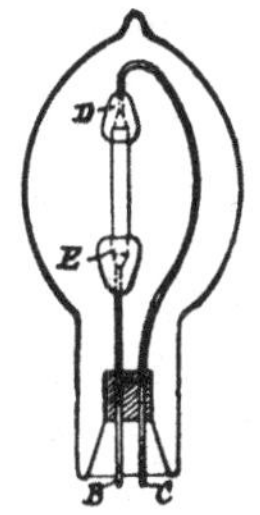

Fig. 477.

Crookes.—The filament is mounted in the first place upon its permanent support, and is then placed with this support in a vessel from which the air can be exhausted. Vapour is then admitted, and the filament is rendered incandescent. The lamp bulb is made in two parts, the filament being mounted on the one part, and enclosed in the chamber described in Specification 3799 of 1881 (see Appendix), to adjust its conductivity, the terminals of conductors being joined to the fixed terminals in the chamber. The two parts of the lamp are then brought together and fused, and the globe is exhausted and sealed as usual. The lamp is then finished in a small machine, where its neck is set with plaster into a wooden cup, the conductors passing through holes in the bottom. The projection on the neck of the lamp, left where the connection with the air pump is made, is loosely enveloped with paper to prevent contact between it and the plaster, and is received in a cavity *b* in the cup bottom. The terminals *c* are then screwed down on to the conductors, which are twisted round their shanks. In Fig. 479 the cup is partly of brass. The lamp foot is screwed into a socket, its terminals coming against horn-like springs, each in the shape of a segment of, say, one-third of an annulus bent upwards from the base to which one end is fixed. Or the lamp may be suspended from a button of insulating material, provided with two metal springs connected to the leads and entering holes in the lamp terminals. The lamp may be suspended directly from conductors passing to the ceiling, and secured to the lamp terminals by small levers having cam-shaped ends working in a slot in the terminal. The conductors are laid in mouldings along the ceiling, having removable caps, and connected to pieces of metal on the side of an insulating cylindrical boss extending out from the centre of the rosette. The conductors from which the lamps hang

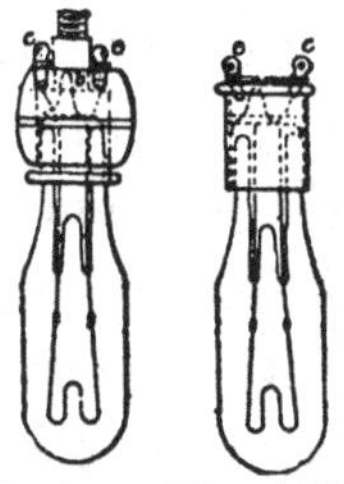

Fig.478. Fig. 479

have their ends inserted into holes in the metal pieces, and a ring or slider is drawn over them. For side wall connections, two separate metal studs are each provided with a slider securing the conductors into a hole in the side of the stud. Each circuit is completed through a switch consisting of two bent conductors fixed to a wooden base. One bent conductor is depressed, and an eye on the other part engages with an arm on the first part. When the part having the eye is depressed by means of a button, the eye is removed and the arm springs back, breaking the circuit.

Defries.—Fig. 480 shows a section of a coupling arranged to carry an incandescence lamp, and Fig. 481 a section of a coupling applied to a swing pendant, or a ball-and-socket joint. The end of each conductor is drawn through a hole in a plug D of insulating material; one end being coiled up into a small cylindrical coil I and let into the centre, or the end is inserted into a short metallic tube passing through the plug D. The other end is bent into a circular coil K (Fig. 481) laid flat on the surface of, or in a groove in, the plug and concentric with the coil formed by the other end, or the end may be connected to a metallic

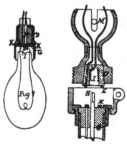

Fig. 480.   Fig. 481.

washer K (Fig. 480) inserted into the plug. The whole may be placed in a casing screwed on to a suitable fixture. A cap or coupling screwed on to fit the other part of the joint, is placed over the casing. The other part of the joint consists of a piece of insulating material G screwed or fitted to the cap, and on the other end is a circular metal plate K of a diameter equal to the flat hoop K and connected to one conductor. A wire or pin H, preferably split and connected to a second conductor, is fixed in the centre of this insulating material G concentric with the metal ring K, and makes contact with the central coil I. As shown in Fig. 481, a pin M on the socket fits into a groove in the ball and prevents the ball revolving in the socket and leaves it free to swing in all directions.

Fig. 482.

Gatehouse and Alabaster.—The filament is fixed in position in the vacuum globe in the position shown in Fig. 482, Fig. 483 being a plan of the same. The slightly thickened ends of the filaments have copper deposited on them, as at B; on this is deposited a layer of other metal, preferably platinum, which admits of the sheathed ends being fused into the glass. The external part D of the filament is picked out of its compound metallic sheathing, the space being filled with cement, or the ends of the wires E may be

Fig. 483.

soldered into the hollow ends. To do away with an outside joint the wires soldered into the hollow ends may be of platinum and pass through into the globe in the neck of which they are fused.

Knowles.—The carbon filament C, Fig. 484, is supported by small carbon cylinders E. It is cemented into cavities in the carbons with a mixture of bichloride of potassium, platinum black or sponge, and sugar, or a mixture of Indian ink and platinum or platinum black or sponge, and water, or water glass. The carbon cylinders are connected to the conductors by cups F soldered to the conductors. The cups F are slit longitudinally, and being of slightly less diameter than the carbon, grip it firmly. The filaments may be connected to the carbon cylinders in various other ways, *e.g.*, the cylinder may be perforated at both ends, one end of the filament being inserted in one end and the conductor in the other, or the carbon cylinder may be slit, the ends of the filament being passed into the slit, or the filament may be supported between divided pieces of carbon set in the cup F. The cup F may be replaced by a spiral. The conductors are passed through an inverted glass cup I and the glass pressed around them, or a cavity may be formed on the top side of the cup, and the conductors sealed by cement. The sides of the cup I are pressed down to the bottom of a second cup J filled with a suitable cement. The lamp is then exhausted through either of the tubes B. The cup J may be formed on the cup I by turning up its sides, or a cylinder of india-rubber may be used. The lamp is set upon an insulating base-piece having on its under side metal sectors, connected to the conductors, and between which is a metal piece. A fixture is provided at its upper side with an insulating disc into which metal sectors connected with the leads are set. The lamp is set on the fixture and held by a pivot secured by a screw K. By turning the lamp round, the upper and lower sectors make contact and complete the circuit, or the filament may be short-circuited by the metal piece between the upper sectors, or a series of metal points.

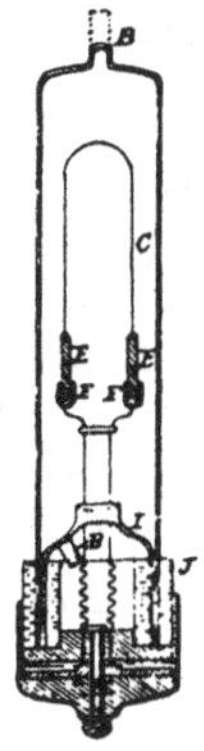

Fig. 484.

Edison.—An inert or azotic gas is admitted to the bulb of the lamp, and the supply is cut off at a definite pressure. The carbon filament, when such is used, is produced by the carbonisation under pressure of any suitable organic material, reduced or not to an amorphous or semi-amorphous condition, as is well understood. Fig. 485 shows a view of the lamp, and Fig. 486 shows it connected with a Sprengel pump D, the mercury enter-

ing at *b* and passing out at C. During the latter part of the operation
the filament is gradually raised to a degree
of incandescence higher than that at which
it is intended to be used. The inert or azotic
gas is then allowed to pass in, the filament
being still incandescent, until the pressure
is sufficient, which is determined by the dis-
appearance of a blue halo formed on the
positive terminal, on the gas being first
admitted. A Geissler spark gauge is pre-

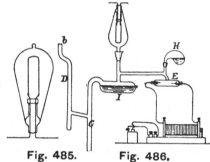

Fig. 485.   Fig. 486.

ferred to determine the exact point at which to seal off the lamp. The
spark gauge is worked by a constant battery and an induction coil. The
pressure may also be determined by a mercurial column, the pressure for
nitrogen being about 29 in. The gas may be introduced from a vessel, or
preferably, and as shown in Fig 486, disengaged by heat from a solid such
as cyanide of mercury or other suitable substance contained in a vessel H.
The chamber I contains the drying apparatus.

Fergusson.—In making the filaments, fibre of bass wood is placed over
the end of a carbon rod, being kept from slipping by pins, and is twisted
about the rod in the form of a double spiral,
and carbonised. If a looped filament is preferred, the fibre is placed over an upper carbon
rod, beneath which its ends are crossed, and
passes under and over two outer rods and down
in a straight line to a suitable attachment.
The filaments are attached to the conductors
by small hollow carbon cylinders. Referring to
Fig 487, the inner periphery of the cylinder *d*
is slightly convex, and the upper end of the
conductor *c* is cut out, and on being pulled

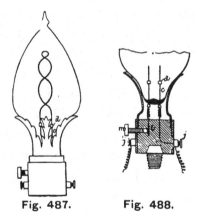

Fig. 487.   Fig. 488.

down firmly wedges the filament between itself and the cylinder. As
shown in Fig. 488, the conductors *c* pass through the glass globe, and are
formed with a loop at their outer ends, locking with wires embedded in
plaster-of-paris, one being connected to the terminal *j*, insulated from the
casing, and the other to a conducting plate *l*. By turning the screw *m*,
which is in connection with the casing, and thus with the other terminal *j*[1],
the circuit is completed. In another arrangement, the plate *l* is permanently
connected to the insulated terminal *j* and the other conductor *c* to *j*[1], and on

turning the screw *m* the circuit is completed without passing through the filament. A resistance equal to the lamp is introduced into the switch circuit. The leads may be arranged within the hollow fittings of the lamp, the holder, on being screwed into place, making connection by means of buttons.

Cherrill.—Chemical means are employed to obtain the vacuum. A gas, such as ammonia, is formed in or passed into the bulb of an incandescence lamp, and before it reaches the pump which is used to assist the process, it is caused to pass through a substance, such as phosphoric anhydride, with which it will combine and be converted into a non-gaseous form. Referring to Fig. 489, A is the bulb, B a chamber containing the phosphoric anhydrid or similar substance, and *d* a pipe leading to the pump.

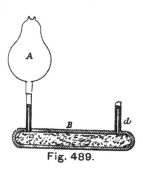

Fig. 489.

Soward.—Two metallic (preferably platinum) conductors are inserted in a globular or other shaped glass vessel at a suitable distance apart. A carbonaceous gas or vapour, such as marsh or coal gas, is introduced into the glass vessel, and is decomposed by the passage of electric sparks from any suitable high tension electric generator. A slight deposit of carbon takes place upon the end of one conductor, and gradually increases until a complete bridge is formed. This bridge forms the filament to be used as the light-emitting conductor. After exhausting and sealing the glass vessel the lamp is ready for use. If the metallic electrodes are made tubular, a suitable conducting paste is injected into them, which, issuing from their conical ends, surrounds the ends of the filament, the external ends of the conductor being then closed. Fig. 490 shows a lamp provided with four conductors *b* and three deposited filaments *c*. The filaments may be formed in a separate vessel by the passage of the electric spark between highly purified graphite, and be attached to conductors and inserted in globular or other shaped vessels. The conductors pass through the top of the glass stem, and are arranged to receive split metallic tubes carrying a projecting tongue to which the filaments are attached. The end of the conductor is flattened, an aperture inserted therein, and coiled into the form of a volute. The inner end is slightly twisted, and the end of the filament inserted in the aperture, and between the coils of the volute; the flattened extremity of the inner end is, when released, pressed firmly against the filament and a con-

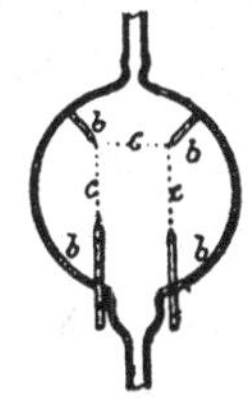

Fig. 490.

F F 2

ducting paste of cellulose and black lead with a metallic powder, is applied to the joint. In the construction of lamps used with filaments prepared from cellulose, a glass pillar is closed at one end, at or near which, and also below it, is provided a ring of glass surrounding the pillar, and held by glass arms. The metallic conductors are bent round the rings, and preferably lie in radial recesses therein, the filaments being attached to these conductors and arranged in series. The one lead passes up to the top of the hollow glass pillar, and the other passes through the funnel-shaped neck of the pillar to the lower ring.

Guest.—This arrangement relates to a method of removing atmospheric air from the bulb of the lamp, and of sealing the glass without using a vacuum pump. Referring to Fig. 491, which is a vertical section, the glass B, containing the carbon filament, is filled with a hydro-carbon or non-inflammable gas by means of a fine tube running up into the globe, and a stopper being applied to the nozzle *f*, the bulb B is suspended within the vessel A, the conductors of the lamp being received in spring holders I formed as spring forceps and attached to slides sliding in the standards G. A window of mica or glass made air-

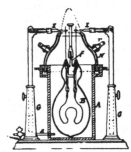

Fig. 491.

tight by a frame and rubber packing, is provided in the side of the vessel A, or it may be made entirely of glass. The vessel A is provided at the top with an annular trough *m* containing mercury serving to seal the lower edge of a removable cover N. A non-combustible or inert gas is then introduced within the vessel A by means of the pipe *s*, and combustible gas by the pipe *r*, to an annular orifice around the cover N, which cover is preferably hinged to the vessel A, and is contracted at its upper end. The gas is ignited at the annular orifice, forming a flame, as shown by the dotted lines, which keeps out the atmospheric air from the vessel A. The plug is then removed from the nipple *f*, and a current passes through the filament by way of the standards G. When the whole of the atmospheric air has been displaced, the nipple *f* is fused by means of a pair of spring tongues carrying platinum electrodes and insulated conductors connected to the electrodes. The spring tongues are inserted through the top of the cover N (the handles being suitably bent away so that they will not be in the flame), and caused to grip the nipple *f*. An electric current is then passed through them, and the electrodes are slightly separated, forming an arc which fuses the nipple.

Harrison.—Metallic pieces of L shape are sealed in the neck of the bulb and support the carbon filament. The legs are connected to the limbs which project through the glass. The filament is carbonised whilst united to an iron staple in a closed vessel containing a preventive of oxidation. In Fig. 492 the neck of the bulb is elliptical in cross section, and a slit extends some distance down each side of its narrow sides. The carbon filament attached to the two L-shaped pieces *c*, is inserted into the bulb, the hooked ends of the pieces *c* resting on the bottom of the slits ; the angles of the hooks are coated with glass before insertion. The glass is then pinched or closed on the L pieces, and a mandrel held in the centre of the neck. The tubular part of the neck is then drawn out to form a small tube for exhaustion. To prepare the carbon filament, a narrow strip of wood pulp paper is bent to a U shape, and its two ends secured to a U-shaped piece of iron wire (which may be made with a coil in each of its limbs) by means of a paper tube and a carbonaceous cement, preferably containing a metallic oxide. The filament is then carbonised in the usual way in a closed vessel supplied with a hydrocarbon gas, or a liquid such as vaseline or mercury. The bow of the iron staple is then cut off, and the ends of the L-shaped pieces inserted into the coil, or the ends of the L-shaped pieces may be slit, and the straight iron wire inserted.

Fig. 492.

Swan.—Several patents have been obtained by Mr. Swan during the past two years in connection with incandescence lighting. The first of these, dated September 15, 1882, is illustrated in Figs. 493 and 494, and refers to the means of attaching the lamps to their sockets. Two peculiarly formed hooks engage with the terminal loops of the lamp, and make firm contact with them independently of springs. Referring to the illustrations, which are elevations at right angles to one another, the wires will be seen, upon which are formed the hooks *b*², under which the terminal loops *c* of the lamp are passed and inserted in the non-conducting body part. The wires are continued in one or more coils, which lie close to the curve of the hook, so that the loops *c* engage between, and are nipped closely by, two or more contiguous windings of the coils ; the other ends are bent at right angles and soldered to the ends of binding screws, to which the leads are attached. When the lamp is not pendent from the holder, spring wires or supports *f* are fixed to the body part and hold the bulb firmly in position. In a modification the body part is formed in two pieces, one carrying the wires with the hook

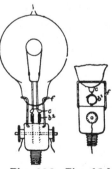

Fig. 493   Fig. 494.

and coils and the supports if required, the wires being connected to conducting plates having projections thereon, and the other being provided with recessed conducting springs connected to the leads. The projections of the one part engage with the recesses of the other part, and complete the connection between the two.

In a second patent dated May 21, 1883, the part A of the holder, Figs. 495 and 496, to which the bulb is directly attached, is provided with metallic clip pieces or jaws *a* which tend to close by the action of springs, but can be pressed apart by a pusher *d*. A peg on one part of each jaw takes into a recess in the other part. The holder is formed to give a firm seating for the base of the bulb, which is attached by pressing the pusher so as to cause the jaws to open, when the loops forming the terminals of the lamp are passed over the pegs, and on releasing the pusher, the springs *e* cause the jaws to close together and grasp the terminal loops. The part A is attached to the part B, which is affixed to the bracket, by means of the springs *b*. These are undercut or formed with projecting parts at their ends, and engage with shoulders or ledges on opposite sides of the upper portion of A, which portion is passed into a recess on the part B, and is then given a partial rotation. The springs *b* are also formed with recesses, and by giving the part A a partial rotation, these recesses make contact with metallic pins projecting from the jaws *a*. The pins *a* may be brought out of contact with the springs, while the shoulders or ledges on A are still engaged with the projections on B.

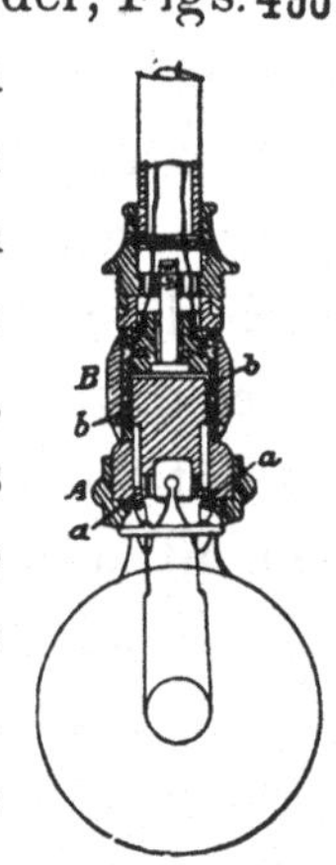

Fig. 495.

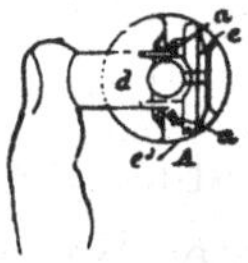

Fig. 496.

Fig. 498.   **Fig. 499.**

A third patent, of August 13, 1883, refers to the manufacture of incandescence lamps. The bulbs, after being blown, are softened at the parts where the holes are to be formed for the passage of the terminal wires, and nipple pieces are blown or drawn at these softened parts. The tops of these nipple pieces are cut off, leaving the requisite passages and quantity of glass for fastening the terminal wires. The shaping of the neck of the bulb, and the contraction of the exhausting tube connected therewith, may be effected either before or after the holes for the terminal wires are

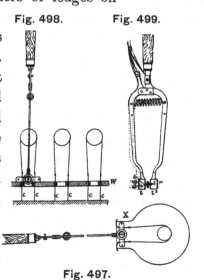

Fig. 497.

made. The neck is then cut, dividing the bulb into two portions, one part X shown in Fig. 497, and another portion consisting of the other part of the neck and the exhausting tubes, the holes for the terminal wires being in one or the other part. The wires are then secured by fusion of the glass in their places, and the two parts of the bulb are sealed or fused together. In a modification perforated cast glass closing pieces, which fit the neck of the bulb, are employed, the holes being produced therein during the casting. Fig. 498 shows a series of carbons attached to their terminal wires *c*, and held in a support W which prevents them being injured. They are removed from this support by a clip or pincers constructed as shown in Fig. 499, and the carbons are introduced as shown in Fig. 497. The gripping parts $t\,t^2$ are arranged upon central pins *s* carried in heads *r*, against which the parts *t t* are kept by springs *r*.

According to a fourth patent dated October 30, 1883, the globe or bulb is formed with grooves, ribs, projections, or depressions, engaging with springs on the holder. As will be seen from Fig. 500, the springs *b* on the holder engage with the recess *a* in the neck part of the bulb. The conductors *c* pass between springs or spring clips *d*. The holder is shown constructed as described in Specification 2528 of 1883 ; the springs *d* will, however, serve to retain the two parts of the holder in position, and the contact parts at *e* may, therefore, be rigid. The springs *b* may be ornamental or concealed by an ornamental part of the holder. The sides of the springs *b* may be arranged to bear against projections on the bulb to prevent it being rotated, or the depressions may be made of a shape to merely receive the springs, instead of as circular grooves. Or springs are attached to the loops, and their other ends secured to a central

Fig. 500.

piece with ear-pieces passing up grooves in the sides of the central hole in the covering part of the holder, into which they are drawn by the springs. If the conductors *c* pass out at the sides of the neck, loops or knobs on their ends engage with recesses in two of the springs *b*, which are arranged to form part of the circuit.

END OF VOLUME II.

INDEX.

——:-:——

	PAGE
Abdank's Arc Lamp	404
Aberration, Chromatic	180
Absolute Capacity of a Condenser	78
,, Measurement in Electricity · and Magnetism, Gray's	97
Absorption, Electric	82
,, Photometer, Stevenson's... ...	144
,, and Waste of Energy in a Circuit	89
Accommodation of the Eye, Power of ...	182
Achromatism	182
Action of Lenses in causing Refraction ...	179
Activity, Electric	85
Adams's Investigation of Selenium ...	137
,, Method of Showing the Characteristic of a Dynamo	93
Adjustment for Distance, Optical	183
Agricultural Society's Dynamometer, The Royal	227
Air Condenser	82
,, Muirhead's	82
Alabaster and Gatehouse's Tests of Ball's Unipolar Dynamo	379
Allan's Automatic Regulator	314
Amos's Dynamometer	196
Anderson and Easton's Dynamometer ...	195
Andouin and Bérard's Investigations of the Carcel Lamp	157
André's Arc Lamp	392
Angular Difference of Deviation in Light Rays	181
,, Displacement of Reflecting Galvanometer	38
Arago's Photometer	121
Arc Lamps, Electric	391
,, ,, Abdank's	404
,, ,, André	391
,, ,, Breguet	416
,, ,, Clark and Bowman ...	412
,, ,, Crompton	399
,, ,, Crompton and Bürgin	402
,, ,, Egger-Kremenetsky...	421
,, ,, Hawkes	395
,, ,, Hochhausen	294
,, ,, Hochhausen (Double Carbon)	298
,, ,, Lever	414
,, ,, Piette and Krizik ...	425
Arc Lamps, Electric, Solignac	408
,, ,, Varley	420
,, ,, Werdermann ...	419
,, ,, Weston	271
,, ,, ,, Brake-Wheel Feed Mechanism ...	275
,, ,, ,, Single Carbon	272
,, ,, Zipernowsky ...	423
Arc Lighting, Hochhausen's Dynamo for ...	282
,, ,, System, Weston ...	261
,, Resistance of the	90
,, Voltaic. Arrangement of Carbons for the	171
Argand's Carcel Lamp	157
Arithmetic, Day's Electric Lighting	97
Armature Coils, Connecting with Commutator; Weston's Dynamo	266
Armature, Crompton's Step-wound	389
,, Ferranti	304
,. Ferranti (Continuous Current Dynamo)	308
,, Ferranti-Thomson	301
,, Gordon 500 Light Dynamo... ...	357
,, Hochhausen	284
,, ,, Core of the	284
,, ,, Winding the	285
,, Hopkinson-Muirhead	384
,, Matthews	383
,, Weston	263
,, ,, Various Modes of Coiling	264
Arrangement of Carbons for the Voltaic Arc .	171
Automatic Cut-out for Incandescence Lamps, Maxim	279
,, ,, Weston	276
,, Regulation of the Chertemps-Dandeu Dynamo	374
,, Regulator, Allan	314
,, ,, Hochhausen	287
,, ,, ,, Motor for	288
Ayrton and Perry's Dynamometer	236
,, ,, Photometer	149
Back Electromotive Force	93
Balance, Raffard's, Dynamometric 200, 202, 205	
Ball's Unipolar Dynamo	376

PAGE

Ball's Unipolar Dynamo, Alabaster and Gate-
house's Tests of ... 379
,, ,, ,, Constructive Details of 378
,, ,, ,, Sabine's Tests of ... 377
Battery Currents 11
,, Electromotive Force of a 10, 73
,, Internal Resistance, Lodge's Method
of Finding ... 65
,, ,, ,, Mance's Method
of Finding ... 65
,, ,, ,, Munro and Muir-
head's Method of Measuring 66
,, Resistance Test, Kempe's 66
,, Superior Limit of a 76
,, Waste of Local Heat in a 91
Becquerel on Standard Candles 156
Beeman's Incandescence Lamps ... 429
Beetz's Experiment with Electrolytic Cells ... 91
Belt Dynamometer, Farcot 219
,, ,, Froude 203, 213
,, ,, Hefner Alteneck ... 222
,, ,, Parsons 221
,, ,, Tatham 214
Belt Dynamometers 212
Bérard and Andouin's Investigations of the
Carcel Lamp 157
Bernstein's Incandescence Lamp 429
Berzelius, Selenium Discovered by 138
Bidwell's Experiments with Selenium ... 137
Bifilar Suspension 40
Blind Spot of the Eye 185
Boltzmann's Deductions on Condensers ... 15
Bougies de l'Etoile 152
Bouguer's Method of Measuring Luminous
Intensity 105
,, Photometer 106
Bourdon's Dynamometer 211
Bourry's Dynamometer 229
Bowman and Clark's Arc Lamp 412
Brackets and Fittings for the Edison System... 335
,, for Incandescence Lamps, Weston's... 281
Brake, Deprez, Dynamo-Electric ... 198, 240
,, Prony 193
,, Prony, HP. Absorbed by Heinrichs'
Dynamo, as shown by 256
,, Prony and Morin Dynamometer, Fric-
tion of Heinrichs' Dynamo Calculated
from 256
,, Prony ; a Testing Installation ... 245
,, and Wheel Feed Mechanism for
Weston's Arc Lamp 275
Bramwell's Dynamometer 198
Brauer's Dynamometer 196
Breguet's Arc Lamp 416
Bridge, The Metre or Slide Wire 53
,, The Wheatstone 50
,, ,, Resistances of ... 50

PAGE

Brightness, Effective and Intrinsic, Ratio of 186
British Association Standard Resistance,
The 5
Brown's Dynamometer 205
Brush's Electric Meter 327
,, Fusible Cut-out... 325
,, Incandescence Lamp 317, 429
,, ,, ,, Fittings 319
,, ,, ,, and Regulator ... 321
,, ,, System... 312
,, Regulator for Theatres 325
Budapest, Electric Lighting of the National
Theatre 365
Bunsen's Photometer 117
,, and Roscoe's Chemical Photometer... 132
,, ,, Photometric Experiments 130
Bürgin and Crompton's Arc Lamp 402

Cable Testing 70
Cables, Submarine, Capacity and Charge of 18
,, ,, Finding the Capacity of 80
,, ,, Measuring the Insulation
Resistance of 69
Calibrating Resistance 56
,, ,, Foster's Method of ... 56
Calorimetric Method, Joule's 87
Candles, Becquerel on Standard 156
,, Composition of Standard 155
,, as Luminous Standards 152
,, Péclet on Standard 156
,, Standard Spermaceti... 155
,, "Star" 152
,, Stearine, Standard 155
Capacity 15
,, and Charge of Submarine Cables ... 18
,, of a Condenser, The Absolute ... 78
,, of a Condenser, Determining the ... 76
,, of a Condenser, Finding the, from
Galvanometer Throw 79
,, Finding, from Rate of Fall of
Potential 81
,, of Plates 15
,, of a Sphere 16
,, of Submarine Cables, Finding the .. 80
,, Two Concentric Spheres 16
Carbon, Resistance of 8
,, Hochhausen Lamp with Single ... 294
,, Weston Lamp with Single 272
Carbons, Positive and Negative 170
,, for the Voltaic Arc, Arrangement of... 171
,, Hochhausen Lamp with Double ... 298
,, Weston Lamp with Double ... 271
Carcel Lamp, The 156
,, Andouin and Bérard, Investiga-
tions of... 157
,, Argand 157
,, Consumption of Oil in 158

PAGE

Carcel Lamp, Dumas and Regnault's Investigation of the 159
  ,,   Lange 157
  ,,   Oil for the 160
  ,,   Perrot 157
  ,,   Wick for the 160
Carcel, Standard, Seven Gas Jet 164
Carpentier's Dynamometer 199
Cascade or Series Condensers 19
Cell, Clark's Standard 9
  ,, The Daniell's Standard 9
  ,,   ,,   ,, Modified 9
  ,, De la Rue's Standard 9
Cells, Electrolytic, Beetz's Experiment with 91
  ,, in Series, Internal Resistance of ... 12
Central Station Indicator, Weston 278
C.G.S., System of Units, The 6
Characteristic of a Dynamo (Adams's Method) 93
  ,,   ,,   ,, (Hopkinson's Method) 92
Charge and Capacity of Submarine Cables ... 18
Chemical, Light, and Heat Rays 187
  ,, Photometer, Eden 137
  ,,   ,, Roscoe and Bunsen ... 132
  ,, Photometers 129
Cherrill's Incandescence Lamp... 435
Chertemps-Dandeu Dynamo, The 370
  ,,   ,,   ,, Automatic Regulation of the ... 374
  ,,   ,,   ,, Coils and Magnets 372
  ,,   ,,   ,, Sabine's Tests of the 373
Chromatic Aberration 180
Chrystal's Electricity and Magnetism ... 96
Circuit, Energy in a, Absorbed and Wasted... 89
  ,,   ,, of, in a, its Development and Distribution 89
  ,, of the Gordon 5000 Light Dynamo ... 359
Clark and Bowman's Arc Lamp 412
Clark's Compensation Method of Measuring Electromotive Force 75
  ,, Potentiometer 76
  ,, Standard Cell 9
  ,, Standard Cells, Electromotive Force of 95
Clerk-Maxwell's Electricity and Magnetism... 96
Coiling the Weston Armature 264
Coils, Armature, connecting with Commutator; Weston's Dynamo 266
  ,, of the 5000 Light Gordon Dynamo ... 358
  ,, Resistance 57
  ,,   ,, Fleming's Method of Winding 57
  ,, Shunt 13
Collector Rings in the Ferranti Dynamo ... 305
Combined Electric and Gas Bracket, Edison... 335
Commutator, Connecting Armature Coils with ; Weston's Dynamo 266
  ,, Ferranti 309

PAGE

Commutator, Hopkinson-Muirhead's Continuous Current Dynamo ... 386
  ,, Weston's Dynamo 263
Comparing the Electromotive Force of two Batteries, Poggendorff's Compensation Method 73
  ,, Small Resistances, Method of ... 63
Comparison of Luminous Standards, Girout's 153
  ,, Resistances 6
Compensation Method, Clark's, of Measuring Electromotive Force 75
  ,,   ,, Poggendorff's, of comparing the Electromotive Force of two Batteries 73
Component, Horizontal, of Galvanometer ... 33
Composition of Standard Candles 155
Compound Shunt Dynamos 92
Concentric Spheres, Capacity of 16
Condensers 15, 17
  ,, The Absolute Capacity of... ... 78
  ,, Air 82
  ,, Air, Muirhead's 82
  ,, Boltzmann's Deductions on ... 15
  ,, Capacity of 15
  ,, Cascade or Series 19
  ,, Determining the Capacity of ... 76
  ,, Finding the Capacity of, from Galvanometer Throw 79
  ,, Measuring the Insulation Resistance of 69
  ,, Method of Measuring Electromotive Force 75
Conductivity 5
  ,, Specific 6
Conductors, Cylindrical, Specific Resistance of 64
  ,, Division of Currents in 12
  ,, Heat Developed by 86
  ,, Joint Resistance of 7
  ,, Measuring the Insulation Resistance of 69
  ,, Metallic, Resistance of 4
Cones and Rods of the Eye 184
Congress, Electrical, on Photometric Standards 167
  ,, of Electricians on Luminous Standards 151
Connecting Armature Coils with Commutator; Weston Dynamo 266
Constant of the Galvanometer... ...30, 48
Consumption of Oil in the Carcel Lamp ... 158
Continuous Current Dynamo, Ferranti ... 306
  ,,   ,,   ,, Ferranti, Armature of the ... 307
  ,,   ,,   ,, Hopkinson-Muirhead ... 385
Core of the Hochhausen Armature 284
Cornu's Photometer 111

PAGE

Cornu and Violle's Platinum Photometric
Standard　　...　　...　　...　　...　　... 167
Coulon's Radiometer Photometer　　...　　... 142
Counter, Electrical Revolution, Kempe and
Ferguson　　...　　...　　...　　...　　... 248
Counter Electromotive Force　...　　...　　... 91
Critical Current　...　　...　　...　　...　　... 92
Crompton's Arc Lamps ...　　...　　...　　... 399
Crompton and Bürgin's Arc Lamp　　...　　... 402
Crompton's Step-wound Armature　　...　　... 389
Crookes's Incandescence Lamp　　...　　... 431
Crova on Gas-Jet Standards　　...　　...　　... 163
　　,,　　Investigations in Spectro-Photometry
by　　...　　...　　...　　...　　... 147
　　,,　on Platinum Photometric Standards ... 166
Crova's Spectro-Photometer　　...　　...　　... 148
Cumming's Introduction to the Theory of
Electricity　　...　　...　　...　　...　　... 96
Current　...　　...　　...　　...　　...　　... 10
　　,,　　Critical　...　　...　　...　　...　　... 92
　　,,　　Distribution, Kirchoff's Laws of　　... 51
　　,,　　Galvanometer, The　　...　　...　　... 46
Currents, Battery　　...　　...　　...　　... 11
　　,,　　Division of, in Conductors　...　　... 12
　　,,　　Induction, Strength of　　...　　... 83
Cut-out, Fusible, Brush　　...　　...　　... 325
　　,,　　,,　　Edison　　...　　...　　... 338
　　,,　　for Incandescence Lamps, Maxim　... 279
　　,,　　Weston's Automatic ...　　...　　... 276
　　,,　　,,　　Fusible Plug　　...　　... 277
Cylindrical Conductor, Specific Resistance of. 64

Damping Magnetic Needle Oscillations, Ratio of 61
Daniell's Standard Cell ...　　...　　...　　... 9
　　,,　　,,　　,,　　Modified　　...　　... 9
Darwin's Dynamometer　　...　　...　　... 237
Day's Electric Lighting Arithmetic　...　　... 97
　　,,　　Examples in Electrical and Magnetic
Measurement ...　　...　　...　　... 97
Dead-beat Galvanometers　　...　　...　　... 62
De la Rue's Standard Cell　　...　　...　　... 9
Decrement, Logarithmic, of a Galvanometer . 61
Deflection of Magnets ...　　...　　...　　... 36
Deflections, Measuring Discharge　　...　　... 71
　　,,　　from Shunted Galvanometer, Dis-
charge　　...　　...　　...　　... 71
Defries's Incandescence Lamp ...　　...　　... 432
Degrand's Photometer ...　　...　　...　　... 108
Deliège and Felu's Dynamometer　　...　　... 201
Deprez's Dynamometer ...　　...　　... 198, 240
Deschanel's Natural Philosophy　　...　　... 96
Determining the Capacity of a Condenser　... 76
Development and Distribution of Energy in a
Circuit ...　　...　　...　　...　　...　　... 89
Deviation in Light Rays, Angular Difference of 181
Device for Starting Second Carbons (Hoch-
hausen Lamp)　　...　　..　　... 299
　　,,　　Weston's Fusible Safety　　..　　... 280

PAGE

Diagram of Dynamo Efficiency (Thompson's) . 94
Difference of Deviation in Light Rays, Angular 181
Discharge Deflections, Measuring　　...　　... 71
　　,,　　,,　　from Shunted Galva-
nometer　　...　　... 71
Displacement, Angular, of Reflecting Galva-
nometer...　　...　　...　　...　　...　　... 38
Distance, Optical Adjustment for　　...　　... 183
Distribution of Currents, Kirchoff's Laws of... 51
　　,,　　and Development of Energy in a
Circuit ...　　...　　...　　...　　... 89
　　,,　　Electrical, Gaulard and Gibbs's
System of　　...　　...　　...　　... 342
Division of Currents in Conductors　　...　　... 12
Double Carbon Arc Lamp, Hochhausen　　... 298
　　,,　　,,　　,,　　Weston　　... 271
Draper's Photometric Experiments　　...　　... 130
　　,,　　Platinum Photometric Standard　... 166
Duboscq's Photometer ...　　...　　...　　... 125
Dumas and Regnault's Investigations of the
Carcel Lamp　　... 159
　　,,　　,,　　Photometer ...　　... 160
Dynamo for Arc Lighting, Hochhausen　　... 283
　　,,　　Ball's Unipolar　　...　　...　　... 376
　　,,　　,,　　,,　　Alabaster and Gate-
house's Test　　... 379
　　,,　　,,　　,,　　Constructive Details
of ...　　...　　... 378
　　,,　　,,　　,,　　Sabine's Test of　... 377
　　,,　　Characteristics of a (Adams's Method) 93
　　,,　　,,　　,,　　(Hopkinson's Me-
thod)　　...　　...　　...　　... 92
　　,,　　Chertemps-Dandeu ...　　...　　... 370
　　,,　　,,　　,,　　Automatic Re-
gulation of... 374
　　,,　　,,　　,,　　Coils and Mag-
nets...　　... 372
　　,,　　,,　　,,　　Sabine's Test of
the...　　...　　...　　...　　... 373
　　,,　　Edison, for Shipboard　　...　　... 330
　　,,　　,,　　1200 Light　　...　　...　　... 340
　　,,　　Efficiency, Thompson's Diagram of　94
　　,,　　Ferranti, Collector Rings for　　... 305
　　,,　　,,　　Continuous Current　　... 306
　　,,　　,,　　Continuous Current, Ar-
mature of the　　... 307
　　,,　　,,　　5000 Lamp　　...　　... 303
　　,,　　Ferranti-Thomson ..　　...　　... 300
　　,,　　,,　　Efficiency of the 302
　　,,　　Friction of Heinrichs', Calculated
from Morin Dynamometer and
Prony Brake　　...　　...　　... 256
　　,,　　Ganz ...　　...　　...　　...　　... 361
　　,,　　,,　　Electrical Details of the　　... 363
　　,,　　,,　　Exciting Machine of the　... 361
　　,,　　Gordon 5000 Light　　...　　...　　... 354
　　,,　　,,　　,,　　Armature of the 357
　　,,　　,,　　,,　　Circuits of the 359

PAGE

Dynamo, Gordon 5000 Light, Coils of the ... 358
,, ,, ,, Frame and Magnets of the ... 357
,, ,, ,, Regulator of the 360
,, Gramme Multipolar 350
,, Heinrichs, Efficiency of 257
,, ,, HP. absorbed, as shown by Prony Brake .. 256
,, ,, Preece's Tests of ... ·256
,, ,, Table of Tests of : Values Calculated 255
,, ,, Table of Tests of : Values Measured 254
,, Hochhausen 282
,, Hopkinson-Edison 332
,, Hopkinson-Muirhead 385
,, ,, ,, Continuous Current 385
,, for Incandescence Lighting, Schuckert Modified 313
,, Matthews Multipolar 381
,, ,, ,, Armatures of the 383
,, Schwerdt 388
,, Weston 261
,, ,, Armature of the 263
,, ,, Commutator of the ... 263
,, ,, Connecting Armature Coils with Commutator ... 266
,, ,, Field Magnets of the ... 267
,, ,, for Incandescence Lamps 267
,, Zipernowsky 368
Dynamometer, Amos 196
,, Ayrton and Perry 236
,, Belt 212
,, Bourdon 211
,, Bourry 229
,, Bramwell 198
,, Brauer 196
,, Brown 205
,, Carpentier 199
,, Darwin 237
,, Deprez 198, 240
,, Easton and Anderson ... 195
,, Electro, The 39
,, Emerson 202, 239
,, Emery 196
,, Farcot 219
,, Felu and Deliège 201
,, Froude 203, 213
,, German 206
,, Hefner Alteneck 222
,, Hirn 211
,, Imray 197
,, King 207, 227
,, Kretz 194

PAGE

Dynamometer, Latchinoff 236
,, Matter 237
,, Megy 229
,, Morin 229
,, Morin and Prony Brake, Friction of Heinrichs' Dynamo calculated from 256
,, Neer ... ... · 236
,, Parsons 221
,, Piobert and Fardy 193
,, Prony 193
,, Raffard ... 200, 202, 205
,, Royal Agricultural Society ... 227
,, Ruddick 233
,, Siemens's Electro 41
,, Silver and Gay 207
,, Smith 208, 239
,, Tatham 214
,, Taurine 234
,, Thiabaud 201
,, Valet 233
,, Weber's Electro 39
,, Weyher and Richmond ... 201
DYNAMOMETERS 193
Dynamos, Edison, Types K, L, Z 331
,, Efficiency of 90
,, Ordinary Shunt 92
DYNAMOS, RECENT, AND LAMPS 259
Dynamos, Self-Regulating 92
,, Shunt Compound 92
,, ,, and Series 92
Dyne, The (Unit Force) 84

Easton's and Anderson's Dynamometer ... 195
Economy and Rapidity of Working Dynamos 94
Eden's Chemical Photometer 137
Edison Dynamo for Shipboard 330
,, . Dynamos, Types K, L, Z 331
,, Electric Light Installation at Milan ... 340
,, ,, Lighting on the s.s. Tarawera and Waihora 329
,, Fittings for Isolated Electric Lighting 336
,, Fusible Cut-out 338
Edison-Hopkinson Dynamo, The 332
,, Incandescence Lamps 433
,, Regulator, The 333
,, System, The 329
,, System, Brackets and Fittings for the 335
,, 1200 Light Dynamo 340
Effective and Intrinsic Brightness, Ratio of ... 186
Efficiency of Ball's Unipolar Dynamo ... 380
,, Dynamos 90
,, ,, Thompson's Diagram of 94
,, the Ferranti-Thomson Dynamo 302
,, the Heinrichs Dynamo ... ... ·257
Egger-Kremenetsky's Arc Lamp 421
Electric Absorption 82

PAGE

Electric Activity.. 85
,,   Arc, The   90
ELECTRIC ARC LAMPS 391
,,   ,,   Abdank   404
,,   ,,   André 391
,,   ,,   Breguet 416
,,   ,,   Clark and Bowman ... 412
,,   ,,   . Crompton 399
,,   ,,   Crompton and Bürgin ... 402
,,   ,,   Egger-Kremenetsky ... 421
,,   ,,   Hawkes 395
,,   ,,   Hochhausen 294
,,   ,,   Hochhausen (Double
                  Carbon) 298
,,   ,,   Lever 414
,,   ,,   Piette and Krizik ... 425
,,   ,,   Solignac 408
,,   ,,   Varley 420
,,   ,,   Werdermann 419
,,   ,,   Weston 271
,,   ,,   Zipernowsky 423
,,   Energy, Storage of 91
,,   and Gas Brackets Combined, Edison 335
,,   Light Installation at Milan, Edison 340
,,   Lighting Arithmetic, Day's ... 97
,,   ,,   The Brush Incandescence
                  System 312
,,   ,,   Edison, on the s.s. Tarewera
                  and Waihora 329
,,   ,,   Ferranti System, The ... 300
,,   ,,   Hochhausen System, The ... 282
,,   ,,   Isolated, Edison's Fittings
                  for 336
,,   ,,   at the National Theatre,
                  Budapest 365
,,   ,,   Weston System, The ... 261
,,   ,,   Zipernowsky System, The... 363
,,   Meter, Brush 327
,,   Motors and High Potentials 95
,,   Regulator, Edison 33
,,   Work 84
Electrical Congress on Photometric Standards 167
,,   Distribution, Gaulard and Gibbs's,
                  System of 342
,,   and Magnetic Measurement, Day's 97
ELECTRICAL MEASUREMENTS 3
,,   Phenomena and Theories, Notes on
                  Tyndall's 96
,,   Resistance 4
,,   Revolution Counter, Kempe and
                  Ferguson 248
,,   Testing, Kempe's Handbook of ... 97
Electricians, Congress of, on Luminous
Standards 151
ELECTRICITY AND ELECTRICAL MEASURE-
MENTS, WORKS ON 96
Chrystal's Electricity and Magnetism.

PAGE

ELECTRICITY AND ELECTRICAL MEASURE-
MENTS, WORKS ON 96
Clerk-Maxwell's Elementary Treatise on
    Electricity.
Clerk-Maxwell's Treatise on Electricity
    and Magnetism.
Cumming's Introduction to the Theory
    of Electricity.
Day's Electric Lighting Arithmetic.
,, Examples in Electrical and Mag-
    netic Measurement.
Deschanel's Natural Philosophy.
Everett's Units and Physical Constants.
Ferguson's Electricity.
Gordon's Physical Treatise on Electricity
    and Magnetism.
Gray's Absolute Measurements in Elec-
    tricity and Magnetism.
Jenkin's (Fleeming) Electricity aud Mag-
    netism.
Kempe's Handbook of Electrical Testing.
Kohlrausch's Introduction to Physical
    Measurements.
Mascart and Joubert's Treatise on Elec-
    tricity and Magnetism.
Thompson's Elementary Lessons in Elec-
    tricity and Magnetism.
Tyndall's Notes on Electrical Phenomena
    and Theories.
Electricity, Maxwell's Elementary Treatise
                  on 97
,,   Theory of, Cumming's Treatise
                  on the 96
Electrification 69
Electro-Chemical Equivalent of Silver (Ray-
                  leigh's) 83
,,   Dynamometer, Siemens 41
,,   ,,   Weber 39
,,   Dynamometers 39
Electrolytic Cells, Beetz's Experiments
with 91
Electrometer, Thomson's Quadrant 21
,,   ,,   ,, Induction
,,   Plate for 24
,,   Trap-Door 24
Electromotive Force 8
,,   ,,   Back 93
,,   ,,   of a Battery ... 10, 73
,,   ,,   in a Battery, Superior
                  Limit of 76
,,   ,,   Clark's Compensation
                  Method of Measuring 75
,,   ,,   of Clark's Standard Cells 95
,,   ,,   Condenser Method of
                  Measuring 73
,,   ,,   Counter 91
,,   ,,   Measuring 75

	PAGE
Electromotive Force, Poggendorff's Method of Measuring	74
,, ,, of Two Batteries, comparing by Poggendorff's Compensation Method	73
,, ,, Weston's Method of Regulating	268
Elementary Lessons in Electricity and Magnetism, Thompson's	96
,, Treatise on Electricity, Maxwell's	97
Emerson's Dynamometer	202, 239
Emery's Dynamometer	196
Energy in a Circuit Absorbed and Wasted	89
,, ,, its Development and Distribution	89
,, Heat	86
,, Storage of Electric	91
Ergometer, Smith	239
Everett's Unit and Physical Constants	98
Examples in Electrical and Magnetic Measurement	97
Exciting Machine of the Ganz Dynamo	361
Experiments, Draper's Photometric	130
,, with Electrolytic Cells, Beetz's	91
,, Fizeau and Foucault's Photometric	129
,, Photometric, Sautter-Lemonnier's	172
,, Roscoe and Bunsen's Photometric	130
,, with Selenium, Bidwell's	137
,, ,, May's	137
,, ,, Obach's	138
,, ,, Siemens's	138
Eye, The Blind Spot of the	185
,, The Four Lenses of the	179
EYE, THE, AS A PHOTOMETRIC INSTRUMENT	174
,, The, Power of Accommodation of the	182
,, Rods and Cones of the	184
,, Siemens's Selenium	138
,, The Structure of the	176
Fall of Potential, Rate of, Finding Capacity from	81
Farcot's Dynamometer	219
Fardy and Piobert's Dynamometer	193
Feed Mechanism, Brake Wheel, for Weston's Arc Lamp	275
Felu and Deliège's Dynamometer	201
Ferguson's Electricity	96
Ferguson and Kempe's Electrical Revolution Counter	248
Fergusson's Incandescence Lamp	434
Ferranti Commutator, The	309
,, Continuous Current Dynamo	306
,, ,, ,, Armature of the	307
,, Dynamo, Collector Rings of the	305
,, ,, Field Magnets of the	305
	PAGE
Ferranti 5000 Lamp Dynamo	303
Ferranti-Thomson Dynamo, The	300
,, ,, ,, Armature of the	301
,, ,, ,, Efficiency of the	302
Field Magnets of the Hopkinson-Muirhead Dynamo	384
,, ,, ,, Weston Dynamo	267
Finding Capacity of Condenser from Galvanometer Throw	79
,, ,, from Rate of Fall of Potential	81
Fittings and Brackets for the Edison System	335
,, Brush Incandescence	319
Five Thousand Light Dynamo, Ferranti	303
,, ,, ,, Gordon	354
,, ,, ,, Gordon, Armature of the	357
,, ,, ,, Gordon, Circuits of the	359
,, ,, ,, Gordon, Coils of the	358
,, ,, ,, Gordon, Frame and Magnets of the	357
,, ,, ,, Gordon, Regulator of the	360
Fleming's Mode of Winding Resistance Coils	57
Focussing Images on the Retina	183
Force, Electromotive	8
,, ,, of a Battery	10, 73
Formulæ, General Photometric	105
Foster's Method of Calibrating Resistances	56
,, ,, Measuring nearly Equal Resistances	55
Foucault and Fizeau's Photometric Experiments	129
Foucault's Photometer	107
,, ,, Degrand's Modification	108
Frame and Magnets of the 5000 Light Gordon Dynamo	357
Friction of Heinrichs' Dynamo, calculated from Morin Dynamometer and Prony Brake	256
Froude's Dynamometer	203, 213
Fundamental Law, Ohm's	3
,, Laws of Photometry	102
Fusible Cut-out, Brush	325
,, ,, Edison's	338
,, ,, Weston's Plug	277
,, Safety Device, Weston's	280
Galvanometer, Constant of the	30, 48
,, The Current	46
,, Dead-beat	62
,, Graduation of the	43
,, Horizontal Component of	33
,, Logarithmic Decrement of a	61
,, The Potential	42

PAGE

Galvanometer, The Reflecting 38
,, Shunted, Discharge Deflections
from 71
,, Shunts 13
,, Siemens's Torsion 46
,, Tangent, The 28
,, ,, Gaugain 28
,, Throw, Finding the Capacity
of a Condenser from ... 79
,, Torsion, Finding the Constant of 49
,, ,, Galvanometer ... 46
Ganz's Dynamo 361
,, ,, Electric Details of 362
,, ,, Exciting Machine of 361
,, System of Transmission 364
Gas and Electric Bracket Combined, Edison 335
,, Jet Standard, Seven Carcel 164
,, ,, Standards, Crova on 163
,, Jets as Luminous Standards 153
Gaugain's Tangent Galvanometer 28
Gaulard and Gibbs's System of Electrical
Distribution 342
,, ,, System on the Metro-
politan Railway 348
Gay and Silver's Dynamometer 207
General Arrangement of Heinrichs' Testing
Installation 242
,, Photometric Formulæ 105
,, Principles of Electrical Measurement 3
Generator, Secondary, Gaulard and Gibbs ... 345
German Dynamometer 206
Girout's Comparison of Luminous Standards 153
Glass-Blowing Machine, Wright and Mackie 323
Gordon 5000 Light Dynamo 354
,, ,, ,, Armature of the . 357
,, ,, ,, Circuit of the ... 359
,, ,, ,, Coils of the ... 358
,, ,, ,, Frame and Magnets
of the... ... 357
,, ,, ,, Regulator of the 360
Gordon's Physical Treatise on Electricity and
Magnetism 96
Graduation of Galvanometers 43
Gramme Multipolar Dynamo 350
Gray's Absolute Measurements in Electricity
and Magnetism 97
Grindle's Socket for Incandescence Lamps ... 320
Guest's Incandescence Lamp 436

Hammerl's Photometer 116
Handbook of Electrical Testing, Kempe's ... 97
Harding's Speed Indicator 250
Harrison's Incandescence Lamp 437
Hawkes's Arc Lamp 395
Heat Developed by a Conductor 86
,, Energy 86
,, Light, and Chemical Rays 187

PAGE

Heat, Local, Waste of, in a Battery 91
,, Unit of 87
Hefner Alteneck's Dynamometer 222
Heinrichs' Dynamo, Efficiency of 257
,, ,, Friction of, Calculated
from Morin Dynamo-
meter and Prony Brake 256
,, ,, H.P. Absorbed, as shown
by Prony Brake ... 256
,, ,, Preece's Tests of ... 256
,, ,, Tables of Tests of; Values
Calculated 255
,, ,, Tables of Tests of; Values
Measured 254
Heinrichs' Mode of Testing 253
,, Testing Installation; General Ar-
rangement 242
High Potentials and Electric Motors... ... 95
,, Resistances, The Measurement of ... 68
,, ,, Phillips's Method of Making 60
Hirn's Dynamometer 211
Hochhausen Arc Lamps, The 294
,, Armature, The 284
,, ,, Core of the 284
,, ,, Winding the 285
,, Automatic Regulator, The ... 287
,, ,, ,, Motor for 288
,, Device for Starting Second Carbon 299
,, Double Carbon Arc Lamp ... 298
,, Dynamo for Arc Lighting ... 282
,, Single Carbon Arc Lamp, Mecha-
nism of the 294
,, System, The 282
Hopkinson-Edison Dynamo, The 332
Hopkinson-Muirhead Dynamo, The ... 385
Hopkinson's Method of Showing the Charac-
teristics of a Dynamo 92
Horizontal Component of Galvanometer ... 33
Horse-Power Absorbed by Heinrichs' Dynamo,
as shown by Prony Brake 256
Horse-Power Absorbed and Wasted in a
Circuit 89
Hot Incandescence Lamp, Measuring the
Resistance of 429

Illumination of Oblique Surfaces 104
Images on the Retina, Focussing 183
Imray's Dynamometer 197
INCANDESCENCE LAMPS :
,, ,, Beeman 429
,, ,, Bernstein 429
,, ,, Brush 317, 429
,, ,, ,, Fittings for ... 319
,, ,, Cherrill 435
,, ,, Crookes 431
,, ,, Defries 432
,, , Edison 433

PAGE

INCANDESCENCE LAMPS—*continued.*

,, ,, Edison Fittings for ... 335
,, ,, Fergusson 434
,, ,, Grindle's Socket for ... 320
,, ,, Guest 436
,, ,, Harrison 437
,, ,, Hot Incandescence, Measuring the Resistance of 68
,, ,, Knowles 433
,, ,, Life of 367
,, ,, Maxim's Automatic Cut-out for 279
,, ,, Müller 429
,, ,, Soward 435
,, ,, Swan 437
,, ,, Weston's 281
,, ,, Woodhouse and Rawson 427
,, Lighting, Schückert's Modified Dynamo for 313
,, System, The Brush ... 312
Indicator, Harding's Speed 250
,, Weston's Central Station ... 278
,, Young's Speed 251
Indices of Refraction for Different Bodies ... 177
Induction Currents, Strength of 83
,, Plate of Thomson's Quadrant Electrometer 24
Inertia Dynamometer, Froude's ... 203
Installation, Electric Light at Milan, Edison's 340
INSTALLATION, A TESTING 242
,, ,, Morin's Dynamometer 243
,, ,, The Prony Brake 245
INSTRUMENT, PHOTOMETRIC, THE EYE AS A ... 174
Insulation Resistance of Condensers, Measuring the 69
,, ,, of Conductors, Measuring the 69
,, ,, Measurement of ... 60
,, ,, of Submarine Cables, Measuring the ... 69
Intensity, Luminous, Bouguer's Method of Measuring 105
Internal Resistance 8
,, ,, of a Battery, Lodge's Method of Finding ... 65
,, ,, of Cells in Series... ... 12
,, ,, Mance's Method of Finding 65
,, ,, Munro and Muirhead's Method of Finding ... 66
,, ,, of a Voltaic Cell 8
Intrinsic and Effective Brightness, Ratio of ... 186
Introduction to the Theory of Electricity, Cumming's 96
Investigation of the Carcel Lamp, Andouin and Bérard's 157

PAGE

Investigation of the Carcel Lamp, Dumas and Regnault's 159
,, of Selenium, Adams's 137
,, in Spectro-Photometry, Crova's 147
Isolated Electric Lighting, Edison's Fittings for 336

Jacobi's Law 94
Jenkin's Electricity and Magnetism 96
Jet Standards 153, 163
,, ,, Crova on Gas 163
Joint Resistance of Conductors 7
Joule, The... 87
Joule's Calorimetric Method 87

Kempe and Ferguson's Electrical Revolution Counter 248
Kempe's Battery Resistance Test 66
,, Handbook of Electrical Testing ... 97
King's Dynamometer 207, 227
Kirchoff's Laws of Current Distribution ... 51
Knowles's Incandescence Lamp 433
Kohlrausch's Law 15
,, Physical Measurement... ... 97
Kretz's Dynamometer 194

Lamp, Argand Carcel 157
,, The Carcel 156
,, ,, Andouin and Bérard's Investigation of 157
,, ,, Consumption of Oil in ... 158
,, ,, Dumas and Regnault's Investigation of 159
,, ,, Lange's 157
,, ,, Oil for the 160
,, ,, Perrot's 157
,, ,, Wick for 160
LAMPS AND DYNAMOS, RECENT 259
LAMPS, ELECTRIC ARC 391
,, ,, ,, Abdank 404
,, ,, ,, André 391
,, ,, ,, Breguet 416
,, ,, ,, Clark and Bowman ... 412
,, ,, ,, Crompton 399
,, ,, ,, Crompton and Bürgin 402
,, ,, ,, Egger-Kremenetsky ... 421
,, ,, ,, Hawkes 395
,, ,, ,, Hochhausen 294
,, ,, ,, Hochhausen (Double Carbon) 298
,, ,, ,, Lever 414
,, ,, ,, Piette and Krizik .. 425
,, ,, ,, Solignac 408
,, ,, ,, Varley 420
,, ,, ,, Werdermann 419
,, ,, ,, Weston 271
,, ,, ,, Zipernowsky 423

PAGE

LAMPS, INCANDESCENCE :
,, ,, Beeman 429
,, ,, Bernstein 429
,, ,, Brush 317
,, ,, ,, Fittings for ... 319
,, ,, Cherrill 435
,, ,, Crookes 431
,, ,, Defries 432
,, ,, Edison 433
,, ,, ,, Fittings for ... 335
,, ,, Fergusson 434
,, ,, Grindle's Socket for ... 320
,, ,, Guest 436
,, ,, Harrison 437
,, ,, Hot Incandescence, Mea-
suring the Resistance of 68
,, ,, Knowles 433
,, ,, Life of 367
,, ,, Maxim's Automatic Cut-
out for 279
,, ,, Müller 429
,, ,, Soward 435
,, ,, Swan 437
,, ,, Weston 281
,, ,, Woodhouse and Rawson 427
Lange's Carcel Lamp 157
Latchinoff's Dynamometer 236
Law, Jacobi's 94
,, Kohlrausch's 15
,, Ohm's Fundamental 3
,, ,, Verification of 4
,, of Sines in Refraction 177
,, of the Squares of Distances ... 103
Laws of Current Distribution, Kirchoff's ... 51
,, Photometry, Fundamental 102
Length of Light Rays, Wave 189
Lenses, Action of, in Causing Refraction ... 179
,, of the Eye, The Four 179
Leslie's Photometer 126
Lever's Arc Lamp 414
Life of Incandescence Lamps 367
Light, Heat, and Chemical Rays 187
,, Rays, Angular Difference in Deviation of 181
,, Wave Lengths of 189
,, Waves, Oscillation per Second ... 190
Limit, Superior, of Electromotive Force in a
Battery 76
Local Heat, Waste of, in a Battery 91
Lodge's Method for Finding the Internal
Resistance of a Battery 65
Logarithmic Decrement of a Galvanometer ... 61
Luminous Intensity, Bouguer's Method of
Measuring 105
Luminous Standards 151
,, ,, Candles 152
,, ,, Congress of Electricians on 151
,, ,, Gas Jets 153

PAGE

Luminous Standards, Girout's Comparison of 153

Machine, Exciting, of the Ganz Dynamo ... 361
,, Glass-Blowing, Wright and Mackie 323
Magnetic Needle Oscillations, Ratio of Damping 61
Magnetism and Electricity, Chrystal's ... 96
,, ,, Clerk-Maxwell ... 96
,, ,, Gray's Absolute
Measurements in 97
,, ,, Jenkins ... 96
,, ,, Mascart and
Joubert ... 97
,, ,, Physical Treatise
on, Gordon ... 96
,, ,, Thompson's Ele-
mentary Lessons
in 96
Magnets and Coils of the Chertemps-Dandeu
Dynamo 372
,, Deflection of 36
,, Field, Weston Dynamo 267
,, and Frame of the 5000 Light Gordon
Dynamo 357
,, Swinging 33
,, Vibrating, Middle Passage of ... 35
,, Vibration of 34
Mance's Method for Finding the Resistance of
Batteries 65
Mascart and Joubert's Electricity and Mag-
netism 97
Masson's Photometer 114
Matter's Dynamometer 237
Matthews's Multipolar Dynamo 381
,, ,, ,, Armature of the 383
Maxim's Automatic Cut-out for Incandescence
Lamps 279
Maxwell's Elementary Treatise on Electricity 97
May's Experiments with Selenium 137
Measurement, Electrical, General Principles of 3
,, of High Resistances, The ... 68
,, of Insulation Resistance ... 69
,, of Nearly Equal Resistances,
Foster's Method 55
,, Photometric 168
,, of Resistances 60
MEASUREMENTS, ELECTRICAL 3
Measuring Discharge Deflections 71
,, Electromotive Force, Clark's Com-
pensation Method of 75
,, Electromotive Force, Condenser
Method of 73
,, Electromotive Force, Poggendorff's
Method of 74
,, Insulation Resistance of Submarine
Cables 69
,, Internal Battery Resistance,
Mance's Method of 65

PAGE

Measuring Internal Battery Resistance, Munro
    and Muirhead's Method of   ...   66
  ,,   Luminous   Intensity, Bouguer's
    Method of...   105
  ,,   the Resistance of a Hot Incan-
    descence Lamp   68
  ,,   Resistance by Telephone   54
Meghom, The   5
Megy's Dynamometer   229
Metallic Conductors, Resistance of   4
Meter, Electric Brush   327
  ,,   The Watt, Siemens   86
Method, Clark's Compensation, of Measuring
    Electromotive Force   75
  ,,   of Comparing Small Resistances   ...   63
  ,,   for Finding the Resistance of Batteries   64
  ,,   Joule's Calorimetric   87
  ,,   Foster's, of Calibrating Resistance   56
  ,,   of Measuring Electromotive Force,
    Condenser   73
  ,,   ,,   Electromotive Force,
    Poggendorff's   ...   74
  ,,   ,,   Internal Battery Re-
    sistance, Lodge's...   65
  ,,   ,,   Internal Battery Re-
    sistance, Mance's...   65
  ,,   ,,   Internal Battery Re-
    sistance,   Munro
    and Muirhead's   ...   66
  ,,   Poggendorff's Compensation, for Com-
    paring the Electromotive Force of
    two Batteries   73
  ,,   of Regulating Electromotive Force,
    Weston's   268
  ,,   of Showing the Characteristic of a
    Dynamo (Adams's)...   93
  ,,   of Showing the Characteristic of a
    Dynamo (Hopkinson's)   92
Metre or Slide Wire Bridge, The   53
Metropolitan Railway, Gaulard and Gibbs's
    System of Distribution on the   348
Microfarad, The   19
Microhm, The   5
Middle Passage of Vibrating Magnets   ...   35
Milan, Edison Electric Light Installation at...   340
Mode of Testing ; Heinrichs' Installation   ...   253
Modification of Foucault's Photometer, De-
    grand's   108
  ,,   Leslie Photometer, Ritchie's   126
  ,,   the Prony Brake   245
Modified Daniell's Standard Cell   9
  ,,   Schuckert Dynamo for Incandescence
Lighting   313
Morin's Dynamometer   229
  ,,   Dynamometer and Prony Brake,
    Friction of Heinrichs' Dynamo, Calculated
from   256

PAGE

Morin's Dynamometer ; a Testing Installation   243
Motor for Hochhausen's Automatic Regulator   288
Motors, Electric, and High Potentials   ...   95
Muirhead's Air Condensers   82
Muirhead-Hopkinson Dynamo, The   384
Muirhead and Munro's Method for Measuring
    Internal Battery Resistance...   66
Müller's Incandescence Lamp   429
Multiple Arc, Resistance in   7
Multiplying Power of Shunt   13
Multipolar Gramme Dynamo   350
  ,,   Dynamo, Matthews   381
  ,,   ,,   ,,   Armature of
  the   383

Nacks's Selenium Photometer   141
Natural Philosophy, Deschanel's   96
Neer's Dynamometer   236
Negative and Positive Carbons   170
Newton's Rings   121
Notes on Electrical Phenomena and Theories,
    Tyndall's   96

Obach's Experiments with Selenium   138
Oblique Surfaces, Illumination of   104
Ohm, The   5
Ohm's Fundamental Law   3
  ,,   Law, Verification of   4
Oil for Carcel Lamp   160
  ,,   Consumption of, in a Carcel Lamp   ...   158
Optical Adjustment for Distance   183
Ordinary Shunt Dynamos   92
Oscillations of Magnetic Needle, Ratio of
    Damping   61
  ,,   per Second of Light Waves   ...   190

Parsons's Dynamometer...   221
Péclet on Standard Candles   156
Perrot's Carcel Lamp   157
Phillips's Method of Making High Resistances   60
Photometer, Arago   121
  ,,   Ayrton and Perry   149
  ,,   Bouguer   106
  ,,   Bunsen   117
  ,,   Cornu   111
  ,,   Coulon Radiometer   142
  ,,   Crova's Spectro-   148
  ,,   Degrand   108
  ,,   Duboscq   125
  ,,   Dumas and Regnault   160
  ,,   Eden's Chemical   137
  ,,   Foucault   107
  ,,   Hammerl   116
  ,,   Leslie   126
  ,,   Masson   114
  ,,   Nacks's Selenium   141
  ,,   Roscoe and Bunsen   132

	PAGE
Photometer, Rumford	116
,, Siemens's Selenium	139
,, Stevenson's Absorption	144
,, Wheatstone	120
,, Wolff	110
,, Chemical	129
,, Selenium	137
Photometric Experiments, Draper's	130
,, ,, Fizeau and Foucault's	129
,, ,, Roscoe and Bunsen's	130
,, ,, Sautter-Lemonnier's	172
,, Formulæ, General	105
PHOTOMETRIC INSTRUMENT, THE EYE AS A ...	174
,, Measurements	168
Photometric Standards, Crova on Platinum...	166
,, ,, Draper's ,,	166
,, ,, Electrical Congress on	167
,, ,, Schwendler's Platinum	166
,, ,, Violle and Cornu's Platinum	167
,, ,, Zöllner's Platinum ...	166
,, Unit, Sautter-Lemonnier's	163
PHOTOMETRY	101
,, Fundamental Laws of	102
Physical Constants and Units, Everett's	98
,, Measurements, Kohlrausch's	97
,, Treatise on Electricity and Magnetism, Gordon's	96
Piette and Krizik's Arc Lamp...	425
Piobert and Fardy's Dynamometer	193
Plates, Capacity of	15
Platinum Photometric Standards, Crova on ...	166
,, ,, ,, Draper's ...	166
,, ,, ,, Schwendler's	166
,, ,, ,, Violle and Cornu's...	167
,, ,, ,, Zöllner's ...	166
Poggendorff's Compensation Method of Comparing Electromotive Force of two Batteries...	73
,, Method of Measuring Electromotive Force	74
,, Rheocord	59
Positive and Negative Carbons	170
Potential	84
,, Galvanometer, The	42
,, Rate of Fall of, Finding Capacity from	81
Potentials, High, and Electric Motors ...	95
Potentiometer, Clark's	76
Power of Accommodation of the Eye ...	182
,, of Shunt, Multiplying	13
,, Unit, The Watt	86
Preece's Tests of Heinrichs' Dynamo ...	256
Principles, General, of Electrical Measurement	3

	PAGE
Prony Brake, HP. Absorbed by Heinrichs' Dynamo, as shown by ...	256
,, ,, and Morin Dynamometer, Friction of Heinrichs' Dynamo, Calculated from	256
,, ,, The ; a Testing Installation ...	245
,, Dynamometer	193
Quadrant Electrometer, Thomson's	21
,, ,, ,, Induction Plate for	24
Quantity	82
Radiometer Photometer, Coulon's	142
Raffard's Dynamometer... ... ...200, 202, 205
Rapidity and Economy of Working Dynamos	94
Rate of Fall of Potential, Finding Capacity from	81
Ratio of Damping Magnetic Needle Oscillations	61
,, Effective and Intrinsic Brightness ...	186
,, Torsion 34, 36
Rayleigh's Electro-Chemical Equivalent of Silver	83
Rays, Angular Difference in Deviation of Light	181
,, Light, Heat, and Chemical	187
RECENT DYNAMOS AND LAMPS...	259
Reflecting Galvanometer, The...	38
Refraction...	177
,, Action of Lenses in Causing ...	179
,, Indices of, for Different Bodies ...	177
,, The Law of Sines	177
Regnault and Dumas's Investigation of the Carcel Lamp ...	159
,, ,, Photometer... ...	160
Regulating Electromotive Force, Weston's Method	268
Regulation, Automatic, of Chertemps and Dandeu's Dynamo	374
Regulator, Automatic, Allan	314
,, ,, Hochhausen	287
,, ,, ,, Motor for	288
,, and Brush Incandescence Lamp ...	321
,, The Edison Electric	333
,, of the Gordon 5000 Light Dynamo	360
,, for Theatres, Brush	325
Replenisher, Thomson	26
Resistance of the Arc	90
,, Box	58
,, Boxes, Warden and Muirhead ...	60
,, Calibrating	56
,, of Carbon	8
,, Coils	57
,, ,, Fleming's Method of Winding	57
,, Electrical	4
,, Foster's Method of Calibrating ...	56
,, of a Hot Incandescence Lamp, Measuring the	68

PAGE

Resistance, Insulation of Conductors, Measuring the 69
,, Internal 8
,, ,, of a Battery, Lodge's Method of Finding... 65
,, ,, of a Battery, Mance's Method of Finding... 65
,, ,, of a Battery, Munro and Muirhead's Method of Measuring 66
,, ,, of Cells in Series ... 12
,, Joint, of Conductors 7
,, Measurement of 60
,, of Metallic Conductors ... 4
,, in Multiple Arc 7
,, of Selenium 8
,, in Series 7
,, of Shunts 14
,, Specific 64
,, ,, of Cylindrical Conductor 64
,, ,, of Wire 64
,, ,, of Tellurium 8
,, Standard, British Association ... 5
,, Test, Kempe's 66
Resistances, Calibrating 56
,, Comparison of 6
,, Foster's Method of Measuring nearly equal 55
,, The Measurement of High ... 68
,, Measurement of, by Telephone... 54
,, Method of Comparing Small ... 63
,, Phillips's Method of Making ... 60
,, of the Wheatstone Bridge ... 50
Retina, Focussing Images on the 183
Revolution Counter, Electrical, Kempe and Ferguson 248
Rheocord, Poggendorff 59
Rheostat, Weston 268
,, Wheatstone 58
Richmond and Weyher's Dynamometer ... 201
Rings, Newton's... 121
Ritchie's Photometer, Modification by Leslie 126
Rods and Cones of the Eye 184
Roscoe and Bunsen's Chemical Photometer ... 132
,, ,, Photometric Experiments 130
Royal Agricultural Society's Dynamometer... 227
Ruddick's Dynamometer 233
Rumford's Photometer 116

Sabine's Tests of Ball's Unipolar Dynamo ... 377
,, ,, the Chertemps-Dandeu Dynamo 373
Safety Device, Fusible, Weston 280
Sautter - Lemonnier's Photometric Experiments 172
,, ,, ,, Unit 163

Schwendler's Platinum Photometric Standard 166
Schwerdt's Dynamo 388
Secondary Generator, Gaulard and Gibbs's ... 345
Selenium, Adams's Investigation of 137
,, Bidwell's Experiments with ... 137
,, Discovered by Berzelius 138
,, Eye, Siemens 138
,, May's Experiments with 137
,, Obach's Experiments with 138
,, Photometer, Nacks 141
,, Photometer, Siemens' 139
,, Photometers 137
,, Resistance of 8
,, Siemens's Experiments with ... 138
Self-Regulating Dynamos 92
Series or Cascade Condensers 19
,, Internal Resistance of Cells in 12
,, Resistance in 7
Seven Carcel Gas Jet Standard 164
Shipboard, Edison Dynamo for 330
Shuckert, Modified Dynamo for Incandescence Lighting 313
Shunt Coils 13
,, Dynamos, Compound 92
,, ,, Ordinary 92
,, Power of Multiplying 13
,, and Series Dynamos 92
Shunted Galvanometers, Discharge Deflection from 71
Shunts in Ordinary Use... 14
,, Resistance of 14
Siemens's Electro-Dynamometer 41
,, Experiments with Selenium... ... 138
,, Selenium Eye 138
,, Selenium Photometer 139
,, Torsion Galvanometer 46
,, Watt Meter 86
Silver and Gay's Dynamometer 207
,, Rayleigh's Electro-Chemical Equivalent of 83
Sines, The Law of, in Refraction 177
Single Carbon Arc Lamp, Mechanism of the Hochhausen ... 296
,, ,, ,, Weston 272
Slide Wire or Metre Bridge, The 53
Small Resistances, Method of Comparing ... 63
Smith's Dynamometer 208, 239
Socket, Grindle's, for Brush Incandescence Lamp 320
Solignac's Arc Lamp 408
Soward's Incandescence Lamp 435
Specific Conductivity 6
,, Resistance 6, 64
,, ,, of Cylindrical Conductor... 64
,, ,, of Wire 64
Spectro-Photometer, Crova 148
,, Photometers 147

	PAGE
Spectro-Photometry, Crova's Investigation in	147
Speed Indicator, Harding	250
,, ,, Young	251
Spermaceti Standard Candles	155
Sphere, Capacity of	16
Spheres, Capacity of two Concentric	16
Spot, The Blind, of the Eye	185
Squares of Distances, Law of the	103
Standard Candles, Becquerel on	156
,, ,, Composition of	155
,, ,, Péclet on	156
,, ,, Spermaceti	155
,, ,, Star	152
,, ,, Stearine	155
,, Cell, De la Rue	9
,, ,, Clark	9
,, ,, Clark's Electromotive Force of	95
,, ,, The Daniell	9
,, ,, Modified	9
,, Photometric, Draper's Platinum ...	166
,, Resistance, The British Association	5
,, Schwendler's Platinum Photometric	166
,, Seven Carcel Gas Jet	164
,, Violle and Cornu's Platinum Photo-	
metric	167
,, Zöllner's Platinum Photometric ...	166
Standards, Crova on Platinum Photometric...	166
,, Gas Jet 153,	163
,, Gas Jets, Crova on	163
,, Luminous	151
,, ,, Congress of Electricians on	151
,, ,, Girout's Comparison of ...	153
,, Photometric, Electrical Congress on	167
"Star" Candles	152
Station, Indicator, Weston's Central	278
Stearine Standard Candles	155
Step-wound Armature, Crompton's	389
Stevenson's Absorption Photometer	144
Storage of Energy, Electric	91
Strength of a Current	10
,, Induction Currents	83
,, Magnets, Verification of... ...	45
Structure of the Eye	176
Submarine Cables, Capacity and Charge of ...	18
,, ,, Finding the Capacity of ...	80
,, ,, Measuring the Insulation	
Resistance of	69
Superior Limit of Electromotive Force in a	
Battery	76
Surfaces, Oblique, The Illumination of ...	104
Suspension, Bifilar	40
Swan Incandescence Lamp	437
Swinging Magnets	33
System, Weston	261
,, Brush Incandescence	312
,, Edison	329
,, ,, Brackets and Fittings for	335

	PAGE
System of Electric Lighting, Weston ...	261
,, of Electrical Distribution, Gaulard	
and Gibbs	342
,, Ferranti	301
,, Hochhausen	282
,, of Transmission, Ganz and Co.'s ...	364
,, of Units, The C.G.S.	6
,, Zipernowsky	363
Tables of Tests of Heinrichs' Dynamo; Values	
Calculated	255
Tables of Tests of Heinrichs' Dynamo; Values	
Measured	254
Tangent Galvanometer, The	28
,, ,, Gaugain	28
Tarawera (s.s.) and Waihora, Edison Electric	
Lighting on the	329
Tatham's Dynamometer...	214
Taurines' Dynamometer	234
Telephone, Measurement of Resistance by ...	54
Tellurium, Resistance of	8
Testing Cables	70
,, Heinrichs' Mode of	253
,, Installation, A	242
,, ,, Morin's Dynamometer...	243
,, ,, The Prony Brake ...	245
Tests, Alabaster and Gatehouse's, of Ball's	
Unipolar Dynamo	379
,, of Heinrichs' Dynamo, Preece's ...	256
,, Kempe's Battery Resistance	66
,, of the Prony Brake	246
,, Sabine's, of Ball's Unipolar Dynamo ...	377
,, ,, of the Chertemps - Dandeu	
Dynamo	373
Theatre, National, Budapest, Electric Light-	
ing of the	365
Theatres, Brush Regulators for	325
Thiabaud's Dynamometer	201
Thompson's Diagram of Dynamo Efficiency ...	94
,, Elementary Lessons in Electricity	
and Magnetism	96
Thomson's Potential Galvanometer	42
,, Quadrant Electrometer	21
,, ,, ,, Induction	
Plate for	24
,, Replenisher	26
Time of Vibration	35
Torsion Dynamometer, Hirn	211
,, Galvanometer, The	46
,, ,, Finding the Constant of	49
,, Ratio of	34
Transmission, Ganz and Co.'s System of ...	364
Trap-Door Electrometer	24
Twelve-Hundred Light Dynamo, Edison ...	340
Tyndall's Notes on Electrical Phenomenon	
and Theories	96
Types, K, L, Z, of Edison Dynamo	331

PAGE

Unipolar Dynamo, Ball 376
,,   ,,   ,, Alabaster and Gate-
house's Tests ... 379
,,   ,,   ,, Constructive Details
of 378
,,   ,,   ,, Sabine's Tests of ... 377
Unit Electromotive Force 9
,, Force, The Dyne 84
,, of Heat 87
,, of Power, The Watt 86
,, Resistance 4
,, Sautter-Lemonnier's Photometric ... 163
Units, The C.G.S. System of 6
,, and Physical Constants, Everett ... 98

Valet's Dynamometer 233
Varley's Arc Lamp 420
Verification of Ohm's Law 4
Verification of Magnet Strength 45
Vibrating Magnets, Middle Passage of ... 35
Vibration of Magnets 34
,, Time of 35
Violle and Cornu's Platinum Photometric
Standards 167
Volt, The 9
Voltaic Arc, Arrangement of Carbons for the 171
,, Cell, Internal Resistance of 8

Waihora (s.s.) and Tarawera (s.s.), Edison
Electric Lighting on the 329
Warden and Muirhead's Resistance Boxes ... 60
Waste in Battery from Local Heat 91
Watt Meter, Siemens 86
,, The, Unit of Power 86
Wave Lengths of Rays 189
Waves of Light, Oscillation per Second of ... 190
Weber's Electro-Dynamometer 39
Werdermann's Arc Lamp 419
Weston's Arc Lamp, Brake Wheel Feed Me-
chanism for 275
,,   ,, Mechanism of the ... 271
,, Arc Lighting System... 269
,, Automatic Cut-out 276
,, Bracket for Incandescence Lamps... 281
,, Central Station Indicator 278
,, Duplex Arc Lamp 271
,, Dynamo 261
,,   ,, Armature of the 261
,,   ,, Commutator of the ... 263
,,   ,, Connecting Armature Coils
with Commutator ... 266
,,   ,, Field Magnets of the ... 267
,,   ,, Incandescence Lamps ... 267
,, Fusible Plug Cut-out 277
,,   ,, Safety Device ... 280
,, Method of Regulating Electromotive
Force 268
,, Rheostat 268

Weston's Single Carbon Lamp 272
,, System of Electric Lighting ... 261
Weyher and Richmond's Dynamometer ... 201
Wheatstone Bridge, The 50
,,   ,, Resistance of the ... 50
Wheatstone's Photometer 120
,, Rheostat 58
Wick for the Carcel Lamp 160
Winding the Hochhausen Armature 285
Winding Resistance Coils 57
,,   ,,   ,, Fleming's Method . 57
Wire, Specific Resistance of 64
Wolff's Photometer 110
Woodhouse and Rawson's Incandescence Lamp 427
Work 83
,, Electric 84
Working Dynamos, Rapidity and Economy
of Working 94
WORKS ON ELECTRICITY AND ELECTRICAL
MEASUREMENT 96
Chrystal's Electricity and Magnetism.
Clerk-Maxwell's Elementary Treatise on
Electricity.
Clerk-Maxwell's Treatise on Electricity
and Magnetism.
Cumming's Introduction to the Theory
of Electricity.
Day's Electric Lighting Arithmetic.
,, Examples in Electrical and Mag-
netic Measurement.
Deschanel's Natural Philosophy.
Everett's Units and Physical Constants.
Ferguson's Electricity.
Gordon's Physical Treatise on Electricity
and Magnetism.
Gray's Absolute Measurements in Elec-
tricity and Magnetism.
Jenkin's (Fleeming) Electricity and Mag-
netism.
Kempe's Handbook of Electrical Testing.
Kohlrausch's Introduction to Physical
Measurements.
Mascart and Joubert's Treatise on Elec-
tricity and Magnetism.
Thompson's Elementary Lessons in Elec-
tricity and Magnetism.
Tyndall's Notes on Electrical Phenomena
and Theories.
Wright and Mackie's Glass Blowing Ma-
chines 323

Young's Speed Indicator 251

Zipernowsky Arc Lamp, The 423
,, Dynamo, The 368
,, System, The 363
Zöllner's Platinum Photometric Standard ... 166

APPENDICES.

—:o:—

Appendix.—Abstracts of Electrical Patents from January 1, 1873, to June 30, 1882.

Appendix A.—Reports on Tests made at the Münich Electrical Exhibition on Dynamos, and Arc and Incandescence Lamps.

NOTE TO APPENDIX.

————:o:————

THE Abstracts of British Patents for electrical inventions, commenced in the first volume of "Electric Illumination," and brought down to the end of 1872, are continued in the following pages, and include notices of all applications filed between the 1st of January, 1873, and the 30th of June, 1882. The number of these specifications is so great, and in many cases the abstracts are necessarily so extended, that it has proved inconvenient to extend the work to the conclusion of 1883, when the new Patent Law came into force. The abstracts of those patents granted between July 1, 1882, and December 31, 1883, will therefore be appended to the third and final volume of "Electric Illumination," which will also include complete abstracts of United States electrical patents. The period now dealt with is the most important in electrical invention, and it is hoped that full justice has been done to inventors, and that the abstracts are in all cases sufficiently complete to serve the purpose for which they are designed.

W. L. W.

November, 1884.

ABSTRACTS OF PATENTS

RELATING TO

ELECTRIC ILLUMINATION.

COMPILED BY W. LLOYD WISE.

(For Abridgments of Patent Specifications prior to 1873 see Appendix to "ELECTRIC ILLUMINATION," Vol I.)

1873.

*65.—J. F. Wiles, London. **Securing Railway Carriage Doors.** 4d. January 6.

MAGNETO - ELECTRIC GENERATOR.—The attraction of an electro-magnet for its armature is used to actuate the catches which secure railway carriage doors. The necessary electric current for exciting the electro-magnet is derived from a magneto-electric generator, driven from one of the carriage axles, and preferably placed in the guard's van. When the train comes to a standstill the current ceases, and the catches are withdrawn, allowing the doors to be opened.

87.—J. R. Chislett, Plymouth. **Applying Electricity for Curative and other Purposes.** 10d. (9 figs.) January 8.

ELECTRO-MAGNETS.—The inventor makes these of conical form, by increasing the number of turns as he proceeds with the winding from end to end. The core is made to consist of a cylindrical bundle of soft iron wires.

91.—S. W. Konn, London and St. Petersburg. (*A. N. Lodighin, St. Petersburg.*) **Producing Heat by Electricity.** 8d. (8 figs.) January 8.

INCANDESCENCE LAMP.—" This invention relates to the production of heat," and presumably light, " by electricity, by the employment of solid stems of carbon within hermetically closed vessels, charged with azotic or other gas, which does not support combustion." The carbon stems are held " by claws on pins, the removal of which breaks the contact when desired." The electric current is provided by an "electro-magnetic apparatus," preferably Wilde's or Noble's. Several applications of this invention are described, in which the incandescent carbon is made to communicate its heat to water or air. The illustration represents one form of this apparatus, *a* being a V shaped carbon bridge, and *b* the external envelope. No mention

is made of adapting this invention to lighting purposes.

*476.—R. Werdermann, London. **Purifying and Refining Metals and Alloys.** 4d. February 10.

MAGNETO - ELECTRIC GENERATOR.—The process consists in applying the currents produced by a magneto-electric generator to the purification of

metallic bodies, whilst they are in a molten state. "To prevent the development of heat" in the magneto-electric generator, "the connecting wires pass through a cooling apparatus."

534.—C. Owen, London. Signalling in Railway Trains. 4d. February 13.

CONDUCTORS. — These are provided with end clips, by which they are coupled up continuously. On each clip are fitted two eyes, and a stout spring pressing on the eyes insures contact. A slip coupling is formed of two telescopic tubes. The inner tube is sprung and provided with a "tipping piece," soldered to the end, which slips into a groove in the outer tube. The two tubes are readily pulled apart in the act of slipping a carriage, but will not separate by the ordinary motion of the train.

556.—F. H. Atkins, London. Filtering Apparatus. 2s. 2d. (12 figs.) February 14.

MAGNETO-ELECTRIC GENERATOR.—For the purpose of precipitating organic or other impurities in water the inventor passes through his filtering apparatus the currents derived from a magneto-electric generator.

618.—H. Wilde, Manchester. Producing and Regulating Electric Light. 10d. (4 figs.) February 19.

ELECTRO - MAGNETIC AND DYNAMO - ELECTRIC GENERATORS.—This relates to improvements in the generator described in Specification No. 842 of 1867. The helices in which the currents are induced, instead of being coiled round the armatures, as described in the above-mentioned patent, "are coiled round the extremities of the iron cores of the electro-magnets, which are prolonged for that purpose, or have iron cores fixed thereto, or an iron wheel or disc (the periphery of which is divided into as many segments as there are electro-magnets in each circle) is made to revolve between the circles of stationary electro-magnets and coils, for the purpose of generating electric currents." The electro-magnets may be excited by a separate generator, or "by the current from one or more of the stationary coils in which the major current is induced through the intervention of a commutator," while the current from the remainder of the coils is employed in the outer circuit.

ARC LAMP.—The carbons are made to approach and to recede from each other by means of a right and left-handed screw connected with the carbon holders. Each of the screws can be actuated independently of the other for the purpose of keeping a fixed focus when the lamp is used in conjunction with any optical apparatus. The regulation is done by hand

HOLOPHOTE. — A lens or parabolic reflector is mounted on a platform, which carries the lamp above described. This platform is made to revolve on a vertical spindle by means of a worm and toothed wheel, and is hinged in front to the metal supporting frame, so as to admit of an up-and-down motion, the requisite adjustment being effected by a screw at the back. The whole apparatus thus retains whatever position is given to it, both horizontally and vertically, without the use of any locking instrument. The inventor claims also the arrangement of handles, whereby the above described adjustments are brought about.

686.—F. J. Bolton and C. E. Webber, London. Obtaining Photometric Measurements. 4d. February 24.

PHOTOMETER. — The inventors state "certain bodies when exposed to light of greater or less intensity become changed in their electrical conductivity." According to this invention the body to be acted upon by the light is placed in an electrical circuit, which also includes an electrometer or other gauge of electrical resistance. The light is then directed on the said body, and the measurement of altered conductivity furnished by the electrometer is taken as an index of the value of the light. The same method may be employed to measure the comparative transparency of various substances.

799.—B. Hunt, London. (A. F. C. Reynoso, Paris.) Extraction of Iodine. 4d. March 5.

ELECTRO-MAGNETIC GENERATOR.—The current derived from an electro-magnetic generator is employed for the purpose of decomposing an acidulated solution of iodates which compose the "mother waters of the treatment of 'caliche.'" The iodine is thereby separated, and appears as scales, which float on the surface of the liquid.

1,000.—C. H. O. H. D'Arras. Applying Electro-Magnetism for Producing Locomotion. 4d. March 18.

ELECTRO-MAGNETIC LOCOMOTIVE.—"This apparatus is so arranged as to cause a wheel to run along a smooth iron rail by the force with which electro-magnets mounted on the wheel are attracted to the rail, which serves as a fixed armature to these magnets." The electro-magnets are placed radially so that their poles pass through the tyre of the wheel at intervals. A suitable commutator distributes the current to the electro-magnets in turn as they approach the rail, and cuts off the current when the electro-magnet pole has reached its lowest position and begins to leave the rail. When two wheels are fixed on one axle their magnets are arranged so as to act in alternating succession.

The source of electrical "power" might be fixed, and the current "might be transmitted by conductors laid along the rails, or by the rails themselves." In these cases the electrical current would be communicated to the magnet coils by conducting rollers arranged to run along the conducting wires, or by the tyres of the wheels having on them insulated conductors spaced to correspond with the spacing of the magnets.

1,178,—H. Highton, London. Electric Telegraphs. 6d. (4 figs.) March 29.

ELECTRO MAGNETS.—These are made with cores of boiler plate three-quarters of an inch thick, eight inches wide, and ten inches long, united in pairs at one end, and "wound with six or more layers of No. 16 copper wire well insulated with silk."

1,180.—J. H. Johnson, London. (*H. Fontaine, Paris.*) Magnets. 6d. (5 figs.) March 31.

MAGNETS.—In lieu of being made with rigid bars, these are formed of a series of very thin flexible blades, united in a bundle, which arrangement "admits of a great amount of magnetism being concentrated in a small quantity of steel." The blades are connected together by pieces of copper or iron. Wires may be used instead of steel blades. This form of magnet may be applied to the generators of Gramme, Wylde, Siemens, Holmes, and others.

1,450.—R. H. Courtenay, London. Magneto-Electric Induction Machine. 4d. April 22.

MAGNETO-ELECTRIC GENERATOR. — The helices are so arranged that the whole of them are connected at one end to a collar of copper, the other ends being connected with a conducting rod, insulated in the centre. "Through the method of induction used, the currents are collected at the terminals without break, leakage, or friction, and do not require a commutator to change them. The current is induced by two or more iron wheels" "revolving past permanent or electro-magnets. On the opposite side a series of helices are arranged forming a half-circle, which receives the magnetism from the revolving wheels or bars. On both sides of the iron wheels or bars, a series of helices are screwed into iron plates, between which plates the wheels or bars revolve, one end of each helix being connected with the collar before mentioned, the continuous induction through the whole series of helices building up a powerful current of electricity." A second part of this invention consists in "collecting from a revolving cylinder of helices, the currents generated, into a box of mercury." A third improvement consists in driving the axle of the generator by means of a lever connected with a flywheel, and to which flywheel is attached a smaller wheel by means of a second lever. "The axle of the wheel is driven by an engine of small power, the resistance to be overcome being only the wheels on the axles, and the restraining powers of the magnets."

1,558.—J. T. Sprague, Birmingham. Galvanometers. 8d. (3 figs.) April 29.

GALVANOMETERS.—These are constructed with several circuits having powers of deflection increasing in definite ratios, as 1, 10, 100, so that the one graduation serves to measure all currents. The dials are graduated to measure definite units, or to measure practical work, such as the weight of metal deposited per hour. The dials are also graduated to indicate directly the resistance of a circuit in ohms, by using a standard battery, and to indicate the E.M.F. of a cell by making the resistance one or more ohms as required, and generally "to convert any arbitrary or indefinite scale of graduation into the corresponding values in definite units." The principle of the construction of these instruments is described. For measuring large currents, instead of placing the needle within the coil, the inventor mounts it outside a single turn, "so as to obtain the action due to the difference between those of the upper and lower parts," and if necessary causes a part of the current to traverse a second circuit, "weakening the effect of the other by being laid in a reverse direction." Another plan is thus described :—"I mount the needle upon the stand, and surround it with a single turn of wire, as in a tangent galvanometer, but I make this turn movable on an axis in the line of the needle's position of rest, causing the current to enter and leave it by means of two springs which form the axis at one end of the frame. As the conductor is inclined from the vertical line, its effect is diminished till it becomes nothing when horizontal." Vertical galvanometers are provided with controlling magnets "to supersede the earth's influence." Tangent galvanometers are graduated by maintaining a steady deflection for a measured time, whilst the current is doing chemical work, such as depositing copper. The results observed are reduced into terms of one chemical equivalent in grains per ten hours, from which the current strength is then calculated.

1,713.—A. M. Clark, London. (*G. Planté, Paris.*) Apparatus for Lighting Lamps, &c. 6d. May 10.

SECONDARY BATTERY.—This is constructed of two lead strips coiled helically, and placed in a glass vessel filled with water, acidulated with ten per cent. of sulphuric acid, and enclosed in a box, the

lid of which is provided with clips for holding platinum electrodes. Silver strips may be used in place of lead. This battery is described as a "store of power," and is applied to an apparatus for lighting lamps, &c.

1,845.—E. Tyer, London. Signalling on Railways. 3s. 4d. (29 figs.) May 21.

MAGNETO-ELECTRIC GENERATOR.—Four coils are fixed on the poles of the permanent magnets, the centres of their cores forming the four equidistant corners of a square. Upon each core is "a very thick and broad iron armature or horn," directed towards the centre of the square, and of such a length that, in describing a circle from the centre of the square, the ends of these armatures are shaped off, following the line of such circle, "at that point where their breadth will exactly equal the intervening space between each in succession, thus dividing the circle into eight equal parts." A rotating armature is mounted in the centre of the circle with its two ends shaped to correspond with the curvature of the armature cores. On each side of this rotating armature is screwed an iron plate of the same width, but extending beyond it, and overlapping the ends of the two opposite armatures of the cores. Motion is imparted to the armature by suitable means.

1,903.—C. W. Cooke, London. Winding Electro-Magnets with Insulated Wires. 10d. (5 figs.) May 27.

ELECTRO-MAGNETS.—This invention relates to an apparatus for winding insulated wires on cores, and consists of a sheave clamped on the core with arms carrying bobbins, and guide pulleys fitted to revolve round the said sheave. The guide pulleys are arranged so as to be in an electrical circuit, with an indicator which gives warning of a defect in the insulation of the wire.

1,970.—M. Evans, Glasgow. Apparatus for Signalling, &c., in Railway Trains. 1s. 6d. (31 figs.) May 31.

MAGNETO - ELECTRIC AND DYNAMO - ELECTRIC GENERATOR.—For signalling in trains the inventor employs steel permanent magnets or soft iron magnets, having slight residual magnetism, which are placed round one or more axles of the carriages, and these acting on soft iron cores, or bobbins covered with insulated wire, generate electricity through the motion of the train. The magnets, instead of being on the axles of the vehicles, may be driven therefrom by suitable gearing. Several forms of this generator are illustrated and described. In the accompanying figure, B B are two of four electro-magnets, and A is a Gramme ring in connection with the studs *a*,

which form the commutator. The current is led off by a spring at each side, pressing against these

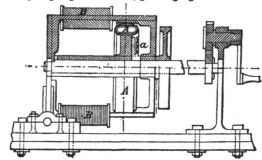

studs. One-half the current is used for exciting the field magnets.

2,006.—C. W. Siemens, London. (*W. Siemens and F. Von H. Alteneck, Berlin.*) Producing and Regulating Electric Currents. 1s. 10d. (17 figs.) June 5.

DYNAMO-ELECTRIC GENERATORS.—"Between the poles of one or more magnets or electro-magnets is fixed an iron cylinder, leaving a space between its periphery and the faces of the magnetic poles, which are hollowed out to a cylindrical form concentric with the said cylinder." In this annular space a cylindrical shell of light metal or other material is caused to revolve by mechanical power around the axis of the cylinder. "On this shell is wound insulated wire, in a direction parallel to the axis, such wire crossing the ends of the shell from the outer periphery thereof on one side, to the outer periphery on the other side. There may be several such wire coils, each covering an arc of the periphery on each side of the shell. The ends of the wires constituting each coil are connected respectively to pieces of metal, rollers, springs, or brushes of conducting material, which are insulated from one another, but which in their rotation with the shell and coils bear successively against two stationary conductors insulated from each other, which constitute the poles of the machine. On causing the shell with its coils to rotate by mechanical force, currents of electricity are generated in the coils as they successively pass the fixed magnetic poles, and by properly arranging the conductors from these several coils in relation to the poles of the magnet, and to the conductors on which they bear in their rotation, these electric currents are transmitted through any conductor connecting the poles of the machine." The currents may be made either continuous, intermittent, or alternating. "The inner iron cylinder may itself be rendered magnetic by coiling on it longitudinally an insulated electric wire in the manner of what is known as Siemens' rotating armature, and in this case the

outer magnets might be dispensed with." The inner iron cylinder might be made to rotate. The separate shell for receiving the wire could then be dispensed with, the wire being coiled longitudinally on the rotating iron cylinder itself. The wire may be coiled on the shell in two layers. A greater number than two magnetic poles may be made to surround the above described cylindrical armature. To avoid sparking, the sections of the commutator are made elastic, so that they can yield to the spring, brush, or roller under which they pass, and thereby remain longer in contact. The field magnets are excited by the current from the generator itself or from any other source. This generator may be employed to produce the electric light in a suitable lamp, and by transmitting electric currents through the generator it becomes an electro-motor. Fig. 1 is a longitudinal section of a generator con-

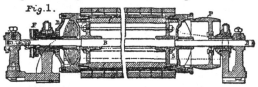

Fig. 1.

structed in accordance with this invention. A A are the field magnets, B is a fixed spindle, which supports the iron cylinder C. S is a shell supported on the discs $s\ s$, bored so that they can revolve freely on the spindle B, and provided with the driving pulley P. On the shell S are wound logitudinally the coils of insulated wire W, the ends of each of which are brought through the journal D and connected to the commutator F. Fig. 2 illustrates diagrammatically a method of connecting

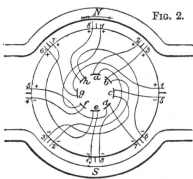

FIG. 2.

the ends of the armature coils, when such coils are in duplicate. a, b, c, d, e, f, g, h are the commutator sections to which the ends of the coils marked respectively 1, 2, 3, &c., and $1^1, 2^1, 3^1$ are attached. The line of connection is $1, f, 4^1, 4\,g, 6^1, 6, h, 8^1, 8, a, 2, 2^1, b, 3, 3^1, c, 5, 5^1, d, 7, 7^1, e, 1^1$ back to 1. A pair of brushes rub one on each side of the commutator when the coils are thus connected.

ARC LAMP.—The carbon points are made to

approach or recede from one another by the action of the electric current itself. The principle of this lamp is the differential action of two magnets, one having a thick wire coil in the main circuit and the other a fine wire coil in a shunt circuit. The lamp is adapted for use with either continuous or alternating currents. The above generator and lamp will be found very fully described in "Electric Illumination," Vol. I., pages 278 and 406. The inventors' claims are:—A generator, in which "a shell coiled longitudinally with insulated conducting wires on its outside is made to revolve in the annular space between fixed external magnetic poles and a fixed internal cylinder of iron which may be independently magnetised." A generator in which the wire "is coiled longitudinally over the external surface of an iron cylinder which is made to revolve within magnetic poles. The use of a rotating shell coiled both externally and internally in combination with fixed external or internal magnets." The method of winding the armature. The use of springs for the commutator sections. The use in an electric lamp of "two electro-magnets working a wheel in opposite directions by means of pawls, such magnets being rendered active or inert according as the electrical resistance varies." The use in electric lamps of "one electro-magnet working by a pawl a wheel in opposition to a weight or spring, the said magnet being rendered active or inert according as the electrical resistance varies."

*2,015.—A. V. Newton, London. (*J. D. Wallace, New York.*) **Electro-Magnetic Engines or Motors.** 4d. June 5.

MOTOR.—This invention consists in the construction of magnets of "broken annular sections," and includes also a "novel operation of a permutator" whereby the motor is controlled. "The fixed magnets are rectangular bars whose breadths should be to their lengths as two to three. Within a circle of these is mounted a similar series of magnets upon flanged plates, secured to a central driving shaft." One end of "the coils of this series is electrically connected with their several cores, and the other ends of the coils are attached to a ring which is itself connected with an insulated rod, centred in the driving shaft." The ends of the coils last mentioned communicate with one section of a "permutating wheel" mounted on the driving shaft, and the other section of this wheel is connected to the opposite ends of the coils through the insulated rod. A pair of "current wheels," mounted on rocking levers, embrace the permutator, and convey to it the current. The levers are actuated by a sector gearing with a worm, the turning of which adjusts the wheels relatively to the permutator. A pole changer is used to reverse the

motor. The course of the current is through the fixed magnets and one current wheel to the permutator, thence to the movable magnets, back to the permutator, and out by the second current wheel. The winding of the magnet coils is arranged so as to give opposite polarities to any two magnets in juxtaposition.

*2,077.—W. Rowett, Liverpool. Electric Telegraph Cables or Ropes. 4d. June 11.

CONDUCTORS.—This relates to the use of manilla or hemp in place of jute for covering insulated electrical conductors, whereby the strength of the cable is increased. It is preferred to adapt the invention to several insulated conductors, laid parallel, and secured with a serving of tape, in the interstices of which the hempen lines are laid. The core thus prepared, may be preserved by "Rowett's well-known" compound, and is further improved by being put together with "Rowett's preserving elastic marine glue." The conductor may be simply enveloped in hemp prepared with the above-named marine glue, without other insulating material. The strength of the cable may be increased by forming the strands of "wrong-way yarn," applied from conical bobbins so as to "keep the turn in the yarn."

2,091.—H. A. Bonneville, Paris and London. (*W. Radde, New York.*) Conductors for Telegraph Wires. 1s. (4 figs.) June 13.

CONDUCTORS.—An insulated bridge, enclosed in a box, forms the connecting link between sections of pipes which enclose one or more conducting wires, in such a manner, that by opening the box, access can be had to the wires. Another part of the invention relates to a pipe containing a series of insulated passages, separated from each other and from the enclosing shell by a layer of cement, bitumen, or other suitable material. Each passage takes a conducting wire.

2,266. — G. Zanni, London. Telegraphic Apparatus. 1s. (8 figs.) June 30.

MAGNETO-ELECTRIC GENERATOR.—This invention relates to a "magneto-electric Morse ink printing telegraphic apparatus." The generator consists of a Siemens armature placed between the poles of a compound circular permanent magnet. The armature is rotated by clockwork.

2,302.—W. R. Lake, London. (*H. T. Brownell, Hartford, Connecticut, U.S.A.*) Plating Metals. 4d. July 2.

CONDUCTORS.—Wires are nickel-plated, and then subjected to a temperature of from 480° to 700°

Fahrenheit which "unites and incorporates the two metals." A permanent inoxidisable surface is thus obtained.

2,618.—F. A Palmer, New York and London. (*C. Gaumé, New York.*) Electro-Magnetic Engines. 6d. (2 figs.) August 2.

MOTOR.—The object of this invention is to avoid the "pull back" or retardation which is such an objection in motors of the usual construction. Two frames are united below by three bars, each of which supports a horseshoe electro-magnet in an inclined position as regards the radial lines drawn from the centre of a shaft, which runs in bearings in the upper part of the side frames. Each of the electro-magnet coils is joined by one end to an insulated terminal, which receives a wire leading to the negative pole of the source of current. The wire from the positive pole is connected with the frame of the motor. A wheel attached to the shaft carries four armatures, which, as the wheel revolves, come close to and between the poles of the fixed magnets. The armatures consist of "a central bar attached at its centre to the face of the wheel, and having crossheads formed upon them, about midway between its centres and ends, said crossheads having short arms or bars formed upon their ends, parallel with the central bar, and the ends of which project to equal distances upon the outer and inner sides of said crossheads. Armatures thus constructed are found to be free from retardation or 'pull back.'" A suitable commutator with contact wheels distributes the currents to the magnets at the right time. The above described armatures only are claimed.

2,656.—W. R. Lake, London. (*F. L. Pope, New Jersey, U.S.A.*) Railway Signal Apparatus. 1s. 10d. (13 figs.) August 7.

MAGNETO-ELECTRIC GENERATOR.—One form of generator consists of an armature of H section, having upon it a coil of insulated wire. Currents are induced in this coil when the armature is made to vibrate between the poles of a compound permanent horseshoe magnet. The other form of generator, described and illustrated, is designed to be driven by a weight and cord, is an ordinary Siemens' armature, rotating between the poles of a battery and of permanent horseshoe magnets.

2,870.—J. B. Stearns, Boston, U.S.A. Electric Telegraph Apparatus. 10d. (10 figs.) September 1.

SECONDARY BATTERY.—The inventor employs a secondary battery "composed of many plates of one metal (and sometimes called Ritter's pile)" as a substitute for a condenser, for the purpose of ob-

viating the disturbances of telegraphic instruments due to induction or static charge. The invention is especially of use in systems of duplex telegraphy.

2,969.—W. Moseley, Manchester. Electrical Signal Apparatus. 1s. 10d. (59 figs.) September 10.

CONDUCTORS.—Wires are sewn or woven in between strong tapes, saturated with a mixture of gum and and boiled linseed oil. The tapes can be secured to a wall surface by nailing, without the use of staples. "Highly insulated wires" are covered preferably with three layers of pure india-rubber, over which cotton or hemp is applied, and the whole is then immersed in a solution of equal parts of gum and boiled oil. When this coating is dry, a wrapping of metal ribbon or tape is added. This last covering should consist, by preference, of a strip of tinned copper, which is drawn over the insulated wire, and passed through molten tin, in order to unite the overlapping edges.

INSULATORS.—These are constructed of an earthenware disc-shaped top portion, to which is fixed cylindrical ebonite extension downwards, forming the bell. The ebonite cylinder is applied warm, in an expanded condition, and is secured by marine glue, or other cement. A variety of forms of insulators for special purposes are illustrated and described.

*3,078.—S. J. Moore and R. H. Courtenay, London. Electro-Motor. 4d. September 19.

MOTOR.—The electro-magnets are constructed each in four parts, the poles of which are movable, working in grooves forming part of the supporting frame. The stationary portions of the magnets have solid iron cores, and the movable portions have tubular poles in which springs are inserted for the purpose of connecting electrically the movable and fixed portions of the magnets. The magnet coils are connected to the grooves of gun-metal, in which the movable parts of the magnets slide in order to complete the circuit, and thereby "effect the transmission of an equivalent amount of magnetism in proportion to their size to the poles, as in the ordinary form of fixed magnets." The sliding tubes of the movable magnets "terminate at the poles of the same, and communicate with suitable keepers divided by an insulator, so as to keep the currents of magnetic forces separate." Insulators are employed "to keep the magnets isolated from deterrent or absorbent forces." The current is transmitted to the motor by "a commutator worked by a spring" or by hand, "so that the whole power generated may be under control and regulated to work with the greatest accuracy."

3,083.—A. Wilkinson, London. Coating and Preserving Telegraph Wires, &c. 4d. September 19.

CONDUCTORS.—The wires are first cleaned and then passed through a bath of Brunswick black or gutta-percha, diluted, after which they are covered with fibres which have been prepared with a mixture of white lead, pitch, japan, shellac, tallow, naphtha, and linseed oil. After being dried, the wires are passed through a bath of the above compound, and then receive a final coating of white lead. Two or more wires thus prepared may be united in a cable, which may be sheathed with metal wires, tube, or strips.

3,086.—J. Fottrell, Dublin. Composition for the Manufacture of Pipes and Conduits. 4d. September 20.

INSULATORS.—These are made from a material consisting of sand or finely powdered stone, mixed with shale oil, Trinidad bitumen, and bituminous rock, in suitable proportions.

3,461.—W. E. Newton, London. (*J. S. Camacho, Havana.*) Electro-Magnets. 6d. (2 figs.) October 24.

ELECTRO-MAGNETS. "This invention consists in constructing an electro-magnet of a series of concentric tubes, all of which are of equal length; each tube being surrounded by one or more coils of insulated conducting wire."

3,736.—W. Darlow and H. Fairfax, London. Magneto-Electric Apparatus for Curative and other Purposes. 6d. (3 figs.) November 18.

MAGNETO-ELECTRIC GENERATOR.—A series of electro-magnets are arranged helically upon a spindle which is rotated by hand or clockwork, in front of a series of fixed permanent magnets. The electro-magnets may be encased in an enamelled cylinder, provided with rubbers, by which frictional electricity may be combined with magneto-electrical currents. It is not quite clear how the currents are led off from the generator.

*3,780.—W. Hooper, London. Telegraph Cables. 4d. November 20.

CONDUCTORS.—An insulating material for coating conductors consists of the "foots" of cotton seed oil, or other vegetable drying oil, which is oxidised by treating it whilst hot with nitric acid, and hard pitch obtained from the distillation of cotton seed or other vegetable oil. This compound is sufficiently fluid for use at a little over 300 deg. Fahr., and it may be applied to conductors direct, or over an

insulating covering of india-rubber. Fibrous yarns used in electrical cables may be saturated with this material to render them non-absorbent.

3,862.—M. Gray, London. Insulating Underground Telegraph Wires. 10d. (2 figs.) November 26.

CONDUCTORS.—To maintain the insulating property of gutta-percha, when it is employed as a covering for underground electrical wires, the inventor proposes to enclose such wires, in a continuous length of cast-iron pipes, capable of containing water sufficient to keep the insulated wires immersed. At given distances testing boxes are provided. These boxes have an elastic diaphragm through which the wires are threaded, and by which means access is gained to their ends when the box cover is removed, without the necessity of drawing off the water contained in the pipes.

3,863.—M. Gray, London. Insulating Underground Telegraph Wires. 10d. (3 figs.) November 26.

CONDUCTORS.—Naked underground wires are supported by stretching them through perforated blocks of insulating material fixed at suitable intervals in a continuous length of pipes, which are made as far as practicable air-tight. Dry air is then passed through such pipes to remove the moisture therefrom.

3,997.—W. Hooper, London, and J. M. Dunlop, Windermere, Westmoreland. Telegraph Cables. 4d. December 4.

CONDUCTORS.—This invention is, in the main, identical with that described in No. 3780 of 1873, but in addition relates to the use of a material termed "ondroic," obtained from the distillation of stearine, for the same purpose for which the soft pitch of cotton seed oil, mentioned in the former patent, was employed.

*4,079.—W. Rowett, Liverpool. Electric Telegraph Cables or Ropes. 4d. December 11.

CONDUCTORS.—This specification is almost, word for word, a repetition of No. 2077 of 1873.

4,167.—C. L. Madsen, Copenhagen and London. Electric Telegraph Cables. 4d. December 18.

CONDUCTORS. — Metallic conductors are surrounded with an insulating compound composed of wax, resin, paraffine, and turpentine. Small proportions of fats may be added to render the material

harder or softer as required. Cables are made thus:—Tapes of fibrous materials saturated with the above compound are applied to the conductor, and over them may be a coating of a compound consisting of Stockholm pitch, wax, and tallow, and over this again may be another layer of tape saturated with the first-named compound. The process of working is described in detail.

4,171.—G. S. de Capanemb, Rio de Janeiro. Insulators. 10d. (17 figs.) December 19.

INSULATORS.—These are made entirely of glass, porcelain, or other suitable non-conducting material, of single or double bell-shape, supported on a central stem, the crown of the bell projecting above such stem without any metal hood. In the upper face of this projecting crown, a chase or groove is formed to receive the wire, such groove extending diametrically across the crown. The middle of the groove is widened out for the reception of a ball or collar on the wire. The groove tapers to a greater width from the central widened part outwards to allow for the swing of the wire when laid in the groove. The wires have collars of cylindrical, spherical, or other shape, cast upon them at the required intervals, an instrument resembling a bullet mould being used for the purpose, and these collars are laid in the widened part of the groove in the insulator, whereby the wire is prevented from being pulled in either direction. In some cases a pin is passed through the crown of the insulator above and at right angles to the wire to further secure it.

*4,193.—J. Rubery, London. Telegraphic Wires or Conductors. 4d. December 20.

CONDUCTORS.—A hardened and tempered steel wire is coated with an electro deposit of copper, and may be subsequently insulated with a covering of india-rubber, gutta-percha, or other material.

4,277.—H. Highton, Putney. Electric Telegraphs. 6d. (5 figs.) December 30.

ELECTRO-MAGNETIC GENERATORS.—In order to "multiply the power of a current at the end of a cable or telegraph line," the inventor makes it circulate round the field magnets of a small generator, having an ordinary Siemens armature, which is rapidly rotated by suitable means, and thereby producing a "secondary current which will act more powerfully than the original line current."

1874.

94.—W. E. Newton, London. (*J. B. Stone, Boonton, New Jersey, U.S.A.*) Electro-Magnets. 10d. (4 figs.) January 7.

MOTORS. — This invention relates to rotary

motors in which the magnetic pull is obtained by consecutive action consequent on the unequal division, as regards number, of the fixed and moving magnets, and consists in constructing the

magnet poles corrugated on their opposed faces, so as to give "largely increased acting surfaces" to the magnets as compared with plain faced ones. The frame for the fixed magnets and the shaft for the moving ones are both "made of magnetic metal, whereby the contiguous, fixed, and moving magnets are changed (when magnetic) and attract each other consecutively into (as regards effect or action) a ⊔ form of magnet with the opposite poles attracting each the other, and constituting a ⊔ magnet with its one pole fixed and its other pole movable." The cores of the fixed magnets are cast in one piece with the side frames, and those of the movable magnets with their hub. The commutator s a disc with contact pieces forming the terminals of the moving magnets, which pieces come in contact with spring "fingers" insulated from the frame of the motor, but electrically connected to the coils of the fixed magnets. The other ends of the fixed magnet coils join a common return wire, and the other ends of the revolving magnet coils make contact with the revolving shaft by which the current enters the motor.

124.—E. T. Truman, London. **Insulated Telegraphic Conductors.** (6d.) January 9.

CONDUCTORS.—This relates to improvements in the machinery for covering conductors with insulating material, described in Specification No. 482 of 1872. In order to impart a uniform rotation to the wire, instead of rotating the carrying and receiving drums, the wire is drawn aside into a loop, and is then caused to rotate around its line of travel. Joints in stranded conductors are made "in the manner in which sailors splice a rope," and soldering is dispensed with. A hinged mould is then applied to the joint, into which plastic gutta-percha is forced. This invention also relates to improvements in the "filling machine," described in Specification No. 78 of 1870. A hopper receives the insulating material, and in this hopper is a screw which delivers the material to another screw, working in a cylinder, provided with a suitable outlet. The combined effect of the two screws is to secure regularity in the delivery of the material. This invention further relates to improvements on the gutta-percha washing machine described in Specification No. 41 of 1870. These consist in providing means for closing or partly closing the openings in the outer cylinder, so as to render the escape of material in the raw state, and in any small pieces impossible until agglomeration has taken place.

265.—E. H. C. Monckton, Fineshade, Northamptonshire. **Magnetic Engines.** 2s. 2d. (17 figs.) January 21.

CONDUCTORS.—A cheap preparation for insulating conductors consists of sulphur, or a mixture of 95 per cent. sulphur, and 5 per cent. of yellow orpiment, or white arsenic, combined with fibrous material, and solutions of shellac and silk, in combination with fibrous materials are also employed. A cable may be made by coating iron wires with "sulphur ointment," and winding round them an insulated copper wire. The conductor thus prepared is surrounded with one of the above described sulphur compositions, applied by preference to a woollen cloth, which is wound on helically. The central iron wire thus becomes an electro-magnet "of enormous length and also of considerable power." There is "no tendency for the electricity to escape," as the "two wires forming the cable are both better conductors than water."

MAGNETS.—The inventor states that "long magnets being comparatively speaking thin, must necessarily have thin poles; these must therefore be secured to large and solid pieces of soft iron." Very long magnets, in order to save room, "may be insulated and buried." Coils of electro-magnets which have to convey high tension currents, are made by "winding the wires in separate lengths in opposite directions, commencing each helix by coiling it from its bottom next the magnet upwards, and leaving either a clear air space," or placing insulating materials between each coil. "The top ends of each pair of coils are united by themselves, and the bottom ends by themselves." By another method of construction, a thin coil surrounds each pole, and a thicker coil the centre of an electro-magnet.

MAGNETO-ELECTRIC GENERATOR.—Two insulated steel rings or cylinders are used. One is fixed and the other is made to revolve within it, but without touching it. The rings are "magnetised in the usual way," and "their peripheries are rendered polar by drawing the poles of the magnet employed in magnetising them radially across them." Insulated copper rings are attached to the sides of both the steel rings, and the latter are made to rotate in contrary directions, or one is made to rotate while the other remains fixed. "The electricity generated by their rapid revolutions is conveyed away by wires attached to springs or rollers fixed to them, made to press on and slide or roll over the insulated copper ring conductors. By the rotation of two insulated steel or malleable iron discs, placed nearly in contact, similarly prepared, and properly magnetised," an electric induced current is generated. "Here the electricity is drawn from the periphery by means of an insulated copper ring, and from the centre of the disc, by means of insulated copper. Cones and hemispheres may also be made to revolve and rotate one within the other in a similar manner." Induced currents are also obtained "by rotating between the poles of a mag-

net, armature magnets made like a wheel with spokes."

ELECTRIC LIGHT.—The inventor proposes to use the electric light in combination with reflectors, to enable medical men to inspect diseased parts of the human body.

COMMUTATORS.—These have, by preference, silver, or electro-plated copper rollers, which are followed by a surface cleaner, made of velvet.

GLOBES, which surround the electric light are prevented from cracking, by being made in pieces united in a frame.

MOTOR.—In "ring engines" the ring is constructed in pieces so as to admit of easy coiling of strong iron wire round the ring in the vacant space between the armatures, and the armatures are made "with double curves rising to the centre so that the motion of the revolving magnet can be reversed." It is preferred "to drive on outside the armature ring another flat ring of soft iron, covered with an iron coil," and in place of having one revolving magnet, the "electro-magnetic wheel" already described is used. A reciprocating electro-magnetic locomotive is described, and the inventor states that he constructs railways with insulated rods, by preference of copper, secured between the rails, along and against which either a single insulated metallic grooved wheel runs, or else these rods are clipped by a pair of such wheels. "The electricity generated from any source along the railway can thus be conveyed to the magnets actuating the engine." For this purpose large fixed steam engines, wind, or water power may be employed. This is a lengthy and somewhat rambling specification of 42 pages.

*293.—T. Walker, London. Transmitting Electric Currents. 4d. January 23.

CONDUCTORS.—This invention relates to a method of utilising for telegraphic purposes an uninsulated or partially insulated conductor. An inexpensive partially insulating material consists of a mixture of shellac, resin, and tar, or resin, tar, and oil, to which may be added pitch or asphaltum. After these compounds have been applied to the conductor it is preferred to cover it with a serving of tape or fibre.

380.—G. T. Bousfield, Sutton, Surrey. (*H. J. Smith, Boston, U.S.A.*) Electrical Machines. 8d. (10 figs.) January 29.

STATIC GENERATOR.—This consists of a shallow box of vulcanite containing a "frame plate," condenser, and generating plate placed parallel to each other and the whole enclosed in an outer wooden case. The generating plate lies between the condenser and frame plate, and is revolved by means of a handle between two cushions, coated with amalgam. The cushions are provided with flaps to prevent the electricity from escaping from the generating plate until the excited portion of its surface reaches the collectors, which are serrated strips of metal. These collectors are in metallic connection with one set of plates of the condenser, whilst the other set is joined to the cushions. "Posts," or discharging terminals, are attached to the condenser, which is so arranged that the forward motion of the crank to generate electricity and charge the condenser, moves it forward through a small arc, whereby its terminals are moved away from the discharging terminals. A retrograde motion of the handle restores the condenser to its original position and discharges it. The vulcanite casing of the generator is made air-tight, the spindle working through a gland. External to this casing is an india-rubber pouch, which further insures the exclusion of moisture.

440.—S. J. Mackie, London. Signalling on Railway Trains. 1s. 6d. (14 figs.) February 3.

CONDUCTORS.—For coupling the ends of conductors used for signalling on railway trains the inventor provides them with draw tubes sliding over each other or with hooked ends dropping into bayonet sockets.

447.—J. Macintosh, London. Insulating Telegraphic Wires, &c. 4d. February 4.

CONDUCTORS.—Cotton seed pitch is mixed with india-rubber by means of rollers until it becomes a "hot plastic heterogeneous mass," in which state it is applied to conducting wires for insulating them.

485.—J. J. Harrop, Manchester. Tin and Lead Piping, &c. 1s. (11 figs.) February 6.

CONDUCTORS.—This invention relates to machinery for the manufacture of tin-lined lead pipes, applicable also for covering electrical cables with a lead protecting envelope.

*515.—W. F. Reynolds, London. Speed Indication for Engines. 4d. February 9.

GENERATOR.—The inventor arranges "an electro-motive apparatus" in connection with some moving part of an engine, and carries a wire to a suitable position within sight of a responsible person, where the wire is attached to a spindle of an index hand free to travel over a dial. The wire is so arranged that as the main shaft rotates the wire becomes heated in proportion to the speed of such shaft, and expands in the direction of its length, thereby operating the index hand on the dial.

862.—J. Imray, London. (*T. J. Waters, Japan.*) Laying Telegraph Cables. 10d. (4 figs.) March 18.

CONDUCTORS.—For laying electrical cables under water or under ground the inventor uses a kind of plough mounted on wheels which is drawn along the bottom of the water or along the soil in which the cable is to be embedded. The plough has a tubular shank extending obliquely downwards behind its wheels, so as to cut a furrow in the soil, and two thin flukes extending horizontally from the shank serve to keep its end down at the required depth below the surface. The cable is passed down the interior of the hollow shank and is thus guided to the bottom of the furrow cut by the plough.

979.—W. E. Prall, Washington, U.S.A. **Compressed Air Apparatus for Signalling on Railways, &c.** 10d. (7 figs.) March 19.

CONDUCTORS.—In connection with a system of signalling on railways by means of compressed air, the inventor proposes to place a small cable containing telegraph conductors in the pipes which convey the compressed air "to protect thereby the telegraph wires from atmospheric disturbances," and to save the expense of the usual supports.

1,159.—W. M. Bullivant, London. **Telegraph Cables.** 4d. April 2.

CONDUCTORS.—Around an insulated conductor is arranged a number of hempen yarns saturated with tar or other bituminous compound. The yarns are immersed in water to shrink them and then dried, and lastly saturated with tar. Without again moistening them they are laid around the insulated conductor in a very long lay, two servings being applied in opposite directions, separated by a layer of pitch containing silica. Over these yarns is sometimes applied a binding of other yarns with a short lay. The whole cable is covered with prepared pitch and silica compound, and is then complete. In this way is produced a very light cable, which, it is claimed, will not shrink and damage the insulated conductor.

1,261.—W. R. Lake, London. (*M. Day, Mansfield, Ohio, U.S.A.*) **Electric Light Apparatus.** 8d. (6 figs.) April 11.

ARC LAMP.—In the illustration the standard is made in two parts, the one A of wood, and the other B of metal. C, C, C, C are metallic arms for supporting the guides L and D, which are respectively of soft iron and brass. Between the lower arms C C an electro-magnet R can attract the soft iron guide L. Below the magnet is a coiled spring K, to the drum of which is attached one end of a cord or ribbon J, preferably of steel, its other end being fastened to the lower end of the guide L. The spring K gives motion to all parts of the lamp and may be replaced by a weight if desired. Upon

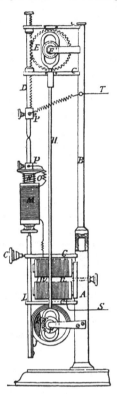

the spring spindle is the insulated pinion and jamb nut F. Between the upper arms C C is the wheel E, gearing into the rack on the guide D, and upon the wheel spindle is another insulated pinion and jamb nut F[1]. H is a connecting rod with slots at each end, geared respectively to opposite sides of the pinions F and F[1]. Upon the upper end of L is a magnet M with core N, spring O, and carbon socket P. A similar socket is attached to the guide D. The set screw c holds the guides in any position when the lamp is not in use. S and T are the terminals of the lamp. The sockets P P can hold one or more carbons parallel to but not touching each other. The operation of the lamp is evident from the figure. The lower magnet R regulates the feed whilst the magnet M draws the arc, both magnets being in the main circuit. When double carbons are used the light is alternately emitted from each pair, the change from carbon to carbon being effected very quickly at the instant that the wear of one pair has gone on sufficiently long to bring the second pair into actual contact. Reversals of current are employed for keeping the arc in a focus. In another form of lamp "an ordinary marine clock movement" is employed to restrain the spring K, and in a third form of lamp the upper carbon-holder is a rack gearing with a wheel on the spring spindle.

1,438.—R. Werdermann, London. **Cutting Rock or Stone, &c.** 8d. (5 figs.) April 24.

ELECTRIC LIGHT.—Part of this invention relates to the use of the electric arc produced between carbon pencils, for the purpose of quarrying porphyry, granite, and like stones. A jet of air is conducted through a tube to the carbon points and carries the arc forwards, causing it to act as a blowpipe.

1,446.—J. H. Johnson, London. (*F. E. de Mersanne, Paris*). **Production of Electric Light, &c.** 3s. 4d. (35 figs.)

ELECTRIC LIGHT.—This invention is based on the principle of obtaining light from "a succession of rupture sparks (obtained by alternately establishing and breaking contacts) between the electrodes of a regulator," and also consists in the employment of revolving circular electrodes of dense carbon. Very complete drawings accompany this specification and clearly explain the details of the somewhat complicated apparatus for carrying the invention into effect. Another part of this invention relates to a current distributing apparatus "having rubbers and boards provided with studs or plates, and so arranged as to effect the division of a current into a number of currents of an intensity equal or proportional to given numbers." Another current distributor with mercury contacts is also described.

*1,554.—H. Wilde, Manchester. **Excavating Coal, &c.** 4d. May 2.

TRANSMISSION OF POWER.—The generator described in Specification No. 842 of 1867 is employed to produce a current of electricity, which is transmitted by suitable conductors to a distance, where it is retransformed into power and employed for the purpose of coal cutting. A reciprocating motor is described, but a rotary motor may also be used, and its motion converted into a reciprocating motion by means of cranks and levers.

*1,766.—M. Eustace, Glasnevin, Dublin. **Jointing and Tightening Wires.** 4d. May 19.

CONDUCTORS.—Wires are joined by inserting their ends, side by side, into a socket piece from opposite directions, after which a taper key is driven through transverse holes in the sides of the socket between the two enclosed wires to secure them. This invention also includes a wire tightening appliance.

1,820.—E. O. W. Whitehouse, London. **Electric Light Apparatus.** 4d. May 22.

ARC LAMP.—Cylindrical carbon electrodes are employed, and these are mounted on axes by preference placed at right angles to each other, and

made to rotate, at the same time being screwed forward axially, "so as to present in turn every part of their surfaces toward each other." Clockwork or other mechanism rotates the carbons, and advances them axially. The current is reversed at intervals to equalise the consumption of carbon, and "to retransfer the carbon from the pole to which it had been carried, back again to that from which it came."

1,855.—G. Zanni, London. **Telegraphic Cables.** 8d. (6 figs.) May 27.

CONDUCTORS.—A central conductor, consisting of one or more copper wires, is surrounded with a series of iron or steel wires, and the whole is united by being passed through a bath of molten tin. An insulating coating of gutta-percha or other material is then applied, over which tinfoil is wrapped "to exclude moisture;" or in place of the tinfoil, the core is covered with a cotton tape and passed through a bath of molten tin, whereby the insulating material is protected from the injurious effects of the water pressure at great depths. Several cores may be combined in one cable. Suitable sheathing may be added, and shore ends may be protected by threading them through lengths of gas piping.

*2,106.—H. Conybeare and G. Naphegyi, London. **Treating the Juice of the Zapote or Chickley Tree.** 4d. June 17.

CONDUCTORS.—This invention relates to the preparation of the juice of the zapote tree (a variety of the ficus elasticus) for use as a substitute for india-rubber or gutta-percha in the insulation of electrical conductors. The inventors call the material "zapotine." It can be vulcanised.

2,625.—T. Slater, London. **Electro-Magnets, &c.** 4d. July 28.

ELECTRO-MAGNETS.—The core upon which the insulated wire is wound, is made of conical form, the diameter of the base of the cone being by preference about twice its height. At the apex of the cone is a screwed stem with a nut for the purpose of securing the electro-magnet to any apparatus of which it is to form part. "By increasing or diminishing the size of the conical core, and surrounding it with a correspondingly thicker or thinner wire, magnets of greater or less power will be produced according, with an equal amount of battery power."

*2,717.—G. Haseltine, London. (*H. C. Cook, Brooklyn, U.S.A.*) **Illuminating Compass Cards, Dials, and Gauges.** 4d. August 5.

ELECTRIC LIGHT.—A vacuum tube, worked by a Rhumkorff's coil and sulphate of mercury battery,

is employed for the illumination of mariners' compasses, water gauges, &c.

*3,006.—H. Highton, Putney. Submarine Telegraphy. 4d. September 2.

CONDUCTORS.—The main feature of this invention is the use of phosphor-bronze in the construction of cables. It may be employed for the conducting wires, or for the sheathing, and in the former case may be combined with copper wires to give increased conductivity.

3,156.—R. Werdermann, London. Magneto-Electric Machines. 1s. 4d. (6 figs.) September 16.

DYNAMO-ELECTRIC GENERATORS.—In the apparatus constituting this invention "the electric currents are induced in such a manner that all those halves of the convolutions of the wire lying on one side of the core are induced by magnetic poles of the same polarity." This principle is termed by the inventor "isopolar induction." The advantages claimed for this mode of constructing generators are facility for regulating the electromotive force, and the collection of currents without a commutator, springs, or rubbing contacts. One form of generator is constructed as follows :—On an annular plate of metal is coiled an insulated wire or ribbon of copper. This plate is fixed like a tyre on a wooden disc, fitted on a shaft. The boss of the disc is surrounded by two metal rings, insulated from each other. These rings are each connected with one end of the wire on the plate, and lie in contact with rubbers connected with the terminals of the generator, or the ends of the wire may be passed in opposite directions through the axis of the shaft, and inserted in mercury cups. The wire coiled on the plate may be divided into sections, and an arrangement of plug commutators may be fixed to the wooden disc, whereby such sections can be joined up parallel or in series, as may be desired. On each side of the annular armature first described is fixed a soft iron ring, such ring forming the polar extension of one or more permanent or electro-magnets, and one ring being of opposite polarity to the other. Instead of a flat armature one of elliptical or circular section may be used, the field magnet poles being suitably shaped to embrace it as closely as possible. Several armatures may be combined in one generator, in which case there will always be one more field magnet than there are armatures. A hollow soft iron cylindrical electro-magnet may afford two unipolar magnetic fields, in each of which an annular armature may revolve, one to act as a "feeding" and the other as a working armature. The armature may be of spherical or ellipsoidal form, coiled meridianally or equatorially with insulated copper wire, in which case the field magnets are made of suitable form to embrace it. The illustration represents another form of generator embodying this invention. The armature consists of a wooden cylinder *b b*, on which is wound helically a rod or ribbon of iron, coiled with insulated copper wire B; *c c* are collecting hoops. The armature is sup-

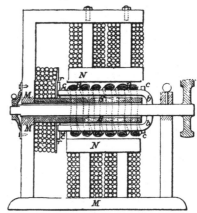

ported by the four rollers *r*. M is an electro-magnet having three branches forming two north poles N and one south pole S. This is called a trifurcated magnet ; magnets with many poles are called polyfurcated. The soft iron rings and polar extensions, as well as the cores of the electro-magnets, are made by preference of electro-deposited iron, and that side of the polar extensions which face the armature are roughened "like a file, whereby the power of magnetic radiation is increased."

3,172.—M. Eustace, Glasnevin, Dublin. Jointing and Tightening Wires for Fences, Telegraphs, &c. 8d. (5 figs.) September 17.

CONDUCTORS.—This invention relates to methods of joining the ends of wires, used as electrical conductors or otherwise, and has already been described in Provisional Specification No. 1,766 of 1874.

*3,381.—C. Goldstone, Southampton ; J. Radcliffe, Retford ; M. Gray, London. Signalling in Railway Trains. 4d. October 3.

CONDUCTORS.—These are joined by a coupling of peculiar construction, the one end being the counterpart of the other, to facilitate making up trains provided with through conductors for signalling purposes. To each end of the conductor to be coupled is secured a block of insulating material to which is fitted a pair of elongated clip springs. The ends of the conducting wires are soldered to these clips. Strips of vulcanite, wedge-shaped at their ends, triangular in cross section,

and tapering inwards, are attached to the inner faces of these springs. On one of their inclined sides a metallic plate is secured, which comes in contact with a similar plate on the counterpart clip. These metal plates each have at at their outer ends a clip, which in the act of coupling up the conductors will be forced apart, but on the couplings being severed, themselves come into contact. The couplings are provided with a waterproof tubular covering of vulcanised india-rubber, to protect them from rain and dust.

3,509　E. H. C. Monckton, Fineshade, Northamptonshire. **Galvanic Batteries.** 1s. 4d. (12 figs.) October 12.

ELECTRIC LIGHT.—This invention includes the application of the electric light to mines and collieries, by means of reflectors, its intensity being subdued by the use of suitable screens or shades of ground glass. Reflectors are also employed to break up the shadows. When there is danger from gas, the light is enclosed in a glass case.

ELECTRO-MAGNETS.—In Patent No. 265 of 1874 it was omitted to state that the ends of the magnets therein described should be made larger in proportion to their lengths, and that the coil placed on these ends should also be large in proportion to their thickness. The copper wire round the great length of the thin part of the magnets need not be coiled close. Another improvement is to case the magnet coils with iron slit cylinders, and, if desired, to add other coils and slit cylinders in succession. The weight of iron in any magnet should be from 2½ to 3 times the weight of the wire in the coils. Magnets may be made by combining several rods or flat thin sheets of iron, and it is preferred to electroplate them with copper or nickel. Permanent circular magnets are made by rotating the rings over the poles of electro-magnets, and continuing the rotation whilst they are being withdrawn. Compound flat electro-magnets are made of thin iron plates, over which separate coils are placed, and their ends are joined to a back plate of iron. Other modifications of these magnets are described.

MOTOR.—This is constructed on principle of the Arago disc. A large metal disc is placed between the poles of a very large horseshoe magnet, such poles being curved so as to embrace the periphery of the disc. The current is led to the magnet and the disc, the connection being made to the disc by means of slotted rings which encircle its circumference, by a spring sliding over the same, or by other suitable means. A modification of the above motor is described, in which two copper discs are placed between two circularly disposed sets of electro-magnets. The discs have their centres and circum-

ferences in connection with the source of electricity. A centrifugal governor is used to regulate the speed of the motor by breaking the circuit as the weights rise, and a flywheel is also applied. A sliding armature, which can be pushed on and off the magnet poles by means of a lever is used to vary the power of the motor. The poles of the magnets are in some cases shaped "to the configuration of the magnetic curve," *i.e.*, the curve taken by iron filing at a magnetic pole. A reciprocating motor is also described.

3,521.—W. H. Preece and **C. Goldstone**, Southampton; **J. Radcliffe**, Retford; **M. Gray**, London. **Signalling in Railway Trains.** 1s. 4d. (12 figs.) October 13.

CONDUCTORS.—A coupling, for electrical conductors extending throughout a railway train, consists of two counterpart spring clips, to which the ends of the conducting cables (each provided with two insulated wires) are attached. These clips are described in Provisional Specification No. 3381 of 1874.

3,771.—A. M. Clark, London. (*H. Lartigue*, *Paris.*) **Electrical Commutator, &c.** 2s. (55 figs.) October 31.

COMMUTATOR.—This invention relates to an apparatus for inverting or interrupting electric currents, and consists of an air-tight box of non-conducting material, with or without partitions, enclosing some mercury, which becomes the medium for connecting the ends of platinum wires which penetrate its sides, and are externally connected with the circuits it is desired to operate upon. The changes of connection are made by tilting or inverting the box. Many applications of the principle of this invention are illustrated and described.

*3,800.—H. Timmins, London. **Joining Wire for Telegraphs and Fencing.** 4d. November 4.

CONDUCTORS.—Joints in conducting wires are made by means of a socket with a right and left handed internal screw thread, into which socket the wire ends are screwed.

*3,974.—L. Sterne, London. **Telegraph Insulators.** 4d. November 18.

INSULATORS.—Two or more (preferably three) lugs or claws are formed at the top of the insulator, under which lugs or claws the wire engages. In this way the wire is secured without the necessity of using binding wire. The lugs are arranged out of a direct line, so that when the wire is engaged it assumes a cranked form, and is thereby more efficiently held.

4,261.—G. F. James, Salford. **Circular Braiding Machines.** 10d. (10 figs.) December 10.

CONDUCTORS.—This invention relates to improvements in braiding machines, which, amongst other purposes, are employed for covering electrical conductors. The improvements consist of a mode of weighting the spindles of circular braiding machines so as to allow the bobbins to be changed without using "dummies ;" in the application of a spring to each governor whereby the number of working spindles can be varied ; and in the application of a creel with movable arms.

4,435.—C. E. Powell, Binfield, Berks. **Electric Light Lamp.** 1s. 10d. (8 figs.) December 24.

ARC LAMP.—The carbon points are placed *in vacuo,* and an automatic arrangement is provided for closing the circuit if from fracture of the lamp or other cause, the vacuum cease. The lamp consists of an air-tight vessel or globe of glass, suitably supported, and the apparatus for adjusting the carbon points. The carbon-holders are fixed to the lamp stock or body, one being insulated therefrom, but connected to a conical centre piece, and having a spring fastened to it which presses against and forms a connection with the other carbon-holder. The lamp body has a conical hole to receive the centre piece, and is fitted with a cock through which the globe is exhausted. The centre piece is bored to receive a movable plug, capable of working up and down air-tight. This plug rests on the spring, short-circuiting the poles of the lamp. The lamp is worked by exhausting the air in the globe ; the plug is then depressed by the atmospheric pressure and pushes the spring away from the poles and the current passes between the carbon points. The containing vessel may take the form of a para-boloid with plane front or any other convenient shape.

4,454.—H. H. Bonneville, Paris and London. (*T. Chutaux, Paris.*) **Electric Motor.** 10d. (3 figs.) December 28.

MOTOR.—This relates to a combination of motors in connection with a single series of frames, driving shaft, and flywheel. The motor illustrated as an example is a combination of four independent motors. A disc carrying armatures on its circumference is fixed in a shaft and revolves in a vertical plane so that the said armatures pass very close to the poles of four electro-magnets, fixed on the frame of the motor, and directed to the centre of the disc. The armatures, which are placed parallel to the axis of the disc, are made up of several pieces of sheet-iron bent to a triangular section. On one end of the driving shaft is mounted a commutator consisting of a circular insulating core, with a number of contact pieces on its circumference which close or open the circuit on striking contact rollers communicating by springs with two of the four electro-magnet coils. A second commutator of similar construction distributes the current to the second pair of electro-magnet coils. The two other motors act respectively by means of a pawl on ratchet teeth formed on the main disc, already referred to, and through the medium of a crank or eccentric attached to the main shaft.

*4,475.—S. J. Neave, London. **Conducting Cores for Telegraph Cables.** 4d. December 30.

CONDUCTORS.—A spiral conducting wire is employed instead of the straight ones ordinarily used, with a view to gaining elasticity, and to obviate danger from breakage or "knuckling through" of the conductor.

1875.

441.—S. A. Kosloff, St. Petersburg. **Production of Electric Light.** 10d. (5 figs.) February 5.

INCANDESENCE LAMP.—The electric light is produced "by means of an electric current heating to a white heat sticks of carbon placed in the circuit, and being enclosed hermetically in a globe filled with nitrogen gas." A carbon rod has its ends inserted into blocks of carbon which are supported between two insulated arms. The current is conveyed to it by means of wires inserted in holes in these carbon blocks. The insulated arms are secured to a suitable standard, the whole being placed on a stand. In order to avoid the inconvenience of preparing nitrogen gas the air within the containing globe is rarefied, and the small quantity of oxygen remaining is then "transformed into oxide of carbon." Several carbons may be combined in one lamp. A certain amount of end play is given to the carbon rods in their supports to prevent them breaking when they expand. The lower part of the globe is provided with valves which are closed by a water joint.

473.—W. Clark, London. (*D. F. Lontin, Paris.*) **Application of Electro-Dynamic Machines for Obtaining Metals, &c.** 4d. February 8.

DYNAMO-ELECTRIC GENERATOR.—This invention

relates to the use of currents derived from dynamo-electric generators for the purpose of decomposing metallic salts. The inventor employs a generator "wherein the whole of the current produced is returned to the primary electro-magnets," in which circuit are placed the "voltameters" or vessels containing the substances to be operated upon. Another application of this invention is the employment in a similar way of a dynamo-electric generator for exciting an alternate current machine for the purpose of producing the electric light.

478.—W. S. Laycock, Sheffield. Railway Brakes. 10d. (4 figs.) February 9.

MAGNETO - EECTRIC GENERATORS. — These are worked from the carriage axles of railway trains, and the electric current produced is employed to actuate an electro-magnetic brake.

792.—W. Lloyd Wise, London. (*T. Masin, Brussels.*) Railway Brakes. 1s. 10d. (8 figs.) March 3.

CONDUCTORS.—A coupling for double conductors is included in this invention, and consists of two metal cylinders, placed on the ends of the conductor but insulated therefrom, one fitting into the other with a telescope joint. The conducting wires communicate with contact springs attached to insulating plugs within the cylinders, such contact springs establishing the desired connection when the metal cylinders are brought together.

*806.—M. A. Wier, London. Igniting Gas. 4d. March 4.

MAGNETO-ELECTRIC GENERATOR.—The current from a magneto-electric generator is used to heat a platinum wire to incandescence, for the purpose indicated in the title.

970.—P. Jensen, London. (*S. V. Konn, St. Petersburg.*) Electric Light Apparatus. 8d. (2 figs.) March 16.

INCANDESCENCE LAMPS.—A glass cylinder with closed top is fixed air-tight to a metallic base, from which pass upwards two copper conductors, one being in connection with the metallic base, and the other passing through, but insulated from it. The latter is joined, in a hollow chamber under the base, to an insulated terminal. The conductors terminate in the upper part of the glass cylinder in holders, between which several stems of graphite are inserted. The upper holder is a perforated plate through which a number of platinum or silver studs, of different lengths, pass, but are insulated therefrom. A hinged plate conveys the current to the highest stud. The lower part of each stud has a socket for the reception of the carbon rod. If any rod should break in use, then the stud which

bears on it drops, and the hinged plate falls on to the next highest stud, and the current is thus transferred to the next carbon. A metal guard is fixed inside the lower part of the glass cylinder to prevent breakage from heated pieces of carbon, and a cock in connection with the interior of the glass cylinder serves for pumping out the air.

*1,487.—R. H. Courtenay, London. Electro-Magnetic Motor. 4d. April 23.

MOTOR.—This invention, which has for its object "the increase or multiplication of magnetic power, by which means electricity is the initial power or prime mover of other powers, such as air and water," consists in improvements in the apparatus described in Provisional Specification No. 2179 of 1874, in which, however, no reference is made to motors, nor does the present specification mention them, except in its title. The invention states: "Having formed a piece of iron into a U shape, I place across the two limbs, and at any convenient distance apart, one or more induction coils, each one being in itself an electro-magnet, and so arranged that similar poles are placed on the same side, the ends of the coils being connected with springs attached to the conductors by insulated binding screws, so as to allow all or any of the coils to be employed at discretion. Each compound magnet, arranged as described, is worked with a quantity arrangement of electricity, one element of the battery to one coil preventing as far as possible counter currents."

1,719.—F. W. Ewen, Manchester, and G. F. James, Salford. Insulating Telegraph Wires, &c. 10d. (5 figs.) May 8.

CONDUCTORS.—Ground glass, fine sand, or other non-conducting powder, is employed for insulating electrical conducting wires. The wire is carried through a hopper containing the ground glass or other powder, and a convenient number of yarns are guided through holes in the top or bottom of the hopper to carry forward the powder, which is surrounded by yarns or wires passing through guides below the hopper. The whole is then braided and passed through hot tar. The machinery for conducting these operations is illustrated and described.

1,800. — J. Faulkner, Manchester. Electrical Appliances. 10d. (11 figs.) May 14.

ELECTRO-MAGNETS.—This invention consists in using an iron tube to cover the outside of the coil of an electro-magnet, the iron core being secured or not to an end base or bottom. This system of making electro-magnets is termed by the inventor, "Faulkner's Altandi Systemæ." A second coil

may be added, and a second tubular iron cover, and so on to any desired extent.

1,938.—F. Field, London, and **R. Talling,** Lostwithiel, Cornwall. **Insulating Compounds.** (4d.) May 27.

CONDUCTORS.—In Patent No. 3,778 of 1869 a process is described for producing an insulating material by combining ozokerit with gutta-percha, india-rubber, or other like substance. The inventor states that the compounds so prepared become brittle, and in order to remedy this defect he combines the ingredients by dissolving them in a suitable solvent, such as coal tar naphtha, or by subjecting them to a process of mastication. The compound thus produced remains perfectly pliable, and may be used as a covering for electrical conducting wires. Fibrous materials such as mica, asbestos, and the like may be incorporated with this compound, which may or may not be vulcanised.

2,043.—A. M. Clark, London. (*C. A. Hussey, New York.*) **Electro-Magnetic Engines.** 10d. (5 figs.) June 3.

MOTOR.—"Induction currents of the magnets and sparks at the commutator are entirely avoided" in this motor. A circular frame carries two stationary magnets with radial arms and ⌈ shaped pole-pieces, which are placed by a commutator on the revolving shaft in connection with the driving current, so that the currents in the magnets are alternately reversed and act "with continuous force" on a central revolving magnet, shaped similarly to the fixed magnets, but with ⊤ shaped pole-pieces. "The commutator revolves with the central magnet, and is made of positive and negative sections with separating insulating layer or bands for reversing, by the alternating contact with the adjustable current transmitting contact wheels, the polarity of the stationary magnets. The formation of induced currents is prevented by the lapping over of the ⊤ shaped pole ends across the space between two adjoining pole ends of the outer magnets," and thus supposing induction currents should be formed, "they would serve by reversing the current in the side poles to assist the magnetic force." The non-formation of induced currents avoids sparking.

2,049.—L. Nelson and **I. E. Anderson,** New York. (*H. M. Paine and E. L. Paine, New Jersey, U.S.A.*) **Electro-Magnetic Motors, &c.** 10d. (3 figs.) June 3.

MOTOR.—Electro-magnets for motors have their cores made of a number of plates combined instead of being in one solid piece. Two such magnets are

set on one shaft "under such conditions that the poles of one limb shall be the opposite poles of the corresponding end of the other limb." The wires are wound around the limbs or cores to the right and left respectively in separate coils, "the non-insulated end of each wire being secured to its limb, while the two opposite ends of the two coils connect in one common union with an insulated conductor which passes through and out of the shaft." With this arrangement of parts the current will simultaneously pass around each limb, and thus secure a "synchronous polarisation of the limbs.' The armature in "rotatory electro-motors" is grooved longitudinally in order to "increase the duty of the magnet" by taking advantage of the "pulsative action" at the moment the edge of the magnet passes the edge of the armature. This pulsative action is said to greatly exceed the "general pull." The speed of motors, constructed in accordance with this invention, is regulated by the insertion of resistances into the circuit, a rheostat being used for the purpose.

2,059.—F. Greening, Plaistow. **Compounds of Soluble Gun-Cotton.** 4d. June 4.

CONDUCTORS.—An insulating material for electrical conductors is made by submitting mixtures of paraffine, shellac, resin, or gum, and also the residues left after distilling the heavier tar oils or mixtures of these substances with soluble guncotton, to the action of creosote obtained from wood tar. When these materials are incorporated they assume a "uniform plastic condition" suitable for application to conducting wires.

2,205.—C. Clamond, Paris. **Electro-Motor Machines.** 4d. June 15.

MOTOR.—Two bundles of electro-magnets, formed of rods of soft iron surrounded by coils of wire, are united electrically by a commutator. One of them is fixed, and remains always magnetised in the same manner. The other rotates, and as it does so "the direction of its magnetisation changes alternately from one electro-magnet to the other." In this way successive attractions and repulsions occur between the poles of the electro-magnets of the two bundles, which result in rotary motion of one. The commutator is a cylinder of wood, provided with grooves in which six copper blades are lodged. These blades "are grouped in two series of three branches," which communicate with each other by means of two copper rings, "so that as soon as the current arrives in one of the branches it is at the same time transmitted to the two others of the same series." The circuit is closed by means of adjustable contact combs, which are fixed to copper plates, but insulated therefrom. With these plates are connected the copper blades by

which the currents (preferably from a thermopile of the inventor's) arrive. To the same blades are soldered bands which pass round the fixed bundle to the coils, from which they distribute the current. In all the coils of the two bundles the wire is wound in the same direction, so that a change of direction of the current suffices to reverse the polarity of the magnet. This is effected by the commutator. The wires of the bobbins are united by a hexagon, the sides of which are connected three with one band and three with the other band. By this arrangement the current is sent in opposite directions in two consecutive bobbins, "thus imparting magnetisation of contrary name to the electro-magnets to oppose the variable poles of the movable bundle."

2,410.—F. M. A. Chauvin, L. H. Goizet, and A. Aubrey, Paris. Electric Submerged Lamp. 8d. (6 figs.) July 3.

INCANDESCENCE LAMP.—The body of the lamp is a glass bulb A, the neck of which is hermetically closed by an india-rubber plug B. Two metal conductors C C separated and insulated from each other, pass through the plug B, and at their ends within the globe are joined by means of the helical platinum wire D. In place of this platinum wire may be substituted two carbon rods, "carried by rods forming springs to bring them into contact in

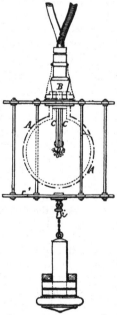

proportion as they are worn away by the combustion" It is stated that in consequence of the strong heat, the air contained within the bulb will be dilated and expelled, producing "a vacuum very favourable to the intensity and radiation of the electric light." The lamp is designed for use under water.

*2,564.—T. A. Hequet, Paris. Electro-Magnets. 4d. July 17.

ELECTRO-MAGNETS.—The object of this invention is to annihilate residual magnetism, which is done by "separating magnetically the two bobbins of the electro-magnet from each other, either by dividing the iron yoke, which joins them, or by introducing non-magnetic material between the bobbins and the said yoke at their points of junction therewith."

2,633.—F. R. Lucas, London. Submarine Cables. 8d. (2 figs.) July 24.

CONDUCTORS.—This relates to the manufacture of deep sea cables, so as to give the greatest possible strength, combined with compactness and low specific gravity. A core, after being served with jute or other yarn, is sheathed with Manilla hemp yarns and wires laid side by side helically around the core. The cable may receive an outside serving of yarn and compound in the usual way.

*2,767.—E. G. Brewer, London. (*S. A. Kosloff, St. Petersburg.*) Electric Light. 4d. August 5.

INCANDESCENCE LAMP.—This provisional specification describes substantially the same invention as that for which a patent was granted to S. A. Kosloff on July 5 of this year, numbered 441.

2,787.—J. H. Johnson, London. (*F. E. de Mersanne, Paris.*) Electric Signals. 2s. 10d. (38 figs.) August 7.

ELECTRIC LIGHT.—This invention relates to the use of electric lights for signalling purposes and consists of apparatus for producing several signal lights simultaneously in different places, by means of a single current derived from a single generator. For this purpose the lamps are constructed to produce a succession of sparks between two carbon or other electrodes, the principle of their construction consisting in making one or both of the electrodes vibrate into or out of contact. The current is distributed to the lamps by apparatus, which is illustrated in detail and very fully described, in such a way that no two lamps ever vibrate together, and so as to produce in each of them sparks which never appear simultaneously, and thus the intensity of the sparks produced is in proportion to "the integral energy of the current of the generator." This specification runs to 23 pages.

*2,946.—C. A. Faure, London. Thermo-Electric Batteries and Electro-Motors. 4d. August 21.

MOTOR.—A thermo-electric battery is constructed around the core of the electro-magnets of electro-motors, that is to say, the coils of the electro-magnets are divided at intervals, and a thermo-electric element is inserted in each division. The helix on each branch of the magnet will thus be continuous, and if the four ends of the two coils be properly connected, the magnet will exercise strong attraction on any iron armature capable of revolving in front of its free poles. By automatically interrupting the circuit, the attraction of the magnet will cease, and the armature will be free to continue its motion under the momentum it has acquired. Another plan is to construct a magnet capable of revolving on a shaft having its axis coincident with that of the magnet. The magnet is formed of steel wires arranged around one or more drums, fixed to the shaft, so as to form a hollow cylinder. Round each pole of this magnet is a copper cap. The current from a thermo-electric battery is then made to "circulate radially" in these caps, when the hollow magnet will be caused to rotate. The caps may be built up of a series of cylinders, grooved and tongued, with the interposition of thermo-electric elements.

*2,996. — W. J. Kilner, London. Magneto-Electric Apparatus. 8d. (5 figs.) August 26.

MAGNETO - ELECTRIC GENERATOR.—The object of this invention is to obtain a continuous current

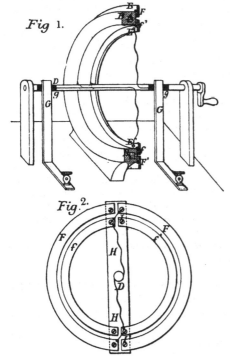

Fig 1.

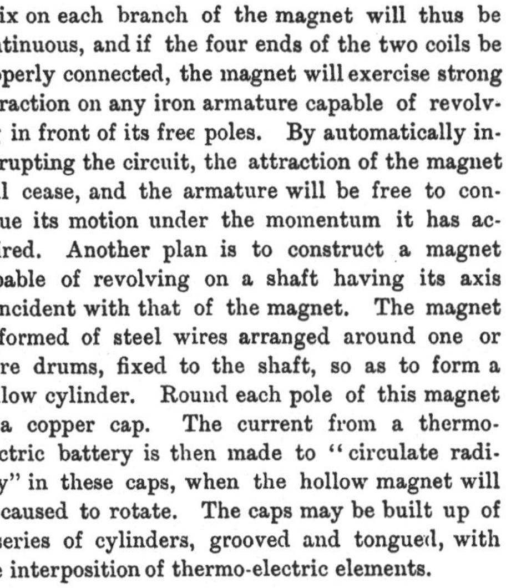

Fig. 2.

of electricity by means of a magnet rotating within

a soft iron ring, on which are placed coils of insulated copper wire. Fig. 1 represents a section of a generator, constructed on this principle. A is the soft iron ring surmounted by the coils of insulated wire B, the ends of which terminate in knots or brushes E E', pressing on the copper segments F F, ff^1. Fig. 2 is another view of these segments showing their connections, and mode of arrangement. They are secured to the axle D by the cross-piece H H, and revolve with it. Fixed on the same axis is a bar magnet (not shown in the illustrations), with its axis placed in a line with the junctions of the copper segments I I', and its ends reaching almost to the inner circumference of the coils B. The segments F F', ff^1 are connected in the manner shown to the copper rings g g^1, against which the terminal springs D D, G G press. The action of the generator is explained as follows: When the N. pole of the magnet approaches any given point in the soft iron ring, this point becomes magnetised with S. polarity. The intensity of this pole becomes greater as the rotating magnet approaches, so that a current of electricity is induced in the coil placed round this point in the iron ring; but when the N. pole is receding, this point loses its magnetism and begins to acquire opposite polarity on the approach of the S. pole of the rotating magnet. The induced current is thus caused to change its direction, but from the arrangement of the collecting segments F F', ff it is evident that each current reaches the terminals in the same direction. More than one rotating magnet may be used, in which case the number of collecting segments must be increased in proportion. A commutator of complicated construction for varying the connections of the coils, and thereby the electromotive force of the generator, is illustrated and described.

3,085.—T. H. Richardson, Leeds, and A. Moffatt, Mirfield, Yorks. Signalling in Railway Trains. 1s. 6d. (13 figs.) September 2.

CONDUCTORS.—A cable for use on railway trains consists of three copper wires "imbedded by preference in india-rubber of good quality." Connections between carriage and carriage are made by short lengths of cable having helical wire conductors enclosed in vulcanised india-rubber tubing, the wires terminating in metal eyes, which take into suitable hooks on the carriages, to which hook the main cables are connected.

3,243.—E. P. Alexander, London. (*E. Bürgin Paris.*) Magneto-Electric Machines, &c. 1s. (19 figs.) September 16.

DYNAMO-ELECTRIC GENERATORS.—The improvements embodied in this invention relate particu-

larly to the armatures of dynamo-electric generators. Four distinct methods of construction are described and illustrated, but in principle they are similar. By one method the armature is built up of a number of elements, each consisting of an electro-magnet of the form represented in Fig. 1,

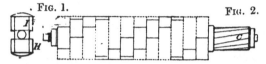

. FIG. 1.   FIG. 2.

having two insulated coils H I, and a perforation in the centre of its length. These elements are threaded upon a shaft, and fixed as shown in Fig. 2, which represents a complete armature. The coils are all joined up in one series, and the points of juncture of those of each element are connected respectively to the several divisions of the commutator C, the sections of which correspond with the number of elements. The lines of division of the commutator are disposed obliquely "in order to obtain a continual contact of the segments with the rubbers." The number of commutator sections may also be double that of the elements, in which case the two coils on the latter are treated separately, each junction having a corresponding commutator plate. By another method of construction the elements are + shaped, and each provided with four coils, one on each arm. These armatures are caused to rotate between the poles of one or more electro-magnets suitably shaped so as to embrace the armature as closely as possible without being in actual contact therewith. To obtain currents of great quantity, the wires of the armature coils "may be replaced by insulated sheet copper ribbons."

MOTORS.—Generators constructed as described above, form efficient electro-motors.

3,364.—W. R. Lake, London. (*J. B. Fuller, Brooklyn, U.S.A., and J. N. Crandall, Norwich, Connecticut, U.S.A.*) **Magneto-Electric Machines.** 8d. (8 figs.) September 25.

DYNAMO-ELECTRIC GENERATOR.—The armature consists of several radial iron arms, wound with coils of insulated wire, arranged at equal distances apart around a horizontal shaft or cylinder. The ends of the coils are connected to form one complete circuit with all the coils in series. The junctions between the wires of the adjacent coils are also connected to a sectional commutator, fixed to the shaft, against which on opposite sides two metal springs press. There are two field magnets placed vertically, one on either side. The pole-pieces of these magnets are curved and tapered, and they together embrace the greater portion of the periphery of the armature. The invention states that

by tapering the magnetic poles "the abrupt falling off of the magnetic force is prevented, and the resulting current rendered more uniform." There may be any number of field magnets, and there should be as many commutator springs for each commutator as there are magnetic poles. The magnets may be excited by a battery, magneto-electric generator, or a suitable number of coils from the generator itself. The arms of the commutator may each have two coils, the currents from one of which would be used to excite the field magnets.

3,416.—J. H. Johnson, London. (*J. S. Camacho, Paris.*) **Electro-Magnetic Engines.** 10d. (8 figs.) October 1.

MOTOR.—A number of armatures are arranged upon the surface of a cylinder, and come in turn under the attractive influence of a series of electro-magnets fixed to a frame, and placed radially with respect to the axis of the cylinder. The armatures are composed of a number of iron plates " arranged normally at the acting surface of the electro-magnets, and magnetically insulated from one another by the interposition of any suitable substance," such as paper, resin, and the like, "which form bad conductors of magnetism. In this manner the magnetisation is produced successively in each of the plates of which the armature is composed, and which in consequence of their small volume are thus brought to a condition of maximum magnetisation as soon as they come within the range of magnetism of one of the electro-magnets, and the travel of the armature is completed in the condition of maximum magnetism," whence it is said "a greater production of force, with the same expenditure of electricity, is accomplished." In the commutator the piece of metal serving as a conductor, instead of being made of rectangular shape, is constructed in the form of a " trapezium or triangle, so that the rubber which acts against it continues in contact therewith during only a fraction of a revolution, varying according to the position in which the contact maker is placed." By adjusting the position of the commutator upon its axis of rotation, the length of time the rubbing surfaces are in contact is regulated, and thereby the speed of the motor. The electro-magnets used in this motor are by preference made of tubular section.

3,466.—W. Prosser, London. **Electric Lamps.** 10d. (14 figs.) October 6.

ARC LAMPS.—This invention relates to electric lamps in which the light is produced either by the combination of a fluid with a solid electrode, or by the meeting of two or more carbons at a point. In

the former type of lamp the inventor claims the use of a "malleable glass" globe to enclose the light, a double glass, having an intervening space to prevent the access of air to the glass immediately enclosing the light, and a plurality of fluid jets as a positive or negative electrode. In the second type of lamp two negative carbon electrodes, the one opposite to or placed at an angle with the other, are kept in contact by a spring or weight, their point of junction being immediately beneath the positive electrode, which is also of carbon. The negative electrode may in some cases be a disc kept slowly rotating. In another form of this lamp two or more carbon electrodes radially abutting against each other may be employed for each pole.

*3,666.—J. F. Lackersteen, London. Telegraph Cables. 4d. October 21.

Conductors.—A copper conducting wire is surrounded by a casing of iron, steel, or other suitable metal by means of a draw plate. The compound wire thus produced may be reduced to any desired gauge, the copper core and iron casing still maintaining the original relative proportion between their sectional areas. It is preferred to insulate such wires with a covering of vulcanised india-rubber, cured after application to the wire. Other materials, however, may be employed, and the whole may be sheathed with a covering of woven or twisted wire. Several conductors prepared in this way may be combined into one cable, which may then receive an additional protective sheathing.

3,762.—T. A. Edison, Newark, New Jersey, U.S.A. Autographic Printing. 1s. 2d. (25 figs.) October 29.

Motor.—This invention relates in the main to an electric pen, or puncturing apparatus, which is actuated preferably by a small rotary electro-motor. This consists of a horseshoe electro-magnet attached to a frame that supports the axis of a flywheel. Across one side of this flywheel is fixed an armature, which as the wheel revolves passes very close to the poles of the electro-magnet. The commutator is a spring acted upon by a notched and flattened disc, so as to make and break the circuit through a contact screw.

*3,795.—G. Zanni, London. Telegraph Signalling and Printing Apparatus. 4d. November 1.

Magneto-Electric Generator.—This invention relates to the application of magneto-electric apparatus to sending telegraphic signals, but no details of such apparatus are given.

*3,798.—S. E. Phillips and W. C. Johnson, Charlton, Kent. Telegraph Cables. 4d. November 2.

Conductors.—Electrical cables are protected by a serving of "Hessian," or strips of canvas, without selvage, applied either internally for the protection of the core, or externally over the sheathing wires, and laid round the cable preferably without overlapping the last turn of the same strip of serving, but making a flush joint with it. For core serving the canvas is applied either unprepared or tanned, but for external use it is saturated with a preservative compound of tar, pitch, and silica. If two or more successive servings of canvas are applied, they are wound right and left handed alternately.

3,999.—A. Browne, London. (*D. F. Kimball, New York.*) Electro-Magnetic Engines and Galvanic Batteries. 8d. (7 figs.) November 18.

Motors.—A shaft is provided with a suitable number of T shaped armatures, the cross-pieces of which are parallel with the axis of the shaft and at equal distances from it. Two electro-magnets are arranged at opposite ends of the base of the frame that supports the rotating shaft, and have T shaped polar extensions corresponding with the armatures. A commutator of the usual construction is provided, by which the circuit through the magnets is intermittently completed, so as to cause them to alternately attract and release the armatures. The inventor considers it an important feature in this motor that the several armatures are entirely isolated from each other except through the common hub, because the residual magnetism in the electro-magnet cores is less when they are deprived of any body to act upon, and the so-called "back pull" is therefore obviated.

4,115.—W. T. Henley, Plaistow, Essex. Submarine Telegraphic Cables, &c. 1s. 2d. (20 figs.) November 26.

Conductors.—The object of this invention is to combine lightness with strength in submarine cables, and at the same time to produce them cheaply. A number of strong cords or strands of Manilla hemp are first laid round the core with a long twist, and around these strands is applied with a short twist a tape or webbing formed of jute, and preferably saturated with marine glue. Steel wires are then laid over this, in the same direction and with the same length of lay as the Manilla strands, so that the wires lie in the interstices formed by such strands, the intervening webbing being pressed in by the wires, but preventing the latter from forcing their way between the cords. As many wires as

cords may be employed, but for a light cable, half the number will serve. After the wires are laid on, the whole is served with a coat of compound webbing and afterwards passed through hot marine glue. Another part of this invention consists in curing india-rubber covered conductors in ozokerit, paraffine or other similar hydro-carbon, instead of in high-pressure steam. The ozokerit or other material is contained in a closed steam-jacketted cylinder, into which the core is introduced. The core may also be cured in the usual way, and when completely dried it may be placed in a vessel of heated ozokerit under pressure.

4,118.—E. L. Paine, Newark, New Jersey, U.S.A., and L. Nelson, New York. (*H. M. Paine, Newark, New Jersey, U.S.A.*) Electro-Magnetic Motors. 1s. 4d. (9 figs.) November 26.

Motors.—This invention relates to motors similar in principle to that described in Specification No. 2,049 of 1875, and consists in improved means for making and breaking contact, in such motors, between one pole of the battery, and the rotating commutator. Around the shaft carrying the revolving electro-magnets, which shaft is made hollow and of non-conducting material, a number of metal rings are arranged. The number of rings corresponds to the number of electro-magnets, and each ring is connected by a wire passing through the hollow shaft to the coil of its corresponding magnet. Around the peripheries of the rings a number of crossbars are arranged parallel to the axis of the shaft. The plates or bars of each ring are equidistant around the circumference, so that between every two plates of one ring the plates of another ring or rings can intervene. The rings and their crossbars are all insulated from each other. To obviate sparking at the commutator, the inventor interposes between the terminal of the motor, and the commutator just described, a bridge-shaped metal shoe, having two bearing parts or edges the same distance apart as the crossbars of each ring, so that when in position, this shoe will bear against two plates at a time, in such a way that the connection between the plates of the ring connected with one magnet is gradually cut off whilst the connection with the plates of the ring of the succeeding magnet is being gradually made.

4,206.—W. R. Lake, London. (*C. L. Van Tenac, Paris.*) Electrical Apparatus for Producing Fire and Light. 2s. (72 figs.) December 4.

Incandescence Lamp.—This is a platinum spiral heated by the current derived from a suitable battery, the object being to provide a portable apparatus for obtaining light.

4,311.—A. M. Clark, London. (*E. Bertin, Paris.*) Dynamo and Magneto-Electric Machines. 6d. (3 figs.) December 11.

Dynamo and Magneto-Electric Generators.—To obviate "certain disadvantages" in the usual forms of commutators, as applied to electrical generators, the inventor places on the shaft of the generator as many collectors, or pairs of contact rings as there are induction coils, by which means a portion only of the current may be used, if desired, and all the coils of the generator may be coupled either for tension or quantity at will. For example, a machine of twenty-eight coils may have seven commutators or pairs of contact rings, the first of which receives the current from a single coil, the second from two coils, the third from three, and so on. In this way any number of combinations may be effected. Both commutators and contact rings may be used in the same machine to produce alternate or continuous currents, as may be desired. The inventor enunciates the shunt principle of connecting up generators as follows: "I take a dynamo-electric machine of any kind having a continuous current and completely close the circuit upon itself, but instead of interposing the work to be done in a break made in the circuit, I interpose it in a derived current taken from the poles of the machine, which furnish the currents to the electro magnets. In this arrangement, which I term a lateral derived current, if the work to be performed is constant as in electro-metallurgy and other operations, the result is very much improved."

4,384.—W. Smith, London. Joining Telegraph Wires. 8d. (12 figs.) December 17.

Conductors.—Joints in insulated conductors are made as follows: A clamp consisting of two plates of vulcanite or other suitable material is provided, each plate being grooved semicircularly from end to end, so that when the two plates are held face to face there is a passage between them somewhat smaller in diameter than the covered wire to be joined. A fine screw thread is cut in the passage at the two ends and the central portion is enlarged to allow sufficient insulating material to be placed round the joint. The conducting wire is joined in the usual way, and the insulating material is applied over the joint. The clamp is then put on and forcibly squeezed together by means of a portable vice.

1876.

367.—F. L. B. Bedwell, London. **Connecting Submarine Telegraph Cables with Lightships, &c.** 6d. (5 figs.) January 29.

CONDUCTORS.—This invention relates to a swivel for the purpose indicated in the title. The swivel has a hollow spindle through which the cable is passed.

386.—A. M. Clark, London. (*D. F. Lontin, Paris.*) **Dynamo-Electric Machines.** 6d. (6 figs.) January 31.

DYNAMO-ELECTRIC GENERATORS.—In constructing dynamo-electric generators : 1st. The field magnets are made to revolve, whilst the armature coils remain stationary. This dispenses with a commutator, and allows the currents from the armature coils to be dealt with separately, or to be coupled up as desired. 2nd. The field magnets are considerably lengthened so as to allow of one or more conducting wires being placed on them from the ends of which currents may be collected. In this way both continuous and alternate currents may be generated in one machine.

557.—E. Tyer, London. **Signalling in Railway Trains.** 6d. (9 figs.) February 11.

CONDUCTORS.—In coupling electrically the ends of conductors used for signalling in railway trains, the inventor employs wire brushes in the following way : The ends which are to be coupled are provided, one with a cylindrical plug, and the other with a tube, and flexible wire brushes are fixed, either upon the outer surface of the former or on the interior of the latter. The object is to obtain a rubbing contact. Another form of this coupling consists of a jointed lever arm on the end of each carriage carrying at its free end a metallic brush. These brushes are brought together when making up a train.

*610.—G. Weathers, London. **Rotatory Electric Light.** 2d.

ARC LAMP.—The chief difficulty attending the use of arc lamps arises in the regulation, which difficulty the inventor overcomes " by causing a central ball, to which the negative current of the battery is conveyed, to be surrounded on all sides by brushes or balls to which the positive current of the battery is transmitted. The opposite electricities are further developed in intensity, and made to act upon one another at a greater distance by the aid of a mechanical contrivance, whereby the positive and negative balls are caused to rotate upon their axes in opposite directions."

*802.—W. J. Kilner, London. **Producing Continuous Currents of Electricity.** 6d. (5 figs.) February 26.

DYNAMO OR MAGNETO-ELECTRIC GENERATORS.—This specification describes an invention substantially similar to that set forth in Provisional Specification No. 2,996 of 1875.

836.—P. Jablochkoff, Paris. **Electro-Magnets.** 6d. (6 figs.) February 29.

ELECTRO-MAGNETS.—The coils of electro-magnets have their conductors brought into metallic contact with the soft iron core. These coils are preferably made of a thin metallic ribbon wound upon the core in such a manner as to have one of its edges in contact with the core, whilst the flat surfaces of successive coils are separated from each other by a thickness of non-conducting material. The core may take the form of an iron disc, on the surface of which a spiral conductor is fixed.

1,412.—E. Maden, Portland. **Machinery for Compressing Air.** 6d. (4 figs.) April 1.

MAGNETO-ELECTRIC GENERATOR.—In connection with air compressing machinery, and for exciting certain electro-magnets which form part of it, the inventor prefers to use " Wilde's or the Gramme Company's " electric generators.

*1,557.—W. R. Lake, London. (*J. B. Fuller, New York, and J. N. Crandall, Norwich, Connecticut, U.S.A.*) **Magneto-Electric Machines.** 2d. April 12.

DYNAMO-ELECTRIC GENERATORS.—This invention relates to generators which have two sets of electro-magnets and two revolving armatures, and which have a connecting plate so placed as to become the " neutral magnetic point in both sets." This plate should be a heavy iron casting, and may be in one piece with the electro-magnet cores. In order that both of these magnets shall produce in the connecting plate " the same polar condition " the magnet coils should be wound " so that some of the poles are similar while the other poles are dissimilar. The polar direction from the south poles will be alike through the bridge towards the north poles. The relative strength of the two magnets will depend upon the location of the bridge and the relative length of each set of cores. Each magnet will act independently of the other, and possess all the characteristics at the poles of separate magnets." Between the magnet poles armatures " of any ordinary construction " may be caused to revolve, the currents of one amature being used for exciting the electro-magnets, while those from the other may be employed for any useful purpose. It is difficult to gather a clear idea of the invention from the description given in this specification.

1,569.—H. R. Newton, London. **Water Supply and Flushing Apparatus.** 6d. (8 figs.) April 13.

CONDUCTORS.—This invention in the main relates to a system of drainage, comprising the construction of a subway beneath the road, and it is in this subway that the inventor proposes to place electrical conducting wires for telegraphic or other purposes.

1,704.—B. Fixsen, St. Petersburg and London. (*L. Danckwerth, St. Petersburg.*) **India-Rubber and Gutta-Percha Compounds.** 4d. April 22.

CONDUCTORS.—New insulating compounds are formed "by the combination or amalgamation of ozocerite, or ozokerite (also known as fossil resin, or fossil wax, or earth wax)" with india-rubber, gutta-percha, or a mixture of india-rubber and gutta-percha, or with other compounds. The proportions in which the ingredients are combined, vary with the uses to which the compound is to be put. "Very useful results" are obtained by kneading together 100 parts of gutta percha and 50 parts of ozocerite. Ozocerite combined with india-rubber produces an easily vulcanised compound, which retains its pliability, and is unaffected by atmospheric influences. It is also proof against acids, and whilst it is not liable to oxidation, "its insulating properties for electric purposes are unimpaired," and its specific gravity is nearly the same as ordinary caoutchouc. The gutta-percha compound has an insulating capacity greater than that of ordinary gutta-percha.

*1,900.—I. L. Pulvermacher, London. **Generating and Applying Electricity.** 2d. May 5.

RHEOSTAT.—"One part of this invention consists in making a resistance coil for measuring and testing the strength of an electric current from bands made of thin German silver wire," spun by a process used by the inventor for making bands of a different kind.

CURRENT METER.—This is a "voltameter or electrolytic current measurer, combined with a variable resistance changer and measurer, and graduator of the strengths of the currents."

1,931.—W. Clark, London. (*L. Bastet, New York, U.S.A.*) **Electro-Magnetic Engines.** 6d. (5 figs.) May 8.

ELECTRO-MAGNETIC MOTOR.—A horizontal rotary shaft or hub is provided with a suitable number of T shaped armatures, consisting of radial arms terminating in rectangular crossheads, each armature being separated from and having no metallic connection with any other armature except what may exist through a common hub. With this arrangement of armatures, the inventor states that he is able to produce a motor free from any "back pull" and therefore very powerful. Four horse-shoe magnets are fixed with their cores vertical, and their poles, past which the revolving armatures move, forming the four corners of a square. The cores of the electro-magnets have T shaped pole-pieces. A cylindrical commutator of ordinary construction is employed to distribute the current to the electro-magnets at such time as any one armature is in a position to be attracted. The inventor considers it a great advantage that the T shaped armatures are "magnetically isolated from each other" except through their common hub.

1,944.—W. T. Henley, Plaistow. **Electric Telegraph Conductors.** 8d. (27 figs.) May 9.

CONDUCTORS.—The main object of this invention is to construct cheap conductors and to more efficiently protect them from injury. When more than one conductor is used, the wire or strand is covered with one or more coats of india-rubber laid on helically, over which is applied a layer of felt. The covered wire is then passed through ozokerit, or other similar material, and the required number are stranded together. The rope of wires thus produced is then coated with vulcanisable india-rubber, which is cured in accordance with Patent No. 4,115 of 1875. Another mode of insulating conductors is to cover the wire with cotton and then saturate it with an insulating compound. To remove all air and moisture from the cotton, the wire is enclosed in an air-tight vessel, surrounded by a steam jacket, and when heated the air is exhausted from the vessel, and the insulating compound allowed to pass in. The wire is afterwards taken out and passed through a die or rollers to make it round and smooth. The compound may be a mixture of gutta-percha, india-rubber and ozokerit, or paraffine, india-rubber, linseed oil, and resin, or india-rubber and cowrie gum. The core described in Specification No. 1,415 of 1875, may be passed through hot ozokerit. Conductors for subterranean use are enclosed in wrought iron or steel troughs of various sections, which are illustrated in the drawings. Various devices are employed for fastening together the top and bottom halves of these troughs, bayonet joints, studs and key-hole perforations, lappings of wire, and wrought-iron clips and wedges. Wood troughing with wrought-iron lids may also be used to protect conductors. In submarine cables, to protect the core from the ravages of boring animals, the inventor proposes to lay small wires in the interstices between the larger wires of the sheathing, so as to effectually close any opening by which such animals could intrude themselves. This patent also includes a collapsible iron pipe

which is closed up by suitable tongs as soon as the conductor has been inserted.

2,107.—**E. P. Alexander,** London. (*A. Lemaire-Douchy, Paris.*) **Apparatus for Testing the Air of Mines.** 6d. (1 fig.) May 18.

MAGNETO-ELECTRIC GENERATORS AND MOTORS.—In connection with apparatus for testing the air of mines, the inventor proposes to work certain cocks "by the action of electro-motors or magneto-electric machines of the Gramme or other type."

2,725.—**P. Protheroe,** Surbiton. **Establishing Telegraphic Communication with Light-ships.** 6d. (6 figs.) July 3.

CONDUCTORS.—This invention relates to a method of establishing electrical communication with light-ships, and consists in securing the cable in such a way that the motion of the ship will not twist or drag it. This is done by bringing the cable to an immovable mooring, sunk as described in Provisional Specification No. 2,468 of 1876. To this mooring a tube is attached by a universal joint, and through this tube the cable is led to the upper part of the buoy, where there is another universal joint, to which is attached an arm, connected by a mooring cable to the lightship. Within a water-tight casing in the buoy are "insulated contact surfaces so arranged as to allow for the horizontal rotary movement of the mooring cable relatively to the buoy and also for its vertical oscillation," one surface being in connection with the electrical conductor and the other with an insulated wire led along the mooring cable to the ship.

*2,750.—**F. Faucher,** Paris. **Electric Brake for Horses.** 2d. July 5.

MAGNETO-ELECTRIC GENERATOR. — To control runaway horses the inventor employs the currents derived from a "Clarke's magneto-electric machine," and applies the same to the horses' bodies by means of suitable wires attached to the harness.

2,759.—**J. Rubery,** London. **Wire for Electric Telegraphs, &c.** 2d. July 6.

CONDUCTORS. — A copper wire of high conductivity is inserted in the channel of an iron or steel wire of U section, and the two are passed through a draw-plate, by which process the iron or steel wire is closed upon, and completely surrounds the copper core. Connecting wires made in this way are light, strong, and of low resistance.

2,821.—**G. Zanni,** London. **Magneto-Electric and Electro-Magnetic Apparatus.** 6d. (4 figs.) July 11.

MAGNETO-ELECTRIC GENERATORS.—This invention relates mainly to a cylindrical armature for

magneto-electric generators. The figure shows a cross section of this armature, which consists of a number

of iron cores of **T** section grouped around a central shaft, each core being wound with insulated wire in a direction parallel with the axis of the armature. An ordinary cylindrical commutator has its conducting strips connected to the armature coils, the opposite coils being coupled together. A modification of this construction consists in making the peripheral portion of the armature an iron cylinder or tube. The armature may be arranged in combination with a permanent magnet or magnets as described in Specification No. 3,245 of 1870. In some cases the magnetism of the permanent magnets is augmented by surrounding them with a copper wire in connection with the commutator springs.

DYNAMO-ELECTRIC GENERATOR.—The above described armature may be used in combination with electro-magnets for generating electric currents.

2,866.—**A. Deiss,** Plaistow, and **R. Scaife,** London. **Treating Gums, Wax, Rubber, &c.** 4d. July 13.

CONDUCTORS.—Insulating compounds for coating electrical conductors are very cheaply produced by the following process : Pure or impure gums, wax, rubber, gutta-percha, old or new, ceraffine, ozokerit, asphalte, sulphur, and other mineral and animal products, are treated with solvents under pressure, and with the assistance of steam heat, in a closed vessel. An agitating apparatus mechanically aids the chemical action of the solvents, and when they have dissolved the whole of the solid bodies, the resulting solution is run through a filter and is then ready for use, or if it be desired to recover the solvents, the solution is evaporated, and the vapours collected in a condenser. "Excellent results" are obtained in some cases by mixing solutions of different gums or bitumen.

*2,886.—**J. Dewar,** Cambridge. **Electrometers, &c.** 2d. July 14.

ELECTROMETERS.—This invention applies to electrometers in which a horizontal capillary tube $\frac{1}{70}$ of an inch in diameter is used in connection with a reservoir at one or both ends, such tube having a globule of electrolytic liquid between two columns of mercury in the reservoirs. In instruments of

this kind, the motion of the globule is very small and the action uncertain. The inventor avoids these difficulties by employing a horizontal tube of $\frac{1}{17}$ of an inch diameter, thereby rendering the apparatus much more sensitive. This modification enables the exact measurement of electromotive forces to be obtained in comparison with a given standard, as the distance traversed by the globule of electrolytic liquid is proportional to the electromotive forces acting thereon. The movement of the globule may be recorded on a cylinder by means of a marking instrument set in motion by a fine wire passing out through a small hole in the horizontal tube, such wire being connected to a movable float worked by the globule. The electromotive force, instead of being measured by the displacement of the globule, may be calculated from the additional weights of mercury added to either reservoir to retain such globule in its normal condition. A second part of this invention relates to the construction of standards of electromotive force. A cup of mercury has at its lower end a capillary orifice dipping into an electrolytic liquid. The height of mercury in the cup must be maintained at a constant level above the capillary orifice, or at the level of the electrolytic liquid. By the dropping of mercury from the capillary orifice into the liquid, a definite electrical potential is obtained depending in amount on the diameter and length of the capillary tube, and the height of mercury above the orifice, which must be kept constant in order to produce an invariable potential. In every case the apparatus must be calibrated by comparison with a standard battery.

3,003.—R. H. Ridout, Monmouth. **Galvanometers.** 2d. July 25.

GALVANOMETERS.—As usually constructed, these instruments have a division in the coil through which the suspension of the needles is accomplished, the result of this method of construction being that "the centre parts of the coil, which should be the most energetic, are really lost." The inventor makes a coil in one solid piece, and suspends a pair of astatic needles, one within it and one below it, the upper needle being pivotted on a needle point placed within the coil on its lower surface. By this arrangement it is possible to suspend from the centre of the lower needle a dependent wire needle or rod, which, passing through a perforated piece of glass or agate, allows the vertical oscillations of the needle to be controlled.

3,099.—H. P. Scott, London. (*W. Strickler, Lebanon, Pennsylvania, U.S.A.*) **Telegraph Wires and Cables.** 6d. (3 figs.) August 3.

CONDUCTORS.—The metallic conductor consists of a wire grooved longitudinally or otherwise, "so ·

that atmospheric presence is secured" between the wire and its covering. The latter consists of sheets of mica which have been heated to a cherry red, cooled in water and then dried, a base of Manilla paper or other suitable fibrous material being used to give them continuity. The covering thus prepared is applied to the grooved conductor in strips wound on helically; over this a serving of cotton is given to the core and then a coating of asphaltum made liquid by the addition of mineral tar. Finally, a helical serving of paper strips is applied.

3,264.—A. M. Clark, London. (*D. F. I. Lontin, Paris.*) **Dynamo-Electric and Magneto-Electric Machines.** 1s. 10d. (60 figs.) August 19.

DYNAMO-ELECTRIC GENERATORS.—To avoid inconvenient heating of the induced helices of dynamo-electric machines the inventor proposes to multiply their number. Sparking at the commutator is said to be due to three causes: "1st, the currents proper; 2nd, the extra current; 3rd, static electricity." The current proper must be retained, but the extra current may be made to disappear by the use of condensers. Static electricity is produced in but small quantity, if the insulation of the helices is not "too perfect." The oxidising effect of sparking at the commutator can be avoided by enclosing it in a bath of non-drying oil. A generator embodying the above improvements is illustrated. P is an iron wheel in the

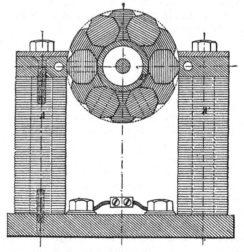

form of a pinion, provided with arms and mounted on a shaft to which rotary motion may be imparted. Each arm of the wheel has coils of insulated wire upon it, in which electric currents are induced when the wheel is rotated in front of the poles of the electro-magnet A A[1]. Part only of the electricity produced is used to excite the

field magnets. A commutator gives a continuous direction to the currents generated. In a modification of this generator there are two "induction wheels" on the same shaft, one being applied to excite the field magnets and the other to outside wires. The coils of the "induction wheels" are joined up so as to form a closed circuit. Several modifications of this generator are described, including alternate current and electrolytic generators, and the inventor specially claims "the principle of sending all the electricity produced by one or more series of coils or induction wheels to the electro-magnets of the dynamo-electric machine," and "of introducing in a break of the circuit the apparatus in which the desired chemical or magnetic result is to be produced, the principle having been laid down in Patent No. 473 of 1875." Another improvement is effected by using flexible metal combs to rub on the commutator. In arranging a number of induction wheels on the same shaft the coils of each wheel are displaced slightly with reference to those of the other wheels. To diminish the rapidity of magnetisation and demagnetisation, and thereby to reduce sparking at the commutator, the poles of the electro-magnets may be extended so as to surround the induction wheels, either partly or wholly. The coils of the induction wheels may be of wire or metallic ribbon. The action of the apparatus may be reversed so as to make the induction wheels the inductors, in which case the currents might be taken from the electro-magnet coils without the use of a commutator. In some cases a separate generator may be used to excite the field magnets. A commutator is described which consists of a toothed wheel, having the teeth insulated from each other, and each tooth connected with one of the coils of the induction wheels. The collectors are also toothed wheels, entirely metallic, gearing with the commutator, and adjustable on a quadrant-shaped bracket.

MAGNETO-ELECTRIC GENERATORS.—The above-described generators may be modified so as to become magneto-electric generators.

ELECTRO-MAGNETIC MOTOR.—The generators included in this invention, when supplied with an electric current, become efficient motors, and they may also be employed as brakes by taking advantage of the immense resistance to rotation which arises immediately the circuit is closed.

3,315.—J. H. Johnson, London. (*F. E. De Mersanne, Paris.*) Electric Light for Signals, &c. 6d. (10 figs.) August 23.

ARC LAMP.—The object of this invention is to produce an arc lamp for signalling purposes which can be adjusted from a distance, the operator being able at will to effect any desired movement of the carbons, either as regards their positions relative to each other, or to the optical focus of the enclosing lantern. The means by which it is proposed to effect this are as follows :—An auxiliary current is employed, in the circuit of which are "certain appliances" contained in the base of the lamp and a manipulator for producing the signals. In closing the circuit of this auxiliary current by suitable contact makers on the manipulator, all the functions of the lamp are instantaneously performed. The lamp case is of cylindrical form, having a dioptric lens at about the centre of its height. The carbon holders are nuts running on vertical screws, and each of the latter has at its lower end a pair of crown ratchet wheels, between the opposed teeth of which a pawl works. The pawl oscillates between two electro-magnets placed one above and one below it, and accordingly as it is attracted by one or the other ratchet wheel. Two horizontally placed electro-magnets also act on the pawl so as to give it a side movement and thereby effect rotation of the ratchet wheels and screws. Each screw being furnished with the above appliances, it can be rotated in either direction by closing the circuit to one or the other of the vertical electro-magnets, and afterwards to the horizontal electro-magnets alternately. A suitable switchboard is arranged so as to facilitate the transmission of the currents as required. A small electro-magnet is applied to the lower carbon-holder and can impart to it a vibratory movement, so as to effect contact of the carbons, and thereby establish the arc. The lamp can be made automatic by the use of clockwork to operate the distribution of the current to the regulating mechanism.

3,412.—J. E. Massey, London. Ships' Logs, &c. 8d. (22 figs.) August 30.

CONDUCTORS.—Insulated conductors are used in connection with ships' logs when it is desired to communicate their indications to the ship by means of electricity.

*3,462.—E. G. Brewer, London. (*E. E. S. Facio, Paris.*) Obtaining Light by Electricity. 2d. September 2.

ELECTRIC LIGHT.—This invention relates to the use of "Gessler" (? Geissler) tubes in connection with a portable battery, for such purposes as seeing the time at night by one's watch, examining thermometers, &c.

3,533.—W. C. Johnson and S. E. Phillips, Charlton. Telegraph Cables. 4d. September 8.

CONDUCTORS.—For serving electrical cables a woven fabric called "Hessian" is used. This is cut into strips, and applied either as a packing

or protection for the core, or externally over the sheathing wires or yarns, with or without the use of a dressing of compound or other preservative. The preparation of this fabric is described, and the inventors claim the process as part of their improvements. The fabric is saturated with a highly silicated compound, and rolled upon a spindle, after which the roll is cut up into discs by a "knife of peculiar form."

3,534.—W. C. Johnson and **S. E. Phillips**, Charlton. **Insulators for Telegraph Wires.** 6d. (4 figs.) September 8.

INSULATORS.—The principle of this invention is the "introduction of a liquid insulator between the line and the earth," which is effected in several ways. An ordinary cup-shaped insulator may be inverted, filled with the liquid, and provided with a cover to exclude rain, or an erect bell-shaped insulator may have its lip turned up inwardly so as to contain the liquid, or, lastly, a bell-shaped insulator may have within it a cup-shaped vessel with lip turned up outwards. The liquid in all cases is by preference a hydro-carbon "which will not support a film of moisture or dust on its surface."

*3,552.—R. Applegarth, London. (*P. Jablochkoff, Paris.*) **Electric Light.** 4d. (8 figs.) September 11.

ELECTRIC CANDLES.—The mechanism generally used in electric lamps is entirely dispensed with. Instead of causing the carbon points to approach each other automatically by means of machinery, in proportion as they are consumed, they are placed side by side with an insulating substance between them, which consumes at the same rate as the carbon points. For this purpose porcelain, brick, magnesia, or other non-conducting substance may be employed. To light the lamp the tips of the carbons are united by a small slip of carbon. Carbons of different thickness are used to insure equal consumption of both. One or both of the carbon rods may be placed in a tube or tubes of insulating material. In order to keep the arc in a focus, clockwork may be employed to raise the candle, such clockwork being controlled by a portion of the electric current. The carbons may be inclined towards each other if desired.

ELECTRODES.—The carbons used for the candles above described may consist of: (1) The ordinary carbons as used for the electric light. (2) Hollow carbons of different sizes. (3) Carbons made of compressed coal.

3,623.—C. W. Harrison, London. **Compounds for Preserving Metals, &c.** 4d. September 16.

CONDUCTORS.—A protecting varnish for outdoor electrical conductors is made by "mixing together and heating at a low boiling point for a short time ¼lb. ozocerit, ¼lb. gutter-percha or india-rubber, 1lb. rectified resin oil, and 2lb. linseed oil varnish." Submarine cables are protected from injury by marine animals by coating them with a compound of equal parts of ozokerit, caoutchouc, gutta-percha, and gum euphorbia, or alkaloid extract.

3,670.—C. A. Faure, Faversham. **Thermo-Electric Generators and Electro-Motors.** 6d. (12 figs.) September 19.

ELECTRO-MAGNETIC MOTOR.—This consists of two horseshoe electro-magnets fixed horizontally and parallel to each other, and so arranged that the poles form the four corners of a rectangle, and also lie in the same vertical plane. The magnet poles are furnished with wrought-iron pieces of U shape, between the branches of which the radii of a star-like armature revolve. A contact breaker is arranged so as to allow the electro-magnets to be excited alternately. The armature is fixed to a spindle running in bearings, and is placed parallel to the electro-magnets. The motor is designed for use with a thermo-electric generator worked by a gas flame, and is provided with a centrifugal governor acting on the supply of gas.

3,782.—I. L. Pulvermacher, London. **Appliances for Generating and Applying Electricity.** 8d. (18 figs.) September 28.

CURRENT METER.—A glass jar is hermetically closed by a stopper through which a tubular glass stem is fitted. The upper end of this stem is graduated, and near the portion where it enters the stopper an index is attached. A graduated band on the lip of the jar shows the amount of rotation which has been given to the tube. The glass stem is continued towards the bottom of the vessel, where it is bent into circular form. It has a thin platinum strip as an electrode through its whole length and with the lower end projecting beyond the end of the glass. The top end of the platinum strip has one of the conducting wires from the source of current affixed to it by a loosely fitting ring. Within the vessel is a separate glass tube also bent in circular form, closed at one end, and of such a diameter that the stem tube may enter into it, and thus reduce the thickness of water within it, the containing jar being filled with water to a given height. The open end of the larger tube has one or more platinum wires attached in ring form, and a conducting wire leading to a terminal screw. The water used is tinted so that its rising in the tubular stem may be easily seen. This rising will be in proportion to the amount of decomposition effected by the passage of

a current, and will therefore indicate the strength of such current. The free end of the bent portion of the glass stem is presented at starting close to the mouth of the larger tube, the index finger being at zero; by causing it to enter the larger tube the resistance is thereby increased, the amount of such increase being indicated upon the graduated band above mentioned. A removable stopper admits of the gases being liberated when the apparatus is wanted for fresh observations.

GALVANOMETER.—This is termed by the inventor "an automatic compensating galvanometer," and it serves as a measurer and graduator of resistance. A hollow cylindrical box has a groove in the bottom of it, charged with oil, in which the "bob" of an index finger, forming part of a galvanometer needle, is caused to move when the needle is deflected. This needle consists of a series of dished plates spread out fan-like at their pole ends, and attached at right angles to the index finger. The edges of the plates are fastened to a segmental metallic band, forming a kind of hood, over the top of which a strip of "conductable paper" is placed, the ends of such paper being in connection with the two battery poles through "fine metallized threads" dipping into mercury cups in the bottom of the containing box. Different resistances are obtained by blackleading the paper on its surface, or by using a metallic powder as paint. Included in the circuit is a fine wire coil which surrounds the top part of the case of the instrument, and has the magnetised needle in its centre. "The current in the coil causes a movement of the needle and a deflection of the index finger in proportion to the strength of the current." The deflection is read off from a graduated scale. As the needle turns on its axis one end of the paper strip is lowered, and the current travels through the black-leaded surface which is away from the hood, thereby offering a resistance in proportion to the quantity of paper unwound. In this way compensation is effected for variation of current strength.

RESISTANCES.—Another part of this invention refers to a substitute for resistance coils, which consists of thin German silver wire "spun in band form," analogous to the voltaic bands devised by the inventor.

4,159.—W. Hibell, Birmingham. Joining Iron and Steel Wire. 2d. October 26.

CONDUCTORS.—This invention relates to joints in iron and steel conducting wires, and consists of scarfing the ends and brazing the junction. The scarf is preferably made by reducing each of the two wire ends to a semicircular section for about an inch from the end, and making the ends wedge-shaped and the shoulders, against which they take,

under-cut. In joining small wires the scarf is a simple incline at a very small angle.

4,222.—G. Zanni, London. Application of Magneto-Electricity, &c., to Baths. 6d. (1 fig.) November 1.

MAGNETO-ELECTRIC GENERATOR.—For applying electricity to curative purposes, the inventor prefers to produce it by apparatus described in Patents No. 2,419 of 1858, and 2,821 of 1876. The former relates to a magneto-electric generator for giving intermittent currents, and the latter to a generator of a continuous current.

4,280.—H. J. Haddan, London. (*E. Weston, Newark, New Jersey, U.S.A.*) Magneto-Electric Machines. 6d. (8 figs.) November 6.

DYNAMO-ELECTRIC GENERATORS.—In the generators described by the inventor the currents generated in a number of revolving armatures are passed through the field magnets, thus "involving a peculiar construction differing widely" from that

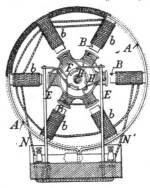

adopted up to the present in generators, which involve the principle of mutual accumulation. The diagram shows an end view of this generator. An iron cylinder A is suitably supported. From the interior of this cylinder a number of radially projecting magnets B B are arranged, all pointing towards a common centre, which magnets are preferably made broad transversely, and short radially. The magnets B B are wrapped with wire or ribbons *b b*, and the cylinder A may also be similarly wrapped. In the central space is a shaft C, carried on bearings D D, and having upon such portion of it as is within the cylinder A a series of armatures E E, all connected to an iron hub F. The armatures are of iron, wrapped with wire or ribbon. When these armatures are revolved past the magnets B B, currents of electricity are generated in the wires surrounding the armatures, and these currents are collected and made continuous by a commutator. This commutator consists of two

pieces fixed to the revolving shaft, but insulated from it, and from each other. Projections on one piece fit into recesses on the other piece, so that the collecting brushes are alternately in connection with each. The brushes are so placed that both are never in connection with the same half of the commutator at one time. On revolving the armatures the currents which are generated in their coils, after being made continuous by the commutator, pass by the wires N N¹ to the coils $b\,b$, and if desired through the coils $a\,a$, surrounding the iron ring A. The current then traverses an outer circuit in which it may be utilised for various purposes. One-half of the armature coils are connected to one part of the commutator, and the remainder to the other part.

AUTOMATIC SWITCH. — To avoid reversal in electro-depositing machines the inventor uses an automatic apparatus constructed as follows : " A cup with internal ribs contains mercury and can be rotated by means of a belt leading from the shaft of the generator. A wire dips into the cup and remains in contact with the mercury as long as the cup is stationary, or only rotating at a slow speed, but when the same is rotated rapidly the mercury is depressed at its centre and contact is broken. By including this switch in the circuit of the dynamo, and connecting the electro - depositing vat to the wire and mercury cup respectively, a means is provided of short-circuiting the vat whenever the speed of the dynamo falls below a certain limit, and in this way the counter current of the vat is prevented from entering the dynamo and effecting a reversal of its poles. Another means of preventing reversal is to make one or more of the field magnet cores partly or entirely of hardened steel.

4,312.—A. M. Clark, London. (*P. E. Smith, S. R. Spruill, and W. R. Wood, Scotland Neck, Halifax, N. Carolina, U.S.A.*) **Electric Light Buoy.** 6d. (1 fig.) November 7.

ELECTRIC LIGHT.—The object of this invention is to provide a waterway night signal for harbours, shoals, and other dangerous places. A buoy carries a glass globe in which a "suitable electrical light apparatus is arranged," and the current is led thereto from the shore by means of a cable.

4,380.—A. O'Neill, Baltimore, U.S.A. **Pipe Joints.** 6d. (29 figs.) November 11.

CONDUCTORS.—Pipes for carrying underground conductors are constructed with lugs or flanges on each end, and a collar encircling the joint, provided with flanges or lugs adapted to fit those on the pipe ends, and to draw the ends together by a rotary movement of the said collar. Each section of pipe has a recessed end for the reception of a gasket and spigot to fit securely on such gasket.

These pipe joints are very fully illustrated in the drawings accompanying the specification.

4,597.—E. H. C. Monckton, Fineshade, Northamptonshire. **Electric Motors.** 8d. (25 figs.) November 28.

ELECTRO - MAGNETIC MOTOR. — Circles of permanent or electro-magnets of horseshoe form are arranged in juxtaposition around two circular frames, "each pole of each magnet being consecutively placed so as to face the opposite pole of the next magnet ; this is effected by either placing the magnets, laid flat, around the peripheries of the frames, or by making them in pieces to be joined together in any desired form so as to be arranged vertically with their poles turned aside." By separating the poles, "one or more plain iron or magnetised steel discs can be introduced, packed side by side almost in contact, which arrangement will also unite in one large magnet the entire circle of smaller ones." These discs can be arranged so as to revolve, and also "to break contact by being slotted from the circumference towards the middle." Another form of motor is constructed by placing a condenser between two electro - magnet cylinders, "each rotating opposite to a similar fixed magnet circle." A commutator is arranged so that the poles of the rotating magnets attract those of the fixed magnets, while simultaneously they repel the poles of the magnets placed opposite to them : just as the poles are passing the centres of attraction a commutator reverses the current so that "the power is applied continuously all round the circle." An intensity coil is sometimes used in addition to the condenser, by which the tension of the current is increased ; "the velocity produced by intensifying the current, being in its own action an element of great power." By placing the circle of rotating magnets with their poles opposite to those of the fixed ones, each magnet acts "doubly," that is, attracts at one side while repelling at the other, thus avoiding back pull and the loss of power.

DYNAMO-ELECTRIC GENERATOR.—By substituting for the disc described, with reference to the first form of motor, an electro-magnet "formed like a reel, with slotted ends," or a "revolving wheel formed of electro-magnets, both wheels being insulated from their axles, and by revolving either in contact with the poles of the magnets arranged in circles, the electro-magnetic circles will be alternately completed and broken, and by a suitable arrangement any current generated can be drawn off." "Gramme rings" are built up by coiling insulated wire on a central hollow tube with thin sheet iron interposed between every successive two or more coils. The form of horseshoe magnets is varied by curving their ends inwards so as to make them face each other, and by applying to them

"novel bowl-shaped poles dished by preference inwardly," by which means the magnetic force is caused to lie in the rims. A magnet of this form is made to revolve with its poles lying within the hemispherical poles of a large magnet of similar construction, presumably to generate electric currents. An improved commutator consists of two metal rings, one on each side of a wooden cylinder, across the periphery of which, and inserted therein so as to be flush with its surface, are placed wires or pins. These wires alternate one with the other in making connection with one or the other ring. The terminals of an electro-magnet coil are "inserted apart with a frame of non-conducting material," which is fixed over the cylinder so that the terminals of the magnets as they rotate come in contact simultaneously with two of the wires, negative and positive, lying across the periphery of the cylinder.

CARBONS.—Carbon points for the electric light are made by uniting pulverised carbon with clay in sufficient quantity to bind it. Cornwall, "or other china clays" are preferred, and the mixture is subjected to hydraulic pressure previously to being baked.

4,636.—P. Moritz, Naval Officer on Board the French Man-of-War "Oriflamme." **Electric Night Signal Apparatus.** 6d. (14 figs.) November 30.

ELECTRIC LIGHT.—This invention relates to an arrangement of apparatus for signalling at sea by means of the electric light, in connection with which it is preferred to employ a "Gramme" generator and "Serrin" lamp.

4,705. H. E. Newton, London. (*H. Menier, Paris.*) **Telegraphic and other Conductors.** 6d. (2 figs.) December 5.

CONDUCTORS.—After the wire has received its insulating coating it is conducted through a vessel containing a freezing mixture which quickly hardens the coating, and allows of the conductor being immediately coiled.

4,805.—R. Werdermann, London. **Electric Lighting Apparatus.** 6d. (11 figs.) December 12.

CANDLES.—Various cross sections are illustrated, in most cases showing two carbons with an interposed thickness of insulating material, which in one candle encircles both carbons. A candle consisting of two carbon tubes, one within the other, but separated from it by insulating material, is also shown. When light is wanted in one direction only, one electrode is a large plate of carbon, from which the other and smaller electrode is separated by a sheet of mica.

ELECTRODES.—To protect carbon electrodes from contact with air they are coated with a composition of two or more of the following materials, viz., calcined magnesia, sulphate of magnesia, borax, boracic acid, sulphate of lime, fire-clay, glass, alumina, or other bodies which "in burning emit a great amount of light."

ARC LAMP.—The carbons are curved circularly. One is fixed, and the other secured to a pivotted arm which may be provided with a balance weight, preferably containing sand, which escapes and diminishes the weight. Clockwork or electro-magnets may be used to automatically regulate the distance of the carbon points from each other.

4,905.—S. A. Varley, Hatfield. **Apparatus for Producing Electric Light, &c.** 10d. (28 figs.) December 19.

DYNAMO-ELECTRIC GENERATORS.—This invention consists in improvements on the generator described in Patent No. 3,394 of 1866, by which "greatly increased magnetic potential in the bobbins" is obtained. The bobbins are divided into groups with a soft iron field magnet to each group, instead of to each bobbin, as in the generator described in the above-mentioned patent. A commutator may be provided for each group of bobbins, or when electricity of a very high potential is required, the number of commutators may correspond with the number of bobbins. The commutators are cylindrical and have their peripheries so divided, and the several sections so connected, that the insulated wires of the bobbins which are joined to the commutator sections, at certain determined positions, are cut out of the electric circuit, and whilst so disconnected "acquire magnetism only," such magnetism being further increased by causing electricity developed in other parts of the generator to pass through the insulated wires of the said bobbins after they have been put again into the electric circuit, but before their magnetic polarity is reversed. The high magnetic potential so acquired is converted into electricity at the next reversal of the magnetic polarity. The generators illustrated have a series of cylindrical helices with iron cores arranged circularly around and parallel to a horizontal shaft. The field magnets are also similarly arranged, either in a larger circle, with pole-pieces turned inwards to cover the ends of the helices, or in two circles of the same diameter, one on either side of the circle of helices. Commutators suitable for either form of generator are fully described and illustrated, and the principle of their action consists in cutting out the bobbins from the electrical circuit at regular intervals by cutting away portions of the discs, the number of divisions having relation to the number of changes which have to be effected at each revolution of the shaft.

The inward motion of the springs which press against the peripheries of the commutators is limited by insulating stops. " Part of the electricity developed by the machine is diverted to maintain the magnetism of the soft iron magnets, and the remainder is used to produce the electric light." This is effected preferably by wrapping the field magnets with two insulated wires, one having larger resistance than the other. The circuit of larger resistance is always closed, and the circuit of less resistance is used for the electric light. "When the construction of the machine admits of it," the field magnets are made of decreasing diameter towards their acting poles.

ARC LAMP.—A curved carbon rod rests upon the periphery of a disc of carbon "usually bevelled on the edge." In some cases the carbon disc is made in segments with interposed pieces of insulating material, over which an arc forms when the lamp is in action. Motion is given to the carbon disc by the electro-motor described below or other suitable means.

ELECTRO-MOTOR. — Hollow helices of insulated wire are arranged end to end, and between them move iron discs, which are mounted on a spindle. The iron discs have teeth on their edges and the helices are encased in iron tubes having teeth at both ends, corresponding in number to those on the discs. Commutators are provided to close and open the circuit through the helices in succession, and the discs are arranged so that their teeth "break joint."

ELECTRO-DYNAMOMETER. — A hollow bobbin is wrapped with soft iron wire, and accurately in its centre a permanent magnet is mounted on pivots. Circular or segmental coils of insulated copper ribbon are mounted above and below the poles of the magnet. An index moving over a divided scale is connected by a spring to the upper pivot of the magnet, and retains it in the centre of the coils. These coils are put into the circuit, and when the current passes the magnet is deflected. "The index may then be turned round until the torsion of the spring is sufficient to bring the magnet back again into the centre of the coils . . . The distance travelled by the index to effect this indicates the amount of electric force which is being developed."

1877.

76.—J. C. Fuller and G. Fuller, Stratford. Essex. Insulating Supports of Telegraph Wires. 6d. (2 figs.) January 8.

INSULATORS.—These are made more or less in the form of an inverted truncated cone, and the electrical resistance is increased by corrugating or undercutting the outer surface of the cone. When an inner cup is provided the two parts are made separable. The usual screw and nut on the insulator stem is dispensed with, and a slot and cotter used instead.

166.—H. Baggeley, London. Firebricks and Tubes. 4d. January 12.

CONDUCTORS.—Vitrified pipes suitable for containing underground conductors are made from a composition consisting of pipe-clay 60 lb., chalk 30 lb., powdered white glass 6 lb., powdered green glass 3 lb., powdered Welsh spar 2 lb., powdered "tinkle" 1 lb., and black oxide of cobalt 1½ lb.

270.—C. F. Varley, London. Apparatus for Generating Electricity. 8d. (16 figs). January 20.

DYNAMO AND MAGNETO-ELECTRIC GENERATORS.— Actual or nearly actual contact is maintained between the armatures and poles of the field magnets. The magnets with the intermediate cores surrounded by helices form a complete iron ring or circuit, and they have their respective N. and S. poles continuously closed, or nearly so, notwithstanding the movement of the armatures, but the armatures "alter the direction of the magnetic conduction through the inducing cores surrounded by electric conductors," the effect of which commutation is to generate electricity in the conductors forming the helices. In a simple form of this generator two horseshoe magnets are employed, placed one opposite to the other, and between their poles are inserted two soft iron cores on which coils of covered wire are wound. The two N. poles are in contact with the two ends of one of the cores, and the two S. poles with the two ends of the other. A reciprocating armature makes contact first with the two poles of one magnet and then with those of the other. The faces of the magnets and of the armature may be grooved to increase the area of the surfaces in contact and the armature may advantageously be made double, and its dimensions such that it may close with one pair of poles before leaving the other pair. The armature may revolve instead of reciprocating. By another arrangement an even number of horseshoe magnets are placed radially around a circle with like poles uppermost. Similar poles are united by soft iron cores wound with covered wire forming the circuit in which the currents are generated. By preference this much of the generator is carried upon a

gun-metal cylinder or ring, within which a corresponding frame, having armatures upon its surface, rotates. As this frame rotates the armatures couple together the poles of each magnet in succession. In some cases motion is given to the armature by a finger key.

ELECTRO-STATIC GENERATOR.—This is an improvment on the generator described in Patent No. 206 of 1860, and preferably consists of a vulcanite disc mounted on an insulated axis, and capable of being rapidly rotated. On the base of the disc radial strips of tin foil are fixed at equal distances apart. In front of the disc are two collecting combs so placed that the tin foil strips pass close to them as the disc is rotated. Near the disc on the other side are two insulated pieces of wood, which serve as conductors. They are of curved form, and each inductor covers about a quadrant of the operative part of the disc. The inductors are so fixed that the leading end of each is immediately opposite to the collecting comb. For the purpose of charging the inductors there are other combs placed similarly to the collecting combs, but removed from them by an angular distance of 90 deg. These combs are in metallic connection with the inductors, each to each, and they carry wire brushes which make contact with studs upon the face of the disc in connection with each of the tin foil strips. This arrangement renders the machine self-charging. The brushes can be lifted out of contact as soon as the machine is at work. This generator is employed in combination with the magneto-electric generator above described in the production of the electric light.

369.—W., E. W., and A. Harvey, North Woolwich. **Manufacture and Application of Soft Steel or Ingot Iron.** 6d. January 29.

CONDUCTORS.—This invention relates to a process by which a " soft steel," suitable for electrical conducting wires or cable sheathing, is produced.

494.—P. Jablochkoff, Paris. **Electric Lamps, &c.** 6d. (12 figs.) February 6.

CANDLES.—Pieces of carbon are placed side by side and separated by an insulating substance, which may be kaolin, glass, the ingredients of glass and porcelain, earths, or silicates. It is preferred to form the candle by ramming the insulating material into an asbestos cartridge case containing the carbons. When the case is filled it may be sealed with silicate of potass. The lower ends of the carbons are inserted into pieces of metal tube separated from each other by asbestos, and the tube ends are then connected by vice jaws to the conducting wires. The whole may be mounted on a non-conducting base and provided with a suitable

glass globe. The heat produced by the electric current fuses the substance between the carbons, and dissipates it. The resulting light may be coloured by introducing into the insulating material suitable substances. Instead of placing the carbons in a case charged with insulating powder, they may be embedded in solid insulating material, or the carbons may be made tubular filled within with the same material which surrounds them outside. To light these candles a piece of carbon is placed on the two points and then withdrawn, or a " pulverulent match " is employed. Several candles may be placed in one circuit, or to guard against interruption of the light each may be worked by relay from the main circuit.

522. J. H. Cordeaux, Birmingham. **Fixing the Insulators of Electric Telegraphs.** 6d. (4 figs.) February 7.

INSULATORS. — The stem is screwed into the insulator, and an elastic washer is interposed between a collar on the stem where the screw terminates and the shoulder inside the cavity of the insulator.

*555. E. W. Beckingsale, Newport, Isle of Wight. **Covering and Insulating Wire.** 2d. February 9.

CONDUCTORS.—The conducting wire is insulated and protected by covering it with paper, or paper pulp applied in a continuous length without seam or lapping. This covering may be applied either to a naked wire or to one which has already been covered.

*687. W. J. Russell and R. Wilson, Croydon. **Fastenings for Gas Syphon Boxes, &c.** 2d. February 20.

CONDUCTORS. — This invention is applicable to road boxes in connection with subterranean conductors, and consists in a method of locking the lids of such boxes by a key actuating bolts which take into holes in the side of the box. When the lid is closed the removal of the key locks it, and, consequently, no lid can remain unfastened.

732. J. H. Lovel, Sunderland. **Electro-Magnetic Engines.** 6d. (13 figs.) February 22.

ELECTRO-MAGNETIC MOTOR.—A soft iron disc is attached to a shaft, one end of which rests in a suitable bearing. The other end is connected to a crank fixed on a vertical shaft working in bearings. Below the disc is a circle of electro-magnets with their poles all in the same horizontal plane. When one electro-magnet is excited by an electric current, the periphery of the disc is attracted, and

the magnets are made to act consecutively by a suitable commutator. An undulating rotary motion is by these means given to the disc, and rotary motion to the crankshaft, from which power may be taken. The advantage claimed for this form of motor is that there is no impact of the armature upon the electro-magnets, but only a rolling contact, and there is, therefore, no danger of the electro-magnets becoming permanently magnetised.

805. W. Boulton, Burslem. Apparatus for Making Articles in Pottery. 6d. (12 figs.) February 28.

INSULATORS. — This invention relates to the manufacture of insulators, and more especially to improved constructions of "jolleys" used for moulding them when in a plastic state, and to methods of mounting and actuating such jolleys.

833. W. T. Henley, Plaistow. Electric Telegraph Apparatus. 6d. (9 figs.) March 1.

CONDUCTORS.—For a cheap insulating material for electrical conductors, the inventor proposes to use a compound of gutta-percha and cotton-seed pitch mixed in equal proportions. India-rubber may also be added to the compound if desired, or it may be applied as a separate coating interposed between successive coatings of the above compound. Among the insulated conductors, described in Specification No. 1,944 of 1876, there is one of triangular section placed in a lower channel-shaped piece with square sides. It is now proposed to make these sides to fit the triangular sides of the conductor.

MAGNETO-ELECTRIC GENERATOR.—This refers to details facilitating the manipulation of the alphabetical telegraph instruments described in Patents No. 734 and 2,464 of 1861.

*855. T. G. Glover, London. Protecting Submarine and Subterranean Telegraph Cables from Insects, &c. 2d. March 3.

CONDUCTORS.—To protect the core of submarine and subterranean cables from the destructive effects of the toredo, white ant, &c., a strip of lead foil is wound helically around the core with the edges overlapping about a quarter of an inch, and over this is placed a serving of hemp saturated with castor oil. The servings of the outer wires of the cables are also saturated with castor oil.

1,264.—C. D. Abel, London. (*W. Winter, Prague.*) Production of Photographic Pictures on Woven Fabrics. 4d. March 31.

ELECTRIC LIGHT.—The inventor claims the use of the electric light, "produced by dynamo-electric apparatus," for obtaining magnified positive photographs on woven fabrics, when such fabrics are impregnated with iodide or chromide of silver.

1,387.—F. F. A. Achard, Paris. Electric Apparatus for Actuating Railway Brakes. 6d. (15 figs.) April 9.

SECONDARY BATTERY.—For working the electric brake described in this specification the inventor employs a Planté battery in conjunction with Daniells' cells, in order to obtain a current of great intensity at the moment it is desired to apply the brake. The inventor states that the Planté cells "play a part analogous to that of Leyden jars in static electricity; they constitute real accumulators."

1,416.—H. A. C. Saunders and A. Jamieson, London. Protecting Telegraph Cables and Wires. 2d. April 11.

CONDUCTORS.—To preseve the cores of submarine or underground cables from the attacks of the toredo and other animals, the servings which cover such cores are saturated with audiroba oil. This oil is also employed to saturate the servings of the iron sheathing wires, thereby preventing the rusting of such wires.

1,693.—G. Pickersgill, Todmorden. Expansion and Contraction Coupling for Signal and Telegraph Wires. 2d. May 1.

CONDUCTORS.—When used for connecting electrical wires this coupling is made of copper or brass, and consists of a tube provided with one or more piston rods, on each of which and within the cylinder are mounted any required number of discs of vulcanised india-rubber. These rods after passing through the covers, provided on each end of the tube, are attached to the wires.

1,829.—R. Werdermann, London. Electro-Magnetic Apparatus for Developing Motive Power, &c. 6d. (4 figs.) May 10.

ELECTRO-MAGNETIC MOTOR.—A cast-iron frame carries two equal-sized discs placed horizontally one above the other, and to each disc is attached a series of electro-magnets arranged circularly, with the poles of the magnets on one disc facing the poles of those on the other disc, and with a space between opposed poles. In this space is placed a disc-shaped armature which is fixed to a shaft supported below in a spherical bearing, and attached above to a crank and vertical shaft. The vertical shaft carries a driving pulley and a commutator by which the electric current is distributed to the electro-magnets. Under the successive attractions of the electro-magnets the armature receives a rolling

motion which is communicated to the vertical shaft through its crank. It is preferred that the electro-magnets of one series should be opposite to the spaces between those of the opposite series. To avoid loss of power by the coercive force of the iron the electro-magnets and armature should be of pure wrought iron, or a feeble reverse current should be sent through the electro-magnets at the moment the circuit is broken. Or for the same purpose the magnet or armature, or both, may be coated with copper or some other suitable material. The electro-magnets used in this motor are constructed as follows : Two iron tubes, one smaller than the other, are coiled externally with copper wire, and the smaller one is placed within and concentrically with the larger one. In the annular space between the two tubes is placed a circle of smaller tubes also coiled with copper wire. These are made of opposite polarity to the two larger tubes, and the latter are always magnetised.

MAGNETO-ELECTRIC GENERATOR. — When the above apparatus is used as a generator permanent magnets surrounded by helices are used instead of the electro-magnets.

*1,910.—F. W. Heinke, Twickenham. Producing Electric Light. 2d. May 16.

ARC LAMPS.—Carbon disc electrodes are made to rotate in contrary directions, so as to work against one another preferably at right angles. It is preferred to make the two discs to rotate synchronously by clockwork. To provide for wear of the discs one of them is advanced at every revolution, the movement being accomplished by suitable mechanism. Sharpening appliances are applied to the edges of the discs to keep them suitably shaped. When bars or rods are used instead of discs they have a reciprocating motion imparted to them.

1,996.—P. Jablochkoff, Paris. Producing and Dividing Electric Light. 6d. (3 figs.) May 18.

INCANDESCENCE LIGHT.—With the candles described in Specification No. 3,552 of 1876 light was produced by the action of the electric current upon the conducting points themselves, but in the present invention it is the result of the action of the current upon refractory bodies, such as kaolin placed between the conductors, whereby these bodies are raised to a sufficient temperature to become luminous. In carrying out this invention a high tension current is necessary, and this is obtained by placing in the main circuit a number of induction coils corresponding to the number of lights to be employed, and having the terminals of their secondary coils attached to the holders which support the slab of kaolin. In putting the lamp in action the current of the induction coil is caused to pass through "a kind of conducting match" placed on the surface of the refractory slab.

When continuous currents are employed the induction coils are provided with interruptors and condensers, but in the case of alternating currents these are dispensed with, and the coils are constructed similar to the electro-magnets described in Specification No. 836 of 1876. The circuit of each coil may be interrupted at several points for the establishment of as many separate lamps. The consumption of kaolin is very small.

2,094.—A. M. Clark, London. (*Lontin and Co., Paris*). Electric Light Apparatus. 8d. (16 figs.) May 29.

ARC LAMP.—The regulation is effected by the employment (in place of electro-magnets) of a simple wire, which becomes heated by the passage of the current, and consequently expands in length. This alteration in the length of wire is applied to control the regulating mechanism. The wire should be of such a gauge that it may be heated to 400 deg. Centigrade by the passage of the current. A fine wire solenoid may be employed to supplement the action of the extensible wire. Such solenoid is attached either "directly or indirectly" to the two carbon-holders, so that if the carbons are in contact or the arc is short, the coil will be inactive, its activity increasing proportionately as the arc lengthens. The action of the coil may be applied to release the feeding mechanism, or to bring into action a commutator designed to effect mechanically the approximation of the two ends of the expanding wire. The invention also consists in using the solenoid alone as a regulator without the expanding wire. In order to guard against the fine wire solenoid receiving too much current about 3 centimetres of very fine wire is inserted in its circuit which, in the above-mentioned contingency, fuses and interrupts the circuit. This feature of the invention may be applied to another purpose. In working several lights in series in one circuit a fusible wire is used to retain a suitable commutator ; but immediately the wire is fused the commutator comes into use, and allows the current to support the other lights while the extinguished light is being readjusted. In some cases the fusible wire may be replaced by a multi-metallic device whose expansion and consequent change of form will actuate the commutators. By suitably proportioning the size and length of the wire of the solenoid a large number of regulators may be introduced into one circuit. These solenoids may be wound with a cable composed of several independent wires which can be connected according to requirements.

2,106.—J. H. W. Biggs, Liverpool. Apparatus for the Manufacture of Common Salt and Carbonate of Soda, and for Packing Salt, &c. 1s. 10d. (83 figs.) May 30.

ELECTRO-MAGNETIC MOTOR.—For giving indicacations of the density of brine the inventor employs an electro-magnetic motor, making the rise and fall of a float regulate the amount of current it receives, and thereby its speed of rotation.

2,527.—W. H. Kerr, Malahide, Dublin. Preparation of Materials to be used in the Manufacture of Porcelain, &c. 2d. June 30.

INSULATORS.—The raw materials are fired separately and then mixed together whilst subjected to the action of heat in a furnace. The materials are then ground and afterwards moulded, cast, thrown, or blown (as a glass). Glazing can be dispensed with.

2,585.—W. Prosser and W. E. Moore, London. Lamps, &c., for Electric Light. 6d. (21 figs.) July 5.

ARC LAMPS.—A strip of carbon is partially enclosed in a metallic case, and towards its exposed surface a carbon rod is fed. The strip of carbon is traversed whilst the rod is uniformly fed by suitable mechanism so as to maintain a focus. Two or more rods may be presented to the same strip if desired. Electrodes arranged radially in circular or other frames or carriers are adopted when it is required to maintain the light without interruption for a considerable length of time, and they are brought into successive action very rapidly. These multiple electrodes may be used in combination with disc or other electrodes, or with drums. When fluid electrodes are employed, they are enclosed in hermetically sealed glass shades in accordance with Patent No. 3,466 of 1875, and such fluid electrodes may consist of two or more intersecting streams of mercury.

CARBONS are formed by compressing powdered boxwood, charcoal, or other suitable material.

2,934.—A. M. Clark, London. (*S. Marcus and B. Egger, Paris.*) Electric Lamps. 8d. (6 figs.) July 31.

ARC LAMPS.—The principle of this invention is the regulation of the carbons directly by a series of consecutive coils, through some only of which the current passes at one time according to the position of a soft iron core, so that the latter is always subject to the maximum influence of the coils whatever may be its position. The lamp described and illustrated is of upright form, and is made focus keeping by gearing the carbon-holders by means of cords and drums of unequal size. The upper carbon-holder is weighted so that there is a tendency for the carbons to approach each other. A long vertical solenoid, consisting of several independent coils placed one above the other, has working within it a soft iron core connected directly to the lower carbon-holder, and through the cord gear to the upper carbon-holder. The wire ends of the separate coils are taken to contact pieces arranged in a vertical line, and over which run contact rollers, depending from the lower carbon-holder. The office of these rollers is to transfer the current at intervals to a higher set of coils, as the carbons consume, and thus to follow up the soft iron core as it rises, always exerting uniform attraction upon it.

2,982.—A. M. Clark, London. (*E. Reynier, Paris.*) Electric Lamps, &c. 8d. (18 figs.) August 3.

ARC LAMPS.—The principal distinctive features of the lamps comprised in this invention consist in rendering the progressive movements of the circular electrodes completely independent the one of the other, and in keeping the arc stationary and unobscured. This latter object is effected by inclining the electrodes to one another at an angle of from 20 deg. to 120 deg., their axes being also inclined to each other but in the same plane, or by making the plane of one electrode nearly a tangent to the circumference of the other. In the lamp described the latter arrangement is adopted. The disc electrodes are rotated independently by clockwork, one being fixed as regards its plane of rotation, and the other controlled by a solenoid and iron core opposed to a helical spring. The spring tends to bring the electrodes together, whilst the action of the solenoid is to part them. The two are so adjusted as to maintain an arc of the required length.

ELECTRODES.—Disc electrodes are made of a composition of powdered graphite 100 parts, powdered sugar 20 parts, and iron filings 5 parts. The mixture is moistened with dilute nitric or hydrochloric acid to form a thick paste, from which the electrodes are moulded. They are then dried at a gentle heat and afterwards stoved for 24 hours at a temperature rising from 40 deg. C. to 100 deg. C.

3,170.—L. Denayrouse, Paris. Kindler for Electric Lights. 4d. (3 figs.) August 21.

CANDLE HOLDER.—Two or more electric candles are mounted on a stand of convenient form, and a key is provided for making and breaking the circuit for each candle. The key is worked by one arm of a lever, the other arm of which has a stud pressed by a spring against the candle which is burning near its lower end. When this candle has burned nearly down so that the stud on the lever is no longer supported by it, the lever is moved by the spring, the key actuated, and the current transferred to the next candle.

3,187.—P. Jablochkoff, Paris. **Magneto-Dynamo Electric Machine.** 6d. (3 figs.) August 22.

DYNAMO-ELECTRIC GENERATOR.—A bobbin of insulated wire of diameter greater than its length is placed between two soft iron discs, which are fixed to a soft iron shaft passing through the centre of the bobbin. The peripheries of the discs are notched so as to form a series of teeth, each tooth of one facing a space between adjoining teeth of the other. The discs are caused to rotate near a series of electro-magnets so arranged that the teeth

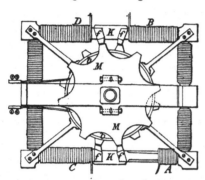

successively pass opposite poles of these magnets. This form of generator produces alternating currents, but by applying a suitable commutator it may be made to give continuous currents. Referring to the diagram, A, B, C, D are the field magnets, *f f* soft iron pole-pieces, K K brass distance pieces; *b* is the insulated wire bobbin, and M M the notched discs.

3,469.—I. L. Pulvermacher, London. **Generating, &c., and Applying Electricity.** 10d. (41 figs.) September 14.

CURRENT REGULATOR.—In the bottom of a rectangular gutta-percha box is placed a resistance coil consisting of a cord of gutta-percha impregnated with graphite, wound helically upon a porcelain cylinder. The upper portion of the gutta-percha box is wound with many turns of fine wire, forming a flat bobbin with vertical axis. A magnetic needle is pivotted so as to be capable of vibrating in a vertical plane within the bobbin, and it carries on one side a metallic comb so constructed that as the needle moves towards a horizontal position the teeth of the comb are successively inserted in a row of mercury cups, each in connection with a portion of the resistance coil. A pointer also attached to the magnetic needle traverses a dial. In proportion as the current which passes through the bobbin increases in strength the needle is deflected, the comb lifted, and more and more resistance inserted till the desired compensation is effected. In another form of this instrument strips of a

graphite composition are attached to the rocking needle which as it is deflected by the current immerses a greater or less number of the strips in a mercury trough and thereby alters the amount of resistance in circuit as before.

3,552.—A. S. Hickley, London. **Constructing and Arranging Magneto-Electric Apparatus.** 6d. (11 figs.) September 21.

MAGNETO-ELECTRIC GENERATOR.—To each pole of a horizontally fixed permanent horseshoe magnet is attached one end of a small soft iron bar, the other end carrying a vertical rod of the same metal, which acts as a core to a coil of insulated wire through which it passes. A bar of soft iron rests on the tops of the two vertical rods and forms an armature. When this armature is vibrated by suitable means electric currents are generated in the coil.

3743.—J. H. Johnson, London. (*Société l'Alliance, J. Miot, Paris.*) **Magneto-Electric Machines.** 6d. (5 figs.) October 9.

MAGNETO-ELECTRIC GENERATOR.—The object of this invention is to subdivide the currents produced by a magneto-electric generator, and this is attained by the addition of an "auxiliary apparatus" to a generator of the ordinary type. The generator illustrated has four series of horseshoe magnets arranged diagonally in a rectangular frame, with their poles lying in the circle described by the revolving bobbins, of which there are three sets of eight, situated so as to alternate with the four series of magnets. The wires are coiled in the same direction on all the bobbins, and the currents generated are collected by connecting them with a common conductor. The "auxiliary apparatus" before referred to consists of one or more copper discs attached to but insulated from the main shaft of the generator, each disc being in connection with an independent set of bobbins. Separate rubbers bearing on these discs constitute the means of distributing the currents to several different circuits. The currents are alternating.

3,839.—R. Jablochkoff, Paris. **Distributing and Increasing with Atmospheric Electricity Currents Proceeding from a Single Source of Electricity for Supplying several Lighting Centres.** 4d. (5 figs.) October 17.

ELECTRIC LIGHTING.—Instead of using currents proceeding from a single source directly, they are first converted into static electricity and then reconverted into dynamic electricity. This is effected by interposing condensers or Leyden jars between the main circuit and branch in which it is desired to maintain a light. By these means the current may

be distributed in several directions, and there is the additional advantage that atmospheric electricity is developed and accumulated in the condensers, "from which it is directed in the form of currents to the illuminating apparatus." The last may be the electric candles described in Specification No. 1,996 of 1877.

3,854.—G. Pickersgill, Todmorden. Expansion and Contraction Coupling for Signal and Telegraph Wires. 6d. (8 figs.) October 18.

CONDUCTORS.—This coupling is substantially the same as that described in Provisional Specification No. 1,693 of this year.

3,981.—M. H. Smith, Halifax. Electric Magnets and Magnetic Motor Engines. 6d. (8 figs.) October 27.

ELECTRO-MAGNETS.—A coil of copper wire rests on an iron base-plate (preferably circular), and an iron case attached to the base-plate surrounds the coil. The armature consists of an iron plate the size of the case, to which is attached an iron core sliding within the coil.

ELECTRO-MAGNETIC MOTOR.—A circle of electro-magnets act successively on a series of armatures attached to a disc, which is thereby caused to roll or "wobble." The motion thus obtained is converted into rotary motion by a crank and shaft.

4,036.—H. A. Dibbin, Hallaton, Leicester. Electric Apparatus for Applying Railway Brakes. 6d. (6 figs.) October 31.

MAGNETO-ELECTRIC GENERATOR.—The currents required for operating a brake electrically in accordance with this invention may be obtained from a magneto-electric generator amongst other sources.

CONDUCTORS.—Conducting wires are united at their ends by a spigot and socket joint formed of split and sprung metal tube, and sheathed with india-rubber.

4,053.—H. Conradi, London. (*Emil André, Ehrenbreitstein, Prussia.*) **Electro-Metallurgical Apparatus.** 4d. November 1.

MAGNETO-ELECTRIC AND DYNAMO-ELECTRIC GENERATORS.—This invention consists in the application of the above generators to the production of metals from impure metals, alloys, ores, rubbish, and scrap.

4,232.—G. Zanni, London. Magnetic Apparatus for Ringing Bells, &c. 6d. (9 figs.) November 13.

MAGNETO-ELECTRIC GENERATOR.—Improvements on apparatus described in Specification No. 2,721 of 1871. The permanent magnets are made of

horseshoe shape instead of circular. An armature (presumably Siemens') is rotated by hand in a cylindrical recess formed between the magnet poles, the required speed of rotation being obtained by the use of toothed or friction gear. In one form of this generator rotation of the armature is obtained from a direct pull by means of a straight rack acting on a train of toothed wheels gearing to a pinion on the armature, and in another form the motive mechanism is clockwork. These generators have a commutator of the usual construction for making the currents generated in the armature continuous.

4,275.—F. W. Heinke, Twickenham. Producing Electric Light. 2d. November 15.

ARC LAMP.—Two carbon discs are caused to rotate near each other, the electric arc forming at their edges. The discs may be enclosed in a glass or metal case with a top and bottom aperture, through which the heat "created by the combustion of the carbon" causes a draught of air to pass. This current is applied to turn a fan "which by suitable intermediate gear gives motion to the carbons, and works the feed." The feed may be worked by clockwork if the carbons are rotated thereby, or by a small engine worked by a boiler placed on top of the carbons. The feed may be altered for different rates of combustion, qualities of carbon, and other circumstances by a series of "eccentric step cams" arranged on one spindle, and any one may be thrown into gear with a movable toothed wheel gearing into a deep tooth wheel on the feed screw spindle.

4,341.—A. G. Bell, London. Coupling Wires for Conveying Electricity. 4d. November 20.

CONDUCTORS.—Relates to using a return wire on telephone circuits in order to eliminate induction. The two wires are insulated and then stranded together.

4,412.—F. H. Ziffer, Manchester. Electric Lamps. 8d. (7 figs.) November 23.

ARC LAMP.—This invention consists: (1) In causing the carbons to approach or recede by means of screws actuated by spur or bevel gear; each carbon may be worked independently if desired. (2) In causing the carbons to revolve as they approach each other. Horizontal and vertical forms of this lamp are illustrated and described in detail.

4,432.—J. Rapieff, London. Production and Application of Electric Currents for Lighting, &c. 3s. 10d. (454 figs.) November 24.

DYNAMO-ELECTRIC GENERATORS.—These improvements may be grouped into three series. 1. A method of constructing a compound bobbin or electro-magnet consisting of several smaller bobbins

united to a pair of pole-plates. 2. The use of a "closed magnetic circuit" in each group of bobbins. 3. Arrangements for avoiding the bad effect of useless induction currents in the magnet cores and coils. To obviate the inconvenience of retarded magnetisation due to mutual induction in systems of bobbins, the cores of such bobbins are made up of a series of thin electro-magnets with their like poles similarly placed. Another plan consists in coiling the bobbin with a ribbon having a warp of cotton or silk, and a woof of conducting wire. Many forms of compound electro-magnets constructed in accordance with the above methods are described, and the inventor states that in all cases the magnetising currents must be intermittent or alternating. For generating induced currents a conductor is coiled with another conductor, both being insulated; when an intermittent current is sent through the inner conductor an induced current appears in the outer coil. If the compound conductor be coiled round a third conductor, and an intermittent current be sent through the latter, the induced current which appears in the exterior coil will be stronger, and its strength will be further increased by coiling the triple system around an iron core. A dynamo-electric generator involving these principles consists of a ring armature "coiled with one or several conductors, which may be wires or ribbons of any shape, insulated or not, and may be wound in one or in several series." The diagram, Fig. 1, shows three systems of connecting the

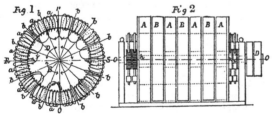

separate sections of the ring coils. A category of ten distinct advantages of this kind of armature are recited. The field magnets may consist (1) of a suitably-shaped arrangement of one or more iron sheets or wires, coiled with wire, or (2) of compound electro-magnets as described at the beginning of this abstract. Fig. 2 shows a generator in which A A A are the inductors and B B B the armatures. The latter are keyed to the shaft O, and are rotated whilst the inductors A A A remain fixed. Several other forms of generator are described and illustrated, which consist of applications of the various arrangements of electro-magnets and armatures referred to above. Sectional commutators of the ordinary kind are provided for giving the currents a constant direction, and the currents are taken off by brushes, springs, rollers, or mercury baths. When the armature is fixed the following arrange-

ment may be adopted : A disc, provided with as many projections as there are poles in the generator, is secured to the rotating shaft, and these projections press successively at determined moments on contact springs which are connected on the one side to the bobbin terminals and on the other side to the field magnet coils.

ARC LAMPS.—Two or more carbons are used for each pole, and they are secured obliquely so that

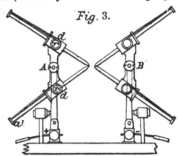

Fig. 3.

the outward motion of each carbon is limited at its ends by its encounter with the other carbon. Fig. 3 shows a lamp on this principle ; $a\,c$, $a^{\text{I}}\,c^{\text{I}}$ are the carbons of one pole, $d\,d^{\text{I}}$ are the holders, and $r\,r$ springs with cords attached for advancing the carbons as they consume. A piece of carbon is placed across the poles when it is desired to establish the electric arc. For obtaining a continuous light of great intensity the poles are made of non-combustible material, and the arc is maintained by feeding between the poles a substance capable of giving off vapour which acts like the oil which feeds an ordinary lamp.

ELECTRODES. — These may consist of carbon coated with metal, or with cores of metallic wire. When intense light is required carbons may be formed of a mixture of coal with a substance which disengages heavy gases under the heat of the arc, or magnesia lime and other such substances may be combined with the coal. This wordy specification, which runs to 42 pages, includes a vast amount of detail which cannot be referred to in this abstract.

4,435.—S. A. Varley, Hatfield. Electrical Apparatus for Lighting, &c. 10d. (13 figs.) November 24.

DYNAMO-ELECTRIC GENERATORS.—This invention has reference chiefly to the class of generator described in Specification No. 4,905 of 1876, and has for its object the more economical production of electric currents, and their distribution or division. The revolving bobbins have each a wooden core surrounded with a number of iron wires arranged parallel to the axis of the core. Convolutions of insulated wire are wrapped over the iron wires, and then a second series of iron wires is added, and a second coil, and so on to any desired extent.

The field magnets may be built up in a similar way to the revolving bobbins. To divide the light "electric division bobbins" are used. These consist of rings of helices like those just described, except that they are wrapped with two or more wires side by side instead of one continuous length of wire. These helices have soft iron pole-pieces. There are as many distinct circuits as there are separate insulated wires on the helices. In an apparatus with four circuits, three may each include an electric lamp, and the remaining circuit should be connected to the reversing commutator of a generator such as that described in Specification No. 4,905 of 1876. When the generator is in action, alternating currents will be produced in the circuits including the lamps. Reversals of the currents in the division bobbins may be obtained by causing them to move between the poles of electro-magnets "polarised in alternate directions," and when this plan is adopted the bobbins are made with hollow cores of soft iron wires, and are mounted so as to move freely in two soft iron tubes having a slot cut through them on one side. These tubes form extensions of the poles of electro-magnets and envelop some of the helices; they also approach one another to within three-quarters of an inch. Metallic contact-pieces may be attached to the ends of the helices and project into the slots in the tubular extensions of the magnet poles; and three sets of contact springs are mounted over such extensions. One set of springs rests on the contact-pieces in the central space separating the slotted tubes, and the other two sets on the contact-pieces near the poles of the electro-magnet, and these are connected to one of the poles of the lamp, the others being joined to the second pole. A rapid to-and-fro motion in the direction of the length of the slotted tubes given to the helices by suitable means produces alternating currents in such helices. The above-described generators are available as motors.

DEVIATOR.—To re-establish the circuit in the event of a carbon breaking in an arc lamp the inventor uses an apparatus consisting of an electro-magnet in the main circuit, which on losing its current releases an armature, and thereby completes the circuit again through a carbon resistance.

*4,464.—L. Simon, Nottingham. (*S. Schuckert, Nürnberg.*) Dynamo - Electric Machines. 4d. (2 figs.) November 27.

DYNAMO-ELECTRIC GENERATOR.—The armature is an iron ring composed of a number of sheet-iron rings insulated "magnetically" from each other, and wound like a Gramme ring. The field magnet poles are deeply recessed to receive the armature, and thus bring an increased portion of its surface under the influence of the magnetic field. An ordinary sectional commutator and brushes are used to collect the currents generated.

4,748.—W. R. Lake, London. (*E. Weston, Newark, New Jersey, U.S.A.* Magneto-Electric Machines. 6d. (4 figs.) December 14.

DYNAMO-ELECTRIC GENERATOR.—Improvements on the generator described in Specification No. 4,280 of 1876. These are as follows: (1) Commutator brushes attached to a disc moving concentrically with the shaft and are adjustable about it. (2) For electro-depositing, an automatic switch operated by centrifugal force short-circuits the electrodes in the vat as soon as the speed falls to a prescribed point. (3) In order to prevent reversal of polarity of a generator "a constantly closed differential circuit of prescribed resistance" is combined with the main circuit. This is effected by a resistance coil placed as a shunt to the main circuit. (4) To keep the generator cool its interior parts are coated with paraffine, and it is placed in a water-tight shell through which water is caused to circulate.

4,824.—B. Hunt, London. (*D. Brooks, Philadelphia.*) Insulating Telegraph Wires, &c. 6d. (5 figs.) December 19.

CONDUCTORS.—Copper wires are wrapped with cotton and then immersed in melted paraffine, after having been kept at a temperature of 320 deg. Fahr. for an hour and a half. For underground or submarine use several of these conductors may be drawn into lengths of wrought-iron tubing, and the tube then filled up with paraffine oil. Stand-pipes are placed at intervals along a line of these conductors by which a head of oil is kept in the tubes.

4,893,—A. M. Clark, London. (*Lontin and Co., Paris.*) Dynamo-Electric and Magneto-Electric Machines. 4d. December 24.

DYNAMO-ELECTRIC GENERATORS.—The object of this invention is to obtain continuous currents without a commutator. If a magnet be made to move parallel to an iron bar on which several small coils are arranged, a current continuous in direction will be produced in each coil. If the bar ends be united so as to form a ring the same effect will be obtained during the passage parallel thereto of a magnet. One or more such rings may be made to turn in front of a number of similar magnetic poles and the currents so produced may be collected and employed to excite the electro-magnets. The ring "may be composed of a series of coils attached one after the other in a circular or polygonal form." The field magnets may be placed either laterally outside or inside the ring, or may envelop its surface more or less.

1878.

*131.—D. Gollner, Batavia. Electrical Cables. 2d. January 10.

CONDUCTORS.—This invention consists in covering the gutta-percha cores of submarine and subterranean conductors with a coating of india-rubber in order to guard them from the ravages of marine and other animals.

162.—W. R. Lake, London. (*J. J. McTighe, Pittsburg, Pennsylvania, U.S.A.*) Producing and Applying Magneto-Electricity. 6d. (8 figs.) January 12.

MAGNETO-ELECTRIC GENERATORS. — A vibratory motion is given to one of the poles of a magnet "having an electrical coil or helix within its field of induction," and thereby "intermittent or pulsatory" electric currents are generated in such coil. This invention is designed for telegraphic and telephonic purposes and takes the form of a horseshoe magnet with one of its poles flattened out to form a diaphragm and its other pole provided with a coil which lies between the two poles.

240.—E. T. Hughes, London. (*W. Wallace, Ansonia, Connecticut, U.S.A.*) Electric Lamps. 4d. (2 figs.) January 18.

ARC LAMP.—This consists of a square wooden base carrying two metallic bars parallel to each other and with a space between. The carbons are rectangular plates and are clamped one to each bar with their edges parallel and close to each other. The electric arc forms between these edges and travels onwards as the carbons consume.

251.—C. W. Siemens, London. Electric Telegraph Conductors, &c. 6d. (4 figs.) January 19.

CONDUCTORS. — After covering the conducting wires with insulating material in any known manner, and to any desired thickness, the core thus formed is drawn through the hollow mandrel of a chuck on which is mounted a bobbin of metallic tape. The chuck being made to revolve, as the conductor is drawn onwards, the metallic tape is wound in helical form on the latter with more or less overlap as may be desired. The conductor is then drawn through a smooth die to close down the convolutions of the taping. The machine for effecting these operations is illustrated and described.

292.—G. Andrews, London. Motive Power. 2d. January 23.

MAGNETIC MOTOR. — "The operation of this machine depends on the property of attraction and repulsion of the negative and positive poles of permanent magnets respectively, the independent force counteracting the gravitation exercised by the weights employed to keep the revolving magnets turning, at the same time overcoming the friction to which every machine is necessarily liable; this force, therefore, has the effect of keeping the weights always wound up, and as the permanent magnets rather increase in power from use, the machine will be self-sustaining."

308.—L. A. Brasseur and S. W. M. de Sussex, London. Galvanometers. 6d. (3 figs). January 23.

GALVANOMETER.—The invention has more especially for its object the construction of a relay for use with long electric conductors. Projecting at right angles from the N. pole of a vertical permanent magnet is a light pivotted iron armature, a soft iron adjustable continuation of the S. pole parallel to and slightly above the armature serving to support it and maintain its stability. Midway between the ends of the armature rise on either side from the N. pole of a second permanent magnet, the soft iron core-pieces of two small electro-magnets, causing the armature to be equally repelled. At the further end and in prolongation of the armature, projects a needle carrying a light pivot descending from which is an arm carrying a small plate, and projecting horizontally from its sides is a lever, having at one extremity a fine platinum point, and at the other an adjustable counterpoise. Two reservoirs containing mercury, respectively connected to the N. and S. poles of a local battery, are mounted immediately under the small plate and fine platinum points.

*311.—H. E. Newton, London. (*Harriet G. Hosmer, Rome.*) Obtaining Motive Power. 2d. January 23.

MAGNETIC MOTOR.—A heavy horizontal balanced metallic wheel, the hollow rim of which is filled with lead, has its boss also hollow, and containing a number of armatures working, as the wheel revolves, in front of a similar number of permanent magnets secured to the framing. On the other side of these magnets are heavier hinged armatures held in place by springs, but removable by tappets attached to the boss. The stationary magnets exerting an attractive force on the armatures in the boss of the wheel cause it to revolve when these have reached a certain point; they require to be relieved from the attraction of the magnets, "otherwise they would be held in contact therewith. This release of the rotating armatures is done by counteracting or destroying their attractive force by allowing the other armatures to come in contact with the opposite sides of the

magnets. On so doing, the magnetic current, or the greater portion thereof, will be diverted from the movable armatures," &c.

312.—H. Clifford, London. **Submarine Telegraph Cables.** 4d. January 23.

CONDUCTORS.—To protect the insulated cores of conductors from the ravages of insects, they are covered by a metallic sheathing prepared as follows : A tape of felt, or other fibrous material, is treated with a mixture of resin and resin oil or marine glue, and while the compound is still in a warm and plastic state, the tape is pressed on one or both sides of thin wire gauze, sheet brass, or other metal, and to insure its perfect adhesion it is pressed between a pair of rollers ; or the metal itself may be treated in like manner. This prepared tape is then laid helically round the insulated core ; each overlap as it passes through the closing rollers of the machine is pressed against the neighbouring underlap, thus forming a compact protecting covering. The usual covering of jute is then served over the prepared tape.

375.—E. T. Truman, London. **Manufacture of Insulated Telegraphic Conductors, &c.** 4d. January 29.

CONDUCTORS.—The conductor is insulated with vulcanite as follows : A tinned conductor is covered with a suitable thickness of a mixture composed of india-rubber, ozokerit (or other wax), and sulphur ; the covered conductor being then inserted into a lead tube, is baked or cured under pressure or confinement, at a proper temperature, for a sufficient length of time. It is asserted that the lead tube will give the conductor the necessary power of bending without the vulcanite cracking or rupturing.

*446.—H. V. Weyde, London. **Illuminating Objects to be Photographed.** 2d. February 2.

REFLECTORS.—A parabolic or other concave reflector is employed in combination with a shield of opaque or semi-translucent material, to intercept the direct rays of light.

*471.—M. Gray, London. (*N. E. Reynier, Paris.*) **Carbon Electrodes.** 2d. February 5.

CARBONS.—The carbon rods having been immersed in a solution of caustic soda or potash are coated with iron or nickel by electro-deposition.

*492.—S. J. Coxeter, London. **Battery Rheostats.** 2d. February 6.

RHEOSTATS.—Holes are drilled through a block of wood, and in these are placed various resistances composed of a mixture of graphite and glass powder. The holes are closed at the upper part by metallic plugs and at the lower part by a metallic plate electrically connecting them all. By means of an arm various resistances may be switched in.

596.—E. B. Bright, London. **Applying Electricity.** 6d. (11 figs.) February 13.

RHEOSTATS.—An arrangement of rheostats and thermostats gives indication on a central instrument of the locality of an outbreak of fire.

611.—A. G. Bell, London. **Electric Telephonic Apparatus.** 6d. (5 figs.) February 14.

MAGNETS.—The magnets are made tubular, and the tube is bent down upon itself so that the outer and larger tube surrounds the inner and smaller tube ; the coil is wound on the inner tube. The vibrating plate may be formed from a portion of the magnet itself.

717.—R. A. Kipling, Paris. **Electric Lamp.** 6d. (8 figs.) February 21.

ARC LAMP.—A lamp adapted to burn four carbons, the two positive being on a higher level, and crossing the two negative at right angles. Each carbon slides in a tubular casing, and is fed forward by a spring or travelling band. A store of carbons may be arranged like the charges of a repeating firearm or a revolver-like action may be attached to the tubular casing. The frame on which the negative carbons are mounted is made to slide, and is pressed upwards by springs; the armature of an electro-magnet being attached to this frame draws it down on the passing of the current and establishes the arc.

759.—F. Lambert, London. **Telegraph Cables.** 6d. (2 figs.) February 23.

CONDUCTORS.—To afford better protection to the insulated core, it is first passed through a bath of pitch, tar, oils, collodion, or other agglutinating material, and is then covered with a layer of slag wool, the whole is then served spirally with a tape impregnated with slag wool.

*767.—H. Conradi, London. (*R. Wiebe, Wolgaste, Prussia.*) **Covering Telegraph Wires, &c.** 2d. February 25.

CONDUCTORS.—The wires have painted on them, layer after layer, a solution of india-rubber and bisulphuret of carbon (CS_2). Chatterton's compound in solution with CS_2 may be similarly employed, or a combination of the two solutions in the proportion of 4 to 1 may be used.

*861.—T. F. Scott, Birmingham. **Apparatus for Producing Electric Light.** 2d. March 2.

CARBONS. — Finely powdered carbon or coke mixed with a thick paste of flour or starch is moulded in dies and dried by exposure to the air, and further hardened by heating. A second and more intense heating is then applied to decompose the flour or starch. Flexible carbons are made by incorporating with the mixture fibres of asbestos or hemp or other mineral or organic fibres. The blocks are made of such a figure that when fixed on a flexible band and passed over a pulley the ends shall interlock, or endless tapes of carbon may be employed. A stream of finely divided carbon is projected on the arc.

902.—N. T. Neale, London. **Transmitting Light and Apparatus therefor.** 6d. (8 figs.) March 5.

REFLECTORS.—The light is condensed by a reflector and passed into a reflective tube through which it is transmitted to the place where it is desired to be utilised, it being there dispersed or condensed by other reflectors. The tube may be in contact with a heat-conducting material.

915.—H. C. Spalding, Bloomfield, N.J., U.S.A. **Transmitting Power by Electric Currents.** 6d. (2 figs.) March 6.

TRANSMISSION OF POWER.—A current generated by any ordinary dynamo actuates a motor constructed as follows : The outer annular casing of soft iron suitably suspended between two metallic rods passing through its edges and secured to the base-plate has its inner edges cut, circumferentially, into a series of ratchet-like steps ; in the illustration six are shown of a long isosceles configuration. Mounted on suitable bearings and concentric to this casing is a revolving shaft having rigidly fixed to it, electro-magnets, whose core-pieces approach sufficiently near to the steps of the annular casing to be alternately attracted by them, an ordinary commutator arrangement being employed to admit the current at the right moment.

953.—J. H. Johnson, London. (*Z. T. Gramme and E. L. C. d'Ivernois, Paris.*) **Electro-Magnetic Machines.** 6d. (2 figs.) (Patent dated March 29, 1878.) March 9.

ELECTRO-MAGNETIC GENERATOR.—A generator for producing alternating currents. Arranged round an annular armature composed of insulated iron wire are a series of coils or bobbins of copper wire, each coil being insulated from the others and coupled up to give 32, 16, 8, or 4 distinct currents as required. Radially mounted on a shaft and revolving in the armature are a number of bar electro-magnets separately and oppositely excited, having their polar armatures extended either way in the path of their rotation.

*1,195.—H. C. Spalding, Bloomfield, N.J., U.S.A. **Electric Conductors.** 2d. March 26.

CONDUCTORS.—Any electrical conducting fluid is put into non-conducting tubes fitted at both ends with metallic connecting plugs.

1,196.—H. C. Spalding, Bloomfield, N.J., U.S.A. **Electric Conductors.** 6d. (9 figs.) March 26.

CONDUCTORS.—To obtain a greater conducting area laminated plates suitably arranged and covered with glass or enamel are used in place of solid conductors, or a glass or enamel tube may be lined with a thin metallic sheet by electro-deposition.

1,197.—H. C. Spalding, Bloomfield, N.J., U.S.A. **Apparatus for Regulating the Motion of Electric Motors.** 6d. (2 figs.) March 26.

GOVERNOR.—An ordinary ball governor driven by the motor actuates by a link a contact lever or switch which increases or decreases the resistance in the circuit as the speed of the motor becomes excessive or is diminished.

1,228.—H. Wilde, Manchester. **Electric Telegraphs.** 6d. (6 figs.) March 28.

MAGNETO-ELECTRIC GENERATOR.—A Pacinotti armature is surrounded at equi-angular distances by a circle of electro-magnets of alternate polarity. The current from one or more coils on the armature, collected at a commutator, is taken to excite the electro-magnets, the remaining current, collected at a pair of rings, is used for external work. The magnets may be separately excited.

1,248.—J. Oppenheimer, Manchester. **Insulators for Telegraphic Purposes.** 4d. (1 fig.) March 29.

INSULATORS.—The pin of a porcelain insulator, open at its top, is in connexion with the earth, and has its upper end pointed and surrounded by a ring of insulating material. A metallic cap, to which the line conductor is attached, covers and is cemented to the insulator and closely approaches the point of the pin ; the whole thus forms a lightning conductor.

1,442.—G. E. Alder and J. A. Clarke, London. **Producing Light for Photographic Purposes.** 4d. April 11.

REFLECTORS.—A suitable reflector, by preference built up of small pieces of silvered glass, in combination with a shield of semi-transparent material, is employed to intercept the direct rays of light.

*1,467.—H. C. Spalding, Bloomfield, N.J., U.S.A. **Originating and Developing Electric Currents.** 2d. April 12.

DYNAMO-ELECTRIC CURRENTS.—Consists in arranging and bringing into simultaneous action a series of dynamo-electric generators " whereby a feeble current produced in the first may be increased or intensified by successively passing through the remaining machines or engines of the series."

*1,587.—I. L. Pulvermacher, London. **Production, Application, and Regulation of Electric Currents.** 4d. April 20.

GALVANOMETER.—A modified form of galvanometer is described as a "compensator," a pivotted "resistance blade" dipping in a trough of mercury being gradually lifted or lowered by the action of the galvanometer. In the construction of a motor the "resistance blades" are used in place of a commutator.

*1,677.—G. Zanni, London. **Electrical Signalling Apparatus.** 4d. April 25.

MOTOR.—A signalling mechanism consisting of a vertical armature mounted between two electro-magnets. Attached to the armature is an escapement device engaging with the teeth of a ratchet wheel which operates a pointer recording the signals on a dial.

*1,770.—C. D. Abel, London. (*N. A. Otto, Deutz.*) **Apparatus for Igniting the Charges of Gas Motor Engines.** 2d. May 2.

MAGNETO-ELECTRIC GENERATOR.—The armature, preferably of the Siemens type, may revolve continuously, its coils and magnets being so adjusted as to give a spark when the gaseous charge has been admitted to the cylinder, or a rapid motion is given the armature only when it is required to ignite the charge. Contact inside the cylinder is made by a movable contact piece, which is separated at the right moment from a fixed insulated contact by suitable mechanism. The current may be passed into a condenser.

1,927.—P. Jensen, London. (*A. J. B. Cance, Paris*). **Electric Apparatus.** 6d. (7 figs.) May 14.

MOTOR.—The magnets of this motor are made by winding on a solid or built-up core a number of turns of insulated wire and then laying on the outermost coil of wire a number of small bars of soft iron, the winding being continued over these bars. This alternate arrangement of wire and iron bars may be repeated as often as desired. The motor described is composed of three such magnets arranged as an inverted triangle, with their axes parallel. A shaft carries at each end five projecting armatures, which pass close to either face of the magnets, and an ordinary commutator admits currents to the magnets as desired.

*1,932.—W. Scantlebury, London. **Galvanic Batteries and Electrical Appliances.** 2d. May 14.

VACUUM TUBES.—This relates to the combination of a condenser, induction coil, and vacuum tube for scenic effects.

*1,937.—W. Abbott, London. **Coating Telegraph Wires.** 2d. May 15.

CONDUCTORS.—The conductor, either bare or previously insulated, has braided over it a cotton or other yarn, or fibre, impregnated with the residium from the distillation of ozokerit in a liquid or heated state.

1,988.—P. la Cour, Copenhagen. **Phonic Wheels, &c.** 6d. (14 figs.) May 17.

MAGNETIC MOTOR.—A small electro-magnet acts by means of the phono-electric current upon the iron teeth of a horizontal mercury-weighted wheel, which first has a spinning motion imparted to it by hand, its rotation being then maintained by the magnet, a tooth passing the pole for each wave of current.

2,003.—H. J. Haddan, London. (*C. Brush, Cleveland, Ohio, U.S.A.*) **Apparatus for the Generation and Application of Electricity for Lighting, Plating, &c.** 1s. (59 Figs.) May 18.

This invention relates to : 1, the general arrangement of the generator ; 2, the novel form of armature ; 3, the construction of the commutator ; 4, the arrangement of the conductors on the field magnets ; 5, an arc lamp ; and 6, encasing or plating carbons.

DYNAMO-ELECTRIC GENERATOR. — The general arrangement will be understood by reference to Fig. 1, a vertical section, through its axis, of the

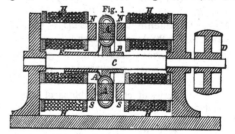

Fig. 1

generator, wherein the ring armature A, of iron or

other magnetic substance, is attached to a limb B of any suitable material, rigidly mounted on the shaft C, which also carries the commutator cylinder E, and is driven by the pulley D. The electro-magnets H H are excited by the whole or a portion of the current; the pole pieces N N, S S, being segmentally extended.

ARMATURE.—A perspective view of this is given in Fig. 2, from which it will be seen to consist of

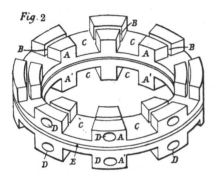

Fig. 2

two parts A A[1], firmly secured together with suitable insulating material E between them. An annular groove B is formed on one or both sides of the armature, while C C are depressions which are wound full of insulated wire, the holes D D, together with the groove B serve to keep the armature cool and destroy any induced currents.

COMMUTATOR. — In Fig. 3, a view in cross-

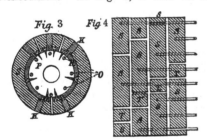

Fig. 3　　*Fig. 4*

section, P is a hub or cylinder of wood or other suitable non-conducting material, to which are attached by suitable screws, separate metallic plates or sub-segments R, having attached to them the bobbin wires O, and on which are mounted the wearing segments S T, the sub-segments corresponding in fashion and location to the segments above them. The copper conducting and insulating segments S S, T T, are secured to the sub-segments R by screws K having the bearing parts of their cylindrically elongated heads near the lower side of the segments. Fig. 4 shows in plan the method of placing and connecting the segments S T. The insulating segments T may be omitted and open spaces kept in their place.

ELECTRO-MAGNETS. — These are arranged horizontally in pairs on either side of the armature and

have their polar faces extended to present a large surface thereto; a fine high resistance wire in comparison with external circuit, or "teaser," is arranged as a shunt, and is wound on the cores to maintain a permanent magnetic field, the main coils being wound on over the "teasers" or otherwise.

ARC LAMP.—A helix of insulated wire in the form of a hollow cylinder A rests on an insulated plate A[1] upheld by a metallic standard E. The soft iron core is partly supported by adjustable

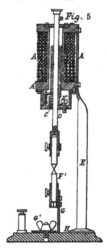

Fig. 5

springs C, through which passes the rod B, moving freely in the helix; a clamping ring D surrounding the rod B has one edge over a lifting tongue C[1], the opposite edge being under the head of an adjustable set screw D[1]. The standard E is fastened to a base H, to which is attached the lower carbon F[1] by means of a support G, having its lower part slotted and bent at right angles, and secured to the base by a set screw G[1]. The upper part carries a carbon-holder composed of a split cylinder, one piece being attached to the support and the other resting in a stirrup provided with a set screw. By suitable link and lever connections the clamping ring is arranged to control a rod operating the feed of the lamp in which both carbons move.

CARBONS.—In coating carbons it is preferred to electro-plate them with copper, nickel, or other suitable material, or a ribbon of any one of the materials may be wrapped spirally round the carbon, or carbon compounded with magnesia, lime, &c., may be rammed into a metallic tube.

*2,008.— J. N. Aronson, London. Scenic Effects, &c. 2d. May 20.

ELECTRIC LIGHT.—The scenes are "dressed" with vacuum tubes so arranged that any group or groups may be partly or totally extinguished. The dresses of the players may be decorated in a similar manner, portable batteries being then used.

*2033.—A. M. Clark, London. (*D. Ward and A. Ball, Berkshire, N.Y., U.S.A.*) Electro-Motors. 2d. May 21.

MAGNETO-ELECTRIC MOTOR.—A wheel carrying a number of magnets alternately wound in opposite directions revolves in a ring to which are attached a corresponding number of permanent magnets. A current changer makes the current alternate, causing the electro-magnets to be attracted to and partly to pass the permanent magnets from which they are then repelled by the reversal of the current.

2,193.—A. F. St. George, Redhill, Surrey. Apparatus for Producing Induced Currents of Electricity, Specially Applicable to Transmitting Signals, &c. 6d. (3 figs.) June 1.

MAGNETS.—Fine insulated wire is coiled on a bar magnet ; a coil of thicker wire is superimposed, having in its circuit a suitable contact break. The induced current of the fine wire causes the diaphragm of the telephone to vibrate, and so give a signal.

2,281.—C. W. Siemens, London. Distributing and Regulating Electric Currents to Work Lamps, &c. 6d. (8 figs.) June 7.

ELECTRIC LIGHT.—Thin strips of metal are employed which elongate or contract according to the current passing, and operate mechanism by which resistance is switched in or out of the circuit, or the movement of the thin strip may directly effect the adjustment of the carbons in a lamp.

2,399. — W. P. Thompson, Liverpool. (*E. Reynier, Paris*). Obtaining Light by Electricity. 6d. (7 figs.) June 17.

ARC LAMP.—A base-plate supports a hollow vertical standard, in the upper part of which slides, in friction rollers, a rod from which projects at right angles the adjustable upper carbon-holder, which, descending by its own gravity, feeds the carbon on to the disc-shaped lower carbon at a line out of its centre, thus furnishing tangential force, which causes the disc to turn, the disc being carried by a bracket-like holder projecting from the vertical standard. Contact with the upper carbon is made in a lateral direction near its point.

*2,401.—C. Dubos, Paris. Apparatus for Producing Electric Light. 2d. June 17.

ARC LAMP.—One carbon is mounted on a soft iron bar or tube sliding loosely through a helix of wire. The bar or tube may be balanced by weights. Where both carbons are required to move, they are connected by a cord arranged over suitable pulleys to regulate their respective rates of travel, or both carbons may be regulated by separate helices.

2,477.—R. Werdermann, London. Apparatus for Electric Lighting. 8d. (15 figs.) June 21.

ARC-INCANDESCENT LAMP.—The upper electrode has a considerably larger sectional area than the lower one, which is fed up to it by weights or other suitable means: "An infinitesimally small arc is produced, to which the incandescence of the carbon is due." An arrangement of lamps in parallel circuit with the means of switching in equivalent resistances is described.

2,640.—E. Leek, Longton, and J. Edwards, Fenton, Staffs. Magnets for Separating Metallic Substances from Materials used in the Manufacture of Pottery, &c. 6d. (20 figs.) July 2.

MAGNETS.—The materials are passed over permanent or electro-magnets of various forms, the magnets being nickeled or silver-plated to prevent their corrosion.

2,816.—G. E. Pritchard, Bishop's Stortford. Magnetising Metal or other Substances, &c. 6d. (12 figs). July 13.

MAGNETS.—Thin plates of hard steel magnetised in the usual way are bent to a parabolic form, and a number of them screwed together, the centre becoming a point of strong polarity.

*2,878.—E. J. Harding and H. Bull, London. Electric Motor. 4d. July 18.

ELECTRIC MOTOR.—Mounted along the shaft are radial "flyers," alternately attracted and repelled by electro-magnets placed under the shaft, a suitable current changer being provided.

*2,930.—H. E. Newton, London. (*Harriet G. Hosmer, Rome.*) Apparatus for Obtaining Motive Power. 2d. July 23.

MAGNETIC MOTOR.—This is identical with No. 311 of 1878, with the exception that the movable armatures are in the form of rollers, and the magnets detached from one another by presenting the same poles.

2,943.—W. R. Lake, London. (*J. W. de Castro, New York, U.S.A.*) Lighting Apparatus for Mines, Tunnels, &c. 6d. (6 figs.) July 24.

ELECTRIC LIGHT.—The light is reflected into the workings, &c., by an arrangement of mirrors, prisms, and tubes. The tubes contain caustic lime or chloride of calcium to keep the air within them dry and clear.

2,962.—**J. H. Nettlefold**, Birmingham. **Joining Wires.** 6d. (18 figs.) July 25.

CONDUCTORS.—In joining conducting wires their ends are doubled back for a few inches, the loop so formed at the end of one wire is passed through the loop of the other, and the whole drawn up tight.

3,016.—**W. Morgan-Brown**, London. (*D. G. Haskins, Boston, Mass., U.S.A.*) **Lifeboat and Signal Buoy.** 6d. (6 figs.) July 30.

ELECTRIC LIGHT.—From "a rotary electrical apparatus, a conducting wire extends upwards" and "passing out terminates in luminous rotating points."

3,134.—**C. H. Siemens**, London. (*E. W. Siemens and F. H. von Alteneck, Berlin*). **Apparatus for the Dynamical Production and Application of Electricity.** 1s. 2d. (32 figs.) August 8.

DYNAMO-ELECTRIC GENERATORS.—The first genetor described is of horizontal construction and has its coils wound on a cylindrical armature, each pair of opposite divisions being coiled with two separate wires ; one of the wires starting from an insulated metallic ring and traversing the length of the armature terminates at one of the commutator plates ; the other wire starting from the directly opposite plate of the commutator terminates at the metallic ring, which thus receives one end of all the wires. Two sets of tangential brushes collect the current. In the modified form shown in Fig. 1,

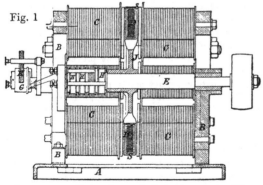
Fig. 1

the armature D is a flat soft iron ring made either solid or built up of iron wire and is attached to the shaft E by arms projecting from the boss J. On the base A are fixed the two end frames B B, secured to which are the electro-magnets C projecting inwards parallel to the axis of the machinery, the iron plates F of trapezoidal shape forming polar fields. The coils S are formed of insulated wire transversely wound on the ring D, and the coils of the magnets are so connecte that the two

polar fields facing each other are always of one polarity, and the two next in order of opposite polarity ; the coils S thus have alternating currents induced in them which may be transmitted separately from the machine or combined in groupings by means of the insulated rings H and tangential brushes K. The electro-magnets having constant polarity the coils take their current from the brushes of a commutator G. In a further modification a non-magnetic disc rotates between the electro-magnets as last described, and has wound in its plane on elongated cones, preferably of wood, a number of coils of insulated wire, the electromagnets having their polar ends correspondingly extended by attached iron plates. The arrangement of insulated rings and commutator last described is used to collect the current. Another form of generator has its electro-magnets arranged radially, the coils of the armature being wound on a cylindrical shell revolving between an inner and outer set of electro-magnets. Fig. 2 shows,

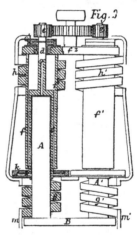
Fig. 2

half in elevation and half in vertical section, a unipolar generator for producing currents of low electromotive force, in which two cores A A¹ secured to a base-plate B, all of soft iron, form a horseshoe magnet. Revolving about the cores are cylindrical copper shells C, on an axis *d* on which is a pinion *e*. Outside the shells C are cylindrical iron shells *f f¹* secured to a plate *f²*, and with it forming a horseshoe magnet. Surrounding the cores A A¹ at their lower parts are helices of thick copper *g g¹*, similar helices *h h¹* surrounding the upper parts of *f f¹* ; these are in electrical connection with the shells C through the shallow vessels *k* and *l* containing mercury, the current going to the external circuit by the terminals *m m¹*. This invention also relates to various methods of coupling up separate generators "so as to produce a chain of dynamo-electric action."

ARC LAMPS.—In the first lamp described, the

two carbons held in suitable holders, incline to each other, and meet at their points, being kept in this position either by springs or their own weight. The point of a rod of refractory material is thrust between the points of the carbons by the action of a solenoid, and separates them, thus forming the arc, the rod being consumed at the same rate as the carbons. In another form of lamp the carbons are placed vertically over each other, the lower carbon-holder being pivotted in the fork end of an armature drawn upwards by a spring, but actuated by an electro-magnet, through which alternating currents pass, thereby giving to the lower carbon rapid vertical oscillations, the upper carbon feeding downwards by its own weight as it is consumed. This arrangement is capable of being inverted, in which case the lower carbon is fed upwards by a float, or a spring in tension.

3,250.—H. Wilde, Manchester. **Producing and Regulating Electric Light.** 6d. (2 Figs.) August 17.

ARC LAMP.—Two carbons, arranged vertically and parallel, slide through two clipping holders; one holder, pivotted at its lower end, has an armature which is acted on by an electro-magnet to part the carbons, an opposing spring keeping them in contact. When the carbons have burnt to a certain depth an occulting ring obscures the light and they have to be raised. A ring of insulation round the bottom of one carbon prevents their lower ends coming in contact.

*3,315.—C. W. Siemens, London. **Apparatus for Electric Lighting.** 2d. August 22.

ARC LAMP.—The carbons are fed through two tubes, inclining downwards and forwards, by rollers actuated by clockwork, the last wheel of the train being controlled by "flyers," on the spindle of which is fixed a friction wheel. Immediately over this wheel is a small weight suspended from the centre of a thin strip of platinum or other metal, having one end connected to each of the troughs; should the carbon points come too close more current passes through the metal strip, which by its consequent expansion lowers the weight on to the friction wheel and so retards the feed movement.

CARBONS.—To provide a continuous supply of carbons the back end of one has a countersunk hollow, and the front end a conical point; the attendant serves the point with a paste made of pasty caoutchouc solution, or carbon powder and adhesive gum, and presses the gummed point into the previous carbon.

DYNAMO-ELECTRIC GENERATORS.—To obtain alternating currents from constant current generators a commutator is placed on the shaft consisting of successive plates connected respectively and alternately to the electrodes of the machine, and having a pair of tangent wire brushes bearing against them, connected respectively to electrodes for the external circuit.

*3,317.—E. A. Obach, Charlton, Kent. **Galvanometers.** 2d. August 22.

GALVANOMETERS.—The meridian ring or coil is mounted on horizontal bearings, one at each side of the needle dial, and is provided with a graduated scale and set screw, permitting of its being adjusted at various degrees of inclination to the axis of the needle, which is capable of movement only in the horizontal plane.

3,338.—J. E. F. Ludeke and A. J. Thorman, London. **Obtaining and Applying Motive Power by Magnets.** 6d. (14 figs.) August 24.

ELECTRO-MAGNETIC MOTOR. — "Batteries" of horseshoe magnets, arranged with the poles alternately so as to form hollow cylinders, are placed at either end of a base-plate; a soft iron shaft running through the centre of the cylinders carries two crosses of soft iron working close to the faces of the "batteries;" above this, and geared with it, is a second shaft carrying a cross formed of two horseshoe magnets with their poles suitably bent outwards; these revolve in the space between the two crosses on the lower shaft and one of sufficient length just to clear the same. Another motor is in the form of a revolving wheel or cylinder, in the interior of which "grooves or curves are formed of peculiar construction" radiating from the centre to the periphery; in each of these grooves is placed an iron ball movable therein, but prevented from bearing at the periphery. Suitably arranged magnets alternately attract and repel the balls. In either of the foregoing electro-magnets may be substituted for the permanent magnets.

3,470.—C. W. Harrison, London. **Obtaining Light by Electricity.** 4d. September 2.

ELECTRODES.—Consists in impregnating the carbons with some of the metals or elementary bodies which exist in the solar spectrum or its atmosphere combining those whose colours are complimentary, *e.g.*, sodium with potassium, zinc, or mercury; this being effected by immersing the carbons in suitable solutions of the metals, or the ingredients may be mixed together in the form of powder and united by pressure. Flexible carbons are made by coating filaments of silk, cotton, &c., previously treated with acid to render it more

combustible with the mixture of carbons and metals.

ELECTRIC LIGHT.—Induced currents are used to maintain the light ; these are obtained from induction coils made of insulated ribbons of iron and copper wound one over the other. When the primary current passes through the copper coil, induced currents are created in the iron. A group of these coils may be arranged in series and made to sustain an increased light.

3,603.—G. J. Stanford, London. (*E. F. Phillips, Providence, R.J., U.S.A.*) Flexible Electric Conductors and Tips therefor. 6d. (11 figs.) September 12.

CONDUCTORS. — This relates to the braiding of telephonic conductors and the attachment of a "tip" wire by means of a thin metal thimble or shell ; the tip wire having been joined to the end of the conductor, by looping or otherwise, the shell is passed over the joint and crimped on to the braided end.

3,622.—W. Smith, London. Submarine Electric Telegraph Cables. 4d. September 13.

CONDUCTORS.—The object is to construct submarine conductors of small specific gravity. The conductor is coated with one or more layers of insulating material, and this is covered with a mixture of gutta-percha and shellac. Over this the iron or steel wires are laid at equal distances apart in long open spirals. To embed them in the core they are heated, or the core itself may be heated, the core being then covered with further coatings of insulating material.

3,656.—A. H. P. Stuart-Wortley, London. Manufacture of Electrodes for Electric Lighting, &c. 4d. September 17.

ELECTRODES.—To prevent the too rapid consumption of the electrodes the points are protected by hoods or caps of asbestos, or the electrodes may be made of the following composition : Gas retort carbon, 10 parts ; plumbago, 2 parts ; asbestos, 1 part ; or steatite, 1½ parts. It is preferred to set the electrodes obliquely so as to give a slight shearing action when they run together.

3,658.—J. H. Johnson, London. (*A. de Meritens, Paris.*) Magneto-Electric and Electro-Magnetic Machines. 6d. (16 figs.) September 17.

ELECTRIC GENERATORS.—This invention relates to direct current and alternating current generators. In the first or magneto-electric generator the armature is composed of a brass ring, around which are coils, with soft iron segmental cores

separated by diamagnetic partitions ; the inducing magnets, of horseshoe form built up of many thin plates, are mounted on two brass rings attached to the base-plate, and are arranged round the circumference of the armature in such a manner that the intervals between them are equal to the solid portions. The current is collected from two insulated rubbers mounted on the shaft. In the second or dynamo-electric generator for supplying direct currents, an alternating current generator has its hollow shaft coupled to the shaft of a second generator constructed as follows : straight inducing electro-magnets, having their cores formed of fine iron wire, are mounted on two brass wheels rigidly attached to a hollow shaft ; at either end or pole of these magnets is a ring of coils (constructed as before described) mounted in brass frames bolted to the base-plate. The alternating currents produced by the first generator pass along the hollow shafts to the straight electro-magnets of the second generator, inducing in the coils of the latter direct currents which are collected from two binding screws without the use of a commutator.

3,676.—H. E. Newton, London. (*Harriet G. Hosmer, Rome.*) Magnets and Armatures for Producing Motive Power, &c. 6d. (2 figs.) (Sealed the 28th of March by order of the Lord Chancellor.) September 17.

MAGNETO-ELECTRIC MOTOR. — This is fully described in Specifications Nos. 311 and 2,930 of this year.

3,837.—A. Longsdon, London. (*A. Krupp, Essen.*) Electric Light Apparatus. 6d. (4 figs.) September 28.

ARC LAMP. — In this lamp, which is focus-keeping, the descent of an upper weighted carbon-holder causes the ascent of the lower holder by means of a small chain connecting them together and passing over suitably proportioned pulleys whose speed of rotation is controlled by a tooth wheel which gears with a fan revolving in a vessel of mercury ; the tooth wheel is fitted with a pawl admitting of the carbon-holders being drawn apart to insert fresh carbons. The feed of the carbons is controlled by a brake actuated by a shunt magnet pressing on the periphery of a friction wheel, the brake lever being so arranged that it grips and draws forward this wheel, slightly separating the carbons should they get too close. The lower carbon-holder may be immovable in lamps where it is not necessary to keep a focus.

*3,912.—W. S. Wilson, Sunderland. Manufacture of Electric Lights, &c. 2d. October 4.

ELECTRODES.—It is proposed " to form the

electric arc by means of a gas or vapour of sufficient conducting power, such as a metallic vapour or hydro-carbon, preferably saturated or mixed with carbon."

3,971.—G. and E. Woods, Warrington. Connecting the Ends of Wires for Continuous Fencing, Electric Telegraphs, &c. 6d. (5 figs.) October 9.

CONDUCTORS.—The ends of conducting wires are joined together : 1, by a split metallic tube having threads cut on its conical outer surface, the two halves being jammed together by means of two nuts ; 2, by laying the ends in taper grooves in a cylindrical piece of metal and fastening them by nuts ; 3, by putting them into a frame holding two roughened eccentric or cam-like jaw-pieces, which jam the ends on any attempt being made to withdraw them ; and, 4, by inserting them in a frame having on either side one roughened cam-piece, the opposite surface being corrugated.

3,976.—M. R. Ward, Brighton. Magneto-Electric Machines. 6d. (18 figs.) October 9.

DYNAMO-ELECTRIC GENERATOR. — The first generator described is arranged similar to Gramme's, but has the core of the ring armature built up of a number of soft iron segments insulated from each other, on which the insulated wire is wound parallel to the axis of the armature, the whole being mounted on a wooden hub. The commutator is built up in the ordinary way of alternate strips of metal and insulating material. In the second generator an armature, as above described, is mounted on a hollow shaft and revolves past two electro-magnetic inductors fixed rigidly on a shaft, and having their outer polar faces cylindrically extended to present a large surface to the interior of the revolving armature. In a generator for producing alternating currents, electro-magnets of alternate polarity rotate inside a series of H-shaped segmental cores bolted together so as to form an outer drum-like casing, insulated wire being wound on these cores as described in the armature of the previous generator. The shaft carries two insulated metallic rollers, to one of which the extreme end of the wire going round the cores of the magnetic inductors is attached, the other end being attached to the second ring ; pressing on these are brushes of the ordinary form. In a modification of this generator a double set of magnetic inductors revolve, the one inside and the other outside a ring of fixed coils, both the inductors and coils being constructed as previously described ; the two inductors which influence simultaneously one coil are of like polarity.

*3,985.—H. W. Tyler, Kt., Colchester. Electric Lighting. 2d. October 9.

ELECTRIC LIGHT.—This invention applies to : 1, the use of platinum or other gaseous liquid, or solid materials to give a light ; 2, a suitable switch and resistances ; and 3, the combination of the above.

3,988.—St. G. L. Fox, London. Electric Lighting, &c. 6d. (6 figs.) October 9.

INCANDESCENCE LAMP.—This lamp has a filament formed of an alloy of platinum and iridium attached at either end to conductors passing through and hermetically sealed in the top of the lamp, or the filament may be passed several times over and depend from a glass bar or tube placed in, and near the top of, the lamp. To distribute the electrical energy one pole of the generator is attached to a system of main conductors and their sub-branches, and the other pole to earth, which is used throughout the system as a return.

CURRENT METER. — The quantity of electric energy used is measured by an electro-magnetic voltameter joined up in a derived circuit from the main conductor, and consisting of an electro-magnet and automatic make and break ; the voltameter being joined up between the points of contact, the extra current formed on breaking contact passes through the voltameter and decomposes the water.

SECONDARY BATTERIES.—A number of secondary batteries joined together in series between the main and earth serve as reservoirs for the electricity.

CONDUCTORS.—The mains may be carried in gas pipes and suitably insulated therefrom. To maintain a constant electromotive force, a quadrant electrometer, placed in the circuit, is so arranged that the deviation of its needle either way will complete one or other of two electric circuits, which acting on an electro-magnet, cause it by suitable means to regulate the supply of steam to the engine working the generator.

3,991.—J. Clark, London. Adapting the Electric Light to Domestic and other Purposes. 4d. October 9.

ELECTRIC LIGHT —To subdivide the main current a bath of mercury is placed at the end of the main conductor, and from this a number of finer conductors, all preferably of uniform length and resistance, lead to the lamps. The lamps are made by enclosing finely powdered charcoal in glass tubes, which are then surrounded by a larger globe. Asbestos, either powdered or in small lumps, or chromate of iron, or bichromate of potash, may be used as a luminous medium.

4,006.—P. Jensen, London. (*S. Marcus, Vienna.*) Electric Lighting Apparatus. 6d. (5 figs.) October 10.

INCANDESCENCE LAMP.—In this lamp an adjustable carbon disc is slowly rotated by the descent of a weighted carbon rod which impinges on its periphery in a line between its centre and circumference; the spindle of the disc is cut with a fine thread, and this gives it a slight horizontal movement. The lamp may be inverted, the carbon rod being then pressed up by a weight and string arranged over a pulley.

*4,016.—J. Munro, London. Electric Lighting, Heating, and Motive Power Apparatus. 4d. October 11.

ELECTRIC LIGHT.—The object of this invention is to divide the current and supply it in separate pulses to each lamp. To accomplish this a divider is used which consists of a rapidly revolving shaft connected to one pole of the generator. The ends of all the separate lamp circuits are arranged radially round this shaft and a brush contact attached to the shaft sweeps the ends as it revolves, thus making contact with every lamp in each revolution, or the ends of the conductors may be arranged alongside the shaft and contact made by cams arranged helically on the shaft, or the shaft may have an oscillating motion, or a vibrating fork may distribute the current which may be passed through a condenser on its way to the divider.

*4,022.—J. W. T. Cadett, London. Arrangements for Electric Lighting. 4d. October 11.

RESISTANCES.—To maintain a steady light an excess of current is generated and resistances are placed between the generator and the lamp, "such resisting medium to be made to deliver the amount of current required for each place." An automatic resistance is made by causing a motor to raise an arm to which is fixed a rod of carbon dipping into a bath of mercury; when the resistance thus caused becomes too high an electro-magnet releases an armature and thus short-circuits the motor. In arc lamps a rod of carbon is arranged to dip into a trough of mercury as the carbons draw apart, so counterbalancing the increased resistance of the arc. An electro-magnet switches in an equivalent resistance in the event of the arc being extinguished.

4,031.—A. M. Clark, London. (*L. Clemandot, Paris.*) Softening or Diffusing Electric and other Lights. 6d. (4 figs.) October 11.

GLOBES.—Double globes are used and the intervening space is filled with materials of a flocculent nature such as powdered glass, glass wool, spun glass, calcined mica, or silver leaf, either separately or in various combinations. The materials and globes may be tinted, or the globes may be built up of lenses or prisms.

*4,041.—W. North, Huddersfield. Electro-Magnetic Engine for Driving Rotary Hair Brushes. 2d. October 12.

ELECTRO-MAGNETIC MOTOR.—An iron spindle has a number of radial electro-magnets mounted on it, and an outer drum-like case carries on its inner side a corresponding number of soft iron armatures and a number of brass contact pins, which make and break the magnet circuits as the case revolves. The rotary brush is on the outer side of the case.

4,043.—St. G. L. Fox, London. Applying Electricity for Lighting and Heating. 6d. (7 figs.) October 12

INCANDESCENCE LAMPS.—This invention relates to covering the filaments in incandescence lamps with a refractory substance such as finely divided asbestos, fireclay, lime, magnesia, or steatite. For heating purposes a coil of fine wire is used. For boiling water, a coil of enamelled wire is placed at the bottom of the vessel. A bad conductor, such as graphite in combination with fireclay, may be made to form the lining of a furnace. Where very high temperatures are required, the coil, placed inside a non-heat conducting furnace, closely surrounds the crucible. To regulate the flow of current, a rod of copper or silver is joined up in the circuit, and by being plunged more or less into a glass containing mercury, increases or diminishes the resistance.

SECONDARY BATTERIES.—A number of secondary batteries, coupled together in series, are each fitted with a commutator enabling them to be coupled in quantity and charged by a current of low electromotive force, and to be capable of giving, by an alteration of the commutators, the full tension.

*4,046.—G. P. Harding, Paris. Obtaining Electric Light. 2d. October 12.

ARC LAMPS.—The weight of the electrode is supported by a column of fluid maintained at a constant head; as the electrode consumes it is carried up by the fluid.

4,047.—G. P. Harding, Paris. Apparatus for Regulating Electric Light. 6d. (4 figs.) October 12.

REGULATOR.—To avoid over-heating of the electrodes, a wire, of less resistance than the lamp, is placed in the main circuit and held in tension by a small lever contact maker and spring. On the current becoming excessive, this wire elongates and completes a shunt circuit, in which suitable resistances are placed, or an electro-magnet may be used

to automatically switch in the resistance. A wire may be caused to heat its casing of lime or other material, and an arc light may be formed in a cavity of a piece of lime, or lime may be carried on the electrodes by the action of the current.

4,066.—J. Imray, London. (*Société Generale d'Electricité, Paris*). **Magneto-Dynamo Electric Machines.** 6d. (12 figs.) October 14.

DYNAMO-ELECTRIC GENERATOR.—In this generator the electro-magnets, preferably of U form, are mounted obliquely on a drum-like boss revolving within a ring of soft iron cores, either flat or of U form, wound with insulated wire and carried on two fixed annular frames of gun-metal, to which they are attached by an extended polar face at each end, entering a notch in the frame, and being secured therein by a set screw; the cores are placed either parallel to, or obliquely with, the axis of the shaft; the core-pieces may have a pole at either end and one in the centre, in which case the outer poles are of the same and the centre one of opposite polarity; the magnet coils are then given increased obliquity, so that each is within the influence of three of the cores. The cores of the U shaped electro-magnets may be coiled by forming the wires into a bundle and winding them zig-zag fashion over the alternate coils of the cores, the end wires of the bundle being connected in such a manner as to make the whole into one circuit. The current is collected at insulated rings mounted on the shaft.

*4,074.—A. Arnaud, London. **Indefinitely Dividing Electric Currents.** 2d. October 14.

DISTRIBUTION OF CURRENTS—To establish an equilibrium of resistance throughout a circuit, both conductors are put in communication with the positive pole, one direct and the other through the work to be done; the current then follows a parallel course in the two conductors passing from the first to the second by the work to be done, and to the end of the circuit before returning to the negative pole. In electric lighting the lamps are put in parallel.

4,079.—I. L. Pulvermacher, London. **Translucent Medium for Reflecting, Refracting, and Diffusing Light, &c.** 6d. (4 figs.) October 14.

REFLECTORS.—A dish-shaped globe of glass or other suitable material has its inner side furnished with a chemically deposited layer of platinum, rendering the surface brilliant and yet retaining a certain transparency; the light is thus diffused in the direction of a suspended re-reflecting arrangement of suitable shape.

4,094.—I. L. Pulvermacher, London. **Producing and Regulating Electric Currents.** 8d. (13 figs). October 15.

GALVANOMETER.—This relates to improvements on the batteries and galvanometer described in Specification No. 3,469 of 1877. In this instrument, the "resistance blade," dipping into the mercury trough, is made of a piece of glass or ebonite, wound spirally with platinum wire; the blade is actuated as previously described.

4,100.—F. H. Varley, London. **Producing Electricity and Electric Light.** 6d. (11 figs.) October 15.

ARC LAMP.—This relates to the invention already described in Specification No. 1,702 of 1877. A stream of finely divided carbon is projected on the lower electrode. An electro-magnet, controlled by the action of the current, regulates the opening in a hopper; in a modification the hopper is replaced by a vertical cylinder, terminating at its lower end in a conical orifice, in which a disc, actuated by means of linkwork connected to an electro-magnet, regulates the supply of carbon powder. The carbon powder may be fed to the lower electrode by an endless band moved by rollers connected to clock-work, or in any other way. Similarly the carbon powder may be fed between the negative and positive electrodes, these being placed either horizontally or vertically, inclined to one another. In an arrangement for working in a vacuum, the hopper has a fine orifice, and a trembling hammer, to maintain a steady flow of carbon powder, is placed over the electrodes which are inclined to one another, and the whole is a placed in a glass or other exhausted chamber.

*4,114.—E. J. C. Welch, Manchester. **Dividing and Distributing Electric Currents for Lighting, &c.** 2d. October 16.

DIVISION OF CURRENTS.—This invention "consists in transmitting the main electric current along each of the circuits of a series consecutively." The apparatus is composed of a rotating piece in connection with one terminal of the generator, having the connections with which it is consecutively to complete a circuit ranged around it, but insulated from it and from each other.

*4,116.—G. Forbes, Glasgow. **Electric Lighting.** 2d. October 16.

ARC LAMP.—The light is exhibited in flashes of extremely short duration by making and breaking contact between two carbon points. "A convenient way of giving a sufficiently rapid motion to the carbon points is by mounting them upon a revolving wheel."

4,132.—A. S. Hickley, London. **Electric Light-ing Apparatus.** 2d. October 17.

ARC-INCANDESCENCE LAMP.—Two carbon rods pressed upwards by springs impinge on the under side of a block of magnesia, lime, or other refractory material, and are each in connection with one pole of a generator; a piece of fine wire or foil passed across the top of the carbons serves to establish the arc; or a platinum or other wire, or piece of carbon, passing through a hole in the block of lime, connects the points of the two carbons and is withdrawn by an electro-magnet on the passage of the current; or the magnet may switch on a current of high tension which "sparks" across the tops of the carbons, the sparks acting as a bridge to convey the primary current, whereupon the electro-magnet shuts off the high tension current; or a constant high tension current may be used. Three parallel carbons may be employed, the two outer ones being negative, and the inner one positive, each being pressed against the block by a spring.

*4,140.—W. Scott, London. **Electrical Apparatus.** 2d. October 18.

AUTOMATIC SHUNT.—The apparatus is constructed of a compound bar of metals differing in their degree of expansion under heat; these are by preference bent in the form of a helix, through which the whole or a portion of the current may pass. This helix is fixed at one end, and attached at the other end to an arm provided with sliding pieces arranged to alter the resistance in two or more rheostats and so vary the current passing through them. In cases where the variation of the current is slight, a simple piece of metal, or a small column of mercury acting on a piston, may be used to actuate the arm. A similar arrangement of shunt may be used to short-circuit or throw off the main circuit when all the branches are thrown off. A compound bar may give motion to an index on a graduated scale to indicate the amount of electricity passing.

4,163.—J. N. Aronson and H. B. Farnie, London. **Electric Lighting.** 4d. October 18.

ELECTRODES.—Osmium tips are used, made from osmium in a powdered or granulated state which is consolidated by heat or pressure. In the first case it is moderately compressed in plaster-of-paris moulds to the size and form required; the mould is then put into a muffle and submitted to a white heat for about an hour. In the second case it is compressed in steel moulds under a pressure of, say, 1,200 tons to the inch. The tips are then placed in holders, preferably of carbon, and placed in the electric circuit. The passage of a powerful current causes the osmium to more or less fuse and become closely consolidated and homogeneous.

*4,180.—I. L. Pulvermacher, London. **Producing Light by Electricity.** 4d. October 19.

INCANDESCENCE LAMPS.—These are formed by winding a thin platinum wire over two larger wires of good conducting material kept apart at the distance of half the length that it is intended to give to the loops, the edges of the platinum wires being soldered to the conducting wires. The band may be used in this form or folded longitudinally, so forming loops of the platinum wires. The band may be rolled into a circular, convolute, or spiral form and inserted in a glass enclosure. In another lamp thin platinum wires are looped through holes in a sheet of mica; all the wires are connected at the back of the mica, and in front they are connected in alternate order with the two conductors. Another lamp is constructed of platinum foil mounted on mica, having elongated clamp contact pieces along the sides.

*4,206.—R. P. Higgs, London. **Magneto or Dynamo-Magneto Electric Machines.** 2d. October 22.

GENERATOR.—It is proposed to construct the generator so that only magnetic poles of like polarity are presented to one conductor. In a generator having the conductor and exciting magnet rotating together as a single whole rotating body, it is preferred to wind a bar of iron, which can rotate on its longitudinal axis, as a straight electromagnet and place collecting springs at the ends and middle from which to convey the current induced. In a generator in which the exciting magnets are fixed, it is preferred to form the periphery of a wheel, of curved electro-magnets; this wheel, mounted on an axle, is caused to rotate in the end of a cylinder formed of electro-magnets; in the other end of the cylinder a second similar wheel rotates.

4,208.—C. W. Siemens, London. **Means and Apparatus for Electric Illumination.** 8d. (7 figs.) October 22.

ARC LAMP.—In the first lamp described the arc is formed between the hemispherical ends of two short tubes of steel or other material kept cool by a current of water playing on their interior surfaces; carbon may be interposed between these tubes, or the negative electrode only may be cooled by water and the other made of carbon, preferably in the form of a disc, which slowly rotates. The hollow negative electrode is fixed at the longer end of a tubular arm pivotted to the lamp frame, and has

on its shorter end an armature actuated by an adjustable electro-magnet, working in opposition to a spring, which regulates its distance from the revolving disc. A small cistern, fitted with an overflow pipe, receives the water after it has passed round the hollow negative electrode, and is so arranged on a leaf spring as to form part of the lamp circuit. When the water ceases flowing the spring, overcomes the weight of the cistern and the circuit is broken. A small spring is provided to " tail off " the current and so prevent the contact pieces being damaged. The axis of the carbon disc rests on the curved arms of a pivotted and counterpoised lever, and is retained by radius yokes pivotted to the fixed framing of the lamp ; the rotation of the axis (caused by its inclination to run down the curved arms and by the preponderance of the half-circumference of the disc which has yet to pass the arc) is controlled by a flyer geared with a train of wheels, a suitably arranged brake, pivotted with, and attached to, the tubular arm of the negative electrode, checks the flyer so long as the lamp burns a normal arc. A modification of the negative electrode consists of a pipe bent round upon itself and fed by a small circulating cistern placed on the top of the lamp frame. The flow and return pipe expose a considerable cooling surface. The distance between the electrodes is maintained by the expansion and contraction of a strip of metal placed in the main circuit.

*4,212.—C. T. Bright, Kt., London. **Lighting by Electricity.** 2d. October 22.

INCANDESCENCE LAMPS.—The light is worked by the secondary coils of an induction apparatus situated at each point where the light is required, the primary coils are coupled up to a main conductor common to them all. The lamps consist of glass vessels containing a rarefied gas such as nitrogen and vapours of phosphorus, mercury, ether, or volatile hydro-carbons ; the current passing through these renders them luminous. The current supplied to each building is regulated by means of rheostats, the amount used being indicated by a series of wheels controlled by an electro-magnet in the circuit, and having indicators upon their axles.

4,226.—T. A. Edison, New Jersey, U.S.A. **Developing Electric Currents and Lighting by Electricity.** 1s. (68 Figs.) October 23.

MAGNETO-ELECTRIC GENERATOR. — This consists of a large tuning-fork, which may be either a permanent or electro-magnet, or have electro-magnets mounted on the outer sides of both its arms ; opposite these arms are fixed electro-magnets, in which the current is induced. The vibrations are maintained by a small air, gas, or steam engine

suitably connected to the end of each arm of the fork.

AUTOMATIC CUT-OUT.—The light emitting spiral of an incandescence lamp is coiled round a vertical rod of metal which expands when the heat of the spiral becomes excessive, and lowers a contact lever attached to its bottom end, which thus cuts the lamp out of circuit ; or the rod of metal may be adjacent to the spiral and reacted on by the radiated heat ; or the air in a closed vessel containing the spiral or luminous strip expands when the heat reaches too high a point, and causes a thin metal diaphragm to make contact with a terminal, and so cut the lamp out of circuit ; or the closed vessel may be in communication with a chamber arranged similarly to an aneroid, and which would actuate the cut-out ; or a bent tube, containing mercury, in which is a movable float attached to a contact lever, may be used as a cut-out. A safety arrangement prevents the circuit being interrupted in the event of either the spiral or rod breaking, a resistance coil, equal to the lamp, being switched in should the lamp break or not be required.

SECONDARY BATTERIES.—These may be used, the current being put through them by the expansion of metal rods.

CURRENT METER.—This consists of a resistance coil proportioned to the number of lamps used, and an electrolytic cell, containing, in a solution of copper, two electrodes of copper, one very thick and the other very thin, the small portion of current passing through the cell carrying over copper and depositing it on the thin plate. Various methods of coupling in lamps, so as to maintain equal resistances in each sub-branch, are shown and described.

REGULATOR.—An index or galvanometer at the central station indicates to the engineer the strength of the current, so that he may regulate the machine, or a governor may be employed to throw in resistance or short-circuit the current.

4,278.—E. J. C. Welch, Manchester. **Apparatus for Electric Lighting.** 6d. (3 figs.) October 25.

CURRENT DIVIDER.—This consists of a cylinder having on its inner circumference an odd number of insulated contact pieces. Mounted on a shaft, revolving concentrically with the cylinder, are an even number of adjustable rubbing contact pieces in connection, through the shaft, with one pole of the generator ; as the shaft revolves the circuit is completed with one of each of the insulated contact pieces in turn by each of the rubbing contacts.

4,283.—J. E. Stokes, Upton Manor, Essex. **Electric Lighting.** 6d. (15 figs.) Oct. 25.

AUTOMATIC CUT-OUT.—The light emitting spiral

surrounds a rod of metal which by its expansion actuates a lever attached to its lower end, which makes contact with a relief circuit, and thus cuts the lamp out when the heat becomes excessive.

*4,287.—F. Verrue, Paris. Magneto-Electric Engines. 2d. October 25.

MAGNETO-ELECTRIC GENERATORS.—Upon a cast-iron frame are fixed two bronze drums on which the coils, "composed of a pair of flat parts, or bars of soft iron, the copper wire being wound round them," are fitted. On a shaft "is keyed a bronze drum having wide flanges on which the magnets are carried; these magnets are placed end to end, so as to form a distinct circle or crown around each of the rings of coils. The wires are united to collectors which collect the currents."

*4,304.—C. E. Shea, Foot's Cray, Kent. Dividing and Distributing the Current Produced by Galvanic Batteries, &c. 6d. (4 figs.) October 26.

DIVISION OF CURRENTS.—Two discs, insulated from each other, are mounted on a revolving shaft, and each connected to one pole of a generator; projecting from their faces they have two studs or contact pieces; the ends of the various circuits are collected together and arranged around the circumference of two similar, but fixed, discs placed so that the projecting contact pieces from the revolving discs bridge them across, and so complete each circuit in rotation.

*4,313.—A. A. Cochrane, London. Obtaining and Applying Electricity. 2d. October 26.

GENERATORS.—It is proposed to obtain electricity from vessels in which water, steam, or vapours or gases may be heated or generated, care being taken to insulate the same from the earth. It is proposed to collect electricity in vessels similar to gasometers, constructed on the principle of Leyden jars. To regulate the intensity of the current through metallic conductors, the conductivity of the conductor is modified.

*4,315.—B. P. Stockman, London. Apparatus for Producing Light by Electricity. 2d. October 26

ARC LAMP.—The lower carbon is placed in a cylinder containing mercury, in which is a second inverted cylinder, through the top of which a piston rod passes, and to this the carbon is secured and fed upwards by flotation, the upper carbon being a fixture; or the carbons may be attached to one arm of a pivotted lever, and the mercury float attached to the other arm.

4,316.—J. W. T. Cadett, London. Electrical Arrangements. 6d. (1 fig.) October 26.

CONDENSERS.—These are placed in the main circuit in series, and deliver electricity for any purpose that may be required. An electro-magnet placed in the circuit operates a rheostat, and so maintains an even resistance between all the condensers.

*4,317.—F. D. Tilleard, London. Lamps for Lighting by Electricity. 2d. October 26.

LANTERN.—Railway carriages, locomotives, steamships, &c., are lighted by any known form of arc lamp, enclosed in a lantern made of iron or other opaque material, in the shape of a box having apertures at the base, from which extend mirrors or reflectors of glass to reflect the light in the direction required.

*4,338.—C. W. Harrison, London. Obtaining Light by Electricity. 2d. October 28.

LAMP.—This is made by placing three insulated tubes vertically, and parallel, the two outer ones being filled with a conducting fluid, preferably mercury, which flows on to a vibrating piston near the top of the centre tube, a series of continuous sparks or flashes being produced. The whole is covered with a glass bottle, from which the vapour is collected and condensed.

4,346.—E. Berthard, Cortaillod, and F. Borel, Bonday, Switzerland. Manufacture of Telegraph Cables. 6d. (2 figs.) October 29.

CONDUCTORS.—These are made by introducing the conducting wire and insulating material into a leaden pipe as it comes from the press. Similarly a double conductor may be made by taking a conducting wire, then a layer of insulating material, enclosing these in a conducting tube, then another layer of insulating material, and finally enveloping the whole in the outer protecting tube of lead or other suitable material.

4,347.—J. S. Wilson, Sunderland. Apparatus for Producing Electric Light. 6d. (2 figs.) October 29.

ARC LAMP.—On a base-plate are fixed two vertical electro-magnets and two uprights; rigidly attached to one of these uprights is a projecting arm carrying at its outer end a carbon held in a suitable holder, to the other upright is pivotted a similar arm carrying the other carbon, approaching the first one at a vertical angle, and held in contact therewith by a spring. Attached to the other and shorter end of the pivotted arm are a second pair of electro-magnets moving at right angles to the axes of, and in a parallel plane with, the fixed

magnets ; these separate the carbons and establish the arc. One carbon-holder is pivotted, and the carbon points are pressed together by a coiled spring.

*4,348.—J. S. Wilson, Sunderland. **Secondary Galvanic Batteries.** 2d. October 29.

SECONDARY BATTERIES.—The cathode and anode are made of iron, and the electrolyte is preferably made of equal portions of saturated solutions of nitrate of soda and sulphate of soda. An electro-magnet with a vibrating spring is fixed between the primary and secondary battery, so that the current from the secondary battery will be approximately constant and alternate, first in the direction of the primary battery, and then in the contrary direction.

*4,388.—S. F. Van Choate, New York. **Producing Electric Light, &c.** 4d. October 31.

INCANDESCENCE LAMP.—The ribbon or spiral of this is formed of asbestos, mica, platinum, or carbon, or any combination of these.

CURRENT REGULATOR.—This consists in fitting near each lamp an automatic shunt, and suitably proportioned resistances in connection therewith.

*4,403.—T. A. Bell, Liverpool. **Apparatus for Dividing and Distributing Currents of Electricity.** 2d. October 31.

CURRENT DISTRIBUTOR.—A wheel mounted on an axis in electrical connection with one pole of a generator has a contact maker mounted on its periphery ; round a frame partly surrounding the wheel is arranged one end of each circuit, the other ends going to the remaining pole of the generator ; the revolving wheel thus completes each circuit in rotation ; or the ends of the wires may be arranged so that spring pistons slightly projecting through the frame may make contact with them when pressed upwards by a stud on the revolving wheel ; or studs may project from a flat disc and make contact with the ends of the wires arranged round a similar flat disc placed in the same plane with them.

*4,407.—A. M. Clark, London. *(C. Davis, Paris.)* **Applying Electricity as a Source of Heat.** 2d. October 31.

HEATING.—A generator is used in connection with wires of platinum or other metals for heating purposes ; or the sparks from a Ruhmkorff coil may be utilized for the same purpose.

4,438.—E. T. Truman, London. **Preparing Gutta-Percha for Insulating Telegraphic Conductors, &c.** 4d. November 2.

CONDUCTORS.—The gutta-percha, after cleansing, and while in the state of a spongy porous mass, is submitted to a twisting or "spinning" action to expel the water and air. In insulating a joint made in a conductor the copper wires are suitably connected, and have placed over them a perforated metallic mould slightly overlapping each of the insulated ends, which is again enclosed in a larger metallic cylinder, and has ends projecting through into this cylinder. Gutta-percha in a heated, semi-fluid state is forced into the mould under pressure, and when the gutta-percha begins to exude through either end, that end, and in turn the other, is drawn within the cylinder, thus securing a good union between the old and new coverings.

4,456.—F. H. W. Higgins, London. **Apparatus for Producing, Maintaining and Subdividing Electric Light.** 6d. (2 figs.) November 4.

ARC LAMP.—In this lamp the lower carbon, as it is consumed, is fed up by a column of mercury against a block of carbon held in an arm pivotted from the stand of the lamp, which permits of its being swung on one side to insert the fresh lower carbon. In a modification the upper carbon may be lowered by a wheel and ratchet or other suitable mechanical arrangement.

*4,462.—H. L. Thomson, London. **Production of Electric Light.** 2d. November 5.

CURRENT DIVIDER.—One pole of the generator is connected to a revolving insulated metallic disc, whose periphery is built up of alternate segments of conducting and insulating material. Copper brushes, placed so that only one will be in metallic contact at a time, are arranged in the plane of the disc, and press on its circumference at different parts ; these lead to the lamps, the return conductors going to the other pole of the generator.

*4,464.—F. B. Buddeley, London. **Applying Railway Brakes.** 2d. November 5.

GENERATORS.—A generator of any known type is used in connection with electro-magnets to apply the brakes.

4,466.—C. Stewart, London. **Distributing Electricity for the Production of Electric Light.** 6d. (2 figs.) November 5.

CURRENT DIVIDER.—This consists of a rotating arm, in connection with one pole of a generator, terminating in a spring contact passing over a number of insulated metallic contact pieces, disposed at equal intervals round the face of a disc, and each in connection with one of the branch wires. The returns go to the other pole of the generator.

4,473.—F. Gye, London. **Apparatus for Obtaining Electric Light.** 2d. November 5.

ARC LAMP.—One electrode is a disc of carbon, which may revolve slowly, mounted on and clamped to a plate by a coned ring. The other electrode is held stationary at a short distance from the face of the disc, which, when the arc forms, wears away in a long spiral line.

*4,476.—G. R. Bodmer, London. **Electric Lighting.** 6d. (3 figs.) November 5.

CURRENT DIVIDER.—This consists of two rotating discs mounted one on either end of a shaft, and insulated from each other, each disc being in connection with one pole of a generator. The discs rotate in two rings, the one having the leads and the other the returns of the branch circuits arranged around their circumferences. A contact-maker on each disc completes the circuits in succession ; or there may be as many discs as there are circuits, a similar number of contact-makers being provided.

ARC LAMP.—A vertical revolving cylinder driven by a coiled spring has mounted on its circumference a number of clips suited to hold "Jablochkoff" candles, each half of the clip being insulated from the other. These clips pass over two contact-makers, which complete the circuit through the "candles." A pin passed through one of the electrodes holds the lamp in contact till the "candle" has burnt down to it, when the coiled spring moves the next "candle" up to the contact pieces.

4,502.—E. G. Brewer, London. (*T. A. Edison, N.J., U.S.A.*) **Lighting by Electricity.** 6d. (3 figs.) November 7.

INCANDESCENCE LAMP.—The "candle" or light-giving device is a tapering hollow cylinder split vertically to near its top, the current passing up one side and down the other. One of the thermal current regulators described in Patent No. 4,226 of 1878 is used to cut out the lamp should its heat become excessive. It is preferred to make the "candles" with a finely powdered metal such as platinum, iridium, or ruthenium, incorporated with oxide of magnesia or zirconium, lime, silicon, or boron, the whole being moulded by pressure. The material may then be saturated with a salt of a metal difficult of fusion.

*4,518. — W. North, Huddersfield. **Electro-Motors.** 2d. November 7.

ELECTRO-MAGNETIC MOTOR. — The two south poles and the two north poles of the electro-magnets are placed adjacent, so as to have only one nodal point in the armature.

4,553. — M. Gray, London. **Manufacturing Carbons for Electrical Purposes.** 4d. November 9.

CARBONS.—These are made by powdering carbon prepared in any approved form, washing it in acid or alkaline solutions, and then mixing it with molasses or other adhesive substance which liquefies with heat. It is afterwards moulded to suitable shapes by forcing it through a moulding press, the carbons as they are delivered from the press being submitted to a high temperature to carbonise the saccharine matter. They are finally immersed in boiling sugar or molasses to fill up the pores, and reheated to again carbonise the saccharine matter.

4,559.—C. Davis, London. [Mechanism, &c., **for Producing Light and Heat by Electricity.** 2d. November 11.

DYNAMO-ELECTRIC GENERATORS.—This relates to driving generators from the wheels or axles of vehicles by toothed wheels or belts, the currents generated being used to light or warm the vehicles.

*4,568.—J. Mackenzie, London. **Electric Light Apparatus.** 2d. November 11.

LAMP FOR CANDLES.—To enable long electrodes to be used they are placed in tubes, side by side or otherwise, into which mercury is admitted from an elevation at a sufficient rate to float the electrodes up as they are consumed.

4,573.—G. Zanni, London. **Apparatus for the Electro-Deposition of Metals and the Production of Light by Electricity.** 6d. (3 figs.) November 11.

DYNAMO-ELECTRIC GENERATOR. — Electro-magnets arranged in a circle are attached to the interior circumference of a case, and their core-pieces of soft iron are made hollow to permit of water flowing through them. The soft iron ring of the armature is made in segments, each segment having a quantity of insulated copper wire coiled upon it. The segments are united by tongues and grooves.

ARC LAMP.—Two carbon rods have a small toothed wheel fitted on one of each of their ends, the wheels engaging with each other, and one being conical to admit of the point of one carbon approaching the other carbon as they are consumed ; the rods are caused to rotate by a spring, and the distance of their points apart is maintained by an electro-magnet.

*4,575.—H. W. Tyler, Kt., Colchester. **Electric Lighting.** 2d. November 12.

INCANDESCENCE LAMP.—In this lamp the incan-

descent substance is enclosed in a refractory casing, such as German glass, rock crystal, glaze, fireclay, or steatite, which may be sufficiently transparent for the light to pass through, or to become incandescent.

4,595.—C. F. Heinrichs, London. **Apparatus for Generating Electric Currents, &c.** 10d. (21 figs.) November 13.

DYNAMO-ELECTRIC GENERATORS.—The armature of the alternating generator shown in Fig. 1 is constructed of eight sections m^1 m^2, m^1 m^2, &c., four sections being wound in one direction, and the others in the reverse direction, one end of the wire from each section being attached to the commutator plate c^1, and the other ends to the plate c^2. The plates are magnetically separated by a brass ring keyed on the shaft s. Surrounding the armature, secured to the iron rings r^1 r^2 and attached to the frame f, are four inducing magnets m n, m n, &c., all wound with insulated wire in one direction, and coupled together in pairs, each of the two wires from each set being connected with the brushes b^1 and b^2 (not shown), and the others with the terminals t^1 t^2, thus giving the inducing magnets the same polarity at one end of the arma-

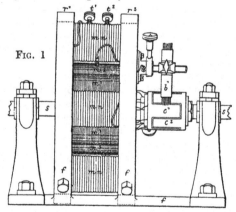

FIG. 1

ture, and the reverse polarity at the other end. In a generator for continuous currents the armature is constructed of eight sections, each constituting a magnet, and all are wound in one direction, but are magnetically connected by an iron ring mounted on the shaft. The commutator consists of eight insulated copper plates, to each of which is attached one end wire of each of two adjacent sections, so that all the sections are, through the commutator plates, in connection with each other. Five inducing magnets are secured to the frame as in the generator first described, and surround the upper half of the armature, currents being thus generated in all the sections of one-half of the armature in one direction, and in the other

half in the reverse direction. Each brush is mounted in a holder sliding on a flat-sided shaft, on which it can be clamped by a set screw; a wormwheel on one end of the shaft is actuated by a worm on the end of a set screw, and a spring acting in opposition to this set screw tends to press the brush on the commutator.

ARC LAMP.—To the frames f f, shown in Fig. 2, are attached two depending brackets carrying at their lower ends pivots, upon which revolve two pulleys projecting from, but attached to which are carbon-holders carrying the **C** shaped carbons. The pulleys are actuated by crossed chains passing

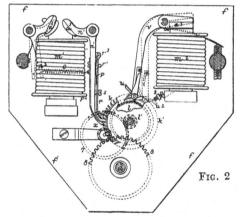

FIG. 2

over a differentially speeded drum keyed on the spindle d^1; on this is also keyed the wheel 8 geared by wheel 7 to the spindle l^1, upon which are fixed the escapement wheel k and ratchet-wheel l. The magnet m^1 with armature a^3 and lever n^1 with the pallets r and x regulate the feed of the carbons through the escapement wheel k; and the magnet m^2 with armature a^5 and lever v with pawl u acts as a separator upon the ratchet-wheel l. The stud 1 serves to lift the pawl out of the teeth of the ratchet-wheel. The armature a^2 is only attracted by a very powerful current when it acts on the armature n^1 by the spring o. In a modification the two circular carbons have a rotating flat carbon disc interposed at right angles between their points; or two pairs of carbons, the upper or positive pair being at right angles to the lower negative pair, may be used; or two parallel circular carbons having their points impinging on a horizontal cylinder of carbon cause the cylinder to revolve by their falling motion as they are consumed.

4,601.—B. Hunt, London. (*M. H. Alberger and S. W. Pettit, Philadelphia, U.S.A.*) **Telegraph Cables, &c.** 6d. (11 figs.) November 13.

CONDUCTORS.—After coating the conductor with glass, it is introduced into an iron pipe, and the

whole is then heated and rolled or drawn to the size required. In making a joint the two ends are rounded off so that the conductor projects slightly beyond the insulating matter, and this again projects beyond the outer tube ; the two ends are then brought forcibly together by an ordinary screw coupling. A number of conductors may be similarly insulated and placed in an outer protecting envelope.

*4,611.—E. Edwards and A. Normandy, London. Producing Light, Heat, and Motive Power by Electricity. 4d. November 13.

ELECTRIC LIGHT.—The main circuit is laid in the ordinary way, and at every point where a lamp is required a coil is put in to induce secondary currents, by which the lamps are worked. The light may be produced by decomposing water and directing the ignited jet of gas against a block of lime, or points of carbon may be placed in glass globes from which the air has been exhausted.

4,626.—St. G. L. Fox, London. Electric Lighting. 8d. (6 figs.) November 14.

CURRENT REGULATOR.—The object of this is to maintain a constant E.M.F. in the branch leads, and it consists of an electro-magnet placed in circuit between the branch and earth which attracts a pivotted counterpoise. This has at its end a vessel containing mercury, electrically connected to the main conductor, into which dip the ends of a fixed resistance forming a connection through the mercury between the main and the lamps. On the E.M.F. increasing, the armature, and by it the vessel of mercury, is drawn down, thus leaving fewer ends of the wire for the passage of the current, and so increasing the resistance. In a second arrangement the counterpoise armature actuates a needle connected to a local battery. The needle moves on a graduated scale, and thus indicates the E.M.F. A sliding piece on the scale, fitted with contact pegs, may be set where required ; the needle making contact with either of these causes the current of the local battery to operate one of two electro-magnets, thus actuating a switch or the steam valve of the engine.

CURRENT METER.—Air under pressure which may be maintained by an electric pump or a head of water is passed through an ordinary gas meter, its passage being regulated by a valve controlled by an electro-magnet placed in the circuit. The quantity of air passing bears a constant ratio to the quantity of electricity passing by the magnet. Water may be used in place of the air, or the registering train of the meter may be driven by mechanical means, in which case the electro-magnet

by attracting its counterpoise armature causes a friction disc, normally resting in the centre of a second revolving disc, driven by clockwork, to approach its circumference, imparting motion to it which it transmits to the registering train by means of a long gearing pulley.

*4,635.—E. J. C. Welch, Manchester. Electric Lighting. 2d. November 15.

CURRENT DIVIDER.—A modification of Patent No. 4,278 of 1878. The inventor employs a double set of revolving contact pieces insulated from each other and each connected to one of the poles of a generator ; the wires of each circuit are connected to two fixed contact pieces with which the revolving pieces make contact.

4,645.—J. S. Sellon, W. Ladd, and H. Edmunds, London. Obtaining Electric Light. 6d. (1 fig.) November 15.

ELECTRODES.—For arc lamps a "point" of pure iridium is fixed in a metallic setting, preferably platinum, and this is attached to a plug of copper of suitable form. For incandescence lamps various alloys of iridium and platinum are used for the filaments.

4,646.—J. S. Sellon, W. Ladd, and H. Edmunds, London. Electric Lamps. 6d. (6 figs.) November 15.

ARC LAMP.—In the lamp first described the carbons are placed in tubes set as a V, and are fed downwards by means of springs. One tube is pivotted near its lower extremity and is separated from the other by an electro-magnet, which is employed to establish the arc. Projecting from near the bottom of each tube is a bracket carrying iridium tipped supports, against which the conical ends of the carbons press. In a vertical lamp the carbons are placed in parallel tubes and fed upwards by springs against the iridium tipped supports. A metallic contact block rests between the points of the carbons and is held there by a spring, from which it is drawn by an electro-magnet when the current passes ; or the tops of the carbons are bridged across by a metallic block.

4,650.—T. Clarke and E. Smith, Torquay. Manufacture of Sulphuric Acid, &c. 8d. (2 figs.) November 16.

ARC LAMPS.—The nitrogen and hydrogen gases for the formation of ammonia are passed through a white heat produced by the electric arc.

4,662.—J. T. Sprague, Birmingham. Producing Electric Light, &c. 6d. (1 fig.) November 16.

INCANDESCENCE LAMP.—The filament is formed of a compound material such as calcium, barium, magnesium, &c., intermingled with finely divided carbon or platinum. Several filaments may be arranged in one tube, and to switch these in according to the strength of the current, a tube containing mercury is enclosed in a coil of wire conveying the current, which draws down a corepiece and presses the mercury into a vessel, thereby making electrical contact with the ends of wires which are arranged in the vessel.

REFLECTORS.—To more evenly distribute the light a reflector of circular trough form and of opal glass or other translucent material, with a smooth upper surface, is suspended below the light, which is thus sent into an upper reflector of suitable shape, whence it is diffused over the desired area.

4,671.—W. L. Scott, Wolverhampton. Production, Regulation, and Distribution of Electric and Electro-Calcic Lights. 6d. (16 figs.) November 18.

CONDENSERS.—These are employed in connection with the lights and have a surface equal to at least one and a quarter times the surface of the conducting wires, and are formed of plates, bands, or ribbons of metal insulated from each other, having between them a layer of iron in the form of plate, wire, or powder. Alternate plates are coupled and the two groups are connected to one of the electrodes. The conducting wires may be replaced by ribbons of metal, several being superimposed, but insulated from each other, thus forming an elongated condenser.

SECONDARY BATTERY.—A secondary battery may be used in connection with these condensers to take up any surplus of current.

INCANDESCENCE LAMP.—The light emitting portion is made of lime, magnesia, alumina, zirconia, or other metallic oxide, mixed or coated with platinum, iridium, silver, copper, tungsten iron, carbon, or graphite, and is held or supported by wires of platinum, iridium, or palladium. This constitutes the "electro-calcic burner," which may be steeped in or have incorporated with it a phosphorescent substance to modify or colour the light. To largely reduce the resistance of the electro-calcic burners, "a large metallic reflector with or without the use of a stream of cold water" is employed. The globes containing the burners may contain suitable vapours. The lower electrode where two are used may be fed up by a mercury float. A jet of mercury is used as one electrode. Continuous, alternating, or intermittent currents are employed.

4,686.—E. N. Parod, Paris. Transmitting Electricity to Great Distances. 8d. (6 figs.) November 19.

CONDUCTOR.—This consists of a large cylindrical condenser of glass or other material, formed of tubes coupled together with suitable collars and carried on insulating supports in earthenware pipes divided longitudinally, the upper and lower halves being cemented together; the inner side of the glass pipes is lined with metal, or water may be used, as the conducting medium. The outer side is covered with rolled tin or other metal. The condenser is charged by two rods insulated from each other connected respectively to the inner and outer metallic casings; similar insulated rods are placed where it is desired to take branches from the condenser. Quicklime is put in the bottom of the earthenware tube to absorb any moisture. When a supply of "quantity" electricity is required, it is taken from the condenser by cylinders of metal, and when a high tension current is required elongated cones are used, connected at their apices with the apparatus to be supplied.

*4,689.—E. J. C. Welch, Manchester. Apparatus for Producing the Electric Light. 2d. November 19.

INCANDESCENCE LAMPS.—A thin "pencil" of carbon, or other material, is held at its upper end in a fixed socket, and has a movable contact sleeve embracing and sliding on its lower part; on an increase of current this sleeve is drawn down by the armature of an electro-magnet placed in the circuit, thus increasing the resistance of the filament. A coil, acting by expansion, may be used to move the sleeve in place of the electro-magnet; or a movable contact spring may be employed.

*4,690.—J. H. Johnson, London. (*A. de Méritens, Paris*). Electric Regulators, Lamps, or Candles. 2d. November 19.

ARC LAMP.—The carbons are burned within a tall glass dome, and are so placed that the arc shall be two-thirds of the way up the dome. An annular space formed between the base and the dome serves to admit the external atmosphere.

*4,693.—A. Reimenschneider and F. S. Christensen, Nestred, Denmark. Apparatus for Dividing Electric Light. 4d. (1 fig.) November 19.

ARC LAMP.—An exhausted glass cylinder is mounted on a suitable metallic base, on which is fixed the lower carbon, and carries on its upper end a metal cylinder, in which works an air-tight piston, whose rod working through an air-tight gland carries the upper carbon. On the passage of the current the air in the glass cylinder is rarefied, and the compressed air under the piston raises it, and so draws the carbons apart. The current

divides between the lamp and an equal resistance, either of which may be cut out by a switch.

4,696.—C. E. Crighton, Newcastle-upon-Tyne. Electric Telegraph Insulators. 6d. (2 figs.) November 19.

INSULATORS.—In the upper part of an ordinary porcelain insulator is a vertical slot, extended at its lower part to form a horizontal aperture of a duplo-conoidal shape, having a ridge at its centre. In the aperture, and resting on the ridge, is placed the conductor, over which, and in the slot, is inserted a wedge, slightly arched on its under side. The wedge is pressed on the conductor by screwing a cap on the top of the insulator.

4,699.—A. Melbads, Ramsgate. Driving Sewing and other Machines by Electro-Magnetism. 6d. (6 figs.) November 19.

ELECTRO - MAGNETIC MOTOR.—Fixed on a revolving spindle are a number of radial armatures, insulated from each other; a number of electro-magnets are so placed as to attract these. A commutator, with twice as many metallic parts as there are armatures and the same number of contact rollers, passes the current to each magnet in turn.

4,705.—F. J. Cheesbrough, Liverpool. (*W. E. Sawyer, New York, and A. Man, Brooklyn, N.Y., U.S.A.*) Distributing, Regulating, and Applying Electric Currents for Lighting, &c. 1s. (8 figs.) November 19.

ELECTRIC LIGHT.—This invention relates to a distributing system, a means of automatically supplying the current to meet the demand, a lamp and light regulating apparatus, and a current meter.

DISTRIBUTING SYSTEM.—In this a revolving eccentric switch makes and breaks contact with the several lamp circuits, condensers being placed near the lamps to maintain a constant current during the intervals.

CURRENT REGULATOR.—A tilting cog-bar slides in a pivotted sleeve, and has its one end projecting over and within the influence of an electro-magnet, the other end being attached by links to the cut-off valve of the engine. On either side of the pivotted sleeve is one of two small toothed rollers which engage with the cog-bar and revolve in opposite directions. A spring mounted in opposition to the magnet tends to keep the cog-bar in gear with one roller, and thus admit more steam, but when, by the increase of current, the magnet preponderates, the cog-bar is drawn into gear with the oppositely revolving roller, and steam is proportionately cut

off. When a normal current flows the cog-bar vibrates over the two rollers.

LIGHT REGULATING APPARATUS.—A suitable metallic holder has a groove in which is a sliding piece prolonged in the form of a flat spring having at its end a cross-piece which, as the slide is moved by a rack and pinion, makes contact successively with sixteen (or more) pins arranged eight on either side. The outward travel of the slide is limited by a stop and its inward travel by contact with an armature lever, of an electro-magnet, bent at right angles and pivotted, the longer arm moving between two contact studs. The current normally flows by one of the studs, through the coils of the magnets to the slide holder, and by the slide, spring, and cross-piece to one of the eight pins and thence to the lamp; should the circuit be interrupted the armature, actuated by an opposing spring, leaves the magnet, makes contact with the other stud, and the current flows by a suitable resistance. The slide is then run in and, striking the armature, forces it into the first position, and the current flows as before. By connecting the pins on one side to suitably arranged resistances in a shunt circuit, the light of the lamps may be regulated at the same time, maintaining the resistance in the circuit constant. The first pin on one side, and the last on the opposite side, are not electrically connected.

CURRENT METER.—The object of this meter is to indicate the total number of lamp-hours during which the current is supplied, and to accomplish this the metallic cylinder D, mounted on the shaft

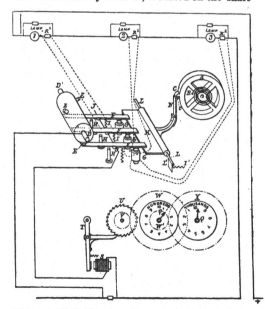

D¹ has set spirally, upon it as many contact pins E as there are lamps; the cylinder is rotated by clock-

work, of which the balance wheel A, driven in one direction by the spring B, is the governor. The armatures F, of electro-magnets G, work on the shaft J, and are opposed by springs I, forcing them on the stops H. Under the levers F is a metal strip K, fixed to shaft L, and forced upward by spring I^1 acting against the arm L^1. To the arm K is secured a light spring N and bent piece O engaging with C. Current enters at the—point, traverses the lamps 1, 2, 3, and a portion is diverted by the resistances R^2, and energises the magnets G, which, attracting the armatures F, depress the plate K, and so release the balance-wheel A. The cylinder D then rotates, its pins making contact with the shaft J, through levers F, and thus energising the magnet S, which by its armature T drives the indicating train W X. Each lamp causes the ratchet wheel U to be moved one tooth in each revolution of the cylinder.

4,707.—W. J. and R. T. Meredith, Bridgenorth. **Apparatus for Securing the Doors of Railway Carriages and Tram Cars.** 6d. (11 figs.) November 20.

GENERATOR.—Any ordinary generator is employed to supply the current in connection with various locking devices actuated by electromagnets.

4,762.—J. T. Sprague, Birmingham. **Apparatus for Generating and Measuring Electric Currents.** 8d. (14 figs.) November 22.

DYNAMO-ELECTRIC GENERATOR.—The core of the electro-magnet is a conductor of electricity, and is preferably made of a sheet of iron of the necessary width, compactly wound with an insulating material between the turns; or strips of insulating material may be inserted at intervals to form air spaces; or two such sheets may be arranged to form the length of the core, having their ends connected by a central rod of iron, or iron discs otherwise insulated from the several coils, so as to form of the whole one continuous conductor in which the current shall flow in the same direction throughout; or the core is constructed of insulated iron wire wound in a coil. The iron core may constitute the magnet, or it may be wound with copper wire in the same direction. The ends of the cores may be open or terminated with iron pole-pieces of various shapes. The inducing system may consist of permanent magnets constructed of bars of thin steel, having tongues at their ends, connected together by pole-pieces of soft iron having corresponding grooves, or the pole-pieces may be used in connection with segments wound with insulated wire. The magnets are so arranged that the pole-pieces act as "consequent" poles. An inducing system composed of radial magnets having their inner ends

united by an iron boss forms a series of horseshoe magnets, to the outer end of which are connected pole-pieces, to which are also connected the polar ends of magnets forming, if straight, a many-sided polygon, or a circle if curved. These are so arranged that the several pole-pieces, while acting as "consequent" poles within the system, collect and direct the whole of the magnetic actions externally as alternate N. and S. poles. The induced system has one of these wheels placed on each side of it with their poles opposed, in which case the pole-pieces may be connected by electro-magnets, so enclosing the induced system. The induced system may be in the form of a ring and revolve within the pole-pieces of the inducing system. In a continuous current machine a complete magnetic circle is constructed of two or more segments united together by means of pole-pieces. Upon the ring so formed coils of wire are placed. When the core is composed of sheets of iron their interlapped ends form the pole-pieces, and permit of the circulation of air. When the core is of disc form, so that the wires cannot entirely cover it, the spaces are filled with pieces of iron. The winding is done in sections, and the connections are as in the Gramme generator.

COMMUTATOR.—The wearing surfaces slide in grooves in insulated carriers, and are retained in place by end plates. The usual brushes are employed. To avoid sparking it is preferred to couple the separate plates of the commutator with a secondary battery capable of absorbing the "extra" current at break, and giving it up afterwards.

GALVANOMETER.—Coils of wire are constructed in pairs, one for each side of the zero line, and form a chamber in which the moving parts of the instrument freely move. Two such pairs are arranged back to back, so as to have reverse magnetic fields. The moving systems have opposite magnetic actions, and being fixed on one axis all the influences tend to move it in one direction. Soft iron may be inserted in the moving coils, or soft iron needles may be mounted on the same axis as the first pair of coils, but placed at a considerable angle to them. Various methods of suspension are employed. In another form of instrument a vertical helix causes an iron core to descend, while a second helix causes its core to ascend. The two cores are connected by a cord passing over a wheel, and give the required indication by means of a pointer; or an electro-magnet attracting an armature forces mercury out of an elastic vessel into a graduated glass tube, the armature being counterpoised by an inverted conical iron weight suspended in a vessel of mercury.

CURRENT METER.—Any suitable chemical decomposition may be employed to record the current

passed in a given time. Thus two balanced plates change the direction of the current by the weight of metal passed from one plate to the other, causing the deflections of a pivotted armature to alternately make contact with two points, each deflection of the armature advancing a ratchet-wheel one tooth, and so recording the quantity of current passed. Or cylinders mounted on axes rotate by the change of weight displacing their centre of gravity, and so actuate a train of recording wheels.

4,774.—I. L. Pulvermacher, London. Producing Light by Electricity. 6d. (9 figs.) November 23.

ARC LAMP.—A rod of carbon, previously insulated, has wound round it in spiral coils a thinner rod of carbon. One end of the rod projects beyond the spiral and has a metallic cap fixed on it. The lamp consists of two metallic tubes separated by a glass tube; the metallic cap is placed in the shorter internal tube and the last coil of the spiral in contact with the top of the outer tube. The arc is formed by a temporary connection at the top of the carbons. In a lamp to maintain the light at a fixed point the "rod spiral" is carried on the top of a deep ferrule sliding in a metallic tube with slight friction, and drawn up by counterweights. A metallic tube passing through the centre of the hollow carbon rod, has at its upper end an inverted conical stop which permits the carbon to be fed up only as it is consumed; a rod passing through this tube, terminating at its top in a small metallic brush inclined downwards, serves to make contact between the carbons and so start the arc.

INCANDESCENCE LAMP.—This is constructed as described in Patent No. 4,180 of 1878.

*4,812.—F. W. C. Vogel, London. (*N. C. Vogel, Rotterdam.*) Machinery for Generating Electricity. 2d. November 26.

DYNAMO-ELECTRIC GENERATOR.—"Both systems" (*i.e.*, the inducing and induced) "are made to revolve in inverse sense, and so when both are revolved at 1000 revolutions a minute, the intensity due to 2000 revolutions is obtained."

*4,821.—R. Sabine, Hampton Wick. Electric Lamps or Regulators. 2d. November 27.

ARC LAMP.—This relates to methods of regulating the distance of the carbons. In the first instance a fan is rotated by the ascent of the heated air and sets in motion a suitably arranged train of wheels; in the second instance, where a top illumination is required, a block of cast iron is substituted for the lower carbon, a portion of this fuses and forms a small pond of molten material; in the third instance, the carbons are placed vertically side by

side, one being pivotted, which allows of its being repelled by the fixed carbon according to the degree of current flowing.

4,844.—I. L. Pulvermacher, London. Dynamo-Electric Machine for Producing Electric Currents. 6d. (9 figs.) November 28.

DYNAMO-ELECTRIC GENERATOR.—This consists of two horizontal cylinders built up of a number of soft iron rings bolted together, but separated from each other by washers, the inner one revolving in the fixed outer one. Longitudinal parallel grooves, to contain insulated copper wire, are cut in the outer surface of the revolving cylinder, and a corresponding number of grooves are cut in the inner surface of the fixed cylinder. The wires are laid in these grooves in zig-zag form from one end to the other alternately. The current of the revolving cylinder is collected at two insulated rings and taken from terminals attached to the fixed cylinder. Inner and outer fixed cylinders may be used, in which case a central revolving cylinder has grooves cut on both its surfaces. The shaft is preferably mounted in bearings having adjustable split bushes.

4,847.—F. J. Cheesbrough, Liverpool. (*W. E. Sawyer, New York, U.S.A. and A. Man, Brooklyn, N.Y., U.S.A.*) Electric Lamps. 8d. (9 figs.) November 28.

SEMI-INCANDESCENCE LAMP. — The lamp is enclosed in a tall glass globe having a flange at its lower end over which is placed a brass ring. A glass stopper ground to the globe is secured by a wooden disc, under which come discs of rubber and brass, drawn up by bolts passing through the brass ring and discs. Passing through the stopper and hermetically sealed therein are two coiled conducting tubes reaching half way up the globe; the one closed at the top has a small hole communicating with the bottom of the lamp, the other is open at the top. On the ends of these tubes are screwed binding posts fitted with washers to make an air-tight joint, the lamp being exhausted and the posts screwed on afterwards. A disc of soapstone is placed at the top of these tubes, to which the carbon-holders are secured. The lower one carries a rounded point of carbon, and the upper one is a rod bent at right angles at the top, and sliding in a short tube, in which it is guided by a pin working in a slot; from the projecting arm depends the carbon, which is guided in a projecting clamp formed of a strip of silver bent round the rod, and having its points protected by two pieces of iridium. A lump of sodium or potassium, and one or two lumps of freshly burned lime, are placed in the globe. A modification of the lamp has a number of upper carbons arranged round a circular

lower carbon, and in a further modification the upper carbon is shortened and the guiding clamp dispensed with.

4,949.—C. W. Siemens, London. (*H. v. Alteneck, Berlin*). **Electric Lamps.** 6d. (7 figs.) December 4.

ARC LAMPS.—In the first lamp described, the carbons are suspended from a framing and meet at an acute angle, the one being fixed and the other attached to a pivotted lever counterweighted so as to tend to keep their points in contact, the passage of the current in opposite directions tending to repel them. The arm of the lever may be at right angles to the carbon, one end being weighted and the other attached to the core of a solenoid; the core may have a dash-pot action, or be made of a group of insulated iron wires to adapt it for alternating currents. In another lamp the carbons are placed vertically one over the other. An iron core in a solenoid is attracted downwards in opposition to a weight or spring, and is connected by a lever to a vertically sliding rod, which is itself connected through a ratchet and pawl to a vertical slide carrying the upper carbon. The rod and slide remain together so long as the stroke of the core can effect the adjustment, but when the carbon is so far wasted that the rod descends to the limit of the stroke, the pawl comes in contact with a fixed stop, and the carbon descends till the pawl engages again and the self-adjusting movement is resumed. The upper carbon-holder may be connected to the lower one, so as to form a focus-keeping lamp. The descent of the carbon when its slide is disengaged from the rod, is regulated by a pendulum and escapement device. The movements of the core are controlled by a piston working in an air cylinder. In a modification, the core connected to the carbon rod as before slides in a solenoid, one part of which is wound with a low resistance and the other part with a high resistance coil, the counteraction of the two coils causing the carbons to adjust themselves to suit the current.

4,960.—A. V. Newton, London. (*Weston Dynamo-Electric Machine Company, Incorporated, Newark, N.J., U.S.A.*) **Apparatus for Generating Electric Currents, and for Producing Electric Light.** 10d. (18 figs.) December 4.

DYNAMO-ELECTRIC GENERATOR.—The horizontal electro-magnets are carried at their outer ends by iron standards in which are parallel slots to provide for the circulation of air between the upper and lower magnets. The pole-pieces connecting the inner ends of the magnets are constructed of a number of wrought-iron plates each curved on their inner edges and spaced apart to corre-

spond and be in line with the spaces between the discs forming the armature. The wrought-iron plates have their ends notched, and are placed in the moulds into which the metal for the core-pieces of the magnets is run. The cylindrical armature is composed of a series of parallel discs corresponding in number and spacing to the wrought-iron plates. Longitudinal grooves are cut on the surface of the cylinder, and in these the insulated wires are laid in a double system of coils in separate divisions, the divisions being looped together and the loops connected to the commutator plates; the loops of the first system are connected to plates of the commutator diametrically opposite to the plates to which the loops of the second system are connected; the free end of one system is connected to the free end of the other, their other ends being connected to one plate of the commutator. A number of separate insulated coils are wound on the cores of the electro-magnets so that each shall have the same length and number of convolutions; one group of the free ends is connected to one of the brushes, the other ends are coupled through the outer circuits to the other brush. The commutator consists of a cylinder provided with eight insulated spirally arranged plates mounted on the projecting end of the shaft and carried in two brass bracing pieces, the flanges of which are bolted to the upper edges of the poles of the stationary magnets. The brushes are supported in slotted studs inserted in slots in an oscillating plate mounted loosely on the outside of the bearing of the shaft, and provided with a set screw. To admit of the accurate setting of the brushes, the hub of the oscillating plate has a pointer behind which a curved scale is fixed; the studs are also fitted with curved wires serving as gauges for the length of brush required. In the second or compound multiple circuit generator the stationary coils traverse the exterior of a skeleton cylinder, the ends of each coil being carried to binding posts. The revolving armature has a number of core-pieces, corresponding to the number of coils on which an insulated wire is wound, forming coils on each core-piece. One end of the wire is taken to an insulated collar on the shaft, and the other end through the shaft to a second collar. The exciting current of a second generator is supplied to these collars by a pair of brushes.

ARC LAMPS.—In the first lamp the two carbons are placed vertically side by side in metallic holders connected with the circuit wires. On the outer side of the positive electrode a cylinder of fine glass is placed affording by volatilisation a conducting vapour which, by its passage across the point of the positive to the point of the negative electrode, defines the path for the current and keeps the arc at the points of the carbons. An "auto-

matic lighter," consisting of a bar of carbon mounted in a tilting holder, rests on the points of the carbons, and is withdrawn from them by an electro-magnet on the passage of the current, or a bar of carbon may be mounted on the end of a rack and caused to descend and make contact between the points of two fixed vertical carbons, from which it is withdrawn by the action of an electro-magnet operating an endless band by means of a rocking clutch armature and ratchet-wheel. In a further modification one or both of the electrodes are mounted in tilting holders worked by the armatures of electro-magnets, which in the latter case are geared together to insure the electrode oscillating in the same plane. The attraction of the armature is in opposition to a spring through a curved lever, so arranged that the force of the spring varies to correspond to the variable attractive force of the magnet upon the armature in the various positions which it occupies as the electrodes diminish in length.

4,961.—H. C. Byshe, London. **Magnetic Appliances for Curative Purposes.** 2d. December 4.

MAGNETS.—The surfaces of magnets are preserved from oxidation by a paint composed of pulverised iron, boiled linseed oil, and turpentine.

*4,988.—S. P. Thompson, Bristol, and W. P. Thompson, Liverpool. **Production of Electric Light.** 2d. December 5.

DYNAMO-ELECTRIC GENERATOR.—Silver wire is used in place of copper wire to decrease the size of the rotating armatures. In Ruhmkorff's coils it is proposed to do away with the copper wire and iron core, substituing for them a spiral of iron wire, preferably a strand of seven silver-plated wires twisted and drawn together so as to squeeze silver into the interstices, the whole being then externally electro-plated and covered with silk.

ARC LAMP.—A pencil of blacklead and pipeclay or other similar material is fed in between two fixed platinum poles and catching fire burns "with a brilliant flame, producing at the same time the well-known electric light."

*5,011.—S. Colme, London. **Electric Candles.** 2d. December 7.

ELECTRODES.—The electrodes are made of ultramarine either pure or mixed with carbon, molasses, kaolin, plaster-of-paris, or with silicate of soda or potash. To increase their conductivity a copper or platinum wire may be placed within the electrodes.

*5,044.—J. H. Johnson, London. (*F. E. de Mersanne and E. Bertin, Paris.*) **Electric Lamps.** 2d. December 9.

SEMI-INCANDESCENCE LAMP.—The lamp is mounted on a wooden stand carrying a brass socket surmounted by a brass ring, and a plug of silicate of magnesia traversed by a metal conducting wire terminating in a finer platinum wire, is placed in the socket. Arranged radially on the upper part of the ring are two or more tubes, each carrying a weighted carbon rod pressing in turn against the platinum wire; the weight on one when it is consumed comes in contact with the catch and lever arrangement of the other. The lever devices may be worked by electro-magnets.

*5,053.—J. H. Johnson, London. (*F. E. de Mersanne and E. Bertin, Paris.*) **Regulating Apparatus for Lighting by Electricity.** 2d. December 10.

CURRENT REGULATOR.—A resistance, preferably of platinum wire, is placed near the generator, and on this slides a contact sleeve. The attendant moves this sleeve so as to interpose more or less resistance as the platinum wire shows greater or less signs of heat.

5,060.—A. V. Newton, London. (*F. E. de Mersanne, Paris.*) **Electric Light Regulators.** 6d. (9 figs.) December 10.

ARC LAMP.—The upper and lower carbons are fed towards each other by clockwork. The lamp standard is a square tube, to the front of which the arm carrying the upper carbon is rigidly attached; the arm carrying the lower carbon is secured to a spring blade mounted on the front of the tube; an electro-magnet attracts this spring blade, and thus draws the carbons apart laterally, to produce the arc. A shunt magnet controls the clockwork. The frames containing the guide rollers for propelling the carbons can be adjusted to any desired angle.

5,076.—J. H. Johnson, London. (*E. Bertin and F. E. de Mersanne, Paris.*) **Electro-Magnetic and Magneto-Electric Machines.** 6d. (17 figs.) December 11.

DYNAMO-ELECTRIC GENERATOR.—A base-plate has a vertical disc at either end, to which are attached coils alternately wound in opposite directions. Carried in bearings supported by the vertical discs is a shaft on which a plate is fixed, having on its circumference a number of electro-magnets presenting alternately opposite poles on each side of the plate, and the axes of their cores corresponding to the axes of the coils attached to the discs. The inducing current passes to the electro-magnets by contact rollers bearing on a disc mounted on the shaft. The cores of the coils may be coupled in pairs by an iron plate so as to

k

form horseshoe magnets. The core-pieces may be solid or built up of iron wires. Any number of electro-magnet systems may be employed, a corresponding number of discs and coils being used.

*5,105.—R. Punshon, Brighton. Carbon Points or Candles for Electric Light. 2d. December 13.

CARBONS.—Finely powdered carbon and asbestos are ground together with an amalgam of mercury and silver; cohesion is secured by the addition of albumen, gum, and treacle, the desired shape being formed by pressure in a mould. A flexible candle is made of a number of short pieces of carbon joined together, end to end, by having a tape previously saturated with soda, gummed to one or both sides of the "candle" lengthwise and continuously, the ends of each portion being covered with an adhesive mixture which causes them to adhere slightly without being rigid. The jointed carbon is rolled round a drum and carried to a disc of carbon by grooved friction wheels. The carbon composition may be used for the filaments of incandescence lamps.

5,110.—J. H. Johnson, London. (*F. E. de Mersanne and E. Bertin, Paris*). Regulators for Electric Light. 6d. (11 figs.) December 13.

ARC LAMP.—A hollow T shaped frame supports a vertical tubular arm at each end of its crossbar; on top of these tubular arms, and at right angles to them, are fixed adjustable frames carrying the horizontal carbons. The latter are propelled by guide rollers to which motion is imparted by a small endless chain passing over pulleys and pressed by spring rollers against a pulley driven by the descent of a frame. On this frame is mounted an electro-magnet and a train of wheels terminating in a fly controlled by the action of the armature of an electro-magnet coupled in as a shunt to the arc. In a modification the electro-magnet is fixed to the frame of the lamp and a weight is employed in its place, the armature, as before, controlling the train of wheels. A second electro-magnet operates by a tension pulley on the endless chain and draws the carbons apart. The driving weight may be composed of two core-pieces bent at right angles at their top and bottom ends, the bottom hinged ends being shaped so as to encircle and slide on a rod; on the lower ends of these core-pieces are low resistance coils in the main circuit, tending to cause the armatures to grip the rod; above these are wound high resistance coils coupled as a shunt, tending to cause the armatures to release their grip, on which the carbons are brought together by the descent of the magnets. The carbons may be fed forward by a screw motion, in which case they are placed in a tube having a longitudinal slot on two of its opposite sides and a revolving helix in its interior; the carbon is laid in the helix and a pin passing through its end engages with the helix and travels in the two slots, or an endless chain may impart motion to the carbon direct by means of a hook attached to the carbon-holder engaging in a link of the chain.

5,123.—G. E. Dering, Lockleys, Herts, Obtaining Light, Heat, and Motive Power. 4d. December 13.

GENERATOR.—A dynamo-electric or other generator is used to decompose water, and the resulting gases are separately collected and stored for future use.

5,139.—W. B. Brain, London. Electric Lighting. 6d. (26 figs.) December 14.

DYNAMO-ELECTRIC GENERATOR.—The armature is composed of a ring of sheet iron or wire on which insulated copper wire is wound, the whole being mounted on a wooden knob and attached thereto by flat radial plates; the exciting magnets are of the Gramme type, and the core-pieces are bridged across so as to nearly envelop the armature. In a modification the shaft has a circular plate mounted on it, to which are attached a series of magnets arranged in sets of alternate polarity, and held at their outer periphery by a brass ring; on either side of the plate are rigidly mounted similar rings of magnets. The currents are collected from a commutator of the ordinary kind.

ARC LAMP.—Three or more carbons are used to form the positive pole; these are placed horizontally and radiate at equal angles. The three negative carbons, placed horizontally and alternately with the positive carbons, bisect these angles. The six carbons are fed to their common centre by any suitable means. In another lamp a vertical carbon is used in connection with horizontal carbons, its feed being regulated by three electro-magnets. A large horizontal magnet has hinged at its outer ends one of the outer ends of each of two small magnets, which, on the current passing, lift the carbon by means of spring clamps. In a lamp having two horizontal and one vertical carbon the frame has two horizontal tubes projecting at its lower end; in these tubes the carbons slide, and are fed together by the descent of the cage containing the rocking magnets and vertical carbon. This cage is connected by suitable cords passing over pulleys to each of the horizontal carbons.

5,152.—G. W. Whyte, Elgin, N.B. Electric Lamps or Lights, &c. 6d. (7 figs.) December 16.

ARC LAMP.—This is a lamp in which the arc is formed between the edges of two carbon discs, mounted vertically one over the other on rotating collar centres, the lower carbon being rotated by means of a small electro-motor, coupled in a shunt circuit and communicating its motion to the upper carbon by means of an endless belt. The spindle of the lower or positive carbon is mounted in slotted bearings in the upper forked end of an inverted bow frame, secured on the top of an insulated core working in opposition to a spring of a regulating magnet attached to a reciprocating rod sliding freely in a hollow guide. The spindle of the upper carbon is carried at right angles to a second reciprocating rod working freely in a hollow guide tube ; both these rods have rack teeth cut at their lower ends, which engage with gear wheels proportioned for the different rates of consumption of the two carbons, so as to nearly balance each other. The gear wheels are actuated by the armature of an electro-magnet whose stroke is controlled by an adjustable spring. The motor, in the first lamp described, consists of a horizontal magnet and armature, on which latter is mounted a spring pawl engaging with the teeth of a ratchet-wheel which communicates motion to a vertical spindle terminating in a square hole carrying the square shaft of a small worm. This worm gears into a wheel mounted on the hollow spindle carrying the collar to which the carbon disc is attached. The square shaft of the worm held in suitable bearings permits of its vertical movement when the arc is struck. In a modification the arc is formed by two electro-magnets, mounted on the armature of two vertical magnets, causing their cores to grip the vertical standard supporting the bow frame, at the same time the vertical magnets draw down their armature and the arc is established. In a modification of the motor, two fixed vertical electro-magnets have rotating over them two electro-magnets fixed on a vertical central spindle carried in a footstep bearing at its lower end, and bush bearing at its top, and communicating motion to the carbon by gearing as before described.

*5,165.—A. M. Clark, London. (*L. Reynard, Paris.*) Electric Lamps. 4d. (4 figs.) December 16.

ARC LAMPS.—In one lamp the carbons have racks formed on them and are fed forward by a toothed wheel ; in another lamp the carbons are formed with a screw thread and are fed by revolving holders having corresponding female threads ; the toothed wheel and holder are actuated by any suitable means.

5,183.—W. R. Lake, London. (*J. B. Fuller, New York, U.S.A.*) Electric Lighting Apparatus. 8d. (15 figs.) December 17.

INDUCTION APPARATUS.—Large magnet cores arranged parallel to each other with their ends magnetically connected, have soft iron discs around their centres, and at a distance from these discs are beads of insulating material. The outer ends of the cores are coiled with thick insulated wire so connected together and to the generator as to produce two consequent magnetic poles. Between the iron beads and the insulated terminations of these coils are wound coils of finer wire from which the induced currents for the lamps are derived. A pivotted iron arm regulates the light by strengthening or weakening the magnetic poles. The invention also relates to a safety device, in which an electro-magnet on an increase of current attracts an armature which is in the circuit and draws it from a contact point leading direct to the lamp, thereby inserting a resistance.

CURRENT METER. — Two horizontal magnets wound so as to present opposite poles are fixed with their axes in line. An armature vibrates between them on the passing of current, and rotates by means of a spring pawl at its upper end a ratchet-wheel geared with suitable indicating mechanism.

*5,197.—H. Wilde, Manchester. Electric Light Apparatus. 2d. December 18.

ARC LAMPS.—This invention relates to improvements on Patent No. 3,250 of 1878, and consists in placing the carbons in straight V grooves in which they are pressed by friction rollers and fed upwards by a spring acting on a piston in the tube in which they are contained, their upward motion being arrested by a block of lime with which the incandescent points come in contact. Compressed air may be used to raise the carbons. The arc may be started by a piece of carbon attached to an armature which is drawn away by an electro-magnet on the passing of the current.

*5,257.—J. H. Johnson, London. (*A. de Méritens, Paris.*) Induction Apparatus. 2d. December 24.

INDUCTION APPARATUS.—An alternating current is caused to pass through the two coils of an inducing electro-magnet having its cores formed of bundles of fine wire encased in a sheath of sheet iron and connected together at one end by a piece of soft iron. A precisely similar magnet, but coiled with finer wire, is placed with its poles opposite the poles of the first electro-magnet, from which it is separated by a thin slip of brass. The currents induced in the fine wire coils are employed to work lamps, &c.

*5,270. — F. R. Lucas, Greenwich. Underground Telegraphs. 2d. December 24.

CONDUCTORS.—These are laid in stoneware cylinders, secured together by a cemented socket or clamp.

*5,281.—A. M. Thompson and H. D. Earl, Crewe. Electric Lamps. 2d. December 27.

ELECTRODES.—To prevent the rapid destruction of the electrodes that portion "intended by its incandescence to produce the light is or may be enclosed in solid glass or porcelain or other suitable non-electro conductors."

5,306.—T. A. Edison, N.J., U.S.A. Developing Magnetism and Electric Currents and Apparatus for Illuminating by Electricity. 6d. (10 figs.) December 28.

INCANDESCENCE LAMP.—The filament is made of a wire of platinum, or platinum-iridium alloy, coated with a "pyroinsulation" of an oxide of metal such as cerium, lime, &c., dissolved in acid. The "pyroinsulated" wire is wound on a cylinder of lime and connected in the circuit, or a helix of flattened wire may be insulated and used as a "burner" in the same way. The expansion and contraction of the burner is used to actuate a safety device as described in Patent No. 4,226, of 1878. The burner may be surrounded by a metal cylinder which becomes incandescent and radiates the light. For insulating the wire a small carriage has mounted on it a sponge containing the solution, and a spirit-lamp; the carriage is drawn, sponge first, under the stretched wire. The burner can be made up of finely divided iridium mixed with an oxide of titanium, iron, or other metal.

CONDUCTORS. — An insulated copper-stranded cable is carried in a tube which serves as the return. An insulated strand and branch pipe leads off to each house. A safety device for cutting off the current to a house consists of an armature working against a spring adjusted to the current required; an excess of current causes an electro-magnet to draw away the armature from a contact and thus break the house circuit, the armature being caught and held by a spring detent.

SECONDARY BATTERIES.—These are described as constructed of lead sheets wound to any convenient form and placed in acidulated water in closed cases; a switch is provided for switching any battery into circuit. The gases developed when the batteries are fully charged pass into a chamber fitted with a flexible diaphragm which acts on a switch and breaks the battery circuit.

*5,307.—J. B. F. Freeman, London. Carbons for Electric Lighting. 2d. December 28.

CARBONS.—These are made in the form of helices of considerable diameter and are mounted to rotate on axes so as to feed their points together.

5,321.—J. S. Pierson, Brooklyn, N.Y., U.S.A. Underground Telegraph Lines. 6d. (2 figs.) December 30.

CONDUCTORS.—These are laid in grooved planks and covered with asphalte. Several planks placed on top of one another are put in a casing which is then filled with asphalte.

1879.

27. — B. A. Raworth, Manchester. Electric Lighting. 6d. (11 figs.) January 3.

DYNAMO-ELECTRIC GENERATOR.—The first part of this invention relates to the cleansing of the com

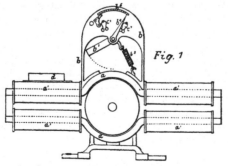

Fig. 1

mutator by means of a rotary brush driven either

direct from the machine or by a separate shaft. The second part of the invention consists in arranging an indicator as shown in cross section in Fig. 1; $a\,a$ are electro-magnets, $a^1\,a^1$ coils, b a frame carrying in pivots a piece of iron b^1 held off the magnets by the spring b^3. When no current is passing the indicator finger b^4 occupies the position shown, and bearing against C^1 closes a battery circuit and rings a bell; should the carbons run together, the finger is moved to C^2, closing the circuit again. In the event of a short circuit, the shunt C^3 will be put in contact, thereby providing a path for the current without its traversing the coils $a^1\,a^1$, thus stopping the action of the generator. The shunt remains in action until readjusted. The third part of the invention refers to the use of a galvanometer and magnetic indicator, to detect any leakage of current. The magnetic attraction

may be shown on the segment b^3, or a needle acting in opposition to a spring may be mounted at d. The galvanometer, placed in the external circuit at a distance from the machine, has its dial marked in unison with the dial of the magnetic indicator; should the latter point to a higher figure than the galvanometer, it will show that part of the current does not leave the machine. A method of dividing electric currents in definite proportions by means of mercurial resistances is described in the fourth part of the invention. Two cylindrical vessels are fitted with discs of vulcanite, having their surfaces cut in **V** shaped concentric grooves, partly filled with mercury and connected so as to form one continuous circuit; a head of water is maintained above the mercury to absorb heat. In the lower side of the discs chambers are cut communicating with the grooves by passages, thin sheets of india-rubber being cemented to the lower sides of the discs, and the spaces between these and two flexible diaphragms are filled with water. These diaphragms are attached to either end of an armature, pivotted in its centre and actuated by two electro-magnets placed one at either end and acting against springs tending to keep the armature in equilibrium. The entering current divides between the electro-magnets, traverses their coils, and thence through the **V** grooves to the outer circuit. When one-half of the divided circuit becomes stronger than the other, its magnet attracts the armature, with the attached diaphragm, thus withdrawing mercury from the **V** grooves of the one disc and forcing more into the grooves of the other.

Arc Lamp.—This, the fifth part of the invention, is illustrated by Fig. 2, a side elevation, and Fig. 3, an enlarged section. Referring to the first figure, l is the box containing the mechanism, and two tubes m^1 and m^2 depend from it, connected respectively to the two poles. The carbon-holders m^3 m^4, sliding on the tubes and having insulating bushes at n n, are supported by the string o, which passes round the snatch-block o, to proportion the travel of the carbons; p is a pillar on which the lamp stands, and is connected to the positive wire; the negative wire is insulated and carried up the pillar to p, where it makes contact with the lamp, the pillar making contact at p^2. The lamp may be suspended by the hooks q^1 q, each of which is in connection with one of the tubes m m. In the longitudinal section the string o, passing over the grooved pulley o^3, has one end fastened to the carbon-holder m^4, the other end being secured to the lever o^4. The pulley o^3 is geared with the copper disc s, revolving between the poles of the electro-magnet s, which, attracting its armature s^2 on the admission of current, separates the carbons. Should the current fail, the spring s^3 overcomes the armature, and by the catch t and ratchet-wheel t^1

sets the train of gear in motion, causing the carbons to approach each other. The finger t^3 is pressed on the disc s by the electro-magnet t^2, and the action of the finger is regulated by the spring t^4 and rod t^5 actuated by the thumb-screw t^6; u is a catch for stopping

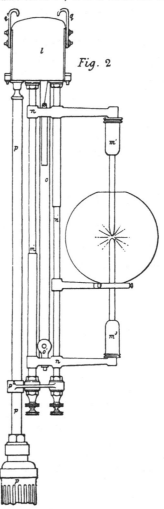

Fig. 2

the lamp. In a modification the grooved pulley and its spurwheel are mounted on the boss of a tumbler plate moved by a link connected to an armature; a pinion carried by a plate mounted on a boss formed on the back of the tumbler plate carries a ratchet and a plain wheel, against which the tumbler bears and acts as a brake. On the entrance of current the armature is attracted and the end of the string lowered; the tumbler is also lowered and bears against the plain wheel, preventing it and its geared train from turning; consequently the grooved pulley is lowered and the carbons are separated. On the current through the magnet weakening the armature lifts, and the upward motion of the tumbler draws the carbons

together, at the same time a pawl gearing with the teeth of the ratchet-wheel tends to put the carbons closer together. Should the circuit be interrupted, the armature makes its full upward stroke, and thereby causes the tumbler to relieve the plain wheel, and the carbons run together. When the friction of the carbon-holders is too

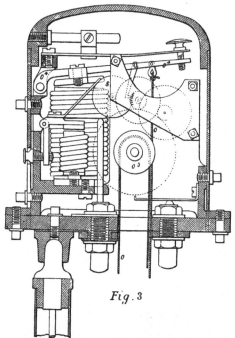

Fig. 3

great for the magnet to overcome, a shunt is employed consisting of a vertical sliding pin of brass, mounted in a small frame carrying a stud of ivory and one of metal, between which a pin projecting from the tumbler plate works with sufficient play to make and break contact without moving the carbon-holders. A finger projecting from the frame makes contact with the magnet coils at any desired point, and the tumbler plate is in connection with the positive pole. When the current decreases, part of the current is shunted, and when the current increases the magnet exerts its full strength.

*51.—E. K. Dutton, Manchester. (*F. Reese, Dortmund.*) **Signalling on Railway Trains, &c.** 2d. January 4.

GENERATOR.—Any ordinary generator is employed in connection with electro-magnets to give a signal to the driver or guard.

*65.—A. M. Clark, London. (*E. Ducretet, Paris.*) **Electric Lamps.** 4d. (1 fig.) January 6.

ARC LAMP.—The lower electrode is placed in a tube filled with mercury and floats up against a stop of refractory material.

81.—K. W. Hedges, London. **Electric Lamps.** 6d. (7 figs.) January 8.

ARC LAMP.—The carbons descend in guide tubes as they waste, their points abutting on a piece of refractory material on the face of which the arc is formed. The distance between the carbons is maintained by an electro-magnet, whose armature is attached to one of the guide tubes. The hollow block supporting the refractory material may be water jacketted.

83.—W. Ladd, London. (*J. Jaspar, Liège.*) **Electric Lamps.** 6d. (4 figs.) January 8.

ARC LAMP.—This is constructed to burn vertical carbons placed one over the other, both movable, and so connected together by cords and pulleys that a counterweight causes them to approach each other and a solenoid separates them. The frame consists of a base and top plate, supported by standards. The lower carbon-holder is a tube free to slide vertically in guides formed of a tube projecting from the base and an opening through the top plate; the lower end of the tube is of iron to form an armature, and is slotted longitudinally to admit a portion of a grooved pulley to which is attached a cord depending therefrom into the tube, the weighted end being fastened thereto. The spindle of the pulley is mounted in a forked bracket fixed to the underside of the top plate and supports a second pulley of about double the diameter of the first, to which is attached a cord connected to the lower end of the upper carbon-holder; the spindle also carries a small barrel connected by a cord to a weighted lever. A solenoid, one end of which is in connection with the frame of the lamp, and the other with the source of electricity, surrounds the tubular guide of the lower carbon. The upper carbon is carried by a holder projecting downwards from an arm attached to a vertical tube sliding in a tubular insulated guide fixed in the top plate of the lamp. The tube passes down through the guide and its end is fastened to a horizontal plate sliding on a fixed guide rod; to this plate is attached the cord from the second pulley. By the side of and parallel with the tubular guide is a mercury tube open at its top to receive a rod projecting downwards from the arm, by which contact is made with the upper carbon. The carbons tend to approach each other by the drag of the weighted lever attached to the small barrel, and the solenoid by drawing down the lower carbon-holder tends to separate them.

144.—W. Morgan-Brown, London. (*W. F. C. McCarty and Baron Sellière, Paris*). **Electric Lighting, &c.** 6d. (5 figs.) January 13.

ARC LAMP.—The lamp consists of a bed-plate

supporting a vertical standard, upon which slides a boss provided with two arms carrying at their outer ends clamping collars which converge upwards at an angle of about 30°; in these clamping collars are placed hard rubber tubes closed at their lower ends by a cap which serves as an abutment for a spring pressing a piston upwards; their upper ends closely approach each other, and are protected by a coating of asbestos. The electrodes, composed of carbon, platinum, iridium, and magnesium, are fed up by the springs, and form the arc across a pellet of platinum-iridium alloy at the top of the tubes, a vertical rod of chemically pure calcium being fed down from a tube held by a bent arm. A double globe, in which water is placed, is used to soften the light.

153.—**J. C. Mewburn**, London. (*Dr. M. C. F. Nitze, Vienna*). **Instruments for Illuminating and Examining Parts or Cavities of the Human or Animal Body.** 6d. (18 figs.) January 14.

INCANDESCENCE LAMP.—A platinum or other wire is made incandescent by an electric current, and the heat is absorbed by a constant flow of water. The lamp so arranged is used in combination with various instruments fitted with lenses for the direct illumination of the part to be examined.

*178.—**A. de Meritens**, Paris. **Obtaining Light by Electricity.** 2d. January 15.

CANDLES.—These are preferably semicircular in section, and are set in holders a sufficient distance apart to admit of a third carbon being placed between them. The tips are connected by a combustible material.

*192.—**G. Remington**, London. **Preventing Rapid Destruction of the Electrodes used in Electric Lighting.** 2d. January 16.

ARC LAMP.—The arc is formed in a globe devoid of oxygen, the globe being charged with hydrogen or any other suitable gas. The carbons may be water jacketted.

211.—**J. Rapieff**, London. **Producing and Applying Electric Currents for Lighting, &c.** 1s. 10d. (232 figs.) January 18.

DYNAMO-ELECTRIC GENERATOR.—To lessen the absorption of magnetism the cores of the electro-magnets are made of diamagnetic material covered on the outside with iron plates, sheets, or wire; or the magnetic parts can be placed alternately with the diamagnetic parts. The non-magnetic cores may be coiled with iron wire, and over them may be placed the inductive coils, or an iron wire of a sufficient size may be substituted for both these coilings. Electro-magnets may be constructed of a solid or compound iron disc provided with two slots, and the coiling put on in such a manner that the wire is always close to the disc and bends round each of its halves from one side only. If more than two poles are required the discs are provided with a corresponding number of slots, and the coiling is laid through them alternately on both sides of the disc, and always in one direction. Two discs are arranged in the form of a cross for four poles, three for six poles, and so on. These discs may be alternated with iron pieces of suitable dimensions. To form a compound magnet a number of the discs have their poles connected alternately in pairs and may be formed into a ring, disc, or cylinder by connecting pieces taking the form of two rings, discs, or cylinders between which the magnets are fixed. By arranging a series of electro-magnets, having alternately their N. and S. poles connected together, as a ring, an endless magnet is obtained having its poles of one polarity on one side, and of opposite polarity on the other. To increase the local magnetic influence the core-pieces are provided with variously shaped projections. To secure the maximum magnetic influence on one side of the electro-magnet the disc core prepared from several pieces of iron is coiled in the usual way, and to it is added a similar disc and the coiling is continued round both discs as if they were one single piece, this process being repeated till the magnet is of the required strength. To increase the action of the magnetic field the armatures are shaped to correspond to the form of the poles. The two branches of electro-magnets have a flat form and are placed side by side, or a polarised fixed armature enclosing a soft magnetising core may be employed. To attract two distinct armatures at different times two parallel magnets of opposite polarity have one of the armatures hinged to one of their poles, the second armature being free. The cores of electro magnets are wound by always coiling the wire in the same direction around and along the axis of the core, a fresh wire being taken for each layer, thus obtaining as many series of conductors as there are layers. The layers may be variously grouped. In arranging the generator the inductors may have their S. poles exactly opposite the N. poles, both polarities being placed side by side or one retreating somewhat behind the others. The arrangement of the induced conductor may also be varied : 1, when the poles of the inductors face each other from both sides of the ring, the coiling is obliquely, and if a second layer of coiling be required its inclination is in the opposite direction; 2, when the poles on one side of the ring face the intervals between the poles on the opposite side, the wire is coiled approximately in the plane passing through the axis of rotation of the disc. In

another arrangement the inductors may be placed with their like poles opposite to each other on both sides of the induced part, in which case they are arranged on each side alternately. The induced conductor is mounted on a ring or disc of iron, copper, brass, or wood, in the form of separate bobbins or zig-zags, but always so that the distance between the branches of one bobbin is approximately equal to the width of the poles. In another form the induced conductor consists of separate bobbins with two long parts and two shorter parts, placed in layers one above the other. The induced conductor may be arranged as a disc, cylinder, or ring, and the iron core may be wound with copper wire, or the conductor itself may be iron. When the bobbins or zig-zags are constructed without any iron parts they are arranged in proper order and connected by a substance which in time becomes hard. A number of discs containing the induced wire are placed alternately between the inductors. The parts may be arranged to give direct or alternating currents. To lessen the friction of the shafts a rotating conical bush is placed on the shaft; where the speed is excessive a number of such bushes placed one within the other are used.

COMMUTATOR.—A commutator for combining the alternate currents of various circuits into one direct current consists of a number of discs, equal to the number of circuits, each having four metallic projections in the form of a cross, the spaces being filled by insulating material. The discs are insulated from each other and from the shaft on which they are mounted, and they are metallically connected to the ends of their respective circuits, and so placed that the retreat of each consecutive disc as regards the preceding one is about one quarter of the intervals between them. The width of the brushes equals the united width of all the discs.

ARC LAMPS.—The current has two paths, one by the carbons to the coils of a permanent magnet, and thence to the return, and the other through an electro-magnet direct to the return; finding the carbons apart it takes the second course, and in doing so causes the movable part of the electro-magnet to be attracted, which by means of a rod lifts the lower carbon-holder and so puts the carbons together; the current then divides and part going by the bobbins fixed to the permanent magnet demagnetises it, and releases an armature which cuts out of circuit the electro-magnet and the carbons fall apart, establishing the arc. The electro-magnet may be cut out of the circuit after the arc is formed by a contact piece attached to the armature, the armature being then retained by a suitable hook or ratchet. Revolving discs of carbon may be used, the current being conveyed

to them by brushes bearing on special washers fixed on their metallic axles. The heat of the arc is employed to bring to incandescence a refractory substance, or a gas or powder is burned by the arc; or the carbons may be enveloped by suitable alloys, which after being melted accumulate in cups and are consumed in the arc. In a modification two carbons placed parallel and insulated from each other are fixed in metallic holders; a piece of conducting wire laid across their ends starts the arc. Two pairs of carbons may be inclined to each other and fed upwards by a weight.

SEMI - INCANDESCENCE LAMP.—Two carbon rods are fed together at an angle, and their points kept together by spring rollers. The carbons may be jacketted with a good conducting material. One of the rods may be replaced by a fixed block of carbon. The feed of the carbons may be controlled by an escapement wheel and tooth, the tooth being actuated by an electro-magnet alternately magnetised and demagnetised; the alternations are controlled by a single electrical arrangement acting on the whole circuit, by an escapement for each separate lamp, by a beam of light acting on a thermopile, or by a small portion of the current itself being periodically interrupted. Liquid poles, such as quicksilver, may be used, in which case two vessels in the form of "a so-called Marriott's jar," each provided with a refractory end and connected to the respective poles of the generator are used; various forms of vessels may be used either in the open or under an air-tight cover.

SECONDARY BATTERIES.—These may be placed in the main circuit, a special commutator being arranged to switch them in or out of the circuit and to make the connection of their elements.

*245.—R. E. B. Crompton and P. W. Willans, London. **Apparatus Employed in Electric Lighting.** 2d. January 21.

ARC LAMP.—A rotating disc electrode is mounted so as to allow of its axis being moved towards the light as it diminishes, and a roller, bearing on its periphery, is caused to advance the axis. Two such electrodes may be used, or one in combination with a stationary electrode.

259.—T. W. Grieve, London. **Lamp or Apparatus for Shading the Electric and other Lights, &c.** 6d. (8 figs.) January 22.

REFLECTORS.—A lantern, fitted with a reflector, has a number of different coloured glass shades, the lower parts of which are hinged so that they may be drawn up under the light. The lantern may be lowered by chains and pulleys to give scenic effects.

277.—S. Colme, London. **Electric Lighting** January 22.

CARBONS.—These are formed of ultramarine in its pure state, or mixed with carbon, kaolin, plaster-of-paris, molasses, or with any powdered metal, the mixture preferred being ultramarine, carbon, powdered copper, and molasses. The candle is baked in a zinc tube, which sublimates and partly enters into the candle.

299.—H. J. Haddan, London. (*E. Molera and J. Cebrian, San Francisco, U.S.A.*) Apparatus for Producing Light by Electricity, &c. 10d. (18 figs.) January 24.

REFLECTORS.—The first part of this invention relates to the diffusion of light by suitable lenses, reflectors, prisms, &c.

ARC LAMP.—This consists of a vertical vessel filled with a suitable liquid, in which are placed side by side, two independent floats working through a guide plate at their upper ends : in these floats the carbons are placed, and slide through contact collars placed near their points. The floats ascend as the carbons are consumed. The carbons may be vertical or inclined. Various modifications of this are shown, such as a large float to which the lower carbon is attached direct, the upper one being carried by a string passing over a pulley, and having its free end fastened to the float, or an electro-magnet, attracting the top of an elastic supplementary cylinder, may be employed to maintain a constant level of liquid. In a flat closed cylinder the bottom plate of soft iron is attached by an elastic ring ; projecting from the top cover are two tubes, bent at right angles and fitted with pistons ; in these the carbons are placed. An electro-magnet tends to repel the bottom plate and force the carbons together.

GENERATORS.—The inventor proposes to maintain the speed of generators constant by driving them from a water motor worked by a constant head of water.

*307.—W. Moseley, Liverpool. Insulated Wires and Cables. 2d. January 24.

CONDUCTORS. — Twisted strands of horsehair saturated with melted resin and oil are wound on the conductor which is then covered longitudinally with a copper ribbon tinned on both sides ; a tin tape is laid over this, having the joint at the opposite side and an outer and thicker copper tape is added, and the whole is then drawn through a die plate heated to 450 deg., which solders the surfaces together. The metal sheathing may be corrugated.

*325.—E. L. Paraine, London. Electrical Light Apparatus. 2d. January 27.

ARC LAMP.—To maintain a constant feed the carbon "is actuated by any element of greater specific gravity than itself."

332.—Baron Elphinstone Musselburg, N.B., and C. W. Vincent, London. Obtaining Currents of Electricity. 6d. (5 figs.) January 27.

DYNAMO-ELECTRIC GENERATOR. — Referring to the illustration, it will be seen that the framing consists of two standards A A, connected by three bridge-pieces similar to B. Mounted on the hollow journals *c c* is the drum C of non-magnetic mate-

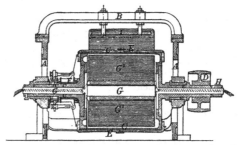

rials, carrying on its periphery the insulated coils E E, laid parallel to its axis, the free ends of each coil being connected to the commutator F. Mounted on the dead shaft G is the electro-magnet G^1 formed of pairs of radial soft iron plates wound with insulated copper wire so that alternate poles are N. and S. The ends of the wires are brought out through longitudinal grooves in the shaft G which is secured to the bridge-piece H to prevent its rotating. Outside the drum C are three electro-magnets I secured to the bridge by the bolts and nuts *b b*, and having their poles opposed to the poles of the magnet G^1. The commutator is of the ordinary kind, the brushes are carried by the insulated ring K, a pair being provided for each independent current required. Permanent magnets may be substituted for the electro-magnets, and the drum may be fixed whilst the magnets revolve.

*350.—F. Young and J. Freeman, London. Increasing and Diffusing Artificial Light. 2d. January 28.

REFLECTORS.—A frame having any number of sides fitted with "bulls-eyes," with a reflector at its top, is used to diffuse the light.

*382.—J. Imray, London. Electric Lamps. 2d. January 30.

ARC LAMP.—The carbons are in the form of hollow cylinders and are rotated and fed towards each other at right angles by clockwork. The arc is formed at the ends of the tubes and consumption takes place in a helical direction. The distance apart of the carbons is regulated by an electro-

magnet or the expansion of a metal strip. The spring giving motion to the feed gear, is arranged to move a piston in a cylinder containing liquid, the outlet being controlled by a valve attached to the armature of the electro-magnet, which thus regulates the mechanism.

*389.—W. R. Lake, London. (*E. B. Watson, Yokohama*). **Reflectors for Electric-Lighting Apparatus.** 2d. January 30.

REFLECTORS.—These are made of silvered glass of suitable shapes. The glass may be coloured.

390.—G. Watson, Castle Eden. **Protecting Coating for Ships, Telegraph Cables, &c.** 2d. January 30.

CONDUCTORS.—The protecting covering is saturated with a mixture of "sene-grease" and naphtha or other suitable vehicle.

*416.—J. D. F. Andrews, Charlton, Kent. **Electric Lamps.** 2d. February 1.

CANDLES.—These are preferably in the form of discs placed face to face, and separated by a layer of plaster-of-paris. The arc presents itself at some part of the periphery and travels round it. The positive disc is made thicker to allow for its more rapid consumption. When the light is to be presented in one direction only, the positive carbon is backed by a metal plate connected by a metal tube to its conductor, the other conductor being connected to the negative carbon by an insulated metal rod passing through the tube, and having at its outer end a screw or nut to clamp together the carbons, with the insulating layer between them.

427.—C. Dubos, Paris. **Electric Lamps.** 4d. (6 figs.) February 3.

ARC LAMP.—Semicircular carbons are held in holders attached to the ends of radius bars fixed to pulleys, round which pass cords connected to a weighted lever tending to keep the points of the carbons in contact, a cord from a second pulley of each radius bar being connected to a soft iron core of a solenoid placed in the lamp circuit which tends to separate the carbons. In a modification the carbons are mounted on counterweighted arms, one of which is carried by a shaft and the other by a hollow journal moving freely on the shaft; both are fitted with pinions which are engaged by a larger pinion actuated by a solenoid tending to separate the carbons.

*454.—R. W. H. P. Higgs, London. **Electric Lighting.** 4d. (3 figs.) February 5.

ARC LAMP.—A vibrating armature actuated by an electro-magnet mounted on a base-plate carries a block of carbon. Attached to the base-plate is a vertical standard carrying a tube in which slides a carbon rod controlled by a counterpoised friction arm and roller. The circuit is completed through the rod, block, armature, and electro-magnet.

*465.—S. W. M. de Sussex, Brussels. **Luminous Electric Buoys, &c.** 2d. February 5.

ELECTRIC LIGHT.—The battery, constructed as described in Patent No. 2,194 of 1877, is connected to a Ruhmkorff's coil and Geissler's tube.

508.—H. S. Weyde, London. **Shades or Lanterns for Electric Lights, &c.** 6d. (1 fig.) February 8.

REFLECTORS.—These consist of an arrangement of slightly concave reflectors placed in parallel zones above and below the level of the luminous focus, the upper set being concave on their lower surface and the lower set on their upper surface. The two sets are separated by an equatorial prism.

523.—E. T. Bousfield, Bedford, and **W. R. Bousfield**, Bristol. **Electric Light.** 6d. (9 figs.) February 10.

DYNAMO-ELECTRIC GENERATOR.—The armature is built up of longitudinal segmental W-shaped bars of soft iron in the form of a drum, insulated coils being laid round the webs of the bars, and consequently parallel to the axis of the drum. The bars are in metallic contact and are secured by two end plates having annular recesses into which their ends fit. One plate is fixed to the axle and the other is forced against the ends of the bars by a nut working on a thread cut on the axle. The drum revolves between the poles of fixed electro-magnets and the currents are collected by means of a commutator and brushes of ordinary type. Flat segmental bars may be used and the insulated wires placed so that the periphery of the drum would present an unbroken surface of coils.

ARC LAMP.—A standard mounted on a base-plate carries a bell-crank lever provided at its forward end with the positive carbon-holder in which the carbon is free to slide, but is retained by a clip spring; the pendent limb of the lever is pressed by a coiled spring tending to bring the carbons in contact. A second smaller lever serves to grip the carbon but is withdrawn by a spring having an armature at its tail end. An electro-magnet placed in the lamp circuit, on the current being established, causes the small lever to grip the carbon and simultaneously lifts the crank lever, thus drawing the arc. In a modification both carbons tend to approach each other and are connected together by a cord passing over pulleys, one end

being fastened to a drum. On the drum spindle is a small toothed pinion gearing into the rack rod of the upper carbon-holder, and mounted on the same spindle is a brake wheel having its periphery fitted with a number of small armatures which come under an electro-magnet placed in the lamp circuit. The electro-magnet acting on this wheel draws up the sliding cage in which it is mounted, thereby separating the carbons and establishing the arc. In a further modification one end of the lifting cage may be pivoted, in which case the arc is formed by the carbons being thrown out of line.

565.—J. Formby, Formby, Lancashire. **Dynamo-Electric Machines.** 6d. (3 figs.) February 13.

Dynamo-Electric Generator.—The inductors consist of a number of straight coils wound on wood and arranged longitudinally in the form of a drum; they are supported in upright frames, and have all the N. poles and all the S. poles grouped together, and opposite their ends are two annular revolving coils, also wound on wood and mounted on a shaft. Outside these revolving coils are two fixed coils attached to the frame. The ring cores may be strengthened with metal. Two concentric rings formed of a number of collectors, are placed on each of the revolving coil frames, each piece being connected to one end of a coil, from which the currents are collected by ordinary brushes. A series of terminals are arranged on the top of the generator for connecting the coils either in quantity or tension, or for making a number of circuits.

644.—J. F. Wiles, London. **Magneto-Electric or Dynamo-Electric Machines.** 6d. (4 figs.) February 18.

Dynamo-Electric Generator.—The field-magnets are mounted on a rotating shaft, and revolve in the opposite direction to the armatures, which are arranged in a ring. Each system has its commutator and brushes of the usual kind.

684.—M. A. Wier, London. **Apparatus for Producing Light by Electricity.** 6d. (18 figs.) February 20.

Arc Lamp.—The electrodes are formed of a number of small bars of copper, placed parallel to, but insulated from, each other, and set in the lamp so as to form a kind of inclined grating, over which is passed a constant stream of finely divided carbon falling from an upper hopper into a lower one, the relative positions of the hoppers being periodically reversed by suitable mechanical means; or the lamp may be contained in a revolving glass cylinder, the inner periphery of which is fitted with steps which carry the powder up and deposit it in the upper hopper; or the cylinder may be plain and rapidly rotated, a fixed stop sweeping the powder off into the hopper as it rises. The lamp may consist of a cylinder closed at its lower end by a disc of glass, on which is laid a quantity of the powdered carbon, the electrodes being vibrated in close proximity to the glass by an electro-magnet. The electrodes may be of conical form.

718.—W. F. Jack, London, and **F. Greening,** Southall. **Production and Application of Substitutes for Collodion.** 4d. February 22.

Conductors.—These may be covered by a substance made from a coal or wood fibre, soluble in alcohol or hydro-carbon oils, and formed into a solution therein either by itself or mixed with oils, gums, resins, baryta, &c. The compound is made of such a consistency as to be capable of being formed into sheets by rolling, or the conductors may be dipped into the compound while it is in a liquid state.

740—F. Mori, C. E. Hallewell, W. Milner, and **W. Griffin,** Leeds. **Electric Lamps.** 6d. (11 figs.) February 24.

Arc Lamps.—A revolving commutator, actuated by a coiled spring, carries several pairs of parallel carbons, and its motion is controlled by a series of stops. A circular disc of insulating material has a central upright pillar with radial arms, to each of which is affixed a carbon; the other carbons are carried by holders pivoted to pedestals fixed to the plate, terminating at their shorter ends in armatures acted on by magnets attached to the underside of the plate. The coils of the magnets have one of their ends connected to contact studs and the other ends to the pivoted holders. A commutator, insulated from the upright pillar, and in permanent contact with one of the terminals, has radial arms carrying contact pieces at their ends so arranged that, on the commutator revolving, they will rub over the faces of the magnet contact studs. The commutator is prevented from revolving by stop-pieces mounted on vertical spindles, through the heads of which pins pass bearing against the carbons. As soon as one pair of carbons burns down to its pin, a coiled spring throws the stop out, and the commutator moves forward to the next pair. When the last pair of candles are consumed, the commutator comes in contact with a stud which completes the circuit.

749.—C. Dubos, Paris. **Galvano-Magneto-Dynamo-Electric Machine.** 6d. (6 figs.) February 25.

Dynamo-Electric Generator. — The electro-

magnets consist of a ring of soft iron divided by segmental collars into a number of intermediate portions, each of which is coiled transversely with insulated wire so as to give alternate polarity to each adjacent pair of coils. A number of these rings are arranged in a frame side by side and may be separately excited. In the spaces between the electro-magnets revolve the armatures, each consisting of a wooden boss carrying as many coils as there are divisions in each of the annular electro-magnets. The cores of the coils are made of tin-plate rolled and soldered to form hollow cylinders. In a modification the electro-magnets consist of a number of cranked bars coiled with insulated wire so as to give the cheeks of the cranks alternate polarity. The coils of the armatures revolve between these cheeks. The currents are collected by an ordinary commutator.

783.—G. P. Harding, Paris. Obtaining Electric Light. 6d. (7 figs.) February 26.

ELECTRODES.—These are coated with lime or other refractory material, or a metallic conductor may be employed in place of the carbon. A sheet or disc of lime is used to reflect the light.

805.—W. P. Thompson, London. (*W. W. Gary, Washington, Columbia, U.S.A.*) Magneto-Electric Apparatus. 6d. (6 figs.) February 28.

MAGNETO-ELECTRIC GENERATOR. — This relates to a generator involving the use of a horseshoe permanent magnet having the armature extended across its two poles "and then arranging the armature to move only between the neutral line and the magnet." The instrument is chiefly applicable to telegraphic purposes.

830.—G. G. Andre, Dorking. Electric Lamps. 8d. (9 figs.) March 1.

ARC LAMP.—This is designed for use in coal mines, &c., and consists of four vertical carbons fed down by weights on to the peripheries of four carbon discs; the carbons are automatically rendered operative one at a time by a commutator, and the whole arrangement is enclosed by a glass cylinder. The discs are eccentrically mounted and an armature actuated by an electro-magnet coupled in the lamp circuit permits a spring to turn and raise the discs should the current fail, thus cleaning their surfaces. The tubes in which the positive carbons slide have each at their lower ends a lever, the outer and heavier end of which serves to press the carbon against the lower part of the tube. The commutator consists of a number of electro-magnets and armatures, one less than the number of sets of carbons to be used, connected together in such a manner that when one of the carbon connections is interrupted, its electro-magnet shall release its armature which sends the current through the electro-magnet of the next carbon, thus causing it to light. In another lamp the points of the pivotted positive carbons incline to a central negative rod and are drawn from it, one at a time, in a similar way. In another arrangement two discs of carbon are a greater distance apart at their centres than at their outer periphery, the space between them being left empty or filled with insulating material, and they are mounted on a vertical stem connected at either end to the respective poles of the generator.

863.—C. D. Abel, London. (*J. C. Jamin, Paris.*) Electric Lamps. 6d. (12 figs.) March 4.

ARC LAMP.—The two carbons are placed parallel to, and a slight distance apart from each other; the current is made to first traverse a wire loop placed parallel to and on each side of the carbons and across their terminals, the current flowing in the same direction in the loop and carbons. In another lamp a flat carbon plate presents its edge to a carbon rod, the circuit wire passing from the upper socket of the rod parallel with the plate towards its lower socket, where it turns off at right angles. The carbon rods may have coils of wire round them which will cause the arc to revolve. The lower carbon may be in the form of a cylinder within which is placed a block of lime. The cylinder is within a solenoid which causes the arc formed between the upper carbon and the block of lime to revolve.

*876.—J. B. Spence, London. Electric Lighting. 2d. March 5.

ARC LAMP.—The ends of the electrodes are formed of platinum fitted in tubes in which water circulates. A gas or other flame passing between the platinum points has the arc formed within it.

*920.—B. G. Holmes, Leytonstone. Globes or Shades for Lamps. 2d. March 8.

REFLECTORS.—The upper part of a globular envelope is constructed of white glazed earthenware, the lower part of clear glass. The parts may be separate and suitably joined together.

925.—K. W. Hedges, London. Electric Lamps. 6d. (3 figs.) March 8.

ARC LAMP.—The carbons, rendered partly magnetic, slide freely towards each other in guide tubes, so that their points meet, or one may be stationary or moved by a spring, while the other descends by its own weight. Near the carbons are electro-magnets, placed in the lamp circuit, which

attract them, and so stop their advance when the current is strong. One guide tube is pivotted, and this is attracted laterally by an electro-magnet opposed by a spring on current being admitted to the lamp; the carbons are thus separated and the arc established.

*927.—R. C. Thompson, London. Electric Light. 2d. March 8.

ELECTRODES.—These consist of powdered carbon placed in tubes, preferably of carbon also, and fed to the arc by a "screw conveyor" or by gravity.

947.—H. J. Haddan, London. (*C. F. Brush, Cleveland, Ohio, U.S.A.*) **Apparatus for Electric Lighting. 8d. (20 figs.) March 11.**

ARC LAMP.—The upper carbon is provided with a mechanism which alternately keeps it stationary, and allows it to drop while the lower carbon is fed, by the resultant force of magnetism caused by the passage of the current through two coils, one in the main circuit and one in a shunt circuit. As will be seen by reference to the illustration, a suitable base K has attached to it a metallic post E, carry-

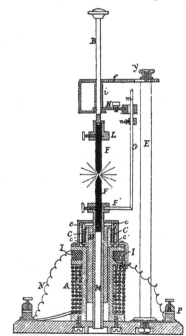

ing an arm x, through which the upper carbon-holder B slides; this holder, or rod, is surrounded by a ring clamp H, supported on projections attached to the arm x; a loose tube l surrounding the rod prevents the clamp being carried up with it. D is an iron core rigidly attached to the tube M, terminating in the carbon-holder L; c are arms which, by means of the spiral springs C, support and

force upwards the core D, and with it the carbon F[1]; c^1 are adjusting screws. The arm G, carried by the carbon-holder L, passes at its upper end through a hole in the clamp H, provided with an insulating piece at m. The adjustable collar n is so placed that the core D, at the limit of its upward movement, slightly raises the clamp H. The coarse wire helix A has one of its ends attached to the binding post N, the other in connection with the lower carbon-holder. The fine wire, or adjusting helix I, is wound in the opposite direction to A, and its ends are connected to the posts P and N. P is a binding post attached to the post E. To adjust the carbons the arm x is provided with a slot, and is clamped on post E by the nut y. When the lamp is not in use, the spring C forces the core D upwards, the collar n on arm G raises the end of the clamp H, and the rod B is released and falls until the carbons F F[1] are in contact. In a modification the arm G is dispensed with, and the fine wire helix arranged as a solenoid actuates a core attached to the ring clamp, or the lower carbon is fixed and the coarse wire helix and core abolished, the fine wire helix being used to actuate the ring clamp. Various methods of combining the coarse and fine wire helices may be used. The invention also relates to a method of shunting the current from any lamp in a circuit when its resistance becomes too high. This is accomplished by a high resistance magnet, which attracts an armature, and so opens another path for the current. In the path so opened is a second low resistance magnet, which serves to maintain the armature in its new position.

959.—H. Haymen, London. Preventing Incrustation of Metallic Surfaces, &c. 4d. March 11.

CONDUCTORS.—A composition for insulating conductors is composed of a mixture of pyroxiline, with an earthy or other pigment, thinned by the addition of a volatile spirit, to which is added glycerine, oil, caoutchouc and gum, resin, or gelatine. The conductor is passed through a bath of the liquid composition.

960. — L. Simon, Nottingham. (*S. Schuckert, Nürnberg.*) **Dynamo - Electric Machines. 6d. (3 figs.) March 11.**

DYNAMO-ELECTRIC GENERATOR. — A flat ring armature is made up of several sheet-iron discs magnetically separated from each other. The coils consist of "copper wire or any other good conductor wound about the iron ring." The pole plates of the horizontal electro-magnets face the ring and cover two-thirds of its entire surface. The "brushes consist of a number of insulated metallic pieces, each one connected with a conducting wire coming from the revolving armature."

*988.—M. Birkhead, Liverpool. **Utilisation of the Fall of Sewage for Motive Power.** 2d. March 13.

ELECTRIC LIGHT.— Motors, preferably ordinary vortex wheels, are placed in the conduits of the sewers. "One main use for these sewer motors will be in working dynamo-electric machines for the electric light."

*1,012.—A. L. Coke, Bristol. **Electric Lighting.** 2d. March 14.

CURRENT DIVIDER.—"The contact may be made and broken in any of the well-known ways aided by machinery of any convenient kind," to enable the whole current "to be fully turned on to each light or series of lights in succession."

*1,110.—G. B. Finch, London. **Cutting Coal.** 2d. March 20.

ELECTRO-DYNAMIC MOTORS.—These are used to drive boring bars, the current is supplied from a generator situated on the surface.

*1,122.—St. G. L. Fox, London. **Electric Light.** 2d. March 20.

INCANDESCENCE LAMP.—The filament is composed of two highly refractory materials, one a conductor such as plumbago, and the other a non-conductor such as magnesia, zirconia, or steatite, ground together to a very fine powder, slightly moistened, and then formed to the shape required under great pressure. The surface may be coated with hard carbon by sending a current through the filament while it is immersed in a heavy hydro-carbon. Each end of the filament is connected to a piece of plumbago which may be held by metallic clips, or supported loosely in metallic supports having sufficient play to allow of the insertion of the "candle" between them, or they may enter passages in a block of steatite. The parts are hermetically enclosed within a glass globe from which all traces of oxygen have been removed.

*1,158.—J. Dove, London. **Motive Power.** 2d. March 24.

MOTIVE POWER.—The rise and fall of a pontoon placed in a tidal stream "can be used for generating electricity."

1,175.—J. Imray, London. (*Société Générale d'Electricité, Procédés Jablochkoff, Paris*). **Electric Candles.** 6d. (10 figs.) March 24.

CANDLES.—Two metallic wires are placed parallel, or slightly inclined to each other, and each is enclosed in an insulating envelope of carbonaceous matter, such as anthracite, moulded to a cylindrical form, with a thin middle rib. Two or more wires may be used for each "wick," and the insulating material may have ingredients which undergo chemical reaction and produce increased light. To provide for the rekindling of the candle metallic particles are mixed with the insulating material. The several candles in one lantern are all connected to the circuit wires, but have different resistances at their points; the one of least resistance lights first, and so on.

1,207.—T. A. Dillon, Dublin. **Producing Electric Lights.** 4d. March 26.

ELECTRIC LIGHT.—An engine serves to compress air which is distributed to the houses by pipes. Each house is provided with a generator driven by the compressed air, and lamps of any known type are used. The waste heat from chimneys, gas lamps, &c., may be utilised to produce the necessary steam. The turnstiles at theatres, or "the rush of crowds up and down stairs may be thus simply utilised."

1,347.—T. A. Dillon, Dublin. **Transmitting Messages and Printed Matter, &c., by Electrical Cables.** 6d. April 4.

ELECTRIC LIGHT.—To record telegraphic signals an electric light is reflected by a mirror galvanometer on to a series of plates rendered photographically sensitive.

*1,387.—A. W. Lake, London. (*N. S. Keith, New York, U.S.A.*) **Dynamo and Magneto-Electric Machines for Lighting, Plating, &c.** 4d. April 8.

DYNAMO-ELECTRIC GENERATOR.—The space in which the armature revolves is of such shape that the points of greatest magnetic attraction are close to the armature, while the points of least attraction are further away, the space approaching the form of an ellipse. Various polygonal forms may also be used. The polar ends of the armature cores are made of not greater, and even of less, sectional area than the body of the core. The "contrary currents" generated in the coils of the electro-magnets are diverted by a rotary circuit maker and breaker into a shunt external to the main circuit. The circuit maker consists of a cylinder fastened on the axis of the armature, upon the periphery of which are attached a number of brass plates equal to double the number of the pole ends of the armature to which they are attached. Alternate plates are electrically connected while the intermediate plates are insulated from each other. Brushes connected to the shunt press on this commutator.

ARC LAMP.—A toothed wheel has a drum on each side of it, and fastened on the same axle, one drum being twice the diameter of the other.

The carbon-holders, one of which is heavier than the other, are connected to these drums by cords, and a rocking lever armature actuated by an electro-magnet opposed by a spring engages in the teeth of the wheel, serving to separate and keep the carbons apart so long as current passes. When the current ceases the spring repels the armature, and the carbons run together by gravity.

*1,445.—J. Whiteley, Rochdale. Obtaining Motive Power. 2d. April 10.

MOTOR.—" Static electricity, positive and negative, may be used in place of magnetism, north and south, but contact must be avoided to prevent the recomposition of the two electricities."

1,476.—H. J. Haddan, London. (*A. M. Rosebrugh, Toronto, Canada.*) Telephones, &c. 6d. (10 figs.) April 16.

MAGNETO-ELECTRIC GENERATOR. — A rotating spindle carries an arm on which are mounted two small electro-magnets revolving in close proximity to the polar faces of a flat horseshoe magnet. The currents are collected from a commutator arranged in the usual manner.

*1,622.—R. C. Thompson, London. Producing Electric Light. 2d. April 24.

ELECTRODES.—Carbon in a plastic state is fed to the arc through tubes, or tubular carbons may have a suitable downward inclination to one another, and powdered carbon may flow down them in a continuous stream.

1,635.—J. Mackenzie, London. Electric Light Apparatus. 6d. (5 figs.) April 25.

ARC LAMP.—The upper carbon is carried in a tubular holder pivotted to the end of a lever moving on a central axis, the other end of the lever being the armature of an electro-magnet. On the opposite side of the tubular holder a small lever is mounted having its outer end depressed by a weight; its shorter end bears against the carbon and securely holds it in the tube. On the failure of the current the magnet releases the lever and the tubular holder falls; the outer end of the small lever is then lifted by an adjustable stop-piece and the upper carbon falls on to the lower one. The lower carbon is fed up by being mounted in a tubular holder having a screw-rod at its lower end, which rod is slotted from end to end and has a pin engaging in the slot to prevent its rotating. The holder is caused to revolve by a ratchet and pawl actuated by a second electro-magnet, thus causing the screw-rod to rise and with it the lower carbon. In a modification the lower carbon is forced upwards by a spiral spring and retained by a cord wound round

a roller controlled by an escape wheel and pallets actuated by an electro-magnet. The same electro-magnet may operate a rod which serves to regulate the feed of the upper carbon.

1,692.—J. S. Sellon and H. Edmunds, London. Regulating Electric Currents. 6d. (5 figs.) April 30.

AUTOMATIC SWITCH.—This consists of a rectangular frame on which is mounted a spring contact piece retained by a fine tension wire through which the current passes. When the current is excessive the tension wire expands and allows contact to be made with a stud, thus cutting out of circuit the lamps. The switch may be placed so as to interrupt the exciting current supplied to a number of generators. A switch may be employed to automatically place more resistance in circuit by making contact with a series of contact pieces to which fine wires leading to the main circuit are attached.

RHEOSTAT.—This is an arrangement of resistance whereby an equivalent amount of resistance is put in circuit on switching out a lamp. The lamps being in the circuit of the induced current and the resistances in the circuit of the exciting current, the generation of the current is so controlled as to limit the production to the requirements of the work.

*1,791.—J. S. Sellon and H. Edmunds, London. Electric Lamps. 2d. May 6.

INCANDESCENCE LAMPS.—Where lamps requiring different currents are to be placed in the same circuit those requiring less current have a number of fine wires connecting the two sides of the space bridged over by the incandescent filament, which wires carry over the surplus current. A spring retained by a tension wire cuts the filament out of circuit should the current become excessive. In a lamp containing two filaments the second is brought into circuit on the fusion of the first. Three posts are employed for carrying the filaments, two fixed and one jointed. The first of the fixed posts is connected with one end of each of the filaments, the other ends being connected to the jointed and second fixed post respectively. The jointed post is fitted with a spring which on the breaking of its filament causes it to make contact with the second post.

*1,949.—J. S. Sellon and H. Edmunds, London. Apparatus for Generating Electric Currents. 2d. May 15.

MAGNETO-ELECTRIC GENERATOR.—Mounted on either side of an annular armature are a number of electro-magnets, the poles of which are shaped to

embrace the armature. The magnets are so coiled that those on one side are all of one polarity and those on the other side are of opposite polarity. The armature is composed of a number of short segment-shaped bobbins of soft iron provided with end flanges and wound with insulated copper wire, the terminal wires being connected either in groups or in series, or in the form of independent circuits. The magnets and armature are made to rotate in opposite directions.

1,959.—J. Hopkinson, London. **Electric Lamps.** 4d. (1 fig.) May 16.

ARC LAMP.—Points of the main circuit on opposite sides of the arc are connected by a fine wire, forming an electro-magnet or solenoid. When the distance between the carbons becomes too great the electro-magnet or solenoid actuates a suitable switch whereby the main magnet of the lamp itself is short-circuited and the carbons are permitted to approach each other.

1,969.—A. Longsdon, London. (*A. Krupp, Essen.*) **Electric Light Apparatus or Lamps.** 6d. (4 figs.) May 16.

ARC LAMP.—The core of a solenoid actuates a lever connected to the lower carbon-holder and also connected by a rod to a clamping device attached to the upper carbon-holder. On completing the circuit the solenoid draws down the lever and with it the lower carbon, clamping at the same time the upper carbon, and thus establishing the arc. In a modification the clamping arrangement is actuated by an electro-magnet placed in the main circuit which releases the upper carbon when the current becomes weakened.

*1,971.—A. M. Clark, London. (*N. E. Reynier, Paris*). **Lighting by Electricity.** 8d. (8 figs.) May 16.

AUTOMATIC SWITCH.—A number of change lamps are provided and switched in to replace any other lamp that may be extinguished. Each of the change lamps is placed in a derivation of the main circuit, and in each of the derivations, except the last one, electro-magnets are placed, which open the circuit of the succeeding lamp when they are magnetised, and close it when they are demagnetised.

*2,000.—J. Cougnet, Brussels. **Electric Lamps.** 2d. May 20.

ARC LAMP.—The lower carbon is fed up by a cord and weight passing over a pulley. A spring-controlled lever armature, attracted by an electro-magnet in the lamp circuit, acts as a break on the periphery of the pulley. The upper pole consists

of a block of carbon let into a hemispherical polished metal holder. For use in mines the lamp is placed in a glass globe, and the lower carbon rests in and is fed up by mercury contained in a cylinder.

2,016.—F. Saunders and **L. Danckwerth,** St. Petersburgh. **Compound for Insulating, &c.** 2d. May 21.

INSULATING COMPOUND.—One-third by weight of a mixture of equal parts of vegetable tar oil and coal-tar oil, is heated in a boiler with another third of hemp-seed oil, at a temperature of about 140 degrees C., until the mass becomes stringy. There is then added the last third by weight of boiled linseed oil. With this composition is mixed five to ten per cent. of ozokerit and a little spermaceti, the mass being again exposed to the same temperature for several hours, and finally about one-twelfth part by weight of sulphur is added.

2,019.—J. Chretien and **C. Felix,** Paris. **Traction Machinery.** 6d. (9 figs.) May 21.

TRANSMISSION OF POWER.—The currents produced by generators are used to work motors adapted to drive the various machinery used in timber yards, railways, mines, agriculture, &c.

2,040.—J. L. Haddan, London. **Traction and Ways.** 10d. (43 figs.) May 22.

ELECTRIC RAILWAYS.—The trains may be driven by a Gramme or other suitable generator used as a motor, or an iron piston may be forced along a diamagnetic tube forming the rail, the poles of electro-magnets fixed to the carriages exerting an attractive influence on the pistons. The current may be generated by a thermopile carried on the train.

2,060.—A. F. Blandy, London. **Apparatus for Electric Lighting.** 6d. (2 figs.) May 23.

REGULATING CURRENTS. — A "governor" for regulating the currents to lamps in the brushes of a divided circuit, is formed of a soft iron core sliding in a tube, round which wires are wound parallel to each other, and connected to the branches, so that the currents shall traverse the coils in opposite directions, thus "making" the circuit in one branch, sets up a momentary induced current in the other, which maintains the arc situate in that branch, while the electrodes of the second lamp touch and separate. The core may be counter-balanced to withdraw it when both branches are at work, the current in either branch attracting it when only one branch is at work.

2,110.—C. W. Siemens, London. **Producing Light and Heat by Electricity.** 10d. (12 figs.) May 27.

ARC LAMP.—The lower terminal is water-cased and consists of an iron wire fed up by a spring through a thin copper tube surrounded by another tube of larger diameter, a constant flow of water being maintained through the annular space between the two tubes. The upper carbon is free to slide down a tube having longitudinal internal projections and is clamped in the tube by its lower end resting against the platinum tip of an abutment pole screwed obliquely through an eye at the lower end of the casing. To this eye is attached a convex facing of porcelain. The upper carbon tube is fixed at the end of a lever pivotted on the frame and is suspended by a thin metal strip forming part of the lamp circuit, which by its expansion and contraction regulates the length of the arc, or the tube may have on its outer surface a casing of soft iron sliding within a high resistance solenoid coupled in across the arc. The tube and carbon are counterbalanced by an adjustable weight on a lever. The top of the tube is attached to a flexible diaphragm forming the cover of an air vessel having a small air-hole which acts as a dash-pot against the action of the solenoid. In a modification both poles are of carbon, the lower one being fed up by a weight against two oblique screws which press it against two chisel edges. The upper carbon has let into a groove on one side of it, a strip of zinc or other easily fused metal which rests against the edge of the hole in the reflecting shield, and which as it fuses allows the carbon to descend. Two weighted levers pressing against the carbon near its point make electrical contact with it. Both carbon-holders are attached to arms of bell-crank levers, the other arms of which are connected by a non-conducting link. One lever carries a weight, and its extended end rests in the loop of a platinum wire in the circuit, the tension of which is adjustable, and which by its expansion and contraction under the heating due to the current regulates the length of the arc. In a further modification the horizontal carbons are fed forward by volute springs, their front ends abutting against adjustable chisel edges. The line wires are flexibly connected to the carbon guide tubes, and these are linked to the vertical arms of counterweighted bell-crank levers, on the horizontal arms of which bears a disc of non-conducting material attached to the core of a high resistance solenoid coupled as a shunt to the arc; the core is prolonged upwards, and has attached to it a piston working in an air cylinder; its upper end terminates in a cup in which shot is placed for adjustment. The arc may be used for heating a crucible.

2,111.—W. R. Lake, London. (*J. Puvilland and T. Raphaël, Paris.*) **Electric Lighting Apparatus.** 6d. (5 figs.) May 27.

ARC LAMP.—Circular carbons are fixed in holders at the ends of two counterpoised radius bars mounted on axes sustained by supports fixed to the frame of the lamp, and a regulating slide permits of the adjustment of the bars according to the radius of the carbon rings. On pulleys attached to the axes metallic chains are wound which pass over guide pulleys to a wooden crossbar and are attached by cords to weights tending to keep the carbons in contact. A magnet is placed in the main circuit, and by means of an armature actuating a balanced lever tends to separate the carbons. The balanced lever is pivotted and provided with spring pawls which engage with ratchet-wheels on the axes of the carbon-holders to separate the carbons but become disengaged on the reverse stroke of the lever.

*2,199.—I. Furstenburgh, Bradford. (*L. Siemens, Charlottenburg*). **Lighting by Electricity.** 2d. June 3.

CARBONS.—These are provided with a "wick" of glass or other vitreous substance.

2,267.—G. Grant and R. Sennett, London. **Illuminating by Electricity.** 2d. June 9.

ELECTRODES.—These are made hollow, and gas or finely divided substances are passed through them to the arc. Hollow platinum electrodes may also be used.

2,301.—R. Werdermann, London. **Apparatus for Electric Lighting.** 6d. (6 figs.) June 10.

SEMI-INCANDESCENCE LAMP.—This relates chiefly to improvements on Patent No. 2,477 of 1878. The counterpoised arm holding the upper carbon disc is pivotted to a sleeve sliding on an upright pillar which is so adjusted that the disc shall always rest on the point of the lower carbon; this is fitted to slide through guides pivotted together so as to be clamped round the carbon rod. The pivotted arm of the carbon disc is attached by a rod to the longer arm of a lever, the shorter arm of which is prolonged as a spring having at its upper part a set screw, bearing against the lower carbon, and so regulating its feed. When the lower carbon is totally consumed this lever makes contact with a stud, thus opening a by-path for the current. In a modification the spring is attached to the pivotted arm direct, and bears through a block of insulating material on the lower carbon. The spring short-circuits the lamp when the lower carbon is consumed. The weight of the pivotted arm may, by means of a link and lever, be made to grip the lower carbon, a short circuit being established as before on the consumption of the lower carbon. The clamping ring of the disc is supported in centres in such a manner that its

weight preponderates on the side in contact with the lower carbon. The spring is mounted on the clamping ring, and bears against the carbon. The pivotted arm is connected to a lever on which are mounted two adjustable spring rollers gripping the lower carbon.

REFLECTORS.—These consist of an opal glass dish placed under the light, and a broad deflector placed above the light.

2,3I7.—C. F. Heinrichs, London. Puncturing or Perforating Materials for Use as Stencils. 6d. (24 figs.) June 11.

ELECTRO-MAGNETIC MOTOR.—A small motor, consisting of two vertical electro-magnets, has a rotating armature fixed to the stylus of the pen; the reciprocating motion is given by a double inclined plane, over which the armature runs on suitable rollers, or the electro-magnets may revolve, the armature being fixed.

2,32I.—J. D. F. Andrews, Charlton. Production and Regulation of Electricity for Illumination. 6d. (16 figs.) June 11.

DYNAMO-ELECTRIC GENERATOR.—The rotating armature consists of a long cylinder having a number of iron cores mounted on its surface, parallel to the axis and wound lengthways with coils of insulated wire. The coils are connected to a suitable commutator, whence the currents are collected in the usual way. Two stationary electro-magnets extend radially on each side of the armature, their outer ends being connected by iron so as to form a complete horseshoe. A commutator for varying the power of the lights consists of a cylinder, part of which is of conducting and part of non-conducting material, and against this press a number of springs, each connected to a separate set of coils on the generator. By turning the cylinder more or less round a greater or less current is put into the circuit. In another form slides are provided, consisting of two metallic plates in connection with the lamp terminals, and insulated from each other, which are pushed between springs connected to separate sources of electricity. In a commutator for altering the connections of a number of wires into wholly or partly parallel circuits or series, a sliding non-conducting plate has a number of metal strips fixed on it. The lamp wires are connected to pieces fixed on the sliding plate, and a number of contact springs connected to separate sources of electricity are arranged so as to make contact with them, and by bringing different combinations of the springs in contact with the strips the current in the line wires is varied as desired.

ARC LAMP.—The carbons are placed vertically one over the other; the upper one is held by friction between rollers, one of which is on the shaft of the armature of a magneto-electric motor having its coil in the lamp circuit. The action of the motor is to separate the carbons, whilst the weight of the upper carbon brings them together. Two horizontal carbon plates meet and form the + electrode, and a third parallel lower plate presents its end to their junction and forms the — electrode. The distance of the lower plate from the others is automatically regulated by an electro-magnet having its coil in the lamp circuit. The — electrode may be at right angles to the two plates. In another lamp the lower carbon is mounted on a lever which has armatures on each side of its fulcrum acted on by two electro-magnets, the second of which has at its lower end an armature making contact, when not attracted, with a stud. When the current traverses the line wire, the first magnet raises the lower carbon and establishes the arc. The current passes through the carbons, and also through the second magnet, which attracts its armature from the stud, and so cuts the first magnet out of circuit. The automatic adjustment of the lower carbon is then carried on by the second magnet, which also maintains the feed by a spring pawl acting on a toothed rack. In another lamp two vertical plates of carbon have a central plate between them; all are insulated from each other, the outer plates being connected to an alternating current generator. A rod of carbon attached to the armature of an electro-magnet in the lamp circuit establishes the arc between the plates. The plates may be duplicated. Two discs may be used, one a little below and parallel to the other, and carried by a rod passing through the tubular support of the first. A few turns of wire about the tube cause the arc to rotate. In a modification the rod is free to move, and is attracted by a magnet, thus bringing the carbons together and putting a second magnet in circuit, which breaks the circuit of the first magnet, and the discs then fall apart.

*2,322.—C. D. Abel, London. (*H. Sedlaczek and F. Wiknlill, Leoben.*) Electric Lamps. 2d. June 11.

ARC LAMPS.—The two electrodes are carried by pistons contained in two closed vessels filled with liquid and communicating with each other, the descent of the upper piston and carbon causing the ascent of the lower one. The flow of liquid is controlled by a valve actuated by a solenoid which checks the flow on any increase of current.

2,339.—A. M. Clark, London. (*A. de Méritens and Co., Paris.*) Obtaining Light by Electricity. 6d. (8 figs.) June 12.

CANDLES.—The arc is formed between a number

of electrodes, the outer pair being of larger diameter than the inner pairs. The current is divided, and resistances are interposed before the smaller carbons, by which the larger portion of the current is caused to traverse the larger carbons.

*2,340.—A. M. Clark, London. (*C. J. P. Desnos, Paris*). Electrodes for Electric Light. 2d. June 12.

ELECTRODES.—These consist of an iron rod surrounded by a cylinder of carbon, which is coated with a metallic sulphide electrically deposited.

2,397. — J. Sheldon, Birmingham. Joining Wires, Strips of Metal, &c. 8d. (40 figs.) June 17.

CONDUCTORS.—A thick hollow disc of metal, perforated at its centre, has two holes of the size of the wires to be joined. The holes lie parallel and close to each other on opposite sides of a diameter of the disc. The ends of the wires are passed through the holes in opposite directions, and are separated and forced into the marginal parts of the hollow disc by a pointed conical tool passed through the central perforation, which is then closed up by a rivet. For joining strips a hollow tube is used, the openings being narrow slots.

2,402.—T. A. Edison, N.J., U.S.A. Electric Lights and Apparatus for Developing Electric Currents, &c. 10d. (18 figs.) June 17.

INCANDESCENCE LAMP.—A spiral of platinum wire enclosed in a glass bulb has its ends passed through and sealed in the glass. The bulb is exhausted of air by a Sprengel pump, and sufficient current is passed through the spiral to heat it to 150 deg. Fahr., at which temperature it is allowed to remain for about 15 minutes, during which time it yields any air or gas it may contain. The pump being kept constantly at work, the temperature of the filament is raised at intervals of 15 minutes, by increasing the current, until the spiral attains vivid incandescence, when the bulb is sealed. The "pyro insulated" burner described in Patent No. 5,306 of 1878 may be substituted for the spiral of platinum.

CURRENT REGULATOR.—A small horseshoe electro-magnet provided with a governor is employed to open the lamp circuit when the strength of the current becomes too great. This magnet is attached to a vertical shaft and revolves in a plane parallel to a similar electro-magnet fixed above it. The shaft is provided with a pair of centrifugal governors attached at their lower ends to a sliding collar on which one arm of a lever is pressed by a spring, the other arm being fitted with a block in

juxtaposition to springs by which the circuit is simultaneously opened in many places when the governors attain a certain velocity. A block carrying the contact springs may be drawn from the lever arm, thus requiring a greater speed of the governors to break the circuit. The commutator is fixed to the shaft, and consists of a ring of insulating material having its periphery faced with platinum, and cut in two parts, one part being connected to one pair of magnets, and the other part to the second pair. An automatic starter consists of a light lever and armature, fitted with a pawl, which engages with a ratchet on the shaft on the admission of the current.

CARBONS.—A press for manufacturing carbons consists of a block of iron with a central tapered hole in which a divided steel mould fits. After filling the mould with the pulverised material, a steel plunger section is placed in it and forced down; other sections are then added and forced down until the mass is of the required density.

DYNAMO-ELECTRIC GENERATOR.—This generator has upright field magnets yoked together at the top, and provided with massive pole-pieces at their lower ends. The armature is built up of two wrought-iron discs, with a wooden cylinder between them, which is wound with iron wire until flush with the edge of the iron discs; over this is then wound, in longitudinal sections, the insulated induction wire. Hard rubber discs of greater diameter than the iron discs are secured on the shaft, and keep the induction wires from contact with them. A polarised magnet, included in the circuit, opens it automatically should a reverse current be sent through the generator. The wire of the parallel induction helix is endless, and is wound to obtain a continuous current. The field magnets are excited by currents supplied from thermopiles, acted on by the heat given off from the steam engines.

CONDUCTORS.—These are carried in pipes lined with ebonite leading to cast-iron service boxes, from which branches are taken to the houses. The mains consist of a number of bare wires, and should have a resistance not exceeding one-tenth the total resistance of all the lamps. The pipes and boxes are air-tight, and a vacuum may be maintained in them or they may be charged with warm air.

2,481.—J. Hopkinson, London. Transmission of Power by Electricity. 6d. (2 figs.) June 21.

REVERSING GEAR FOR MOTORS.—Mounted on opposite sides of the axis of the armature are two insulated pins coupled together by a link to admit of their being turned simultaneously in the same direction. Each pin has two collecting brushes, so placed that when one of the brushes carried by

each pin is in contact with the commutator, the armature will rotate in one direction, and when the other brushes are in contact the armature will rotate in the reverse direction.

2,543.—F. J. de Hamel, London. Manufacture of Carbon Candles or Points for Electric Lighting, &c. 4d. June 25.

CARBONS.—These are made of a paste formed of pulverised carbon mixed with molasses or oil and forced through die-plates at the end of a cylinder by a propeller working in the cylinder. The carbon rods have metallic wire "wicks." Hollow carbons are made by moulding the carbons on a rod which is afterwards withdrawn. Wires may be embedded in the surface while the carbons are in a humid state, or plate carbons may have metallic wires or sheets laid between them, the whole being then rolled together under pressure.

2,629.—A. Blondot and J. Bourdin, Paris. Submarine Telegraph Cable. 4d. (2 figs.) June 30.

CONDUCTORS.—This invention consists of a single copper conductor having a first covering of gutta-percha, a second of caoutchouc, a third of gutta-percha, a fourth of caoutchouc, and a fifth of gutta-percha mineralised by a mercurial salt. Over this is a further covering of tarred hemp. In making joints the two ends of the conductor are bevelled and soldered with silver solder. A copper sleeve, internally silver soldered, covers the bevels. The sleeve is externally tapered to make the joint more flexible.

2,652.—C. W. Siemens, London. (*F. and H. von Alteneck, Berlin*). **Electric Light.** 6d. (10 figs.) July 1.

ARC LAMP.—This refers to improvements on patents Nos. 2,006 of 1873, and 4,949 of 1878, and consists in dividing the conductor at some part of its course into two branches, one leading through the coil of a solenoid or electro-magnet of high resistance directly back to the external circuit, and the other through a solenoid or electro-magnet of low resistance back to the external circuit through the arc. The armatures of the two electro-magnets are connected by a lever to one of the carbon-holders in such a manner that the attraction of the high resistance coil causes the carbons to approach, and that of the low resistance coil tends to separate them ; thus, when both are excited, the carbons take a position dependent on the resultant of the two forces acting on them. The feed arrangement is the same as that described in the latter specification above referred to.

2,706.—J. B. Spence, London. Combining Metallic Sulphides with Sulphur, &c. 4d. July 3.

CONDUCTORS.—These are insulated by drawing them slowly through a bath containing a mixture of the natural sulphides, such as of iron or copper, ground to an impalpable powder and combined with about 30 per cent. of sulphur while at its melting point.

2,744.—G. Whyte, Elgin, N.B. Electric Lamps. 6d. (6 figs.) July 5.

ARC LAMP.—This relates to improvements on Patent No. 5,152 of 1878. The feed mechanism is operated by a solenoid placed in the lamp circuit which replaces the small motor described in the previous specification.

*2,769.—H. E. Newton, London. (*H. W. Cook, Niederdorf, Tyrol.*) **Electric Lighting.** 2d. July 7.

DIVIDING CURRENTS.—To maintain several lights a commutator is employed to distribute the current to the various lamps in rapid succession.

INCANDESCENCE LAMPS.—These are arranged in districts preferably having a small area. A separate wire is taken from the main conductor to each lamp, the other wire from which returns to a pilot lamp at the central station. The radiated heat of the pilot light acting on a thermo-electrometer shows by its needle when the lamps are at their proper state. The electrometer also regulates the current, which may be sent intermittently to the various groups of lamps, by lifting or lowering an iron cone in a bath of mercury, thus interposing more or less resistance as may be required.

*2,772.—R. Allison and W. J. Hunter, Ipswich. Fishing Apparatus. 2d. July 7.

ELECTRIC LIGHT.—Geissler's tubes are suspended in the neighbourhood of fishing lines or nets, the currents being supplied from any suitable generators on board the vessels.

2,821. — G. Zanni, London. Automatically Opening and Closing Electrical Circuits, &c. 6d. (2 figs.) July 10.

COMMUTATOR.—This is designed to prevent the reversal of polarity of generators used in electroplating, and consists of an electro-magnet having its coils in the circuit of a second small generator geared to the first. The main or bath circuit passes through the frame and insulated armature of the electro-magnet, which is attracted in opposition to a spring as soon as current is developed. The spring repels the armature and breaks the bath circuit when the generator stops.

*2,823.—R. Bossomaier and F. Schwegler, London. Producing Wire Ropes, Cords, &c., with Coloured Surfaces. 2d. July 10.

CONDUCTORS.—Metallic wire cords are silvered by electro-plating and then passed through a heated tube, colouring matter being imparted to them by means of rotating brushes. They are then passed through a second heated tube to set the colour. "A new article of manufacture is thus produced."

2,962.—A. M. Clark, London. (*L. A. Hermann, Paris.*) Joints of Pipes for Containing Underground Telegraph Wires. 8d. (2 figs.) July 21.

CONDUCTORS.—The joints of pipes to contain underground conductors are made with spigot and socket. The socket is enlarged and formed with a circumferential groove to contain lead packing. In making a joint between the sleeve and the tubes which it connects, cast-iron rings each provided with two annular spaces of diameters corresponding to the sleeve and tube are used. The lead packing is placed in the annular spaces and set up to make a tight joint.

*3,001.—W. Moseley, Liverpool. Insulated Wires and Cables, &c. 2d. July 23.

CONDUCTORS.—The inventor applies the process described in Patent No. 2,926 of 1873, to "line wires."

3,023.—J. H. Johnson, London. (*E. Wheeler, Philadelphia, U.S.A.*) Telegraph Wires. 4d. (3 figs.) July 25.

CONDUCTORS.—For telegraphic purposes these are made of a core of steel enclosed in a covering of wrought iron.

3,085.—B. J. B. Mills, London. (*F. Million, Lyons*). Regulators for the Electric Light. 6d. (4 figs.) July 30.

ARC LAMP.—The carbons are carried by supports sliding vertically and geared by rack teeth to a pinion having a larger toothed wheel fixed upon its axis. This wheel is caused to rotate by a tooth fixed by a spring to a cross-piece projecting from a pendulum. Two electro-magnets placed either in the main or a shunt circuit, cause the pendulum to oscillate according to the intensity of the current. The current for the electro-magnets may be supplied by a battery and controlled by a central regulator, or each regulator may be operated by a secondary current derived from the main current.

*3,272.—W. R. Lake, London. (*L. Concornotti, Crémone, Italy*). Electric Lighting Apparatus. 2d. August 13.

ARC LAMP.—This consists of two parallel insulated metallic rods having forks at their upper ends in which metallic or carbon rollers are pivotted. A bar of carbon, preferably double concave in transverse section, rests on a portion of the peripheries of the two rollers, and descends as it is consumed. The carbon is held by a nipping holder, which slides in a guide held by a prismatic socket fixed to the frame of the lamp. The whole arrangement may be hermetically enclosed in a glass globe.

3,355.—W. Morgan-Brown, London. (*P. A. Portier, Paris*). Manufacture of Electrodes for the Electric Light. 6d. (3 figs.) August 20.

ELECTRODES. — These are formed of metallic tubes, preferably iron or steel, into which is forced a paste consisting of oxide of magnesia and a little water. Or lime, alumina, carbonates of lime, and magnesia, or baryta, and strontia, or mixtures of any of these may be used.

*3,368.—T. Young, Cessford, N.B. Joinings for Wire Fences, Telegraph Wires, &c. 2d. August 21.

CONDUCTORS.—Joints in conducting wires are made by inserting their ends into a coupling piece having a rectangular slot in the direction of its length for the wires to lie in. The ends are brought through holes at right angles to the slot, and then flattened to prevent their withdrawal.

3,425.—G. Pitt, Sutton. (*Messrs. Mangin and L. S. Lemonnier and Co., Paris*). Apparatus for Producing Parallel and Divergent Pencils of Light Suitable for Signalling, &c. 6d. (17 figs.) August 25.

HOLOPHOTES.—The dispersing apparatus is composed of two systems of lenses, the one plano-convex and the other plano-concave, so that when in contact they produce the same effect as a glass with parallel faces, and by separating them a variable divergence is obtained. One lens is fixed and the other is moved on guides by a screw, The reflector is of silvered glass, having spherical surfaces of different curvatures. The arc lamp may be fed automatically or by hand, and the axis of the carbons is placed obliquely to the axis of the light directing arrangement. The cast-iron cylindrical lantern is vertically and horizontally adjustable by screw gearing, and is mounted on a turning carriage fitted on top of a wrought-iron trelliswork standard, capable of being set at different heights. The lantern may be mounted on two axes, movement being imparted by a tangent screw and toothed segment. For use on a launch the axes of the projector are carried by a forked arm.

3,472.—A. Wilkinson, London. Compounds for Insulating Telegraph Wires. 4d. August 28.

CONDUCTORS.—These are first coated with a mixture consisting of 1 lb. of beeswax, 3 lb. of white lead, "a proportion" of Stockholm tar, and a small quantity of tallow. The wires are drawn through a bath of the heated liquid, and then through a die-plate to regulate the thickness, and before the first coating is set the operation is repeated. They are then braided with a fibrous material, and an outer coating, composed of pitch, ground cork, and bisulphide of carbon is applied. The protected wires may be coated with lead. Submarine cables may be made hollow, and gas may be conveyed through them to light buoys. A second insulating composition consists of 1 lb. each of wheat flour and linseed oil, 6 oz. of white lead, and 3 oz. of gutta-percha, sufficient gold size being added to render the mixture flexible.

3,509.—R. E. B. Crompton, London. Regulating Mechanism for Electric Lamps. 6d. (13 figs.) September 2.

ARC LAMP.—In this lamp the regulating mechanism is controlled by the action of magnets upon an armature on which is pivotted a stop piece or detent. The arc is formed by the magnets drawing down the lower carbon and simultaneously checking the descent of the upper carbon. In the figures the stroke of the lower carbon-holder B[1] is

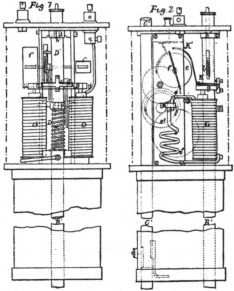

regulated by an adjustable collar b^1, and is maintained at its highest position by the spring D acting on the collar d. The rack rod C[1] carries the upper carbon and engages with the last of a series

of wheels e^1, e^2, e^3, e^4, the first, e^1, being the flywheel which carries two vanes f. The current passes through the controlling electro-magnets G G, which draw down the armature g, and with it the soft iron piece h turning on a horizontal pin h^1, to which is fixed the curved slip k k^1 so shaped as to rest on the edge of the wheel e^1 and prevent its rotation when the armature g is attracted. When current diminishes through the arc exceeding its normal length the spring l overcomes the attraction of the armature g upon the piece h and lifts the slip k off the wheel e^1, which permits the carbon to descend until the strength of the current is restored. A detent wheel may be used in place of, or in combination with, the flywheel.

*3,543.—R. H. Courteney, London. Electro-Magnetic Induction Apparatus. 2d. September 3.

INDUCTION APPARATUS.—Four soft iron bars arranged in the form of a square, with their ends separated, have wound on each of them two primary and two secondary coils. A contact breaker of any known form is used to make and break the primary circuit.

3,565.—W. Elmore, London. Dynamo-Electric Machines. 8d. (17 figs.) September 4.

DYNAMO-ELECTRIC GENERATOR.—This generator has an armature of disc form, the magnets of which are constructed of two soft iron sectors joined together by iron core-pieces of the same shape as, but smaller than, the sectors. One sector serves as a flange for bolting the magnets to the faces of the revolving disc; the other acts as the polar face of the magnet. Two rings of inducing magnets, similarly constructed and corresponding in number and relative position to those on the disc, are bolted to the inner faces of two circular frames fastened to the base-plate. The armature shaft runs in bearings carried by these frames. A bush forming the centre of the armature disc has holes lined with insulating material through which the ends of the coils are taken to a commutator of ordinary construction, the currents being collected by brushes. An alarum to indicate reversal of polarity, when the generator is used for electro-plating, consists of a polarised armature poised between two electro-magnets placed in the main circuit, which completes a bell circuit on the reversal of current. The amount of current passing is indicated by an electro-magnet acting on a pivotted armature, the longer end of which is graduated and fitted with a sliding weight.

3,587.—F. J. Cheesbrough, Liverpool. (*E. Sawyer, New York, U.S.A.*) Electric Lamps and Switches therefor. 6d. (5 figs.) September 6.

INCANDESCENCE LAMP.—This lamp is in general construction similar to that described in Patent No. 4,847 of 1878. The carbon rod is fed up between two carbon rollers by a wire passing over pulleys, and having one end fastened to a spindle actuated by a ratchet and spring pawl attached to an armature of an electro-magnet, which may act automatically; or a band switch may be used to admit the current when it is desired to feed the carbon up.

3,679.—G. F. Redfern, London. (*J. H. Bloomfield, Concordia, Argentine Republic*). Insulators. 6d. (5 figs.) September 13.

INSULATORS.—These are of the vertical type and are made of porcelain. The lower end is concave, and the upper convex end carries two downwardly projecting cup flanges covering the shank around which the wire is fastened. The cup flanges are arranged one within the other, the inner one being slightly the shorter. A securing screw, having a conical head with a square hole to receive a tightening key, passes through a longitudinal perforation in the insulator countersunk at its lower end. A rubber ring is jammed into the countersunk part by the conical head of the screw, and a similar ring is pressed into the upper end of the space between the screw and insulator and the crossbar, which is concave on its under side to receive the upper part of the insulator.

3,697.—A. M. Clark, London. (*J. M. A. Gérard-Lescuyer, Paris.*) Electric Lamps or Regulators. 1s. (9 figs.) September 15.

ARC LAMPS.—Two cylinders, filled with liquid, are connected at their lower parts and contain pistons, one of which feeds up the lower carbon, the other being weighted, by its descent forces the liquid into the adjoining cylinder. The lower carbon abuts against a disc of carbon, the arc being maintained by the repulsion due to the passage of the current. The disc is mounted on a lever free to oscillate on a quadrant, on which it is retained by four metal straps. The weight of the disc is sustained by an adjustable spiral spring. The lower carbon is guided at its top by a clip, one jaw of which is jointed and has a projecting arm engaging in a hook formed on the lever. The disc normally rests on the top of the carbon, releasing the jaw. The passage of the current repels the disc which by its lever then clips the lower carbon. The lower carbon may be fed up by a spiral spring, an electro-magnet in the lamp circuit serving to clamp it on the current passing. In a suspended lamp a carbon rod descends in an iron tube, and is controlled by an electro-magnet formed like a piston and sliding in the iron tube; the admission of current checks its descent, and the disc, mounted

on a thin leaf spring attached to, but insulated from, the tube, is repelled. The descent of the upper carbon rod may be controlled by an electro-magnet acting on the periphery of a drum on which is wound a cord suspending the weighted carbon-holder. The parts of the lamp may be duplicated where two upper carbons are required. The ascent of the lower carbon rod may be caused by a coiled spring within a drum on which is wound a cord passing over guide pulleys to the carbon-holder, an electro-magnet controls the rotation of the drum, and the upper carbon is lifted by a second electro-magnet acting on a lever armature, to one arm of which the carbon-holder is attached.

3,750.—H. J. Haddan, London. (*C. F. Brush, Cleveland, Ohio, U.S.A.*) Electric Lamps, &c. 8d. (10 figs.) September 18.

ARC LAMPS.—Two or more sets of carbons are used, the movable holders of which are operated by ring clamps, to which a tilting and lifting motion is imparted by a lifter actuated by electro-magnets, or by the expansive action of heat generated by the current upon any suitable apparatus connected with the lamp mechanism. The lifter is so formed and balanced as to lift one of the carbons slightly in advance of the other on the passage of the current. A tubular support surrounds one of the carbon-holders, and is of such length that when one set of carbons is sufficiently consumed a stop on the carbon-holder shall rest on the support, so that the weight of the carbon-holder and support shall always be borne by the lifter. The lamp is adjusted so that the magnets shall carry a definite load, and to insure a steady motion a dash-pot, having the cylinder movable and the plunger fixed, is employed. An automatic shunt consists of a fine and coarse wire helix, the fine wire being in the lamp circuit; on an excess of current passing, it attracts an armature which completes the circuit of the coarse wire, and at the same time interposes a suitable resistance.

3,771.—J. Brockie, London. Electric Lamps. 6d. (8 figs.) September 19.

ARC LAMPS.—These are regulated periodically by a magnet, the current of which is momentarily broken at pre-arranged intervals by a commutator. The upper carbon is carried by a rod having a cross-head which slides in guide bars. The feed of the rod is controlled by a gripping clutch, which is operated at intervals by the armature of a magnet placed in a branch circuit in connection with the commutator. The stroke of the armature is opposed by an adjustable spring. The regulation may be effected by magnets placed in the main circuit, a smaller magnet serving to periodically short-circuit them. When a number of lamps are placed

in circuit each is provided with a small wheel, which is partly rotated by the small magnet at each impulse, and permits the main magnets to be short-circuited only at a definite point in its revolution. By setting the wheels in the various lamps at different points the carbons are adjusted at different times. For alternating currents the carbons are placed at an angle to each other; both are free to move, and are adjusted by a magnet and clutches as before described. In a lamp having a defined feed as well as a periodical readjustment, the carbon-holder passes through a tight-fitting sleeve, and is fed forward by a pawl actuated by an electro-magnet, the sleeve being lifted by a clutch at determined intervals. The pawl is made to engage with the rack when the magnet is active by means of a fixed pin. The clutch has a weight at one end and an armature at the other, and a central hole just large enough to admit the sleeve. In a double carbon lamp a swinging stop prevents one pair meeting by locking its pawl; a pin on the carbon-holder comes in contact with the swinging stop, when its carbons are consumed, and liberates the second pair, at the same time locking the pawl of the first pair.

3,778.—W. R. Lake, London. (*P. Arbogast and T. J. McTighe, Pittsburg, Penn., U.S.A.*) Manufacture of Telegraphic Cables. 6d. (30 figs.) September 19.

CONDUCTORS.—The wires are covered by drawing them through a molten mass of glass or other vitreous material, or by laying them between two portions of the material and applying heat. A coupling box, each half of which is flanged, embraces the ends of the sections, which are joined together in the usual way and insulated.

*3,793.—J. Harrop, Sheffield. Apparatus and Application thereof for Producing and Subdividing the Electric Light. 2d. September 20.

ARC LAMP.—The inventor places "the carbon points or other substitutes therefor in metallic holders in juxtaposition on the same electrode." A drum, constructed partly of copper and partly of insulating material, and a copper comb bearing on it, form part of the circuit. The drum revolves, "and thereby partly or wholly subdues the aforesaid electric light without breaking the continuity of the aforesaid electrode or circuit."

*3,831.—P. Jensen, London. (*W. E. Prall and H. Obrick, New York, U.S.A.*) Protecting and Insulating Underground Telegraph Wires. 2d. September 23.

CONDUCTORS.—These are enclosed in a metallic protecting pipe fitted with an interior insulating tube of glass, asbestos, &c., made in sections. The metallic pipe is dipped in coal-tar and the tube in paraffine or wax. Air-tight test boxes having stuffing-boxes at each end to allow for expansion and contraction, are provided at suitable intervals, and outlets, with stuffing-boxes, are fitted on the test boxes.

3,843.—A. M. Clark, London. (*W. Hadden, New York, U.S.A.*) Electrical Signalling Apparatus. 6d. (3 figs.) September 24.

RHEOSTAT.—This is designed for use with telegraphic instruments, and consists of a glass tube filled with powdered carbon, having one end of the conductor fixed in one of its ends, and the other end sliding in or out according to the amount of resistance it is desired to interpose.

3,875.—C. W. Harrison, London. Electric Lighting. 6d. (14 figs.) September 25.

ARC LAMPS.—The feed of the carbons is regulated by a brake-wheel having on its spindle two drums of different diameters to which the two carbon-holders are attached by flexible cords. The upper holder descends by gravity. Each holder is guided by three pulleys, one, backed by a spring, serving to keep them in electrical contact with the other two by which the currents are conveyed. The carbons are separated and the arc formed by an electro-magnet drawing down an armature, to which is fixed a string passing over a pulley to one of the drums. The passage of the current deflects a magnetic needle which presses a block against the brake-wheel, and so maintains the arc. The drums may be made conical to compensate for the variation of driving weight caused by the combustion of the carbons. In a second lamp a series of flashes is made by a number of carbons arranged as a ring on a copper plate. Concentrically to the circle formed by the carbons is a conical mirror attached at its centre to a revolving spindle. The arms carrying the mirror at its centre and circumference are diametrically continued beyond its periphery, and form two contact pieces which make and break contact with the carbons as the mirror revolves. The arms have also attached at their outer extremities an annular mirror. A long feather on the spindle permits the bush carrying the rotating arms to be fed up by a spring as the carbons consume. In another lamp two discs of carbon are fixed at the required distance from the opposite edges of a cylindrical ring of carbon. The current passes by two brushes which bear opposite each other on the outer edges of the discs. The arcs are started by two carbon rods pressed forward by springs, but withdrawn by electro-magnets on the current passing. The discs and ring revolve. In a street lamp-post the light is situated in the base,

and is projected upwards on to an inverted conical reflector which diffuses it. In lighting a building a tube is used in which are placed reflectors to diffuse the light where required.

ELECTRODES.—These are formed of pure lamp-black and an equal part, by measure, of wood pulp freed from impurities and finely powdered. Naphthaline, pitch, or resin is dissolved in boiling tar and maintained at 250 deg. Fahr., until the more volatile hydro-carbons are expelled and the whole is plastic. To this is added, while still hot, a sufficient quantity of the lamp-black, &c., to form a dry plastic mass. The whole is then ground together and moulded under pressure. The negative electrodes for incandescence and contact lights are formed of lamp-black impregnated with iron by immersion in a neutral solution, and this is mixed with the plastic material above described.

RHEOSTAT.—A number of varying resistance pieces, insulated from each other, are arranged in a metal box radially round a vertical insulated metallic spindle on which is mounted a contact piece. One binding post is connected to the spindle and the other to the metal box.

3,885.—E. A. Bridges, Berlin. (*L. Heyer, Berlin*). **Treating Used or Waste Vulcanised or other Caoutchouc, &c.** 4d. September 26.

CONDUCTORS.—A boiler is provided having three compartments separated by wire sieves. In the bottom compartment is water, in the centre the waste rubber, and on the top the fire, which serves to melt the rubber and consume the volatile ingredients; the melted portions drop into the water, and are collected and boiled until sufficiently fluid to run through a sieve which separates any extraneous matter. The rubber is then mixed with "siccative" and oil of turpentine, and used for insulating purposes.

*4,087.—J. H. Johnson, London. (*H. M. A. Berthaud, Paris*). **Manufacture of Aluminium and Magnesium.** 2d. October 9.

GENERATOR.—Any dynamo-electric generator is employed for the electro-deposition of the above metals.

4,189.—C. E. Crighton, Newcastle-on-Tyne. **Electric Telegraph Terminal and Shackle Insulators.** 6d. (3 figs.) October 16.

INSULATORS.—A horizontal arm carried by the post has a hollow conical insulator attached to it by a metal encircling strap and bolts. A second insulating body or "shed" covers and is fixed to the conical one by cement. Through the centre of these passes a conducting bolt for connecting the subterraneous and aërial wires, and for carrying the strain of the latter. This bolt is screw-threaded at one end, and fitted with a conical plug adapted to the aperture in the central insulator, the extremity of which is closed by a screw cap. The conical plug is provided with india-rubber washers, which form a cushion under the strain of the conducting wire. The central insulator is undulated in form to prevent the accumulation of water.

*4,306.—T. Wilkins, London. **Automatic Contact Breakers for Electric Circuits.** 2d. October 22.

AUTOMATIC CONTACT BREAKER.—A vacuum bellows is provided in connection with the contact pieces, and is so arranged that the destruction of the vacuum shall separate the contacts. This is effected either by small pin holes, or by making the body of the bellows of a porous material, or the bellows may be separated by springs, the tension of which is adjustable. Clockwork may also be employed to separate the contact pieces. The contacts may be placed in a vessel which is filled with mercury and provided with an escape. An arrangement similar to the hour-glass may be used with the points of contact near the neck, the powder or liquid being a good conductor. The whole may be mounted on axes so as to turn automatically by the alteration of its centre of gravity.

4,354.—A. S. Hickley, Catford Bridge, Kent. **Electric Lamps.** 6d. (4 figs.) October 25.

ARC LAMPS.—A vertical plate has two horizontally projecting arms, the upper one of which carries a disc of carbon. The extremity of the lower arm carries a gripping carbon-holder, which is acted on by a "presser" terminating in an armature at one end and having a non-conducting roller press on one part of the carbon-holder through which a carbon rod is fed up by weights. The disc electrode is carried by an axle turning in centres at the extremity of a forked lever pivotted on the upper arm, and the inner end of the lever carries an armature, the stroke of which is controlled by a set screw. To the axle is fitted a ratchet-wheel gearing with a pawl pivotted to the fixed arm. On the passage of the current the electro-magnets supported on the lower fixed arm attract the armatures, thus gripping the lower carbon, and raising the disc by the forked lever, at the same time turning the ratchet-wheel one tooth. In a modification a spring arm attached to the fork lever grips the lower carbon.

4,400.—W. R. Lake, London. (*E. I. Houston and E. Thomson, Philadelphia, Penn., U.S.A.*) **Electric Apparatus.** 1s. (22 figs.) October 29.

DYNAMO-ELECTRIC GENERATORS.—An iron base carries vertical supports to which the cores of the field magnets are attached. These cores are built up of plates, rods, or rolled sheets of iron. The pole-pieces are serrated on their edges, and nearly enclose the armature, the gaps between them being in a plane inclined in the direction of rotation of the armature. The coils of the field magnets are wound conically, having the base of the core nearest the armature. Accessory coils serve to cover the gaps between the pole-pieces. The armature core consists of one or more pieces of iron of circular or polygonal shape, having projections extending radially outward on which coils are wound longitudinally, the remaining spaces or "flutings" being filled by winding coils over the ends of the cylinder. The coils in the flutings have the two ends respectively of each single coil connected to opposite strips of the commutator. It is proposed to connect the coils on the projections to two separate commutators, the two ends respectively of a single coil being joined each to a strip in one of the cylinders, and those of the next adjacent coil to the next adjacent pair of strips, and so on. The collecting brushes extend over and rest in contact with all the strips connected to coils generating currents in the same direction.

ARC LAMPS.—To a base is secured a solenoid surrounding a hollow iron core, in which loosely slides a metallic rod carrying the lower carbon-holder. The core is drawn down in opposition to a spring. Attached to the core by an extensible spring is a movable platform through which the rod passes. The descent of the rod is prevented by clamping stops attached to the top of the platform, which at the same time allow of its upward feed movement. The core is fitted with stops acting in the same manner. The upper electrode is carried by a rod sliding in the arm of a standard. A spring interposed between the rod and arm serves to make its support elastic. The current enters the solenoid and passes by a sliding contact to the rod carrying the lower carbon, hence by the upper carbon and arm to the terminal. The arc is established by the descent of the core, and with it the platform and carbon-holder. In a second lamp the lower carbon-holder is attached to an armature opposed by an adjustable spring actuated by magnets placed in the main circuit, and carried on a crossbar fixed to the lower ends of rods depending from the frame. The rod carrying the upper carbon-holder passes between friction rollers, one of which is attached to a spring. The descent of the rod imparts motion through one of the rollers to a second toothed wheel, the rotation of which is prevented by a projection from an armature held against the wheel by a spring and acted on by an electro-magnet placed in a shunt circuit of high resistance. On the current entering the lamp the lower carbon is drawn down by its electro-magnet, the upper carbon remaining locked until the resistance of the arc becomes too great, when the shunt magnet acts. The lamp may be arranged to keep focus by suitably gearing the carbon-holders together.

SECONDARY BATTERIES.—In a vessel of glass are placed two horizontal plates, preferably of copper, one near the top and the other near the bottom of the vessel, each being connected to an insulated binding post. The vessel is nearly filled with a saturated solution of zinc sulphate, and hermetically sealed. The space between the plates may be filled with sawdust saturated with the same solution. The passage of the current decomposes the solution, metallic zinc being deposited on one of the plates, and metallic copper being taken from the other plate, and converted into copper sulphate. A number of the plates may be combined to form a battery. An automatic switch cuts the battery out of circuit when the E.M.F. of the generator is insufficient to overcome that of the battery at the same time that it introduces an equivalent resistance. A sliding contact piece is used in connection with a number of studs to put into circuit a variable resistance. The batteries may be charged in separate sections by providing the conducting wires of each section with metallic terminals, with which the poles of the generator are brought successively into contact.

4,405.—A. V. Newton, London. (*F. Tommasi, Paris*). Electric Lamps. 6d. (6 figs.) October 29.

ARC LAMPS.—A number of carbon rods, circularly disposed, abut against a block of carbon carried by a central rod. The carbon rods float in insulated mercury tubes carried by brackets secured to the lamp pedestal. The rods are switched into the circuit in pairs by clockwork mechanism actuating a pair of insulated arms, which pass over contact studs connected to the pairs of carbons and make contact between them and two concentric rings connected with opposite poles of the generator. In a lamp for submarine use the whole is inserted in a glass cylinder closed at its ends by discs held in place by a rod terminating at its upper end in a ring to which is attached the lowering cord. Separate conductors lead to each carbon, and the commutator is carried in the boat. The tubes may be connected with reservoirs containing mercury to maintain a constant pressure.

4,428.—E. G. Brewer, London. (*E. A. Chambrier, Paris*). Electrical Apparatus. 6d. (14 figs.) October 30.

ELECTRO-MAGNETS.—Part of the invention con-

sists of "combined new arrangements of electro magnets whereby the apparatus can be regulated in such manner as to find the most favourable point for obtaining the maximum attractive power;" another part consists in "changing the form of the ordinary plane surfaces of the armature so as to increase the dimensions of the surface presented."

*4,456.—J. T. King, Liverpool. (*C. Linford, Pittsburg, Penn., U.S.A.*) **Apparatus for Containing and Protecting Telegraph Wires.** 2d. November 1.

CONDUCTORS.—A "container" is made of metal "rolled, cast, or otherwise formed," with grooves enamelled with porcelain, in which the insulated conductors are laid. The outer surfaces are tinned or japanned to prevent rusting. Covers fit over the containers and coupling pieces surround the joints. The joints are made with cement run into grooves formed in the covers and coupling pieces. Vitreous pipes are fitted where it is desired to lead a conductor to the surface.

4,534.—C. W. Siemens, London. **Electric Machines.** 6d. (4 figs.) November 6.

DYNAMO-ELECTRIC GENERATORS.—The coils of the electro-magnets are wound so as to have a resistance of not less than ten times that of the armature coils, which is done by greatly increasing the number of convolutions without reducing the sectional area of the wire. The coils are arranged so that the increased coils form portions of a branch circuit connected to the external circuit of the generator beyond the commutator. An increase in the external resistance causes a larger proportion of the current to pass through the electro-magnets, which are thereby more powerfully excited, and conversely when the external resistance is lessened. The same arrangement may be applied to electro-dynamic motors, the regulation producing more power when it is required.

4,549.—J. Bell, London. **Insulated Electric Conductors.** 6d. (8 figs.) November 8.

CONDUCTORS.—These are covered with asbestos, which is wrapped on spirally in the form of paper or braided on as a thread, or the conductor is drawn through pulped asbestos so as to leave a coating. In any case the conductors constructed as above are coated with silicate of soda to bind the whole together and to withstand moisture.

4,555.—J. Bell and G. Scarlett, Liverpool. **Electro-Magnets.** 2d. November 8.

ELECTRO-MAGNETS.—In place of winding the coils with a single continuous wire, two or more independent wires, insulated from each other, are

used. All the wires from one end of the coils are connected to one terminal and those from the other ends to a second terminal.

4,576.—T. A. Edison, New Jersey, U.S.A. **Electric Lamps.** 6d. (3 figs.) November 10.

INCANDESCENCE LAMP.—The high-resisting carbon spiral a is preferably made of a fibrous vegetable substance, coated with a plastic compound of lamp-black and tar, and is held by the platinum wires $d\,d^1$, cemented in the thickened ends $c\,c^1$, and fastened to the leading wires $x\,x$ by the clamps n;

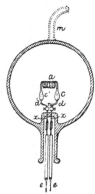

$e\,e$ are copper wires for connecting to the mains. The tube m, shown in dotted lines, leads to the vacuum pump, and is removed after sealing the globe.

4,589.—C. F. Heinrichs, London. **Apparatus for Generating Electric Currents, &c.** 2s. (62 figs.) November 11.

DYNAMO-ELECTRIC GENERATORS.—The soft iron core of the ring armature is ∪ shaped in cross section, and may be solid or be made of sheet iron, and may have iron wire wound round it as shown at r in the figure, the sheet iron being turned at s, to

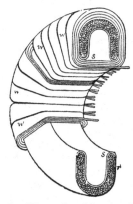

form a guard. The coils of insulated conducting wires are wound as shown at w. In a continuous

current generator the armature coils are connected to the commutator slips in the usual way. Upon a base-plate are mounted two standards fitted with bearings in which the shaft revolves. Two bosses fixed on the shaft have radial arms to which the ring armature is attached. The inducing electromagnets surround the entire outer surface of the armature in section, and about two-thirds in circumference. Two or more generators may be mounted on one shaft. In an alternating current generator the ring armature is fixed, and the inducing magnets rotate. These magnets are excited by a separate generator, constructed as previously described. The armature is fixed to a ring frame by twelve iron holders, the rotating inducing magnets being mounted on bosses keyed to the shaft. Each section of the armature is connected to one of the iron holders to which are fitted binding posts. The exciting current traverses the inducing magnets by means of two metal rings mounted on, but insulated from, the shaft on which bear suitable brushes. In a modification of this generator the ring armatures of both the exciter and multiple circuit generators are fixed to a suitable framing. The shaft carries an iron plate, on one side of which are fixed the inducing electro-magnets of the smaller ring armature, and on the other side those of the larger armature. Both armatures are fixed to their respective side frames and to their supports by iron holders, the side frames being fixed to a strong ring frame attached to the base. The sections of the smaller armature are connected to each other continuously through the plates of a fixed commutator. The brushes revolve with the shaft. The continuous current produced is employed to excite the magnets of both generators, each section of the larger armature being connected to an insulated terminal fixed to the frame. When used as a motor, double brushes are employed, one pair or the other being brought into action.

ARC LAMPS.—This relates to improvements on Patent No. 4,595 of 1878. The two semicircular carbons are carried in holders fitted in the outer ends of two radial arms attached to spindles having a common centre in brackets depending from the lamp frame. To the arms are fixed two pinions, which gear by racks with two other pinions, on the spindle of which is mounted a wormwheel gearing with a screw. On the spindle of the screw are two ratchet-wheels, with their teeth set in reverse directions. A second spindle carries a lever and a double cam-shaped armature actuated by an electromagnet. The lever has two pallets, one of which turns one of the ratchet-wheels, the other turning the second ratchet-wheel in the reverse direction. An adjustable guard surrounds a portion of the circumference of the ratchet-wheels, and prevents the pallets acting unnecessarily on the wheels. The guard is so constructed that if the lever be placed at its centre both pallets are lifted on it. A spring acts in opposition to the armature, and tends to close the carbons together. A small shaft pressed in from the outside lifts the pallets off the ratchet-wheels, and by a bevel-wheel, engaging with the screw spindle, the carbon holders can be separated by hand. Screws may be used in place of racks to connect the carbon-holders with the regulating mechanism. In another lamp the negative semi-circular carbon is about half the diameter of the positive one, the feed is effected by an electro-magnet acting as a brake upon the circumference of a segment-shaped armature, the separation of the carbons being effected as previously described. The carbon-holder arms may be balanced by the rack, the pivots on which they turn being placed at angles to each other. A thick rod of carbon may be placed above the arc to act as a reflector. In a further modification two pairs of semicircular carbons, crossing each other at right angles, are employed. Each arm is fixed at one end to a bevelled pinion, and each pair is geared to a second bevelled pinion, which maintains equality of feed. The upper or positive carbons are carried on pinions, mounted on a movable frame, which is lifted on the passage of current and forms the arc. In another lamp two semicircular carbons and one straight carbon rod are used. A curved metal electrode, through which water passes, may be used in place of the negative carbons.

CARBONS.—The discharge passage from the cylinder, out of which the carbon preparation is pressed by the piston, is made in a curved form. It is preferred to make the carbons oval in section, and to press them in dies to the required size and section.

ELECTRO-MAGNETS.—The fourth part of the invention relates to winding electro-magnets with insulated conducting tubes in place of solid wires, thus providing for the passage of air, water, &c., round the cores.

4,590.—G. P. Harding, Paris. Electric Lighting. 6d. (11 figs.) November 11.

ARC LAMPS.—The carbons are placed horizontally, preferably one above the other, in two tubes, and are held at one end in a cap which is fed forward by a weight. The descent of the weight is controlled by a cam actuated by an electro-magnet gripping the suspending cord. One of the tubes is pivotted and is momentarily attracted by an electro-magnet in a shunt circuit to establish the arc, and is then repelled by a spring. In a modification the controlling apparatus is an expansible spiral, placed near the arc, pressing a crank which releases the gripping cam. Fusible stops may be employed for the same purpose, a number

being mounted radially on a disc which turns a portion of a revolution as each one is consumed and thus frees the feed gear during the time of its movement.

4,653.—J. Hopkinson, London. Apparatus for the Transmission of Power by Electricity. 6d. (6 figs.) November 14.

Electro - Magnetic Motor. — This invention relates to methods of reversing the direction of rotation of the armature, and has reference to improvements on Patent No. 2,481 of 1879. The brushes are carried by a cross-piece, movable about the axis of the armature. The electrical connections are made by spring contact pieces arranged in relation to fixed metal pieces so as to be connected with either the positive or negative source of electricity only when the brushes are in their correct positions.

4,696.—A. M. Clark, London. (*J. S. Lamar, Augusta, Georgia, U.S.A.*) Electro - Magnetic Engines. 6d. (5 figs.) November 18.

Electro-Magnetic Motor.—A shaft mounted in bearings carries a wheel having a series of electro-magnets attached to its periphery with their axes parallel to the shaft and equidistant from each other. The cores are wound so as to be alternately N. and S., and are furnished at their ends with soft iron elliptical pole-pieces having their longer sides tangential to the circle in which they revolve. On a wooden frame surrounding the wheel are secured an equal number of permanent magnets arranged on both sides of the frame in alternate polarity. The commutator, preferably mounted on a separate shaft driven by gearing, consists of a number of interlocking tongues, insulated from the shaft, on which two springs connected with the poles of the generator bear. Two contact springs also bear on the commutator and on two insulated rings, mounted on the shaft, to one of which the initial and to the other the terminal ends of the helices are connected. Two or more wheels carrying electro-magnets may be mounted on the same shaft, the arrangement of the permanent magnets then being such that the magnets of the one series will face the space between two magnets of the next series. Each set of electro-magnets is provided with a separate commutator. A mechanical starter consists of a crown-wheel and lever so arranged that when its longer arm is moved its shorter arm will engage one of the teeth of the crown-wheel. A wheel reverser and brake consists of a wooden block fitted with metallic ends projecting below the wood and resting on two parallel electrical conductors connected with the commutator springs ; the block

turns on an insulated bolt, and the metallic ends kept in contact by a coiled spring, are constantly pressed by springs connected to the poles of the generator. The current is reversed by moving the ends from one of the underlying conductors to the other. A "substituter" is employed to switch in other batteries in place of those exhausted. A centrifugal governor opens the circuit when the speed becomes excessive.

*4,718.—W. B. Godfrey, London. Apparatus for Holding the Electrodes for Use in Electric Lighting. 2d. November 20.

Carbon-Holders.—These are arranged so that when one pair of carbons has burnt to a given point, another pair shall automatically succeed them. The action depends on a wire attached to the carbons at the desired point, which is arrested by a stop so long as the wire is intact, but which, when the wire is fused, puts the apparatus in motion.

4,796.—T. E. Gatehouse, London. Electric Lighting Apparatus. 8d. (12 figs.) November 25.

Arc Lamp.—A horizontal axle, mounted between two standards, carries a metal disc, on the periphery of which are fixed radially a number of carbon-holders, each oscillating on a pivot. Bolted to the metal disc is a disc of ebonite, to which are fixed a corresponding number of carbon-holders. The discs are rotated by a spring. To each of the fixed carbon-holders is attached a rod, from which a wire stop is carried to the carbon; these stops, catch against an adjustable insulated detent mounted on one of the standards. Attached to the detent is a contact spring which establishes connection between the respective separate carbon-holders and a terminal on the base of the lamp. The points of the carbons are held together by a fusible composition. Or a magnet in a shunt circuit tilts the movable carbon-holder by means of an armature and rod, and brings the points of the carbons into momentary contact. The carbons may be disposed in a straight row or in a circle. In an arrangement for carrying Jablochkoff candles a high resistance shunt magnet lifts a catch lever on the extinction of the arc, and permits the next candle to come to the points of contact. The contact piece may be rotated round a centre by a spring which is controlled by a fusible stop-piece fixed to the same centre as the contact piece. The carbon-holders are set each in succession nearer the centre, so that the second pair shall check the stop-piece after its end has been fused off by the first pair and so on. In a lamp for burning semicircular carbons an axle

carried in suitable bearings has its two holders insulated from each other, and carries at its centre a hollow boss of insulating material, into which are fixed at a little distance apart two arms fitted at their outer ends with carbon-holders. The carbons are kept equidistant by pieces of porcelain. Attached to the axle are discs of metal, each carrying radiating arms, through holes in the end of which are placed fusible stops. Two arms make contact with the carbons at any given point. The axle is rotated by a coiled spring, and is held in position by an arm catching against the fusible stops. After fusing the last stop the lamp short-circuits itself.

4,821.—W. Elmore, London. Production of Alloys for Shipbuilding, &c. 4d. November 25.

GENERATORS. — An alloy consisting of about 85 per cent. of nickel and 15 per cent. of magnesia, is added to the iron of which the electro-magnet cores of dynamo-electric generators are made; about 10 per cent. of the alloy is used.

*4,846.—E. Mourdat, Paris. Substitute for Gutta-Percha. 2d. November 27.

INSULATING MATERIAL.—This consists of the residue left after distilling the bark of the birch.

*4,916.—J. C. Mewburn, London. (*J. Heins, Philadelphia, U.S.A.*) Compositions for Insulating Telegraph Wires, Cables, &c. 2d. December 1.

CONDUCTORS. — Three insulating compositions are described. The first is composed of about 50 lb. of natural asphaltum, 30 lb. of resin, 10 lb. of beeswax, 10 lb. of caoutchouc, 10 gallons of linseed oil. For driers, a mixture is made of about 12 lb. of litharge, 15 lb. of oxide of manganese, 5 lb. of borax, and 5 lb. of lime. For protection from the influence of heat a mixture of 40 lb. of powdered asbestos and 18 lb. of alum are used; the wire is covered with a fibrous material and then boiled in a bath of the solution. The third composition is for filling in the pipes in which the conductors are carried, and consists of about 60 gallons of dead oil, 40 lb. of caoutchouc, 1,120 lb. of natural asphaltum, 320 lb. of powdered asbestos, and 150 lb. of powdered alum. The whole is combined by heat, and introduced into the pipe in a fluid state.

5,056.—M. Gray, London. Supplying Plastic Compounds of India-Rubber and Gutta-Percha to Moulding or Shaping Dies. 6d. (3 figs.) December 10.

CONDUCTORS.—The insulating compound is fed under pressure into a cylinder, fitted with a propelling screw, by steam-heated rollers. The compound leaves the rollers in the form of a sheet and is contracted in breadth by end guides which convert it into the form of a cord, in which shape it enters the cylinder. The object is to free the mixture from all air and moisture.

5,085.—W. Lloyd Wise, London. (*E. Bürgin. Basle, Switzerland.*) Dynamo-Electric Machines and Electromotive Engines. 8d. (12 figs.) December 11.

DYNAMO-ELECTRIC GENERATOR. — A cylindrical armature consisting of a number of polygonal-shaped rings on the sides of which the coils, each set slightly in advance of those on the ring next behind it, are wound, revolves in a tunnel bored in the central pole-pieces of the electro-magnets. The frame is formed of two horizontal electro-magnets each consisting of a single piece of cast-iron, attached firmly to each other at their outer ends. The coils are wound so as to produce N. and S. poles at the middle of the length of each magnet. The general construction of the generator is shown in Fig. 1,

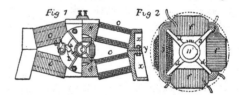

in which $1^* 1^*$ are the core-pieces, $c\,c$ the coils, Z one of the brushes, 2, 2 terminal enlargements of the core-pieces by means of which they are bolted together, N and S the cylindrically-bored pole-pieces in which the armature revolves. One of the armature rings is shown in detail at Fig. 2, and consists of a bush m, from which arms d of non-magnetic material radiate. Mounted on these arms is the polygonal iron ring j on the sides of which the coils of insulated copper wire $f\,f$ are wound; The whole is mounted on the shaft D^1. The armature consists of a number of these rings arranged as above described. The coils are wound in the same direction on each side of the polygonal core and are connected successively with each other, thus the end of the first coil is joined to the beginning of the second coil and so on, the end of the last being joined to the commencement of the first so as to form a closed circuit. Each of the joints is connected with a corresponding plate of the commutator, which is composed of the same number of plates as the cylinder has coils. Each plate is insulated from the others and corresponds with the coil lying in the same radius. The brushes bear on opposite plates of the commutator and are connected to the coils of the electro-magnets, or

the electro-magnets may be separately excited or be excited by part only of the current generated, or be replaced by permanent magnets. The wires of the armature coils may be replaced by insulated copper ribbons. In a generator for producing alternating currents the stationary frame is made of wrought or cast-iron and consists of two or more core-pieces. The coils are wound so as to form poles of like name on diametrically opposite sides of the shaft, on which are mounted a number of rings or "elements" constructed similarly to those previously described, but having the wire wound in different directions on alternate sides of the rings; thus each ring forms a closed circuit and the two ends of its wire are connected to the commutator rings. The polygonal core of each ring has its edges equal in number to the pole-pieces of the electro-magnets. The commutator G, as shown at Figs. 3 and 4, is constructed with a sleeve insulated from and secured on the shaft D^1 and is composed of a series of segments S^1, S^2, S^3, S^4, all

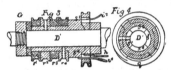

insulated from each other and corresponding in number to the "elements" of the armature. To these segments are connected the wires i leading from the elements, the wire of the first element being connected to the segments S^1 and so on. The commutator comprises also the rings r^1, r^2, r^3, r^4, insulated from each other and secured to the segments, with one of which each ring is connected. The ring q is insulated from all parts of the commutator, but is connected with each of the elements by the wires L L, &c. The currents are taken from brushes bearing on the rings r^1, r^2, and return by a brush bearing on the ring q. The ring q and its brush may be omitted by connecting the wires b with the shaft. The electro-magnets are separately excited by a continuous current. In a self-exciting alternating-current generator the armature is composed of four elements constructed as last described but all set in the same position and not spirally on the shaft, so that the points on which the two ends of the wires come are divided equally on the circumference of the cylinder. The terminal wires of the armature coils are connected to the commutator as before described, and the coils of the electro-magnets are connected to a brush bearing on the ring q and another bearing on the segments s.

5,127.—T. A. Edison, New Jersey, U.S.A. Electric Lamps, &c. 6d. (6 figs.) December 15.

INCANDESCENCE LAMPS.—This relates to an improved method of making the carbon filament and

attaching it to the conductors of the lamp described in Patent No. 4,576 of 1879. The filaments are made of paper, preferably "Bristol board." A number of suitably-shaped narrow strips with slightly enlarged ends are cut out and laid in an iron mould, a sheet of tissue paper being interposed between each strip, and a light weight made of a piece of gas retort carbon is then laid on the top. The mould is closed with a cover and the whole gradually raised to 600 deg. Fahr. and allowed to cool, after which it is raised to white heat. The clamps for connecting the filament with the conductors are made of a steel spring, tipped at its ends with platinum and bent into a bow, having its ends crossed and turned back towards each other, similar to the figure 8, the opening being at the upper end.

5,156.—S. Pitt, Sutton, Surrey. (*C. E. Scribner, Chicago, U.S.A.*) Electric Light Regulators. 8d. (3 figs.) December 16.

ARC LAMPS.—The upper carbon-holder is attached to a rod held by a friction clip carried upon the armature of an electro-magnet in the main circuit. This magnet separates the carbons on the passage of the current. The feed is controlled by a magnet, placed in a shunt circuit, the armature of which is suspended by springs. During its downward movement the armature carries with it the carbon-rod by means of a clutching device. The return motion is effected by breaking the shunt circuit. A small slide mounted in the frame is capable of moving in the same direction as the armature, and rests, when in its highest position, against a spring contact point which completes the shunt circuit. The slide carries two adjustable screws, between the ends of which a finger attached to the armature plays. In the downward movement of the armature the finger comes against the lower screw, and moves the slide from the spring-contact point, and so breaks the shunt circuit. The armature, is then lifted by its suspending springs and the circuit completed again. The clutching device consists of two springs of unequal strength mounted one at each end of the armature, through the centre of which the rod slides, which springs cause the armature to take a skew position when attracted, and grip the rod.

5,157.—H. F. Joel, London. Magneto-Electric Machines and Electric Light Apparatus. 10d. (31 figs.) December 16.

DYNAMO-ELECTRIC AND MAGNETO-ELECTRIC GENERATORS. — Two circular-shaped standards of dia-magnetic material mounted on a base-plate support in insulated bearings a compound axle made up of wood, and having two end prolonga-

tions of iron. On this axle are mounted the iron cores of the inducing electro-magnets of quadrant-shaped outline, the spokes and rims of which are wound with insulated wire or tape, and connected so that the centres of the rims shall form the pole terminals, adjoining poles being reversed. On the rims are fitted iron tooth-shaped projections, between which the wire is wound lengthwise. Supported in the annular rims of the standards are circular-shaped electro-magnets, which have toothed iron projections corresponding in number and position to those in the armature. The ends of the cores are prolonged and turned up, so as to bring only the centres of the coils under the influence of the inducing electro-magnets. The electro-magnets surround the armature, and are wound lengthwise with insulated wire laid between the iron projecting teeth in a direction opposite to the winding of the armature. The coils are connected across the end, the coils of one set to the first coils of the other set. A commutator is used to change the alternating currents of the outside coils into continuous currents. The long coils may be divided into separate smaller coils, all having the same polarity. In a modification an axle carrying radial permanent horseshoe magnets revolves within the circular electro-magnets. Attached circumferentially between the ends of the permanent magnets are electro-magnets provided with toothed shaped iron projections corresponding to those in the circular electro-magnets. In a continuous current generator the induced coils revolve between permanent magnets, fixed with their like poles all on one side of the ring. The cores of the induced coils are connected together by centre pieces, the wires on each being wound in opposite directions. The centre pieces extend beyond the coiling on the cores, and have each a separate coil of wire on them. In a modification the magnetic action is obtained by induction, the centre pieces not being in actual contact with the core-pieces. Either the inducing or induced systems may rotate. The commutator is movable on the shaft to which it is attached by a strong spiral spring, and has a rim of soft iron fastened to it; soft iron prolongations of the induced electro-magnets project down in front of the rim, and retard the movement of the commutator during the magnetisation of the induced electro-magnets, the spring assisting its forward movement during the period of rest or demagnetisation.

SEMI-INCANDESCENCE LAMPS.—Two contact clips of bell-crank form are pivotted to a nipple, and have their lower arms horizontal and the upper ones inclined so as to meet. Passing through the nipple is a tube having a pivotted flange at the top and a collar and rollers near the bottom. A weight suspended from the rollers by a cord presses, through the flanged tube, on the

lower arms of the clip, so closing together the upper ones. The weight also presses the lower electrode upwards against a small block of carbon. When the lower electrode is completely lowered, the enlarged end of the carbon-holder touches a contact and short-circuits the lamp.

ARC LAMPS.—The lower carbon slides through two contact jaws, which are supported on two links forming the armatures to two electro-magnets placed in the lamp circuit. On the passage of current the magnets attract the contact jaws, pressing out springs, and by the parallel movement of the links force the lower carbon down, and by cords and pulleys attached thereto simultaneously lift the upper carbon. Part of the electro-magnets may be coiled as a shunt. A pulley having its edges milled is free to turn on the end of a pivotted arm; an adjustable screw is placed above the pulley round which the cord is passed, and should this be jerked by the abrupt fall of the upper weighted carbon, the pulley is carried upward against the screw and checked. In a modification an electro-magnet in the lamp circuit actuates a pivotted armature, to which at the end of an extended piece is loosely jointed a link, to the lower end of which are pivotted two cross links, these again being jointed to two contact jaws, which are carried on pivots on a cross-piece. A second electro-magnet "of considerable resistance, connected to a distinct and auxiliary circuit wire leading from the machine" has its armature connected to the cross-piece by a lever and connecting link, and at intervals by means of a make-and-break contact causes the jaws to release the carbon. The lower carbon is fixed. Another arrangement of contact jaws consists in mounting on the ends of armatures rollers supported and kept in contact by springs, or the rollers may be mounted on one end of pivotted levers, the other ends of which are weighted. The lamps are constructed in such a way that the act of fixing them in their supports makes the necessary contacts. The movement of the two carbons together may be controlled by a system of parallel links, the centres of which slide in a slotted groove. An equivalent resistance and switch is arranged so that the switching out of the lamp inserts the resistance.

5,206.—G. G. Andre, Dorking, Surrey. Electric Lamps. 6d. (8 figs.) December 19.

ARC LAMP.—This relates to Patent No. 830 of 1879, and consists in surrounding the tube in which the upper carbon rod slides by an outer tube from which it is insulated. The carbon is fed down by a weight, and passing out through a steatite ferrule, rests on the negative electrode of copper carried by an adjustable bracket connected to the outer tube to which the base-plate, formed with a screwed

ring on its circumference, is attached. To the base-plate is fixed a flanged glass globe, which may be surrounded by an outer globe. The steatite ferrule is attached to a contact piece pivotted to the lower end of the inner tube, and by its own weight presses against the carbon a bent rod of copper of segmental form. The lower electrode may be formed with a sloping recess, the sliding of the carbon on the incline throwing it against the upper contact piece ; or the carbon may pass through an eye formed in the lower electrode and rest on an incline below the eye.

COMMUTATOR.—The second part of this invention relates to a commutator for reducing the resistance of an electro-magnet coil, by shunting a portion of the current after the armature is attracted. A frame is provided with two standards, to one of which an armature is pivotted. The outer end of the armature covers a pillar fitted with an insulating ring pressed upward by an adjustable helical spring. A resistance coil, proportioned to take the desired amount of current from the magnet, has one of its ends connected to the pillar and the other to the wire leading to the lamp, which is coiled round a core and passes to one of the standards to which the armature is pivotted. Round a second core is coiled a thinner wire in a direction contrary to the first, so as to neutralise its effects, and this has one end connected to the second standard, which is fitted with a projection making contact with the armature when it is released ; the other end is fitted to an insulated stop, with which the armature makes contact when attracted. One end of the main conductor is attached to the first-mentioned standard, and the other end to the second standard. The current entering the first divides itself between the thick or lamp coil and the armature which is in contact with the second standard. The thick wire core then draws down the armature, and forms contact with the pillar which shunts the current through the resistance coil ; simultaneously the armature makes contact with the insulated stop whereby the thin wire is thrown into circuit, thus partly neutralising the effect of the thicker coil. An increase of resistance in the lamp causes more current to pass through the thin coil, thereby neutralising the effect of the thick coil and releasing the armature.

1880.

18.—J. W. Swan, Newcastle-on-Tyne. Electric Lamps. 2d. January 2.

INCANDESCENCE LAMP.—To prevent the carbon gradually wasting, and the interior surface of the globe being coated with a carbonaceous deposit when the globe is nearly exhausted by a "Sprengel" pump, the carbon is rendered incandescent by the passage of an electric current, and the exhaustion of the globe is then completed.

33.—T. A. Edison, New Jersey, U.S.A. Developing Electric Currents for Electric Lights. 4d. January 3.

ELECTRIC LIGHT.—A high resistance dynamometer, "similar to that of Webber," is used to indicate the rise or fall of E.M.F. in the external circuit. Two instruments are used connected across the conductors in multiple arc. The movements of the dynamometer are transferred to large dials by gearing. Each engine and its generators are arranged so as to form a complete system. Two large copper rods, and a number of connecting levers, afford a means of coupling the generators in multiple arc. The field magnet of each generator is so wound, or the pulleys are so proportioned, that strengthening the magnet or increasing the speed shall adjust the E.M.F. of induction to the same amount in all the generators. Sets of subsidiary generators, driven by two or more of the engines, supply current to excite the field magnets of all the generators. A switch provides for changing a subsidiary generator from one set to the other ; these generators are also arranged in multiple arc. The field magnets of the main generators are connected in rows "of, say, ten" in series, and these rows are connected in multiple arc. Switches are provided for disconnecting the rows of field magnets when the armature has been disconnected from the mains. The field magnets of the subsidiary generators are separately excited, and in this exciting circuit is an adjustable resistance operated by a switch, by which means the output of current may be controlled. Each generator has in its armature circuit a short wire of bismuth, or other fusible metal, to prevent the admission of current from the mains should the generator from any cause be stopped while the others continue running. Each generator is fitted with fast-and-loose pulleys.

ELECTRO-MAGNETIC MOTORS.—These have the field magnets of higher resistance than the armature, and are connected to the mains in multiple arc. They are run at high speeds, using a small pulley on the motor. When used for pumping water, a float serves to make and break the circuit.

The speed may be regulated by an ordinary governor actuating a lever which breaks the circuit when the speed becomes excessive.

75.—A. M. Clark, London. (*L. Raynier, Paris.*) Dividing and Regulating the Electric Light. 8d. (4 figs.) January 7.

Arc Lamp.—The current is sent through two branches, in one of which is the lamp and in the other a resistance, preferably consisting of a carbon of small diameter ; the branches are re-united and pass on as a single wire to the next lamp. The carbons are fed together by right and left-handed screws, connected by an insulated coupling and actuated by spring barrels. A solenoid placed in the main circuit throws into gear either one or other of the springs, and so causes the carbons to recede from or approach each other. In a modification the solenoid is placed in a divided circuit, and has its current regulated automatically by a second solenoid, placed in the main circuit, having its core fitted with a metallic crosshead, to which are attached two rods of carbon terminating in pistons dipping in vessels containing mercury, which interpose more or less resistance as required. In another arrangement each lamp is fitted with an electro-magnet, placed in the direct circuit, the armature of which falls and closes a metallic circuit. Should any of the lamps become extinguished, the armature is raised and the current passes by a divided circuit, having a resistance equal to the lamp.

79.—R. Werdermann, London. Producing and Utilising Electric Currents, &c. 6d. (4 figs.) January 8.

Dynamo-Electric Generator.—The armature consists of a drum or disc of wood fixed to a shaft by bosses. On the periphery of this disc are mounted the coils, which are wound on an iron core, of ring shape, in a direction parallel to its axis. The two ends of each coil are attached to two insulated conductors fitted one on each side of the disc, so that there are as many insulated conductors on each side of the disc as there are coils. Insulated metal brushes fixed on the polar extensions of the field magnets collect the currents. The field magnets nearly surround the armature, and may be permanent or electro-magnets. The coils may be joined up in series or for quantity, and they may be wound all in one direction for continuous currents, or alternately in opposite directions for alternating currents.

Incandescence Lamp.—The glass containing globe is in one piece, with double walls and an intervening space. This inner space is filled with carbon or other conductor, or is plated with silver.

The outer globe is encircled externally with spiral conductors, and the space between the two globes is filled with a gas, preferably oxygen or siliciuretted hydrogen. The carbon of the inner and the spiral conductor of the outer globe are each connected to one of the terminals of an induction coil. The inner glass globe may be replaced by a globe of platinum. In a modification, a glass globe, charged with gas as before, has a cylindrical projection on which is mounted a suitably wound bobbin, lined internally with silk or other rubbing material. A bracket, carrying inducing magnets, is provided with a socket on which is mounted the bearing for a spindle attached to the bobbin. A current passing through the bobbin causes it to revolve, and the frictional electricity developed on the glass renders the enclosed gas luminous.

*144.—V. P. Lambert and L. E. Iverneau, Paris. Application of the Tides as Motive Power. 2d. January 13.

Electric Light.—A reservoir is so constructed that it will be filled at high tide. The accumulated water works a turbine which in turn drives a generator for producing the electric light.

203.—J. Clark, London. Developing Electric Light. 2d. January 16.

Incandescence Lamp.—This refers to improvements on Patent No. 3,991 of 1879. The filament consists of asbestos mixed with paper, the pulped compound being made into sheets from which the carbons are cut. These are " charred " in suitable cases. The filament is enclosed in a globe charged with carbonic acid gas or water.

*219.—J. L. Corbett and W. Lockhead, Glasgow. Propelling and Steering Steam Ships, &c. 4d. January 17.

Electric Light.—A peculiarly formed propeller may be used as a motor " for the driving of electro-magnetic machines."

*231.—F. W. Heinke and G. Long, London. Submarine Electric Lanterns. 2d. January 19.

Lanterns. — The bottom of a lamp for submarine use is made with a flanged plate, preferably of brass, to withstand the action of the water. To sink the lantern and steady it when standing, legs of lead are attached to it. The conductors are connected to unions secured to the plate, and lined internally for a certain length with vulcanite, sufficient space being left for the reception of copper discs, to which the wires leading to the two poles of the lamp are connected. On the upper surface of the plate is an india-rubber disc which is bolted

between it and a second plate; a glass cylinder rests on a second india-rubber disc. Similar plates and discs form the lantern top, and are secured to the upper end of the cylinder by tie-rods, passing outside the cylinder. A metal rod passes through a stuffing-box in the top of the lantern and extends down to the thumb-screw regulator of the lamp, which is preferably of the "Siemens" type.

250.—J. W. Swan, Newcastle-on-Tyne. Electric Lamps. 6d. (8 figs.) January 20.

INCANDESCENCE LAMPS. — To prevent cracking and leakage, two caps of platinum are attached to the glass globe, or to tubes proceeding from it, at the points where the conducting wires enter the lamp enclosure. The caps and wires are soldered together, or are in firm contact with each other. To prevent the rupture of the filament it is formed of cardboard or parchment paper, produced by immersing bibulous paper in dilute sulphuric acid, and washing with water till the whole of the acid is eliminated, after which it is dried and bent into the form of a loop. The loop is then enclosed in a vessel containing powdered charcoal, and the whole submitted to a white heat. The filaments are held in the globe by clips of platinum or platinum iridium. The third part of the invention relates to coating the wires with glass or enamel for preventing the evolution of gas occluded by the conducting wires.

315.—W. R. Lake, London. (*E. J. Houston and E. Thomson, Philadelphia, Penn., U.S.A.*) Apparatus for Generating, Controlling, and Utilising Electricity for Lighting, &c. 1s. (33 figs.) January 23.

DYNAMO-ELECTRIC GENERATORS.—The field magnets consist of a hollow rectangular iron frame, upon the two vertical sides of which coils are wound in such a manner as to cause the remaining sides of the frame to become of opposite polarity. The frame is made in symmetrical halves, bolted together. Pole-pieces, shaped to conform to the circular outline of the armature, overlap the junction of the two halves of the frame, and have accessory coils wound on the necks by which they project from the frames. These are wound in the same plane as the coils on the field magnets, and intensify the polarity of the pole-pieces. The sections of the frame may be slightly apart, or holes may be made to permit of an air circulation. The armature is mounted on a shaft running in bearings, supported by standards attached to prolongations of the frame, and consists of two discs of iron placed centrally on the shaft, between which, but insulated from them, are secured separate iron sections having a central rib extending from disc to disc. Projecting from the rib in a direction coinciding with the circum-

ference of the armature core, is a series of elongated teeth, fitting loosely into the spaces between similar teeth on the adjoining sections, which are sufficient in number to complete the outline of the armature. The coils pass diametrically across the ends of the armature core, and longitudinally in diametrically opposite spaces parallel to the axis. Bands of brass encircling the armature keep the coils in place. The armature may have three coils, each consisting of a single length of wire, in which case there will be six terminals, half connected to each other, or to a metallic ring fitted on the shaft, and the remainder to the segments of the commutator. Instead of connecting the terminals of the coils together, they may be connected to a second commutator, placed adjacent to the first, one end of one of the coils being connected to a segment of the first, while the other end is connected to a diametrically opposite segment of the second commutator. A wide pair of brushes, making contact with both commutators, is then used. When a greater number of parallel wires are used, a proportionate increase in the number of commutators is made, a pair of brushes being fitted to each commutator. In one arrangement of circuits two commutators and a ring are used. The two positive brushes of the commutators are connected to form one terminal of the external circuit, the two negative brushes forming the other terminal of the circuit in which the field magnets are placed. A common connection of the armature coils is made to a ring mounted on, but insulated from, the shaft, and this is joined to the middle of the field-magnet circuit, thus maintaining a connection of the equipotential points of the armature coils, the operative circuit, and the magnet coils; or each brush may form one terminal of a distinct circuit, the returns being brought to the ring. The form of the armature core sections may be modified to permit of their being wound separately and afterwards put together. When six coils are wound alternately the opposite coils have one pair of ends connected to each other; of the six remaining terminals the three that are alternate are coupled together, and the remaining three are carried to the commutator. When finer wire is used, one terminal of each of the opposite coils are joined together, so that the direction of winding is reversed in passing across the axis. The joints are connected by a wire leading to one segment of the armature; the remaining coils are joined in pairs respectively, and are connected to the two remaining commutator segments in a similar manner. The remaining six ends are connected in two sets of three, alternate ends being connected in a set. The commutator consists of three segments, the separating slots being at an oblique angle to the axis. Each segment has two metal rods passing through it parallel to the axis; the

ends of the rods fit in plates, from which they are insulated, carried on the shaft. Metal sleeves, placed on the rods, serve to keep the segments centrally between the plates, and have the armature coils connected to them. The internal diameter of the ring is greater than that of the shaft, thus providing for insulation and circulation of air. Two pairs of brushes are preferably used. To adjust the brushes automatically they are mounted on a support capable of rotation on its axis, and a flexible cord attached at one end to the support passes under a pulley running in the forked end of a lever forming the armature of an electro-magnet placed in a shunt circuit. The other end of the cord is attached to a lever, the position of which is regulated by centrifugal governors. A spring tends to move the brushes in a direction opposite to that of the revolution of the commutator. To dispense with the centrifugal regulation, and to impart an amount of motion to the brushes directly dependent on the amount of current received by the brushes, small accessory brushes are provided, which receive their current from the segments after they have passed out of contact with the main collecting brushes. The accessory current so received, causes an electro-magnet, placed in a shunt circuit, to throw in or out of action any suitable motor device attached to the collecting brushes, for their adjustment. When this arrangement is applied to a commutator of the "Pacinotti" form, it is preferred to polarise the armature, so that reversal of the current will repel the armature and so break the motor circuit. The polarised armature is also used with generators, having the coils of their armatures connected as a closed circuit.

Arc Lamp.—The intervals of passage of the main current through a motor controlling the feed is regulated by a contact lever operated by an electro-magnet placed in a shunt circuit. The lower carbon is fixed, and the upper carbon rod is held laterally between two grooved rollers, one of which is adjustable towards the rod. The spindle of the second roller has upon it a toothed wheel gearing with a pinion on the shaft of a "small electromotor of any suitable construction," "adapted to move in both directions in a magnetic field." The armature of an electro-magnet, placed in a shunt circuit, moves a lever provided with a contact piece which makes or breaks the motor circuit. To prevent too great a separation of the carbons a small platform surrounding the upper rod is provided with a catch, and permits of a limited downward motion of the rod only. In a modification the following arrangement is substituted for the motor. An electro-magnet, placed in the main circuit, attracts an armature attached to the end of a cord passing over a grooved pulley placed on the same axis as, and substituted for, the toothed wheel.

The other end of the cord is attached to an adjustable spring. The shunt magnet diverts the main current from the controlling electro-magnets when the arc becomes too long. The feed of the upper carbon may be insured by mounting one of the rollers on one arm of a pivotted lever, the other arm of which forms an armature attracted by the shunt magnet should the arc resistance become too great. The descending rod is preferably narrowed at its upper end, which permits of its falling through the rollers when the carbons are consumed, "and thereby gives notice that a renewal of carbons is needed."

347.—W. G. White, Paris, and J. N. Sears, London. Kerbways, 4d. (4 figs.) January 27.

Conductors. — These are carried in a metal trough forming the base of the kerb and also in circular apertures formed in a series of metal plates which, when placed together, form the kerb itself. The top plate has a ribbed or corrugated surface.

*350. — J. P. C. de Puydt and J. Cougnet, Brussels. Electric Lamps. 2d. January 27.

Arc Lamp.—"Whatever be the form of apparatus or arrangement used, the illuminating source is an electric arc of light produced in a vacuum or in a partial vacuum in which is a gas."

351. — J. Davis, Birmingham. (*E. Marx, F. Aklem and J. Kayser, New York, and A. G. Tisdel, Ilion, New York, U.S.A.*) Telephones, &c. 6d. (6 figs.) January 27.

Induction Coil.—Two layers of the primary wire are first wound on the core, the secondary and primary wire are then wound on together for two or more layers, and are then covered with paper. The operation is repeated until the coil is complete.

455.—T. H. Blamires, Huddersfield. Lamps for Electric Lighting. 6d. (7 figs.) February 2.

Semi-Incandescence Lamps. — The lamp consists of a globe terminating at its lower end in a tube, and having at its upper extremity an orifice ground flat. This orifice is closed by a metallic stopper held down by a capped washer and nut. The upper carbon is fixed to the stopper and the lower carbon is held by a glass support resting on a narrowed portion of the tube. The lamp is inverted and filled with mercury, the end of the tube is then placed in a vessel containing mercury and the lamp brought to a vertical position. The head of mercury serves to maintain a vacuum and keep the lower carbon in contact with the

upper one. The connections are made to the metallic stopper and to the lower vessel containing the mercury. A small reservoir of water serves to prevent the expansion of the globe. The vessel is closed to prevent any variation of atmospheric pressure altering the height of the column of mercury. The globe may be fitted with a soft metal pipe through which it is filled. The vacuum is produced as before and the tube is nipped together so as to perfectly close it. In a lamp for burning Jablochkoff candles the globe has a ground neck, into which a hollow stopper having two tubes fits. The vacuum is produced as before, a vessel containing mercury being provided for each tube. The connections are made through the vessels. Modifications of this lamp are described and illustrated.

*469.—F. J. Brougham, London. (*C. Sabatou, Paris.*) **Globes or Lanterns for Lamps, &c.** 2d. February 2.

LANTERNS.—Two hollow globes of suitable shape are arranged one within the other, the space between them being filled with water or other liquid, such as refined oil or glycerine. In the liquid is dissolved a silver, or other analogous salt. The solution may be tinted. When applied to incandescence lights the inner globe is exhausted of air.

*478. —T. Morgan, London. (*N. Glouchoff, Moscow*). **Electro-Magnetic and Dynamo-Electric Machines.** 4d. February 3.

MAGNETO-ELECTRIC GENERATOR.—This consists of a flat ring of iron wire, covered with a copper wire, wound in a spiral coil. The ring is maintained in a fixed position, and an axis placed within it carries one or more permanent horseshoe magnets arranged with their poles on each side of the ring, like poles being on the same side. In another arrangement the steel magnets have their poles of like denomination placed on opposite sides of the ring. The magnets may have segment-shaped pole-pieces. The iron ring may have teeth formed on each of its sides, the intervals between the teeth being filled with coils of insulated wire. Permanent or electro-magnets may be placed on each side of the ring, as previously described. In a modification an iron plate has an insulated iron wire coiled in a spiral in the form of a ring of the same dimensions as the iron ring, and in such a manner that the direction of the current is always perpendicular to the wire and to the direction of the iron ring which is in motion. Concentric grooves are formed in the plate, and these are filled with a continuous spiral coil of insulated wire. Two of these electro-magnets are employed with their coils turned towards the ring which revolves between them.

ELECTRO-MAGNETIC MOTOR.—By passing a current through the coils on the ring either of the above generators will become a motor. The commutators consist of contiguous metal discs placed in communication with the inducing and induced circuits in the case of dynamo-electric generators or motors, and with the induced circuits in the case of magneto-electric generators or motors.

GALVANOMETERS.—Either of the before described arrangements may be employed to denote the intensity or quantity of the current, which may be expressed in measures of weight corresponding to the work performed.

553.—G. W. Wigner, London. **Regulators or Lamps for the Electric Light.** 8d. (4 figs.) February 9.

ARC LAMP.—The regulator is actuated by the pressure of an intermittent supply of water from

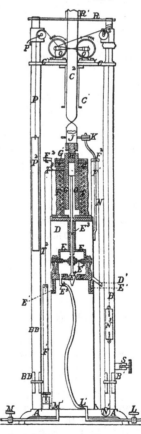

a higher external source, acting on a piston or bellows, and admitted through a valve controlled by an electro-magnet, or solenoid, placed in the main circuit. The lower carbon-holder is fixed to the core of the solenoid, which is carried on a hollow piston working in a cylinder. The water is

conducted by a flexible tube into the piston at full pressure, and by means of a diaphragm raises a conical valve communicating with the lower part of the cylinder. When the normal current is flowing the solenoid closes the valve against the pressure of the water, and the motion of the piston ceases until the current begins to fail, when the valve is again opened and the piston rises. The motion of the upper electrode is regulated by a cord attached to the piston. The drawing represents a lamp, having vertical carbons. The same arrangement is applicable to lamps with horizontal or oblique carbons. A is the base-plate, B columns which support the plate C, carrying gripping wheels for upper electrode. The columns are insulated at B^1, B, B^1; D is a metal cylinder with openings L^1 M^1 through bottom, E hollow piston, in which is a flexible diaphragm E^2, E^3 rod carrying valve E^4, F electro-magnet, F^1 trunk round magnet, G iron core loaded by collar and weights G^1, H holder into which the lower electrode J is fitted, and which passes through guide K, L tap for water supply communicating with pipe L^1 leading to the cylinder D and connected with the pipe E^5, M tap communicating by tube M^1 with interior of cylinder, N cord, O gripping wheels, P cord passing in the reverse direction to cord N carried over guiding pulley and attached to weight P_2, R cover to gripping wheels, S screw connected electrically with the carbon C_2, T screw insulated from the base of the lamp connected to the coil of the magnet F.

578.—T. A. Edison, New Jersey, U.S.A. **Electric Light.** 8d. (13 figs.) February 10.

INCANDESCENCE LAMP.—The conductors to which the filament is attached, are passed through the closed head of a tube, which is fitted into a neck formed as a prolongation of the lamp bulb. The part of the head where the platinum wires pass through may be of white enamelled glass. The filament is secured to the upper ends of the platinum wire by clamps, and the tube is passed into the neck of the lamp and fused therein. The lamp is exhausted through a small tube which is afterwards broken off. The supporting clamps and all the metallic parts within the vacuum are of aluminium or platinum, which is previously prepared and treated by heat in a vacuum. The inventor has "discovered that if the resistance of the light-giving body be increased, its radiating surface remaining the same, the same amount of light will be produced, but the conductors may be diminished in size proportionally to the increase of resistance," less current being required. To increase the lamp resistance, the filament is made

longer and doubled upon itself, or several filaments may be connected together in series. The filaments are preferably made of a single fibre of bleached Manilla hemp; the ends are enlarged by wrapping them with tissue paper previous to carbonising the fibre. The lamp socket as shown in Fig. 2, consists of a cylinder g of insulating material, provided on opposite sides with the metallic contact pieces 13, 14. The clamps h^1, h^1 connect the filament a to the conductors 1, 2, which lead to the

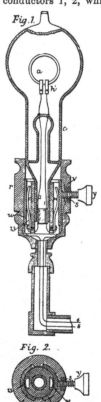

Fig. 1.

Fig. 2.

contact pieces 13, 14. The socket r of suitable insulating material, receives the neck of the lamp c, and has a screw-threaded aperture in its base, which permits of its being fastened to a bracket. On the interior of the socket are the plates u, v, insulated from each other; a metallic nut s is let into the exterior, and the metallic screw y passes through it. A conductor 5 leads to the plate u, and a conductor 6 to the nut s. Contact is made by turning the screw y inwards.

583.—C. H. Siemens, Berlin. (*E. W. Siemens, Berlin.*) **Electric Railways, &c.** 6d. (8 figs.) February 10.

DYNAMO-ELECTRIC MOTORS.—The currents generated by stationary dyanamo-electric generators are conveyed along conductors, which may consist

partly or wholly of the rails, to suitable motors. The continuous electrical connection is maintained wholly by the wheels, or partly by the wheels, and partly by rollers, springs, or brushes.

602.—T. A. Edison, New Jersey, U.S.A. Utilization of Electricity for Light, Heat, and Power. 8d. (4 figs.) February 11.

ELECTRIC LIGHT.—Each district has a central station provided with a prime motor, suitable generators, and means for determining the amount of current generated and supplied. The magnets of the generators supplying the current to the working circuit are separately excited by other generators having their field-magnet coils coupled in series, and excited either by a galvanic battery or by their own current; while the generating coils of the supply generators are coupled in multiple arc to the main conductors. All the translating devices are arranged on the multiple arc system, each device being in its own divided circuit. To indicate a variation of resistance in the outside circuit, galvanometers are placed across the main conductors at the central station, and these are frequently tested by standard cells; test lights are also placed in the central station. To maintain a constant electromotive force the battery feeding the field magnets of the exciting generator is arranged in connection with a series of resistances, so that the energy of the battery current may be varied, causing in turn a variation in current of the supply generators. It is preferred to use conductors placed in insulated pipes, made watertight and carried underground, provision being made for house connections. A pair of conductors are preferably laid along each side of the street. Each house is provided with a suitable meter to measure the amount of current passed. Where motors are used, each is fitted with a governor, which breaks the circuit on an excess of speed.

*624.—G. F. James, Salford. Manufacture of Engine Packing, Insulated Telegraph Wires, &c. 2d. February 12.

CONDUCTORS.—The funnel of the braiding machine has an outer casing, through which steam passes to keep the contents of the funnel in a heated state. A series of stirrers mounted on a toothed wheel serve to agitate the contents of the funnel. The finished cable is wound on to a reel or cylinder.

*630.—Hon. R. T. D. Brougham, London. Lighting by Electricity. 2d. February 13.

CARBONS.—To reduce the rate of consumption of the carbons, they are combined with a liquid carbonaceous substance either before or during use in the lamp.

636.—A. M. Clark, London. (*C. L. Pilleux, Paris.*) Producing the Electric Light. 6d. (6 figs.) February 13.

ARC LAMP.—This invention consists in interposing between two metallic conductors the extremity or extremities of one or more carbons which descend or are fed up as they are consumed. A lamp is described, consisting of two iron tubes, set at an angle and fitted with springs, which feed the carbons upwards. The conductors form helices round these tubes, and are wound in reverse directions to each other, thus causing the carbons to repel one another immediately the current passes. The conductors terminate in claw-shaped ends, which touch the cone formed by the combustion of the carbon, and act as stops.

695.—W. R. Lake, London. (*C. E. Chinnock, Brooklyn, and J. de Hart Harrison, Newark, U.S.A.*) Telegraph Lines. 6d. (7 figs.) February 17.

CONDUCTORS.—The object is to prevent currents on one line wire being conveyed to another, and to carry off all external currents caused by escape or induction. Around each line wire is wound an insulating covering. The external electric conductor made of lead, tin, or other foil is wound over the covering as shown, and is connected with a wire, and the outside currents are thence conveyed to the ground.

699.—W. R. Lake, London. (*C. E. Chinnock, New York, and J. de Hart Harrison, Newark, New Jersey, U.S.A.*) Apparatus for Preventing the Injurious Effect of Induced or Escaping Electrical Currents in Telephonic or Telegraphic Lines. 6d. (1 fig.) February 17.

CONDUCTORS.—To prevent induced or escaping currents from reaching line wires or conductors, a wire extends along the tops of the posts, supported by insulators, and connected with the ground at suitable distances by a ground wire. The line wires are carried by insulators, around the shanks of which metal foil is placed forming an electric conductor. Other wires connect the conductors with the top wire, and the escaping currents are conducted to the ground.

725.—J. Imray, London. (*La Société Generale d'Electricité — Procédés Jablochkoff — Paris.*) Distributing Currents for Electric Lamps or Candles. 6d. (11 figs.) February 18.

COMMUTATOR.—As will be seen from the illustration, the object is to cut one of a group of Jablochkoff candles out of the circuit when its carbons are nearly consumed; and to accomplish

this, a strip D, consisting of two metals having different degrees of expansion when heated, is bent to an inverted ∪ form, and secured by one of its limbs to one of the clamps which hold the candle. While the strip remains cold the current passes through the candle and the connections M, N; but when the arc approaches the bottom, its heat causes the strip to partially unbend and release a spring P, which makes contact with the screw *o*, establishing a circuit of low resistance from M to O, the latter being common to all the group. In place of the spring, the outer arm of the strip may be prolonged and come in contact with a fixed stop. Fig. 2 shows a commutating apparatus for bringing a number of lights successively into circuit. A

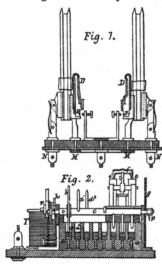

Fig. 1.

Fig. 2.

vessel *a*, of ebonite, is divided into a number of compartments, a^1, a^2, &c., there being two more compartments than there are lamps, one terminal of each lamp being connected to one of the compartments. Each compartment contains mercury, and in the first is partly immersed a metal disc *b*, fixed on the horizontal shaft *c*, on which are also fixed metal pins d^1, d^2, &c., one for each light, arranged helically round the shaft, so that as one pin is leaving the mercury in its compartment, the pin immediately following it, is entering the mercury in its compartment. A star wheel *i*, fixed at one end of the shaft, is worked by an anchor *h, h*, from the armature R of an electro-magnet T. A second electro-magnet, B, through the coils of which the current normally passes, has on its armature lever *l*, a fork *f*, the prongs of which are immediately over two cells *e, e¹*, containing mercury. When, by the extinction of one of the candles, the circuit is broken, the magnet B ceases to attract its armature, and the fork *f* dips into the mercury in *e, e¹*; the current then passes to the magnet T, which, attracting its armature,

moves the anchor *h*, and causes the shaft *c* to partly rotate the pin next in order, becomes immersed in the mercury, and the next candle is brought into circuit. An index *p*, fixed on the shaft, shows on a dial *q* which lamp is in circuit.

751.—F. J. Cheesbrough, Liverpool. (*A. K. Eaton, Brooklyn, New York, U.S.A.*) **Telegraph Wires.** 6d. (3 figs.) February 20.

CONDUCTORS.—The wire is covered with a fibrous or spongy material, and encased in a tube of ductile metal, either perforated or plain, such covering being put on in an apparatus provided with a nipple and die through which the wire passes, and a cylinder so arranged that an annular space is left, and by the action of a plunger hot lead or its equivalent is caused to issue in form of a tube through the die. The pores and interstices of the intervening covering are then filled with a liquid insulating material, either with or without the aid of a pump.

*756.—A. W. L. Reddie, London. (*H. Menier, Paris.*) **Electrical Conducting Wires.** 2d. February 20.

CONDUCTORS.—The inventor employs a metallic coating in contact with the surfaces of insulating coverings of electric wires to modify their electrical properties. The metal is used in very thin sheets, or is powdered and caused to adhere to the di-electric.

814.—J. H. Johnson, London. (*F. A. Gower, Paris.*) **Telephone Wires.** 6d. (4 figs.) February 24.

CONDUCTORS.—Wire is wound in helical coils upon the insulating material covering the conduct-

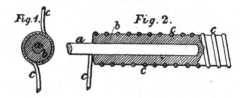

Fig. 1.   *Fig. 2.*

ing wire. In Figs. 1 and 2, *a* is the conducting wire, *b* insulating material, *c* wire coiled helically round the whole and communicating with the earth. By this invention the inductive action of adjacent wires is prevented.

832.—R. T. D. Brougham, Electric Lamp. 6d. (5 figs.) February 25.

LANTERN.—Air is excluded from the lamp by surrounding it wholly or partly by a liquid contained between two vessels or envelopes. The figure shows a vertical section through an André

incandescence lamp. The globe *h* is cemented to a

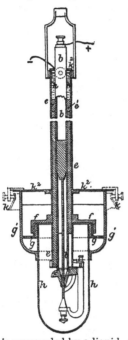

ring *g*, and is surrounded by a liquid, contained in the saucer *g*¹, forming a liquid joint.

842.—A. M. Clark. (*L. J. Bouteilloux and W. Laing, Paris.*) **Producing Electric Light.** 6d. (8 figs.) February 25.

ARC LAMP.—The electric arc is always maintained of the same length by allowing one of the carbons to be pressed by its own weight, or by special mechanism, against a non-conducting or an insulated core inserted in the centre of the other, and which is gradually dissipated during the con-

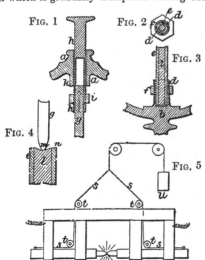

sumption of the carbons. The electrodes are held in split tubes, screwed on the outside and compressed by a nut. Figs. 1, 2, and 3 are sections of a carbon-holder, in which *g* is the carbon, *h* the split tube, and *l* the nut. Fig. 4 is a section of the carbons, in which *e* is a stick of carbon bored to receive the non-conducting core *l*, and *g* is the opposite carbon resting on it. Fig. 5 is an elevation of a lamp with horizontal carbons, arranged to maintain the luminous focus in one position. The carbons are fed together by the descent of the weight *u*.

849.—H. J. Haddan, London. (*C. F. Brush, Cleveland, Ohio, U.S.A.*) **Dynamo-Electric Machines.** 6d. (8 figs.) February 26.

DYNAMO-ELECTRIC GENERATORS. — This invention has for its object the adaptation of dynamo-electric generators to variable external conditions of resistance, without altering their speed, but by varying the intensity of the magnetic field according to the requirements of the working circuit, either by shunting from one or more of the inducing magnets a portion of the current, or by diverting it from a suitable number of their convolutions. Fig. 1 represents a portion of a dynamo-electric generator, provided with a resistance adjustable by hand round the field magnet circuit; Fig. 2 represents a modification of the same arranged to work automatically; Figs. 3, 4, and 5 show other similar modifications; Fig. 6 represents a field magnet coil adapted to have various portions of itself short-circuited; Fig. 7 a modification of the same; Fig. 8 a helix adapted to have various

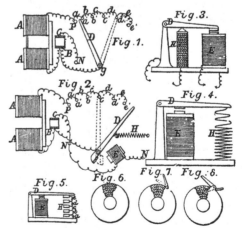

portions of itself cut out of circuit. In Fig. 1, A A are the magnets, B C the commutator brushes, P N the generator terminals; *a*, *b*, *c*, *d*, *e* resistance coils connected together in series, and to the metallic studs *a*¹, *b*¹, *c*¹, *d*¹, *e*¹; D is an arm electrically connected to the brush B, and capable of

being joined to any of the studs a^1, b^1, c^1, d^1, e^1, whenever it is found necessary to shunt a large or small quantity of the current from the magnets. In Fig. 2 the shunt arm D is operated automatically by the electro-magnet E, excited by the working current. Fig. 3 shows the resistance coils replaced by carbon discs, whose resistance varies under the varying pressure of a magnet armature, and Figs. 4 and 5 are modifications thereof. The field magnets are excited wholly by the main current, any of the above arrangements being used for varying the magnetising effect of the exciting current according to requirement, either by shunting or diverting the current from a portion of the "field of force" coils.

872.—D. G. Fitzgerald, London. Magneto-Electric and Dynamo-Electric Machines. 6d. (10 figs.) February 28.

GENERATORS.—This invention relates chiefly to that class of electric generators in which there is a ring or armature of soft iron carrying coils of wire, and aims at the substitution for the large field magnets, at present in use, of comparatively small electro or permanent magnets arranged in the plane of the ring, and nearly surrounding it and its coils both "longitudinally and transversely." Fig. 1 is a sectional elevation of an electric generator, in which A is the ring, carried by a disc on the shaft F. The coils of wire B have their extremities

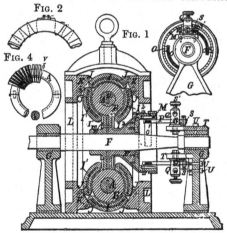

FIG. 2   FIG. 1   FIG. 4

brought out and connected to the commutator; I, I^1, I^2 are the "saddle-back" field magnets, and their essential characteristic is that they are always curved in the direction transverse to the axis of the magnet, whilst they are also sometimes curved in the direction of this axis, and, in the case of electro-magnets, are surrounded by coils of wire, every turn of which is partly convex and partly concave. The winding of the wire is reversed at two points diametrically opposite (see

Fig. 2), to produce opposing poles on each side of the machine. Owing to the form of the ring A, interstices are often left between the different coils at their outside edge, and these are filled up "with considerable advantage" by soft iron wedges or blocks V (Fig. 4) cast or slipped on to the ring. The ends of the wire of the coils are attached to plates M (Fig. 1), and the currents are collected by by springs o o^1, fixed to the bar P carried in the terminal Q. Rings and magnets of various sections, besides those illustrated herewith, are illustrated.

*885. — J. Graddon, London. Producing Motive Power and Light by Electricity. 2d. February 28.

MOTORS. — The object of this invention is to produce motive power from wheels wound with conductors, and surrounded with metallic cases having other conductors connected with an electric battery. To this arrangement may be added pumps to compress fluids to "assist in creating power."

*886.—C. W. Harrison. Apparatus for Obtaining Electricity. 2d. February 28.

GENERATOR.—A bar, or series of bars arranged as an endless chain, has a continuous motion within a series of helices of conducting material, thereby generating electric currents in them, which are collected in the usual manner. The motion of the bars may be vibratory or they may be stationary, and motion may be imparted to the helices.

*905.—J. C. L. Loeffler, London. Covering Wire with Insulating Material, &c. 2d. March 1.

CONDUCTORS.—The object of the invention is to apply several layers of insulating plastic material to electric conductors at one operation. The wire is drawn through a succession of dies of gradually increasing size, and receives a covering in each. The dies may be all in connexion with one forcing cylinder, or each may have a separate supply; the apparatus may be used to complete the coverings at one operation, or at several.

925.—J. H. Guest, Brooklyn, New York, U.S.A. Electric Lamps. 6d. (3 figs.) March 2.

INCANDESCENCE LAMP.—To maintain an air-tight joint around the conducting wires that enter incandescence vacuum lamps, enlarged conducting wires are twisted into a strand so as to render them elastic, and sealed after the glass has been melted into their interstices, by mercury contained in cups. In the illustration, a is the lamp globe, h the incandescent carbon hung in the loops f g of

the conducting wires, around the upper part of which are the cups *b c* containing mercury. The

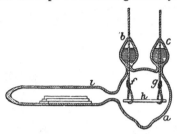

chamber *i* forms a magazine for spare carbons, one of which can be shaken into the loops *f g* when required; but such chamber may be dispensed with.

*1,046.—V. Poulet and Commelin, Hautmont, France. Compressing Air by Electricity for Obtaining Motive Power. 2d. March 10.

TRANSMISSION OF POWER.—Two "Grammes" are used, one as a generator, and the other as a motor, to drive air-compressing machinery.

*1,049.—C. Groombridge, London. Magneto-Electric Brakes, &c. March 10.

ELECTRO-MAGNETS.—A series of electro-magnets are employed as brakes for railway trains. The magnets "bite the rails" when "electric contact is established."

1,109.—H. W. Cook, Stondon Massey, Essex. Motive-Power Engines, Regulators, and Feed-Water Heaters. 6d. (15 figs.) March 15.

DYNAMO-ELECTRIC GENERATOR.—"A convenient form of dynamo-electric machine" is used in connection with an electro-magnetic arrangement to operate the throttle or expansion valve of the engine.

1,125.—J. C. Mewburn, London. (*T. A. Burtis Putnam, New York, U.S.A.*) Electric Railway Signals, &c. 8d. (19 figs.) March 16.

CONDUCTOR.—An electrical conductor is arranged in a cavity on the road bed, preferably between the rails, and adapted to be elevated or protruded by

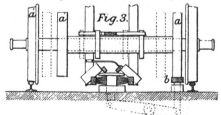

the pressure of an extra wheel *a* on the loco-

motive axle acting through levers. This conductor, when protruded, is swept by metallic brush conductors on the locomotive, by which means a circuit is closed. Wires lead from the protruding conductor to any point on the line, and are caused, through suitable mechanism, to make and break a circuit when required.

*1,136.—J. H. Johnson, London. (*A. de Méritens, Paris.*) Magneto-Electric and Dynamo-Electric Machines. 4d. March 16.

DYNAMO-ELECTRIC GENERATOR.—The object is to excite the field magnets of an alternate current machine without the aid of special auxiliary coils. Two additional commutators are employed, each of which has every alternate plate replaced by a non-conductor. The coil on each leg of a field magnet is connected to the brush of one commutator, and the coil on the other to the brush of the second commutator. In operation one coil only is excited at a time, the other one having its current cut off by the commutator brush resting upon an insulated plate, from which it follows that all the positive impulses traverse one coil, and all the negative impulses the other, and thus maintain the magnet in a uniform polar condition.

1,178.—J. Perry, London. Dynamo or Magneto-Electric Machine. 6d. (5 figs.) March 18.

DYNAMO-ELECTRIC GENERATOR.—This invention relates (1) to the use of coils on a movable armature or ring, the plane of each winding making an angle between 90 deg. and zero with the direction of motion; (2) to placing magnetic poles of opposite polarity so as to break joint on the opposite sides of a coiled armature or ring, whereby

FIG. 1.   FIG. 2.

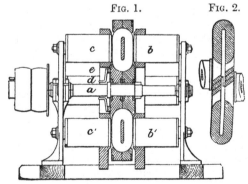

nearly all the lines of force make an angle of from 20 deg. to 70 deg. with the direction of motion; (3) to the use of an annular armature of coils wound at any angle to the direction of motion when soft iron is not used as a core; (4) to a particular form of armature, and to a new commutator; and (5) to the use of dynamo-electric generators

with a greater internal resistance than 20 ohms, for telegraph work. The illustration shows one of the many forms that generators constructed in accordance with this invention may assume. The ring is placed close to the field magnets, and does not require the aid of a soft iron core, as used in the Siemens and Gramme machines, and which at high speeds can no longer help in intensifying the magnetic field. The core is made of steel or phosphorbronze, and in some cases is dispensed with entirely. The rings are first wound on a peculiarly shaped bobbin, then soaked with a stiffening insulating material, and when dry are taken from the bobbin and slipped on the core "at a decided angle" to it. Referring to the illustration, the bobbins and core are carried on a wood centre fixed on the shaft a, and are further held by two wood clip rings screwed together on the outside. The field magnets $b\,b^1$, $c\,c^1$ are carried at their outer ends in cast-iron frames, and at their inner ends in wooden rings, shown in section. The opposite magnets are not upon the same centre line, one system being turned about ten degrees in advance of the other, so that a line joining the centres of their poles makes an angle of from 20 deg. to 70 deg. with the direction of motion of the armature. The commutator is formed of cranked or curved rods d attached to an insulating ring e, and separated by air spaces. "The brushes may be few or many, depending on the way in which the coils of the armature are connected with one another." A tubular form of electro-magnet is described, consisting of a hollow core, on which a sufficient quantity of insulated wire is wound, the whole being surrounded by an outer metallic tube. An armature revolves in the annular space between the ends of the tubes.

1,184.—J. T. King, Liverpool. (*C. Linfold, Pittsburgh, U.S.A.*) **Apparatus for Insulating, Containing, and Protecting Surface and Underground Telegraph Wires.** 6d. (8 figs.) March 19.

CONDUCTORS.—The object is to protect surface and underground conductors from growths of fungus, admission of moisture, and to render them accessible for repairs. Fig. 1 is a perspective view of a telegraph cable and container. The cable a consists of cotton-covered wires held together by a strip of cotton wrapped spirally round them; the interstices are filled with melted paraffine or tar. The cloth wrapper is coated with an elastic impervious material, and, if desired, another strip of cloth is wound round in an opposite direction, and also coated with material. Fig. 2 shows a section of the cable. Fig. 3 shows a container with an iron box or bed, protected from rust by tinning or other treatment. The interior is fluted at b to

form receptacles for the wires, and the whole interior surfaces of them are coated with porcelain enamel, as at c, to prevent the passage of moisture to the interior. A cover d fits within the flanges e, and rests upon the shoulders f. The sections of the container are united by a coupling box g, which

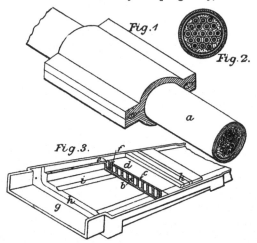

Fig. 1. Fig. 2. Fig. 3.

has a transverse groove h and longitudinal grooves i. The form of the coupling box is adapted to the shape of the bed, and is attached by screws. When the wires are placed in the grooves, the cover d is cemented down, the cement being fitted into the groove k and between flanges e. Cement is also poured into the coupling box, and fills up the grooves and recesses.

1,217.—W. H. Lake, London. (*E. J. Molera and J. C. Cebrian, San Francisco, U.S.A.*) **Apparatus for Examining Microscopic Photographs.** 6d. (8 figs.) March 20.

ELECTRIC LIGHT.—A small generator, driven by a treadle, is used in conjunction with "an electric light" to illuminate the object under examination.

*1,239.—E. Quin, Leyland, Lanc. **Composition as Substitute for India-Rubber.** 2d. March 23.

CONDUCTORS.—These may be coated with a compound consisting of chloride of sulphur mixed with bisulphide of carbon and naphtha, or other volatile neutral solvent. To these is added rape or other good bodied oil, and the whole allowed to stand until the volatile part is nearly evaporated. The result is a light-coloured substance suitable to take the place of india-rubber.

1,244.—J. H. Johnson, London. (*W. W. Griscom, Philadelphia, U.S.A.*) **Regulating Electro-Magnetic Motors, &c.** 6d. (2 figs.) March 23.

MAGNETO-ELECTRIC MOTOR.—A motor is driven by a battery whose plates are so arranged that, to regulate the power of the motor, they can be lifted to a greater or smaller extent out of the exciting liquid by mechanism attached to a treadle or lever.

1,259.—J. H. Johnson, London. (*W. W. Griscom, Philadelphia, U.S.A.*) **Electro-Magnetic Motors and Dynamo-Electric Machines.** 6d. (11 figs.) March 24.

MAGNETO-ELECTRIC MOTOR AND DYNAMO-ELECTRIC GENERATOR.—This relates to constructing electro-magnetic motors or dynamo-electric generators with the entire body of the field magnets enveloping a portion of the armature in the direction of its rotation so as to utilise the free magnetism distributed over the magnet as well as that concentrated at the poles. The figure illustrates the invention ; *a* is a rotating armature of any convenient construction, *b* is a bar of soft iron of rectangular or other section, bent into the form shown,

and wound with insulated wire to form an electro-magnet whose poles are at the opposite sides of the armature, and whose length surrounds it as far as possible. Two such magnets may be combined pole to pole in a manner similar to that adopted in the Siemens generator, or they may be provided with heavy polar faces, or may be replaced by permanent steel magnets of similar form. The core of the armatures may be of the ordinary Siemens double **T** section, or of a **Z** section, to provide a gradually increasing attraction on the part of the armature for the magnet. If, when used as a motor, the ends of the wires surrounding the field magnets are joined, powerful induced currents will pass through the coils as the armature revolves, and will "thereby increase the power of the motor." An ordinary two-part commutator is used, and in either case the armature or the field magnets may revolve.

1,295.—E. M. Allen, London. (*A. Knudson and F. L. Kane, Brooklyn, New York, U.S.A.*) **Improvements in Insulating Electrical Conductors.** 6d. (2 figs.) March 30.

CONDUCTORS.—Conductors, more especially such as are designed for telegraphic purposes, are coated with an insulating fibrous material covered or saturated, or both, with a waterproof flexible adhesive compound composed of Trinidad or other natural asphaltum, mixed with non-drying and softening material, such as paraffine oil, petroleum residuum, &c.

The proportion of native bitumen and non-drying material depends upon the nature and character of the bitumen and the place in which the conductor is to be used, but a suitable compound for ordinary purposes may be made of 65 parts of Trinidad mixed with 14 parts of petroleum residuum. Fig. 1 represents a conductor coated with a continuous

covering of fibrous material applied in any convenient manner. The wire is then saturated or coated with the compound by passing it through a quantity of the material kept by heat in a liquid condition. In covering a joint between two conductors, the ends are stripped and the joint is then wrapped with narrow strips of loosely woven cloth, or tape, previously saturated with the compound, as shown in Fig. 2.

1,333.—G. F. James, Salford. **Machinery for the Manufacture of Engine Packing, Insulating Telegraph Wires, &c.** 6d. (2 figs.) March 31.

CONDUCTORS.—The invention has for its object to combine the processes of lapping and braiding or plaiting any desired number of threads, strands, &c., round a central core in one machine ; the core, after it is lapped, being taken through one or more series of braiding spindles. The figure shows a front view of a portion of a combined lapping and braiding machine : *a* is the lapping spindle with bobbin *b* and flyer *c*, both revolving round a central

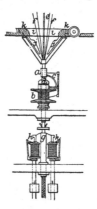

hollow spindle *d*. Through this hollow spindle passes the core *e* surrounded by a series of yarns *f*, which are lapped together by the material carried upon the bobbin, and drawn off by the flyer in the usual manner. The lapped material then passes through the hollow spindle *g*, and is braided with yarn, &c., from the braiding spindles *h*. The

specification relates (2) to an improved hopper *i* provided with an outer casing, through which steam, or other heating medium, may be circulated to keep its contents in a heated, and, if required, in a liquid state ; *j j* are knives or stirrers mounted upon a wormwheel *k*, to which motion is given by suitable apparatus ; (3) to an improved drawing-off arrangement not shown in the illustration.

*1,339.—J. C. Mewburn. (*W. de Fonvielle, Paris.*) **Rotary Induction, or Electric Motor.** 2d. April 1.

MAGNETO-ELECTRIC MOTOR. — This consists in employing the rotary action directly engendered by the passage of an interrupted current in the presence of a piece of magnetic metal capable of moving round an axis. This principle is applied either in the form of a motor or scientific instrument. In the latter case the motive part may be enclosed in a glass or other vessel full of air or liquid, or exhausted of air. The motor is composed (1) of a galvanometric frame on which are wound a number of turns of insulated electric wire, (2) of the motor proper "formed of a piece of magnetizable material, either solid or hollow," and mounted on a pivot in proximity to the said frame, (3) of magnets grouped in suitable position around the motor.

1,385.—T. A. Edison, Menlo Park, New Jersey, U.S.A. **Electric Machines and Motors.** 6d. (5 figs.) April 5.

GENERATORS AND MOTORS.—To obviate the induced current which circulates in the body of rotating armatures made of a solid mass of metal, or of large rings of metal, the armature is made of discs or rings $\frac{1}{32}$ in. to $\frac{1}{64}$ in. thick, secured together, and insulated from each other by sheets of tissue paper ; and, to avoid sparking at the commutator, the brushes are arranged so that they bear obliquely upon the face of the commutator, preferably at an

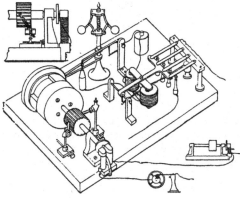

angle of about 30 deg. with the axis. To insure

even wear over the commutator, a disc forming the armature of an electro-magnet attached to the frame of the machine is fixed to the end of the shaft. When the magnet is charged, its armature together with the shaft is attracted, a spring forcing them back when the magnet is inactive, or an iron pulley fixed to the shaft in the vicinity of the field magnets may take the place of the spring, the attraction between it and the magnets serving for the recoil of the shaft. The magnet is alternately charged and discharged by means of a contact-breaker consisting of a spring bearing upon the periphery of a rotating disc, one part of the periphery being of insulating and the other part of conducting material. Fluctuations in speed consequent upon variations of load or work are corrected by a centrifugal governor driven by the machine shaft, the governor acting upon an adjustable circuit lever which acts by either making or breaking the circuit to the motor. To prevent too sudden fluctuation, the main shaft is provided with a heavy flywheel, and in order to get rid of the large spark occurring when the main circuit is broken in one place only, the circuit is broken at several places simultaneously by an arrangement of contact-breaking levers, operated by the governor either directly or by means of an auxiliary electro-magnet and armature which the governor connects with a local battery. The disposition of the various parts is clearly shown in the accompanying illustrations. In using electro-motors the speed of the actuated machine is controlled by the application of a brake to the main driven wheel, the brake being pivotted in such a manner as to be capable of bearing, by pressure from a treadle or by hand, upon the wheel, but is kept normally therefrom by a spring. In order that the actuated machine may be driven at a less or greater speed than the motor, as desired, a differential motion, consisting of friction gearing, or other well-known arrangement, is interposed between the two, and worked in combination with the above described brake.

1,392.—W. R. Lake, London. (*H. S. Maxim, Brooklyn, New York, U.S.A.*) **Dynamo-Electric Machines.** 8d. (10 figs.) April 5.

DYNAMO-ELECTRIC GENERATORS.—This invention has for its object to provide a regulator for a dynamo-electric generator, which shall automatically control the generation of the current, so as to make it equal at all times to the amount required for use without appreciable variation in the electromotive force. This is effected by using an auxiliary generator to excite the field magnets of the main generator, and by automatically altering the position of the brushes of the auxiliary generator with reference to the neutral points of the commutator, the position being controlled by a sensitive electro-

magnet in a shunt circuit. The specification relates also to an improved armature, in which both the iron core and the coils are so divided by air passages, without breaking up the continuity of the mass of iron or wire at any place in the armature, that a rapid circulation of air is maintained through the passages for cooling the machine; and further to a commutator arranged to prevent the sparking which occurs when the plates are made straight, and the shrinkage of the insulating material between the plates when they are made spiral in one direction. The commutator plates are so made that they run parallel to the axis of the generator on the inside, and form on the outside a double

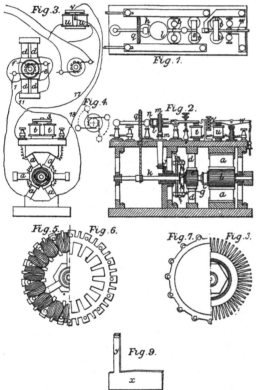

spiral. In Figs. 1, 2, and 3, *a a a* are the field magnets of the main generator, *b* is the armature, and *c* the commutator, *d d* are the field magnets of the auxiliary generator, *e* the armature, *f* the commutator, *g g* the brush holders, *h h* the brushes, *i i* are arms carrying the brush holders and affixed to a sleeve surrounding the shaft of the generator, *j* a toothed sector affixed to the same sleeve and engaging the bevel pinion *k*; *l l* is a double disc mounted on shaft *m*, which carries the wheel *k*. The double disc turns with the shaft, but is free to move vertically upon it for a short distance; *n* is a wheel mounted on shaft *q* and revolving between the edges of the disc *l l*, so as to engage with

either as required; *r* is a lever controlling the vertical position of the disc *l l*; *t* is a strong electro-magnet in the circuit of the auxiliary generator, and *s* is its armature; *u* is an electro-magnet of high resistance placed in a shunt of the main circuit, and *v* is its armature connected to the lever *w*. The current from the auxiliary generator passes through the main field magnets, through shunts 17 and 18, or the magnet *t*, according as the shunt is open or closed. When the shunt is open the path of the current is from 1 to 16, but when the shunt is closed the magnet *t* is cut out, and the current follows 17 and 18 instead of 5, 6, and 7. The magnet *u* opens and closes the shunt, the position of which obviously determines the magnetic condition of *t* and consequently the position of the armature *s*. By means of the lever *r*, one of the discs *l l*, according to the position of the armature *s*, is brought into contact with the revolving friction disc *n*, which through the medium of the intervening gear moves the brushes of the auxiliary generator round their commutator. The two extreme positions of the brushes are shown in Fig. 4. Figs. 5, 6, and 7 show the construction of the improved revolving armature, the body of which consists of thin plates of best annealed iron having "lodial" projections serving as polar extensions, and with other and longer projections for holding the wire of the coils in position. The plates are separated from one another by washers, and are fastened together by rods of dia-magnetic material passing through holes at the circumference. The armature is attached to the shaft by means of a hub with spokes, leaving the space between open for the ingress of air. Figs. 8 and 9 represent the commutator, Fig. 9 showing a separate plate, *x* being the commutator plate, and *y* a projection for connecting it to the armature coils.

1,397.—C. D. Abel, London. (*F. Krizik and L. Piette, both of Pilsen, Austria.*) **Electric Lamps.** 6d. (11 figs.) April 6.

ARC LAMPS.—The essential feature of novelty in this invention consists in the use of solenoid cores, having their metallic mass lessened towards each end; and combining with them two differential solenoid coils, with their axes in prolongation of each other, one being placed in the main circuit, and the other as a shunt across the arc for the regulation of the electric lamp. Figs. 1, 2, 3, and 4 show the principle on which the invention is founded. In a core of uniform section (Fig. 1) placed within solenoid coils S S¹ of equal magnetic force, the attraction on the core will depend on its position relatively to the coils, hence when in the position shown it will be more strongly attracted by S¹ than S. By making the core as shown in

Figs. 2, 3, 4, notwithstanding the different positions occupied by the core, it remains equally attracted by the two coils S S[1]. Fig. 5 shows the core F attached to the socket of the upper carbon, the whole being balanced by a weight attached to a cord passing over a pulley R. When the carbons are too near, the attractive power of S being increased, and that of S[1] diminished, the core F is attracted upwards. When the carbons are too far apart, S loses force, and S[1] gains force, so that

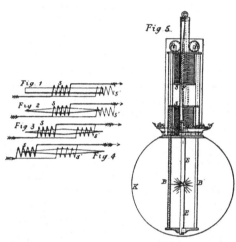

the core F is attracted downwards, and causes the carbons to approach. In a lamp to burn horizontal carbons the core is carried on rollers. The core may be reduced towards the ends either by tapering it from the middle, or by boring it from each end with a tapering pole, or by reducing it by steps from the centre towards the ends. In a focus-keeping lamp, each of the carbon-holders has attached to it a half-taper or step core, each half-core being free to move within one of the solenoids, the carbon-holders are then connected together in the usual way by gearing. "It is of advantage to combine with the pump an automatic shunt of any known construction."

1,407.—O. Heaviside, London. **Electrical Conductors, &c.** 6d. (8 figs.) April 6.

CONDUCTORS.—To render the circuit independent of inductive influences, two insulated conductors are used and placed one within the other. The main conducting wire is surrounded with insulating material, which is covered with the second conductor in the form of brass or other metallic tape or foil; this is then sheathed with an outer covering of insulating material. Should only a part of the line be exposed to induction, that part only need be protected by the metallic tape, the current returning the rest of the distance by earth.

1,475.—F. Wirth, Frankfort. (*M. M. Manly, R. P. Manly, and W. J. Philips, Philadelphia, U.S.A.*) **Telegraphic Cables or Conductors.** 8d. (15 figs.) April 10.

CONDUCTORS.—The tube and wire to be insulated are fixed at their lower ends and strained constantly in an upward direction, the molten insulating material being forced into the lower end. The wire and tube are kept under a yielding tensile strain while being cooled, and columns of the molten insulating material are applied at both ends of the tube, thus compensating for any contraction which may occur. When a number of wires are to be used they are introduced into the tube by threading the ends through perforations in two plugs, one of which passes through the tube and the other is fastened in the end. The ends of the wires are secured to the travelling plug, and are carried through the tube, the other plug being prevented from moving inwardly. The cable is twisted by being clamped to a rotating flywheel. The apparatus for filling the tube consists of a fixed lower head bored centrally and flared downwardly, and a movable upper head similarly bored, but flared in the opposite direction. Each head has a lateral passage connected to a supply pipe. The lead tube is passed through the two central orifices, and its ends are flared out to secure it in the heads. The top head is moved by a weighted cord passing over a pulley, the wire being held in tension by a similar device. A constant head of insulating material is maintained from a hopper, and the supply pipe leading to the lower head, together with the hopper, is steam-jacketted. A second tube of greater diameter serves as a water jacket to the inner lead tube, which is pierced to coincide with the passages in the upper and lower heads respectively, through which the insulating material is forced. In a modification the insulating material is admitted through the lower head at the same time that water is admitted to the water jacket, the insulating material being supplied in excess and allowed to flow out at the top. The joints are made by passing a close-fitting sleeve with flaring ends on to one of the tubes; the ends of the wires are then passed through hard rubber templates and soldered; the joint is then placed in a mould, and the cavity filled with insulating material, which, when it has set, is coated with a non-hardening insulator, such as balsam of fir; the sleeve is then drawn over the joint and the ends compressed, care being taken to preserve the flared ends, over and about which plumbers' wipe-joints are made.

1,507.—G. G. Andre, Dorking. **Electric Lamps.** 6d. (8 figs.) April 13.

LANTERNS.—This relates to an electric light lan-

tern by which a definite quantity of air is supplied to the incandescent carbons, while the lamp is burning, an automatic ventilation being produced when the current ceases. The illustration shows one of the inventor's lamps, to which the present invention has been applied ; a is the small hole for the ingress of the air, and it may, if desired, be made adjustable in size by any suitable means. The tube b for the carbon is provided at the top with a small aperture b. As long as the

current is passing, the core tube c is drawn down by the action of the solenoid d, and the leather washer c then covers the aperture b, preventing the escape of the air. When the lamp goes out, and the current ceases, the valve c rises, and allows the air to escape, thus producing automatic ventilation. Further drawings represent the adaptation of the invention to other lamps. In all cases a solenoid, or electro-magnet, is used to operate a valve.

*1,510.—G. Wells, London, and **A. Gilbert,** Worcester Park, Surrey. **Insulators.** 4d. (5 figs.) April 13.

INSULATORS.—The object is to secure telegraph

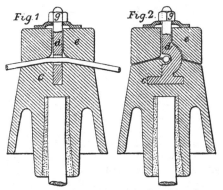

wire, so that in the event of its breaking between any two posts, the other parts of the wire on the other sides of the posts shall remain securely held. The lower part of an ordinary vertical insulator has imbedded in it a screw bolt d. The wire, after being strained, is laid in the groove made for its reception in the lower part c, and a cap e, having a corresponding groove, is placed thereon, and fastened down by a nut g, so that the wire is gripped between the upper and lower groove in the manner shown. The bolt d is bent to allow of the wire being readily replaced or removed.

1,552.—A. M. Clark, London. (*J. M. A. Gérard-Lescuyer, Paris.*) **Electric Lamps.** 6d. (2 figs.) April 15.

ARC LAMP.—The chief improvement consists in the arrangement of the electrodes, which are each formed of two carbons inclined towards each other in the form of a v, and so shaped at the points as to touch each other in a vertical plane. The lamp is composed of two pairs of tubular carbon-holders $a\ a^1$, through which the two sets of carbons $b\ b^1$ slide. The whole of the mechanism is mounted on a plate c, to which the holders a are permanently fixed, the other holders a^1 being hinged to it at d ;

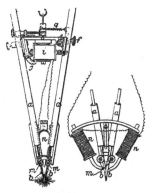

i is an electro-magnet placed in a shunt circuit, and serves to bring the carbons together to establish the arc, the spring g separating them to the extent allowed by the adjusting screw f ; $m\ m$ are gunmetal rollers, which are used both to guide the carbons and to make electric contact between the holders and the carbons near their points ; $n\ n$ is a fan-shaped electro-magnet, through which the main current passes, curved prolongations of the cores being provided, as shown, to localise the arc.

1,553.—C. D. Abel, London. (*Cie. Générale d' Eclairage Electrique, Paris.*) **Electric Lamps.** 6d. (6 figs.) April 16.

ARC LAMP.—This refers to Specification No. 863 of 1879, wherein the carbons were encircled in the direction of their length by an "electrical

loop," to localise the arc. The present lamp has the "electrical loop," composed of a number of turns of wire carried in an iron sheath, which becomes magnetised on the passage of the lamp current, and attracts a spring armature, a stud on which separates the carbons, and establishes the arc. For this purpose one of each pair of carbons is held in a socket pivotted to the frame, and so loaded that its point bears against the point of the other fixed carbon. Both carbons may be advanced by clockwork so as to keep the arc at one point. In a lamp for burning a number of pairs of carbons in succession, one carbon of each pair is pressed by a spring against a fusible stud near its root, the other carbon is jointed to a lever connected to the armature of an electro-magnet, which may be part of the sheath covering the "electrical loop," in the lamp circuit. When the circuit is completed, one or other of the pairs of carbons becomes kindled, and all the carbons are separated; when the arc reaches the fusible stud the circuit is broken, and the armature drops, bringing the points into contact again, and the operation is repeated. When several lamps are placed on the same circuit, an automatic cut-out and resistance is used. The main conductor divides into two branches at each lamp; one branch passes through the coil of an electro-magnet, the lamp, and to the main conductor beyond the magnet. The other branch passes by the armature of the electro-magnet, a contact, and a resistance equal to the lamp, back to the main conductor. Should the lamp be extinguished, the armature falls and makes contact, shunting the current through the resistance to the main conductor.

1,563.—J. H. Johnson, London. (*T. W. Stanford and S. Milligan, Melbourne, Australia.*) **Railway Brakes.** 6d. (2 figs.) April 16.

GENERATOR.—An "electrical machine" is used to supply current to the brakes, which are electro-magnets "shod with iron shoes made to fit the periphery of the wheel."

1,580.—E. P. Alexander, London. (*Karl Zipernowsky, Buda-Pesth.*) **Dynamo-Electric Machines.** 6d. (4 figs.) April 17.

DYNAMO-ELECTRIC GENERATOR.—In a dynamo-electric generator for producing currents of great tension a cylindrical armature, wound lengthwise with coils of insulated wire, is caused to revolve within a ring-shaped electro-magnet consisting of an iron ring wound over sections of its circumference with coils of insulated wire, so as to produce a number of successive polar fields of alternate polarity, and having inwardly projecting polar extensions. The coils of the armature

are wound in the form of rectangles having their long sides parallel to the axis, the width of each coil being such that its two sides are at the same

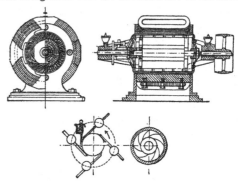

time approaching and leaving two of the polar fields. The core of the armature, for very strong currents, is made of wrought-iron plates whose continuity is broken; and for producing continuous uniform currents it is formed of spiral-shaped rings of insulated iron wire wound round a cylinder of non-magnetic material. The currents are collected by a cylindrical commutator, which, in order to diminish wear and tear of the plates, is lubricated with pure petroleum, drawn from a tank placed above the commutator, through a porous wick which extends from the tank in such a manner as to bear tangentially against the commutator. The armature is made hollow, and has openings in its ends and periphery, so that the air put in motion by internal oblique fans may keep the machine cool. The drawings without further reference will show the disposition of the various parts.

1,585.—G. Scarlett, Liverpool. Coils for Electro-Magnets, &c. 6d. (4 figs.) April 17.

ELECTRO-MAGNETS.—The coils are wound with two or more insulated wires in multiple arc, the core having as many divisions as there are wires, the wire in each division being "of the same resistance as the line in telegraphy, or of the pro-

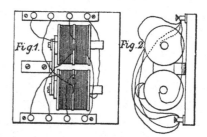

ducer of electricity in electro-magnets." If preferred, the insulated wires may be wound together upon the bobbin. The main point to be observed in these coils is that the depth of the coil should not be less

than the diameter of the core. Fig. 1 shows a horseshoe magnet wound with eight independent wires, and Fig. 2 a bobbin wound for telegraphic purposes with two such wires.

1,649.—W. R. Lake, London. (*H. S. Maxim, Brooklyn, New York, U.S.A.*) **Electric Lighting Apparatus, &c.** 8d. (12 figs.) April 21.

ARC LAMP.—This invention relates firstly to a sensitive electric regulator having the mechanism above the luminous focus, and in which "all friction is removed from the regulating device by suspending the working parts upon springs, which at the same time support the upper carbon and its feeding mechanism free from contact with all other parts of the lamp, and serve as frictionless conductors of the current to the carbon-holder." The upper carbon-holder is at the end of a vertical rack, guided in the movable core of a solenoid or axial magnet, and supported by the teeth of a pinion, which forms one of a train of wheels mounted upon a non-magnetic extension of the core of the solenoid, and consequently capable of a vertical oscillating motion by which the fingers of a star wheel fitted with vanes, in the train, may be brought into or out of contact with a fixed detent on the lamp case. The core, together with the train of wheels and the rack, is suspended by two fixed and one adjustable spring, and is raised and

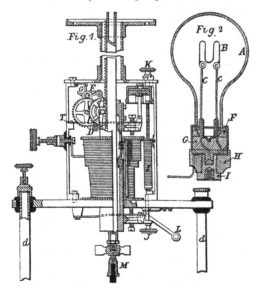

lowered within certain limits as the strength of the current circulating in the solenoid increases or diminishes. When the arc lengthens the core gradually sinks until the star wheel is clear of the detent, and then the wheels slowly revolve and allow the rack to run down, until the increased

current raises the core and stops the wheelwork. In the illustration M is the carbon-holder, with a ball-and-socket attachment to the rack rod. The double spring T, fixed at either end to the case, supports the core D, as does also the spiral I, which can be adjusted by the screws J and K. G is the fixed stop, and F the star wheel affixed to an arm carried on the solenoid core. L is an arm for locking the lamp. The globe for protecting the carbons is carried from the lower ends of the rods *d d*, which are telescopic to allow of it being easily lowered, and are locked in position by suitable catches.

INCANDESCENCE LAMP.—This invention relates secondly to an incandescence lamp, and refers (*a*) to the method of attaching the carbons, (*b*) to the formation of the vacuum, (*c*) to the sealing of the globe. The carbons and the platinum supports have enlarged flattened ends, and are fixed together by platinum screws and soft carbon washers, placed at each side of the hard carbon filament. In forming the vacuum, the air is first exhausted from the globe by a pump, and an atmosphere of hydrocarbon admitted to replace it. This vapour is then exhausted, and another supply admitted until nearly every trace of oxygen is removed, and nothing left but a slight residuum of hydro-carbon vapour which will be precipitated on the hottest part of the filament, when the latter is in operation. The method of sealing is illustrated in Fig. 2, where A is the globe, B the carbon filament, C C platinum supports, F plaster-of-paris, G shellac or copal, H vulcanite base, I a metallic core. The conductor S^2 is connected to C^2 by a metallic ring R^2, the conductor C is in connection with the metallic core I. This invention further relates to an apparatus and a process for carbonising substances for making the filaments of incandescence lamps, by exposing wood, the inner bark of trees, cardboard, &c., bent to the required shape, to a high temperature in the presence of hydro-carbon gas or vapour.

1,704.—H. J. Haddan, London. (*A. Bureau, Gand.*) **Electric Lamps.** 6d. (3 figs.) April 26.

ARC LAMP.—The extremities of the two carbons C C are embedded in a block M of refractory material, such as marble, which is cut out so as to form a vault A directed towards the side which is to be illuminated, there being at the top of the vault two orifices which lead to the points of the carbon, thereby forcing the arc to take a prescribed path. It is stated that by this means the consumption of the carbons is lessened by the protection from the air which the marble block affords them, and that by the high temperature of the refractory material a disaggregation of the carbons,

favourable to the formation of the arc, is secured, thus permitting the use of currents, either alternating or continuous, of small quantity. The lighting of the lamp is effected by joining the poles with a small strip of carbon or other conductor. The

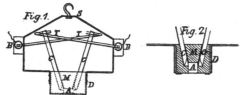

carbons may be placed at any desired angle or in different planes, or one of the carbons may be replaced by a metallic electrode; the current is conducted to them by connections T turning on centres B, and resting in grooves at the top ends of the carbons. The lamp is enclosed in a frame D provided with a means of suspension S.

*1,720.—J. R. Edwards, Liverpool. Attachment of Telegraph Wires to Insulators. 2d. April 27.

INSULATORS.—Instead of attaching the line wire to insulators by binding wire, a steel wire in the shape of a ring welded at the joint is used so as to tightly hold the line wire slightly bent round the insulator. In applying the clip or ring it is doubled until the two halves nearly touch, and the line wire is placed between the two loops formed by the double ring, which is then placed over the top of the insulator and forced into the groove.

*1,824.—R. C. Anderson, Wood Green, Middlesex. Electric Telegraphs. 2d. May 4.

CONDUCTORS.—In lieu of the ordinary metallic conductor, a fluid conductor such as sea-water is used, contained in a tube of suitable material. The resistance to the passage of the current being much greater than obtains with the ordinary wire, currents of high tension such as induced currents are necessary.

1,826.—J. E. H. Gordon, Dorking, Surrey. Apparatus for Producing Electric Light. 6d. (1 fig.) May 4.

ARC LAMP.—The light is produced by rapidly alternating currents of high tension discharged between knobs of refractory metal which are thereby caused to become heated and to emit light. The source of electricity may be either an induction coil or a magneto or dynamo-electric generator, provided that the currents are rapidly alternating, and are of high tension. In the illustration G G are wires of platinum or iridium ending in balls of the same metal, the outer ones being connected to the source of the electricity which

springs from one to the other. A is a glass globe and K an insulating plate on which the terminals

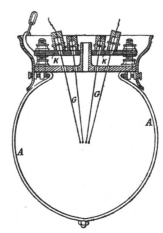

are secured. The central aperture serves as a chimney. To start the lamp the knobs are brought together either " by hand, by the use of suitable levers, or automatically."

1,840.—W. R. Lake, London. (*T. L. Clingman, North Carolina, U.S.A.*) Production of Electric Light. 6d. (4 figs.) May 5.

INCANDESCENCE LAMP.—The incandescing portion consists of a mass (in the form of a sphere or spheroid or other shape) composed of the oxide of zirconia, or a mixture of zirconia and plumbago or other carbon, or of zirconia and alumina, or magnesia, lime, silica, or mixtures with one or more of these substances. The mass is coated with a glazing of zirconia or one of the other substances above named. This mass is supported on each side by wires or pins of platinum, brass, iron, or other metal connected with copper electrodes.

The pins may be connected at several points with the mass or only at one point, or may be expanded at the point of contact. The outer coating of zirconia assists in completely excluding the air from the carbon, which is therefore protected from combustion. The incandescence is produced by the passage of the current through the mass, and the carbon being protected from combustion no feed mechanism is necessary. In the illustration A is the mass, B the platinum wires or pins, and C the electrodes. In some cases openings are made through the mass in different directions to admit air to affect the amount of heat at different points.

1,958.—J. H. Johnson, London. (*G. L. Anders and T. A. Watson, Everett, Mass., U.S.A.*) **Telephonic Exchange Systems.** 1s. 8d. (38 figs.) May 12.

MAGNETO-ELECTRIC GENERATOR.—A magneto-electric generator is used to transmit an induced current for signalling purposes, and consists of a permanent horseshoe magnet having a pair of coils fixed on its polar extremities with their axes at right angles to the plane of the magnet. The armature of the coils is fixed to a lever pivotted on a suitable support, and is removed from the poles by depressing one end of the lever with a sliding push bar. The magnetic make and break induces the current. In another generator two rotating armatures are used, and are so arranged that the currents produced by one is of minimum while that of the other is of maximum intensity. A permanent magnet is provided with pole-pieces between which revolve two armatures of the Siemens type, carried in suitable bearings and rotated by a gear wheel, engaging pinions on the armature shafts. The currents are collected by springs bearing on two part commutators, and the connections are made to enable either high or low tension, or continuous or alternating currents to be sent through the circuit.

*1,960.—G. Wells, London, and **A. Gilbert**, Worcester Park, Surrey. **Insulators for Telegraph Wires, &c.** 2d. May 13.

INSULATORS.—The insulator is surmounted by a head, shaped to receive one or more bolts, which have a small groove on the under side of their heads to grip the wire against corresponding grooves in the insulators. In a second arrangement the head of the insulator is formed with one flat side, over which a band of iron passes and grips the wire between a groove in its inner side and the insulator, and is tightened by a screw passing through its opposite side, and abutting against the insulator.

1,998.—W. R. Lake, London. (*C. A. Seeley, New York, U.S.A.*) **Magneto-Electric Machines.** 6d. (6 figs.) May 14.

DYNAMO-ELECTRIC GENERATOR.—The armature consists of a flat disc made up of a number of sectors of insulated wire, wound so as to be in the plane of revolution, and secured at their centre and periphery by iron plates, by which they are attached to the shaft. The faces of the field-magnet poles are set radially to conform to the armature, and those on either side of the disc are set in the same plane. The sectors may be coupled in series or in multiple arc; in either case commutators of the ordinary construction are used. The radial wire only is comprised within the magnetic field, the connecting wires being entirely beyond it, and serving as a support for the armature structure.

2,037.—W. Clark, London. (*P. Drew, Brooklyn, New York, U.S.A.*) **Electric Lamps.** 6d. (5 figs.) May 19.

INCANDESCENCE LAMPS.—The glass vacuum chamber or globe A is formed with a tubular extension a, and inner concentric cylinder a^1, closed at the top, and having the sealing tubes $b\ b$, formed with bulbs o, for containing mercury to seal the conducting wires c, which pass freely through the tubes. The tubes may be closed at their lower ends around the wires, in which case the bulbs o are dispensed with, or they may be extended upward from the closed lower end of the extension a,

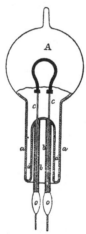

and closed at their upper ends around the wires. The filament is shrunk on the conducting wire during the process of carbonisation, and is made of greater thickness at the point of connection. A secondary air-chamber, either above or below the globe, and opening into it, is used. "The residual air will be forced, by being heated when the lamp is in use, into the secondary chamber, and remain until the current is cut off."

*2,068.—F. H. Warlich, London. **Production of Pictures by the Agency of Light.** 2d. May 21.

ELECTRIC LIGHT.—"The electric or other light" is employed for the more speedy and accurate production of enlarged photographs.

2,121.—E. Mourlot, Paris. **Substance for Replacing the Gutta-Percha, &c.** 2d. May 25.

CONDUCTORS.—An insulating material is made by extracting the gummy substances contained in

the external skin or bark of birch trees and mixing it either with gutta-percha or india-rubber, without restriction as to the proportions. By adding a small proportion of the product to ordinary gutta-percha or india-rubber, it imparts to it the property of remaining unchangeable, prevents cracking or breaking, and adds "a power of resistance" which neither india-rubber nor gutta-percha possesses.

*2,147.—W. Lloyd Wise, London. (*J. A. Mandon, Paris.*) **Electric Light Apparatus.** 2d. May 26.

ARC LAMP.—The carbons are curved, so as to form arcs of circles. Each carbon is carried by an adjustable balance lever, supported on an insulated axis, and is connected by a circularly curved arm to a float immersed in an insulated bath of mercury or other fluid, which is contained in a stationary vessel and traversed by the current. This vessel is shaped to an arc of a circle, of which the balance lever axis is the centre, and is situated on the same side as the carbon. The curved arm carrying the float is so proportioned that, as the carbon is consumed, it will gradually ascend in the mercury in such wise that the loss of carbon will be correctly compensated by the upward movement thereof, the two carbons being so balanced that the points are maintained at the required distance apart. It is stated that by causing the current to pass through mercury all unpleasant noise and vibration is obviated.

2,179.—A, Specht, Hamburg. (*O. Schumann, Hamburg.*) **Hollow Glass Reflectors for Lamps, &c.** 6d. (10 figs.) May 28.

REFLECTORS. — These consist of hollow glass cupolas coated inside with silver, and are adapted for reflecting the light either downward or sideways. Several forms of reflectors are shown, in which the chief feature is the formation of a neck *b*

through both walls of glass, for the lamp cylinder, and a bulb *a*. The silvering of the glass is performed in the usual manner; but to give the silver solution a permanent hold on the glass and to protect it from the influence of heat and the atmosphere, a thin coating of asphalte varnish is applied.

*2,236.—S. Cohne, London. **Electric Lamp, &c.** 2d. June 1.

ARC LAMP.—The carbon-holders are carried on the arms of a parallelogram frame actuated by a

vessel submerged in a cylinder containing water or other fluid, and arranged so that, as the consumption of the carbons takes place, the lower one rises and the upper one descends. A syphon may be substituted for the cylinder, the vessel carrying the carbons floating upon the fluid in the long leg. On the short leg is a cover containing a small aperture through which the fluid passes at the rate required by the consumption of the carbons.

2,252.—G. G. André, Dorking, and **E. Easton,** London. **Electric Lamps.** 6d. (3 figs.) June 2.

SEMI-INCANDESCENCE LAMP.—The light is produced by the incandescence of a carbon electrode in contact with a copper electrode, and the novelty consists in feeding a carbon or carbons placed end to end in a holder so as to slide therein, by means of a cord and weight, which, when the negative electrode approaches the positive, are brought into action by means of a regulator or brake controlled by the electric current. The electrode C^2 slides freely in the bracket C, and has a contact made with it by the spring L^3, carried by the rod C^1 passing at its lower end through a solenoid of high resistance, which serves to prevent sparking when contact is broken at L^3. The electrode C^2 follows the wasting

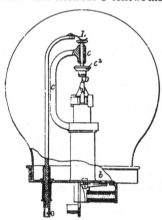

carbon D until it has reached the end of its travel, when a further consumption of carbon will break the circuit, and diminish or extinguish the light. The carbons D are placed end to end in a tube, and are forced up by an arrangement resembling a carriage lamp with a weight replacing the usual spring. The weight is controlled by being attached to a cord passing round a brake pulley held or released by the action of a solenoid (not shown) wound differentially so that when the current is flowing normally it is out of action, but when the resistance of the lamp is increased, it lifts the brake lever and allows the weight to feed on the carbon; *b* is a ventilating hole controlled by apparatus described in Specification

No. 1,507 of 1880. K is a contact point on to which the upper pole falls to short-circuit the lamp when the carbon is totally consumed. In a lamp in which the upper carbon descends, the holder is made of sufficient weight to force the carbon down, the lower electrode being forced up by a counter-weight.

2,271.—J. J. W. Watson, St. Marychurch, South Devon. **Artificial Illumination, &c.** 8d. (17 figs.) June 3.

INCANDESCENCE LAMPS.—The object of the invention is to produce increased illumination by discharging currents of induced electricity through gas and other flames. Several modes are proposed. Among others, three "fish-tail" or "batswing" burners are inclined upwards to produce a "cocked-hat" flame, this form being well adapted for use with the electric arc proper, or with an incandescent platinum coil. The platinum coils may be wound on spools of asbestos, and placed either vertically or arched across the flames, or they may be of "pine-apple" or double "bee-hive" form.

ARC LAMP.—From a suitable base rise two hollow standards of glass, surmounted by hollow curved arms, carrying at their centre a ring out of which jets of gas issue. The two carbons are inclined together, and are fed upwards, so that the arc shall be formed in the focus of the gas jets. The carbons are guided near their points by grooved rollers, and have their lower ends held in holders fitted to a crossbar attached to a rod provided at its lower end with a piston. This piston fits in a cylinder in which a pressure of water is maintained by an annular piston depressed by a spring fitting in a second larger cylinder surrounding the first one. The annular piston is lifted by a rack and pinion, the flow of liquid being controlled by a taper plug valve. A solenoid carries on the top of its core a bevelled block of steatite which separates the points of the carbons on the admission of the current which should be alternating.

CONDUCTORS. — These are carried in glazed earthenware pipes fitted with earthenware diaphragms placed at each joint. The mains should be of tapering section. The negative main, where direct currents are used, is an uninsulated iron rod offering the same resistance as the thickest section of the positive main. Automatic cut-outs and resistances are used to equalise the current. The gas supply cock may also be part of an electrically controlled switch, which simultaneously closes the circuit and urns on the gas.

CURRENT METER.—The gas meter is fitted with an extra set of dials worked by the gas when thrown into gear by an electro-magnet placed in the main circuit.

2,272.—T. Slater, London. **Obtaining, Increasing, and Employing Currents of Electricity, &c.** 6d. (5 figs.) June 3.

DYNAMO-ELECTRIC GENERATOR.—This consists of a generator with field magnets having their opposing poles N N, S S, as shown, separated by strips

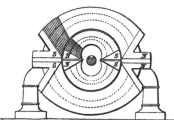

of dia-magnetic material. Two sections of the rotating armature are illustrated, in each case composed of two similar electro-magnets with their like poles opposed and separated by dia-magnetic material. A compound armature consists of "two or more lengths on the same axle." The commutator and connections are as usual. Each part of the compound armature may be of different diameter, the current passing from the smaller to the larger. The large part may have a primary and secondary coil, one coil receiving the current from a galvanic battery "and the other coil receiving its current from the permanent or fixed magnets of its own machine."

2,453.—J. C. Mewburn, London. (*F. A. Achard, Paris.*) **Electric Apparatus for Working Railway Brakes.** 6d. (17 figs.) June 17.

DYNAMO-ELECTRIC GENERATORS.—This refers to improvements in the electrical apparatus described in Patent No. 1,387 of 1877. The inventor states that "instead of employing piles" (preferably of the Planté type) "as the source of electricity for operating the brakes, magneto-electric and dynamo-electric induction machines can be employed, worked by the wheels of the vans in which they are placed." A commutator and relay are also described.

2,665.—W. R. Lake, London. (*R. B. Lamb, Camden, New Jersey, U.S.A.*) **Protective and Insulating Casings for Underground Telegraphic Wires, &c.** 6d. (3 figs.) June 29.

CONDUCTORS.—The casings are made of cylindrical blocks of terra-cotta or other suitable substance, and have passages or channels lined with india-rubber or other equivalent substance, through which the conductors are carried. No joint is mentioned.

2,764.—G. G. Andre, Dorking, Surrey. **Electric Lamps.** 8d. (21 figs.) July 6.

ARC LAMP.—The regulation of the arc is effected

by a small electro-motor. The motor armature is wound with a double set of wires, each set having its own commutator, so that two currents of opposite directions may be passed through the coils at the same time, causing the armature to revolve in the direction determined by the predominating current. The main current is passed round the electro-magnets of the motor, a branch circuit of high resistance running through one set of coils on the armature. A shunt circuit, having a resistance equal to that of the branch and the arc, runs from the positive to the negative pole of the lamp, and passes through the other set of coils of the armature, so that the currents flowing through the armature are equal to the shunt current flowing in the opposite direction to that of the branch. Instead of winding the armature ring differentially, two separate rings may be used on the same axis, or the electro-motor may be arranged to work on the principle of the Wheatstone bridge. In this case the electro-magnet coils are wound in two

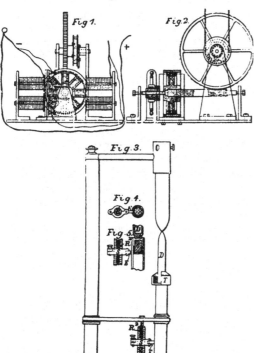

Fig 1.   Fig.2.

Fig.3.   Fig 4.   Fig.5.

separate circuits, one of high and the other low resistance. The current divides on entering the motor, one part passing through the thick coils and through the arc, the other part flowing through an extra resistance and through the thin coils. The helix of the motor forms the bridge between the junction of the two members in each branch. Consequently any variation in the resistance of the arc causes the helix to revolve and move the posi-

tion of the carbon in the direction required. Figs. 1 to 5 represent the motor and the manner in which its revolving helix is connected with the carbon, Fig. 2 showing the method adopted where the carbons are long, in which case an endless cord is used to actuate the feed mechanism. Figs. 3 to 5 show an arrangement suitable for short carbons. In this the motor is placed upright and caused to work by a system of gearing the screw R^2, which by means of the nut S and arm E, raises or lowers the carbon D. The nut S is made in halves and held together by a peg. T is a cup containing copper-coated shot, forming contact for the carbon. The motor instead of being worked by the difference of two currents continually flowing through it, may be actuated by a current thrown into it only when required either by the differential action of two electro-magnets or solenoids upon an armature, causing a portion of the main current to pass through one or other of the helix coils as required, or by a similar differential action upon a magnetic needle the motion of which is caused to make it break the circuit through the motor.

DYNAMO-ELECTRIC GENERATOR. — When the motor is to be used as a generator, the parts are enlarged and the radial arms doubled in the axial direction, the boss and rim being proportionately enlarged as required. The rim is divided into a number of contiguous segments, each comprising a group of arms and coils which are coupled up for quantity or tension as desired.

SEMI-INCANDESCENCE LAMP.—The carbon-holder is arranged as described in Patent No. 2,252 of 1880. The negative electrode is so hinged to a portion of the lamp that it descends upon the positive carbon as the latter burns away. When it has descended a minute distance, a spring on the end of the arm carrying the electrode comes in contact with a stud, and shunts a portion of the current through, and starts the electro-motor. In this case the armature is wound with a single coil, and the motor is only required to turn in one direction. A short-circuiting switch is provided to prevent the circuit being broken should the lamp become extinguished.

*2,835.—G. W. von Nawrocki, Berlin. (*T. Balukiewicz, St. Petersburg.*) Electric Signalling and Controlling Apparatus for Trains. 4d. July 9.

DYNAMO-ELECTRIC GENERATOR. — A dynamo-electric, or other generator, is used to operate the signals on the engine and in the guard's van.

2,888.—H. J. Haddan. (*C. Cummings, Virginia City, Nevada, U.S.A.*) Signalling Apparatus for Mines. 6d. (7 figs.) July 13.

CONDUCTORS.—These are carried down the shaft

and are connected to the source of electricity by a brush mounted on the cage. The completion of the circuit causes an alarm to be given at the surface.

2,893.—Baron Elphinstone, Musselburg, and C. W. Vincent, London. Apparatus for Generating Electric Currents. 8d. (11 figs.) July 13.

DYNAMO-ELECTRIC GENERATOR.—This invention is an improvement on the generator embodied in Patent No. 332 of 1879. To obtain a steady and uniform current, the coils are arranged diagonally upon the armature drum. A further improvement consists in causing a definite proportion of the induced current to traverse the electro-magnets. To insure this the armature is provided with two layers of diagonally arranged coils, one set of which is connected with the field magnets, through a commutator, and the ends of the other set are likewise connected with a commutator, which may be arranged to give either a continuous or an alternating current. Or the wires of the two sets of coils may be laid side by side, their ends being carried to separate commutators as before, or the coils may be arranged in three or more layers, and connected either for tension or quantity, one or more sets being employed to feed the magnets, and the remaining sets for supplying the outside circuit ; or again one set of coils only may be used, the wire on the rotating

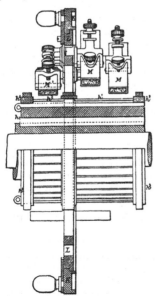

drum in this case being so proportioned in area and length to that on the field magnets, that the latter cannot carry the whole of the electricity which the revolving coil is capable of producing

when its maximum speed of rotation is reached. Thus the drum coils will act as if they were in two layers, the field magnets taking only their proportion of the current. In general construction this generator resembles the one previously illustrated in the abstract of No. 332 of 1879, with the exception of the commutator, which as shown in the figure consists of a cylinder of vulcanite h grooved longitudinally to receive two sets of brass flanged plates h^1 h^2, the flanges of the former being double the length of the latter ; the plates are held in place by the binding hoops h^3. The plates correspond in number to the coils on the armature. The brushes are fitted in holders carried by a ring L, capable of axial motion on its annular frame K. The holders M M^1 are divided into two sets and mounted on opposite sides of the ring L, from which they are insulated. The brushes M bear on the slips h^1 and take off the current for the outside circuit, the other brushes collect the current for the field magnets. In case one set of coils only is used, the commutator is made with one set of slips and brushes only.

2,980.—A. M. Clark, London. (*J. H. Guest, Brooklyn, New York, U.S.A.*) Regulators for Electric Lamps. 2d. July 20.

ARC LAMP.—A "thermoscopic rod" is employed to automatically regulate the length of the arc. The lineal expansion of the rod, according to the intensity of the current, is multiplied by levers which by means of clamps separate the carbons. A section of carbon or metal of low conductivity is combined with the expansion rod "whereby the heat due to resistance is rapidly generated and dissipated." A shunt is mentioned for dividing the current when the carbons are separated to a definite point.

3,025.—P. Jensen, London. (*M. Avenarius, Kiev, Russia.*) Electric Lighting. 6d. (8 figs.) July 22.

ELECTRIC LIGHT.—This invention relates to a system of electric lighting in which a series of lights C_1 C_2 C_3 (Fig. 1) are produced by a single alternate current generator M, voltameters $V_1 V_2 V_3$ being inserted in shunt circuits as shown. The electric burners C may be either lamps or Jablochkoff candles. One form of voltameter suitable for this purpose is shown at Fig. 2, the metal plate A forming one pole of the instrument, and the plate F the other pole. The plates $F_1 F_2 F_3$, &c., all fitting closely the glass cylinder H, are spaced upon a glass rod E_1, forming a continuation of the metal rod E. The polarisation of the apparatus is determined by the number of metallic discs F employed, and its resistance by the distance of these discs

apart, and by the distance of the lowest disc from the bottom. Modifications are described in which the discs F are placed in a vertical position to allow of the escape of the gases from the acidu-

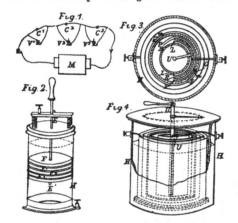

lated water upon the passing of the current. Another suitable form of voltameter is represented by Figs. 3 and 4, in which I and K are glass cylinders, and R a half-cylinder of glass capable of turning upon the axis U, and of interposing any portion of its surface between the metal plates L P. The inner plate P and the outer plate F^m are connected together by the metallic strip o.

*3,041.—C. G. Gumpel, London. Producing and Directing Electric Currents, &c. 2d. July 24.

DYNAMO-ELECTRIC GENERATOR.—The inventor "concentrates the polarity as much as possible into a narrow field" by distributing "an even number, preferably not less than four of such intensely polar fields, uniformly round the circumference of the circle in which the armature coils move, the polarity of these fields being alternated." The armature coils are wound either as in the Gramme or Siemens generator, and may be connected in any desired combination, either to insulated rings or to the slips of a commutator. When a ring armature is used the ring is constructed of several parallel "tyres" of wood, on which is fixed a "sheet iron casing" or "bars of iron side by side, but not in magnetic contact with each other." The wires of the brushes slide freely in a casing, and are pressed against the periphery of the commutator by a caoutchouc spring. When the brush has to extend over an arc of the periphery thin plates are used, placed side by side, and their ends are hollowed to conform to the commutator surface.

INDUCTION COILS.—Instead of superposing the secondary coil, the primary and secondary wires are wound on together.

ELECTRO-MAGNETIC MOTOR.—The above-described generator may be used as a motor for steering gear, in which case a commutating apparatus is arranged so that the direction of the currents through the field magnets shall remain constant while those through the armature may be changed at will.

3,310.—E. T. Truman, London. Insulating Telegraph Conductors, &c. 4d. August 7.

CONDUCTORS.—The principal object of the invention is to make the stranding and covering of the wire one continuous operation. This is effected, in combination with apparatus described in former Specifications No. 878 of 1870, No. 482 of 1872, and No. 124 of 1874, by a lay plate or tube, which is caused to rotate, and the strand, finding no stationary point until it has passed the lay plate, the separate wires are laid and covered each with its compound. Another part of the invention relates to a material or compound with which the strand is coated before being covered with gutta-percha, consisting of ozokerit, 10 parts; pitch, resin, or gutta percha, 5 parts; Venice turpentine, 5 to 10 parts.

*3,324.—C. G. Gumpel, London. Electric Machines. 2d. August 16.

DYNAMO-ELECTRIC GENERATORS.—This refers to Specification No. 3,041 of 1880, and describes various ways of winding an armature in the form of a hollow open-ended shell having a shape approaching that of an ellipsoid. The ends of the shell are divided by projecting fins which are continued as ribs along the outer surface of the shell, thus forming a number of sections for the reception of the insulated wire. Assuming that there are four magnetic fields, then the number of divisions is a multiple of four by any odd number. Three methods of winding the wires in the sections are given. The currents are collected by an ordinary commutator. This method of winding is applicable to motors.

*3,416.—C. H. Frome and G. C. Gibbs, London. Apparatus for Illuminating and Producing Theatrical Effects. 2d. August 23.

ELECTRIC LIGHT.—To give a flashing scenic effect, the ends of the conductors are attached at intervals to either side of a rope from which they are insulated. The rope is stretched across the stage, and a small carriage provided with metal plates completes the various circuits as it is drawn over the rope. The lamps are made of glass cylinders having different coloured sections, and may be secured to the head of the performer in such a manner that by "pulling a wire or string the cylinder may be turned."

3,424.—G. Barker, Birmingham. (*W. W. Jacques, Boston, Mass., U.S.A.*) Electric Conductors, &c. 4d. August 24.

CONDUCTORS.—These are coated with a composition consisting of india-rubber mixed with a "preserving compound" composed of Venice turpentine and beeswax in about equal parts. This is placed in a vessel arranged to be heated and exhausted of air. When at about 80 deg. Cent., the rubber is added and the heat raised to 100 deg. Cent., the air being gradually exhausted. The air is now admitted and the mass allowed to cool down to 80 deg. Cent. again. The impregnated rubber is now mixed on warm rolls with sulphur in the usual way. An insulating composition composed of the following ingredients may be used : Rubber, 30 parts ; preserving compound, 7 parts ; sulphur, 3 parts ; oxide of zinc, 10 parts ; steatite, 10 parts ; asphalte, 40 parts. The wires are covered with the composition in the usual way, and are finally packed in powdered soapstone and vulcanised.

*3,438.—A. E. Gilbert, Edinburgh. Securing Telegraph Wires to their Insulators. 2d. August 25.

INSULATORS. — This comprises a bow-shaped fastener or clip having its free ends hooked or turned round so as to grip the conductor some inches in advance of and beyond the insulator, and to pull it firmly into the groove on one side by the insertion of the strong hollow central breast of the clip into the groove of the insulator on the other side, the curve of the clip being such as to pull or slightly bend the conductor on each side of the insulator towards the line through its centre, so that the straining of the wire automatically secures it and its clamp within the groove on each opposite side of the insulator.

3,494.— St. G. L. Fox, London. Electric Lamps, &c. 6d. (13 figs.) August 28.

INCANDESCENCE LAMP.—This refers to Specifications Nos. 3,988, 4,043, and 4,626, all of 1878. In Fig. 1, A is the globe, *a* the luminous filament, *b* a block of steatite, *c* a pair of steel clips cemented to a block *d* of porcelain by fused borax ; *e e* are platinum wires fused into glass tubes *f f* filled with mercury to prevent leakage ; *g* is an india-rubber stopper covered with a layer of mercury *i* and coated with marine glue *j*. The filament is made of vulcanised fibre which has been baked at a white heat. Several pieces of fibre, cut to the shape shown in Fig. 2, are arranged in a spring clip K and baked in a crucible (Fig. 3) filled with powdered charcoal. The baked strips are carbonised by treating them at a white heat in benzole vapour. Fig. 4 shows the apparatus for exhausting the lamps. A glass

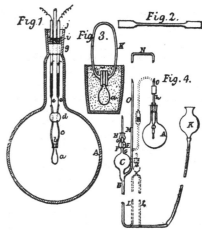

tube B, 30 in. long, terminates in a bulb C, which has a ground neck to receive at times the lower end of a glass rod E. This neck opens into another bulb F above, through the neck G of which the glass rod also passes air-tight. L L is an air trap that may or may not be used. In operation the vessel K is filled with mercury when in a lowered position, and is then raised to such a height that while it becomes nearly empty, the mercury rises into the two bulbs C and F, filling the same up to the cup H. The neck G of the upper bulb is then closed by the rod E, the neck of the lower one being left empty. The open vessel K, which is connected by a flexible tube, is then lowered about 36 in. in order that the mercury may fall well below the point where the tube M communicates with the tube B. The consequent fall of mercury in the bulb produces a vacuum which is filled by air in the lamp, forcing its way through the exhausting tube *h*, *o*, N M. The open vessel is then again raised, the neck G of the upper bulb being opened to allow the air to escape. The two bulbs C and F again become filled with mercury, the neck G of the upper is again closed, the open vessel lowered, and so on until the lamp is almost exhausted. The pumping action is then modified, so as to exhaust the last traces of air.

*3,496.—C. W. Harrison, London. Apparatus for Obtaining Electricity. 2d. August 28.

MAGNETO-ELECTRIC GENERATOR. — An endless chain of bars passes through a number of helices of insulated wire. The motion of the bars may be vibratory instead of continuous, or the helices may have motion imparted to them. The currents may be taken from each coil or from a group or groups of them. In generators producing alternating currents, the currents are passed "through one or more coils of conducting material, the reacting

currents of which coils are made by a suitable commutator placed on the axis of the machine to act in direction with the initial current."

*3,509. — J. Hopkinson, London. Electric Lamps. 2d. August 30.

ARC LAMP.—Each or either carbon electrode is carried by the piston of a hydraulic cylinder, the movement of which controls the advance of the carbons. The passage of fluid into or from the cylinder, or from one cylinder to the other, is controlled by an electro-magnet placed either in the principal circuit, or in a shunt circuit of high resistance. A third cylinder communicating with each of the others may be used to adjust the position of the arc. A second electro-magnet placed either in the main or a shunt circuit is employed to establish the arc.

3,637.—P. M. Justice, London. (*H. C. Spalding, Boston, U.S.A.*) Lighting Cities by Electricity. 6d. (5 figs.) September 7.

ELECTRIC LIGHT. — The inventor proposes to " flood the atmosphere with light irradiated in many directions" by electric lights situated on high towers arranged in triangular groups, and each fitted with a reflecting lantern.

3,765.—E. G. Brewer, London. (*T. A. Edison, New Jersey, U.S.A.*) Electric Lamps, &c. 1s. 2d. (41 figs.) September 16.

INCANDESCENCE LAMPS.—The filaments should have a resistance of not less than 100 ohms, and are preferably made from a single fibre of "monkey bast." Several filaments are bound together and soaked in a solution of sugar, after which the wooden clamps for attaching them to the conducting wires are affixed, and the whole carbonised. The platinum wires are inserted in the clamps previous to the carbonisation. When made from bamboo the cane is split, and the pith removed; the slips are then passed before a knife, which cuts them to a uniform thickness. After this they are placed between the two halves of a mould which determines their shape and size, the projecting portions of the slips being removed with a cutting instrument. The filaments when shaped have enlarged ends for attachment to the conducting wires. In making the filaments of wood, a block is shaped to the form of a web with enlarged ends from which slices are cut off, softened by moisture, and bent to the required shape; or the block may be shaped to the form of an arch before the slices are detached to obviate the bending process. The carbonising of fibrous filaments is done in a nickel box, shaped to the form of the filaments. The clamps are carbonised in vacuo to remove occluded gases. Wood filaments with enlarged ends are

laid in a suitably shaped groove formed in a metal plate, a number of plates are placed one over the other, and then subjected to heat in a closed nickel case. Several forms of plates and grooves are described, in all cases the length of the groove being such as to allow for contraction of the material. The carbonising stove is of nickel, preferably in two parts, and gaseous fuel is used. To prevent oxidation of the filaments, a pipe conducts into the stove an oxygen-absorbing element. Where it is desired to use lamps of lower lighting power than the standard lamp, they are placed in a divided circuit, each being provided with a filament of proportionate resistance and radiating power, in connection with a single circuit breaker to insure their turning off or on simultaneously. In lamps for producing large lights, and where large conductors have to be used, the filament is placed in a Torricellian vacuum formed at the head of a U tube, and the conductors pass through the mercury. This lamp is supported on a frame provided with levelling screws. To detect any flaw in the filaments, they are subjected to a preliminary heating in vacuo in a temporary lamp. Where it is desired to fit the lamps with double filaments, the inner end of one of the lamp conductors fits into a Y shaped clamp; one end of a filament is fastened in each limb of the clamp, the other ends being fastened in individual clamps, each having its own conductor. A circuit-controlling device is so constructed that the circuit through either filament may be closed, or both may be on either in series or multiple arc. To prevent the separation of the glass and wire at the point where the conductors enter the globe, a small glass bulb is blown of slightly larger diameter than the distance apart of the conductors. The conductors are passed through and each sealed in this globe at two places, which is then placed in an aperture left in the larger, and sealed thereto, at about the meridian line of the smaller globe. Both the globe and bulb are simultaneously exhausted. In a lamp which may be taken apart, the filament is supported upon a tube closed at one end, the conductors passing through the tube being sealed at the top closed end; the enclosing globe is made with a neck somewhat larger than the tube. A rubber packing is interposed between the outer walls of the tube and the inside of the neck, securing the two together, the end of the tube projecting beyond the end of the neck, and passing through a rubber cup, the base of the neck resting upon the bottom of the interior of the cup. A rubber packing fills the space between the exterior of the neck and the interior of the cup. The space in the cup is filled with mercury which is retained in place by the rubber packing. The lamp is exhausted and sealed in the usual manner.

*3,808. — F. G. Willatt, London. Dynamo Machines. 2d. September 20.

DYNAMO-ELECTRIC GENERATOR. — This relates principally to an armature cast in one solid piece with radiating blades narrowing towards the centre. These blades are wound longitudinally and are made to revolve between the north and south poles of a stationary electro-magnet, preferably circular in form so as to encompass the armature. The blades of the armature are arranged right and left of its longitudinal axis on the centre boss, and may be wound and coupled up as a whole or separately, the commutator being of the ordinary description.

3,809.—J. B. Rogers, London. Dividing and Subdividing the Electric Current for Lighting. 6d. (11 figs.) September 20.

To divide a current the inventor connects the leading wire from a dynamo-electric generator to the apex of a metallic cone or conoidal body, and leads off the various line wires from points in the circumference of the base. To intensify a current he couples several generators to the circumference of the cone, and draws one powerful current from the apex.

3,832.—W. Elmore, London. Dynamo-Electric Machines. 6d. (8 figs.) September 22.

DYNAMO-ELECTRIC GENERATOR.—This invention relates principally to improvements in the subject of Letters Patent No. 3,565 of 1879, whereby a more thorough cooling of the revolving armature and its magnet is effected. The armature cylinder, boss or disc, and its magnet cores, are constructed with a continuous internal chamber connected with conduits formed in the shaft as shown in Fig. 1, so that a constant stream of water or other refrigerating liquid can be maintained through them. A special commutator is also provided allowing a

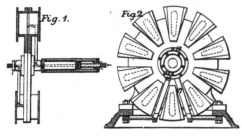

Fig. 1.   Fig. 2.

greater facility of connection, and avoiding the secretion of metallic dust under the brushes. A metal barrel piece is mounted on, and insulated from, the shaft and runs in an insulated bearing. The barrel carries on its inner end terminals to which the ends of half the armature coils are attached. The outer end forms half the sections of the commutator which stand away from the shaft, and are secured thereto by insulated nuts, and divided from the remaining half of the sections, which are connected through the shaft to the coils on the other half of the commutator. The whole is held in place by a lock nut. The air circulates in the recesses formed between the shaft and the commutator slips. Ring connectors 16 and 17, Fig. 2, are employed for joining up the wires from the stationary magnets to the brushes and terminals of the machine.

3,880.—P. Jensen, London. (*T. A. Edison, New Jersey, U.S.A.*) Systems of Conductors for the Distribution of Electricity, &c. 6d. (20 figs.) September 24.

DISTRIBUTING CURRENTS.—The patentee states that when many lamps or motors are arranged upon the multiple arc system it is essential that an equal electromotive force be maintained in all parts of the system. When, however, each set of conductors is run out from the central station in a straight circuit containing a number of lamps, the electromotive force is apt to be the greatest near the station, and to diminish gradually towards the end of the conductors. The object of this invention is to obviate such irregularity by using the conductors as feeding conductors only, and by placing the lamps on service circuits connected to the main conductor in such a way that all the lamps are electrically equidistant from the source of the current. When only a few lamps near the central station are used they are placed upon a direct circuit therefrom with sufficient resistance to equalise the electromotive force to that of the more distant circuits. Separate feeders are laid to buildings using a number of lights. Earth may be used as half of the circuit. Where there are several central stations the feeding circuits of all the stations are connected. One method of laying the conductors consists in placing each service set (positive and negative) in squares around a central station, and having feeding circuits connected to each set at several points ; the service sets being connected together. In another method the positive feeding conductor is connected to the service conductor on the side of the square opposite to that at which the negative is connected, so that the same mass of conductor intervenes between the terminals of all the house circuits.

CONDUCTORS.—Where a conductor varying in sectional area is used it is preferably composed of several single uninsulated wires of different lengths, a few of which extend the whole length, the others ending at various points. These are bound together at intervals. Where the current is used for power as well as light, two branch circuits run from the

mains into the house, and a meter placed in each enables the currents to be charged for at different rates.

3,894.—P. Jensen, London. (*T. A. Edison, New Jersey, U.S.A.*) Electro - Magnetic Railroads, &c. 1s. 6d. (36 figs.) September 25.

TRANSMISSION OF POWER.—This invention consists in a complete electro-magnetic railway system embracing the generation, distribution, and utilisation of electric currents for motive power. Fig. 1 shows the general arrangement of a central station and the rail connections. M T, M T^1, and M T^{11} are the main rails. A portion M T of sufficient length to accommodate one or more trains is electrically separated form the remainder of the sections, each section having its central station. The parts M T^1

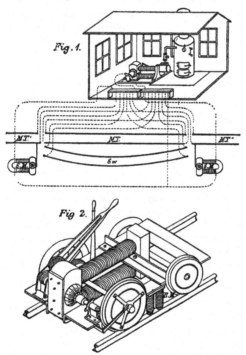

and M T^{11} are movable, so as to constitute switches by which trains may be shunted on to the siding S W. The motors S M serve to pull over the points. Fig. 2 is a perspective view of a motor car. The rails form the conductors, contact being made between them and the motor by means of brushes. The motion is conveyed from the armature spindle to the driving wheels by means of frictional or toothed gearing. Two magnets, one of which is shown at E M, exert an attractive force upon the rails, and thereby increase the grip of the driving wheels. The motor, of any ordinary construction, may be fitted with a central gripping wheel, which is

thrown into gear when ascending an incline, or " creepers " may be used consisting of a casing fitted with eccentrically pivotted rollers, which allow the casing to slide along the rail, but prevent retrograde movement by gripping the rail. The " creepers " are so geared to the motor as to draw the carriage forward. An electrometer in the circuit gives indication of the position of a train on the section. Electro-magnets are used as brakes. Each car is fitted with collecting brushes so as to make a contact with the rails in many places, and conductors, running the whole length of the train, convey the current to the commutators. When it is desired to use the rails of an ordinary railroad they are insulated from the sleepers by suitable means. All the devices using current are arranged in multiple arc, the rails forming the main conductors. Incandescence lamps are used for lighting the carriages and for head lamps. The specification contains seven sheets of drawings and thirty-four claims, and is too long for condensation within our limits.

3,928.—W. R. Lake, London. (*E. Thompson, New Britain, Conn., U.S.A.*) Apparatus for Generating and Utilising Electricity. 8d. (16 figs.) September 28.

DYNAMO-ELECTRIC GENERATOR.—The object of this invention is to increase the effective power of magneto and dynamo-electric generators, and to provide an armature for the latter, the magnetising effect upon which shall not be limited to the point of magnetic saturation of the field magnets, but shall be further reinforced by the direct action upon the iron core of the said armature, and the insulated wire surrounding the same, of the currents traversing the whole of the field magnet coils, which coils are made to enclose

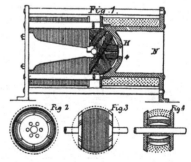

and surround the revolving armature, so that the iron core of the armature becomes practically a movable extension of the field magnets themselves. The field magnets N are cylindrical in form and the ends H are hemispherical, in order to embrace the armature, which is spherical. Rings of metal are provided to keep the coils, which project beyond

the pole-pieces, in place. Figs. 2, 3, and 4 show how the armature is built up of two metal plates and coils of iron wire previously to the insulated wire being wrapped round it, the dotted line in Fig. 3 showing the ultimate outline. The coils may be wound and coupled up as required, but a special arrangement for preserving uniformity of outline consists in winding three coils of insulated wire intersecting one another, at the axis, at angles of about 60 deg. One half of one coil is wound on, then one half of the second coil, then the whole of the third coil is wound on; next the remaining half of the second coil, and finally the remaining half of the first coil. The commutator consists of two flanges fixed on the shaft with a segmental ring mounted between them. Each segment is provided with two rods which pass through, but are insulated from, holes in the flanges. The segments are mounted at some distance from the flanges, and their inner diameter is larger than the shaft, thus providing an air space. The insulated wires from the armature coils pass through the shaft, and are connected to the ends of one of the rods of each segment. The gap between any two segments is preferably made obliquely.

3,936.—J. W. Fletcher, Stockport. Shackle and Terminal Insulators for Telegraph Wires. 6d. (20 figs.) September 28.

INSULATORS.—This insulator consists of a cylinder of earthenware around the outside of which one or more grooves are formed to receive the telegraph wires, which may be held in place by straps or other suitable means. The cylinder is pierced with a central hole to allow the supporting bolt to pass, and two annular recesses are made within the cylinder, one on each side. The specification gives various modified forms to meet special circumstances, all of which are similar to the figure, and consist essentially of two No. 8 insulators joined head to head. The upper cup may be provided with a mushroom head.

3,964.—P. Jensen, London. (*T. A. Edison, New Jersey, U.S.A.*) Magneto-Electric Machines, &c. 10d. (11 figs.) September 30.

DYNAMO-ELECTRIC GENERATORS.—The first part of this invention relates to the mode of holding the commutator brushes, which is clearly shown in the accompanying illustrations. To maintain a uniform speed the engine, provided with a governor and automatic variable cut-off, is connected direct to the axle of the armature, preferably by a crank-pin on a balanced disc. "Generators of very great capacity" are constructed with a series of field magnets, fixed together with the engine on a self-contained base-plate fitted with non-magnetic supports as shown in Fig. 4, the arma-

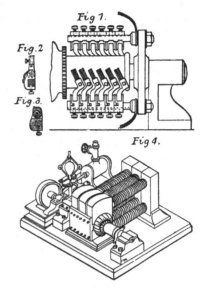

ture being proportioned to act as a flywheel. To maintain a uniform electromotive force a small

auxiliary generator is used to excite the field magnets.

MOTORS.—The current is regulated by arranging that the circuit is broken at every revolution of the governor. Connected with the ball arms, so as to be moved thereby, is a sleeve composed of an insulating and a conducting portion, their line of union being a diagonal, which if developed would be V shaped. A spring bears on the sleeve, and the circuit to the motor passes through the spring and sleeve; thus the circuit is broken at regular intervals, the motor running by momentum until the circuit is again completed. In communicating power from the motor to the driven mechanism, the rotary motion of the armature is first converted into an oscillatory motion, and then into a continuous rotary motion by pawl friction gear. When it is desired to generate currents of very high potential, and to use small main conductors, "a battery" of separate and distinct generators, having all their armatures mounted on one shaft, and connected up in series, is used. Motors are arranged in the same way. Where a slow speed water-wheel is the source of power for driving generators, it is employed to pump water to a height, and the water is then used to drive a quick-speed wheel geared direct to the "battery." The field magnets of the "battery" of motors are arranged as a shunt, and are excited by a portion only of the current. To enable any coil to be removed from the armature singly, only that portion of the coil which is upon the operative face is made of wire, the wires of the coil being connected at the end by metallic plates fastened to an insulating base,

and insulated from each other. These plates project at the proper points above the general surface of the core, at which points the wires are secured to them. At one end each plate is connected to the commutator block.

*3,971.—A. M. Clark, London. (*A. Niaudet, and E. Reynier, Paris*). **Dynamo-Electric Machines, &c.** 10d. (22 figs.) September 30.

DYNAMO - ELECTRIC GENERATORS. — The fixed electro-magnets are in line with the axis of rotation of the coils. The ends of the cores of the coils are connected by thin circles of iron; or a thin iron ring may be placed on the poles of the fixed magnets. The rotating coils may be parallel or radial to their axis of rotation. The commutator may be constructed of radial bars with intervening spaces, against the ends of which springs bear, or of metallic segments fixed on a disc of wood. The fixed electro-magnets may be either in the main circuit or in a derivation therefrom.

4,005.—E. G. Brewer, London. (*A. J. B. Cance, Paris*). **Electric Machines.** 6d. (11 figs.) October 2.

DYNAMO-ELECTRIC GENERATOR.—This refers to Specification No. 1,927 of 1878, and consists essentially of one or several induction crowns or rings, formed of bobbins "horizontally independent," and placed circularly on a support in the form of a wheel mounted on the shaft of the generator. Whilst rotating, the poles of the induced bobbins pass successively before the poles of a certain number of "inducting or conducting" bobbins, placed two and two on the same diameter of the apparatus, preferably four bobbins placed two and two on the same vertical diameter. These generators give at will continuous or alternate currents. To obtain continuous currents the wires of the induced system are arranged in such manner as to collect by means of a four-part "collector," the currents developed in the system, and to make them pass through the inducing system. To ob-

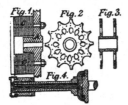

tain alternate currents, an electric current coming from an intermediate source is caused to pass in the field magnets, and by means of two rubbers bearing on insulated rings the currents developed in the induced system are collected. Fig. 1 illus-

trates the coupling of the induction bobbins, Figs. 2 and 3 show the bronze inductor carrier for supporting the induction crowns, and Fig. 4 is a longitudinal section of the shaft and collector. Each end plate A carries four bobbins C C, between which the induced bobbins fixed on the discs (Figs. 2 and 3) rotate. The free poles of the magnets C C are united by a "magnetic crown," as shown, to intensify their action. The induced bobbins are arranged in two concentric circles, of which those of the first row lie in the recesses shown in the periphery of the disc, and those of the second row between those of the first row. Every alternate bobbin in each circle is of opposite polarity and the magnetic poles of the field magnets "are opposite." Fig. 4 shows how the wires are led from the bobbins to the "collector."

4,007.—G. Zanni, London. **Magneto-Electric Apparatus for Railway Signalling.** 6d. (4 figs.) October 2.

MAGNETO-ELECTRIC GENERATOR.—The generator described in Patent No. 4,232 of 1877 is used to operate railway signals.

4,009.—J. G. Lorrain, Edinburgh. (*G. Trouvé, Paris*). **Obtaining and Applying Motive Power, &c.** 6d. (7 figs.) October 2.

MOTORS.—This relates to electro-motors, and consists of a Siemens armature revolving in the magnetic field of an electro-magnet. To avoid the dead point or "angle of indifference to motion," which in machines where the armature revolves concentrically with the magnets, amounts, it is stated, to nearly 25 deg., the polar surfaces of the armature, or preferably of the magnets within which the armature revolves, are made eccentric to the axis of rotation, and are by preference either oval or volute. The arrangement may be applied to generators.

4,049.—A. W. L. Reddie, London. (*A. Biloret and C. Mora, Paris*). **Electric Machines.** 6d. (4 figs.) October 5.

DYNAMO-ELECTRIC GENERATOR.—This consists of a wheel on a horizontal shaft carrying a number of bobbins of insulated wire retained between iron cheeks. This part of the arrangement is somewhat similar to a Siemens alternate current machine. The field magnets are four vertical coils with iron cores and polar extensions carried out towards the bobbins on the rotating wheel. It is stated that "by this invention the induced current is passed round all the poles of the field magnets at once, and a current from each pole where the wire of that pole comes out can be utilised." The commutator is of the ordinary type and rotary metallic wire

brushes are used. The result is that a Gramme or a Siemens machine constructed thus could burn four lights at once.

4,081.—W. R. Lake, London. (*A. Lemaire and E. Lebrun, Paris.*) **Electrical Signal Apparatus for Railways.** 6d. (7 figs.) October 7.

SWITCHES.—Two switches are described, one an ordinary two-way switch and the other a push, which in its normal position maintains contact with one line, and when depressed completes a second circuit.

4,191.—G. P. Harding, London. **Electric Lamps.** 6d. (11 figs.) October 15.

ARC LAMPS.—The illustration shows the mechanism of a non-focussing lamp. The electro-magnets a a are upon a shunt circuit, and when excited attract the armature b, connected to the eccentrics c c. Assuming the parts to be in the positions shown, with the upper carbon out of contact with the lower one, the first effect of a current is to cause the armature b to be attracted by the electro-magnets, in opposition to springs, and by its motion to

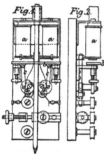

rotate the eccentrics d d, thereby allowing the upper carbon to fall on to the lower one. By this means a part of the current is diverted from the coils and their armature recedes, consequently the upper electrode is again gripped and slightly raised to establish the arc. In a modification the upper carbon or carbon-holder is sustained between fingers or tongs held together by springs. A lamp is also described, having converging carbons which are pressed outwards by springs, against fixed stops acting as brakes. The arc is formed by the magnets closing the points of the carbons together, and releasing them on the passage of current. An automatic shunt is used, consisting of two electro-magnets and an armature, which on the extinction of the lamp makes contact with a "butting screw," and shunts the current through an iron wire resistance.

4,192.—G. P. Harding, London. **Electric Lamps for Locomotive Engines.** 4d. October 15.

ELECTRIC LIGHT.—One or more lamps "of any convenient construction" are placed on the front of the locomotive to illuminate the track, and they are so arranged that each lamp illuminates a definite portion of the road. The generator is driven either by a small auxiliary engine or direct from the locomotive axle.

*4,220.—H. G. Hosmer, Rome. **Apparatus for Obtaining Motive Power.** 2d. October 16.

MAGNETO-ELECTRIC MOTOR.—This refers to Specification No. 3,676 of 1878, and consists of a ring carrying vertical permanent magnets, electro-magnets being mounted on a horizontal axis "placed centrally of the ring." "A current of electricity will be generated in the coils of the armatures as the latter come opposite the magnets, which will have the effect of magnetising the armatures, and thus converting them for the moment into electro-magnets."

*4,227.—A. Budenberg, Manchester. (*C. F. Budenberg and B. A. Schäffer, Buckau*). **Apparatus for Preventing Explosions in Coal Mines, &c.** 4d. October 16.

COMMUTATOR.—A cylinder provided with contact studs completes in succession the circuit through a number of spirals of fine platinum wire placed in various parts of the mine to explode the gases liberated.

4,254.—W. A. Benson, London. **Apparatus for the Distribution of Artificial Light.** 4d. (3 figs.) October 19.

REFLECTORS.—The direct rays are screened from the room by a saucer-like reflector, which deflects them upwards on to a second reflector, by which they are diffused as required.

4,265.—W. R. Lake, London. (*C. A. Hussey and A. S. Dodd, New York, U.S.A.*) **Dynamo-Electric Machines.** 6d. (7 figs.) October 19.

DYNAMO-ELECTRIC GENERATOR.—The object of

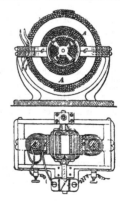

the invention is to produce a compact, powerful, and effective generator. Referring to the illustration, A A are the field magnets which may be formed either as a ring or a cylinder. They have polar extensions C C partly surrounding the rotating armature, which is of the section shown and is provided with air passages b, b, b. It may be wound and coupled up in various ways, as in the Siemens or Gramme armature, or the coils may surround each of the ribs $b\,b$. The commutator is of the ordinary type.

4,391.—P. Jensen, London. (*T. A. Edison, New Jersey, U.S.A.*) **Measuring the Amount of Electrical Current Flowing through a Circuit.** 6d. (2 figs.) October 27.

CURRENT METER.—In this apparatus the indications are obtained by using balanced plates in a depositing cell, so arranged that the deposition of metal upon one plate causes an overbalancing, which operates registering devices and also contrivances for reversing the circuit through the cell. The plan of operation is as follows : The current is

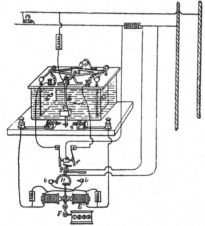

flowing from plate C to C_1, and as metal is deposited upon C_1 it gradually overbalances C and brings the arm d, fitted with an adjustable weight to determine the excess of weight of one plate over the other, against screw q, whereupon the circuit through magnet coil E^1 is closed. E^1 attracts arm F and operates the register. The upper portion of arm F strikes the horseshoe-shaped lever D, causing it to break circuit at l, and carrying it over against the stop l^1. At the same time the arm g is carried in the reverse direction, leaving stop f^1, which thereupon comes in contact with stop e, from which the arm g carries the stop f. The current is then free to pass through the cell in the opposite direction, when a similar cycle of operations is performed.

4,393.—W. R. Lake, London. (*H. S. Maxim, Brooklyn, New York, U.S.A.*) **Electric Lighting Apparatus.** 6d. (16 figs.) October 27.

INCANDESCENCE LAMP.—The object of this invention is to so construct the globe and other parts of the lamp that a conductor of any required resistance may be used, and that the lamp may be readily taken apart for the insertion of a new filament, and yet be air-tight. For this purpose the globe is made with a neck sufficiently wide to admit the filament. This neck is fitted with a glass stopper ground into its place and serving as an entrance for the conductors, which in the case of powerful currents are divided into several branches and are fused into the glass. The lamp may be exhausted by making the stopper tubular and fitting on the top of the tube a glass cap with a ground joint. The cap has a tubular branch on its side, for attachment to the air pump, whose opening corresponds with an opening in the side of the stopper, so that as the cap is turned round it acts as a valve. The space above the stopper is filled with gum or wax.

*4,408.—G. W. von Nawrocki, Berlin. (*W. S. Limbeck, Langendreer, Germany*). **Photometers.** 2d. October 28.

PHOTOMETER.—The ray of light is thrown on to a plate of selenium, the altered electrical resistance of the latter becoming a measure of the light.

*4,428.—J. H. Johnson, London. (*A. Berjot, Paris.*) **Electric Lamps.** 4d. October 29.

ARC LAMP.—" The essential part of the apparatus is composed of a parallelogram" having two horizontal bars, carrying the mechanism for actuating the carbons, joined to two vertical bars by blade springs. The upper carbon is fed down by a rack controlled by four wheels and pinions, which in turn are controlled and the arc formed by an electro-magnet placed in the lamp circuit. The current is automatically switched to a fresh pair of carbons by a pivotted lever provided at one end with a tooth which engages with the rack of the carbons to be next consumed. The other smooth end of the lever falls into a recess on the top of the rack of the burning carbons, and thus frees the second rod. Stops are used to regulate the travel of the racks. A spring bridges across the lamp terminals on the consumption of the last pair of carbons. The loss of weight of each carbon is compensated for by a ball of lead on the end of a horizontal lever suitably pivotted and resting at one extremity on a bracket attached to the stationary part of the parallelogram. The descent of each rack draws a lever down and the ball falls

on to the movable part of the parallelogram. The carbon-holders consist of copper cylinders fitted with internal springs.

4,482.—E. George and J. B. Morgan, Liverpool. **Cables for Telephones.** 6d. (6 figs.) November 3.

CONDUCTORS.—The inventors propose to eliminate induction currents in wires suspended near each other by insulating the separate wires and connecting the outsides of the coatings to earth at frequent intervals.

4,495.—W. R. Lake, London. (*J. V. Nichols, Brooklyn, New York, U.S.A.*) **Electric Lighting Apparatus.** 6d. (1 fig.) November 3.

INCANDESCENCE LAMP.—This is an incandescence

lamp with conductor A of high resistance enclosed

by a glass globe B in which a vacuum has been formed. Copper conducting wires C are used instead of platinum wire, and the novelty of the invention consists chiefly in the material and method employed for securing these conductors to the base of the lamp. That part of the globe immediately surrounding the wires is formed of a "metallo-vitreous" cement made by mixing various metallic substances with silica and potash, as, for example, in the following proportions: oxide of lead 58 parts, silica 17 parts, oxide of iron 10 parts, oxide of copper 10 parts, and potash or soda 5 parts.

4,608.—C. F. Heinrichs, London. **Generating, Sub-Dividing, and Transmitting Electric Currents, &c.** 1s. 2d. (45 figs.) November 9.

DYNAMO-ELECTRIC GENERATOR.—This invention relates to improvements on Patents No. 4,595 of 1878 and No. 4,589 of 1879. The illustration shows the generator in cross section. The armature, of annular form, is wound with D shaped helices of insulated wire forming a closed circuit. On the shaft s is keyed the boss $a\,a$, carrying the armature core constructed by coiling upon the thin cast-iron shell $r\,r$, of U section, a quantity of soft iron wire. The armature revolves within the pole-pieces of the electro-magnets $m^1\,m^2$, &c., excited by the rectangular coils $w^1\,w^1$, forming part of

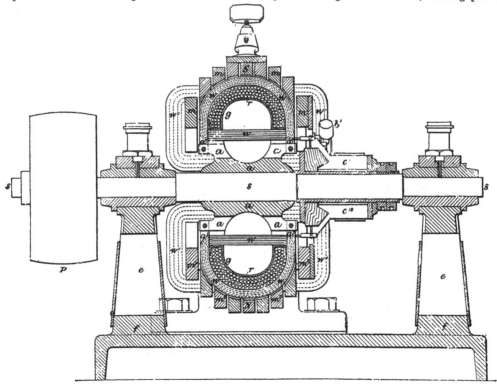

the main circuit. A commutator, the segments of which are preferably arranged obliquely, and brushes of the ordinary type are used. The ring is attached to the boss by brass loops, of which two are shown marked $a^1 a^1$, $a^3 a^3$. These loops pass over the iron wire core, and their ends are secured by screws to the boss $a a$, so as to form ventilating openings. Several modified forms of armature ring are shown, the coils of which may be coupled up to form one or many independent circuits, in the latter case a pair of brushes being provided for each circuit. The exciting magnets may be permanent steel ones. Each commutator plate may consist of two parts insulated from each other, having a wire of high resistance interposed between them to prevent interruption of the current when several sets of brushes are used.

ARC LAMP.—This is a lamp for burning carbons

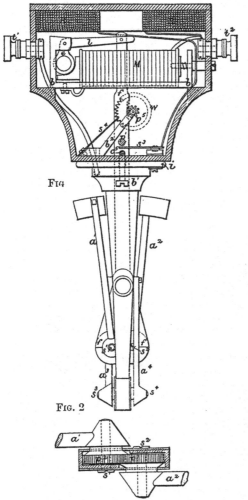

FIG. 2

of circular form. Referring to the figures, it will be seen that the upper positive pair of carbon-holders

$a^1 a^2$ are pivotted at the points s^1 and s^2 to two small equal-sized pinions $p^1 p^2$, shown in detail in Fig. 2, by which the swing of the arms $a^1 a^2$ is rendered equal on each side of the vertical axis of the lamp. A similar method of pivotting is adapted to the lower carbon-holder arms $a^3 a^4$. The geared pivots of the upper pair are attached to the rod f suspended from the long arm of the bell-crank lever l, of which the short arm is the armature e of an electro-magnet M, which on the passage of current attracts the armature e, lifting the rod f, and with it the upper carbons, and thus establishing the arc. To prevent a jerking movement a small rack forming part of the rod f gears with the pinion p^5 on which is a toothed wheel w, against which presses a click spring s^4. The lower carbons are in electrical connection with the frame b^1, which, while insulated from the case of the lamp and positive terminal l^2 by the ebonite washer i, is connected to the negative terminal l^1. A coil of German silver wire R R, having a resistance equal to the arc, is substituted for the latter by means of a projecting pin p^1 on the rod f, which, when the rod drops, brings the spring s^3 against the contact screw c^5, which is connected to b^1, and by the wire b, to the negative terminal. The pinions $p^1 p^2$ are rigidly attached to the radial arms $a^1 a^2$, turning on the pins $s^1 s^2$, fixed to opposite sides of the frame. In a modification, motion is communicated to the two carbons by racks gearing with pinions having their fulcrums in brackets depending and insulated from the lamp. The racks gear at their upper ends with a pinion controlled by an electro-magnet, actuating by its armature a pawl engaging with teeth on the periphery of the pinion. A second electro-magnet, placed in another circuit, disengages the pawl and permits the carbons to approach each other at predetermined intervals. A block of fireclay may be placed above the carbon points as a reflector. In a further modification the carbons, fitted with projecting pins near their lower ends, are placed within spirals contained in tubes, which spirals are rotated by clockwork and suitable gearing, the pins sliding in grooves in the tubes. The top ends of the carbons press against platinum points arranged on levers, causing the lower ends of the levers to grip the carbon. The grippers also insure contact with the electrodes. The arrangement of the spiral and slotted tube is also applied to a semi-incandescence lamp in which the carbon rod is propelled against a block of refractory material. The circular carbons are made by compressing the material in horizontal dies.

4,614.—C. W. Siemens, London. **Electric Lamps.** 6d. (3 figs.) November 10.

ARC LAMP.—Referring to the illustration, the

core H of a high resistance solenoid coil E, situated upon a shunt circuit, is balanced by a small spring and surmounted by a cup in which shot is placed for regulation. To the core are attached rods K L, supported at their other extremities in angular beds $k\ l$, and fitted with ratchet teeth, which ordinarily stand just clear of the ratchet wheels C D upon the axes of the carbon feed rollers $a\ b$. In operation, as the distance increases between the carbons A and B, the resistance to the direct circuit of the lamp is augmented, and consequently a larger portion of the current passes through the shunt circuit of the solenoid coil; the core H is thus attracted upwards, and by its ascent draws up the two bars K and L. These, as they retire from their rests k and l, bear against the ratchet wheels

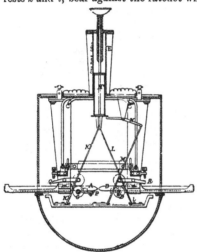

C and D, and partly turn them round in the direction of the arrows, thus causing the carbons to approach each other. Should the carbons be too near, the solenoid will have little power, and descending will move the tappet lever N, raising the bar M, and reversing the motion of the feed roller B. The lamp is focussed by the carbon A being fed rather faster than the carbon B, and being provided with an abutment stop P. When either carbon is consumed the upper feed roller drops and short-circuits the current through the contact screw R or Q. The same arrangement may be applied to vertical or inclined carbons provided with suitable counterweights. A method of making the carbon supply continuous consists in forming the pencils with male and female ends that can be joined by cement and a wire pin.

4,621.—E. G. Brewer, London. (*E. H. Johnson and T. A. Edison, New Jersey, U.S.A.*) **Magneto-Electric Signal Apparatus.** 6d. (5 figs.) November 10.

Magneto-Electric Generator.—To enable the

generator to be cut out of circuit when not in use, the handle is connected to the shaft by a pin taking into a slot with inclined faces, and the first effect of rotating the handle is to slide it endways on the shaft and break a shunt circuit. A commutator is provided for sending either continuous or intermittent currents according as the generator revolves in one direction or the other. This is accomplished by providing the commutator with two surfaces so arranged that when turning one way a collecting spring is in contact with one of them, and when turning the opposite way the spring is in contact with the other surface.

*4,674.—R. Kendal, London. **Telegraph Cables.** 2d. November 13.

Conductors.—To protect cables against "boring marine animals" the core has helically wound around it a ribbon of sheet zinc, a second similar ribbon being wound on over the first in the reverse direction.

*4,745.—J. E. H. Gordon, Dorking. **Electric Lamps.** 2d. November 17.

Incandescence Lamps. — When incandescent iridium is the source of light a current of air is passed through the lamp, and carried through a passage in which the iridium dissipated by the heat is condensed.

4,755.—J. A. Berly and D. Hulett, London. **Electric Lamps, &c.** 6d. (10 figs.) November 18.

Lanterns.—To enable the lamp to be drawn up and down, it is suspended to a gallery attached to the lower end of a rod, down either side of which the insulated conductors are brought to the lamp terminals, the rod and gallery being attached to an arm and counterweighted. The upper end of the rod has a block of insulating material fitted to slide in a tube provided with insulated contact slips extending nearly to the bottom of it. Spring contact pieces, projecting from the ends of the conductors, and embedded in the block, complete the circuit. In a modification, the lamp terminals are connected to the lower ends of two vertical tubes, containing mercury, carried on an arm and counterweight. The conductors are attached to an insulating block and dip into the mercury tubes.

4,779.—F. Harmant, Paris. **Electro-Magnetic Apparatus for Table Services, Offices, and Warehouses.** 4d. (3 figs.) November 19.

Magneto-Electric Motor.—A small electromagnetic motor of ordinary construction is used in connection with a current carrying railway for domestic and other use.

*4,825.—C. Kesseler, Berlin. (*E. Kuhlo, Stettin*). Dynamo-Electric Motors. 4d. November 22.

MAGNETO-ELECTRIC MOTOR.—This consists of three electro-magnets arranged around a wheel upon which are placed, preferably, eleven armatures or keepers. A commutator directs the current into the electro-magnets alternately, so as to maintain a continual attraction on the armatures in one direction.

4,833.—E. H. T. Liveing, London. Apparatus for Detecting and Measuring Small Quantities of Inflammable Gas in Coal Mines, &c. 6d. (6 figs.) November 22.

MAGNETO-ELECTRIC GENERATOR.—A "magneto-electrical machine" is used in conjunction with platinum spirals to indicate the presence of gases. To prevent sparking, one end of the armature coil is connected to its core, the bearings being used as one terminal. The other end of the coil terminates in a stout insulated wire passing through the hollow shaft; the end of this wire is brought to a smooth rounded point, and presses against a polished steel spring forming the other terminal.

*4,851.—H. Law, London. Current Meters. 2d. November 23.

CURRENT METER.—The meter is so constructed that the "axle which is put into motion by the current, imparts its motion to a vertical axle enclosed in a tube closed at its upper extremity." The axle is carried to the upper part of the tube, and "at one point in its revolution it completes an electric circuit which gives a signal to the observer."

4,866.—W. R. Lake, London. (*H. S. Maxim, Brooklyn, U.S.A.*) Electric Lighting Apparatus. 6d. (6 figs.) November 23.

ARC LAMP.—The upper carbon descends by gravity driving a train of gearing combined with an electro-magnet and armature lever operating a detent connected with the gearing, and also a carrier for the lower carbon movable vertically and connected with the armature lever. A dash-pot fixed to the armature or core of the electro-magnet, and surrounding its head, controls the action of the magnet. Thin flat supports are used for the carrier and globe, placed edge to edge upon the same side of, and in the same vertical plane with, the focus. The illustration shows a vertical section of the lamp. A is a wood case, C a thin flat bar of metal supporting the globe, and moving freely up and down in slots in the arms *a a*; D the upper carbon carrier, E the lower carbon carrier. The two sides *d d* are held together by the screw *b*; G is a flat

bar supporting lower carbon carrier; H is a link pivotted to bars F and G; J the armature lever; L the axial magnet placed in the lamp circuit; N an adjusting screw for limiting the movement of the core; O a train of gearing; and Q a rod passing

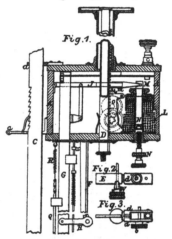

Fig. 1.　Fig. 2.　Fig. 3.

through the lower carbon carrier and suspended by the spring R. The operation of the lamp is as follows: The forked head W is drawn down and turned until its legs stand upon the lower surface of the lower carrier E, when the spring R raises the bar G and lever J until the detent *m* clears the escapement wheel. The upper carbon carrier then descends by its own weight until the carbon points come in contact, when the current excites the coil L and draws down its core M, at the same time depressing the lower carbon and arresting the descent of the upper carbon. When the arc lengthens the coil L becomes too weak to overcome the tension of the spring R, so that the lower carbon is gradually raised and the detent *m* withdrawn, allowing the upper carbon to descend until equilibrium is again restored. The length of the arc is regulated by adjusting the tension spring R, and the lamp may be stopped by turning the head screw W until it straddles the lower carrier E, when the spring V, obtaining a bearing point against the case A at the shoulder *x*, overcomes the tension of spring R, and keeps the lower carbon depressed.

4,886. — J. Hopkinson and A. Muirhead, London. Dynamo-Electric Machines. 8d. (10 figs.) November 24.

DYNAMO - ELECTRIC GENERATOR.—To decrease the sparks between the collecting brush and the commutator the brush is divided into parts, and resistance is interposed between the part that is last in contact with the commutator, and the leading wire or conductor. The rotating armature consists of a pulley, on the periphery of which is coiled a series of layers of sheet iron insulated

from each other. On one or both faces of the ring thus formed, radial slots are cut to admit coils of insulated wire, which lie in planes parallel to the plane of revolution. When a continuous current is required, coils are placed on both faces of the armature, and are arranged alternately so that when one coil is at its dead point those next to it on the opposite face are producing their maximum effect. The field magnets are made in a similar manner to the armature, but preferably with solid wrought-iron core-pieces. Both in the armature and the magnets each coil is packed close to its neighbour, and is of a tapered quadrilateral form.

4,914.—W. Lloyd Wise, London. (*J. A. Mandon, Paris*). **Electric Light Apparatus.** 8d. (3 figs.) November 25.

ARC LAMP.—In the lamp shown in the illustration there are two carbons, each formed like a quadrant, and balanced to turn freely round a centre of rotation with a counterweight and a curved tube L containing mercury, and receiving a float J attached to a stem I, such float and stem being so formed that as the carbons

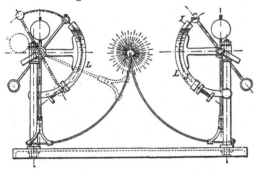

waste the float and stem rise a proportionate distance through the mercury, whereby the arc is maintained at one point. A lamp is also described having straight carbons rising out of inclined tubes of mercury, and meeting at a point where they are guided by friction rollers carried by arms jointed to an insulated axis having a horizontal armature and counterweight. An electro-magnet, placed in the lamp circuit, attracts the armature and separates the electrodes.

4,933. — J. W. Swan, Newcastle - on - Tyne. **Electric Lamps.** 6d. (4 figs.) November 27.

INCANDESCENCE LAMPS.—The filament is formed from cotton thread, soaked in a solution of sulphuric acid, and carbonised in the usual way. The ends of the filament are thickened where they are held by the platinum clamps by wrappings of cotton thread which are fixed by the action of the acid. The lamp is built up as described in Patent No. 250 of 1880.

*4,952.—A. M. Clark, London. (*J. Garnier, Paris.*) **Furnaces for Metallurgy of Copper, &c.** 2d. November 27.

CONDUCTORS.—To improve the conductibility of the copper "the last trace of oxygen" is eliminated by adding phosphides of copper to the mass while in a furnace lined with a basic lining. To prevent the galvanic action which is usually set up in copper sheathing, phosphorus is added, but not sufficient to effect its malleability.

4,988.—K. W. Hedges, London. **Electric Lamps.** 6d. (13 figs.) November 30.

ARC LAMPS.—This refers to Patents Nos. 81 and 925 of 1879, and consists in using two positive and one negative electrode, and a refractory abutment. In this lamp two rectangular troughs are attached to the supports, and are inclined to each other in V form; two of the carbons slide freely in these troughs, and meet at a point below. The carbon carrier is rigidly attached to the framing of the lamp, the trough carrying the negative carbon being pivotted to the frame, and its motion controlled by the armature of an electro-magnet placed in the lamp circuit. The passage of the current attracts the armature, thus separating the carbons and establishing the arc. The main positive carbon is prevented from sliding through the trough by the pressure of the smaller carbon sliding in a trough so placed that the point of contact is sufficiently near the arc to insure the gradual consumption of the smaller carbon. The negative carbon-holder is provided with an adjustable platinum stop, which presses against the carbon and controls its feed. Electrical contact is made by providing each carbon-holder with a contact piece hinged to the trough at its upper end, and carrying at its lower end a brass block grooved out to fit the cylindrical carbon against which it presses. The length of the arc is determined by a screw controlling the movement of the armature. In a modified form of the lamp, designed for burning in series with others, the carbons are separated by a solenoid having its core attached (by a non-magnetic connection) to the armature of an electro-magnet, the high resistance coils of which form a shunt across the arc, and are so disposed with respect to the armature as to influence it in an opposite direction to that of the solenoid. When the circuit is completed the carbons are drawn apart by the solenoid, the distance to which they are separated being determined by the difference of attractive force on the armature of the solenoid and magnet. The descent of the carbons and electrical contact therewith is as described in the first lamp. When alternating currents are employed, only one positive carbon is used.

SEMI-INCANDESCENCE LAMP.—The carbon de-

scends against the periphery of a chisel-edged metal disc, and is pressed on laterally by a bent lever loaded with an adjustable weight. The cover of the lamp is provided with a safety valve in case of explosion of combustible gas produced within the glass. Such explosions may generally be prevented by introducing into the lamp glass a little ammonia or other volatile substance.

5,004. — J. W. Swan, Newcastle - on - Tyne. **Measuring and Recording Electric Currents, &c.** 6d. (2 figs.) December 1.

CURRENT METER.—The apparatus produces a record of the time during which the current has passed, by means of counting mechanism brought into action by, and during the passage of, the current. Each translating device requires a separate apparatus, but several can be arranged in one case, as in the illustration, where it will be seen that a cylinder A, rotating at an uniform rate, is provided with pegs *a a*, which, each time they

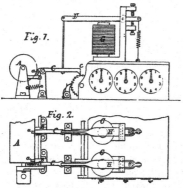

meet the sway lever C, cause it to rotate the wheel *c*, one tooth thus actuating the counting mechanism. When no current is passing the sway lever is held out of gear with the wheel *c* by a spring acting on the shorter end of the pivotted armature H. When the current passes in any given circuit the electro-magnet G of that circuit attracts its armature H and lowers its sway lever into gear with the toothed wheel belonging to its particular set of counting mechanism.

*5,008. — H. Wilde, Manchester. **Electro-Magnetic Induction Machines.** 2d. December 1.

DYNAMO - ELECTRIC GENERATOR. — To avoid counter currents the discs in which the armatures are mounted are slit in a radial direction towards the centre. The armatures are also similarly slit and are fixed in the discs in such a position that he slits coincide. The ends of the electro-magnets are terminated by overlapping pole-pieces which are also slit.

5,014. — J. W. Swan, Newcastle - on - Tyne. **Electric Lamps.** 6d. (18 figs.) December 2.

INCANDESCENCE LAMPS. — The filaments are formed from cotton threads, treated with sulphuric acid, and carbonised by heat in a crucible containing powdered carbon. The ends are thickened by wrapping of bibulous paper, muslin, or other material susceptible of being parchmentised. The lamp, which may contain two or more filaments, is constructed in the ordinary manner, the conducting wires being coated as described in Patent No. 250 of 1880.

*5,033. — J. H. Johnson, London. (*A. de Meritens, Paris*). **Electric Lamps.** 2d. December 3.

INCANDESCENCE LAMP.—Currents of high potential are discharged across the space between two iron balls, mounted on the ends of platinum conductors and enclosed in an exhausted chamber. The balls gradually become incandescent, but are protected from oxidation by the nitric and hyponitric acids formed under the influence of the sparks.

5,137.—W. T. Henley, Plaistow, Essex. **Electric Machinery and Apparatus for the Production of Light and Heat, &c.** 2s. 8d. (45 figs.) December 9.

MAGNETO-ELECTRIC GENERATOR.—A wheel carrying permanent magnets revolves between two sets of electro-magnets having their poles placed intermediate between each other. Referring to the illustration it will be seen that the base-plate carries the cast-iron standards A A, to which the electro-magnets *a a*, *b b*, having their cores formed of wrought-iron oval tubes filled with soft iron wire, are bolted. The brass wheel C, fixed on the

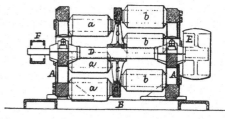

shaft D, carries six pieces of soft iron $c^1 c^1$, which as they revolve bring the magnets on one standard alternately in magnetic connection with the magnets on the other standard. The electro-magnets are wound so as to be alternately N. and S., and the ends of the coils are carried to a commutator, from which the currents are collected in the ordinary way.

DYNAMO-ELECTRIC GENERATORS.—The pieces of soft iron carried by the brass wheel may be wound

so as to form electro-magnets. The commutator and winding is arranged so that the fixed electro-magnets have similar poles opposite each other ; or the currents may alternate in the fixed magnets. The wheel may have attached to its periphery two semicircular pieces of iron provided with indentations for the reception of the coils. Two or more wheels may be mounted on the same shaft, so as to revolve between the upright magnets, which are made to move in slides by means of nuts engaging on a bar provided with a right-handed screw on one half and a left-handed screw on the other half. The electro-magnets may be excited by the current from a voltaic battery. The field magnets, arranged parallel to the shaft, may be mounted with their outer ends surrounding the peripheries of the armatures when two are mounted on one shaft. The field magnets may be arranged vertically around a circular base, the induced magnets being carried by and depending from a connecting ring keyed on a vertical shaft driven by bevel gearing. The shaft carries two commutators, each mounted on a separate brass bush, and insulated therefrom. The second commutator is replaced by metallic rings should alternating currents be required for the outside circuit. The currents are collected by brass springs. In winding the electro-magnets, small wire is used and wound in separate layers all in one direction, the starting ends being all connected together at one end, and the terminals at the other end. The commutator consists of metal rods moved in a vertical direction by cams on the shaft, terminating in forks which dip into vessels of mercury through which the circuit connections are made. Each fork has a long and short leg, the contact being made and broken by the short leg. The inventor states : " In conveying the current from electro-magnets on a revolving shaft, I propose to use an insulated metal disc turned thin at the edge, revolving in a cup of mercury on a pillar insulated from the base of the machine." A modification consists of a brass bush adjustable on the shaft, and carrying two insulated copper rings provided with inwardly projecting pieces, the pieces of one ring projecting into the spaces of the other without touching. The currents are collected by copper springs.

ELECTRO-MAGNETIC MOTOR.—By connecting the coils on one of the armatures with the field magnets in such a way that the current shall flow in one direction in the armature, and by means of a commutator be alternately reversed in the field magnets, this part of the generator, on the passage of current from a voltaic battery, becomes a motor, capable of driving another current producing armature, wound to produce high tension currents.

ARC LAMPS.—Two discs of carbon are inclined towards each other, and are carried on spindles fitted in the upper ends of jointed levers. A disc of insulating material, which burns away equally with the carbons, is mounted on a horizontal axis and rotates by clockwork. The carbon discs receive their motion by being pressed against the rotating disc by springs. Cone pulleys are provided to compensate for decreased speed of the disc peripheries. An electro-magnet, placed in the main or a shunt circuit, on the passage of current attracts its armature, and forces up the insulating disc, thus separating the carbons and establishing the arc. On the current failing the disc drops and allows the carbons to come together, a small arm at the same time checking the clockwork. The current passes to the discs through mercury cups, into which dip small discs mounted on the spindles. In a modification the carbons are mounted on axes parallel to each other, the disc of insulating material being pressed against their peripheries. The disc is pressed up by a cam driven by the clockwork employed also to rotate the disc. In a further modification two discs of carbon are rotated by clockwork, and are drawn together by a spring in opposition to a cam driven by the clockwork, which controls their rate of approach. The arc is formed by an electro-magnet placed in the main circuit. Carbon rods may be substituted for the carbon discs, and fed up by clockwork against an insulated disc. An electro-magnet serves to separate the carbons and form the arc, and "a coil and permanent magnet suspended to retard the train should the pencils be fed forward too fast" is used.

INCANDESCENCE LAMP.—The open end of the glass globe is cemented into a metal collar formed with a flange and screwed to receive a metal cup having a flange and india-rubber washer ; " there is a valve and stop-cock to attach to a syringe or air pump for the purpose of exhausting the air when necessary, and this may be done every night before lighting." Two brass pillars pass through an ebonite block fixed in the metal collar, and serve as the conductors to a cross-piece of refractory material.

5,211.—P. Adie, London. Lighting Mines. 2d. December 13.

INCANDESCENCE LAMPS.—The lamps are so connected to the leading wires that the fracture of the filament breaks the circuit. (Provisional protection not allowed).

5,237.—W. T. Glover and G. F. James, Salford. Machinery for Braiding, Lapping, or Covering Telegraph Wires, Crinoline Steel, Engine Packings, &c. 6d. (3 figs.) December 14.

CONDUCTORS.—To avoid injury to the finished conductor, the core is pushed by the feeding rollers

into the machine instead of being drawn through by the part that is completed.

5,268.—A. W. L. Reddie, London. (*J. André, Paris.*) Transmitting Drawings, Characters, and Writing by Electricity, &c. 6d. (7 figs.) December 15.

CONDUCTORS.—These consist of a band of composite web, formed by grouping a number of metallic conducting wires, which are insulated from each other by a weft of textile material, the warp threads being the conductors themselves. The web so formed may be used for the brushes of generators.

*5,275.—D. G. Fitz-Gerald, Brixton. Electric Lighting, &c. 2d. December 16.

INCANDESCENCE LAMPS.—The filament is formed of carbonised vegetable substance and is surrounded by an atmosphere of hydro-carbon, or other vapour. The ends of the filaments are enlarged with any suitable metal by electro-deposition, extended if desirable over the whole surface of the conductor. When a metallic filament is employed it is embedded in an infusible substance.

5,352.—S. Pitt, Sutton. (*O. Lugo, New York, U.S.A.*) Dynamo-Electric Telegraphy. 8d. (4 figs.) December 21.

DYNAMO-ELECTRIC GENERATOR.—A continuous current generator is used and is so arranged that the resistance of a portion of the outside circuit is approximately equal to that of the generator, and the exterior circuit being kept practically constant in resistance by means of a rheostat, composed of a bar divided into sections and connected by plugs, which sections are connected to resistance coils.

*5,358.—F. M. Lyte, London. Protecting Iron and Steel Ships from Corrosion. 2d. December 21.

DYNAMO-ELECTRIC GENERATOR.—The negative pole is connected to the parts to be protected while the positive pole is immersed in the electrolytic solution, in this case the sea water.

5,404.—M. C. and T. J. Denne, Redhill, Surrey. Locking and Unlocking Railway Signal and Point Levers. 6d. (4 figs.) December 23.

CONDUCTORS.—The various parts of the apparatus are "connected by wires in the ordinary way," the circuit being completed by a "bridge piece" attached to the engine.

5,498.—J. J. Shedlock, Uxbridge. Treatment of Articles of Cast Iron. 4d. December 30.

INSULATING MATERIAL.—Articles of cast iron which are to be used for electrical apparatus are covered with an insulating compound by exposing the surface to the action of a dissolving agent, and then enclosing them in an air-tight box in which a vacuum is maintained while the casting is heated. The compound consisting of pitch, resin, india-rubber, or gutta percha is then admitted and forced into the pores of the metal by the atmospheric pressure.

1881.

39.—H. J. Haddan, London. (*J. D. Townsend, New York, U.S.A.*) Street Curbs and Gutters for the Reception of Telegraph Wires. 4d. (6 figs.) January 4.

CONDUCTORS. — The conduit for telegraph and other conductors consists of a flanged case or chamber forming a street curb. A parallel flanged chamber forms the gutter, the chambers being suitably connected together at the sides and ends, and fitted with a removable cover forming a step and gutter piece.

48.—W. R. Lake, London. (*E. Etéve, Paris.*) Apparatus for Generating and Utilising Electricity for Lighting, &c. 6d. (9 figs.) January 4.

DYNAMO - ELECTRIC GENERATOR. — The field magnets, of horseshoe form, are built up of a number of separate plates, and are provided with coils on each leg. The armature consists of a "coil mounted on an axle and actuated by a pair of toothed wheels." The two-part commutator is of ordinary construction. A cam-shaped plate and spring are used, "the caloric effect of which" is utilised for igniting, say, a spirit lamp.

ARC LAMP.—The base, of glass, carries two uprights supporting two parallel spindles geared together, on the ends of which are secured discs of carbon. Motion is imparted by a falling weight.

65.—P. M. Justice, London. Electric Lighting. (*H. C. Spalding, Boston, U.S.A.*) 6d. (1 fig.) January 6.

ELECTRIC LIGHT.—This consists of a system of lighting towns by groups of electric lights massed together in elevated positions, the groups being

subdivided into lesser groups, each sub-group being in a circuit distinct from and independent of the others, and the lights composing each sub-group being so arranged as to be interspersed with the lights of other sub-groups.

78.—J. E. H. Gordon, Dorking. Dynamo-Electric Machines for Electric Lighting. 6d. (8 figs.) January 6.

DYNAMO-ELECTRIC GENERATOR.—The armature consists of a number of wheels mounted upon a shaft, carrying at their peripheries cylindrical bobbins of insulated wire arranged with their axes parallel to and equidistant from the axis of the generator. The bobbins, wound on solid or tubular iron cores, fit in recesses in the wheels in which they are secured by screws overlapping their ends. They are separately excited by a continuous current, and are wound so that the poles on each side of a given core and the one opposite to it are of contrary polarity to itself. The moving parts of the exciter are preferably mounted on the armature shaft. The field magnets consist of a number of cylindrical coils of insulated wire, carried in recesses formed on the peripheries of brass stationary brackets placed between and outside the wheels composing the armature. The coils are secured in the recesses by brass bands. The cores, consisting preferably of bundles of iron wires, have their axes parallel to the axis of the generator. These bobbins are so arranged that the ends of the cores of the rotating bobbins pass close to their ends. Each of the stationary magnets may supply a separate circuit or they may be coupled up as required. The fixed bobbins are each divided into a number of sections by discs of insulating material. The bobbins adjacent to the outer faces of the armature are preferably half the length of the others, and are coupled together in pairs, their cores being connected by iron armatures. Two insulated wheels fixed on the shaft convey the current from the exciter to the armature coils. No commutator is described.

93. — J. Imray, London. (*C. Herz, Paris.*) Telephonic Apparatus and Conductors. 6d. (10 figs.) January 8.

DYNAMO-ELECTRIC GENERATOR. — A dynamo-electric or magneto-electric generator is employed to supply the current to the instrument.

129. — J. H. Johnson, London. (*C. Faure, Paris*). Galvanic Polarisation Batteries or Magazines of Electricity. 6d. (6 figs.) January 11.

SECONDARY BATTERIES. — The elements of the batteries are prepared by mechanically coating, electro-plating, or depositing upon them a layer of spongy or porous lead of suitable thickness. The plates, which are by preference of lead, are coated with metallic salts or oxides mixed with inert substances, as for example sulphate of lead mixed with pulverised coke. When the two elements thus prepared are immersed in dilute sulphuric acid and connected to a source of electricity, the salt on one of the plates will be reduced and form an adhesive porous coating. On the current being reversed the other plate will be similarly affected, and the first plate will be reoxidised or peroxidised, and these two will constitute a galvanic couple. Fig. 1 is a battery constructed with flat plates, Fig. 2 a battery with circular plates, and

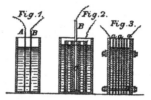

Fig. 3 a battery of flat plates forming cells. A and B, Fig. 1, are thin lead plates; B, Fig. 2, is a central rod of lead or carbon enclosed in a cylinder C of porous materials; the annular space between C and the central element is filled in the first instance with sulphate of lead and coke. In Fig. 3, $a\ a\ a$ are plates of lead, arranged parallel to each other and secured by strips of india-rubber. The specification describes the application of the invention to driving tramcars and boats. A regulating switch is employed for putting in, or removing from, the circuit, sections of the battery.

153.—A. Muirhead and J. Hopkinson, Westminster. Electric Lamps. 6d. (5 figs.) January 12.

ARC LAMP.—Figs. 1 and 2 are elevation and plan of the lamp. Fig. 3 shows a modified form of brake gear. The upper carbon C is carried by the holder A, upon which is cut a screw-thread B, engaging a similar thread in the boss of the wheel F. D is a key-way or "screw of infinite pitch," which keeps the holder from revolving. In operation the arc is formed by an electro-magnet acting on either the upper or lower carbon-holder, and as the upper carbon falls, the wheel F is caused to revolve by the screw A, and rotates the pinion G and the brake-wheel H. The brake-wheel is controlled by an electro-magnet on a shunt circuit acting against a spring, which, when the resistance of the arc increases, is overpowered and drawn away from the wheel. As shown in Fig. 3, the toothed wheels may be replaced by a brake disc provided with paddles working in fluids, the amount of immersion

being controlled by an electro-magnet placed on a shunt circuit.

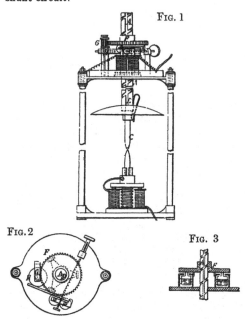

FIG. 1

FIG. 2   FIG. 3

INCANDESCENCE LAMP. — The carbon filaments are prepared from silk, horn-beam, gumostic fibre, Manilla paper, bone, cartilage, or assein, freed from earthy matter, by embedding strips of the material in lampblack in a porcelain crucible, through which the vapour of carbon bisulphide is passed. The crucible is then closed and heated to redness for one hour. The looped ends of the filaments are secured to metallic conductors by clips, and are then cemented into long closely fitting glass tubes which are passed through holes in the globe accurately ground to receive them. The globes are filled with vapour of carbon sulphide and hermetically sealed. The filaments may be platinised by dipping them in a solution of ammonio chloride of platinum and then heating them to redness.

200.—J. Imray, London. (*J. E. Cabanellas, Paris*). Transforming, Conveying, and Applying Power by Means of Electricity, &c. 1s. (32 figs.) January 15.

DISTRIBUTING CURRENTS.—This invention relates to the means of distributing the current over a single conducting system in such a manner that the varying local consumption may be automatically regulated at the central station. This is attained by a system of "electric cocks" and automatic regulators. The "electric cock" is a combination of an electric motor and a generator, having their movable parts fixed on the same spindle. The motor is driven by the main current, the generator supplying the current to the local circuit. The regulation of this generator is effected by a centrifugal governor working a lever which short-circuits it when the velocity becomes too great. The "electric cocks" are placed at intervals in the main circuit forming the connections between it and the various local circuits. The main generator is caused to send out the exact amount of current required by the action of an automatic "sentinel." If the elements of the main generator be "hydro-electric, thermo-electric, or of such kinds," they are arranged in series, their poles being connected separately to insulated springs arranged in a circle, over which passes a "brush" connected to the return wire, the motion of the "brush" in either direction putting in circuit more or less of the elements. The "sentinel," or relay, is actuated by variations in the main current, on an increase in which it puts in circuit the "sentinel battery" so as to turn an electric motor in one direction, and on a diminution in the other; this motor turns the "brush," until the main current is of the requisite strength, when the "sentinel" breaks contact. In a modification a differentially wound motor is employed. "If the elements of the producer be of dynamo-electric kind . . . . the corrections intermediate between what would result from the suppression or addition of one element is effected by a variation of speed." In this case the electric motor worked by the "sentinel battery," besides turning the correcting brush, also rotates a centrifugal governor which regulates the speed of the engine. A more exact regulation is obtained by winding the field magnets of one of the "elements" with separate coils connected to a commutator similar to that employed for cutting out the "elements" whose brush moves in the same manner. This serves to produce fractional parts of the requisite current. When the electro-magnets of the generators are separately excited, an "electro-magnetic gauge" is used, which opens the exciting circuit when the predetermined intensity is exceeded. Water-wheels, turbines, windmills, &c., are used to generate the main current, the hollow axles of the wheels carrying the moving parts and a stationary spindle the fixed parts of a generator. When annular magnetic inductors, of ring form, and actuating a single armature of the Gramme type, are used in the "electric cocks," both the inducing and induced parts being respectively connected to their proper commutators, the two sets of brushes, to one of which the exciting current conductors are connected, are caused to revolve together round their respective commutators. Various combinations of the inducing and induced coils are described. The main current may be transformed into alternating currents by suppressing the induced collector and sub-

stituting for the continuous coil of the ring a series of independent coils, each of a length equal to that between the points of the inducing coils corresponding to two consecutive positions of the inducing brushes. The inventor states "that with the concentric or side-by-side arrangements described for the inducing and induced coils, and by the use of iron wire in bundles, the cock can be made for general use in the form of a straight tube that can be sold by the yard." It is further stated that "the rotation of the brushes only causes a very limited consumption of local energy for the passive expenditure of the friction of the brushes in the medium and on the collectors," their speed being controlled by clockwork, or by a small electric motor, the speed of which is in turn regulated by an "electric ratchet." This "consists in the combination of two distinct induced circuits operating successively, and causing each in its turn the movement of the ratchet one division." The description of the "electric ratchet" is too vague to permit of the method of construction being understood. Mathematical formulæ are given to obtain "the radius of the circle round which the rectilineal commutator should be wound," and also for the construction of the commutator. The ratchets may be worked by continuous or alternating currents, and their direction of rotation reversed by bringing into action a supplementary set of brushes.

218.—J. E. H. Gordon, Dorking. Apparatus for Producing Electric Light. 6d. (7 figs.) January 17.

INCANDESCENCE LAMPS.—This relates to modifications of a former Patent, No. 1,826 of 1880. The

light is produced by the discharge of high tension alternating currents between balls of refractory metal. A represents the two balls at the end of thin stems held or guided by the glass cross-piece and spiral C. The globe contains hydrogen or nitrogen at a pressure of one-third to one-half an atmosphere.

225.—St. G. L. Fox, London. Electric Lamps, &c. 6d. (8 figs.) January 18.

INCANDESCENCE LAMPS.—This refers to Patents Nos. 3,988, 4,043, and 4,626 of 1878, and No. 3,494

of 1880. According to this invention the filaments of incandescence lamps are formed of vegetable threads wound round a block of refractory material of a pear-shaped section provided with a steel cutting edge at the pointed end. The whole is heated in a crucible when the threads break over the cutting edge. The baked threads are removed from the block and placed in suitable numbers in clips and immersed in an atmosphere of coal gas, while an electric current is passed through them so as to raise them to a white heat. The resistance of the partially carbonised threads is overcome by the employment of secondary currents produced by momentarily short-circuiting a dynamo machine, until the filaments attain a condition when a moderate current will traverse them. To thicken the ends the main part of the filament is short-circuited by a copper wire and the current is continued until a sufficient amount of carbon has been deposited upon them from the surrounding gas. Contact between the ends of the filament and the platinum holders is made by smearing them with Indian or Chinese ink. To extinguish street lamps, without affecting the lamps in adjoining houses fed from the same main, there are connected in series by a line wire a number of electro-magnets, one to each street lamp. Vibrating between the poles of these electro-magnets are permanent magnets connected to the main. These permanent magnets, according to the direction of the current sent through the line wire, make contact with one of two stops, thereby switching in or out the street lamp. In a modification a high resistance electro-magnet, through which a feeble current always passes, has a spring armature balanced to the normal electromotive force of the main. The armature on a momentary increase or decrease closes the circuits of other electro-magnets, one of which turns the lamp circuit on, and the other turns it off.

229.—H. H. Lake, London. (*H. A. Clark, Boston, U.S.A.*) Restoring Waste Vulcanised India-Rubber or Gutta-Percha, &c. 6d. (12 figs.) January 18.

CONDUCTORS.—The waste rubber is ground into small particles and boiled with a little water until the sulphur is nearly removed. Two to ten per cent. of palm oil is then added, and the whole mixed and heated, after which the mass is subjected to the vapours of turpentine or camphine until in a plastic condition fit for working. Before being made up two to ten per cent. of resinous matter is added to the desulphurised material. When used for coating electrical conductors it is wound on the outside with a string of tinfoil.

245.—C. L. Clarke and J. Leigh, Manchester. Apparatus for Lighting Gas. 4d. January 20.

INDUCTION COILS.—To have both terminals of the wire at the outer ends of the coil the core is divided into two halves by a washer of ebonite. One end of a wire is placed through a hole in the washer near the core, and one half the coil wound with this wire to the required diameter; one end of a second wire is then soldered to the end protruding through the washer, and with this the second half of the coil is wound.

253.—C. G. Gumpel, London. Apparatus for Producing Electric Currents, &c. 10d. (31 figs.) January 20.

DYNAMO-ELECTRIC GENERATOR.—The inventor's object is to "concentrate the polarity" of the field magnets "as much as possible into a narrow field." To this end an even number, preferably not less than four, of electro or permanent magnets having alternate polarity are distributed about the circumference of the armature. The pole-pieces of these magnets are formed so as to converge towards each other. The rotating armature, in shape, approaches a hollow open-ended barrel, being larger in diameter at its centre than at its ends. The core may be constructed of an iron barrel, slotted to impede longitudinal induction and supported at each end by non-magnetic bosses fixed on the shaft; or it may be constructed of a number of insulated iron rings supported on a wooden centre and clamped together by two end bosses; or it may consist of iron wire wound on a wooden core. In all cases the surface is divided by longitudinal ribs in the spaces between which the insulated copper wire coils are wound. Three methods of winding the coils are described. The commutator consists of a cylindrical core having mounted on it, side by side, a number of insulated metallic rings, each connected with one or more coils of the armature. On the outer peripheries of these rings, and parallel with the axis of the armature, are fixed the insulated commutator slips. The brushes, of "chain" form, are constructed of short links of metal, the edges of the links overlapping the commutator to the desired arc. The ends of the "chains" are jointed to two arms hinged together and pressed towards the commutator by an adjustable spring. The linked chain may be replaced by a number of flexible metal strips pressed on the commutator by a bow spring. When brushes of the ordinary type are used the commutator rings are fitted with segments, each extending over the desired arc, the circumferential spaces being filled in with insulating material. When the commutator has to serve for either continuous or alternating currents, the rings are made broader than the projecting segments, the brushes being arranged to bear either on segments or continuously on the lower surfaces presented by the rings.

ARC LAMP.—The upper carbon is held in a socket tube having on its side a toothed rack gearing with a pinion, on the axis of which is fixed a ratchet wheel. The ratchet wheel rests on a shoe hinged to the end of the armature of an electro-magnet having its coil in the lamp circuit. A spring pawl rack gears with the ratchet wheel, which, along with the pinion, is mounted in a movable frame having its descent limited by an adjustable stop. On the admission of current the magnet lifts its armature and with it the shoe, wheels, and frame to an extent determined by a stop. On the arc resistance becoming too great the armature falls, and the ratchet wheel, held by the spring pawl rack, turns and feeds down the carbon. In a modification a spring friction pawl operated by an electro-magnet of high resistance placed in a branch circuit, is substituted for the pawl rack. In another arrangement a solenoid having its coil in a by-pass circuit attracts downwards its core when the arc resistance is too great, thereby releasing the shoe from the ratchet wheel. The wheels are supported on an arm hinged to the main frame instead of being carried in the small frame. In a further modification a spring friction pawl is worked through a lever by the solenoid which operates the shoe.

TRANSMISSION OF POWER.—A continuous current generator supplies current to the armature of an electric motor having its axis geared with, say, the rudder arm of a ship. A commutator is arranged so as to change the direction of the current through the motor armature, by which the rudder is put to port or starboard as required. The commutator is operated by moving a steering handle and so arranged as to make its full stroke even when the handle is moved only a short distance. The steering handle segment carries a spring contact, travelling over a number of insulated plates, each connected to one of a number of contact points traversed by an arm fixed to the rudder shaft. This arm and also the segment of the handle forms a by-pass to the main circuit, and in this is placed an electro-magnet, the spring armature of which is linked to a sliding contact spring sliding on two insulated contact plates having interposed between them two or more plates, each connected to one of the contact plates through a resistance. By these means the rudder is caused to take a position corresponding to that of the steering handle.

264.—A. Apps, London. Apparatus for Producing Light by Means of Electric Currents, &c. 6d. (5 figs.) January 21.

CURRENT METER.—To obtain a correct measurement of the current passing through an induction coil or other apparatus, a dynamometer is employed having a spring, similar to a Salter's balance, which indicates the force exercised by the magnetic core of the apparatus upon a second core or armature. The invention further relates to a multitubular form of magnet and conductor contained within the inductor, "securing better insulation than heretofore, whether solid or fluid, both of the magnet proper and its surroundings and internally contained currents or charges which is also applied vertically or at right angles to the first order of parts when the currents pass in annular planes alternately with dividing facilities, and in multiple current in secondary, tertiary, or further formations, whether the effect be continuous, or vibratory, or intermittent."

INCANDESCENCE LAMP.—The currents are separated on the parts of the conductors in or about the induced field of the multitubular magnet, and the discharges utilised to produce light in rods or filaments of carbon.

ARC LAMP.—This is a fixed upright rod, with a second rod descending by gravity and suitably guided in its descent. The descending rod makes, by preference, a fluid contact in an iron tube. The lamp is supported on a glass base and contained within an air-tight glass globe.

288. — J. Richardson, Lincoln. Controlling and Regulating the Speed of Engines Employed for Driving Dynamo-Electric Machines. 8d. (7 figs.) January 22.

CURRENT REGULATOR. — The governor of the motor engine is driven by a small dynamo motor C impelled by the current from the main electric

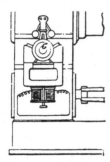

generator, and running faster or slower as the current rises or falls below the normal amount. If the current fails altogether a weight, usually sustained by an electro-magnet E, falls and closes the throttle valve. The current may be made to actuate the throttle valve by an electro-magnet, placed in the main or a shunt circuit, and having its armature connected to the valve spindle.

304.—R. Werdermann, London. Manufacture of Carbon and Graphite, &c. 6d. (1 fig.) January 22.

CARBONS.—These are made from sugar, raw cotton, wood pulp, or similar materials, either alone or mixed with essences, tar, &c. After carbonisation the product is mixed with a solution of pure sugar, or tar, or oil, and moulded to the required shape. When dry the rods are placed in a vessel containing a solution of sugar, tar, oil, or other carbonaceous liquid and the air is exhausted, after which the rods are again dried and are raised to a red heat by a current of electricity.

455. — F. Barr, New York, U.S.A. Signal Buoys. 6d. (4 figs.) February 3.

DYNAMO-ELECTRIC GENERATOR.—A "dynamo-electric engine" supplies the current for an "electric lamp or lantern" placed upon the top of the buoy.

473.—E. P. Alexander, London. (*J. C. White and H. H. Hayden, New York, U.S.A.*) Conveying Money, Documents, &c., in Mercantile Establishments, &c. 8d. (21 figs.) February 4.

MOTOR.—The articles are placed in small carriers which "are rendered self moving under the action of any suitable motor, mechanical or electric."

497.—H. Wilde, Manchester. Electro - Magnetic Induction Machines. 6d. (2 figs.) February 5.

DYNAMO-ELECTRIC GENERATORS.—*a a* are wheels or discs made of brass or other non-magnetic metal, and slit in a radial direction; *b b* are the armatures fitted with pole-pieces *b¹ b¹*, the slits in which coincide with the slits in the wheels. The pole-pieces are fitted in recesses in

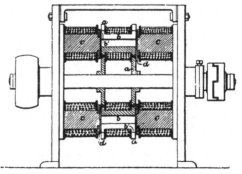

the brass wheels *a*; *c c* are the electro-magnets with pole-pieces *d* slit radially and about one-half larger in diameter than the cores, while the pole-pieces *b¹ b¹* are about one-sixth larger in diameter

than the armature cores. The field magnets may be separately excited or the current of one or more coils of the armature may be used for exciting them. In order to prevent the destructive spark on the commutator when the circuit is interrupted, an electro-magnet, excited by the main current, is employed to short-circuit the currents through the machine till the external circuit is re-established. The commutator is of ordinary construction.

539.—E. G. Brewer, London. (*T. A. Edison, New Jersey, U.S.A.*) **Electric Lamps.** 8d. (16 figs.) February 8.

INCANDESCENCE LAMP.—In incandescence lamps in which the filament is straight, and is not a bow or curl, as has been hitherto the case, the filament is held extended upwards by a vertical glass arm carrying a conductor. In order to prevent heat from the incandescent material being carried down to the point where the conducting wires are fused into the glass, a long interval is interposed between the clamps of the carbon and globe, and the conducting wires are supported by a glass pillar. Lamps to be suspended may be cheaply constructed of a filament, a sealed glass sphere, and protruding conductors. Figs. 1 and 2 show an arrangement whereby the intensity of a lamp may be modulated at will. The base B contains the resistance ring of heavy carbon, shown in Fig. 2. One main conductor is

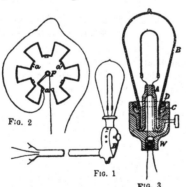

Fig. 2

Fig. 1

Fig. 3

attached to the stud in the centre, and the current flows through the metal arm F to any of the contact pieces *a a*, to which the arm may be applied by means of its external knob. In the illustration the arm rests on the terminal of the second leading wire, and the carbon resistance is out of circuit. Several forms of glass supports to prevent the bending over of the fibre are illustrated, and the form of the glass stem through which the wires pass is modified to withstand the air pressure better. Fig. 3 illustrates a new method of sealing the lamp; W is a wooden base, A a glass socket fitted therein, of a form resembling a **W**

in cross section. The globe B is fitted into the socket A by grinding, and the joint is further secured by the mercury C, and rubber ring D. The small screw at the side of the wooden base serves to switch in or out the lamp. The conducting wires are sealed at their entrance into the globe, and also again at their exit from the chamber formed under their place of entrance.

562.—P. Jensen, London. (*T. A. Edison, New Jersey, U.S.A.*) **Manufacture of Carbon Burners.** 8d. (12 figs.) February 9.

INCANDESCENCE LAMPS.—Flexible carbon sheets or filaments for incandescence lamps are made by precipitating carbon from a gas or vapour on to sheets of nickel or cobalt heated to a high temperature. The metal is removed from the carbon so precipitated by being dissolved in acid. Blocks or

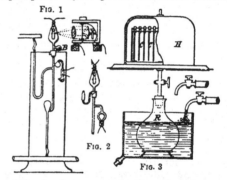

Fig. 1

Fig. 2

Fig. 3

crucibles of carbon are prepared by carbonising paper or pulp between metal dies. The hydrogen that is not removed by the heat is taken up by passing chlorine through the retort. After the filament is placed in the globe the latter is exhausted to carry off the occluded gases while the current is passing, but as the heat does not extend to the clamps their temperature is raised by rays projected by a mirror and lens (see Fig. 1) from an arc light. After the lamp has been sealed off from the pump a small amount of gas is left which is absorbed by cocoa-nut charcoal contained in the globe B, which is heated to a high temperature, and left upon the stem for four-and-twenty hours. In Fig. 2 the charcoal containing bulb is heated by a current of electricity. Fig. 3 shows a method of filling up minute cracks in filaments by the deposition of carbon upon or in them from hydro-carbon vapour. C C are the filaments, R a reservoir of hydro-carbon in a tank of hot water, H a flask heated in any convenient manner. The clamping ends of the filaments are electro-deposited. The filaments are preferably made of fibrous vegetable substances, the conducting powers of their enlarged ends being increased by placing them in a vessel through which is passed a hydro-carbon vapour.

639.—W. R. Lake, London. (*H. S. Maxim, Brooklyn, U.S.A.*) Preparing Carbon and other Conductors for Electric Lighting. 8d. (1 fig.) February 15.

INCANDESCENCE LAMP. —In practice the unfinished filament, made in any ordinary manner, is treated in an exhausted receiver provided with proper electrical connections; a current of electricity is passed through it to expel the occluded gases, and these are pumped from the receiver, after which pure hydro-carbon or other carbonaceous gas is admitted, of a pressure about equal to 1 in. of mercury. The current is then increased, carrying the temperature up to a high point; the heat disassociates the elements of the gas and the carbon set free is deposited in a state of fine subdivision in the pores and upon the exterior of the carbon, gradually reducing its resistance and changing its illuminating capacity. The process is used for manufacturing lamps of a standard illuminating power, the process being carried on until the light emitted exactly balances the light of a lamp whose terminals are maintained at the same potential.

696.—C. W. Siemens, London, and **A. C. Boothby,** Kirkcaldy, Fife. Apparatus for Working the Brakes of Railway Trains. 6d. (5 figs.) February 17.

MOTORS.—The electric currents from a dynamo or other generator placed on the engine are conveyed by wires along the train to electric motors attached to each carriage which operate the brakes.

*715.—J. G. Tongue, London. (*A. Lacomme, Paris*). Electric Lamps, &c. 2d. February 18.

ARC LAMP.—The object is to bring a succession of carbons into the circuit as they are burnt out. The carbon rods are arranged around the periphery of a wheel, but the method of operating is not clearly described.

768.—E. G. Brewer, London. (*T. A. Edison, New Jersey, U.S.A.*) Connecting the Carbon Ends to Conducting Wires in Electric Lamps. 6d. (1 fig.) February 23.

INCANDESCENCE LAMP.—This relates to the connection of the wires and the carbon filament contained in the hermetically sealed glass globe of an incandescence lamp, the object being to dispense with the delicate clamps hitherto used. The method of uniting consists in first attaching the carbon filament to the wire by mechanical means, and then completing the attachment by plating, the deposited metal securing good electrical connection between the wire and the carbon. The

wire can be mechanically attached by flattening and splitting the end, and inserting the carbon in the forked opening, the sides of which are pressed down on the carbon, or the wire may simply have a turn or two round the carbon, or be bound to it by fine wire. The illustration shows an electro-plating cell, which is also claimed as part of the invention, having a rubber stopper F to close the

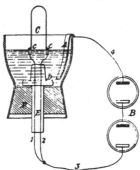

bottom. E is a supporting tube to receive the conducting wires 1 and 2, which are sealed therein, the tube having an enlargement D for sealing into the neck of the enclosing globe of the incandescence lamp. At the upper ends the wires 1 and 2 are secured to the filament C; the tube E is passed through an aperture in F until the point of union is just covered by the plating solution in G, connections 3 and 4 are then made with a battery B, the plate A forming the anode, and the ends c c the cathode of the depositing cells. The parts of the wire not required to be plated are coated with varnish or wax.

*774.—J. Fyfe, London. Electric Lamps. 2d. February 23.

ARC LAMPS.—This relates more particularly to the Krizik and Piette ("Pilsen") type of lamp, and consists in placing the regulating solenoids beneath the lamp, a "thin non-conductor" leading up to the crosshead of the lower carbon being attached to it at its upper end, and to the core piece at its lower end. The upper and lower carbon-holders are suitably geared together.

783.—J. Perry and **W. E. Ayrton,** London. Electrical Conductors. 8d. (23 figs.) February 24.

CONDUCTORS.—This is designed to supply a means of insulating the conductor that supplies the current to electrically driven trains. To this end the train does not make direct contact with the main conductor, but with short successive lengths of a secondary conductor that is put in communication with the main conductor as the train passes over it. The main conductor is in-

sulated in any well-known manner, as by india-
rubber coverings, or is carried upon posts and
insulators, and is connected to the secondary conductors by branch wires. In the drawing, A B is a
copper rod in short lengths fastened to corrugated
steel discs D D upon an ebonite ring E E. The
ring is secured to a cast-iron box carried on the
sleeper. The rod and disc have sufficient elasticity
for two or more of the latter to be simultaneously
depressed by an insulating collecting brush or
roller carried on one or all the carriages of the
train, thereby bringing the stud E into contact
with the stud G, which is electrically connected
by the severed wire H to the main conductor. By

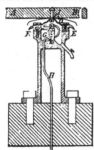

this arrangement leakage from wet and dirt is confined to the length of rod depressed by the train.
Several modifications of this part of the invention
are described and illustrated, in all cases the depression of a flexible disc being employed to complete
the main circuit. A somewhat similar arrangement may be employed to give the signalmen
graphic indications of the progress of a train driven
by steam or electricity. A main conductor, in
which flows a continuous current, is put to earth
by the contacts such as E and G, its resistance
constantly diminishing or increasing as the train
progresses. In the cabin is a galvanoscope which
points out on a diagram a position corresponding
to the resistance in circuit and consequently to the
distance of the train from the cabin. Further, the
specification, which is very long, describes an absolute block system in accordance with which a
train on entering a blocked section of the line is,
independent of any action on the part of the driver
or guard, deprived of all motive power and powerfully braked.

785. — **W. E. Ayrton,** London. **Covering
Wire for Electrical Purposes.** 4d. February 24.

CONDUCTORS.—The conductor is formed of a
wire warp interwoven with a non-conducting weft,
or a non-conducting weft woven with a wire warp.
The web is applicable, among other things, to
resistance coils, galvanometers, electro-magnets,
condensers, and artificial submarine cables.

792. — **P. Jensen,** London. (*T. A. Edison,
New Jersey, U.S.A.*) **Electric Lamps, &c.**
6d. (1 fig.) February 24.

INCANDESCENCE LAMPS. — This consists in so
arranging incandescence lamps in various combinations of series and multiple arc that the total
resistance on any one circuit is constant, whether
it contains one, two, or more lamps. This is effected
by increasing the density and conducting power of
the filaments, preferably made of graphite carbon,
correspondingly with increasing their radiating
surface.

803.—**R. Waller,** Leeds. **Dynamic Apparatus
and Motors, &c.** 10d. (25 figs.) February 25.

DYNAMO-ELECTRIC GENERATORS. — The first arrangement of "dynamic electro machine" has
field magnets "composed of two hollow and one
solid magnet on each side of the armature;" and
these are wound so that "there would only be the
poles uncovered." The armature core is hollow,
and is wound in six sections. A modification has
two or more armatures geared together. In a
second modification the electro-magnets radiate
inwardly from a circular frame of wood, the armature being mounted on a shaft, the axis of which
is placed eccentrically to the common centre of
the field magnets. In other modifications, electro-
magnets are arranged to approach to and recede
from each other by means of levers.

CARBONS.—These are made of zig-zag or spiral
form.

859.—**J. W. Fletcher,** Stockport, Cheshire.
**Apparatus for Erecting and Repairing
Overhead Telegraph Wires.** 6d. (2 figs.)
March 1.

DRAW VICES.—This applies to what are called
draw vices, and consists in making the winding
pulley in halves, by which means, when the pulley
is filled, the wire can be removed by opening

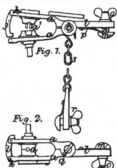

the hinged frame *a a*[1]: in combining the pulley
with a gripping device which is formed by prolonging the arm *a* to form one jaw *l* of a vice, *m*

being the other jaw made with a stem n passing through l; o is a thumb nut, and p a spring for causing m to move away from l when nut o is unscrewed : in a locking latch or bolt j pivotted at k to arm a for keeping the two arms a a^1 in the position shown in Fig. 2 : and in forming an eye q attached to joint b and a draw vice r connected to the eye by the chain s.

879.—A. Shippey, Bayswater. Electric Light Signalling. 4d. March 1.

ELECTRIC LIGHT.—This relates to means of electric light signalling for military purposes, or for saving life at sea, and consists essentially in employing a balloon or kite, to which is suspended an electric lamp connected by conducting wires, the return wire being attached to a float to be thrown into the sea.

894.—J. J. Sachs, Sunbury, Middlesex. Electric Lamps. 6d. (3 figs.) March 2.

SEMI-INCANDESCENCE LAMPS.—The two carbons are placed in tubes b b, and rest upon a plate c of

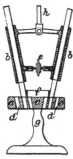

asbestos. The plate c is perforated with a series of holes d of smaller diameter than the carbons, which are adjusted in position over the holes by means of the screw e; a strip of graphite f serves to connect the carbons : g is a stand for supporting the plate c, the tubes b being suspended by h. In another lamp a number of carbons are arranged round a plate in a similar manner. The perforations in the plate allow the emission of the light whilst preventing the carbons from passing through.

913.—W. T. Glover and G. F. James, Manchester. Machinery for Twisting and Lapping Wire or other Materials in the Manufacture of Ropes and Cables. 6d. (6 figs.) March 3.

CONDUCTORS.—The wire is wound on a bobbin having on one of its ends a series of pins, and mounted on a horizontal stud, carried by an arm on a hollow vertical spindle. An escapement is also mounted on a horizontal stud carried by the

vertical spindle, and is provided with a catch designed to come in contact with each of the pins upon the bobbin in turn. This escapement also carries a roller over which the wire passes on its way to a slot in the hollow spindle, in which it runs under a pulley attached to a weighted drop rod. So long as the drop rod is supported by the wire, the roller keeps the escapement in gear with the pins, and prevents the bobbin from revolving more than one tooth at a time. Should the wire break or run out, the rod falls and stops the machine.

922.—J. Elmore, London. Galvanising Iron. 4d. March 3.

DYNAMO-ELECTRIC GENERATOR.—The iron after pickling and cleansing is plunged into a solution of salts of zinc and connected by a copper rod with the cathode pole of a powerful dynamo-electric generator. In the same solution are sheets of zinc which are connected to the anode pole of the dynamo-electric machine.

1,016.—E. G. Brewer, London. (*T. A. Edison, New Jersey, U.S.A.*) Weber Meters. 6d. (7 figs.) March 9.

CURRENT METERS. — The invention relates to the class of apparatus described in Specification No. 4,391 of 1880. In one arrangement a definite proportion of the current to be measured is shunted through an open-bottomed vessel partially filled with water and floating upright in the same liquid. The water in the vessel is decomposed by the current, and the gases formed displace a portion of the water in the vessel, lessening its specific gravity and causing it to rise. The rising of the vessel actuates a lever connected with a registering mechanism, so that the amount of gases produced being proportional to the current through the vessel, and this in turn being proportional to the main current, the latter may be indicated by the registers. An arrangement is provided whereby, when a definite amount of decomposition has taken place, the vessel will rise to a sufficient height to close an electrical circuit, which will heat a platinum wire and explode the gases, so that they will be recomposed into water and the vessel will fall. These operations are continuously repeated so long as the current passes. According to another form of the invention an electro-depositing cell has a metal lining which forms one of the electrodes thereof, connected in any suitable manner with one of the leading wires. The other electrode is a piece of metal suspended within the cell by a spring, the stress of the spring being so adjusted that the metal plate will be sustained at the top of the cell when there is a minimum of deposit thereon, while the maximum of deposit will cause it to sink to the bottom.

An index and a scale are attached to the springs, and the electrodes are also connected with a registering apparatus and circuit reversers, which change the direction of the current through the cell, so that the metal is deposited on and dissolved from each of the electrodes alternately. A third form of measuring apparatus consists of a motor to which is given a certain definite amount of resistance to overcome, combined with registering devices; the rapidity of its motion is proportional to the current. The field and armature coils may both be in the main circuit, or one or both may be in a shunt circuit. The resistance may be fan blades or paddles working in liquid.

*1,027.—J. A. Berly, London. **Electric Lamps.** 4d. March 10.

CANDLES.—This relates to a carbon formed of two separate "candles," and to means of establishing the arc. An automatic switch for changing the candles, and a switch operated by compressed air, are also included in the specification, but in the absence of drawings the arrangements cannot be well understood.

1,040.—A. A. Common and **H. F. Joel,** London. **Electric Lamps.** 6d. (17 figs.) March 10.

ARC LAMP.—This relates to a device for establishing and controlling the arc and other devices for feeding and regulating the length of arc. Fig. 1 is a sectional elevation and Fig. 2 a plan of one form of lamp somewhat similar to that described in Specification No. 368 of 1870. C is the core of a differential solenoid M, taking hold of a lever L, carried by two arms W on the shaft P. The lever has a toe which, when the lever is raised, bites against the wheel D, and acts as a pawl to turn the said wheel. E is the upper carbon and E¹ the lower. When the current is sent through the thick coil of the solenoid and through the carbons, the core C is drawn upwards, lifting the lever L and turning the disc D, and with it the pinion F. The motion of the pinion raises the rack R and the upper carbon, establishing the arc. When the current begins to fall the shunt coil forces down the core and lever L, thereby releasing the wheel D, and allowing the upper carbon-holder to slide down as fast as the fluid in the cylinder P¹ can escape through a fine hole drilled through the piston P. When the arc is at its normal length, the lever is again raised, gripping and retaining the wheel D. Fig. 3 is a modification in which two coils mm are combined with one double core C, the two sets of discs D, the gripping levers L being as before. In this arrangement a forward movement is given to one of the levers, so that it never releases its carbon until the other pair of

carbons is consumed. Fig. 4 is a sectional view of a similar lamp, but with two clamps H H¹ gripping a rod E, the adjustment being made by screws S S¹. Fig. 5 shows an additional device for effecting a more delicate feed than that of the lamp in Fig. 1. Here an adjustable spring K is fitted to

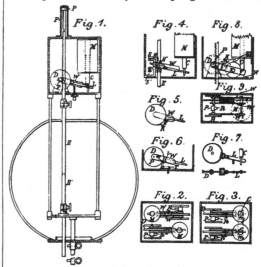

Fig. 1.   Fig. 4.   Fig. 8.   Fig. 5.   Fig. 9.   Fig. 6.   Fig. 7.   Fig. 2.   Fig. 3.

the lever L, so as to control the revolution of the disc when the lever falls. Fig. 6 is a modification of Fig. 1, in which the tripping of the bell-crank lever T is so arranged that the arm W rests by means of a pin w upon the top of the vertical arm T; any additional fall of L causes the projection at its end, under the pivot l, to press against the vertical arm T and push it away, thus releasing W and momentarily restoring the gripping lever L to its former position. This allows the feed to take place with a very small movement of the core, and yet provides for the full action at starting. Fig. 7 shows another form of the gripping lever L. Figs. 8 and 9 give sectional elevation and a plan of a lamp with wheel gearing to regulate the rise and fall of the rack. On the axle of P is hinged a frame W, W¹, W², the other end of which is attached to the core C. The upward movement of the core lifts the frame and gear, and first slightly raises the rack before the train revolves. A further upward movement brings the rim of the end wheel D² into contact with the adjustable spring K and brakes the whole train, the rack R being then lifted as in the previous lamps. The reverse action takes place upon the core falling, and the movement of the carbon is thus controlled. The inventors propose using this as a controlling lamp for regulating any desired number of simple and inexpensive lamps; thus when the frame W is at its lowest point it can be caused to make contact momentarily and complete the circuit of a wire connected to the electro-magnets of any number of

lamps, in such a manner that at the moment when an electric current is sent into the coils of the electro-magnets, it is made to release all the top carbons, allowing them to drop, and on the cessation of the current, to be again raised.

SEMI-INCANDESCENCE LAMPS.—The invention further relates to semi-incandescence lamps of the Werdermann type, such as were described in Specification No. 5,157 of 1879, and consists principally in means for maintaining the carbon in contact with the metal pole. This is accomplished by using an electro-magnet of high resistance, connected as a shunt across the burning carbon, the armature of which serves to separate the jaws through which the carbon passes. In a modification, the body of the lamp consists of a split tube, one half fixed to the nipple and block at its top, the other half being hinged and in contact with a side arm carrying the lower electrode. Each half of the tube is a distinct conductor. The hinged half is secured by means of a button and catch. Before the tube can be opened this catch piece must be turned and brought into contact with the other half, thus short-circuiting the lamp. A chandelier arrangement is described and illustrated in which the suspension cords form the conductors.

1,097.—J. H. Johnson, London. (*La Société Anonyme la Force et la Lumière, Société Générale d'Electricité, Paris.*) Voltaic Batteries, &c. 6d. (11 figs.) March 14.

SECONDARY BATTERIES.—The metallic cells are formed as hollow truncated cones and are packed one within the other with their smaller ends downwards. Metallic contact is prevented by non-conducting packing pieces, and, when two liquids are employed, by felt envelopes.

1,107.—W. R. Lake, London. (*B. Barda, Vienna*). Electrical Bath Apparatus. 6d. (4 figs.) March 14.

GENERATOR.—An "electric machine" is used to supply the currents to plates placed in various parts of the bath.

1,119.—S. Pitt, Sutton, Surrey. (*O. Lugo, New York, U.S.A.*) Telegraphy. 10d. (6 figs.) March 15.

CONDUCTORS.—To obviate the effects of induction the return wire is wound helically round the other conductor, and is of such a section that its total resistance is equal to that of the line wire. For mechanical reasons the return conductor is composed of several wires.

*1,232.—H. E. M. D. C. Upton, Newcastle-on-Tyne. Electric Lamps. 2d. March 21.

INCANDESCENCE LAMP.—The inventor "rests the end of a vertical carbon upon two studs or rollers, which are the terminals of the conductors to and from the lamp."

1,235.—G. A. Tambourin, Marseilles, France. Apparatus for Electric Lighting. 6d. (2 figs.) March 21.

ELECTRIC LIGHT.—This relates to the separate and direct production of electric lights at different and distant points by means of compressed air or other fluids, which are compressed at a central station, and conducted through conduits at a proper pressure to the interiors of the lamp-posts,

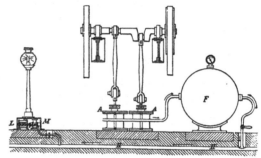

where these fluids work small motors of any suitable construction attached to dynamo-electric generators. A A are the pumps, F is the receiver, H the pressure pipe, M the motor, and L the dynamo-electric generator situated in the base of the lamp-post.

1,236.—J. A. Berly, London. Electric Lamps, &c. 8d. (21 figs.) March 21.

ARC LAMP.—The figures illustrate, in elevation and plan, the principle of the invention. *a* is a fixed carbon-holder, and *b* a movable holder oscillating on a spring hinge, and connected to the armature *d* of the high resistance electro-magnet *m*, which is situated on a shunt circuit. When the

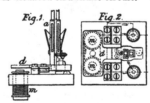

arc is not established, the current passes round the magnet, which attracts its armature and moves the one carbon into contact with the other, thus opening a direct path for the greater part of the current, which practically leaves the magnet, allowing the carbon to resume its vertical attitude.

CANDLES.—The specification illustrates the application of the above principle to a similar

lamp for burning several sets of candles, also a Jamin and Wilde lamp each controlled in the same way. Further, it describes a method of automatically switching a fresh candle into circuit in lamps of the Jablochkoff-Gadot description by connecting one pole of each of the candle holders in parallel arc to the leading wire, and fixing a return wire to each remaining carbon in such a way, that when a candle is nearly consumed the wire melts off it, and the current is free to choose a fresh path through one of the unlighted candles. As, however, this result is not always attained, in consequence of the current dividing, an alternate method is provided comprising a rotating armature actuated at intervals by air, water, or electricity from the central station.

1,240.—E. G. Brewer, London. (*T. A. Edison, New Jersey, U.S.A.*) **Armatures for Electric Machines.** 6d. (3 figs.) March 21.

DYNAMO-ELECTRIC GENERATOR.—The object is to furnish means, additional to those of Patent No. 3,964 of 1880, for easily removing and replacing one or more coils of the armatures of dynamo-electric generators. For the generative portions of the armature, coils of wire or naked bars are employed, electrical connection from each bar to the corresponding bar on the opposite side of the armature being made through discs. If each longitudinal set of wires, or each bar, be considered as one coil, a series of discs equal in number to half the coils is used at each end of the armature. Referring to the illustrations, which are diagrammatic in their character, A (Fig. 1) represents a disc at the commutator end of the armature, and B a disc at the other end. Each disc A has a hole punched out of the centre, and a

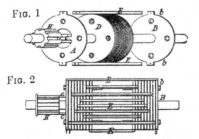

FIG. 1

FIG. 2

finger *c* bent out at right angles to the commutator block H. In Fig. 2 two only of such fingers are shown, for sake of clearness, connected to discs Nos. 1 and 3. D is a core, not shown in Fig. 2, of insulated discs of soft iron. Supposing the upper bar E (Fig. 2) to be moving through the polar space of the magnetic field, a current would traverse it, say, from left to right, enter the disc 2 by the lug *b*, cross the disc, and return by a bar not shown,

but immediately behind the lowest bar E to the disc 3, and then two paths would be open to it, one by the tongue *c* to the commutator and the other through the second bar E attached to the disc 3. The same would occur in each pair of bars and discs, the course of the current being similar to that in a Siemens low-tension generator for plating.

1,257.—S. Pitt, Sutton, Surrey. (*W. B. Espeut, Jamaica*). **Laying Telegraph Wires.** 6d. (4 figs.) March 22.

CONDUCTORS.—Underground wires are carried naked in a tube of porcelain, either laid directly

in the soil or enclosed within iron tubing. The various lengths are securely jointed together, and T pieces are provided at intervals to permit of branches being led out. The illustration is a cross-section of the tube, A A being the wires.

1,358.—R. Harrison and **C. Blagburn,** Newcastle-on-Tyne. **Electric Lamps.** 6d. (4 figs.) March 26.

SEMI-INCANDESCENCE LAMP.—A carbon rod is kept continuously in contact with a block of refractory material towards which it is fed upwards between two grooved rollers near its point. The rollers are caused to turn and press against the carbon by cords attached to a descending weight. When the carbon rod is consumed the lamp is short-circuited by the carbon-holder making contact with a pin metallically connected to the other pole.

1,384.—W. R. Lake, London. (*A. G. Holcombe, Danielsonville, Connecticut, U.S.A.*) **Electric Lighting Apparatus.** 6d. (6 figs.) March 29.

ARC LAMP.—The peculiarity of this lamp consists in its containing no electro-magnetic apparatus. The regulation is controlled by the attraction or repulsion of electric currents traversing insulated conductors. The upper carbon *c* runs down by gravity as soon as the brake wheel *e* is released from contact with a block carried on the lever *f*, which is actuated by the attraction of two flat coils $f^2 f^3$, one fixed and the other movable, the stroke of the movable coil being limited by an adjustable stop. The arc is established by the solenoids *m* and *m*[1]. The carbons and solenoids form one circuit, and the flat coils $f^2 f^3$ a second or shunt circuit acting on the well-known differential principle. The ends

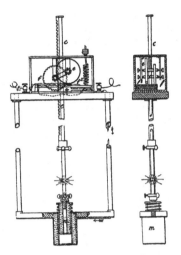

of the carbons are combined with pyrotechnic composition to give notice when they are nearly consumed.

*1,393.—J. H. Thomson, Shoeburyness. **Electric Drills.** 2d. March 29.

MOTOR.—Two concentric drills revolving in opposite directions are impelled by a small electromotor engaging with bevel gearing placed upon their spindles.

1,412.—J. Scott and W. H. Akester, Newcastle-on-Tyne. **Manufacture of Carbons for Incandescent Electric Lamps.** 2d. March 31.

INCANDESCENCE LAMPS.—The filament is made from brushmaker's fibre, called "kitool," the piasiva fibre being preferably used. This is drawn through a conical tube having a knife edge at the smaller end. A number of fibres so treated are placed together and bent to the desired shape, and baked to set them to this shape, after which they are carbonised in the usual manner.

1,422.—W. Crookes, London. **Production of Electric Light.** 4d. March 31.

INCANDESCENCE LAMP.—The inorganic constituents of the cellulose is removed from carbon filaments by exposing them, in the form of thin rods, either before or after carbonisation, to the action of dilute hydrochloric or other suitable acid, or by heating the carbon to whiteness in free chlorine. The residual gas is removed by enclosing in the bulb, or in a chamber connected to the bulb, a gas-absorbing substance, such as thorina, heated to a degree below redness, or the residual gas may be one that is readily taken up by some special absorbent.

1,447.—C. W. Siemens, Westminster. (*Siemens and Halske, Berlin*). **Electric Machines.** 6d. (7 figs.) April 1.

DYNAMO-ELECTRIC GENERATORS.—The essential feature of this invention consists in combining together to form a continuous current separate electrical impulses that are generated immediately after one another at different parts of the generator, whereas in generators of present construction such combination can only be effected with impulses generated in consecutive magnetic fields. Figs. 1 and 2 show a generator for the production of intermittent currents of like or of alternate direction. On the fixed frames B B are an even number (ten) of electro-magnets C. The pole of each magnet is of opposite name to the pole of the magnet facing it and to poles on each side of it, so that between each pair of facing poles there is a powerful magnetic field. Through these magnetic fields are caused to revolve bobbins S S of insulated wire wound on wooden cores of elongated shape fixed on a wheel. If the number of the coils S be the same as that of

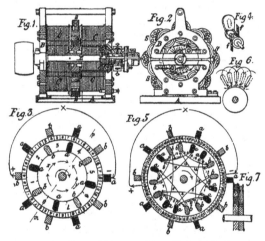

the magnets C, as hitherto, then all the coils have currents induced in them simultaneously as they approach and leave the successive magnetic fields. According to the present invention, however, the number of coils S is made to differ by an even number from that of the fields G through which they pass; in the illustration there are eight coils and ten fields. The effect of this is that every pair of diametrically opposite coils is always in the same phase of induction, each successive pair being subject to a greater or less inductive influence, as the case may be. Thus the current attains its maximum or minimum strength successively in each pair of diametrically opposite coils. The wires of all the coils are connected to form a continuous circuit, whilst they are wound alternately in opposite directions, the impulses in all the coils thus becoming

added together. The axis carries a commutator *t* of ordinary construction, consisting of forty insulated plates divided into five groups of eight, each plate being separated from the next of its own group by seven intermediate plates. This grouping is effected by eight insulated rings *r* fixed on the axis, to each of which five of the commutator plates are connected by five wires *d*, and as the rings *r* are connected each to the wire joining two successive coils, the successive groups of commutator plates correspond in order with the successive connections of the coils. The diagram, Fig. 3, illustrates the action of the generator. The outer squares *a b* represent ten stationary fields of alternate polarity, and the inner ellipses *a b* represent eight coils that revolve through them. The divided circle between these represents the commutator, having its forty plates arranged in five groups, each numbered 1 to 8. The numbers 1, 2, 3, &c., between the ellipses indicate the wires that lead from the connecting wire of each pair of adjoining coils to the rings *r*, and thence to the commutator plates. Thus wire 1 is connected through its rings to all the plates marked 3, wire 2 to all the plates marked 8, and so on. The arrows marked respectively + and − indicate the brushes. All the coils approaching the magnetic fields are marked with the same letters, *a* approaching *a* and *b* approaching *b*, have currents set up in them in one direction indicated by the arrows *x*, and all the coils approaching fields marked by different letters, *a* approaching *b* or *b* approaching *a*, have currents set up of opposite direction, indicated by arrows *y*. There will always therefore be in every situation of the coils as they revolve a diametrically dividing line, such as *p m*, towards one end of which there will be an accumulation of positive and towards the other end of negative electricity. As shown, this line passes through 4 and 8, and as the brushes rub on the corresponding plates 4 and 8 of the commutator they take off the accumulated electricity. The imaginary line *p m* always moves round in a direction opposite to that of the commutator, and with such greater velocity that the brushes are always in the proper position to receive the currents. The number of coils might be doubled by arranging them in two places overlapping each other, as shown by Figs. 6 and 7, the connections being in that case arranged as in Fig. 5, in which the arrows indicate the directions of the currents to and from the several coils, and the numbers marked on the lines connecting the coils indicate the connections through the rings to the respective commutator plates, which in this case are eighty in number. Instead of collecting the currents from a number of coils together, each separate coil might be arranged to deliver its current, and these currents, which would be alternating, might be collected into a curre of constant direction, in the manner shown in Fig. 4. This invention applies to other forms of generator than the one shown, and also to magneto-electric and electro-dynamic motors, the characteristic feature being the combining with an even number of stationary magnetic fields of alternate polarity, a number of revolving alternately oppositely wound coils differing by two or other even number from the number of fields, the wires of the coils being all connected to each other and to the commutator.

*1,474.—J. C. Mewburn, London. (*J. Bourdin and S. A. de Maltzoff, Paris.*) **Apparatus for Laying Underground Telegraphic or Telephonic Conductors.** 2d. April 5.

Conductors.—A plough carries a reel on which the conductor is wound, and as it is unwound by the action of the plough it passes through a hole in a share into the bed which the nose of the share prepares for it, a trailing roller replacing the sod.

1,526.—J. D. F. Andrews, Westminster. **Electric Lamps, &c.** 6d. (12 figs.) April 7.

Arc Lamps.—This refers to Patent No. 2,321 of 1879, which related to lamps having discs or plates of carbon fixed at a little distance apart parallel to each other, the voltaic arc being presented at their edges. Fig. 1 shows a disc lamp according to the present invention. The lower disc has a countersunk central hole *c* and is connected to the conductors at its outer edge, while the other disc, which has no hole, is connected to the conductor at its centre or at its edge. The arc presents itself at the edge of the countersunk hole and moves round it as the

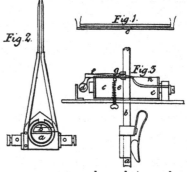

carbon consumes, or may be made to revolve round it by the inductive action of a neighbouring coil or magnet. Also both carbons may have central countersunk holes and be connected to the conductors at their outer edges. When parallel plates of carbon are used they are connected to the conductors at their opposite edges, one extending beyond the other. When two plates are employed, the one fixed and the other movable, the lamp

takes the form shown in Fig. 2. The current traverses a magnet of which the core is divided longitudinally into two parts a and b; these are mutually repelled when the current passes, and one of them being pivotted and carrying the movable carbon, the arc is established. This form of magnetic core is applicable to other purposes than regulating lamps. Fig. 3 shows a lamp for feeding ordinary cylindrical carbons. The upper carbon a is held in a guide tube b; an electro-magnet c, placed in the lamp circuit, has facing it an armature d, which is kept away from it by an adjustable spring e. The armature has two arms f, between which is pivotted a semicircular bush g and a U shaped yoke. A spring on the armature lever bears against the tail of the yoke, tending to cant the yoke into an inclined position, its canting movement being limited by a stop. When the lamp is put in circuit the armature is attracted and the yoke canted so as to pinch the tube and raise it to form the arc. As the current falls off the grip of the yoke is released and the carbon slides down. In a modification a double clutch feed is employed, and also an arrangement whereby a resistance is inserted into the circuit when the lamp is extinguished. This is accomplished by a high resistance magnet attracting its armature in opposition to a spring, and so completing a by-pass circuit in which the resistance is placed.

1,536.—J. L. A. Dupont-Auberville, Paris. (*V. Delaye, Paris.*) **Electric Lamps.** 6d. (3 figs.) April 8.

ARC LAMP.—Each carbon pencil i is pressed by a weight and cord q q against a refractory non-

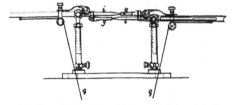

conducting abutment y. The arc is formed between the two electrodes, which are gradually forced towards the abutments as they waste.

1,543.—St. G. Lane Fox, London. 6d. (1 fig.) **Electric Lamps.** April 8.

INCANDESCENCE LAMP.—The illustration shows in vertical section a lamp constructed according to this invention. a is the globe, b b mercury tubes with enlargement b^1 b^1 to facilitate the escape of air when the mercury is poured in, d is cotton wool, e plaster-of-paris, ff conducting wires, gg platinum wires fused into the bottoms $n n$ of the tubes b,

and extending into the mercury, and r r blocks of plumbago or carbon drilled with holes to receive

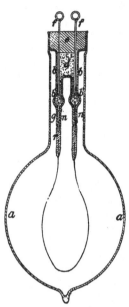

the wires g and the ends of the filament, the contacts in each case being made with Indian ink.

1,596.—A. W. L. Reddie, London. (*H. Sedlaczek and F. Winklill, Leoben, Austria*). **Electric Lamps.** 6d. (16 figs.) April 12.

ARC LAMPS.—This invention relates to the feed mechanism of lamps having their carbon-holders supported on fluid in closed vessels communicating with each other at their lower parts. The regulation may be effected by the direct action of a centrifugal governor driven by the engine, or by an electro-magnet or solenoid. In the former case the arrangement is based on the fact that as the resistance of the arc increases so does the speed of the engine, and *vice versâ*. A lamp on this principle is shown in Figs. 3 and 4, in which each of the electrodes is carried by a piston 3 or 4. In the base of the apparatus and between the two cylinders is a socket a with a brass plug b, the two being provided with ports n n^1 and c c^1. The plug is operated by a disc g and a lever connected to the sleeve of a centrifugal governor. As soon as the engine is set in motion the disc g will be caused to turn on its axis by the rod, and with it the plug will move, carrying the opening c gradually past the corresponding opening n of the box a until it has closed the same, and thereupon the communication of the two cylinders on the side next the positive pole will be cut off, the hollow plug b being, however, during its further rotation still in communication with the negative cylinder 2 through the large opening c^1

and the passage n^1. As the rotation continues, wings on the plug, sliding upwards on correspondingly inclined surfaces on the case, will lift the plug (Fig. 3). The rising of the plug causes the negative electrode to fall to a corresponding extent, whereby the voltaic arc is formed. When the velocity of the engine, in consequence of the consumption of the carbon points, reaches such a degree that the small slot of the plug between c and c^1 comes in

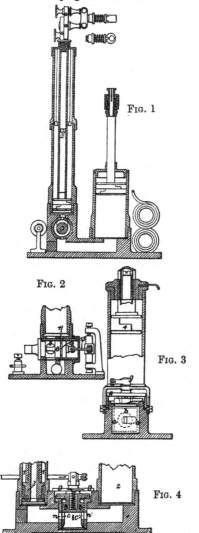

FIG. 1

FIG. 2

FIG. 3

FIG. 4

front of the passage n, and thereby re-establishes communication between the two cylinders 1 and 2, the carbon points will then approach each other until the rotation of the engine becomes slower, and the plug is moved to break the communication. A further set of inclines is provided to come into action if the carbons break and a very rapid feed is required. Figs. 1 and 2 illustrate a lamp governed

by an electro-magnet with an armature having a small range of motion balanced by a short spring acting through a long lever. As soon as a sufficiently powerful current passes through the coil of the electro-magnet and the carbons (then in contact), the piston 6, by means of the link 13, will be drawn backwards, and the holes 11 of the cock will be closed, while the space 8 at the rear of the piston will be increased, causing the negative carbon to fall and produce the arc. As the current decreases the piston returns, establishing communication between the two cylinders through the port 11, the groove 12, and the pipe below. When several lamps are used in series the electro-magnet is replaced by differential solenoids.

1,636.—St. G. Lane Fox, London. Apparatus for Producing Motion by Electricity, &c. 6d. (2 figs.) April 14.

CURRENT REGULATOR.—This invention refers to apparatus in which an electrometer or electro-dynamometer connected between the main conductors and the earth, such, for example, as those described in Specifications Nos. 3,988 and 4,626 of 1878, is employed to actuate a motor which is caused to control the commutators of secondary batteries, or the actions of rheostats, or for other purposes in distributing electricity. $m\,m$ are two electro-magnets, the circuits of which are respec-

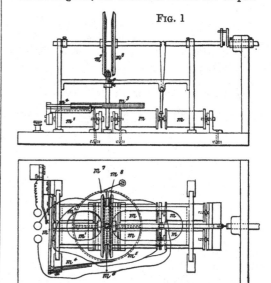

FIG. 1

FIG. 2.

tively completed by an electrometer or electro-dynamometer when the electromotive force in the mains rises or falls beyond a given limit. These two coils are connected in joint circuit, Fig. 2, with a third electro-magnet m^1, having a trembling

armature provided with a pawl m^4, gearing into the ratchet wheel m^5. When the circuit of either one of the electro-magnets $m\ m$ is completed, the armature of m^1 is set in action, and one or other of the wheels m^7 or m^8 is put in gear with pinion m^6, until the necessary adjustment has been effected.

1,653.—J. H. Johnson, London. (*Société la Force et la Lumière, Société Générale d'Electricité, Brussels*). **Electric Lamps.** 8d. (22 figs.) April 14.

INCANDESCENCE LAMP.—To obtain sufficient resistance in an incandescence lamp without the use of an attenuated conductor, the inventor divides the filament in several places, letting the parts lie in contact with each other, the "augmentation of the resistance being directly in proportion to the number of divisions." In Fig. 1, $a\ a$ is the filament of carbon, iridium, or other suitable material held between platinum terminals A A, and divided into sections as shown. Fig. 2 shows such a filament set horizontally in an exhausted globe between two spring arms. Fig. 3 is an example of a semi-incan-

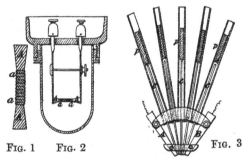

FIG. 1   FIG. 2   FIG. 3

descence lamp to burn in the open air; $c\ c\ c$ are carbon rods impelled towards a common point by the weight $p\ p$; A B are the two terminals of the lamp between which the current passes, making its way transversely across the carbon rods and heating them to the necessary degree. Various forms and adaptations of the filament are described and illustrated.

1,670.—G. S. Grimstone, New Cross, Kent. Electric Lamps. 6d. (5 figs.) April 16.

ARC LAMP.—Fig. 3 is a sectional plan on lines X X of the lamp shown in elevation in Figs. 1 and 2. To the soft iron core A, moving in the solenoids of high and low resistance B C, is pivotted one end of a lever D, which has its fulcrum on the fixed bracket E, and has pivotted to its other end the suspended frame F, having two loops $F^1\ F^2$, through which passes freely the carbon-holder G. The holder is held in the said loops by the action of an eccentric locking piece H, pivotted

to the frame and weighted by an arm H^1, so as to press the carbon-holder against the sides of the loops $F^1\ F^2$. Thus, as the core A rises, owing to the increased attraction of the solenoid of high resistance C, consequent on the burning away of the carbons S, the frame F, and with it the carbon-holder and upper carbon, will descend until the arm H^1 of the locking piece comes in contact with the stop K, whereupon a slight further descent of the frame will cause the eccentric H to be turned by its arm H^1, so as to free the carbon-

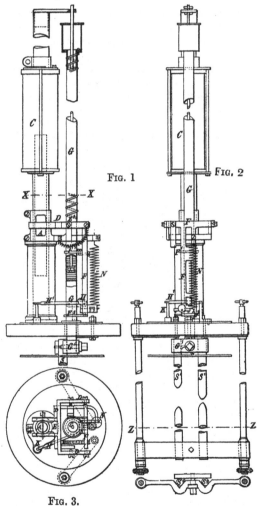

FIG. 1   FIG. 2

FIG. 3.

holder, which will then descend by gravity through the loops $F^1\ F^2$, its descent being regulated by the action of the loosely fitting piston L in the hollow carbon-holder filled with glycerine. When the arc is sufficiently reduced in length, the solenoid B will raise the frame and bring the arm H^1 off the stop K. N is a spring to balance the excess of weight of the solenoid. In some cases it is

attached to a cord which unwinds from a barrel as the carbon consumes. The carbon-holder G has a crosshead G^1, in which are secured two upper carbons $S^1 S^1$. On first starting the lamp the current will pass through one pair of carbons, and as that pair burns away the downward motion of the holder will cause the second pair to approach nearer than the first pair, the current will then be transferred to the second pair. Thus the current will alternately pass by each pair of carbons until both pairs are consumed.

1,676.—J. H. Johnson, London. (*C. A. Faure, Paris*). **Secondary Electric or Galvanic Batteries.** 6d. (10 figs.) April 16.

SECONDARY BATTERIES.—This invention relates to means for supporting the active substance against the surface of the element in the battery described in Patents No. 129 and 1 097 of 1881. Referring to the illustrations, Fig. 1 represents a cell in which a lead plate A, prepared by receiving a layer B of suitable composition, such as oxide of lead, is connected to a sheet C of permeable material, pre-

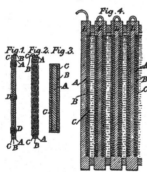

ferably felt, by means of clips or rivets D of lead. In Fig. 2 the sheet C is secured to the plate A by stitching with thread or twine. When narrow plates, prepared on both sides, as in Fig. 3, are used, the composition is supported by folding the sheet C round the plate A having the layers B interposed, the edges of the sheet being sewn together. Fig. 4 illustrates the invention applied to a battery of flat lead plates arranged parallel to each other. They are covered with the active substance on one side and applied to the partitions L to impart to them the necessary rigidity. The partitions are inserted in a trough and divide it into cells. The invention further describes a battery of pyramidal-shaped cells packed within each other.

1,683.—A. M. Clark, London. (*A. d'Auriac, St. Fleur, France.*) **Electrical Apparatus for Signalling on Railways.** 8d. (5 figs.) April 16.

CONDUCTOR.—A metallic rod laid between the rails forms a conductor, by which signals can be exchanged between the train and other trains or stations.

1,685.—A. M. Clark, London. (*J. M. A. Gerard-Lescuyer, Paris.*) **Electric Lamps or Regulators.** 8d. (9 figs.) April 18.

ARC LAMP.—This invention relates to various arrangements of electric lamps in which a small arc is produced, thus effecting the division of the electric light. In Figs. 1 and 2, A is a guide upon which slides a frame B C carrying a solenoid D of fine wire, through which passes freely the brass carbon-holder E, carrying at its upper end a soft iron armature F. The carbon-holder E is suspended by a spring G from a small bracket H carried on the plate C, so that the armature is a few millimetres above the upper end of the solenoid. A second armature K has a jaw L, which is caused by the spring M to bind against the guide A. The upper carbon is fixed in the holder E, and passes through guide rollers. The lower is carried by the bracket S. The current passing through the shunt circuit of the solenoid causes the armatures K F to be attracted, at the same time releasing the grip of the jaw L and allowing the bracket B to descend until the carbons meet. The current is thus diverted from the solenoid, and the two armatures are withdrawn by their springs, the one causing the jaw L to again bind against the guide A, and

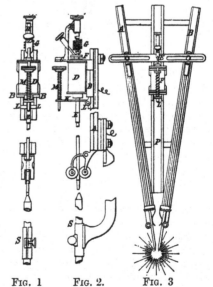

FIG. 1   FIG. 2.   FIG. 3

thus arrest the descent of the bracket B, and the other causing the upper electrode to be raised to produce the arc. A modified form of this lamp is also described in which the carbon-holder is retained by the friction of a screw point forced against it by

a spring and released at intervals by an electro-magnet. Fig. 3 shows another modification in which the carbons slide in two guides A B. Between them is another guide in which slides a solenoid F, having at its lower end an armature L adapted to bind against the guide P, and at its upper end an armature for effecting the separation of the carbons. When the current traverses the shunt coil it is released from the guide and drops down the slide carrying the carbons before it until they meet; the coil then becomes inert, and the lower armature grasps the guide, while the upper one raises the bar E and separates the carbons. In a still further modification the descent of the upper carbon is controlled by a tightly fitting piston to which the carbon is attached, sliding in a cylinder, a partial vacuum being formed by the descent of the carbon and piston.

1,696.—S. Pitt, Sutton, Surrey. (*O. Lugo, New York, U.S.A.*) Telegraphy. 6d. (2 figs.) April 19.

CONDUCTORS.—To enable two independent circuits to be worked simultaneously through the same "compound conductor," two insulated parallel conductors forming one circuit have the two insulated conductors forming the second circuit wound helically around them.

*1,699.—J. Wetter, London. (*R. S. Jennings, Baltimore, Mass., U.S.A.*) Block Signalling for Railways, &c. 2d. April 19.

ELECTRIC LIGHT.—The passage of a train over a predetermined spot switches on an electric lamp.

*1,745.—C. D. Abel, London. (*P. Jablochkoff, Paris.*) Electrical Batteries. (2d.) April 22.

SECONDARY BATTERIES.—In these secondary batteries the acting electrodes are covered with an oily or resinous body. They may be made of spirally wound plates of metal immersed in oil, or of baskets of perforated metal filled with coke and immersed in oil.

1,762.—J. A. Fleming, Cambridge. Preparation of Materials for Electric Insulation. 6d. (1 fig.) April 23.

INSULATING MATERIAL.—The material is made from wood deprived of its moisture in any ordinary way, and impregnated under pressure with paraffin wax or with a mixture of paraffin wax and resin.

1,783.—E. G. Brewer, London. (*T. A. Edison, New Jersey, U.S.A.*) Measuring Electric Currents. 6d. (4 figs.) April 25.

CURRENT METER.—In his Patent No. 4,391 of 1880 the inventor described a weber meter consist-

ing of an electro-depositing cell placed in a shunt to the main circuit, so that a definite proportion of the current passed through the cell, and left a record of its quantity in the amount of metal deposited upon one plate of the cell. In use, the resistance of the cell and consequently the amount of current passing through it decreases with an increase of temperature, and to obviate this defect the shunt is formed of two parts, the resistance of one of which increases as that of the other decreases under changes of temperature. Supposing the second resistance to be a coil of copper wire and the first to be a 20 per cent. solution of sulphate of copper, then the variations of the two would balance each other if the resistance of the wire were three times as great as that of the solution. Two cells are used, in one of which the deposition takes place with greater rapidity than in the other.

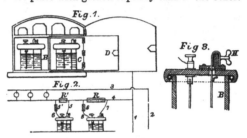

Fig. 1 is a front elevation of a box or case containing two cells which constitute a meter; Fig. 2 shows two cells with diagram of circuit connection; Fig. 3 is a detailed view in section of a portion of a cell. In Fig. 1 the inner door D covering the cell C is kept locked, while the outer door is always open for inspection. In Fig. 2, 1 and 2 are the main conductors from which a derived circuit 3, 4 leads into the house in which the meter B C is placed for measuring the current passing through the house. Resistances R R[1] deflect portions of the current into the shunt circuits, 5, 6 and 7, 8, the resistance R[1] being less than R, so that a greater current passes through B than C. S S[1] are the compensating resistances of fine copper wire. Fig. 3 shows how one of the plates may be withdrawn to be weighed by slackening the thumbscrew H and raising the plate through the slot *f*.

1,787.—A. M. Clark, London. (*H. J. Müller and A. Levett, New York, U.S.A.*) Dynamo-Electric Machines. 1s. 4d. (13 figs.) April 25.

DYNAMO-ELECTRIC GENERATOR.—The object is (1) to provide a generator so constructed that separate currents may be produced, one of which may be used to excite the field magnets, and at the same time to perform work in the external circuits,

whereas the other currents perform work in the external circuits only, and these different currents from the same generator may be of various degrees of strength. (2) To facilitate the separation of the armature coils from the fixed magnet coils, and to provide a form of wheel that will produce a current of air to prevent overheating of the coils. Referring to the illustration, the armature rotates between two sets of field magnets, of which there are six in each set. The magnet cores are connected by brass plates O^1 to O^6, and are coupled together in one circuit by means of a

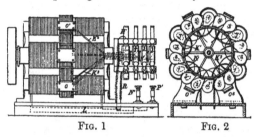

FIG. 1   FIG. 2

FIG. 3

wire L, which proceeds from the terminal N^1, traverses the whole of the magnets in succession, and goes in the form of a coil R to the lower brush of the first commutator I^1. From the other brush H^1 of the same commutator a wire extends to the terminal P. The armature wheel has eighteen coils designated J^1 to J^3. These coils are divided into three groups of six each, that is, as many as there are field magnets, in the following manner: A wire K^1 connects all the coils $J^1 J^1 J^1$ for each group, while the coils J^2 for each group are connected by a wire K^2, and the coils J^3 are connected by a wire K^3. These wires pass to the respective commutators $I^1 I^2 I^3$. Each commutator consists of six segments, arranged concentrically within a flanged ring a (Fig. 3); the segments 1, 3, and 5 are connected with each other, and the segments, 2, 4, and 6 are also connected with each other. The wire K^1 connects the segments 1, 3, and 5 of the commutator I^1, and the coils $J^1 J^1 J^1$ of an armature coil group with each other, and passes back to the commutator, and connects the segments 2, 4, and 6 of the commutator I^1. In like manner the wire K^2 connects the segments 1, 3, and 5 of the commutator I^2, the coils $J^2 J^2 J^2$, and the segments 2, 4, and 6 of the commutator I^2, and

so on. The armature coils are so arranged that the moment they leave one field magnet one-eighth of their cross section covers the next adjoining magnet, as shown in Fig. 2. The specification further describes a generator provided with a relay and resistance for causing the reverse or secondary current orginating in the solution of a plating bath to pass through the magnets in the same direction as the main current.

1,802.—P. Jensen, London. (*T. A. Edison, New Jersey, U.S.A.*) **Electric Lights, &c.** 10d. (22 figs.) April 26.

INCANDESCENCE LAMPS.—This invention relates to lamps so arranged that the act of placing them into their sockets completes the circuit of the irrespective terminals and conductors. Fig. 1 is an example of one form of lamp and socket. Two metal rings $a\,b$ are fixed on the neck B, one of them, a, being formed into a male screw thread; from these rings the wires 3, 4 lead to the clamps $c\,c$ of the filament. Upon the interior of the socket is an internally screwed metal band connected to one of the main conductors, the other conductor being connected to the contact breaker c. Instead of forming one of the rings into a screw thread it may be made concave and held by spring finger pieces in the socket; or the insulated neck

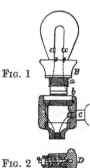

FIG. 1

FIG. 2

may be slightly bevelled and have two concavities, one on either side, in the bottoms of which are metal plates connected to the filament. Spring fingers attached to the socket enter these concavities and make the necessary contacts while they secure the lamp. Each lamp is provided with a contact breaker C, by which it may be turned in or out as desired. In this appliance, which is shown in section in Fig. 2, is a bush screwed into the lamp socket, carrying the spindle x ending in the contact point u. This spindle is constantly pressed forwards by a spring, its motion being restrained by a pin y, which, as the handle D is rotated, drops into one or other of two cross grooves, one only of which is sufficiently deep to allow the point of the spindle to come in contact

with the terminal of the lamp. The invention further relates to inserting a piece of lead wire in the lamp circuit so that in case of an abnormal current the lead melts and destroys the connection: also to methods of suspending shades so that the fittings shall not cast a shadow, and of constructing brackets and chandeliers for electric lamps similar in general design to those already employed for gas.

1,809.—W. R. Lake, London. (*P. B. Delany, New York, U.S.A.*) **Telegraphic Cables.** 8d. (18 figs.) April 26.

CONDUCTORS.—The illustrations show one form of cable made according to this invention. The wires C C are strained through insulating distance pieces in a tube, the space between them being filled up with powdered mica. In some cases an inner tube serves for the circulation of a heating medium to

dissipate any moisture which may have condensed on the wires; in other cases the wires pass through the inner tube, and the annular space serves for the passage of the heating medium. The specification describes various methods of laying, carrying, and ramming such pipes. The tubes may be placed in wooden troughs above ground or in subterranean channels. The insulating compound is forced into the tubes by air pressure.

1,835.—H. J. Haddan, London. (*C. F. Brush, Cleveland, Ohio, U.S.A.*) **Current-Governors for Dynamo-Electric Machines.** 6d. (3 figs.) April 28.

DYNAMO-ELECTRIC GENERATORS. — The object of this invention is to produce constant currents from dynamo-electric generators working under variable external conditions without varying the speed of their armatures. This is effected by exciting the field magnets by a shunt current from the main conductors, and interposing a variable shunt across the magnet coils. Fig. 1 shows the governor in section, while Fig. 2 shows it applied to a generator. The operation of the device is as follows : the binding posts $p\,n$, forming the terminals of the shunt circuit, are connected with the field magnets of the generator. The course of the shunt circuit is then from the post p, through the piles H H, formed of carbon discs, and the outer portion of the solenoid coil E to the post n. The posts P N are put in the working circuit of the generator in such a manner that the main current shall pass through the helix E in the same direction as the shunt current. The

weight c is so adjusted that when the generator is working to its full capacity, and the normal current is passing through its portion of the helix E, the magnetism of the core is just sufficient to sustain the lever D without subjecting the piles to pressure. If the resistance of the external circuit be now lessened the current will be increased in the helix E, and the core drawn upwards compres-

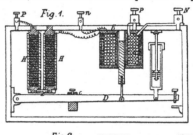

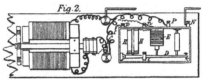

sing the piles H H. Current will then be shunted from the field magnets until the main current falls to its original strength. Some increase of current strength is, however, required to maintain the pressure, but to render this as little as possible the shunted current is made to pass through a portion of the helix E. The piston of the dash-pot has a valve in it which allows the lever D to fall suddenly but retards its rising.

1,852.—H. J. Haddan, London. (*L. Somzée, Brussels*). **Electric Lighting Apparatus.** 2d. April 29.

ARC LAMP.—"The voltaic arc is produced in any convenient mode, and the powdered excitant" (*i.e.* charcoal, &c.) "supplied to it, as and when required." This relates also to passing an electrical current through a gas flame.

1,856.—A. W. L. Reddie, London. (*J. Hunebelle, Paris.*) **Construction of Railways.** 6d. (4 figs.) April 29.

CONDUCTORS.—This invention relates to a composite rail, especially adapted for the reception within it of electrical conductors. A continuous chair is used consisting of a base-plate having two vertical checks, the upper part of the space between which is occupied by the web of the rail, the lower part being utilised for the reception of the conductors.

1,873.—W. T. Henley, Plaistow. **Telegraph Cables, &c.** 6d. (13 figs.) April 30.

CONDUCTORS. — The conducting wires are em-

bedded in bands or belts of insulating material and
may be packed in troughs underground, or be made

in the form of coils and carried overhead. In the
latter case steel wires are introduced in connection
with the copper ones to give the necessary strength
to the whole.

1,918.—E. G. Brewer, London. (*T. A. Edison,
New Jersey, U.S.A.*) **Manufacture of Car-
bon Conductors for Incandescent Electric
Lamps.** 4d. May 3.

INCANDESCENCE LAMPS.—In Patents Nos. 4,576
and 5,127 of 1879, 578 and 3,765 of 1880, and 562 of
1881, are described various methods of manufacturing incandescing carbon filaments with the
qualities of flexibility and high resistance. The present invention is to furnish a method by which such
filaments can be made of graphite or plumbago,
graphitoidal silicon, boron, zirconium, and the
like. The material used is reduced to an impalpable powder and pressed into sheets in a die box.
The density is varied according to the pressure
employed, and the resistance regulated by the
amount of pulverised carbon mingled with the
"basic material." When the materials have little
or no cohesion they are made into a paste by a
fluid having in solution a hydro-carbon, the men-
struum being one that is readily evaporable. From
sheets thus prepared the filaments are stamped,
after they have been heated to incandescence in
hydro-carbon vapour.

1,922.—J. B. Rogers, London. **Electric Lamp.**
6d. (10 figs.) May 3.

SEMI-INCANDESCENCE LAMP. — In lamps constructed according to this invention, the carbons
are caused to adjust themselves and to bring their
points into electrical contact either by their own or

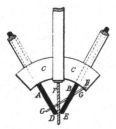

by an applied weight, the points bearing against
an asbestos or other non-combustible stem. In the
illustration, which shows a typical form of lamp,

the two carbons A B, are set at an angle in sockets
C C, and permitted to slide until their points D E
come into contact with the strip of asbestos F.
"The sockets or tubular portions C C are in electrical circuit when the strip of metal G, which is
articulated at H, rests across and in contact with
the two carbous A B."

1,942.—J. Brockie, London. **Electric Arc
Lamps.** 6d. (10 figs.) May 4.

ARC LAMP.—Two magnets or solenoids are placed
in each lamp, one in the main and the other in a
shunt circuit. The second magnet releases, retards,
or stops the feeding train, while the first magnet
separates them, and at the same time, by its motion,
tightens or slackens the spring against which the
second solenoid acts, so that when a strong current
exists in the main circuit the first magnet, in
addition to directly lengthening the arc, will
tighten the said spring. By this arrangement the
lamp becomes self regulating, and may be worked

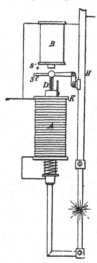

with currents of various powers without adjustment or regulation. When the solenoids are
arranged as in the figure, and the main current
from any cause becomes diminished, the solenoid
A will reduce the tension of the feeding spring D,
and thus compensate for the simultaneous weakening of the influence of the shunt magnet B, while at
the same time it brings the lower carbon a little
nearer the upper one, reducing the length of the
arc and adjusting the feeding gear to the decreased
strength of current. The core of the solenoid A is
of iron in the upper half only, and is pulled downwards against a spring not shown. The shunt
solenoid B actuates a brake K upon the upper
holder H. The lever of this brake plays between
two stops $S^1 S^2$. the lower of which forms an abut-

ment which allows the lower solenoid core as it moves up and down to diminish or increase the tension of the feeding spring D. The descent of the carbon is retarded by a dash-pot or a train of wheels. In a modification the lower carbon-holder forms the core piece of a solenoid placed in the main circuit. The specification illustrates the invention as applied to lamps with various forms of regulating gear and with more than one set of carbons.

1,943.—E. G. Brewer, London. (*T. A. Edison, New Jersey, U.S.A.*) Electric Lighting. 4d. (1 fig.) May 4.

INCANDESCENCE LAMPS.—When a number of lamps are fed by currents from a main kept at a constant potential, and one lamp is taken as a standard, the others may be constructed to give a smaller light by diminishing their radiating surfaces and increasing their resistances in definite proportions. Thus supposing a lamp to give 16-candle power under the normal potential, then, if a

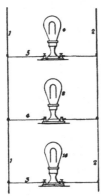

lamp having half its radiating surface and twice its resistance be substituted for it, a light of eight-candle power will be given, and so on. In the figure, 1, 2 are the main conductors with lamps 4, 8, 16 coupled in multiple arc.

1,961.—P. Higgs, New York, U.S.A. Magneto-Electric Machines. 10d. (14 Figs.) May 5.

DYNAMO-ELECTRIC GENERATORS.—Fig. 1 is a perspective view of the generator, and Fig. 2 is a detail view explanatory of the construction of the armature, which is of the Gramme type. Each of the spools E E is wound separately and then the whole are built up into a continuous ring by connecting their end flanges to the arms or spokes projecting from the boss. Plates of mica *c*, Fig. 2, are interposed between the bobbins to prevent the circulation of induction currents. The field magnets are formed of soft iron bars F F, around which the exciting

coils are wound; these coils may be extended around the parallel pieces *d* connecting the cores with the common pole-pieces *f*. At each of the ends of the pole-pieces is a termination of peculiar

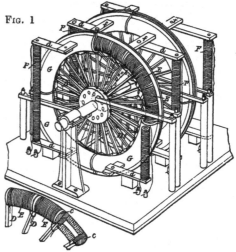

FIG. 1

FIG. 2

form marked G, such that when four of these are placed together they constitute a hollow circular tube in which the armature revolves, a narrow slot being left for the passage of the spokes D. The armature coils may be wound in series or in multiple arc and be connected to a single commutator, or they may be wound so as to produce two or more distinct currents, and be connected to two or more commutators to supply currents circulating in independent fields. The commutator is a cylinder of insulating material pierced with holes for rods of copper which, as they wear under the destructive action of the spark, can be set up afresh to the face of the cylinder. The specification illustrates by diagram several ways of connecting the field magnets in series and parallel to each other, and also methods of connecting up the field magnets when one generator is driven by the current from another, and also when a number of lamps are supplied from one machine. The earth is preferably used as a return. The radiation of heat from the line wire is assisted by attaching metallic plates having large surfaces. The conducting wires are made from an alloy of steel, aluminium, and silver.

1,968.—W. R. Lake, London. (*N. Bouliguine, Paris*). Electric Light Apparatus. 6d. (4 figs.) May 5.

ARC LAMP.—This lamp has six pair of carbons, the arc being formed in the first instance between any one pair, and after burning there for a time moving to another pair, and so on. Referring to

the drawings the current traverses the coils of the electro-magnet *d*, and then goes by the spindle B to the carbon-holders E, of which there is one at the end of each of the six arms of the wheel K. After crossing the arc it proceeds to one of the fixed carbon-holders F, and away to the conductor. At the commencement the spring *p* rotates the spindle B and brings each of the carbons *l* into contact with one of the carbons *m*. On the passage of the current the electro-magnet tends to cause the armature to place its axial line parallel to the axial line of the magnet, thus rotating the spindle and moving

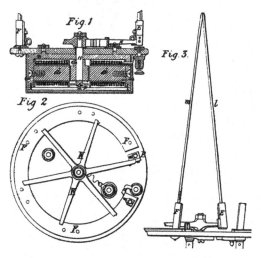

the carbons out of contact. The arc will thus be established between one pair, and as they burn they will gradually approach each other until those of another pair touch, when the arc will leave the first pair, and be established between the second pair and so on. It is stated " that the last pair of carbons will burn for twice as long a time as the preceding ones, as it will be the longest ; thus the consumption or wear of the pairs of carbons will be almost loxodromic."

*2,013.—A. Masson, Bordeaux. **Apparatus for Generating and Conveying Electricity.** 2d. May 9.

DYNAMO-ELECTRIC GENERATOR.—" The application of an earth current to a dynamo machine for regulating the path of electricity generated, combined with the application of dualised wires," is stated to be the nature of the improvement. " An insulating medium composed of cement mixed with powdered iron " is used. The generator consists of two sets of metal discs, one set fixed to a support and the other mounted on a revolving plate, " all such discs being furnished with projecting magnets made to interlace without touching one another."

2,038.—H. J. Haddan, London. (*R. J. Gülcher, Bielitz-Biala, Austria*). **Electric Lighting Apparatus.** 6d. (5 figs.) May 10.

ARC LAMP.—The feed of the carbons is regulated, and the arc established by the same electro-magnet, which acts directly on the upper carbon-holder, and holds or releases it without the intervention of wheels, catches, or grippers. The electro-magnet is carried in trunnions, around which it can oscillate to strike the arc. Referring to the illustration, the current enters the magnet D at the trunnions, traverses the coil, and escapes partly through the spring E to the negative terminal, and partly to the upper carbon-holder, and across the arc. As soon as the magnet is excited it attracts the iron stem F of the carbon-holder, while at the same time its other end moves towards the fixed armature H.

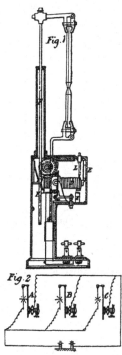

Each end of the magnet is rounded to the arc of a circle whose centre is at the trunnions, consequently, when it moves, its left-hand face rolls over the stem F, raising or lowering it as if the two constituted a rack and pinion. J acts as a magnetic brake to control the speed of the oscillations. As the carbons burn away the magnet gradually rises at its right-hand end until it reaches the stop L, after which any further decrease of current causes the attraction by which the stem F is held to be lessened until it slips past the face of the magnet, and so reduces the arc. A thin covering of brass is interposed between the magnet and the surfaces

that slide over it. Several of these lamps may be placed in parallel circuit and burnt from one generator as shown in Fig. 2. Lamp A is first lighted and after it has become steady the current is turned through B, and again after a short interval through C, and so on, the lamps mutually aiding in the regulation of each other. In order to have the resistance equal throughout the circuit the area of the lower leading wire between A and B is only two-thirds that of the main lead, while between B and C it is only one-third. In the upper lead these conditions are reversed. The specification gives a diagram showing this arrangement extended to twelve lamps in three parallel circuits of four each.

*2,079.—C. H. Gimingham, London. Electric Lamps, &c. 2d. May 12.

INCANDESCENCE LAMPS.—This relates to the method of inserting the conductors carrying the filament. This is done by passing the conductors through a short conical tubular glass stopper ground to fit the stem of the glass bulb, and filling up the space with "vacuum cement" made of a mixture of beeswax and resin. The conductors are severed and then twisted together, the twisted parts being embedded in the cement.

2,198.—C. D. Abel, London. (*W. Tschikoleff and H. Kleiber, St. Petersburg*). Electric Lamps. 6d. (6 figs.) May 19.

ARC LAMP.—The arc is established and the carbons continuously adjusted by two small electric

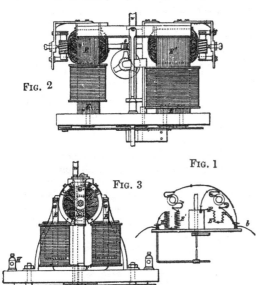

FIG. 2

FIG. 1

FIG. 3

motors combined with the usual mechanism of a lamp. These motors oppose each other, one tend-

ing to increase and the other to decrease the arc, the former being placed in the main circuit, and the latter in a shunt circuit between the lamp terminals. Figs. 2 and 3 are two elevations of the mechanism. E and E¹ are two small Siemens electro-motors, each having two field magnets with extended pole-pieces surrounding its armature. Both armatures are fixed on the same spindle on either side of a worm e, which, through intermediate wheels raises and lowers the rack W attached to the upper carbon-holder. The leading wire b, as shown in Fig. 1, is divided into three branches, the first of which traverses the field magnet coils, and the second the armature coils of E, both meeting at P and going through the carbons and across the arc to the other lamp terminal H. The third branch leads in succession through the armature and magnet coils of E¹ to the same terminal H, the connections being so made that the two armatures tend to rotate in opposite directions. When direct currents are employed one motor only need be used, with the high and low resistance coils wound side by side on its field magnet cores. The carbons are kept in contact by springs when the lamp is inactive.

*2,212.—C. A. Barlow, Manchester. (*A. de Meritens, Paris*). Dynamo-Electric Machines. 2d. May 20.

DYNAMO-ELECTRIC GENERATOR.—This invention is based on the discovery "that the pole of a straight magnet becomes more intense if a mass of metal is applied to its other pole." The generator is composed of two magnetised clusters of horseshoe form, each cluster having two branches with sixty-four plates. These branches are arranged in a curve so that their inner surface is equidistant from the periphery of a ring that rotates between them, each of the sixty-four plates being laid on a cylindrical surface formed in an opening cast in an annular magnet carrier, and in a recess cast in a second annular magnet carrier. The magnet carrier is of bronze and the arms of the magnets pass through the openings. The second magnet carrier is of cast iron, and it forms the base-plate or magnetic yoke common to all the clusters. Outside the bronze magnet carrier on the driving shaft is a ring which contains sixteen bobbins, each composed of eighty plates of soft iron divided so that each bobbin contains in itself four separate small bobbins. The sixteen large bobbins are fixed to the small ring, and the sixty-four small bobbins are covered with suitable wire. The details of the commutators are described, but in the absence of drawings they cannot well be understood. The inductors of the generator are made of two clusters of permanent horseshoe magnets, the four poles of which form an almost complete circle.

2,215.—P. R. Allen, London. **Couplings for Electrical Conductors.** 6d. (21 figs.) May 20.

CONDUCTORS.—To couple the wires used as conductors of electricity in railway trains, the inventor adapts the existing brake couplings to carry the terminals of the electrical conductor attached to each carriage, the contact being made by the bodies of the couplings themselves or by spring plungers, or other devices. In order that the contacts may be kept constantly clean and free from dirt, scouring devices are so arranged as to be brought into operation by the act of coupling. The illustrations are elevation and plan of an electrical connection with the Westinghouse brake coupling. The electrical contact is made by a pair of plungers B B in connection with the conductors C C which extend along the train ; when the brake couplings are locked together the plungers are pressed into contact by spiral springs. Within each half of the brake coupling there is one of these plungers and also a piece of cleaning or scouring material A. In the act of locking the coupling each half turns a portion of a revolution on the other, and the plungers before coming into contact are caused to

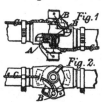

pass over the cleaning or scouring material A A, and thus the dirt is removed and good contact insured. The specification illustrates the invention in connection with the Smith and Clayton couplings, and also with other contact devices in place of the plungers B B. It also describes a cable coupling consisting of male and female parts, the male part having trunnions and a spring plunger, and the female part hooks and a recess for the reception of the plunger.

2,217.—W. R. Lake, London. (*P. B. Delany, New York, U.S.A.*) **Electrical Cables and Conductors.** 6d. (8 figs.) May 20.

CONDUCTORS.—A number of discs, pierced with holes, are threaded on a wire rope. The conducting wires are passed through these holes, the intermediate parts of the discs acting as distance pieces to keep the wires apart.

2,256.—W. R. Lake, London. (*W. C. Allison, Philadelphia, U.S.A.*) **Supporting Structures for Electric Wires or Conductors, &c.** 6d. (8 figs.) May 24.

CONDUCTORS.—These are carried on light metal arches thrown across the streets and furnished with insulators.

2,263.—J. C. Cuff, London. **Apparatus for Effecting Electrical Measurements.** 6d. (6 figs.) May 24.

RHEOSTAT.—The object of the invention is to simplify and improve resistance boxes. The bobbins are wound with a single wire helically, the direction of winding being reversed in each layer. A hollow plug is fitted into the end of each bobbin to remove it from the ebonite baseplate. Each bobbin is supported by means of a brass rod which passes through it, and is suitably fixed in the base-plate. The upper end of this rod forms a contact stud, and the lower end is flattened to receive a cross-pin, which retains in place a block of vulcanite; this block is pressed against the pin by a spiral spring mounted on the rod, inside the bobbin, the cross-pin being connected to the coil of the next bobbin. Over the brass rods metal bars are arranged with knobs sliding upon them, which can be put in contact with the rods to connect them to the bars. The box is provided with a Wheatstone bridge and the connections to the terminals are made in the usual way.

*2,264.—W. C. Barney, London. **Cables or Conductors for Electrical Circuits.** 4d. May 24.

CONDUCTORS.—To prevent induced currents from neighbouring conductors, an insulated conductor has a metallic covering which is insulated from the earth, or the conductor may have upon it insulated metallic sleeves about fifty feet in length, and separated by intervals of ten feet. The insulated conductor and its metallic covering may be encased in insulating material, and the conductor and metallic sleeves may be similarly encased.

2,272.—J. W. Swan, Newcastle-on-Tyne. **Secondary Batteries or Apparatus for Storing and Conserving Electricity.** 4d. (6 Figs.) May 24.

SECONDARY BATTERIES.—To facilitate the construction of secondary voltaic piles, plates are employed having a cellular, corrugated, or grooved surface or surfaces, the object being to obtain an interstitial construction of plate capable of affording a very large amount of acting surface in a small

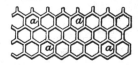

compass, and to prevent the coating of oxide or spongy lead from falling away from the plate, as it

would from a plain vertical surface unless held in position by some material external to the said coating. The illustration shows a plate with carriers *a a* for the reception of the spongy lead; the outer surfaces may also be covered with the same material.

*2,304.—W. Crookes, London. Manufacture of Apparatus for the Exhibition of Electric Light. 2d. May 25.

INCANDESCENCE LAMPS.—To insure good contact between the carbon filament and its connecting wires, metal is deposited by the electrotype process on to the filament and connecting wires at and around the connections.

*2,323.—J. H. Johnson, London. (*La Société La Force et la Lumière, Société Générale d'Electricité, Brussels*). Secondary Electric Batteries. 2d. May 26.

SECONDARY BATTERIES.—This invention relates to methods of partitioning or arranging the electrodes for the purpose of retaining the products of the process of electrolysis against the sides of the electrodes. For this purpose a porous vessel is described, similar to that mentioned in Patent No. 1,630 of 1881, and formed by folding a sheet of felt or other material in any suitable number of radiating folds. In each compartment of this porous vessel is placed a sheet of metal capable of forming a negative electrode. On the exterior of the vessel are placed plates capable of acting as the positive electrode. All the internal plates are electrically connected and likewise all the external plates. In a modification the external plates are placed in a second porous vessel. The electrodes may be ropes of lead or zinc wire, the partitions then consisting of a gimp of woollen thread.

2,344.—P. L. M. Gadot, Paris. Electrical Lighting Apparatus. 8d. (13 Figs.) May 27.

ARC LAMP.—This invention relates to a lamp for holding candles of the Jablochkoff type, and for automatically effecting the lighting of each in succession as the preceding one is consumed. Figs. 1 and 2 show an ordinary cruciform lamp with four candles, each of which is held between a couple of jaws. One jaw of each couple is connected to the same terminal B; the other jaws have each a separate terminal *b*. To each candle-holder there is added (1) an expanding spring D, consisting of two or more metals that are expanded differentially by heat; (2) a contact device C having springs *r r* in electric communication with the next holder; (3) an extinguisher E placed, in this instance, in the centre of the lamp in communication with the central terminal B. The working is as follows:

The spring D keeps the piece C out of contact with the inner jaw, but when the candle is on the point of going out the heat of the arc causes the spring to expand, and in expanding it ceases to hold the clip C, which, acted upon by its springs *r r*, bears strongly against the inner jaw, thus placing the next inner jaw and consequently the next candle in the electric

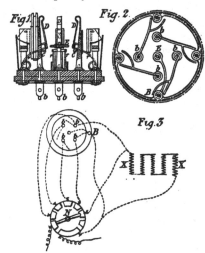

circuit. If the arc does not then leave the first candle and establish itself on the second, the spring continues to expand until it touches the extinguisher E, thus putting the first inner jaw in communication with the central terminal and extinguishing the candle. It is not necessary that the current should traverse the springs *r r*, and if desirable, flexible tongues can be arranged to provide a path for the current. When it is necessary to be able to switch the current upon the breakage of a carbon, a "cross circuit" (Fig. 3) is employed. In case of accidents the movable contact piece of the commutator N is placed upon the contact piece corresponding to the candle which follows the broken one. X X represents a resistance coil that is thrown into circuit when the last candle is consumed. The specification further illustrates the invention as applied to lamps in which each pair of jaws has two independent terminals, and it also includes a modified form of extinguisher, in which the current is caused to traverse a bath of mercury.

2,369.—S. Cohne, London. Electric Lamp. 4d. May 30.

ARC LAMP.—The lamp is "based on hydrostatic principles." Within a cylindrical vessel containing glycerine there is placed a suitable float, which is submerged, and by its buoyancy supplies the motive power for the movement of the carbons. It supports two double racks secured to a top and bottom crossbar forming together a frame

"worked by vertical pinions." One electrode is fixed in each crossbar in such manner that they are brought opposite to each other point to point. On each of the two crossbars is fixed a telescopic sliding tube to enable the current to pass through the circuit without interfering with the motion of the electrodes. In the base of the submerged vessel is fixed a "ratch" and opposite to it a pawl, "representing the feeder," controlled by an electro-magnet. "The pawl consists of a wheel actuated by a cam attached to an armature of an electro-magnet. The wheel projects so as to throw into gear the ratch of the float. When the current enters the electro-magnet it attracts the armature and with it the cam, pushing the wheel carrying the ratch downwards, separating the two carbons. As soon as the current breaks, the armature becomes released, the wheel is free to revolve, and allows the ratch to move up and feed the carbons." The arrangement is difficult of comprehension, as the specification contains no drawings.

2,375.—**H. E. Newton**, London. (*C. A. Hussey and A. S. Dodd, New York, U.S.A.*) **Magneto-Electric Machines, Magnets, and Telephones.** 6d. (8 figs.) May 31.

DYNAMO-ELECTRIC GENERATOR. — A permanent magnet M of ring form is enclosed in a case in which it is embedded in plaster-of-paris. The armature A is composed of a number of plates arranged side by side, "and having alternate inward and outward projections." In the recesses between the outward projections the longitudinal coils of insulated wire B[1] are wound, the coils being connected together, and to the commutator plates C. The magnet

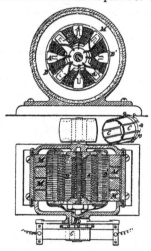

is provided with poles or consequent points, and "the armature travels not only before the poles, but also before the whole length of the magnet, and through the entire field of force; hence, electric currents

are generated throughout its entire revolution." The magnet sections are magnetised by being placed in contact with the poles of a powerful stationary electro-magnet. The parts of the ring "where north and south poles or consequent points are desired to be formed" are placed opposite the south and north poles of the said electro-magnet. A second "electro-magnet, having its poles furnished with inward extensions, is arranged so that its poles will correspond with those of the first electro-magnet, and it is then repeatedly moved over the side of the magnet section from one portion of the periphery towards the stationary electro-magnet, and from a diametrically opposite portion of the periphery towards the first electro-magnet." The section is then reversed, and the opposite side treated in the same way. As shown in the drawings, the electro-magnets are each wound so that both their poles are of the same name. The commutator plates are turned up at one end, and provided with poles in which the ends of the coils are inserted and secured by screws.

2,394.—**S. Pitt**, Sutton, Surrey. (*O. Lugo, New York, U.S.A.*) **Electric Circuits.** 6d. (6 figs.) May 31.

CONDUCTORS.—The inventor refers to Patent No. 1,119 of 1881, and claims the electric conductor shown in the drawing, consisting of two or more

solenoids connected together in series, and having the helical conductor of each solenoid joined to the axial conductor of the next solenoid in each direction and *vice versâ*. The object is to neutralise induction.

2,398.—**R. M.** and **W. V. O. Lockwood**, New York, U.S.A. **Telephones.** 6d. (3 figs.) May 31.

ELECTRO-MAGNETS.—To enable connections to be made at three places on the coil, a short length wire is joined on after one third of the coil has been wound, the three ends being brought through perforations in one of the end flanges.

2,402.—**G. Hawkes**, London, and **R. Bowman**, Ipswich. **Electric Lamps.** 8d. (8 figs.) May 31.

ARC LAMP.—The lower carbon is fed up by mercury, and checked in its rise by an oscillating gripping piece operated by an electro-magnet. Fig. 1 is a sectional plan to an enlarged scale of Fig. 2. B is a tube lined with glass, and con-

taining mercury, in which the lower carbon-holder is immersed, its movements being steadied by the perforated leather piston C. The carbon-holder rises between the arms of an electro-magnet E, and also through a gripping piece F attached to the armature E¹ of the magnet. On the passage of current the arc is formed by the upper carbon holder plate G¹ being attracted by the magnet G placed in the lamp circuit. In order to cause the electro-magnet E to regulate the feed with great accuracy the following contrivance is adopted : H is an electro-magnet in the lamp circuit, between one pole of the generator and the lower carbon. Opposed to this is a second magnet H¹, and between the two is an oscillating armature I. "The armature H¹ connects with the electro-magnet E, and is intended to impart rapidly intermittent currents thereto for the purpose of enabling it to actuate the oscillating gripper F, and allow the carbon C to rise gradually at an almost imperceptible rate." One terminal of E is connected to the plate A, and through the wire 2 and magnet H

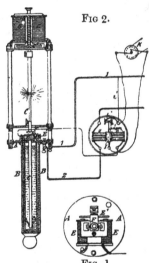

FIG 2.

FIG. 1

to the generator ; the other terminal of E connects with one terminal of H¹, the other terminal of which is joined by the elastic coil i to the oscillating armature I. An extension of I bears against a fixed stop i connected by a wire i to the leading wire I. As the arc grows longer the current in the lamp circuit and the magnet H becomes less, and that in the shunt circuit and magnets E and H¹ greater. At a certain point the armature I will be attracted by H¹, breaking the shunt circuit and rendering inert the two magnets in it. This will momentarily release the carbon-holder from the gripper P and leave it free to rise until the magnet H draws the armature I back against the contact piece again. The arma-

ture will continue to vibrate until the arc is so reduced that the magnet H¹ can no longer draw it away from H. To insure greater delicacy of adjustment, it is proposed to couple together the oscillating armature I and fixed stop I¹, inserting between them the revolving contact breaker K. When there is only one lamp on a circuit the electro-magnet 4 is replaced by a spring. If it be desired to employ two sets of carbons in one lamp, the second set is held out of action by a double-ended spring lever, which takes at one end into a notch in the second carbon-holder, and bears at the other end on the plain part of the first carbon-holder. When the first pair of pencils is nearly consumed a notch on the first carbon-holder comes opposite the end of the aforesaid lever, which immediately enters it, and at the same time leaves the notch in the second carbon-holder, which is then free. The specifiation contains a diagram of connections for lamps in parallel circuit, and in series. In the latter case the magnet H of the controlling device is replaced by a spring which holds the armature I in contact with the fixed stop i.

2,416.—**F. Wolff**, Copenhagen. (*C. P. Jurgensen and L. V. Lorenz, Copenhagen*). **Magneto-Electric Machines.** 6d. (3 figs.) June 1.

DYNAMO-ELECTRIC GENERATOR.—This generator is, on the whole, similar to the Gramme, but has a field magnet of horseshoe form, and a central stationary electro-magnet acting as an inner field magnet, inside the annular armature, and formed as two bar magnets placed crosswise, with their north poles and south poles respectively united by pole-pieces. This interior magnet is carried on, or is united to, a stud projecting from a bracket on the frame, and has its poles opposite to like poles of the outer field magnet. The armature core is built up of rings insulated from each other, and held together by bolts. At each end it is closed by a disc of non-magnetic material ; one of these discs is fixed to the end of a short shaft that runs in a bearing and carries the driving pulley at its further end, and the other is bored to take a bearing and revolves upon the stud that carries the inner stationary magnet. The field magnets may be excited in any usual way, or the generator may work as a dynamo machine.

2,437.—**E. Edmonds**, London. (*G. M. Mowbray, North Adams, Mass., U.S.A.*) **Metallic Circuits for Electrical Transmission.** 6d. (9 figs.) June 2.

CONDUCTORS.—Two conductors are combined in close proximity to each other, one uninsulated, say, of iron, with or without ground connections, and the other insulated, the pair forming a complete

metallic circuit. The illustration shows one method of carrying out the invention. B ,B, B, B are four insulated copper conductors embedded in

grooves in the iron rod A, the edges of which are subsequently closed round them by means of rollers.

2,449.—C. V. Boys, Wing, Rutland. **Apparatus for Measuring Mechanical and Electrical Power.** 6d. (10 figs.) June 3.

POWER METER.—The instrument is designed to measure the work done by the current between two points of a conductor, that is to say, the product of the main current by the difference of the potentials at the two points. An integrating cylinder is mounted in a yoke frame which is caused by a "mangle motion" pinion to travel to and fro along a fixed rod, and also to swing so as to bring the cylinder into contact with each of two discs alternately, as it makes its to-and-fro strokes. The mangle motion is driven by clockwork. The swivel frame in which the discs are mounted is fixed on the axis of a beam carrying at one end a weight and at the other the movable core of a fixed solenoid, wound with an outer and inner coil, both of large wire and in the same direction and placed in the main circuit. The core has an upper and lower coil of fine wire wound in opposite directions, these coils forming a by-pass circuit connected to the two points of the main circuit between which points the work expended is to be measured. As the core is more or less drawn down into the solenoid in opposition to an adjustable pendulum weight the discs are more or less inclined to the axis of the integrating cylinder, causing it, as it reciprocates, to turn more or less round. The number of revolutions of the cylinder in a given time is shown by a counting arrangement. When the instrument is used for a weak current the solenoid coils are surrounded by iron. A detent worked by the spring armature of an electro-magnet stops the clockwork on the cessation of current. The integrating cylinder may carry a band of paper on which a diagram would be drawn by a fixed tracing point.

2,482.—E. G. Brewer, London. (*T. A. Edison, New Jersey, U.S.A.*) **Electric Machines, &c.** 8d. (9 figs.) June 7.

DISTRIBUTING CURRENTS.—When a number of "translating devices" (motors, lamps, &c.) are in

cluded in a circuit fed from one source, it is necessary that the current should be so regulated that each of such devices will receive its proper amount whether part or all of them be in operation at a given moment. Figs. 1, 2, 3, and 4 illustrate one method of accomplishing this. The field magnets of the generator or generators are excited from a separate generator driven by an independent steam engine having a governor (Fig. 2) provided with an adjustable weight, and a handwheel and screw whereby its position can be regulated at pleasure. B, C, D, E (Fig. 1) are four generators coupled in parallel circuit : A is the generator for exciting

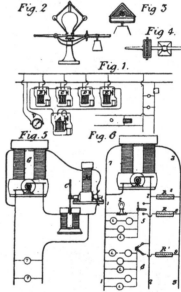

their field magnet coils, and V is an electro-dynamometer in the main circuit. As the current rises or falls, the attendant moves the governor weight out or in, and by so doing varies the strength of the magnetic field. The specification also states that the weight may be moved by an electro-magnet, but does not explain the method. When self-exciting generators are employed the governor is applied to the engine. Another method of attaining the same result consists in varying the electromotive force of a magneto-electric generator, used to oppose the current in the field circuit of the generator, the counter electromotive force having the function of a resistance. An electro-motor M (Fig. 5) of small resistance is included in the circuit of the field magnets of the generator. Upon its shaft is mounted a copper disc C, which rotates between the polar extensions of an electro-magnet included in the field circuit of the generator G, and also in the circuit leading to the translating devices 7, 7. If no lamps are in circuit the entire resistance of

the exterior circuit through A and M is comparatively large, and the magnet A is feebly excited, consequently the disc C encounters little resistance in cutting through the lines of force between the poles, and the motor running at a high speed, throws a strong counter electromotive force into the magnet circuit, by which means the main current is diminished. Another portion of the invention relates to appliances whereby, when one or a number of lamps are thrown into circuit, shall automatically put in operation means of proportionately increasing the generative capacity of the generators. In Fig. 6, the generator helix is connected to the main conductors 1, 2, between which are the groups of multiple arcs 5, 6, containing lamps L. The coils of the field magnets are connected on one side by a wire 7 directly to one of the main conductors. Upon the other a conductor 3 leads to the most remote group, and branch conductors run to each group at 8 and 9, each branch having a resistance R or R¹. If the switch for the circuit 5 be closed, simultaneously the field magnets will receive the proper amount of exciting current. When the second group is closed the total resistance of the field circuit will be again reduced, and the current circulating round the magnets increased. The upper switch is shown open, and the lower one closed. In the former case the wire 8 is not in connection with the wire 2. R² is a resistance through which sufficient current will flow to insure a certain definite amount of current passing through the field magnet coils at all times. When a system contains both lamps and motors, the motion of the latter throws an opposing current into the mains, causing a temporary disturbance of the lights. According to this invention, a resistance equal to that of each motor when in motion, is placed beside it, and when the motor is started a centrifugal governor on its axis draws a finger along the resistance gradually cutting it out of circuit as the speed increases. Another method of regulating the excitation of field magnets is to drive the exciter by an electromotor situated on a shunt circuit from the main leads. Adjustable resistances can be placed either in the motor circuit or in the field circuit. To prevent injury arising from any momentary current having a higher E.M.F. than that normally used, condensers are employed, placed in multiple arc in each house or section.

*2,484.—W. P. Thompson, London. (*F. Van Rysselberghe, Brussels.*) **Neutralisation of Currents in Telegraph and Telephone Lines.** 2d. June 8.

CONDUCTORS. — "The induced currents in the lines are practically neutralised by giving a re-

sistance to the reception coil enormously greater than the line resistance, instead of the two being equalised as is now the practice."

2,492.—P. Jensen, London. (*T. A. Edison, New Jersey, U.S.A.*) **Electric Lamps.** 8d. (17 figs.) June 8.

INCANDESCENCE LAMPS.—To make the globe, the blower takes the glass directly from the melting pot and shapes it to the form shown in Fig. 1, leaving an aperture large enough for the introduction of the carbon and its support. Upon the other end of the globe a small exhaust tube is formed. The leading-in wires are laid in a tube, Fig. 2, in which an enlargement has been blown, and one end is brought to a welding heat and firmly and hermetically sealed on the wires. The wires are made in three parts, a central piece of platinum for sealing into the glass, and two copper extremities, the upper of which is of sheet copper, or copper wire with a flattened end, to wrap round the end of the carbon filament. The wires being sealed in the tube and connected to the filament, the whole is introduced into the globe, and welded in position as shown in Fig. 3. "When carbon filaments are used for lighting by incandescence a phenomenon is found, to which may be applied the term 'electrical carrying.' This is an absolute carrying or moving of the carbon itself from the negative to the positive end of the carbon." The amount depends on "the resistance of the filaments, the degree of incandescence, the electromotive force between the clamping electrodes, and

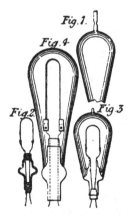

Fig. 1. Fig. 4. Fig. 2. Fig. 3.

the state of the vacuum." In view of this the carbons are arranged so that the portion having the least resistance shall be at the negative clamp, or carbons are manufactured having a less resistance at one terminal than at the other, as in Fig. 4, where one end of the carbon is thicker than the other. "In the present state of the art, however,

it is impossible to manufacture a carbon which will not eventually be destroyed." In order to avoid a more than momentary darkness, two lamps may be placed side by side, and the current automatically switched into the second upon the failure of the first. For this purpose a small electro-magnet is placed in the circuit of the lamp and provided with an armature, which, so long as the current passes, is held closely to it, but if the current ceases it is withdrawn by a spring and completes the circuit of the second lamp. Any number of lamps may thus be placed together and lighted and used successively. The specification illustrates three methods of carrying out this part of the invention; in each case an electro-magnet and armature serves to divert the current. Fig. 5 is an arrangement to

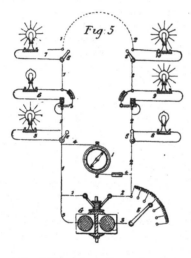

Fig. 5

prevent the failure of one lamp of a series extinguishing the remainder. G is a generator, 1 and 2 the main circuit, 3 the field magnet circuit with an adjustable resistance. I is an electro-dynamometer in the derived circuit 4; 5, 6, 7, 8, 9, 10 are branch circuits. In 5, 7, 8, 10 switches *s s* are shown to short-circuit the lamps: at 6 and 9 are illustrated means for preserving the circuit complete, notwithstanding injury to a lamp. When there is no current the parts stand as in 6. When the circuit is closed, the current divides between the lamp circuit and a small resistance shunt circuit, and, exciting the magnet, lifts the armature and breaks the resistance circuit. When the lamp fails the armature drops on to the lower stop and the current follows the other circuit. A further object of the invention is to obviate the ill effects of the blackening of the globes by the deposition of carbon thereon, by attaching to each lamp a short magnet, presenting such a polarity to the positive side of the carbon filament that it will attract the highly electrified carbon vapour downwards to the clamps instead of

permitting it to be deposited on the glass. In place of a magnet, a solenoid may be used.

2,495.—E. G. Brewer, London. (*T. A. Edison, New Jersey, U.S.A.*) **Electric Arc Lamps.** 6d. (1 fig.) June 8.

ARC LAMP.—Either or both of the carbons are rotated around their longitudinal axes at a speed, by preference, of two to three thousand revolutions a minute. A motor of the Pacinotti or other type in the lamp circuit, or in a shunt circuit, or clockwork, may be used to effect the rotation. Any form of feeding or regulating mechanism may be used, but the one shown in the illustration is well adapted to be combined with rotating carbons. The upper carbon *c* is connected with the rod by the ball-and-socket joint *a*, and is guided near its point by the guide *b*. E is the electro-magnet of the motor, the coils of which are in the shunt circuit 3—4. The revolving armature is supported

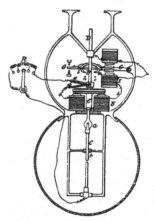

by a sleeve on the top of the frame, through wh slides the rod D, the two being connected by feather; *d e* are commutator springs touching a revolving circuit breaker. An adjustable resistance R is placed in the shunt circuit for the purpose of regulating the speed of the motor. G is a horizontal armature pivotted on the frame of the motor and playing in the fields of the two electro-magnets, of which the upper is in the lamp circuit and the lower in the shunt circuit. The lever G has two spring pawls that clamp downwardly on the rod D, and have arms *l m*, and stops *n o* are located at such points on the frame that the pawls are thrown upwardly away from rod D, when the armature lever reaches a certain point in its downward movement.

2,532.—G. Gouraud, London. (*P. B. Delany and C. H. Johnson, New York, U.S.A.*) **Electric Cables.** 6d. (6 figs.) June 10.

CONDUCTORS.—These are separately insulated and arranged side by side to form a flat cable, and are imbedded in soft metal such as lead. By this means their inductive action is mainly expended on the sheathing; and to further prevent induced currents the conductors are crossed at each joint or

coupling, as shown in Fig. 3, so that each wire in turn occupies every one of the six available positions. The metal cover may be flat, or grooved as in Fig. 1. Fig. 2 illustrates the method of manufacture. The whole is run between rolls which force the soft metal to completely fill the interstices.

2,538.—M. R. Ward, London. **Electric and Magnetic Brakes.** 6d. (1 fig.) June 10.

ELECTRIC LIGHT.—Each vehicle carries an electric generator driven from the axle. On closing the circuit resistance is offered to the movement of the train, and the current so produced is stored in secondary batteries and utilised for lighting and heating purposes.

2,542.—S. J. Mackie, London. **Electrical Insulated Wires and Conductors.** 4d. June 11.

CONDUCTORS.—This invention consists in insulating wire or core by flat servings of yarn impregnated with an insulating compound. Above the yarn other tapings or servings may be wound, and the whole may be finally braided. The coverings may be waterproof in themselves, or be saturated with waterproof materials and dusted with powdered glass, or be waterproofed by any means, such as gelatine and bichromate of potash. For the protection of insulated conductors they may be placed in troughs or tubes, which are filled with compositions of powdered glass and sulphur, cement, plaster-of-paris, or bitumen.

2,563.—G. G. André, Dorking. **Electric Lamps.** 6d. (10 figs.) June 13.

ARC LAMP.—In this lamp the upper carbon is suspended from an iron rod B, which is held up by magneto-frictional contact with pole-pieces of a magnet suspended from a pair of electro-magnets D D, so long as the arc is maintained within the determined limit of length. The greater part of the entering current is passed through the magnet D, which is attached to the plate E. A branch I¹, from the main wire I, connects to a smaller electro-magnet K of greater resistance, having two coils wound in opposite directions; the pole-pieces C¹

are carried out to the iron rod B, and are shaped to embrace it. "A third electro-magnet R, having similar pole-pieces C² C³, is arranged on the other side of the rod B, and is wound with a coil connected to one of the coils of the electro-magnet K, and forming a second branch from the main wire I, and so that the resistance of this coil and the coil on the magnet R are together equal to that of the other coil on the electro-magnet K; the exit end of the wire is connected to a bracket M, which is insulated from the plate G and extends over an armature N, which may form electrical connection with it. The plate G carries a fourth electro-magnet O, the coils of which are wound with wire of very high resistance, and form a shunt circuit past the arc. The weight of the magnet R, or rather of its core, is supported by a light spring, and its upper pole-piece is in close proximity to a peg or stop under the plate E." The action is as follows: "When the carbons are together and the current turned on, the magnets D, K, and O become

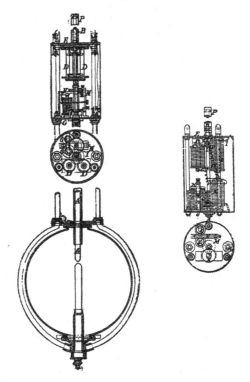

excited, the main current passing through D and K; the pole-pieces C will then adhere to the rod E, and the electro-magnet D will draw up the electro-magnet K; the arc will thus be established. When the arc becomes in the least lengthened, the magnet O closes the circuit through that coil of the magnet K which is connected to the bracket M, and through the coil of the magnet R by means of

the armature N, and thus neutralises the adhesive action of the pole-pieces C, and excites the pole-pieces of the magnet R, the spring of which will be slightly compressed by the superior weight of the rod B and carbon A, and will drop on to the peg, whereby the current through the magnet O is decreased, and the circuit through one of the coils of the magnet K and through the magnet R broken. The magnet K will then resume its action, and the pole-pieces C adhere to the rod B." In a modification the electro-magnet R is omitted, and the tube P on the top of the lamp is surrounded by a cylinder containing glycerine and having an annular weight or easily fitting piston with two holes covered with a valve ring. The rod B is counterbalanced by a weight. In another arrangement, where a lower carbon is fed upwards, and which is also applicable to incandescence lamps having such a carbon rod, the lower carbon is clamped to an iron cylinder provided with a long loosely fitting float piston supported in a cylinder of mercury. In this case the magnet K has its coils wound in opposite directions, the inner being arranged to form a branch from the main conductor, and the outer coils, of high resistance, to form a shunt past the arc. Another pair of electro-magnets with coils of lower resistance, and corresponding with the magnets D in the illustrations, are arranged underneath the first-named magnets, and have their wire coils connected, one to the positive terminal, and the other to the carbon supporting rod. These electro-magnets serve the purpose of separating the carbons when first started, by drawing down the first-named, or feed magnets, as in the previous arrangements. As the arc lengthens the pole-pieces of the magnet K are demagnetised so far as to allow the rod to slide past them, impelled by the float. When the arc is adjusted the rod is again held. Instead of the branch and shunt connections it is possible to use the arrangements above described for the purpose of exciting the feed magnets. The whole of the framing of the lamp may consist of a tube, the upper end of the carbon only extending above it, and having a protecting casing around the magnet. In all the preceding cases the negative carbon is supposed to be fixed. Where a weight and endless cord are employed for feeding a carbon in an incandescence or arc lamp, as in Patent No. 2,252 of 1880, the retarding action of a piston and liquid is employed, and for this purpose the lower pulley, over which the endless band passes, is provided with a crank-pin for working a perforated piston in a small cylinder of liquid. This pulley is also provided with an iron disc wheel to which a brake is applied by magnetic friction, regulated, as above described, according to the length of the arc.

2,572.—H. E. NEWTON, London. (*C. A. Hussey and A. M. Dodd, New York, U.S.A.*) **Electric Lamps.** 6d. (7 figs.) June 13.

INCANDESCENCE LAMP. — A translucent globe, shown in Figs. 1 and 2, has prolonged ends, through which extend holders B B[1] for carbons C, one of which can be heated to incandescence at a time. The holders B are all connected to the metal band D, and the holders B[1] to insulated plates E on a commutator, the upper one of which is then in contact with spring G, which forms one of the lamp terminals. The current entering the lamp at D, traverses one carbon and leaves at E. When that carbon breaks or becomes inefficient the

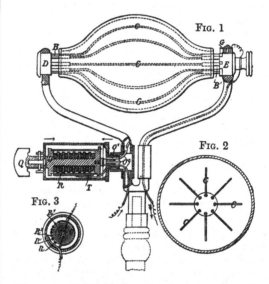

FIG. 1

FIG. 2

FIG. 3

lamp can be partially rotated, and a second carbon brought into circuit. An electro-magnetic appliance for effecting the rotation automatically, on the failure of the current, consists of an electro-magnet placed in the lamp circuit having an armature fitted with a pawl, which engages with the teeth of a ratchet wheel fixed to the longitudinal shaft of the lamp. The current can be regulated by resistance coils upon a spool O actuated by the handle Q. Outside the coils are metal bars R, R[1], R[2], &c. (Fig. 3), which are connected, at short distances apart, to the flanges of the spool O, and are thereby insulated from each other. The bars are severally coupled to the wires of the respective coils. One wire of the circuit is joined to a loose sleeve O[1], on a metallic stem O[2], extending from the spool, so that the wire will not be turned when the spool turns. The bar R is in electrical communication with the stem O[2]. The leading wire goes to the fixed contact spring T, which bears on the bars R[1], R[2], &c., when the spool is rotated. When T rests on R the coils are out of

circuit, and when it rests on R¹ the first coil is in circuit, and so on.

2,573.—H. E. Newton, London. (*O. A. Hussey and A. S. Dodd, New York, U.S.A.*) Means for Supporting and Protecting Electrical Wires, &c. 6d. June 14.

CONDUCTORS.—This invention consists in combining with the kerb of the sidewalk a conduit for receiving and protecting electrical conductors. Fig. 1 illustrates a number of buildings and streets having wires for electric street lamps and houses conducted to them, and Fig. 2 is a horizontal section of a sidewalk. Around the kerb of each block extends a conduit H. On the block 9 is a station I, where the generators are arranged, one for the street lamps and one for private users. The street lamps are marked L and are supported on crosspieces extending from block to block. O is another station for the supply of the houses on the faces of the blocks 6, 5, and 4,

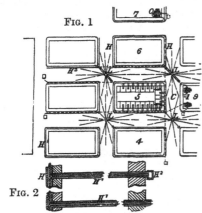

FIG. 1

FIG. 2

that are on the street C. The conduits are of segmental form, provided with flat backs to rest against the kerb. H¹ (Fig. 2) is a branch into a house; H² is the same branch stopped by a cap when not required for the passage of a conductor.

2,592.—W. R. Lake, London. (*H. A. Clark, Boston, U.S.A.*) Manufacture of Telegraphic or Telephonic Conductors, &c. 6d. (10 figs.) June 14.

CONDUCTORS.—The object of the invention is to adapt gutta-percha covering machines firstly to coat

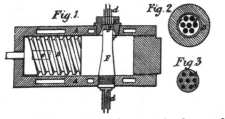

Fig. 1. Fig. 2.

Fig. 3

a wire which has or has not been previously coated,

and is of a nature to yield either in thickness or length, or both ; and, secondly, to produce a compound conductor with a series of parallel wires inclosed in one mass of insulating material. The apparatus consists of a guide for the wire, which at and along its end at which the wire passes into the die opening has parallel sides projecting into the die opening. The illustration shows the invention as applied to forming a compound wire, Figs. 2 and 3 being enlarged sections of the die F (Fig. 1). A is a chambered cylinder, B the screw, e the hole opening to the said screw, F the guide for the wires d, E the die having an opening f. The wire guide F is provided with a series of tubes g, each adapted for the passage of a wire, and separated by a free space from its neighbours, so that the insulating material can circulate between them before reaching the wires issuing from their ends.

*2,606.—A. Muirhead, Westminster. Electric Accumulators. 2d. June 15.

SECONDARY BATTERIES.—The battery is made of thin sheets of metal separated by a permeable material and arranged like a condenser. The sheets are coated with sesquioxide and red oxide of lead, or with red oxide of lead and peroxide of manganese, or alternately with a mixture of red oxide of lead and peroxide of manganese and red oxide of lead only ; or there is interposed between the sheets acetate or nitrate of lead.

2,612.—W. Crookes, London. Electric Lamps, &c. 4d. June 15.

INCANDESCENCE LAMPS. — In constructing the filaments cellular tissue or cellulose, such as paper, linen, or cotton thread is purified from silica by hydrofluoric acid, and treated with a solution of cupro-ammonia. By this means the structure of the cellulose is wholly or partially destroyed, and, after evaporation of the ammonia and extraction of the copper, it is obtained in a dense form. The air may be expelled from the fibre by a previous soaking in strong ammonia. The copper can be extracted by dilute acid or by the passage of a strong electric current through the carbonised filament. Another way is to entirely dissolve a loose form of cellulose, such as cotton wool or Swedish blotting paper, and then by evaporation to produce a thin film upon the bottom of the dish. Partially dissolved threads may also be steeped or dipped in a thick solution of cellulose to fill up their pores. It is not necessary to extract the copper from the filament at its junction with the clamps, as its presence there is advantageous. These junctions may be coated with a thick solution of cellulose. The filaments are stamped out of sheets and twisted into the desired shape while still damp, and they are carbonised in any ordinary

manner. In case the filament after carbonisation does not possess the right resistance, or is imperfect, it is heated electrically in a vacuous vessel which contains a hydro-carbon compound, of which the boiling point is high, and the vapour tension at ordinary temperatures very low. Naphthaline or zylol or chloroform are suitable hydro-carbons for the purpose.

*2,618.—J. Jamieson, Newcastle-on-Tyne. Governing Electric Machines. 2d. June 16.

GOVERNOR.—Solenoids or magnets are employed to throw mechanical appliances into gear, to control the current in a main circuit, or to control the magnetism of the field-magnets of electrical generators.

*2,635.—W. C. Johnson and **S. E. Phillips**, Charlton, Kent. Floating Apparatus for Generating and Conveying Electricity for Production of Electric Light and Transmission of Power. 2d. June 16.

ELECTRIC LIGHT.—The necessary machinery is mounted on a suitable barge or vessel, so as to be easily moved from place to place.

2,703.—J. Richardson, Lincoln. Electric Appliances for Moving Tramcars, &c. 2d. June 20.

MOTOR.—The car is driven by a motor actuated by the current from a secondary battery, which is charged at intervals from a stationary generator. The momentum of the car in descending an incline is utilised for partially re-charging the battery.

2,711.—T. A. B. Putnam, New York, U.S.A. Electric Danger Alarms or Signals for Railways, &c. 8d. (10 figs.) June 21.

DYNAMO-ELECTRIC GENERATOR.—A constantly working dynamo-electric generator is mounted on the locomotive. The circuit is completed through the rails, and includes the coil of an electro-magnet. So long as the circuit remains complete no signal is given, but when it it is broken the armature of the electro-magnet is released and an alarm sounds.

2,739.—H. E. Newton, London. (*A. Gravier, Paris*.) Distributing Electricity. 8d. (8 figs.) June 22.

DISTRIBUTING CURRENTS. — This specification appears to relate to an arrangement for feeding a number of arc lamps in parallel circuits from one generator, and contains a long mathematical investigation of the conditions to be fulfilled, and upon which the invention is based. The invention consists, it is stated, in a novel application of derived circuits calculated according to the laws of Joule and Ohm, and arranged so as to (1) distribute the

electricity in determined proportions; (2) to preserve to each apparatus the proper intensity; (3) to assure the independence of the various apparatus; (4) to keep the production of electricity always equal to the requirements. The invention comprises a return wire, a regulator of production, a regulator of consumption, a commutator (see illustration), an equaliser of light, and a distributor which may be either a secondary battery or a mass of metal. When three lights are to be fed, one leading wire is connected to the equalisers of light, from which currents pass to the lamp, and the other directly to the lamp. "The conductors should normally form reservoirs of electricity, but as the consumption augments or diminishes, the conductors pass from the condition of reservoirs to that of channels of conveyance, and to cause them to return to their normal condition the production of electricity should vary." This variation is brought about by a small motor that acts on the steam supply of the engine. The distributor comprises a prism P carrying contact pieces. In one angular position it connects the terminals A B and wires L N, and when

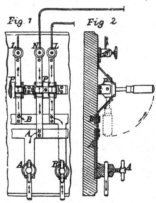

turned into another position it breaks the connection. The "equaliser of light" is an arm moving over the terminals of a number of resistances so as to include more or less of them in circuit.

2,761.—L. A. Groth, London. (*D. Lachaussée, Liège*.) Electro-Magnetic Induction Machine for Dividing a Direct Current into Alternate Currents. 6d. (2 figs.) June 24.

DYNAMO-ELECTRIC GENERATOR.—This generator is so constructed that all the bobbins are independent of each other, and are so arranged that each may be put into and taken out of place without interfering with the others. As shown in the illustrations, the rotating plates A A carry the field-magnet bobbins B B, the two sets of which are separated by a fixed drum D, in which is placed any suitable number of metallic spirals E, of small thickness, but of a diameter sensibly equal

to that of the electro-magnet bobbins. The figure shows the construction of the drum D, which is made of wood; one bobbin E is shown in its recess, while the other recesses are empty. "The

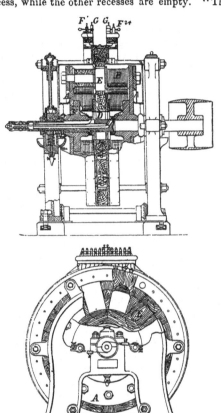

spirals are wound round a very subdivided magnetic material, or are even wound simultaneously on the wires or strips." All their ends are connected to small terminals F^1 to F^{24}, parallel to which are placed large terminals G and G^1, which serve to connect these to one or more return wires, and "allow of the grouping of the currents in tension or in quantity, even when the machine is working."

*2,770.—R. H. Courteney, London. Electrical Resisting Mediums. 2d. June 24.

INCANDESCENCE LAMPS.—The filaments of these are made of silk thread rubbed with saccharine matter, by preference the fruit of the date, then with plumbago, saccharine matter, and asbestos. Another method consists in carbonising date fruit and stones, and cloth, and mixing the resultant mass with asbestos and cotton carbon powder. The filament is mounted in the usual way in an exhausted globe.

2,782.—H. E. Newton, London. (*Société Universelle d'Electricité Tommasi, Paris.*) Secondary Batteries. 6d. (5 figs.) June 25.

SECONDARY BATTERIES.—As will be seen from Fig. 1, the electrodes A A^1 are formed of lead plates cast with a series of shelves inclining upwards ; in the cells so formed sheets of lead foil d d^1 are laid. The lead foil may be dispensed with, and the plates with their shelves made of an alloy of lead and tin. In this case the shelves are formed closer together. The tin is extracted by electrolytic action, so as to

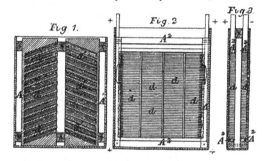

leave the lead porous. The spaces between the shelves are then filled with pure lead filings, or with lead precipitated from one of its salts, the preparation known as "Saturn's tree" being preferably employed. In the modification shown in Figs. 2 and 3 the shelves are dispensed with, sheet lead being cut into strips which are laid one above another on the bottom flange of the plates. The pile of strips is then soldered to its supporting frame, the vertical wall of the plate, as well as the shelves, being in this case dispensed with. Two of these frames A^2, containing strips of lead d, are placed in a vulcanite vessel about ⅜ in. apart, and rest upon blocks of wood.

2,788.—B. J. B. Mills, London. (*F. Million, Lyons, France.*) Apparatus for Obtaining Light by Electricity. 6d. (5 figs.) June 25.

ARC LAMP.—Each of the two carbons is mounted in a carriage a a^1, and is impelled towards the other by cords wound round one of the windlasses H H^1, which are turned by cords attached to the counterweight J. The travel of the carbons is controlled by cords wound on the barrel F, and gradually let off according to the consumption. Fig. 2 shows the regulating apparatus developed in one plane, and Fig. 3 is a section of a modified form of carbon-holder, in which the current passes through the spring d and roller b. Upon the axis of the barrel F is a ratchet wheel K, "furnished with a spur M and operated by a lever N, having at one end a counterpoise O, and at the other a fork, to which are attached two bars P P^1 passing through two solenoids Q Q^1; a stop R serves to work the click

L." The core of the resistance coil Y is formed of a bundle of iron wire. This construction " no longer requires long lengths of conductor wound thereon, provided that it forms two currents of induction and of contrary direction." Q' is "called the solenoid of substitution, because at the time of lighting the lamp it is substituted for the solenoid of regulation, and permits the lighting without oscillation of the arc." Z is a solenoid whose core dips into an insulated vessel of mercury. Assuming the carbons to be apart and the circuit to be completed, the current entering at r (Fig. 2) will follow the conductor s, and will pass between the contacts 3 and 4 through the mercury, which is raised by the core I lying in it. It will then traverse the resistance coil Y and the solenoid Q' to the opposite

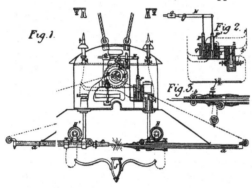

Fig. 1.   *Fig 2.*   *Fig. 3.*

conductor, and in so doing "will magnetise the bar P', and attract it with so much power that all the current will pass by this solenoid," and will draw down the lever N. "This last will draw back the click L, which will be raised by the friction of the spur M on the stop R." The divider will thus be rotated and the carbons moved together; whereupon the current will partly traverse the solenoid Z, and drawing up the cores will allow the mercury to fall away from the contacts 3, 4, and thus cut the solenoid Q' out of circuit. The counterweight O will then raise the cores P and P', and at the same time the click L will turn the ratchet wheel and thus separate the carbons to establish the arc, after which its regulation will be effected by the shunt solenoid Q. Should it totally fail, a circuit is again opened through the mercury. The lamp circuit is completed by hauling it up, when the conical-topped terminals of the lamp come in contact with the correspondingly recessed circuit terminals as shown in Fig. 1.

*2,807.—A. C. Ranyard, London, and **J. A. Fleming**, Cambridge. **Telegraph Cables, &c.** 2d. June 27.

CONDUCTORS.—Manilla or other fibre is twisted or plaited round the conducting wire, and impreg-

nated with "paraffin butter," by forcing the same into the fibre while the conductor is in an exhausted chamber.

*2,823.—A. P. Laurie, Edinburgh. **Secondary Batteries.** 2d. June 28.

SECONDARY BATTERIES.—Electrodes of zinc and carbon are immersed in a solution of iodide of zinc in water.

*2,833.—G. G. André and E. Easton, London. **Electric Incandescent Lamps, &c.** 2d. June 28.

INCANDESCENCE LAMPS.—The filament is made from natural fibre, without alteration of form or structure other than results from carbonising, and is placed in its straight or natural form between the metallic conductors. The globe is exhausted in the ordinary way, the conductors entering at opposite sides or ends.

2,848.—J. G. Lorrain, London. **Treatment of Carbon for Electric Lighting.** 4d. June 29.

CARBONS.—Carbon rods, filaments, &c., are impregnated with carbon by employing a substance containing carbon in conjunction with a substance that will act upon, or be acted upon, by the former, so that decomposition shall take place and the carbons shall be deposited in and upon the body under treatment. As an example, carbon rods soaked in turpentine are treated with gaseous chlorine at gradually increasing temperatures up to bright red. Another method is to boil them in syrup and carbonise the occluded sugar by sulphuric acid.

2,851.—W. R. Lake, London. (*J. J. Wood, New York, U.S.A.*) **Electric Lighting Apparatus.** 6d. (3 figs.) June 29.

ARC LAMP.—This invention relates to a means of affording a path round a faulty lamp when its

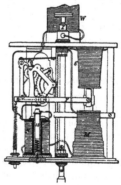

resistance rises above a determined amount. Re-

ferring to the illustration, M and S are differential coils, the latter being in a shunt circuit. These act conjointly on a lever F pivotted in the frame, and carrying a small cage in which is mounted a train of wheels, the first of which gears with the carbon-holder rack, and the last drives a fly. This fly is controlled by a spring detent *b* fixed on a connecting spring which acts also as a connecting bar to give the cage an approximately parallel motion. The solenoid K, in the lamp circuit, attracts its core 15, situate under a contact piece, in opposition to a spring. Should the feed gear stick or the carbons break, the solenoid K will release its core and make a short circuit for the current, from the lamp frame through the resistance W (equal to the arc) to the insulated terminal I. Should the carbons again meet, the current will divide between them and the resistance W and the core 15 will be drawn from the contact piece.

2,879.—E. Tyer, London. **Railway Signalling, &c.** 10d. (16 figs.) July 5.

Electric Light.—The circuit of an incandescence lamp is completed by the passage of a train over one part of the line, and broken so soon as the train leaves that portion of the line.

2,930.—E. P. Ward, London. **Electric Lamps.** 2d. July 5.

Incandescence Lamps.—Iron or steel wires are used in place of the usual platinum conducting wires, each being first coated with glass, the two coatings are then connected together by fusion at a suitable distance from one end, and one afterwards fused to the stem of the lamp bulb.

2,944.—J. P. Hooper, London. **Manufacture of Electric Telegraph Wire Ropes or Cables.** 6d. (1 fig.) July 5.

Conductors.—To prevent the springiness and tendency to kinking of a cable, the sheathing is made with strands having alternately a right-hand and left-hand lay ; as shown in the illustration A is

the core, B a covering of hemp, and C C¹ the alternate strands forming the outer sheathing. Two machines, revolving in opposite directions, are used in the manufacture, the right-hand strands being laid on by one and the left-hand strands by the other.

2,954.—P. Jensen, London. (*T. A. Edison, New Jersey, U.S.A.*) **Electric Machines.** 6d. (8 figs.) July 6.

Dynamo-Electric Generator.—The invention relates firstly to constructing the armature without an iron core, and, secondly, to constructing a generator so that it will generate a continuous current of high E.M.F. in the same direction without the use of " pole changers," all the inductive portions of the armature being constantly in circuit, and the internal resistance very low. This is accomplished by constructing the armature as a disc divided into radial sections 1 to 16 (Fig. 1). These sections are bare copper bars joined edgewise by insulating

Fig. 1

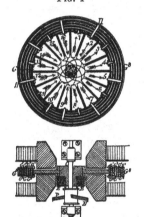

Fig. 2

material, shown by black radial lines in the drawing. The radial bars are turned outwardly at their inner ends (Fig. 2), and are each separately connected to one of the insulated circular plates E surrounding the hub. Each plate has a tongue, turned outwardly and secured in a groove of the commutator cylinder. " The bars nearest to diametrically opposite positions are in this way connected together in pairs, and with a commutator between them," as diagramatically shown in Fig. 1. The bars revolve between magnetic poles (Fig. 2), and outside these are surrounded by concentric insulated copper rings G¹ to G⁸. The number of exterior rings and of plates on the boss is severally half that of the radial bars. Each ring is connected to two of the metal bars, as shown by white lines in Fig. 1, and connects two radial bars, the terminal bar of one opposite pair being thereby connected to the initial bar of another pair, so as to make a continually closed circuit through all the bars. The method of connecting the bars and rings together and to the commutator is described. The concentric rings being outside the polar field generate no current. The neutral line extends vertically through the centre of the armature, the brushes P N making contact at the ends of a horizontal diameter ; the bars next the neutral line

a a

are connected with the central side commutator bars as shown. To strengthen the disc a supplementary disc may be added formed by winding a thin strip of iron around the hub in spiral convolutions with a strip of paper interposed. Around this disc is shrunk an iron ring, and radial bolts are passed through holes in the disc, and screwed into the hub. From its spiral shape this disc does not cut the lines of force, but becomes a detached portion of the magnet. Or the copper bars may be made double with a strengthening disc between them. Fig. 4 illustrates the application of concentric rings for making the multiple

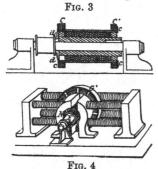

FIG. 3

FIG. 4

arc connections of the copper bars in a generator or motor having an armature consisting of a cage of copper bars united at their ends in pairs by discs. At each end of the cylindrical armature is arranged a series of insulated copper rings C C¹, placed outside the polar extensions of the field magnets. Each ring is provided with two projecting fingers c, which are turned inwardly and connected with the ends of the proper longitudinal bars. Midway between the fingers, at the commutator end of the armature, each ring is connected with a bar of the commutator cylinder by means of an angular piece d.

2,989.—J. Hopkinson, London. **Apparatus for Transmission of Power by Electricity.** 6d. (7 figs.) July 7.

MOTORS.—This invention relates to the use of a spring in combination with the reversing arrangement described in Patents Nos. 2,481 and 4,653 of 1879, for throwing the brushes into a neutral position when the load is removed; and to a method of varying the speed of vehicles, &c., propelled by electricity, by using two or more motors having two or more independent circuits and coupling them in parallel circuit or in series for high or low speeds respectively.

3,010.—H. Gardner, London. (*E. Roe, Perth, West Australia.*) **Signalling Code and Apparatus.** 8d. (5 figs.) July 8.

ELECTRIC LIGHT.—The "electric light may be employed for transmitting the signals."

*3,015.—W. R. Lake, London. (*J. J. C. W. Greb, Frankfort - on - the - Maine.*) **Electric Lighting Apparatus.** 4d. July 8.

ARC LAMP.—This provisional specification gives a long description of an arc lamp, but does not contain any drawings. The lamp is intended to burn in series with other like lamps, and includes a carbon resistance that is put in circuit when the lamp fails. Although the details of the mechanism are minutely described, in the absence of drawings the action cannot well be gathered from the specification.

3,032.—Sir W. Thomson, Glasgow. **Regulating Electric Currents, &c.** 10d. 9 figs.) July 9.

SECONDARY BATTERIES.—This invention relates to, firstly, an automatic appliance for regulating by means of electro-magnetic force, the work of an electric accumulator by increasing or diminishing the number of working cells. One appliance gives a constant strength of current and another constant electromotive force. In the former case the electro-magnetic force due to the whole current flowing through a short thick coil is used, in the latter case the force of a branch current through a fine coil connecting the working electrodes is used. A (Fig. 1) is the first plate of the first cell of the battery, B the connected second plate of the first cell and first plate of the second cell; C, D, E, &c., represent the other plates; a^1, b^2, b^1 c^2, &c., are insulated studs. The studs a^1 to h^1 are con-

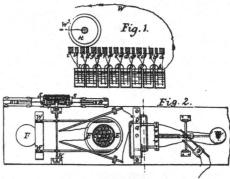

nected to the plates A, B, C, &c., by short thick wires; b^2 to l^2 by wires having considerable resistance, each doubled on itself to minimise the electro-magnetic inertia. The plate A is connected by the wire W to the working circuit or to the charging circuit; R is a roller by which one or two of the studs are put in connection with the working or charging circuit by the wire W¹. The connection may also be made by letting the roller run in a **V** groove of which one of the sides is composed of the flat inclined faces of the said studs, and the other side of a straight continuous metal bar to

which the wire W[1] is connected. This plan is shown in Fig. 2. The invention relates, secondly, to keeping the strength of an electric current or the electric force between two conductors nearly constant when changes are made in the work, and in cases where a simpler appliance has not the requisite power to effect the regulation. It consists in applying a small electric motor to do the work needed to diminish the steam power for driving the main generator, or to diminish the battery or accumulator. An electro-magnetic make-and-break is used to move the contacts for throwing on and off the small motor. When it is the strength of the current that is to be regulated, the poles of the motor are connected to points of the main circuit on the two sides of a contact, which is broken when the current is too strong, and made again when the current is not too strong. When it is the electromotive force that is to be regulated, one pole of the motor is kept in connection with one of the main conductors. The other is put in connection with and removed from the main conductor according to the electromotive force. In each case the instantaneous effect of throwing the coils of the motor into circuit is to check suddenly the electric action. In the former case this is done by augmentation of resistance in the main circuit, and in the latter by shunting part of the current through the motor coils. When the current or electromotive force becomes too weak, a corresponding automatic arrangement directs an electromotor to augment the exciting power by a reverse action to that described above. The same motor may serve both purposes by being arranged to run in either direction. Any type of motor may be used. That shown in Fig. 2 is a steel magnet surrounded by a movable coil, which does its work by a single stroke. E is the movable coil; W is a counterpoise to it, and M the actuating steel magnet in the regulating motor; *e*, *w*, and *m* represent the similar parts made smaller in the primary regulator or double-acting relay as it may be called; F is a pan to receive adjustment weights when the regulating motor is used without relay; R is the contact roller similar to that in Fig. 1; S S are the studs forming one side of a V groove in which R rolls; V is a continuous bar forming the other side of the groove; B B is a guiding carriage for the roller. The roller and groove are at a considerable distance below the level of the motor, and the former is moved by a long arm depending from the shaft K K, according to the current passing through the coil; Q *q* represent contacts made by the double-acting relay so as to send currents in opposite directions through E according as *e* tilts *w* up or down. A resistance coil is illustrated in the specification made from a coil of iron wire, the rings of which can be successively thrown in

and out of contact by a roller and groove such as Fig. 1. The third part of the invention relates to an automatic make-and-break, by which the charging circuit of an accumulator is made or broken according as the electromotive force does or does not suffice to add to the charge in the accumulator, and is also broken when the current is stronger than desired. In this apparatus a balance beam carries at one end a fine and at the other a thick wire coil, each surrounding a vertical magnet such as M (Fig. 2). The ends of the fine coil are connected to two cups of mercury, and the ends of the thick coil dip into the said cups, which are so placed that when the beam tilts Q up, one or both of its ends are lifted out of the mercury. One mercury cup is connected to one pole of the generator, and the other cup to the corresponding pole of the accumulator. The other poles of the source and accumulator are permanently connected. When a current passes through the fine coil in the proper direction the beam is tilted, and a circuit is made through the mercury cups and the thick wire coil to keep it down. Should the charging current at any time cease, or be reversed, the beam rises and the mercury contacts are broken. The fourth part of the invention consists of arrangements for applying the accumulator or parts of it to the source when recharging is required, and freeing the accumulator or parts of it, when ready for work, and applying them to the work when wanted. Briefly this is effected by separating the battery into parts by additions of cells to one end of a series, and taking off cells from the other end of the same series and joining the detached parts, whether in series or parallel, so that while some of the parts are doing work others may be receiving a fresh charge. To do this automatically two circuits are arranged in the form of neighbouring circles. Either of these circuits may be broken at any point; the first circle is permanently broken at a point, and the two charging electrodes are applied to the two contact pieces thus separated. Similarly the second circuit is broken, and the electrodes leading to the external work are applied to its separated contact pieces. "The accumulator is arranged after the manner of Volta's crown of cups in the neighbourhood of these circles." The connections are made by a shaft rotated at the proper speed by gearing from the dynamo and fitted with a weight having proper projections to act upon the contact pieces. This weight rotates with the shaft, and is lifted gradually and let fall suddenly. A modification consists of a small truck which carries contact springs, and is caused to traverse the circle in a predetermined time.

3,053.—L. Jacobson, Berlin. Battery Telephones, &c. 6d. (14 figs.) July 12.

INDUCTION COIL.—The core is covered with one coil of the primary wire, over this the coils of the secondary wire are laid, and over this again the remaining coils of the primary wire are wound.

*3,073.—D. Graham, Glasgow. Lamp Casings or Holders for Containing and Protecting Electric Lights, &c. 2d. July 14.

LANTERN.—A glass globe is elastically cushioned between an upper and lower metallic portion, and is protected by wires shaped to the contour of the glass. The upper and lower parts are locked together by a padlock. When contact points are used in mines they are covered with oil or water, or are situated within a sealed tube.

3,122. — St. G. L. Fox, London. Electric Bridges for Lamps. 2d. July 18.

INCANDESCENCE LAMP—The filaments are made from fibre, preferably the roots of an Italian grass (Andropogon Ischœmum, or Chrysopogon Gryllus). The fibres are boiled in a solution of caustic soda or potash, their outer skins removed, and they are then boiled in water. The fibres are straightened while still damp and are then bound together and stretched on a block of carbon, buried in plumbago, or hermetically seated in a crucible and raised to a white heat in the usual manner.

*3,146. — G. K. Winter, London. Effecting Intercommunication in the Lighting and the Working of the Brake Mechanism of Railway Trains by Electricity. 4d. July 19.

ELECTRIC LIGHT.—A generator is placed on the locomotive and supplies current to an accumulator carried in the goods van. The invention also comprises a modification of the coupling hooks described in Patent No. 2,637 of 1874, a combination of the means of intercommunication, and an electric brake system.

3,166.—W. Morgan-Brown, London. (*G. P. Harding, Paris.*) Electric Lamps. 6d. (5 figs.) July 20.

ARC LAMP.—As will be seen in Fig. 1, the upper carbon is suspended by a chain, its descent being regulated by an escapement B, which is actuated by an automatic make-and-break placed in a shunt circuit. The escapement has a recoil action, arranged so that the recoil is equal to any desired amount of the feed, and which "serves to produce the arc by giving falls of precise measure and separation of the carbons after contact." The telescopic carbon-holder D may be replaced by a chain and guides. "The action of the current is to pass through the coil and attract the armature, which moving the escapement allows one tooth to

pass and the carbon to fall. Contact is then made by platinum point and the armature feed, when escapement moves back and recoil is obtained. This takes place until carbons are in contact, and on separation by recoil the lamp is lighted." A clutch-lamp is shown in Fig. 2. The holder passes

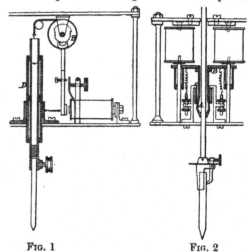

FIG. 1   FIG. 2

through the spring-clip A, fixed on the armature B. On the failure of current the armature and clip descend, and the holder is released, until the increased current raises the clip, and drawing the jaws within the ring E compresses them and lifts the carbon. In a third form of lamp two pivotted armatures close upon the holder when attracted, after the manner of "devil's claws."

3,169.—A. M. Clark, London. (*W. Laing, Paris.*) Electric Machines. 4d. July 20.

COMMUTATORS.—To obviate the heating of the commutator when used with an alternating current generator, it is arranged to work in a box filled with water.

*3,177.—T. J. Mayall, Reading, U.S.A. Insulated Coatings and Covers, or Cases for Telegraphic Wires and Cables, &c. 4d. July 21.

CONDUCTORS.—Copper wires, laid parallel with each other, are enclosed between two wide continuous bands of india-rubber, which are then cemented together and vulcanised. Such strips are packed edgewise in underground troughs.

3,187.—W. R. Lake, London. (*J. V. Nichols, Brooklyn, U.S.A.*) Electric Lamps, &c. 6d. (7 figs.) July 21.

INCANDESCENCE LAMPS.—The conducting wires are first embedded in a small circular disc of vitreous cement, such as a composition of potash, silica, and oxide of iron and copper; this disc is

sealed directly into the opening of the main globe, and the exterior surfaces are smoothed over. In the illustrations B is the disc of cement, with extensions upwards around the conductors. The globe is made

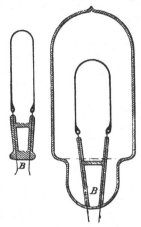

in the form of a cylinder with hemispherical ends. The copper conductors are passed through holes in the ends of the filament and clamped over, the connections being afterwards electro-plated.

3,189.—W. R. Lake, London. (*H. S. Maxim, Brooklyn, New York, U.S.A.*) **Electric Lamps, &c.** 8d. (11 figs.) July 21.

INCANDESCENCE LAMPS. — Several methods of manufacturing incandescing conductors are described. The first consists in cutting the blank from fibrous material, carbonising the same, bending the carbonised strips into the desired shape, and then electrically heating them to give them a permanent set; the second in cutting the strips from previously carbonised sheets, and then bending and heating them; the third in carbonising and electrically heating in a carbonaceous vapour narrow sheets, and cutting the filaments from them. B (Fig. 1) is a stopper moulded with two indentations on each side. By means of a drill these indentations are extended through the glass, and through the holes thus obtained slightly tapering wires D D are drawn, covered with powdered gum copal. The whole is then heated, and the wires drawn further into the holes. E E are two conductors united at the upper end to a block of the same or other material. Another method of inserting the wires D D is to make them in the form of taper steel plugs and grind them into their places. A lamp with such connections is shown having two or more filaments. The lamp can be rotated about a central plug, which makes contact with one end of each filament, thus bringing the filaments successively in contact with the line wires. Another method of sealing the conductors into the glass stopper is shown in Figs. 2 and 3. Each conductor has

grooves or threads formed around part of it (Fig. 3), and on such portion a small quantity of vitreous cement, or a composition of potash, silica, and the oxides of iron and copper is caused to adhere while hot, about which when hard a second layer of the material, but containing a smaller proportion of metal, is formed, and then over this a third layer of nearly pure glass. The wires with their adhering mass of cement are then laid in a mould, and a glass stopper is formed about them, which is ground into

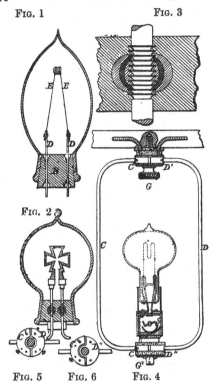

FIG. 1   FIG. 3

FIG. 2

FIG. 5   FIG. 6   FIG. 4

the open neck of the lamp globe. Fig. 4 is an elevation of a lamp and bracket. The bars C and D are insulated from each other, and are attached at their ends to circular segmental plates C' D', C'' D'' (Figs. 5 and 6). These are bound together by clamping screws between washers of insulating material. The screw G serves to fix the bracket to its support; the other screw G' clamps the frame to the lamp holder. S is a circuit breaker.

3,190.—R. H. Hughes, London. **Fittings for Electric Lamps, &c.** 6d. (17 figs.) July 21.

FITTINGS.—The fittings may be specially constructed, or gas fittings adapted. At Fig. 1 is shown one method of making and breaking contact between the conductors 1 and 3, by means of a metal plug in which is a piece of insulating material shown in solid block. In Fig. 2 the parts of the apparatus carrying the contacts are moved towards

each other by a rack and pinion. Several modifications of a similar kind are illustrated. The globe of the lamp is mounted on a sliding tube, capable,

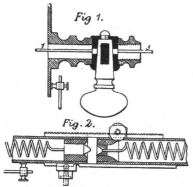

Fig 1.

Fig. 2.

when raised, of being locked in position. When the wires will not pass through the centre of the fittings, an outside groove is cut in which they are embedded.

3,211.—H. H. Lake, London. (*L. Létrange, Paris.*) **Obtaining Zinc from its Ore, &c.** 8d. (2 figs.) July 22.

DYNAMO-ELECTRIC GENERATOR.—A "Gramme" or other suitable generator is employed to precipitate the metallic zinc.

3,214.—A. M. Clark, London. (*L. J. Bouteilloux and W. Laing, Paris.*) **Electric Lamps.** 8d. (15 figs.) July 22.

ARC LAMPS.—In this lamp one of the carbon electrodes has a central core of insulating material for the other to bear against. In the first lamp described a vertical solid carbon rod abuts against a non-conducting core within a hollow carbon cylinder. The arc is formed between the rod and the cylinder, the feeding being effected by gravity, the upper electrode descending as the non-conducting material fuses. One electrode may be constructed of three concentric cylinders, viz., a central rod of carbon, an intermediate cylinder of kaolin, and an external cylinder of carbon. A second lamp has the electrode arranged horizontally, the weight of the lamp acting as the propelling power. In a third lamp an arc is drawn while the non-conducting core is also retained. The solid electrode is forced upwards by a spring, its rate of travel being regulated by an electro-magnet placed in a shunt circuit, and acting on a spring brake. This lamp may be arranged to establish its arc by the brake-shoe tilting the carbon sideways until it comes in contact with one edge of the cylindrical outer carbon. The lower carbon may be arranged to move by a counterweight, and have its motion checked by a

train of wheels. In another form of lamp, shown in the illustrations, two inclined carbons meet at a point below a tubulous carbon having a non-conducting core, and sliding within a tube *f*, provided with a contracted orifice to prevent the cylinder falling out. The tubular electrode is

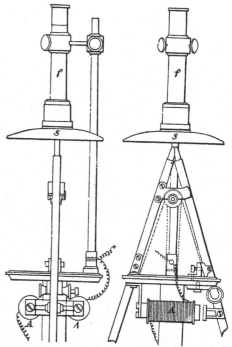

closed by a plug of carbon to facilitate the first lighting. The inclined carbons are fed upwards by a chain and weight, and when extinguished are separated by the electro-magnet A (placed in a shunt circuit) to allow them to rise and make contact with the upper one. S is a reflector. The cores may be made of various materials when it is desired to give colour to the light.

3,231.—E. G. Brewer, London. (*T. A. Edison, New Jersey, U.S.A.*) **Commutators for Dynamo or Magneto-Electric Machines, &c.** 6d. (3 figs.) July 23.

COMMUTATORS.—To minimise the sparking it is

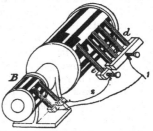

subdivided, since "the spark at each part is re-

duced about as the square of the number of points at which the circuit is broken." The insulation, shown black in the illustration, is widened, and the conducting bars narrowed at one end of the commutator, and upon this portion a single brush *e* is arranged to bear noticeably behind the ends of the main brushes *d d*. This " isolated brush " is not connected with the main brushes directly but with a series of breaking points on a " breaking cylinder " B, provided with conducting bars and insulated spaces corresponding with those on which the " isolated brush " bears. In working, the local circuit between two bars, and a portion of the main circuit, are continued through each " isolated brush " after the main brushes have left each commutator bar, so that no spark is produced at the points of the main brushes. When each " isolated brush " leaves a commutator bar, the current passing through it is also broken at a number of points on the breaking cylinder simultaneously with the breaking of the current on the commutator cylinder. The connections are clearly shown on the drawing, 1 being the main conductor. There is a corresponding set of brushes on the opposite side, but these are not shown.

3,240.—T. E. Gatehouse, London. Obtaining Electric Light, &c. 6d. (10 figs.) July 25.

INCANDESCENCE LAMPS.—To regulate the amount of current passing through the platinum filament, a carbon filament is employed, and coupled in parallel circuit with the platinum, the resistance of the platinum increasing and that of the carbon decreasing as the temperature rises. Fig. 1 shows the arrangement diagrammatically : P is the platinum filament, C a carbon shunt adjustable by the sliding contact S : W W¹ are the leads. Fig. 2

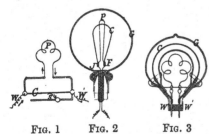

FIG. 1. FIG. 2. FIG. 3.

shows an actual lamp, the same letters indicating similar parts. The connections J F are bound with iron wire, which combines with any oxygen that may leak into the globe. The conducting wires are passed through a mass of "Chatterton's compound," which slightly melts and forms a seal. Fig. 3 "represents a lamp in which the carbon filament C is alone placed inside a crescent-shaped exhausted globe G, the platinum incandescent

filament being placed outside." In a fourth lamp the carbon shunt is arranged in the base or standard. In a fifth both filament and shunt are carried on a spiral spring. One or more lamps may be regulated by a single shunt, the length of which is adjusted by a contact roller moved by a solenoid in the main circuit. A resistance, as a substitute for the lamp when not burning, may be combined with each lamp base. When it is desired to modify the light by hand, an adjustable contact brush, connected to one terminal of the circuit, is arranged to slide over a slip of carbon connected to the other terminal.

3,254.—H. H. Lake, London. (*D. Brooks, Philadelphia.*) Cables or Conductors for Telegraphic Purposes, &c. 6d. (5 figs.) July 26.

CONDUCTORS.—Each wire is wound with fibrous material and over this with a wire. Several such wires are held together with a metal ribbon, and enclosed within a lead pipe into which insulating material is injected from a closed receiver, the impelling force being the elasticity of the vapour

of a volatilised hydro-carbon, "the wrappings being in contact with each other, and with the enclosing pipe form an electrical communication between all the metallic wrappings inside the pipe and the pipe itself, whereby the induced currents are carried off to the exterior pipe, and the inductive effect on the other wires lessened."

3,274.—D. Graham, Glasgow. Lamp Casings or Holders for Containing and Protecting Electric Lights, &c. 10d. (15 figs.) July 26.

LANTERNS.—When applied to an incandescence lamp the upper part is hollow, to receive the stem and contact pieces of the lamp, and is connected with the lantern by bayonet or other joints. The lantern consists of upper and lower metallic portions, embracing a globe between them. For use in houses, &c., the lamp casing is made as a flat or curved or conical reflector. Electric lamps are also made according to this invention, so that when hung upon a properly constructed support, suitably connected to the leading wires, they instantly light, cutting a resistance coil, that replaces them, out of circuit. In mines, where it is dangerous to have a spark, the contact is made and broken in mercury covered with water.

3,283.—S. Pitt, Sutton, Surrey. (*S. J. M. Bear, Mitchell, Iowa, U.S.A.*) Electric Generators. 8d. (12 figs.) July 26.

DYNAMO-ELECTRIC GENERATORS.—The object of

this invention is to permit the poles of the field magnets and the armature to be brought into contact with each other, which is effected by so com-

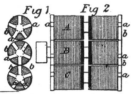

Fig 1   Fig 2

bining the two that the armature shall roll upon the field magnets. The illustrations show one method of carrying out the invention, in which A C are the field magnets and B the armature. Each core piece has five polar extensions $a\,a$ with insulating material b interposed between them. The three are geared together and roll in contact. The meetings of the polar extensions of the field magnets and those of the armature cause electrical impulses which are collected by any suitable commutator. In a second generator a Gramme ring rolls between two Siemens armatures. In a third the cylinder B is replaced by a number of bobbins set radially upon an axis with their polar extensions suitably arranged. In a fourth, two Siemens armatures roll epicycloidically within a cylindrical magnetic field. In a fifth, a bobbin, such as B, with starlike polar extensions, has an epicycloidal movement in a cylindrical magnetic field. In a sixth, a Siemens armature revolves against a cylinder formed of ten electro-magnets set radially, the cylinder being five times the diameter of the armature. In a seventh, a shaft provided with two coils has cranks at its extremities, on which are discs rolling within magnetic fields.

*3,305.—Sir C. T. Bright, London. **Apparatus for Displaying and Regulating Electric Light.** 2d. July 28.

ARC LAMP.—In this lamp the action of the heavier parts is made more sensitive to minute differences of current by the employment of a relay regulator. The carbons are separated by an electro-magnet, and their feed regulated by a second electro-magnet, which comes into action when the relay circuit is completed. A rack rod and train of wheels are used, the last wheel being an escapement either controlled by an electro-magnet, or rotating in a trough of mercury.

3,349.—A. W. L. Reddie, London. *(D. A. Chertemps, Paris.)* **Electric Lamps.** 6d. (2 figs.) August 2.

ARC LAMP.—This invention relates especially to the carbon-holders, and consists in providing the lower one with a bracket e, which carries it clear of the rod and so permits a long length to be used. The upper holder is jointed in two directions, and

is so arranged that by turning the screw f^1 the point of the electrode is moved to or from the rod k, and

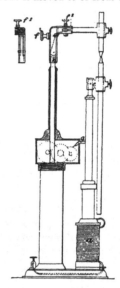

by turning the screw f^2 it is moved sideways. The catch g stops the wheelwork when the solenoid a is excited. The feed mechanism is of the Serrin type.

3,362.—Dr. J. Hopkinson, London. **Electric Lamps, &c.** 6d. (7 figs.) August 3.

ARC LAMPS.—This invention relates to a combination of means for utilising fluid friction in regulating the feed of the carbons, without retarding the motion of the armature in establishing the arc, with an appliance for holding and releasing the carbon holder. In Fig. 1, b is the armature of the

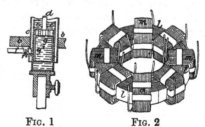

FIG. 1   FIG. 2

controlling magnet of the lamp; in it is fixed a vertical tube c, rather longer than the acting length of the carbon, and closed at the top by a cap from which depends a rod d, ending in a perforated piston g and a short tube g^2. Between the piston and the outer tube is a cylinder e filled with viscous fluid, and acting at its lower end as the carbon-holder; f is a bent lever provided at its outer end, which stands over or rests upon an adjustable stop, with a counterweight. It has a toe-piece f^2, which bears against the cylinder e, and jams it when the free end is clear of the stop. When the current

passes, the electro-magnet attracts the armature in opposition to a spring, lifting it, the outer and inner tubes, and the piston, and thus establishes the arc. As the carbons waste, the armature falls and the long arm of the lever f coming in contact with its stop, the pressure on the toe-piece f^2 is reduced, and the cylinder e descends slowly, relatively to the tube c, and the piston g, and adjusts the arc. In a modification two magnets are employed, one in the lamp circuit to establish the arc, and one in a shunt circuit to regulate the feed.

INDUCTION APPARATUS.—Fig. 2 represents a coil of wire "with a considerable co-efficient of self-induction, and so arranged that this co-efficient may be varied. By its means the current, from an alternating current generator, may be varied without interfering with the generator, or if an arc lamp be used, without altering the length of the arc. The current may also be divided and arc lamps burnt in parallel circuit. The annulus l is formed of a coiled ribbon of sheet iron, round which several coils $m\ m$ of insulated wire are wound. These coils can be coupled by plugs like an ordinary resistance box. Another form of coil is made like an ordinary electro-magnet with a U shaped core that can be more or less intruded into the bobbins. The current to the lamp is diminished by putting more coils in circuit as required. The specification contains two diagrams, the first showing the annular coil placed in a circuit with several incandescence lamps arranged in parallel arc, and the second illustrating four arc lamps in parallel circuit with a coil of the second form in each lamp circuit, in either case the light being varied by bringing more coils, or a greater portion of each coil, into operation.

3,369.—K. W. Hedges, London. **Working Gear and Appliances used in Electric Lighting.** 6d. (8 figs.) August 3.

SWITCH.—The illustrations show a switch, in which the novelty appears to consist in the contact cylinders D D being loose, so that they can be rotated on their axes to insure good contact with

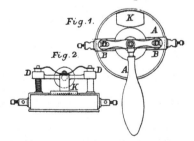

the terminals B B. When the handle is turned the cylinders slide on to the non-conducting faces A A. K is a stop. The invention also includes an instrument consisting of terminals, some of

which are connected to the generators, and some to the leads. By means of taper plugs one generator may be thrown out of circuit and a fresh one substituted for it.

DYNAMO-ELECTRIC GENERATORS.—In connection with each generator there is an alarum, consisting of an electro-magnet and a lever, which falls when the current ceases. The method of driving the generators is as follows: The pulley runs in contact with and between the driving pulley and an idle pulley, the centre of which latter is either above or below a line drawn through the centres of the other two. A belt encircles the whole of the pulleys nipping the idle pulley, which is carried on a movable arm and is held tightly against the driven pulley.

3,380.—W. P. Thompson, London. (*J. A. Pel, Liége.*) **Apparatus for Detecting and Recording the Passage or Stoppage of an Electric Current through a Conductor.** 6d. (2 figs.) August 4.

CURRENT METER.—The object of this invention is to provide apparatus that will make a mark upon a paper dial each time the line wire is traversed by a current. The dial (Fig. 1) carries a paper disc divided radially into twelve divisions representing the hours, and concentrically into fourteen rings

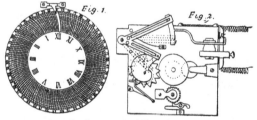

representing the days and nights of one week. The disc is continuously rotated by clockwork under the pointer I, which is steadily withdrawn from the central ring to the outer one by fourteen successive movements at twelve hour intervals. When a current is passing the line the electro-magnet (K^1 Fig. 2) is excited, and brings the pointer against the paper either steadily or intermittently, and so leaves a record of the time and duration of the passage of the current.

3,388.—C. Detrick, Philadelphia, U.S.A. **Continuous Underground Pipes.** 8d. (44 figs.) August 4.

CONDUCTORS.—The earth is excavated to the desired shape, and the pipes formed of plastic material on formers or cones, which are progressively advanced. A series of hollow formers of paper are carried by partitions, and the outer spaces filled in with a plastic material, the con-

ductors passing through the paper tubes. Numerous shapes and methods of construction are described and illustrated.

3,394.—St. G. L. Fox, London. **Generating Electric Currents, &c.** 6d. (12 figs.) August 5.

DYNAMO-ELECTRIC GENERATORS. — This generator is shown in Fig. 1 in plan, Fig. 2 being an end elevation. There are four field magnets on

FIG. 1

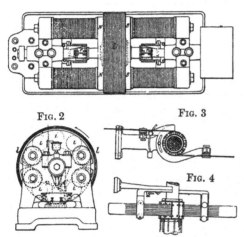

FIG. 2   FIG. 3

FIG. 4

each side of the armature. Each pair, with poles of like name, are fitted with a common pole-piece N N or S S, and consequently there are two pole-pieces of opposite signs (as shown in dotted lines in Fig 2) at each side. The armature is made of a disc of hard wood bolted to a metal nave fixed on the shaft, and provided with coils *i i* wound on iron wire cores, and seated in recesses in its periphery; they are held in place by wedges, and the whole is bound by strong string *l*. The two ends of each coil are carried to separate commutators *r r*, which may be coupled up in parallel circuit or in series. Figs. 3 and 4 show the construction of the commutator. Upon the shaft is a cylinder of baked wood, and upon this are secured segmental brass strips. The working faces of the strips are covered with light removable copper plates to take the wear. The field magnets are preferably excited by "a shunted portion of the current."

3,400.—J. H. Johnson, London. (*J. B. V. Mignon and S. H. Rouart, Paris.*) **Electric Machines.** 6d. (7 figs.) August 5.

DYNAMO-ELECTRIC GENERATORS.—The illustrations show a generator, in which a number of radial electro-magnets are fastened in front of a rotating metal disc having a coil *a a* of insulated wire on its periphery, "the ends of the said wire

being brought up to the axis so as to enable the currents to be discharged by means of an ordinary commutator." To utilise the inductive action of the earth six insulated wires are wound helically and parallel to each other around a rotating

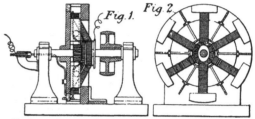

cylinder whose axis is parallel to the dipping needle, the currents being collected at an ordinary commutator. The several forms of frictional generators shown in the illustrations are also described.

*3,401.—J. H. Johnson, London. (*G. Tissandier, Paris.*) **Propelling or Navigating Aerostats or Aerial Machines.** 2d. August 5.

MOTOR.—An electro-motor is made to drive a screw propeller, and is supplied with current from a secondary battery.

3,402.—J. H. Johnson, London. (*J. B. V. Mignon and S. H. Rouart, Paris.*) **Electric Lamps and Manufacture of Carbons for same.** 6d. (2 figs.) August 5.

CARBONS.—Fig. 1 shows an apparatus for making hollow or cored carbons. The core composi

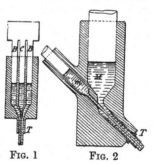

FIG. 1   FIG. 2

tion is placed in the central tube, and the covering composition in the annular outer chamber. The two are forced out together into the nozzle T, the central part by the ram C, and the external part by the tube-like ram B. Fig. 2 shows another apparatus for a similar purpose, in which M is the outer, and *m* the core composition. The two rams can be geared together to produce cores of different diameter. By a modification hollow carbons may be produced. As a protection against atmospheric influences the carbons are covered with enamel or glass.

ARC LAMP.—If convex or concave cups of glass be arranged above and below the arc "a remarkable increase in the light is obtained by a peculiar incandescent effect."

3,404.—E. G. Brewer, London. (*A. G. Desquiens, Paris.*) **Apparatus or Appliances for Automatically Lighting Electric Candles.** 2d. August 5.

ARC LAMP.—This invention relates to an automatic appliance for putting "candles" successively into circuit as the previous ones are consumed. The expansion of a spring by the heat of the arc effects the change, but the specification does not sufficiently explain the mode of operation.

3,409.—G. Westinghouse, Jun., London. **Regulating the Dynamical Production of Electricity.** 6d. (6 figs.) August 6.

GENERATORS.—This invention relates to controlling the supply of actuating fluid to the motor driving an electric generator by means of the influence of the current itself. A soft iron core, within a solenoid traversed by the current, is suspended from a spring and operates a valve attached to it by a rod. This valve controls the inlet and exhaust passages of a cylinder in which is a piston connected to the throttle valve of the engine. An excess of current causes the core to descend and admit fluid under pressure to the regulating cylinder, while a deficiency of current raises the core and allows the cylinder to exhaust. The fluid pressure may be obtained from a small rotary pump combined with the apparatus and driven from the engine.

3,423.—J. G. Dudley, Carmarthen. **Actuating Musical Boxes, &c.** 6d. (2 figs.) August 8.

MOTOR.—A small electro-motor is employed to actuate the instrument.

*3,435.—F. Wright, London. **Incandescent Electric Lamps.** 2d. August 8.

INCANDESCENCE LAMP. — The platinum conductors are filed in the form of a wedge at the ends, being thickest at the extremities. A carbon collar is slipped over the wire, and the end of the filament is inserted within the same collar. When the collar is pushed up the wedge-shaped part locks or binds the filament and conductor together.

*3,437.—F. Wright, London. **Incandescent Electric Lamps.** 2d. August 8.

INCANDESCENCE LAMP. — The filament is made from bass fibre, boiled in weak sulphuric acid, and carbonised in a crucible. It is placed in an atmosphere of hydro-carbon, and is successively dipped in oil and heated by a current until it attains the required conductivity. This part of the operation may be arrested automatically by an electro-magnet, which when the desired degree of conductibility is attained breaks the circuit of the fibre. The finished filament is placed in a globe that is exhausted through a fine hole, which is finally closed by a screw point.

3,441.—R. R. Moffatt and S. Chichester, Brooklyn, New York, U.S.A. **Apparatus for Generating Electricity, &c.** 9d. (8 figs.) August 9.

DYNAMO-ELECTRIC GENERATOR.—The invention consists firstly of a rotating cylindrical armature having its greatest length of cross section parallel with its axis, and arranged in such manner with relation to the poles of the field magnets that they will envelop a large portion of both the inner and outer surfaces of the armature cylinder. Referring to Figs. 1 to 4, the armature

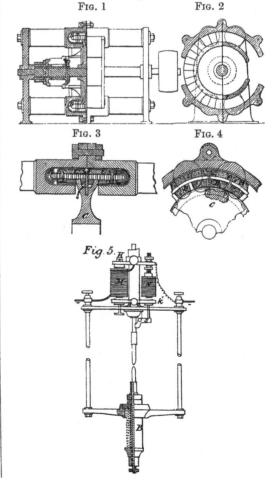

FIG. 1   FIG. 2

FIG. 3   FIG. 4

Fig. 5.

core is formed of pieces B grouped together to constitute a ring (Fig. 4). C is the hub provided with arms of non-magnetic material; i i are small lugs, ribs, or fillets of iron which are secured between the bobbins; they are in metallic contact with the core-pieces B, and serve to convey the magnetic force to them from the field magnets. The poles of the field magnets extend entirely across the outer surface and also around a large portion of the inner surface of the armature (Fig. 3). The wires are wound round the core-pieces of sheet iron as shown, the ends being brought out to the commutator.

ARC LAMP.—Fig. 5 shows an elevation of a lamp in which the arc is established by the solenoid B drawing down the lower carbon. The descent of the upper carbon-holder, which is counterweighted, is regulated by the brake lever s. M is an electro-magnet in the lamp circuit, and N a magnet in a shunt circuit. Both magnets are carried by brackets not shown in the drawings, and have common armatures k and k^1 above and below them. These armatures are pivotted in the centre, so that they can rise and fall like scale beams. When the arc is established and the current is flowing both armatures are attracted by the magnet M, and the end of the lower one holds the brake s to its work. As the resistance increases the shunt magnet obtains increased power, and neutralises the effect of the other, thus releasing the carbon-holder.

*3,455.—W. E. Hubble, London. (*A. F. W. Partz, Philadelphia, U.S.A.*) Electric Lighting. 2d. August 9.

DIFFUSING LIGHT.—The source of light is situated underground, and the light is reflected up a tube. This tube is provided at its top with a reflector, consisting of a conical ring and an inverted cone by which the light is dispersed.

*3,456.—W. R. Lake, London. (*A. L. Arey, Cleveland, Ohio, U.S.A.*) Electrical Apparatus for Lighting, &c. 2d. August 9.

DYNAMO-ELECTRIC GENERATOR.—The field magnets consist of a pair of helices with opposing poles of the same name. A piece of soft iron encases each helix and connects them both in two places above and below the armature. Upon the shaft is a rotating armature "divided into four posts." A fixed armature is carried on a stationary hollow shaft. Four brushes are used, two to supply the field magnets, and two the exterior circuit.

ARC LAMP.—The armatures of an electro-magnet are "so arranged in relation to the regulating mechanism that the ratio between the rate of motion of the positive carbons and the rate of motion of the armatures . . . shall equal the ratio between the given changes in the voltaic arc and

the corresponding change in the position of the said armatures," &c. In the absence of drawings the description is scarcely intelligible.

3,463. — F. R. Lucas, London. Chain for Protecting Submarine Telegraphs, &c. 6d. (5 figs.) August 10.

CONDUCTORS.—The cable is drawn through the centre of a double-link chain, and protected from chafing by the surrounding links. To allow of the swinging of the ship the cable has a swivel in it, in which the conductor passes through stuffing-boxes, and is divided and connected by spring contacts.

3,464.—S. Von Sawiczeski, Paris. Electric Brake. 6d. (2 figs.) August 10.

GENERATOR.—A generator is employed to supply the necessary current for actuating an electro-magnetic brake for railway trains.

3,472.—E. J. Harling and E. Hartmann, London. Dynamo-Electric Machines. 8d. (9 figs.) August 11.

DYNAMO-ELECTRIC GENERATOR.—In this generator a ring armature has a number of projecting pieces of soft iron as shown in the illustrations, which are attached to the sides, or to the peripheries, as the case may be, of a number of parallel soft iron rings. Between the projections P P the coils C are wound. The magnetic field is so arranged that the point of greatest magnetic intensity is not situated in the middle of the field, but near the end where the

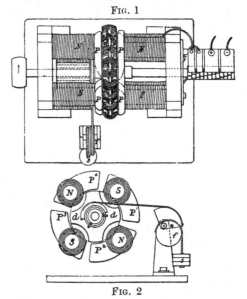

FIG. 1

FIG. 2

coils leave, so that when one of the above-mentioned projections and its generating coil passes

from one neutral point to another it is withdrawn suddenly from the magnetic influence. N S represent the magnets, and P^1 to P^4 the pole-pieces. The strength of the magnetic field is varied by a regulator or keeper consisting of two soft iron poles of opposite polarity arranged to attract a bar of soft iron moving in close proximity with the faces of the field magnets, but never coming into actual contact therewith. The more the regulator is attracted the more it will cover the faces of the magnets, and so diminish their effect on the rotating coils; d (Fig. 2) is one form of the regulator. Any further attraction of d in the direction of the arrow will lessen the space between d and P, the attraction being exercised in opposition to the weight s suspended from the eccentric f. The circuits are arranged as follows: Considering four diametrically opposite coils, the first end of coil 1 is connected to the first end of coil 2, the last end of the latter to the last end of coil 3, and the first end of coil 3 to the first end of coil 4; the last ends of coils 1 and 4 are connected to their respective segments of the commutator in such a manner that each end is in contact with four diametrically opposite segments, which do not, however, form part of the same cylinder. The segments 1, 3, 5, and 7 of the commutator are all connected with coil 1, while the segments between the same are connected to coil 4. When the coil is under the influence of the first portion of the pole-pieces P, its current is conducted through the field magnet coils, but as soon as it approaches the cores of the magnet the current is sent through the working circuit.

3,473.—E. H. Harling and E. Hartmann, London. Electric Lamps, &c. 6d. (13 figs.) August 11.

ARC LAMP.—The feed mechanism is allowed to revolve with greater or less velocity, or is entirely arrested by a brake, which consists of a rotating keeper revolving between the poles of two electromagnets. One of the magnets is in the lamp circuit, and when the normal current is flowing its attraction is sufficient to stop the rotation of the keeper. The other magnet is in a shunt circuit, and its power, which increases as the arc lengthens, tends to neutralise that of the first magnet and leave the keeper free to move. In some cases a frictional magnetic brake presses against the periphery of the keeper, its pressure varying with the force of the magnetism induced in the said keeper. Referring to the illustrations it will be seen that the two carbon-holders are suspended by cords wound in opposite directions round drums of different diameters on the same shaft. The weight of the upper holder preponderating raises the lower one and rotates the feeding mechanism. The cord

from the lower holder is led round a sheave r on the end of a lever which is attracted downwards when the circuit is established, thus dropping the lower holder a sufficient distance to establish the arc. Upon the drum shaft c is a toothed wheel t, which gears with a pinion p upon a shaft which carries the disc keeper A. g is a stop that comes down upon the disc when the current is abnor-

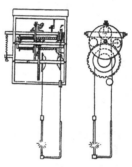

mally strong. The regulation of the lamp may be effected by a sliding piece that can be set to bridge the magnetic poles. The specification also illustrates a modification in which the keeper rotates within polar extensions resembling those in use on a Gramme generator, and in which worm gear is substituted for the spurwheel and pinion. The lower carbon may be fixed, in which case only one drum is used, and the cord attached to the upper carbon-holder is then carried over a pulley and fixed to an armature so as to be drawn down on the passage of current, and thereby raise the upper carbon a sufficient distance to establish the arc.

3,483.—E. G. Brewer, London. (*T. A. Edison, New Jersey, U.S.A.*) Electric Conductors, &c. 6d. (16 figs.) August 11.

CONDUCTORS.—In the distribution of electricity in streets and blocks, junction boxes are placed at the intersection of the streets, and the positive and negative mains are respectively connected together, where they cross, to equalise their potential. In the face of each block "a safety catch-box is located," in which the current is run through a

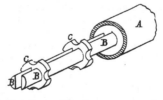

metal bridge that will melt upon an undue increase of current. The illustration shows the conductors employed and the method of laying them. B B are two strips of copper held apart by washers

C C of Manilla pasteboard and enclosed within the iron tube A, the vacant space being filled with asphaltum. Service boxes are placed in front of each house and a service box, safety catch, and current meter is arranged on each floor. A solid rod may be used in place of the two strips.

3,539.—W. R. Lake, London. (*C. Williams, F. W. Harrington, and T. W. Lane, Massachusetts, U.S.A.*) **Electrical Apparatus.** 8d. (11 figs.) August 15.

MAGNETO-ELECTRIC GENERATORS. — A small "Siemens" armature is driven by gearing, and revolves between the poles of three permanent horseshoe magnets placed parallel to each other and with their limbs in a vertical plane. The currents are collected by springs bearing on each end of the armature spindle.

3,559.—C. W. Harrison, London. **Electric Lighting, &c.** 6d. (14 figs.) August 16.

ARC LAMP.—In order to intensify the light, the arc is partially surrounded by a reflector of uncrystallised oolite, known as roestone. This may be used alone in slabs, or as a base ground- to powder and mixed with gypsum. The powders are brought to a liquid condition by solutions of

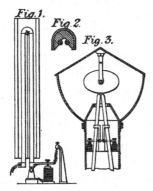

Fig. 1. Fig. 2. Fig. 3.

soda, or compounds thereof, either singly or in conjunction with liquid sulphates of elementary bodies "which exist in the solar spectrum." Figs. 1 and 2 represent in elevation and section an electric candle of the Wilde type surrounded by a reflecting refractory block, and Fig. 3 shows a candle having a reflecting mass above the arc.

3,599.—C. Lever, Bowden, Cheshire. **Electric Lamps.** 8d. (8 figs.) August 18.

ARC LAMP.—The illustration shows all the improvements combined in one lamp but they may be used separately. The upper carbon-holder is clipped between two eccentrically pivotted bow springs S S¹, the outer ends of which are united by the armatures a a^1. On the passage of current

through the electro-magnet coils E E¹, placed in the lamp circuit, the large iron thumb-screws passing through their cores become polarised and

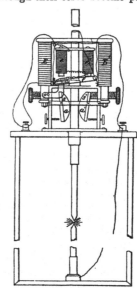

attract the armatures, and the holder is gripped and raised. The high resistance shunt magnet e e provides a passage for the current when the arc resistance becomes too great. Should the lamp circuit be broken the armature f is attracted by the shunt magnet and drawn against a contact piece, thus providing a short circuit around the lamp.

*3,635.—T. Zubini, London. **Electric Light.** 2d. August 20.

ARC LAMP.—A stream of carbonaceous matter, such as carburetted hydrogen or hydro-carbon oil, is fed across the arc from one electrode to the other, the electrodes being made tubular for the purpose.

*3,650.—G. Pfannkuche, London. **Electric Lamps.** 2d. August 22.

INCANDESCENCE LAMPS.—The filaments, whether of carbon or platinum, are coated with china clay and then fused into a solid block of glass. The glass may be of any desired colour, and formed, or ground and polished, like a precious stone.

3,652.—C. L. Clarke and J. Leigh, Manchester. **Coiling Machine.** 6d. (4 figs.) August 22.

CONDUCTORS.—This invention refers to improvements on Patent No. 2,229 of 1880, and relates to a machine for coiling wire and thread side by side in forming induction coils, which may, by a modification, be adapted for forming induction coils of covered wire only. The apparatus has two head-

stocks similar to those of lathes, one having a spindle which can be revolved in either direction. The wire and thread are wound on separate bobbins and are carried through tension devices similar to those of a sewing machine. When the first coil is completed a layer of paraffine paper is placed around it, the machine reversed, and the succeeding coil wound in the opposite direction.

3,655.—R. E. Dunstan, Donhead-Saint-Mary, Wilts, and G. Pfannkuche, Westminster. **Division and Regulation of Electric Currents.** 6d. (3 figs.) August 22.

CURRENT REGULATOR.—Referring to the illustrations, A and B are two solenoids whose cores are connected by a brass rod formed at its central portion E into a double rack. Into these racks gear two segments pivotted at H and H¹, and provided with arms which extend over the terminals G G¹ of a set of resistance coils. Supposing a current to be divided between two circuits, in each of which one of the solenoids is included, and sup-

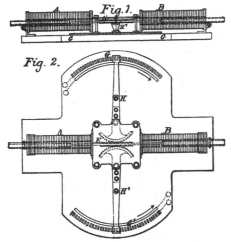

posing the A circuit to have the greater resistance, the core will then be sucked into the solenoid B, and the arms H H¹ moved in the direction of the arrows, throwing resistances into the B circuit and cutting them out of the A circuit. By winding the solenoids suitably different proportions of current may be distributed to the two circuits. When a current is to be divided into four parts it is first halved by such an apparatus, and then each moiety is again divided by similar apparatus. The regulation of the magnet current in shunt-wound dynamo-electric generators is another use to which the appliance may be adapted.

3,668.—W. R. Lake, London. (*T. A. Connolly, Washington, U.S.A.*) **Electric Lighting Apparatus.** 6d. (1 fig.) August 23.

ARC LAMP.—The lamp is intended to be burned in series with others, and its chief feature is the means for short-circuiting it if the arc fails. The current enters at the terminal *c* and divides itself, part flowing through the magnet coils and arc, and part through the lever *e*, screw *f*, and frame A to the terminal K. If the carbons are in contact, or the arc is established, the armature on the lever *e* is attracted, the connection with the screw *f* broken, and the entire current passes through the carbons. If, however, from any cause, the lamp does not

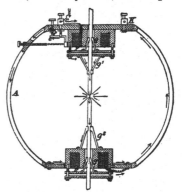

work properly the current goes round it, and it is left out of circuit. When the magnets are excited both the armatures F are raised, forcing the clips *g g* together and raising both carbons. In consequence of the adjustment the upper carbon is raised twice as far as the lower one, and the arc appears between them. There is no provision for regulation, the feeding of the carbons occurring on the extinction of the light only. *g¹ g²* are spring conductors.

3,679.—S. Pitt, Sutton, Surrey. (*S. J. Burrell, Brooklyn, New York, U.S.A.*) **Electric Light Regulators.** 6d. (3 figs.) August 23.

ARC LAMP.—The feed of the upper carbon is controlled by a gripping arrangement. In Fig. 1, A and B represent a coarse and fine wire helix respectively, the two hollow cylindrical cores of which, C D and G, are united by a brass tube H. The piece D is pivotted to two lugs *c c* on C (Fig. 2), and is capable of rocking thereon. When the solenoid A is traversed by a current, C and D are magnetised, and both being of the same polarity the upper end of D is repelled, forcing the lower end inward and binding it against the carbon-holder F, which otherwise slides freely through the cores and is guided at the upper ends by the contact rollers M M (Fig. 3). The weight of the core is balanced by a lever and adjustable spring K, and its upward and downward stroke is controlled by discs and set screws. Assuming the carbons to be in contact and the circuit to be com-

pleted, the piece D is first pressed upon the rod F, and then the two are lifted together to establish the arc. A similar clutch arrangement may be

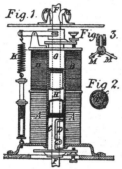

Fig. 1.   *Fig. 3.*   *Fig 2.*

applied to the lower carbon. "The coil B offers a resistance nearly equal to that of the lamp itself, including the arc, and acts as a shunt circuit for maintaining a constant resistance."

*3,711.—F. H. F. Engel, Hamburg. (*C. H. F. Müller, Hamburg.*) **Electric Lamps.** 2d. August 25.

INCANDESCENCE LAMPS.—The conductors which are fused into the globes are made in the form of plates instead of wires.

3,768.—W. M. Brown, London. (*O. Gassett and J. Fisher, Pittsburgh, Penn., U.S.A*). **Automatic Electric Railway Signalling Apparatus.** 10d. (22 figs.) August 30.

CONDUCTORS.—The rails suitably insulated are used as the conductors, and are divided into convenient signal sections.

3,790.—W. R. Lake, London. (*S. D. Strohm, Philadelphia, U.S.A.*) **Covering, Protection, and Insulation of Electrical Conductors or Cables for Telegraphic Purposes, &c.** 6d. (11 figs.) August 31.

CONDUCTORS.—The object of this invention is to provide conduits which can be easily constructed, and can be readily put together and taken apart, and which, when in position, shall form tight and effective insulators for electrical conductors. The

F   *E*

illustration is a section of such a conduit. It is composed of two parts fastened together by longitudinal ribs and key bolts, and each section ends in a diaphragm having apertures to support the

tubes C C in which the conductors are carried. The branch circuits enter through pipes E E. A modified conduit of rectangular section is also described.

3,799.—W. Crookes, London. **Electric Lamps.** 6d. (8 figs.) August 31.

INCANDESCENCE LAMP.—A cylinder of glass is blown into the shape shown at 1, and its end doubled inwards by a two-pronged tool 2, till it takes the configuration illustrated in elevation and plan in 3. At this stage an exhausting tube (shown in dotted lines) is fused to the globe and the neck is divided and the upper end rounded off and closed as at 4. The two hollow points at the top of the projection B are opened and conducting wires passed into them. If platinum wires are used they are sealed directly to the glass, but if other metal is employed, the wires are coated with white enamel or arsenic glass and passed through short cylinders of glass into which the enamel

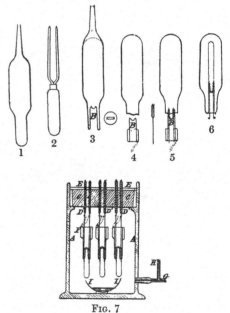

1   2   3   4   5   6

FIG. 7

is melted. Such a conductor and cylinder are shown. The wire is then passed through the open point of the glass projection until the glass cylinder on the wire rests against the end of the glass projection, and the two are then sealed together in the blow-pipe either with or without the interposition of arsenic glass or enamel. The parts now have the appearance shown at 5. After the carbon filament has been connected to the wires the lamp neck can be joined, 6, and the exhaustion be effected. The difficulty of making a good connection between conducting wires other than platinum and the glass can be avoided by

using a compound wire with a core of copper or silver and a platinum sheathing. Fig. 7 shows an apparatus for adjusting the resistances of the filaments. A is a glass vessel, C a plate of vulcanised fibre, D D¹ glass-covered wires, E a mercury seal, G the exhausting pipe, H tube leading to a vessel containing chloroform or other substance. The chloroform may also be placed in the cup I. The resistance of each filament is measured, and if it be too great, a current is sent through it until a sufficient quantity of carbon is deposited on it to augment the conductivity. In order to increase the rigidity of the carbon filaments, they are made of a flat section and receive a twist of a quarter of a circle in being placed in the terminals. The filaments are made as described in Patent No. 2,612 of 1881.

3,804.—P. Jensen, London. (*T. A. Edison, New Jersey, U.S.A.*) **Commutators for Dynamo or Magneto-Electric Machines, &c.** 4d. September 1.

GENERATORS.—To reduce the resistance and prevent sparking between the points of contact, the inventor amalgamates the commutator and brushes.

3,821.—A. L. Fyfe, London, and J. Main, Brixton. **Electric Lamps or Regulators.** 6d. (2 figs.) September 2.

ARC LAMP.—As will be seen from the illustrations a rack on the upper carbon-holder B gears with the first wheel of a train carried in the frame

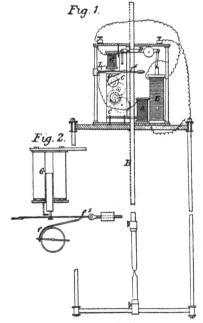

Fig. 1.

Fig. 2.

C, suspended from the lever D and capable of a

slight vertical motion. The last wheel of the train is a brake drum, and is situated beneath the brake shoe f^1 on the balanced lever f^5. When the circuit is completed the solenoid E draws down its core, and by means of the lever D raises the frame C and with it the upper carbon, establishing the arc. As the carbons consume, the power of the shunt magnet G increases, and it attracts the lever f^5, relieving the brake wheel from the weight of the shoe, and allowing the train to move and the upper holder to run down by gravity. L is an adjustable stop which receives the end of the brake lever when the power of the solenoid no longer suffices to uphold the frame. R is a resistance in the shunt circuit. Fig. 2 shows a modification in which a solenoid is substituted for the magnet G. When the lamp is intended for use on a single circuit, the solenoid G is traversed by the main current, and acts against a spring which tends to raise the brake lever.

*3,822.—A. Tubini, London. **Electric Lamp or Regulator.** 2d. September 2.

ARC LAMP.—The carbons are placed side by side in a slightly inclined position, and one of them is covered with kaolin. The core of a solenoid placed in the main circuit is attached to one of the holders and separates the ends of the carbons to form the arc. The movable carbon-holder may be attached to the core of a solenoid placed in the main circuit and opposed by a spring. The passage of current draws in the core and so separates the carbons.

*3,832.—W. R. Lake, London. (*R. Haase and J. P. Recker, Indianapolis, U.S.A.*) **Machinery for Winding or Coiling Wire upon Annular Armatures for Electro-Magnetic or Magneto-Electric Apparatus,** 6d. September 2.

CONDUCTORS.—The machine has a clamp adapted to be securely fastened to, and surrounding the rim of the armature, a rim adapted to surround and revolve upon the clamp arms, one of which carries a spool or bobbin containing the wire and the other a counterbalancing weight and fastening, guiding, and controlling mechanism. In the absence of drawings, which are not filed, the action of the machine cannot be understood.

*3,857.—C. Le Snew, Paris. **Perpetual Electro-Magnetic Motor.** 2d. September 5.

MAGNETO - MOTOR.—"A number of horseshoe magnets are attached to the inside of a fixed ring, in the centre of which is mounted a movable shaft perpendicular to the plane of the ring and carrying a number of arms provided at their extremities with horseshoe magnets."

3,858.—J. Wetter, London. (*W. Wheeler, Massachusetts, U.S.A.*) Apparatus for Diffusing Light. 8d. (34 figs.) September 5.

DIFFUSING LIGHT.—This invention relates to a system of lighting and to apparatus whereby the electric light is maintained in the focus of the reflector, and is transmitted to any number of places which it is desired to illuminate by optical conduction, division, and dispersion.

*3,871.—H. A. Harborow, London. Dynamo-Electric Machines. 2d. September 6.

DYNAMO-ELECTRIC GENERATOR. — An armature wound longitudinally is caused to revolve, "by suitable mechanism," within and in the opposite direction to a second armature "wound with segments of wire." Enclosing both armatures are the field magnets, which may be excited by either or both the armatures or separately.

3,880.—W. R. Lake, London. (*C. Dion, New York, U.S.A.*) Manufactures of Revolving Armatures and other Parts of Electrical Apparatus, &c. 10d. (12 figs.) September 7.

ARMATURES. — This invention relates to the method of construction of an armature, and also to the mechanism for producing it. Fig. 1 is a section of the armature, and Fig. 2 a view explanatory of the method of manufacture. A is a core of soft iron wire, and B a coil of copper ribbon wound edgewise upon the core. The convolutions are insulated by being treated with gelatine mixed with bichromate of potash. To lay the conductor in place the core is divided, and the ends sprung apart to

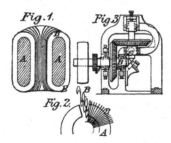

Fig. 1.   Fig. 3.

Fig. 2.

form an opening through which the convolutions can be introduced. Fig. 3 shows a machine for bending the ribbon which is passed by means of feed rollers between the conical ends of the two shafts B C. These ends are not perfect cones, but are cut away for a portion of their circumferences, while the feed rollers are similarly treated so as not to form perfect cylinders. The ribbon is first pushed forward by the feed rollers, and passes between the cones, which are adjusted so as not to grip it at this moment. When a length equal to

the straight side of the core has passed, the feed rollers lose their hold, and the conical rollers begin to nip. In consequence, however, of their form they force one side of the ribbon forward much faster than the other, and so bend it into a curve to pass round the end of the core. The alternate actions of the rollers and the cones produce a helix of the section shown in Fig. 1. A modification of the apparatus is adapted for producing cylindrical helices.

3,890.—D. G. Fitz-Gerald, London. Electric Lamps. 6d. (8 figs.) September 8.

INCANDESCENCE LAMPS.—The carbon filament is prepared by treating unsized paper or vegetable fibre with a cold concentrated solution of zinc chloride, the ends being made thicker by doubling them over. When the material is sufficiently gelatinised it is thoroughly washed in dilute hydrochloric acid, ammonia, and water successively, and after being brought to the required shape is carbonised in the usual manner. To establish contact between the ends of the filament and the conductors, the former are cemented into short metal tubes (see illustration) by a mixture of finely divided carbon and platinum black, made into a

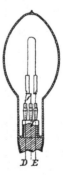

D E

paste with a solution of sugar and water. When the cement is dry it is baked at a temperature below the melting point of the metal, the conducting wires D E are then inserted into the other ends of the tubes, and secured by bending the lip of the tube into nicks cut to receive it. I is a glass stopper, into which three wires are sealed, situated over a bulb filled with cement. The central conducting wire is connected to a supplementary filament of carbon, oxidisable wire, or a combination of an oxidisable and an unoxidisable metal. When a current is sent through this supplementary filament it combines with any residual gas in the globe. To indicate the degree of rarefaction a manometer is used consisting of a length of capillary tube closed at one end and containing a globule of mercury. The inverted end terminates in a bulb and is included in the lamp.

3,893.—**W. R. Lake**, London. (*W. S. Hill, Boston, U.S.A.*) **Electric Lighting Apparatus.** 8d. (11 figs.) September 8.

ARC LAMP. — The first part of the invention relates to a shunt for cutting a lamp out of circuit when the carbons have burnt down as low as desired. The illustration shows this applied to a Weston lamp. From the terminal w^3 a wire extends to the lever f, and from the terminal w another wire extends to the plate g. When the collar a^1 on the rod a descends

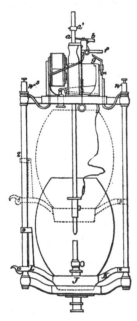

on to the trigger b the lever f is released from a catch and falls on to the plate, making direct connection between the positive and negative binding posts of the lamps. The invention further relates to an improved device for holding a lamp globe as shown in the same figure, and comprises a ring J with arms K L on either side, and an eccentric cam lever to clamp the rod 2. According to the third part of the invention a second set of carbons is put into action when the first set is burnt out, the two being controlled by the same regulating mechanism. This is effected by holding the second pair apart by a grip lever, which is released when a collar on the first carbon-holder descends on to a trigger.

3,911.—**J. Wetter**, London. (*W. Wheeler, Massachusetts, U.S.A.*) **Reflectors.** 8d. (16 figs.) September 9.

REFLECTORS.—Three forms of reflectors especially applicable to "the electric arc" are described. (1) A reflector having a surface such as would be generated by the revolution of a conic sectional curve about a line in the plane of the said curve, and meeting its axis perpendicularly in some point other than the focus. (2) A reflector having a surface generated by the revolution of a conic sectional curve about two or more axes of revolution successively, the said axes intersecting each other in some point of the principal axis other than the focus. (3) A reflector having a surface generated by the movement of a conic sectional curve about one or more lines intersecting its axis at right angles, the said curve moving 360 deg. about the said line or lines.

3,926.—**J. S. Sellon**, London. **Secondary Batteries or Magazines of Electricity.** 6d. (11 figs.) September 10.

SECONDARY BATTERIES.—This invention relates to "the use in the construction of secondary batteries of perforated plates or sheets roughened, serrated, or indented, composed of lead, platinum, or carbon, upon, in, or against which plates spongy or finely divided lead, or other salts or compounds of lead, or other suitable substances or compounds are, or may be, held or retained." In the illustrations Fig. 1 represents a perspective view of a perforated battery plate, formed with corrugations of a dovetail section; Fig. 2 shows a section of a perforated plate formed with angular projections or grooves. This plate may be bent into a rectangular or cylindrical form. Fig. 3 shows an irregular section of a compound battery plate formed of two or more plates which may have

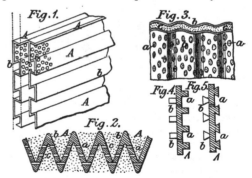

flat or irregular surfaces. Figs. 4 and 5 illustrate a plate cast with slits and projections, the latter of which are flattened or rivetted over during manufacture to cause the retention of the metallic oxide. A A are sheets or plates of lead, platinum, or other material so formed that a large quantity of spongy or finely divided lead may be retained in between them, or against them, by parchmented paper or other similar material, in such a manner as to be readily acted upon by the current. The plates may be formed of corrugated lead, or of

lead cast with holes a, either plain or with flutes, corrugations, indentions, or projections b, in or on which the material c can be packed. In Fig. 3 the oxides are placed between the sheets which are rivetted or soldered together.

3,929.—E. J. P. Mercadier, Paris. Multiplex and Self-Revertible "Teleradiophone." 10d. (13 figs.) September 10.

MOTOR.—A small motor is employed to rotate a plate having radial perforations by which a beam of light is periodically intercepted.

3,932.—P. Jensen, London. (*T. A. Edison, New Jersey, U.S.A.*) Dynamo or Magneto-Electric Machines. 6d. (3 figs.) September 10.

DYNAMO-ELECTRIC GENERATOR.—This invention relates to generators in which inductive bars are employed running lengthwise of the armatures, as described in Specification No. 1,240 of 1881. The induction bars B B¹, Figs. 1 and 2, are now widened to close up the spaces between them, and the copper bars are provided with T shaped ends

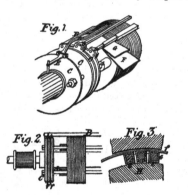

Fig. 1.

Fig. 2.   *Fig. 3.*

secured to the copper discs C which connect the bars on opposite sides of the armature. The alternate bars are curved outwardly, so as to pass over the first line of ears and inwardly again to make connection with the second line of ears. This construction provides a larger contact between the bars and the discs. To lessen the resistance and prevent oxidation, the surfaces are plated with gold or silver, or amalgamated with mercury. For commutator connections the tongues extending from the open centres of the discs, as shown in the former specification, are dispensed with, and exterior rods E are substituted for them. To place the copper discs out of the magnetic field they are packed away from the armature by distance pieces $d\,d$. To prevent "electrical creeping" over the edges of the discs they are bevelled at $f\,f$, and the insulating paper discs are pressed down between them. The

copper bars are insulated by wrappings of parchment japanned, or oiled paper F, and sheets of mica G. To keep the armature cool the spaces between the polar extensions of the exciting magnets are closed by brass plates, and into the box thus formed air is driven by a fan actuated by the armature spindle or otherwise. To allow circulation of the air currents the induction bars are packed off the core H (Fig. 3) by blocks or projections $e\,e$.

3,975.—J. W. Smith, Edinburgh. Carrying or Laying Electric Wires. 6d. (73 figs.) September 14.

CONDUCTORS.—These are carried in the kerbstones or the causeway blocks of streets, which are formed as hollow channels. Special claim is made to fitting insulated brackets in the kerbs for carrying the wires, service branch pipes for leading out individual wires, fitting water-tight covers and joints, and providing means of drainage.

3,976.—P. Jensen, London. (*A. J. B. Cance, Paris*). Electric Arc Lamps. 6d. (5 figs.) September 14.

ARC LAMPS.—In Fig. 1, which shows the general construction of the lamp, the carbons $a\,b$ are fixed to the holders $c\,d$, formed of two frames from each of which a pair of rods rise vertically. The rods from the lower carbon-holder are provided at their tops with two small pulleys $e\,e$, around which pass the cords f, one extremity of each being fixed to the top frame g. The other end of each cord passes round the pulleys $h\,h$, and is fastened to the top frame d of the upper carbon-holder. This frame carries a driving weight h loaded with shot and guided by the rods $i\,i$. The weight acts as a nut, and fits on the central screw j, fixed laterally, but moving freely, on its axis pivotted at the top plate g and the lower platform k. The descent of the weight turns the screw and tends always to keep the carbons in contact. The two platforms g and k, connected by the four guiding columns i, form the fixed frame of the lamp. On the lower platform of this frame are two plates l, carrying the coils m, inside which moves the hollow core n. The upper part of this core is extended in the same form by the copper core o, the length of which is proportioned to the desired magnetic intensity. From the lower end of the screw j depends a copper rod p passing through the axis of the core. The weight of the core may be neutralised by springs. The coil, placed in the main circuit, regulates the lamp by means of the crosspieces $q\,r$ rising and falling with it and guided by the rods $i\,i$. A screw s serves to regulate the position of the core in the coil, and the position of the escapement wheel, which checks the movement ; a

second regulating screw limits the length of the

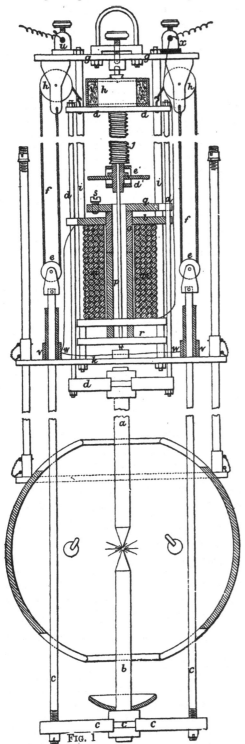

arc. The action is as follows : The current enters

by the terminal u, passes through the frame to the upper carbon-holder and carbon, flows through the lower carbon and the two insulated rods of its holder which slide easily in the sleeves v, electrically connected by the conductor w, and communicating with one end of the coil m. The other end of the coil goes to the binding screw x. The

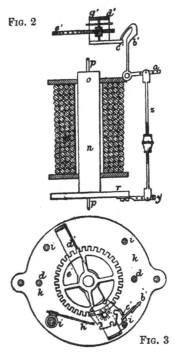

passage of the current magnetises the core, which lifts and separates the carbons by the following means. The lower plate of the core carries an arm y (see Fig. 2) to which is attached a connecting rod z, the length of which is adjustable, and the upper end of which is linked to the arm a^1 of a cranked lever pivotted on the frame. The arm b^1 of the lever terminates in a curved path engaging a finger c^1, fast on the frame d^1, which is mounted loosely on the extension p of the screw j. Mounted loosely within the frame d^1 and fast on the rod p is the wheel e^1, gearing with the pinion f^1 pivotted in the frame. The spindle g^1 of the pinion carries also the escapement shown in plan in Fig. 3, which comes in contact with, or moves clear of, the spring h^1, according as the carbons approach or recede. This escapement acts as either a brake or stop. Each recess in the wheel is separated by a tooth on which the spring h^1 bears with sufficient power to stop the movement when the magnetic variation, due to the approach of the carbons, has not been sufficient to bring the recess opposite the spring. On completing the circuit the core n is lifted, raising with it the connecting rod z which tilts the

cranked lever a^1; the curved end of b^1 then pushes the frame d^1, which draws the escapement wheel against the spring b^1, and prevents this wheel and the pinion f^1 being turned by the wheel e^1; then the pinion gearing in the wheel e^1 forces the latter, by its travel with the frame d^1, to move backwards and with it the screw j to which it is fastened. This raises the weight and separates the carbons. As the magnetic intensity of the coil becomes reduced the core tends to fall, and the escapement wheel, retiring from the spring h^1 until it clears it, allows the carbons to approach each other.

3,987.—J. S. Sellon, London. Secondary Batteries or Magazines of Electricity. 6d. (3 figs.) September 15.

SECONDARY BATTERIES.—This invention relates, firstly, to the use in secondary batteries of plates, elements, or supports composed of alloys of lead

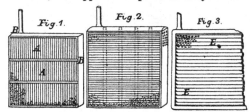

with antimony; secondly, to the employment of plates or elements either constructed as described in Patent No. 3,926 of 1881 or composed of perforated strips, tubes, pieces, or woven fabrics, of lead, or of the above alloy, either separately or combined and affixed to, supported by, or strung upon rods, bars, or pieces of carbon, lead, or other suitable metal. Fig. 1 represents a side view of a plate of a secondary battery composed of a number of perforated lead strips having suitable surfaces, or of metal fabric. Fig. 2 is a modified arrangement of the above, in which the perforated lead strips or metallic fabric are strung or affixed to carbon rods or other supports. Fig. 3 illustrates a further modification, in which the plate is composed of a number of perforated lead tubes, inside which the material is packed. A A are lead strips, or lead fabric indented, corrugated, or roughened to retain the material to be packed therein, and carried by supports B.

*4,005. — J. S. Sellon, London. Storing Electricity. 2d. September 16.

SECONDARY BATTERIES.—The electrodes are constructed of thin sheets or gauze of platinised copper, or pieces of carbon, and are immersed in a solution of acetate of lead.

4,011.—B. Hunt, London. (*A. E. Brown, Cleveland, Ohio, U.S.A.*) Electric Lamps. 8d. (6 figs.) September 17.

ARC LAMP.—In this lamp the electrodes have a constant feed, less than the speed of their consumption. When the arc sensibly lengthens, the feed is temporarily increased to bring the carbons again within the proper distance and then it falls back to its normal rate until the limit of arc resistance is again exceeded. Referring to the illustration, F is one of two solenoids on a shunt circuit, each of which has a movable core H, attached to the balance beam G, which is pivotted at a. From a yoke on this beam depend two rods g attached at their lower ends to a long piston fitting fluid tight, or thereabouts, in the tubular extension of the upper carbon-holder. L is a second balance pivotted at c to the beam G. From L there depends a rod, situated between the two rods $g\,g$, and carrying a small plunger or cylindrical valve that fits in an axial cylindrical cavity in the piston, and controls ports or openings that form a communication between the fluid (glycerine) above and below the

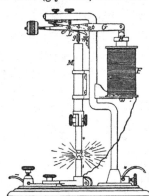

said piston in the tubular extension M. When the lamp is burning, the upper carbon is supported by the atmospheric pressure acting on an area equal to the large piston. A slow leakage of fluid takes place through the piston, and consequently there is a steady descent of the upper electrode, which, however, is not equal to its consumption. When the arc lengthens, the increased current through the coils draws down the coil cores, lengthening the arc still further and pressing the end of the balance beam L against the stop m. This has the effect of tilting the beam with respect to the beam G, and raising the valve, the lower part of which is tapered, so as to allow an increased flow of fluid through the main piston, and the more rapid descent of the carbon-holder. As the resistance of the arc decreases, the cores rise and shut off or diminish the direct current of fluid. In a modification, a single solenoid is employed and has its core working within a cylinder partly filled with liquid. In a further modification, the arc is established by a solenoid in the main circuit, the valve being controlled by a solenoid placed in

a shunt circuit. In a focussing lamp the descent of the upper carbon is caused by suitable gearing to raise the lower carbon.

4,017.—S. Hallet, London. Electric Lamps, &c. 4d. September 17.

INCANDESCENCE LAMPS.—The filament is supported between two platinised carbon discs and is protected by a covering of silicon. In a modification two filaments are provided, one being intended to combine with the residual oxygen contained in the globe. Strips of carbon may also be let into a block of lime to serve as conductors thereto, and be coated with silicon.

*4,024.—W. M. Brown, London. (*E. M. Fox, New York, U.S.A.*) **Electric Lamps.** 2d. September 19.

INCANDESCENCE LAMP.—An apertured stop plug is used by which the globe is exhausted, the opening being closed by turning the plug, which also serves to support the filament and its conductors. The holders are fixed on the filaments and attached to the conductors by clamping devices.

*4,026.—E. de Pass, London. (*La Société Anonyme des Câbles Electriques, Système Berthoud, Borel, et Cie., Paris.*) **Dynamo-Electric and Magneto-Electric Machines.** 2d. September 19.

MOTOR.—The armature "represents in section a double T with curved flanges or shoes" and revolves within "the fixed outer coil." The armature receives a continuous current while the outer coil is traversed at each half-revolution by reverse currents. The commutator is composed of two parts mounted on an insulating bush.

4,034.—P. Jensen, London. (*T. A. Edison, New Jersey, U.S.A.*) **Dynamo or Magneto-Electric Machines and Electro-Motors.** 6d. (8 figs.) September 19.

DISTRIBUTING CURRENTS.—When a battery of generators, arranged in parallel circuit, supplies current to a district, the excitation of the field magnets needs to be varied to adjust the current produced to the demands made upon it at any moment. This is effected by increasing or diminishing the resistance included in the field magnet circuit and may be done by hand or automatically. The latter method is illustrated in Fig. 1, where four generators are shown with their commutator brushes connected to two main conductors 1 and 2, and their field magnet coils in derived circuits from the same main conductors. The exciting current flows along the wire 4 to the movable arm L, and then along one of the wires 5,

6, 7, 8, or 9, after which it divides into four parts and passes through the four sets of field magnet coils to the other main wire 2. R R are resistances. When the point of the lever L is in its highest position no one of these resistances is in the circuit, whilst when it is in its lowest position there are four resistances in each magnet circuit. The position of the lever is determined by the speed of the centrifugal governor G, driven by an electro-motor on a derived circuit as shown. When

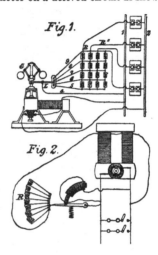

little work is being done in the district the speed of the motor increases and more resistance is thrown into the magnet circuits to lessen the intensity of the magnet, and *vice versâ*. Fig. 2 shows an arrangement in which a curved solenoid and core are substituted for the electro-motor. When lamps and motors are fed from the same circuit the throwing into action of the motor draws away current from the lamps until the speed sets up an opposing electromotive force which acts as a resistance and equalises the distribution of current. To prevent this, each motor is provided with a set of resistances and a lever to throw them out of circuit, one by one, as the speed increases. When this is done automatically the resistances are connected with movable contact blocks, located in line with each other, and forced together in succession by the movement of a centrifugal governor.

4,037.—W. Clark, London. Secondary Batteries. 6d. (11 figs.) September 19.

SECONDARY BATTERIES.—The object of this invention is to produce Planté battery plates upon which thick coats of lead oxide may be formed without liability to flake off. In the simplest form a composite electrode is formed of thin sheets or laminæ of lead, folded around a thicker sheet forming a support, and the whole enveloped in a sheet of parchment paper. The sheets are either plain or

corrugated as in Figs. 1 and 2, and are first treated with dilute sulphuric acid to give the plates a covering of lead sulphate. Another form of electrode is made by packing thin lead plates, alter-

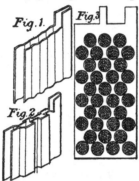

nately plain and corrugated or crimped, between stout lead covers and binding the whole together by india-rubber bands. Such an electrode may be enclosed in a sheet or case pierced with holes as in Fig. 3.

4,057.—H. E. Newton, London. (*Société Universelle d'Electricité Tommasi, Paris.*) **Production and Employment of Continuous Electric Currents in Railway Carriages, &c.** 6d. (8 figs.) September 20.

DYNAMO-ELECTRIC GENERATOR.—In a train in which electric lights are fed by a dynamo-electric generator driven from the axle of one of the

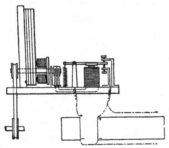

vehicles, and in which a secondary battery is provided to maintain the light during stoppages, the current from the battery when in use divides itself between the lamp circuit and the generator, and is to a considerable extent wasted. To prevent this, when the speed of the generator falls below a certain point, it is cut out of the circuit. One method of effecting this is by a centrifugal governor, driven from the same source as the generator, and so arranged that, when its arms fall below a certain angle, it shall separate two springs forming part of the generator circuit. Another device for the same object is shown in the illustration, and consists of a small magneto generator

driven by a strap from the axle, and sending a current through an electro-magnet. When the speed is sufficiently great the magnet attracts its armature and completes the circuit; this brings the generator, which hitherto has revolved idly, into action.

4,058.—W. R. Lake, London. (*J. B. Henck, Boston, Mass., U.S.A.*) **Electrical Cables.** 6d. (21 figs.) September 20.

CONDUCTORS.—To diminish the effects of induction the conductor is divided into sections which overlap each other. The wires are covered with hemp impregnated with "Chatterton's compound" and enclosed in a metal tube, which is sufficiently thin to be slightly corrugated by the outer protecting wire as it is laid on. Several forms of overlapping conductors are shown and described.

4,060.—A. M. Clark, London. (*N. de Kabath, Paris.*) **Regulating the Discharge of Secondary Batteries.** 6d. (3 figs.) September 20.

SECONDARY BATTERIES.—To obtain a uniform current, notwithstanding the variations of the battery or of the external circuit, the height of the liquid in the battery is varied, and consequently the area of the plates included in the active circuit. Fig. 1 illustrates the idea, and diagrammatically shows the plates but slightly immersed as they would be at the commencement of their action. Fig. 2 shows the manner of regulation; a is the battery; B C two tanks for liquid, situated

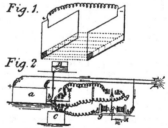

respectively above and below; b is a three-way cock, by which liquid from B can be fed to a, or liquid from a discharged to C. This cock is controlled by two electro-magnets set in action by a relay; g is a shunt circuit. When the current in the main circuit decreases, the spring m^1 of the relay moves the index to put one of the electro-magnets f into action, and opens communication between the tank B and the battery. On the other hand, if the current increase beyond the determined amount, the spring m^1 is overpowered by the magnetic attraction, and the opposite magnet f turns the cock to allow part of the liquid in the battery to flow out.

4,069.—W. P. Thompson, London. (*W. W. Gary, Boston, Mass., U.S.A.*) **Indicating Apparatus or Signals for Railway Switches or Points.** 6d. (4 figs.) September 21.

MAGNETO-ELECTRIC GENERATOR. — A magneto-electric generator and switch arrangement is used to move the signals.

4,070.—W. P. Thompson, London. (*W. W. Gary, Boston, Mass., U.S.A.*) **Indicating Apparatus or Signals for Railway Switches or Points.** 6d. (1 fig.) September 21.

MAGNETO-ELECTRIC GENERATOR. — A magneto-electric generator, receiving motion through the passage of the trains, is employed to actuate the signals.

*4,093.—E. G. Brewer, London. (*P. B. Delany and E. H. Johnson, New York, U.S.A.*) **Electric Cables.** 2d. September 22.

CONDUCTORS.—A series of insulated wires are arranged in a flat lead pipe and separated by walls of lead. After the wires are placed in the partitions the pipe is compressed round them so as to exclude the air.

4,107.—F. E. Fahrig, Southampton. **Dynamo-Electric Machine.** 6d. (11 figs.) September 23.

DYNAMO-ELECTRIC GENERATOR. — In this generator the armature is constructed so that the coils may be readily removed and replaced by others according to the current required. Referring to Fig. 1, the magnetic field is produced by four or more pairs of electro-magnets, the ends of each upper and each lower pair being connected by semicircular pole-pieces N S, N S. The armature

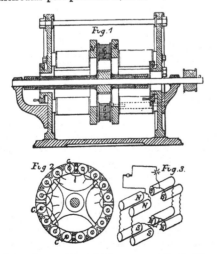

consists of a number of transverse coils C (Fig. 2),

arranged in a circle, and inserted in holes in pairs of segment-shaped plates, of which the annular frame of the armature is built. These segments are rebated at the joints and united by counter-sunk screws, and are carried by the arms of spider frames fixed upon a boss keyed on the shaft. The coils may be coupled in any desired combination, and are connected to the commutator strips in the ordinary manner. In a modification, the coils are duplicated, and two commutators used. "The field magnets are connected as shown in perspective in the diagram (Fig. 3) so as to have the opposite polarities indicated by the letters N S. This is only one illustration of the way the magnetic field may be coupled, which may be varied in different ways." The commutator strips, of L shaped section, are embedded around a boss of hard wood, and provided each with a suitable binding screw, by which the terminals of the coils are connected to the strips.

4,128.—J. Imray, London. (*M. Deprez and J. Carpentier, Paris*). **Distributing and Regulating the Transmission of Electrical Power.** 6d. (12 figs.) September 24.

DISTRIBUTING CURRENTS.—The consumers have each two secondary batteries, into which current is conveyed from the mains by a branch wire. An "automatic galvanometer" actuates a switch by which each battery is alternately changed from the charging to the working circuit. The field magnets of the main generator are excited partly from the main current and partly from a separate source. The speed of a motor is controlled by a centrifugal governor which breaks contact when the speed becomes excessive. In order to vary the force of the combined current produced by two alternate current generators connected to the same axle, "the fixed parts" of the first generator are shifted relatively to those of the second generator. To maintain a constant current through a number of motors placed on the same circuit an "indicator motor" is included in the circuit. This indicator motor actuates a rheostat which throws more or less resistance into the exciting circuit of the main generator. In a second regulator an electro-magnet, placed in a branch circuit, attracts its armature in opposition to a spring. A second derivation supplies a small motor. Should the resistance in the main circuit increase, the electro-magnet releases its armature, which, acting as a brake, stops the small motor; this consequently offers less resistance to the current, so that the generator becomes further excited. To convey a quantity current to a distance, with a minimum loss, it is first converted into a high tension current by means of an induction coil, and conveyed to the distant

d d

station, where it is re-converted into a quantity current by a second induction coil.

4,168.—W. P. Thompson, Liverpool. (*J. W. Langley, City of Ann Arbor, Michigan, U.S.A.*) **Governing Apparatus for Electric Machines.** 6d. (2 figs.) September 27.

Dynamo - Electric Generator. — A piece of magnetic metal is suspended in such relation to the poles of the field magnets in a dynamo-electric generator, that when such generator produces more current than is required, the attraction of the poles upon the piece of magnetic metal will cause it to approach them, and will tend to establish magnetic communication between them, thereby reducing the current generated in the armature. In the illustration, A A are the field magnets of a dynamo-electric generator; P P are the pole-pieces nearly surrounding the armature B, which may be of any construction; C C¹ are pieces of magnetic metal

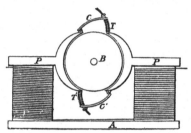

each pivotted to one pole-piece, and guided by curved rods T T¹, and bearing on springs of different lengths, so as to oppose an increasing force to the movement of the pieces C C¹ as they approach the poles. As long as the resistance of the external circuit is constant the free end of C will remain stationary; if, now, its resistance be diminished, as for instance by extinguishing part of the lamps upon it, or if the speed of the armature be increased, the attraction of the pole P upon the free end of C will increase so as to overcome the resistance of the springs and to draw it nearer to the pole P, thus diminishing the free magnetism in A, and weakening the current produced by the generator. When the normal conditions again obtain the piece C moves back to its former position. The invention is also applicable for governing the speed of electro-motors.

4,174.—E. G. Brewer, London. (*T. A. Edison, New Jersey, U.S.A.*) **Electric Lamps, &c.** 8d. (14 figs.) September 27.

Incandescence Lamps.—The collar for attaching the lamp to the socket serves to make connection with the conductors. This is accomplished, as shown in Fig. 1, by dropping the metal rings *f g* into a suitable mould D, wires being first soldered

to the inner surfaces of the rings, so as to project upwardly on opposite sides of the mould. The wires of the lamp are bent up on opposite sides of the same, and it is placed in the centre of the mould, and held upright by a spring holder B. The wire ends of the lamp and collar rings are twisted together and turned down into the mould, after which the mould is filled with plaster-of-paris Fig. 2 shows the finished lamp. The invention secondly relates to lamps in which the glass supporting the conductors is held in a tapering soft rubber stopper forced into the neck of the globe, and held there by atmospheric pressure. Such a lamp can be taken to pieces when the filament is destroyed, and the parts used again. Another method

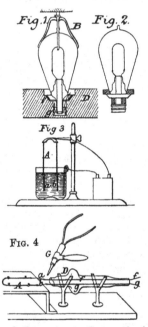

is to seal a platinum ring to the mouth of the globe, and another ring to the glass of the conductor support, and to solder the two rings together when the lamp is being completed. Figs. 3 and 4 show a method of uniting the carbon filament with the conductors. The filament A, which has short copper wires *a b* mechanically secured to its ends, is suspended in an electro-plating cell, so that the junctions become coated with metal. In Fig. 4, D is the glass for supporting the conductors *f g*. G is a blow-pipe for uniting the copper wires *a b* to the platinum wires *f g*. When a shorter lamp is necessary the conductors which support the carbon are sealed directly in the lower end of the bulb, the usual glass support being dispensed with. The fittings of incandescence lamps terminate in hooks, which fit on corresponding hooks forming the circuit terminals. In mines and other explosive places

both the lamps and hook-like connections are submerged in water.

4,193.—C. H. Gimingham, Newcastle-upon-Tyne.
Electric Lamps. 4d. (3 figs.) September 29.

INCANDESCENCE LAMPS.—In connecting the carbon filament to the terminals the ends of the latter are flattened, and the flattened ends are formed into a tube by drawing through a wire plate or otherwise. The filament is mounted directly in these tubes;

FIG. 2   FIG. 1   FIG. 3

Fig. 1 represents a terminal wire with its end flattened out into a thin plate, which is bent into tubular form as shown in Fig. 2, the ends of the filament being placed in the tubular parts as shown in Fig. 3. The combined spring of the carbon and metal holds the filament firmly in place.

ends of the filament are clamped in their places, it is inserted in an electrotype bath, and a coating of copper is deposited all over it, except immediately against the clamps. After removal from the bath the filament is immersed in hydro-carbon liquid or vapour, and a current is sent through it which raises the uncoated portions to redness, and causes a deposit thereon of solid carbon, which welds together the filament and the socket. A similar result may be obtained by immersing the greater part of the filament in mercury. In order to produce uniformity in the light-giving power of incandescence lamps, there is arranged near to them, during manufacture, a thermo-electric contact breaker, which interrupts the current when a standard temperature is attained. The invention also includes a device for switching the current through a second filament when the first fails, but this is not claimed as novel.

4,207.—C. A. Barlow, Manchester. (*A. de Meritens, Paris.*) **Dynamo-Electric Machines.** 6d. (4 figs.) September 29.

DYNAMO-ELECTRIC GENERATOR.—In this generator (Fig. 1) the magnetic fields are produced by

FIG. 1

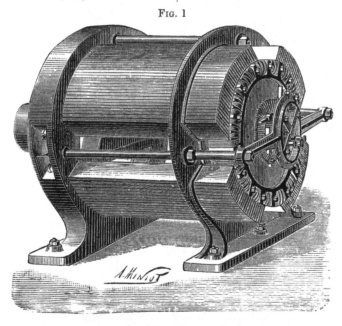

4,202.—J. W. Swan, Newcastle-upon-Tyne.
Incandescent Electric Lamps. 4d. September 29.

INCANDESCENCE LAMPS.—This invention relates to a method of producing by means of local electrical heating a coating or deposit of carbon over the junction of the metallic socket and the carbon filament inserted in it. For this purpose, after the

four compound permanent magnets E E¹ (Fig. 2) arranged with alternating poles around the two circular frames of the generator, so as to lie parallel with its axis, and to form part of a cylindrical surface within which the armature G rotates. Each of these compound magnets is built up of sixty-four steel laminæ which are held together, and to the brass framing of the generator,

by three bolts, indicated by the dotted lines. Fig. 3 is a diagrammatic view of the armature ring, showing also the commutator and brushes. Fig. 4 is a section of the armature in which its

FIG. 2

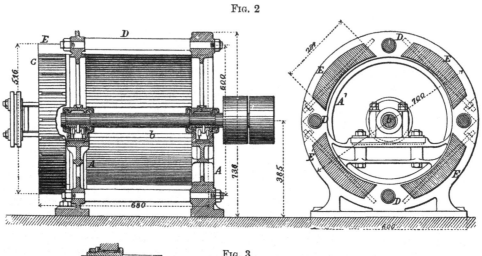

FIG. 5

FIG. 3

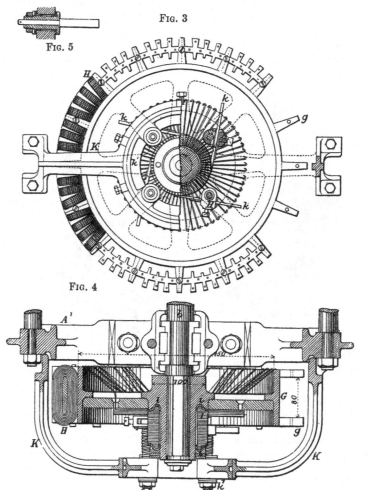

FIG. 4

position relative to the framework and the commutator is shown. The armature consists of a ring built up of sixty-four bobbins H H, each of which is wound upon a compound iron core H, made up of eighty laminæ of soft iron. Each core is divided into four parts by projecting pieces, and in the recesses are wound the coils of insulated copper wire, the outer surfaces of which are flush with the outer surfaces of the separating teeth. Each compound bobbin is fitted on to the brass wheel between the projecting pieces g, and held in place by bolts passing through the pieces g, and through grooves at the end of the bobbin cores. The coils of the armature are connected in series as in the Gramme armatures. The commutator I (Fig. 3) is built up of sixty-four strips of copper, insulated from each other by layers of silk, and forming a hollow cylinder which is attached to the shaft. The junction between each contiguous pair of coils is connected to its corresponding section on the commutator. Two pairs of brushes $k k$, $k k$ (Fig. 3) formed of strips of hard elastic brass are used. The alternate brushes are connected together to form the poles. Each brush is mounted on an insulated pin (shown in detail in Fig. 5) and maintained in contact with the commuator by spiral springs. The four pins are attached to a brass ring capable of rotation about the axis of the commutator.

4,255.—A. Watt, Liverpool. Secondary Batteries. 4d. October 1.

SECONDARY BATTERIES.—The first form of battery consists of lead rolled into sheets and coated with a mixture of calcium hydrate, manganese dioxide, and sodium or calcium chloride, or preferably with "Weldon mud" mixed with an equal volume of powdered coke. Plates of carbon may take the place of the lead plates. Also carbon plates in a solution of sulphate of manganese can be employed in one cell, as the sulphate decomposes and covers the carbon with peroxide of manganese. The second form of battery consists of alternate layers of carbon and granulated manganese dioxide mixed with granulated carbon, preferably in equal volumes, or of lead and manganese dioxide or layers arranged as follows :

$$\mathrm{Mn\ O_2\ Pb} \ \Big|\ \mathrm{Mn\ O_2\ C} \ \Big|\ \mathrm{Mn\ O_2} \ \Big|\ \mathrm{Pb\ Mn\ O_2\ C}$$
$$+\ \mathrm{C} \qquad +\ \mathrm{C} \qquad +\ \mathrm{C} \qquad +\ \mathrm{C}$$

and so on, or

$$\mathrm{Pb} \ \Big|\ \mathrm{Mn\ O_2\ C} \ \Big|\ \mathrm{Pb} \ \Big|\ \mathrm{Mn\ O_2\ C}$$
$$+\ \mathrm{C} \qquad\qquad +\ \mathrm{C}$$

the vertical bars being porous diaphragms. The exciting liquid is sal-ammoniac or common salt, or dilute sulphuric acid. A convenient form of cell consists of an ordinary battery tank divided down the centre by a porous diaphragm. The charged porous pots, fresh or exhausted, of Leclanché cells may even be used without further preparation than immersing them in one of the aforesaid solutions. If one pot be fresh and the other exhausted the couple is charged and ready to work. Another modification consists of a porous pot containing a plate of carbon and charged with a mixture of manganese dioxide and carbon, or coke, in a state of coarse powder, and placed in a leaden vessel containing dilute sulphuric acid, the space between the pot and the leaden vessel being filled with a mixture of manganese dioxide and carbon or coke, or simply with granulated lead.

4,271.—W. R. Lake, London. (*A. D. Maikoff and M. de Kabath, Paris.*) Electro-Magnetic Apparatus. 6d. (5 figs.) October 1.

ELECTRO-MAGNETS.—According to this invention the core surrounded by the wire coil acts, not as heretofore upon an exterior armature, but upon an armature placed within the coil. The action is characterised by the fact that it is not produced by the attraction exerted by the current upon the

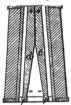

armature, but by the repulsion caused by similar polarity being imparted both to the coil and to the armature. The illustration shows one method of putting the invention into practice. When a current passes through the coil the two pieces d and e pivotted at their lower ends, are similarly magnetised and repel each other.

4,294.—A. G. Schaeffer, Newcastle-on-Tyne. Manufacture of Incandescent Electric Lamps. 2d. October 4.

INCANDESCENCE LAMPS.—The globe is exhausted through a small metallic pipe, preferably of platinum, which acts as one of the conductors. After exhaustion the pipe is nipped together, severed, and the end soldered over.

*4,296.—C. D. Abel, London. (*L. A. Brasseur and O. Dejaer, Brussels.*) Conduits for Telegraphic or Telephonic Conductors Laid in Streets or Roadways. 2d. October 4.

CONDUCTORS.—These are carried in a trough having a flanged cover either separate from or hinged to it. The cover forms part of the roadway.

4,304.—H. Aylesbury, Bristol. Dynamo Machines for the Production and Distribution of Electric Currents. 6d. (3 figs.) October 4.

DYNAMO-ELECTRIC GENERATOR.—This generator appears to be an aggregation of a number (twelve) of dynamo generators in one framework, and its advantage is stated to consist in the facility it affords of producing a number of independent currents and of combining these currents so as to vary

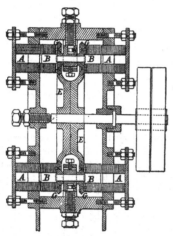

the strength and E.M.F. of the resultant. Considering one element of the machine (say the right-hand lower one), A is an electro-magnet, wound on two **D** shaped cores, so as to present two poles at its free end, and capable of axial adjustment; and B is a circle of ten sector-shaped electro-magnets packed together to form a cylinder, which rotates around its axis in face of the magnet A. Each of the magnets B is fixed on a disc G, which is carried on a pin that runs in an adjustable bearing attached to the framework. E is a rotating disc, provided with two **V** grooves in its periphery, which gear with **V** projections on the edges of the discs G and cause them to revolve. The specification does not explain how the currents are collected or distributed.

*4,305.—**H. J. Haddan**, London. (*L. Somzée, Brussels.*) **Electric Lamps.** 2d. October 4.

SEMI-INCANDESCENCE LAMP.—The principle of this invention consists chiefly in constructing electric lamps to emit light by incandescence as well as by the voltaic arc. A thin rod (preferably of carbon) is used, which may be fixed or movable on the second electrode of large section, the arc being formed at the circumference of a rod of refractory material. The specification contains no drawings, and the arrangements cannot well be understood.

4,310.—**A. P. Laurie**, Edinburgh. **Secondary Batteries.** 4d. October 4.

SECONDARY BATTERIES.—The essential feature of this invention is the storing of chlorine, for use as a source of electricity, by combining it with copper to form cuprous chloride. The cell is constructed as follows: A copper plate is coated with a paste of cuprous chloride, and is wrapped in parchment paper. It is then dipped into a solution of chloride of zinc along with a zinc plate. The zinc and copper plates are then put in metallic connection, and an electric current flows until all the cuprous chloride is reduced to a mass of spongy copper. The zinc plate is then replaced by a copper plate and the cell is ready for charging. The action of the chlorine on the spongy copper can be regulated by means of copper gauze placed over the spongy copper and in metallic communication with the copper plate. The cells may consist of vertical plates, or of shallow copper trays placed one within the other.

4,311.—**J. H. Johnson**, London. (*C. A. Faure, Paris.*) **Electric Lamps.** 6d. (4 figs.) October 4.

INCANDESCENCE LAMPS.—This invention relates to (1) the holders for the filaments; (2) to means whereby the filament may be removed without destroying the body of the lamp; (3) to improved means of exhausting the lamp; and (4) to the production of the filament itself. The globe has a tubular stem, to which a metallic collar or stem b is attached by fusing, or by electro-plating the stem and fixing

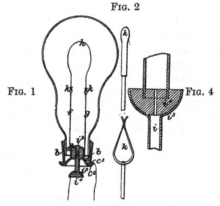

FIG. 2

FIG. 1

FIG. 4

FIG. 3

the collar to it by solder, or by fusible metal that will expand in cooling. The filament holder consists of a metal tube c closed by a plug a of glass, porcelain, or other non-conductor. The said tube c is splayed out into a trough c^2 to contain a metal c^3 which melts at a low temperature and receives the outer edge of the socket b so as to make an air-tight joint. The two conductors $f\,g$, upon which is fixed the filament k, are arranged as follows: The one electrode is formed by a piece of wire g soldered to the tube c, and the other holder is formed by a small tube passing through the plug d and

through the plate i^2 to which the wire f is fastened. The filament is secured by two spring clips k of metal, which are made by bending the same into an elongated loop and then flattening it. Figs. 2 and 3 are two views of such a clip. To effect the final sealing the metal i^4 in the cup i^3 at the end of the tube i^1 is perforated with a very fine hole (Fig. 4). After exhaustion the fusible metal is heated and the hole thereby closed. The filament is manufactured by drawing or cutting graphite into small strips, which are heated and bent into loops or other forms whilst hot.

4,383.—St. G. Lane-Fox, Westminster. Electric Bridges for Incandescent Lamps, &c. 6d. (4 figs.) October 8.

INCANDESCENCE LAMPS.—This invention relates (1) to the employment, for the purpose of reducing filaments for incandescence electric lamps to a definite or specified resistance, of a series of bottles or vessels charged with hydro-carbon gas or vapour in which the filaments are immersed, the said vessels having connections so arranged that the filament in any of such vessels can be connected at will with a source of electricity, and to a measuring apparatus employed to test the resistance of the filament. (2) To reducing the filaments in the first

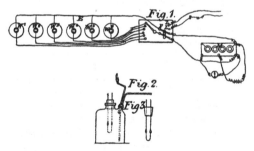

place to a resistance which is somewhat greater than that required in the lamps, and then, after they have been mounted in their holders and before hermetically sealing them in the lamps, reducing or adjusting their resistance to the exact degree required. Each of the bottles W^1 to W^6 (Figs. 1 and 2) contains a filament, and is charged with coal-gas or other hydro-carbon vapour. They are all connected to a common lead E at one end, and each has a separate lead at the other end. By means of a key F the current can be sent into each bridge in succession from the terminals LL'. When the process of building up a bridge is nearly complete the current from the generator is cut off, and its resistance is measured by the apparatus M. If the original source of electricity have a constant electromotive force a galvanometer may be included in the main circuit.

*4,396.—J. James and J. C. F. Lee, London. Manufacture of Carbons for Electric Lamps, &c. 2d. October 10.

CARBONS.—These are made from carbonaceous materials and compressed in moulds between upper and lower dies, by hydraulic pressure.

*4,398.—A. W. L. Reddie, London. (*E. Volckmar, Paris.*) Secondary Batteries or Electrical Accumulators. 2d. October 10.

SECONDARY BATTERIES.—The usual porous pot is replaced by two perforated plates of lead having the perforations filled with granules of pure lead. The plates are subjected to pressure and placed together, with strips of insulating material between them, in a suitable vessel containing acidulated water.

4,405.—A. M. Clark, London. (*Josephine de Changy, Paris.*) Producing Electric Light, &c. 10d. (10 figs.) October 10.

SEMI - INCANDESCENCE AND INCANDESCENCE LAMPS.—The carbon conductor is rendered incandescent "without being traversed by the current" by being placed within a platinum spiral through which the current passes or by being placed between two electrodes, as in the figure, where $p\,p$ are two rotating discs of platinum connected to the poles of a generator and moving in contact with a carbon pencil. The carbons employed in these lamps, and the filaments of incandescence lamps, are formed by causing porous substances of vegetable or mineral origin to absorb matters capable of yielding carbon by calcination in a closed vessel or *in vacuo*. The materials, such as yarns, pumice stone,

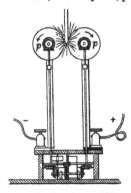

&c., are immersed in a boiling hydro-carbon holding in solution, resin, tar, or other bodies rich in carbon, and when thoroughly impregnated are brought to the desired form for the pencils, and afterwards are alternately baked and dipped in syrup. Such carbons may have a metallic core. Carbons may also be made from piassavei, sorghum, alfa, rattan, rattan pith, or couch grass, by soaking these materials

in sulphuric acid, washing, drying, and immersing them in a bath of glucose, sugar, and water; also from long-fibred wood free from knots, and from bone and ivory, by cutting the same into slips and immersing them in sulphuric acid, or boiling them in turpenpentine before carbonisation. The invention also includes an incandescence lamp globe, the tubular neck of which is formed with a swelling for engaging with a metallic clip which serves to form a contact.

4,409.—W. O. Callender, London. **Manufacture of Telegraph Conductors, &c.** 4d. October 11.

CONDUCTORS.—This invention relates to a compound of 35 parts of bitumen and 35 parts of oil residue, the latter obtained by subjecting oil to distillation in such manner as to leave the residue in an elastic condition with a specific gravity of 1.00. In place of the oil residue the substance known as elastikon may be used. The bitumen and residue are melted separately, three or four parts of vegetable oil being added in each case, and mixed at a temperature of 220 deg. to 250 deg. Fahr.; the temperature is then raised to 280 deg. or 290 deg. Fahr., and if the material require to be vulcanised, flowers of sulphur are added, the heat again being raised to 340 deg.; the mixture is finally stirred for about forty-five minutes, and kept at the last-named temperature for about two hours.

4,439.—J. Jameson, Newcastle-on-Tyne. **Incandescent Electric Lamps.** 6d. (11 figs.) October 12.

INCANDESCENCE LAMPS.—This invention relates to a method of sealing up relays of carbon filaments in incandescence lamps, so that when one fails it may be removed and a fresh one be substituted without access to the interior of the lamps. The filaments are mounted on an endless band passing

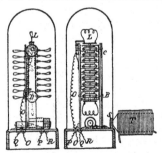

over rollers, arranged to bring each into position successively. They are each connected to a bar having insulating material between the points at which the ends of the filaments are attached. The band passes over two pulleys B C, driven by worm gear actuated by an electro-magnet and ratchet

mechanism, the conducting wires being connected to the top roller, so that only the filament L, occupying the highest position, is in circuit. The two wires O P convey the current to the electro-magnetic mechanism. The movement of the filaments may be rendered automatic by placing an electromagnet in the circuit Q R, and so arranging it that when excited it will break the circuit O P. When the former circuit fails the current traverses the latter, and by a make-and-break device operates the mechanism until a second filament is moved into position. To clean the interior of the globe a little bag of iron filings S is drawn up against the glass by the attraction of a magnet T. By moving the magnet or the globe the bag can be drawn all over the interior surface. Other methods of effecting this result, but all substantially the same in principle, are described and illustrated.

4,454.—J. T. Sprague, Birmingham. **Instruments for Measuring Electric Currents, &c.** 6d. (15 figs.) October 12.

CURRENT METER.—This refers to Patent No. 4,762 of 1878, in which a galvanometer having two soft iron needles was employed, and consists in so arranging the circuit as to obtain more perfect magnetism in the needles. This is accomplished by inserting within the ordinary frame containing the coil by which the needle is deflected, a second frame, the wire on which is wound at right angles to that on the ordinary frame. Or a semicircular frame may be used, wound so that in whatever position the needle may be the wire crosses it at right angles. In a quantity meter a moving electrode floats or sinks in the liquid of a cell according to the amount of metal deposited on it; the electrode is made hollow and fitted with a stem on the principle of the hydrometer. When "large currents" are used the floating electrode has fitted to its lower surface a number of parallel plates connected together so as to act as an electrode of large surface, and arranged to move in the spaces of a similar system acting as a fixed electrode, provision being made to keep the opposite surfaces close together and to prevent contact. The circuit of the moving electrode is completed by a bent wire sliding freely in a tube of mercury. An index fixed to the moving parts traverses a scale and indicates the "degrees of deposit effected." The electrode may be suspended from an axle having a pointer which traverses an insulated metallic dial, a spring clip, adjustable on the dial, conveying currents to any connected apparatus on the pointer making contact with it. A current reverser may be put in action by the pointer making contact with insulated stops attached to the dial at the beginning and end of the scale The electrodes, when a pair is used, may be

caused to actuate a rocking frame which in turn moves a train of recording dials. Or a pair of electrodes may have mounted between them, upon suitable bearings, an induced electrode in the form of a cylinder, which rotates when the deposit on one side alters the position of its centre of gravity. To obviate the necessity of occasionally changing the + and — electrodes, an external cylinder is used, composed of a number of insulated copper strips, and provided with suitable contacts for passing the currents into those plates which face opposite sides of the induced electrode. "As metals and liquids vary in their resistance in opposite ratios, they can give correct results only at the temperature of adjustment, if they are used as shunts to each other;" the inventor therefore forms the shunt circuit of graphite, and if necessary adds thereto a tube of mercury or a German silver wire. A warming apparatus to prevent the meter freezing consists of a thin platinum wire wound on a glass tube and surrounded by a second glass tube; these are open at their ends and are placed in the liquid causing convection currents to be set up therein. A "thermo-regulator" consists of a bulb and U tube with a neck for the insertion of the + and — wires, one of which is in contact with the mercury whilst the other is insulated except at its point. The contraction of the air by cold causes the mercury to rise and complete the circuit.

4,455.—J. W. Swan, Newcastle-upon-Tyne. **Secondary Batteries, or Apparatus for Effecting Electrical Storage.** 4d. October 13.

SECONDARY BATTERIES.—The surface of the lead plate is extended by cutting, or scraping, or equivalently acting upon the same, and thus facilitating the storing action. The whole extent of the plate is not so treated, but a portion is left solid to support the weakened part. The cutting may be effected in various ways, such as by tools in a shaping machine, or by engraved, indented, or roughened dies.

4,472.—C. V. Boys, Oakham, Rutland. **Electric Meter.** 6d. (13 figs.) October 13.

CURRENT METER.—To measure the current passing through a conductor, clockwork is provided with an escapement governed by a pendulum or balance, the oscillations of which are determined by the force of an electro-magnet or solenoid in the circuit, so that as a less or greater quantity of electricity passes the amount of movement permitted to the clockwork is less or greater. In order to insure accuracy the escapement should be of the dead-beat kind, incapable of giving such influence to the balance as would cause it to oscillate. The clockwork may be driven by a weight or spring wound up from time to

time by hand. It is preferred, however, to render it self-acting by combining it with remontoir apparatus, which is acted on at intervals by an electro-magnet in the circuit. A counter connected to the clockwork shows by suitable dials the quantity of current passed. Fig. 1 is a front view and Fig. 2 an inverted plan of one form of meter. B B are two limbs of an electro-magnet the coils of which are traversed by the current to be measured. These limbs terminate in polar extensions b b, between which there oscillates an armature on a vertical axis C. The lower part of the axis constitutes the cylinder of an escapement engaging with an escapement wheel D, connected to an arbor E driven by clockwork. The remontoir apparatus is arranged as follows: On the fixed axis f there are mounted (1) a toothed wheel F, which gears with a pinion e, and has a groove cut in its circumference to receive

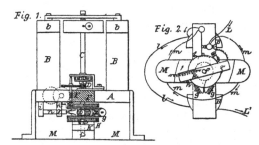

Fig. 1.　　　*Fig. 2.*

a driving band; (2) a barrel containing a volute spring; (3) a ratchet wheel H, grooved like F, to receive the same driving band, which passes over guide pulleys g g mounted on spring brackets projecting from the barrel; and (4) an armature K, which can oscillate between the poles M M of an electro-magnet fixed to the base A; this armature is drawn by a spring k against a stop pin k^1, and it carries a spring pawl h engaging with the teeth of the ratchet wheel H. The coils of the electro-magnet M M are in a by-pass circuit m m, branching off from the main circuit L L^1 at the terminals. These terminals are usually connected by a spring n, bearing against a contact screw n^1, and in that case the main current passes direct from L by the wires l l, through the coils of the electro-magnet B B to L^1, the magnet M M being then too feebly excited to effect displacement of the armature K. When, however, the stud g^1 pushes the spring n from the screw n^1 the current passes through the magnet M M, and the armature makes its stroke, winding up the spring. The specification also describes two modified forms of remontoir apparatus, in the first of which each complete oscillation of the electric balance gives a slight impulse to the winding mechanism; in the second the winding is effected only when the movement of the balance becomes so far reduced as to demand a fresh impulse.

4,478.—R. Harrison, Newcastle-upon-Tyne. Electric Lamps. 6d. (4 figs.) October 14.

INCANDESCENCE LAMPS.—The carbon filaments for incandescence lamps are prepared by punching them out of thick paper, in the shape C or C¹, by a

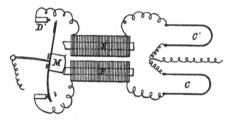

punch having in its face a **V**-shaped hollow so as to present two cutting edges and an intermediate recess in which the paper is compressed. The filament has two wide ends into each of which an eyelet is fixed to receive the conducting wires. In order to arrange for one lamp taking the place of another in case of extinction, the filaments are provided in duplicate and connected to the shunting apparatus shown in the illustration. E E¹ are two electro-magnets, one in the circuit of each filament, and M is an armature provided with contact pieces. One magnet is made more powerful than the other, so that as long as the filaments are intact the armature is attracted in one direction and the opposite circuit is broken at D¹. When the first filament fails its magnet becomes inoperative, and the position of the armature is reversed, thereby switching the current into the other filament.

*4,496.—J. H. Johnson, London. (*La Société Anonyme la Force et la Lumière, Société Générale d'Electricité, Brussels.*) Regulators for Electric Motors. 4d. October 15.

RHEOSTAT.—The essential part of this regulator is an ordinary metallic chain. The chain presents a series of points of contact, each of which offers a resistance dependent on the pressure upon it. When it is tightly stretched the resistance is comparatively small, but when it is slackened, or allowed to hang loosely, the resistance is considerably increased. This provisional specification also describes several ways of including a chain in the field magnet circuit of an electric generator so as to vary the intensity of the current.

4,504.—J. Brockie, London. Electric Arc Lamps. 6d. (4 figs.) October 15.

ARC LAMP.—Referring to the illustration, the two upper carbon-holders A¹ A² fall freely by their own weight when they are not retained by the cam levers C¹ C². These levers grip the holders when their inner ends are raised by the armature C of the electro-magnet M M. When a current is

passed through the circuit, the shunt magnet M attracts the armature, lifting it and the two levers and carbons. The current forms an arc between one pair of carbons, the other pair remaining inactive. When the magnet M is momentarily cut out by an arbitrary commutator or interrupter, or momentarily short-circuited, or reversed, by a regulating coil E,

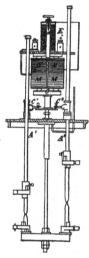

the armature will retire from the magnet allowing the clutches to relax their grip on the holders. The two pairs of carbons will then fall together and will be instantly raised, the arc appearing only between the pair that is incandescent. The clutch magnet M has a coil at its lower part in the main circuit, and a coil N at its upper part in a shunt circuit; this latter coil, when put in circuit by the regulating shunt coil E, reverses the polarity of the clutch magnet, and thereby releases the armature and the clutches then fall away; but as this has the immediate effect of reducing the strength of the regulating coil E, the reversing circuit is at once broken, and the main magnet, being no longer reversed, lifts up the armature smartly, and the arc is re-established; thus, whenever the arc becomes too long, the main magnet is momentarily reversed and the readjustment effected. In a modification, a third magnet, wound in the same circuit as the reversing coil N, is placed below the armature, so that if the reversing coil N should prove too weak or too strong, this additional magnet will pull the armature C from the poles of the clutch magnet M. Reference is made to a prior Specification No. 3,071 of 1879.

*4,507.—A. E. Gilbert, London. Insulators for Telegraph Wires, &c. 2d. October 15.

INSULATORS.—An earthenware insulator has an oblique lateral channel, the sides of which are under-cut and provided with lips. The strained horizontal conductor enters the opposite end of

each of the two recesses and is housed behind the projecting lips, which prevent it escaping sideways.

*4,508.—J. H. Johnson, London. (*E. W. Parod, Paris.*) **Production, Collection, or Storage and Distribution of Electricity, &c.** 4d. October 15.

DISTRIBUTING CURRENTS.—This invention relates (1) to the combination with the condensing conductors described in Patent No. 4,686 of 1878, of secondary batteries " to collect and store up the electricity generated in excess ;" (2) to automatic apparatus for placing the batteries in communication with the conductors ; and (3) to automatic apparatus for regulating and controlling the intensity of the distributed currents, and for throwing the generators in or out of action, and for operating the controlling valves of steam engines. Under the first head it is stated that "each of the metallic armatures of the said condensers may be connected with a second armature in direct contact with the first, and consisting of water acidulated or saturated by any metallic salt," to which any suitable substance is added to form a permeable mass in which the excess of current is stored. Under the second head, a disc, provided with an armature and contact pieces, is attracted equally by a permanent and an electro-magnet so long as the current remains normal ; should the tension however increase, the electro-magnet causes the disc to insert uncharged accumulators into the circuit, and should it decrease, the permanent magnet causes it to insert charged accumulators into the circuit. Under the third head, an armature, placed between a permanent and an electro-magnet, is moved according to their relative power, and in moving changes or reverses the current.

4,518.—W. R. Lake, London. (*J. J. Journaux, Paris.*) **Apparatus for Driving or Operating Sewing Machines by Electricity.** 2d. October 17.

MOTOR.—An electric motor is geared with the machine by elastic and variable friction appliances.

4,533.—R. R. Gibbs, Liverpool. **Electric Lamps.** 6d. (3 figs.) October 18.

ARC LAMP.—This lamp is arranged to automatically bring a fresh " candle " into action as the preceding one is consumed. The action is as follows : The current passes to the fixed socket b, thence through the carbons $h\ h^1$, to the movable sockets c ; it then passes through the lever d, spring f, lever d^1, spring f^1, and so on to the spring f^3 and return wire. When the " candle " $h\ h^1$ has been consumed below the lower edge of the insulating

strip i, the two carbons are free from each other, and the spring k moves the lever d until the sockets c and b are in contact. The spring f is at the same time moved into contact with the socket of the next candle through which the current then

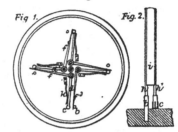

Fig 1.   Fig. 2.

passes. When continuous currents are used the carbons are insulated on three sides, leaving the adjacent sides bare, or the carbons are caused to revolve in opposite directions. The same arrangements are also applicable when alternating currents are used.

4,541.—R. Kennedy, Paisley. **Generation, Collection, and Distribution of Electro-Magnetic Currents, &c.** 6d. (9 figs.) October 18.

DYNAMO-ELECTRIC GENERATOR.—In this generator the field magnets $a^1\ a^1$ consist of two broad horseshoe shaped cores, suitably wound, having pole-pieces within which the armature, consisting of a number of rings $e\ e$ mounted upon the axis b, rotates. Each ring carries a number of coils wound upon an annulus of iron wire supported by arms, the separate

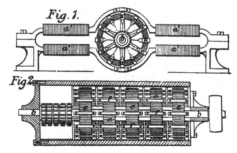

Fig. 1.

Fig 2.

coils being divided from each other by iron clips $f\ f$ that serve to convey the inductive force from the field magnets to the armature core. Each armature ring has its own commutator with the necessary brushes, and one ring is used solely to excite the field magnets. The currents may be distributed in various ways, and the coils coupled in series or parallel arc according to the exigencies of the installation.

4,552.—P. Jensen, London. (*T. A. Edison, New Jersey, U.S.A.*) **Dynamo or Magneto-Electric Machines.** 6d. (4 figs.) October 18.

DYNAMO-ELECTRIC GENERATOR.—This invention relates to means of regulating the generative power of a generator. "For this purpose in one case the principle is made use of that the power of an electro-magnet can be weakened by diminishing the mass of the yoke connecting the cores, and that such power can be strengthened by increasing the mass of the said yoke until a maximum is gained." The illustration shows an automatic method of keeping the electromotive force, developed by a generator, constant. The yoke C C, connecting the two branches of the field magnets, has a central conical opening F, in which plays the conical block G, forming the magnetic circuit regulator. The stem of this block may be screwed and be adjusted by hand,

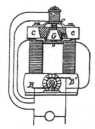

or it may form the core of a solenoid. This solenoid "is used to draw the cone into the yoke, the same being retracted by a spring f and by its own weight. The coil of the magnet is in the main or consumption circuit." The same result may be accomplished in a second way by the use of levers or bars of iron adapted to be adjusted closer to or farther away from the pole ends and yoke of the magnet, or the pole ends alone, so as "to partially shunt the magnetic currents or lines of force away from or around the field in which moves the revolving armature." Two methods of arranging the levers are shown in the specification ; one in which a lever connects B and C and also B¹ and C, and another in which B and B¹ are connected.

4,553.—P. Jensen, London. (*T. A. Edison, New Jersey, U.S.A.*) **Charging and Using Secondary Batteries.** 6d. (4 figs.) October 18.

SECONDARY BATTERIES.—This invention relates to means for maintaining the electromotive force of a current from secondary batteries constant, by reinforcing the primary current from time to time, so as to keep up the electromotive force in the main or consumption circuit. Fig. 1 is a diagram representing the method of charging the secondary cells, and Fig. 2 a diagram of one method of using the cells and maintaining the electromotive force of the current. 1, 2 (Fig. 2) are the main conductors of a distribution system, of which A represents the lamps connected in multiple arc. B represents secondary cells arranged in two series C D, the

cells of each series being coupled together and connected to the positive main conductor 2 by switches a. At the other end the two series are connected together by a wire, from which the wire 9 leads to the positive commutator brush b of a dynamo or magneto-electric generator E. The negative commutator brush c is connected by the wire 5 with 1. The field circuit 6, 7, of E, has a series of resistances and switch d. Each series C D has the right

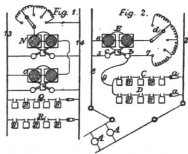

number of cells to give the desired electromotive force. Both series may be connected by a with 2 at the start, to give the primary current, or only one series may be so connected, and then the other can be thrown in when the electromotive force of the primary current drops. The generator E is not operated until the current due to both C and D commences to drop in its pressure. Then the switch d is moved to the upper contact, and the engine that drives E is started. This throws a reinforcing current in multiple arc through the secondary cells. At each subsequent drop in the electromotive force the switch d is moved inwardly to the next contact, and the standard electromotive force maintained. Two other methods of arranging rows of cells are described with suitable connections for throwing them into circuit successively. For charging the cells two or more dynamo or magneto-electric generators N O are connected in multiple arc with conductors 13 and 14, Fig. 1. The field coils of the generators are connected in series. The cells are arranged in series Q R, connected to the conductors 13 and 14. When the cells are fully charged they are disconnected from 13, 14, and conveyed to the points of consumption.

4,559.—F. M. Newton, Taunton. **Apparatus for Generating and Utilising Electricity.** 6d. (13 figs.) October 19.

DYNAMO-ELECTRIC GENERATOR.—This invention relates to a method of constructing the armatures of dynamo-electric generators, and to the construction of an arc lamp. Fig. 1 is a vertical longitudinal section of the armature and commutator of an electric generator. Fig. 2 is a cross section of the same, with part of the coils removed, and Fig. 3 is an end view. Upon a shaft B are threaded a

number of notched or toothed discs A, separated from each other by intervening air spaces, and so arranged that the notches together form longitudinal grooves. The external air has access to all the spaces through the helical grooves F F, and is thrown out through the coils by the centrifugal force and the action of the blades or wings E E. The method of winding the armature is as follows : The wire is laid down groove 1, then across the end of the armature, as a chord of the circle, and back down groove 6. Crossing the other end

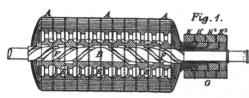

Fig. 1.

of the core, it again goes down groove 1, and so on until 1 and 6 are filled. Grooves 9 and 14, 2 and 13, and 5 and 10 are then wound in a similar manner. At this stage eight of the sixteen grooves are filled, and the portions of the coils which appear at each end of the core enclose a square. The remaining grooves are then wound in the order of 4 and 15, 7 and 12, 11 and 16, and 8 and 3. One end of each coil is coupled to the appropriate end of the coil that lies in a parallel plane with it, and the remaining end of each coil is connected to the commutator. Thus the eight coils form four circuits. The commutator G consists of four metallic rings on a wood boss. There is one ring for each circuit on the armature, and each ring is divided into two parts or segments, of which one constitutes five-eighths and the other three-eighths of the whole circumference. One end of each circuit is attached to one part of a ring, and from the construction it follows that when the circuit is in the neutral part of the field, one end is insulated, and the two brushes are connected through the major segment of the ring. Two modified forms of commutator are also described ; in one the two ends of the coils are connected together, as well as the two brushes at the neutral point, and in the other both ends of the coil and brushes are insulated. By aid of the commutator the coils may all be connected in parallel arc, or they may be joined in series by connecting the positive brush of H to the negative of H¹, and the positive of H¹ to the negative of H², and so on.

ARC LAMP.—Fig. 5 is a section of an arc lamp in which the feature of novelty lies in the method of releasing and retaining the upper carbon. The carbon A passes through a socket G (see detail view), in which is an inclined slot or gate containing a cylinder or roller O. This roller jams the

carbon against the opposite side and holds it so long as the lamp is burning with the normal arc. As the current begins to fail, the core of the solenoid

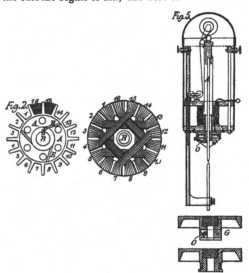

Fig. 5.

descends, lowering the socket and carbon until the roller O comes in contact with the fixed finger below it, and is forced up the slot, when the carbon is freed and descends to adjust the arc.

4,571.—E. G. Brewer, London. (*T. A. Edison, New Jersey, U.S.A.*) **Measurement of Electricity in Distribution Systems.** 6d. (3 figs.) October 19.

CURRENT METERS.—The object of this invention is to provide check meters at the central station to measure the total quantity of electricity sent forth, and to check it, if necessary, against the aggregate results of the consumers' meters ; also to determine the loss from leakage by coupling the negative conductor to earth and measuring the flow into it. This flow will exactly equal the leakage ; 1, 2 are the main feeding conductors, and 3, 4, 5, 6, 7, 8 the consumption circuits ; from the latter

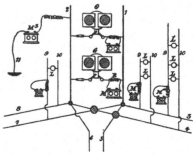

extend the house conductors 9, 10, supplying the lamps L ; G represents the generators connected in multiple arc ; M M are the current meters arranged

in shunts between G and 1. These meters can be situated, if desired, as shunts to the conductors 3, 5, and 7 of the consumption circuits. A resistance R is employed to shunt a definite proportion of the current through the meters. M^3 is the leakage meter in line 11 from earth to 2. Any form of meter may be employed.

4,576.—E. G. Brewer, London. (*T. A. Edison, New Jersey, U.S.A.*) Meters for Measuring Electric Currents. 6d. (4 figs.) October 19.

CURRENT METERS.—This invention relates to an electro-depositing meter to indicate a correct result with a smaller current than the copper cell previously employed for the same purpose. This is accomplished by using amalgamated zinc electrodes in a solution of sulphate of zinc. The plates have a heavy coating of zinc deposited on them before being placed in the meter, "which zinc will be thoroughly amalgamated while being deposited." The depositing cell M is arranged in a shunt from one of the

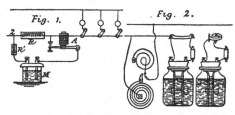

Fig. 1.   Fig. 2.

conductors 2 of a house or other consumption circuit, a resistance R being placed in the line to shunt a definite small portion of the current through the meter. A wire resistance R^1 is placed in the meter circuit to compensate for the effects of change of temperature in the resistance of the cell. To prevent the establishing of a counter current, and the recomposition of the solution when no current is flowing through the house, an electro-magnet A is provided for automatically breaking the shunt when the current ceases. To prevent the temperature in the cell falling too low, a carbon or metal resistance is placed beside it (Fig. 2), and at a certain temperature a coiled spring makes contact and diverts a definite portion of the current through the resistance, thereby generating heat. The meter may comprise two cells, one with a much greater resistance than the other, to be used as a check. In place of glass vessels and metal plates, two concentric copper cylinders may be used.

*4,582.—A. M. Clark, London. (*J. de Changy, Paris.*) Lighting of Railway Carriages, &c., by Electricity. 2d. October 19.

ELECTRIC LIGHT.—A generator, driven by one of the axles, supplies its current to a series of

secondary batteries from which "the current is conducted to lamps on the incandescence system."

*4,591.—H. J. Haddan, London. (*G. Dessaigne, Villefranche, France.*) Generating Electricity. 2d. October 20.

GENERATORS.—The object of this invention is to utilise the power of flywheels by attaching to them magnets, "and placing bobbins or electro-magnets around such magnets," &c. "Electricity is obtained in this way without any additional working expense."

4,592.—A. Millar, Glasgow. Apparatus for Generating Electricity, &c. 8d. (13 figs.) October 20.

MAGNETO-ELECTRIC GENERATOR.—Referring to the illustration, A is a plate of bronze resting upon the edges of an open cistern B filled with water. C C C are permanent magnets, and E E E are soft iron bars wound with coils of insulated wire. A series of soft iron rollers G G, in number one less than the number of magnets, are driven in a circular path on the surface of the plate A, and come in contact with the ends of C and E simultaneously. When the shaft H is made to rotate in either direction the magnetism of the magnets is communicated to the bars E E by the rollers, and currents of electricity are generated at the moment

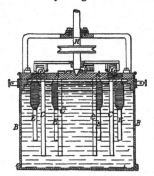

of "magnetisation and demagnetisation." The heat which is produced is absorbed by the liquid in the tank. When water is used, the coils must be insulated by japan or other waterproof material. A modification of this generator has horseshoe in place of bar magnets. In order to lessen the force required to separate the rollers from the magnets, these latter have pole-pieces that gradually taper in one direction.

MOTOR.—The invention also relates to an electro-motor. This consists two parallel discs, the first having electro-magnets on the inside of its rim like the teeth of an internal gear wheel. The second has electro-magnets on its periphery like the teeth of a spurwheel, and is placed within the other, so

that the teeth on one side of it gear with the teeth on one side of the other disc, in the same way that a solid pinion and an internally toothed wheel usually gear. The number of magnets on one disc is greater than that on the other in the ratio of twelve to ten, and on account of this difference of numbers, and also owing to the eccentricity of the shafts, "the poles of one pair of magnets are at a considerable distance apart, the next pair nearer together, the third pair still nearer, and the fourth pair are passing each other. A current of electricity sent through the coils of the magnets causes the poles to be attracted, and as they can approach each other only when the two discs are made to rotate simultaneously in one direction, a motion of rotation is established."

4,607.—H. F. Joel, London. Magneto-Electric Machines. 6d. (25 figs.) October 21.

DYNAMO-ELECTRIC GENERATORS.—The improvements "are essentially in the construction of the frame, the field magnets, the revolving armature, the commutator, and the brushes and frames; and, further, in the novel disposition of the magnets relatively to the armature, and in the method of connection between the armature coils through the commutator." The illustrations are two elevations of one form of generator. There are four field magnets with poles N S, N^1 S^1 at their centres, and each pole carries a pole-piece n n^1. The revolving armature has nine coils, and is wound on a hollow annular core, carried on a non-magnetic boss provided with arms. The construction of this centre or core can be partly seen in Fig. 2. It consists of a boss with arms, which support two rings of wrought iron, each made of two layers of overlapping plates, and separated by an air space. Around these two

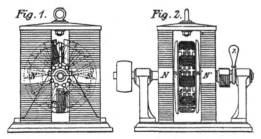

Fig. 1.　　　　*Fig. 2.*

rings the coils are wound, after the manner that obtains in the Brush armature, and between them iron cheek pieces e e are fixed. These cheeks have hooks that pass over the edge and down between the rings, which are magnetically insulated from each other. It has been "found that the strongest action obtainable from ring armatures with projecting cheeks, of this or other construction, occurs when the cheeks make and break from the ends of the extending pole-pieces." To intensify this

action the pairs of pole plates are arranged "to alternately overlap and be in advance of one another. In this case the coil is at the moment of its passing across the centre of the pole plates comparatively neutral, whilst at the same moment, in consequence of the odd numbers, there is no neutral coil in front of the centre of the other poles." The pole plates may be modified by affixing thereto slight internal radial projections, like broad chipping pieces. In such an arrangement the neutral coils are cut out between the pole plates as in ordinary ring armatures, and not in the centre, "and the magnetic make and break takes place at or near the centres of the pole plates." By this construction of pole plates there is gained "the advantage of the currents from the induction action being added to the currents from the magnetic make and break." In Fig. 2 there is shown one of Varley's commutators, consisting of a number of rings, corresponding to the single or double coils on the armature, with insulating spaces for reversing and cutting out the bobbins as desired. The brushes and holders can be adjusted angularly by the handle Z, and each brush can be moved in and out by a rack and pinion. Another form of commutator is made with segmental plates, each plate being stepped in the middle, so that the brush is in contact alternately with one plate and two plates. In conjunction with this commutator there may be constructed on the same axle a short circulating device, consisting of an indented metal wheel with the indentations filled with insulating material. "This device enables one through a subsidiary brush or brushes to short-circuit from the commutator through the exposed metallic portions at the intervals of the insulating material occurring at the time of the increase consequent on the cutting out of the neutral coil." This device may be used to connect up in a shunt circuit supplementary coils of wire coiled round the field magnets in a similar method to that advocated by Wheatstone, or the current may be led through a separate leading wire to be utilised as may be desired. A second type of generator has its field magnets arranged as in the figures, but the opposite poles are of unlike name. The two rims of the armature core are magnetically connected, and each is wound with independent coils. The connection between the coils and the commutator plates is made by wires, each of which leads from one commutator plate and is branched to the similar poles of a pair of coils, one in each ring, but one coil in advance of the other. In a third generator the special feature is that the ends of the pole-pieces are bevelled off, so that the armature coils emerge gradually from their influence. In the fourth and last generator there are four polar fields around an armature like the one shown in the figures, but with eight coils instead of nine.

These are divided into two distinct circuits, and, from the construction of the armature, each system of coils is at one time alone in connection with the leading wires through the brushes, and at another time the two systems are combined in parallel circuit.

4,617.—A. M. Clark, London. (*H. B. Sheridan, Cleveland, Ohio, U.S.A.*) **Electric Lamps.** 6d. (6 figs.) October 21.

ARC LAMP.—This invention relates to an arc lamp of which Fig. 1 is a sectional elevation, Fig. 3 is a similar view of the regulating mechanism to a larger scale, and Fig. 2 is a section at right angles to Fig. 3. The lower carbon is raised by a cord passing over guide pulleys in the framing and wound on the groove 1 of the pulley H. The upper carbon-holder depends from a cord wound in the opposite direction on the groove 2. The pulley H rides loose on a pin, and is turned in one direction, to establish the arc, by the end of the spring i acting as a pawl, and taking into teeth on the flange, and in the other direction by the pawl x on the lever w. The adjustable spring i is depressed by an arm h on the lever d, which is operated by the differential coils D and E, of which D is in the

FIG. 1.   FIG. 2.

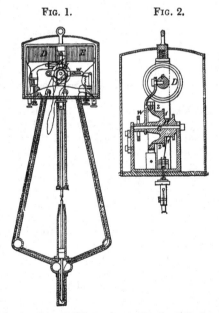

main circuit, and E in a shunt circuit. The lever w is centred on the arm v, and slotted out to pass the pin G ; it is moved by a driving pin on an arm projecting horizontally from the lever d. At its outer end it carries an insulated plate m, electrically connected to one terminal, which is put in contact with a similar plate o, connected to the other lamp terminal, when the carbon-holders have reached

the limit of their travel and the arc fails. Fig. 1 shows the position of the parts when the lamp is ready for lighting. Upon the passage of the current the core e moves into the coil D, and the lever d presses the spring i on to the toothed wheel, turning the pulley in the opposite direction to the hands of a watch and separating the carbons. As the arc lengthens the core e moves back into

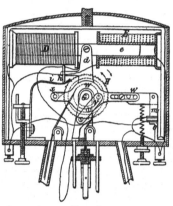

FIG. 3.

the coil E, and the levers d allow the spring i to rise, at the same time by means of the lever w and pawl x it turns the toothed wheel in the opposite direction, and causes the carbons to approach each other. As the levers d, w have different fulcra, but are connected together, the pawl x will travel faster and further than the holding plate h, and the pulley will make its movement before being fully released from the spring i. The outer end of each coil is closed by a "vibrating plate" like a reed, to act as a dash-pot.

*4,632.—J. S. Sellon, London. Secondary Batteries. 2d. October 22.

SECONDARY BATTERIES.—These are made by packing peroxide of lead round pieces of metal or carbon. The metal, such as platinum, is used in the form of gauze, sheet, or foil, plain or corrugated, and is fixed to wooden or other frames. The oxide is enclosed in porous materials to form cells. Sheets of lead attached to one or both sides of the cells form the second electrode. On dipping into acidulated water a sufficient quantity is absorbed to put the battery in action without the use of a free liquid.

4,654.—G. G. Andre, Dorking. Electric Incandescent Lamps. 6d. (6 figs.) October 24.

INCANDESCENCE LAMPS.—To form the filament a vegetable fibre, such as rattan, is immersed in a mixture of one part nitric and four parts sulphuric acid, and, after washing, is steeped in a

solution of nitro-cellulose in ether, or other solvent, where it is allowed to remain until it assumes a semi-dissolved state. The material thus prepared, when dry, is formed into the shapes required, which are then carbonised in the ordinary way and treated in ether, tar, boiled linseed, or other drying oil, sugar syrup, solution of starch, dextrine, or the like, to fill up the pores. The filaments are then calcined to carbonise the absorbed substances; or the pores may be filled up by heating the carbon in a vessel of hydro-carbon gas; or the filament may be steeped in drying oil, and the oil allowed to dry before calcining; or burnt linseed oil mixed with lampblack may be rubbed into the fibre. The ends of the filament may be inserted into sockets at the ends of the conductors and be fixed by dissolved nitro-cellulose, wood pulp, or burnt boiled oil, and carbonised by the current. This is shown in Fig. 1, where *b b* are carbon blocks or metallic sockets, attached to the conducting wires *c c*, *a* being the filament, and *d d* wires connected to each of the

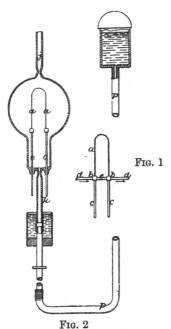

Fig. 1

Fig. 2

blocks, which are united by a wire *e* to complete the circuit during the carbonisation of the contacts. Another method is to couple the filament to the wires by the aid of a clip of mica or steatite. Fig. 2 shows the exhausting apparatus. At each end of the bulb there is fixed a contracted tube, and to the lower of these there is coupled a vessel of mercury by means of a flexible tube *p*. When the vessel is raised the globe is filled with mercury, the greater part of the air being expelled through

the tube *j*, which is filled with a soft stopping. The vessel is then lowered more than 30 in. to create a barometrical vacuum in the globe. This liberates the occluded gases of the filament, which on again raising the vessel are forced out of the tube *j*. This is unstopped and then sealed, the tube *k* being sealed when the vessel is again lowered.

4,659.—R. H. Courtenay, London. Lighting by Gas and Electricity Combined. 6d. (4 figs.) October 25.

ELECTRIC LIGHT.—Four methods of carrying out this invention are illustrated. In the first, a curved platino-iridium wire, rendered more or less incandescent by an electric current, is placed in the upper fringe of the flame of a batswing burner. In the second, a vertical wire, rendered incandescent, passes down the centre of a Bunsen burner. In the third, a Jablochkoff candle has a passage for gas made in the "colombin." In the last, an ordinary arc light has an Argand gas burner surrounding the lower carbon.

*4,664.—J. Imray, London. (*J. Carpentier, Paris.*) Electrometers. 2d. October 25.

CURRENT METER.—The instrument consists of two tangent galvanometers arranged at right angles to one another, and set to act upon the same needle. Or one coil may be placed above the other, and each may have its own needle, provided that the two needles be rigidly connected. The instrument is designed for measuring the relative intensities of two currents.

4,684.—F. C. Kinnear, London. Balloons for Illumination. 6d. (3 figs.) October 26.

ELECTRIC LIGHT.—This invention refers chiefly to captive balloons, the exteriors of which are illuminated by electric light, the current being supplied by a generator on the earth. In free balloons the current is stored in secondary batteries carried by the balloon itself.

*4,771.—C. Crastin, London. Lighting by Gas, &c. 2d. November 1.

INCANDESCENCE LAMPS.—A burner constructed of "any suitable heat-resisting, non-conducting material formed with holes to receive the ends of a number of loops and rings of platinum . . . . may be employed for electric lighting."

4,775.—H. A. Bonneville, London. (*L. Daft, Brooklyn, U.S.A.*) Electric Lamps. 6d. (6 figs.) November 1.

ARC LAMP.—In its simplest form, as shown in Fig. 1, this lamp has no regulating mechanism

whatever. The upper carbon slides down a tube F, until arrested by the claws $b\,b$, which only allow the part that has wasted in diameter under the action of the current to pass through them. The lower carbon is similarly regulated, the lifting power resulting from its being immersed in mercury. The arc is established by the solenoid A. The core is attached firmly to a prolongation of the carbon-holder, and the weight of the two is carried by a

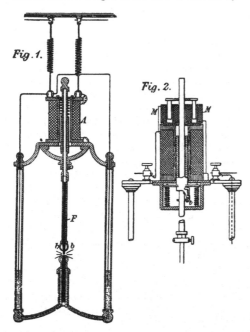

Fig. 1.

Fig. 2.

spiral spring. As soon as the circuit is established, the core is lifted and the arc appears. The lamp is suspended by springs to prevent the vibrations of the support affecting the light. When the lamps are burnt in series a fine differential coil M is added to each (Fig. 2), and it is so arranged that its force is opposed to that of the solenoid. Fig. 2 also shows a regulating appliance that may be used in place of the claws b to feed down the upper carbon-holder. It consists of a bell-crank lever t pivotted to the core. The horizontal arm is a loop embracing the carbon-holder, which can slide through the core. When the vertical arm is strongly attracted to the core the horizontal arm clips the holder and retains it.

*4,777.—E. R. Prentice, Stowmarket. **Electric Lamps.** 2d. November 1.

ARC LAMP.—The upper carbon is fixed to a piston which moves in a cylinder filled with glycerine. Communication is made between the two ends of the cylinder by a passage governed by a solenoid core acting as a valve. When the

circuit is completed the core moves downwards, compressing the fluid in the lower part of the cylinder and raising the carbon-holder piston. As the carbons waste the core rises and allows the glycerine to pass from the lower to the upper side of the piston.

*4,778.—F. Wright, London. **Electric Lamps.** 2d. November 1.

INCANDESCENCE LAMPS.—The globe has a " bottle neck," clamped at its upper end over two short platinum conductors ; these are connected to two copper wires which at their other ends are connected to the filament by two other short platinum wires. The globe is exhausted through a tube projecting obliquely from the neck.

4,780.—A. T. Woodward, New York, U.S.A. **Insulating Electric Conductors, &c.** 6d. (2 figs.) November 1.

CONDUCTORS.—This invention relates to the mode of rendering the whole of the conductors and test-boxes water and air-proof, and at the same time of permitting easy access to the test-boxes. The

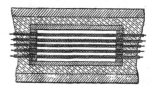

conductors are laid in a tube and run in with a non-conducting composition composed of silica, 66 parts ; resin or pitch, 34 parts ; wax, 26 parts ; and oil, 3 parts. The mixture does not penetrate the testing-boxes, although it surrounds them on all sides.

4,792.—W. E. Hubble, London. (*J. M. A. Gérard-Lescuyer, Paris.*) **Apparatus for Closing the Circuit on the Extinction of an Electric Lamp, &c.** 6d. (8 figs.) November 2.

AUTOMATIC CUT-OUT.—The figure illustrates one device according to this invention. Two metal cups D D¹ are placed in the circuit, at opposite sides of the lamp, underneath two metal rods C C¹, connected by a crossbar. The cups are partly filled with mercury in such manner that during the normal working of the lamp one rod is always in the mercury, while the other is slightly above it. The two rods are suspended by a platinum helix A, offering a greater resistance than the lamp, and held by a binding screw G, which communicates by a shunt circuit with a binding screw H, connected with the main circuit on

the side of the lamp corresponding to the rod C¹, which is not in contact with the mercury. The circuit wires x to and from the lamp are connected respectively to the cups D and D¹, so that during the normal working of the lamp the current passes in the manner indicated by the full line arrows; but when from any cause the lamp goes out the current passes through the mercury and up the platinum helix. The helix then becomes heated and

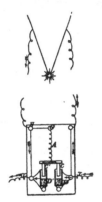

allows both rods C C¹ to descend and form a short circuit between the two mercury cups. Instead of the helix, fusible material or a thin rod of carbon may be employed. Several modifications are described. In the first the platinum helix is replaced by a permanent magnet surrounded by a coil. The passage of a current demagnetises the coil and allows the rods to drop. In the second an electro-magnet in a shunt circuit tilts a lever and thus makes the connection.

4,797.—C. L. Gore, New York. **Telegraphic or Telephonic Cables and Conductors.** 6d. (5 figs.) November 2.

CONDUCTORS.—Each wire is insulated by collars or washers of insulating material threaded on to it, and when laid underground the wires are packed together as shown in the figure. The intervening

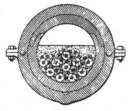

space is filled with semi-liquid material, such as water glass. The inner tube is of earthenware and the outer one of cast iron, the lower half of which is provided with a suitable recess to contain a strip of india-rubber, the upper half being provided with a corresponding projection.

*4,812.—H. J. Haddan, London. (*W. de Busscher, Brussels.*) **Apparatus for Using Electric Lamps on Locomotives, &c.** 2d. November 3.

ELECTRIC LIGHT. — "An electric lamp with movable reflector" is used. The current is supplied by a generator driven by friction gearing.

4,819.—W. Lloyd Wise, London. (*E. Bürgin, Basle, Switzerland.*) **Motors Actuated by Electricity.** 6d. (3 figs.) November 3.

MOTOR.—Referring to the illustration, A is a hollow sphere of thin brass divided into two hemispheres B and C, whose junction forms a horizontal great circle. These hemispheres are wound with insulated conducting wires D, coiled in convolutions parallel to the horizontal plane of junction. On the shaft F is fixed a core G wound with insulated wire into the form of a sphere. J is the commutator formed of three insulated parts (I., II., III.). The extremities of the conductor on the core G are

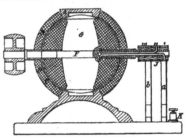

connected to the segments I. and II., whilst those of the conductor on the stationary part are connected to a pair of brushes of which one only, viz. b, is shown in the drawing. The current enters at K, which is connected to the brush a and the segment I., and then traverses the layers of the conductor on the core and returns to II. In this manner it always traverses the bobbin in the same direction. From the segment II. the current passes through the brush b into the outer conductor, and after circulating through it comes out again in the segment III., whence it passes through the other brush to the negative terminal. The core G is of non-magnetic material, but it may be soft iron or a permanent magnet. In the latter case it need not be provided with coils.

4,820.—W. Lloyd Wise, London. (*E. Bürgin, Basle, Switzerland.*) **Electric Lamps or Regulators.** 6d. (11 figs.) November 3.

ARC LAMP.—In the illustrations Fig. 1 is a side elevation of a lamp provided with two sets of carbons, one set of which comes into action when the first set is consumed. Fig. 2 is a sectional plan. N S is an electro-magnet through which the current passes before arriving at the electrodes;

an armature *i* is suspended before the magnet and forms the side of a parallelogram, of which the other sides are the connecting links *k l* and the casing. To this armature is fixed a bracket carrying a horizontal shaft, on which are two pinions in gear with racks on the carbon-holders sliding in guides in the bracket. The pinions are free to turn on the shaft in one direction, but are held in the opposite direction by pawls and ratchet wheels. R is a brake wheel fast to the shaft and situated below a brake shoe W attached to the casing. The method of maintaining one pair of electrodes out of action until the other is consumed is as follows : A lever is pivotted on a stud placed between the two carbon-holders. The pivot is much nearer one holder than the other, whence it

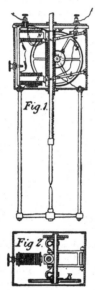

Fig.1

Fig 2.

follows that if the two ends of the lever be forced into contact with the two holders one will exert a much greater pressure than the other. If this pressure be suitably regulated one rod can slide down while the other will be jammed until the end of the first holder has dropped below the lever, when the second holder will be free. When the current traverses the lamp the electro-magnet attracts the armature *i*, raising it and the bracket with the parts attached to it. By this movement the brake wheel comes in contact with the shoe and it is held. At the same time the upper carbons are raised and the arc is established between the free pair. When the arc lengthens the armature descends and the brake wheel is released, consequently the upper electrode moves downwards. In the illustration the armature is shown wrapped with a coil. This coil may either be in the lamp circuit to increase the power of the magnetic attraction, or it may be in a shunt circuit and be

opposed to the regulating magnet. The specification also describes a simpler form of lamp with only one pair of electrodes. In this lamp the carbon-holder is suspended by a cord and descends through a long tubular guide attached to the armature. The cord passes over a pulley at the top and is wound on a drum on the horizontal shaft. There is no toothed gearing. The carbon-holder is formed of a split nozzle forced, by a weight, into a conical mouthpiece.

4,824.—C. H. Carus-Wilson, London. Electric Current Meters. 6d. (7 figs.) November 3.

CURRENT METER.—The object of this invention is to register the amount of electricity which passes through a wire or conductor. It consists mainly of a peculiar method of obtaining a variable rotation of one shaft from another shaft rotating with a constant angular velocity ; and, secondly, of means by which the piece connecting the one shaft with the other is moved in accordance with variations in the current to be measured. If the pulley A be driven at a uniform velocity from an external source, and if the pulley A^1 be driven from it by the intermediate friction pulley C, it follows that if the angular position of the pulley C be varied according to the intensity of the current, the revolutions of A^1 will be a measure of the current. The variations of the position of C can be effected in many ways. According to one its journals run in a fork connected to an electric dynamometer, which turns it in one direction or the other. Another method for moving the wheel C is partly illustrated in Fig. 2, "and is specially important as it enables a meter to be made which practi-

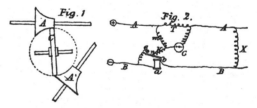

Fig. 1   *Fig. 2.*

cally measures the amount of resistance in any circuit so that the sum is not affected by variation in the electromotive force in the mains." A A, B B represent the two mains from the generating machine ; T is a small resistance, a fraction of an ohm ; X is the resistance in the circuit, consisting of a certain number of lamps ; G is a galvanometer joined up with resistances T and X, and two other resistances *m* and *n*, so as to form a Wheatstone's bridge ; *a* is an electro-magnet in the main circuit, and directly any current passes through X the lever *b* is attracted and strikes against the stop *e*, thereby putting *n*, *m*, and G in circuit between A A and B B. When the galvano-

meter is in equilibrium n must be the same function of X that m is of T ; n is constructed in the form of a box of resistances, and automatic appliances are provided for putting these resistances in and out of circuit. When the needle of the galvanometer moves to either side it acts as a relay, and puts a motor in operation which alters the resistance in n until the needle returns to zero, and at the same time moves the wheel C (Fig. 1) to cause the meter to give indications corresponding to the current.

4,825.—C. H. Carus-Wilson, London. Regulating Dynamo - Electric Machines, &c. 6d. (7 figs.) November 3.

CURRENT REGULATOR.—The objects of this invention are, firstly, to provide a regulator for dynamo-electric generators to proportion the supply of current to the demand ; secondly, to prevent the current exceeding a given maximum ; and, thirdly, to provide a governor for electric motors to limit their velocity. Under the first head a centrifugal governor is driven by an electro-motor arranged as a shunt between the main leads. When the arms of the governor rise above a determined position it short-circuits a portion of the convolutions of the field magnet coils of the generator to be controlled. The illustration shows another method of accomplishing the same result. The shaft a is driven by the motor and rotates the two bevel wheels b^4 b in opposite direc-

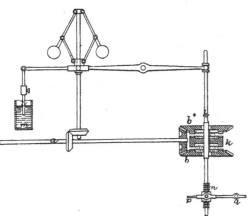

tions. Each of these carries a clutch on its inner side, which usually is free of the double clutch k, fast on the shaft h. In the field magnet circuit of the generator there is included the carbon rod l and the mercury bath m. When the governor commences to rise it first lifts the carbon rod partly out of the mercury, and thus increases the resistance of the circuit. Should it rise still further it will put the clutch on b in gear with the clutch k, when the shaft h will be rotated, and the nut n will be raised by the screw. The lever p q is connected at

its other end to a system of resistances in the magnet circuit ; thus a great increase of electromotive force will cause the insertion of permanent resistances, while small variations will be compensated for by the carbon rod. The lever p q may also be attached to the throttle valve of the steam engine to control its position. If it be necessary to vary the electromotive force in the system at will, change gear or cone pulleys are interposed between the motor and the shaft a, so that the speed of the governor may be varied. The same effect may be produced by including a set of resistances, variable by hand, in the magnet circuit of the motor. To prevent the speed of an electro-motor rising above a determined maximum it is caused to drive a governor that will short-circuit its terminals when a given velocity is attained. If it be desirable that the maximum shall be variable at will the governor is driven by a pair of speed cones and an intermediate friction wheel, as illustrated in the preceding abridgment.

4,850.—C. J. Allport, London, and R. Punshon, Brighton. Carbons for Electric Lighting. 2d. November 5.

INCANDESCENCE LAMPS.—The filaments of incandescence lamps are prepared, according to this invention, by soaking a thread or fibre of asbestos in a solution of sugar or syrup, and carbonising it. The process is repeated until the required amount of carbon is deposited on the thread.

CARBONS.—Asbestos may also be used as a binding agent in conjunction with carbon pencils to prevent disintegration.

4,851.—D. T. Piot, London. Electromotive Engines. 8d. (9 figs.) November 5.

MOTOR.—The cores of the fixed field magnet coils E are bent down to form long radial polar extensions a. The cores of the armature coils G are also bent down to form the extensions b, rotating in close proximity to a. The distribution of

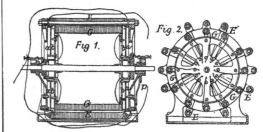

the current is effected by a commutator composed of segmental plates running under rollers connected to the generative source. The handle p serves to adjust the rollers so that the rotation of the armature may be stopped or its direction reversed.

*4,854.—J. B. Rogers, London. Production, Storage, and Utilisation of Electricity for Lighting or Power Purposes. 2d. November 5.

SECONDARY BATTERIES.—The current produced by any suitable generator is stored in secondary batteries. These are composed of a rectangular box open at its top and containing a similar shaped box whose open side is lowermost. The inner box is provided with racks having sheet lead arranged in zig-zag order between them. The outer box contains an acidulated liquid.

4,855. — J. B. Rogers, London. Electric Lamps. 6d. (10 figs.) November 5.

SEMI-INCANDESCENCE LAMP. — This invention consists in (1) arranging carbons so that the arc or incandescence light can be produced in the same lamp, and by the same carbons, at will, by a simple movement; (2) a lamp with two separate carbons, adjustable by weights, in contact with a refractory body; (3) a special method of making and breaking the circuit of suspended lamps with

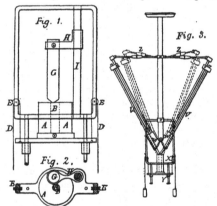

adjustable fittings, so as to dispense with hanging wires; (4) fitting three carbons, two to a lower frame and one in an upper frame resting upon the two lower ones, and capable of being shifted to one or the other, or on to both. Figs. 1 and 2 illustrate the first part of this invention. The lower carbon is made with a central core B of refractory material and has an upward motion under the action of a weight through the cords D and pulleys E. The upper carbon G is carried by an arm H, free to move on a pillar I, so that an arc or incandescence light is produced accordingly as it rests on the central core or edge of the lower carbon respectively. In the second part of this invention a movable lower frame carries a carbon rod and a strip of refractory material, which are made to abut against an upper rod of refractory material and carbon respectively. The rods are parallel and near to each other. The arc is es-

tablished by means of a metallic bar, which is pushed across so as to touch both carbons. In carrying out the third head it is proposed to fix contact buttons to the ceiling in the case of pendent fittings, so that as the lamp is lowered the circuit is broken. With brackets the make and break is established by a sliding action. Fig. 3 shows a lamp constructed in accordance with the fourth part of this invention. Two carbons V V are fixed in the upper frame W, and one, X, in the lower frame Y in the position shown. Each upper carbon can be moved to right or left by means of the screwed rods Z, so as to produce an arc or incandescence light, as described with reference to Figs. 1 and 2. The upper carbons V V may be connected to opposite poles so as to produce light between themselves, the requisite motions being obtained by means of weights. A lamp on this principle with upright as well as inverted carbons is described and illustrated.

*4,885.—W. C. Johnson and S. E. Phillips, London. Insulating and Protecting Underground Electric Lighting Wires, &c. 2d. November 8.

CONDUCTORS.—Each strand is separately insulated by dried fibrous material, several layers being put on in opposite directions. The mains are carried in continuous iron pipes, and suitable junction boxes are provided for the branches, which pass through stuffing-boxes in the lids. The pipes and boxes before being laid are heated and then plunged into oil, and are subsequently occasionally flushed with oil.

4,939.—A. F. St. George, London. Apparatus for Producing Light by Means of Electricity. 6d. (1 fig.) November 11.

INCANDESCENCE LAMPS.—To render the usual exhausted globe unnecessary, the carbon filament a is embedded in a composition b consisting of powdered flint mixed with silicate of sodium, or other metallic silicate or silicic, and sufficient

glycerine to form a paste. The enclosing tube c is made from powdered flint and water, vitrified in a furnace upon a cylindrical core. Powdered carbon or metal may be mixed with the composition a. When a metallic filament is used it may be sealed directly into a block of glass.

4,942.—S. Pitt, Sutton, Surrey. (*L. Gaulard and J. D. Gibbs, Paris.*) Applying Electric Currents in the Production of Light, &c. 6d. (1 fig.) November 11.

INCANDESCENCE LAMPS.—The inventors explain the fact of incandescence lamps giving a much less return in light for the power expended in the generator by suggesting that dynamo-electric generators produce a "quantity of electricity determined by the volume of the inducing wire, and under a tension dependent on the resistance of this inductor," and that "the arc lamp utilises both the quantity of electricity and its tension, whilst the incandescence lamp, on the contrary, transforms into light only

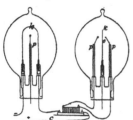

the volume of the current produced." The lamp illustrated herewith is intended to avoid this difficulty. A carbon thread *k*, the extremities of which have been previously coppered, is "rendered incandescent by the quantity of the current of the circuit with which it is connected." Two insulated platinum conductors, terminating in small spheres, are placed parallel to and in a plane at right angles to that of the wires *k k*, their extremities facing each other. The conductors receive the current of a Ruhmkorf bobbin placed upon the circuit *c*. "In these conditions the static charge only of the conductor *c* acting by influence upon the fine wire of the bobbin will determine in the latter an induced current, which in the form of a spark will manifest itself at *o* concurrently with the incandescence; thus the total quantity of electricity produced by the generator is utilised."

4,948.—G. G. André, Dorking. Electric Lamps. 1s. (17 figs.) November 11.

ARC LAMP.—In this lamp the feed current is separate from, and may be independent of, the main or lighting current, and is only momentarily established at intervals; and although the feed current is sent into the lamp at certain intervals, it has no effect on the feed mechanism till the automatic action of the lamp, dependent on the length of the arc, makes the necessary connections; no two lamps on the same circuit necessarily feed at the same time. Referring to the illustration, when the lamp current is turned on the arc is established by the electro-magnet E, which is in the main circuit, the current flowing to the lower carbon through the rollers D⁴. When the rod is drawn down, the ring E⁵ is caused to grip the holder F and raise it. As the arc lengthens the resistance in the main circuit is increased, and

a greater proportion of the current passes through the shunt circuit on the feed controlling magnet, which is situated in the base of the lamp, and is not shown in the figures. The armature of this

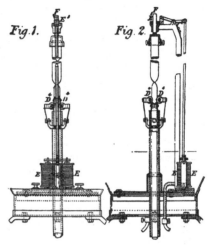

magnet acts as a relay, and when attracted it sets the connections in such a way that when the periodic feed current arrives it passes through an electro-magnet which raises the lower carbon by a step-by-step action, or it may be fed up as described in Patent No. 2,563 of 1881. If the feed current arrive when the relay is set the opposite way no action results. An electro-motor in the engine-house switches on the periodic current at frequent intervals to all the lamps fed from that centre. In order to indicate changes in the intensity of the current due to extinction of lamps or to variation in speed of the generator, an indicator is put in the circuit. This consists of a semaphore, which is worked by a current sent into it by a relay.

4,966.—W. R. Lake, London. (*J. B. Johnson, Boston, Mass., U.S.A.*) Electrical Signalling Apparatus for Railways. 6d. (1 fig.) November 12.

MAGNETO-ELECTRIC GENERATOR.—Mechanism, put in motion by the wheels of a passing train, rotates the armature of a magneto-electric generator, the current produced serving to actuate the signals.

*5,002.—S. Vyle, Middlesbrough, Yorks. Dynamo-Electric Circuits. 2d. November 15.

DISTRIBUTING CURRENTS.—In lighting systems the usual return wire is dispensed with, and the earth employed for the return circuit.

*5,006.—F. Wright, London, and F. A. Ormiston, Twickenham, Surrey. Regulating the Production of Electricity by Dynamo-Electric Machines. 2d. November 15.

REGULATING CURRENTS.—A greater or less resistance is introduced into the circuit of the field magnets by the action of a solenoid on a system of rollers, a spring being thereby caused to move against a surface which presents contacts in connection with resistances.

5,032.—S. Brear and A. Hudson, Bradford. Apparatus for Working Railway Signals, &c. 6d. (7 figs.) November 17.

GENERATOR.—A "generator" is used to supply current to an accumulator which in turn transmits the current to electro-magnets forming part of the signalling mechanism.

5,080.—R. E. B. Crompton, London. Apparatus for the Conduction and Distribution of Electric Currents. 6d. (4 figs.) November 21.

CONDUCTORS. — In the illustrations A A are enamelled copper mains or conductors placed in a

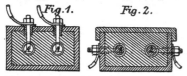

Fig. 1.   *Fig. 2.*

case or tube and run in with purified vitreous furnace slag. Connections are made to them by copper studs that are brought out through insulating bushes in the lid. The ends of the conductors are jumped up and screwed and united by screw couplings. The threads on the adjoining ends of the conductors are of slightly different pitch, so that the act of coupling forces them into good metallic contact. Fig. 2 is a modification suitable for being laid under foot-paths.

5,096.—W. R. Lake, London. (*F. Blake, Weston, Mass., U.S.A.*) Electrical Commutators. 6d. (2 figs.) November 22.

COMMUTATOR.—This is an instrument for connect-

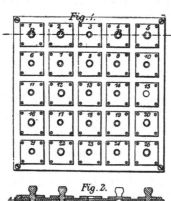

Fig. 1.

Fig. 2.

ing up electrical circuits, such, for instance, as the wires running into a telephone exchange, in any desired order. Each of the plates 1—24 is connected to a separate line wire. The plates A, B, C, D, E are coupling plates, and are separated from each other by insulating material. Supposing it be desired to couple lines 1 and 4, a metal peg with two collars is inserted in each of the plates 1 and 4. The upper collar fits and makes contact with the numbered plates, while the lower collar fits the hole in the plate B. The current now goes from plate 1 through peg B to plate B, and through the second peg B to the plate 4. Similarly 2 and 5 may be coupled through plate E, and so on.

5,104.—A. M. Clark, London. (*G. Fournier, Paris.*) Electric Batteries. 4d. November 22.

SECONDARY BATTERIES.—This invention relates to employing in secondary batteries a mixture of glycerine or glyceric acid, and of metallic oxides capable of forming therewith a solid compound insoluble in water. In carrying out the invention lead oxide, for example, is taken in powder and mixed with glycerine. The paste is run into moulds and allowed to set, and the solid plates produced are employed as one of the elements of a battery. When the other element is zinc the hydrogen reduces the lead as fast as it is evolved, and no polarisation is set up. Greater effect may be obtained by the addition to the mixture of peroxides. A plate, composed of glycerine, lead oxide, and lead peroxide, plunged in a dilute solution of sulphuric acid containing a plate of copper, constitutes a cell in which the acid attacks the copper and the hydrogen reduces the oxide. If a current be sent through the cell after the acid is exhausted, the elements will be returned to their original combinations. A secondary battery may have one plate of oxide, peroxide, and glycerine, and the other of lead oxide, metallic lead in powder, and glycerine.

5,159.—R. E. B. Crompton, London, and D. G. Fitzgerald, Brixton. Galvanic Batteries. 4d. November 25.

SECONDARY BATTERIES.—This invention relates, firstly, to the method of obtaining in an electrode a porous structure, offering large surface, by the removal through solution, chemical action, or heat, of a component originally brought into a condition of intimate admixture or combination with the permanent component of the electrode. Secondly, to the preparation of porous electrodes by the removal, through electrolysis, of sodium, potassium, zinc, cadmium, iron, antimony, copper, or silver from mixtures or alloys of one or more of these metals with lead. Thirdly, the construction of electrodes

by the compression of mixtures containing lead in a subdivided state, together with a component or components to be removed by solution or otherwise.

5,185.—E. G. Brewer, London. (*A. G. Water-house, New York, U.S.A.*) **Electric Lamps.** 1s. (33 figs.) November 28.

ARC LAMP.—In the illustrations, Fig. 1 is an elevation of the regulating mechanism of the lamp; Fig. 2 is a plan; and Fig. 3 is a separate view of the clutch, which is of the Brush type. B is the carbon-holder, passing between the two horizontal magnets E H, of which E is in a shunt circuit and H in the main circuit. The magnet E has pole-pieces G outside the carrying brackets D; the core of the magnet H is carried on pivots I, and is free to rotate about its axis. It has pole-pieces K which extend towards G. The power of the magnet H is normally so great that by induction it can overcome the polarity of E, which is opposed to it, and the two sets of pole-pieces are mutually attracted. When the arc lengthens, however, this preponderance is lessened, and the pieces K fall, and in so doing actuate the feed mechanism. As shown in the drawings this consists of a clip, or

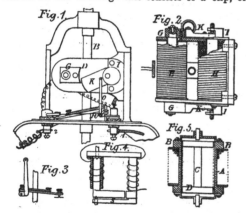

clamp, through which the holder passes freely when it is nearly horizontal, but which grips the holder when it is raised into an inclined position. To render the grip more certain the clip carries a number of steel plates, each of which is perforated to obtain a separate hold. Fig. 4 shows a modification of the controlling magnet, in which one movable pole-piece only is employed; it is supported on pivots, of which one is carried in a bracket S of non-magnetic material. Fig. 5 is a second modification. A is a cylinder of iron upon which the main circuit coils are wound, and which is provided with pole-pieces at B. C is an iron core pivotted on its longitudinal axis, and wound with shunt circuit coils. It has polar extensions D. When no current is passing, the poles D are held away from the poles B by suitable devices, but when the circuit is com-

plete the action is similar to that already described with reference to Fig. 1. A shunting appliance is provided which cuts the lamp out of circuit if the current cease, and cuts the arc out of circuit if the fine coils cease to act. The spring R (Fig. 1) is connected to the negative terminal, and its free end is controlled by the link O. If it be greatly depressed it forms a direct connection between the two terminals 1 and 2, while if it be raised beyond a certain point it forms a short circuit from the end of the main coil to the terminal 2. The next part of the invention consists of a combination with a carbon-holding rod, fed by gravity or by other means, of two controlling clutches, so arranged that the rod is at all times positively engaged by one or other of them, to the end that over-feeding may be prevented. In practice, two clutches are provided, both operated by the controlling magnet, and one is arranged so as to clamp and tend to lift the carbon rod when the strength of the magnet increases, while the other is set to engage with the rod, and by a connection with suitable mechanism to retard its downward movement whenever the magnet releases the first clutch. Three pieces of apparatus for carrying out the invention are illustrated. In the first two the retarding apparatus is a dash-pot, and in the third a fly. To enable two sets of carbons to be used in the same lamp, it is proposed to support the two clutches, which engage with the separate carbon-holders, at opposite ends of a lever pivotted at one side of its centre, and itself supported and moved upwards and downwards bodily by the regulating electro-magnet. In the operation of this device, the clutch whose end is supported by the longer end of the lever, will overbalance the clutch at the shorter end of the lever, raising it, and with it the carbon rod, so as to keep the electrodes separated. The clutch at the longer end works in the usual way. Should the acting carbon rod stick, its weight will be taken off the lever, and then the opposite carbon-holder will tilt the lever, and its clutch will be put in operation. The invention further refers to devices to be employed in place of a clamp for allowing a retarded feed of the carbon-supporting rod or carrier. Many forms of apparatus are illustrated. They are (1) two kinds of verge escapement; (2) balance escapement; (3) escapement and dash-pot; (4) verge wheel and pallet; (5) escapement and paddle; (6) ball governor; (7) a variable pendulum. Several forms of carbon clamps are described. The first is tightened by a wedge. In the second two jaws are nipped to the carbon by a screw; the third is provided with devices to prevent the rod from turning about its axis, but at the same time to allow it to move bodily in any direction so as to permit of proper setting of the carbons.

5,198.—C. H. W. Biggs and **W. W. Beaumont,** London. Secondary Batteries for the Production, &c., of Electric Currents. 6d. (7 figs.) November 28.

SECONDARY BATTERIES.—In the usual form of secondary batteries the products of electric decomposition appearing at the poles are collected and retained in the charged battery itself, but according to this invention these products are collected in separate receivers, and are recombined as required. Instead of obtaining the hydrogen by electrolysis it may be generated by chemical means, and, in some cases, the oxygen may be prepared as peroxide

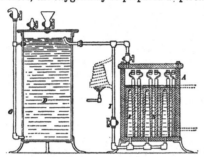

of lead in a similar way. Referring to the illustration, A is a vessel containing the electrodes N N[1], and connected to the holder D, into which any excess of hydrogen evolved during the charging of the battery is conveyed. A pipe I conveys the collected hydrogen into chambers formed in the porous lead electrodes ; and a coil, heated by a gas jet, is provided to intensify the action of the gas when required. The pipe G maintains a constant head of water in the receiver D. The electrodes are made from finely divided lead, preferably that obtained by reducing sugar of lead with zinc, placed in a suitable mould, and caused to cohere by pressure.

5,216.—J. E. Liardet and **T. Donnithorne,** London. Hydraulic Motors, &c. 8d. (12 figs.) November 29.

ELECTRIC LIGHT.—This invention relates to the use of "spiral hydraulic motors" specially adapted for driving generators for "the production of electricity to be used in the electric light."

5,226.—A. W. Brewtnall, London. Joining Branch to Main Conducting Wires for Electrical Purposes, &c. 6d. (11 figs.) November 29.

CONDUCTORS.—The coupling, as shown in Fig. 1, consists of a hook-shaped jaw to receive the main wire c, having a hole e, intersecting the jaw, for the insertion of the branch wire, which is pressed into contact with the main by the screw d. Fig. 2 shows a coupling in which the branch wire is in-

sulated, the connection to it being made by the

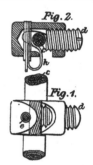

fusible wire h, which is melted when the current exceeds a determined amount.

*5,229.—W. R. Lake, London. (*J. S. Williams, Riverton, New Jersey, U.S.A.*) Utilisation of Electricity for Lighting, &c. 4d. November 30.

INCANDESCENCE LAMPS.—"The electric current is caused to pass through pulverised or granulated particles of conducting materials," preferably carbon. The material is enclosed "in a transparent tube or case" which may be sealed and exhausted. In order to compensate for the waste of material, a reservoir is provided, from which "the supply is constant, so that there cannot take place an interruption of the light." In a second lamp a liquid conductor is used. For distributing heat a conductor, contained in an exhausted vessel, is placed in a second vessel through which a circulation of water is maintained.

5,233.—W. R. Lake, London. (*J. S. Williams, Riverton, New Jersey, U.S.A.*) Utilisation of Electricity for Lighting, &c. 1s. (72 figs.) November 30.

INCANDESCENCE LAMPS.—This invention relates to the utilisation of electricity for lighting and heating, and consists, firstly, in the construction of a lamp as illustrated in Fig. 1, in which there is shown a conductor having its two main portions connected by a series of transverse filaments, which form so many passages for the current. The ends of the main parts of the filaments are held between spring clamps of metal. In a modified lamp the bars a[1] are arranged vertically. Secondly, a number of independent filaments are included in one globe, and are furnished with contact pieces operated by hand, so that one or more may be placed in circuit at one time. Thirdly, a number of platino-iridium wires are arranged in an exhausted globe, upon a form, by which they are bent into a balloon-like, fan-like, or other shape. Fourthly, an incandescing filament is made of a number of thin laminæ laid together to the end that a rupture may not

extend across the whole of them. Fifthly, the light-emitting carbon is encased within a hollow piece of glass and surmounted by a reflector *p* (Fig. 2). Numerous modifications of this idea are described and illustrated. Sixthly, the invention

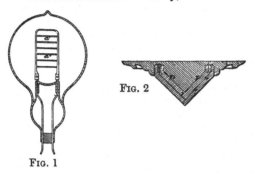

Fig. 2

Fig. 1

relates to apparatus wherein the electric current is caused to develop heat. Many methods of applying this are illustrated, in which are employed terracotta, water, and air for the diffusing media, the temperature being controlled by hand, or by the expansion of fluids or metals.

5,261.—H. E. Newton, London. (*E. Volckmar, Paris.*) Secondary Batteries. 6d. (2 figs.) December 1.

SECONDARY BATTERIES.—A certain number of secondary couples are arranged in a single gas-tight vessel horizontally one above the other. There are as many elements as there are plates, less one, each plate being at the same time the positive electrode of one couple and the negative electrode of the following couple, with the exception of the first and last plates, which form the poles of the battery. The plates are charged in the usual way with spongy lead or equivalent material, and are im-

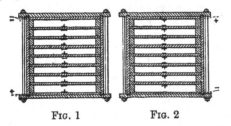

Fig. 1   Fig. 2

mersed in the exciting liquid in a vessel capable of being turned over to reverse the positions of the plates. Under the influence of the charging current the positive electrode of a secondary battery disengages hydrogen, which is carried to the negative electrode, the negative electrode disengaging oxygen, which stores itself at the positive electrode. In discharging, the operations are reversed. In a battery arranged as in Fig. 1, the hydrogen disengaged from the positive electrode will rise to the negative element, while the oxygen will descend from the negative to the positive in accordance with the law of densities. If, now, the battery be turned upside down, as in Fig. 2, and the charged battery be set to work, the gases will still flow in their natural direction.

5,272.—W. F. King and A. B. Brown, Edinburgh. Electric Lamps. 6d. (3 figs.) December 2.

ARC LAMP.—The upper carbon-holder 9 is connected to the rod of a piston 14, which works in a cylinder filled with oil. There is an external passage 15, leading from the lower to the upper end of the cylinder, and this passage is controlled by a valve 18. The weight of the piston and carbon-holder tends to force the liquid through this passage when the position of the valve permits it. When the valve is closed, no flow can take place, and con-

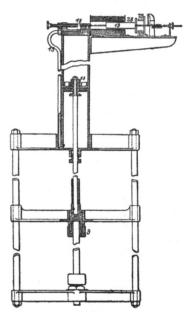

sequently the upper carbon-holder cannot move. The valve is controlled by the solenoid core 19, which is moved in opposition to an adjustable spring. When the current is strong, the core is drawn into the coil and the valve closed, but when the arc lengthens, and the current diminishes, the spring draws back the core and opens the valve. 29 and 30 are two contact pieces connected to the two terminals of the lamp, and when the piece 28 is drawn against them the lamp is cut out of circuit. The specification does not explain how the arc is established.

5,286.—A. R. Sennett, Worthing. **Production of Electric Light, &c.** 6d. (9 Figs.) December 3.

INCANDESCENCE LAMPS.—This invention relates to the manufacture of carbon electrodes and filaments. These are made by reducing cellulose by mechanical means to a state of fine division, and depositing it on a permeable diaphragm by withdrawing the water in which it is floating from beneath it. The film is raised by placing a second diaphragm above it and exhausting the air from the back. The sheet of material thus obtained is compressed *in vacuo* between steel plates, and is then stamped into the required forms and carbonised in an exhausted chamber, an inert gas being injected before heat is applied. The ends of the conductors are suitably enlarged for the attachment of the platinum connections, and sockets are formed in an inner glass shell in the globe of the lamp to receive the enlargements. The invention further relates to a mode of hastening the action of the mercury pump, by employing two sets of reservoirs, so arranged, that while the air is being expelled from one set it is expanding into the other set.

5,295.—H. E. Newton, London. (*A. J. Gravier, Paris.*) **Mechanism for Regulating the Feed of Electrodes in Electric Lamps.** 6d. (6 figs.) December 3.

ARC LAMPS.—For the purpose of unlocking the wheel train of an arc lamp, when the difference of potential at the two terminals exceeds a given amount, "there are provided two bobbins *a b* (Fig. 1), two soft iron cores A B of the same size and weight, a base C, and an armature *d*. The armature is very short and light, and is formed of several elements placed in juxtaposition . . . . and soldered together with tin solder." Its movement on either side is limited by small screws and it is fitted with a brass spring *s*. "The bobbins may be placed parallel or otherwise. The bobbin *a* is

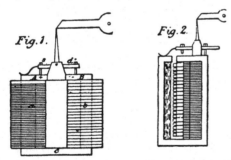

Fig. 1.   *Fig. 2.*

wound with a fine wire of certain determined resistance, comparatively great. The bobbin *b* carries a thicker wire of small resistance." *a* is in a derived and *b* in the main circuit. "At Fig. 2 another form of differential unlocking electro-magnet is shown. It is composed of a bobbin mounted upon a soft iron core and wound with two coils, one of

thick wire and the other of thin, these being arranged so as to be traversed by currents in opposite directions." Another form of apparatus includes a Desprez fish-bone galvanometer, acting as a relay, to send a current into an unlocking magnet. The resistances of the coils and the tension of the spring are determined mathematically, and it is stated that "a variation of less than .01 of a weber will suffice to produce the movement of the armature."

5,309.—J. A. Fleming, Nottingham. **Preparation of Materials for Electric Insulation.** 4d. December 5.

INSULATING MATERIAL.—Wood, flour, sawdust, bran, straw, cotton, &c., are taken in a finely divided state, and after desiccation by a current of hot air are mixed with paraffin and resin and moulded into the required form under pressure.

5,316.—R. Laybourne, Newport, Mon. **Apparatus for Lighting Railway Carriages, &c., by Electricity.** 6d. (6 figs.) December 5.

ELECTRIC LIGHT.—An electric generator, driven by friction or belting from one of the axles of the train, supplies current to a secondary battery from which the current passes to "the electric lamps" in various parts of the train.

*5,322.—J. Imray, London. (*J. Carpentier and Dr. O. de Pezzer, Paris.*) **Electric Accumulators.** 2d. December 6.

SECONDARY BATTERIES.—To lighten and simplify the accumulator, the negative plate is made very thin, and the positive plate about twice the thickness and half the area of the negative plate, each plate consisting of several blades bent double and arranged side by side in a porous cell. A number of these cells are placed in a water-tight box charged with acidulated water.

5,338.—D. G. Fitzgerald, C. H. W. Biggs, and **W. W. Beaumont,** London. **Secondary Batteries.** 6d. (9 figs.) December 6.

SECONDARY BATTERIES.—Porous electrodes are made by minutely puncturing one or both surfaces of sheets of lead. The invention also relates to the manufacture of such plates by minutely puncturing both their surfaces, then treating the plates by immersion in dilute sulphuric acid in a vessel from which air may be withdrawn, so as to produce a deposit of lead sulphate on them, then rolling the plates so as to extend their surface by reduction of thickness, and finally repuncturing them. These porous electrodes may be made by the electro-deposition of lead on a fine net fabric, which is preferably folded to make up any required

thickness, or by the application of heat and pressure to lead granules, with or without oxide or sulphate of lead, in suitable moulds, or they may be constructed from fabrics of woven lead wire, chemically treated, folded, and pressed. The specification illustrates a machine in which a number of needles, suitably held, and actuated by a crank driven in opposition to a spring, are used to puncture the plates.

5,354.—P. Cardew, Chatham. Apparatus for Indicating the Speed of Revolution of Shafts. 6d. (3 figs.) December 7.

GENERATOR.—The apparatus consists of an electric generator and an instrument for measuring the intensity and direction of the currents. Figs. 1 and 2 show two elevations of a magneto generator with a Siemens armature, and Fig. 3 is a diagram showing the general arrangement. The currents are sent through two galvanometers, one in the

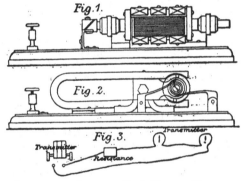

Fig. 1.

Fig. 2.

Fig. 3.

engine-room and one on the bridge, and by means of resistances the indications of these galvanometers are made to coincide with the speed of the main shaft. When it is not desired to show the direction of rotation, alternating currents may be used, and the heating effect of the current may be employed to put in operation a thermometric indicator.

*5,367.—W. R. Lake, London. (*H. S. Maxim, Brooklyn, U.S.A.*) Machine for Coating Insulated Electrical Conductors with Lead. 6d. (3 figs.) December 8.

CONDUCTORS.—The conductor is drawn through a die into which fluid lead is pumped. The nozzle of the die has a water jacket to partly cool the lead, so that it shall require a pressure of about 5000 lb. on the square inch to force it forwards. The water valve, and the supply valve of the fuel for melting the lead, are controlled by a rod subject to the pressure in the cylinder.

*5,368.—J. D. Mucklow and J. B. Spurge, London. Photometer, &c. 2d. December 8.

PHOTOMETER.—This consists of a number of tubes closed at the top by perforated opaque plates. The bottom of each tube is covered by a stencil plate between which and a cover is placed the film to be tested. The inventors state that the intensities of different lights may be compared with this instrument but do not state how.

*5,396.—C. F. and F. H. Varley, London. Electric Lamps. 6d. (14 figs.) December 9.

ARC LAMP.—Fig. 1 shows a general view of the regulating mechanism and part of the upper carbon. The lower carbon is carried by the arm E, of which Fig. 2 is a separate view. Its holder is made to swivel (Fig 3) in order that it may be turned out of the way when the upper electrode is inserted in its place. The motive power of the lamp is derived from the helical spring C, which is forced to the upper end of the case A by the insertion of the upper carbon. The reflex action of the spring is retarded by the piston O, which is composed of

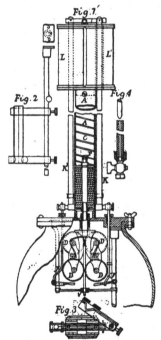

Fig. 1.

Fig. 2.

Fig. 4.

Fig. 3.

discs of felt separated by layers of paper and provided with an adjustable air valve P. The felt is turned up in a lathe by a red-hot tool to secure a good fit. The feed mechanism is controlled by two solenoids L L¹, the former in a shunt circuit and the latter in the lamp circuit. The solenoid cores $K K^2$ are connected by two rods J to two detent arms as shown. These arms are capable of engaging with teeth on the lower rollers D D. There are two pairs of these rollers, between which the upper carbon passes, and they are supported in a frame,

which can rise and fall. Each pair, considered vertically, is free to swivel about the centre R, and also to move bodily, so as to "enable the carbon to move in any direction, and consequently the poles of the carbon, by the action of the guide V (although the carbon may be crooked) are always directed to the right spot." In starting the lamp the upper carbon is passed through the guide V, then between the two pairs of rollers D, and up the case A, pushing the piston and spring before it. When the circuit is completed through the second carbon the rod K is drawn upwards "forcing the rod K² downwards until the regulation arms are forced into the teeth of the rollers D D, forcing them back and carrying along with them the carbon rod B, establishing the arc." When the arc increases in length the coil L balances the coil L¹, and drawing up its core withdraws the arms J J from the rollers. The carbon then slides down until again checked by the arms. A modified form of lamp has two cylinders and pistons, one opposite the other, each with rollers and detent arms. The rollers are constructed to give the proper proportionate feeds to the two carbons. The arc is enclosed in a globe filled with carbonic anhydride, or carbonic oxide, or hydrogen, or nitrogen, dried in its passage to the globe. To direct the light where it is most useful small parabolic reflectors are employed, or rings or coils of glass of prismatic section, coated with silver, platinum, or other reflecting metal. In place of these, surfaces covered with bicarbonate of soda, sulphate of baryta, or other crystalline substance may be used. If the carbons be saturated with wax, or hydro-carbon, their combustion will be retarded. The same effect may be produced by conducting a jet of gas into the arc through or around one of the electrodes, as in Fig. 4.

*5,400.—T. Rowan, London. Electric Light Lamps or Lanterns. 2d. December 9.

REFLECTORS.—A series of prismatic reflectors is used to diffuse the light.

*5,407.—W. Abbott and F. Field, London. Manufacture of Compounds for Electrical Insulation. 2d. December 10.

CONDUCTORS. — The conductor is first covered with a compound of ozokerit and india-rubber as described in Patent No. 1,938 of 1875, and surrounded with a layer of india-rubber, with which an excessive quantity of sulphur has been incorporated; the compound covering is then vulcanised.

5,418.—J. E. Liardet and T. Donnithorne, London. Arrangement of Electrical Apparatus on Railway Trains. 6d. (5 figs.) December 10.

GENERATORS. — Electric generators are driven from the axle of one or more of the vehicles, and the currents produced are stored in secondary batteries. Fig. 1 shows one method of driving. The axle carries two double bevel wheels Q¹, driving each

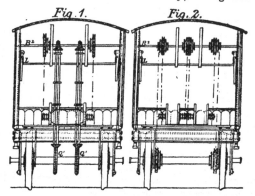

two vertical shafts, which in turn drive a horizontal shaft B². From this shaft belts convey the power to the armature of the generator. In Fig. 2 the transmitting media are belts and pulleys. The shaft B² is carried in a swing frame which can be raised and lowered by the cranked lever L, so as to tighten or slacken all the belts simultaneously.

5,451.—J. Pitkin, London. Secondary Batteries. 6d. (2 figs.) December 13.

SECONDARY BATTERIES.—The illustration shows one of the elements. A is an open frame of wood; B is a mass of very thin shavings or turnings of lead, or strips or pieces of lead foil, or of other

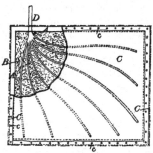

very thin sheets of lead, packed into the frame A, and retained therein by coverings C C of felt or flannel stretched over each side and secured by the wooden pegs c. The terminals are formed of rods of lead D with branches as shown.

5,468. — J. Imray, London. (*J. M. Stearns, Brooklyn, U.S.A.*) Telegraph or Telephone Conductors. 4d. (5 figs.) December 14.

CONDUCTORS. — This invention relates to the combination with a number of grouped insulated conductors of a metal sheet, folded with and among

them, so as to interpose itself between those that are adjacent to each other. This sheet is connected to earth so as to carry off induced currents and to prevent interferences arising from induction. Fig. 1

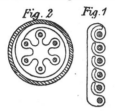

Fig. 2   *Fig. 1*

shows a number of conductors arranged in a vertical plane, as along a wall, with a sheet of metal wound in among them. Fig. 2 illustrates the invention applied to conductors in a pipe.

5,477.—W. R. Lake, London. (*C. F. de la Roche, Paris.*) **Electric Lamps.** 6d. (8 figs.) December 14.

ARC LAMP.—The arc is developed in the interior of a chamber or space in a block of refractory material, having in its walls openings in the form of a frustrum of a cone, so as to allow the light to emerge in determined directions. The illustrations are sections at right angles to each other of one form of the refractory block ; *a a* are

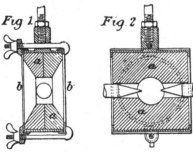

Fig 1.   *Fig. 2*

two pieces of compressed magnesia held together by screws and pierced with holes for the admission of the electrodes. The conical openings are closed by plates of mica *b* to prevent the setting up of air currents. Other forms of blocks are shown in the specification, one having four conical openings, and another only one such opening.

5,481.—D. G. Fitz-Gerald, London. **Secondary Batteries.** 4d. December 14.

SECONDARY BATTERIES.—Carbon in conjunction with lead is used for the electrodes. The carbon is used in the form of fragments in direct contact with each other, the interstices being filled with lead in a state of fine division, or with an oxide or sulphate of lead. The carbon fragments may be first impregnated with lead by immersing them in a solution of acetate of lead, and subsequently re-

ducing the metal either by heat or by electrolysis. In constructing a cell the fragments are divided into two portions, each of which is in contact with a plate of carbon, the two portions being separated by a porous partition. The cells are built up in wooden vessels rendered waterproof and coated internally with marine glue, the electrolyte being introduced through a glass tube cemented into the upper part of the wooden vessel. In a modification the carbon and lead fragments are placed within a perforated tube of carbon or lead, the perforations being filled or covered with a porous material.

5,490.—W. R. Lake, London. (*J. A. Mondos, Neuilly, France.*) **Electric Lamps.** 6d. (6 figs.) December 15.

ARC LAMP.—Upon a base-plate there is fixed a standard B, to which is pivotted a bent lever C, carrying at one end a counterweight Q and at the other end a core moving within the solenoids E E, which are on a shunt circuit. The lever C is connected to the tubular carbon-holder T, within which slides the carbon. This tube carries a bracket H, which serves as the fulcrum of a second lever L, carrying an armature *a*, subject to the attraction of the magnet coils E. Below this lever is fixed a tongue piece or cam *l*, which enters

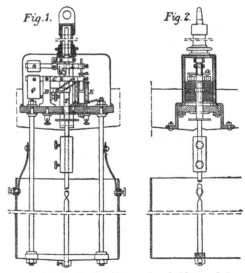

Fig. 1.   *Fig. 2.*

an opening in the tubular carbon-holder and jams the carbon, preventing it from descending. When the circuit is first completed the total current passes through the coils, which draw in their cores, and cause the carbon-holder to descend. At the same time the armature *a* being attracted the cam *l* is withdrawn and the carbon allowed to slide through the holder until it meets the lower electrode. Immediately the greater part of the current is diverted from the coils, the lower counter-

weight raises the carbon-holder, and the upper weight forces the cam into contact with the electrode and causes the two to rise together to establish the arc. As the arc lengthens the upper carbon is allowed to drop slightly from time to time by the pressure of the cam being relaxed. The invention also refers to a lamp, in which a thin pencil falls by gravity towards a carbon block, carried on one arm of a bell-crank lever, which is subject to the attraction of an electromagnet. The other arm carries a brake shoe pressing against the thin pencil. When the circuit is completed the magnet draws down the block slightly and presses the brake against the other electrode, establishing a very small arc. The specification further illustrates a lamp on the principle of that first described, with two sets of carbons, the upper ones being both connected by a yoke to a rod passing through the hollow holder. The arcs are established by drawing down the lower electrodes, the holders of which are attached to the cores of two solenoids carried on the lower part of the lamp frame.

5,494.—J. W. Swan, Newcastle-upon-Tyne. **Secondary Voltaic Cells.** 2d. December 15.

SECONDARY BATTERIES.—The surfaces of the lead plates, preferably constructed as described in Patent No. 2,272 of 1881, are converted into white lead by the combined action of acetic acid, carbonic acid, and atmospheric air, and are then reduced to a metallic state by the action of electrolytic hydrogen. The plates so produced are arranged and used as in the Planté cells.

5,499.—J. W. Swan, Newcastle-upon-Tyne. **Measuring and Recording Electric Currents.** 4d. December 16.

CURRENT METER.—In the inventor's former Specification, No. 5,004 of 1880, there is described an electric meter so constructed that it is necessary to have a pair of wires between it and each lamp. The object of this invention is, in part, to avoid the use of these wires. The main conducting wire, or a shunt from it, is now taken round the several electro-magnets which determine the engagement of the pawls with the counting mechanism, and the springs which counteract the pull of these electro-magnets are so adjusted that when the current from one lamp only is passing one pawl only is in action. When two lamps are burning two pawls will be in gear, and so on. The motor which drives the recording mechanism comes into motion when the first lamp is lighted. In a modified meter a single solenoid is substituted for all the electro-magnets, and the depth to which its core is drawn in determines the number of

pawls that come into operation. In a voltametric meter the electrolytic apparatus and an automatically variable resistance are connected in series and are placed between the main conductors in the same way that the lamps are. The action of a solenoid in the main conductor causes a plunger to rise or fall in a tube of mercury in proportion to the current passing, so as to make the rise or fall of the mercury short-circuit or disconnect branches taken from different lengths of the resistance coil, these lengths being so adjusted that the resistance of the coil shall diminish or increase in steps corresponding to lamp units of currents in the main conductor; the amount of gas evolved in the voltameter consequently is proportional to the number of lamps lit. The gas may be made to cause the oscillation of a beam connected to counting mechanism.

5,521.—G. Grout and W. H. Jones, London. **Secondary or Polarisation Batteries for Storage of Electric Energy.** 4d. December 17.

SECONDARY BATTERIES.—This invention relates, firstly, to the combination of lead or other metal with carbon; and, secondly, to the application of lead dust in secondary batteries. Flour, starch, or meal is mixed with an oxide or salt of lead, and sufficient syrup is added to make it plastic. This material is moulded into sheets or blocks and exposed to sufficient heat in a closed vessel to oxidise the organic matter and reduce the metallic salt. In constructing some of the elements it is advantageous to take red oxide of lead and combine it with flour and syrup to form a paste. This is applied by a spreader to the surface of paper, or other carbonisable material, which is rolled or folded to the desired form, and then carbonised with a solid lead core to form an attachment. If the moulded block is to be of a very porous nature it is mixed with yeast and fermented, as if it were bread, and then carbonised. For some purposes metallised carbon is employed in the powdered state to surround the conductor. The powder is prepared by adding a salt of the desired metal to pulverised charcoal, or to a carbonaceous substance, and exposing it to a reducing heat. For the construction of elements according to the second part of this invention, lead or a plate of carbon is taken, and dusty or granular lead is heaped upon it. This dust is prepared by mixing powdered charcoal with molten lead and stirring it until it becomes cold.

5,524.—R. Kennedy, Glasgow. **Electric Lamps, &c.** 8d. (7 figs.) December 17.

INCANDESCENCE LAMPS.—This invention relates, first, to a mercurial air pump (Fig. 1). The lamp

globe to be exhausted is shown at the top, and is connected to the pump by two india-rubber washers and a mercurial seal. In operation the three-way cock *d* is turned to place the pipes *e* and *c* in communication, and mercury is poured down the former until it rises to the valve *b*, driving all the air before it. The cock is then turned to connect *c* and J, and the mercury allowed to run out. This is repeated

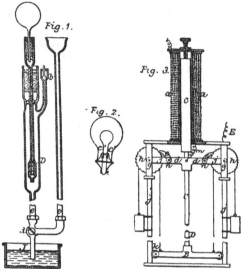

as often as necessary, the valve D preventing the mercury rising into the globe, and the valve *b* preventing inlet of air. Secondly, to connect the platinum conductors and the filament, in an incandescence lamp, the former have helices *c' c'* (Fig. 2) formed on their ends, and are raised to a red heat by a current in a carbonaceous atmosphere, by which means carbon is deposited on and fills the interstices of the helices. The filament is made by preference from a fibre of the aloe-tree.

Arc Lamp.—The invention also relates to the arc lamp, shown in Fig. 3. "The crossbar *d* carries at its ends the pawls *e e*, by pins *n n*, which pass through holes in the crossbar *d*, pawls *e e*, and pawl levers *j j*, the latter being mounted at their outer ends upon the axes *g g*, which also carry the ratchet pulleys *h h*." "When the lamp is put into circuit the current passes through the solenoid *a*, thence through the wire *m* to the upper carbon C, and from this across the arc to the lower carbon D, which is not insulated from its carrier, thence to the return wire E, through the metallic connections formed by the rollers *k k* and the framing *j*. When the current passing through the solenoid *a* is of less than the proper strength the solenoid core *c* descends, and at the same time releases the pulleys *h h*, so that the carrier B and the lower carbon rise, while at the same time the upper carbon descends. This shortens the arc and allows a greater quantity

of current to pass; the solenoid then draws up the core *c* and thus checks the pulleys *h h*." This description does not appear to correspond exactly with the drawings. In a modified lamp the upper and lower carbon-holders are both suspended by cords passing round the pulleys *h h*, and tend to counterweight each other, the upper being slightly the heavier. The bar *d* merely serves to operate the pawls.

5,525.—W. H. Akester, Glasgow. Dynamo-Electric Machines. 6d. (7 figs.) December 17.

Dynamo-Electric Generator.—This invention relates principally to the construction and method of winding the compound armatures used in this generator. Referring to the illustrations, Fig. 1 is a sectional side elevation, Fig. 2 a plan, Figs. 3 and 4 transverse vertical sections, Fig. 5 a diagram showing the mode of connecting the coils, and

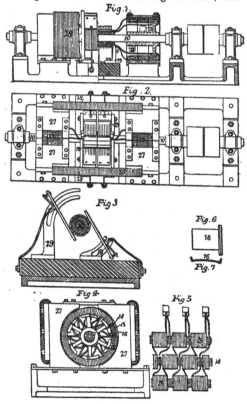

Figs. 6 and 7 are face view and section of a soft iron piece held in the armatures by the insulated copper wire coils. The shaft 10 has fixed on it, near each bearing, a boss formed with spokes 13, having flanges at their outer ends on which are fixed the armature rings. In the drawings two armatures, each made up of three rings, are shown. The rings are fixed by thin wedges of wood driven

in between the flanges of the spokes and the thin metal plates protecting the coils. Each ring consists of an annulus 14 (Fig. 4) of soft iron wire wound with a number of coils of insulated copper wire 15, which each bind down a soft iron piece 16 (Figs. 4, 6, and 7), the flange of which projects outwards, and has its outer edge flush with the coils. The connections for continuous currents are sufficiently indicated in Fig. 5, the commutator being of ordinary construction, as shown in Fig. 3. The brushes are fixed to slotted brackets 19 bolted to a wooden block and connected by wires to the terminals. The field magnets consist of two horizontal bars placed parallel to and on opposite sides of the shaft 10, and connected by pole-pieces 27 formed with concave cylindrical surfaces. The field magnets are so excited by the coils 28 that the N. and S. poles of one respectively face the S. and N. poles of the other. The pole-pieces 27 (Fig. 2) are formed so that the exciting coils 28 may be carried to some extent behind them. Two or more pairs of magnets may be arranged equidistantly round the shaft 10, the poles being placed alternately. The pieces 16 may have projecting parts at both ends.

5,536.—J. E. H. Gordon, London. **Dynamo-Electric Machines.** 6d. (8 figs.) December 17.

DYNAMO-ELECTRIC GENERATOR.—This invention relates to improvements on the generator described in Specification No. 78 of 1881. The generator therein described comprised two revolving wheels carrying electro-magnets and one stationary wheel provided with helices, there being as many electromagnets on the wheel as there were stationary

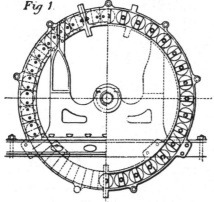

Fig 1.

helices. To prevent the coils in adjoining helices interfering with each other, the number of helices of the stationary ring are in the present invention double in number those of the electro-magnets on one of the rotating rings. The magnets thus act alternately on the alternate sets of coils,

and the intervening coils are practically idle and shield the others from mutual action. The cores of the helices are lengthened beyond what is actually required for the wire, and the plate to which they are attached is set back into a field of greatly diminished inductive action. The space between the wire and the plate may be left empty or filled with wood. Fig. 1 is a side elevation of

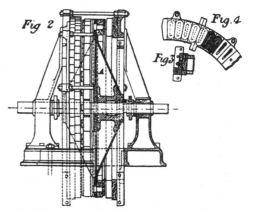

the generator, Fig. 2 an end elevation, Fig. 3 a section of one of the fixed rings and one of the inducing helices, and Fig. 4 is an elevation of part of one of the fixed rings. The field magnets are carried on a rotating wheel A, the construction of which can be seen in Fig. 2. Each magnet is provided with a pole-piece at each end.

5,551.—J. H. Johnson, London. (*W. W. Griscom, Philadelphia, U.S.A.*) **Armatures for Magneto-Electric and Dynamo-Electric Machines, &c.** 6d. (2 figs.) December 19.

ARMATURES.—This refers especially to armatures of the Siemens type, and "consists in dividing the bi-polar extremities of such armatures into two or more sections concentric with each other, one or more of such sections being at such a distance as to utilise the magnetic lines of force which will

fill the space between the two poles of the field magnets, whilst the other section or sections are arranged so as to pass in close proximity to the poles of the field magnets." As will be seen on reference to the illustration, the armature A has its bi-polar extremities divided into sections concentric with

each other, the sections *m* revolving at such a distance within the field magnets B and C as to fully utilise the magnetic force. The projections *p* are longitudinal ribs extending beyond the periphery of the sections *m*.

*5,566.—A. Millar, Glasgow. Producing Electric Currents. 2d. December 20.

MAGNETO - ELECTRIC GENERATOR. — This generator is of the "exploder" type, and has a soft iron armature which rotates before the poles of a permanent magnet coiled with insulated wire. Several arrangements are mentioned but not described as regards their detail.

5,593.—L. S. Powell, London. (*J. M. A. Gérard-Lescuyer, Paris*). **Dynamo-Electric Machines.** 6d. (3 figs.) December 21.

DYNAMO - ELECTRIC GENERATOR.—The general construction of the generator is clearly shown in the drawings. C C are straight electro-magnets fixed upon the circumference of a rotating brass wheel, with their poles arranged alternately upon either side, which wheel runs between two circles

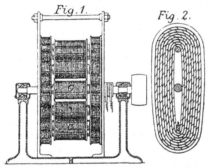

of fixed electro-magnets. Fig. 2 is a side view of one of these electro-magnets with one of its covering plates removed. A thin sheet of metal is interposed between each layer of wire in the coils, to assist in the inducing action and to prevent the formation of extra currents. As the induced currents are alternate in direction, a separate generator is necessary to excite the field magnets.

5,599.—W. Smith, London. Insulated Conductors for Telegraphic Uses, &c. 2d. December 21.

CONDUCTORS. — The dielectric is made from gutta-percha and zinc white, which is incorporated with the gutta-percha while it is being masticated in an ordinary masticating machine.

5,600.—S. Pitt, Sutton, Surrey. (*E. T. Starr, Philadelphia, U.S.A.*) **Electric Lighting Apparatus for Railway Trains, &c.** 6d. (4 figs.) December 21.

GENERATOR.—A generator, driven by one of the axles, feeds a secondary battery in addition to a head and tail light. Distinctive signals may be given by means of a circuit breaker driven from the axle so as to indicate the speed of the train.

5,614.—J. T. Jones and J. H. Wild, Leeds. Apparatus for Boring Rocks. 8d. (6 figs.) December 22.

MOTOR.—The drills are mounted on an adjustable travelling frame and may be driven by a "dynamo-electric machine."

5,615.—J. N. Culbertson, Antwerp, and J. W. Brown, London. Cables for Telephonic and Telegraphic Communication. 6d. (5 figs.) December 22.

CONDUCTORS.—These are threaded through insulating perforated diaphragms in pipes laid under ground. The pipes and wires are prepared in lengths and coupled in position, the wires being connected by short spiral springs.

5,618.—D. Graham, Glasgow. Holder for Incandescent Electric Lights. 6d. (3 figs.) December 22.

INCANDESCENCE LAMPS.—The illustration shows the holder in section. A is a block of non-conducting material made to screw on to the nipple of an ordinary gas bracket. Around this is an elastic metal clipping piece, which embraces the neck of the lamp globe; it is divided into two

parts *b b*[1], insulated from each other and connected respectively to the two leading wires. Each part carries a screw that can be put in contact with one of the lamp terminals. E E are brackets for an external shade. The holder is provided with a contact breaker not shown in the illustration.

*5,623.—C. A. Carus-Wilson, London. Measuring Electric Currents. 4d. December 23.

CURRENT METERS.—This invention relates essentially to the employment, for the purpose of measuring electric currents, of a secondary battery or batteries, which, or each of which, is first charged or acted on by the current to be measured, and is then allowed to discharge or run down, or has its current reversed, these alternate actions being repeated and made to communicate motion to, or to

regulate, the action of any given counting or integrating apparatus. Seven modes of carrying out the invention are briefly described, but the methods of reversing the connections and actuating the counting mechanism are not explained.

5,631. — J. S. Sellon, London. Secondary Batteries. 2d, December 23.

SECONDARY BATTERIES.—This invention relates to constructing the " terminals, plates, supports, retainers, or frames employed in secondary batteries of a material or materials not readily subjected to the destructive influence of oxidation." "Carbon, either in a solid form or amalgamated with other substances, or asbestos, wood, papier-maché, cellulose, either alone or in combination with metal not easily oxidised," may be used for the purpose.

*5,632.—J. S. Sellon, London. Construction of Incandescence Lamps. 2d. December 23.

INCANDESCENCE LAMPS.—Opal glass is employed in place of transparent glass in making the globes.

5,651.—St. G. Lane Fox, London. Electric Current Meters. 6d. (4 figs.) December 24.

CURRENT METER.—This invention has reference to meters of the class described in Patent No. 4,626 of 1878. Referring to the illustrations, C C is an electric motor acting upon a horizontal sway beam which makes and breaks the circuit every time it touches the contact springs F F. The spring I serves to insure the beams stopping in a right position for starting when the current passes. The

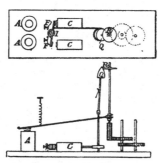

motion of the beam is communicated by a pawl to a wheel on a vertical shaft jointed in the middle and carrying at its upper end a conoidal friction drum Q, which runs in contact with a disc R. This disc and its shaft is raised or lowered by the magnet A according to the current passing, and therefore rotates at speeds that are proportionate to the consumption of the electric energy. The lower end of the shaft drives a counter.

5,656.—E. G. Brewer, London. (*C. E. Ball, Newhaven, Conn., U.S.A.*) Apparatus for Working Electric Clocks, &c. 8d. (8 figs.) December 24.

SECONDARY BATTERIES.—The use of "local batteries" is dispensed with, and a secondary battery employed to maintain a constantly charged main line circuit.

5,660.—L. S. Powell, London. (*J. M. A. Gérard-Lescuyer, Paris*). Electric Lamps. 6d. (9 figs.) December 24.

ARC LAMPS.—Two elevations of a lamp designed to be worked by alternating currents are shown in the illustrations. The upper carbon-holder is connected to a piston working in a copper cylinder surrounded by a coil of fine wire E in a shunt circuit. Between the cylinder and the coil is a tube of iron acting as the core of the coil. The lower carbon-holder is carried by a movable frame consisting of two rods A A, which can move endwise, and are supported by adjustable springs. The separation of the carbons is effected by means of an iron armature O, broken away in Fig. 2, connected with the two rods A A, and situated underneath the lower extremity of the magnet E. When the magnet is excited the armature rises and carries the lower carbon with it. The descent of the upper carbon is arrested when required by a brake device worked by the upper extremity of the same magnet. This brake consists of two vertical rods, connected together at their upper extremities by an armature T, and jointed at their lower extremities to two bent claws U capable of oscillating on pivots V. That part of each claw which faces the upper carbon-holder terminates in a sector provided with two projections capable of clasping the carbon-holder. Two helical springs W serve to counterbalance the weights of the rods and brake. Before lighting the lamp the carbons are separated so that the current passes entirely round the coil, which becomes excited and attracts both top and bottom armatures. The attraction on the upper one has the effect of drawing down the rods S S and opening the claws U ; thus the upper carbon-holder is freed and descends slowly, at a rate dependent on the opening of an air valve. The lower armature is also attracted and raises its carbon to meet the upper one. When the two are in contact the magnet loses its power, the lower carbon drops to establish the arc, and the upper one is locked by the brake device. In a modified lamp of this type, adapted for use with either direct or alternate currents, a second coil in the lamp circuit is placed below the armature O, and acts upon it in opposition to the shunt coil above. In another form of lamp the descent of the upper carbon-holder is regulated alone by the

action of air admitted to a cylinder, such as H, by a valve. This lamp is provided with a fine and a thick coil, the former being the uppermost. A copper tube is passed through both coils, and encloses another copper tube sliding within it for a short distance. The upper end of the inner tube carries a piston and the lower end an armature. That part of the tubes which is surrounded by the lower coil is enclosed in an iron sheath, which constitutes the acting core. The valve for controlling the descent of the upper carbon-

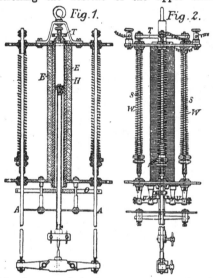

Fig. 1.   *Fig. 2.*

holder is suspended within the upper coil, which acts as a solenoid, and is of the piston-valve type, covering a port in the side of the tube in which it moves. This valve, being of iron, is drawn down when the coil which surrounds it is excited, in consequence of an abnormal resistance in the arc, and air is admitted to the cylinder, in which the piston carrying the upper carbon-holder moves. The lower carbon is supported as shown in the illustration. The specification also describes a slightly modified arrangement with a conical valve.

*5,661.—J. H. Johnson, London. (*C. Labye and L. de Locht-Labye, Paris.*) **Pipes for Containing Electrical Conductors.** 4d. December 24.

CONDUCTORS.—The conductors are carried in cast-iron jointed troughs covered with lids. The wires are separated by sea-sand, powdered glass, or by a mixture of asphalte and resin, the horizontal layers being covered with sheets of glass.

5,665.—S. A. Varley, Islington. **Apparatus for Producing and Regulating Electric Light, &c.** 10d. (19 figs.) December 24.

DYNAMO-ELECTRIC GENERATORS.—This invention relates, firstly, to an improved dynamo-electric generator. Upon a central axis there are fixed a number of wheels of non-magnetic material (Fig. 2), having blocks of iron let into their peripheries as shown. In each of these blocks a rectangular slot is cut, and an iron bar A″, seen in end view in Fig. 2 and elevation in Fig. 1, is threaded into each row of blocks. Between each pair of blocks a coil is wound on the bar, so that each coil has an iron centre and an iron cheek or flange at each end. A thin sheet of iron is bent round the coil between each layer of wire, the edges of the sheet not quite touching each other. In the generator illustrated there are four coils on each bar and sixteen bars, or sixty-four coils in all.

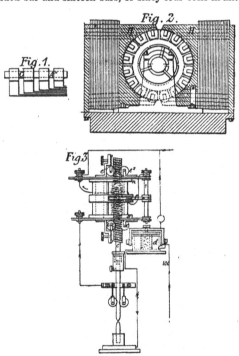

Fig. 2.

Fig. 1.

Fig. 3.

The wheels, coils, and bars, together constitute the armature which rotates between the field magnets H H. These magnets are plates wound with coils and set on edge, each plate having its own coil. The width of the plates is equal to that of the magnetic blocks in the wheels, that is, to the space C (Fig. 1) between the coils, and there are as many magnets on each side as there are wheels. The poles of adjoining magnets are alternately north and south, and the poles of opposite magnets are of opposite name. From this it follows that the blocks or cheeks of a coil that is passing through the magnetic field are magnetised in opposite directions. As indicated in Fig. 2, the magnets are not all on the same level, but are

stepped, so that the coils on a given bar enter and leave the magnetic field in succession. Each coil is connected to a separate commutator plate and two commutators are used for convenience. When alternating currents are desired, the field magnets are made half the depth of those shown, and four poles are arranged round the circle. The invention relates, secondly, to an apparatus for controlling the speed of the steam engine in accordance with the current required. This comprises two parts, a relay and a magnetic centrifugal governor. The relay consists of a cylindrical electro-magnet moving within the tubular poles of another electro-magnet and weighted so that it does not move until the current reaches a determined amount. When this happens, the current, or a part of it, is sent to the coils on the governor. The spindle of the governor drives the arms through a pawl and ratchet. Each arm is in two parts, a solid cylindrical portion jointed to the central collar, and a tubular portion affixed to the ball or weight. The solid portion carries an electro-magnetic coil, and when this is traversed by a current it draws up the hollow portion, thereby greatly shortening the length of the arm. "The effect of this is to reduce the diameter of the orbit through which the governor balls revolve, and by a well-known law it results that their rate of revolution becomes immediately increased and consequently they travel round faster than the shaft which communicated motion to them; thus the balls become raised by the centrifugal force, and the supply of actuating fluid to the motor is arrested by closing the throttle valve in the usual way."

ARC LAMP.—The invention relates, thirdly, to two forms of electric arc lamp. In the first the lower carbon is stationary while the upper is connected to an iron rod *c*, Fig. 3, wrapped with naked wire, the convolutions of which are insulated from the rod and from each other. This rod moves within hollow tubular polar extensions *n s* of a fixed electro-magnet not shown very clearly in the illustration. The current enters the lamp as indicated by the arrow 100, passes through the short-circuiting magnet *d*, and thence to the lower carbon. It then crosses the arc, leaves the upper carbon by the rollers shown, and, dividing, is led to the contact rollers *e e*, *e*[1] *e*[2], by which it enters the coil on C, and, flowing upwards and downwards, escapes by the contact springs *f* to the coils of the fixed electro-magnet and thence to the negative terminal of the lamp. By this arrangement a south pole is formed in the rod C at the place where the contact spring *f* touches it, and as the pole is just between the north and south poles of the fixed magnet it is impelled upwards by the magnetic force. But however much

the rod may move, the magnetic pole remains stationary at the point where the current leaves the coil. When the circuit is completed through the lamp, the rod and upper carbon are lifted until the resistance of the arc decreases the current to a point at which the lifting power of the magnet and the force of gravity exactly neutralise each other. The lamp is regulated by counterweighting to a greater or less extent the weight of the rod C. In a second form of lamp the heated air and gas rising from the arc are made to drive a fan, and thus supply the motive power for feeding the electrodes. Three bevel wheels, thrown in and out of gear by an electro-magnet, determine the direction of the motion.

5,667.—S. A. Varley, London. **Collection and Distribution of Electric Currents.** 6d. (8 figs.) December 24.

DISTRIBUTING CURRENTS. — According to this invention the contact pieces that are in contact with the insulated segments of the commutators of electric generators move over them with a rolling friction, and have a motion of rotation slightly greater or slightly less than the circumferential velocity of the segments. The illustration shows one method of carrying out the invention. The commutator is made in the form of a lantern pinion, all the teeth being insulated from one another. The collectors are toothed wheels D D[1] gearing with the lantern pinion, and carried in

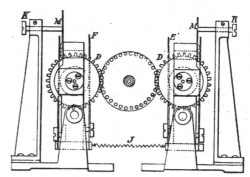

pivotted frames, which can be forced towards the pinion by the elastic pressure of the springs M and set screws K. The spring J tends to move the frames in the opposite direction. F E are brake springs which bear against the toothed wheels to prevent backlash. The specification illustrates another form of commutator with plain insulated segments running in contact with grooved cylinders that are driven positively by gearing on the central axis of the armature at a circumferential speed slightly greater or less than that of the commutator.

5,668.—Sir William **Thomson**, Glasgow. **Dy-namo-Electric Machines, &c.** 1s. 10d. (59 figs.) December 26.

Dynamo-Electric Generator.—This invention relates, firstly, to revolving metallic brushes used in place of the fixed brushes usually in contact with the commutator bars. These brushes are made partly elastic, or the commutator bars are elastically connected to the rotating armature and the brushes made rigid. The rotating brushes are made of cups packed into each other, each alternate cup being cut radially into a number of divisions, and the whole fixed upon a spindle. Fixed springs bear upon the outer cups to lead off the current. Instead of the cups, discs of copper, radially split, may be employed, and when fixed together they may have a cylindrical or conical form; in the latter case they may run against a disc or face commutator. For very high speeds the commutator bars are made elastic by fixing copper brushes to them, with the wires standing out like the fibres of a nail brush. When it is desired to make contact on several con-secutive commutator bars, the bars are made in the form of brushes and run against fixed brushes of con-siderable width. Figs. 1 and 2 are sectional diagrams of an improved dynamo-electric generator, which consists of a drum or cylinder B, formed of insulated copper bars built together like the staves of a barrel. These bars are bound together externally by cord S or by hoops of non-magnetic metal. A plug of wood inserted in one end of the drum resists the pressures of the roller brushes, shown by dotted circles C C, and the antifriction bearing rollers, partly shown by segmental dotted lines A¹, on both of which the drum may be supported. The trans-mission of power to or from the drum may be effected by the tangential force of one or more of the bearing rollers, or by means of an axle fixed into the mouth of the drum at the end remote from the brushes. In the latter case, the axle must be hollow to allow a sup-port for the interior part of the electro-magnet to pass through it. The magnetic fields K K are produced by means of an electro-magnet of circular form in two segments. The smaller of these has its soft iron core M' wound with a conductor of any suitable kind, while the larger is wound with strips suitably disposed to fill the whole available space with copper conductor. The lines G H, G K, show the form of one of the coils before being slipped on to the core. The thickness of the strip of copper forming the coils is greater at the point H than nearer to the magnetic fields K K, the number of convolutions at each place being inversely as the thickness. The commutator bars are merely pro-longations of the main bars, but the more usual arrangements may be adopted in this particular. The other end of each of the main bars is metalli-cally connected to the diametrically opposite bar,

but when great electromotive force is required the Hefner Alteneck method of connection may be adopted. Figs. 2 and 3 show another form of generator. This consists of a wooden disc A with projecting wood teeth *t*, mounted on an axle B. A flat or square strip of wire C, bent round the teeth, with its ends carried to the axis, is fastened to the disc in the fol-lowing manner: A sufficient number of notches are cut across the edge of the disc to receive the bends of the conductor. This being placed

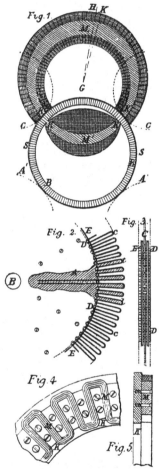

in position, wooden pins D are then laid in the notches over the conductor, and a wire E is passed two or three times round the edge of the disc to press the wooden pin into the notch. This genera-tor is adapted for producing alternating currents, the magnetic fields by which the radial portions of the moving conductors are excited being produced either by permanent or electro-magnets. When electro-magnets are used they are arranged as shown in Figs. 4 and 5, so that each straight radial portion of the conductor serves to excite the soft

iron of two contiguous electro-magnets M M. A circular cast-iron ring K is provided having pieces of soft iron M bolted to it, and surrounded by zig-zag conductors. A separate exciter must be used with this generator. The arrangement of dynamo shown in Figs. 1 and 2 has the advantage that there is no moving iron in it and no change of polarity. An alternative plan substitutes for the binding cords a composite cylinder formed of thin flat iron rings insulated from each other and bolted together. The interior magnet, which corresponds to M', Fig. 1, is flat instead of circular, and is situated diametrically within the copper bars. The part that corresponds to the exterior magnet is merely a heavy iron armature, or keeper, with two pole-pieces running longitudinally of the cylinder opposite to the poles of the internal magnet. The cylinder or drum may be supported on antifriction pulleys at different parts of its length, and be driven by a belt applied directly to it. Sometimes the cylinder is arranged vertically and its weight is supported on a turntable.

CURRENT METERS.—For measuring the difference of potential between any two points in a circuit in which a current is flowing constantly in one direction, an instrument is employed consisting of an annular circular coil of insulated wire, and a system of magnetic needles, supported by a jewelled cap, as described in Specification No. 1,339 of 1876, and carried in a suitable frame, which is also provided with a light index moving over a horizontal scale. The needle and index are enclosed in a box which can be moved along parallel to the axis of the coil to vary the sensibility of the instrument. The scale is so divided that the numbers are proportional to the tangents of the angles of deflection. Over the box there is placed a semicircular permanent magnet, in such a position that the line joining its poles is horizontal and parallel to the plane of the coil, and passes through and is bisected by the magnetic axis of the needles. When the current would produce a deflection greater than the range of the instrument, a counteracting force is applied by means of a bar magnet with its axis in the same plane as the axis of the coil. When currents are to be measured, and not potentials, a coarse coil is substituted for the fine one. The specification describes such a galvanometer and also two other forms. To determine the quantity of electricity that has flowed through a circuit in a given interval, a rectangular coil, pivotted on knife edges, is provided with a lever, which carries, at its lower end, a globe situated between an inclined rotating disc and a cylinder placed with its axis horizontal and parallel to the disc, similar to Professor James Thomson's disc, globe, and cylinder integrator. The disc is driven by clockwork, and imparts more or less of its motion to the cylinder, according to

the position of the globe. To insure the prime mover, which drives the dynamo, receiving the proper amount of actuating fluid, a weighted pulley is placed in a frame, so that it rides on the driving belt, and according to the tightness or slackness of the belt the pulley rises and falls, and in so doing opens or closes the throttle valve of the motor. For measuring the speed of dynamos, one or more cantilever springs are placed on the shaft parallel to the axis, and under the influence of centrifugal force these springs bend outwards, and their positions are read off on a fixed dial marked experimentally. Another method is to arrange three cylindrical glass vessels to rotate on their vertical axes and to partly fill them with liquid. The liquid assumes a paraboloidal form, and the distance to which its upper edge is raised is a measure of the speed. The three vessels are geared at different speeds, so that the instrument will indicate a large range of velocities.

5,681.—J. Richardson, Lincoln. Construction and Arrangement of Dynamo-Electric Machines. 6d. (4 figs.) December 27.

DYNAMO-ELECTRIC GENERATOR.—A series of armatures A B, at right angles to each other, are so arranged in connection with field magnets C D, that one set is at the point of maximum intensity, while the other is at the point of minimum intensity, and as one increases in intensity the other

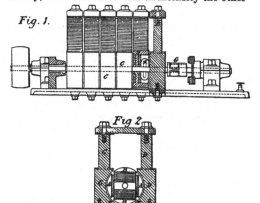

Fig. 1.

Fig 2.

diminishes in the same ratio, the sum of the two producing a constant strength of current. Six pairs of armatures with their corresponding field magnets are shown in the drawings. G is a commutator of ordinary construction. The armature coils may be coupled up as desired.

5,687.—C. A. Carus-Wilson, London. Controlling and Regulating the Production and Distribution of Electricity. 8d. (18 figs.) December 27.

DISTRIBUTING CURRENTS.—This invention relates, firstly, to a means of supplementing the action of the governor of a steam engine driving electric generators, so as to keep a constant electromotive force in the mains. A weight, arranged to act upon the governor as it rotates, is under the control of two electro-magnets or solenoids, one to apply the force and another to take it off. The magnets are brought into action by currents which are established or interrupted by a galvanometer needle. When the electromotive force is of the desired degree, the needle is stationary between two stops, the weight keeping the governor down a little, so that when the electromotive force rises, the needle is deflected against one of the stops, making circuit through one of the magnets and raising the governor balls. If the electromotive force fall, the opposite will happen. a is a solenoid having a core b, of which the shaded portion is of iron; c is a pointer connected to the core and vibrating between two stop pins. When the pointer touches one of the stops, it make a connection through one of two solenoids, which tends to raise or depress the governor balls as the case may

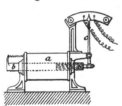

be. Instead of acting on the throttle valve, the solenoids may throw wheels in or out of gear to vary the expansion gear of the engine. The invention relates, secondly, to regulating gas engines, by causing the action to which the impulse is due, to put a certain resistance in the circuit, or in the generator, so that while the speed is above the normal, owing to the sudden impulse, a resistance is in circuit which keeps the electromotive force in the main down, and, as the extra speed diminishes, this resistance is withdrawn. One method of effecting this is by a rotating porcelain cup, containing mercury, with a carbon-rod projecting down its centre line into the mercury. The rotation causes the mercury to assume a paraboloidal form and consequently more or less of the length of the carbon rod is put in circuit. Besides employing a resistance, an opposing electromotive force may be brought into play by the increased current which is the result of a variation in speed. In the field magnet circuit of the generator a small electric motor is placed, so that the whole of the exciting current goes through it. This motor drives a fan or brake "set for a certain number of revolutions." "If from any cause the electromotive force of the machine rises, it drives the motor faster, thereby increasing

the opposing electromotive force and keeping down the exciting current. This method may be employed also to regulate the electromotive force in the circuit of a single lamp, or a series of lamps, and also to rotate the lamps themselves. Furthermore, in order to obtain a constant electromotive force in the main circuit, secondary batteries are so arranged that when charged to a certain degree the circuit is broken, so that the generating machine ceases to charge the batteries, and no current can flow back from the batteries to the generating machine. When circuit is again made through the batteries, the current will flow at the required electromotive force, and the generating machine will recommence charging the batteries. By such means apparatus can be constructed analogous to a hydraulic accumulator, or to a gasometer. Batteries are charged through a kind of electric valve, so that the current can pass forwards but not backwards. In one apparatus, a commutator or switch is turned by a solenoid core, working against a spring, so as to make and break the generator circuit according to the condition of the secondary batteries. To show to what extent the secondary batteries are charged, a small secondary battery has its two plates plunged into two cells separated by a porous diaphragm, and connected by an external pipe filled with the fluid element, and containing at some point a screw or paddle wheel provided with counting mechanism. When a cell is being charged, the liquid in one tends to rise with a force proportional to the strength of the current, but since the two partitions are connected by an external pipe, the liquid flows through the pipe and actuates the counting mechanism. As the battery is discharged, the action is reversed, so that the counter always indicates the amount of any charge remaining. The invention also includes apparatus for enabling the engineer to make and break any circuit, at any position, without having the whole current brought to a commutator close by him. This is accomplished by a solenoid whose core is pivotted to a spring steel pawl-shaped armature, which engages with teeth attached to a brass wheel, having four segments of insulating material, and against which the ends of the conductor bear by means of roller contact. A switch is so arranged as to cause the solenoid to move the brass wheel a certain portion of its revolution at each movement of the switch.

5,702. — J. W. Swan, Newcastle - on - Tyne. Sockets or Holders for Electric Lamps. 6d. (33 figs.) December 28.

INCANDESCENCE LAMPS.—This invention relates to sockets or holders for incandescence electric lamps, whereby a lamp, without any external fittings, can be readily attached to the conducting

wires and make good contact with them. In the first form, the holder is provided with two hooks insulated from each other and terminating in binding screws. The connection between the lamp and the socket is effected by engaging the two hooks with the two eye-like terminations of the lamp conductors, firm contact being insured by a compressed spring. In the second, the binding screws for attachment to the conductors are replaced by circular concentric contact pieces. In lieu of the helical spring enveloping the lamp stem, it may be caused to act upon a tube adapted to the said stem. An example of this modification is shown in Figs. 1 and

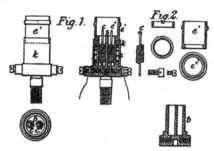

2, in which the hooks c c^1 are connected to binding screws on opposite sides of the holder; e^1 is the tube adapted to the lamp stem, and fitted between the body b and an external tube k, and acted upon by a spring e. In the example, the hooks c c^1 are also acted upon by springs l^1 l^2 in a state of tension, but these may be dispensed with if desired. As a further modification, the spring e may be dispensed with, the body of the holder being formed with a screw thread, and the tube e^1 with a corresponding internal thread, so that, by the act of screwing, the required tension is placed upon the hooks and eyes, and efficient electrical contact is secured.

5,738.—J. G. Lorrain, Westminster. **Electric Lamps.** 4d. December 31.

INCANDESCENCE LAMPS.—This invention relates to silvering, coating, or covering a portion of the globe of an incandescence lamp with a silvering or reflecting material.

5,743.—G. Pfannkuche and R. E. Dunstan, London. **Electrical Resistances.** 6d. (7 figs.) December 31.

RESISTANCES.—This invention relates to "improved electrical resistances in which the conductor is embedded in glass, plaster-of-paris, or other solid insulating material, which aids in diffusing the heat resulting from the passage of the current." The figure illustrates the invention applicable to a lamp; a is a cylinder of glass, b a platinum spiral

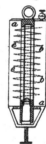

traversed by the current, with contact pieces e e communicating with the central cylindrical cavity. Into this cavity mercury can be forced to any height, by a screw, so as to short-circuit any desired number of the coils. The specification also illustrates the invention applied for heating purposes to an urn and to a railway carriage.

5,751.—W. R. Lake, London. (*A. L. Duwelius, L. W. Goss, P. Higgs, F. R. Merrell, H. D. Peck, and H. Walter, U.S.A.*) **Operating or Controlling Railway Brakes by Electricity.** 6d. (9 figs.) December 31.

GENERATOR.—The locomotive carries a magneto or dynamo-electric generator and a secondary battery, and forms one end of a system of mains running the length of the train, having derived circuits, between which the electro-magnetic appliances for controlling the brakes are placed.

1882.

14.—A. Mackie, London. **Apparatus for Electric Lighting.** 6d. (7 figs.) January 2.

AUTOMATIC SWITCHES.—The object is to provide means whereby an excess of current, supplied to a lamp, may be shunted to a second, and even to a third lamp or conductor. Fig. 1 illustrates the invention applied to lamps arranged in parallel arc. The positive conductor is connected to one end of a solenoid or magnet a, having an armature b, whose motion is controlled by a spring e; the other end of the solenoid is connected to a stud d^1 always in contact with the lever c. The positive conductors of the lamps are connected, respectively, to the studs d^1 d^2 d^3, the negative conductors being connected to a common return. When the circuit is completed, the lever c makes contact, successively,

with the studs $d^2 d^3$, as the current increases in strength, and the lamps $g^2 g^3$ are successively included in the circuit. In a modification, the positive conductors of the lamps are connected to a conductor which includes the magnet a, the negative conductors being connected to the studs $d^1 d^2 d^3$, and the stud d^1 to the return lead. Fig. 2 shows the lamps arranged in series. The construction of the magnet and lever is similar to that in Fig. 1. The solenoid a is in the main circuit, its inner terminal being connected to the positive terminal of the lamp g^1. The negative terminal of the lamp g^1 is connected to the positive terminal of the lamp g^2, and also to the contact stud d^1, and similarly throughout the series. All the studs $d^1 d^2 d^3 d^4$ are in contact with the lever c; thus the last three lamps are normally short-circuited, until the movement of the lever, by an excess of current,

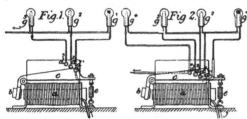

breaks the contacts. According to another arrangement, the magnet, armature, and lever are applied for throwing extra lamps into circuit when the current is increased. The device is applicable to street lamps, so that every other one can be extinguished by lessening the supply of electricity from the central station. An arrangement is also shown in which there are five lamps, arranged in parallel arc, and four contact levers, successively attracted as the current increases, by a double electro-magnet. The main current passes through the electro-magnet to a contact stud, connected to the first lamp, and also to the pivots of the four contact levers, to which the positive terminals of the remaining four lamps are respectively connected; the negative terminals of the lamps are connected to contact studs in communication with the negative conductor.

*29.—D. G. FitzGerald, C. H. W. Biggs, and W. W. Beaumont, London. Secondary Batteries. 2d. January 3.

SECONDARY BATTERIES. — The electrodes are obtained by utilising the local action occurring between lead and a metal electro-negative to lead. A solution of plumbic sulphate in ammonic sulphate is used as an electrolyte. The electrodes may be prepared by blowing air, alone or charged with other materials, into molten lead. The surface of the plate may be corrugated, perforated,

or otherwise increased, as described in Specification No. 5,338 of 1881.

49.—J. Hopkinson, London. Measuring and Recording Quantity of Electricity, &c. 6d. (8 figs.) January 4.

ELECTRICITY METERS.—The velocity of a centrifugal governor is controlled by electrical apparatus. The governor is driven by an electro-motor arranged as a shunt to the current to be measured. The derived current passes first round the electro-magnets a, then by a brush to the armature, and thence by brushes d and e to an insulated ring, and by a wire to an insulated ring fixed within the core of an electro-magnet k, through the coils of which the current to be measured passes. The governor balls $n\ n$ tend to raise the iron core i, which revolves with and slides on the motor shaft. This motion is resisted by the attraction of the core m, fixed on the shaft l, and the electro-magnet k on the core i. The electro-magnet k is surrounded by a cylinder of

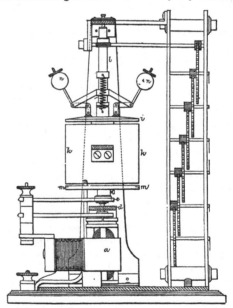

iron supported by the frame. The shunt circuit is completed through the core i, which, if the motor should revolve too fast, will be raised and break the circuit, reducing the velocity of the motor, and keeping it proportional to the current passing through the coil k. In modifications, iron pieces suspended by springs are attracted in opposition to centrifugal force by the core i, which in this case is fixed to the shaft of the motor, the iron pieces serving also to break the circuit of the motor, or the motor may be replaced by a clockwork arrangement, the speed of which is controlled by the flange of the core i coming in contact with brake screws. In another arrangement, the motor

drives a vertical shaft carrying three or more vertical tubes, connected with each other, and containing mercury. A movable coil, turning in bearings and carrying a contact piece, is operated by a stationary coil, the current to be measured passing through the two coils. If the speed of the motor increase too much, the mercury is depressed in the central tube and leaves the contact piece, and thus lessens the speed, and on any increase of current both the contact piece and the mercury are depressed. In all cases the motor operates the counting mechanism by a worm gearing as shown.

55.—J. Perry, London. Apparatus Used in the Distribution of Electrical Energy. 4d. January 5.

DISTRIBUTING CURRENTS. — The object is to supply each consumer with a current of electricity which is quite independent of the demand from and supply to other consumers. Supposing (1) that the translating devices are arranged in parallel arc; it is then necessary to maintain a constant difference of potential between the main and return conductors. The current, in this case, is furnished to the mains by a "series" dynamo, and a magneto-electric generator, or other apparatus producing a constant E.M.F. equal to the E.M.F. which should exist between the consumer's supply and return conductor. The two generators are placed in series, and the speed of the dynamo is such that its E.M.F. is proportional to the current circulating in the coils of its field magnets. Instead of using a "series" dynamo, a "shunt" dynamo may be employed, in which case the magneto generator will be introduced, either into the shunt, which includes the field magnets, or into the part which contains the wire of the revolving armature, but preferably it is stationed in the main circuit outside the machine, so that the current passing through it goes direct to the supply conductor. Supposing (2) that the translating devices are arranged in series; it is then necessary that they should all receive a constant current of electricity, however the resistance of one or more of them may vary. In this case, one or more "shunt" dynamos are combined with one or more magneto generators (or other generator producing a constant E.M.F.) in such a way that either the circuits of the field magnets, or the armature, or both, contains a magneto generator. The dynamo should be driven at such a speed that the E.M.F. produced in its armature, by rotation, is equal to the current in the circuit of its field magnets, multiplied by the sum of the resistance of its armature and field magnets, the constant current resulting being that which the magneto generator would produce in a circuit whose total resistance is equal to the resistance of the armature of the dynamo. When charging accumulators,

coupled partly parallel and partly in series, by a generator arranged according to the first system, the number in parallel circuits is diminished, and in the second system the number in series is diminished. This mode of distribution is further explained, in the specification, by examples and algebraic formulæ.

64.— L. A. Groth, London. (*R. J. Gülcher, Biala, Austria.*) Magneto and Dynamo-Electric Machines. 6d. (4 figs.) January 5.

DYNAMO - ELECTRIC GENERATOR.—In order to reduce the heating of the wire coils, the armature is constructed of a ring M, formed of laminated insulated soft iron rings, either of a wedge-shape in cross section (Fig. 4), or rectangular at its outer part and wedge-shaped at its inner part

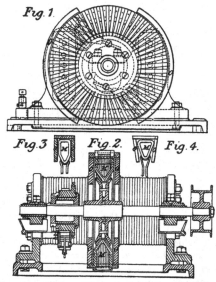

(Figs. 2 and 3). The poles O of the inducing magnets have a channel or U shape cross section, and the sides are of a depth equal to, or greater than, that of the rotating ring, plus the coils of wire thereon. When the ring is of triangular section, the sides of the channel incline inwards (Fig. 4). By this construction, a current of air is drawn in at one end of the pole-piece and driven out at the other.

69.—E. H. T. Liveing, London, and C. V. Boys, Oakham. Manufacture of Incandescent Lamps. 4d. January 6.

INCANDESCENCE LAMPS.—This relates to methods of manufacturing incandescence lamps, in which the electrical connection between the outside wires and the carbon filaments is made by suitably insulated conducting wires, passing through a narrow tube attached to one end of the globe, into which entrance of the air is prevented by partly filling the

tube with well-boiled pitch, or other resinous cement, which will not give off sufficient vapour to injure the lamp. By this method the use of platinum wires is avoided. Undue heating of the cement is prevented by employing a considerable length of twisted conducting wire between the filament and the cement, and to these wires are sometimes fixed radiators, to assist the passage of the heat out of the lamp. For small lamps, capillary coke is made use of for the conducting element, thickened or not by heating in a hydro-carbon, or in any well-known way. In another form of lamp, where the filaments are replaceable, the globe is made with a neck sufficiently wide to admit of the introduction of the filament, and it is cemented by pitch to a support forming one terminal of the lamp, the other terminal being insulated and passing through it.

*72.—R. Kennedy, Glasgow. Secondary or Reversible Batteries. 2d. January 6.

SECONDARY BATTERIES.—A plate of clean lead is coated with a layer of lead peroxide or dioxide, and is laid in a depositing bath filled with dilute sulphuric acid. Over this another sheet of lead is suspended, and the current sent through the two, so as to form a close-grained deposit of spongy lead on the lower plate, which is then cut up to suit the size of the cells of the secondary battery, and is painted with lead dioxide made into a paste with water and glycerine. The plates cut from the upper sheet or plate are painted with a paste of water, glycerine, lead dioxide, and lead sulphate, and constitute the other elements of the couples. The plates forming each element are placed with their prepared sides facing each other in a separate cell, with a sheet of non-conducting cloth or paper on them. Plates of tin, iron, or other suitable metal or conducting material, immersed in a suitable acid, may be used in lieu of the lead plates.

84.—W. R. Lake, London. (*C. E. Ball, Philadelphia, U.S.A.*) Dynamo-Electric Machines. 6d. (2 figs.) January 6.

DYNAMO-ELECTRIC GENERATOR.—Each armature is arranged to rotate within the inductive influence of only one pole of a field magnet, the armatures and field magnets being in the same circuit. Referring to the illustrations, the two parallel bars C C^1 are sustained in the uprights B, which are secured to the metal bed-plate, and constitute the cores of the electro-magnets. The two pole-pieces C^2 C^3, of opposite polarity, are fastened to the insides of the armatures, and rotate in opposite directions, each in the inductive field of *one* pole-piece only. The direction of the current is as

follows: Leaving the commutator F it passes in succession through the left-hand coil and the middle coil of the bar C, through the coil c^2, through the external circuit, through the coils

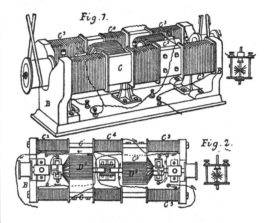

c^3 c^4 c^5, by a brush to the commutator F^1, through the armature D^3 by the other brush of the commutator F^1, to the brush of the commutator F, and through the armature D^2 to the starting-point. In a modification, the poles are of the same kind, and the armatures rotate in the same direction on the same, or on separate shafts.

85.—W. R. Lake, London. (*J. S. Williams, Riverton, N.J., U.S.A.*) Machines and Apparatus for Generating and Utilising Electricity for Lighting, &c. 1s. 4d. (62 figs.) January 6.

DYNAMO-ELECTRIC GENERATORS. — The generators are constructed with duplicate armatures, or duplicate induced faces, one for positive and the other for negative currents, so as to produce continuous currents without reversal of the currents in the parts of the generator connected to the external circuit. Two rings of electro-magnets are suitably supported upon a base-plate, and are wound in such a manner that the poles of each ring adjacent to the armature are of the same kind, the two sets being of the same or of opposite polarity. Between the two rings of electro-magnets are mounted, on a shaft rotating in suitable bearings secured to the base-plate, two discs or rings, which carry the armature coils, each disc or ring facing one set of electro-magnets. On the shaft between the two discs may be mounted a coil, wound to produce on each disc a pole of opposite kind to that of the adjacent ring of electro-magnets. In another arrangement, each coil of the armature projects through two plates, from which it is preferably insulated, and faces the two rings of electro-magnets; or each coil may be divided into two parts which come respectively under the influence

of one or other of the rings of electro-magnets. In another generator, the armature coils, ribbons, or bars, are moulded upon the faces of a disc, which rotates between two rings of field magnets, or two such armatures may each rotate within grooves in pole-pieces of the same polarity, the pole-pieces of one armature being of opposite polarity to those of the other. Instead of coils or ribbons, each disc may be provided with a continuous conducting surface. In another generator, the field magnets are arranged round the periphery of the armature. The circle of poles adjacent to the armature are of the same polarity, or a circle of poles at one end of the armature may be of opposite polarity to those at the other. The armature consists of bars, strips, or ribbons extending lengthwise of the shaft. If the poles adjacent to the armature be all of the same polarity, the bars or strips are united to a common conductor at each end of the armature, or are connected alternately at their ends to form a single continuous conductor in each section. If the pole-pieces at opposite ends of the armature are of opposite polarity, the bar or strips at each end are connected by a conductor; or the sections at one end of the armature are united with diametrically opposite sections at the other end of the armature; or one end of each section may be joined to one end of the adjacent section on the other end of the armature. The conducting portions may be built up upon a magnetic, non-magnetic, or insulating base. The coils of the armature may be connected in series, or those of one disc or set may be connected in series alternately with those of the other disc or set, and may or may not include the coils of a central electro-magnet, if such be used in their circuit. In another method, the ends of the different coils are taken to a switch box and may be connected up as required. The current is collected by brushes bearing on solid conducting rings at the ends of the armature shaft. Several methods of forming the armature are described. The bars or strips of the armature may be arranged in a mould and secured together by pouring into the mould a liquid insulating material that will solidify. The surface of the conductors adjacent to the field magnets may be covered or not at pleasure. The bars may be supported in the moulds by plates which are afterwards removed. In another method, the armature core is placed in a mould having projections which come in close contact with the core, or with insulating material thereon. The insulating material is then poured in, and the coils or bars are placed in the grooves formed by the projections on the core. The conductors may be formed in a mould so as to be of uniform section and of the desired shape. In another method, the core to be wound is mounted

upon a shaft and enclosed in an air-tight box, heated so as to render glass or other insulating substance sufficiently pliable to be wound with the wire strip or ribbons which it serves to insulate, or the conductor may be passed through a bath of vitreous pyro-insulating or other material. The conductor is guided by fireclay, zirconia, or other suitable rollers. The insulating material employed may be a compound of salts and oxide of zinc. For controlling generators, clockwork or other mechanism, having combined with it "circuit makers," interrupting or reversing the currents, in or through batteries, at intervals, is employed, and may be combined with a motor, galvanometer, or measuring apparatus, so that the position of the circuit maker is governed by the strength of the current. The electricity generated within one battery is led by a suitable conductor to a storage battery, or is otherwise utilised, the circuit makers reversing the current as required. The first-mentioned batteries may be so arranged that the mechanism, when the E.M.F. falls below a given point, reverses the current.

95.—W. J. Mackenzie, Glasgow. Electric Lamps. 6d. (8 figs.) January 7.

ARC LAMPS.—The carbons are fed by forcing them forward through a ring provided with internal projections of a not readily fusible material, such as agate, iron, or platinum, which projections are fitted against the pencils, at points equidistant around the circumference, and at that part where it passes from the cylindrical to the conoidal shape. Referring to the illustrations, it will be seen that each carbon is supported in a long tube, and is held at its outer ends by a metal piston, sliding in the tubes and urging the carbons, by means of a spring,

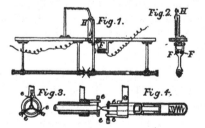

against three studs *e* (Figs. 3 and 4) fixed on a ring. The studs of one ring are fitted intermediately between those of the other. One tube or carbon-holder is suspended from pivotted links, so that it can be moved back to establish the arc. To effect this an electro-magnet F is provided, acting upon an armature on one of the links. The magnetic attraction is opposed by a spring and lever H, the latter being so shaped that the more nearly the armature approaches the magnet the greater the leverage through which the spring acts. The current passes

by one of the links to the ring supporting the studs, and thence across the arc to the other ring and the electro-magnet F, or *vice versâ*. When this method of regulation is not sufficiently delicate, another is employed, in which a small electro-motor is caused to rotate in one direction, or the other, by a part of the current used to produce the light, or by a battery, the current being directed by a relay arrangement. The motor is connected either direct, or through gearing, to one of the carbon-holders. In order to reverse the current through the armature of the motor, the brushes are respectively connected by conductors to a central stop, and to a piece having contact stops on either side of the central stop. Two arms, carried by the link, and operated by the magnets F, are connected respectively to the conductors on either side of the arc, and make contact with the central stop and one outside stop, thus operating the motor in one direction, or with the central stop and the other outside stop, operating the motor in the reverse direction, or they completely interrupt the current. In another arrangement, the arc is struck by the action of an electro-magnet in the main circuit on one carbon-holder, and normally the electromagnet F causes the link to make a contact and short-circuit the motor, but on the lengthening of the arc the contact is broken, and the motor, being thus brought into the circuit, advances the other carbon-holder.

120.—J. E. Liardet, Brockley, and **T. Donni-thorne**, London. **Storing Electrical Energy.** 8d. (4 figs.) January 9.

SECONDARY BATTERIES.—The inventors employ : (1) Cells packed with lead shot. (2) Cells filled with hollow spheres of lead prepared by coating balls of soluble substances, such as salt, chalk, &c., with lead, and dissolving out the cores. (3) Cells filled with broken charcoal, coke, asbestos, &c., coated with metallic lead. (4) Lead, deposited from a salt of lead, such as the acetate, by means of zinc, and coated, or not, with any oxide of lead, or sulphate of lead. (5) Lead plates attached to compressed plates or cakes of red lead. Other cakes, obtained by mixing red oxide, or other suitable salt of lead, with plaster-of-paris, cement, china clay, or powdered glass, may be employed. (6) Finely divided porous lead, obtained by filling a cell, having two compartments divided by a porous wall, with acetate of lead solution, and immersing therein, as electrodes, two plates of zinc, or one plate of zinc and one plate of lead, encased in oxide or sulphate of lead, in suitable jackets, and as soon as the zinc is dissolved and replaced by metallic lead, drawing off the solution and replacing it by acidified water. In cases where the cells are started with

metallic lead plates, peroxides of hydrogen are added to the acid to assist in the formation of peroxide of lead. In Fig. 1, the terminals A and B are composed of thin plates of lead or carbon, surrounded by a coating of finely divided porous lead, overlaid with a lead salt and plaster-of-paris, in the proportion of eight parts of the former to one of the latter. Fig. 2 illustrates a battery having one terminal A made of sheet lead, encased in a mass of finely divided porous lead D, which is kept in place by thin folds of lead foil. The other electrode B is composed of carbon surrounded by red lead and plaster-of-paris. Porous lead may be obtained direct from the lead ore by treating the latter with an acid such

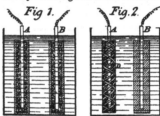

as hydrochloric acid, the chloride of lead thus obtained being washed, and then dissolved in boiling water, from which it is precipitated by scraps of zinc; or galena is first roasted, then treated with hydrochloric acid, and afterwards with a solution of brine, from which the dissolved lead is precipitated by zinc. The porous lead so prepared is moulded by the hands around a rod or plate of some conductor, which is then placed upright in a suitable mould into which is poured a mixture of a salt of lead and plaster-of-paris. The conducting terminals of the cells may be protected by gutta-percha or other material. In the provisional specification it is stated that plates, or other forms of metal, may be substituted for the corresponding forms of lead, and that they may be coated with an oxide of tin.

129.—W. H. Preece, Wimbledon, and **J. James**, London. **Electrically Lighting Railway Trains.** 6d. (6 figs.) January 10.

ELECTRIC LIGHT.—The running axles of the train work a compressing pump, so as to charge reservoirs with compressed air. This air is employed

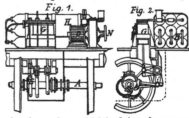

to work air engines, which drive dynamo-electric generators to supply the lamps in the train. In

order to maintain the lights during stoppage, the pumping power and reservoirs are made of suitable size to keep the generators in action after the axles come to rest. A is the axle carrying a sleeve E, connected to it by a clutch. Upon this are fixed eccentrics which drive the air pumps G and deliver air to the reservoirs H. N is the air engine and O the dynamo.

130.—W. T. Henley, Plaistow. Machinery for Obtaining, Transmitting, and Applying Electric Currents for Lighting, &c. 1s. 6d. (20 figs.) January 10.

DYNAMO-ELECTRIC GENERATORS.—This invention consists in the construction of electrical generators, in which the field magnets are straight, and are fixed parallel to, and symmetrically about, a rotating shaft. The number of such field magnets may be two or more, and they are carried in frames of non-magnetisable metal attached to the bed-plate of the generator. The armatures, of which there are usually two, are rings of iron "having recesses for the reception of coils of insulated wire," and are

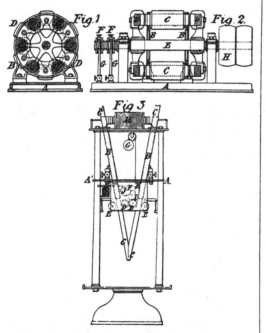

fixed one at each end of the rotating shaft, so as to revolve in close proximity to the faces of the field magnets. "These rings and coils act in quite a different way to the machines known as the 'Gramme' or 'Brush' machines, as they form distinct electro-magnets revolving at each end of the fixed magnets." The currents from the coils on one of these rings are passed through the field magnets, a commutator being used to make them

continuous, and the currents from the other ring are employed in the outer circuit. The coils of the latter are connected up in series, and the currents generated within them are used as alternate currents, or are changed into a continuous current by means of a commutator. As many independent circuits as there are coils may be supplied, in which case the commutator has the desired number of contact pieces and pairs of springs. Figs. 1 and 2 represent a dynamo-electric generator with six field magnets, constructed in accordance with this invention. A is the foundation plate, B the standards supporting the field magnets, C D are the revolving rings with their coils, E the shaft carrying the rings, F the commutator, G G the pillars and springs for conveying the currents, and H H the fast and loose pulleys. The cores of the field magnets may be of solid iron, or of iron tubes filled with soft iron wires, or of cast iron with the pole-pieces cast on, or of thin plates of soft iron. The armature coils may be wound on cores of like construction. The rings are supported by arms of non-magnetic material such as brass, and may be solid or laminated. In modifications, each ring is made with two projections of soft iron, on one side, which rotate before two rectangular field magnets. The spaces between the projections are wound with coils of insulated wire, or projecting pieces are attached to the outer sides and edge of the ring, the pieces being opposed to the poles of the field magnets, which poles extend and surround the ring; or horseshoe magnets and a single ring may be employed. In another arrangement the field magnets and the armature coils are fixed, a bar of iron attached to shaft by non-magnetic material rotating before the poles of both. The bar of iron may be wound with insulated wire, and the current produced be commutated to excite the field magnets.

ARC LAMP.—Fig. 3 shows an arc lamp in which the carbons are inclined at an angle to each other, and are fed downwards, meeting below. The stand A, with the pillars and two platforms A^1, carries the two tubes B, through which the carbons C are fed downwards by the rollers D pressing them against the non-conducting rollers E; the rollers D are geared together by two wheels, one of which gears with the wheel I on the same axis as the barrel F, round which a cord is wound actuated by the weight G. To prevent the carbons running down too quickly, the wheel I gears with a pinion on the axis of the wheel J, and drives a fly K. A small electro-magnet in the main circuit is employed to release the fly and allow the carbons to be fed forward. An electro-magnet M on the upper platform causes the carbons to separate at their lower ends, and establishes the arc. The tubes and carbons move on a fulcrum formed by the rollers D and E. Two other lamps of this type,

but with the carbons meeting in an upward direction, are illustrated and described, as well as another type of lamp in which the carbons are in the same vertical line, and their holders are attached to racks whose motion is governed by a train of wheels, fly, and electro-magnets similar to those employed in the lamp first described.

144.—H. J. Haddan, London.   (*E. Boettcher, Leipzig, Saxony.*)   **Secondary Galvanic Batteries.** 2d. January 11.

SECONDARY BATTERIES.—A very thin plate of sheet zinc is immersed in a solution of sulphate of zinc (about 1 part of the sulphate to 3 parts of water) forming the cathode. Thin lead foil folded lengthwise, previously dipped into an aqueous emulsion of pure litharge and coated with a thick layer of the litharge, forms the anode. On the passage of a current, pure zinc is deposited on the zinc plate, and the litharge is oxidised to peroxide of lead. By closing the circuit of the cell, the deposited zinc is re-dissolved, and the peroxide, and also a layer of the oxide of lead, is reduced to metallic lead. By alternately charging and discharging the cell for a few times, the cell will be rendered ready for use.

157.—G. Hawkes, London. **Apparatus for Electric Lighting.** 8d. (11 figs.) January 11.

ARC LAMPS.—The arc is struck by means of an electro-magnet arranged in the main circuit and actuating the upper carbon-holder. The lower carbon-holder is fed upwards by weights, its motion being controlled by a piston working in a cylinder containing a suitable fluid, and by a clutch arrangement operated by an electro-magnet placed in a shunt across the arc. Fig. 1 is a vertical section, and Fig. 2 a sectional plan taken immediately below the plate A^1. The frame of the lamp consists of three metal base-plates A, A^1, and A^2, rigidly connected by vertical stems A^3, A^4, and A^5. The stem A^3, which is in connection by the terminal below with the positive conductor from the generator, is insulated from these metal plates, and the stem A^4, which is in connection with the negative conductor through the plates A, A^1, is insulated from the upper plate A^2 only. This plate supports the mechanism for striking the arc and regulating the advance of the upper carbon C and holder C^1, while the plate A supports the mechanism for regulating the advance of the lower carbon D and its holder D^1. This carbon traverses a long tube B, pendent from the bottom of the lamp, and prolonged at B^1 by a reservoir of liquid into which a piston descends. When the current is switched on the lamp, it passes by the stem A^3 and terminal g to the electro-magnet G, and thence by another terminal to the plate A^2, which is in metallic contact with the car-

bon C^1. It then traverses the carbons and complete its circuit through the lower holder D^1 and the plates A^1 A. In traversing the double-coiled electro-magnet G, it attracts the armature G^1, and thereby

FIG. 1.

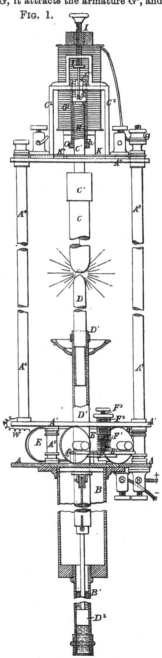

separates the upper carbon from the lower and forms the arc. In order to give a regular feed to the upper carbon, a coiled spring H surrounds the upper part of the carbon-holder C^1 and is retained between two

collars h h^1, which also surround the carbon-holder. The lower collar h is suspended from the guide frame C^2 by means of pendent rods which pass freely through the head of the guide frame, and are each provided with a head to take a bearing on the guide frames. The upper collar h^1 is attached to a pair of guide rods, which, after passing up through the guide frame C^2, are connected together by a crosshead, upon which bears a screw I, acting upon a spring, which determines the position at which the upper carbon will come to rest. On the strength of the main current decreasing through a lengthening of the arc, the spring comes into action, and forces the armature and carbon-holder downwards, until its further expansion is checked by the pendent collar h. After this, the armature falls by its own weight, and the upper carbon drops into contact with the lower one, thereby closing the main circuit. The electro-magnet then comes into full play again and re-establishes the arc as before. The lower carbon is advanced to maintain the arc by the pull of two counterweights, the cords from which pass over two pulleys and are attached to D^1. This advance is controlled by a clutch device. The carbon-holder D^1 passes through a hole in the plates A^1 A, and between the poles of an electro-magnet E, which is in a derived circuit as shown by the wires W W. Attached to the armature E^1 of this electro-magnet is a gripping piece F, which is carried by pivot pins f, and has an opening through it to allow of the free passage upwards of the carbon-holder when the armature E^1 is attracted by its magnet, and the gripping piece raised. Should, however, the strength of the current in the electro-magnet become less, owing to a too rapid rise of the carbon, the reacting force of the spring F^1 presses down the armature, and the gripping piece arrests the upward motion of the carbon-holder, until the arc attains its proper length, and the derived current in the electro-magnet E its normal strength. The lower carbon is then unlocked by the electro-magnet, and the carbon advances as before. The action of the spring F^1 is regulated so as to limit the upward play of the armature E^1, and to prevent its being drawn to the poles of the magnet. To serve this double purpose a hollow regulating screw F^2 is provided. On passing through the plate A^1 this screw bears on the coiled spring F^1, which is made fast to the armature E^1. The hollow screw is tapped to receive an elongated screw F^3, which projects down through the coiled spring so as to form a stop to the rising armature. By the adjustment of the screw F^2 the force of the spring is regulated so as to counteract the energy of the electro-magnet to any desired extent, and the upward range of the armature is properly limited. The piston D^2 moving in a well B^1 of

viscous liquid, such as glycerine and water, regulates the speed of advance of the lower carbon. The device employed to cut a lamp out of circuit is as follows : Insulated from the top plate A^2, but resting on it, is a strip of metal K, which is in metallic connection with the stem A^3, and underlies one end of the armature G^1. A similar strip of insulated metal K^1 is connected to the other stem A^4. The armature G^1, at the parts which overlie the strips K K^1, is fitted with platinum points. Whilst the lamp is in action, the armature, being held up by the attraction of its electro-magnet, is clear of these insulated metal pieces, but as soon as an undue resistance is offered to the main current, the power of the electro-magnet is overcome by the spring H, and the armature drops down and short-circuits the currents by the strips K K^1. In a lamp having two pairs of carbons one of the lower carbon-holders is locked out of action. The armature of the electro-magnet in the main circuit is made of a cruciform shape and carries a pair of carbon-holders. The armature of the electro-magnet

Fig. 2.

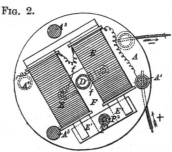

in the derived circuit is formed in two parts, to each of which is attached a gripping piece. The lower carbon-holders are mounted independently, and between them is a gripping lever free to rock vertically on a pin carried by a bracket bolted to the underside of the lower plate. One end of this lever grips one of the carbon-holders, and the other end is made elastic and projects into the path of an adjustable stud on the other carbon-holder. This lever is held in operation by means of a rod passing through guide holes in the two lower plates, but releases the carbon-holder when the adjustable stud comes in contact with its elastic end. In another arrangement, a light spring attached to the lower plate holds down the gripping-piece by pressing against the edge of the armature attached to it, it being released by the attraction on the armature when the first pair of carbons is burnt out.

169.—H. S. Raison, London. **Electro-Motors and Dynamo-Electric Machines, &c.** 8d. (2 figs.) January 12.

MOTORS.—The field magnets and armature are caused to rotate in opposite directions, the one in-

side the other, and each have a globular or spherical form. The armature is wound on a cast-iron core of oval form a, having pole-pieces a^1. The outer coil is built up of segments n^1 n^1, which may be wound separately before they are fixed in position. The coils of wire surround the cores over its entire surface and the axis of the coils is perpendicular to the axis of rotation. A number of openings are left in the cores, both of the field magnets and of the armature, which allow of the circulation of air and lessen

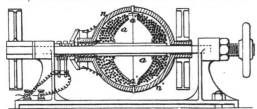

the heating, or the core of the armature may be made of some non-conductor. The commutator consists of two complete brass rings, conveying the current to and from the armature coils, and two half rings conveying the current from the armature to the field magnets. As shown, the current in the field magnets is reversed every time their poles come opposite those of the armature. The poles a^1 may be placed at any other angle to the axis of rotation.

DYNAMO-ELECTRIC GENERATOR. — The inventor prefers "to make the portion framing the outer hemisphere of a number of coils arranged around its circumference with provision for fixing if required; the commutator would form a number of divisions, and the brushes or collectors would be made to embrace nearly the whole. When the said outer part is fixed, the currents may be taken off without the intervention of a commutator, in which case they are alternating."

*185.—H. J. Haddan, London. (*A. Morel, Roubaix, France.*) **Electric Accumulators.** 2d. January 13.

SECONDARY BATTERIES.—Two metallic blades or good conductors are inserted in a series of receptacles containing water or other suitable liquid. If an electric current be passed through these elements, the receptacles will be filled with gas, and when this current is broken, the gases will recombine and give out an electric current. The gases which feed or generate the electricity may be produced in any other known way, and may be employed with increased or diminished pressure.

224.—W. R. Lake, London. (*T. S. Williams, Riverton, N.J., U.S.A.*) **Electric Lighting Apparatus.** 6d. January 16.

INCANDESCENCE LAMPS. — For the filaments of incandescence lamps, metal, or a metallic alloy, is deposited upon a base of zirconia, gypsum, lime, fireclay, asbestos, or other insulating and comparatively indestructible material, or upon graphite, carbon, or the like, such base being subsequently removed if desirable. The base may consist of a non-conducting support coated partially or entirely with a metallic or carbonaceous solution; or it may consist of wire gauze, on which is deposited a refractory metal or carbon by immersing it in a bath containing in solution platinum, iridium, ruthenium, manganese, or a combination of these, which are electrolytically deposed upon it. If the support is to be afterwards removed, the metal is deposited on one side only, the base, which may consist of sheet tin or metallic foil, stamped out to the required shape, being protected on the other side by a suitable material. The anode in the bath may be a plate of the deposited metal, or iron. The filament is united to the conductors by fusion, electro-deposition, or by means of a carbon deposit. The filament, according to another method, is formed of highly refractory metal, or alloy, compressed between rolls, or by stamps or dies, which give it the desired form, the metal being heated by the passage of a current prior to or during the process, and the terminals of the filaments being formed of increased section. The filaments are mounted within an exhausted or other globe in the usual manner. In the provisional specification it is stated that the electrodes may be coated with glass, or other like material, by compression between rolls, or by drawing through stamps or dies, and bent to the required form whilst hot; the terminals project beyond the glass, which is united to the filament by the application of heat.

CANDLES. — The provisional specification also states that a metallic cement is employed as an insulating base in electric candles, the candles being placed in a transparent chamber having an outlet at the top, the inlet of air being prevented by a valve.

ARC LAMPS.—These are described in the provisional specification only. The electrodes are mounted one above the other, the bottom electrode being provided with an extension, into which the upper one fits. The upper electrode will fuse, and feed upon, and build up the lower electrodes, which may afterwards serve as an upper electrode. The electrodes may be placed in a gauze, or grating, which will become incandescent. A separator, provided with a central aperture which becomes highly incandescent, may be provided between the electrodes, and pulverised or granulated electrodes may be fed to the arc by a hopper.

232.—H. R. Meyer, Liverpool. **Permanent Way for Electric and Telephonic Coductors.** 6d. (10 figs.) January 17.

CONDUCTORS.—The naked ribbons, wires, or rods, are embedded in superposed grooved blocks, slabs, or channels A of earthenware, rendered impermeable by immersion in hot asphalte, pitch, tar, or varnish. The blocks are tongued and grooved, or otherwise connected end to end, and covered with a lid; the naked conductors are laid in the grooves and run in with pitch or cement to insulate them. The upper surface of each block is formed with side ridges on which are grooves; tongues on the inside of the ridges take into the grooves. The blocks A are arranged to break joint with each other, and the covers to lap-joint with each other, and break joint

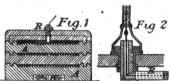

with the upper block. Connections are made to the outside by means of an arm R, which passes through the lid or side, and can be turned to point to any particular conductor; it is formed of vulcanite, and a tight joint is made round it by means of a lead bush. The joint in the conductors is soldered, carefully covered with india-rubber, and protected by a funnel slipped over it, and resting on the wooden washer. The whole is mounted on a line of wooden sleepers K, or on a cement bed. In another arrangement, the blocks A are placed one above the other in a box.

234.—W. R. Lake, London. (*C. A. Hussey and A. S. Dodd, New York, U.S.A.*) **Dynamo-Electric Machines.** 6d. (3 figs.) January 17.
DYNAMO-ELECTRIC GENERATORS AND MOTORS.—The core A of the field magnet is formed in one piece and wound with coils *b*. The core is deeply indented on the side adjacent to the armature, the coils being wound into the indentations, and the intermediate parts a^1 a^2 forming the poles. The coils are connected together in series (as shown), the terminals from the first and last coils being brought to the binding posts N N, and, from the way in which the coils are coupled, consequent or alternate, poles are formed between them. The armature is similar to the field magnet, but is built up of thin plates, and connected by insulated wires *d* extending from the first and last coils to two insulated metallic rings D D^1, which rotate in contact with the brushes E E^1. The wires *a* extend beyond the rings D D^1, one to each of the two metallic plates of a commutator H, to which the brushes E E^1 can be transferred. The two plates each consist of a metallic band having a number of strips equal to half the number of coils in the armature; the strips of one extend between those of the other, and the

brushes bear upon these strips. The machine may be adapted to produce either alternating or direct currents, at will, as follows: From the brushes E E wires *e* e^1 extend to switches K K^1, which control communication between these wires *e* e^1 and the wires *f* f^1, leading to binding posts L L^1. From the binding post B B wires *g* g^1 extend to switches M M^1; the switch M controls communication between the wire *g* and a wire *h*, which leads to the brush E, when it is adjusted to bear on H. The other switch M^1 controls communication between

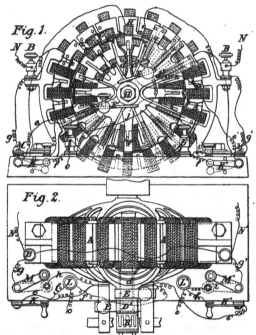

the wire g^1 and the wire *i*, which leads to the binding post L^1. A wire *j* leads from the brush E^1 to the binding post L. Wires N lead to the binding posts B B and wires *o* to the posts L L^1. The wires *o o* may be regarded as an outside circuit in contradistinction to the machine circuit. When switches M M^1 are closed, and K K^1 open, the coils of the armature are in circuit with the field magnet. When the brushes are shifted to the rings D D^1, the switches M M^1 are opened and K K^1 are closed, and the field magnets are excited by a current through N N, and alternating currents will be produced.

245.—W. R. Lake, London. (*A. de Khotinsky, Paris.*) **Apparatus for Regulating Electric Currents.** 6d. (6 figs.) January 17.
CURRENT REGULATORS.—The object is to provide for the automatic regulation of the current (1) by inserting as a shunt to the field magnets of the exciter and generators (or of the generator alone if self-exciting) a conductor of variable resistance;

or (2) by placing a second variable resistance in the circuit of the current which excites the field magnets. Referring to the illustration, A is the exciter, and B B the generators. The exciting current leaving the brush a divides into two portions; the one passing through the field magnets of the exciter and also of the generators, enters the part f of the rheostat C at c, and, traversing the resisting wire and conductor e, returns to the exciter through the brush b, and the second portion passes through the other part of the rheostat, through the resisting conductor coiled on the insulating cylinder g, and returns by the conductor e and brush b to the exciter. The terminals of the resisting conductor are connected respectively to the axes of the insulating cylinders f and g, and it slides through a contact plate e, situated between the cylinders, and connected to the brush b, and serves to increase the resistance in the exciting circuit and diminish that of the shunt circuit, or *vice versâ*. The cylinders f and g are rotated in opposite directions by the pinion h, which is con-

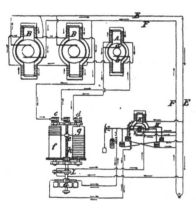

nected to the electro-motor G. The current from the generators B passes into the main heads E F, and a small motor K is arranged in parallel arc to these conductors. The current passing through the motor tends to turn its armature against the resistance of a spring or weight. To the armature shaft are attached two insulated levers H, whose ends respectively make or break contact with four cups connected as shown. These levers are connected to the brushes of the motor G. If the current be too strong, the contacts are made at the two cups shown on the right of the motor K, and the current passing to the motor G in a certain direction, it rotates it and increases the resistance on the cylinder f, and decreases the resistance on the cylinder g; if the current be too weak, the weight predominates, and the contacts are made with two cups, shown on the left of the motor G, and the current passing to the motor G in the opposite direction causes it to rotate in the opposite direction and decrease the resistance on the cylinder f,

and increase the resistance on the cylinder g. On the revolution of the shaft of the motor K, a circuit is completed to the electro-magnet L, which is made in the form of a horseshoe, having its two arms hinged together, and carrying on their inner sides blocks bearing on the shaft of the motor G, thus serving as a brake; or the blocks may be fixed, and the other attached to the armature of a horseshoe electro-magnet. In other arrangements, the rheostat is dispensed with, the shaft of the motor k carrying a lever to which is attached, at either end, a bar dipping more or less into acidulated water, and thus acting in a similar manner to the rheostat. A glycerine pump, with a small aperture in its piston, may be employed to regulate the movements of the lever.

252.—H. H. Lake, London. (*La Société Universelle d'Electricité Tommasi, Paris.*) **Electrical Accumulators.** 6d. (5 figs.) January 18.

SECONDARY BATTERIES.—Each element is formed of two continuous sheets of lead $a\,b$, folded several times, and placed in a vessel in such a manner that a double fold of the one is always contained within a double fold of the other. In order to augment the surface, the sheets of lead are grooved in the

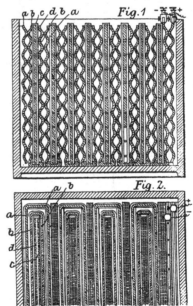

direction of their length, and there is placed between the two adjacent sides of a fold of the same plate a piece of cloth or millboard, or another plate of lead c, enveloped in lead wire d, and secured thereto by an autogenous welding at the lower part. All the plates are perforated to allow of circulation of the liquid element.

SWITCH.—The invention also comprises a switch, which permits of connecting, as desired, the accumulator only, or the accumulator and the apparatus with the source of electricity, in such a manner that, in the first case, the accumulator is charged, and in the second, the source of electricity and the accumulator, connected in series, both act upon the apparatus.

289.—J. Humphrys, London. Secondary Batteries for Storage of Electricity.　4d. (19 figs.)　January 20.

SECONDARY BATTERIES.—The invention consists in making the plates of an open lattice form, that is to say, the sides of the framework of each plate are connected with bars either of a diamond or oval section, these bars being arranged at an angle on the principle of a Venetian blind, serve to retain

Fig. 1.

Fig. 2.

the spongy lead or active composition in place, thereby exposing a maximum area of the spongy lead to the action of the liquids. In Fig. 1 plates of oval section are shown, and in Fig. 2 diamond-shaped plates. In other forms, the bars are wedge-shaped, or tapered, or they may be arranged in a spiral frame, or one above the other, or may extend diagonally across the frame.

305.—J. N. Aronson, London. Electric Lamps. 6d. (8 figs.)　January 21.

INCANDESCENCE LAMPS.—The globes of incandescence lamps are constructed with reflectors formed in or on the material of the globe itself, or with prisms or lenses which form part of the globe. The reflective surface may be produced by silvering, painting, gilding, or enamelling on the inside or out-

side of the globe. The specification illustrates several forms of reflective globes suitable for throwing the light downwards, as for domestic illumination, or to one side for use in theatres. The figure shows a globe formed with circular prisms, the rays being refracted in a downward direction.

319.—J. S. Sellon, London. Construction of Secondary Batteries.　2d.　January 21.

SECONDARY BATTERIES.—The supports, retainers, or frames, in, or on which, the active material is deposited, are constructed partially or entirely of pieces of metal, carbon, or other suitable material, which either is conductive in itself or can be rendered so, or on to which strips of lead or other metal can be fixed. The terminal plates may be constructed as above, or as described in Patents Nos. 3,926, 3,987 or 5,631, all of 1881.

333.—T. J. Handford, London.　(*P. B. Delany, New York, U.S.A.*)　Electric Cables.　6d. (4 figs.)　January 23.

CONDUCTORS.—It is the object of this invention to interpose a continuous conductor into the meshes between the wires of a plaited or braided cable, and thus eliminate from this class of cable the last vestige of induction. The insulated wires are woven loosely into a flat braid *a a*, so that they

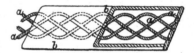

will cross and recross, leaving open meshes between them. This plait is introduced in a flat pipe *b*, of lead or other ductile metal, and the tube is then passed between rollers to force the inner surfaces of the tube together through the meshes between the wires.

339.—E. de Pass, London.　(*B. Abdank, Paris.*)　Regulating Electric Lamps.　6d. (4 figs.)　January 23.

ARC LAMPS.—Electric lamps are regulated by means of apparatus termed "resistance balances," which may be placed at any convenient distance from the lamps. The figure represents one arrangement. The current from the dynamo M flows into the low resistance coil A, and thence to the lower carbon of the lamp L, passing through the lamp and returning to the other terminal of the machine by the route shown. The high resistance coil B is in a shunt circuit between Q and S, this circuit also including a number of resistance coils R, a part only of which is shown. The coils A B are adjustable in a graduated slide. An indicator Z, travelling over a graduated scale, inserts more or less of the resistance R into the derived circuit according to its position upon the insulated rod O O[1]. The resistance introduced into the main circuit can be measured by means of either of the indicators Z or the movable coils. By increasing the resistance R the core C C is attracted towards F, while by diminishing the resistance the core moves in the

direction of E. In practice, the index Z is set to show the number of ohms on its scale corresponding to the resistance of the lamp. Should the resistance of the arc increase, the core is moved towards E, and striking the spring i, breaks contact at i^1. The lamp-regulating bobbin is by this means introduced into the derived circuit, and causes the

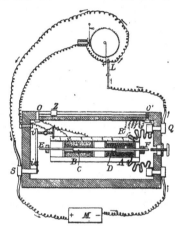

release of the upper carbon. By giving a very slow movement to the carbons, an almost continuous regulation can be obtained. In a modification, the core, acted upon by the two solenoids, is fixed to a rod movable on an axis and kept in its normal position by two springs. Its free end moves over a graduated scale, and, at certain limits, a contact is made and the carbon released.

346.—R. E. B. Crompton, London. Electric Lamps, &c. 8d. (4 figs.) January 24.

ELECTRIC ARC LAMPS.—Referring to Figs. 1 and 2, $a\,a$ is a rack rod acting as the upper carbon-holder and gearing with the first wheel c^1 of a train of three wheels, of which the last is a smooth-surfaced brake wheel, fitted with tangential spring flies $d\,d$, which move outwards against the brake e when the speed is excessive. This gearing is carried in a frame which can be moved upwards and downwards within the main frame for a short distance. The brake lever e is fitted between the cheeks of the gearing frame at F, and carries an armature e^1, which is attracted by the feed magnet C against the regulating spring f. This magnet C is in a shunt circuit, while the main magnet G, which establishes the arc, is in the lamp or arc circuit. When the circuit is completed, the magnet G attracts and raises the gearing frame, and with it the rack rod and carbon, and so establishes the arc, and it also brings the brake lever within the attractive influence of the magnet C. The tension of the regulating spring is so adjusted that while the arc is of the normal resistance the brake

wheel is kept stationary. As this resistance increases, the spring is overpowered by the magnet, and the wheel train revolves until the condition of equilibrium is again attained. Fig. 3 shows a modification, in which the main magnet G, which establishes the arc, is on an alternative or cut-out circuit, the general arrangement of the lamp being otherwise the same; N N is a separate magnet called a cut-out magnet, having an armature, which, when attracted, breaks the direct circuit through the magnet G. The latter, when in action, draws down the gearing frame so as to bring the carbons into contact, and when its circuit is broken, the frame is forced upwards, by a spring, a distance

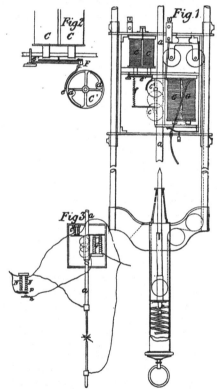

equal to the length of the arc. When the circuit is completed, to start the lamp, the armature p of the magnet N is in contact with the piece n, and the current can flow through the magnet G direct. At this time it cannot pass through N, because the arc is still open; immediately the magnet G brings the carbons together, so as to complete the circuit through N, the armature p cuts the magnet G out of circuit, and the spring O lifts the frame and establishes the arc. The lower carbon is lifted by a cord and guide pulleys, as shown, when the lamp is intended to be focussing, or if alternating currents are employed, the cord is attached directly to the lower carbon. The specification describes

the alterations in the framing that must be made when the lamp is intended to stand on a table, as in lighthouse illumination, the mechanism in this case being placed below the light. Another part of the invention relates to a device for preventing electric lighting circuits acquiring a static charge of high potential. " It is probable that this static charge is due to the fact that where there is an electric discharge through gaseous media, there is an absence of symmetry in the behaviour of the current, and the amounts of free electricity escaping into the atmosphere are not the same for positive and negative electricity, so that there is a continual tendency to accumulate a charge of that kind of electricity which escapes less readily than the other." The remedy for this consists in connecting the circuit by a wire of high resistance (say 10,000 ohms) to earth.

359.—J. N. Aronson, London. Electric Lamps. 8d. (24 figs.) January 24.

INCANDESCENCE LAMPS.—This relates to means for bringing a second incandescing conductor into circuit upon the failure of the first. This may be effected either by hand or automatically. In the former case, each filament in a globe is provided with an independent terminal, and the lamp socket has contact pieces connected to the positive and negative conductors. Each terminal may be enclosed in a glass tube fused around it, the tubes being then fused into a mass and fixed in the globe, or the tubes may pass through the closed end of a hollow cylinder, or the conductors may be passed through a block of glass. By inserting the globe neck into the socket in different angular positions, or to different depths, any particular conductor may be placed in circuit, or in some cases several, or all of the filaments, may be included in the path of the current. In another arrangement, an end of each filament is connected to a common terminal, and the other end to a separate insulated contact piece on the outside of the lamp, situated below a spring in connection with one conductor. A ring, provided with slots, forces one spring into connection with the contact plate below, the other entering the slots in the ring. When automatic apparatus is used to effect the change, several modifications may be adopted. If the lamp have but two filaments, an electro-magnet wound with two coils in opposite directions, one of small resistance in the lamp circuit and the other of high resistance as a shunt to the filament, may be employed to attract an armature when one filament gives way, and in so doing to complete the circuit to the second filament, or to cut the low resistance coil out of circuit. One coil may be replaced by a spring, or by a polarised armature, or by both. When there are more than two filaments, the magnet may be wound differentially,

and work a circular contact breaker, after the manner of a make-and-break device. If the magnet be not powerful enough for this, a coiled spring may be added, and be controlled by the magnet acting on an escapement. When several lamps are arranged in series, a differential system is adopted. There is combined with one terminal of each filament an armature, provided with a spring capable of holding it in two positions, like the blade of a pocket knife, either in contact with, or a short distance from, the magnet. The first filament is connected directly to the conductors, and a shunt circuit is made round it, which includes the second filament and a magnet coil of high resistance. When the first filament fails, an increased current passes through the shunt, and causes the magnet to attract its armature, which comes up against a contact piece, and, cutting the coil out of circuit, affords a direct path from the current to the filament. The third, fourth, and fifth filaments are similarly respectively connected to the second, third, and fourth filaments. Another method is to connect the positive terminal of the first filament to the leading wire, and to put it in communi-

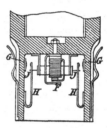

cation with a spring forming part of a shunt circuit through the second filament. Between the spring and the terminal of the second filament there is interposed a high resistance, of such a nature that, when the first filament fails, and the greater part of the current traverses the said high resistance, it will melt, and allow the spring to move and make good metallic contact with the terminal. The spring may be made of two metals of different expansibility. The third terminal is connected to the second, and the fourth to the third, in a similar way. The illustration shows the socket of a lamp arranged according to this plan. The positive end of each filament is connected to a spring H, which has a tendency to press against the contact G, but is, unless the filament has come into action, held back by a tie of fusible metal, connected at the other end to the conductor J. When a filament fails, the armature of the electro-magnet comes away and makes contact at F, by which means the current is sent through one of the ties J, and melting it, a new contact is established at G, putting a fresh filament into circuit, and attracting the armature afresh.

361.—W. R. Lake, London. (*H. A. Clark, Boston, U.S.A.*) **Electrical Conductors or Cables.** 6d. (11 figs.) January 24.

CONDUCTORS.—One object of this invention is to prevent inductive action between the different conductors of a series lying adjacent to each other. The insulated conducting wires are laid side by side upon a sheet of conducting material; or, if preferred, the conducting material, with the series of insulated conductors, may be rolled up into a tubular form, when the external conducting material will also serve to protect the insulated conductors from injury. Several concentric layers may be employed to form a compact cable, in which case the tubular conducting material on both sides of the wires will

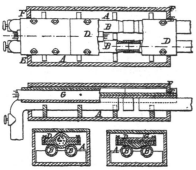

co-operate to divert the inductive effect of the currents flowing through them. To still further neutralise inductive action, the wires are crossed at intervals. Another object is to maintain the central position of the conductor, in insulated wires, during the process of vulcanising. To effect this, a vulcanising chamber A is employed, having steam pipes B B running through it. D D are moulds, each in two parts, one shaped to receive a round cable, and the other a cable in the form of a flat belt or band, G G. The moulds run in the direction of the length of the oven, and pass through an opening E in each end F, and project as shown. The vulcanising proceeds in the usual manner, successive portions of the cable being brought under the action of the oven.

377.—Sir C. T. Bright, London. **Electric Lamps.** 6d. (6 figs.) January 25.

ARC LAMPS.—Two electro-magnets are employed for separating and regulating the carbons. One end of the first strikes the arc, and its other end actuates an armature, by which the second electro-magnet is short-circuited, or its effect otherwise neutralised for a short interval of time. This second electro-magnet regulates the feed by means of clips, which hold, or release, the upper carbon-holder when necessary. The first electro-magnet is placed by preference in a shunt circuit. The detail of this lamp will be understood from the illustrations. In Fig. 1

a a are the poles of an electro-magnet in the main circuit, wound with large wire; *b* is a ring-shaped armature fixed to the slide rods *c c*, which pass through the frame of the upper part of the lamp, and carry at their lower ends the plates *d d*, which, in turn, support the pivotted axles *e e* of the two levers *f*. These levers carry at their other ends two armatures *g*; *h h* are electro-magnets of high resistance, comprised in a circuit shunting the arc. When the current passes through both circuits, *b* is attracted and the sliding frame is raised; at the same time *g*, being attracted by the poles of *h h*, one of which is shown at *i*, grips the carbon-holder *j j* by the jaws of *f*, and rises with it; *k* is an armature fixed to the contact lever *l* (see Fig. 2); *m m* are two poles of the electro-magnets *h h*, which attract the armature *k*, and thereby make contact between *l* and *n*, whenever the resistance of the arc circuit increases, and more current in consequence passes through *h h*; *l* and *n* are respectively connected to the ends of the coils *h h*, which, thus being cut out of circuit, cease to attract *g*, and the carbon-holder is then free to fall. The lever *l*, however,

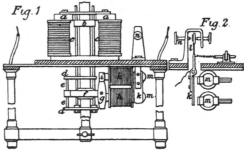

Fig. 1   Fig. 2.

immediately returns to its place of rest, and the jaws again grip the carbon-holder, this intermittent action being repeated until the necessary feed is effected; adjustable soft iron pole-pieces enable the action of the lever *l* to be regulated to correspond with the best conditions of resistance for the arc. The upper carbon-holder is hollow, and contains glycerine, or other suitable fluid, and a piston attached to a suitable guide works within it. Where extreme regularity is needed, the fluid preferred is mercury, which is also made to effect the electrical connection to the carbon electrode, at the same time constituting an adjustable resistance, which compensates for the diminishing resistance of the carbons as they consume. A braided copper wire is used to connect the carbon-holder when a non-conducting fluid is used. A rack, gearing into clockwork mechanism, may be substituted for the fluid. The relay, by which the carbons are separated, may be placed at one end of the coil which regulates the feed. Solenoids may replace the electro-magnets. The separation of the carbons, in this case, is effected by the lower carbon being drawn down

through the lower solenoid, the lower end of which actuates a movable armature similar to k; or both solenoids are placed in the upper part of the lamp, the upper carbon-holder being surrounded by an iron sleeve, passing into the first solenoid, and connected by non-magnetic metal to a soft iron rod entering the second solenoid; or one electro-magnet or solenoid only may be used.

386.—W. T. Henley, Plaistow. **Cores for Telegraph Cables, &c.** 6d. (1 fig.) January 26.

CONDUCTORS.—These improvements refer to the construction of a core with multiple conductors, which may be used at will for a number of circuits, or for a few, or as one only. As shown in the drawing, the conductors are first insulated separately, and are then all enclosed in another coat of insulating material, and sheathed with yarn and wires, or

surrounded with strong Manilla or Russian hemp, or other fibrous material, laid round the core in long spirals, and then sheathed in galvanised steel wire, and finally "covered with tape or yarn, and compound in the ordinary way." All the separate cores may be enclosed in one exterior coat of insulating material.

392.—W. P. Thompson, London (*Union Electric Manufacturing Company, New York, U.S.A.*) **Obtaining Light by Electricity, &c.** 6d. (2 figs.) January 26.

ARC LAMP.—The upper carbon C^2 is carried by a rack, which gears with a pinion D^1 upon a shaft to which is also fixed the hollow drum D^2. Upon

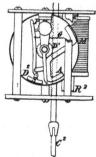

the same shaft, an arm is loosely pivotted, carrying the bent armature A at one side, and the clamp shoe S at the other. M is a differential magnet,

with pole-pieces R^1 R^2. When the circuit is completed, the magnet attracts the armature A, and causes it to rotate, and at the same time the shoe S is carried round and outwards, until it jams itself against the inside of the drum D^2, causing it to rotate, and raise the upper carbon by the rack and pinion, thus establishing the arc. As the resistance of the arc increases, the armature moves back gradually, until it attains a point where the shoe loses its hold, and then the drum rotates and the rack runs down until the arc is readjusted. The lamp is adjusted by means of the stop P projecting radially from an axle, and against which the lever is pulled by the spring G, the stop P determining the position at which the shoe S will act.

441.—C. F. Varley, London, and **W. Judd**, Penang, Straits Settlements. **Electric Railways.** 4d. January 28.

CONDUCTORS.—The conductor is laid in a groove in sleepers impregnated with paraffine, wax, ozokerit, bitumen, or other suitable substance, in the ordinary manner. The sleepers are partially covered by guard plates. The rails may be used as the conductor, in which case they, as well as the fishplates and bolts, are galvanised. The groove in the sleeper is closed by a waterproof insulating band. "Through this band there are conducting bolts which connect the conductor with the contact roller or wheel, when the wheel traverses over and depresses the band. The space between the insulating band and the sleeper is sometimes filled with petroleum, or other insulating material." The metals, rollers, or wheels, are insulated at the sides, and are mounted on a jointed axle. "A reversing commutator in the car enables the driver to go either forward, or if in motion to stop, or go backwards."

454.—G. and E. Ashworth, Manchester. **Metallic Brushes, &c.** 4d. January 30.

GENERATOR.—The pins or wires of a hair brush are connected to an electric generator.

*493.—C. J. Allport, London. **Preparation of Asbestos as an Insulating Material.** 2d. February 1.

INSULATING MATERIAL.—The asbestos is incorporated with black wax obtained from ozokerit, solid paraffine, or other suitable hydro-carbon, and is pressed into heated moulds, and then, if necessary, passed through hot rollers. Sulphur, or other suitable hardening material, may be introduced.

497.—G. Little, Passaic, N.J., U.S.A. **Electro-Magnets and Armatures, &c.** 6d. (8 figs.) February 1.

ELECTRO-MAGNETIC MOTOR.—The object of this invention is to utilise magnetic force, without destroying contact between the magnetic pole and armature, and this is effected by giving the latter a rolling motion " whereby the metallic surfaces of the poles are brought into actual contact for obtaining the maximum magnetic force, and the magnetism is neutralised, or partially or entirely reversed, as the rolling armature passes the place of greatest magnetic action, so as to prevent the sticking of the armature to the pole." The figures are respectively an elevation and section of "a revolving magnet with rotating armature rollers." "The magnet is composed of two heads upon a metal tube, with a helix wound in between the heads," each head being divided up into " pole-faces " of a single or double **V** shape. These are shown at A A. B is a cylindrical armature, capable of rotation on an axis running in bearings. The magnet poles are placed so as to operate progressively, the pole-pieces of the magnets 1, 2, and 3 coming into action in succession. A commutator f, with contact rollers $e\ g$, is employed to distribute the currents, and to reverse the direction of rotation ; k is a wheel which may be driven by this motor, thus adapting it for

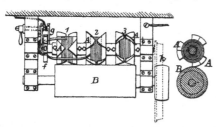

use as a locomotive. When operated as a stationary engine, this motor may, with advantage, be worked " in a shunt circuit." Another form of motor is illustrated and described, in which three cylindrical armatures, attached to a disc, roll upon the poles of a fixed central set of electro-magnets, and thereby effect a rotation of the disc. Rings of non-magnetisable metal are introduced into peripheral grooves in the pole faces supporting the rolling armatures, to guide the same. A tubular shaft may be provided with passages for air to be forced through to keep the parts cool. The helices of the electro-magnets are wound in the usual manner, except that a connection is made from about the middle of each helix to the negative binding post, and the extreme ends are respectively connected to two spring arms. The positive terminal is connected to a spring, tending to bear against the spring arm connected to the exterior end of the coils, but acted upon by a wheel, which has as many projections as there are pairs of pole faces, so as to make contact with the other spring arm just before the armatures reach the pole faces, and with the

post, for a moment only, when the armatures reach the middle of the pole faces.

513.—C. V. Boys, Wing, Rutland. **Electric Meters.** 6d. (3 figs.) February 2.

CURRENT METER.—Figs. 1 and 2 are respectively the side and back views of this instrument. A balance wheel A is suspended by a long flat spring B, which passes through a slit in the lever arm C, which is pivotted at c, counterbalanced by a weight C^1, and drawn upwards by an adjustable spring C^2. On this arm, near its pivot, is fixed a cylindrical armature D, near the poles of an electro-magnet E, whose coils are in the main circuit. D is attracted proportionally to the current passing through E, the arm C being thereby more or less depressed, and shortening to a greater or less extent the torsive length of the spring B below the slit in C, and thus more or less accelerating the oscillations of the balance A. The number of oscillations of A, in a given time, is recorded by a counter operated from its axis, and will measure the quantity of electricity that has passed through E in a given time. These oscillations may be maintained by suitable means, as

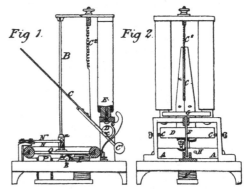

Fig 1.   Fig 2.

described in Specification No. 4,472 of 1881. In the instrument illustrated, this is effected by the electro-magnet P acting on the armature Q. The coils of this electro-magnet are occasionally connected to the main circuit by the action of a grooved stud on the hanging finger n, causing contact of the spring N N[1] ; this connection is effected only when the excursion of the stud k is so much reduced that the finger n is caught and held in its groove. To insure the starting of the balance, it is always stopped in such a position that there is torsion on the spring B. This is effected by the detent lever F having an armature G, which is released when no current passes, thus allowing the other end of the lever to engage the ratchet stud H, and thereby to hold the balance. The number of oscillations is recorded by a counter, worked by a spring pawl, attached to the bar Q, acting on a ratchet wheel R geared to the counter. The integrating apparatus described in

Patent No. 2,449 of 1881 might be applied to this electric meter, by making the axis of the lever C to serve as axis on which swivels a disc pressing against a reciprocating cylinder, as described in the above-mentioned patent.

538.—W. R. Lake, London (*J. J. Barrier and F. T. de la Vernède, Paris.*) **Electrical Accumulators.** 6d. (14 figs.) February 3.

SECONDARY BATTERIES.—Lead ribbon, shaped as shown in Figs. 1, 2 and 3, is wound upon itself in various forms, Fig. 4 being one in which a circular arrangement is chosen. The surfaces of these ribbons are separated by a cement, consisting of one part platinised charcoal, one part litharge, and one part glycerine. The objects of employing the first ingredient are (1) to prevent the lead bands touching each other ; and (2) to absorb hydrogen. Each element is made to consist of four discs, arranged as follows : " Two discs are united with the cement to form a half element. Each element is placed in a compartment of ebonite, or other suitable substance, and is connected to the other for quantity or tension as required." The elements are coupled by means of a tail of lead

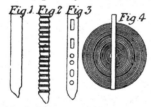

soldered on one of the faces of each. Each half element is separated from its neighbour by a layer of liquid, or by a porous plate, and in the latter case it is coated with a layer of platinised charcoal. The elements thus constructed are "prepared " by means of a bath of potash, or a bath saturated with a neutral acetate of lead. The troughs containing the elements are divided into perfectly tight cells by means of the elements employed. Corrugated lead sheets are used in another form of secondary battery, the corrugations being filled in with the cement already described, and various arrangements of the discs or elements are mentioned and illustrated. Platinised pumice or lead, or any oxide or salt of platinum, may be substituted for the platinised charcoal, and any suitable metallic filings may be added to the cement. A solution of acetate, or other salt of lead, may be used in lieu of sulphuric acid.

540.—J. D. F. Andrews, Glasgow. **Dynamo-Electric and Electro-Dynamic Machines.** 6d. (5 figs.) February 3.

DYNAMO-ELECTRIC GENERATOR.—A wire rope,

wound with coils E, of insulated wire, is twisted helically around a wooden cylinder A, and rotates between the field magnets N S. By suitably proportioning the width of the coils to the diameter of the barrel, each coil of one convolution of the rope is made to enter and leave each field, before the coil on the next convolution, each coil being connected to the commutator plates. "When the commutator is of the usual construction, consisting of plates insulated from each other, there is added to the pair of diametrically opposite brushes, usually employed, additional brushes, rubbing on other points of the circumference, so as to collect and distribute the electricity while the revolving coils are entering and leaving the magnetic fields. In like manner, when the machine has several sets of poles, there are employed two commutator brushes for each

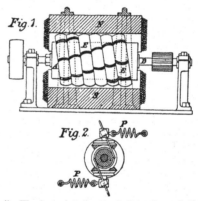

pole." The barrel A is attached to the axis B by a spring, through which the turning power is transmitted. When this power increases, the spring is deflected, and the coils are angularly displaced as regards the commutator plates, or the armature may be fixed, and the driving pulley rigidly connected to the commutator, or may be connected to the shaft through a spring. According to another method, illustrated in Fig. 2, a similar result is obtained by driving through a differential motion consisting of four mitre wheels. The brushes are connected to the frame of the two intermediate wheels, which is maintained in position against the strain of the driving power by springs P P. As the power increases, the springs are elongated, and the frame and brushes slightly rotated.

MOTOR.—These improvements are applicable to motors.

*542.—W. R. Lake, London. (*M. Levy, Paris.*) **Regulating the Transmission of Electrical Energy.** 4d. February 3.

DISTRIBUTING CURRENTS. — This provisional specification, which is divided into twelve sections, describes, with the aid of a great number of algebraic

formulæ, apparatus for regulating the transmission of electrical energy, and also capable of being used as a speed regulator for machinery. If the translating devices be arranged in a single circuit, it is necessary to regulate only the intensity of the current. This is effected by placing on the shaft of the admission valve of the motor which drives the generators, two small electric motors, one driven by a constant current (*e. g.* a Daniell battery) and tending to open the valve, the other being driven by the current to be regulated, and tending to close the valve. The first motor may be replaced by a spring or weight, or any constant force tending to open the valve, and exactly balancing the second motor, when the current is at the required intensity. If the translating devices be arranged in multiple arc, the governor is placed in one of the circuits, which may also be used to excite the field magnets of the generators. If it be required to keep the speed of the generators constant, they are placed in two circuits connecting two points, one on the positive and the other on the negative lead. Another set of generators is arranged with their field magnets in the first circuit, and their armatures in the second circuit. By regulating the speed of one of the generators, and keeping the current in the first circuit constant, or by removing the generator from that circuit, the difference of potential between the two points connected by the circuits will be constant.

560.—J. S. Williams, Riverton, New Jersey, U.S.A. **Boilers, Condensers, &c.** 4d. February 4.

ELECTRIC LIGHT.—The inventor describes several methods of coating boilers, condensers, &c., with a non-oxidisable metal or alloy, and in his provisional specification a method of coating a ship. A metal, or a metallic salt, is arranged as one pole of a battery, and the iron, or metal of the ship's side, as the other pole, the electrolyte being the salt water. The metal from the metallic salt will be deposited on the sides of the ship, and a current generated, which may be employed for lighting or heating, either directly or by the aid of secondary batteries. By passing a current in the reverse direction, the deposited metal is recovered. Floating batteries may be constructed in a similar manner.

563.—A. J. Jarman, London. **Arc Electric Lamps.** 6d. (7 figs.) February 6.

ARC LAMPS.—A solenoid A, Fig. 1 (or electromagnet), is mounted in such a manner as to be adjustable with regard to its core D, so that the desired amount of attraction may be produced with varying currents. The base of the core is provided with lugs E, carrying pins or rollers, upon which slide crossbars, pinned at their centres, and carrying

the gripping pieces H H¹, drawn together by a spring and accurately fitted for embracing the upper carbon-holder. The current passes from the solenoid A to blocks, which exert such pressure against the rod as shall prevent its too sudden descent when the current ceases, and insure good electrical contact. Upon a current flowing through the coil, the core is drawn upwards against the resistance of two springs fixed to the pillars B B¹, first causing the crossbars to straighten, and to clutch the carbon-holder, then to lift it and establish

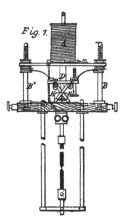

Fig. 1.

the arc, the length of which can be regulated by lowering the solenoid A on the pillars B B¹, and by a screw regulating the rise of the armature. By means of the crossbars, a parallel motion is imparted to the clutch pieces. When two or more lamps are arranged in series, the shunt coil or regulator shown in Fig. 2 is employed. A represents the solenoid of the lamp, to the left of which is a smaller solenoid, wound for high resistance, and forming a shunt to the lamp circuit. Upon the resistance of the arc increasing, the high resistance

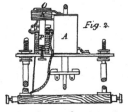

Fig. 2.

coil comes into action, and, attracting the core P, diverts the current through the coil K. This, in its turn, attracts the disc-shaped end O of the core P, and thus holds the bar F firmly against the contact piece J, thus short-circuiting the carbons. In a modification, a spring forces the lower carbon against the upper, and on the passage of the current, the crossarms force the clutch pieces against the rod, drawing it down and establishing the arc.

578.—B. J. B. Mills, London. (*W. M. Thomas, Cincinnati, Ohio, U.S.A.*) **Electric Lamps.** 6d. (2 figs.) February 7.

ARC LAMP.—The lower electrode is fixed and is composed of copper with an iridium tip J. The upper one is suspended by a cord passing over a pulley, and carrying a counterweight at the other end, by which the speed of fall can be regulated. D is a solenoid wound on a split brass bobbin, with one coil of wire, and furnished with a soft iron core K, to which the brass or copper rod L, carrying the upper carbon-holder, is attached. The wire of the solenoid has the covering removed from it along

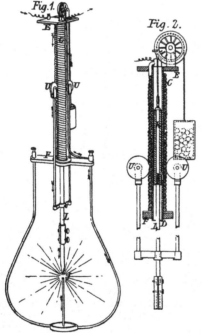

two lines traversed by the wheels U, which are connected to the spring arms of an adjustable yoke on the carbon-holder. The current enters at the top of the coil, circulates in its convolutions until it reaches the wheel U, where it leaves, and proceeds through the electrodes, and the pendent frame, back to the negative binding post, as indicated by the arrows. As the carbon consumes, the rollers U descend, and in so doing increase the length of the active part of the coil, so that the core is always maintained at a constant distance from the lower end of it.

621.—J. B. Rogers, London. **Effecting and Maintaining the Continuity of Divided and Sub-Divided Electric Currents for Lighting.** 6d. (1 fig.) February 8.

DISTRIBUTING CURRENTS.—Negative and positive currents are distributed from a main station, radially, to intermediate stations, each of which may be a generating and storage station, and from these to distant stations, putting all the intermediate and distant stations into connection with each other for maintaining continuity throughout the system.

626.—A. A. Common, Ealing. **Electric Lamps.** 6d. (11 figs.) February 9.

ARC LAMPS.—The upper carbon-holder is carried by a steel tube D, free to revolve in a sleeve H, to which is connected by a pin joint the toe-piece J, whose outer end rests on a plate lever C, working on the top of two rounded plates B, and thus having a variable fulcrum. The upper end of the tube D contains mercury, and serves as a cylinder for the piston Q, which, together with its rod, is a fixture; the rod passes through a cover screwed to the top of a chamber S (forming an enlarged extension of the tube D), and provided with a long collar, which prevents the escape of mercury around the piston, should the lamp be laid down or turned up. The piston Q is provided with several small holes, covered

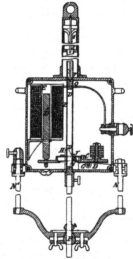

by a platinum valve V, which, being heavier than the mercury, always rests upon the holes, and allows the tube D to move freely upwards but controls its downward motion. A metal ring, held by screws, in any given position on the lever C, is provided with two extensions, bent down so as to stand one on either side of the piece J, which can thus be carried round to any desired point, so as to vary and determine the lift of the carbon. The plate lever C, actuated by the soft iron core D^2 of the solenoid E, is provided with adjustable weights to partly balance the core D^2, and is kept in position, should the lamp be turned over, by a pin G. The lower carbon is carried in the ball-holder P, so that it can be adjusted to be exactly in alignment with the upper carbon. The conductors are firmly fastened to lugs on the casing,

and connected to the terminals. The action is as follows : The carbons being together, the current passes through the coil E, the tube D, the carbons, and the rod N to the next lamp. The core is immediately raised, and pivots on its fulcrum the plate lever C, which carries with it the outer end of the toe-piece J, which by its other end and the sleeve H instantly grips the tube D. The further lift of the plate lever, whose motion gradually increases in proportion to that of the core, owing to the rounded form of its fulcrum, now carries with it the tube D, and thus establishes the arc. As the resistance increases, due to the consumption of the electrodes, the lever C tilts back and releases the grip of the toe-piece J, thus allowing the tube to drop gradually against the resistance of the mercury. In a modification adapted for street lighting, the upper carbon is fixed, and the lower carbon is movable, the piston Q being made larger, so as to act also as a float, and assist in raising the lower carbon, the framework also being slightly modified.

700.—J. S. Williams, Riverton, N.J., U.S.A. Generation, Storage, and Utilisation of Electricity for Lighting, &c. 10d. February 13.

DYNAMO-ELECTRIC GENERATOR.—Preferably, two armatures are employed, each being rotated in the "field" of a single pole only, of a field magnet, so as to generate currents of low E.M.F.

DISTRIBUTING CURRENTS.—The generators are so combined with secondary batteries, and current-making devices, that the generators are charging a portion of the batteries at low tension, whilst other portions are being discharged into the main conductors, at such a higher tension as may be required. The circuit-making devices may consist of a series of reciprocating arms on a shaft, operated by an electric motor, or otherwise, and carrying a series of insulated contact pieces, or weighted contact makers ; or the displacement of mercury, or other conducting liquid, operated by blades on the shaft, may be employed. An automatic cut-out is arranged in the circuit of the generators, and the secondary batteries, to avoid reversal of the current, and it may be caused to break the primary circuit and complete a shunt circuit. This may consist of a switch, operated by an electric governor, or a motor, solenoid, or electro-magnet, in the circuit. Two electro-magnets may be used, one in the charging, and the other in a shunt circuit. An independent secondary battery may be employed for operating these devices. By using secondary batteries which increase the E.M.F. of the current, a smaller generator and a smaller sectional area of the conductors is required. Auxiliary secondary batteries may be placed at different points of the

main conductors, and may be arranged to supply the current to the translating devices, at a reduced, or the required E.M.F. As applied to electric railways, the circuit to the motor on the engine may be controlled from a fixed station by means of the mechanism employed to actuate the signals or points. The electric energy may be utilised for heating metal passing through rolls, for heating or melting metallic ores, for heating rooms, &c.

SECONDARY BATTERIES.—In the floating batteries described in the Specification 560 of 1882, free admission is provided for salt water. The sides of the vessel may be coated with graphite or a metallic paint. One electrode may be of copper or iron, and the other of zinc, and the electrolyte may be acidulated or salt water. A rotating or reciprocating contact maker, operated by the generator, or by a motor, thermostat, or electro-magnet and armature, arranged in the circuit, may be employed to produce alternating currents from a continuous current generator. The contacts may be made with mercury.

730.—C. A. Faure, London. Apparatus for Measuring and Registering Electric Currents. 6d. (3 figs.) February 15.

CURRENT METER.—This invention is based upon the well-known philosophical apparatus, in which the rotation of a conductor, carrying a current, round a pole is demonstrated. B is a core of soft iron, wrapped with an insulated conductor traversed by the current to be measured, and provided with an extension C, surrounded by a non-conducting envelope. In this extension is a footstep for the spindle H, which has two arms dipping into mer-

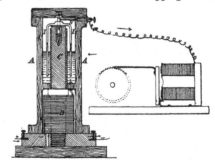

cury in the annular space between C and the wooden case A. The current enters at +, circulates through the magnet coil, and then, passing through the mercury, rises up the arms I I and escapes through the magnet core to the terminal —. The moving force is directly proportional to the square of the current, and the resistance to the square of the velocity ; whence it follows that the speed of rotation is a measure of the current. At each revolution, a projection on the spindle comes in contact with the spring N, and sends a momentary

current into a counting apparatus, which may be of convenient form, such as an electro-magnet with a step-by-step motion, as shown in the illustration. The specification also describes a modified instrument, in which the arms rotate between a hollow outer pole and an inner one. Through the latter passes an adjustable pillar, supporting by a step the lower end of the pivot H, the principle of action being the same as in the first described instrument.

GALVANOMETER. — In order to indicate temporarily the strength of a current, the rotation of the conductor is resisted by a spiral or other spring, and the amount of deviation of the conductor from a fiducial mark gives directly the value of the current. The electro-magnet may be replaced by a permanent magnet, the square roots of the dial readings indicating the current.

740.—A. M. Clark, London. (*Solignac et Cie., Paris.*) **Electric Lamps, &c.** 6d. (4 figs.) February 15.

ARC LAMP.—In this lamp, the feed of the carbons is regulated by the fusion or softening of a tube or stick of glass. The lamp is shown in elevation in Fig. 1, Fig. 2 representing the point of one carbon on a large scale. The carbons F are fed together by cords E, attached at one end to the lamp frame A, passing round sheaves D on the rear ends of the carbon-holders, and winding on spring barrels K.

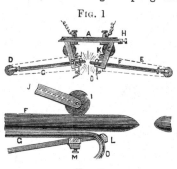

FIG. 1

FIG. 2

G are rods of glass extending parallel to the whole length of each carbon F. The rods G abut against stops of copper L, and their ends O, bending downwards, under the influence of the heat generated by an arc of given length, permit of the carbons being progressively fed towards each other. In a modification, this principle of regulation is applied to one carbon only, the other being controlled by a solenoid, the evident object of which is only to separate the carbons and so establish the arc.

SEMI-INCANDESCENCE LAMP.—In a semi-incandescence lamp, a glass rod is fixed to, and parallel with, one of the carbons, and abutting against the other carbon, and is consumed immediately an arc is

formed, thus causing the contact of the carbons to be renewed.

DYNAMO-ELECTRIC GENERATOR. — The rotating field magnets, semicircular in cross section, are placed together upon an axis, so as to form a cylinder, which revolves within the armature, formed of coils wound upon iron cores, arranged in a circle parallel to the axis of the field magnets. The coils of the field magnets constitute a shunt to the main circuit, a stationary collector distributing the current to two brushes carried by the rotating frame of the field magnets. The inventor does not describe the method of winding the coils, nor their connections.

756.—J. Brockie, London. **Machines for Producing Electric Currents.** 6d. (2 figs.) February 16.

DYNAMO-ELECTRIC GENERATOR.—The revolving armature is built up of two iron rings C D, mounted upon a magnetic boss B, and transverse compound iron bars E, fixed between the rings. Upon the two extremities of these bars coils of insulated wire F are wound, the ends of the coils being led to a suitable commutator. The armature revolves between the poles of four field magnets, whose polarity is so arranged that the two rings C D shall be of opposite polarity in any given length of the crossbars. These

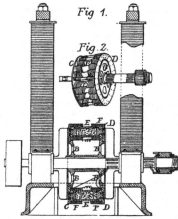

Fig 1.

Fig. 2.

crossbars, or magnets of the armature, are constructed of several pieces of iron fastened together, or of bundles of iron wire, or of hollow cases filled with iron filings, and are wound with insulated wire as shown at F, and are arranged to lift out from between the two iron rings. The field magnets are separately excited, or intermediate coils may be wound on the central part of the bars, and the current from them be used to excite the field magnets.

760.—C. W. Siemens, London. (*E. W. Siemens, Berlin.*) **Dynamo-Electric or Electro-Dynamic Machine.** 6d. (3 figs.) February 16.

Motor.—The same coils of insulated wire serve both for the induction of electrical currents, and also for induction of magnetism. Referring to the illustrations, a is a hollow rotating cylinder, formed by bending a plate of soft iron into an arc of a circle of about 270 deg., and filling up the remainder with a segment of brass. It is contained within a stationary case, made of two brass plates k k^1, and two ebonite end rings, supported by the screws m from the main frame, and is caused to rotate, by a ring of teeth cut on its left-hand end gearing with pinions e on three external shafts, driven by a large wheel h on the central axis. The cylinder is further supported, and guided, by three plain rollers d, and three angular rollers d^1, the latter running in a groove to limit the end play. Around the case k k^1 are wound lengthwise a number of stationary coils of insulated wire, 18 being shown in the figures, disposed in three groups

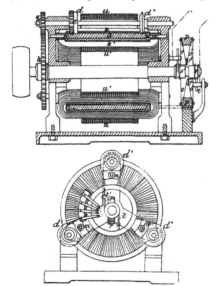

of six each, and arranged between the blank spaces left for the rollers and pinions. Each coil has two ends ; consequently there are 36 wires, connected in the following manner : A ring p of non-conducting material has fixed on each of its sloping sides 18 springs o and o^1. These springs extend inwards radially, and are, by preference, inclined in the direction of rotation of the shaft. The springs o on one side of the ring p, and the springs o^1 on the other side, incline towards each other, so that they meet ; thus there are 18 pairs of springs in all. One end of the wire of each coil is connected to one of the springs o, and the other end of the same wire is connected to one of the springs o^1, not to that which meets the former, but to the spring next in order, so that the outgoing end of each coil becomes connected, through the meeting of a

pair of springs, with the ingoing end of the coil next in order, and if all the 18 pairs of springs were simultaneously in contact there would be a closed circuit through all 18 coils. On the axis there are fixed two knives or separators r, and r^1, which, as the shaft revolves, pass between the pairs of springs o and o^1, separating each pair successively. Each knife has on one side a facing of non-conducting material, so arranged that r makes electrical contact with the springs o, and r^1 with springs o^1. The knife r is connected to the ring s, insulated from the axis, and the knife r^1 to a similar ring s^1, and against these rings rub the brushes t t^1, connected to the wires of the exterior circuit.

Dynamo-Electric Generator.—When used as a generator, the cylinder a, having certain residual magnetism, is rotated through the coils, inducing currents in them, and the currents so produced react on the cylinder, increasing its magnetism until the generator reaches the limit of its power. The circuit of the coils being broken at successive points, by the knives r r^1 separating the successive springs, the currents are directed into the external circuit. The power of the generator may be increased by external and internal shells u u^1 of soft iron. If the generator is intended to run in either direction, the knives have two metallic sides, separated by a non-conducting layer, and connected respectively to two rings, each provided with brushes, one ring of each pair only, however, being connected to the circuit at a time.

*761.—C. J. Chubb, Gloucester. **Dynamo-Electric Machines.** 2d. February 16.
Dynamo - Electric Generator. — The field magnets revolve in the opposite direction to the armature, or "a conducting material, connected magnetically with such magnets, revolves in an opposite direction to the armature."

766.—J. S. Williams, Riverton, N.J., U.S.A. **Generation, Storage, Utilisation, &c., of Electricity.** 4d. February 16.
Distributing Currents. — A current of low E.M.F. at the generating station is transformed into a current of high E.M.F., by the aid of secondary batteries. This current, or a current of high E.M.F. direct from a generator, is transmitted by conductors of relatively small section to the places where it is to be utilised, and, by the aid of secondary batteries, is again transformed into a current of low E.M.F. If independent secondary batteries be employed with the translating devices, these may be supplied with currents of different E.M.F. according to their requirements. The batteries are so arranged, that whilst at work a portion is being charged, whilst other portions are being discharged.

SECONDARY BATTERIES.—The plates may be prepared by depositing a suitable metal upon both sides, and over one end of an insulating sheet; or thin layers of metallic salts, or oxides, or iron coated with zinc, may be employed. An iron support may be coated with zinc, and immersed in salt water. The elements may be placed in a vessel of iron or steel, lined with an insulating material, such as cement, in which the edges of the plates may be embedded, or they may be separated by insulating strips. Sheets of lead foil, or other metal, separated by thin sheets of paper, parchment, or asbestos, may be closely packed together, and enclosed in a chamber containing acidulated water. The elements are provided with terminals, so that they can be coupled in tension, or otherwise, as required.

REFLECTORS.—In the provisional specification it is stated that, with electric lighting apparatus, a reflector, having a series of reflecting faces, each at an angle of 45 deg., is employed, and that the light is maintained at a certain fixed point, through the medium of an incandescent conductor, or an adjustable electrode, forced against a fixed electrode of refractory material.

ELECTRODES.—An alloy of manganese, iridium, or platinum, with other suitable metals, or a liquid conductor, or metallic oxide, enclosed in a suitable transparent material, may be employed.

*774.—J. C. Mewburn, London. (*A. M. J. Jeune, Paris.*) **Protecting Cables or Wires for Conducting Electricity.** 2d. February 17.

CONDUCTORS.—A braided or woven-wire sheathing is employed as an external protecting envelope.

819.—S. Pitt, Sutton, Surrey. (*E. T. Starr, Philadelphia, U.S.A.*) **Electric Lighting Apparatus for Railway Trains, &c.** 8d. (8 figs.) February 20.

DYNAMO-ELECTRIC GENERATOR.—This invention relates to improvements on Patent No. 5,600 of 1881. Referring to the illustration, which is diagrammatic in its character, *e* represents the spindle of a dynamo-electric generator, carrying a friction pulley G, to

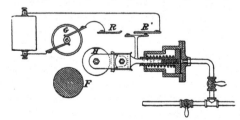

be driven from a car axle F. The two are connected by an intermediate wheel H, which can be made to gear with both of them, by means of a cylinder, into which steam, or compressed air, can be directed.

The current is conducted to a secondary battery P. The connection at R R¹ is broken, and the battery is cut out of circuit, as soon as the generator is stopped. The current from the battery is utilised in the head and tail lights, and also for actuating the brakes, being led to the lights by a system of commutators, which produce a flashing red light as the train moves forward, a flashing green light as it moves backwards, and a white light when it is stationary.

831.—J. Rapieff, London. **Electric Lamps, &c.** 6d. (18 figs.) February 21.

ARC LAMPS.—"The actuating force used in this mechanism is the force of gravity, or springs, or hydrostatic pressure. As it is necessary to obtain two distinct functions of the lamp—viz., the bringing together, and the separating of the carbons, it is also necessary to have two distinct regulating parts of the mechanism." Fig. 1 shows an arrangement in which two weights are used, and Fig. 2 one with a single weight. Referring to Fig. 1, "*q* shows the weight, and *r* a smaller weight. *a*, *b*, *c*, *d* are pulleys on shafts fixed to a suitable frame. *n* and *m* are also pulleys attached to the weight *r*; an endless cord, or chain, *g h* passes round these pulleys, in such a way that it cannot slip round the pulleys *a* or *b* without revolving them; the weight *q* is attached

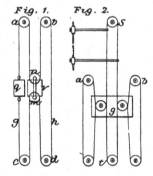

to the cord by means of a clutch, in such a way, that it can be easily raised by hand during the time of charging the lamp. For the same purpose the pulleys *a* and *b* are provided with ratchet wheels and pawls. The pulleys *a* and *b* are also provided either directly, or indirectly, or by means of a train of wheels or other suitable gear, with a brake, or escapement, actuated by an electro-magnetic appliance, whereby the rotation is checked, or impeded. So long as the pulleys *a* and *b* are held stationary, both weights remain suspended. By releasing the pulley *b*, the weight *r* is allowed to descend, but so soon as the pulley *b* is checked, and the pulley *a* released, the weight *q* will descend, at the same time raising the weight *r*." Now, if this weight be connected with one or both

of the carbons of the lamp, the carbons will be brought together, or separated, as the case may be. "Fig. 2 shows an arrangement in which the problem is solved by means of one weight g only; a and b are similarly constructed pulleys, for the purpose of regulating the descent of the weight g, so that, by releasing the pulley a, the weight g will descend, pulling after it the cord, or chain, which will cause the pulleys s and t to rotate in one direction. By checking the rotation of the pulley a, and releasing the pulley b, the direction of rotation of the pulleys s and t will be reversed." If the carbon-holders be connected to the branches of the cord, or to the pulleys, they will be made to approach and separate as desired. The specification contains ten diagrams illustrating various modifications of the regulating mechanism, and six diagrams wherein fluid pressure takes the place of the chains or cords. In the latter case, the weights are pistons moving in cylinders filled with liquid. Each end of each cylinder is connected to each end of each other cylinder, and the communication is controlled by valves actuated by electro-magnetic appliances, the coils of which are differentially wound.

834.—W. R. Lake, London. (*B. Lande, New York, U.S.A.*) **Electric Lamps.** 6d. (4 figs.) February 21.

Arc Lamp.—In this lamp, the upper carbon rod f passes between two grooved pulleys $l\,l$, which nip it gently. When the pulleys are free, the rod runs down between them, rotating them as it goes. Under each pulley is a partly circular brake arm n, attached to the magnet core d. So long as

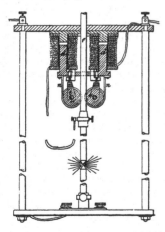

the normal current is flowing, the brakes hold the pulleys, but when the arc increases, the pulleys, brakes, and rod descend until the pulleys are arrested by a stop, when a further descent of the brakes releases the rod, and allows it to run down to feed the carbon.

*837.—I. L. Pulvermacher, London. **Apparatus for Collecting and Storing Electric Currents.** 2d. February 21.

Secondary Batteries.—Platinum cases, filled with platinum sponge, are placed in a vessel charged with a conducting liquid. "Electrodes receive the current set in conduction with the liquid." Loose helices of platinum, wound upon a cylinder, may have the spaces between the turns filled with platinum sponge. Each layer is enveloped in a porous material, and the ends of the various helices are coupled up as desired.

838.—W. R. Lake, London. (*B. Lande, New York, U.S.A.*) **Dynamo-Electric Machines.** 6d. (7 figs.) February 21.

Dynamo-Electric Generator.—This invention relates principally to means of inducing a circulation of air through the armature, to keep down its temperature. The armature consists of a flat ring, or hollow cylinder, composed alternately of projecting parts b, and recessed parts c, wound with

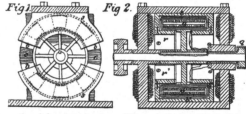

coils of insulated wire. The whole is attached by arms to a revolving shaft, provided with a commutator q. Air passages, or grooves, $g\,g$ are formed in the body of the ring, and have outlets whereby air may be projected through the coils. Inlets are made in the solid parts $b\,b$. The field magnet poles consist of curved and grooved castings $s\,s$, united at the outside by the curved pieces $t\,t$, and carrying on the inside the curved plates $v\,v^1$.

849.—F. H. F. Engel, Hamburg. (*G. Franke, Hamburg.*) **Glass Reflector for Gas and other Lights.** 6d. (8 figs.) February 21.

Reflectors.—This invention relates to the construction of single-walled glass reflectors, in which the silver is protected against outer influences. This is effected by applying coatings of shellac varnish to the silvered surface.

856.—J. S. Williams, Riverton, N.J., U.S.A. **Generation, Storage, &c., of Electricity.** 2s. 4d. (48 figs.) February 22.

Distributing Currents.—One or more generators are connected to the main conductors, either directly, or, if it be desired to increase or regulate the E.M.F. of the current, through the agency of

secondary batteries. In connection with these conductors, and at suitable intervals, are arranged independent sets of secondary batteries, each connected to one or more translating devices, and provided with one or more independent switches, operated by electro-magnets, or other devices, so as to place the batteries in circuit as required. Any one or more of these batteries may be arranged to be independently cut out, or put into circuit with the charging, or discharging, conductor. In order to maintain the activity of the secondary batteries at a practical uniformity, the insulated terminals of the conductors are mounted on a lever arm, which is caused to be moved at certain times, or at a determined rate of speed, in proportion to the generating capacity of the generator, and the discharge from the batteries. A "cut-off," operated by an electric current, renders any part of the apparatus independent of the other parts. In an arrangement, adapted : (1) for a railway, for operating the engines, switches or signals : (2) for lighting or heating : or (3) for other similar purposes, the current from a low tension generator is conducted to secondary batteries, the elements of which can be charged at a low tension, and be discharged at a high tension, into conductors, in connection with which, at suitable intervals, are auxiliary secondary batteries, which may be arranged to reduce the E.M.F. of the current, according to the requirements of the translating devices. One portion of each auxiliary battery can be charged whilst other portions are being discharged into a conductor connected (in the case of a railway, through a sliding shoe, or friction roller) either directly, or through a series of secondary batteries, to the motors on the engine, or to other translating devices. The generators on the engine of a train may charge a series of secondary batteries thereon, from which the circuit is completed to motors, on the engine, by circuit-controlling switches, worked by the engine-driver, or guard, or both, or from certain fixed stations on the line. Auxiliary conductors may be led from the batteries on the engine to furnish energy for the development of light or heat, or for operating movable mechanism along the line. The supply of energy to the secondary batteries on the engine may be controlled by the engine-driver, or guard, and be supplied on a principle like that by which water is supplied to locomotive water tanks. A similar system of distribution may be arranged to actuate motors, electro-magnets, &c., and to operate the "points" or signals, the currents being controlled by switches. A motor may be made to compress air for giving audible signals by means of a fog-horn. In order to light or heat the vehicles of the train, to operate motors on the axles of the vehicles, or motors operating the

brakes, conductors are arranged along the train, and are in connection with secondary batteries on one, or more, of the vehicles. The connection between the stationary conductors, and the engine, is made by hinged shoes, or rollers, the contact being broken when they are raised. A liquid lubricant may be employed at the contact surfaces, for making a perfect contact, and lessening the friction. If a return insulated conductor be used, the two conductors may be placed in channels, or grooves, in an insulating support.

MOTORS.—An insulating support for the armature is fixed upon the axles, or upon the wheels, of the engine, and insulated wires, or ribbons, are secured to the support, and covered by one or more metallic pieces. The field magnets, are insulated from the axle and driving wheels, and are arranged so that the armatures revolve in close proximity to their polar extensions. An annular electro-magnet, having semicircular polar extensions, may be employed. The collectors, or circuit makers, are placed at the ends of the axle. In large motors, the armatures may be constructed in sections, secured to the driving wheels, and the field magnets may be placed between the wheels, supported so as to provide for the variation in height of the body of the vehicle due to the springs. Contact makers govern the number of the series of elements employed at a given time to actuate the motor, or the independent sections of the armature may be included in, or withdrawn from, the circuit. The motors may be transformed into generators when desired, to retard the speed of the train, and are connected with circuit-making devices, to enable them to charge secondary batteries on the engine. The motors may also be arranged to actuate the driving wheels by means of friction gear. The power may be distributed throughout the train, and be controlled from the guard's van, by devices, which will reverse the current through the motors. The brakes of the vehicle may also be operated by an electro-motor.

SWITCHES.—The circuit maker may consist of a plunger, which causes the displacement of a conducting liquid, making contact successively with conductors connected to the different sections of the armature. A similar contact maker may be employed for reversing the current through the motor. Switches, suitable for charging and discharging the secondary batteries, are arranged at intervals along the conductors, and consist of two V-shaped pieces, making contact with V-grooved-blocks, and attached to opposite ends of a weighted lever, carrying the armature of an electro-magnet placed in a shunt circuit to the conductors, or in an independent circuit. In a circuit-breaking device, one of the contacts may be omitted. A switch, for enabling one part of a secondary

battery to be charged, whilst another portion is being discharged, consists of a shaft operated by time mechanism, or by the generator, or by the motors, &c., and carrying contact pieces.

RHEOSTATS.—The circuit-maker may consist of a movable and a fixed wedge (or other suitably shaped piece) of carbon, lead, or other conductor, arranged in an insulating casing, so as to vary the distance between the two pieces, or to vary the height of a conducting liquid, which makes contact between them. In another arrangement, studs, on a movable plate, can be brought into contact with one, or more, conductors, communicating with one, or more, independent translating devices. The plate may be actuated by a lever, button, or electrical apparatus. A registering device shows the number of conductors in circuit, and thus indicates the resistance. The contacts may be so disposed that on continuing the rotation of the plate, the current to the translating devices will be reversed, a central contact being connected to one conductor, and the other conductor being connected to contacts on either side of the central contact.

SECONDARY BATTERIES.—One pole of the battery is a liquid, which, when heated, gives off a gas, and the other pole is the metal of the containing vessel. Air, or gas under pressure, may be employed for storing electrical energy, conductors provided with an extended surface being placed in the air or gas chambers. The conductors may consist of alternate sections of metals and metallic oxides. Secondary batteries may be provided within the stand of an electric lamp, so that it can be disconnected from the "leads" and used as a hand lamp. This specification covers 34 pages and has ten sheets of drawings. There are 14 claims.

*866.—J. C. Mewburn, London. (*La Société Alamagny et Oriel, Paris.*) Cables for Telegraphic Purposes, &c. 2d. February 22.

CONDUCTORS.—The lead pipe, in which the wires are enclosed, is of larger diameter than the cable to be protected, and is closed upon it by drawing through a diminishing series of rolls, or draw plates.

869.—C. E. Spagnoletti, London. Dynamo-Electric Machines, &c. 6d. (8 figs.) February 22.

DYNAMO-ELECTRIC GENERATOR.—The field magnets, of which there may be several, are large single magnets A, capped in such a manner as to collect the magnetic lines of force at each end, and conduct the same to a given central point. Between these caps is a wheel C (Fig. 1), having at the end of each spoke a coil of wire D, the ends of which are led to the commutator. F F are

collectors, placed so that only one set at a time is in contact with the wires, or commutator bars, "and when, by revolving, the said bars are not touching the first two, they are in contact with the second two. These collectors F F connect one, or more, pairs of the coils to the lamps, or outer circuit, and the other to the field magnets. The arrangement is so made, that when one or two of the revolving coils are within the influence of the field magnets, the current is taken to the main circuit, and when leaving or nearing the other end of the electro-magnet, the contrary current is thrown into the field magnets, and keeps up their power, and by this arrangement, both the field magnets, and the main circuit, are electrically charged." "The revolving coils of any kind may be placed in such a manner, and made to revolve in the surplus currents of electricity that are carried by the rush of air from this dynamo machine, or any other system, as an auxiliary wheel (as represented by dotted lines in Fig. 1), and thus the present wasted

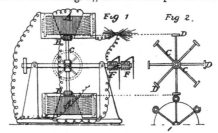

power may be utilised and the number of lamps increased." The specification also describes an electric motor, in which the noticeable feature is that the armature rocks on the poles of the magnet, and never leaves it. Dynamo-electric generators are also made "of two or more electro-magnets with coils, one divided, either throughout, or at the ends only, by insulating material, or metal for magnetism, so as to distribute the lines of force, and on the core of the other magnet is fixed a jaw which revolves round the other magnet's core."

CONDUCTORS.—A coupling for electric conductors consists of a nut, tapped with right and left-handed threads, and engaging with corresponding screws on the ends of the conductors.

ARC LAMPS.—The feed is regulated by allowing mercury to pass out of a cylinder, from under a piston, by a valve operated by a differentially wound magnet. The arc is established by the carbon-holders moving on trunnions, which permit them to fall apart. The carbons may be placed one above the other, in which case the mercury that flows from the top tube is used to raise the lower carbon.

898. — J. Brockie, London. Electric Arc Lamps. 8d. (3 figs.) February 24.

ARC LAMPS.—This invention relates, firstly, to

mechanical details, and, secondly, to modifications in the principle of the step-by-step, or independent feed lamps described in patent No. 3,771 of 1879. (1) In order to accomplish the adjustment of the arc, after a certain number of impulses of the feeding magnet, which may have been greater or less than the actual consumption of the carbons, the arc-striking magnet was formerly short-circuited, but now it is proposed to reverse its polarity by the extra coil S. This coil is a shunt across the arc, and is momentarily in circuit at the readjusting periods. Sometimes another magnet, in the same circuit, is placed below the armature, to aid in pulling it away smartly. The contrivance for putting the magnet S into circuit, at intervals, is as follows : The magnet F, which gives the step-by-

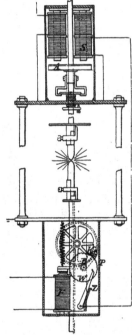

step movement to the lower carbon-holder, also turns a ratchet wheel W ; this wheel is furnished with a projecting insulated boss B, upon which a spring lever rests. An insulated pin P is placed near the outside of the wheel, so that, as the wheel revolves, the pin P will lift the lever off the boss, and at a certain part of each revolution will allow the spring to drop suddenly upon the boss B again. As the lever falls, it rubs against the spring K, and in so doing completes the circuit of the reversing coil S ; the armature A then drops, the two carbons come into contact, and are again lifted the determined length of the arc. The object of the second part of the invention is to dispense with the arbitrary readjustment of the arc, after a definite number of impulses of the feeding holder, and in

its stead to cause each lamp in a circuit to adjust its own arc, from time to time, according to its individual requirements, even although the step-by-step feed should continue a constant quantity. In the first modification, the upper carbon-holder is carried by a solenoid with a long range of motion. If the step-by-step feed of the lower carbon be too quick, the solenoid will cause the upper carbon to retreat before it, while, on the contrary, if it be too slow, the upper carbon will descend a distance equal to the deficit. In a second modification, the solenoid, in addition to the duties already described, inserts, or withdraws, a taper stop into, or from, the path of the step-by-step armature, to bring its throw to the requirements of the consumption of the carbon. In a third modification, the rise of the solenoid core short-circuits the step-by-step magnet, so as to arrest the feeding from time to time, if it be too rapid. In a fourth modification, the regular feed is set to be too slow, and the upper carbon is fed by an arrangement similar in principle to the Brush feed. The feed may be worked by clockwork, instead of by magnetic power, and the stroke of the pawl may be regulated, in either case, by a separate magnet ; the adjusting solenoid tightens or slackens the spring against which the magnet acts, according to the strength of the current, or it moves the magnet farther away from its armature.

905.—J. W. Swan, Newcastle-on-Tyne. Secondary Batteries. 4d. February 24.

SECONDARY BATTERIES.—The electrodes are made of finely divided lead, obtained by submitting galena, or sulphide of lead, to the action of electrolytic hydrogen, or to the chemical action of metallic zinc. The decomposition of the sulphide may be effected, after its application to the electrodes, by connecting it to the hydrogen pole of an electrolytic bath. Non-corrodible pole plates are made from compressed plumbago.

917.—H. J. Haddan, London. (*W. Wheeler, Mass., U.S.A.*) Reflectors, &c. 6d. (13 figs.) February 25.

REFLECTORS.—The reflecting surface is generated by the revolution about its principal axis of a curve, which is constantly variable throughout the revolution. Several modifications, and also ears and clamps for attaching the reflectors to electric lamps, are described and illustrated.

931.—A. M. Clark, London. (*H. B. Sheridan, Cleveland, Ohio, U.S.A.*) Dynamo-Electric Machines. 6d. (8 figs.) February 25.

DYNAMO-ELECTRIC GENERATOR.—The object of this invention is to induce the current in the armature without wide breaks, and nearly continuously,

and also to increase the efficiency of the generator. "The magnet cores E stand with their axes in a piral line, and present an oblong field of force to the armature, the length of the field depending on the pitch of the cores, and their inclination to the axis. The magnet cores are arranged with the longest diameters of their fields of force in the direction in which the armature revolves, by which arrangement the projections of the armature will remain longer in the fields of force, and the generated currents will be considerably increased. One pole of the magnet cores E, with the surrounding helix, can be one pole of an electro-magnet, or two or more of the said cores can be connected on their faces by a plate." The armature core D is made in the form of a hollow iron ring, and is nearly rectangular in cross section. The sides slightly converge as shown. In the inner and outer peripheries of the core, holes H and I are formed for ventilation. Upon the sides of the armature there are angular

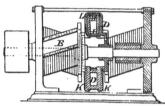

projections K K, which serve to convey the induction to the core, and between these projections the coils are wound Gramme fashion. When the armature is to revolve inside the magnet fields, as in the Gramme generator, and not between them as shown, rows of cylindrical magnet cores are placed in spiral lines radiating from the polar extensions. In a modification, the magnet coils are wound on flat bars, which are arranged in inclined positions with reference to each other, so that they nearly overlap each other laterally. "With this arrangement the face of the pole of each magnet is arranged diagonally across the path of the armature ring in the plane of its rotation. This construction brings the magnets into such positions, that, each section coil L of the armature will pass upon the pole of each magnet, at the instant that it leaves the pole of the preceding magnet, so that the current induced in the armature will be be nearly continuous. With this construction the armature can be revolved in either direction."

941.—J. Richardson, London. **Electrically Controlling, &c., the Speed of Engines Employed for Driving Electric Machines.** 6d. (9 figs.) February 27.

CURRENT REGULATOR.—This invention relates to improvements on Patent No. 288 of 1881. F is a solenoid, traversed by the current from the gene-

rator, driven by the engine which it is desired to control. The core D is connected by a lever H to the rod G of the throttle valve, so that the rod of the valve is in tension, and consequently can be made very small so as to have but little friction in the stuffing-boxes. The valve is raised by the weight of the core, and the cheese-weights J acting through the lever I and link C. If the current ceases, the core falls, and would open the valve wide, were it not that a collar on an inter-

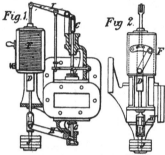

mediate rod, pivotted to the lever I, comes against a second lever pivotted to the valve spindle, and so closes the valve. The invention can be adapted to governing an ordinary throttle valve, by attaching the solenoid core to an arm on the spindle of the valve. The solenoid can be regulated by having its wire wound in sections that can be put successively in and out of circuit. When the currents are small, two solenoids, side by side, may be employed. Instead of operating the throttle valve, the solenoid core may be connected to the expansion slide.

943.—H. E. Newton, London. (*A. I. Gravier, Paris.*) **Machinery for Obtaining Continuous or Alternating Currents of Electricity.** 6d. (18 figs.) February 27.

DYNAMO-ELECTRIC GENERATOR.—This invention relates to "machinery for producing continuous or alternating currents of electricity, without the use of rubbers, collectors, brushes, or commutators, in which an inductor, either with magnetic or electromagnetic commutation, is arranged in the magnetic field of a permanent or an electro-magnet." "Fig. 1 is a transverse section of the machine, showing the arrangement adopted for the magnetic commutator, inductor, or rotary armature, and the fixed magnet. Fig. 2 is a longitudinal section, and Fig. 3 a side elevation of the same. The induction cylinder, or revolving armature, is shown as composed of sixteen bars of soft iron a a, b b, slightly separated from each other by pieces of paper. These bars are mounted upon, and secured at their extremities to, two discs of soft iron D D, by means of screws, and the whole is attached to an axle o turning in suitable bearings. The stationary field magnet A A[1] A[11], B B[1] B[11] is composed of eight cast-iron

rings." " Fig. 1 shows the configuration and the mode of coiling the field magnets. The eight rings are separated by insulating material, and are clamped together by means of bolts, two exterior frames of cast-iron, provided with plummer-blocks, completing the iron part of the fixed field magnet, and being solidly clamped to the eight rings by the same bolts. The winding of the wire, in actual practice, is divided in such a manner as to form forty circuits, the extremities of each of which terminate in a grouping table, to allow of the circuits being connected together in any way that may be found desirable, either in quantity, tension, or series. On starting the rotary armature, it will be polarised by the residual magnetism of the exterior magnet, and an electrical current will be induced

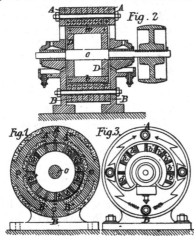

in the coils of the exterior magnet. This current will strengthen the rotating armature and induce a current in the coils thereof. A reciprocal action being thus set up, the currents of electricity will flow from the terminals of the coils of the external magnet, which may be used without employing a commutator and collectors. As a modification of the above, two rotary armatures, of different dimensions, may be mounted on the same axle, the larger one being coiled so as to produce two consequent poles, and the coils connected by metal plates with the coils of the adjacent smaller armature. Both armatures will be provided with field magnets, the larger field magnet being coiled so as to produce two or more currents, one or more of which will circulate in the coils of the field magnet."

986.—W. H. Akester and **T. B. Barnes**, Glasgow. Dynamo-Electric Machines. 6d. (6 figs.) March 1.

DYNAMO-ELECTRIC GENERATOR.—Referring to the illustrations, Figs. 1 and 2 are plan and sectional side elevation, respectively, of a dynamo-electric generator, constructed according to this invention,

and Fig. 3 is a sectional and end elevation of the same. The axle has on it two compound armatures, and between them three commutators 14. There are two parallel horizontal field magnets 15, one at each side of the generator, provided with exciting coils 16. Each magnet has two pole-pieces 17, arranged to oppose those of the opposite magnet. The armatures consist of a number of wheel-like sections, fixed on a brass boss, keyed on the shaft 10. Each wheel piece has a number of cylindrical cases 19, having within them radial spindles of soft iron, wound with coils, whose ends are led to the commutator. The several wheels are arranged on the shaft, so that the spokes form a helix, and enter and leave the magnetic fields successively. Three commutators are shown, to allow of the electricity being

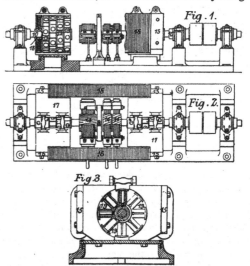

collected in three currents, one corresponding to the whole of one compound rotating armature, and the other two each to about one-half of the other rotating armature. One of the smaller currents is intended for exciting the field magnets. The brushes are insulated from each other, and can be removed from the commutator by a hand lever. A modified generator has the boss made of iron, and brass zones are inserted in the middle of the cases, the object being to interpose a non-magnetic material between the outer and inner ends of the cases. In a second modification, the external diameter of the wheel piece is doubled, the number of field magnets being increased and arranged equidistantly around the armatures.

994.—J. Fielding, Gloucester. **Gas Motor Engines.** 10d. (24 figs.) March 1.

DYNAMO-ELECTRIC GENERATOR. — The gaseous charge is exploded by a make-and-break arrangement, placed in the combustion chamber, the current being supplied by a dynamo-electric generator.

*1,010.—G. M. Minchin and **L. H. Despeissis,** Staines. **Railways.** 2d. March 2.

CONDUCTORS.—To enable the guards of trains to communicate with one another, when near together, a conductor, broken at intervals, is laid between the rails.

1,017.—J. S. Lewis, Birkenhead. **Insulating Apparatus for Overhead Telegraph Lines, &c.** 6d. (8 figs.) March 3.

INSULATORS.—A horseshoe-shaped clip, with two hooked ends, is slipped on to the conductor, and a screw-threaded portion of the insulator is screwed into the loop thus formed.

1,023.—T. J. Handford, London. (*T. A. Edison, New Jersey, U.S.A.*) **Indicating and Regulating the Current of Electric Generators.** 8d. (5 figs.) March 3.

CURRENT REGULATOR.—This invention relates to means for indicating to the engineer in charge of an electric lighting station, when the lights are above, or below, the desired limit of candle power. Three ways of accomplishing this are described in the specification, two being illustrated in Fig. 1, and the third in Fig. 2. The bell shown to the left of Fig. 1 is set in operation whenever the difference of potential at the poles of the generators exceeds, or falls below, a given amount, the action of the bell being stronger in the former case than in the latter. A is the generator, 1, 2 the main leads, and 3, 4 a derived circuit. This circuit contains a resistance R, and an electro-magnet B ; 5, 6 is a shunt circuit around the resistance R, and including the resistance R' and the bell *e*. The armature of the electro-magnet B is upon a lever C, the free end of which can make contact with either of the terminals *c c'*, or can stand between the two, where it will be held steady by the movable stop C', which is drawn by its spring against the pin *b*. When the difference of potential between the leads is too great, there is an increased current in the magnet B, which attracts its armature, moving its free end against the stop *c*. A circuit is thus completed through 3, 5, C, *c*, *d*, *g*, R', 6, and 4, and the bell rings. If, on the other hand, the difference of potential be too small, the spring *b* overpowers the magnet, and the lever C moves towards the contact *c'*, and then the circuit is through 3, 5, C, *c'*, R², *d*, *g*, R¹, 6, and 4, and the bell is rung, but with less force than before, on account of the decreased current and of the extra resistance R². After the alarm is given, the engineer varies the resistance of the field magnet circuit by the lever *r* and resistance coils R¹. To prevent injury to the generators by too great current, a safety-catch wire is inserted in one of the mains, and a shunt, including a bell *l*, is formed round it.

When a current is generated, that is greater than the carrying capacity of the wire loop, it will fuse the catch, and the bell will then come into action, the current at the same time falling to a safe amount on account of the increased resistance in its circuit. Fig. 2 illustrates an automatic appliance for increasing, or diminishing, the resistance of the field magnet circuit, according to the requirements of the system. A is the generator, 1, 2 the main leads, 3, 4 a derived circuit, which includes the magnet coils, and the resistance B ; 5, 6 a second derived circuit, comprising the resistance D and magnet C. The office of this magnet is to hold the lever E midway between the two contacts *c d*, when the difference of potential is normal, and

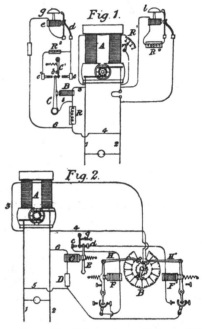

Fig. 1.

Fig. 2.

to move it, or allow it to move, against one of them, when the regulation has to be effected. F F¹ are two electro-magnets, each with an armature lever, that works like the hammer of a trembling bell when a current is flowing round the magnet. When the lever E comes against the contact *d*, the circuit of the magnet F¹ is completed, and the pawl H¹ rotates the ratchet wheel *r¹*, and arm *a*, to cut some of the resistances B out of the magnet circuit. On the contrary, if the lever E makes contact with *c*, the pawl H produces the contrary effect. The movable stop *g* insures a certain amount of stability to the system, and prevents the apparatus being in a constant state of reciprocation.

1,029.—F. Wright and **M. W. W. Mackie,** London. **Incandescent Electric Lamps.** 6d. (5 figs.) March 3.

INCANDESCENCE LAMPS.—This invention relates chiefly to the construction of the bulbs, and to means for attaching the filaments to their conductors. Fig. 1 shows the glass tube A, on which are fixed the conducting wires B B with the carbon filament C, by means of spun glass; this is inserted into the throat D¹ of the bulb D, Fig. 2, which is then heated so as to collapse on the tube A, thus closing the wires in hermetically. While the throat D¹ is soft, it is pushed inward somewhat, by means of the tube A, so as to assume the shape shown at Fig. 2, the

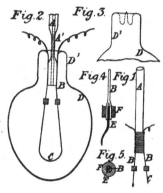

tube A being also heated and drawn out to a small bore at A¹. After exhausting the air through the tube, it is hermetically closed, by melting the part at A¹, the projecting end of which is protected by the throat D¹ of the bulb, as shown at Fig. 3. Figs. 4 and 5 show the mode of securing the carbon filament to the conductors. The end of the platinum conductor B is split and bent into tubular form, and into it is inserted the thickened end E of the filament, which is secured by a ring F of steatite.

1,031.—F. Wright and M. W. W. Mackie, London. Vacuum Pumps for Exhausting Bulbs of Electric Lamps, &c. 6d. (2 figs.) March 3.

INCANDESCENCE LAMPS.—The bulb to be exhausted is connected to a pipe that ascends vertically more than 30 in., and then descends into a vessel of mercury below the level of the liquid, the joint, where it enters the vessel, being securely sealed. This vessel has a valve opening outwards, and is connected by a flexible tube with a second movable vessel of mercury. When the second vessel is raised, the mercury flows into the first, forcing out the air through the valve. When the second is again lowered, the mercury falls, and its place is taken by air from the bulb, the operation being repeated until the globe is sufficiently exhausted. Any suitable hygroscopic substance may be used to dry the air.

1,036.—H. Liepmann and P. S. Looker, London. Manufacture of Carbons for Electrical Purposes. 6d. (3 figs.) March 3.

CARBONS.—The carbons are made from the waste of Corrosso, or vegetable ivory, after it has been submitted to destructive distillation, or to carbonisation by heat, acid, or caustic alkali. The carbon obtained is ground to a fine powder, and boiled in a vacuum, with a saccharine solution, to give it binding qualities. Thus prepared, it is compressed in a mould, and, after drying, the pencils are

boiled in a vacuum in a solution of sugar, and are finally carbonised at the lowest possible temperature, in a partial vacuum. The figure shows a mould for making three carbons at a time. The lower half has removable dovetail ends; B is the upper half forming part of the ram of a hydraulic press. In making filaments for incandescence lamps, the shell only of the Corrosso nut is used.

1,079.—W. Crookes, London. Incandescent Electric Lamps. 4d. March 6.

INCANDESCENCE LAMPS.—The filaments are produced from animal fibres, such as silk, hair, wool, silkworm gut, horn, gelatine, or parchment, by treating such materials with cuprammonia, or other solvents, and then carbonising them in the ordinary manner; or the filaments may be made by carbonising cellulose, precipitated from solution in cuprammonia by evaporation, or by an acid. The junction between the filament and the conducting wires, is formed of a cement of graphitoidal carbon and water, to which a little sugar has been added. The conductors may have cup-shaped terminals, in which the ends of the filament are embedded by the cement. The ends of the conductors may be bent upon themselves, so as to form a clip, in which the filament is held by a light pressure at three or more points, the connection being completed by a carbonaceous cement. The conductors may terminate within a glass cup, into which the end of the filament is loosely placed, the connection being made by a carbonaceous cement. The junction may be made by electro-plating the extremities of the filaments, and inserting them, with or without a cement, into metallic tubular terminations of the conductors; or the terminations of the filaments and conductors may be nipped in metal socket-pieces. The filaments may be strengthened by de-

positing carbon on them, from either chloride or bromide of carbon. The resistance of the filament is adjusted by exhausting the globe, and then filling it with carbonaceous vapour obtained by the application of heat to a hydro-carbon body, such as india-rubber, that is non-volatile at ordinary temperatures. To equalise and measure the resistance of the filaments, at one operation, they are arranged in series, and coupled with resistances in a Wheatstone bridge arrangement, in which is a galvanometer that shows when the filaments have been brought to the required standard of resistance. This is accomplished by raising the filaments to incandescence whilst in contact with a carbonaceous vapour, the current being continued till a balance is obtained.

1,094.—E. H. Johnson, London. Holders for Electric Lamps. 6d. (6 figs.) March 7.

FITTINGS.—The socket is composed of a base A, a contact carrier B, a hollow metallic screw or junction piece C, and an enclosing case D. One terminal of the lamp is connected to the central

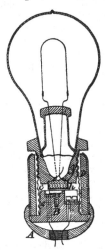

piece *h* and the other to the screw *g*. The main leads are connected to the central screw, and to the peripheral screw *e*. The act of screwing the lamp into the socket puts it into circuit.

*1,132.—G. Smith, Bradford. Earthenware Pipes for the Conveyance of Gas, Water, &c. 2d. March 8.

CONDUCTORS.—The wires are carried by earthenware blocks, perforated with a series of holes, and fixed in a trench lined with cement.

1,139.—T. J. Handford, London. (*T. A. Edison, New Jersey, U.S.A.*) **Dynamo or Magneto-Electric Machines.** 6d. (3 figs.) March 9.

DYNAMO-ELECTRIC GENERATOR.—Generators of the Pacinotti type, constructed according to this invention, have the inactive portions of the coils of a greater cross sectional area, and a lower resistance per unit of length, than the active portions, thus lowering the internal resistance of the generator. This is accomplished by forming the bobbin coils of bars, which extend on the outside of the ring, or angular core, parallel with the axis of rotation, and of wide plates on the inside of the ring, the bars and plates being connected, to produce a continuous bobbin, by radial end bars. Fig. 1 is a perspective

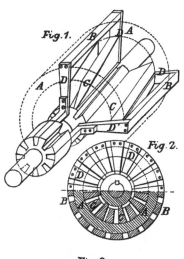

Fig. 1.

Fig. 2.

Fig. 3

view of a portion of the armature of a generator, the annular core being shown in dotted lines. Fig. 2 is a half end elevation of the armature, with the commutator cylinder and its connections removed, and a half cross-section of the armature. Fig. 3 is a plan of the armature, with some of the active bars removed, and the annular core shown partly broken away and in section. A is the annular core, on the outside of which are the active bars B, parallel with the axis of rotation, whilst on the inside are wide interior plates C, having greater cross-sectional area and lower resistance than the active bars. The bars B, and plates C, are connected by radial end bars D D[1]. The plates are set obliquely to the axis of rotation, in order to form with the active bars B, and radial end bars D D[1], a continuous bobbin.

1,142.—T. J. Handford, London (*T. A. Edison, New Jersey, U.S.A.*) **Regulating the Generative Capacity of Dynamo or Magneto-Electric Machines.** 6d. (2 figs.) March 9.

CURRENT REGULATOR.—The generator is regulated by moving the brushes round the commutator towards and from the points of greatest potential. The apparatus is intended for use in cases where the lamps, or other translating devices, are arranged in parallel arc between two mains, and is set in operation, when the difference of potential between the terminals of the said two leads falls above, or below, a certain amount. The arms carrying the commutator brushes are mounted on a yoke, pivotted on the bearing next to the commutator. The yoke carries a wormwheel e, which engages with a worm e^1. Secured to this wormwheel are two ratchet wheels, with teeth respectively leading in opposite directions. With these two wheels, two pawl bars $g\,g$, carried by armature levers $h\,h^1$, engage. The lower ends of the armature levers engage with circuit makers and

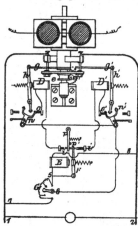

breakers, which are Y-shaped pieces playing between contact screws, and making their forward contacts when the armature levers are retracted, and their back contacts when such levers are drawn forward by the electro-magnets D D¹. In a multiple arc circuit 7, 8 is a controlling electro-magnet E and resistance G. The adjustable shunt 5, 6 runs to the armature lever F, and to the Y-shaped circuit controllers $n\,n^1$. The front and back contacts $i\,i^1$ of F are connected with D D¹, which are also connected with the front contacts $o\,o^1$ of the circuit controllers $n\,n^1$. The spring arm p, and stop p^1, determine the central position of F. The action is as follows : When the electromotive force of the current in 1, 2 is normal, and the candle power of the lamps at the desired point, the lever F will be held in a central position, breaking both divisions of the shunt 5, 6, as shown in the drawings. An

increase in the electromotive force will strengthen the magnet E, and draw the lever F against i, closing the circuit through D. The lever h is then caused to vibrate like a trembling bell, and through the intermediary of the pawl and ratchet wheel, to move the brushes round the commutator away from the line of maximum generation. When the electromotive force falls, the other magnet D¹ is put in action with the contrary result. In a modified arrangement, the power for moving the brushes is derived from a small electric motor. This can be driven in either direction, by reversing the current in its magnet coils. A derived circuit leads from one of the mains through the armature of this motor to a fixed stop. Two movable stops, which are the terminals of the field magnet circuit, normally rest against the fixed stop, and consequently there is no current in the magnet coils. When the electromotive force rises, or falls, beyond a given limit, the armature lever of a controlling magnet, such as E, forces one of the movable stops away from the fixed stop. A circuit is thus opened from one main, through the armature coils, to the fixed stop, then from one movable stop, through the magnet coils, to the other movable stop, and through the armature lever F to the other main. When the armature lever moves over in the opposite direction, and reaches the other movable stop, the direction of the current in the magnet coils is reversed.

1,153.—M. Zingler, London. **Substitute for Gutta-Percha.** 4d. March 9.

INSULATING MATERIAL.—Any one of the hard gums is dissolved by heat in turpentine, to which is added a small quantity of camphor, and sufficient sulphur to give the necessary hardness. The boiling is continued, and a small quantity of albumen added. To the gum solution is then added tannin, and the whole is boiled until it assumes the desired consistency.

1,162.—W. R. Lake, London. (*H. S. Maxim, New York, U.S.A.*) **Apparatus for the Distribution and Regulation of Electric Currents.** 6d. (1 fig.) March 10.

DISTRIBUTING CURRENTS.—The object of this invention is to maintain the generative capacity of the generators equal to the requirements of the circuit, by such means, that the adjustment may not be upset by the resistance of the conductors, which is a constant quantity, and does not vary with the number of lamps in use. To this end, two wires C C¹ are run from the centre of the group of lamps, and are connected to an indicator G, and a regulator R. This latter may be of any known construction, but preferably arranged to alter the position of the brushes on the commutator. By this device the

electromotive force is maintained constant at the

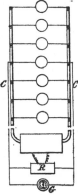

point where the current is utilised, and not at the poles of the generator as is usually the case.

1,163.—W. R. Lake, London. (*E. Weston, Newark, N.J., U.S.A.*) **Electric Lighting Apparatus.** 6d. (4 figs.) March 10.

ARC LAMP.—This invention relates to means for bringing a second set of carbons into circuit, on the onsumption of the first, in lamps having their feed

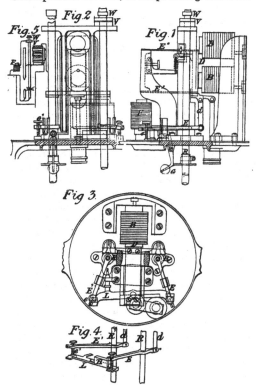

controlled by clutch mechanism. This is done by packing up the abutment of one of the clamping levers, until the other carbons are consumed, and then suddenly removing the packing piece, and

allowing the lever to drop into a position where it becomes operative. The lamp shown in the illustration is of the Weston type, but the invention is applicable to other forms of lamps. B B is a differentially wound controlling magnet, acting on the armature D, which is carried on flat springs E* E*. From this armature depend two rods *d d* (Figs. 2 and 4) pivotted to the two clamping levers E E¹, which seize, in succession, the two carbon-holders R R¹. As shown in Figs. 2 and 5, the lever E¹ is out of action, being packed up on the raised piece *e¹* of the bar L, the carbon carrier R¹ being meanwhile supported by the catch G (Fig. 1). To put the second set of carbons into circuit, the lever L must be rotated through a small angle, thereby releasing the end of the lever E¹, and raising the lever E. This is done by means of the electromagnet F and armature H. When the first pair of carbons is nearly burnt out, the collar W descends on the collar V (Figs. 1 and 5), and completes the circuit through the magnet F, which immediately attracts its armature, and, in so doing, moves the bar L, raising the lever E, and dropping the lever E¹. At the same time, a projection on the bar L releases the catch *b*, and allows the weight of the second carbon to come on to its clamping lever.

1,171.—A. Graham, London. **Mechanism for Regulating the Burning of Carbon or other Electrodes in Electric Lighting Apparatus.** 8d. (12 figs.) March 10.

ARC LAMP.—The essential feature of this invention is the use of a verge escapement for controlling the feed of the carbons. This consists of a crown wheel, on a horizontal axis, driving a pallet shaft set vertically. The motive power is derived from the weight of one or both carbons, and is applied to the crown wheel by means of a grooved pulley

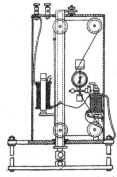

and a cord, or by a rack and pinion. In the accompanying drawing, I represents the crown wheel in front elevation. The pallet shaft carries a balance or fly lever, one end of which is provided with a weight, and the other with a circular rack. Immediately over this rack is a detent finger, operated by a differentially wound electro-magnet.

When the normal current is flowing across the arc, the magnet attracts its armature, and draws the detent into the rack, thereby stopping the feed, and locking the carbons. When the resistance of the arc is such that the armature rises, the balance lever becomes free, and allows the pallet shaft to vibrate and the verge wheel to rotate, paying out cord and allowing the carbons to approach each other. The arc is established by the magnet G, which raises the sheath or tube in which the carbon-holder slides. The specification also illustrates a lamp with the carbons above the regulating mechanism, two lamps with inclined carbons, and a lamp with horizontal carbons, all of which are controlled in the same way. It further shows a lamp in which the regulating mechanism comprises only a magnet, armature, friction clutch, and detent. When the armature descends, it carries the carbon-holder with it, but when, on the arc being shortened, the armature ascends, the detent holds the carbon rod, and the clutch slides over it.

1,172.—J. Wauthier, London. Incandescence Electric Lamps, &c. 6d. (5 figs.) March 10.

INCANDESCENCE LAMPS.—The bulb has an inner neck *a*, sealed into its throat, and within this neck the platinum wires and the exhaustion tube are situated. The method of manufacture is illustrated at Fig. 2. A tube of glass has two nipples ff^1

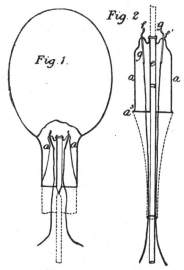

Fig. 2.

Fig. 1.

blown on to it, for the passage of the platinum wires *g g*, and has a tube *e* sealed into its centre. The superfluous part, shown in dotted lines, is cut off at a^3, and the whole is sealed into the globe neck. The globe is exhausted through the tube *e*, which is afterwards sealed and broken off as shown in Fig. 1.

*1,173.—J. H. Johnson, London (*A. de Méritens, Paris*). **Electric Accumulators.** 2d. March 10.

SECONDARY BATTERIES.—The lead plates are separated from each other by frames of wood, preferably covered with india-rubber or other insulating material. The end plates are covered with slabs of wood, projecting above the plates, and through the projecting parts bolts are passed, to clamp the whole together.

1,174.—J. S. Williams, Riverton, N.J., U.S.A. Generation, Distribution, Storing, &c., of Electricity, and Apparatus therefor. 1s. 8d. (20 figs.) March 10.

DISTRIBUTING CURRENTS.—The flow, or head of water, in the pipes or mains, the reservoirs of a water supply system, the drainage from land, or the flow of matter in sewers, may be utilised to drive turbines, or water-wheels, with which are combined dynamo-electric generators. The currents so produced are conveyed into a prime storage system, and thence to auxiliary accumulators, placed at or near the various translating devices, or the prime system may be dispensed with, in which case the generators discharge their currents into the consumption circuits direct. The translating devices are placed in branches and sub-branches of the consumption circuit. To prevent overcharging the conductors, they are connected, by automatic cut-offs, with secondary batteries, so proportioned as to provide for the storage of all the currents that can be produced by the generators. The speed of the turbines is controlled by cut-off gates, which are opened or closed by an electromagnet, and a small electric motor.

DYNAMO-ELECTRIC GENERATOR.—To prevent heating the bearings, the axles have enlargements, which fit in corresponding cavities formed in the bearings, communicating with oil reservoirs, these again communicating with other reservoirs fitted with check valves. A thermostat and circuit closer are connected with the bearings, which, on their becoming heated, close the circuit of an electric motor, and thereby lower the cut-off gate, and so check the turbine. The field magnets are preferably excited by the current from a secondary battery, but may be placed in a shunt from the main circuit. The armatures are made in sections, of box-like shape, which contain the coils, and are so formed as to fit upon, and be secured "to the periphery of the body of the armature." Two armatures are used, and the coils are either coupled up independently, or the coils of one armature are connected to those of the other. The commutator is enclosed in a case, containing at its lower part an absorbent material and a lubricant, which serve to prevent sparking. The brushes are placed vertically, and are pressed

up by springs. The inducing faces of the armatures are arranged perpendicularly to the axles, and revolve in proximity to the outer faces of the field magnets ; or they may be arranged to revolve between the field magnets.

MOTORS.—These, for ordinary purposes, are constructed in the same manner as the generators. When used for propelling vehicles, on rail or tramways, the armatures are built up in sections, and are secured to the inner faces of the wheels. It is preferred to fit one carriage with sufficiently powerful motors to draw a train.

CURRENT METER.—A vibrating armature, pivotted to a suitable support, carries a spring pawl, which engages, on its forward movement only, with a ratchet wheel, combined with recording mechanism, through the medium of speed-reducing gear. The armature is spring mounted, and vibrates before a magnet arranged in a shunt circuit, which is made and broken by a suitable contact maker attached to the armature. A solenoid, included in the shunt circuit, controls the movement of the vibrating armature, through the medium of its core, and a cam-ended link, whose movement is checked by a dash-pot. In a modification, the vibrating armature is somewhat differently pivotted, and the dashpot is replaced by a spring. To prevent tampering with the meter by magnets on its exterior, and to prevent wear of the parts, the movable mechanism is immersed in an insulating liquid.

SECONDARY BATTERIES.—The elements consist of cells of suitable material, such as lead, into which is introduced granulated carbon and lead filings, with or without metallic oxides. The terminals are placed in the granulated mass. The electrolyte is an acidulated or alkaline solution. Lead plates only may be used, separated by granulated glass, or pumice stone.

1,191.—T. J. Handford, London. (*T. A. Edison, New Jersey, U.S.A.*) Regulating the Regenerative Capacity of Dynamo or Magneto - Electric Machines. 6d. (7 figs.) March 11.

CURRENT REGULATOR.—This invention relates to means for automatically regulating the generative capacity of dynamo-electric generators, to accommodate the addition or removal of lamps in a multiple arc system. It is an improvement upon the method described in Letters Patent No. 2,482 of 1881. The figure represents groups of lamps, arranged in multiple arc upon conductors branching from the mains 1 and 2. In one of each pair of branches is placed an electro-magnet B, provided with an armature f. The field circuit is a shunt from the main conductors, and it embraces a set of resistances R, R^1, and R^2, one for each branch circuit. Around the resistances are shunt circuits b, e, d, to the

armature levers f, and their front contact stops i, by the opening or closing of which, by the electromagnets B, the power of the field magnet is diminished or increased. The apparatus operates in the following manner : When there are too few lamps in circuit for the magnets to attract their armatures, the minimum current will flow through the field circuit, but after a certain number of lamps have been added to any particular group, (for example, to circuits 7 and 8), the magnet of that group will be sufficiently excited to attract its armature, and close the circuit. This will cut the resistance R^1 out of circuit, and strengthen the field magnet, whereby the electromotive force of the generator will be increased to the proper degree for supplying all the lamps of that group. As the lamp circuits are broken, the reverse action takes

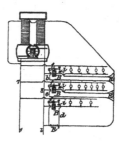

place. In lieu of cutting out and throwing in resistance, the electro-magnets may be used to close and open different circuits around the field magnets. Another part of the invention relates to the use of graduated or successively acting contrivances, for the same purpose. Electro-magnets of graduated power, arranged in a variety of ways, in connection with the main conductors, operate so as to cut out resistances in the field circuits, in succession, as a greater electromotive force is required. To prevent undue incandescence in the lamps, owing to an increase of speed of the motor, and consequent augmentation of electromotive force, an electro - magnet is placed in multiple arc from the main leads, and opens and closes a circuit, through a resistance in one of the mains, or around the graduated magnets, in such a manner as to weaken the power of the field circuit.

1,199.—R. Kennedy, Glasgow. Electric Lamp. 6d. (4 figs.) March 13.

ARC LAMPS.—This is a differential lamp, having the fine wire coil wound on the movable core of the thick wire coil, and travelling up and down with it. The current is conveyed in and out of the fine wire coil, by the chains from which it is suspended. Referring to the figures, the thick wire coil B, in the arc circuit, tends to raise the upper carbon, while the fine coil C, in the shunt circuit, tends to depress it. The latter is

wound on the soft iron tube, which forms the core of the coil B. The upper carbon is connected directly to this core, and passes through a guide o^1. The lower carbon is suspended by chains that pass over pulley v g^1, and are connected to the core, consequently the two carbons move simultaneously in opposite directions. One edge of the pulley g^1 is provided with ratchet teeth, engaging

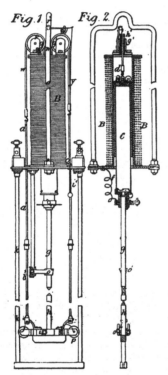

Fig. 1.   Fig. 2.

with a pawl h^1, to prevent the carbons drawing apart while the lamp is in action. A string y lifts out the pawl when fresh carbons are to be put in. The specification illustrates two modifications of the lamp, both embodying the same principle. One is arranged to stand on a table, and the other has both coils movable in relation to each other, the thick wire coil in the latter being suspended between suitable guides.

*1,201.—R. Matthews, Hyde. Dynamo-Electric or Electro-Dynamo Machines. 2d. March 13.

DYNAMO-ELECTRIC GENERATOR.— "Any number of electro-magnets are arranged to present any number of magnetic fields round a circle, presenting a number of successive fields of like polarity. A number of coils of insulated wire, or bars, may be wound or fixed, either on cores of iron, or of wooden material; but, by preference, the core is made of thin sheet-iron strips, arranged in lines of force, or parallel

thereto, or to which is fixed one, two, or three layers of the coils, and then another covering of iron, covered by another coil of insulated wire, and when all the insulated wire is coiled on, the two poles are coupled externally of the coil, so that the induced current commences in the cores before the coils come opposite the magnetic fields." Alternating generators are made, according to this invention, with an odd number of armature coils, and an even number of magnet fields. The commutators are made of rings, having teeth, the number of rings being equal to twice the number of armature coils. This invention appears to be included in Specification No. 3,334 of 1882.

1,211.—H. E. Newton, London. (*A. I. Gravier, Paris.*) Machinery for Obtaining Electric Currents. 6d. (4 figs.) March 13.

DYNAMO-ELECTRIC GENERATOR. — Referring to the illustrations, Fig. 1 is intended to illustrate the principle upon which the generator is constructed. A small helix h, surrounding the magnet A B, is moved successively from O to B, and from B to O^1, from O^1 to A, and back again to O, which movement gives rise to induced currents in the helix, flowing in the direction of the arrows. Fig. 2 represents diagrammatically a generator on this system, for

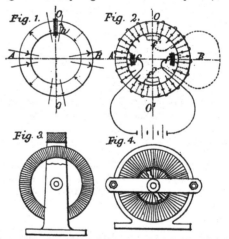

Fig. 1.   Fig. 2.

Fig. 3.   Fig. 4.

the production of continuous currents. The ring is made of iron, as in the Gramme generator, and is surrounded by a series of coils of insulated copper wire. The ends of the coils are connected up in series, and from each junction a wire is led to a plate of a suitable commutator, carried upon the spindle of the generator. In the event of a current being passed into the coils by the aid of two brushes at ff, an annular electro-magnet would be produced with consequent poles at A B, and neutral points at O O^1. During rotation of the ring, the poles and neutral points would not change their position in relation to the generator, but if the ring be fixed,

and the brushes revolve, the poles would travel around the centre. The ring is surrounded by a second series of insulated wire coils, arranged and coupled in a similar manner to the first series, the junctions being connected to commutator plates situated upon the opposite side of the generator. This commutator is provided with a pair of brushes $f^1 f^1$ placed at right angles to the first pair. On causing the composite ring to revolve, the successive passages of the secondary coils, over the consequent poles A and B, will induce currents in the coils, which may be collected by the brushes $f^1 f^1$. In place of causing the ring to rotate, it may, in some cases, be found convenient to make it stationary and to make the brushes $f f$ revolve. It may also be advisable to place masses of iron in front of the magnet poles, and these may form part of the framing, as shown in Figs. 3 and 4. The secondary coils are formed of either thick or thin wire, as compared with the primary coils, according to the character of the currents required. By causing the two secondary brushes $f^1 f^1$ to revolve with the ring, and by connecting them with a pair of insulated metal rings attached to the spindle, alternate currents may be collected. Instead of being supplied with an electric current from an external source, the primary coils may be fed, either partially or wholly, by the current induced in the secondary coils. In the latter case, however, it is necessary that the generator should be separately excited, at the commencement, to enable it to start.

1,225.—M. Evans, Wemyss Bay, Renfrew, N.B. Apparatus for Gauging Carbons, Filaments, &c. 6d. (10 figs.) March 14.

INCANDESCENCE LAMPS. — The filament to be gauged is placed between a fixed and a movable jaw. The position of the jaws regulates the angle of a mirror, which reflects a beam of light upon a scale, the particular point of the scale upon which the beam is thrown indicating the thickness of the filament.

1,249.—L. Levey and E. Lumley, New York, U.S.A. Armatures for Magneto-Electric Machines, &c. 6d. (10 figs.) March 15.

DYNAMO-ELECTRIC GENERATOR.—The core of a Gramme armature is formed of a number of alternate plates of sheet metal and non-conducting material, all the plates being cut in the form of a wheel, with hubs, spokes, and rim, and either with or without projecting pieces in line with the spokes. The plates are threaded on a shaft, and insulated wire is wound between the spokes in the usual manner. In some cases it is only the non-magnetic plates that have spokes, and in others ventilating spaces are left between the iron rings.

ARC LAMP.—This specification also describes an arc lamp, having two differentially wound magnets, acting on armatures on the opposite ends of a balance beam fixed on a rocking shaft. An arm on this shaft is connected by a rod to a clutch washer, which controls the feed of the upper carbon.

*1,274.—F. Wright and M. W. W. Mackie, London. Incandescent Electric Lamps. 2d. March 16.

INCANDESCENCE LAMPS.—The globes are made with a tubular stem for carrying the conducting wires, and the tube of the vacuum pump. The filaments are prepared from the fibres of a plant specially grown in distilled water, or other medium not containing inorganic matter.

1,288.—J. B. Rogers, London. Incandescent Lamps, &c. 6d. (14 figs.) March 16.

INCANDESCENCE LAMPS.—The platinum conducting wires are sealed directly into the globe, without the use of leg-pieces, a portion of the loops forming the external terminations of the conductors being sealed into the glass. The sockets are provided with spring loops, or fingers, tending to force the lamp out, and so insuring good contact between the platinum loops and the external conductors. A switch has a plug of insulating material, with a conducting portion let into it in such a manner, that, on turning the plug a quarter-turn, the circuit is broken. A testing apparatus for filaments acts on the principle of the Wheatstone bridge. The pump is of the Sprengel type. Neither of these two latter arrangements are included in the provisional specification, nor are their points of novelty specified.

1,302.—Hon. R. Brougham, London. Electrolier. 6d. (7 figs.) March 17.

FITTINGS.—The electrolier consists of a tubular stem fixed in any suitable manner. There is a collar upon the stem, and beyond this a flanged nut, screwed upon the stem. Two metal plates of ornamental form are clamped between the collar and the nut, and are separated by a thick disc of vulcanite, and thinner discs of vulcanite serve to insulate the plates from the collar and nut upon the stem. The parts of the plates projecting over the separating disc are formed with radial grooves on their inner surfaces. Each lamp is inserted between the two plates carried by the stem, and is provided with two contact springs, diametrically opposite each other, fixed to the exterior of the socket which encloses the neck of the lamp. The springs press against the plates when the lamp is in place, and the plates are connected to the conductors passing down the centre of the stem and through the thick vulcanite.

1,324.—J. D. F. Andrews, Glasgow. Electric Lamps. 6d. (9 figs.) March 18.

ARC LAMP.—The regulation is effected by a brake, taken off by the core of a solenoid in a shunt circuit, when the arc resistance increases. The solenoid core is attached to the brake lever by a slotted connection, so that the brake is not affected by minute variations of current. Referring to Fig. 1, the upper carbon is carried by a vertical tube, sliding in guides over a fixed piston H. The tube and piston together form a dash-pot, which modifies the motion of the electrode. As the arc resistance increases, the shunt current in the solenoid raises the lower electrode and its core, until the brake D is raised, when the upper electrode slides down. "Or instead of the dash-pot, a second clutch may be employed, worked by a third arm of the core lever G, the arms being so arranged that one clutch becomes released just before the other. Thus the clutch

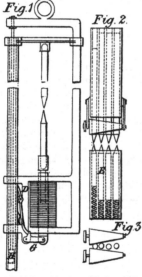

Fig. 1   *Fig. 2*   *Fig. 3*

first released, being of the oblique plate" (Brush washer) "kind, and the other, being a friction clutch, the carbon can descend slowly when the first is released, but cannot descend rapidly until both are released." Figs. 2 and 3 show an arrangement for storing several pairs of carbons, and bringing them successively into action. The lower carbons E are each forced upwards by a strong spring, but are retained within their case by an internal lip that runs along each edge. The upper ones are kept in position by friction, with the exception of one that is grasped between a pair of oblique or conical rollers driven from the controlling mechanism. When the first upper carbon is burnt out, the stump falls from between the rollers, which are immediately pressed nearer together by springs and grasp the second one, and so on.

INCANDESCENCE LAMP.—An incandescence lamp, according to this invention, has a carbon plate, or a series of plates, in lieu of filaments. These are clamped at the opposite edges between loops of the conducting wires. The carbon plate may be divided into sections, which are arranged so that one or more may be brought into circuit, according to the illuminating power required.

1,327.—L. J. Crossley, Halifax, J. F. Harrison, Bradford, and W. Emmot, Halifax. Transmission of Electric Currents of High Tension. 6d. (1 fig.) March 18.

DISTRIBUTING CURRENTS. — This invention relates to certain arrangements by means of which "telephone and other lines of suitable resistance may be used for the conveyance of powerful currents of electricity for electric lighting, electro-plating, the transmission of power, and other purposes where electricity of high tension is employed, without danger to the usual telephonic, telegraphic, and other instruments." In this specification "tension" appears to be employed to signify strength of current. The main leads, coming from the dynamo-electric generators A, situated in the central exchange, are passed through a switch B to the lines, and in each line is interposed an indicator C, wound with thick wire. Each subscriber is also provided with a

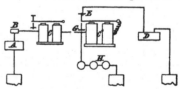

safety indicator C^1, wound with thick wire, and connected to line, the telephone apparatus D being connected to the repose contact E of this indicator. A current of "high tension," flowing through the indicators, will work them both, and the armature F of the indicator C^1, coming in contact with the stud G, will automatically switch the line to the lamps H, or to a secondary battery, or elsewhere, while the currents used in telephonic signalling will not affect the armature, and thus, under ordinary circumstances, the line will be switched to the telephone. When the subscriber wishes to communicate to the central office, he breaks the circuit, and the communicator C indicates the fact. A secondary battery may be used to maintain the lights while the line is in telephonic operation.

1,336.—A. J. Boult, London. (*J. W. Thomas and L. F. Requa, New York.*) **Manufacture of Covered or Insulated Wire.** 8d. (5 figs.) March 18.

CONDUCTORS.—A series of wires, and, at the same time, each wire independently, is covered with two separate longitudinal strips of rubber, each strip having a width equal to half the circumference of the wire; these strips are curved round the wire and joined together at the edges by passing between multigrooved rollers. Two coatings of this kind are applied to the wire, with an intermediate winding of tarred tape between the two rubbers. The covering apparatus consists of two series of grooved rollers, with separate bobbins, carrying the narrow rubber strips, and mounted in operative relation to the separate grooves of the rollers, and with tape-winding devices between the two sets of rollers, whereby the wires are drawn through the first set of rolls and are covered with one rubber coat, which is then wound with the intermediate tape, and finally covered with a second rubber coating in passing through the second set of rollers.

*1,345.—D. Bremner, London. Coupling for Joining Lengths of Wire. 2d. March 20.

CONDUCTORS. — The coupling consists of two metallic parts, identical in shape, and capable of interlocking by means of inclined surfaces, or tongues, or grooves. Before, or after being placed in the coupling piece, the extremity of each wire is bent at an angle, to prevent it slipping through the coupling.

1,347.—S. E. Phillips, London. (*Partly W. C. Johnson, Mentone.*) Machine for Generating Electric Currents. 4d. March 20.

DYNAMO-ELECTRIC GENERATOR.—The object is to wind a given weight of generating conductor into a smaller space than by the usual methods. On the outer edge of a disc, preferably of metal, either a single or double circle of coils is fixed, one circle on each side of the disc. Or a single circle of coils is made to drop into openings cut in the disc, in addition to the other coils. The coils are preferably made from square wire. The field magnets and commutator are of the ordinary construction.

1,363.—F. Maxwell-Lyte, London. Secondary Batteries. 4d. March 21.

SECONDARY BATTERIES.—The electrodes are made of insoluble, or only slightly soluble, lead salts, such as the haloid salts of lead, fused and cast into plates, which are then "formed" in a solution of sulphuric acid in the usual way. The electrodes may be cast upon central wire cores, or be run into a network of wires, or porous carbon. Soluble lead salts are also run into any suitable porous material, and converted, within the pores, into an insoluble

salt, by immersion in a saline solution. The electrodes are built into batteries with separating diaphragms of porous paper, or other suitable material.

1,368.—A. E. Dolbear, London. Electrical Cables. 6d. (3 figs.) March 21.

CONDUCTORS.—To minimise the extent of contact surface between a conductor and its insulating covering, the conductor is supported centrally in a non-conducting pipe, as shown in the figures. Around the conductor *a* is wound spirally a cord of cotton, or other suitable material, preferably hard twisted, its function being to prevent the collapse

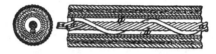

of the insulating pipe *d*. When the pipe is sufficiently strong, the spiral is omitted, and the conductor allowed to lie in direct contact with the pipe. The outer sheathing may be of any approved construction. The pipe *d* is preferably formed of several thicknesses of paper, rendered waterproof, and cemented by india-rubber cement.

1,390.—J. B. Rogers, London. Driving, Regulating, and Indicating Speed of Dynamo Machines. 6d. (11 figs.) March 22.

DYNAMO-ELECTRIC GENERATOR.—This invention relates to apparatus for driving a dynamo-electric generator from the axle of a railway carriage, in such a manner, that its speed is maintained at a uniform rate at whatever speed the train be moving. This is effected by cone pulleys, and a belt-striking gear, under the control of an attendant. The apparatus is provided with reversing gear. The invention, secondly, relates to speed-indicating apparatus, in which a vertical spindle is rotated from one of the drums of the driving mechanism. The spindle is hollow, and is surmounted by a cap, from which two arms project at right angles. These arms carry weights, connected by chains to a crossbar, which is free to rise within slots formed in the spindle. The ends of the chains are attached to a spring encircling the spindle, which opposes the centrifugal force of the weights. The cross-bar carries a rack, into which a pinion, on the back of a dial plate, gears, and to which the indicating finger is attached. It is proposed to use secondary batteries to maintain the lights during the stoppages of the train.

*1,392.—W. Graham and H. J. Smith, Glasgow. Incandescent Electric Lamps. 2d. March 22.

INCANDESCENCE LAMPS.—The filaments are end-

less bands of carbon, the two conductors being connected at points on the circumference equidistant from each other. In a modified arrangement, small carbons are used in place of endless bands.

1,400.—T. E. Gatehouse, London. Incandescent Electric Lamps. 2d. March 23.

INCANDESCENCE LAMPS.—In place of employing the compound filament of platinum and carbon in two parallel circuits, as described in Patent No. 3,420 of 1881, the two materials are arranged to form a continuous filament. A convenient arrangement is to attach to the conducting wires, hermetically sealed in the glass globe, short lengths of platinum wire, connected by a loop of carbon filament. To obtain the best effect, each part of the conductor should have such dimensions, that the resistance of the platinum, at ordinary temperatures, should be to that of the carbon, in about the proportion of 3 or 5 to 100. Such a compound conductor may be introduced into the main conductor of a series of lamps, so as to act as a regulator.

1,412.—O. E. Woodhouse and F. L. Rawson, London. Electric Lighting. 6d. (32 figs.) March 23.

INCANDESCENCE LAMPS.—One part B, of the lamp globe, shown at Fig. 1, is rendered reflective, by the application of silver, or tin foil, either inside the globe before exhaustion, or preferably outside after exhaustion. When externally applied, the silver is protected by a coating of copper, and that, again, by a coat of thin metal, or varnish. Several modifications are described for accomplish-

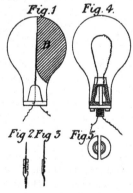

Fig. 1   Fig. 4.

Fig 2 Fig 3   Fig. 5.

ing this object. The filament may be formed of carbon and a metallic substance, the latter being extended beyond the carbon, so as to form a reflector. A modification retains the compound form of filament, but dispenses with its reflective portion. Figs. 2 and 3 represent a method of connecting the filaments to the conductors. The conductors ter-

minate in thin copper, or platinum plates, which are notched, as shown. The pieces notched out are turned up in such a manner as to hold the ends of the filament in a slightly bent form. Or the conductors may terminate in a spiral, bent out of line. The so formed spring is closed together, and the end of the filament inserted, the spring is then allowed to extend, and so holds the filament. Various modifications for connecting the conductors and filaments are described and illustrated. Figs. 4 and 5 show an undulatory junction spring clip, for attaching the lamp to a stand, so as to make a secure joint, and good contact. The stand consists of a split metallic cylinder, each half insulated from the other, and connected each to one of the conductors. The semi-cylindrical clips are undulated in the direction of their length, and fit similar undulations in the neck of the lamp. The ends of the platinum conductors are brought through the upper part of the neck, and are led along the interior of the undulatory plug. When the lamp is pressed on to the cylindrical joint, the latter opens by its elasticity, and so makes contact, the clip serving to hold the lamp.

1,437.—S. Cohne, London. Electric Accumulator. 4d. March 25.

SECONDARY BATTERIES. — A sheet of lead is covered with a layer of mercuric sulphide (Hg S), in the form of cinnabar, or vermilion. The coated lead sheet is bent into the form of a box, or of a spiral, and is used as the positive electrode of the battery. A second sheet is prepared in the same way for the negative electrode. Both sheets are perforated, and are placed in a cell of dilute sulphuric acid. "As soon as the current enters, hydrogen is liberated, which reduces the mercuric sulphide, gradually causing a deposition of metallic mercury on the surface of the lead. Immediately the effects of polarisation are manifested, local action disappears, then the amalgam, formed without being destructive to the lead, decomposes the water, and in that way hydrogen is always conserved, and travels backwards and forwards as the cells are charged and discharged, and peroxide of lead is thus formed, the precipitated sulphur acting as a resistance." Sulphate of mercury ($Hg^2 O S O^3$) may be used in place of mercuric sulphide.

1,444.—R. Werdermann, London. Electric Incandescent Lighting Apparatus. (3 figs.) 6d. March 25.

SEMI-INCANDESCENCE LAMP. — In the lamp represented in Fig. 1, a carbon pencil a is kept in contact with a carbon or metal electrode of larger cross sectional area b, by the pressure of mercury, glycerine, or other suitable liquid. To

prevent incandescence of the carbon pencil below the desired distance, upon the disc d is fixed a collar e of steatite, or magnesia, in such a way that the pencil can slide freely through it. The disc d has a rim for the reception of the glass cylinder f, which is hermetically closed at its upper end by the upper electrode, and at its lower end by the liquid in the tube c. When the carbon pencil becomes incandescent the small amount of

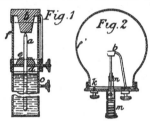

oxygen in the lamp is quickly consumed, and the pencil is then not exposed to further oxidation. Fig. 2 shows a modification in which the carbon pencil and the upper electrode are kept in contact with each other by means of a helical spring m. Upon the disc k, to which the globe f_1 is hermetically attached, is fixed a bracket n, carrying a small grooved metal roller to guide the pencil a. The globe may be exhausted if found desirable.

*1,455.—G. Molloy, Dublin. **Secondary Batteries for the Storage of Electricity.** 2d. March 27.

SECONDARY BATTERIES.—Each plate, consisting of a thin sheet of lead, suitably folded, is fitted in a frame. The plates are placed side by side in a rectangular box lined with lead, the alternate plates being connected together. Each plate, after being fitted in its frame, is covered over with a layer of red oxide of lead, and immersed in dilute sulphuric acid, and left therein for three or four days, during which no current is sent through the acid. The battery is then ready to be charged by a dynamo-electric generator, or otherwise.

1,462.—S. Waters, London. **Electric Lamps.** 2d. March 27.

INCANDESCENCE LAMP.—A portion of the globe of an incandescence lamp is formed into a concave mirror, by coating the glass externally with silver by a chemical process. Tartrate of silver is mixed with ammonia until a milky solution is formed, in which is plunged the part of the lamp to be coated. By passing a current of electricity through the lamp, the solution becomes heated, and metallic silver is deposited upon the immersed portion of the globe. To prevent a dark shadow being thrown by the coated portion, the deposit is made sufficiently thin to be semi-transparent.

1,464.—F. de Lalande, Paris (*Partly G. Chaperon, Mines d'Aloisno, Spain.*) **Electric Piles or Batteries.** 4d. March 27.

SECONDARY BATTERIES.—The principal feature of this battery is the employment of copper oxide as the depolarising material. In one form of construction, the positive electrode consists of a plate of gas retort carbon, contained in a porous cell, and surrounded by fragments of carbons, the interstices being filled with oxide of copper. The negative electrode is formed preferably of zinc in an exciting solution containing from 10 to 30 per cent. of caustic soda or potash. Palladium, platinum, or iron may be used for the negative electrode.

1,465.—A. Smith, London. **Carbons for Electric Lamps.** 4d. March 27.

INCANDESCENCE LAMPS. — For producing high-resistance filaments for incandescence lamps, an excess of hydrochloric acid is passed through the liquid furfurol or fucusol. The reaction produces a black liquid, which is enclosed between two glass plates, separated to a distance equal to the desired thickness of the filament. After the liquid has set the plates are wedged apart, and that plate to which the film adheres is plunged into cold water for the purpose of detaching the film, which is then floated on to a flat surface, and the excess of moisture removed by blotting paper. The film is cut into strips, and these are bent between moulds of suitable shape, made of plaster-of-paris, and exposed to a temperature of 212 deg. When removed from the moulds, the filaments are subjected to a high temperature in closely covered crucibles surrounded by carbon powder, or in porcelain tubes partly filled with carbon powder, through which a current of coal gas is passed during the process. Another method is described, in which furfurol or fucusol is treated with commercial sulphuric acid of specific gravity about 1.84.

CARBONS.—Carbons for arc lamps are produced in the usual way, from finely divided carbon, by compression, or by forcing through a die, and they are afterwards treated with furfurol or fucusol and hydrochloric acid.

*1,483.—C. L. Clarke and J. Leigh, Manchester. **Coiling Machine.** 2d. March 28.

CONDUCTORS.—This invention relates to so fitting the machine described in Patent No. 3,652 of 1881, for making induction or resistance coils, that it shall automatically stop and reverse the direction of the guide at the end of each traverse, so as to allow of the paraffined paper being more readily put in place between the coils.

1,496.—T. J. Handford, London. (*T. A. Edison, New Jersey, U.S.A.*) **Dynamo, or Magneto-Electric Machines.** 6d. (7 figs.) March 28.

CURRENT REGULATOR.—This invention relates to means for automatically regulating the generation of current in dynamo-electric generators. This is accomplished by the use of a rapidly vibrating circuit controller, automatically operated by the current, and serving to regulate the power of the field magnet, by rapidly and successively opening and closing the field magnet circuit. The controller is caused to make and break circuit at a number of points simultaneously, in order to reduce sparking. It is preferably vibrated by an electromagnet, placed in a shunt circuit from the main conductors, an additional resistance being, by preference, placed in the same circuit. The armature lever of this magnet is provided with an adjustable retractor spring, by means of which the power of

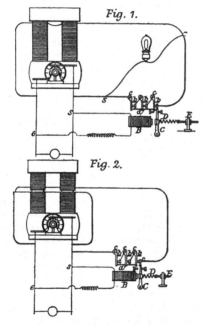

Fig. 1.

Fig. 2.

the field magnets, and, consequently, the candle-power of the lamps, can be regulated. Figs. 1 and 2 represent the field-magnet circuit adapted to be broken, and made, at a number of points simultaneously, by levers c, retracted by springs b against stops, the levers and stops being connected to form a complete circuit in the retracted position. Insulating pins d are carried by all the levers, with the exception of the front one, and they press against the levers immediately in front of them. This mechanism is operated by an electro-magnet B, placed in a multiple arc circuit 5, 6, with or without an additional resistance. Its armature lever C is drawn back by a spring D, which is connected by an adjusting arrangement with the standard E, and when acting under the attraction

of B, its forward movement forces the levers c from the screws, thus breaking the circuit. In order to arrest sparking at the points of the controller, the shunt circuit 7 s is employed, including a resistance, which may be of wire, or an incandescence lamp (Fig. 1), or a portion of the coils of the field magnet (Fig. 2). The working of the apparatus is as follows : When the electromotive force is normal, the magnet B will vibrate the lever C with a regular rapid movement, keeping the main field circuit open a certain portion of the time. A decrease in the electromotive force, produced by the addition of lamps, or a decrease in the speed of the engine, will tend to weaken the magnet B, and its vibrating armature will play more upon its back contact, changing the proportion of the time of the make and break, and allowing the field magnet circuit to remain closed for a greater portion of the time. This will increase the power of the field magnet, and reproduce the normal electromotive force in the main circuit. An increase in the electromotive force will have the opposite effect. In the position of the parts illustrated in the figures, the lever c rests against its back contact, the circuit being closed through the controlling apparatus. In a modification, a second electro-magnet is placed in the same, or in another shunt circuit, and serves to open and close a shunt circuit round the first electro-magnet, by means of an armature acting as a circuit controller, or to make and break directly the circuit of the first electro-magnet. According to another modification, the spark-arresting resistance is divided into a number of fixed or constant parts, around each of which is a shunt circuit, controlled by an electro-magnet, placed by preference in a shunt around a resistance inserted in one of the main conductors. By this means the resistance in the spark-arresting circuit is automatically varied to suit the number of translating devices in circuit.

1,516.—J. Imray, London. (*La Société Anonyme des Cables Electriques, Paris.*) **Material for Electric Insulation.** 4d. March 29.

INSULATING MATERIAL.—Linseed oil, either with or without an admixture of colophonium, is converted into a solid elastic body by the application of heat, and is then used for insulating purposes.

1,548.—W. B. Brain, Cinderford, Gloucestershire. **Secondary Batteries.** 6d. (18 figs.) March 30.

SECONDARY BATTERIES. — Chambers, bags, or closed or partly closed envelopes, enclosing the active agent, are formed of thin sheet lead, or other suitable metal, and combined with such chambers, whether perforated, corrugated, or solid, are pieces

of asbestos wick, felt, spongy india-rubber, flannel, &c., or other non-metallic, porous, and by preference elastic material. Fig. 1 represents an envelope or bag in the course of construction, A being a strip of lead, and C the porous insulating material. Figs.

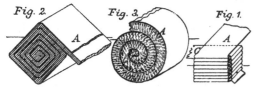

Fig. 2. *Fig. 3.* *Fig. 1.*

2 and 3 illustrate methods of wrapping a band of lead, and a band of porous material, coated with red lead on both sides, so as to form a series of continuous chambers. Two, or any number of these envelopes, may be grouped to form one battery.

1,556.—J. S. Williams, Riverton, N.J., U.S.A. Generation, Storage, and Utilisation of Electricity. 1s. 4d. (35 figs.) March 30.

DISTRIBUTING CURRENTS. — The patentee explains his system of distribution, by the aid of three sheets of elaborate drawings, and a description extending over six pages of the Blue Book. A water-wheel is used to drive a dynamo-electric generator, which supplies its current to a series of secondary batteries. The generator and batteries are connected together by switching devices, so arranged as to provide for charging one portion of the batteries while another portion is supplying current to the discharging circuit. A "converting" device regulates the tension of the current discharged. A meter serves to record the current passed. A "reserve" device stores the current near the point where it is to be used; this reserve may be employed to drive motors, in which the speed is regulated by governors, or otherwise. It is proposed to use light conductors, and currents of very high tension. In driving vehicles on rails, the current is supplied to secondary batteries, carried on the vehicles, an automatic cut-off preventing any back flow of current from the batteries which are mounted on the vehicles. The circuit to the motors may be completed through the wheels, or by special circuit makers. The signals, switches, brakes, and crossings may be actuated by the current. The following specifications are referred to : Nos. 5,233 and 5,742 of 1881, and Nos. 85, 224, 700, 766, 856 and 1,174 of 1882.

DYNAMO-ELECTRIC GENERATOR.—The armature may consist of one wheel, or of a series of wheels, covered with insulating material, and provided with "a coating of conducting material, such as copper," by the ordinary process of electro-deposition. The wheel has a continuous, or sectional band of metal, arranged upon or near its periphery. The currents are collected in the usual

manner. The field magnets are in the form of boxes, filled with scraps of magnetic metal; they may be employed as "reservoirs," in which case a suitable copper plate is used "to induce the currents from the copper wheel." When two armatures are used, they are so arranged that one will generate positive and the other negative electricity. The inductive action of the earth may be utilised in generating currents. The polar extensions of the field magnets may cover the periphery and the sides of the inducing portion of the armature, or the sides only, the winding being such as to insure all the poles on one side being of one denomination, and all on the other side of the opposite denomination. Several drum armatures are illustrated, some of which are provided with projections designed to revolve in recesses formed in the polar extensions of the field magnets. The projections may extend either laterally or from the periphery. The field magnets may be separately excited, or they may form a shunt to the main circuit.

1,570.—W. Jeffery, London. Electric Arc Lamps. 6d. (8 figs.) March 31.

ARC LAMPS.—In the lamp first described, a ring *l*, fitted with an elastic lining, is used as a clutch for controlling the feed of the upper carbon. The ring is attached to the connecting link *i*, by pins or clips, and is held by the adjustable arm *l¹* upon the opposite side of the carbon. The connecting link *i* is actuated, in opposition to an adjustable spring, through a second link and beam, by the core of a

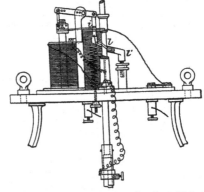

solenoid placed in the main circuit. This lamp may be adapted for burning two pairs of carbons, either together or singly. A lamp is described in which two upper carbons, inclined together at their points, feed down by gravity to a lower carbon, which is controlled by the elastic ring mechanism, and fed upwards by a weight. A shunt circuit, of approximately the same resistance as the arc, is closed automatically by the core of the regulating solenoid, upon the lamp circuit being broken.

1,580.—Sir D. Salomons, Tunbridge Wells. **Electric Lamps.** 6d. (8 figs.) March 31.

INCANDESCENCE LAMPS.—The "filament" is constructed, as shown in Fig. 1, of a block or stick of baked pipeclay *b*, coated with a composition *c*, about one millimetre in thickness, formed of equal weights of plumbago, charcoal powder, and sugar. The whole is subjected to a white heat in a closed vessel for about six hours, and is then sealed, by means of platinum wires *d*, in a glass envelope *a*, which is then exhausted in the usual manner. Several other forms of composite filaments are described and illustrated. In one form, a stock of baked pipeclay is enclosed in a thin tube of

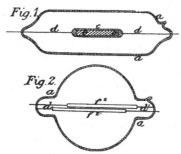

platinum, the platinum connecting wires being kept in position in the globe by means of corks. In another, a stick of opal glass is surrounded by a carbon cylinder, whose internal diameter is about two millimetres greater than the diameter of the glass stick. Fig. 2 represents an arrangement in which two sticks of carbon $f^2 f^2$ are connected to platinum wires *d* and *d'* in an exhausted globe *a*.

SEMI-INCANDESCENCE LAMP.—In a similar arrangement, a small arc is formed in an exhausted glass globe, between a short thick carbon pencil and a carbon ball.

1,587.—A. Tribe, London. **Secondary Batteries.** 4d. April 1.

SECONDARY BATTERIES.—The negative element consists of a rectangular frame, having three sides of prepared wood, slate, or other insulating material, the fourth side being constructed of a suitable conducting material. Into, or upon, this frame, peroxide of lead is placed, and the whole is ensheathed in parchment, felt, or other porous material. The positive element and the electrolytic liquid are of the usual description.

*1,611.—W. R. Lake, London. (*E. Weston, Newark, N.J., U.S.A.*) **Electro-Magnetic and Regulating Apparatus for Electric Lamps, &c.** 2d. April 3.

ELECTRO-MAGNETS.—To obtain great power, and to economise space, these are constructed as follows:

A number of convolutions of thick copper wire are wound upon a hollow metallic spool, provided with flanges at either end. An iron core, working within the hollow spool, forms part of an iron shell which completely encloses the top and sides of the spool and its helix.

1,614.—W. R. Lake, London. (*E. Weston, Newark, N.J., U.S.A.*) **Magnet› or Dynamo-Electric Machines.** 6d. (3 figs.) April 3.

DYNAMO-ELECTRIC GENERATORS.—This invention relates to the class of electric generators which have armatures wound with a plurality of coils, the ends of which are united to form an endless wire, and from which loops are taken off at the junction of the original coils, and connected to separate insulated commutator plates. To avoid short-circuiting, the armature is wound with two or more independent endless wires, which are connected by loops to the commutator plates D, so as to avoid having adjacent commutator plates connected to loops from adjacent coils of the same wire. The illustration

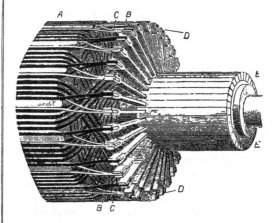

represents the end of a cylindrical armature wound with two wires B and C, shown in white and black respectively, and from this it will be seen that the black and white wires are connected to alternate plates of the commutator. In a generator provided with an armature of this type, the brushes, as they are left by one pair of segments of the commutator, will bear upon another pair of segments with which there is no tendency to short-circuit; and, should two adjacent segments become accidentally short-circuited, it is claimed that no injury can result to the generator, and that no loss of useful effect will occur.

*1,616.—W. B. Brain, Cinderford, Gloucestershire. **Production of Electric Currents.** 2d. April 3.

GENERATORS.—It is proposed to dispense with

the commutator, by so winding the armature coils that they shall remain stationary, while the armature core revolves within them.

*1,618.—J. B. Rogers, London. Electric Lamps. 6d. (3 figs.) April 3.

INCANDESCENCE LAMPS.—The conducting wires projecting beyond the base of the lamp are plain, bent, cup-shaped, or otherwise modified, so as to be readily put in contact with the leads, which are arranged in or on ledges, shelves, brackets, or other suitable supports. The main leads from the generator are attached to vessels, from which secondary leads are taken to the supports for the lamps. The lamps can thus be moved about from one room or bracket to another. A socket is described in the final specification, consisting of a socket piece through which the ends of the lamp wires are carried, and then bent so as to form " springy holdfasts" in a bracket piece containing the terminals of the conducting wires.

1,619.—W. R. Lake, London. (*H. S. Maxim, New York, U.S.A.*) Carbon Conductors for Electric Lamps. 6d. (5 figs.) April 3.

INCANDESCENCE LAMPS.—Continuous lengths of carbon filament for incandescence lamps are formed in a press, shown in the illustration. A quantity of plastic carbonisable compound, such as finely powdered retort carbon and coal-tar, is introduced under the plunger of the lower chamber, and is

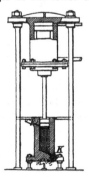

forced through the nozzle K, which is adjustable either by hand or automatically. The adjustments of the nozzle take place periodically, and are so timed, that enlargements shall occur at equal distances along the continuous strip. When the strip is divided in lengths, the enlargements form the ends of the separate filaments. Provision is made for opening and shutting the die gradually, so that the enlargements and contractions do not take place suddenly.

1,626.—J. Munro, London. Electric Light and Power Apparatus. 8d. (17 figs.) April 4.

DYNAMO-ELECTRIC GENERATOR.—In the generator shown in Fig. 1, the armature coils, each of which supplies its current to a separate circuit, are stationary, while the field magnets revolve. The "inducing disc" may be of brass, and have either segmental slabs of soft iron, or electro-magnets bolted on to it. The inventor "may use as a generator" earth currents, or the currents obtained from lightning-rods, these being collected and stored in secondary batteries.

DISTRIBUTING CURRENTS. — A rotary current distributor, illustrated in Fig. 2, has a rotating arm B, which is in electrical contact with the generator, and distributes the current successively to the contacts c, which in turn supply the current to a series of secondary batteries. When the battery is sufficiently charged, the lamps,

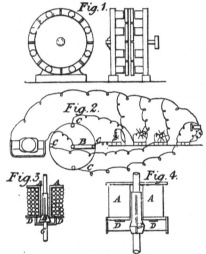

forming a derived circuit from it, are switched in, and the feeding current splits up between battery and lamps. To avoid undue sparking at the contacts, it is proposed to employ condensers, or "a vacuum rotating contact make and break."

SECONDARY BATTERIES.—To prevent a reverse current entering a storage battery, a contact breaker, having a needle mounted similarly to an ordinary galvanometer needle, placed within a high resistance coil forming a shunt to the main circuit, is so arranged, as to break the circuit on the reversal of the current. The temperature of the batteries is raised during the time of charging.

ARC LAMP.—The feed is controlled by the electromagnets A A, Figs. 3 and 4, lifting the conical armature B, when the arc resistance becomes greater than that of the magnets. The springs D D cause the armature to grip the carbon-holder C, when it is in its lower position. In another lamp, two low resistance solenoids, in circuit with the arc, attract or repel, according to the arc's resistance, a high

resistance shunt solenoid, connected by suitable gearing with the upper carbon. Another method of regulation consists of an open spiral, which expands or contracts according to the current passing. The arc is preferably maintained in a vacuum, and is enclosed in a ferrule of lime. The globes for use in foggy weather are preferably made of uranium glass, or a tint may be given to the arc by supplying certain gases to it. The carbons are preferably built up on a core by chemical deposition of pure carbon.

RHEOSTAT.—Elastic carbon filaments, such as are used in incandescence lamps, are adjusted to given resistances and arranged in sets.

SWITCHES.—The hard insulating substance usually placed between the contact plates, is replaced by a soft substance, preferably india-rubber, by which attrition of the moving metallic surface is avoided.

CURRENT METER.—A disc of copper, mounted on an axle, is surrounded by a horseshoe magnet. When a current passes by the axle, through the wheel, into mercury, in which the edge of the disc dips, the disc rotates, and by a suitably graduated counter, registers the current passed. The whole apparatus is enclosed in a thick case of soft iron. In driving vehicles, it is proposed to lay a conductor between the rails, and pass the current at intervals into secondary batteries carried on the vehicle, these supplying the current continuously to any suitable motor.

1,640. — R. Kennedy, Glasgow. Dynamo-Electric Machine. 6d. (3 figs.) April 4.

DYNAMO-ELECTRIC GENERATOR.—The object is principally to obtain several currents of electricity from one continuous rotating wire. The figure represents the main feature of the invention. The armature consists of an inner coil of iron wire,

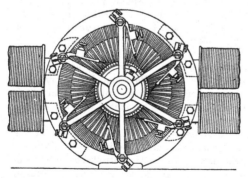

wound continuously in one direction. The generating coil of insulated wire is wound over the iron wire coil at right angles to it. The connections to the commutator are so made that the coil shall be endless. The inner and outer coils are held between two discs of wood, provided with iron plates

on their outer surfaces, by which they are secured to the shaft. The commutator is of the usual multiple plate pattern, while the electro-magnets are arranged to form two independent poles of one polarity above, and two independent poles of the opposite polarity below. Currents are taken off the commutator by six brushes. The field magnet poles may be an even number greater than four, but in any case the number of collecting brushes is greater by two than that of the magnet poles.

1,642.—W. H. Akester, Glasgow. Incandescent Electric Lamps. 6d. (5 figs.) April 5.

INCANDESCENCE LAMPS.—The object of this invention is to facilitate the manufacture of the lamp and the fixing of the filament. Fig. 1 represents the glass globe ready for the insertion of the filament. The exhaustion is designed to be effected through the tube 7, although, if desired, it may take place through the neck 8, the former tube being then dispensed with. Fig. 2 shows the carbon filament 10, with its platinum wire con-

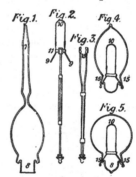

ductors 9, which are fused into the opposite ends of a glass cross-piece 11, held by a pair of grippers, ready for being inserted into the globe. Fig. 3 is a side view of the grippers. Figs. 4 and 5 illustrate completely formed lamps, the platinum wires being fused into the horns 15, and terminated in loops. In the former figure, the lamp is represented as having been exhausted through the tube 7, while in the latter the exhaustion has been effected through the neck 8.

1,647.—St. G. Lane-Fox, London. Manufacture of Incandescent Electric Lamps. 6d. (3 figs.) April 5.

INCANDESCENCE LAMPS.—This invention relates to a method of thickening the carbon filaments at their junction with the connecting wires, and is an improvement upon the methods described in specification of Letters Patent No. 3,494 of 1880, and No. 225 of 1881. Referring to the figures, in order to cause the filament to adhere to the platinum spirals

b, a cement is used, consisting of Chinese ink and plumbago, or other suitable mixture, and the whole is then immersed in an atmosphere of coal gas or carbon compound. By means of the loop *i*, contact is made between the filament and one pole of an electric source, the other pole being connected successively with the carbon pencils *l l*. By this means a small arc is formed between the pencil and the spiral junction, and carbon is deposited, partly from

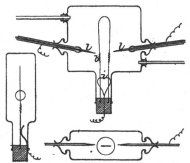

the gas, and partly from the pencil, upon the platinum and filament at the point of contact. The filament is preferably thickened by this method for a distance of about a quarter of an inch from the holder, and the deposits should taper towards the loop of the filament. The arrangement adopted in carrying out this invention is clearly shown by the accompanying illustrations.

1,649.—A. J. Boult, London. (*J. D. Thomas, New York, U.S.A.*) **Underground Conduits for Electric Wires, &c.** 6d. (4 figs.) April 5.

CONDUCTORS.—The conductors are carried in conduits made of baked pottery clay, or equivalent earthenware, formed in hollow longitudinal sections, laid end to end, and secured together by Portland cement. Each longitudinal section consists of an upper and lower half, the latter being trough-shaped with one or more compartments for the reception of the wires. The upper half, or lid, is provided with marginal lips overlapping the lips of the

trough, to which they are afterwards cemented. The ends of each section of trough and lid break joint with the next section, by means of over and underlying ledges, which are made fast together by cement. Prizing notches are formed on one or both sides of the cover, or of the trough, to facilitate their separation when required. For the purpose of insulating the wires, they are preferably coated with india-rubber, according to a process described in an

application for Letters Patent No. 1,336 of 1882. At certain intervals along the conduit, elbow lids are introduced for the purpose of "tapping" the conductors. The lengths are connected together by screw coupling sleeve pieces, which screw on to the threaded ends of the conductors.

1,670.—J. Jameson, Newcastle-upon-Tyne. **Incandescent Electric Lamps.** 4d. April 6.

INCANDESCENCE LAMPS.—The carbon filaments are produced by depositing carbon from hydrocarbon gas of not too rich a character, under a small pressure and a high temperature, around a matrix of glazed porcelain, or other refractory material. To give endurance to the hollow carbon cylinder, during the process of its removal from the matrix, and subsequent treatment, the cylinder is filled with a tenacious cement, such as resin or shellac, and while thus supported a series of discs are sawn or ground off, by means of a hoop-iron saw, with sand and water, or by other suitable means.

1,689.—G. S. Young and **R. J. Hatton,** London. **Electric Lamps.** 6d. (4 figs.) April 6.

ARC LAMP.—The regulation of the upper carbon is effected by a brake, or feed appliance, under the control of an electro-magnet or solenoid. As will be seen from the illustrations, the brake consists of a chamber B, surrounding the upper carbon-holder A, and containing a number of balls C, which jamb between the inclined inner surface of the

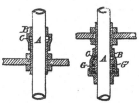

chamber and the holder, and prevent the latter from sliding. When the length of the arc requires adjusting, the magnet releases the chamber B, which moves downwards until the balls come in contact with a stop, when the carbon-holder will be relieved from the grip of the balls, and be free to drop. To keep the balls in proper position, one or more springs G are employed with a sleeve G¹.

1,692.—D. T. Piot, London. **Dynamo-Electric or Magneto-Electric Machines.** 6d. (3 figs.) April 6.

DYNAMO-ELECTRIC GENERATOR.—As will be seen from the illustrations, there is keyed upon the shaft C a boss D, carrying a disc E, preferably of wood, to which are attached the armature bobbins F_2. Discs are also fixed upon this shaft, carrying bobbins F_1, which form supplementary arma-

tures. The field magnet bobbins J, between the pole-pieces of which the armature coils successively revolve, are preferably of horseshoe form. Fig. 3

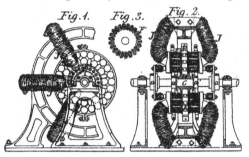

Fig. 1.   Fig. 3.   Fig. 2.

illustrates the elliptical form given to the coils F¹. The commutators $a\,a$ and $a^1\,a^1$ may be of the ordinary construction, and the coils coupled up as desired.

1,697.—The Hon. R. Brougham and F. A. Ormiston, London. Incandescent Electric Lamps. 6d. (10 figs.) April 8.

INCANDESCENCE LAMPS.—By this system of manufacture, the small tube used for exhausting the globe is attached to the closed end of the cylindrical stem, instead of to the bulb, and the appearance of the lamp thereby improved. In Fig. 1, two conducting wires C, which have each been coated for part of their length with a covering of glass C¹, are passed into the oval end of a glass tube B, and their glass covering is made to adhere to the end of the tube, which is then heated and closed. It will then appear as shown in Fig. 2, when a small glass

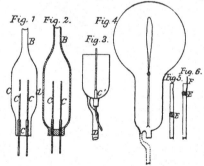

Fig. 1   Fig. 2.   Fig. 4.   Fig. 3.

exhausting tube is sealed to the rounded portion of the end, as in Fig. 3. At this stage the tube is cut off at $c\,d$ (Fig. 1), a filament inserted, and the lamp finished off as illustrated in Fig. 4. The next operation is to exhaust the air through the small tube D, in the ordinary way, and to seal off the tube when the operation has been carried on for a sufficient time. Figs. 5 and 6 represent the method adopted for connecting the ends of a carbon filament to the ends of the conducting wires. Each wire, near its end, is filed away

on one side, and a small coil of thin wire E is slipped upon the conducting wire, as in Fig. 5. An end F of the filament is then held against the side of the wire, and the small coil slipped over it, as indicated in Fig. 6. The small coil E is coated with a composition consisting of a mixture of iron filings, sulphur, and sal-ammoniac, for the purpose of filling the interstices and making better contact.

1,713. — J. Brockie, London. Electric Arc Lamps. 6d. (3 figs.) April 11.

ARC LAMP.—The principal feature of the invention is illustrated by Fig. 1. The armature B of the regulating solenoid A has not, as is usual in electric lamps, a positive connection with the brake or escapement lever C, but, on the contrary, operates it either by mechanical or magnetic friction. By this arrangement, the solenoid core is enabled to take up various positions with regard to C, and no alteration will be required in the tension of the spring S to suit variations in current, unless they be very great. The oscillations of C are limited by the stops D D.

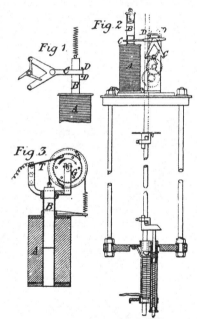

Fig. 2   Fig. 1   Fig. 3.

The application of this device to an electric lamp is clearly shown in Fig. 2. Another mode of regulation is represented in Fig. 3. The oscillations of the solenoid core B, through the medium of a ratchet and pawl, cause the wheel G to revolve. During its rotation contact is made between pins placed around its circumference and the spring T, the successive makes and breaks serving to close and open a branch circuit, which, on completion, reverses the polarity of the " arc striking " magnet, and causes an intermittent feed of the carbon.

1,727.—W. Fisher, Birmingham. Automatic Current Director for Electric Machines. 6d. (9 figs.) April 12.

AUTOMATIC SWITCH. — This instrument is designed for automatically sending the current in the right direction should a reversal of poles take place in the generator. One arrangement is shown in the figures, where B, C, E, F are electro-magnets, operating respectively armatures A and D, which are capable of vibrating between stops Z Z¹ and O P. The conductors from the generator are connected to terminals J and K. M¹ and N¹ are connected to any apparatus through which it is necessary to send a continuous current in one

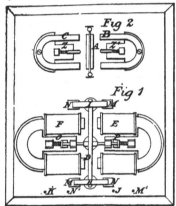

direction. Suppose, for example, that the current enters at J, the suspended magnet A will be attracted to the magnet B, making contact with the stop Z¹, and throwing the magnet E into circuit. This brings H into contact with N, and I with M, causing the current to leave the director by M¹. Should the poles of the machine reverse, the current would enter at K and the magnet A be attracted by C to stop Z. This puts magnet F in circuit, changing the contacts of H and I respectively to M and N, and causing the current still to leave the director by the terminal M¹. Modifications of this arrangement are suggested.

1,747.—D. A. Chertemps and L. Dandeu, Paris. Dynamo-Electric Machines. 6d. (13 figs.) April 12.

DYNAMO-ELECTRIC GENERATOR. — The illustrations refer to a self-exciting alternating current generator, in which the armature is stationary, and the field magnets revolve. The armature bobbin, whose function it is to excite the latter, is wound with wire of a size corresponding to the gauge of the field magnet wire, and its weight is made to correspond with the weight of the other armature coils. By this arrangement, it is stated that exciting currents are proportionate to the requirements of the generating bobbins. The commutator, for trans-

forming the alternating currents furnished by the exciting bobbin, and supplying them continuously in one direction to the field magnets, is shown at h h, Fig. 1. It is constructed from two hollow cylinders, each having one closed end, their peripheries being cut away, so that they interlock the whole of their width in the manner illustrated. The number of interlocking plates is the same as the number of pairs of field magnet bobbins. By adjusting the commutator, so that the collecting brushes pass the middle steps of the plates, at the

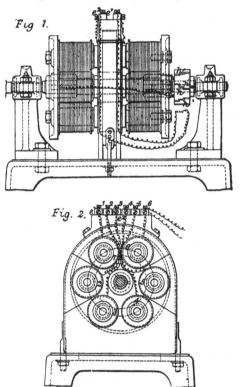

moment that the field magnets are passing the neutral spaces between the armature bobbins, liability to spark will be obviated. Connection wires are led from the separate bobbins of the armature to terminals e¹ e², &c., and a forked sliding piece, not shown, which underruns the screws No. 1, No. 2, &c., is made use of to couple up the terminals as required. In a shunt from each lamp circuit is introduced a small electro-magnet, whose function is to short-circuit a bobbin when its lamp circuit is accidentally broken.

1,760.—J. B. Rogers, London. Dynamo or Electric Current Producing Machines. 6d. (11 figs.) April 13.

DYNAMO-ELECTRIC GENERATOR.—This invention relates to an alternating current generator, the

stationary armature of which consists of a ring, or disc, carrying a number of coils wound in a transverse direction, the rotating field magnets consisting of circular rows of electro-magnets placed as in the Siemens generator. Four different methods of coupling up the armature coils are described and illustrated.

1,769.—J. H. Johnson, London. (*C. A. Faure, Paris.*) Secondary Batteries. 6d. (1 fig.) April 13.

SECONDARY BATTERIES.—The vessels, tanks, or cisterns used to hold the electrolytic liquid are constructed of iron, or other convenient metal, thickly coated on the inside and outside in the following manner: The surfaces of the tank are first coated with varnish made of pitch, linseed oil, paraffine, and tar, applied hot, and afterwards with sheets of asbestos, felt, or canvas, soaked to saturation in the same varnish, several such coatings being successively applied, if necessary. Instead of iron, Portland cement may be employed for the shell of the tank, and in place of the coating described, the inside of the tank may be lined with a surface of glazed bricks, or blocks of wood soaked in the varnish, and cemented together with

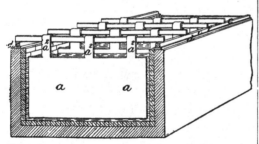

either the varnish or molten sulphur. An arrangement for suspending the electrodes in the tank is shown in the figure. Each electrode a is formed with strips of metal a^2 upon its upper edge, which strips are hooked over flat iron bars covered with sheet lead, stretching across the tank. Along the series of cisterns, which constitute a set of batteries, two large copper bars c are laid upon insulating supports d. Upon one copper bar rest, in electrical contact, all the positive crossbars, and upon the other all the negative crossbars, insulating wedges f separating the positive copper bar from the negative electrodes, and the negative crossbar from the positive electrodes.

*1,774.—A. Muirhead, Westminster. (*J. A. Briggs and F. Kinsman, Bombay.*) Electrical Circuits, &c. 6d. (10 figs.) April 14.

CONDUCTORS.—The objects of this invention are to obviate the disturbing effects of induction, split

currents, earth currents, &c., so affording means of utilising the same trough, cable, or line of supports for carrying two or more electrical circuits close together. This may be accomplished by certain methods of suspending the circuits, by interposing "like" and "unlike helices" in the circuit, or by an "induction guard." The invention is illustrated with ten diagrams.

*1,787. — B. H. Antill, London. Dynamo-Electric and Electro-Dynamic Machines. 2d. April 14.

DYNAMO-ELECTRIC GENERATOR.—The currents are generated by the movement of coils of insulated wire through magnetic fields without changes of polarity, or by the movement of magnetic fields of constant polarity relatively to coils of wire.

1,794.—E. L. Voice, London. Apparatus for Generating Currents of Electricity. 10d. (18 figs.) April 14.

DYNAMO-ELECTRIC GENERATORS.—The principal feature of this invention consists in the generation of currents by the rotation of ring armatures, such as shown in Figs. 2 and 3, to the iron cores of which a magnetic polarity is given, in an unchanging polar field. A typical form of generator is shown, partly

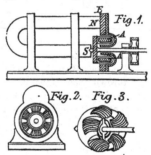

in section, in Fig. 1. The core of the armature A forms a polar extension of the magnet core S, while the armature itself is made to revolve through the constant field formed by the polar extension E, of the core N. Numerous obvious modifications, all embracing the same principle, are described and illustrated; for instance, the armatures may be duplicated, a double set of commutators being then employed.

1,803.—A. R. Leask, London. Manufacturing Incandescent Lamps. 6d. (6 figs.) April 15.

INCANDESCENCE LAMPS.—This invention relates to a gauge for the formation of wire terminals for incandescence lamps. The gauge is shown at Fig. 1, P^1 and P^2 being lengths of steel wire held firmly by binding screws B^1 and B^2. One end of a coil of platinum wire, having been wound helically round

P¹, a length of the wire is measured off from X to Y, and a second helical coil of double the number of turns formed round P². The wire is then released from the gauge by pressing B¹, thus withdrawing P¹ into the body of the gauge. The second made helix is then removed from P² and placed upon P¹, a second length of wire stretched across the gauge, and a third twist taken round P². These operations are continued for the whole length of the coil, which is afterwards separated in lengths through the centre of the intermediate spirals. The separate portions are then annealed and bent

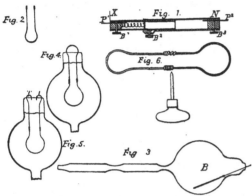

into the shape shown by Fig. 2. The next process consists in inserting the ends of the carbon filaments into the spirals, and in causing the latter to take a firm grip by opening out the separate turns with a blunt tool, as illustrated in Fig. 6, or in otherwise contracting their internal diameter. Figs. 3, 4, and 5 represent the following processes of introducing the conductor into the globe B of the lamp, of securing it there, and of forming the attachment loops T. The lamp is then ready for exhaustion, and is finished off in the usual manner.

1,821.—J. C. Mewburn, London. (*L. Weiller, Angoulême, France.*) Manufacture of Silicious Copper and Silicious Bronze. 4d. April 17.

CONDUCTORS.—A mixture consisting of fluo-silicate of potash, pounded glass, chloride of sodium, carbonate of soda, carbonate of lime, and chloride of calcium, of each one-twentieth part by weight, is introduced into melted copper, or bronze, and renders the mass particularly suited for making electrical conductors.

1,822.—A. S. Church, London. (*J. B. King, Brooklyn, U.S.A.*) Electric Lamps. 6d. (3 figs.) April 17.

CARBONS.—To prevent flickering, and to intensify the combustion of the carbons, they are made hollow, as shown at A in the figure, and have, in the interior, a spiral metallic wire C, designed to increase

the conductivity. D is a tube containing a piece of sponge E, which is covered by a disc of wire gauze.

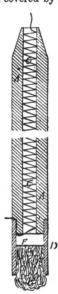

F. The sponge is kept moist, and air passes through it to the arc.

1,830.—Fleeming Jenkin, Edinburgh. Mechansm for Transporting Goods and Passengers by Electricity. 6d. (7 figs.) April 17.

CONDUCTORS.—This invention relates principally o "telpherage," or the transmission of vehicles by electricity to a distance independently of any control exercised from the vehicles. The system comprises the use of strained conductors, A₁, A₂, A₃, &c., Fig. 1, which serve both to suspend the load and to convey the electric energy. In the most simple arrangement, at each support there is a break in the electrical continuity of the conductor, and the separate sections A₁, A₂, A₃, &c., are insulated from each other and from the earth. They are, however, electrically coupled together by movable coupling pieces B. D indicates a train upon the line, its length being such that it always spans one at least of the intervals between the sections. As it enters upon a section, it removes the bridge-piece B, which connected this section with the section in rear, thereby compelling the current to flow through one or more motors carried by and propelling the train. Before the tail of the train leaves the preceding section, it replaces the bridge-piece, and opens the bridge-piece in advance. The ormer operation is effected by means of the circuit-closer E, brought into action by means of electromagnets, which are connected by insulated wires F o the distant bridge-pieces B. The same movement of the train which removes the bridge-piece, connects the wire F with the main circuit, and the

magnet, being excited, closes the bridge-piece with which it is connected. A number of trains may be operated in series upon a single conductor by increasing the energy supplied to the line from the station C, and one train is hindered from overtaking a preceding train by the action of the circuit-closer, which maintains electrical connection between the sections for a certain distance behind a train. A following train, entering upon a part of the line in which the sections are thus connected, will be checked, as the application of power to it depends upon there being a break between the two sections momentarily supporting it.

Motor.—Figs. 2 and 3 represent a suitable locomotive. It consists of two electromotors J J, keyed upon the shafts, which have pinions gearing into the toothed wheels K^2 K^2, which gear with each other across the wire rope A. On the spindles

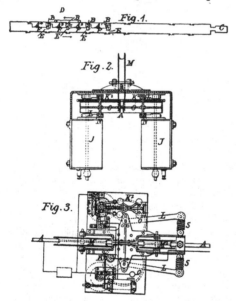

N N, carrying the latter, are keyed the gripping pulleys O O, running in bearings in the arms of the double-armed levers L L, one arm of each lever being centred concentrically with the motor shafts, while the other arm is drawn inwards by the helical springs S. The weight of the locomotive is carried by the pulleys M_1 M_2. Each train is provided with a governor (in part the subject of Letters Patent No. 3,007 of 1882) which, when the speed is sufficient, closes a shunt circuit and allows the current to pass without traversing the coils of the motors. The governor further acts directly to check the train, if the speed becomes excessive. The specification also includes a system of "telpherage," in which two conductors are divided into separate sections, and are provided with connections from one to the other, controlled

by passing trains. The inventor refers to Specification No. 783 of 1881.

*1,850.—R. D. Smillie, Glasgow. Negativing or Destroying the Effects of Induced Currents in Telephonic Lines. 2d. April 18.

Conductors.—The circuits terminate in a pair, or more, of coils, surrounding a central iron core, through which the induced currents pass in a certain order.

1,851.—C. Curtoys, London. Insulated Supports for the Communicating Wires of Telephones. 4d. (4 figs.) April 18.

Insulators.—The end of the arm *a*, not seen, is roughened, and is cemented into a roughened

hole in the insulator B. The other end of the arm may be screw threaded, or fitted with a plate having holes for screws or nails.

1,862.—T. J. Handford, London. (*T. A. Edison, New Jersey, U.S.A.*) Electrical Railways or Tramways. 8d. (15 figs.) April 18.

Electric Railways.—This invention is divided into several parts, having different objects: (1) To provide means for retarding the speed of a train of carriages without the use of brakes. This is accomplished by the insertion, in the electrical circuit, of an adjustable resistance, operated by circuit-controlling and current-reversing mechanism. (2) To provide means for maintaining the strength of the field magnets of the electromotor, when the latter is running at a low rate of speed. To effect this, the field magnets are wound with two sets of coils, one of finer wire than the other. The fine wire coils form the permanent field circuit, while the coarse wire coils are switched in and out of circuit, as required. (3) To provide electrical connections for switches from the main line to sidings, to supply the latter with current, and to obviate a short circuit during shunting. The switch rails are connected by wooden tie-bars, and their ends slide on metal plates in connection with the circuit. (4) To provide similar connections for turntables, which will not become short-circuited by the movement of the latter. (5) To secure good electrical contact throughout the lines

of rails, and to prevent leakage across the ground. The ends of the rails and the fishplates are nickel-plated, and copper strips are used between rails and fishplates at each joint. Also, except at their top surface, the rails are japanned, and the ties are dipped in a boiling insulating compound.

1,867.—A. B. Brown, Edinburgh. Electric Arc Lamps. 6d. (3 figs.) April 19.

ARC LAMPS.—The feed movement of the carbon is effected by the pressure of a piston upon a liquid, the egress of which from a cylinder is controlled by a valve, as described in Letters Patent No. 5,272 of 1881. Fig. 1 represents one modification of a lamp constructed according to this invention. The upper carbon 10 is attached through its holder 12 to the piston 24, the downward movement of which is controlled by the operation of the valve 16 by the solenoid 20. The electro-magnet 33 is employed for striking the arc; its armature is attached to the lever frame 26 of the gripping device 30, which

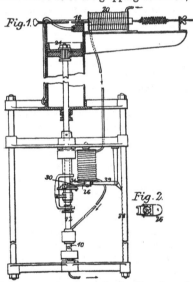

is shown to a larger scale in Fig. 2, and which in its upward stroke clutches and lifts the carbon. To prevent the piston from falling, when the electro-magnets are not excited, the armature 34 is attached to one arm 39 of a lever, whose other arm engages ratchet teeth formed in the guide arm 38. In another modification, the feeding mechanism is carried by the descending piston, and is provided with toggle levers, which hinder the movement of the piston when no current is flowing through the lamp.

,875.—D. G. Fitzgerald, C. H. W. Biggs, and W. W. Beaumont, London. Secondary Batteries. 6d. (14 figs.) April 19.

SECONDARY BATTERIES.—Finely divided crystal-line lead is deposited electrolytically either within the moulds, in which the material is afterwards compressed, or in other moulds of similar form. The figures represent two descriptions of depositing apparatus B, in which A is the anode and C the cathode, the latter being covered with a non-conducting coating, excepting that portion surrounded by the mould M. When in action, lead is dissolved at A, and thrown down upon C, within

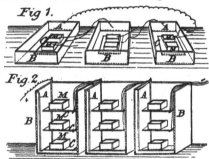

the mould, which, when filled to the desired extent, is removed, and the mass of deposited lead subjected to a sufficient pressure (preferably from 400 lb. to 1,000 lb. per square inch) to constitute it a firmly coherent but porous plate. The process of compression is preferably preceded by a partial oxidation, by air or otherwise. In some cases a contact piece, with spreading fangs, is embedded in the mass before compression.

1,878.—J. H. Johnson, London. (J. M. A. Gerard-Lescuyer, Paris.) Dynamo-Electric Machines. 6d. (9 figs.) April 19.

DYNAMO-ELECTRIC GENERATOR.—The chief feature of this invention is a peculiar method of disposing the armature coils, which consists in winding two coils *e e* upon the armature *f f*, so as to form

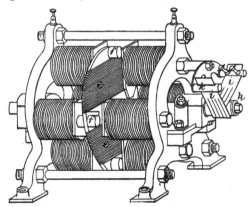

one right-handed and one left-handed helix. The field magnets are of alternate polarity, while those immediately opposite each other are of like polarity. The commutator is arranged outside the framing,

and consists of a cylinder *h*, in two parts, portions of the sides of each being cut away, so as to leave two segments on each part of a length and breadth about equal to the space between them. When the two parts are brought together, the projecting portions of the one fit the recesses of the other, insulating material being interposed between the two. The brushes are arranged as shown at *i i*, the coupling up of the armature and field magnet coils being effected as required.

1,884.—W. R. Lake, London. (*E. Marchese, Turin*). Separating Metals and Metalloids from their Ores. 4d. April 19.

DYNAMO-ELECTRIC GENERATOR. — The current from a dynamo-electric generator is caused to traverse the metals in an electrolytic bath.

1,895.—P. M. Justice, London (*A. Cruto, Piossasco, Italy*). Electric Lighting and Incandescent Lamps. 6d. (7 figs.) April 20.

INCANDESCENCE LAMPS.—Carbon for lamp filaments is produced by the decomposition, by heat, of hydro-carbons. In Fig. 1, carburetted hydrogen is led into a porcelain tube A, with glazed interior surface, passing vertically through a furnace in the manner shown. Carbon is deposited within the tube, and the residual gas is drawn away through

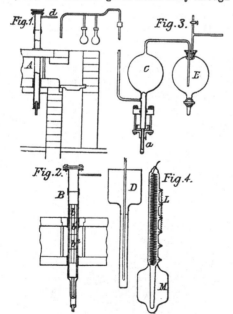

Fig.1. Fig.3. Fig.2. Fig.4.

the tube *d*. Fig. 2 represents a similar arrangement, where the tube B is enclosed in a muffle of refractory material, and contains open porcelain cylinders *b b¹*, &c., upon the surface of which the carbon is deposited. For the production of thread filaments, carbon is deposited upon a hot

platinum wire, which is afterwards volatilised. Fig. 3 illustrates a vacuum pump. Sulphuric acid is introduced into the vessels D and E, and the exhausting machine, not shown, which is connected with E through the tube *a*, is put in operation until the maximum vacuum it can produce is obtained. The receptacle D is then raised, when air will be driven from C into E, whence it will be received by the exhauster. This operation is repeated until the lamps on the left of the figure are sufficiently exhausted.

REGULATING CURRENTS.—In Fig. 4 is represented an automatic current regulator. The tube L contains a spiral of wire, through which the current passes. The end of the tube dips into mercury, which partially fills the vessel M, and rises to a greater height in the tube L, the superior elevation being counterbalanced by a gaseous pressure above the mercury in M. In the upper part of L is a volatile liquid, such as alcohol, which, upon the temperature of the wire increasing, expands and drives the mercury down the tube, exposing a greater length of wire to the passage of the current. In a modification, the movement of the mercury is used to complete the circuit of an electro-magnet, which switches in more or less resistance, according to the amount of current traversing its helix.

1,915.—W. T. Whiteman, London. (*M. Bauer and Co., Paris*.) Electric Lamps. 6d. (8 figs.) April 22.

ARC LAMP.—This relates partly to improvements on the invention described in Letters Patent No. 2,038 of 1881. The improvements consist chiefly in the use of an oscillating horseshoe electro-magnet, shown in plan and elevation in Figs. 3 and 4, with limbs of unequal length, for actuating and controlling the movement of the carbons in an arc lamp. The pole of the shorter limb *d³*, which is in contact with the iron rod *b¹* carrying the upper carbon, is cut to the shape shown in Fig. 4, *d⁴ d⁴* being coverings of brass, leaving the horizontal diameter of the pole alone exposed. The pole *d³* of the longer limb is extended at *d⁶*, and is curved to a radius *d¹ d⁵*. When the current is conveyed to the lamp, the magnet turns upon its pivots, its free end rising to the attraction between the longer limb and the fixed iron block E. The attraction between the shorter limb and the rod *b¹* causes the upper carbon to rise and form the arc. As the carbon points are consumed, the magnet gradually descends until it is arrested by the stop *h*, in which position it remains until the current is weakened sufficiently to allow the rod *b¹* to slip. F, *f¹*, *f²*, *f³* are parts of a magnetic brake. By screwing the iron block *f²* nearer to the strip of non-magnetic metal F, carrying the block, greater brake pressure will be applied to

the face of the pole. The invention also includes a means of attaching a globe to an electric lamp, in which the telescopic tubes N, n^1, n^2, n^3 serve to connect the globe to the base of the lamp, the tubes

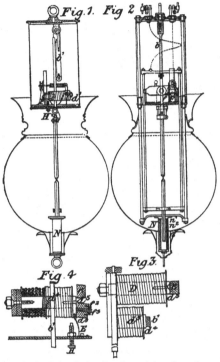

being locked in their closed position by a bayonet joint, or the globe may be suspended by chains and weights.

1,919.—J. Lea, London. **Electric Arc Lamps.** 6d. (3 figs.) April 22.

ARC LAMP.—The upper carbon, shown in dotted lines in Fig. 1, is held between grooved rollers a and a^1, and the jockey roller a^2. Should the carbons

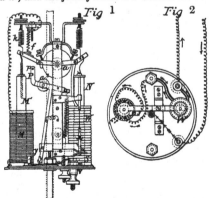

be apart at the time of starting, the current enters at the positive terminal, and, traversing the shunt coils M, attracts the core M^1, until the gripping

lever d comes in contact with the stud p on arm P, and, displacing the jockey roller, causes the upper carbon to fall upon the lower one. The current then passes through the thick wire coil w included in the main circuit, and the core of the shunt coil is withdrawn by the spring k, and the core N pulled down upon the iron base S. The arm x displaces the gripping lever d^1, after turning the roller a in the proper direction, and the arc is established. As the arc lengthens, and a greater proportion of current flows through M, M^1 is attracted until the contact between f and d is broken, when a second coil of greater resistance, either wound round M or upon a separate bobbin, is thrown into the shunt circuit. The effect of this is to release the core M^1, the recoil spring of which, during its retraction, draws back the gripping lever d ready for the next feed. A short-circuiting arrangement is provided upon the frame of the jockey roller, the surface y coming into contact with y^1, when the upper carbon is consumed, or when it accidentally breaks.

1,946.—C. V. Boys, Oakham. **Secondary Batteries.** 6d. (1 fig.) April 25.

SECONDARY BATTERIES.—This invention relates chiefly to the use of very finely divided lead for the electrodes. Lead dust is produced by the agitation of molten lead in a covered vessel, shown in the figure, which also represents the mode of

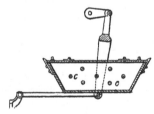

agitating it. c c are crossbars to aid the process of disintegration. The powder thus produced is amalgamated with a weak solution of a salt of mercury, and then compressed to form plates.

1,999.—J. B. Rogers, London. **Accumulating and Storing Electric Currents, &c.** 6d. (4 figs.) April 27.

SECONDARY BATTERIES.—The batteries are constructed of suitably shaped sheets of lead, interposed with layers of felt, or other insulating material, placed in a jar or vessel. Two sets of batteries, one large and the other small, are placed in circuit between the generator and the lamps. The large batteries are charged by the generator, while the lamps are fed from them through the small batteries.

*2,020.—J. C. Asten, London. **Obtaining Electric Light.** 2d. April 28.

CARBONS.—The positive carbon is tubular and contains the negative one, the two being separated by a non-conductor.

2,030.—The Hon. R. Brougham, London. Electrical Switches or Changers. 6d. (3 figs.) April 28.

SWITCHES.—The end of the lever *f*, and the vertical faces of the arcs *g g*, are curved circularly, but the latter are set somewhat eccentrically to the

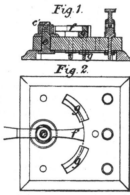

Fig. 1.

Fig. 2.

former, in the manner shown, so that the lever shall be thrust against its fulcrum *e*, when making contact with either of the pieces *g g*.

*2,037.—A. L. Jousselin, Paris. Manufacture of Electric Incandescent Lights in the Vacuum. 2d. April 29.

INCANDESCENCE LAMPS.—The glass globe is closed by a metallic fitting, secured to it by a fusible enamel, composed of 100 parts of sand, 250 of minium, and 150 of protoxide of copper. The conductors pass through two metallic tubes. The filaments are made from strips of woven fabric impregnated with cellulose. These are washed in water and ammonia, dried, and steeped in a solution of sugar, coated with pure graphite, and dried gently. The filaments are cemented to the conductors with a paste composed of pure graphite and pure sugar. Carbonisation is effected so as to obtain an abundant precipitation of carbon on the filament.

2,044.—The Hon. R. Brougham, London. Dynamo-Electrical Machines. 6d. (8 Figs.) April 29.

DYNAMO-ELECTRIC GENERATOR.—The armature coils are made of insulated metal tape, instead of wire, the tape having a width equal to the width of the coil to be formed. These coils B (Fig. 2) are wound on an iron core A, and their ends D are twisted through 90 deg. and led to the commutator. A binding of iron wire K is wrapped over the coils, and dressed to an even cylindrical surface.

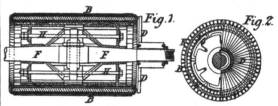

Fig. 1.   *Fig. 2.*

Within the armature so formed, wooden staves I are laid, and supported on centres B, formed in segments and expanded to the required size by being drawn up the tapered portions F F of the spindle by the bolts H H.

2,052.—T. J. Handford, London. (*T. A. Edison, New Jersey, U.S.A.*) Electric Generators and Engines, &c. 6d. (9 figs.) May 1.

DYNAMO-ELECTRIC GENERATOR.—The armature of this generator is built up of a number of copper discs. Fig. 1 shows the generator in elevation, Fig. 2 being a vertical central section, and Fig. 3 an end view of the armature and commutator. The polar extensions N S of the field magnet A are hollowed out on their inner faces to form chambers, which enclose opposite portions of the armature, and their sides are bevelled, as

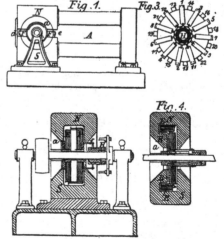

Fig. 1.   *Fig. 3.*

Fig. 4.

shown at *a*, to reduce the attractive force acting directly across between the polar extensions. The armature is composed of an iron core and two sets of copper discs E F. These discs are placed on opposite sides of the iron core, and are insulated from each other by paper. Each disc has two projections *d e* on its periphery, and is joined, by cross connections, to another disc. The connections are made as follows: Assuming that 11 discs are employed in each set, and that the discs of the set E are numbered in the direction approaching the core 1, 3, ... 21, and that the discs of the set F are similarly numbered 2, 4, ... 22, the

outer disc 1, of the set E, is connected to the outer disc 2, of the set F, the disc 2 to the disc 3, and so on throughout the two sets, 21 and 22 being finally connected together. The discs are connected to the commutator H by plates *f*, the alternate connections of the discs, with each other, being connected with the bars of the commutator. In the modification shown in Fig. 4, there is only one set of discs and a number of insulated copper rings or segments of rings E are supported by the rim of the plate B, and connect the copper discs in a continuous series, in a similar manner to that in which the discs of the two sets E F are connected as above described. The invention further relates to regulating a dynamo-electric generator by winding the field magnets with several independent coils, and reversing the flow of the current in one or more of the coils when the current becomes excessive.

2,068.—C. H. Cathcart, Sutton, Surrey. **Secondary Batteries.** 4d. May 2.

SECONDARY BATTERIES.—The negative electrode is made of sheet lead, coated with oxide of lead, while the positive electrode is coated electrolytically with zinc, and well amalgamated with mercury. The solution contains sulphate of zinc acidulated with sulphuric acid.

2,072.—T. J. Handford, London. (*T. A. Edison, New Jersey, U.S.A.*) **Electric Lights.** 6d. (4 figs.) May 2.

ARC LAMP.—The feed of the upper carbon is controlled by a clutch, actuated by two differential magnets M M[1] (Fig. 1) of approximately equal resistance, but having different weights of metal in their coils, M being placed in the main circuit, and M[1] in a shunt circuit. A small resistance R is placed in a derived circuit M. When the carbons are in contact, M predominates, and lifts, by

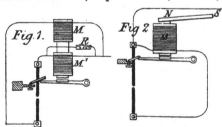

the armature lever and clutch, the upper one, so establishing the arc, M[1] counteracting the sustaining effect of M as the arc resistance increases. Fig. 2 illustrates a modification, in which the electro-magnet M is replaced by a permanent magnet N S. In a further modification, a heating coil, placed in a shunt circuit, and enclosed in an expansible chamber, is used as one of the opposing elements, the move-

ment of the chamber being communicated to the armature lever by a link. In working the lamps in multiple arc, all the shunt coils are connected in series in one of the main conductors.

2,092.—C. Lever, Bowden, Cheshire. **Electric Light Apparatus.** 8d. (13 figs.) May 3.

ARC LAMP.—The coils of the electro-magnets F F, as shown at Fig. 1, and diagrammatically at Fig. 2, are placed in a shunt circuit, and their attraction upon an armature A is counterbalanced by a spring D. When no current is passing, the carbons are held apart by this spring acting through the projection L upon the washer clip B, encircling

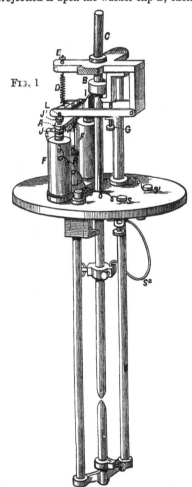

the upper carbon-holder C. Upon the passage of the current the carbons are momentarily drawn together, and then separated by the spring to form the arc, when the current passes by the flexible connection S[2]. The spring D may be replaced by fluid pressure acting on a piston. In a modification, two springs impinge at their curved ends on the

sides of the carbon-holder, and normally raise it to the height required for the production of the arc. The springs have soft iron pieces attached to their under sides, which serve as armatures for two shunt electro-magnets placed immediately under them. The springs may be actuated by a solenoid, in which case a pivotted lever spanning the carbon-holder, and provided with two knife edges, bears on the upper sides of the springs, the free end of the lever being connected to the solenoid core: or the two springs may be connected to an armature common to both of them, as shown at Fig. 3. In a further modification, a spring may be caused to tilt the washer clip, and maintain it in its highest position, a shunt magnet serving to draw down the spring, and so release the clip when the arc resistance becomes abnormal. To prevent an arc forming between the carbon-holder and the lower carbon, should one of the carbons fall out, or to automatically switch out one set of carbons, and

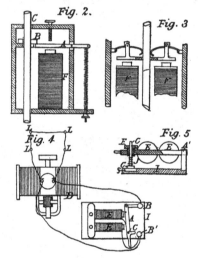

switch in another, in a double lamp, the descent of the carbon-holder is made to release a trigger arrangement, which either cuts the carbon-holder out of circuit, or switches in the second pair of carbons, as the case may be. The methods of regulation before described are applicable to the invention described in Patent No. 3,599 of 1881.

DYNAMO-ELECTRIC GENERATORS.— To enable a self-exciting dynamo-electric generator to start, when a great external resistance exists, the field magnets are first excited by a shunt circuit, which is then broken, either automatically or by hand. Figs. 4 and 5 illustrate the automatic apparatus. From the terminals of the generator two circuits are run, one containing four lamps L L, and the other the shunt apparatus. This comprises an electro-magnet, whose coils E E have a resistance equal to the lamps, a pivotted armature A, and a

sliding contact C, shown to a larger scale in Fig. 5. At first the course of the current is through the magnet coils to the terminal B¹, and fixed portion of the contact, and then across the armature to the terminal B, and back to the generator. As soon, however, as the armature is drawn up to the magnet, one part C of the contact piece slides off the other part S, and the circuit is broken, the whole current then flowing through the lamps.

2,128.—W. Arthur, London. Regulating and Utilising Electric Currents. 6d. (1 fig.) May 5.

SECONDARY BATTERIES.—The circuits are so arranged, that, when any group of lamps not required is switched out, the current, ordinarily passing through them, is diverted through secondary batteries of approximately equal resistance.

2,135.—T. Cuttriss, Leeds. (*Partly C. Cuttriss, Duxbury, Mass., U.S.A.*) Forming Lead for Secondary Batteries, &c. 4d. May 6.

SECONDARY BATTERIES. — The lead plate to be peroxidised is placed in a porous pot in a depositing cell, and is surrounded by a bichromate solution (preferably potash or soda) containing a small quantity of nitric and sulphuric acid. The lead plate forms the anode, while a carbon or metal plate immersed in sulphuric acid forms the cathode. To increase their durability, the plates are made thicker at the upper than at the lower parts.

2,136.—J. Rapieff, London. Incandescent Lamps. 4d. May 6.

INCANDESCENCE LAMPS. — The filaments are formed out of carbon deposited in a crystalline state, either by electricity, heat, or by both, from non-hydrogenic compounds of carbon with iodine, chlorine, bromine, sulphur, selenium, &c., regularity in the resistance of the filaments and in the grain of the carbon being attained by an automatic switch, whereby the depositing current is diverted into a channel of equal resistance when the filament has acquired its proper resistance. Single or multiple filaments, of a spiral or zigzag form, may be prepared from paste, or cotton wool, or flax, hemp, &c., by a process of carbonisation. The conducting wires are secured in a glass thimble, which in turn is secured in the tubular part of the globe. The thimble may be extended for some distance between the conductors, so as to form an insulating bridge between them. The globe is charged with an inert and insulating medium, such as the vapour of chlorine.

*2,138.—A. Millar, Glasgow. Apparatus for Producing Electric Currents, &c. 2d. May 6.

GENERATOR.—The poles of a number of electric or permanent magnets form circles concentric with each other. The currents are produced by rotating a Gramme armature "within the circular space between the limbs of the horseshoe magnets."

2,144.—J. H. Johnson, London. (*J. M. A. Gerard-Lescuyer, Paris.*) **Electric Lamps.** 6d. (6 figs.) May 6.

ARC LAMP.—In this lamp, two parallel carbons are used, the repulsion, due to the current traversing the carbons, furnishing the power which causes one carbon to recede from the other, and so establish the arc. The carbon a is fixed in a stationary holder, while b is carried by a two-legged yoke, provided with pointed feet standing in mercury cups. When no current is passing, the carbon b

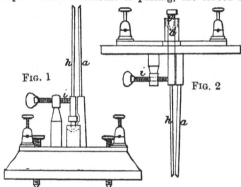

FIG. 1   FIG. 2

leans against a, but as soon as the circuit is completed, b is forced away, and recedes until stopped by the adjustable screw i. Fig. 2 shows the same arrangement inverted; in this the details of the cup d and feet g are more clearly shown. In place of the two equal carbons, a solid electrode may be placed within a hollow one, or one carbon may stand symmetrically within a group of three or more.

*2,184.—C. F. Varley, Bexley Heath, Kent. **Electro-Magnetic and Magneto-Electric Engines.** 2d. May 9.

GENERATOR.—A hollow iron cylinder has another cylinder within, and united to it. This compound cylinder can be rotated within a suitable arrangement of magnets, or within a helix surrounding it, so as to give the double cylinder one magnetic pole. In the annular space between the two cylinders is another cylinder, coiled with wire parallel to its axis. This cylinder forms the second magnetic pole, and is preferably in section like "half a double convex lens."

*2,185.—C. F. Varley, Bexley Heath, Kent. **Electro-Magnetic and Magneto-Electric Engines.** 2d. May 9.

The object of this invention is to obtain rotary motion from electric currents, or *vice versâ*, and to obviate burning at "places of contact." In the method first described, two or more bar magnets are placed end to end, and are free to rotate axially in either direction. Near to, or surrounding them, is the hollow pole of a magnet. Each alternate bar may be a fixture. On passing a current through the bars, rotation is produced. If the bars be rotated, electricity is generated. In a second arrangement, a long bar magnet has magnetic discs upon it, around which discs is a "magnet ring," and the bar between the discs passes through a hole in a magnet. If the N pole of the current be presented to the discs, the S pole is presented to the bar between the discs. Rotation is produced when an electric current passes from end to end of the bar. In a third arrangement, a series of magnetic discs is placed on a revolving bar magnet, each alternate disc being deeply grooved, and each other alternate disc being hollowed in the centre. These run inside holes in stationary bar magnets. In a further arrangement the bar magnet is made in separate insulated segments, and each segment is joined to an insulated ring at the ends of the bar, against which springs press. These springs are so arranged, that the segments may be coupled up parallel, or in series. The rods are made tubular, and the discs are perforated perpendicularly to the axis of the tube. When it is deemed advisable to nickel, or platinise the bars, or segments, it is preferred to do it by electro-deposition. Mercury is used as a lubricator and contact maker. The fixed magnets may be replaced by coils of insulated wire surrounded by the movable bars, and the apparatus can be self-magnetising.

2,186.—H. Lea, Birmingham. **Incandescent Electrical Lamps.** 6d. (15 figs.) May 9.

INCANDESCENCE LAMPS.—The glass globe of an incandescence lamp is surmounted by a hollow neck B, the upper end of which takes a transverse T

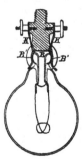

shaped form, whence projects two looped conducting wires $B^1 B^1$. The holder consists of a non-conducting body, two metallic springs K K, and two binding screws, the springs being bent into such

a form as to securely grip the head of the globe when it is forced up between them, and at the same time to form metallic contact with the platinum wires upon which they press.

2,192.—C. J. Allport, London. Manufacture of Bridges or Loops for Incandescence Electric Lamps. 4d. May 10.

INCANDESCENCE LAMPS.—The filaments are made from asbestos fibres, in combination with carbon, the two being united by pressure, interlacing, or otherwise ; or asbestos fibres may be combined with carbon filaments obtained from bodies containing carbon and hydrogen, such as sugar, or paraffine. Asbestos fibres are also combined with carbon filaments derived from hydro-carbons in accordance with Patent No. 4,850, of 1881 ; or combined with hydro-carbons they may have finely divided carbon incorporated with them, after which they are carbonised. Compound filaments, as above described, are carbonised by (*a*) submitting them to direct contact with a hydro-carbon flame ; (*b*) dipping the filament in a hydro-carbon, and igniting ; (*c*) placing it in a crucible containing finely divided carbon, and raising the whole to a white heat. Carbonisation is effected by means of strong sulphuric acid, when saccharine matters are used in the construction of the filament. The compound filaments are compressed by suitable means, such as hydraulic pressure, in order to consolidate them.

*2,207.—C. F. Varley, Bexley Heath, Kent. Electro-Magnetic and Magneto-Electric Engines. 2d. May 10.

The object of this invention is " to get a conductor moving between the poles of the magnets so as to generate electricity." One moving spindle may be used, but it is preferred to employ a multiple of two. In an apparatus having four spindles, on the further end of the left-hand spindle is an insulated conducting disc, and on the remaining portion of that spindle are insulated conducting reels. On the intermediate spindles are insulated discs, and on the right-hand spindle are insulated reels, and at its near end an insulated disc. The discs and reels are either geared together with teeth, or are made to overlap and touch, or they may rotate between fixed pieces of conducting material suitably curved. When combined they form a continuous conductor. The discs are slotted to prevent "false conduction." The slots may be filled with insulating material. In order to diminish resistance at the junctions, jets of mercury are made to play on the discs, or they may dip in mercury cups. Between the discs are placed the poles or magnets, so that the rotating discs cut the magnetic ring, and produce one continuous current through the generator. A second

method of carrying out this invention consists of forming a moving sheet of mercury in the magnetic field. Insulated conductors are in contact with the ends of this sheet to collect the currents generated. The mercury is supplied to the cistern by pumps.

*2,225.—T. Floyd and T. Kirkland, Junr., London. Dynamo-Electric Machine. 2d. May 11.

DYNAMO-ELECTRIC GENERATOR.—The field magnets and armature revolve in opposite directions. The former are constructed in sections, and coupled up in series, so as to obtain consequent points forming N and S poles at extremities of a diameter. A space left between each section permits an air circulation, and so allows of smaller wire being used. The poles of the field magnets almost surround the armature, the neutral points being connected by brass strips. The end frames are insulated from and bolted to the field magnets, and carry the bearings for the armature shaft. The circumference of the armature is divided into twenty-four spaces, between each of which occurs one of the cast-iron radial bars of the box-frame, which is keyed on to the shaft. In each space is a bundle of insulated copper wires, having their ends joined together and brazed to brass caps having tail-pieces, so as to form a connection with the commutator and the next section of wires. The sections are insulated from each other by asbestos paper. The commutator bars are inclined to the shaft, to insure the collecting rollers always touching two of the bars. The collecting rollers have their axes inclined to that of the shaft, so that the centrifugal force shall maintain good contact with the commutator. In a modification, the armature is made to embrace the field magnets both inside and out.

*2,226.—T. Floyd and J. Probert, London. Incandescent Electric Lamp. 2d. May 11.

INCANDESCENCE LAMP.—One or more filaments are enclosed in an exhausted globe, at equal distances from each other, and from the centre of the globe. The filaments, made of cocoa-nut fibre, briza minima, or esparto, are attached at each end to the conducting wires passing through the stem of the lamp. A reflecting ball may be enclosed within the globe.

2,232.—J. M. Stuart, London. Apparatus for Generating Electric Currents. 6d. (2 figs.) May 11.

DYNAMO-ELECTRIC GENERATOR.—This invention relates to the use of duplicate armatures, which revolve in opposite directions. The armature coils are wound longitudinally between soft iron bars A A₁ A, secured to end plates carried by the shafts. These

bars are separated from each other in the same circle by distance pieces D D, D. More than two armatures can be employed if desirable. The field

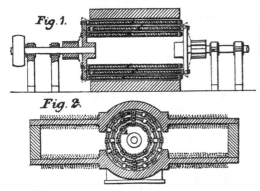

Fig. 1.

Fig. 2.

magnets and commutators are of the ordinary description, and the coils are coupled up as required.

2,233.—J. M. Stuart, London. Electric Lamps.
    4d. May 11.

INCANDESCENCE LAMPS.—A stopper or plug fitting the neck of the globe, carries two glass tubes of about ⅛ in. in diameter, within which are sealed the platinum conducting wires. The wires do not extend the full length of the tubes, but terminate at a distance of ¼ in. to ½ in. from their interior ends, while the tubes themselves extend approximately to the centre of the globe. The air in the globe is rarefied by the use of ether, or other suitable substance, and is then exhausted. A small quantity of carbonic acid is afterwards introduced, and the globe again exhausted as perfectly as possible, and sealed. The carbon filament is prepared from animal carbon, such as horsehair, carbonised and bent in the usual manner. Instead of using animal carbon alone, a combination of animal and vegetable carbon, such as a cotton fibre, around which horsehair is wound, or mineral carbon, may be employed, or a combination of animal, vegetable, and mineral carbon, such as asbestos soaked in melted sugar and wound with horsehair, and subsequently carbonised. The globes are preferably made of iridescent glass.

2,248.—T. Varley, Walthamstow, and H. B. Greenwood, London. Apparatus for Measuring Electric Currents. 6d. (2 figs.) May 12.

CURRENT METER.—The object of this invention is to provide means for measuring the quantity of electricity supplied from any source. It is essential that the speed of the indicating mechanism shall vary with the current passing through the circuit, and that such speed shall not be sensibly affected by friction, and also that the current shall be con-

tinuous. The apparatus is constructed as follows: A disc is arranged so as to be rotated by an electro-motor, or other means, its speed being maintained uniform by an escapement, or other suitable device. In combination with this disc is a roller, carried in a pivotted frame, so arranged as to be oscillated by the varying attractive force of a solenoid included in the circuit. The shaft of the roller is connected with registering mechanism. Fig. 1 is a front elevation of one of these meters, and Fig. 2 is a modification of the same; a is the base-plate, b an electro-motor, c the disc, d the spindle connecting the same to the motor, d^1 a train of wheels and escapement; e is the roller, to the axis of which is fixed the pointer f, and the gear wheel f^1 of the indicating mechanism; g is the supporting frame pivotted at g^1 to the lever h. The spring j keeps the roller and disc in frictional contact. The cord k passes over an eccentric l, carried by a shaft m, and having fixed to it a piece o, which forms the armature of the electro-magnet p. This electro-magnet is included

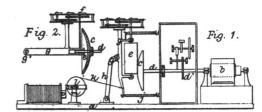

Fig. 2.   Fig. 1.

in the circuit of the current to be measured. The action of the apparatus is as follows: When no current is passing, the roller has a point on its surface bearing on the centre of the disc, and it therefore receives no motion from the disc, but when a current passes through the circuit, the armature is attracted to the pole p of the electro-magnet, thereby altering the position of the roller with respect to the disc. By suitably proportioning the cam, levers, and spring, such motion may be made to be proportional to the variations in the current, and consequently the indicating mechanism in like manner, as the point of contact between roller and disc alters its distance from the centre of the latter. In the modification of this apparatus shown in Fig. 2, the disc c is concave, and gears with a small wheel e, attached to the indicating mechanism. The frame g is pivotted at g^1, and is arranged so that the wheel e can be moved from the centre of the disc to its circumference, in a similar way to that above described.

*2,256.—H. Wilde, Manchester. Regulating and Directing Electric Light. 2d. May 13.

ARC LAMP.—To localise the arc, when used in a projector, the lamp described in Patent No. 618 of 1873, has combined with it one or more electro-

magnetic coils, arranged near to the arc, and included in the main circuit.

2,263.—A. Tribe, London. **Secondary Batteries.**
4d. May 13.

SECONDARY BATTERIES.—Loss of energy, and destruction of the foundation plates, is obviated by exalting the electro-negative character of the plates. This is done by employing plates of lead, more or less converted into sulphides, oxides, arsenides, phosphides, or other electro-negative compounds, by adding to the molten metal sulphur, arsenic, &c., as desired. A positive element is prepared by compressing precipitated lead.

2,286.—R. Kennedy, Glasgow. **Electric Lamps.**
2d. May 16.

ARC LAMP.—For the soft iron tube, referred to in Specification No. 1,199 of 1882, is substituted a tube or core of material other than soft iron, such as brass, copper, wood, or vulcanite. The tube or core may be of conical, or other form. The solenoids may be of iron or other metallic wire, which acts also as the magnetic core.

2,288.—E. L. Voice, London. **Electric Lamps.**
6d. (5 figs.) May 16.

ARC LAMP.—This is a clutch lamp, depending for regulation on the principle, that, if two rods of iron be placed in a hollow bobbin, in which a current of electricity is flowing, they will tend to separate with a force proportional to such current. As will be seen from the illustration, the iron rods C C,

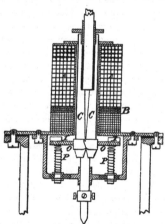

forming the core of the main solenoid A and shunt solenoid B, are hinged together at or near their outer ends. Extension pieces o o, preferably of brass, with blocks of iron I I at their ends, are rigidly fixed to the bars C C. The posts P P are surrounded by spiral springs, and are adjustable transversely in slots made in the frame F, enabling the "biting

power" of the iron rods constituting the core to be regulated. "If the lamp is placed in circuit, the iron rods move apart at the top, and consequently bite the carbon-holder at the bottom, at the same time forming the arc." If the lamps are working in series, the varying resistance of the circuit is compensated for as follows : As soon as one arc gets long, its increased resistance causes the shunt coil B to convey more current, and therefore, to attract the blocks I I, while the core has a tendency to drop, but that cannot happen without first releasing the electrode, by reason of the combined effect of the upturned pressure of the springs, and the increased influence which the shunt coil B has on the blocks I I.

*2,293.—A. Shippey, London, and **R. Punshon,** Brighton. **Insulating, Covering, and Coating Wires used for Electric Lighting, &c.**
2d. May 16.

CONDUCTORS.—Coatings of equal parts of powdered glass and soda silicate, mixed intimately together, are applied moist to the wires. The conductors, when dry, may be covered with thin tape, or asbestos paper soaked in melted paraffine, ozokerit, &c., and finally covered with a solution of india-rubber dissolved in benzole.

2,295. — B. H. Chameroy, Maisons Lafitte, France. **Compensating Dynamo-Electric Machines.** 6d. (5 figs.) May 16.

DYNAMO-ELECTRIC GENERATORS.—The effects of variation of the resistance in circuit are compensated for, by causing the pole-pieces B of the field magnets D to approach or recede from the armature A. For this purpose, the field magnets are pivoted at C, and being free to move in a direction perpendicular to the axis of the armature, an increase in their magnetic intensity, setting up an increased attraction between the pole-pieces and the iron projections F, causes the former to move away from the armature. Their movement is limited by the screw I acting upon the springs H. The armature is composed of coils M, threaded upon an iron wire core, wound helically round the central shaft, the ends of the core being secured in holes in the shaft. The coils are connected up in series, the junctions being connected to a commutator. Soft iron pieces may be placed between the coils, or the iron wire core with the coils may be wound in a helical groove in an iron cylinder. In a modification, the field magnets are arranged around the axis of the armature, which is carried by a sleeve capable of sliding freely upon the shaft. Connected with the sleeve is a ball governor, which, as the speed of the armature increases, causes the armature to approach the field magnets, the return move-

ment being produced by a spring, on a diminution of the speed of revolution The armature may
consist of iron wire rings, covered with coils, and
fitting one within the other; or the coils may be
wound upon a number of rods, each bent spirally
around a central shaft or cylinder, and having their
ends inserted in diametrically opposite openings in
two discs. The commutator bars have a straight
part and an oblique part, upon which latter

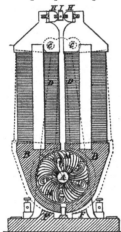

the brushes bear. Thus, when the armature and
commutator are moved along the shaft, the brushes
will still bear on the neutral line. In another
commutator, the bars are arranged in a circle, and
the brushes by their contact with the surface of the
circle "receive the maximum of fluid without the
loss by sparking produced by tangential friction
of other constructions."

2,311.—**J. Anderson** and **W. C. Johnson,**
London. **Submarine Cable Grapnels, &c.**
6d. (2 figs.) May 17.

CONDUCTORS.—The grappling rope contains one or
more insulated electrical conductors, the inboard
ends of which are connected to suitable batteries,
telephones, &c., and which conductors are suitably
arranged within or near that portion of the grapnel
which comes in contact with the cable. When the

cable is received within the prongs of the grapnel,
the conductors are punctured, cut, or otherwise
acted upon mechanically, or electrically, by conduction, or otherwise, so that the presence of the
cables may be indicated on board the ship, or the
presence of the grapnel may be made evident at

the shore end of the cable by means of induced
currents passed from the vessel to the cable.
Swivels are provided in the grapnel rope, as required, to allow the "turn" to circulate. These
are made with contacts to keep up electrical connections, and the internal parts are preferably
enclosed in a suitable oil, or like substance. The
figure shows a section of the cable lying in the
grapnel, in close contact with the insulated coil.
A swivel placed in the grapnel rope, with rubbing
contact for electrical connection of the leading
wires, is illustrated.

*2,318.—**J. A. Cumine,** London. **Electric
Motors.** 2d. May 17.

MOTORS.—The fixed and revolving portions of
the motor are in the same electric circuit. The
electro-magnets are of the bar type, arranged in
conjunction with circular discs, so that two bars
form a horseshoe magnet. A simple form of this
motor consists of three pairs of electro-magnets
bolted three on each side of an iron disc. The
disc is bolted to the frame of the machine, and is
open in the centre to allow a revolving shaft to
pass through it. On this shaft, which runs in
suitable bearings, are two discs, and each disc
carries a set of magnets equal in number to those
on each side of the central disc. The ends of the
fixed and movable magnets are very close together,
and the magnets are so coupled up, that every
time a revolving magnet passes a fixed magnet, its
polarity is changed, and the current in the fixed
magnet is momentarily interrupted, the necessary
changes being effected by a commutator. The
magnets may be coupled either in series or parallel.

DYNAMO - ELECTRIC GENERATOR. — The above
motor may be modified so as to be used as a dynamo-
electric generator, in which case the centre disc is
arranged to revolve, the outer disc being stationary.
The magnets on the end discs are so arranged that
one-half (say the upper) are of one polarity, and the
other half of the opposite polarity. The revolving
disc carries any desired number of magnets "connected to a collector in the same manner as are the
coils on the bobbin of the machine known as the
continuous current Gramme." The arrangement
of fixed magnets is applicable to any generator of
the continuous core type.

2,335.—**C. Defries,** London. **Fittings for
Electric Lamps.** 6d. (6 figs.) May 18.

INCANDESCENCE LAMPS.—Projections on an insulated axle in the lamp-holder are brought, by the
partial revolution of the axle, into or out of contact
with flexible pieces, whereby the electric circuit is
made or broken. As will be seen from the illustrations, a socket is preferably fixed upon the stem W

of the lamp, and is fitted with a pin arranged to enter a bayonet catch in an outer socket T. The inner socket has upon its under face a metal ring N, and also a metal bush O. An insulating block B is secured in a metal casing A, and is provided with springs to retain the socket T. A spring rod, or contact piece F, passes through the block B, and at its upper end is connected to a spiral spring M, or to a metallic ring supported on spiral or other springs, which ring serves to make contact with ring N on the lamp socket. An insulating block C is secured to the lower end of the casing A, and supports a rod G, the upper end of which forms a spring contact, and the lower end of which is designed to make direct contact with one of the main conductors. An insulating axle D, carrying a

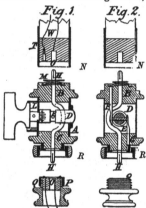

Fig. 1.   Fig. 2.

stud or pin E, passes through the casing A, and between the contacts F and G. A conductor H, passing through the blocks B and C, and split at both ends, makes contact with the bush O in the lamp-holder, and with a bush in connection with the second main conductor. The axle D is provided with a handle and with stops for limiting its motion. By turning the axle D, the circuit is made through the pin E. On reversing the motion of D, the circuit is broken, the spring L insuring that the pin is completely out of contact with the pieces F and G. The holder is connected to an ordinary bracket by means of the screw collar Q and cap R. In a modification, two fixed contacts extend downwards from the upper block B, and are overlapped by the flexible contacts extending upwards from the lower block C. On turning the axle D, projecting studs thereon force the flexible contact against the fixed contacts. This switch may be applied to arc, or other electric lamps.

2,336.—T. J. Handford, London. (*W. A. Stern and H. M. Byllesby, New York, U.S.A.*) **Dynamo or Magneto-Electric Machines and Apparatus for Lighting Railway Carriages, &c.** 6d. (8 figs.) May 18.

DISTRIBUTING CURRENTS.—This invention relates to a system of lighting railway carriages, by the combination of secondary batteries and switches with dynamo-electric generators, worked from the axles of such carriages, provision being made to obviate the polar reversal of the dynamo, and to cause the direction of the current to be the same, whatever may be the direction of rotation of the armature. The secondary batteries are situated between the generator and the lights, and a circuit changer is provided, to connect the generator with one set of cells, while the lights are being supplied from a second set, and then the circuits are reversed so that one battery is being charged whilst the other is running the lights. This reversal is made periodically. An automatic circuit breaker is placed between the generator and secondary battery, so that the current from the latter will not react on the generator. Fig. 1 is a diagram of connections, and Fig. 2 an elevation (partly in section) of the current director. A is the generator, which the current leaves by the communicator springs d d, which are bent where they rest on the commutator, so that the armature may revolve in either direction. The said armature is driven by gearing, or a belt, from one of the axles x of the carriages. In order to avoid reversal of polarity in, and current from, the generator, the current director represented in Fig. 2, and diagramatically in Fig. 1 is interposed in the circuit. Upon the armature shaft c there is a screw thread 3, with a plain portion each side. The nut i fits on the shaft c, and is guided by the rod 4, around which are the two springs 5 and 6. According to the direction of rotation of the shaft, so the nut i will advance towards the springs 5 or 6, thereby bringing into contact the circuit-closing points 17 and 10 with 20, 12 and 15 with 21, or 11 and 14 with 18, and 13 and 16 with 19 respectively. The various connections are clearly shown in Fig. 1, from which will be seen the action of the current director just described. The circuit wires 25 and 24, after passing through the field magnet helices, continue to the circuit changers 31, 32, 33, 34, taking in their circuit the secondary batteries p. The incandescence lamps are placed in multiple arc on the main conductors n o. In the wire 25 is a switch m, and an electro-magnet l is used in a branch circuit x. When the generator is at rest, the current ceases to flow through l, and the switch is released, thereby obviating a reactionary current through the generator from the secondary battery. In modifications, the current director is operated by friction, the contacts being attached to the arms of a friction clamp on the armature shaft. Stops, one on each side, limit the movement of the clamp, which comes in contact with one or the other, according to the direction of

rotation of the armature. Or the contacts may be operated from a sector whose edge is in contact with the armature shaft, and which consequently rolls in one or the other direction. The circuit changers, referred to above, consist of four commutators, as shown in Fig. 1, rotated or turned periodically a half revolution by suitable means, so as to alternately bring into the circuit of the generator, first one, and then the other of the secondary batteries, whilst that one thrown out is joined to the main conductors feeding the lamps. Half of each commutator is of metal and half of

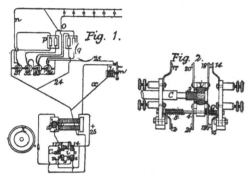

Fig. 1.

Fig. 2.

wood. Each metal half of one corresponds to the wooden half of the next. The metal serves to connect two springs which bear upon it. The negative poles of the batteries p are connected each to the negative lead from the generator, and respectively to two springs bearing on the alternate commutator, the other springs of which are connected to the lead n. The positive poles of the batteries are connected to the lead o, and respectively to two springs bearing on the remaining two commutators, the other springs of which are connected to the positive lead from the generator.

2,340.—**C. W. Vincent**, London. (*Partly Lord Elphinstone, Canada.*) **Dynamo-Electric Machines.** 10d. 19 Figs. May 18.

DYNAMO-ELECTRIC GENERATORS.—This relates, firstly, to a mode of facilitating the manufacture of armature coils constructed according to Letters Patent No. 332 of 1879. The cotton-covered wire to be formed into a coil is passed through a bath of insulating material, such as asphalte dissolved in liquid hydro-carbon, and while still wet is slowly coiled by hand upon a "former" into a parallelogrammic shape, which is maintained by being bound at intervals with pieces of tape. When the insulating material has hardened, the coil is removed from the "former," and it is then moulded to fit the armature drum by pressure, it having been first rendered soft by the application of an electric current. The illustrations represent an

improved armature drum, in which the ends of the coils e overlap each other, and in which the sides of the coils are kept at a uniform distance from the axis, and cover the whole periphery of the drum E. The ends of the coils, instead of being pressed down over the drum-ends, ride over each other upon extensions of the drum as shown, and are maintained in position by cap rings e^1, e^1. The heads are connected to the barrel of the drum (which is of diamagnetic material) by screws. The outer field magnets are carried by side frames, are V-shaped in cross section, and are set concentrically around the drum. The commutator brushes

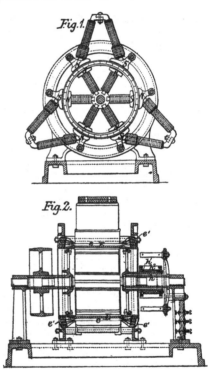

Fig. 1.

Fig. 2.

are carried by an annular frame capable of rotation upon the fixed shaft. The generator illustrated has eighteen armature coils, each composed of two wires, and the commutator has thirty-six plates h^2, placed in longitudinal grooves in a cylinder of vulcanite h^1, carried by a brass cylinder h. The plates are V-shaped in section, and are provided with a bevelled recess at each end, fitting correspondingly bevelled ends on the brass cylinder, from which, however, they are insulated by vulcanite cones. The armature wires are joined in pairs, each pair being connected with a commutator plate. The number of coils may be doubled, each coil being composed of a single wire. The armature thus formed is rotated between an internal and an external set of stationary field magnets, the

polarity of which is alternately N and S. The arrangement is clearly shown in the illustrations. "In some cases the inventor proposes to have two layers of coils thus disposed, and thereby obtains the advantage of causing two currents running in opposite directions to meet at the same collecting brush." In order to facilitate access to the interior of the armature, it is proposed to form the drum in sections, secured together by bolts, &c., passing through flanges at radial divisions of the end plates, and to admit of the separation of these sections, the armature coils are arranged in groups upon each section. The invention further relates to the application of similar coils to disc armatures, the coils being arranged radially and overlapping each other obliquely around the axis of the armature, in the form of a ring, and being clamped between a pair of wooden discs. This armature is rotated between two sets of field magnets with opposing poles, composed of bobbins clamped between fixed boards drilled with a ring of holes for the cores of the bobbins. In the specification above mentioned, the generator was described as fitted with a commutator and six collecting brushes. In this invention but one pair of brushes is employed, and the commutator bars are connected together in groups of three, by means of removable metal coupling plates insulated from each other. By this means, it is designed to reduce the friction of the brushes on the commutator, and the sparking due to extra currents.

2,348.—S. H. Emmens, London. Incandescent Electric Lamps. 6d. (33 figs.) May 18.

INCANDESCENCE LAMPS.—A number of filaments, arranged either in series or in multiple arc, and composed of carbonised ivory, of films of Indian ink, or other carbonaceous material, dried upon the surface of non-conducting material, or a carbonaceous material in the form of a perforated disc, or other flat shape, perforated by holes or slots,

are employed within a single globe. The holders for the filaments are made of platinum, or other suitable wire or ribbon, and are fixed in the glass so that their projecting external portions can be connected to the leads. The holders terminate in loops, engaging with hooks in connection with the leads, or in wires or rods passing into holes in

plugs, and brought into connection with leads by set screws. To effect the regulation of the light in a lamp, the outside terminations of the filament holders are connected to set screws or switches, whereby the current to one or more of the incandescing filaments may be turned off. As will be seen from the illustration, the lamp is supported by a collar, embracing its throat, and either fitting on the plug or socket, or connected therewith by two or more legs working in springs. Firm contacts between the terminals of the lamp and those of the holders are provided for by means of the spiral springs. The lamp may be kept in its place by a set screw, or by one or more springs. The one conductor is connected to the filaments through a central hole in the plug or socket, electrical contact being made by a set screw. The socket can thus be turned freely on the end of the conductor to screw it to brackets.

2,349.—S. H. Emmens, London. Electrical Apparatus. 6d. (12 figs.) May 18.

ELECTRO-MAGNETS.—Straight insulated copper wires are laid side by side, and transversely over them are laid iron wires. This double layer is then rolled up in a direction parallel to the iron wires. The copper wires may be spaced somewhat apart to allow the inner ends to be passed out sideways through the substance of the coil, or they may be placed close together, the inner ends being brought out through the axis of the coil. The whole of the inner and outer ends may be connected together, or the inner and outer extremities may be alternately connected together.

DYNAMO-ELECTRIC GENERATORS. — In applying these electro-magnets to dynamo-electric generators, they are used both in the armatures and for the field magnets, but in the armature the iron wires are made continuous, so as to form a ring, on which the usual coils are wound. Fig. 1 shows a generator constructed in accordance with this invention; C C are the field magnets fitted with pole-pieces D D,

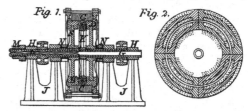

which embrace the periphery of the armature E. All N poles are on one side and S poles on the other. The field magnets are mounted on the frames F F, which are keyed to the hollow shaft G, carried in the bearings H H on the frame J. The armature E is similarly mounted on the frame K and shaft L. The magnet ring and armature may either of them

be revolved, or both in opposite directions. M is a commutator, to which wires are led from the armature through the hollow shaft L. The field magnets may be excited by part, or all of the main current, passed by a brush contact with the insulated ring N N. Fig. 2 is a transverse section of a generator for electrolytic purposes. The relative arrangement of the electro-magnetic field and armature, both revolving in opposite directions, remains as before described, the distinctive feature being in their form, the magnet ring and also the armature ring consisting of plates of iron, wound longitudinally with insulated copper wire. In a modification, suitable for hand power, constructed as in Fig. 1, the pile magnets and armature are made of a larger diameter than is usual, and the outer coils of thicker copper wire, or of coiled insulated ribbon.

MOTOR.—The ring armature is placed in a plane parallel to that of the axis of the field magnets, and is partially embraced by radial pole plates. The field magnets can be secured by a catch, if constructed so as to be capable of revolving.

ARC LAMPS.—The improved magnets are also employed in regulating arc lights. The carbon electrodes are in a glass receiver, either vacuous or filled with a non-oxidising vapour. A magnet is so fixed that the arc is "caused to assume greater curvature with consequent increase of radiating surface and steadiness, and a prevention of circular motion around the carbon points."

SECONDARY BATTERIES.—These "are composed of magnesic or carbon and magnesic electrodes, and electrolytes of sulphate of manganese." They are charged in quantity and discharged for intensity.

2,364.—R. Werdermann, London. Dynamo-Electric Machines. 6d. (2 figs.) May 19.

DYNAMO-ELECTRIC GENERATOR.—The armature, consisting of a magnetic core, carrying a series of coils, rotates in the field of a series of electro-magnets, whose poles, or polar extensions, are fixed on a circular core having the coils between the poles, and are alternately N and S. The coils of the armature are coupled in series, and each junction is connected to a metallic conductor; these conductors are all insulated from each other, and arranged in an insulating disc at the end of the armature. Fixed brushes bear on the conductors, the insulating disc being adjustable. Referring to the illustrations, $a\,a$ are eight polar extensions, arranged upon a circular core composed of rings or hoops, so wound that the poles are alternately N and S. The armature has the general shape of a ring or cylinder, and the coils are wound in a direction parallel to its axis. The bars that form the cores of the bobbins are composed of an alloy of iron 92 parts, and nickel 8 parts, or of iron 92 parts, nickel 7.10 parts, and

cobalt 0.90 parts. The central portion of the armature has a hub, upon which are fixed wooden discs b^2 and segments b, retained between metal cheeks. The bobbins are wound with insulated wire, the ends of adjoining coils being soldered together and connected to a strip of the commutator. This is of disc form, and is rubbed by as many brushes as there are poles, in this case eight.

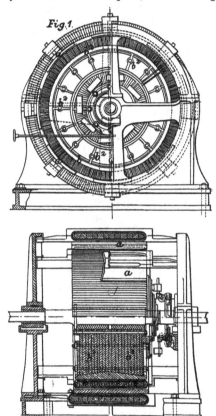

Fig. 1.

The brushes are all mounted on a ring, that can be rotated by a worm, to adjust the brushes to the position where the sparking is least. Each brush is connected to a terminal on a keyboard, and the circuits can be coupled up parallel, or in series. The bobbins of the armature may be wound alternately in reverse directions, in which case there will be a commutator disc at each end of the armature, the armature comprising two closed circuits.

2,370.—J. Brockie, London. Electric Arc Lamps. 8d. (15 figs.) May 19.

ARC LAMPS.—The lower carbon-holder is regulated directly by a differential solenoid, which also controls the feed of the upper carbon, by working an anchor or detent, thus permitting an escapement wheel to move a well-defined distance at each oscillation of the anchor or detent. The top holder

is racked, and gears directly, or by an intermediate train, with a pinion on the axis of the escapement wheel. Each pair of carbons, in an arc lamp having two pairs of carbons, is provided with a magnet or solenoid to strike the arc. The magnets or solenoids are in parallel branches of the same circuit. A locking lever prevents the second pair of carbons coming together until the first pair is consumed. The figure represents a lamp with two sets of carbons, the lower carbons C^2 of each set being fixed. The upper carbons C^1 are attached to racks, which gear into the first wheels of trains under the control of the two arc-striking solenoids A, placed in parallel branches of the main circuit. The last wheels W of the trains are under the control of the brakes D, operated by the shunt coil B. In starting the lamp, one pair of carbons alone is allowed to come together, the other pair being

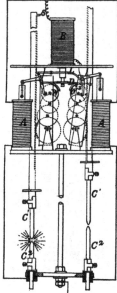

held apart by means of the lever L, which engages the teeth of the brake wheel W^2. When this pair is consumed, the stop S, carried by the holder, trips the lever L, allowing the second pair of carbons to come together. In a modification, a differential solenoid, or magnet, works directly both the lower carbons. The feed of the upper carbons is controlled by a shunt or differential solenoid, the carbon-holder of the second pair being held as in the lamp illustrated, or the regulating coil may be arranged to lift the brakes off the feeding trains in succession. A rod, attached to the frame carrying the lower carbon-holders, is arranged to lift off the brakes when no current exists in the circuit, and it may alter the tension of the regulating spring, according to the pull of the arc-striking solenoid. In arc lamps in which wheel-

work is used to retard the feed of the carbons, the carbons are separated by lifting the first wheel of the train, together with the upper or movable carbon-holder. The first wheel is mounted on a horizontal radius bar, rocking on a pivot in line with the pivot of the second wheel. In a modification, the whole train of wheels is rocked on a pivot on the same horizontal line with the axis of the first wheel, or on a convenient centre, the first wheel of the train gearing into the teeth of the carbon-holder and lifting it when oscillated. The train and framework may form a counterbalance for the carbon-holder. In other forms of lamps, the descending carbon-holders pull up a piston, or counterpoise, in a cylinder, which may or may not contain a liquid. The regulation of the lamp is controlled by a differential solenoid, operating a double spring pawl upon the peripheries of the wheels, over which the suspending chain or cord passes; or the holders may be operated by a lifting lever, or eccentric friction pawl, pressing the carbons against a friction roller. The face of the lever or pawl is covered with an elastic substance, or the pivot upon which it oscillates is flexible. In arc lamps, in which wheelwork is employed, a central brass tube, fixed into the bottom plate of the lamp, encloses the rack or carbon-holder. The teeth of the first wheel of the train pass through a small slot in the tube. This tube forms the mainstay of the lamp, the usual pillars or frames being dispensed with, and protects the wheelwork. In order to automatically adjust the regulating mechanism to compensate for the gradually reduced resistance of the carbons, the tension spring, balance weight, or leverage of the regulating coil is altered. The adjustment is effected by means of a long inclined plane on the side of the carbon-holder, which plane, as it descends, presses back a lever that alters the tension of the spring, the leverage of the regulating coil, the position of a core in the regulating solenoid, or, by means of a cord, the position of the balance weight; or a worm or screw on the axis of one of the wheels, and a travelling nut may be employed to effect the adjustment.

2,375.—C. H. Gimingham, Newcastle-on-Tyne. Air Pumps. 6d. (1 fig.) May 19.

VACUUM PUMP.—The exhaust tube, or passage from the vessel to be exhausted, is sealed in the bottom of an outer vessel open at the top, and is carried up some considerable distance within the outer vessel. Sliding in the outer vessel is a hollow plunger, furnished with leather washers or rings at its lower end to fit as a piston in the outer vessel. The upper end of the hollow plunger is formed with a valve seating, upon which rests a valve, connected by a flexible connection with a valve resting on a

seating on the upper end of the inner tube. A spring tends to pull this latter valve to its seat. The outer vessel contains mercury, which, when the hollow plunger is in its lowest position, rises partly into the plunger, but not level with the valve seating on the central tube. The hollow plunger is reciprocated by convenient means. The air is drawn past the valve into the hollow plunger, and thence forced past the valve in the plunger.

*2,390.—G. Binswanger, London. Apparatus for Lighting Lamps in Railway Carriages. 2d. May 20.

ELECTRIC LIGHT.—The gas lamps of trains are ignited by means of electric sparks, obtained by the aid of an induction coil and voltaic battery. Or the gas, or other lamps, may be lighted by the incandescence of a spiral of platinum wire, through which the current is passed. An arrangement is provided to complete or break the circuit, as required.

2,391.—J. Pitkin, London. Secondary Batteries, &c. 6d. (5 figs.) May 20.

SECONDARY BATTERIES.—The holder, or frame, containing the turnings, or other shreds of lead forming the electrode, is constructed either with louvre strips B, as shown in the figure, or of thin plates of wood, or ebonite, with downwardly and

inwardly inclined apertures, so that the electrolytic liquid has free access to the element, and at the same time the lead peroxide, or spongy lead, resulting from the action of the cell, is retained in its place. The strips B may have their ends inserted in inclined slits in the side bars of the frame. The plates of wood are fixed to opposite sides of a frame.

*2,409.—H. H. Lake, London. (*H. Lory, Paris.*) Electric Accumulators or Secondary Batteries. 2d. May 22.

SECONDARY BATTERIES.—The space separating the opposed plates, in a Planté battery, is divided into two equal parts by a porous partition, and the two compartments thus formed are filled with lead turnings, shavings, filings, or with granulated lead, the object being to extend the operating sur-

faces. The vessel is then filled with acidulated water. Each electrode is composed of one solid plate, with the compressed lead turnings or shavings on either side. Bands of india-rubber insure contact of these three parts, and hold the porous partitions of parchment-paper, felt, or the like, in place; projections on the sides, bottoms, and tops of the partitions bear against the electrodes. The solid plates of the electrodes are fixed to a wooden plate, and are alternately in contact with two lead plates fixed to the cover.

2,414.—J. A. Fleming, London. Insulating Materials. 4d. May 22.

INSULATING MATERIALS.—1. Solid wood (preferably English poplar) is thoroughly desiccated, either in a vacuum, or by superheated steam, or otherwise, and afterwards impregnated under pressure with a mixture of melted bitumen, or asphalte, and one or more substances of the paraffine, anthracene, or resin type, either separately or in conjunction. Among suitable substances of the paraffine type are paraffine proper, hatchettine or mineral tallow, and stearite. Among those of the anthracene type are anthracene, naphthaline, pyrene, and chrysene. The resins which may be used are resins either fossil or recent. The material thus formed is shaped in any ordinary manner into the required form. 2. In lieu of employing solid wood, wood in a finely divided condition, or other vegetable fibrous material such as wood-flour, bran, straw, cotton, jute, hemp, papier-maché, in a finely-divided condition, or asbestos, is treated in a similar manner and moulded into the desired shape under pressure. The following proportions of the various impregnating materials give good results. Bitumen or asphalte about two parts by weight, and substances of the paraffine, anthracene, or resin type from one to two parts. When paraffine wax and resin are used, paraffine wax about three parts by weight, resin about one part. The patentee refers to his former Patents Nos. 1,762 and 5,309 of 1881.

2,419.—W. H. Akester, Glasgow. Electric Arc Lamps. 6d. (7 figs.) May 23.

ARC LAMPS.—The holder of the upper carbon is fixed to a spindle, formed with a screw thread working in an internally screwed block, fitted to turn freely in a guide. The screwed block when raised by the action of a solenoid is prevented from turning. Referring to the illustrations, the upper carbon 13 is attached by a clamp to a holder 16, cut with one or more quickly speeded threads, the holder being hindered from turning by guide rods 15. Surrounding the holder is an elongated nut 17, the head 21 of which is provided with small teeth, engaged by the knife edges 22, carried by the lever 23. The nut 17 is fitted to

move vertically and to turn in a tubular guide, the downward movement being limited by a collar 19, and the upward movement by an adjustable ring. The lever 23 is actuated by the solenoid 26 and core 25, which, in the lamp illustrated, is placed in the main circuit. In operation, upon the starting of the current, the upper carbon is lifted, and the arc formed by the attraction upon the solenoid core, the engagement of the knife edges with the top of the nut preventing the nut from turning under the weight of the carbon-holder. As the carbon consumes, the solenoid releases its core, and the upper carbon-holder and nut descend together, until a point is reached when the descent of the nut is arrested by the collar 19, and the knife edges are taken by the descending core out of contact with the fine teeth

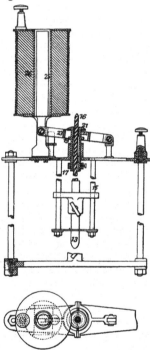

on the upper part of the nut. The nut then is free to turn, and the carbon-holder descends by its own weight until the arc becomes normal, when its motion is hindered by an upward movement of the solenoid core, which once more causes the knife edges to engage the small teeth on the nut. In a modification, the solenoid is placed centrally above the carbons, and a pair of levers is arranged on diametrically opposite sides of, and with their free ends engaging with the head of, the nut 17. The levers are acted on by arms fixed to the bottom of the core 25. In another modification, the lower end of the core engages directly with the head of the nut 17, the core being prevented from turning. The knife edges and teeth may be replaced by mere

frictional contact, the nut 17 being made with a collar having notches upon it, which, when the nut is raised, engage with fixed knife edges, the outer guide for the screw tube being dispensed with. The solenoid may be arranged for working in a shunt circuit, or it may be wound with two coils, one in the main and the other in a shunt circuit.

*2,421.—J. Hickisson, London. Apparatus for Exhibiting Advertisements. 2d. May 23.

ELECTRIC LIGHT.—An electric lamp, over which is a powerful reflector, is employed to light up the surfaces of tablets or plates for advertising purposes. The current is supplied to the carbon pencils or bridges of the lamp from primary or secondary batteries, charged from a dynamo-electric generator, the lamp being provided with the usual connections.

2,425.—J. J. Barrier and F. T. de Lavernede, Paris. Incandescent Electric Lamps, &c. 6d. (6 figs.) May 23.

INCANDESCENCE LAMPS.—The lamps, which are either horizontal or vertical, are fitted with a plurality of filaments, one or more of which may be used at a time. As will be seen, the filaments $x\,x^1$, &c., are attached to platinum conductors $f\,f^1$, &c., which are supported by the perforated glass plates k, and by the glass tubes t, in which they are sealed with mastic or other cement. These tubes are carried by plugs, which are maintained at an invariable distance apart by a glass rod p. The plugs screw into rings d, cemented into the ends of the glass envelope a. The plugs are further

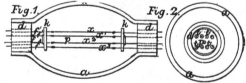

secured by cementing up the ends of the necks. In a vertical lamp, the single neck receives a screw-threaded ring, in which a plug is screwed, and carries glass tubes for the platinum threads supporting the arched filaments. When mounted, a portion of the carbon filament is coated with a composition consisting of equal parts of amorphous phosphorus and pure sulphur, united by a small quantity of gum. Upon the carbons becoming incandescent, the coating burns, furnishing phosphoric acid and sulphurous acid, thereby absorbing the oxygen of enclosed air, and producing a compressed atmosphere in which the carbons do not waste.

2,432.—G. G. Andre, Dorking. Incandescent Electric Lamps. 6d. (3 figs.) May 23.

INCANDESCENCE LAMPS.—Filaments for incandes-

cence lamps are prepared by a process resembling that described in Letters Patent No. 4,654 of 1881. Fibres of an absorbent nature are steeped in linseed or other drying oil, which is afterwards oxidised at a temperature of about 140 deg. Fahr., the process being repeated until sufficient oil has been taken up. As an alternative, the oil may be oxidised in thin layers upon sheets of glass, and then dissolved in naphtha. The fibres are then dipped into the solution and the solvent allowed to evaporate. The filament is made of an inverted **V** shape, the curve at the angle being of very short radius and made more conductive by coating it with carbon or copper. The filaments are preferably carbonised under mercury, in the apparatus shown in Fig. 1, in which a is the muffle containing the fibres, surrounded by a quantity of mercury, which is gradually raised to boiling point, the mercuric vapour given off being condensed in the vessel c

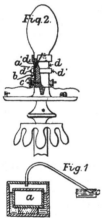

Fig. 2.

Fig. 1

containing water. The carbonised material is then removed from the bath, and subjected to a white heat for about twenty minutes in a suitable furnace. The surface of the filament is improved by the deposition of graphitoidal carbon upon it from a hydro-carbon vapour, and it is further made dense by being heated by an electric current to a bright red heat, and then suddenly immersed in mercury or a mineral oil. This may be done in the lamp bulbs while making the Torricellian vacuum. The filament, during the process of depositing carbon upon it, or when in use, is gradually raised to incandescence by means of a switch containing a connected set of resistances, with each of which contact is made in succession. Fig. 2 represents a lamp-holder, consisting of a short cylindrical piece d, of insulating material, provided with two holes a^1, which receive the lamp wires b, and with side springs c, in communication with the terminals. Upon these springs the wires are pressed, and held by an insulating ring d^1, slipped over them. In the provisional specification, the

patentee states that hollow copper wires, flattened at the parts where they pass through the glass, are substituted for the ordinary platinum conductors ; also that an easily fusible wire, contained within a metallic tube, and having a terminal screw at each end, is placed in the circuit connection.

2,452.—J. Wetter, London (*L. Nothomb, Brussels.*) Incandescent Electric Lamps. 6d. (1 fig.) May 24.

INCANDESCENCE LAMPS.—The filaments may be made, either by compressing strips of Spanish cane between heated steel moulds, and completing the carbonisation by the electric current, or by treating animal parchment in the same way. The ends of the filament are inserted in short coils C C' on the conductors D D', and are fixed there by cement, after which they are heated in an

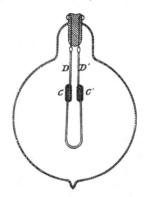

atmosphere of hydro-carbon, which may be obtained by coating the filament with a solution of purified india-rubber in ether. The lamp may contain two or more independent filaments. Two formulæ for the cement are given : they are (1) graphite, 50 parts ; calcined borax, 17 parts ; linseed oil varnish, 13 parts ; refractory clay, 20 parts ; and (2) spongy platinum, 5 parts ; sugar 80 parts ; and graphite 15 parts.

*2,456.—J. Swalwell, London. Apparatus for Driving Dynamo-Electric Machines. 2d. May 24.

GENERATORS. — A standard on the generator frame forms a bearing for the shaft of the armature, on the end of which is keyed a friction pulley. The standard supports also a number of arbors, arranged concentrically around the friction pulleys, and each arbor carries a loose friction pulley, the periphery of which is provided with a suitable elastic medium. The loose friction pulleys are driven by the hoop, or flange, of a wheel keyed on the motor shaft, and drive the armature shaft. The loose friction pulleys may be carried by a

wheel on the motor shaft, and the flange, or hoop, by the fixed frame. Toothed wheels and pinions may be substituted for the friction wheels. The arbors of the loose friction pulleys may be provided with two friction wheels of different diameter, one being driven by the flange, and the other driving the armature shaft. By making these arbors drivers, a series of dynamo-electric generators, out of the axial line of the motor, may also be driven.

2,480.—F. Field, Beckenham, Kent. **Compound for Electrical Insulation.** 2d. May 25.

INSULATING MATERIAL. — "Black wax" (the residue of the distillation of ozokerite), or a compound of this material with india-rubber, produced as described in Patent No. 3,778 of 1869, or No. 1,938 of 1875, is mixed with woody fibre, or cellulose, in a state of fine division. The compound is then subjected to pressure, and formed into blocks. Sulphur may be added to the mixture, and the compound vulcanised. Charcoal, or carbonised cellulose, may be employed with, or substituted for, the woody fibre.

*2,491.—C. W. Vincent, London. (*W. B. F. Elphinstone, Canada.*) **Secondary Batteries.** 2d. May 25.

SECONDARY BATTERIES.—Metallic plates, or cylinders, are formed from finely deposited metals containing a large percentage of hydrogen, such plates being caused to retain their form by being subjected to a suitable pressure, thus obviating the use of supports. Lead, deposited from its solution by electrolysis, or chemically, is preferably employed. In a modification, the metal is deposited on, or otherwise applied to, carbon plates, pressure being then applied. Pyrophoric lead, which after compression is converted into hydride of lead, may be employed.

2,501.—B. Rhodes and G. Binswanger, London. **Materials for Electrical Insulation.** 4d. May 26.

INSULATING MATERIAL.—An insulating compound, suitable for the formation or lining of vessels, is composed of the following materials: Asbestos, 40 parts; shellac, 5 parts; resin, 3 parts; sulphur, 15 parts, all finely powdered and free from grit; india-rubber, 35 parts; and gutta-percha, 2 parts. The whole is thoroughly incorporated together by being passed continuously between heated rollers. The resultant mass is rolled out into sheets, which can be moulded or pressed into any desired form, and also be vulcanised, if required.

*2,512.—E. W. Beckingsale, London. **Incandescent Electric Lamps.** 2d. May 26.

INCANDESCENCE LAMPS.—Two or more globes, or other transparent vessels, are employed, one within the other, enclosing the incandescent conductor, the globes being made perfectly air-tight and exhausted. The object is to prevent radiation of heat, and so increase the light from the lamp.

*2,516.—G. S. Page, Stanley, Jersey, U.S.A. **Materials for Electric Insulation, &c.** 2d. May 26.

INSULATING MATERIALS. — Finely divided dry organic material and a mineral substance are mixed with an organic substance or substances. The mineral substances include metallic oxides, silicates, phosphates, aluminates, borates, carbonates, sulphates, sulphides, tungstates, &c., or carbon in various forms. The organic substances may be obtained from, and by the preparation of, coal tar, coal tar oil or pitch, mineral or shale oil, naphthaline, anthracene, pyrine, chrycine, bitumen, asphalte, resin, gum, mineral wax, gutta-percha, caoutchouc, cellulose, &c., or any so-called waste organic substances.

*2,518.—G. S. Page, Stanley, Jersey, U.S.A. **Compounds for Electrical Insulation, &c.** 2d. May 26.

INSULATING MATERIALS.—These are essentially similar to those described in Specification No. 2,516 of 1882. The organic substance is employed in a liquid or semi-liquid state, and the material is preferably subjected to hydraulic pressure where hardness and elasticity are required.

2,519.—W. H. Akester, Glasgow. **Air Exhaustion Apparatus to be Used in Preparing Incandescent Electric Lamps.** 6d. (4 figs.) May 27.

INCANDESCENCE LAMPS.—The illustration represents a vacuum pump, connected to a single exhauster, though provision is made, by an extension of the delivery and discharge pipes 13 and 14, for the attachment of several exhausters. The lamp bulb to be exhausted is attached to the upper part of the glass tube 5, the mercury being at the level of the dotted lines in the vessel 16; the three-way cock 10 is turned to admit mercury into the pipe 8, from which it drives the air through the valve 9; and the cock is then turned to discharge the mercury through tube 14 into the pump well 22, whence it is raised through the tube 33 into the delivery vessel, any overflow from the latter taking place through the pipe 18. The connection between the delivery pipe 33 and the pump is through the hollow plunger 28, the foot of which supports the

delivery valve. During the discharge of the mercury, fresh air is drawn into the tube 8, and is expelled past the valve 9 on the mercury being again admitted. These operations are continued until the globe is sufficiently exhausted. The syphon bend 21 is designed to prevent the whole of the mercury being drawn from the discharge tube during exhaustion. The tube 5 extends at least 32 inches above the level of the mercury in the vessel 16. An air pipe 24 prevents syphon action in the event of air having access to the main 14, a by-pass pipe and stop-cock being provided to allow the mercury to be run off when desired. The mercury is admitted to the vessel 16 from the delivery vessel, through a stop-cock controlled by a float in the vessel 16, the level of mercury in

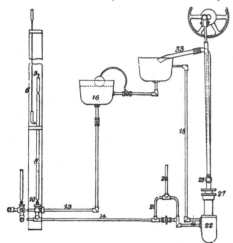

which is preferably a little above the seat of the valve 9. A modified pump for raising the mercury is described, in which the use of the stuffing-box 27 is dispensed with. The plunger, which is hollow, is provided with a foot valve dipping into the pump well, and it works around a central stationary stem in a barrel connecting the well to the delivery vessel, which, in this modification, surrounds the upper part of the pump. The mercury, which has entered, during the descent of the plunger, between the foot valve and the central stem, is forced, during its ascent, through grooves cut in the stem, into the delivery vessel, from which the overflow is conducted to the pump well through the annular space between the plunger and the barrel. The mercury may be raised by an inclined hollow Archimedean screw, or an endless chain of buckets.

2,526.—**W. R. Lake**, London. (*J. J. Wood, Brooklyn, U.S.A.*) **Dynamo or Magneto-Electric Machines.** 6d. (4 figs.) May 27.

CURRENT REGULATORS.—This invention relates to a device for shifting the brushes of a dynamo-electric generator to a more or less favourable position around the commutator. The brushes *h h* are carried on a frame *i* having a toothed segment, as shown, into which gears a pinion *m* worked by a handwheel *n*. The rim of the handwheel is graduated or marked (by experiment), as shown in Fig. 3, and is set in the position corresponding to the number of lights required. Fig. 4 represents, to an enlarged scale, the catch bolt *u* for holding the handwheel at the desired adjustment.

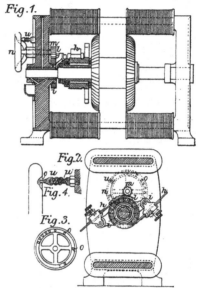

Fig. 1.

Fig. 2.

Fig. 4.

Fig. 3.

The movable bolt part has the form of a grooved sleeve and slides over a hollow stud *u¹*, screwed into the side frame of the machine. A spring enclosed within the hollow of the sleeve, sliding on a stud, tends to press it into engagement with the notches *o* in the rim of the wheel. By pressing the sleeve back, its point may be disengaged from one notch, the wheel turned round to the desired position, and then fixed by allowing the sleeve to spring back into the notch opposite to it.

2,531.—**W. R. Lake**, London. (*J. J. Wood, Brooklyn, U.S.A.*) **Armatures for Electric Machines.** 6d. (5 figs.) May 27.

DYNAMO-ELECTRIC GENERATORS. — The object of this invention is to secure a Gramme armature firmly to its hub. *d* is the wire core wound with coils *e*, connected to the commutator *f* ; *b* is a wooden hub provided with thin projections or plates *g g*, lying in radial slots in the hub, and projecting between the coils on the core. These plates are of non-magnetic material, made in halves, and afterwards soldered together, and act as driver or carrier arms, and maintain the core, coils, and hub in their respective positions. Thin sheets of insulating material are placed between

the plates and the coils. One end of the hub abuts against a collar fixed on the armature shaft, and is formed with a flange at its peripheral edge, against which one end of the ring of coils fits, the coils being slid along the hub and screwed thereto. The armature is wound with three parallel bands of brass, soldered to the edges of the plates *g*, or to a transverse copper strip placed under them.

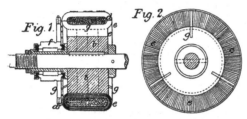

The bands are insulated by tape from the armature. In a modification, the hub is made of brass, with radial blades, which are embedded between the sections of the armature. The outer ends of these spokes are divided in a line with the inner periphery of the armature core, but connected with the inner portion of the spokes by screws, to permit of the attachment and detachment of the ring.

*2,532.—E. W. Beckingsale, London. Treating Certain Materials to Render them Dielectrical. 2d. May 27.

INSULATING MATERIAL.—Porous materials, such as brick, stoneware, clay, earthenware supports for aërial lines, &c., are injected with hydro-carbons, such as petroleum oil, paraffine oil or wax, alone or combined with tar, resin, or shellac, sufficiently to expel the moisture from their pores. The hydro carbon is injected in a liquid or gaseous state, preferably under considerable pressure. Uninsulated conductors may be laid within porous pipes treated as above. The bricks may be employed as a foundation for dynamo-electric generators. Wires in coils may be insulated by a "grouting" of the material under pressure.

*2,557.—M. A. Wier, London. Illumination of Railway Carriages. 2d. May 30.

ELECTRIC LIGHT.—The light from an electric, ox-hydrogen, or other source of light, is distributed by means of lenses, reflectors, and refractors, fixed to side and roof of each compartment throughout the train.

2,558.—J. S. Williams, Riverton, N.J., U.S.A. Generation, Storage, &c., of Electricity. 1s. (83 figs.) May 30.

INCANDESCENCE LAMPS.—The filament is constructed of a metal or a metallic alloy, placed in a suitable chamber or holder, so that it is supported even when heated to liquefaction. The metal is distributed in a thin layer, so as to provide an extended surface, and may be mercury, or an alloy having a low melting point, or a metal, mixed with pulverescent materials, such as lime, zirconia, indestructible materials, glass, plumbago, carbon, mica, oxides of magnesium, or metallic salts. The chamber or holder is preferably transparent, and is constructed so as to permit of the necessary expansion of the metal, the space not occupied by the metal being exhausted, for instance by sealing off the chamber while the metal, which fills the chamber, is heated by the passage of the electric current. Enlargements may be formed in the chamber at the terminals of the conductors leading into it, so that there is an increased body of the metal. In one form of lamp, the metal (or material) is placed in a glass tube, in one end of which a conductor has been sealed so as to extend into the liquid metal; this tube is heated until the metal melts, and a plunger of glass, carrying the other terminal, is forced in, so as to distribute the material in a thin layer or film. The metal may be distributed in fine filaments or thin layers of flat, curved, or conical form, and be supported in chambers of circular, horseshoe, conical, or spherical form. For heating apparatus, the chamber containing the filament is arranged within a casing containing inlet and outlet pipes for water or other liquid, and of transparent or semi-transparent material, so that light, as well as heat, is utilised. The temperature may be controlled as described in Specification No. 5,233 of 1881. Filaments for the lamps may be in the form of plates, ribbons, discs, cones, spheres, or thin layers or films, thin flat loops, with or without flat circular heads, thin flat rings or flat conductors, in the form of a V or M, a double cone, or a series of flat ribbons lying in the same plane at an angle to each other, or conductors with projecting rings, presenting in cross section the form of a cross, the light-giving power being regulated by increasing or decreasing the width. They may be of platinum, iridium, manganese, nickel, steel, iron, or alloys, and be formed by being poured or compressed within suitable moulds, the terminals being of increased section, but not necessarily of the same width. The metal may be rolled into thin sheets or layers, and stamped, or be heated by the electric current (a regulator, resistance or cut-off being in circuit), or otherwise, and poured into a mould; or the metal may be melted in the mould. Carbon, mixed or not with other materials, may be distributed in thin layers, by being compressed between moulds or thin sheets of paper, and be carbonised in the usual manner. A metal, metallic alloy, or carbon may be deposited (*e.g.* as described in Specification No. 224 of 1882) over an extended surface, in the form of a flat ring, loop, or strip, or of hemispherical, con-

cave, convex, angular or other shape, upon non-conducting indestructive material, such as zirconia or asbestos, or upon a transparent substance, or upon a suitably shaped base, which is afterwards removed by heat, chemical agents, or other means, the terminals being subsequently increased in size and connected with the conductors by being coiled around them in a mould, and the contact being secured by molten metal, or by the direct application of heat produced by an electric current. Several forms of the filaments are shown in the drawings.

SECONDARY BATTERIES.—The casing of the battery is constructed so as to be gas-tight, and may be of iron, lined on the inside, and, if desirable, on the outside, with insulating material, which may be composed of bricks and cement, pitch, or asbestos, or of a vitreous substance which will prevent oxidation, or of lead, or an alloy of lead, and antimony or manganese. The casing is divided into two parts or cells by a gas-tight partition, for containing the separate gases forming the positive and negative elements. The gases may be oxygen and hydrogen, or carbonic or sulphurous acid and air. The lower portions of the cells contain acidulated water, or other solution, which prevents the gases from passing through perforations in the lower part of the partition, and mixing, and serves to connect the cells ; they may also be connected by a metallic conductor. The electrodes, which may be coiled, lapped, or folded upon themselves, are supported within the cells, and are composed of sheets, plates, discs, or extended surfaces of metallic fabric, or of surfaces coated with, or formed of, carbon, or of thin sheets of copper coated or impregnated with lead, platinum, manganese, antimony, or other suitable metal or alloy. The electrodes may be coated with carbon by burning turpentine or oil, so that the fumes will permeate the chamber which contains the articles to be coated. They may be coated with metals or metallic salts in a similar manner, the metals being heated by electrical or other means. The interior surface of the cells may be coated in this manner, or lead or other metal, or alloy, may be cast around suitably formed castings of iron or steel. The charging and discharging electrodes may be separate, so as to charge in tension and discharge in quantity, or *vice versâ*, in which case a continuous charge may be given through one set, and a continuous discharge through the other, without the use of switches. Each cell is provided with a safety or pressure valve, or the gases may force out acidulated water or other solution with which the cells are filled, the apparatus being arranged to act upon a switch, which prevents the further charging, and, it may be, completes the circuit to another battery. The connection between the batteries may be completed at the sides, bottom, or top of the casing ; for example, by plates immersed in liquid solution, or covered with a bad conductor, for the purpose of regulating the resistance. The battery is charged by passing the current from a generator across the electrodes, the liquid being decomposed, and the oxygen and hydrogen, or other gases given off, being confined within their respective cells, or it is charged with gases from reservoirs, with which they are connected by pipes, the gases being formed by electrical, chemical, or other means. A "condenser," for regulating the tension of the current from a battery, consists of a metallic chamber, insulated on the inside and outside, and within which is the charging conductor, provided with an extended surface. Plates, discs, or ribbons may be secured upon the conductor within the chamber, and the intervening spaces may be filled with metallic sheets, filings, or scraps, the chamber being then filled with a non-conducting liquid, such as oil or paraffine, and fitted with a tight cover. The metallic chamber is connected to earth. The metallic sheets, filings, or scraps may be replaced by metallic salts or oxides, or finely divided metals or metallic substances, mixed or not with saw-dust, the vessel being filled with sulphuric acid, or an alkaline or other solution. The discharge is regulated by increasing the distance between the negative and positive elements, or between the earth and the chamber. Two such condensers, or their equivalent, may be arranged in connection, one as the positive and the other as the negative.

DISTRIBUTING CURRENTS.—Secondary batteries at a central station are connected by conductors to pairs, or a series, of secondary batteries, at or near the point of consumption, and by regulating the tension of the current, by coupling up the batteries at the central station in tension whilst discharging, the sectional area of the conductors may be considerably diminished. The tension of the current from these batteries may be regulated by "condensers," induction apparatus, or other means, which are so connected with the batteries that parts are being discharged whilst others are being charged. The switches for directing the current may be operated by an electro-magnet controlled from a central station. A meter may be placed in each circuit.

DYNAMO-ELECTRIC GENERATOR.—The provisional specification describes a machine applicable as a generator or motor. The "inducing faces" are constructed of continuous plates, scraps, or coils of metal, say copper, iron, or steel. Insulated strips, wires, coils, or ribbons of metal, the terminals of which extend to opposite ends of the axle of the armature, where they are each connected with a ring, are placed under, within, or between the inducing surfaces. These rings fit in journals or boxes from which the current is taken off. The

entire surface of this armature may be surrounded or covered by any suitable arrangement of magnets, either of horseshoe, annular, or other form. Or a single insulating strip is, or a series are, wound or placed under the continuous metal face, so as to provide for conducting the electricity with which the metal will become charged, the terminals being carried to rings, or to a commutator. The section of the coil is so proportioned as to give the greatest economy. The patentee refers to his previous Patents: Nos. 5,233 of 1881, and 85, 224, 700, 766, 856, and 1,174 of 1882, in conjunction with which the present may be worked.

2,560.—S. Hallet, London. Electric Lamps. 4d. May 31.

CARBONS.—Carbons for arc and incandescence lamps are prepared by mixing finely divided carbon, or ligneous or vegetable pulp (treated or not with dilute mineral acids) of a creamy consistency with water and gelatine, size, or glue, and heating to 150 deg. Fahr.; tannic acid is then added, and the heat maintained till the mass becomes plastic, when it is pressed, moulded, dried, and carbonised. The mould for carbonising consists of three plates of nickel or steel, the central one having grooves corresponding to the shape of the carbons, cut right through it. The carbons are placed in the grooves, and the plates are bolted together so as to make the mould air-tight, and the mould in then heated gradually to nearly a white heat in a carbonising furnace. In order to adjust the resistance of the carbons, they are immersed in a hydro-carbon oil, previously heated to boiling point, and brought to a bright red heat by means of an electric current, at which heat they are kept until they have received a sufficient deposit of carbon.

INCANDESCENCE LAMPS.—Carbons for incandescence lamps are cut from cocoa-nut shell, dried, submerged in a heated bath of cocoa-nut or other oil, and bent to the required shape, after which they are carbonised in a metallic mould. A glass socket, corresponding with an opening in the globe, has two platinum wires fused into it and projecting both ways. On the outside these form the usual connections, and on the inside two flat pieces of metal are hard-soldered thereto. The thickened ends of the carbon are secured to these plates by hard-soldering, by small metallic clips, or by a luting of the plastic carbon described above, the mounted carbons being then dried, and the junction slightly heated by a blowpipe flame. The glass socket is then fused into the globe, which is exhausted in the usual manner. A metal plug is shaped, and may be cemented to the base of the globe; one of the platinum conductors is suitably connected to this plug, and the other, which is insulated, passes

through it, and to its end is soldered a thick copper wire, which is platinised. The lower end of the plug is conical, and fits into a corresponding socket in a bracket or holder, to which it may be further secured by means of a small screw. The copper wire passes between two spring contact pieces, soldered to an insulated conductor passing through the hollow tube of the bracket, the substance of which is used for the return. The switch is made like an ordinary gas cock. The conical plug has a copper wire passing through, but insulated from it, and this wire makes or breaks contact with platinum pieces attached to the adjacent ends of the insulated conductor passing through the tube. A stop-piece limits the motion of the plug.

2,563.—W. R. Lake, London. (*J. J. Wood, Brooklyn, U.S.A.*) Electric Lamps or Lighting Apparatus. 6d. (3 figs.) May 31.

ARC LAMP.—This has reference to devices, in arc lamps, for cutting a faulty one out of circuit when its arc fails or becomes abnormally long by the failure of the feeding mechanism, &c. The cut-out contrivance is combined with the regulating magnet of the lamp, the abnormal attraction of which upon its armature operates to trip a catch permanently cutting out the faulty lamp. As will be seen from the illustrations, the rack of the upper carbon-holder

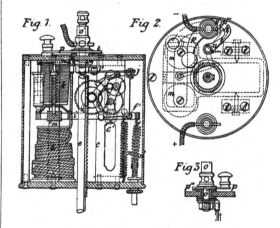

e engages with one of a train of wheels carried in the rocking frame. This frame is pivotted at f^1 to the armature lever i, and is connected to the flat spring j, which is secured at its inner end to a pillar j^1. The armature lever is an open T shaped skeleton frame, pivotted between the standards $c\,c$ on trunnion screws c^1, at a point below the centre of the rocking frame. The magnets $k\,l$ are of the ordinary high and low resistance forms respectively, and the former is placed in a shunt to the lamp circuit. When the lamp is out of circuit, the springs $n\,n^1$ will depress the frame, thus withdrawing the escape

wheel of the train from the stop tooth h, carried by spring j, and allowing the carbon-holder to descend. Upon the current flowing, the direct magnet l will be excited, and depressing the armature m, will move the lever i, and bodily lift the train of gearing and carbon-holder, separating the carbons and bringing the escape wheel against the stop tooth h. The feed is effected by the shunt coils k, which act in opposition to the direct coils l. The length of the arc is adjusted by means of the spring n^1. p is a hand switch, by which the lamp may be thrown out of circuit at will. It is mounted on the stem o^{11} of the negative binding screw o^1, as shown in Fig. 3, so as to turn freely thereon while remaining in electrical connection therewith. When closed, its free end rests upon a metallic seat, which is electrically connected with the body of the lamp, and consequently with the positive terminal; and the lamp is thrown into action by opening the switch as shown in full lines in Fig. 2. The short arm p^{11} of the switch lever has a broad contact surface p^{111}, corresponding to a similar surface on the automatic switch t, resembling a pistol hammer. This switch arm is mounted on a short vertical spindle passing through the top of the lamp and carrying on its lower end a spring u, which tends to rotate it, and a trigger finger, which is tripped by a short horizontal lever pivotted to the underside of the lamp top, shown in Fig. 2. This trip lever projects over a vertical plunger arranged between the legs of the upper magnet, and carried by the armature m. If this plunger be forced up, it causes the trip lever to oscillate, and releases the trigger, whereupon the spring rotates the vertical spindle, and forces the contact piece t^1 against the piece p^{11}, thus making direct contact between the body of the lamp and the negative terminal. Special provision is made that the trigger shall not be tripped without due cause. The armature is loosely connected to its lever, and the latter is furnished with stops that arrest its play before it operates the plunger. To effect this latter result, it is necessary that the attraction of the magnet should increase until it lifts the armature partly out of the lever, and so forces up the plunger. In order to regulate the action to a given length of arc, the weight of the armature is partly carried by an adjustable coiled spring round the plunger. As will be seen from Fig. 2, when the hand switch is closed it automatically sets the safety switch in its open position.

2,565.—A. J. Jarman, London. Dynamo-Electric Machines. 6d. (8 figs.) May 31.

DYNAMO-ELECTRIC GENERATORS.—The armature core consists of an outer cylinder enclosing an inner divided cylinder, the two cylinders being separated by an annular air space. Referring to the illustration, surrounding a thin iron cylinder A^1, divided

in the centre as shown, is another cylinder A, upon which are rivetted the longitudinal ribs B, forming grooves wherein the various armature coils are wound. Small tubes passing through the ribs connect the space between the inner and outer cylinder with the air, to assist the refrigerating effect. This latter is said to be enhanced by covering the opposed iron parts of the armature and field magnets with thin strips of a good heat-conducting metal, such as copper. The ribs are fastened at their ends to

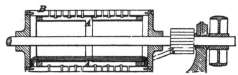

non-magnetic discs, or if the parts are cast, they are insulated from the armature axle by non-conducting bushes. The ribs may be in several pieces, with air spaces between them. Armatures of the disc type may be made in a similar manner. The coils may be connected in series. Two armatures may be arranged together, being connected to supply a tension or quantity current, or both; or a portion of the current from one armature may be employed to excite the field magnets.

2,569.—T. E. Gatehouse, London, and H. R. Kempe, Barnet, Middlesex. Electric Lamps. 6d. (5 figs.) May 31.

ARC LAMPS.—The upper carbon K is supported by two gripping wheels R R^1 and a jockey roller. Upon the axes of the former are keyed two ratchet wheels W W, which are put in motion by the reci-

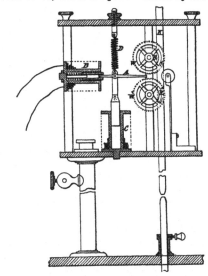

procating pawl A, jointed to the core of the solenoid B. The arm A is also jointed to a link, terminating at one end in the core of the vertical solenoid C,

and at the other end in the recoil spring D. The coil of B is connected through any convenient make and brake arrangement with the lamp, or other circuit, in such a way that the solenoid is intermittently excited, attracting its core in opposition to a spring, and giving a reciprocating motion to the arm A. The coil of the solenoid C, whose function is to deflect the arm A and bring the reciprocating pawl into gear with one or the other of the ratchet wheels W W, according as the resistance of the arc decreases or increases, may form part of the lamp circuit. In this case, a shunt of some kind is provided, to automatically maintain the general circuit in case of the extinction of the lamp. It is preferred, however, to arrange the deflecting solenoid in a shunt circuit of considerable resistance, including a carbon filament in a vacuous bulb. As the electrical resistance of the filament is lessened on becoming hot, the resistance of the shunt decreases as the arc lengthens, and thus, by drawing to itself an increased proportion of current during the lengthening of the arc, causes the deflecting solenoid to be rapidly and forcibly excited. The use of the carbon filament, for the purpose of rapidly and considerably varying the resistance of a circuit, is applicable, generally, to electric lamps in which there is a shunt circuit, either including or not including a solenoid, or magnet, for the regulation of the arc. In a modification, two small rollers press the carbon against a single roller R, on the axis of which is fixed a toothed wheel. The reciprocating arm A is made in the form of an escapement anchor, which causes the wheel to turn in one or the other direction. Or the carbons may be fixed in a tube having rack teeth on its side, engaging with a pinion on the axis of the ratchet wheel. In a self-focussing lamp, the carbons and holders are counterbalanced by weights, or are so connected by cords passing over pulleys that one counterbalances and moves faster than the other.

2,570.—W. R. Lake, London. (*J. J. Wood, Brooklyn, U.S.A.*) **Electric Lamps.** 6d. (7 figs.) May 31.

ARC LAMPS.—This relates to lamps with two or more sets of carbons, and provides simple apparatus, whereby, when the first set is in action, the strain of the second set is supported so as not to be borne by the regulating mechanism. The general arrangement of the lamp is similar to that described in Specification No. 2,563 of 1882. The racks are enclosed at their upper ends by tubes, screwed into the upper head of the lamp. A cap, formed with two sockets fitting the top of the tubes, and secured by a set screw, has a central loop, by which the lamp may be suspended. Referring to the illustration, the right-hand carbon is supposed to be still in action, although nearly consumed,

and the left is ready to be put into circuit, it being meanwhile held up by the lever t and spring h^1, until such time as the projection z shall trip the lever and release it. Both pinions are connected to the spindle by pawls and ratchets, so that when the right-hand pinion rotates to feed its carbon, the spindle can run round idly in the left-hand pinion. In addition to this motion, however, both pinions and spindle have to be raised to establish the arc, in which case the left-hand pinion must roll over the rack in the direction in which the pawl holds. To effect this, the two pawl discs $v\,v^1$, instead of being keyed fast to the spindle, are coupled to it by slots and pins, that allow them a certain amount of play corresponding to the arc,

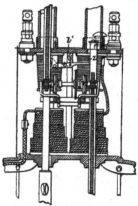

and these slots are so arranged, with relation to each other, that the carbon which is out of action is not disturbed by raising the gear train, the play being always taken up at the active, and always free at the idle clutch, to allow of the backward rolling of the pinion on the idle rack when the gearing is raised. The arrangement of the springs $n\,n^1$, shown in the illustration annexed to our abridgment of Specification No. 2,563 of 1882, is claimed, the fixed spring and the weights of the core and lever being balanced. The spring n^1 is adjusted by removing it from the hook, and turning it in one or the other direction, and replacing it on the hook, the lamp being always ready for action, even when the spring n^1 is detached.

2,571.—W. A. Phillips, London, and S. E. Phillips, Charlton. **Making the Insulating Bodies of Electric Light Conducting Wires Non-Inflammable.** (4d.) May 31.

CONDUCTORS.—These, preferably insulated with cotton, silk, or india-rubber, are passed through a solution of tungstate of soda, or other acids, salts, or oxides, capable of rendering pitch, india-rubber, gutta-percha, and other bodies non-inflammable ; or the non-inflammable salts may be mixed directly with the insulating or mechanical coverings of

electrical conductors. This invention also relates to a method of making an oily solution of the above compound, which may be mixed with any material used to cover wires. A concentrated aqueous solution of tungstate of soda is poured into boiled linseed, or other oil, or into gum, and stirred until the water has evaporated. This is then mixed with the insulating compound.

2,573.—S. Hallett, London. Dynamo-Electric Machine. 6d. (35 figs.) May 31.

DYNAMO ELECTRIC GENERATOR.—Upon a shaft A are fixed bosses B, standing out from which in radial directions are the cores C of the electro-magnets. The shaft A is carried in bearings in non-magnetic standards, one bearing, or standard, being insulated from the bed-plate. In large machines, or in duplex machines, the driving pulley may be placed on the centre part of the shaft, and a commutator at each end, both bearings being insulated. Surrounding these armatures are the field magnets E, whose cores a terminate in discs b, which are provided with pole-pieces c, corresponding with similar pole-pieces attached to the radial armature cores. The cores extend from one standard to the other, and the winding may be continuous, or be divided by a ring of insulating

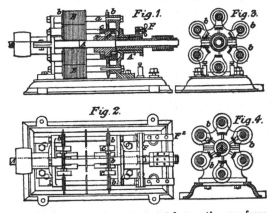

material, or the cores may not be continuous from side to side. As shown in Figs. 1 and 2, the field magnets and armatures are duplicated. The commutator segments correspond to the number of armature sections. Alternate segments of the commutator are formed of one cylindrical piece, with a number of longitudinal slits corresponding to half the number of segments. The movable segments have a ring projecting from their inner surface, and these rings are slipped on to the armature shaft, and are separated from each other by insulating rings, the movable segments lying in the slits of the cylindrical piece, and being insulated therefrom by plugs of insulating material, screwed into tapped holes bored longitudinally between two segments.

Insulating rings fit over projections from the ends of the segments. The brush-holder support F surrounds the bearing A¹, and contains a curved slot, which allows the brush-holders F² to be adjusted round the commutator within suitable limits. The stems of the brush-holders fit in insulated sockets attached to the support F. The brushes are held to the commutator by springs, and may be four in number to each commutator, for better contact, and to avoid sparking. In a modification, the field magnets are arranged radially, their inner ends terminating in pole-pieces attached to a ring, within which revolves the armature, consisting of an annular iron core, upon which are threaded wire coils and iron polar extensions alternately, the whole being driven tightly upon a pulley-shaped core. The field magnets of the two sets, in a duplex machine, may alternate in position, instead of being opposite one another, "so as to counteract as much as possible the resistance to the driving of the armature shaft set up by the magnetism." Permanent magnets may be arranged side by side with the electro-magnets forming the field.

*2,595.—W. Boggett, London. Materials for Use in Secondary Electric Batteries. 2d. June 1.

SECONDARY BATTERIES.—A suitable tool, pressed against a disc of lead secured to a revolving shaft, detaches small particles; or very thin sheets of lead are cut into narrow strips, by a machine similar to a paper-cutting machine, the strips being then placed so as to be cut across their length. The particles of lead are employed in the manufacture of electrodes for secondary batteries.

2,602.—Sir C. T. Bright, London. Secondary Batteries. 4d. June 2.

SECONDARY BATTERIES.—Each cell is divided into two equal parts by a porous diaphragm, and each division is filled with a number of leaden granules, and dilute sulphuric acid. Electrical connection between the cells is made by means of leaden plates, one of which is placed in each cell division. The surface of lead employed in secondary batteries is covered with dioxide of lead, by exposing it to the warm vapours of acetic and carbonic acids, and afterwards treating it with chlorine. Dioxide of lead may be mixed with the lead granules above mentioned, to fill up the interstices between them. The provisional specification describes a process of converting the surface of lead into a protoxide of lead, by exposing it to heat and the air, and afterwards to chlorine gas.

2,604.—F. Des Voeux, Derby. (*A. Bernstein, Boston, Mass, U.S.A.*) Manufacture of Incandescent Electric Lamps. 6d. (3 figs.) June 2.

INCANDESCENCE LAMPS.—The incandescing portion is composed of a substratum of a phosphorescent substance, such as calcined egg or oyster shells, chalk, lime, fluor-spar, &c., upon which is deposited, from hydro-carbon vapour, a film of carbon, covering either the whole or only a portion of its surface. The part not so covered glows with a soft phosphorescent light, aiding the illumination given by the incandescent carbon film. The phosphorescent substance may be covered with a film of carbon, by placing it in a pipe through which hydro-carbon gas is passed, and heating the pipe in a furnace, or the substance may be heated in a vessel containing hydro-carbon vapour, by a spiral of thin wire, through

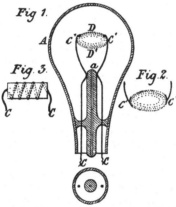

Fig 1.

Fig. 3.

Fig. 2.

which an electric current is passed. The illustrations clearly show the form of the lamp, D D¹ being the incandescing portion. In Fig. 3 the carbon is shown as deposited in a spiral form. The conductors C C are sealed in an annular collar of the stopper, arranged at some distance below the upper end of the central upwardly extending portion of the stopper, so as not to be affected by heat; they may also be sealed to the uppermost portion of the stopper. To increase the endurance of the lamps, the platinum connecting wires C have enlarged ends C¹, where they are attached to incandescing portion D D¹. The lamp is exhausted through the tube *a*.

2,613.—W. E. Ayrton and J. Perry, London. Electric Lamps. 6d. (8 figs.) June 3.

ARC LAMPS.—In these lamps the upper carbon-holder passes through the centre of the regulating electro-magnet, which may consist of one or more coils, but preferably of a single hollow coil, with or without a hollow wrought-iron core. In the illustration two coils are shown, possessing polar extensions P R Q and U S T, whose object is to produce an intense magnetic field in the space R S. Within the polar extensions works the magnet core L M, the upper part of which, L N, is of iron, while the lower part, N M, is of brass, or other non-magnetic

material. To this core, at M, is fixed a frame C C, at one end of which are hinged, at D and F, the ends of two levers D E and F G. Each lever has a projection in the middle of its length, which projections form a pair of grippers supporting either the carbon-holder, or the carbon itself. The core frame and levers are hung from the framing by spiral springs (not shown), and in the normal condition of the lamp, when the arc has been formed, the downward attraction of the core, together with the weight of the movable parts, just

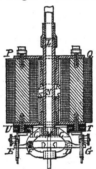

balances the upward pull of the springs. The magnet coils being in a shunt from the main circuit, the tendency of the core, and the frame carrying the carbons, is to descend upon a lengthening of the arc. The frame, however, cannot fall more than a very short distance, before the free ends of the two levers come in contact with stops at E and G, and this diminishes their grip upon the carbon, and lets it slide independently of the frame. The lower carbon is stationary, and is carried in any convenient manner. The frame C, and levers, may be carried on the upper side of the solenoid.

2,618.—R. E. B. Crompton, London. Dynamo-Electric Machines. 6d. (10 figs.) June 3.

DYNAMO-ELECTRIC GENERATORS.—This relates to method of winding disc or ring armatures, so that the whole outer circumference may be covered with the coils. The disc is divided into segments, equal in number to the intended separate coils, and these segments are wound with as many equal and parallel turns of wire as the length of the inner circumferential arc will admit of; the winding is then continued through a series of holes pierced through the disc, or steps, cut in the segment in such manner and order that each successive turn of wire, or groups of turns of wire, is rather shorter than the one preceding it. In this manner the otherwise unoccupied triangles are filled up by winding in a series of turns, or groups of turns, arranged stepwise, so that the whole of the wedge-shaped segments are completely covered. Figs. 1 and 2 illustrate one method of carrying out the invention. In this, the winding of one of the segments is commenced at the oblong

hole a, and is continued, as above described, until the inner surface of the hole a is filled with wire. After this the winding is continued through the holes $c_1 c_2 c_3$, until the full length of the outer bounding arc of the segment is filled. The next segment is then commenced at the hole b, and the winding continued all the way round in a similar manner, each section being formed of as many layers as may be convenient. The holes $a\, b\, c$, &c., are lined with insulating material. Figs. 3 and 4 show another method of winding, applicable to a disc core built up of segments. Each segment is of malleable cast iron, or other magnetisable material, and is formed stepwise along one edge, as indicated at a.

copper bolts, passing through holes in the disc, which may be made either of several thicknesses of iron plate rivetted together, or of a rim of non-magnetisable metal covered with iron plates as in Fig. 7. Figs. 9 and 10 show a modification, in which the holes a and b, corresponding to the holes a and b in Fig. 1, are connected to radial slots extending to the inner circumference of the plates. This affords facility in winding, as by this device the wire can be wound on a bobbin and passed through the central hole of the disc.

*2,619.—R. E. B. Crompton, London. Electric Lighting. 2d. June 3.

Fig. 1. Fig. 2.   Fig. 3.   Fig. 4.

Fig. 9.   Fig. 10.

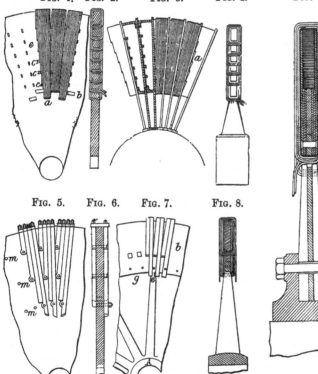

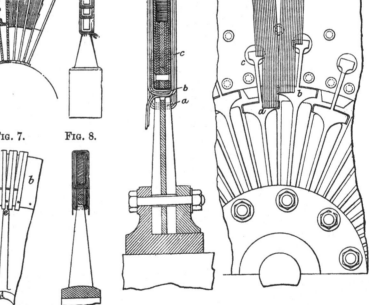

Fig. 5.   Fig. 6.   Fig. 7.   Fig. 8.

The segments are secured together in pairs so that they form a rigid compound segment, capable of withstanding the tangential strain caused by the rotation of the disc. By this arrangement, the respective segments can be wound separately, and built up into a disc, and, if necessary, be subsequently removed and replaced, without disturbing the other segments. The winding is similar to that shown in Fig. 1, and each segment can be coiled before it is bolted to the next. Figs. 5, 6, 7, and 8 show two disc armatures, in which the conductors are thick copper bars $m\, m$, each of which consists of a stirrup of rectangular section. The connections between adjacent turns are made by

ARC LAMP.—The two carbons are attached to sliding holders, so as to approach each other in the usual manner, and they are attached one to the other by means of flexible connections (which may be of regulated weight) passing over pulleys, and so arranged that the travel of one carbon controls the travel of the other, as the rate of wasting away of the point of the first is to that of the second. The length of the arc is controlled by means of a solenoid, or helix, suitably placed, the core of which is attached to the negative carbon at or near the point on its incandescent cone, where it wastes away. This attachment may take the form of a ring pressing on the cone, or of a tube

having one or more claws or pins projecting inwardly. If the lamp is to be burned singly the solenoid can be wound with a coil of thick wire only, coupled up in circuit with the arc, or if more than one lamp is burnt in the circuit, the solenoid may be wound differentially in the usual manner, or according to the arrangement described in Specification No. 346 of 1882. The point of the upper carbon is passed through an insulated guide attached to the ring or tube.

2,623.—W. R. Lake, London. (*J. J. Wood, Brooklyn, U.S.A.*) Devices for Coupling the Armatures and Commutators of Electric Machines. 6d. (6 figs.) June 3.

DYNAMO-ELECTRIC GENERATORS. — This relates more especially to the "Gramme" form of armature and commutator, and consists in a device for connecting the ends of the armature coils to the commutator

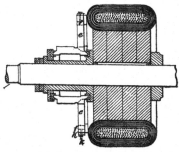

strips ; *e* represents a commutator strip, on the tip of which is shown a small metal block *f*, having on its underside a slot in which *e* is soldered or rivetted. This block has either a level top, or is indented with two parallel grooves, in which the ends of two armature wires are secured by the cap *h* and screw *k*. The underside of the cap has two marginal grooves, or lips, corresponding to the grooves in the block *f*.

*2,628.—H. Defty, Middlesbrough. Collecting and Transmitting Electric Fluid. 2d. June 5.

GENERATOR.—Making use of the inventor's own description, "this invention consists as follows : first there is an arrangement of magnetised metals, of straight or curved bars, built in a consecutive order, forming one combined fagot or pile, to which is fitted at the rear a series of induction wires or rods, leading from a bunker or store of ferruginous ores, to assist in sustaining the magnetic charge from the decomposing mines. Immediately over the positive ends of the fagot or pile, there is attached a steel crossbar, which I name a 'shoe-accumulator,' this being made in such form as to clamp or gather, and then conduct all fluid passing through the fagot or pile into the core of the shoe

accumulator, where there is fixed in contact an arrangement of one or more circuit discs, working on pivots, the above-named discs being fitted with a number of copper wire coils, which I name the 'compounder,' and so arranged as to swing or rotate in order to collect and reconcile within the circuit, the magnetic waves or fluid to one central electro-motive force, discharging one continuous stream up the disc points, where the current is received by union conducting bars, and then transmitted through wires free from variations or broken undulations of current." "The motive power . . . . is by means of a fan wheel driven by the draught of a chimney, or by a system of water-wheels with an arrangement of connecting gear of belts or shafting."

ARC LAMP.—"The object to work machinery or burner for illumination of which the instrument consists, is a small octagon-headed needle, made of a reflecting substance to return the inside ray of light which would otherwise be shaded and lost by the coils or plates as in ordinary use. For carbon burners, the aforesaid carbon being finely powdered and put into a reservoir furnished with a small vent hole, in central position of the poles, which causes a continuous discharge or supply of the carbon upon the electric current, producing a beautiful white or other coloured light according to the chemical mixture with the carbon."

*2,629.—R. Kennedy, Glasgow. Intensifying Fluorescent or Phosphorescent Electric Lighting. 2d. June 5.

ELECTRIC LAMP.—An electric current of high tension is discharged between two conductors in a vacuum, which conductors may be partly or wholly coated with various fluorescent or phosphorescent substances, or fluorescent or phosphorescent vapours may be admitted into an exhausted globe of fluorescent glass, such as uranium glass. The conductor may be of carbon or of metal. A conducting point of metal, or carbon, may be fixed immediately in front of a disc of thin paper, wood, cork, or glass, or metal coated over on one or both sides with the fluorescent or phosphorescent substance, or the fluorescent or phosphorescent substance may be formed into a ball and placed in front of the conducting point. The conductors conveying the current to the lamp are sealed into the glass.

DYNAMO-ELECTRIC GENERATOR.—A generator, suitable for the above described lamp, consists of iron masses revolving before the faces of fixed magnets, the masses being fixed to the ends of revolving iron shafts passing through fixed solenoids, in which the currents are generated. Permanent steel magnets or electro-magnets may be employed.

2,630.—A. J. Jarman, London. Dynamo-Electric Machines. 6d. (9 figs.) June 5.

DYNAMO-ELECTRIC GENERATORS. — This refers principally to a mode of regulating the current generated by a dynamo machine, whereby the pole-pieces A A are caused to approach or recede from the revolving armature. In the illustration, this is accomplished by turning the screw C, which is provided with right and left-handed threads as shown, or the pole-pieces can be regulated by means of set

screws, the pole-pieces sliding within the magnets of which they form part ; or the legs of the magnet, in the Edison machine, may be hinged or pivotted together, and the pole-pieces adjusted by set screws, or screws and wedges ; or the side frames carrying the magnets can be made to slide upon or within each other.

MOTORS.—These may be also regulated by the adjustment of the pole-pieces.

2,632.—W. R. Lake, London. (J. J. Wood, Brooklyn, U.S.A.) Electric Lamps. 6d. (6 figs.) June 5.

ARC LAMP.—The main feature of the invention consists in a simple connection between the negative carbon-holder, which in this case is movable, and the spindle of the wheel train, which is geared with the positive carbon-holder. This connection consists of a rotary spindle, mounted in stationary bearings, and geared on one side with the spindle of the vibrating train, and on the other side with the negative carbon-holder, so that as the train is vibrated up or down, by the usual regulating movements, both carbon-holders will be moved simultaneously in opposite directions, to or from each other, and at a proper speed, thus producing a correct focussing action. The spindle is divided at one point, and the two parts connected by an insulating coupling, the corresponding bearing of the spindle being insulated from the lamp frame. The general arrangement of this lamp is similar to those described in Specifications Nos. 2,563 and 2,570 of 1882. As shown in the drawing annexed to this specification, the mechanism is placed below the carbons, and the lower end of the positive holder works in a closed tube depending from the lower head of the clamp frame, while the main portion of the holder is extended upwardly through a long tube, and from its top projects a lateral overhanging arm, provided with a pendent angularly adjustable clamp for the positive

carbon. The negative carbon is held in the clamp of the negative holder, which rises in line with the positive carbon out of a guide tube. A small point, projecting from the guide of the positive holder, indicates the proper focussing point for the carbons. The standard of the lamp is adjusted by means of a nut screwing on a non-rotative stem, sliding in a pedestal, so as to adjust the focus of the carbons to the focus of the reflector with which the lamp may be provided. The retractile spring of the armature lever is adjusted by means of an L shaped lever, to the horizontal arm of which one end of the spring is secured ; a set screw passes through the vertical arm and bears upon the side of the standard to which the lever is pivotted.

2,636. — A. L. Fyfe and J. Main, London. Dynamo-Electric, Magneto-Electric, and Electro-Magnetic Machines. 8d. (11 figs.) June 5.

DYNAMO-ELECTRIC GENERATOR.—This is a disc armature generator, in which a number of coils, 1, 2, 3, &c., wound upon non-magnetic cores, and provided with iron cheeks *a a* on each side, rivetted to a non-magnetic ring *c*, revolve between two sets of opposing field magnets, one set of which, with its pole-pieces of alternate polarity, is shown in Fig. 3. The chief feature of the machine lies in the method adopted of connecting the revolving

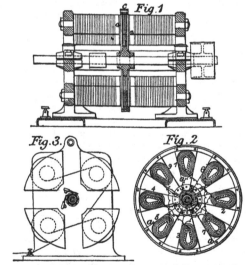

coils to the commutator plates, and in connecting up the plates among themselves, for the purpose of cutting out the idle coils. As shown in Fig. 2, the coils are arranged in two groups, there being eight coils and eight commutator plates. Tracing the connections from one plate, the circuit runs through coils Nos. 1, 2, 3, and 4 in succession, thence to the plate next but one behind the first plate. The second circuit joins together the two plates on either

side of the first plate, through the coils Nos. 6, 7, 8, and 9. Lastly, each plate is connected directly to the one diametrically opposite to it on the commutator. The collecting brushes are arranged to touch the commutator as shown in Fig. 3. In one form of machine the commutator is made stationary, and each section of the armature carries a brush forming part of a spring. When the armature coils arrive at a certain position between the poles, the brushes bear against a separate part of the commutator, and are thereby cut out of circuit during portions of the revolution. Two commutators, placed side by side, are sometimes employed, and their lateral adjustment is provided for by a clutch motion. Fig. 1 represents a similar machine with eight pairs of field magnets.

2,642.—W. E. Ayrton and J. Perry, London. **Registering the Amount of Work Given Electrically to any Part of an Electric Circuit in a Given Time.** 4d. June 5.

POWER METER.—The apparatus consists of "a small machine shaped like a dynamo-electric machine." The movable field magnet and the fixed armatures are wound respectively with a coil, or coils, of small resistance, through which the main current, or a portion of it, passes, and a coil or coils, of high resistance, placed in a shunt across the main circuit, or *vice versâ*. The average force acting between the movable and fixed parts is proportional to the product of the currents flowing in them, and this is proportional to the horse-power being expended in the main current, hence the horse-power so expended, multiplied by the velocity of rotation, is proportional to the mechanical power given out by the motor. If the only work done by the motor is done in overcoming fluid friction, or other resistance which is proportional to the velocity of rotation, the number of revolutions made by the motor, in any time, is proportional to the electrical energy given to the circuit in that time. The number of revolutions is registered in a way similar to that employed in gas or water meters. Steel or cast iron must not be used in the motor, and it is preferable not to use any metallic core in the moving part. Another method of registering the energy given electrically to any circuit, consists in placing on a pendulum or balance of a time-keeper a coil of wire, and in fixing another coil in the neighbourhood. One of these coils is of high resistance, and connects the extremities of the circuit whose energy it is wished to record. Through the other circuit the main current passes, and the coils are so placed, that, in the absence of gravity or springs, the attractions and repulsions between them would cause a very slow vibration of the regulator. The effect of this arrangement is to shorten or quicken the timekeeping,

and this variation, under these conditions, is proportional to the total energy given to the circuit in any time.

*2,643.—H. Woodward, London. **Secondary Batteries, &c.** 2d. June 6.

SECONDARY BATTERIES.—The box, or case, is of any suitable shape (preferably in the shape of a beehive, or dome, or bridge), the sides, top, and bottom being corrugated or fluted. The electrodes are placed inside, and are built up of plates, sheets, or strips of lead, or other metal, also corrugated or fluted, and they may be formed with projections or knobs. The box or case is placed in a shallow trough containing suitable liquid, cotton wick being placed between the plates entering the liquid to keep them moist. The generator and motive power engine are fixed to a frame that can be drawn from place to place.

2,644. — L. Varicas, London. (*G. Richardson, Philadelphia, U.S.A.*) **Underground Electrical Conductors.** 10d. (34 figs.) June 6.

CONDUCTORS.—The bare conductors are encased directly in hydraulic cement. In laying a row of conductors, a wooden board, having a wave-like contour, is embedded in the bottom of a trench, so as to be flush therewith, its width being less than the width of the foundation, but reaching to the edge thereof. The foundation, which may consist of brick or broken stone ground with cement, is made so as to alternately overlap the inwardly curved portions of the wooden board, and to extend over the adjacent earth, thus leaving supports at intervals should the board rot. The foundation being levelled, the bare conductors are stretched over it, so as to be about an inch from the top of the foundation, and hydraulic cement is then poured around and over the conductors, and thoroughly trowelled to make it solid. When the cement has set, the trench is filled with earth. In order to provide for access to the conductors, vertical circular tubes, reaching from the top of the cement to the surface, and provided with covers or caps, may be employed, the conductors being exposed at these points, where they are preferably bent upwards, and galvanised and protected by some non-conducting fluid. In some cases, the conductors are passed over a porous block resting on the earth at the bottom of the trench, and grooved or not ; this block is made of hydraulic cement mixed with sand or gravel, so as to be porous and retain more less moisture, and it serves " to discharge surplus electricity due to atmospheric or earth currents." The block may rest on the cement and have lateral projections extending into the earth. In order to provide for additional rows of conductors, a shelf is

formed around each man-hole, and rests upon an extension of the foundation, the second row being laid around instead of through the man-holes, and being provided with separate man-holes. In constructing pipes for the accommodation of insulated conductors, the flues of the pipes are made tapering, and the larger end of each flue abuts and encircles the smaller end of the flue of the next adjacent pipe, the conductors being drawn through from the larger to the narrower end. In laying single flue pipes, they are preferably square (with round flues), so that a second can be laid above the first, the pipes of one row breaking joint with those of the other. Each pipe has a socket at the larger end, and a tenon at the smaller end. Pipes having a number of flues are preferably cylindrical, and are connected together by short couplings of somewhat larger diameter, an annular recess being formed in each end of the coupling, and the central web being flat or concave with the rim of the coupling projecting beyond this part. The pipes have a rim at each end, to fit the annular recess, and a central web, either flat or convex, to fit the web of the coupling. The flat surfaces should have a layer of cement spread over them. The inner side of the rim of the coupling is bevelled, and the pipes are constructed with soles, so as to make the bottom side flush with the coupling. To enable a circular group of conductors to pass an obstruction, the couplings are constructed at one end to fit the pipes, and are flattened towards the other end to the required width for ranging the flues in a row. Suitable man-holes are provided at intervals. In another method of laying and insulating conductors, the conductors are drawn through pipes, preferably of carbonised stone cement, vaults being constructed at suitable intervals and provided with as many man-holes as groups of conductors can be laid in the conduit. The conductors are carried by a lateral arch into the man-hole, and are passed upwards over a transverse block (of porous material capable of conducting away the surplus electricity), back through the arch to the next section of the conduit. The conductors are stretched so as to hold them about an inch from the bottom of the conduit, and hydraulic cement in a plastic condition is introduced. This is preferably effected by means of sectional troughs, run into the conduit from each end until they meet at the centre. An endless belt, arranged in each sectional trough, and constructed with transverse paddles, is driven by suitable means, and as the troughs are gradually withdrawn the plastic cement is fed into the conduit and distributed. Three blocks, "swivelingly connected," are then drawn through the section; the first is comparatively light, and is provided with scrapers at its end to scrape the sides

and top of the conduit, and serves to smooth the surface of the cement; the second block is much heavier, and is provided with a series of trowels at the bottom (a space being left between the bottom and the trowels about equal to the thickness of the coat of cement required above the conductors); a lateral notch is cut in each trowel near where it is joined to the block, and each conductor passes into the notch of a corresponding trowel, the trowels lifting them slightly and serving to compact the cement under them; the third block is made heavy and serves to smooth the surface of the cement. The forward end of each block is suitably bevelled. This method of insulation is continued in the vault, and the corresponding man-hole is filled up with a suitable removable insulating material. The second group of conductors is then laid and insulated in a similar manner. In supplying buildings with electricity, the branch conduit is composed of pipes erected to reach the highest point of the building, and the conductors branch off laterally at the desired points, these branches only being exposed. The conductors are stretched, whilst being insulated, by means of a sliding sash provided with a series of fixed clamps, constructed so that they can grasp a conductor sideways at any point of its length. The sliding sash is operated by screws working in nuts fixed to the sash. In order to be able to stretch any particular conductor, after the conductors have been stretched collectively, a series of similar movable clamps are arranged on the sash, and in line with the fixed clamps; or one laterally adjustable clamp may be employed. These clamps terminate in a screw threaded rod, on which is a nut, constructed with ratchet teeth, and operated by a hand lever and pawl.

2,654.—R. J. Hatton, Stratford, Essex, and A. L. Paul, London. **Electric Lamps.** 6d. (7 figs.) June 6.

ARC LAMPS.—In Fig. 1 the upper carbon-holder is geared with a pinion c_1, carried by a lever which is pivotted at o^1. The end of the long arm of the lever carries the spring pawl i, and is connected through link c^2 with the solenoid core f. Upon the axis of the pinion c^1 is fixed a ratchet wheel, which, when the coil g is excited, is rotated by the downward movement of the pawl, and causes the upper carbon-holder to rise and establish the arc. When the current becomes weakened by an increasing length of arc, the recoil spring k lifts the pawl out of gear with the ratchet wheel, and the carbon-holder descends. A train of wheels and a fly regulate the speed of the descending carbon. Fig. 2 shows a modification of the arrangement, in which the solenoid is replaced by an electro-magnet, and the lamp is made focus-

keeping, by attaching the lower carbon to a rack gearing with a smaller pinion placed upon the same axis as the pinion c'. The lever e may be dispensed with, the pinion c' and ratchet wheel d being fixed, and the spring pawl i being attached directly to

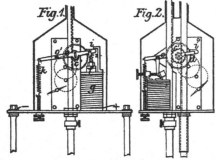

Fig.1.   *Fig.2.*

the solenoid core. The ratchet and pawl may be replaced by various forms of brake wheels and straps, the strap being first caused, by the solenoid or magnet, to grip the brake wheel, and then to slightly rotate it to form the arc.

2,658.—A. Muirhead, London. Secondary Batteries. 4d. June 6.

SECONDARY BATTERIES.—The formation of the active surfaces of the plates is effected by decomposing, electrically, a solution containing lead, whilst the solution is flowing in a stream past the plates. A metallic deposit is preferably obtained on all the plates in the first instance. The plates are connected to one pole of a generator, and are arranged within a vat against its two opposite sides, and on opposite sides of a large plate connected to the other pole. A stream of the electrolytic liquid, preferably protoxide of lead (Pb O) dissolved in caustic potash, is caused to flow through the vat, and on a current being passed, metallic lead is deposited on the plates. The plates on the opposite sides of the vat are then disconnected and connected to the poles of the generator, "a solution of sulphate of lead in dilute sulphuric acid" being caused to flow through the vat. The cistern, into which the liquid flows, contains oxide of lead to replenish the electrolytic liquid, and can be raised so as to cause it to flow back in the reverse direction through the vat. The provisional specification states that a small quantity of phosphoric or tartaric acid may be added to the liquid, to increase the coherency of the precipitated lead.

*2,659.—W. B. Brain, Cinderford, Gloucestershire. Primary and Secondary Batteries. 2d. June 6.

SECONDARY BATTERIES.—The electrodes described in the Provisional Specification No. 1,548 of 1882, or lead, or lead oxide, or other oxide, or chloride, taken from an electrolytic bath, are subjected to a considerable pressure, and pressed into a body of great solidity and density. The primary charging may be in dilute sulphuric acid, with or without chloroform, turpentine, alcoholic liquids, enchlorine, or chloride of ammonia. Plug switch or shunt appliances are used with batteries formed of these plates, by which they can be coupled up so as to vary the electromotive force, while drawing from each cell the same quantity of current at the same time. The contacts may be solid or liquid, such as mercury, the surface of which is protected from the atmosphere.

2,660.—J. Wetter, London. (*W. Stanley, Bergen, N.J., U.S.A.*) Carbon Burners for Electric Lamps. 4d. June 7.

CARBONS.—Carbon filaments for incandescence lamps are made from the hair of various animals. Human hair is considered to be the most suitable, and especially that of Chinamen, as being coarser, straighter, and more uniform in section, than that of other races. The greasy and fatty matter is removed by treatment with alkalies, or weak acids, or by heat. The filaments are then dried upon a straight smooth surface under a slight tension, in order to prevent them from buckling during the operation of carbonisation. After drying they are placed in suitable moulds, provided with grooves, in which the hair is brought into the desired form, and are then carbonised in the ordinary way.

2,661.—J. Blyth, Glasgow, and D. B. Peebles, Edinburgh. Producing and Measuring Electric Currents. 6d. (6 figs.) June 7.

DYNAMO-ELECTRIC GENERATOR.—This is an annular armature generator, with separate external and internal field magnets. The construction is clearly shown by Figs. 1 and 2. Upon the armature shaft 11 is a gun-metal disc 15, with a cylindrical extension 16, to which is attached the armature core 17 (consisting preferably of iron wire) by the links 18, in the manner shown. In the spaces between the links coils of insulated wire are wound upon the iron wire core, one of them being indicated in section at 19, Fig. 2. The external field magnets 9, 10, have pole-pieces 20, between which, and the corresponding pole-pieces 23 of the internal magnets, the armature coils are caused to revolve. The ends of the internal magnets are supported by an iron disc 24, in the centre of which is a brass bush within which the shaft revolves. In a modification, the internal magnets take an H form, with a central tubular boss carried loosely by the shaft, the parts between the boss and pole-pieces being wound with the magnetising coils; the boss is kept from revolving upon the shaft by a convenient form of stop.

CURRENT METER.—The current measuring apparatus, Fig. 3, comprises an inverted glass vessel 27, through the plug in which are inserted two inverted syphons 28, 29, the former for maintaining a supply of acidulated water in the vessel, and the latter communicating with a gas meter 31. Immersed in the acidulated water are two platinum electrodes, formed of corrugated plates, and these are part of a high resistance shunt to the main circuit, the resistance of which bears a definite relation to that part of the main circuit which is shunted, and thus the gas evolved in the vessel will also

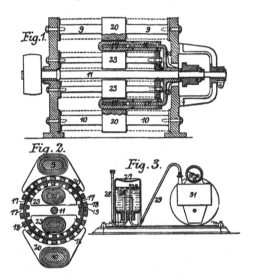

Fig. 1.

Fig. 2.

Fig. 3.

bear a definite relation to the total current flowing. This gas, as it is formed, passes through the meter, from the indications of which the strength of the current can be found. The water in the vessel 27 is slightly acidulated, and a small hydrometer, or specific gravity bead, indicates the acidity. For keeping a sufficiently constant level of water in the meter, any suitable contrivance may be adopted, or a hydro-carbon, or other oil of a non-volatile character, may be used instead of water. The oxygen and hydrogen evolved may be kept separate, and they may be passed through separate drums on the same axle, one drum being twice the size of the other ; or one only of the gases may be measured.

*2,674.—E. De Pass, London. (*J. Gloker, Paris.*) Electric Lamps, 4d. June 7.

ARC LAMPS.—Two solenoids are employed, whose "casings" resemble "the frame carrying the wire of a multiplying galvanometer." The outside of the casing is covered with several coils of wire, and within it moves an iron bar, or plate, mounted on an arbor which carries a piece of soft iron, or armature, fixed almost normally to the bar or plate.

To one end of the armature is attached a spring (for maintaining it in position), whose other end abuts against a screw for regulating its tension ; or the spring may be dispensed with. The inner end of the armature is fitted in a frame, covered at the top and bottom by coils forming two other solenoids, the four solenoids being connected in one circuit so "as to have the same pole at a suitable point." The stem of the upper carbon-holder passes within the second pair of solenoids, and a spring, pressing against it, retards its descent, and conveys the current to it. On the current being passed, the bar or plate tends to assume a position perpendicular to the coils of the first-mentioned solenoids, and the armature describes an arc of a circle, until, entering the field of the stem of the upper carbon-holder, which is magnetised by the current, it adheres to the stem, and continuing its movement, raises the upper carbon and strikes the arc. In a modification, instead of the second solenoids and armature, two electro-magnets are mounted directly on the arbor. An arm, abutting against a stop, determines the extent of motion of the magnets and their armatures, and is attached to one end of a spring. A disc, serving as the armature, carries a socket, through which passes an arbor supporting a roller, forcing the upper carbon against a second roller pressed against the carbon by a spring. On the passage of the current, the bar tends to assume a position perpendicular to the coils of the solenoids, and the electro-magnets attract their armature, thereby causing the rollers to raise the upper carbon-holder.

2,676.—A. M. Clark, London. (*J. M. A. Gérard-Lescuyer, Paris.*) Preparing Electrodes for Secondary Batteries. 4d. June 7.

SECONDARY BATTERIES. — The pair of plates to form the battery are connected to the positive pole of a generator, and immersed in an alkaline solution of a suitable metallic salt (such as a manganate, ferrate, alkaline chromate, zincate, stannate, antimoniate, but preferably plumbate or plumbite, prepared by dissolving massicot or litharge (Pb O) in a boiling alkaline lye), contained in a vessel, the sides of which preferably constitute the negative pole. If the solution is at a temperature between 15 deg. and 70 deg. Cent., minium (Pb$_3$ O$_4$) is formed on the plates, and plumbic acid (Pb O$_2$) if the temperature falls below 15 deg. Cent. ; the hot method is the most rapid. When the coating of oxide is sufficient, the plates are withdrawn and washed, and are immersed in acidulated water and charged in the usual way. Iron, copper, or other metals, or carbon, gilt or platinised, if necessary, in the form of plates, sticks, granules, or gauze, may be coated in a similar manner with spongy metallic lead, or with oxide of lead. The plates may be

rolled up in spiral form, and separated by wood, caoutchouc, or felt.

2,686.—M. A. Wier, London. Electric Lamps. 2d. June 8.

ARC LAMPS.—A small magneto, or dynamo-electric or other motor, rotates the electrodes at great speed. The weight of the carbons and holders may be taken off their pivots by the attraction of an electro-magnet. One, or both electrodes, may be made to revolve, or a central electrode may have others placed round it, and either it, or they, may be arranged to work in pairs alternately, the rotation causing the current to be made and broken by means of a suitable commutator and springs.

*2,688.—C. G. Gumpel, London. Voltaic Batteries. 2d. June 8.

SECONDARY BATTERIES.—The plates, of metal, carbon, or other material, are made in the form of discs, and are placed in proper order in a caoutchouc tube, rings of vulcanite or other suitable substance being interposed between the pairs of discs, so as to form spaces for the exciting substance, and prevent the discs being pressed together.

2,694.—W. R. Lake, London. (*E. Weston, Newark, N.J., U.S.A.*) Dynamo or Magneto-Electric Machines. 6d. (5 figs.) June 8.

DYNAMO-ELECTRIC GENERATORS.—The armatures heretofore constructed by the inventor are successively wound with two systems of coils, one overlying the other, each system being wound in diametrically separate divisions, and each division, or coil, being connected to the adjoining division by a loop from which connection is made to one of the commutator bars; or each system of coils is wound upon itself, so that the two systems lie alongside of each other in the divisions of the armature, and equidistant from the centre of the same. By this latter arrangement, the two coils of each armature division are practically of the same length. According to the present invention, the several divisions of the two systems of coils are alternately superposed. The armature core is preferably composed of a number of plates or discs of magnetic metal, mounted on a shaft, and held at short distances apart by collars or rings of insulating material of considerably less diameter than the discs. The cylinder, when built up, is provided with longitudinal grooves, or recesses, in which the coils are wound. Referring to Fig. 1, in any two diametrically opposite recesses of the armature an insulated conductor is wound, until one-fourth of the recess is filled with a coil a, the two ends of which are temporarily secured. The armature is then turned half round, and the coil b is wound upon the coil a. When the same number of layers is reached, a loop b^1 is formed, and the winding is continued alongside of the coil a, until the coil c is formed. The wire is then cut off, forming a free end, which is temporarily secured. The armature is then turned back, and the coils d and e wound in a similar manner; these operations are repeated until all the recesses are filled, each with four coils. The free ends left after completing the winding are joined to form a continuous conductor, and from the loops so formed, and those originally existing, connection is made to the commutator. In another armature, suitable for currents of great quantity, the core is wound with a double system of continuous coils in separate

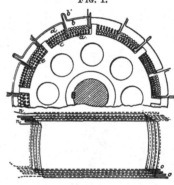

FIG. 1.

FIG. 2.

recesses. Heretofore loops have been taken to the commutator only at the junction of each coil with the next coil, there being one loop to each armature division. According to the present invention, two loops are taken off, one at the junctions between the several coils, and the other from the middle of each coil, each coil being formed with an odd number of turns, so that the loops of each coil will be at diametrically opposite points of the armature, as clearly shown at n and o, Fig. 2. These two loops are respectively connected to separate commutators at either end of the armature, the brushes of which are connected in multiple arc, or in series, as required. When heavy conductors are employed, instead of winding them back on themselves, they are cut, and the coils formed successively and separately.

2,701.—A. R. Leask, London. Exhausting Apparatus. 6d. (6 figs.) June 9.

INCANDESCENCE LAMPS.—This invention refers mainly to a vacuum pump. The framework is constructed of wood, braced by cross-bars, some of which are made hollow to enclose conductors from a generator. Several pumps are attached to the front and back of the framework. The pumps each consist of a large bulb, connected

directly by a flexible tube to the bottom of the pump bulb, which is supported upon brackets. A small pipe extends laterally from the neck of the pump bulb, and then rises vertically to the height of the centre of a smaller bulb connected by a neck to the pump bulb. This pipe communicates through a syphon tube to a tube of larger diameter, arranged below the small bulb, and to which are attached, by small necks, the bulbs to be exhausted. The syphon is connected to the small pipe by an ordinary mercury seal. The neck between the pump bulb and the small bulb is fitted with a ground-glass stopper. To work the pump, the large bulb is hung up the stopper removed, and mercury poured in until it reaches the middle of the small bulb; the stopper is then inserted, and the large bulb is lowered, when the mercury recedes from, and leaves a more or less perfect vacuum in the pump bulb, and air from the bulbs to be exhausted rushes in; the large bulb is then raised, and the mercury rises in the pump bulb and cuts off the return of the air, and the stopper is removed to allow it to escape, the same operation being repeated as often as necessary. The syphon tubes are carried up to a sufficient height to prevent the mercury passing over the bend, and are secured to the frame by clips, slits in which fit over screws or projections on the frame, so that the clips can readily be removed. These tubes may be made of iron, but preferably the shorter leg is of glass, so that the height of the mercury can be seen. The lamp is held during exhaustion in a divided clip of spring wire, held together by a movable sleeve. The connections from the generator to the lamps are made by branch conductors leading from the main conductors, which are enclosed in hollow cross-bars. A switch is provided in the main conductor.

2,712.—W. R. Lake, London. (*F. Krizik and L. Piette, Pilsen, Austria.*) Electric Lamps. 6d. (2 figs.) June 9.

ARC LAMPS.—The regulation of the arc is controlled by the attraction of a solenoid upon a long tapering core C, the lower extremity of which serves as the holder for the upper carbon K, the weight of the latter and the core being partly counterbalanced by that of the lower carbon and its support, as shown, by means of cords attached to the lower part of a brass tube guided by rollers, carrying the upper carbon-holder, and having within it the taper iron core; the lower carbon is carried by a frame, suspended from cords which pass over suitable pulleys. In order to increase the delicacy of the action of the solenoid, its coil is wound, step by step, in increasing diameter upwards, as represented at Z, Fig. 2, and the spaces formed

by the steps are wound with an opposing coil Y, of

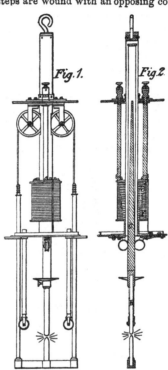

Fig. 1.   *Fig. 2.*

high resistance, placed in a shunt to the lamp circuit.

2,722. — A. P. Price, London. Secondary Batteries. 2d. June 9.

SECONDARY BATTERIES.—The leaden electrodes (both or either of them) are encased, enveloped, enclosed, or otherwise protected by, or in, spongy india-rubber, or india-rubber sponge. The spongy india-rubber may be utilised as a buffer between the elements.

2,723. — C. G. Gumpel, London. Electric Lamps. 6d. (5 figs.) June 9.

ARC LAMPS.—The upper carbon-holder passes freely through the centre of the main solenoid B, and is attached to a guided frame. The holder, frame, and carbon are partly counterbalanced by a weight (shown as surrounding the holder), connected to it by cords passing over pulleys. A light non-magnetic tube D, carrying the sliding frame *d*, with the rollers *e* and *f*, slides freely within the solenoid B and around the hollow core G. The roller *e* is pivotted upon one arm of a bell-crank supported by the sliding frame, the other arm being jointed to the solenoid core. Upon the axis of the roller *f*, which is mounted in the frame *d*, is fixed a feed wheel F actuated by a pawl *h*, or equivalent device, put in motion by the

core H of the high resistance solenoid K. The coil of the solenoid B is in the direct lamp circuit, and when its core is attracted, the jockey roller e grips the carbon-holder, lifting the frame d and the upper carbon-holder, and forming the arc. When this becomes too long the wheel F is turned so as to feed the upper carbon down, by the attraction of the solenoid K upon its core. The travel of the frame d downwards is limited by the stops b. A spring acts as a brake to the feed wheel F. In a modification, the rollers e and f are mounted on a kind of toggle joint pivotted to the frame d, and

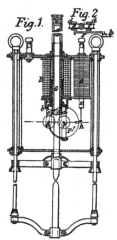

Fig. 1.   Fig. 2.

the core G is divided into two parts, the upper being connected to the tube D, and the two parts being attracted towards each other. In another modification, the rollers are mounted in a toggle joint, linked to a lever carrying the armature of an electro-magnet in the main circuit. The feeding solenoid may have below it a coil in the main circuit, or a lever carrying the armature of the main electro-magnet is linked to a bell-crank, which carries one gripping roller with the feed wheel, and is pivotted on a frame carrying two gripping rollers.

2,734.—J. Mathieson, Stratford, Essex. **Governing the Feed of Electric Arc Lamps.** 6d. (3 figs.) June 10.

ARC LAMPS.—The feature of novelty consists in the means for cutting out the shunt electro-magnet without breaking the circuit. The feed of the carbons is regulated by a train of wheels, of which the last member is a ratchet wheel. A pawl upon the end of a lever takes into this wheel, and when depressed stops its rotation. Upon the other end of the lever is the armature f, situated over the electro-magnet F, placed in a shunt circuit round the arc. The same lever carries the long bracket e^2, and is attached to a spring, by

which it is raised in opposition to the pull of the magnet. When the parts are in the position shown, the shunt current passes from the armature lever through the coils of the magnet F, and, by the contact piece G, through the resistance coil H; but as soon as the power of the magnet becomes sufficient to draw down the armature, a new path is opened for the current through the armature lever, down the

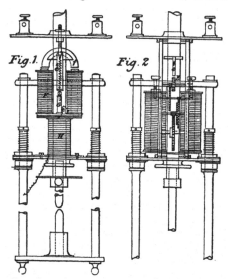

Fig. 1.   Fig. 2.

bracket e^2 to the contact piece G, and thus the magnet momentarily loses its power, the lever rises, thrusting the detent in the path of the ratchet wheel, and stopping the carbons. The armature is again attracted, and the process of alternately stopping and permitting the feed of the carbons continued until equilibrium is effected between the main and shunt circuit, and the armature stands clear of the contact piece.

*2,740.—G. Zanni, London. **Electric Lamps.** 2d. June 10.

SEMI-INCANDESCENCE LAMP.—The positive electrode is formed of two or more rods mounted in the same holder, which touch, or nearly touch, each other, and are fed into contact with the negative electrode by means of weights. The negative electrode may consist of a disc, rod, or block of carbon, copper, magnesium, or lime.

2,741.—G. Zanni, London. **Illuminating Conductors for Incandescent Electric Lamps.** 4d. June 10.

ELECTRODES. — Low resistance filaments are made from a combination of carbonised material with platinum, iridium, or a similar metal. By one method, a film of metal is deposited, either by electricity or from a solution, upon a carbonised thread, and the filament thus formed is coated with

carbon. By another method, a thin metal wire is inserted in a fine Tuscan straw, which is afterwards bent to the shape and carbonised, or the metal may be coated with carbon by means of gas, or in any other suitable manner. The bulbs, in which these filaments are mounted, need not necessarily be exhausted.

2,744.—J. Imray, London. *(J. J. and T. J. McTighe, Pittsburgh, U.S.A.)* **Dynamo-Electric Machines and Electric Motors.** 6d. (9 figs.) June 10.

DYNAMO-ELECTRIC GENERATOR.—The armature is wound Gramme fashion, and has a length about equal to its diameter. Its core is composed of a number of rings a a, of angular or rectangular cross section, each with a piece about the width of a coil cut out of it. The rings are placed together to form a cylinder, with all their slots in line, and the coils, previously wound on a mould, are slipped through the opening and threaded, one by one, on the core until it is filled, as shown in Fig. 2. The rings a are then worked round until no two of the openings are in line, but are all distributed circumferentially at equal intervals and are secured by bolts e e, passing through

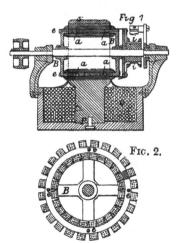

FIG. 2.

them, and the non-magnetic end pieces B B. The commutator is formed of a block of vulcanite with metal bars i passed through it endwise, and is afterwards turned down to an even face. The bars are bent up at their inner ends for connection with the wires, and are clamped in position by means of an insulating flange, to which the bent ends are secured by screws. The field magnet has only one coil Q. One end of the core is provided with a curved pole-piece, and the other end is connected to an iron flange or head, forming the base of an iron cylinder surrounding the coils, and ex-

tending upwardly from both sides to a junction, and terminating in the curved pole-piece s.

2,752.—J. Lane, London. Electric Lamps. 6d. (12 figs.) June 12.

INCANDESCENCE LAMPS.—In this lamp the globe a is removable, the filament d with its connecting wires h h^1 being carried by a plug c, between which and the globe is a ring or collar b making an air-tight joint. The plug c is sometimes made of brass, or other good conductor, one of the wires h being carried through it in an insulating tube. Through c is a passage for exhausting purposes, and upon a seat in this passage is placed the valve e, which during exhaustion is kept open by an electro-magnet controlled by a spring, or by a spring controlled by an electro-magnet. In the former case, when the proper vacuum is reached, the current through the magnet is cut off when the spring closes the valve. In the latter case, the reverse

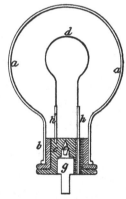

action takes place. A second electro-magnet may replace the spring. After the valve is closed, cement or wax is run into the space and a plug g inserted, by which the lamp may be attached to its bracket or holder. A packing of compressible material may be arranged between the globe and the ring b. The ring b may extend around the neck of the globe and be filled with a suitable material to render the lamp air-tight. In another lamp, two or more conductors are carried through the two ends of the bulb, one end having a removable plug. The shape of the bulb may be hemispherical, the filament being passed in from the side and in a plane parallel with the base of the hemisphere.

2,755.—W. Chadburn, Liverpool. Electric Lamps. 8d. (10 figs.) June 12.

ARC LAMP.—This consists principally of an improved form of gripping device. Three vertical levers a of the first order are pivotted at a^1 to a ring b, suspended by adjustable stops c from the top plate D. The lower ends of the levers

are made to grip the carbon when the upper ends are expanded by a cone *d*, which is moved by the primary solenoid E and the lever F, the other end of which is connected to the shunt solenoid. The cores of the solenoids are made to

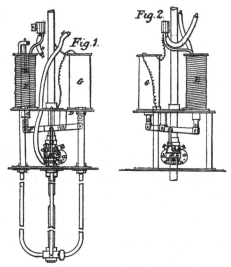

act as dash-pots, being hollow and filled with liquid into which a stationary piston *e* dips. When the lamp is in action, the position of the lever is regulated by the differential action of the two coils, the grip levers being forced inwards or outwards according to the state of the arc.

DYNAMO-ELECTRIC GENERATOR. — The field magnets are vertical and are arranged in two pairs. The common pole-piece of the two inner magnets partly encircles the upper portion of the armature, and the pole-piece of the outer magnets encircles the lower part of the armature and is mounted on, but insulated from, the base of the machine. The other poles of the magnets on either side of the generator are connected by massive pieces of iron. The core of the armature is composed of a coil of iron ribbons, wound round at intervals with copper wire, and forming a spiral of several convolutions, occupying a space equal to the width of the poles. The connections from the copper coils of the armature may be led inside the coils of the spiral to the commutator. The terminal points of each ring of the iron ribbon coil, being the core, are magnetically connected to each other so as to complete the circuit. The spiral coil may lie on a wooden drum.

2,756. — C. G. Gumpel, London. Voltaic Batteries. 6d. (6 figs.) June 12.

SECONDARY BATTERIES.—These batteries are constructed so as to be portable. The electrodes are in the form of discs, with interposed insulating rings, and are placed within a caoutchouc tube, which embraces them with sufficient tightness to prevent leakage between the cells. Referring to the illustrations, the electrodes A are separated by insulating rings and are enclosed within a tube. The exciting liquid occupies the spaces between the plates. The whole is enclosed in a casing fixed in a socket of insulating material. This battery may be combined with an incandescence lamp, as shown, the lowest plate being connected through the metal framing to one terminal, and the highest plate through a switch to the other terminal. The highest plate is also connected to a metal plug

FIG. 1.

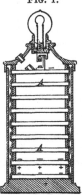

FIG. 2.

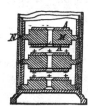

passing through an insulated hole in the framing, for convenience in charging. Fig. 2 shows the plates duplicated, with insulating material M between each pair which are connected together by rivets. In this arrangement, the electrodes are brought nearer together, and the caoutchouc tube is pressed by cords or wires N into grooves in the insulating discs M. The discs M may be made with rims projected upwards, and caoutchouc rings may be placed between the edge of the rim and the bottom of the next higher disc, the caoutchouc tube being dispensed with, and the casing having a screw cap for forcing the plates together.

2,759.—H. H. Lake, London. (*R. R. Moffatt, Brooklyn, U.S.A.*) Electric Lamps. 6d. (8 figs.) June 12.

ARC LAMPS.—The upper carbon rod C is provided with a toothed rack D[1], gearing with a chain

of wheels, the last member of which is the ratchet wheel D⁹. The pawl E is pivotted to a standard erected on one of the levers F², which are at one end pivotted to the frame, and at the other end carry the armature F¹ of a shunt magnet F. The movement of this armature is resisted by a spring F⁴, the tension of which is adjustable. The armature of the main magnet G is mounted on two levers G², which are arranged between the levers F², and are similarly pivotted to the case. These levers are united at their free ends, and the carbon-holder passes between them. A catch F³, consisting of a bell-crank lever pivotted at one end to a rod *b*, which extends between the levers F², is arranged so that its angular portion can act upon the carbon rod. The other end is acted upon by the spring *c*. This catch serves to maintain the carbon rod in any position to which it may be adjusted, but its grip is not sufficiently strong to prevent it being forced down by the pawl E and train of wheels. When the electro-magnet G attracts the armature G¹, this catch raises the rod and establishes the arc. When the magnet loses its power, the tail of the

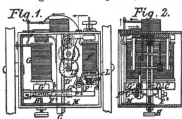

Fig. 1.   *Fig. 2.*

catch meets the stop H, and the other end is lifted clear of the rod, which then descends. The connection to the shunt magnet is through a movable lever L. When the lever F² rises to the top of its stroke, it breaks this connection and demagnetises the magnet; the armature then descends, and at the end of its descent completes the connection, and so on. A projection on the lever G² may work between two stops on the lever L, or the lever L may be provided with a spring contact. If a carbon breaks, the lever G² falls on the contact M, and short-circuits the lamp through a resistance equal to the arc. If the carbon should again come in contact, the lever G² is attracted, and the cutout circuit opened. In a modification, the levers G² are arranged to lift a sliding frame carrying a pinion engaging with the rack and a ratchet wheel, on the same axis as the pinion. An escapement on one of the levers G² controls the rotation of the ratchet wheel, and is connected by an arm to the lever F². The action of this lamp is similar to that already described.

*2,760.—H. H. Lake, London. (*J. G. Richard, Paris.*) **Posts or Supports for Telegraph Wires, &c.** 2d. June 12.

INSULATORS.—The lower part of the post is coated with india-rubber, or enclosed in fabrics impregnated with greasy and isolating substances; or preferably, the lower parts of the posts are inserted in impermeable stone sockets, and fixed therein by hydraulic cement, lead, &c.

*2,763.—D. G. Fitz-Gerald, London. **Manufacture of Peroxide of Lead.** 2d. June 12.

SECONDARY BATTERIES.—Certain salts and insoluble compounds of lead, in admixture with an electrolyte, are subjected to electrolysis in a tank containing suitable electrodes, peroxide of lead being produced at the anode, whilst a metal is reduced at the cathode; or oxide of lead may be first treated with a solution of an alkaline chloride, and a current of chlorine gas passed through the magma formed by adding caustic lime to the alkaline mixture.

2,769. — J. Imray, London. (*P. Jablochkoff, Paris.*) 6d. (4 figs.) June 13.

DYNAMO-ELECTRIC GENERATOR. — A magnetic coiled bobbin, fixed obliquely on an axis, revolves between or within polar fields, so as to present its opposite edges to opposite fields. Fig. 1 shows a

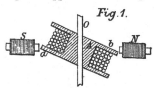

Fig. 1.

simple form of generator: the core or bobbin A, having cheeks *a b* of soft iron, and wound with a coil of insulated wire, is fixed obliquely on the axis O, and revolves between the poles of the electro-

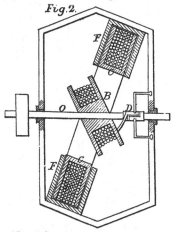

Fig. 2.

magnets N and S. The obliquity of the coil is such that, in each revolution, it presents the edges of *a* and *b* alternately to the poles of N and S, and alternating electric currents are set up in the coil

of A. Fig. 2 shows a generator in which the coil B, fixed obliquely on the axis O, revolves within an oblique core or bobbin C, which is wound with an insulated coil, and has an iron sheath F presenting interior polar edges towards the edges of B. The electric currents set up in the coil of B are collected and converted into currents of uniform direction,

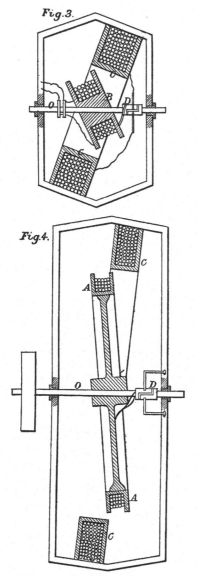

Fig. 3.

Fig. 4.

by means of a commutator D of ordinary construction. In Fig. 3, the bobbin C is of soft iron. The commutator D may be applied, as shown, to alternate the currents in the coil of C, those in the coil of B being constant in direction, and collected in the usual way by rubbers bearing on the rings E. In this case the internal core or bobbin B need not be

of soft iron. When the machine is of large diameter, the interior core or bobbin A may be, as shown in Fig. 4, merely a ring of iron fixed on a wheel of non-magnetic material.

MOTOR.—This invention is also applicable to motors.

2,771.—J. Farquharson, London. Dynamo-Electric Machines. 4d. (6 figs.) June 13.

DYNAMO-ELECTRIC GENERATORS. — The central soft iron core of each field magnet is wound with two or more layers of the insulated conductor; an iron cylinder slit lengthwise is then slipped over the coils, or the coils are covered with iron bars which are again covered with two or more layers of the insulated conductor, the process being continued so that the electro-magnet is built up of the desired number of alternating layers, with iron layers outermost and innermost. The central core is shown as a cylinder closed at one end, the other end forming with the ends of the iron layers an annular pole, slit from the centre to the circumference. The armature is of ring form and rotates between two annular series of these magnets. The iron ring forming the core of the armature is U-shaped in cross-section. The conductor is in the form of a narrow ribbon, insulated on one side, and then folded backwards and forwards so as to form a bar, equal in thickness to the width of the ribbon, and in length to the circumference of the ring. This bar is placed within the armature ring and secured on its inner edge by a removable ring. Wedge-shaped pieces of wood, whose outer ends project beyond the conductor, take the inward pressure of a band for keeping the ribbon firmly in its seating. When the generator is self-exciting, a section of the ribbon is separated, the ends being led to a suitable commutator.

2,776.—F. H. Varley, London. Manufacturing Carbons Applicable for Electric Candles, &c. 6d. (19 figs.) June 13.

CARBONS.—Vegetable or other fibres are twisted into string, or are plaited, knitted, or felted together. This string is stretched on a frame, treated with a caustic alkali, preferably in a solution of high temperature, rinsed, dried, treated with sulphuric acid, again rinsed and dried, immersed in a hot solution of petroleum and tar or other suitable hydro-carbon, and carbonised in an iron or porcelain chamber provided with a cover, hinged or otherwise fitted so as to be air-tight, and with inlet and outlet channels for the circulation of coal or hydro-carbon gas whilst the chamber is heated. The carbons are submitted to a succession of immersions and carbonisations until they become solid. The chamber, if of wrought iron, is

coated inside and outside with fireclay, or a mixture of clay and plumbago.

SECONDARY BATTERIES.—The plates are prepared from solid or thick woven fabrics as described above. Preferably four bands are rolled into a coil, or folded backward and forwards and subjected to the above described treatment. By taking the first and third coil, or the second and fourth, and placing one within the other, a space will be left for the polarising liquid, which may be (1) sulphate of mercury and sulphate of zinc, dissolved in dilute sulphuric acid; (2) sulphate of manganese, sulphate of zinc and sulphate of mercury, dissolved in dilute sulphuric acid; (3) sulphate of manganese dissolved in dilute sulphuric acid; (4) sulphate of zinc and sulphate of manganese, dissolved in dilute sulphuric acid; (5) sulphate of manganese dissolved in water; (6) sulphate of zinc dissolved in water. If the carbon coils require to be elastic, they are immersed in the paraffines or turpenes, but if rigid, in tar, resin, or dissolved treacle or sugar, which may be mixed with fine carbon or farinaceous substances into a thick viscous liquid. Porous cells and carbons for employment as resistances may be made in a similar manner.

SEMI - INCANDESCENCE LAMP. — Two coils or volutes of carbon, prepared as above, are mounted on rotary axes, and the free ends of the coils are forced against each other, the margins of the outer layers bearing on stops so as to guide the ends into contact. The axes are connected to the conductors from a generator.

ARC LAMP.—"The ends of the coils may be made to open by means of regulating mechanism controlled by the current."

2,781.—W. R. Lake, London. (*C. F. de la Roche, Paris.*) Electric Lighting Apparatus. 6d. (5 figs.) June 13.

ARC LAMPS. — The carbons are each forced through a tube or envelope of metal, carrying at its extremity a ring or washer, which serves as an abutment, and in such manner as to leave a free space between the ring or washer and the end of

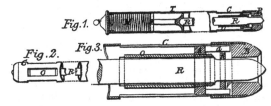

the tube, to permit the carbon to be segregated by the action of the air and heat. Referring to the illustrations, a tube O, of suitable metal, has upon its extremity a screwed sleeve A, provided with

projecting claws *d*, which support a ring K, the internal diameter of which is slightly less than the diameter of the carbon R. A spiral spring M placed in the tube T moves forward the carbon. Between the extremity of the tube O and the washer K (both preferably made of platinum) is a free space, the size of which is regulated by screwing the sleeve A to a greater or less extent upon the tube O. The external tube T is provided with apertures C, the sleeve A being also cut away, to allow of a free circulation of air. The ring B is formed of magnesia, with a central opening, which is made so much larger than the diameter of the electrode as to allow free access of the air. Both carbons may be forced forward by spiral springs, or one carbon may be fed by levers, racks, or pinions, whose movements are controlled by a single spring.

2,803. — F. L. Willard, London. Dynamo-Electric Machines. 6d. (9 figs.) June 14.

DYNAMO-ELECTRIC GENERATOR.—The armature is composed of a series of rings, supported by cheek plates and bolts, and revolves between the poles of an internal and external field magnet. As will be seen from the illustrations, the armature rings C C, carried by the arms D D, revolve in channels formed partly in the horizontal pole-pieces of the external field magnets A A, and partly in the corresponding and opposing pole-pieces of the internal

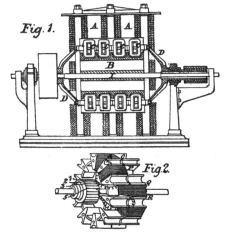

magnet B, this latter being wound in the form of a Siemens armature. Sometimes, instead of being supported by the fixed spindle I, the internal magnet is attached to a disc fixed to the external pole-pieces, and having at its centre one of the bearings of the armature spindle, the rings being bolted to spokes on a boss carried by the armature spindle. The rings C are formed with, or have shrunk on them, collars with their sides parallel to the radial line passing through the centre of the

adjacent coil, *i.e.*, they are V-shaped, as shown in Fig. 2. The rings are united by bolts passing through lugs on the rings, or through the V-shaped collars, and are then bolted to a non-magnetic cheek-plate, or to the arms D. The driving pulley is secured to one cheek-plate, the other being lengthened out into a sleeve carrying the ebonite cylinder on which the commutator segments are mounted. The conductors connecting the armature coils to the commutator pass within the sleeve. The commutator may be fixed inside the bearings. The method adopted in the connections is to join the inner wire of section O to the outer wire of section P, the outer wire of O being joined to plate 1 and the inner end of P to plate 3 of the commutator. Sections Q and R are connected together and to plates 2 and 5, in a similar manner, and so on for the remainder of the coils. The commutator is preferably built up of V-shaped plates, arranged with their narrow and wide ends alternately round the periphery of a cylinder, the brushes being placed at an angle with respect to the direction of rotation, so as to span the gap between two commutator plates. Each ring may be coupled up to a separate commutator, and one or more may be used to excite the field magnets, or the field magnets may be excited by the whole current, or by a shunt.

2,804.—W. R. Lake, London. (*F. Van Rysselberghe, Schaerbeck, Belgium.*) **Employing Electricity for Telegraphic and Telephonic Purposes.** 6d. (2 figs.) June 14.

CONDUCTORS. — This consists in employing for each communication, either two parallel conductors, or two conductors one of which surrounds or envelopes the other, these conductors being insulated from one another and from surrounding bodies. One end of each conductor is insulated, and the other ends are in communication with the receiver and the transmitter respectively.

2,807. — L. Epstein, London. **Secondary Batteries.** 4d. June 14.

SECONDARY BATTERIES. — The electrodes are formed from metallic lead, by the aid of an electric current, and of a permanganate or analogous oxidising agent present in the cell; methylated or other like spirit may also be employed in the cell. To form a cylindrical battery, two sheets of lead are placed one over the other with a water cushion between, and are then rolled into a scroll, the water let out of the cushion, and the cushion withdrawn. To form a rectangular battery, a number of flat plates are placed parallel to one another and packed with charcoal powder between them, the alternate plates being connected together; or a wooden vessel is lined with lead, and has an inter-

mediate or partition plate of lead, and the other electrode is doubled upon itself and placed within the vessel. The current during the process of forming is reversed from time to time.

2,818. — J. S. Sellon, London. **Secondary Batteries.** 4d. (23 figs.) June 15.

SECONDARY BATTERIES.—This consists in making perforations, or interstices, in the plates of secondary batteries of such a form that the material with which the perforations, or interstices, are

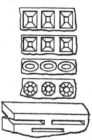

packed will be firmly retained in position by the shape of the perforations or interstices. The figures represent two forms of perforations. Other forms consist of single or double cones, truncated or not.

2,823.—C. Westphal, Berlin. **Generating and Storing Electric Energy.** 8d. (21 figs.) June 15.

SECONDARY BATTERIES.—This specification first describes a method of producing water gas by chemical means, which is afterwards used in conjunction with atmospheric air, or oxygen, in gas batteries to generate electric currents, the batteries being applicable also as secondary batteries. The gases from a reservoir are caught by a series of metallic buckets, attached to the outer side of an endless band passing over pulleys at the top and bottom of a wooden box. The buckets are connected by leaden rivets to an endless strip of lead, situated on the inside of the endless band, and making contact with pulleys which are made of a conducting material not attacked by the liquid in the box. One series of buckets carries the gas, which is conveyed by tubing to a hood surrounding the upper pulley, and the other series carries atmospheric air, or oxygen, down into the liquid in the box, the buckets being kept at a constant distance apart by means of lugs guiding the endless bands; these two gases combine and produce a current which is taken off from the pulleys. If the combined gases form a gaseous product, the discharge of the same is effected by leading the bands upwards in an inclined plane over guide pulleys, so that the gases do not pass into the hood. The buckets may be honey-combed,

or be formed of combs, brushes, or sieves, covered with spongy platinum, or lead, to increase their active surface. The gases may be led under pressure beneath the bottom end of each of a series of superposed metallic gutters, slightly inclined, in a fluid, and with their openings placed downwards, the number of openings being proportional to the height of the water above them. An excess of gas led into a gutter escapes over one edge (which is made shorter than the other) and is caught by the superposed gutter. The slowly rising gutter may be formed by a wire wound helically round a metal cylinder, the gas being forced through minute openings in the cylinder under the wire. In another method, a row of concentrically arranged porous cells is sealed at both ends, so as to form a series of annular spaces. The alternate annular spaces are filled with dilute sulphuric acid, and the remaining spaces are alternately connected together by pipes, and oxygen and hydrogen, or other suitable gases, respectively passed through the two series ; the walls of the cells in contact with the gases are provided with metallic weft, sieves, or brushes. The liquid passing through the walls is conveyed away by pipes, the ends of which dip under the water. The gases may be forced through the porous cells, and in contact with metal sieves, or brushes, secured to the surfaces of the cells turned towards the acid ; or the gases under water may be made to stream out in the form of small bubbles and rise along the surface of a metal plate slightly inclined to the vertical and provided with inequalities ; or the gases may be conducted through porous pipes, or chambers, filled with granular conducting material ; or the gases may be absorbed by suitable liquids which are conducted along the electrodes. When used as secondary batteries, the surfaces on which oxygen is generated by the passage of the current are of a non-oxidisable conductor, such as coal, platinum, gold, or lead. The gases given off are collected in reservoirs and utilised as described above. The bands, in the first described apparatus, are driven in opposite directions respectively when charging and discharging.

2,826.—F. J. Cheesbrough, Liverpool. (*A. K. Eaton, Brooklyn.*) Manufacture of Lead in the Form of Threadlike Fibre. 6d. (2 figs.) June 15.

SECONDARY BATTERIES.—The apparatus consists of a hydraulic cylinder A, a ram B, suitably packed as at C and D, and a lead-retaining cylinder F, fitted with a finely grooved plate G, shown in plan at Fig. 2. The cylinder being charged with hot lead, upon the application of pressure through the ram B, the lead will spin out through the grooves X like a shower bath, as shown by H,

in the form of fine hair or thread-like fibre. The employment of these fibres in secondary batteries

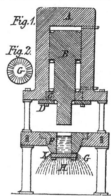

is mentioned in a subsequent patent of this inventor's.

2,830.—W. E. Ayrton and J. Perry, London. Construction and Government of Electro-Motors. 8d. (21 figs.) June 15.

MOTORS.—The magnetic moments of the movable and fixed parts are arranged to be equal, or nearly equal to each another. Referring to the illustrations, the bobbin A, of Siemens type, rotates inside a Pacinotti ring. In the present case, one end of the wire on the bobbin is fastened to the iron, and the other end passes through the spindle to one of the brushes T. The Pacinotti ring is preferably built of two end parts of soft iron E E and a considerable number of plates of iron formed as shown in Fig. 2, bolted together with separating washers Parts of the ring project inwards, so as nearly to touch the surface of the rotating part ; or the ring may be solid, of cast malleable, or wrought iron, with deep circular grooves in the form of narrow screw threads on the inside of the ring, the pitch of the screw being small. The dotted lines show how the wire coils are wound in the groove of the ring, so that they shall be clear of the rotating parts ; their ends are joined up as in the 'Gramme' ring. In Fig. 2 the commutator is shown so that the rubbing surface is plain, the twelve pieces C being fastened to the disc of wood Z. This disc has also on its surface a continuous ring of metal D, to which is connected one terminal B of the machine. The current entering at B passes to the rings D, across the broad brush to one of the commutator pieces C, and to the second narrower brush T, thence through the hollow spindle to the coils of the rotating bobbin, and then to the metal of the bobbin and frame of the machine. A relative motion may be given to the commutator and the Pacinotti ring by means of a handle, a constant electromotive force, or current, being obtained between its termi-

nals as described in Specification No. 55 of 1882. The disc Z may be adjustable to some extent for the purpose of varying the lead. The commutator may be cylindrical instead of disc-shaped. In Fig. 2 the brush-holder is a disc of wood, or other insulating material, fastened to the spindle by means of a set screw passing out at the circumference of the disc. In order to adjust the position of the brushes when the machine is in motion, a clutch is constrained, by spiral feathers on the spindle, to have a relative motion round it, which is communicated to the brushes. A total relative motion of 180 deg. may be given so that the motor can be reversed. The bobbin A may be formed of two semi-cylindrical magnets, joined together to form a cylinder, and one central magnet, passing diametrically through the cylinder; or two or more central magnets may be employed in addition to the two side magnets, the like poles all coming together. In this case, the Pacinotti ring is placed with its plane at right angles to the axis of rotation, and there may be one at each end of the cylinder; or the cylinder may be cut in two halves and be stationary, the Pacinotti

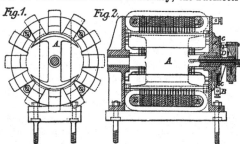

Fig. 1.   Fig. 2.

ring being rotated between the two halves. Two Pacinotti rings may be employed, one revolving within the other, or two "Burgin rotating parts," one being fixed and the other rotating within it. Several arrangements are described and illustrated for regulating the speed of the motor by a governor. When there is a constant electromotive force, a resistance may be introduced and taken out periodically. In one arrangement, a cam, or eccentric, periodically moves a contact block, brush, or roller over studs, so that contact is made during a portion of the periodic time of revolution of the cam; a lateral shifting of the contact block, or of the cam, effected by the governors, places it on studs which are longer or shorter, in the direction of motion of the block, than when the block was in its previous position. Or a contact piece slides between two blocks carried respectively by the motor shaft and the framework, and is guided radially and driven round by the block on the shaft; as the speed increases, the piece is forced outwards by the centrifugal force, against the resistance of a spring, and makes contact between the revolving and the fixed disc, which has

part of its surface of insulating material so situated that the duration of the contact diminishes as the speed increases; the two discs are connected through a resistance. Or the current may be passed to the motor by means of a brush bearing on a commutator, having two plates connected by a resistance; the brush is moved along the cylinder of the governor and makes contact during a greater or lesser part of each revolution with one or the other plate. Or the disc commutator (Fig. 2) may be formed of two series of plates, connected through a resistance, and the broad brush formed of two parts, one bearing on the continuous ring, and the other bearing, more or less, on the two series of plates, and moved outwards against the resistance of a spring by centrifugal force. If the commutator is cylindrical, a separate governor is employed to move the brush. Or the governor may be arranged to alter the lead of the brushes, to alter the positions of the field magnets, or of the pole-pieces. Or several circuits may be employed in the field magnets, more or less of which are cut out by the governors. In other methods, a current passing round both the armatures and a permanent field magnet may diminish the magnetism of the latter; or a magneto-motor, gearing into a very much smaller dynamo-motor, both in the same circuit, or a motor having its field magnet in a shunt circuit, producing a permanent magnetic effect, and also wound with a coil in series with the armature, and weakening the magnetism produced by the shunt circuit, may be employed. When the motors are arranged in a series circuit, the above described arrangements are, when necessary, suitably modified. In one arrangement, the speed is regulated by causing a governor to shunt a considerable part of the current past the motor as the speed increases. Or the permanent magnetism of the field magnets may be opposed by a shunt coil, or the field magnets wound with two coils, one in series with the armature circuit, and the other a shunt to the armature only; or a shunt motor may be combined with a magneto-motor, or with a motor in a shunt circuit. A dynamo motor, having its field magnets excited by the whole constant current, may be employed in lieu of the magneto-motor. The methods of governing by means of differential winding is explained by the aid of algebraic formulæ.

*2,845.—A. Pfannkuche, London. Incandescent Lamps. 2d. June 16.

INCANDESCENCE LAMPS.—The carbon filament is made hollow, and the ends of the conducting wires are inserted to a considerable distance; or if the filament is made of strands, the strands are twisted around the conductor. The filament is preferably made of a hollow plaited or woven fabric, of jute, hemp, cotton, or other suitable fibrous material,

the ends of the wire being inserted before further treatment. The wire ends may be pointed and rough, and cement may be used. A number of such filaments are wound round blocks and covered with graphite, and, after carbonising, left to cool very slowly, the platinum, iron, steel, copper, or other metallic conductors being thereby thoroughly annealed. The conductors may be of platinum only at the point passing through the glass.

2,871.—J. E. H. Gordon, London. Dynamo-Electric Machines. 6d. (10 figs.) June 17.

DYNAMO-ELECTRIC GENERATORS.—This relates to the type of generators described in Specifications Nos. 78 and 5,536 of 1881. The revolving wheel is made of two discs A, of wrought iron, mounted on the axis, and kept apart from one another, at the axis, by a cast-iron distance piece, and at the rim, by a wrought-iron or steel ring E, which carries the electro-magnets. On each side of the discs, and keyed to the shaft, is a cast-iron boss F, with a conical end, to which is secured the apex of a wrought-iron or a steel cone B, the base of which is attached to one of the discs at a small distance from the circumference, the outer extremity being thus left flat for the magnets G to be fixed to it. The cones, discs, ring, distance piece (which is of larger diameter than the bosses, so that the heads of the rivets will not interfere with the bosses), and bosses are all firmly rivetted together and strengthened by

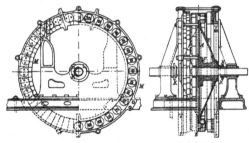

angle pieces. A large gap is made in the sole-plate, through which a portion of the wheel dips into a pit below the generator. The end thrust of the shaft is taken by two loose iron collars K, placed on the shaft, and pressed gently against the inside ends of the journals by means of set screws projecting from the ends of the cast-iron bosses, and these collars carry the contact rings, preferably of phosphor-bronze, and separated from the collars by split insulating rings for conveying the exciting currents to the magnets. Each magnet consists of a core passing through a hole in the flat portion of the revolving discs, which core projects equally on either side. Two brass bobbins are slipped on to the core, one on either side, and radial pole-pieces, rather shorter and of less width than the fixed coils, are fixed to the extremities of the core, and hold the bobbins

in place. The fixed coils are carried by rings M, of cast iron, bolted to the inside of the gap in the sole-plate, and by four struts from the top of low standards on the sole-plate, and by tie-rods parallel to the axis. Each of the fixed rings is made in three portions, bolted together, one being much smaller than either of the other two, and placed preferably at the top, so as to allow of access to the wheel. The cores of the fixed coils are formed of a boiler plate bent upon itself, so that the angle forms the thin end of a wedge, a T-shaped piece with a wedge-shaped head and a screwed movable or fixed stem, which is inserted into a hole in the fixed ring and secured by nuts, being inserted into one end of the folded plate and welded to it. A wooden block slipped on to the core, and secured by pins, forms the back flange. A German silver flange is rivetted to a shoulder cut on the end of the core near the revolving wheel, and may project beyond the coils and be secured to the wood by screws and distance-pieces after the coils are wound. If slits are made in the flange, a cut is made passing through into the opening of the core. Projecting lugs on the flange pass into the opening in the core and prevent the slit in the flange from opening. The field magnets are excited by a separate generator, or by a current from one or more of the coils, the coils being so connected that the opposite pole-pieces, and also the adjacent pole-pieces, are of opposite polarity, there being twice the number of fixed coils on each ring as there are field magnets G. The following appears only in the provisional specification: The commutator consists of two metallic rings mounted on an insulating cylinder, and having metallic strips projecting, parallel with the axis, from their inner sides. The strips of the two rings are together equal in number to the number of the magnets G, and those projecting from one ring are opposite those from the other ring, but do not meet them. A metallic ring is mounted on the centre of the insulating cylinder, and has strips on each side, lying in the spaces between the strips of the two rings. Two brushes bear upon the strips on one side of the central ring, and two upon the strips on the other side. One brush of each pair is connected to one end of each of the alternate coils, and the other to the other ends of the coils. The remaining coils are connected in a similar manner to the other pair of brushes, one pair of brushes being such a distance in advance of the other pair as corresponds to the advance of one set of coils on the other set. The direct currents are taken off by brushes bearing on the end rings, the two sets of coils being thus connected in series; or an extra brush may bear on the central ring, so that either half of the generator may be used independently. The brushes consist of a solid brass plate pressed down by a spiral spring, and having on the under side a number

of strips of thin sheet copper. The frame carrying the four brushes is slightly adjustable. Part of the coils may be connected to a commutator, and the other part to a collector, or to a separate commutator. In another commutator, a number of strips, corresponding to the number of coils, are mounted on an insulating cylinder, on each end of which is a metallic ring, the alternate strips being connected to one ring, and the rest to the other ring. In this case, the four brushes are arranged as above, and bear on the strips, and two collecting brushes bear on the end rings, the series being thus connected in quantity. The commutator is mounted on a thin shaft coupled to the main shaft, and, in order to prevent sparking, it may run in a trough filled with a non-conducting liquid, or have a blast of cold air blown on it at the points of contact of the brushes. The generator may be employed during the day time with its coils arranged in series, and the current be commutated to charge secondary batteries, and during the night with the coils arranged for quantity and to produce alternate currents for lamps.

2,875.—R. J. Gulcher, London. **Gas Batteries and Apparatus for Producing Hydrogen and Oxygen by Electricity.** 6d. (3 figs.) June 17.

SECONDARY BATTERIES.—Three separate vessels, or three compartments f^1, f^2, f^3 in one vessel, communicate with each other by the holes e^3, and are arranged so that the capacity of f^2 is double that of f^1, and that of f^3 is equal to the joint capacity of f^1 and f^2. The compartment f^3 is lined

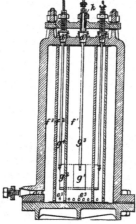

with india-rubber, and the cylinders $f^1 f^2$ are shown as kept in place by the compression of the lining. The platinum, or platinised carbon, or other suitable electrodes $g^1 g^2$ are suspended in the compartments $f^1 f^2$ by means of rods $g^3 g^4$, the rod g^3 being insulated by the hook k from the body of the vessel.

The compartments having been filled with slightly acidulated water, and the three stoppers screwed down, the contents of the exterior chamber are allowed to flow out through the orifice in the plug a^6, which is then screwed down. The rod g^3 is connected to the positive pole, and one of the rods g^4 to the negative pole of an electric generator, and the water in the two compartments $f^1 f^2$ is decomposed, until a sufficient quantity of gas has been generated to force the level of the liquid to the lower edge of the electrodes. During the process the liquid is forced into the chamber f^3, compressing the air in the upper part, and thus maintaining the generated gases under pressure.

2,885.—J. A. Berly, London. (*F. V. Maquaire, Paris.*) **Electric Machines.** 6d. (12 figs.) June 19.

DYNAMO-ELECTRIC GENERATOR. — Fig. 1 illustrates the general idea upon which the invention is based. There are two rows of field magnets, so arranged that the adjacent and opposing poles are of opposite polarity, and between these poles there are two cores upon which the conductor is wound. These cores are built up of isolated pieces of soft iron, connected by non-magnetic material, and their polarities at a given moment are of opposite name. In other arrangements, the opposing poles are of like polarity; or the adjacent poles are of like polarity, in which latter case the poles of one row are either opposite or interspaced with those of the other row; or all the poles of the two rows may be of like polarity, the systems being duplicated. On passing the cores, with their coils, longitudinally between the rows, a current is generated. It is claimed by these arrangements that two sets of currents are obtained, viz., those due to cutting the lines of magnetic force shown in Fig. 1, and those caused by the reciprocal action of the currents in the two conductors. The specification appears to be a literal translation from the French, and is not very clear. According to the illustrations, the armature rotates between two rings of poles arranged as described above. Fig. 2 is a part elevation of the armature. This consists of two similar parts placed side by side. Each part comprises a core divided into sections, as shown in Fig. 2, magnetically insulated from each other. The length of each section in the direction of motion is as short as possible, being, preferably, equal to the space between two adjacent poles, the object being to prevent the tendency to the formation of "consequent points" in the core. Spokes, fixed at their inner extremities into a copper or brass ring, have their other extremities forked to support two other brass rings secured to the frame of the generator. The iron segments of

the cores are each secured at their extremities between the inner and one of the outer brass rings, by means of pieces of sheet iron, against which they terminate as shown in Fig. 2. The segments are joined together by non-magnetic materials. The coils are wound as shown in Fig. 2, the number of layers being the same for the smaller radius as for the larger, and the thickness of the conductor being increased as its breadth diminishes. The exterior sides of each coil facing the opposing poles are parallel, the corresponding interior sides being tapered; the sides of the core being parallel, a free space is left between the coils and the core allowing free circulation of the air. The movable field magnets are wound "so as to regulate at will the quantity of air set in motion . . . . . and in

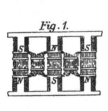

Fig. 1.

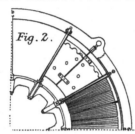

Fig. 2.

such a manner that their profiles are figures of revolution." There are as many commutator-bars as poles in one ring, the sections of the armature being coupled up so as to form one or more circuits, each having a separate commutator. The coils of the field magnets may be arranged in a corresponding number of separate circuits, each receiving current from one section of the armature, or one or more of the armature circuits may be employed for exciting the field magnets; or the armature of one system may be employed to excite the field magnets of another system, whose armature is fixed on the same shaft as the armature of the first

Fig. 3.

system, the object being to obtain a maximum of excitation of the field magnets when each segment of the armature is in the best position for induction. The field magnets may be in a shunt circuit. The coils of the field magnets are of elliptical section, presenting enlarged poles "terminating laterally," so as to more definitely influence the armature coils when approaching or receding from them. In order to regulate the current generated, the armature coils are arranged to be adjustable radially, the adjustment being effected by hand, or automatically, by an electro-motor under the action of a regulator of current. This regulation is effected during the working of the generator. Fig. 3 is a section through half of a generator,

the parts of which are duplicated. Each armature rotates between two rings of poles, the poles being all of like polarity and of opposite polarity to the poles between which the other armature rotates. The inner poles of the two systems are magnetically connected as shown, the outer poles being connected through the iron axle.

2,896.—C. T. Howard, Providence, R.I., U.S.A. Shunts or Switches for Protecting Electrical Instruments from the Effects of Excessively Powerful Currents. 6d. (3 figs.) June 19.

Switch. — The illustrations show a shunt or switch, which is arranged to cut off all electric currents from the interior of buildings, whenever it is desired on account of thunder-storms, &c. Fig. 1 is an elevation showing the current shunted from the telephone, and Fig. 2 a section of the same. The arrows on Fig. 1 show the path open to the current. If the insulating arm, carrying at its ends the

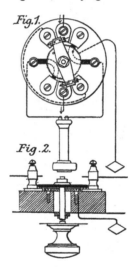

Fig. 1.

Fig. 2.

contacts, be turned slightly in the direction of the hands of a watch, the current would pass through the telephone. The central plate of each of the series of three plates is provided with a serrated edge, as also may be the other plates, the object being to carry off the current from the lines through the serrated central plate, which is connected to earth as shown. The connections between the contact plates and the terminals of the instrument (shown in thick dark lines) is preferably made with a leaden or other safety fuze.

2,898.—A. Swan, Gateshead, Durham. Incandescent Electric Lamps. 6d. (23 figs.) June 19.

Incandescence Lamps.—This relates principally

to methods of preparing the conductors passing through the glass of the bulb. Fig. 1 is a table upon which a conductor is cut into the desired length, and bent into a U form by means of the cutting lever c and the bending lever f, both levers being fixed to the spindle d and operated by the same movement. The bending lever has three graduations, as shown, to accommodate three different thicknesses of conductor. The conductor is laid in a rest g, with one end against an abutment ledge b, and when the cutter c is depressed to cut it off to the desired length, the bending lever f descends and passes into a groove in the bed-plate, bending the conductor, at exactly

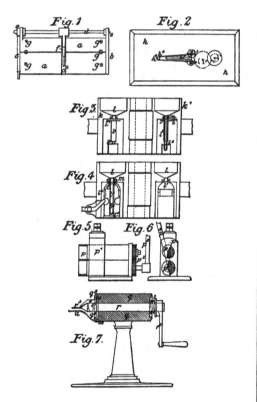

its middle, into a U shape. Fig. 2 represents a contrivance for producing the terminal loops and tags. The curved portion of the conductor is laid against the stop h^3, under the catch h^4, and its ends are received in slots in the ends of spindles projecting above the bed-plate h, and rotated in opposite directions by the gearing shown in interrupted lines, arranged beneath the bed-plate, until further rotation is prevented by the stop piece h^2; the loops are thus formed and the extreme ends or tags of the conductor doubled back against the legs. The distance between the axes of the spindles is equal to that required between the centres of the loops.

Fig. 3 is a mould, opened out, for surrounding the conductor, thus far prepared, in a glass envelope. The conductor j is supported upon pegs m m on one half of the mould, and fitting into recesses in the other half. The mould has an opening l leading into the portion l^1, where the flange of the stem is formed. The portion l' opens into the portion l^2, in which the stem proper is formed. The glass is introduced into the part l, and is pressed by a lever and plug, its downward flow being limited by the plug l^3. Fig. 4 illustrates a mould in which the flat circular recess l^1 is replaced by a dome-shaped recess l^1. By means of a lever the core-piece l^4 is raised from the mould immediately after the casting, to obviate risk of fracture of the stem. This core-piece may be made in halves, the terminal wires passing through a slit l^5. To facilitate adhesion between the conductor and the glass, portions of the metal forming the mould are insulated, and the conductor is raised, by means of an electric current, to a white heat at the moment the molten glass is being pressed around it. In Figs. 5 and 6 is shown a pair of rollers for flattening out the ends of the conductor after the curved portion has been cut away. The rollers p p, which are geared together and move simultaneously by the operation of the handle p^3, have portions of their surfaces removed, as shown, to allow the insertion of the conductors to the full extent left bare by the stem of glass. The rollers are slightly eccentric and flatten the conductor gradually. The flattened ends are then cut to a tapering figure. Fig. 7 is a machine for twisting the flattened ends of the conductor into helical coils for the reception of the lamp filaments. The spindle r is provided at its pointed end with a needle r^3, and with a catch s, which is pivotted in a slot in the spindle at s^2, and the tail s^3 of which lies behind the flange q^3. By pressing the spindle forward, the catch is canted, and allows the flattened end of the conductor to be passed between it and the needle, and when the spindle is released and forced back by the spring t, the tip of the end is held upon the needle. Upon rotating the spindle the coil is made. u is a rest. In a modification, the spring t is dispensed with, the pressure of the catch upon the needle being produced by a spring acting directly upon itself. The end of the spindle is screw-threaded, and works in a screw in the tubular piece q, the number of turns of the screw corresponding with the number of coils to be given to the ends of the wires. The needle is mounted in a stock-piece screwed into the fore end of the spindle. The sealing of the carbon filament into, and over the spiral socket, is completed by the deposition of carbon. In the provisional specification, it is stated that the glass bulb is formed by blowing and rotating the molten metal in a mould, composed of finely pul-

verised coke and plumbago, mixed to a paste with tar or cement, pressed to shape and baked, or it may be formed of steatite. The mould is of such form as to shape the neck, or conceal the line of junction between the bulb and stem.

2,902.—J. T. Sprague, Birmingham. Electric Meters. 4d. June 19.

ELECTRIC METERS.—The gas arising from the decomposition in a galvanic cell, or ordinary voltameter, is caused to re-combine after being measured in a gas battery, and to regenerate the electromotive force expended in the decomposition; or the gas may be delivered as a fine jet into a receiver, and ignited by a heated wire, or by an electric spark, with admission of air if necessary, the heat given off being directed against the face of a thermopile. The measurement of the gases is effected by causing them to rise into balanced receivers, one receiver being preferably balanced by the other. The receivers are fitted with tubes through which the gases escape alternately, at intervals, controlled by the apparatus itself, the reversing commutator and recorder described in Specifications No. 4,762 of 1878, and 4,454 of 1881, being employed: or two, or preferably, three alternately acting receivers are connected to a cranked axle, which is rotated and actuates the commutator and valves. When the gas from one electrode only is measured, the other electrode may be common to all the voltameters, or each may have a separate electrode, and they may be of lead, which will absorb oxygen, or of palladium, which will absorb hydrogen. Two receivers may act alternately, one for generating the gas, which is passed to an electrode in the other. Hydrogen may be generated at one electrode and passed through a gas meter to the electrode of a gas battery, both electrodes being arranged in the same reservoir of liquid, with a common electrode between them composed of spongy lead, or of plates of lead placed parallel to the current. The variation of temperature in measuring the gases is compensated for by altering the capacity of the receiver, or varying the action of the recording train. This may be effected by a spring made of two differently expansible metals, or by a glass bulb floating in liquid, which enters more or less into the stem, or by a bulb which rises or sinks in the liquid as the gas expands or contracts, or the stem may be formed into a U tube containing a liquid which alters its level in the outer arm. This motion may alter a compensation rod, thereby controlling the position of a conical wheel, by which the recording wheelwork is driven, or it may alter the point (as in the case of metallic deposit meters), at which the reversal or contact takes place. The outer arm of the U stem may be graduated and the "reading" utilised for calculating the compensation. The weighing apparatus in metallic deposit meters weighs the metal deposited, or the loss of metal dissolved. The cathode is placed horizontally at the lower part of the cell, and above this is the anode, consisting of an open-ended cylinder arranged vertically, or of a horizontal plate kept at a uniform distance from the cathode by stops. These electrodes may be rendered buoyant so as to be pressed up against stops, or may be counterpoised. If the weight added to the cathode is to be registered, the cathode rests upon three insulating blocks, one of which is hollow and contains mercury, by which connection is made to the cathode when it rests upon the blocks, but is broken as soon as it is raised. The cathode is suspended by non-absorbent cords, between which the anode moves freely to the weighing apparatus, such as a spring balance raised by a lever or screw. The anode also rests on three insulating supports, and can be raised when weighing the cathode, the current being conveyed to it by a conductor bent over to the outside of the cell and dipping into a mercury cup. This wire may carry a bridge between the two cups, which will cut off the whole meter from the current during weighing. If the anode is to be weighed, the cathode is fixed, and the lifting arrangement is required only for the anode. When the anode is buoyant, its upward pressure will afford a means of measurement, or when it is a cylinder, arranged so that a small motion will not materially affect the resistance, it is made buoyant so as nearly to float, and may be attached to a weighing apparatus such as will largely multiply the motion and give constant indications. The two electrodes may be suspended as counterpoises over an axle carrying an indicator. A reverser, arranged either to reverse the action of the meter, or to send a large reverse current through it, is controlled from the outside, and accessible only to the proper authorities. The cathode may be constructed of a layer of mercury, into which the deposited metal is received, causing the mercury to overflow into a receiver, which is graduated or attached to a balance. In order to dissipate the heat in the electrolytic liquid caused by the passage of the current, and utilise it in circulating the liquid, the sides of the containing vessel are constructed with an inner side or plate (forming, or not, one electrode), not reaching to either the top or bottom of the liquid, thus enclosing a film of liquid, which being cooled by the external air will sink; or a tube formed into a helix may form an external connection between the upper and lower strata, and water may be supplied to cool the surface; or the top and bottom may be connected, a pump being arranged in connection and worked by the current, or mechanically.

*2,910.—C. E. Kelway, London. Apparatus for Generating and Utilising Electricity. 2d. June 20.

DYNAMO AND MAGNETO-ELECTRIC GENERATORS.— Two concentric shafts, mounted and arranged to revolve in separate bearings in a line with each other, or arranged so that one shaft revolves within the other, carry respectively the armature and field magnets, and are caused to rotate in opposite directions, at the same or at different speeds. The field magnets may be permanent or electro-magnets. Commutators collect or distribute in the proper direction, "not only the electricity passing through the insulated wire round the armature or armatures, but also that which passes through the insulated wire round the field magnets," when they are electro-magnets.

*2,911.—J. Kincaid, London. Carrying Electric Wires through Streets. 2d. June 20.

CONDUCTORS.— Hollow blocks of iron, stone, wood, asphalte, &c., serve the double purpose of carrying electric conductors, and of being used for the kerbs of footpaths, or for the paving of streets. These blocks may be constructed in various ways, with or without movable tops, having a roughened surface, or a lining of pitch, and a hole or holes for the branch lines. Those blocks at the base may be horizontal, or may slope downwards, so as to carry the conductors horizontally, or sloping downwards at the crossings of streets, or at a junction between the roadway and the kerb. A course of granite or other paving blocks is laid on each side of the blocks in crossing a roadway. The ends are formed so that they can be fastened together to form a continuous tube.

2,912.—S. H. Emmens, London. Apparatus for the Regulation of Electric Currents. 6d. (5 figs.) June 20.

SECONDARY BATTERIES.—The illustration represents a device for automatically charging one set of secondary batteries by a generator, whilst another is discharged by the service circuit, and *vice versâ*, alternately. In the illustrations, the discs F are caused to rotate in opposite directions by the worm E, driven by motor A. The discs are each divided into two insulated halves O O and R R, and have central conducting rings N N and P P, of which N N are always connected with O O, and with the leading and return wires L + and L— of the service circuit, whilst P P are always connected with R R, and the leading and return wires of the generator B. S S and T T are the stationary poles of the batteries C and D, with suitable contact pieces for maintaining electrical connection with the discs F. At every half-revolution of the discs F, the bat-

teries exchange circuits. The changing the coupling up of the batteries simultaneously with the exchange of circuits, by which they may be charged in parallel arc and discharged in series, is effected by the rising and falling of two movable frames carrying the contact pieces, and actuated by an electro-magnet or cams, operated by the discs F, or by the motor A, or by hand. The frames are attached to opposite ends of a lever, and one, the heavier, carries a series of contact pieces, connected together two and two, except the first and last, and the other a series of contact pieces, in which all the positive and also all the negative pieces are respectively connected together. The lighter frame carries a core of a solenoid, or the armature of a magnet in the generator circuit. These contact pieces, when the corresponding frame is depressed,

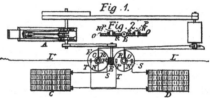

make contact with bars attached in rows to a fixed frame and carrying the electrodes, the batteries being thus connected in series, unless the electro-magnet is excited, when they will be connected parallel. In a modification, the electrodes are connected to jointed levers falling between pairs of insulated springs arranged on fixed bars placed in the service circuit, and making connection between the alternate elements of the battery. A movable bar beneath the levers is provided with uprights corresponding to the electrodes of the battery, and carrying pairs of insulated springs. These uprights are of two altitudes, arranged alternately, so that the levers, when the bar is raised, are lifted in two sets corresponding to the + and — electrodes of the battery; or the same object may be attained by a similar arrangement of the connections along the bar. Conductors make connection between these sets and the generator.

2,913.—S. H. Emmens, London. Secondary Batteries. 6d. (11 figs.) June 20.

SECONDARY BATTERIES.—Strips of sheet lead, or lead foil, are rolled up in the form of short cylinders, and are arranged one over the other, each pair being separated by an insulating diaphragm B, and the elements of each pair by a porous diaphragm. The elements are connected in series as shown. The strips of lead may be wound in cylindrical spirals (notched, if necessary, at the outer circumference to facilitate coiling), or flat rings may be superposed and form two concentric cylinders, the inner rings being either hollow or solid. The two

cylinders form the two poles of the battery. The batteries may be charged with metallic lead and

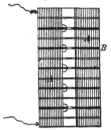

peroxide of lead, either mechanically, or by using a suitable salt of lead as an electrolyte and passing an electric current.

2,914. — S. H. Emmens, London. Electric Lamps. 6d. (4 figs.) June 20.

Arc Lamps —Fig. 1 represents one arrangement for causing the carbons to approach each other on the diminution or cessation of the current. A pair of levers A, jointed scissor-wise, have, at one pair of extremities, arms B to hold the carbons. The normal attraction of the electro-magnets C on the levers A, is counterbalanced by the weights D, or by springs, and on any cessation of current the two carbons are caused to approach each other. In a modification, the cross levers are replaced by bell-cranks, having their horizontal arms acted on by the magnets C, and the ends of their vertical arms connected by a spring which acts in lieu of

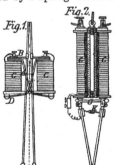

the weight. The arms B are attached at or near the ends of the vertical arms. The carbons rise by their own weight, and are held in position by combustible material, placed at intervals, and attached to a peg which is too large to pass through a hole in the bed-plate. The carbons are fed forward as the material is burnt. Fig. 2 shows another arrangement of automatic feed, the carbons being placed in holders pivotted so as to assume varying degrees of inclination as the candle is consumed. The holders are inclined by the tension of the spring K, and are caused to diverge by the armature O of the magnet C. The current, after passing the arc, may be caused to traverse an outer coil of wire on the magnet C, and oppose the action of the main coil. The carbons may be arranged in line with each other, and one, or both, be fed forward by weights or springs, and be retained in position by contact pieces (pivotted remotely from the carbons and actuated by an electro-magnet), which grip and sustain the carbons as long as the normal current is passing. Any number of carbons may be thus automatically fed forward by putting them in a continuous line.

*2,917.—T. Parker, Coalbrookdale, Salop, and P. B. Elwell, Wolverhampton. Dynamo-Electric Machines. 4d. June 20.

Dynamo-Electric Generator.—The armature is formed of square or oblong section, similar to the rim of a flywheel, exposing the two sides and face to the field magnets, the core of the rim being of a non-magnetic substance, covered with an iron casing. The inductor is a long copper bar (conveniently made in sections), forming a continuous spiral round the rim, strips of iron being placed between each spiral turn, and in contact with the iron casing. The armature is carried on a disc or cylinder of non-conducting material. Several armatures may be mounted on the same shaft. A copper ring or cylinder may replace the spiral bar when currents of low electromotive force are required. Two rollers, each formed of two discs pressed apart by a spring and running in a groove formed on the commutator cylinder, bear with their outer edges against the side of the groove and collect the current. Their opposite edges may also be guided in grooves in solid rollers. Or the current may be collected by two discs on opposite sides of the armature, which bear upon the copper bars when pressed sideways by an arm. One disc is mounted upon a ball-and-socket joint sliding on that part of the armature shaft which passes through the centre of the armature (the boss of the wheel carrying the armature being set sideways), and is connected to the other disc by distance pieces passing through spaces in the boss of wheel, but is insulated from it. The first disc has an arm attached to its centre, extending outwards horizontally in line with the armature shaft, so that when pressed out of line with the spindle, each of the two discs come in contact with the copper spiral upon exactly opposite sides of the spindle, the distance between their inside edges being slightly greater than the thickness of the armature; the discs are thereby rotated by roller friction.

Motors.—An electro-magnet is caused to rotate between the poles of a fixed coil having no iron core, and the current of the internal magnet is reversed at each point where its lines correspond to the lines of the outside coils. The revolving magnet forms a complete disc, so as to present a minimum of resistance to the atmosphere.

SECONDARY BATTERIES. — These are formed of strips of lead, coiled or joined in zig-zag form, with the interstices filled up with oxide or sulphate of lead. The end of each strip is carried to the top to form a pole, and the plates so formed are placed in a skeleton box, made of strips of dried wood soaked in paraffine and dovetailed together.

2,934.— A. W. Brewtnall, Warrington. Suspending or Mounting Electroliers, &c. 6d. (6 figs.) June 20.

CONDUCTORS.— The electroliers are suspended from, or mounted on, a ball and socket joint. The ball and socket are both constructed in segments of metal separated by segments of insulating materials, the metallic segments of the ball corresponding to, and being in contact with, a sufficient surface of those of the socket to permit of free motion without breaking connection between the segments. The ball is constructed of three metallic zones: the uppermost is attached to a metallic stem leading to the neck of the ball, and connected to a conductor going to the lamp: the intermediate zone is connected to a metallic tube leading through the neck of the ball, and connected to the return conductors: the third zone takes the wear, and rests in a gland on the socket, and serves for the attachment of the main stem. The insulating segments are also carried by the neck of the ball, and surround the stem and tube. A central stud on the socket, connected to a conductor from the generator, is pressed by a spring against the uppermost segment; an annular metallic lining exactly corresponds to the intermediate segment, and is connected to the return conductor from the generator. In a modification, the return conductor is the metal of the electrolier, the intermediate zone being dispensed with.

2,943.—H. Aron, Berlin. Primary and Secondary Galvanic Batteries and Cells, &c. 4d. June 21.

SECONDARY BATTERIES.—A material, suitable for use as battery plates, is formed of a mixture of collodion, or other solution of gun-cotton, with salts soluble in alcohol. Thicknesses of textile material are saturated with the mixture and placed in a solution of a metallic salt. The resultant of the action of the metal, and the salt held by the collodion, is here termed "metallodion." A substance called "metallodium" is produced by the reduction of "metallodion," the metallic oxide being converted into metal, and the nitrogen remaining in the derivative of the cellulose being replaced by hydrogen. The metallodium compound may also be produced by mixing metal in a fine state of subdivision with collodion or other solutions of cellu-

lose. A salt, or oxide of metallic silver, platinum, manganese, or preferably lead, may be employed. A plate of conducting material is coated with a fresh emulsion of metallodion, which may be mechanically mixed with carbon, and allowed to dry; or the textile material described above is wrapped round a plate of lead. The cellulose, uniformly deposited between the metallic particles, carries the liquid into the plates by capillary action

2,945.—C. Sorley, London. Plates for Secondary or Storage Batteries. 4d. (2 figs.) June 21.

SECONDARY BATTERIES.—The plates have, in or upon them, a number of grooves, so cut or formed that the ridges between the grooves terminate in thin edges, rough, toothed, or plain. The bottoms of the grooves on one side of the plate preferably alternate with those upon the other side, the grooves being horizontal.

*2,954.—C. A. Carus-Wilson, London. Apparatus for Measuring Electric Currents. 2d. June 21.

CURRENT METER.—Two compartments in a vessel are separated by a porous partition or membrane, and are joined by a syphon, and filled with a suitable liquid, in which the electrodes are plunged. The amount of liquid which, under the action of endosmose produced by the passage of an electric current, flows from one vessel to the other, is registered by a suitable apparatus, arranged to measure the current in the syphon tube, which is directly proportional to the intensity of the current. The electrodes may be constructed so as to absorb the gases given off, and may take the form of secondary battery plates. Suitable gear may be provided for interchanging the positive and negative electrodes at intervals, and to provide against the action of the meter being reversed. A resistance is arranged in the syphon to diminish the electric current passing through it. If the syphon is applied between the two electrodes of a secondary battery, the electrodes being separated by a porous partition, the meter shows the amount of current stored at any time.

*2,962.— M. Volk, Brighton. Incandescent Electric Light Lamps. 2d. June 22.

INCANDESCENCE LAMPS.—To one of the terminals of the ordinary holder a spring is attached, and through the body of the holder is passed a metal stem, which also passes through a clip attached to one of the lamp connections. The stem is provided with an insulated knob. The current passes through this mechanism to the lamp, but when the knob is placed in a certain position, the spring presses against it and the current does not pass to the

lamp; but if the knob be rotated, the stem is brought into metallic connection with the spring, and the current passes.

*2,974.—O. G. Pritchard, Penge, Surrey. **Producing Electric Light.** 2d. June 22.

ELECTRIC LIGHT.—One electrode forms also a gas-burner, the gas providing a path for the current, and the other a feed tube for finely divided carbon, supplied to the arc in a constant stream, to impart by its incandescence the desired intensity to the light. The electrodes consist of infusible and incombustible materials, such as platinum. The relative arrangement of the electrodes may be reversed, and the carbon be mixed with the gas, or pass down a tube surrounding the gas burner.

2,990.—J. H. Johnson, London. (*La Cie. Electrique, Paris.*) **Machinery for Generating, Controlling, and Utilizing Electric Currents.** 6d. (6 figs.) June 23.

DYNAMO-ELECTRIC GENERATORS.—Figs. 1 and 2 are respectively end elevation and horizontal section of the generator. The electro-magnets A are provided with symmetrically arranged multiple polar armatures, and completely surround the armature, which is a Gramme ring B, carried by two external crossbars C. The machines are supported so as to be capable of oscillating.

MOTOR.—Figs. 3 and 4 are respectively vertical and horizontal sections of an electric motor to be employed in combination with the above generator. The magnets A¹ are attached to the frame D, with the axes parallel to the main shaft G, and are provided with large polar armatures E. The shaft G is supported in the frame D, and also rests upon a brass bearing H connected to the polar armatures E. The Gramme ring B¹ is placed externally, so that its collectors or commutator brushes are readily accessible. The brush-holders consist of a flexible blade *a* (Fig. 1), carrying a socket *b* for the reception of the brushes *b*¹, and are fixed to an oscillating block *c*. The pressure exerted by the brushes is regulated by screws *e f*. The coils are preferably covered with hempen cord, to counteract the effects of centrifugal force. Two machines may be arranged end to end with a pulley between the two, and the same machine may be employed either as a generator or as a motor. The cores and pole-pieces of the electro-magnets of the generator are formed of soft iron or steel, or cobalt, &c., while those of the motor are formed of cast-iron or hard steel, so as to retain a considerable coercive force, and prevent the motor from "running away." Fig. 5 illustrates an apparatus for regulating by hand the speed of a motor. The current enters at *a*, passes to the central pin, along arm *m*, to one of the contact pieces *b, c, d,* &c., through the whole, or a

portion, or none, of the resistances to *x* and to the motor. The contact piece N, insulated from the arm *l*, connects the two plates *y* and *z*, and is used in case of a second motor being worked from the same generator. Fig. 6 illustrates an automatic arrangement. If the speed of the motor increases, the current decreases (on account of the difference in the coercive force of the cores of the generator and motor), and the springs R over-

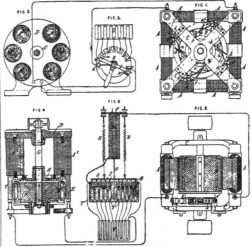

come the action of the solenoid M, and introduce a proportionate resistance, by means of a number of rods S of different lengths, suspended over and capable of dipping into the mercury troughs T, or by means of a commutator, in connection with a rheostat V. The rods may be of equal length, and arranged over troughs of mercury placed at different levels, or a plunger may raise the level of mercury in a vessel.

2,992.—W. R. Lake, London. (*J. M. A. Gérard-Lescuyer, Paris.*) **Apparatus for Regulating the Action of Electric Arc Lamps, &c.** 8d. (6 figs.) June 23.

ARC LAMPS.—This relates principally to a clamping device for regulating the position of the upper carbon. Fig. 1 is a front elevation, partly in section, and Fig. 2 a side elevation of the lamp. The body of the lamp is a hollow tube or chamber connected by two crossbars A B to two rods, which are insulated from the body of the lamp, and carry the lower carbon. The upper carbon, whose upper end is provided with a "Giffard" piston, slides within this tube, a screw at the top of the chamber allowing the air to enter more or less rapidly, and proportionately retard the descent of the carbon. Two solenoids F are fixed upon the crossbar B, one portion (the top) of each being wound with a thick wire in the main circuit, and the other being wound in the opposite direction with a fine wire in a shunt

circuit; the two portions are separated by a disc J. The two cores L, connected by a cross-piece M, slide within the closed tubes, and are provided at their upper ends with cotton or leather discs, regulating the compression of air during upward and exhaustion during downward movements of the cores. A metal disc R serves as a reflector, preserving the lamp from heated currents of air, and acting as an abutment for the brake. The carbon-holder is pressed firmly against the guide pulley U by two springs. The clamping device comprises two metallic pieces N N¹, crossing each other in the form of the letter X, connected to the crossbar M by links O, and provided with screws P P¹ at the other end. Each piece N is provided with a recess or curved part (approximately U-shaped) slightly larger than and partially surrounding the carbon-holder, or it may be formed with projections passing on opposite sides

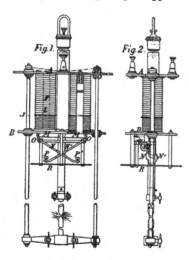

of the holder. As the crossbar M rises, the pieces N N¹ grip the carbon-holder and raise it to strike the arc; when the crossbar descends, the screws P P¹ come in contact with the disc R, and release the carbon-holder, permitting it to descend. In a modification, the carbon-holder passes through a single tubular electro-magnet, the thick wire coils of which act on a lower armature or crossbar, causing the clamping device to grip and raise the carbon-holder, and the thin wire coils of which act on an armature arranged above the magnet, tending to lower the clamping device and release the carbon-holder. In another modification, the clamping device is normally maintained in a raised position by springs, and is lowered by the attraction of an electro-magnet, wound entirely in a shunt circuit on the upper crossbar, the lower crossbar being of non-magnetic material. The arrangement may be reversed, the magnet being placed in the main circuit.

3,002.—P. Jensen, London. (*D. A. Schuyler and F. G. Waterhouse, New York.*) **Dynamo-Electric Machines.** 8d. (13 figs.) June 24.

DYNAMO-ELECTRIC GENERATOR.—The armature coils are divided into sets of four, those in each set and the sets also being disposed symmetrically with relation to one another. Each set is provided with a separate commutator, and one end of each helix is connected to a separate segment of the commutator, the remaining ends being in permanent electrical connection with each other. Fig. 1 is an elevation of a generator, showing a ring or drum armature, constructed according to this invention, rotating between two pole-pieces. Fig. 2 illustrates the manner of dividing and connecting the bobbins to each other and to the commutator, and Fig. 3 shows the bobbins of a single set connected to each other and to the commutator, O representing the commutators, and T and W metallic rings to which the other ends of each set are respectively connected. The brushes of the commutator can be joined either for quantity or intensity, and are so applied, or the segments of

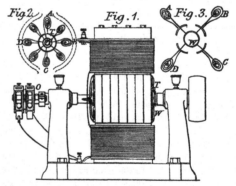

the commutator are of such length, that each coil is kept in circuit preferably for 135 deg. as it passes from one neutral point to the other. As shown, each commutator segment extends over an arc of 90 deg., and two brushes, connected together, bear on the commutator at points which subtend an angle of 45 deg. at the axis. Each coil is thus in circuit alone for an arc of 45 deg. at the centre of the arc of 135 deg., and all the coils are in circuit for the arc of 45 deg. on each side of the central arc. When the brushes of the two or more commutators are connected in parallel arc, they may be made broad enough to bear on both or all the commutators. This invention relates also to generators in which the armature core, moving in the field, induces currents in stationary coils. In one mode of construction, a single coil is wound upon a non-magnetic support in a plane perpendicular to the line joining the two poles of the field magnets, and passing through the neutral points of the fields. The coil is thus in a

plane parallel to the shaft. The armature core consists of two soft iron pieces connected to the shaft by spokes and a hub of non-magnetic material, and revolves before the pole-pieces and within the armature coil, inducing alternating currents. The single coil may be divided into several sections, and a larger or smaller number of field magnets may be employed, each coil or set of coils being wound in the spaces on either side of a pole-piece, or in any other convenient manner, there being half as many coils as there are field magnet poles. The number of soft iron pieces in the armature core corresponds to the number of field magnet poles. In order to commutate the currents produced, the free ends of the coils are connected to stationary brushes bearing upon insulated discs on the shaft. These discs are connected to the commutator proper, which is provided with the usual collecting brushes. In order to induce a continuous current in the coils, the forward ends of the soft iron pieces of the armature core are gradually curved inwards, so that they are introduced gradually within the stationary coils, and are withdrawn suddenly.

MAGNETO - ELECTRIC GENERATOR. — The field magnets and the armature shaft are supported by non-magnetic standards, the magnets being arranged parallel to each other and to the shaft, and provided at each end with inwardly projecting pole-pieces, symmetrically arranged around the shaft, and of alternating polarity. The armature consists of a core forming the shaft of the generator, wound with coils of wire, the ends of which are carried through the shaft to an ordinary commutator. The core is provided at each end with radial projections corresponding in number to half the number of poles before which they rotate. The field magnets thus tend, at any instant, to induce one polarity at one end of the armature core, and the opposite polarity at the other end. Although it is stated that this is a magneto-electric generator, the field magnets are shown wound with coils of wire.

3,003.—A. Wilkinson, London. **Telephonic Wires, Insulating Compounds, and Transmitter and Receiver Diaphragms.** 4d. June 24.

CONDUCTORS.—The wire is wound round a spindle, passed through a hot solution consisting of tin and lead, and drawn on a mandrel; or the tin and lead may be dispensed with, and a filament of fibrous material plaited on. The hollow conductor thus formed is then coated with an insulating compound composed of india-rubber, gutta-percha, tallow, naphtha, ground cork, and bituminous matter. The hollow conductors, insulated or not, may be passed through a braiding machine, and be afterwards coated with the above-described insulating material. Other conductors may be arranged on the exterior of the hollow conductor. The following appears only in the final specification: A number of insulated conductors may be plaited or braided round a hollow core of india-rubber, gutta-percha, or other fibrous material, and the cable so formed passed through an ordinary braiding machine, the threads being steeped in an insulating material such as described above. A hose may be plaited or taped in a machine, and a number of conductors may be placed along the hose either spirally or longitudinally, and finally coated with india-rubber or gutta-percha, the hose being used for the purpose of supplying gaslight to ships, &c.

RESISTANCES.—A conductor may be coiled as above around mandrels or spindles, in order to increase its resistance. A wire is coiled round a spindle to form a hollow conductor, and coated with a compound composed of carbonised ground bones, to which is added, whilst in a heated condition, common stearic or stearine, Stockholm tar, benzoline, and bisulphide of carbon, the whole being well mixed and left in an air-tight vessel for forty-eight hours. These ingredients may be mixed with asbestos.

CARBONS AND SECONDARY BATTERIES. —When carbons are to be produced for electric lamps or batteries, these substances are moulded to the desired form and subjected to hydraulic pressure.

3,007.—F. Jenkin, Edinburgh. **Machinery for the Regulation of Speed in Machinery Driven by Electricity, &c.** 6d. (2 Figs.) June 24.

SPEED REGULATOR.—A governor, driven by the motor, controls the circuit of a relay, which in turn controls that of the motor, the power driving the motor being in excess of that required if its action were continuous. Fig. 1 represents one arrangement of governor and governing relay. A pair of bevelled wheels R, driven by the electro-motor, rotates the balanced centrifugal governor W. The slider U has a wide flange, which, at certain times as the speed increases, comes in contact with the springs 1 and 4. The spring 1 carries a piece of vulcanite, which bears constantly on a spring 2. The spring 2 is separated by a short interval from the spring 3. The circuit wires, from the source of the current, are connected to an insulated binding screw H and to the frame of the machine respectively. The current normally passes through the binding screw H, lever 5, metal bracket 6, and through the motor to the frame. The bracket 6 is connected to one terminal of the motor and to the spring 4. If the speed increases, the slide U presses against the spring 1, and causes contact to be made between 2 and 3, and a portion of the current flows round the electro-magnet 9, and attracts the armature 8 fastened to the lever 5, and the lever

z z

makes contact with the bracket 16, which is connected through the resistance coil 14 to the frame, and thus the whole current divides itself between the two coils 14 and 9; the lever 5 completes contact at 16 before breaking it at 5. The resistance coil 14 may be dispensed with, its object being to reduce the current passing through the springs 2 and 3. If the speed should increase still further, the slider U makes contact with the spring 4, and the circuit of the electrometer is closed, one terminal of the motor being connected with the frame, and thus with the governor and its slide, and the other with the spring 4. An arrangement is shown in Fig. 2 for use with very powerful currents. The main current passes through the shunt lever C, and, according to whether it is in contact with the block C^1 or C^2, it passes through the motor A or not. If the speed becomes too great or too small, two metallic blocks $5a$ and B^b on the lever 5 are connected respectively with the main leads, and come in contact with either the two

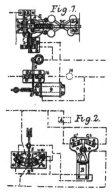

Fig. 1.

Fig. 2.

blocks x^a and x^b, or y^a and y^b, respectively; the shunt current in the two cases passes in opposite direction round the relay motor B, which, by means of a pinion, actuates the shunt lever C. The lever 5 may be worked by the governor directly, or electrically, springs 1, 2, 3 being provided on each side of the slide U (Fig. 1), the spring being dispensed with, and another relay magnet 9 taking its place. Each time the relay breaks contact, it may work an ordinary Morse instrument, or other recorder, which, when the motor is being driven, gives a continuous line, and when cut out, a space. The relay motor (Fig. 2) may control the speed by putting on a mechanical brake, by diminishing or increasing the work done by the machine, *e.g.*, by means of the rate of feed or depth of cut, &c., by controlling a resistance in some part of the circuit, by shifting the position of the brushes on the commutator of the transmitting generator, or by controlling the resistance of the field magnet circuit of the generator. The controlling spring of the governor is adjustable whilst in action.

3,010.—W. E. Debenham, London. Electric Lamps with Incandescent Conductors, &c. 4d. June 26.

INCANDESCENCE LAMPS.—The glass, through which the platinum conductors pass, is made of a mixture of two glasses, one having its coefficient of expansion higher, and the other lower, than that of the platinum, in the proportions indicated by these coefficients. The conductors just inside the glass are made in the form of discs, semi-discs, or extended surfaces lying in contact with the glass and imparting their heat to it. The filament is connected to the margins of these surfaces, which are gradually turned up. The cotton or other thread, to form the filament of the lamp, is soaked in an alcoholic solution of gluten or gelatine, and the joint of the filament with the conductor is covered with flour or other carbonaceous paste, and when dry the alcoholic solution of gluten or gelatine is applied several times, and the joint is then carbonised. Two projections, blown on to the lamp bulb, fit into grooves in a socket and form bayonet joints. The conductors are left projecting, and make contact with metallic spring plates or surfaces in the socket. The pressure of the springs should be strong enough when the lamp is turned round to lay the projecting conductors flat, and give greater contact surface. The projections on the bulb may be welded on the glass, or be on a ring fitted to the glass, or they may be metallic and in connection with the conductors. The springs may press the studs downwards into contact with plates at the bottom of the lamp socket, or the conductors may pass out of the stem of the lamp and bear against connections in the side of the socket, the pressure being gradually tightened as the lamp is turned.

3,025.—E. A. Sperry, Cortland, N.Y., U.S.A. Dynamo-Electric Machines, &c. 10d. (30 figs.) June 27.

DYNAMO-ELECTRIC GENERATORS.—The poles of the field magnets are constructed so as to present extension pieces both interiorly and exteriorly to a transversely wound annular armature. Referring to the illustrations, the cylindrical ring armature with soft iron core a revolves between the pole pieces f, f^1, f^2, and f^3; f and f^1 being of one polarity, while f^2 and f^3 are both of the opposite polarity. Each of the two pairs may be in one casting and magnetised by one or more helices. The armature is constructed so as to have its greatest dimension of cross section parallel with its axis and presents its interior and exterior surfaces, together with one extremity to the poles. The core of the armature comprises two cast-iron end rings, these being originally cast together, and having in cross section the form of an elongated

circle, which after being split, forms two rings, with a semicircular cross section. Each ring has, say, twenty-four equidistant semicircular projections, every fourth one being larger, and on one (the outer) ring left with a square face for attachment to the supporting arms. Six rods are firmly screwed into the six projections on the inner ring, and are then insulated. Segments of sheet iron, having five projections, two large and three small, corresponding to those of the rings, the large projections being perforated for the reception of the rods, are slipped on the rods so as to fill up the alternate spaces, and then the three others are added, filling up the vacant spaces and overlapping the ends of those already in place. A piece of insulating material corresponding in shape to the large projections, is placed over each of the rods. Six more segments are then applied in a similar manner, and so on, the cylindrical core being completed by the outer ring, and the whole being firmly fixed

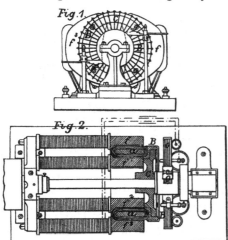

Fig. 1.

Fig. 2.

to a non-magnetic disc C by six bolts. The armature coils are connected either in series or in multiple arc, the ends of the coils being brought to radial wires B attached to a commutator of the usual type, from which the current is collected by the brushes O. The outer ring of the armature core may be of non-magnetic material. Two coils are wound between each two projections (the adjacent sides of the projections being parallel to the radial line midway between them) and "revolve before four equidistant fields." The commutator comprises a cylindrical casting, fitting loosely on the shaft, and having at one end a flange, and a thread at the other. The surface of the cylinder is insulated, and against the inner surface of the flange, and of a nut screwing on the thread, are placed rings of insulating material, in which are cut radial grooves carrying the commutator sections, which are connected to the flexible radial conductors B. The

commutator brushes are composed of flexible conductors arranged in clamps supported by standards. Or the brushes are supported by a collar fitting a boss projecting from the bearing of the shaft, the collar carrying two diametrically opposite slotted projections supporting rods on which the brush clamps are loosely mounted. Springs carried by the clamps, and having their free ends resting in slots in the rods, regulate the pressure of the brushes. The angular position of the brushes is adjusted by right and left-hand threaded screws connected by a nut, one being fixed and the other connected to the collar. In order to adjust the lead of the brushes automatically as the speed increases, a short cylindrical casting is secured on the armature shaft, and has pivotted near its outer periphery two levers, which extend beyond and on opposite sides of the shaft. The free ends of the levers carry weights which by the centrifugal force fly outwards against the resistance of adjustable springs. The levers are connected by chains passing over pulleys to a yoke working through a slot in a short rod, which is thus moved endwise on the shaft, which is bored to receive it, and slotted to receive the yoke. This short rod is forced outwards by a spring, and has at its extremity a groove in which are placed the forks of a lever with a fulcrum at one end, and connected by a rod to a bell-crank lever attached to the collar supporting the brush-holders, or to the face of the nut of the loose commutator. In order to regulate the current, a worm on the armature shaft drives a wormwheel mounted on a shaft carrying a cam, operating one end of a lever, the lower end of which is forked; two pawls are pivotted to the forked ends of the lever and are supported above two ratchet wheels mounted on a shaft, by links connected to a lever on opposite sides of its fulcrum. An electro-magnet in a shunt circuit, or in a portion of the main circuit, attracts an armature on this lever, and according as to whether the current passing through it is greater or less than normal, one or other of the pawls is thrown in gear with its ratchet wheel and rotates the ratchet shaft in one or the other direction. This shaft is screw-threaded and operates a nut having a projection carrying a fork, fitted to a groove in a rod sliding in the end of the shaft, and secured to a cross-piece sliding in a slot in the shaft. Two flexible chains are connected to the cross-piece and pass over guide pulleys, and are secured to the nut on the commutator, which is provided with a retractile spring.

ARC LAMP. — The upper portion of the lamp frame is provided with a hollow cylindrical upward projection, into which is screwed the upper end of a flanged diamagnetic tube carrying the solenoid coils. The lower end of a protecting tube

is also screwed on to the cylindrical projection. The single circuit of the solenoid is divided into four separate parts by conductors, any one of which can be connected to the lamp frame according as strong or weak currents are employed. The core of the solenoid consists of a flanged tube of soft iron, loosely fitting the carbon-holder, and wound with fine insulated wire connected by flexible conductors with the terminals of the lamp, the current traversing the main coils and the shunt coil in opposite directions. A lever, with a fulcrum at one end, and attached at the other end to an adjustable spring and the plunger of a dash-pot, is fixed to the lower extremity of the core, and is connected by two links to the ends of two levers with fulcrums at their other ends. These two levers carry between them a casting, bored loosely to fit the carbon-holder, which has two hooked levers pivotted to its extremities, the hooked ends passing around the holder a short distance below the casting and serving as a clutch. The hooked ends of the levers bear on an adjustable stop, through which the carbon-holder passes, or the ends of the levers are prolonged and bear on adjustable stops. The current passes from one terminal of the lamp, through the solenoid, to the lamp frame and the carbon-holder, across the arc and other carbon to the other terminal. The clutch mechanism may be directly connected to the core. The negative carbon clamp is laterally adjustable. The cut-out attachment comprises a casting vibrating in a horizontal plane on two points of an insulated frame connected with one terminal of the lamp. The casting fits round, but is prevented from touching the core by an insulated stop, and is provided with a contact point, forced into contact with a similar point on an insulated piece of metal, connected through a resistance (equal to the normal resistance of the lamp) to the other terminal by a retractile spring on the pivotted casting. The casting is attracted by the magnetism of the core, and breaks the "cut-out circuit" on the passage of the current through the carbons.

3,033.—F. S. Isaac, London. (*Sir Julius Vogel, at Sea.*) Production of Carbons for Incandescent Electrical Illumination. 2d. June 27.

CARBONS.—The fibres of Rubus Australis Parsona Lygodium (Supple Jack), after having been freed from impurity, are carbonised by any of the known methods.

3,036.—W. E. Ayrton and J. Perry, London. Dynamo-Electric Machines. 6d. (13 figs.) June 27.

DYNAMO-ELECTRIC GENERATORS.—The cores of the field magnets are made of a combination of steel and iron; Figs. 1 and 2 show an arrangement of alternate plates or tubes of steel and wrought iron. By a proper arrangement of winding, at a certain speed it is possible to obtain a constant electromotive force between the two ends of the outside circuit, and with other arrangements of winding and with other speeds it is possible to obtain a constant current. The similar poles of the magnets may be strengthened by joining them with pieces of metal G. The coils are placed on an iron ring made in two pieces, joined together so that the screws are flush with the surface of the ring. The coils are slipped on these halves, a plate of wrought iron, which is cut as shown at C, Fig. 2, being placed between every two coils, which can thus be machine wound, except perhaps the last two, which are wound by hand in place. The iron plates are sometimes U-shaped. The coils may be made and baked as described in Specification No. 1,178 of 1880. By another

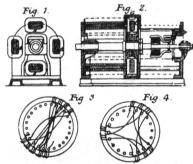

method, the iron ring has iron screws standing out in radial rows, or radial iron pieces, serving to keep the coils, which are wound by hand, in place, increasing the magnetic force, and being especially useful for oblique windings; or a number of radial plates of iron may fit partly around a non-conducting ring, the coils of wire being wound usually very obliquely and by hand. When the coils are thus wound, the pole-pieces are much smaller radially than those shown. The armature is attached to its spindle by a wooden centre, as described in the above-mentioned specification, or by a metallic boss and arms screwed more or less at their outer ends into the iron separating pieces, or these pieces when very thin may pass through saw cuts in the centre, in which they are insulated and secured. If the core is non-conducting, the plates may be soldered to the cuts in the metallic centres. The joining up of the coils when there are four fields is illustrated in Fig. 3. The number of coils is obtained from the formula $\frac{N\,n}{2} - 1$, where N is the number of fields and n any integer. Suppose there are nine coils, their similar ends being respectively a, c, e, g, i, k, m, o, q, and b, d, f, h,

j, l, n, p, r and A, C, E, G, I, K, M, O, Q the commutator pieces ; *a* and *j* are joined to A, *c* and *l* to C, *e* and *n* to E, and so on, the currents being collected by two brushes. Fig. 4 represents the joining up when there are six fields. The core of the armature may be made of phosphor-bronze, or brass, or steel, or other strong material with a great number of soft iron rivets passing through it, the rivets being preferably small in size and very large in number. The rivets may stand out on each side of the core and serve to keep the coils in place.

3,039.—C. P. Nezeraux, Paris.   **Galvanic Batteries.** 6d. (7 figs.) June 28.

SECONDARY BATTERIES.—This invention relates principally to a secondary battery. As will be seen from the illustrations, A and B, Fig. 1, are two plates of lead entirely covered with an adherent coat of ebonite (caoutchouc durci) with the exception of their operative surfaces *m n* : the plates may therefore be said to be encased in a rigid ebonite frame. Each element carries upon the upper edge projections, not shown, which rest upon small metallic strips fixed to the containing vessel R, having a lining V of gutta-percha or lead. These conducting strips serve to attach the plates to the terminals. The contacts may be made by small mercury cups. To render the cells operative, the cavities C D are filled with reduced and peroxidised

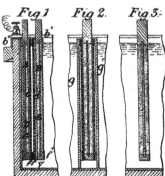

compounds of lead respectively, the former mixed with a solution of an alkaline salt such as potassic cyanide, and the latter simply with water. The plates are then drained, and over each layer of powder is placed a woollen cloth, felt, or similar material *d*, and a rigid perforated plate *e*, connected to each element by an india-rubber band *f*. The couple is then plunged into a vessel containing dilute sulphuric acid, or other suitable excitant. When it is required to recharge the cell, the exhausted material is scraped off the lead plates, and replaced by fresh material. It is then removed to be revivified. By alternative methods, the operative mixtures are formed from an amalgam of lead,

or from finely rasped lead, which is placed in the cavities C and D, and exposed to the usual "forming" action of the electric current, its direction being reversed several times during the process. Fig. 2 represents a double element constructed in a similar manner, *g g*[1] being plates of porous earthenware placed between the couples to prevent polarisation by hydrogen. Fig. 3 also illustrates a double element, but without felts and perforated plates, its cavities being filled with lead powder amalgamated to saturation. In one form of cell the vessel usually containing the exciting fluid is suppressed, two ebonite frames being joined face to face, leaving an unoccupied cavity between the elements, which is filled with fine sand saturated with acidulated water. Two batteries, one having its elements arranged for tension, and the other for quantity, have two of their poles connected together, the remaining two being connected to the motor. After more or less discharge of the elements in tension, they are connected for quantity and the other elements for tension, the object being to obtain the best possible yield from the batteries.

3,042.—F. L. Willard, London.   **Incandescent Electric Lamps.** 6d. (4 figs.) June 28.

INCANDESCENCE LAMPS.—The filaments are made from thin strips of "white lines," or ivory, which are carbonised by baking out of contact with the air, and carbon is deposited upon them by holding them in the naked flame of a lamp, or by other suitable methods. The filaments may be treated with acids or alkalies to destroy all traces of organic matter, and the woody substance may be reduced to pulp by suitable chemicals, and rolled

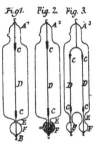

in sheets and stamped and carbonised. They may be immersed in a solution containing platinum or iridium, which is decomposed by an electric current. The filaments are secured to the platinum wires by a cement composed of cannel coke dust and coal tar, the ends being thickened and the whole raised to a red heat. The lamp, of whatever shape, has a small bulb or bulbs formed at its lower end, and filled with granules of pure tin, copper, or iron, or with mercury. The conductors are fused into the glass, the upper one A being attached to the filament at C as described, and the lower one

B inserted into the small bulb F so as to make a good connection with the metal therein. The conductor attached to the lower end of the filament passes through the narrow neck E, which serves as a guide, into the bulb F, and makes a connection with the metal therein and is free to expand and contract.

3,047. — W. Spence, London. (*M. Kotyra, Paris.*) Telephone Receivers. 6d. (13 figs.) June 28.

PERMANENT MAGNETS.—These are composed of multiple bars, separately magnetised and solidly bound together; the polar extremities carry soft iron pieces, the terminals of which are brought close together.

3,048. — G. Macaulay-Cruikshank, Glasgow. (*W. E. Banta, Springfield, Ohio, U.S.A.*) Insulating and Protecting Telegraph Wires. 6d. (3 figs.) June 28.

CONDUCTORS.—The protection consists of a series of insulating terra-cotta or artificial stone troughs A, hermetically connected, and hermetically closed by a coping B of granite, or other rock, or cast iron. At intervals along the trough are brass feet C supporting pillars D. Fitted upon D are washers and glass spools E, with circumferential grooves, to which the conductors are attached by fine binding wire G. The troughs may take the form of a hollow pavement kerb, and be provided with openings for the passage of branch conductors. The ends are preferably interlocked and luted with

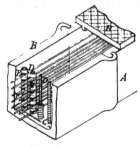

litharge, or other suitable material. The pillars D may be bolted to wooden shoes cemented in a transverse groove in the bottom of the trough, or they may be secured whilst casting or moulding the trough. The washers may be omitted and the spools screw-threaded. The troughs are sunk so that their top surfaces are level with the street or pavement. The conductors may be supported by a number of terra-cotta blocks, retained vertically by posts or uprights, or they may be passed through holes or slots in brass plates, glass, or other insulating thimbles preventing contact between the conductors, and the supporting plates,

which have eyes on their vertical edges, may be secured by pins. The trough may be filled with sand.

*3,054.—J. C. Mewburn, London. (*F. Rigaud, Paris.*) Apparatus for Regulating the Production of Electricity. 2d. June 28.

CURRENT REGULATOR. — The transmission of motion from the engine to the generator is effected by frictional contact of a disc, driven at a constant speed by the engine, against a roller, keyed on the shaft of the generator and movable along this shaft, which roller approaches the centre of the disc as the current increases; or by means of a strap passing over two cones on parallel axes, moved by a fork; or by means of a spherical roller running on a revolving surface. The roller, fork, or other device is controlled by a rack, worked by a pinion keyed on a shaft which is common to two bevel wheels, and between these bevel wheels is another wheel fixed on a shaft at right angles to the other shaft. This second shaft is kept in equilibrium by a spring or counterweight, and the attraction of an electro-magnet in the main or in a shunt circuit, and according to the current passing, is either midway between the two bevel wheels, or in gear with one of them, and thus adjusts the position of the roller or fork; or the magnet may be made to move the two bevel wheels. In another arrangement, a rod attached to the roller or fork may be placed between two rollers rotating in the same direction, and have its position controlled by an electro-magnet; or a contact may be controlled by a ball governor turning under the action of an electro-motor placed in the main circuit.

3,070.—E. de Pass, London. (*C. Roosevelt and B. Abdank, Paris.*) Electric Arc Lamps. 6d. (6 figs.) June 29.

ARC LAMPS.—The upper carbon B is fixed in a clamp P at the lower end of the toothed rod T, which passes through the centre of the fixed iron core *a* of the solenoid A, electric continuity being maintained between the carbon and the solenoid coils by the flexible connection *t*. Resting upon the frame *f* is the soft iron armature D (Figs. 1 and 2), which is provided with a tube sliding freely in the socket *m*. Attached to the tube is a pawl *h* which, when no current passes through the lamp, does not engage the teeth of the rod T. Upon the excitation of the solenoid coils, the armature becomes attracted upwards, and the pawl being brought in contact with the jockey roller *g*, or other equivalent contrivance, engages the teeth of the rod, which it raises and forms the arc. In a modification, one side of the armature is made higher than the other, and the rod T is plain; the higher

portion of the armature being attracted more quickly than the lower portion, the armature is canted and grips and raises the rod T. A counterweight may be provided on the lower part of the armature. In Fig. 3 the armature D is fixed, while the iron core of the solenoid moves, and the pawl *h* is replaced by the lever *l*, which is pivotted to the core at *x*, and carries at its lower end an iron block L. Upon a current flowing through the solenoid coils, L is attracted towards the rod T, which is gripped between the projection *t* upon the centre of the lever and the two screws *v* and *v*¹. At the same time the core moves towards its armature, carrying the carbon-holder and carbon with it. The whole of the mechanism described above is carried by the rack C, the descent of which is regulated by any ordinary mechanism ; or the solenoid A is also wound with fine wire in a shunt circuit, the terminals of which are connected to the microphonic contact of the "resistance balance" described in Specification No. 339 of 1882, the cur-

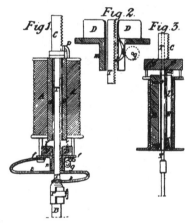

Fig 1. *Fig 2.* *Fig 3.*

rent being passed through the shunt circuit in a contrary direction when the arc becomes too long, causing the armature to descend. In a modification of the resistance balance, the light rod of the balance breaks a contact, causing the shunt current to flow through an electro-magnet, and to the fine wire of the solenoid of the lamp ; the electromagnet attracts an armature, causing the contact to remain broken until the armature of the lamp solenoid descends and breaks the circuit, when the contact will be closed. A spring may be substituted for the main solenoid of the balance. In an arrangement for separating the carbons on their accidentally approaching each other too nearly, one bobbin of the balance is arranged in a shunt circuit, and the other partly in a shunt and partly in the main, and in such a manner that normally the light rod is on the side of the shunt bobbin, but when the arc is too long it is attracted to the side of the compound bobbin and opens a contact,

causing the current to pass through the fine wire of the compound bobbin and to oppose the action of the main wire, causing the bar to suddenly return to its normal position. At the same time that the contact is broken, a current is sent to the lamp and causes the carbons to separate.

*3,073.—H. Binko, London. **Electric Railways and Tramways.** 2d. June 29.

CONDUCTORS. — The electromotive force of the dynamo or motor is increased by arranging the wheels of the additional vehicle or vehicles hauled, so that each or some of them shall collect from one of the rails, or other conductor, an additional quantity of electric current, to be conveyed to the electric motor. When suspended conductors are employed, two or more contact brushes collect the additional current. The current to the electromotor is made and broken through "a peculiarly arranged resistance" enabling the current to be cut off or supplied gradually. The resistance is submerged in water contained in a closed box. India-rubber or other insulating material is interposed between the rails and sleepers, the holding spikes being driven into the sleeper without being in actual contact with the rail, and a washer being interposed between the head of the spike and the insulating material. If chains are used they are made of insulating material, or are reversed, insulating material being inserted into the recesses to support the rail. The ends of the rails, and the parts forming the joints or fastenings, are galvanised or suitably coated to obviate rust and consequent increased resistance.

*3,079.—J. H. Johnson, London. (*L. Bardon, Paris.*) **Electric Lamps.** 2d. June 30.

ARC LAMPS.—The upper carbon is supported by a rod provided with a helix, or quick screw thread, acted upon by a lever which works in combination with a shunt solenoid or electro-magnet, the screw thread preventing the carbon from falling suddenly. The rod passes through guides, the lower guide being provided with an internal helix acting as a nut to the rod. The guides are attached to a bracket carrying the shunt solenoid. To the under side of the threaded guide is pivotted a bent lever, whose short arm bears against the side of the rod, whilst the long arm carries the armature of the solenoid, which is suspended by a spring. The bearing surface of the short arm is concave, and is covered with india-rubber or felt.

*3,097. — A. Watt, Liverpool. **Secondary Batteries.** 2d. June 30.

SECONDARY BATTERIES.—An alloy or mixture of lead and zinc, or other metal electro-positive to

lead, is produced, in a granulated form, by pouring the molten alloy into water, or subjecting it to a jet of high-pressure steam or air, and is pressed into slabs or other solid forms. These slabs are surrounded by a framework of lead cast on them, or are placed in a perforated box of insulating material. The connections are made by strips of lead interlaced or embedded in the mass, as described in Specification No. 4,255 of 1881. The slabs, as above prepared, are immersed in a solution of a salt of lead, the zinc being replaced by spongy lead, or the zinc may be simply removed by an acid: they are then washed, pressed, and placed together in a bath of dilute sulphuric, phosphoric, or boracic acids, or a mixture of these, and separated from each other by insulating material. The negative plates of the batteries described in the above-mentioned specification are formed of a carbon plate surrounded with a mixture of granulated peroxide of manganese and granulated carbon, enclosed between perforated insulating plates, mercury being introduced into a groove in the upper edge of the carbon to make good connection, and retained in position by the cover of the trough; or they are formed of a mixture of granulated oxide of manganese, and granulated or fibrous lead, enclosed in a perforated casing.

*3,099.—A. R. Leask and F. P. Smith, London. Preparation of Carbon Filaments. 2d. June 30.

INCANDESCENCE LAMPS.—The filament is placed in a glass vessel through which is passed carburetted hydrogen gas; a current passed through the filament decomposes the gas, and a deposit of carbon takes place upon the filament; a regulator is provided to control the supply of gas in the vessel. The tool for holding the filament consists of two rods, having a spring catch to each somewhat like the keys of a flute, and adjusting springs. Each rod carries a projection, which presses upon and makes contact with a spring forming the end of the lead or return conductor. A switching arrangement operated by a pedal is used, by means of which contact can be made either with the

dynamo machine, or with the battery system and galvanometer for testing the filament. The filament is observed through a piece of coloured glass. A contact breaker consists of a plunger pressed into an iron vessel containing water, against the resistance of a spring, the circuit being completed through the water. " An electro-magnet is placed upon dynamo return wire."

3,101. — R. H. Courtenay, London. Electric Lamp. 6d. (4 figs.) June 30.

ARC LAMP.—Figs. 1 and 2 are side and front views respectively of the lamp, and Figs. 3 and 4 a vertical section and plan respectively of the gripping device, drawn to an enlarged scale. The iron cores D D, fixed to the plate E, work within the two solenoids C C. The plate E has a wedge-shaped slot F, which carries the gripping blocks G G. The blocks G G, when down, rest upon a

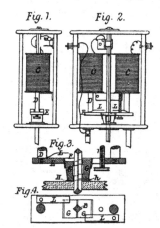

Fig. 1.   Fig. 2.

Fig. 3.

Fig. 4.

plate H, which has an extra thickness *h* on one side to prevent the blocks from jamming. I is an adjusting screw, and L L two springs to retain the blocks in position. When the current passes to the lamp, the solenoids draw up the cores D D, plate E, and blocks G G, and establish the arc, the blocks gripping the carbon rod B so as to retain it at the required height.

ADDENDA TO ABSTRACTS OF PATENTS.

1877.

4,903. — **W. R. Lake**, London. (*E. Weston, Newark, N.J., U.S.A.*) **Magneto-Electric Machines.** 1s. 2d. (22 figs.) December 27.

DYNAMO-ELECTRIC GENERATORS. — The upper ends of the two vertical cores of the stationary electro-magnets, mounted on a bed-plate, are bolted to a cross-piece, and, at the centre of their length, the cores are provided with laterally projecting pole-pieces of opposite polarity. The armature is cylindrical, and its core consists of a hollow elongated sphere, provided with a series of equidistant ribs parallel with its axis and projecting from its ends, by means of which it is secured to, but insulated from, the edges of two cup-shaped discs, each provided with long hubs through which passes the armature shaft. This shaft carries an internal magnet. The commutator is fixed to one hub and the driving pulley to the other. One end of the elongated sphere is removable, being held in place by screws passing radially into the ribs. The screws securing the core to the discs are inserted through plugs of insulating material in the discs. The core may be bisected longitudinally, the two parts being insulated from each other. The number of distinct and similar systems of armature coils corresponds to half the number of ribs, and the two ends of each coil are connected with diametrically opposite plates of a commutator, the number of insulated plates being equal to twice the number of coils. The ends of the elongated sphere are perforated, and radial holes are bored through the ribs, the pole-pieces being provided with corresponding grooves, to allow the air to escape from the interior of the core, the rapid circulation of air tending to cool the structure. The commutator is of disc form, the plates on opposite sides of a fixed diameter being of opposite polarity. Each brush overlaps two or more of the plates. The current passes from one brush, through the outer field magnets, outside circuit, inner field magnet, and shaft to the other brush. The armature may be provided with two commutators, each composed of half the number of plates, in which case, one circuit includes the outer field magnets and the other the internal field magnet. The latter may be rotated in the opposite direction to the armature. The core of the armature may be of paper, rivetted to ribs of metal. The method of mounting the core is stated to be especially convenient when the coils are made of flat strips. In another form of generator, an internal field magnet rotates within a stationary cylindrical armature. The armature core, which may be of non-magnetic material, consists of two or more rings, joined together by equidistant ribs parallel with the axis of the cylinder, which serve to divide the periphery. Each coil is supported in diametrically opposite segments. The outermost surfaces of the ribs are curved to fit the interior of a cylindrical iron shell, to which they are secured by screws, and which is open at the ends and provided with foot-pieces for securing it to the bed. The internal magnet is supported on a shaft revolving in bearings in horizontal arms extending across each end of the iron shell. The shaft is kept from end movement by means of a collar and the hub of the driving pulley. Radial arms, extending inwards from the outer rings of the armature core, support sleeves surrounding the shaft, and thus prevent the portions of the coils crossing the ends of the cylinder from coming in contact with the shaft. The poles of the internal magnet subtend three parts of the armature. The commutator consists of two concentrical strips, which extend laterally from the opposing faces of two hubs, and overlap each other on opposite sides of the shaft. The outer hub is insulated, and is elongated to form a bearing for a brush connected to one terminal of the generator. Outside this hub is an insulated metallic collar upon which a brush, connected to the other terminal, bears. The inner face of the collar is connected, by a conductor enclosed in an insulating tube passing through the hub, to an insulated collar, connected by a similar conductor passing through the enlarged journal of the shaft, to one terminal of the internal field magnet, the other terminal being connected to the shaft, and thus to the inner hub, which is mounted directly on the shaft. The ends of the coils are connected to separate brushes bearing

a a a

on the strips from the hubs, and are secured in slotted stems projecting laterally from the face of an annular cup-shaped disc of insulating material supported on the crossbar. Air, drawn in by the internal magnet, is thrown out through the interstices between the coils and the ribs, and through radial holes in the ribs and iron shell. In order to render the electromotive force of the current constant for the different relative positions of the coils and field magnets, the terminals of a generator are connected to a condenser of ordinary construction. In one form of armature of the cylindrical-ring or hollow-drum type, the core is made in sections, each section comprising a curved portion upon which the coil is wound, and having one end bent radially outward to form a pole, and the other end bent inward to form a contact with the central shaft, the segments being bound together by clamping rings. Two diametrically opposite coils are connected in series and to opposite plates of a commutator. A cylindrical armature, which may be substituted for the armature and internal magnet first described, is constructed of a series of separate discs, each of which is perforated centrally to receive the main shaft, collars being interposed between the discs. Each disc is slit radially and is provided with projecting tongues forming the poles. Two or more holes may be bored through each disc, and they may be mounted on a hollow shaft provided with openings on its periphery corresponding with similar radial holes in the collars ; the hollow shaft may be surrounded at each end by a tubular shield having a wide cup-shaped flange perforated transversely and secured to the end of the armature, which is recessed to form an air chamber. An air supply pipe may be affixed by a swivelling coupling

to the end of the shaft, and air forced through the armature. The coils may be connected to the commutator by conductors passing through insulating tubes in an enlarged journal, or by insulated conductors carried through the hollow shaft. The armature core may be built up of insulated soft iron wire, or strips wound upon the periphery of a hollow shaft or cylinder between radially projecting parallel guides, the ends of the wire or strips being insulated from each other. The provisional specification also describes the following : A mechanical circuit closer and breaker is employed to open and close the main circuit from the generator or to close a short circuit between the terminals, coincidentally with predetermined variations in the strength of the current resulting from changes in the speed. The field magnets are partially wound with a conductor in "a differential circuit," containing a resistance coil, which is constantly closed and provides for a continuous circulation of the current through the coils of the field magnets, even when the main circuit is broken, and thus prevents a reversal of polarity. The circuit breaker comprises a disc driven by the generator, or by the motor driving the generator, mounted on a standard connected with one of the circuit conductors. A cylindrical hub is attached to, but insulated from, the disc. Two blocks slide in opposite radial slots in the disc, and their sides project over the hub, and are forced to bear upon its periphery by means of adjustable spiral springs, except when the centrifugal force overpowers the springs. The hub is connected with the circuit by means of a standard supporting a brush bearing on the rotating hub, the circuit being completed through the disc and hub by means of the sliding blocks.

1881.

190.—E. Edmonds, London. (*G. M. Mowbray, North Adams, Mass., U.S.A.*) **Treating Caoutchouc and Gutta - Percha.** 4d. January 14.

CONDUCTORS. — To prevent the decomposition which takes place on exposing these gums to the air, they are treated with melted naphthaline, at a low temperature, until perfectly combined, every particle of the excipient being subsequently removed by spontaneous evaporation.

1,272.—W. R. Lake, London. (*H. Splitdorf, New York, U.S.A.*) 8d. (6 figs.) March 22.

CONDUCTORS.—The wire is first coated with shellac, which is heated, and then covered with "slivers" or

rovings of cotton, which have their twist taken out of them as they are laid on the wires. This is accomplished by using a machine having a sliver-carrier, provided with feed rollers, so constructed as to draw the sliver off its bobbin as the carrier rotates, and pass the sliver to a pair of rollers, which untwist it and pass it on to a spreading device, which forms it into a flat ribbon. The spread fibres are laid on the wire by a pair of rollers, or guides, set close to it.

*1,577. — J. Hopkinson and **A. Muirhead,** London. **Electric Telegraphs, &c.** 2d. April 11.

DYNAMO-ELECTRIC GENERATOR. —The inventors

use a secondary battery, or, in some cases, a condenser, in combination with a dynamo-electric generator, for telegraphic purposes.

1,834.—H. J. Haddan, London. (*C. F. Brush, Cleveland, Ohio, U.S.A.*) **Reflectors.** 6d. (14 figs.) April 28.

REFLECTORS.—These are concave, and are formed of a flat sheet bent into the form of a parabola. One or more cut-off plates are provided, for the purpose of intercepting the direct rays of light, without interfering with the reflected rays. In another arrangement, a cut-off tube, having a non-reflecting inner surface, is used for the same purpose.

1,897.—W. A. Barlow, London. **Obtaining, Storing, and Utilizing Gas for Lighting and other Purposes.** 6d. (6 figs.) May 2.

DYNAMO-ELECTRIC GENERATOR.—The gases are produced by electrolytic decomposition, a dynamo-electric generator being employed to supply the necessary current.

2,414.—J. A. Sparling, London. **Manufacture of Milanaise, &c.** 6d. (10 figs.) June 1.

CONDUCTORS.—An insulating material for conductors has a cotton core covered with silk, and is laid on the wire by a machine having a hollow spindle, and a bobbin and flyer.

2,564.—J. R. Wigham, Dublin. **Locomotive Engines for Tramways, Railways, &c.** 6d. June 16.

MOTOR.—"Dynamo-electric engines may be employed" to supply the motive power.

2,658.—A. D. Roth, London. **Apparatus for Saving Life and Property at Sea.** 6d. (8 figs.) June 17.

ELECTRIC LIGHT.—A buoy is used, containing a reservoir for storing electricity, and surmounted by an "electric light."

3,049.—F. W. Haddan, London. (*L. G. Woolley, Mendon, Mich., U.S.A.*) 6d. (8 figs.) July 12.

ARC LAMP.—The whole of the regulating mechanism is placed inside a dash-pot, and is kept constantly immersed in glycerine, or other lubricant. Fig. 1 shows the upper part of a lamp; Fig. 2 is an enlarged view of the dash-pot; and Fig. 3 is an elevation of the carbons and clamps. D is an axial solenoid, the core of which is connected by the rod O to the lifting lever P. This lever is pivotted in a groove Q, and when the core is raised, it bites into the inner surface of the carbon-holder, and draws it up. When the core falls, the lever and holder drop

with it, and come in contact with the piston J, depending from the rod I, the frictional connection is then broken, and the carbon-holder descends by gravity until an upward motion of the core

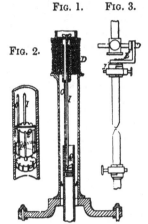

FIG. 1.   FIG. 3.

FIG. 2.

causes the lever again to bite. If glycerine be found objectionable, a spring between Q and J may be substituted for it. The clamp I' of the upper carbon-holder (Fig. 3) is secured to the adjustable bar D'' by solder, which melts when the arc approaches too closely. An insulated chain L serves to catch the clamp as it falls.

3,113.—E. Eteve and C. C. Lallement, Paris. **Motive-Power Engine Operated by Hydro-carburetted Air.** 6d. (5 figs.) July 16.

MAGNETO-ELECTRIC GENERATOR.—A small magneto-electric generator, driven by the engine, is employed to explode the gaseous mixture..

4,458.—W. H. Akester, Glasgow. **Vacuum Pumps.** 6d. (2 figs.) October 13.

INCANDESCENCE LAMPS.—This invention relates to the employment, in vacuum pumps for exhausting incandescence lamps bulbs, of an internal tube and stopper, or closing valves, arranged so that the internal tubes carrying the exhausted lamp can be removed from the other parts of the pump, without destroying the vacuum in the lamps.

4,918.—J. Stanfield and J. L. Clark, London. **Raising Sunken Vessels, &c.** 8d. (21 figs.) November 9.

MOTOR.—A "dynamo-electric machine," enclosed in an air-tight case, is used to actuate the necessary drills, &c., required for piercing the sides of a submerged vessel.

5,000.—C. H. Stearn, Newcastle-on-Tyne. **Mercurial Air Pumps.** 6d. (3 figs.) November 15.

INCANDESCENCE LAMPS.—This invention relates to an automatic feed, by which the mercury employed in a Sprengel pump is periodically returned to its source. Referring to the illustration, A is a compound mercurial tube, which may be of less height than the barometrical column; at its upper end it communicates with the receiver C, and at its lower end with the receiver D. The upper receiver communicates, by a cock and flexible tubes, with an exhausting apparatus, and as soon as a sufficient vacuum is obtained the cock is closed. The lower receiver is supported by a spring e, attached to one arm of a lever centred on a fixed point, and carrying, at its other end, a valve E ground to fit a seating in the globe D. This valve is maintained closed, so long as the weight of the receiver and the

accumulated mercury are insufficient to disturb the spring. The joint G is flexible, so as to permit the motion of the receiver. Communication is made by the passages H with another mechanical exhauster. The cock I cuts off communication between the main part of the apparatus and the receiver, and when it is opened the external atmosphere enters the lower receiver by the valve E, and causes the elastic diaphragm E² to distend towards the left (as in the dotted lines), forcing the mercury through the tubes d and K, up the small tube J, and into the upper receiver C, whence it descends by the tubular connection L, and passes by the tube l to the compound fall tube A, and acts there in the usual manner. When the lower receiver is lightened,

it rises and shuts the valve E; the pump connected to H is then brought into action, and the diaphragm drawn back to the right-hand dotted line, ready for the reception of more mercury. The spring may be replaced by a counterweight, or an electrically controlled contact.

5,548.—L. A. Groth, London. (*H. Goebel and J. W. Kulenkamp, New York, U.S.A.*) Vacuum Pumps. 6d. (1 fig.) December 19.

INCANDESCENCE LAMPS.—The pump is charged by raising the mercury reservoir D above the level of the upper part of the pump C, until the mercury passes through the valve d and the communicating pipe, back to the reservoir. The reservoir is then lowered, and the pump frame tilted to the left, by the handle b, and its connecting rod a. The play of

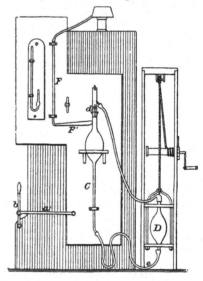

the pump will then begin in the usual manner, exhausting the incandescence lamp globes sealed to the upper end of the tube F, to which the pump is connected by a trough-shaped portion F¹, increasing in depth as it approaches the neck of the pump, and extending nearly at right angles to the main part of the tubes. It is made widest close to the point of connection, so as to throw the mercury towards the angular connection, and form there a seal.

5,742.—J. S. Williams, Riverton, N.J., U.S.A. Heating and Rolling or Shaping Metals. 6d. December 31.

GENERATORS.—The current from a magneto or dynamo-electric generator is supplied to a system of secondary batteries, from which it is utilised for heating metal, during its passage through rolls or dies.

1882.

1,652.—J. J. Wheeler, Chelsea. Curbs of Roads and Footpaths. 4d. April 5.

CONDUCTORS. — The curb is moulded of any cement or other suitable substance, and a channel, or channels, of any suitable section is, or are, left along its entire length, along which may be laid conductors or tubes. At certain intervals the curb is made in two parts, an upper and a lower, held in place by dowels inserted in corresponding holes in the two parts. The curb may be formed of granite, bored out, or of iron, cast to shape.

NOTE.

The following description, by an oversight, was omitted from the abridgment of the specification of J. D. F. Andrews, page lxxxii., No. 2,321 of 1879.

CURRENT REGULATOR.—A solenoid having an iron core is introduced into the external circuit of a dynamo-electric generator driven by a steam or other engine. The core is connected to a throttle or other regulating valve of the engine, and regulates its speed according to the current passing through it. Instead of applying the apparatus to the regulating valve, it may be applied to vary the outlet of a fluid pumped by the engine to act on a piston connected to the regulating valve. This is applicable when the coils of the generator do not consist of parallel wires.

ERRATA.

No.	Year					No.	Year					No.	Year			
4,171	1873	For	Capanemb	read	Capanema.	325	1879	For	Paraine	read	Paraire.	1,217	1880	For	*Cebrian* read	*Celrian.*
4,346	1878	,,	Berthard	,,	Berthoud.	2,016	1879	,,	Saunders	,,	Sanders.	1,596	1881	,,	*Winklill* ,,	*Wiknlill.*
4,464	1878	,,	Buddeley	,,	Bulkeley.	4,087	1879	,,	*Berthaud*	,,	*Berthaut.*	2,618	1881	,,	Jamieson ,,	Jameeson.
4,699	1878	,,	Melbads	,,	Melhado.	4,846	1879	,,	Mourdat	,,	Mourlot.	3,635	1881	,,	Zubini ,,	Tubini.
5,011	1878	,,	Colme	,,	Cohné.	75	1880	,,	Reynier	,,	Roguier.	3,857	1881	,,	Snew ,,	Sueur.
277	1879	,,	,,	,,	,,	231	1880	,,	Long.	,,	Lang.	4,037	1881	insert	(*N. de Kabath, Paris.*)	
299	1879	,,	*Cebrian*	,,	*Celrian.*	1,184	1880	,,	Linfold	,,	Linford.	5,656	1881	for	*Ball* read	*Buell.*

INDEX.

—:-:—

NAMES OF PATENTEES.

Abstracts marked * will be found in the Addenda.

Name	No.	Year
Abbott, W	1,937	1878
Abbott, W., and Field, F.	5,407	1881
Abdank, B.	339	1882
Abdank, B., and Roosevelt, C.	3,070	1882
Abel (Brasseur, L. A., and Dejaer, O.)	4,296	1881
,, (Cie. Générale d'Eclairage Electrique)	1,553	1880
,, (Jablochkoff, P.)	1,745	1881
,, (Jamin, J. C.)	863	1879
,, (Krizik, F., and Piette, L.)	1,397	1880
,, (Otto, N. A.)	1,770	1878
,, (Sedlaczek, H., and Wiknlill, F.)	2,322	1879
,, (Tschikoleff, W., and Kleiber, H.)	2,198	1881
,, (Winter, W.)	1,264	1877
Achard, F. A.	1,387	1877
Achard, F. A.	2,453	1880
Adie, P.	5,211	1880
Akester, W. H.	5,525	1881
,,	1,642	1882
,,	2,419	1882
,,	2,519	1882
,,	*4,458	1881
Akester, W. H., and Barnes, T. B.	986	1882
Akester, W. H., and Scott, J.	1,412	1881
Aklem, F., Kayser, J., Tisdel, A. G., and Marx, E.	351	1880
Alamagny, La Société, et Oriel	866	1882
Alberger, M. H., and Pettit, S. W.	4,601	1878
Alder, G. E., and Clarke, J. A.	1,442	1878
Alexander (Bürgin, E.)	3,243	1875
,, (Lemaire-Douchy, A.)	2,107	1876
,, (White, J. C., and Hayden, H. H.	473	1881
,, (Zipernowsky, K.)	1,580	1880
Allen (Knudson, A., and Kane, F. L.)	1,295	1880
Allen, P. R.	2,215	1881
Alliance, Société l', J. Miot..	3,743	1877
Allison, R., and Hunter, W. J.	2,772	1879
Allison, W. C.	2,256	1881
Allport, C. J.	493	1882
,,	2,192	1882
Allport, C. J., and Punshon, R.	4,850	1881
Alteneck, F. H. von, and Siemens, W.	2,006	1873
Alteneck, F. H. von, and Siemens, E. W.	3,134	1878
Alteneck, H. von	4,949	1878
Alteneck, H., and F. von	2,652	1879
Anders, G. L., and Watson, T. A.	1,958	1880
Anderson, J., and Johnson, W. C.	2,311	1882
Anderson, J. E., and Nelson, L. (Paine, H. M., and E. L.)	2,049	1875
Anderson, R. C.	1,824	1880
André, E.	4,053	1877
André, G. G.	830	1879
,,	5,206	1879
,,	1,507	1880
,,	2,764	1880
,,	2,563	1881
,,	4,654	1881
,,	4,948	1881
,,	2,432	1882
André, G. G., and Easton, E.	2,833	1881
,, ,,	2,252	1882
André, J.	5,268	1880
Andrews, G.	292	1878
Andrews, J. D. F.	416	1879
,,	2,321	1879
,,	1,526	1881
,,	540	1882
,,	1,324	1882
,,	*2,321	1879
Antill, B. H.	1,787	1878
Applegarth (Jablochkoff, P.)	3,552	1876
Apps, A.	264	1881
Arbogast, P., and McTighe, T. J.	3,778	1879
Arey, A L.	3,456	1881
Arnaud, A.	4,074	1878
Aron, H.	2,943	1882
Aronson, J. N.	359	1882
,,	2,008	1878
,,	305	1882
Aronson, J. N., and Farnie, H. B.	4,163	1878
Arras, C. H. O. H. D'.	1,000	1873
Arthur, W.	2,128	1882
Ashworth, E. and G.	454	1882
Asten, J. C.	2,020	1882
Atkins, F. H.	556	1873
Auberville, Dupont- (Delaye, V.)	1,536	1881
Aubrey, A., Chauvin, F. M. A., and Goizet, L. H.	2,410	1875
Auriac, A. d'	1,683	1881
Avenarius, M.	3,025	1880
Aylesbury, H.	4,304	1881
Ayrton, W. E.	785	1881
Ayrton, W. E., and Perry, J.	783	1881
,, ,, ,,	2,613	1882
,, ,, ,,	2,642	1882
,, ,, ,,	2,830	1882
,, ,, ,,	3,036	1882
Baggeley, H.	166	1877
Ball, A., and Ward, D.	2,033	1878
Ball, C. E.	84	1882
Balukiewicz, T.	2,835	1880
Banta, W. E.	3,048	1882
Barda, B.	1,107	1881
Bardon, L.	3,079	1882
Barker (Jacques, W. W.)	3,424	1882
Barlow (Meritens, A. de)	2,212	1881
,,	4,207	1881
Barlow, W. A.	*1,897	1881
Barnes, T. B., and Akester, W. H.	986	1882
Barney, W. C.	2,264	1881
Barr, F.	455	1881
Barrier, J. J., and Lavernede, F. T. de	2,425	1882
Barrier, J. J., and Vernede, F. T. de la	538	1882
Bastet, L.	1,931	1876
Bauer, M., and Co.	1,915	1882
Bear, S. J. M.	3,283	1881
Beaumont, W. W., and Biggs, C. H. F.	5,198	1881
Beaumont, W. W., Fitzgerald, D. G., and Biggs, C. H. F...	5,338 / 29 / 1,875	1881 / 1882 / 1882
Beckingsale, E. W.	555	1877
,,	2,512	1882
,,	2,532	1882
Bedwell, F. L. B.	367	1876
Bell, A. G.	4,341	1877
,,	611	1878
Bell, J.	4,549	1879
Bell, J., and Scarlett, G.	4,555	1879
Bell, T. A.	4,403	1878
Benson, W. A.	4,254	1880
Berjot, A.	4,428	1880
Berly, J. A.	1,027	1881
,,	1,236	1881
Berly, J. A., and Hulett, D...	4,755	1880
,, (Maquare, F. V.)	2,885	1882
Bernstein, A.	2,604	1882
Berthaut, H. M. A.	4,087	1879
Berthoud, E., and Borel, F.	4,346	1878
Berthoud, Borel, et Cie.	4,026	1881
Bertin, E.	4,311	1875
Bertin, E., and Mersanne, F. E. de	5,044 / 5,053 / 5,076 / 5,110	1878 / 1878 / 1878 / 1878
Biggs, J. H. W.	2,106	1877
Biggs, C. H. F., and Beaumont, W. W.	5,198	1881
Biggs, C. H. F., Beaumont, W. W., and Fitzgerald, D. G.	5,338 / 29 / 1,875	1881 / 1882 / 1882
Biloret, A., and Mora, C.	4,049	1880
Binswanger, G.	2,390	1882
Binswanger, G., and Rhodes, B.	2,501	1882
Binko, H.	3,073	1882
Birkhead, M.	988	1879
Blagburn, C., and Harrison, R.	1,358	1881
Blake, F.	5,096	1881
Blamires, T. H.	455	1880
Blandy, A. F.	2,060	1879
Blondot, A., and Bourdin, J.	2,629	1879
Bloomfield, J. H.	3,679	1879
Blyth, J., and Peebles, D. B	2,661	1882
Bodmer, G. R.	4,476	1878
Boettcher, E.	144	1882
Boggett, W.	2,505	1882

	No.	Year
Bolton, F. J., and Webber, C. E.	686	1873
Bonneville (*Chutaux, T.*)	4,454	1874
,, (*Daft, L.*)	4,775	1881
,, (*Radde, W.*)	2,091	1873
Boothby, A. C., and Siemens, C. W.	696	1881
Borel, F., and Berthoud, E.	4,346	1878
Borel, Berthoud et Cie.	4,026	1881
Bossomaier, R., and Schwegler, F.	2,823	1879
Bouliguine, N.	1,968	1881
Boult (*Thomas, J. D.*)	1,649	1882
,, (*Thomas, J. W., and Requa, L. F.*)	1,336	1882
Boulton, W.	805	1877
Bourdin, J., and Blondot, A.	2,629	1879
Bourdin, J., and Maltzoff, S. A. de	1,474	1881
Bousfield (*Smith, H. J.*)	380	1874
Bousfield, W. R. and E. T.	523	1879
Bouteilloux, L. J., and Laing, W.	842	1880
,, ,,	3,214	1881
Bowman, R., and Hawkes, G.	2,402	1881
Boys, C. V.	2,449	1881
,,	4,472	1881
,,	513	1882
,,	1,946	1882
Boys, C. V., and Liveing, E. H. T.	69	1882
Brain, W. B.	5,139	1878
,,	1,548	1882
,,	1,616	1882
,,	2,659	1882
Brasseur, L. A., and Dejaer, O.	4,296	1881
Brasseur, L. A., and Sussex, S. W. M. de	308	1878
Brear, S., and Hudson, A.	5,032	1881
Bremner, D.	1,345	1882
Brewer (*Buell, C. E.*)	5,656	1881
,, (*Cance, A. J. B.*)	4,005	1880
,, (*Chambrier, E. A.*)	4,428	1879
,, (*Delany, P. B., and Johnson, E. H.*)	4,093	1881
,, (*Desquiens, A. G.*)	3,404	1881
,, (*Edison, T. A.*)	4,502	1878
,, ,,	3,765	1880
,, ,,	539	1881
,, ,,	768	1881
,, ,,	1,016	1881
,, ,,	1,240	1881
,, ,,	1,783	1881
,, ,,	1,918	1881
,, ,,	1,943	1881
,, ,,	2,482	1881
,, ,,	2,495	1881
,, ,,	3,231	1881
,, ,,	3,483	1881
,, ,,	4,174	1881
,, ,,	4,571	1881
,, ,,	4,576	1881
,, (*Facio, E. E. S.*)	3,462	1876
,, (*Johnson, E. H., and Edison, T. A.*)	4,621	1880
,, (*Kosloff, S. A.*)	2,767	1875
,, (*Waterhouse, A. G.*)	5,185	1881
Brewtnall, A. W.	5,226	1881
,,	2,934	1882
Bridges (*Heyer, L.*)	3,885	1879
Briggs, J. A., and Kinsman, F.	1,774	1882
Bright, Sir C. T.	4,212	1878
,,	3,305	1881
,,	377	1882
,,	2,602	1882
Bright, E. B.	596	1878
Brockie, J.	3,771	1879
,,	1,942	1881
,,	4,504	1881
,,	756	1882
,,	898	1882
,,	1,713	1882
,,	2,370	1882
Brooks, D.	4,824	1877
,,	3,254	1881
Brougham (*Sabatou, C.*)	469	1880
Brougham, Hon. R. T. D.	630	1880
,, ,,	832	1880
,, ,,	1,302	1882
,, ,,	2,030	1882

	No.	Year
Brougham, Hon. R. T. D.	2,044	1882
Brougham, Hon R., and Ormiston, F. A.	1,697	1882
Brown, A. B.	1,867	1882
Brown, A. B., and King, W. F.	5,272	1881
Brown, A. E.	4,011	1881
Brown, J. W., and Culbertson, J. N.	5,615	1881
Brown, W. Morgan (See Morgan Brown, W.)		
Browne (*Kimball, D. F.*)	3,999	1875
Brownell, H. T.	2,302	1873
Brush, C. F.	2,003	1878
,,	947	1879
,,	3,750	1879
,,	849	1880
,,	*1,834	1881
Budenberg (*Budenberg, C. F., and Schäffer, B. A.*)	4,227	1880
Budenberg, C. F., and Schäffer, B. A.	4,227	1880
Buell, C. E.	5,656	1881
Bulkeley, F. B.	4,464	1878
Bull, H., and Harding, E. J.	2,878	1878
Bullivant, W. M.	1,159	1874
Bürgin, E.	3,243	1881
,,	5,085	1879
,,	4,819	1881
,,	4,820	1881
Bureau, A.	1,704	1880
Burrell, S. J.	3,679	1881
Busscher, W. de	4,812	1881
Byllesby, H. M., and Stern, W. A.	2,336	1882
Byshe, H. C.	4,961	1878
Cabanellas, J. E.	200	1881
Cables Electriques, Société Anonyme des	1,496	1882
Cadett, J. W. T.	4,022	1878
,,	4,316	1878
Callender, W. O.	4,409	1881
Camacho, J. S.	3,461	1873
,,	3,416	1875
Cance, A. J. B.	1,927	1878
,,	4,005	1880
,,	3,976	1881
Capanema, G. S. de	4,171	1873
Cardew, P.	5,354	1881
Carpentier, J.	4,664	1881
Carpentier, J., and Deprez, M.	4,128	1881
Carpentier, C., and Pezzer, O. de	5,322	1881
Carus Wilson, C. A.	5,623	1881
,,	5,687	1881
,,	4,824	1882
,,	4,825	1882
,,	2,954	1882
Castro, J. W. de	2,943	1878
Cathcart, C. H.	2,068	1882
Celrian, J., and Molera, E.	299	1879
Celrian, J. C., and Molera, E. J.	1,217	1880
Chadburn, W.	2,755	1882
Chambrier, E. A.	4,428	1879
Chameroy, B. H.	2,295	1882
Changy, J. de	4,405	1881
,,	4,582	1881
Chaperon, G. (Partly)	1,462	1882
Chauvin, F. M. A., Goizet, L. H., and Aubrey, A.	2,410	1875
Cheesbrough (*Eaton, A. K.*)	751	1880
,,	2,826	1882
,, (*Sawyer, E.*)	3,587	1879
,, (*Sawyer, E., and Man, A.*)	4,705	1878
,, (*Sawyer, E., and Man, A.*)	4,847	1878
Chertemps, D. A.	3,349	1881
Chertemps, D. A., and Dandeu, L.	1,747	1882
Chichester, S., and Moffatt, R. R.	3,441	1881
Chinnock, C. E., and Harrison, J. de H.	695	1880
,, ,,	699	1880
Chislett, J. R.	87	1873
Choate, S. F. van	4,388	1878
Chretien, J., and Felix, C.	2,019	1879
Christensen, F. S., and Reimenschneider, A.	4,693	1878

	No.	Year
Chubb, C. J.	761	1882
Church (*King, J. B.*)	1,822	1882
Chutaux, T.	4,454	1874
Clamond, C.	2,205	1875
Clark (*Auriac, A. d'*)	1,683	1881
,, (*Bastet, L.*)	1,931	1876
,, (*Bertin, E.*)	4,311	1875
,, (*Bouteilloux, L. J., and Laing, W.*)	842	1880
,, (*Bouteilloux, L. J., and Laing, W.*)	3,214	1881
,, (*Changy, J. de*)	4,405	1881
,, ,,	4,582	1881
,, (*Clemandot, L.*)	4,031	1878
,, (*Davis, C.*)	4,407	1878
,, (*Desnos, C. J. P.*)	2,340	1879
,, (*Drew, P.*)	2,037	1880
,, (*Ducretet, E.*)	65	1879
,, (*Fournier, G.*)	5,104	1881
,, (*Garnier, J.*)	4,952	1880
,, (*Gérard-Lescuyer, (J. M. A.*)	3,697	1879
,, ,, ,,	1,552	1880
,, ,, ,,	1,685	1881
,, ,, ,,	2,676	1882
,, (*Guest, J. H.*)	2,980	1880
,, (*Haddan, W.*)	3,843	1879
,, (*Hermann, L. A.*)	2,962	1879
,, (*Hussey, C. A.*)	2,043	1875
,, (*Kabath, N. de*)	4,037	1881
,,	4,060	1881
,, (*Laing, W.*)	3,169	1881
,, (*Lamar, J. S.*)	4,696	1879
,, (*Lartigue, H.*)	3,771	1874
,, (*Lontin, D. F.*)	473	1875
,,	386	1876
,,	3,264	1876
,, (*Lontin & Co.*)	2,094	1877
,,	4,893	1877
,, (*Marcus, S., and Egger, B.*)	2,934	1877
,, (*Meritens, A. de, & Co.*)	2,339	1879
,, (*Müller, H. J., and Levett, A.*)	1,787	1881
,, (*Niaudet, A., and Reynier, E.*)	3,971	1880
,, (*Pilleux, C. L.*)	636	1880
,, (*Planté, G.*)	1,713	1873
,, (*Sheridan, H.B.*)	4,617	1881
,,	931	1882
,, (*Smith, P. E., Spruit, S. R., and Wood, W. R.*)	4,312	1876
,, (*Solignac et Cie.*)	740	1882
,, (*Reynier, E.*)	2,982	1877
,, (*Reynier, N. E.*)	1,971	1879
,, (*Reynard, L.*)	5,165	1878
,, (*Roguier, L.*)	75	1880
,, (*Ward, D., and Ball, A.*)	2,033	1878
Clark, H. A.	229	1881
,,	2,592	1881
,,	361	1882
Clark, J.	3,991	1878
,,	203	1880
Clark, J. L., and Stanfield, J.	*4,918	1881
Clarke, C. L., and Leigh, J.	245	1881
,, ,,	3,652	1881
,, ,,	1,483	1882
Clarke, J. A., and Alder, G. E.	1,442	1878
Clarke, T., and Smith E.	4,650	1878
Clemandot, L.	4,031	1878
Clifford, H.	312	1878
Clingman, T. L.	1,840	1880
Cochrane, A. A.	4,313	1878
Coke, A. L.	1,012	1879
Cohne, S.	5,011	1878
,,	277	1879
,,	2,236	1880
,,	2,369	1881
,,	1,437	1882
Commelin, E., and Poulet, V.	1,046	1880
Common, A. A.	626	1882
Common, A. A., and Joel, H. F.	1,040	1881
Compagnie Electrique, La	2,990	1882
Concornotti, L.	3,272	1879
Connolly, T. A.	3,668	1881
Conradi (*André, E.*)	4,053	1877
,, (*Wiebe, R.*)	767	1878
Conybeare, H, and Naphegyi, G.	2,106	1874
Cook, H. C.	2,717	1874
Cook, H. W.	2,769	1879

	No.	Year
Cook, H. W.	1,109	1880
Cooke, C. W.	1,903	1873
Corbett, J. L., and Lockhead, W.	219	1880
Cordeaux, J. H.	522	1877
Cougnet, J.	2,000	1879
Cougnet, J., and Puydt, J. P. C. de	350	1880
Cour, P. la	1,988	1878
Courtenay, R. H.	1,450	1873
,,	1,487	1875
,,	3,543	1879
,,	2,770	1881
,,	4,659	1881
,,	3,101	1882
Courtenay, R. H., and Moore, S. J.	3,078	1873
Coxeter, S. J.	492	1878
Crandall, J. N., and Fuller, J. B.	3,364	1875
	1,557	1876
Crastin, C.	4,771	1881
Crighton, C. E.	4,696	1878
,,	4,189	1879
Crompton, R. E. B.	3,509	1879
,,	5,080	1881
,,	346	1882
,,	2,618	1882
,,	2,619	1882
Crompton, R. E. B., and Fitzgerald, D. G.	5,159	1881
Crompton, R. E. B., and Willans, P. W.	245	1879
Crookes, W.	1,422	1881
,,	2,612	1881
,,	3,799	1881
,,	2,304	1881
,,	1,079	1882
Crossley, L. J., Harrison, J. F., and Emmott, W.	1,327	1882
Cruto, A.	1,895	1882
Cuff, J. C.	2,263	1881
Culbertson, J. M., and Brown, J. W.	5,615	1881
Cumine, J. A.	2,318	1882
Cummings, C.	2,880	1880
Curtoys, C.	1,851	1882
Cuttris (*Cuttriss, C.*)	2,135	1882
Cuttris, C. (Partly)	2,135	1882
Daft, L.	4,775	1881
Danckwerth, L.	1,704	1876
Danckwerth, L., and Sanders, F.	2,016	1879
Dandeu, L., and Chertemps, D. A.	1,747	1882
Darlow, W., and Fairfax, H.	3,736	1873
D'Arras, C. H. O. H.	1,000	1873
Davis (*Marx, E., Aklem, F., Kayser, J., and Tisdel, A. G.*)	351	1880
Davis, C.	4,407	1878
Davis, C.	4,559	1878
Day, M.	1,261	1874
Debenham, W. E.	3,010	1882
Defries, C.	2,335	1882
Defty, H.	2,628	1882
Deiss, A., and Scaife, R.	2,866	1876
Dejaer, O., and Brasseur, L. A.	4,296	1881
Delany, P. B.	1,809	1881
,,	2,217	1881
,,	333	1882
Delany, P. B., and Johnson, C. H.	2,532	1881
	4,093	1881
Delaye, V.	1,536	1881
Denayrouse, L.	3,170	1877
Denne, T. J. and M. C.	5,404	1880
Deprez, M., and Carpentier, J.	4,128	1881
Dering, G. E.	5,123	1878
Desnos, C. J. P.	2,340	1879
Despeissis, L. H., and Minchin, G. M.	1,010	1882
Desquiens, A. G.	3,404	1881
Dessaigne, G.	4,591	1881
Detrick, C.	3,388	1881
Dewar, J.	2,886	1876
Dibbin, H. A.	4,036	1877
Dillon, T. A.	1,207	1879
,,	1,347	1879
Dion, C.	3,880	1881
Dodd, A. S., and Hussey, C. A.	4,265	1880
,, ,,	2,375	1881
,, ,,	2,572	1881
,, ,,	2,573	1881
Dodd, A. S., and Hussey, C. A.	234	1882
Dolbear, A. E.	1,368	1882
Donnithorne, T., and Liardet, J. E.	5,216	1881
	5,418	1881
	120	1882
Douchy, A. Lemaire-	2,107	1876
Dove, J.	1,158	1879
Drew, P.	2,037	1880
Dubos, C.	2,401	1878
,,	427	1879
,,	749	1879
Ducretet, E.	65	1879
Dudley, J. G.	3,423	1881
Dunlop, J. M., and Hooper, W.	3,997	1873
Dunstan, R. E., and Pfannkuche, G.	3,655	1881
	5,743	1881
Dupont-Auberville (*V. Delaye*)	1,536	1881
Dutton (*Reese, F.*)	51	1879
Duwelius, A. L., Goss, L. W., Higgs, P., Merrell, F. R., Peck, H. D., and Walter H.	5,751	1881
Earl, H. D., and Thompson, A. M.	5,281	1878
Easton, E., and André, G. G.	2,252	1880
,, ,,	2,833	1881
Eaton, A. K.	751	1880
,,	2,826	1882
Eclairage Electrique, Cie. Générale d'	1,553	1880
Edison, T. A.	3,762	1875
,,	4,226	1878
Edison, T. A.	4,502	1878
Edison, T. A.	5,306	1878
,,	2,402	1879
,,	4,576	1879
,,	5,127	1879
,,	33	1880
,,	578	1880
,,	602	1880
,,	1,385	1880
Edison, T. A.	3,765	1880
,,	3,880	1880
,,	3,894	1880
,,	3,964	1880
,,	4,391	1880
,,	539	1881
,,	562	1881
,,	768	1881
,,	792	1881
,,	1,016	1881
,,	1,240	1881
,,	1,783	1881
,,	1,802	1881
,,	1,918	1881
,,	1,943	1881
,,	2,482	1881
,,	2,492	1881
,,	2,495	1881
,,	2,954	1881
,,	3,231	1881
,,	3,483	1881
,,	3,804	1881
,,	3,932	1881
,,	4,034	1881
,,	4,174	1881
,,	4,552	1881
,,	4,553	1881
,,	4,571	1881
,,	4,576	1881
,,	1,023	1882
,,	1,139	1882
,,	1,142	1882
,,	1,191	1882
,,	1,496	1882
,,	1,862	1882
,,	2,052	1882
,,	2,072	1882
Edison, T. A., and Johnson, E. H.	4,621	1880
Edmonds (*Mowbray, G. M.*)	2,437	1881
,, ,,	*190	1881
Edmunds, H., Sellon, J. S., and Ladd, W.	4,645	1878
	4,646	1878
Edmunds, H., and Sellon, J. S.	1,692	1879
,, ,, ,,	1,791	1879
,, ,, ,,	1,949	1879
Edwards, E., and Normandy, A.	4,611	1878
Edwards, J., and Leek, E.	2,640	1878
Edwards, J. R.	1,720	1880
Egger, B., and Marcus, S.	2,934	1877
Electric Manufacturing Co., Union	392	1882
Electricité, Société Générale d'	4,066	1878
,, ,, ,,	1,097	1881
,, ,, ,,	1,653	1881
,, ,, ,,	2,323	1881
,, ,, ,,	4,496	1881
Electricité, Société Générale d', Procédés Jablochkoff	1,175	1879
	725	1880
Electricité, Société Universelle d', Tommasi	2,782	1881
	4,057	1881
	252	1882
Electrique, La Cie.	2,990	1882
Electrique, La Cie. Générale d'Eclairage	1,553	1880
Electriques, La Société Anonyme des Cables, Système Berthoud, Borel et Cie,	4,026	1881
Electriques, Société Anonyme des Câbles	1,496	1882
Elmore, J.	922	1881
Elmore, W.	3,565	1879
,,	4,321	1879
,,	3,832	1880
Elphinstone, Baron, and Vincent, C. W.	332	1879
	2,893	1880
Elphinstone, Lord (Partly)	2,340	1882
Elphinstone, W. B. F.	2,491	1882
Elwell, P. B., and Parker, T.	2,917	1882
Emmens, S. H.	2,348	1882
,,	2,349	1882
,,	2,912	1882
,,	2,913	1882
,,	2,914	1882
Emmott, W., Crossley, L. J., and Harrison, J. F.	1,327	1882
Engel (*Franke, G.*)	849	1882
,, (*Müller, C. H. F.*)	3,711	1881
Epstein, L.	2,807	1882
Espeut, W. B.	1,257	1881
Eteve, E.	48	1881
Eteve, E., and Lallement, C.	*3,113	1881
Eustace, M.	1,766	1874
,,	3,172	1874
Evans, M.	1,970	1873
,,	1,225	1882
Ewen, F. W., and James, G. F.	1,719	1875
Facio, E. E. S.	3,462	1876
Fahrig, F. E.	4,107	1881
Fairfax, H., and Darlow, W.	3,736	1873
Farnie, H. B., and Aronson, J. N.	4,163	1878
Farquharson, J.	2,771	1882
Faucher, F.	2,750	1876
Faulkner, J.	1,800	1875
Faure, C. A.	2,946	1875
,,	3,670	1876
Faure, C. A.	129	1881
,,	1,676	1881
,,	4,311	1881
,,	730	1882
,,	1,769	1882
Felix, C., and Chretien, J.	2,019	1879
Field, F.	2,480	1882
Field, F., and Abbott, W.	5,407	1881
Field, F., and Talling, R.	1,938	1875
Fielding, J.	994	1882
Finch, G. B.	1,110	1879
Fisher, J., and Gassett, O.	3,768	1881
Fisher, W.	1,727	1882
Fitzgerald, D. G.	872	1880
,,	5,275	1880
,,	3,890	1881
,,	5,481	1881
,,	2,763	1882
Fitzgerald, D. G., and Crompton, R. E. B.	5,159	1881
Fitzgerald, D. G., Biggs, C. H. W., and Beaumont, W. W.	5,338	1881
	29	1882
	1,875	1882
Fixsen (*Danckwerth, L.*)	1,704	1876
Fleming, J. A.	1,762	1881
,,	5,309	1881
,,	2,414	1882
Fleming, J. A., Ranyard, A. C.	2,807	1881
Fletcher, J. W.	3,936	1880
,,	859	1881
Floyd, T., and Kirkland, T.	2,225	1882

	No.	Year
Floyd, T., and Probert, J.	2,226	1882
Fontaine, H.	1,180	1873
Fonvielle, W. de	1,339	1880
Forbes, G.	4,116	1878
„	1,097	1881
Force, Société Anonyme la,	1,653	1881
et la Lumière	2,323	1881
	4,496	1881
Formby, J.	565	1879
Fottrell, J.	3,086	1873
Fournier, G.	5,104	1882
Fox, E. M.	4,024	1881
„	4,383	1881
Fox, St. G. L.	3,988	1878
„	4,043	1878
„	4,626	1878
„	1,122	1879
„	3,494	1880
„	225	1881
„	1,543	1881
„	1,636	1881
„	3,122	1881
„	3,394	1881
„	5,651	1881
„	1,647	1882
Franke, G.	849	1882
Freeman, J. B. F.	5,307	1878
Freeman, J., and Young, F.	350	1879
Frome, C. H., and Gibbs, G. C.	3,416	1880
Fuller, J. B.	5,183	1878
Fuller, J. B., and Crandall,	1,557	1876
J. N.	3,364	1875
Fuller, J. C. and G.	76	1877
Furstenburgh (Siemens, L.)	2,199	1879
Fyfe, A. L., and Main, J.	3,821	1881
„	2,636	1882
Fyfe, J. „	774	1881
Gadot, P. L. M.	2,314	1881
Gardner (Roe, E.)	3,010	1881
Garnier, J.	4,952	1880
Gary, W. W.	805	1879
„	4,069	1881
„	4,070	1881
Gassett, O., and Fisher, J.	3,768	1881
Gatehouse, T. E.	4,796	1879
„	3,240	1881
„	1,400	1882
Gatehouse, T. E., and Kempe,		
H. R.	2,569	1882
Gaulard, L., and Gibbs, J. D.	4,942	1881
Gaumé, C.	2,618	1873
George, A. F. St.	2,193	1878
„	4,989	1881
George, E., and Morgan, J. B.	4,482	1880
Gerard-Lescuyer, J. M. A.	3,697	1879
„ „	1,552	1880
„ „	1,685	1881
„ „	4,792	1881
„ „	5,593	1881
„ „	5,660	1881
„ „	1,878	1882
„ „	2,144	1882
„ „	2,676	1882
„ „	2,992	1882
Gibbs, G. C., and Frome, C. H.	3,416	1880
Gibbs, J. D., and Gaulard, L.	4,942	1881
Gibbs, R. R.	4,533	1881
Gilbert, A., and Wells, G.	1,510	1880
„	1,960	1881
Gilbert, A. E.	3,438	1880
„	4,507	1881
Gimingham, C. H.	2,079	1881
„	4,193	1881
„	2,375	1882
Gloker, J.	2,674	1882
Glouchoff, N.	478	1880
Glover, T. G.	855	1877
Glover, W. T., and James, G. F	5,237	1880
„ „	913	1881
Godfrey, W. B.	4,718	1879
Goebel, H., and Kulenkamp		
J. W.	*5,548	1881
Goizet, L. H., Aubrey, A., and		
Chauvin, F. M. A.	2,410	1875
Goldstone, G., Radcliffe, J.,		
and Gray, M.	3,381	1874
Goldstone, C., Radcliffe, J.,		
Gray, M., and Preece, W. H	3521	1874
Gollner, D.	131	1878
Gordon, J. E. H.	1,826	1880
„	4,745	1880

	No.	Year
Gordon, J. E. H.	76	1881
„	218	1881
„	5,536	1881
„	2,871	1882
Gore, C. L.	4,797	1881
Goss, L. W. Higgs, P., Mer-		
rell, F. R., Peck, H. D.,		
Walter, H., and Duwelius,		
A. L.	5,751	1881
Gouraud (Delany, P. B., and		
Johnson, C. H.)	2,532	1881
Gower, F. A.	814	1880
Graddon, J.	885	1880
Graham, A.	1,171	1882
Graham, D.	3,073	1881
„	3,274	1881
„	5,618	1881
Graham, W., and Smith, H. J.	1,392	1882
Gramme, Z. T., and Ivernois,		
E. L. C. d'	953	1878
Grant, G., and Sennett, R.	2,267	1879
Gravier, A.	2,739	1881
„	5,295	1881
„	943	1882
„	1,211	1882
Gray (Reynier, N. E.)	471	1881
Gray, M.	3,862	1873
„	3,863	1873
„	4,553	1878
„	5,056	1879
Gray, M., Goldstone, G., and		
Radcliffe, J.	3,381	1874
Gray, M., Preece, W. H., Gold-		
stone, C., and Radcliffe, J.	3,521	1874
Greb, J. J. C. W.	3,015	1881
Greening, F.	2,059	1875
Greening, F., and Jack, W. F.	718	1879
Greenwood, H. B., and Varley,		
T.	2,248	1882
Grimstone, G. S.	1,670	1881
Griscom, W. W.	1,244	1880
„	1,259	1880
„	5,551	1881
Grieve, T. W.	259	1879
Griffin, W., Mori, F., Halle-		
well, C. E., and Milner, W.	740	1879
Groombridge, C.	1,049	1880
Groth (*Goebel, H., and Kulen-*		
kamp, W. W.)	*5,548	1881
„ (*Gülcher, R. J.*)	64	1882
„ (*Lachaussée, D.*)	2,761	1881
Grout, G., and Jones, W. H.	5,521	1881
Guest, J. H.	925	1880
Guest, J. H.	2,980	1880
Gülcher, R. J.	2,038	1881
„	64	1882
Gülcher, R. J.	2,875	1882
Gumpel, C. G.	3,041	1880
„	3,324	1880
„	253	1881
„	2,688	1882
„	2,723	1882
„	2,756	1882
Gye, F.	4,473	1878
Haase, R., and Recker, J. P.	3,832	1881
Haddan (*Boettcher, E.*)	144	1882
„ (*Brush, C.*)	2,003	1878
„ (*Brush, C. F.*)	947	1879
„ „	3,750	1879
„ „	849	1881
„ „	1,835	1881
„ „	*1,834	1881
„ (*Bureau, A.*)	1,704	1880
„ (*Busscher, W. de*)	4,812	1881
„ (*Cummings, C.*)	2,888	1880
„ (*Dessaigne, G.*)	4,591	1881
„ (*Gülcher, R. J.*)	2,038	1881
„ (*Molera, E., and*		
Cebrian, J.)	299	1879
„ (*Morel, A.*)	185	1882
„ (*Rosebrugh, A. N.*)	1,476	1879
„ (*Somzée, L*).	1,852	1881
„ „	4,305	1881
„ (*Townsend, J. D.*)	39	1881
„ (*Weston, E.*)	4,280	1876
„ (*Wheeler, W.*)	917	1882
„ (*Woolley, L. G.*)	*3,049	1881
Haddan, J. L.	2,040	1879
Hadden, W.	3,843	1879
Hallet, S.	4,017	1881
„	2,560	1882

	No.	Year
Hallet, S.	2,573	1882
Hallewell, C. E., Milner, W.,		
Griffin, W., and Mori, F.	740	1879
Halske and Siemens	1,447	1881
Hamel, F. J. de	2,543	1879
Handford (*Delany, P. B.*)	333	1882
„ (*Edison, T. A.*)	1,191	1882
„ „	1,496	1882
„ „	1,862	1882
„ „	2,052	1882
„ „	2,072	1882
„ „	1,023	1882
„ „	1,139	1882
„ „	1,142	1882
„ (*Stern, W. A., and*		
Byllesby, H. M.)	2,336	1882
Harborow, H. A.	3,871	1881
Harding, E. J., and Bull, H.	2,878	1878
Harding, G. P.	4,046	1878
„	4,047	1878
„	4,590	1879
„	783	1879
„	4,191	1880
„	4,192	1880
Harding, G. P.	3,166	1881
Harling, E. J., and Hart-	3,472	1881
mann, E.	3,473	1881
Harmant, F.	4,779	1880
Harrington, F. W., Lane,		
T. W., and Williams, C.	3,539	1881
Harrison, C. W.	3,623	1876
„	3,470	1878
„	4,338	1878
„	3,875	1879
„	886	1880
„	3,496	1880
„	3,559	1881
Harrison, J. F., Crossley, L.		
J., and Emmott, W.	1,327	1882
Harrison, J. de H., and	695	1880
Chinnock, C. E.	699	1880
Harrison, R.	4,478	1881
Harrison, R., and Blagburn, C.	1,358	1881
Harrop, J. J.	485	1874
„	3,793	1879
Hartmann, E., and Harling,	3,472	1881
E. J.	3,473	1881
Harvey, E. W., W., and A.	369	1877
Haseltine (Cook, H. C.)	2,717	1874
Haskins, D. G.	3,016	1878
Hatton, R. J., and Young,		
G. S	1,689	1882
Hatton, R. J., and Paul, A. L.	2,654	1882
Hawkes, G.	157	1881
Hawkes, G., and Bowman, R.	2,402	1881
Hayden, H. H., and White,		
J. C.	473	1881
Haymen, H.	959	1879
Heaviside, O.	1,407	1880
Hedges, K. W.	81	1879
„	925	1879
„	4,988	1880
„	3,369	1881
Heinke, F. W.	1,910	1877
„	4,275	1877
Heinke, F. W., and Lang, G.	231	1880
Heinrichs, C. F.	4,595	1878
„	2,317	1879
„	4,589	1879
„	4,608	1880
Heins, J.	4,916	1879
Henck, J. B.	4,058	1881
Henley, W. T.	4,115	1875
„	1,944	1876
„	833	1877
„	5,137	1880
„	1,873	1881
„	130	1882
„	386	1882
Hequet, T. A.	2,564	1875
Hermann, L. A.	2,962	1879
Herz, C.	93	1881
Heyer, L.	3,885	1879
Hibell, W.	4,159	1876
Hickisson, J.	2,421	1882
Hickley, A. S.	3,552	1877
„	4,132	1878
„	4,354	1879
Higgins, F. H. W.	4,456	1878
Higgs, R. P.	4,206	1878
Higgs, R. W. H. P.	454	1879
Higgs, P.	1,961	1881

Name	No.	Year
Higgs, P., Merrell, F. R., Peck, H. D, Walter, H., Duwelius, A. L., and Goss, L. W.	5,751	1881
Highton, H.	1,178	1873
"	4,277	1873
"	3,006	1874
Hill, W. S.	3,893	1881
Holcombe, A. G.	1,384	1881
Holmes, B. G.	920	1879
Hooper, J. P.	2,944	1881
Hooper, W.	3,780	1873
Hooper, W., and Dunlop, J. M.	3,997	1873
Hopkinson, J.	1,959	1879
"	2,481	1879
"	4,653	1879
"	3,509	1880
"	2,989	1881
"	3,362	1881
"	49	1882
Hopkinson, J., and Muirhead, A.	4,886	1880
	153	1881
	1,577*	1881
Hosmer, H. G.	311	1878
"	2,930	1878
"	3,676	1878
Hosmer, H. G.	4,220	1880
Houston, E. J., and Thomson, E.	4,400	1879
	315	1880
Howard, C. T.	2,896	1882
Hubble (Gerard-Lescuyer, J. M.A.)	4,792	1881
" (Partz, A. F. W.)	3,455	1881
Hudson, A., and Brear, S.	5,032	1881
Hughes, R. H.	3,190	1881
Hughes (Wallace, W.)	240	1878
Hulett, D., and Berly, J. A.	4,755	1880
Humphrys, J.	289	1882
Hunebelle, J.	1,856	1881
Hunt (Alberger, M. H., and Pettit, S. W.)	4,601	1878
, (Brooks, D.)	4,824	1877
" (Brown, A. E.)	4,011	1881
" (Reynoso, A. F. C.)	799	1873
Hunter, W. J., and Allison, R.	2,772	1879
Hussey, C. A.	2,043	1875
Hussey, C.A., and Dodd, A.S.	4,265	1880
" "	2,375	1881
" "	2,572	1881
" "	2,573	1881
" "	234	1882
Imray, J.	382	1879
Imray (Cabanellas, J. E.)	200	1881
" (Carpentier, J.)	4,664	1881
" (Carpentier, C., and Pezzer, O. de)	5,322	1881
" (Deprez, M., and Carpentier, J.)	4,128	1881
" (Herz, C.)	93	1881
" (Jablochkoff, P.)	2,769	1882
" (La Société Générale d'Electricité)	4,066	1878
" (La Société Anonyme des Cables Electriques)	1,496	1882
" (La Société Générale d' Electricité Procédés Jablochkoff)	1,175	1879
	725	1880
" (McTighe, J.J. and T.J.)	2,744	1882
" (Stearns, J. M.)	5,468	1881
" (Waters, T. J.)	862	1874
Isaac (Vogel, Sir J.)	3,033	1882
Iverneau, L. E., and Lambert, V. P.	144	1880
Ivernois, E. L. C. d', and Gramme, Z. T.	953	1878
Jablochkoff, P.	836	1876
"	3,552	1876
"	494	1877
"	1,996	1877
"	3,187	1877
"	3,839	1877
"	1,745	1881
"	2,769	1882
Jablochkoff, P., La Société Générale d'Electricité	1,175	1879
	725	1880
Jack, W. F., and Greening, F.	718	1879
Jacobson, L.	3,053	1881
Jacques, W. W.	3,424	1880
James, G. F.	4,261	1874
"	624	1880

Name	No.	Year
James, G. F.	1,333	1880
James, G. F., and Ewen, F.W.	1,719	1875
James, G. F., and Glover, W. T.	5,237	1880
"	913	1881
James, J., and Lee, J. C. F.	4,396	1881
James, J., and Preece, W. H.	129	1882
Jameson, J.	2,618	1881
"	4,439	1881
"	1,670	1882
Jamieson, A., and Saunders, H. A. C.	1,416	1877
Jamin, J. C.	863	1879
Jarman, A. J.	563	1882
"	2,565	1882
"	2,630	1882
Jaspar, J.	83	1879
Jeffery, W.	1,570	1882
Jenkin, F.	1,830	1882
"	3,007	1882
Jennings, R. S.	1,699	1881
Jensen (Avenarius, M.)	3,025	1880
" (Cance, A. J. B.)	1,927	1878
"	3,976	1881
" (Edison, T. A.)	3,880	1880
"	3,894	1880
"	3,964	1880
"	4,391	1880
"	562	1881
"	792	1881
"	1,802	1881
"	2,492	1881
"	2,954	1881
"	3,804	1881
"	3,932	1881
"	4,034	1881
"	4,552	1881
"	4,553	1881
" (Konn, S. V.)	970	1875
" (Marcus, S.)	4,006	1878
" (Prall, W. E., and Obrick, H.)	3,831	1879
" (Schuyler, D.A., and Waterhouse, F. G.)	3,002	1882
Jeune, A. M. J.	774	1882
Joel, H. F.	5,157	1879
"	4,607	1881
Joel, H. F., and Common, A.A.	1,040	1881
Johnson (Anders, G. L., and Watson, T. A.)	1,958	1880
" (Bardon, L.)	3,079	1882
" (Berjot, A)	4,428	1880
" (Berthaut, H. M. A.)	4,087	1881
" (Bertin, E., and Mersanne, F. E. de).	5,076	1878
	5,110	1878
" (Camacho, J. S.)	3,416	1875
" (Faure, C. A.)	129	1881
"	1,676	1881
"	4,311	1881
"	1,769	1882
" (Fontaine, H.)	1,180	1873
" (Gerard-Lescuyer, J. M. A.)	1,878	1882
"	2,144	1882
" (Gramme, Z. T., and Ivernois, E. L.C. d')	953	1878
" (Gower, F. A.)	814	1881
" (Griscom, W. W.)	1,244	1880
"	1,259	1880
"	5,551	1881
" (Labye, C., and Locht-Labye, C.)	5,661	1881
" (La Cie. Electrique)	2,990	1882
" (La Société Anonyme la Force et la Lumière, Société Générale d' Electricité)	1,097	1881
	1,653	1881
	2,323	1881
	4,496	1881
" (La Société l'Alliance, J. Miot)	3,743	1877
" (Meritens, A. de)	3,658	1878
"	4,690	1878
"	5,044	1878
"	5,257	1878
"	1,136	1880
"	5,033	1880
"	1,173	1882
" (Mersanne, F. E. de)	1,446	1874
"	2,787	1875
"	3,315	1876
" (Mersanne, F. E. de, and Bertin, E.)	5,053	1878
" (Mignon, J. B. V., and Rouart, S. H.)	3,400	1881
	3,402	1881

Name	No.	Year
Johnson (Parod, E. W.)	4,508	1881
" (Stanford, T.W., and Milligan, S.)	1,563	1880
" (Tissandier, G.)	3,401	1881
" (Wheeler, E.)	3,023	1879
Johnson, E. H.	1,094	1882
Johnson, E.H., and Delany, P.B.	2,532	1881
"	4,093	1881
Johnson, E. H., and Edison, T. A.	4,621	1880
Johnson, J. B.	4,966	1881
Johnson, W. C. (Partly)	1,347	1882
Johnson, W. C., and Anderson, J.	2,311	1882
Johnson,W.C.,and Phillips,E.E.	3,798	1875
" "	3,533	1876
" "	3,534	1876
" "	2,635	1881
" "	4,885	1881
Jones, J. T., and Wild, J. H.	5,614	1881
Jones, W. H., and Grout, G.	5,521	1881
Journaux, J. J.	4,518	1881
Jousselin, A. L.	2,037	1882
Judd, W., and Varley, C. F.	441	1882
Jurgensen, C. P., and Lorenz, L.V.	2,416	1881
Justice (Cruto, A.)	1,895	1882
" (Spalding, H. C.)	3,637	1880
" "	65	1881
Kabath, N. de	4,037	1881
"	4,060	1881
Kabath, N. de, and Maikoff, A. D.	4,271	1881
Kane, F. L., and Knudson, A.	1,295	1880
Kayser, J., Tisdel, A. G., Marx, E., and Aklem, F.	351	1880
Keith, N S.	1,387	1879
Kelway, C. E.	2,910	1882
Kempe, H. R., and Gatehouse, T. E.	2,569	1882
Kendal, R.	4,674	1880
Kennedy, R.	4,541	1881
"	5,524	1881
"	72	1882
"	1,199	1882
"	1,640	1882
"	2,286	1882
"	2,629	1882
Kerr, W. H.	2,527	1877
Kesseler (Kuhlo, E.)	4,825	1880
Khotinsky, A. de	245	1882
Kilner, W. J.	2,996	1875
"	802	1876
Kimball, D. F.	3,999	1875
Kincaid, J.	2,911	1882
King (Linford, C.)	4,456	1879
"	1,184	1880
King, J. B.	1,822	1882
King, W. F., and Brown, A. B.	5,272	1881
Kinnear, F. C.	4,684	1881
Kinsman, F., and Briggs, J. A.	1,774	1882
Kipling, R. A.	717	1881
Kirkland, T., and Floyd, T.	2,225	1882
Kleiber, H., and Tschikoleff, W.	2,198	1881
Knudson, A., and Kane, F. L.	1,295	1880
Konn (Lodighin, A. N.)	91	1873
Konn, S. V.	970	1875
Kosloff, S. A.	441	1875
Kosloff, S. A.	2,767	1875
Kotyra, M.	3,047	1882
Krizik, F., and Piette, L.	1,397	1880
" "	2,712	1882
Krupp, A.	3,837	1878
"	1,969	1879
Kuhlo, E.	4,825	1880
Kulankamp, J. W., and Goebel, H.	*5,548	1882
Labye, C., and Locht-Labye, L. de	5,661	1881
Lachaussée, D.	2,761	1881
Lackersteen, J. F.	3,666	1875
Lacomme, A.	715	1881
Ladd (Jasper, J.)	83	1879
Ladd, W., Edmunds, H., and Sellon, J. S.	4,645	1878
	4,646	1878
Laing, W.	3,169	1881
Laing, W., and Bouteilloux, L. J.	842	1880
	3,214	1881
Lake (Allison, W. C.	2,256	1881

	No.	Year
Lake (Arbogast, P., and Mc-Tighe, T. J.) ..	3,778	1879
" (Arey, A. L.) ..	3,456	1881
" (Ball, C. E.) ..	84	1882
" (Barda, B.) ..	1,107	1881
" (Barrier, J. J., and Vernede, F. T. de la)	538	1882
" (Blake, F.) ..	5,096	1881
" (Bouliguine, N.) ..	1,968	1881
" (Brooks, D.) ..	3,254	1881
" (Brownell, H. T.) ..	2,302	1873
" (Castro, J. W. de) ..	2,943	1878
" (Chinnock, C. E., and {	695	1880
Harrison, J. de H... {	699	1880
" (Clark, H. A.) ..	229	1881
" " ..	2,592	1881
" ..	361	1882
" (Clingman, T. L.) ..	1,840	1881
" (Concornotti, L.) ..	3,272	1879
" (Connolly, T. A.) ..	3,668	1881
" (Day, M.) ..	1,261	1874
" (Delany, P. B.) ..	1,809	1881
" " ..	2,217	1881
" (Dion, C.) ..	3,880	1881
" (Duvelius, A. L., Goss, L. W., Higgs, P., Merrell, F. R., Peck, H. D., and Waller, H.)	5,751	1881
" (Eteve, E.) ..	48	1881
" (Fuller, J. B.) ..	5,183	1878
" (Fuller, J. B., and {	3,364	1875
Crandall, J. N.) {	1,557	1876
" (Gerard-Lescuyer, J. M. A.) ..	2,992	1882
" (Greb, J. J. C. W.) ..	3,015	1881
" (Haase, R., and Recker, J. P.) ..	3,832	1881
" (Henck, J. B.) ..	4,058	1881
" (Hill, W. S.) ..	3,893	1881
" (Holcombe, A. G.) ..	1,384	1881
" (Houston, E. J., and {	4,400	1879
Thomson, E.) {	315	1881
" (Hussey, C. A., and {	4,265	1880
Dodd, A. S.) {	234	1882
" (Johnson, J. B.) ..	4,966	1881
" (Journaux, J. J.) ..	4,518	1881
" (Keith, N. S.) ..	1,387	1879
" (Khotinsky, A. de) ..	245	1882
" (Krizik, F., and Piette, L.) ..	2,712	1882
" (Lamb, R. B.) ..	2,665	1881
" (Lande, B.) ..	834	1882
..	838	1882
" (La Société Universelle d'Electricité Tommasi	252	1882
" (Lemaire, L., and Lebrun, E.) ..	4,081	1880
" (Letrange, L.) ..	3,211	1881
" (Levy, M.) ..	542	1882
" (Lory, H.) ..	2,409	1882
" (McTighe, J. J.) ..	162	1878
" (Maikoff, A. D., and Kabath, M. de)	4,271	1881
" (Marchese, E.) ..	1,884	1882
" (Maxim, H. S.)..	1,392	1880
" " ..	1,649	1880
" " ..	4,393	1880
" " ..	4,866	1880
" " ..	639	1881
" " ..	3,189	1881
" " ..	5,367	1881
" " ..	1,162	1882
" " ..	1,619	1882
" (Moffatt, R. R.)..	2,759	1882
" (Molera, E. J., and Celrian, J. C.)	1,217	1880
" (Mondos, J. A.) ..	5,490	1881
" (Nichols, J. V.) ..	4,495	1880
..	3,187	1881
" (Pope, F. L.) ..	2,656	1873
" (Puvilland, J., and Raphaël, T. ..	2,111	1882
" (Richards, J. G.) ..	2,760	1882
" (Roche, C. F. de la ..	5,477	1881
..	2,781	1882
" (Rysselberge, F. v. ..	2,804	1882
" (Seeley, C. A.) ..	1,998	1880
" (Splitdorf, H.) ..	*1,272	1881
" (Strohm, S. D.) ..	3,790	1881
" (Tenac, C. L. Van) ..	4,206	1875
" (Thompson, E.) ..	3,928	1880
" (Watson, E. B.) ..	389	1879
Lake (Weston, E.) ..	4,748	1877
" " ..	1,163	1882
" " ..	1,611	1882
" " ..	1,614	1882
" " ..	2,694	1882
" " ..	*4,903	1877
" (Williams, C., Harrington, T. W., and Lane, F. W.)..	3,539	1881
" (Williams, J. S.) ..	5,229	1881
" " ..	5,233	1881
" " ..	85	1882
" " ..	224	1882
" (Wood J. J.) ..	2,851	1881
" " ..	2,526	1882
" " ..	2,531	1882
" " ..	2,5?3	1882
" " ..	2,570	1882
" " ..	2,623	1882
" " ..	2,632	1882
Lalande (Partly G. Chaperon)	1,464	1882
Lallement, C. C., and Etève, C. C. ..	*3,113	1881
Lamar, J. S. ..	4,696	1879
Lamb, R. B. ..	2,665	1880
Lambert, F. ..	759	1878
Lambert, V. P., and Iverneau, L. E. ..	144	1880
Lande, B. ..	834	1882
..	838	1882
Lane, J... ..	2,752	1882
Lane, T. W., Williams, C., and Harrington, F. W.	3,539	1881
Lang, G., and Heinke, F. W.	231	1880
Langley, J. W... ..	4,168	1881
Lartigue, H. ..	3,771	1874
Laurie, A. P. ..	2,823	1881
..	4,310	1818
Lavernède, F. T. de, and {	538	1882
Barrier, J. J. {	2,425	1882
Law, H... ..	4,851	1880
Laybourne, R... ..	5,316	1881
Laycock, W. S. ..	478	1875
Lea, J. ..	1,919	1882
Lea, H. ..	2,186	1882
Leask, A. R. ..	1,803	1882
..	2,701	1882
Leask, A. R., and Smith, F. P.	3,099	1882
Lebrun, E., and Lemaire, A.,	4,081	1881
Lee, J. C. F., and James, J.	4,396	1881
Leek, E., and Edwards, J. ..	2,640	1878
Leigh, J., and Clark, C. L. ..	245	1881
" " ..	3,652	1881
" " ..	1,483	1882
Leipmann, H., and Looker, P. S. ..	1,036	1882
Lemaire, A., and Lebrun, E.	4,081	1880
Lemaire-Douchy, A., ..	2,107	1876
Lemonnier, L S., Mangin and Co. ..	3,425	1879
Lescuyer, J. M. A. Gerard- (See Gerard-Lescuyer, J. M. A.)		
Letrange, L. ..	3,211	1881
Lever, C. ..	3,599	1881
" ..	2,092	1882
Levett, A., and Muller, H. J.	1,787	1881
Levcy, L., and Lumley, E. ..	1,249	1882
Levy, M. ..	542	1882
Lewis, J. S. ..	1,017	1882
Liardet, J. E., and Donni- {	5,216	1881
thorne, T. {	5,418	1881
{	120	1882
Limbeck, W. S. ..	4,408	1880
Linford, C. ..	4,456	1879
..	1,184	1882
Little, G. ..	497	1882
Liveing, E. H. T. ..	4,833	1880
Liveing, E. H. T., and Boys, C. V. ..	69	1882
Locht-Labye, L. de, and Labye, C. ..	5,661	1881
Lockhead W., and Corbett, J. L. ..	219	1880
Lockwood, W. V. O. and R. M.	2,398	1881
Lodighin, A. N. ..	91	1873
Loeffler, J. C. L. ..	905	1880
Longsdon (Krupp, A.) ..	3,837	1878
..	1,969	1879
Lontin, D. F. ..	473	1875
" ..	386	1876
" ..	3,264	1876
Lontin and Co... ..	2,094	1877
Lontin and Co... ..	4,893	1877
Looker, P. S., and Leipmann, H. ..	1,036	1882
Lorenz, L. V., and Jurgensen, C. P. ..	2,416	1881
Loarain (Trouvé, G.) ..	4,009	1880
Lorrain, J. G. ..	2,848	1881
..	5,738	1881
Lory, H. ..	2,409	1882
Lovel, J. H. ..	732	1877
Lucas, F. R. ..	2,633	1875
..	5,270	1878
..	3,463	1881
Ludeke, J. E. F., and Thorman, A. J. ..	3,338	1878
Lugo, O... ..	5,352	1880
..	1,119	1881
" ..	1,696	1881
..	2,394	1881
..	1,097	1881
Lumière, Société Anonyme la {	1,653	1881
Force et la {	2,323	1881
{	4,496	1881
Lumley, E., and Levey, L. ..	1,249	1882
Lyte, F. M. ..	5,358	1880
" ..	1,363	1882
Macaulay-Cruikshank (Banta, W. E.) ..	3,048	1882
McCarty, W. F. C., and Sellière, Baron.. ..	144	1879
Macintosh, J. ..	447	1874
Mackenzie, J. ..	4,568	1878
..	1,635	1879
Mackenzie, W. J. ..	95	1882
Mackie, A. ..	14	1882
Mackie, M. W. W., and {	1,029	1882
Wright, F. {	1,031	1882
{	1,274	1882
Mackie, S. J. ..	440	1874
{	2,542	1881
McTighe, J. J. ..	162	1878
McTighe, J. J. and T. J. ..	2,744	1882
McTighe, T. J., and Arbogast, P. ..	3,778	1879
Maden, E. ..	1,412	1876
Madsen, C. L. ..	4,167	1873
Maikoff, A. D., and Kabath, M. de ..	4,271	1881
Main, J., and Fyfe, A. L. ..	3,821	1881
..	2,636	1882
Maltzoff, S. A. de, and Bourdin, J. ..	1,474	1881
Man, A., and Sawyer, E. ..	4,705	1878
" ..	4,847	1878
Mandon, J. A. " ..	2,147	1880
..	4,914	1880
Mangin, Lemonnier, and Co.	3,425	1879
Manly, R. P. and M. M., and Philips, W. J. ..	1,475	1880
Maquire, F. V. ..	2,885	1882
Marchese, E. ..	1,884	1882
Marcus, S. ..	4,006	1878
Marcus, S., and Egger, B...	2,934	1877
Marx, E., Aklem, F. Kayser, J., and Tisdel, A. G. ..	351	1881
Masin, T. ..	792	1875
Massey, J. E. ..	3,412	1876
Masson, A. ..	2,013	1881
Mathieson, J. ..	2,734	1882
Matthews, R. ..	1,201	1882
Maxim, H. S. ..	1,392	1880
" ..	1,649	1880
" ..	4,393	1880
" ..	4,866	1880
" ..	639	1881
" ..	3,189	1881
" ..	5,367	1881
" ..	1,162	1882
" ..	1,619	1882
Maxwell-Lyte, F. (See Lyte, F. M.)		
Mayal, T. J. ..	3,177	1881
Melhado, A. ..	4,699	1878
Menier, H. ..	4,705	1876
..	756	1880
Mercadier, E. J. P. ..	3,929	1881
Meredith, R. T. and W. J.	4,707	1878
Meritens, A. de ..	3,658	1878
" ..	4,690	1878
" ..	5,257	1878
" ..	178	1879

	No.	Year
Meritens, A. de	2,339	1879
,,	1,136	1880
,,	5,033	1880
,,	2,212	1881
,,	4,207	1881
,,	1,173	1882
Mersanne, F. E. de	1,446	1874
,, ,,	2,787	1875
,, ,,	3,315	1876
,, ,,	5,060	1878
Mersanne, F. E. de, and Bertin, E.	5,044	1878
	5,053	1878
	5,076	1878
	5,110	1878
Merrell, F. R., Peck, H. D., Walter, H., Duwelius, A. L., Goss, L. W., Higgs, P.	5,751	1881
Mewburn (Achard, F. A.)	2,453	1880
,, (Bourdin, J., and Maltzoff, S. A. de)	1,474	1881
,, (Fonvielle, W. de	1,339	1880
,, (Heins, J.)	4,916	1879
,, (Jeune, A. M. J.)	774	1882
,, (La Société Alamagny et Oriel)	866	1882
,, (Nitze, M. C. F.)	153	1879
,, (Putnam, T. A. B.)	1,125	1880
,, (Rigaud, F.)	3,054	1882
,, (Weiller, L.)	1,821	1882
Meyer, H. R.	232	1882
Mignon, J. B. V., and Rouart, S. H.	3,400	1881
	3,402	1881
Millar, A.	4,592	1881
,,	5,566	1881
,,	2,138	1882
Milligan, S., and Stanford, T. W.	1,563	1880
Million, F.	3,085	1879
	2,788	1881
Mills (Million, F.)	3,085	1879
	2,788	1881
,, (Thomas, W. M.)	578	1882
Milner, W., Griffin, W., Mori, F., and Hallewell, C. E.	740	1879
Minchin, G. M., and Despeissis, J. H.	1,010	1882
Miot, J., Société l'Alliance	3,743	1877
Moffatt, R. R.	2,759	1882
Moffatt, A., and Richardson, T. H.	3,085	1875
Moffatt, R. R., and Chichester, S.	3,441	1881
Molera, E., and Celrian, J.	299	1879
Molera, E. J., and Celrian, J. C.	1,217	1880
Molloy, G.	1,455	1882
Monckton, E. H. O	265	1874
,,	3,509	1874
,,	4,597	1876
Mondos, J. A	5,490	1881
Moore, S. J., and Courtenay, R. H.	3,078	1873
Moore, W. E., and Prosser, W.	2,585	1877
Mora, C., and Biloret, A.	4,049	1880
Morel, A.	185	1882
Morgan, J. B., and George, E.	4,482	1880
Morgan, T. (Glouchoff, N.)	478	1880
Morgan-Brown (Fox, E. M.)	4,024	1881
,, (Gassett, O. and Fisher, J.)	3,768	1881
,, (Harding, G. P.)	3,166	1881
,, (Haskins, D. G.)	3,016	1878
,, (McCarty, W. F. C., and Sellière, Baron)	144	1879
,, (Portier, P. A.)	3,355	1879
Mori, F., Hallewell, C. E., Milner, W., and Griffin, W.	740	1879
Moritz, P.	4,636	1876
Moseley, W.	2,969	1873
,,	307	1879
,,	3,001	1879
Mourlot, E.	4,846	1879
,,	2,121	1880
Mowbray, G. M.	2,437	1881
,,	*190	1881
Mucklow, J. D., and Spurge J. B.	5,368	1881

	No.	Year
Muirhead (Briggs, J. A., and Kinsman, F.)	1,774	1882
Muirhead, A.	2,606	1881
,,	2,658	1882
,,	4,886	1880
Muirhead, A., and Hopkinson, J.	153	1881
	1,577*	1881
Müller, C. H. F.	3,711	1881
Müller, H. J., and Levett, A.	1,787	1881
Munro, J.	4,016	1878
,,	1,626	1882
Naphegyi, G., and Conybeare, H.	2,106	1874
Nawrocki (Balnkiewicz, T.)	2,835	1880
,, (Limbeck, W. S.)	4,408	1880
Neale, N. T.	902	1878
Neave, S. J.	4,475	1874
Neill, A. O'.	4,380	1876
Nelson, L., and Anderson, J. E. (Paine, H. M. and E. L.)	2,049	1875
Nelson, L., and Paine, E. L. (Paine, H. M.)	4,118	1875
Nettlefold, J. H.	2,962	1878
Newton (Camacho, J. S.)	3,461	1873
,, (Cook, H. W.)	2,769	1879
,, (Gravier, A.)	2,739	1881
,, (Gravier, A. J.)	5,295	1881
,, ,,	943	1882
,, ,,	1,211	1882
,, (Hosmer, H. G.)	311	1878
,, ,,	2,930	1878
,, ,,	3,676	1878
,, (Hussey, C. A., and Dodd, A. M.)	2,375	1881
	2,572	1881
	2,573	1881
,, (Menier, H.)	4,705	1876
,, (Mersanne, F. E. de)	5,060	1878
,, (Société Universelle d'Electricité Tommasi	2,782	1881
	4,057	1881
,, (Stone, J. B.)	94	1874
,, (Tommasi, F.)	4,405	1877
,, (Volckmar, E.	5,261	1881
,, (Wallace, J. D.)	2,015	1873
,, (Weston Dynamo-Electric Machine Co.)	4,960	1878
Newton, F. M.	4,559	1881
Newton, H. R.	1,569	1876
Nezeraux, C. P.	3,039	1882
Niaudet, A., and Reyner, E.	3,971	1881
Nichols, J. V.	4,495	1880
	3,187	1881
Nitze, M. C. F.	153	1879
Normandy, A., and Edwards, E.	4,611	1878
North, W.	4,041	1878
,,	4,518	1878
Nothomb, L.	2,452	1882
Obach, E. A.	3,317	1878
Obrick, H., and Prall, W. E.	3,831	1879
O'Neill, A.	4,380	1876
Oppenheimer, J.	1,248	1878
Oriel, La Société Alamagny et	866	1882
Ormiston, F. A., and Wright, F.	5,006	1881
Ormiston, F. A., and Brougham, Hon. H.	1,697	1882
Otto, N. A.	1,770	1778
Owen, C.	534	1873
Page, G. S.	2,516	1882
,,	2,518	1882
Paine, H. M.	4,118	1875
Paine, H. M., and E. L.	2,049	1875
Paine, E. L. and Nelson, L. (Paine, H. M.)	4,118	1875
Palmer (Gaumé, C.)	2,618	1873
Paraire, E. L.	325	1879
Parker, T., and Elwell, P. B.	2,917	1882
Parod, E. N.	4,686	1878
Parod, E. W.	4,508	1878
Partz, A. F. W.	3,455	1881
Pass (Abdank, B.)	339	1882
,, (Gloker, J.)	2,674	1882

	No.	Year
Pass (La Société Anonyme des Câbles Electriques, Système Berthoud, Borel, et Cie.)	4,026	1881
,, (Roosevelt, C., and Abdank, B.)	3,070	1882
Paul, A. L., and Hatton, R. J.	2,654	1882
Peck, H. D., Walter, H., Duwelius, A. L., Goss, L. W., Higgs, P., Merrell, F. R.	5,751	1881
Peebles, D. B., and Blyth, J.	2,661	1882
Pel, J. A.	3,380	1881
Perry, J.	1,178	1880
,,	55	1882
Perry, J., and Ayrton, W. E.	783	1881
,, ,,	2,613	1882
,, ,,	2,642	1882
,, ,,	3,036	1882
Pettit, S. W., and Alberger, M. H.	4,601	1878
Pezzer, O. de, and Carpentier, C.	5,322	1881
Pfannkuche, A.	2,845	1882
,,	3,650	1881
Pfannkuche, G., and Dunstan, R. E.	3,655	1881
	5,743	1881
Phillips, E. F.	3,603	1878
Phillips, S. E., and Johnson, W. E.	3,798	1875
	3,533	1876
	3,534	1876
	2,635	1881
	4,885	1881
Phillips, S. E. (Partly Johnson, W. C.)	1,347	1882
Phillips, S. E., and Phillips, W. A.	2,571	1882
Philips, W. J., and Manly, M. M. and R. P.	1,475	1880
Pickersgill, G	1,693	1877
,,	3,854	1877
Pierson, J. S.	5,321	1878
Piette, L., and Krizik, F.	1,397	1880
,, ,,	2,712	1882
Pilleux, C. L.	636	1880
Pilsen	1,397	1880
,,	2,712	1882
Piot, D. T.	4,851	1881
,,	1,692	1882
Pitkin, J.	5,451	1881
,,	2,391	1882
Pitt (Bear, S. J. M.)	3,283	1881
,, (Burrell, S. J.)	3,679	1881
,, (Espeut, W. B.)	1,257	1881
,, (Gaulard, L., and Gibbs, J. D.)	4,942	1881
,, (Lugo, O.)	5,352	1880
,, ,,	1,119	1881
,, ,,	1,696	1881
,, ,,	2,394	1881
,, (Mangin, Lemonnier, and Co.)	3,425	1879
,, (Scribner, C. E.)	5,156	1879
,, (Starr, E. T.	5,600	1881
,, ,,	819	1882
Planté, G.	1,713	1873
Pope, F. L.	2,656	1873
Portier, P. A.	3,355	1879
Poulet, V., and Commelin	1,046	1880
Powell, C. E.	4,435	1874
Powell, L. A. (Gerard-Lescuyer, J. M. A.)	5,593	1881
	5,660	1881
Prall, W. E.	979	1874
Prall, W. E, and Obrick, H.	3,831	1879
Preece, W. H. Goldstone, C., Radcliffe, J., Gray, M.	3,521	1874
Preece, W. H., and James, J.	129	1874
Prentice, E. R.	4,777	1881
Price, A. P.	2,722	1882
Pritchard, G. E.	2,816	1878
Pritchard, O. G.	2,974	1882
Probert, J., and Floyd, T	2,226	1882
Prosser, W.	3,466	1875
Prosser, W., and Moore, W. E.	2,585	1877
Protheroe, P.	2,725	1876
Pulvermacher, I. L.	1,900	1876
,,	3,782	1876
,,	3,469	1877
,,	1,587	1878
,,	4,079	1878

Index.

	No.	Year
Pulvermacher, I. L.	4,094	1878
,, ...	4,180	1878
,, ...	4,774	1878
,, ...	4,844	1878
,, ...	837	1882
Punshon, R.	5,105	1878
Punshon, R., and Allport, C. J.	4,850	1881
Punshon, R., and Shippey, A.	2,293	1882
Putnam, T. A. B. ..	2,711	1881
Putnam, T. A. B. ..	1,125	1880
Puvilland, J., and Raphaël, T. ..	2,111	1879
Puydt, J. P. C. de, and J. Cougnet ..	350	1880
Quin, E.	1,239	1880
Radcliffe, J., Goldstone, G., and Gray, M. ..	3,381	1874
Radcliffe, J., Gray, M., Preece, W. H., Goldstone, C.	3,521	1874
Radde, W.	2,091	1873
Raison, H. S. ..	169	1882
Ranyard, A. C., and Fleming, J. A. ..	2,807	1881
Raphaël, T., and Puvilland, J. ..	2,111	1879
Rapieff, J. ..	831	1882
,, ..	4,432	1877
,, ..	2,136	1882
,, ..	211	1879
Rath, A. W. ..	*2,658	1881
Raworth, B. A. ..	27	1879
Rawson, F. L., and Woodhouse, O. E. ..	1,412	1882
Recker, J. P., and Haase, R. ..	3,832	1881
Reddie (André, J.) ..	5,268	1880
,, (Biloret, A., and Mora, C.)	4,049	1880
,, (Chertemps, D. A.)	3,349	1881
,, (Hunebelle, J.)	1,856	1881
,, (Menier, H.)	756	1880
,, (Sedlaczek, H., and Wiknlill, F.)	1,596	1881
,, (Volckmar, E.)	4,398	1881
Redfern (Bloomfield, J. H.)..	3,679	1879
Reese, F...	51	1879
Reimenschneider, A., and Christensen, F. S. ..	4,693	1878
Remington, G...	192	1879
Requa, L. F., and Thomas, J. W. ..	1,336	1882
Reynard, L. ..	5,165	1878
Reynier, E. ..	2,399	1878
,, ..	2,982	1877
Reynier, E., and Niaudet, A.	3,971	1880
Reynier, N. E. ..	471	1878
,, ..	1,971	1879
Reynolds, W. F. ..	515	1874
Reynoso, A. F. C. ..	799	1873
Rhodes, B., and Bingswanger, G. ..	2,501	1882
Richards, J. G...	2,760	1882
Richardson, J...	288	1881
,, ..	2,703	1881
,, ..	5,681	1881
,, ..	941	1882
Richardson, G. ..	2,644	1882
Richardson, T. H., and Moffatt, A. ..	3,085	1875
Ridout, R. H. ..	3,003	1876
Rigaud, F. ..	3,054	1882
Roche, C. F. de la ..	5,477	1881
,, ..	2,781	1882
Roe, E. ..	3,010	1881
Rogers, J. B. ..	3,809	1880
,, ..	1,922	1881
,, ..	4,854	1881
,, ..	4,855	1881
,, ..	621	1882
,, ..	1,288	1882
,, ..	1,390	1882
,, ..	1,618	1882
,, ..	1,760	1882
,, ..	1,999	1882
Roguier, L. ..	75	1880
Roosevelt, C., and Abdank, B. ..	3,070	1882
Rosebrugh, A. M. ..	1,476	1879
Rouart, S. H., and Mignon, J. B. V. ..	{3,400	1881
	3,402	1881}

	No.	Year
Rowan, T. ..	5,400	1881
Rowatt, W. ..	2,077	1873
,, ..	4,079	1873
Rubery, J. ..	4,193	1873
,, ..	2,759	1876
Russell, W. J., and Wilson, R. ..	687	1877
Rysselberghe, F. van ..	2,484	1881
,, ,, ..	2,804	1882
Sabatou, C. ..	469	1880
Sabine, R. ..	4,821	1878
Sachs, J. J. ..	894	1881
Salomons, Sir D. ..	1,580	1882
Sanders, F., and Danckwerth, L. ..	2,016	1879
Saunders, H. A. C., and Jamieson, A. ..	1,416	1877
Sawiczeski, S. von ..	3,464	1881
Sawyer, E. ..	3,587	1879
Sawyer, E., and Man, A. ..	{4,705	1878
	4,847	1878}
Scaife, R., and Deiss, A. ..	2,366	1876
Scantlebury, W. ..	1,932	1878
Scarlett, G. ..	1,585	1880
Scarlett, G., and Bell, J. ..	4,555	1879
Schäffer, B. A., and Budenberg, C. F. ..	4,227	1878
Schaeffer, A. G. ..	4,294	1881
Schuckert, S. ..	4,464	1877
,, ..	960	1879
Schumann, O. ..	2,179	1880
Schuyler, D. A., and Waterhouse, F. G. ..	3,002	1882
Schwegler, F., and Bossomaier, R. ..	2,823	1879
Scott, J., and Akester, W. H. ..	1,412	1881
Scott (Strickler, W.) ..	3,099	1876
Scott, T. F. ..	861	1878
Scott, W. ..	4,140	1878
Scott, W. L. ..	4,671	1878
Scribner, C. E. ..	5,156	1879
Sears, J. N., and White, W. G. ..	347	1880
Sedlaczek, H., and Wiknlill, F. ..	{2,322	1879
	1,596	1881}
Seeley, C. A. ..	1,998	1880
Sellière, Baron, and McCarty, W. F. C. ..	144	1879
Sellon, J. S. ..	3,926	1881
,, ..	3,987	1881
,, ..	4,005	1881
,, ..	4,632	1881
,, ..	5,631	1881
,, ..	5,632	1881
,, ..	319	1882
,, ..	2,818	1882
Sellon, J. S., Ladd, W., and Edmunds, H. ..	{4,645	1878
	4,646	1878}
Sellon, J. S., and Edmunds, H.	1,692	1879
,, ,,	1,791	1879
,, ,,	1,949	1879
Sennett, A. R. ..	5,286	1881
Sennett, R., and Grant, G...	2,267	1879
Shea, C. E. ..	4,304	1878
Shedlock, J. J...	5,498	1880
Sheldon, J ..	2,397	1879
Sheridan, H. B. ..	4,617	1881
,, ..	931	1882
Shippey, A. ..	879	1881
Shippey, A., and Punshon, R. ..	2,293	1882
Siemens (Alteneck, H. von ..	4,949	1878
,, (Von Alteneck, F. and H.)	2,652	1879
,, (Siemens and Halske)	1,447	1881
,, (Siemens, E. W.)	760	1882
,, (Siemens, W., and Alteneck, F. H. von)	2,006	1873
,, (Siemens, E.W., and Alteneck, F. H von)	3,134	1878
,, (Siemens, E. W.	583	1880
,, and Halske ..	1,447	1881
Siemens, C. W. ..	251	1878
,, ..	2,281	1878
,, ..	3,315	1878
,, ..	4,208	1878
,, ..	2,110	1879
,, ..	4,534	1879
,, ..	4,614	1880

	No.	Year
Siemens, C. W., and Boothby, A. C. ..	696	1881
Siemens, E. W. ..	583	1880
,, ..	760	1882
Siemens, W., and Alteneck, F. H. von ..	2,006	1873
Siemens, E. W., and Alteneck, F. H. von ..	3,134	1878
Siemens, L. ..	2,199	1879
Simon (Schuckert, S.) ..	4,464	1877
,, ..	960	1879
Slater, T. ..	2,625	1874
,, ..	2,272	1880
Smillie, R. D. ..	1,850	1882
Smith, A. ..	1,465	1882
Smith, E., and Clarke, T. ..	4,650	1878
Smith, F. P., and Leask, A. R. ..	3,099	1882
Smith, G. ..	1,132	1882
Smith, H. J. ..	380	1874
Smith, H. J., and Graham, W. ..	1,392	1882
Smith, J. W. ..	3,975	1881
Smith, M. H. ..	3,981	1877
Smith, P. E., Spruill, S. R., and Wood, W. R. ..	4,312	1876
Smith, W. ..	4,384	1875
,, ..	3,622	1878
,, ..	5,599	1881
Société Alamagny et Oriel ..	8?6	1882
Société Anonyme des Cables Electriques ..	1,496	1882
Société Anonyme la Force et La Lumière, Société Générale d'Electricité ..	{1,097	1881
	1,653	1881
	2,323	1881
	4,496	1881}
Société Générale d'Electricité	4,066	1878
Société Générale d'Electricité Procédés Jablochkoff ..	{1,175	1879
	725	1880}
Société l'Alliance, J. Miot ..	3,743	1877
Société Universelle d'Electricité, Tommasi ..	{2,782	1881
	4,057	1881
	252	1882}
Solignac et Cie ..	740	1882
Somzée, L. ..	1,852	1881
,, ..	4,305	1881
Sorley, C. ..	2,945	1882
Spagnoletti, C. E. ..	869	1882
Spalding, H. C. ..	915	1878
,, ..	1,195	1878
,, ..	1,196	1878
,, ..	1,197	1878
,, ..	1,467	1878
Spalding, H. C. ..	3,637	1880
,, ..	65	1881
Sparling, J. A. ..	*2,414	1881
Specht (Schumann, O.) ..	2,179	1880
Spence (Kotyra, M.) ..	3,047	1882
Spence, J. B. ..	876	1879
,, ..	2,706	1879
Sperry, E. A. ..	3,025	1882
Splitdorf, H. ..	*1,272	1881
Sprague, J. T. ..	1,558	1873
,, ..	4,662	1878
,, ..	4,762	1878
,, ..	4,454	1881
,, ..	4,902	1882
Spruill, S. R., Wood, W. R., Smith, P. E. ..	4,312	1876
Spurge, J. B., and Mucklow, J. D. ..	5,318	1881
Stanfield, J., and Clark, J. L. ..	*4,918	1881
Stanford (Phillips, E. F.) ..	3,603	1878
Stanford, T. W., and Milligan, S. ..	1,563	1880
Stanley, W. ..	2,660	1882
Starr, E. T. ..	5,601	1881
,, ..	819	1882
Stearn, C. H. ..	*5,000	1881
Stearns, J. B. ..	2,870	1873
Stearns, J. M. ..	5,468	1881
Stern, W. A., and Byllesby, H. M. ..	2,336	1882
Sterne, L. ..	3,974	1874
Stewart, C. ..	4,466	1878
Stockman, B. P. ..	4,315	1878
Stokes, J. E. ..	4,283	1878
Stone, J. B. ..	94	1874
Strickler, W. ..	3,099	1876
Strohm, S. D. ..	3,790	1881
Stuart, J. M. ..	2,232	1882
,, ..	2,233	1882

Name	No.	Year
Stuart-Wortley, A. H. P. ..	3,656	1878
Swalwell, J.	2,456	1882
Swan, A.	2,898	1880
Swan, J. W.	18	1880
,, ...	250	1880
,, ...	4,933	1880
,, ...	5,004	1880
,, ...	5,014	1880
,, ...	2,272	1881
,, ...	4,202	1881
,, ...	4,455	1881
,, ...	5,494	1881
,, ...	5,499	1881
,, ...	5,702	1882
,, ...	905	1882
Sussex, S. W. M. de ..	465	1879
Sussex, S. W. M. de, and Brasseur, L. A. ..	308	1878
Sueur, C. le ..	3,857	1881
Talling, R., and Field, F. ..	1,938	1875
Tambourin, G. A. ..	1,235	1880
Tenac, C. L. van ..	4,206	1875
Thomas J. D... ..	1,649	1882
Thomas, J. W., and Requa, L. F. ..	1,326	1882
Thomas, W. M. ..	578	1882
Thompson (Gary, W. W.) ..	805	1879
,,	4,069	1881
,,	4,070	1881
,, (Langley, J. W.) ..	4,168	1881
,, (Pel, J. A.) ..	3,380	1881
,, (Reynier, E.) ..	2,399	1878
,, (Rysselberghe, F. van) ..	2,484	1881
,, (Union Electric Manufacturing Co.) ..	392	1882
Thompson, A. M., and Earl, H. D. ..	5,281	1878
Thompson, R. C. ..	927	1879
,,	1,622	1879
Thompson, S. P., and W. P. ..	4,988	1878
Thompson, E. ..	3,928	1880
Thomson, E. and Houston, E. J. {	4,400	1879
	315	1880
Thomson, H. L. ..	4,462	1878
Thomson, J. H. ..	1,393	1881
Thomson, Sir W. ..	3,082	1881
,,	5,668	1881
Thorman, A. J., Ludeke, J. E. F. ..	3,338	1878
Tilleard, F. D. ..	4,817	1878
Timmins, H. ..	3,800	1874
Tisdel, A. G., Marx, E., Aklem, F., and Kayser, J.	351	1880
Tissandier, G. ..	3,401	1881
Tommasi, F. ..	4,405	1879
Tommasi, Société Universelle d'Electricité .. {	2,782	1881
	4,057	1882
	252	1881
Tongue (Lacomme, A.) ..	715	1881
Townsend, J. D. ..	39	1881
Tribe, A. ..	1,587	1882
,,	2,263	1882
Trouvé, G. ..	4,009	1880
Truman, E. T. ..	124	1874
,,	375	1878
,,	4,438	1878
,,	3,810	1880
Tschikoleff, W., and Kleiber, H. ..	2,198	1881
Tubini, T. ..	3,635	1881
Tubini, A ..	3,822	1881
Tyer, E... ..	1,845	1873
,,	557	1876
,,	2,879	1881
Tyler, H. W. ..	3,985	1878
,,	4,575	1878
Union Electric Manufactur-Co. ..	392	1882
Upton, H. E. M. D. C. ..	1,232	1881
Varicas (Richardson, G.) ..	2,644	1882
Varley, C. F. ..	270	1877
,,	2,184	1882
,,	2,185	1882
,,	2,207	1882
Varley, C. F. and F. H. ..	5,396	1881
Varley, C. F., and Judd, W. ..	441	1882
Varley, F. H. ..	4,100	1878

Name	No.	Year
Varley, F. H. ..	2,776	1882
Varley, F. H. and C. F. ..	5,396	1881
Varley, S. A. ..	4,905	1876
,,	4,435	1877
,,	5,665	1881
,,	5,667	1881
Varley, T., and Greenwood, H. B ..	2,248	1882
Vernede, F. T de la, and Barrier, J. J. {	538	1882
	2,425	1882
Verrue, F. ..	4,287	1878
Vincent, C. W. and Elphinstone, Baron.. {	332	1879
	2,898	1880
Vincent (Elphinstone, W. B. F.) ..	2,491	1882
,,	684	1879
,,	2,557	1882
Vincent (Elphinstone, Lord, Partly) ..	2,340	1882
Voeux (Bernstein, A.) ..	2,604	1882
Vogel (Vogel, N. C.) ..	4,812	1878
Vogel, N. C. ..	4,812	1878
Vogel, Sir J. ..	3,033	1882
Voice, E. L. ..	1,794	1882
	2,288	1882
Volckmar, E. ..	4,398	1880
,,	5,261	1881
Volk, M... ..	2,962	1882
Vyle, S. ..	5,002	1881
Walker ..	293	1874
Wallace, J. D... ..	2,015	1873
Wallace, W. ..	240	1878
Waller, R. ..	803	1881
Walter, H., Duwelius, A. L., Goss, L. W. Higgs, P., Merrell, F. R., Peck, H. D. ..	5,751	1881
Ward, E. P. ..	2,930	1881
Ward, M. R. ..	3,976	1878
Ward, D., and Ball, A. ..	2,033	1878
Ward, M. R. ..	2,538	1881
Warlich, F. H... ..	2,068	1880
Waterhouse, A. G. ..	5,185	1881
Waterhouse, F. G., and Schuyler, D. A. ..	3,002	1882
Waters, T. J. ..	862	1874
Waters, S. ..	1,462	1882
Watson, E. B. ..	389	1879
Watson, G. ..	390	1879
Watson, J. J. W. ..	2,271	1880
Watson, T. A., and Anders, G. L. ..	1,958	1880
Watt, A. ..	4,255	1881
,,	3,097	1882
Wauthier, J. ..	1,172	1882
Weathers, J. ..	610	1876
Webber, C. E., and Bolton, F. J. ..	686	1873
Weiller, L. ..	1,821	1882
Welch, E. J. C. ..	4,114	1878
,,	4,278	1878
,,	4,685	1878
,,	4,689	1878
Wells, G., and Gilbert, A. ..	1,510	1880
	1,960	1880
Werdermann, R. ..	476	1873
,,	1,438	1874
,,	3,156	1874
,,	4,805	1876
,,	1,829	1877
,,	2,477	1878
,,	2,301	1879
,,	79	1880
,,	304	1881
,,	1,444	1882
,,	2,364	1882
Westphal, C. ..	2,823	1882
Westinghouse, G. ..	3,409	1881
Weston, E. ..	4,280	1876
,,	4,748	1877
,,	1,163	1882
,,	1,611	1882
,,	1,614	1882
,,	2,694	1882
,,	*4,908	1877
Weston Dynamo Electric Machine Co. ..	4,960	1878
Wetter (Jennings, R. S.) ..	1,699	1881
,, (Nothombe, L.) ..	2,452	1882
,, (Stanley, W.) ..	2,660	1882
,, (Wheeler, W.) ..	3,858	1881
,,	3,911	1881

Name	No.	Year
Weyde, H. V. ..	446	1878
Weyde, H. S. V. ..	508	1879
Wheeler, E. ..	3,023	1879
Wheeler, J. J. ..	1,652	1882
Wheeler, W. ..	3,858	1881
,,	3,911	1881
,,	917	1882
White, J. C., and Hayden, H. H. ..	473	1881
White, W. G., and Sears, J. N. ..	347	1880
Whiteley, J. ..	1,445	1879
Whitehouse, E. O. W... ..	1,820	1874
Whiteman (Bauer, M., and Co. ..	1 915	1882
Whyte, G. W. ..	5,152	1878
Whyte, G. ..	2,744	1879
Wiebe, R. ..	767	1878
Wier, M. A. ..	806	1875
,,	2,686	1882
Wigham, J. R... ..	*2,564	1881
Wigner, G. W. ..	553	1880
Wiknlill, F., and Sedlaczek, {	2,322	1879
	1,596	1881
Wild, J. H., and Jones, J. T. ..	5,614	1881
Wilde, H. ..	618	1873
,,	1,554	1874
,,	1,228	1878
,,	3,260	1878
,,	5,197	1878
,,	5,008	1880
,,	497	1881
,,	2,256	1882
Wiles, J. F. ..	65	1873
,,	644	1879
Wilkins, T. ..	4,306	1879
Wilkinson, A. ..	3,083	1873
,,	3,472	1881
,,	3,003	1882
Willans, P. W., and Crompton, R. E. B. ..	245	1879
Willard, F. L. ..	2,803	1882
Willatt, F. G. ..	3,042	1882
Williams, C., Harrington, F. W., and Lane, T. W. ..	3,539	1881
Williams, J. S. ..	5,229	1881
,,	5,233	1881
,,	84	1882
,,	224	1882
Williams, J. S. ..	560	1882
,,	700	1882
,,	766	1882
,,	856	1882
,,	1,174	1882
,,	1,556	1882
,,	2,558	1882
,,	*5,742	1881
Wilson, C. H. Carus (See Carus Wilson)		
Wilson, J. S. ..	4,347	1878
,,	4,348	1878
Wilson, R., and Russell, W. J. ..	687	1877
Wilson, W. S. ..	3,912	1878
Winter, G. K. ..	3,146	1881
Winter, W. ..	1,264	1877
Wirth (Manly, M. M. and R. P., and Philips, W. J.) ..	1,475	1880
Wise, Lloyd (Bürgin, E.) ..	5,085	1879
,, ,,	4,819	1881
,, ,,	4,820	1881
,, (Mandon, J. A.) ..	2,147	1880
,,	4,914	1880
,, (Masin, T.) ..	792	1875
Wolff (Jurgensen, C. P., and Lorenz, L. V.) ..	2,416	1881
Wood, J. J. ..	2,851	1881
,,	2,526	1882
,,	2,531	1882
,,	2,563	1882
,,	2,570	1882
,,	2,623	1882
,,	2,632	1882
Wood, W. R., Smith, P. E., and Spruill, S. R. ..	4,312	1876
Woodhouse, O. E., and Rawson, F. L. ..	1,412	1882
Woods, E. and G. ..	3,917	1878
Woodward, A. T. ..	4,780	1881
Woodward, H. ..	2,642	1882
Woolley, L. G. ..	*3,049	1881
Wortley, A. H. P. Stuart- ..	3,656	1878

	No.	Year		No.	Year		No.	Year
Wright, F.	3,435	1881	Young, G. S., and Hatton, R. J.	1,689	1882	Zanni, G.	4,232	1877
,,	3,437	1881	Young, T.	3,368	1879	,,	1,677	1878
,,	4,778	1881				,,	4,573	1878
Wright, F., and Mackie, M. W. W. {	1,029	1882				,,	2,821	1879
	1,031	1882	Zanni, G.	2,266	1873	,,	4,007	1880
	1,274	1882	,,	1,855	1874	,,	2,740	1882
Wright, F., and Ormiston, F. A.	5,006	1881	,,	3,795	1875	,,	2,741	1882
			,,	2,821	1876	Ziffer, F. H.	4,412	1877
Young, F., and Freeman, J.	350	1879	,,	4,222	1876	Zingler, M.	1,158	1882
						Zipernowsky, K.	1,580	1880

INDEX.

———:-:———

SUBJECT MATTER.

Abstracts marked * will be found in the Addenda.

	No.	Year
ACCUMULATORS (See " Batteries, Secondary.")		
Advertisements, illuminating, Hickisson	2,421	1882
Ammonia, producing, Clarke and Smith	4,650	1878
Arago disc, Monckton ..	3,509	1874
ARC LAMPS — (See also "Candles, Electric;" "Arc-Incandescence Lamps;" "Semi-Incandescence Lamps;" "Carbons;" and "Electrodes.")		
De Mersanne	2,787	1875
Prosser	3,466	1875
Weathers	610	1876
Prosser and Moore ..	2,585	1877
Rapieff	4,432	1877
Brush	2,003	1878
Siemens and Alteneck ..	3,134	1878
Wilde	3,250	1878
Siemens	3,315	1878
Krupp	3,837	1878
Harding	4,047	1878
Cadett	4,022	1878
Siemens	4,208	1878
De Meritens	4,690	1878
Pulvermacher	4,774	1878
Sabine	4,821	1878
Thompson	4,988	1878
Whyte	2,744	1879
Harrop	3,793	1879
Eclairage Electrique ..	1,553	1880
Gordon	1,826	1880
Apps	264	1881
Berly	1,027	1881
Cohne	2,369	1881
André	2,563	1881
Million	2,788	1881
Greb	3,015	1881
Woolley	*3,049	1881
Bouteilloux and Laing ..	3,214	1881
Arey	3,456	1881
Rogers	4,855	1881
Mondos	5,490	1881
Varley	5,665	1881
Williams	224	1882
Brockie	898	1882
Munro	1,626	1882
Kennedy	2,286	1882
Defty	2,628	1882
Gloker	2,674	1882
Roosevelt and Abdank ..	3,070	1882
Actuated by weights, Rapieff	831	1882
Adjusted at a distance, De Mersanne	3,315	1876
Adjusting position of arc, Hopkinson..	3,509	1880
Aërostatic, Gerard-Lescuyer	5,660	1881
Alignment of carbons, Common	626	1882
Applied to producing ammonia, Clarke and Smith	4,650	1878
Applied to signalling, De Mersanne	3,315	1876

	No.	Year
ARC LAMPS—		
Arc formed by repulsion due to current, Gerard-Lescuyer	3,697	1879
Arc striking mechanism—		
Siemens	4,208	1878
Wilson	4,347	1878
Heinrichs	4,595	1878
Sellon, Ladd, and Edmunds	4,646	1878
Reimenschneider and Christensen ..	4,693	1878
De Mersanne	5,060	1878
" and Bertin	5,110	1878
Whyte	5,152	1878
Rapieff	211	1879
Bousfield	523	1879
Hedges	925	1879
Brush	947	1879
Mackenzie.. ..	1,635	1879
Krupp	1,969	1879
Puvilland and Raphaël	2,111	1879
Andrews	2,321	1879
Crompton	3,509	1879
Houston and Thomson ..	4,400	1879
Harding	4,590	1879
Gatehouse	4,796	1879
Scribner	5,156	1879
Joel	5,157	1879
Eclairage Electrique ..	1,553	1880
Watson	2,271	1880
Heinrichs	4,608	1880
Maxim	4,866	1880
Hedges	4,988	1880
Henley	5,137	1880
Million	2,788	1881
Bouteilloux and Laing..	3,214	1881
Harling and Hartmann ..	3,473	1881
Daft..	4,775	1881
André	4,948	1881
Gerard-Lescuyer.. ..	5,660	1881
Mackenzie.. ..	95	1882
Henley	130	1882
Hawkes	157	1882
Brown	1,867	1882
Lea	1,919	1882
Brockie	2,370	1882
Akester	2,419	1882
Hatton	2,654	1882
Moffatt	2,759	1882
Attaching globes of, Bauer and Co.	1,915	1882
Automatic lighter—		
Weston Co.	4,960	1878
Wilde	5,197	1878
Automatically putting electrodes successively in circuit, Desquiens ..	3,404	1881
Automatically short-circuiting main magnets of, Brockie	3,771	1879
Automatically switching in second pair of electrodes, Godfrey	4,718	1879
Automatically switching into circuit, Reynier	1,971	1879

	No.	Year
ARC LAMPS—		
Carbons (See " Carbons.")		
Carbon coils for, Varley ..	2,776	1882
Carbon holders, (See "Carbons, Holders for.")		
Clutch—		
Brush	2,003	1878
De Mersanne and Bertin	5,110	1878
Brain	5,139	1878
Boucfield	523	1879
Brush	947	1879
Krupp	1,969	1879
Gerard-Lescuyer.. ..	3,697	1879
Brush	3,750	1879
Brockie	3,771	1879
Hickley	4,354	1879
Houston and Thomson..	4,400	1879
Harding	4,590	1879
Scribner	5,156	1879
Joel	5,157	1879
Houston and Thompson	315	1880
Harding	4,191	1880
Heinrichs	4,608	1880
Common and Joel ..	1,040	1881
Andrews	1,526	1881
Grimstone	1,670	1881
Gerard-Lescuyer ..	1,685	1881
Gülcher	2,038	1881
Hawkes and Bowman ..	2,402	1881
Moffatt and Chichester ..	2,441	1881
Edison	2,495	1881
André	2,563	1881
Harding	3,166	1881
Hopkinson.. ..	3,362	1881
Lever	3,599	1881
Connolly	3,668	1881
Burrell	3,679	1881
Brockie	4,504	1881
Newton	4,559	1881
Daft	4,775	1881
Waterhouse	5,185	1881
Mondos	5,490	1881
Gerard-Lescuyer.. ..	5,660	1881
Hawkes	157	1882
Bright	377	1882
Jarman	563	1882
Common	626	1882
Graham	1,171	1882
Levey and Lumley ..	1,249	1882
Jeffery	1,570	1882
Munro	1,626	1882
Young and Hatton ..	1,680	1882
Lea	1,919	1882
Edison	2,072	1882
Lever	2,092	1882
Voice	2,288	1882
Ayrton and Perry ..	2,613	1882
Gumpel	2,723	1882
Chadburn	2,755	1882
Emmens	2,914	1882
Gerard-Lescuyer.. ..	2,992	1882
Sperry	3,025	1882
Roosevelt and Abdank ..	3,070	1882
Courtenay	3,101	1882
Combined with gas—		
Spence	876	1879

ARC LAMPS combined with gas—	No.	Year
Grant and Sennett ..	2,267	1879
Watson	2,271	1880
Courtenay.. ..	4,659	1881
Varley	5,396	1881
Pritchard	2,974	1882
Compensating for burnt carbon, Berjot ..	4,428	1880
Compensating for reduced resistance of carbon, Brockie	2,370	1882
Compensating for variation of driving weight, Harrison	3,875	1879
Completing circuit to, Million ..	2,788	1881
Connecting to leads, Joel ..	5,157	1879
Controlling number of, Common and Joel	1,040	1881
Conveying current to carbons of, André	2,764	1880
Cut-outs, automatic (See also "Cut-Outs and Switches, Automatic.")		
Powell ..	4,435	1874
Sawyer and Man ..	4,705	1878
Mori, Hallewell, Milner, and Griffin	740	1879
Brush	947	1879
"	3,750	1879
Gatehouse	4,796	1879
Roquier	75	1880
Eclairage-Electrique	1,553	1880
Harding	4,191	1880
Berjot	4,428	1880
Heinrichs	4,608	1880
Siemens	4,614	1880
Andrews	1,526	1881
Wood	2,851	1881
Lever	3,599	1881
Connolly	3,668	1881
Hill	3,893	1881
Sheridan	4,617	1881
Waterhouse	5,185	1881
King and Brown ..	5,272	1881
Hawkes	157	1882
Jarman	563	1882
Jeffery	1,570	1882
Lever	2,092	1882
Wood	2,563	1882
Moffatt	2,759	1882
Sperry	3,025	1882
Dash-pot arrangement for—		
Von Alteneck ..	4,949	1878
Siemens	2,810	1879
Brush	3,750	1879
Maxim	4,866	1880
Common and Joel ..	1,040	1881
Grimstone	1,670	1881
Gerard-Lescuyer ..	1,685	1881
André	2,563	1881
Woolley	*3,049	1881
Hopkinson	3,362	1881
Sheridan	4,617	1881
Gerard-Lescuyer ..	5,660	1881
Hawkes	157	1882
Bright	377	1882
Common	626	1882
Andrews	1,324	1882
Brockie	2,370	1882
Gerard-Lescuyer ..	2,992	1882
Sperry	3,025	1882
Defining point of arc, Weston Co.	4,960	1878
Differential, Siemens and Alteneck ..	2,006	1873
Double water globe for, M'Carty and Sellière	144	1879
Driven by air current, Heinke	4,275	1877
Driven by clockwork, { Heinke	4,275	1877
Heinke	3,315	1878
Driven by engine, Heinke	4,275	1877
Duplex (See also "Multiplex.")		
Day	1,261	1874
Brockie	3,771	1879
Common and Joel ..	1,040	1881
Hawkes and Bowman	2,402	1881
Hill	3,893	1881
Bürgin	4,820	1881
Waterhouse	5,185	1881
Mondos	5,490	1881
Hawkes	157	1882
Weston	1,163	1882

ARC LAMPS—	No.	Year
Duplex, Brockie ..	2,370	1882
" clutch, Brockie ..	4,504	1881
" with electrodes alternately consumed, Grimstone ..	1,670	1881
Duplex, with movable gearing frame, Wood ..	2,570	1882
Electrode floats up against refractory stop, Ducretet	65	1879
Electrode formed by two inclined carbons—		
Bouteilloux and Laing ..	3,214	1881
Rapieff ..	4,432	1877
Electro-magnet for, Emmens	5,206	1879
	2,349	1882
Encircled by "electrical loop," Eclairage-Electrique {	863	1879
	1,553	1880
Escapements for, Waterhouse	5,185	1881
Feeding pulverised substances to, Williams ..	224	1882
Governed by magnet, Day	1,261	1874
Globes for—		
Holmes	920	1879
André	5,206	1879
Munro	1,626	1882
Hand regulated, Wilde ..	618	1873
Hermetically enclosed, Concornotti	3,272	1879
Holding globe of, Hill	3,893	1881
Hydrostatic—		
Harding	4,046	1878
Stockman	4,315	1878
Higgins	4,456	1878
Molera and Celrian	299	1879
Paraire	325	1879
Imray	382	1879
Sedlaczek and Wiknlill	2,322	1879
Gerard-Lescuyer..	3,697	1879
Wigner	553	1880
Cohne	2,236	1880
Hopkinson.. ..	3,509	1880
Sedlaczek and Wiknlill	1,596	1881
Hopkinson.. ..	3,362	1881
Brown	4,011	1881
Prentice	4,777	1881
King and Brown ..	5,272	1881
Rapieff	831	1882
Spagnoletti ..	869	1882
Brown	1,867	1882
Increasing light from—		
Mignon and Rouart ..	3,402	1881
Harrison	3,559	1881
Williams	224	1882
Emmens	2,349	1882
In vacuo—		
Powell	4,435	1874
Varley	4,100	1878
Eclairage Electrique	863	1879
"	1,553	1880
Puydt and Cougnet	350	1880
Munro	1,626	1882
Localising arc, Wilde ..	z,256	1882
Magnets for, Emmens	2,349	1882
Magnets for unlocking wheel train of, Gravier ..	5,295	1881
Maintaining distance between parallel carbons of, Hedges	81	1879
Mercury contact for, Jaspar	83	1879
Mining, André ..	830	1879
" Cougnet ..	2,000	1879
Multiplex (See also "Duplex.")		
Prosser and Moore ..	2,585	1877
Kepling	717	1878
Bodmer	4,476	1878
Mori, Hallewell, Griffin, and Milner	740	1879
André	830	1879
Jablochkoff	1,175	1879
Brush	3,750	1879
Gatehouse	4,796	1879
Eclairage Electrique	1,553	1880
Bergot	4,428	1880
Lacomme	715	1881
Berly	1,236	1881
Brockie	1,942	1881
Bouliguine	1,968	1881
Gadot	2,344	1881
Gibbs	4,533	1881
Andrews	1,324	1882

ARC LAMPS—	No.	Year
Multiplex, switch for, Jablochkoff ..	725	1880
Optical apparatus for (See also "Optical Apparatus, Light Diffusing Reflectors and Lanterns.")		
Wilde	618	1873
De Mersanne ..	3,315	1876
Regulated by air admitted to cylinder, Gerard-Lescuyer ..	5,660	1881
Regulated by easily fused metallic strip, Siemens ..	2,110	1879
Regulated by expansion—		
Lontin	2,094	1877
Siemens	2,281	1878
"	3,315	1878
"	2,110	1879
Harding	4,590	1879
Guest	2,980	1880
Regulated by governors driven by engine, Sedlaczek and Wiknlill	1,596	1881
Regulated by paddle-wheel, Krupp ..	3,837	1878
Regulated periodically, Brockie ..	3,771	1879
Regulating, Heincke ..	4,275	1877
Regulating distance between carbons, Imray ..	382	1879
Regulating feed by softening of glass, Solignac ..	740	1882
Regulating length of arc, { Wood	2,563	1882
	2,570	1882
Regulating from a distance, Abdank ..	339	1882
Regulator for, Million ..	3,085	1879
Resistance balance for, Roosevelt and Abdank ..	3,070	1882
Reversing polarity of arc striking magnet, Brockie	898	1882
Self-focussing (See also "Arc Lamps, with Semicircular and with Inclined Carbons and with Cords.")		
Day	1,261	1874
Marcus and Egger ..	2,234	1877
Pulvermacher ..	4,774	1878
Von Alteneck ..	4,949	1878
De Mersanne ..	5,660	1878
De Mersanne and Bertin	5,110	1878
Brain	5,139	1878
Whyte	5,152	1878
Jaspar	83	1879
Keith	1,387	1879
Harrison	3,875	1879
Houston and Thomson..	4,400	1879
Bouteilloux and Laing ..	842	1880
Krizik and Piette ..	1,397	1880
Eclairage Electrique	1,553	1880
Siemens	4,614	1880
Mandon	4,914	1880
Fyfe	774	1881
Sedlaczek and Wiknlill..	1,596	1881
Harling and Hartmann..	3,473	1881
Cance	3,976	1881
Brown	4,011	1881
Sheridan	4,617	1881
Varley	5,396	1881
Kennedy	5,524	1881
Mackenzie.. ..	95	1882
Henley	130	1882
Crompton	346	1882
Spagnoletti ..	869	1882
Kennedy	1,199	1882
Gatehouse and Kempe ..	2,569	1882
Crompton	2,619	1882
Wood	2,632	1882
Self-regulating, Brockie	1,942	1881
Short-circuiting magnet of, Rapieff ..	211	1879
Short-circuiting main, solenoid of, Hopkinson	1,959	1879
Solenoid core for, Krizik and Piette..	1,397	1880
Stopping action of, Maxim	4,866	1880
Street, Harrison ..	3,875	1879
Submarine, Tommasi ..	4,405	1879
" Heinke and Long	231	1880
Supplied with finely divided carbon and gas, Pritchard	2,674	1882

Arc Lamps—	No.	Year
Supplied with powdered charcoal, &c.,		
Rapieff	4,432	1877
Varley	4,100	1878
McCarty and Sellière ..	144	1879
Rapieff	211	1879
Grant and Sennett ..	2,267	1879
Somzee	1,852	1881
Williams	224	1882
Supplying currents to, *Blandy*	2,060	1879
Supporting globes of, *Maxim*	4,866	1880
Suspended by springs, *Daft*	4,775	1881
Switching in pairs of rods, *Tommasi*	4,405	1879
Unlocking wheel train of, *Gravier*	5,295	1881
Arc Lamps with—		
Abutments (*See also "Arc Lamps with Refractory Electrodes."*)		
Ducretet	65	1879
Siemens	2,110	1879
Harding	4,191	1880
Hedges	4,988	1880
Roche	2,781	1882
Adjustable fulcrumed lever, Common	626	1882
Arc striking solenoid automatically cut out, Andrews	2,321	1879
Automatic cut-out for solenoid, *Million* ..	2,788	1881
Automatic make and break, Harding	3,166	1881
Bevel gearing, Varley ..	5,665	1881
Brake, direct—		
Bauer and Co. ..	1,915	1882
Brockie	1,942	1881
Brake operated by shunt solenoid, Andrews ..	1,324	1882
Carbon abutment block, Higgs	454	1879
Carbon and cast-iron block, Sabine	4,821	1878
Carbon disc and rod, *Gerard-Lescuyer*	3,697	1879
Carbon disc and rod, Hickley	4,354	1879
,, dies, André ..	830	1879
,, fed through hollow carbon, Thompson ..	1,622	1879
Carbon holders supported on fluid in closed vessels, *Sedlaczek and Wiknlill* ..	1,596	1881
Carbon passing through steatite ferrule, André ..	5,206	1879
Carbon placed within rotating spirals, Heinrichs ..	4,603	1880
Carbon resistance, *Greb* ..	3,015	1881
Carbon resting on platinum tip, Siemens	2,110	1879
Carbon rod and block, Cougnet	2,000	1879
Carbon rod and plate, Jamin	863	1879
Carbon rod fed against insulated disc, Henley ..	5,137	1880
Carbon rods floating in mercury tubes, *Tommasi*	4,405	1879
Carbon rod pressed against insulating core of tubular carbon—		
Bouteilloux and Laing ..	842	1880
,, ,, ,,	3,214	1881
Rogers ,, ,,	4,855	1881
Carbon rod resting on copper block, André ..	5,206	1879
Carbon rollers and rod, *Concornotti*	3,272	1879
Carbons abutting against chisel edges, Siemens ..	2,110	1879
Carbons, balanced, Kennedy	1,199	1882
Carbons C-shaped, Heinrichs	4,595	1878
Carbons at right angles—		
Whitehouse	1,820	1874
Heinrichs	4,595	1878
Imray	382	1879

Arc Lamps with—	No.	Year
Carbons, circular (*See also "Arc Lamps with Curved and with Semicircular Carbons."*)		
Heinrichs	4,595	1878
,, ...	4,608	1880
Carbons, circular and mercury float, Mandon {	2,147	1880
	4,914	1880
Carbons, circular and radius, arms, Mandon ..	4,914	1880
Carbons concentric, Asten	2,020	1882
Carbons fed by weights, Hawkes	157	1882
Carbons, fed to common centre, Brain ..	5,139	1878
Carbons held by combustible material, Emmens	2,914	1882
Carbons, helical—		
Freeman	5,307	1878
Pilleux	636	1880
Carbons normally held apart, Lever ..	2,092	1882
Carbons passing through orifices in refractory block, *Bureau* ..	1,704	1880
Carbons, vertical and horizontal, Brain ..	5,139	1878
Circular series of rods abutting on carbon block, *Tommasi*	4,405	1879
Clutch regulating descent of weight, Harding ..	4,590	1879
Clockwork—		
Day	1,261	1874
Whitehouse	1,820	1874
De Mersanne ..	3,315	1876
Werdermann ..	4,805	1876
Heinke ..	1,910	1877
Reynier ..	2,987	1877
De Mersanne ..	5,060	1878
Imray ..	382	1879
Brockie ..	1,713	1882
Coiled spring, *Day* ..	1,261	1874
Compound solenoid, *Marcus and Egger* ..	2,934	1877
Contact maker, *Marcus and Egger* ..	2,934	1877
Converging carbons and abutments, Harding ..	4,191	1880
Cords—		
Jaspar	83	1879
Bousfield	523	1879
Keith	1,387	1879
Joel	5,157	1879
Crompton	346	1882
Rapieff	831	1882
Crompton	2,619	1882
Cord attached to spring and main magnet armature, *Houston and Thomson*	315	1880
Cords and escapement, *Harding* ..	3,166	1881
Cords and gearing, Harling and Hartmann ..	3,473	1881
Cords and pulleys, *Bürgin*	4,820	1881
Cords and ratchet—		
Sheridan	4,617	1881
Cougnet	2,000	1879
Harrison	3,875	1879
Gerard-Lescuyer ..	3,697	1879
Cord, roller, and escapement, Mackenzie ..	1,635	1879
Cords and weight—		
Million	2,788	1881
Cance	3,976	1881
André	2,563	1881
Cords, weights, ratchets, and pawls, Kennedy ..	5,524	1881
Curved carbons and mercury float, Mandon ..	2,147	1880
Curved electrodes (*See also "Arc Lamps with Semicircular Carbons."*)		
Werdermann ..	4,805	1876
Varley	4,905	1876
Curved metallic water-cased electrode, Heinrichs ..	4,589	1879
Differential motor—		
André	2,764	1880
Tschikoleff and Kleiber ..	2,198	1881
Differential solenoid—		
Von Alteneck	4,949	1878

Arc Lamps with—	No.	Year
Differential solenoid—		
Kennedy	1,199	1882
Differential solenoid and special core, *Krizik and Piette* ..	1,397	1880
Definite feed and periodic adjustment, Brockie	3,771	1879
Disc electrode (*See also "Arc Lamps, with Rotating Disc Electrode."*)		
Prosser	3,466	1875
Prosser and Moore ..	2,585	1877
Reynier	2,399	1878
André	830	1879
Andrews	2,321	1879
Double electrode, Prosser ..	3,466	1875
Electrodes in sealed glass receiver, Emmens ..	2,349	1882
Electrodes, tubular, Zubini	3,635	1881
Endless cord and weight, André	2,563	1881
Escapement—		
Cance	3,976	1881
Graham	1,117	1882
Expansible metallic spiral, Harding ..	4,590	1879
Fan driven by heated air, Sabine	4,821	1878
Feed magnet automatically short-circuited, Bright ..	377	1882
Feed regulated by clutch, Moffatt and Chichester ..	2,441	1881
Feed roller, *Houston and Thomson* ..	315	1880
Feed wheel and ratchet, Grumpel	2,723	1882
Float—		
Cohne	2,236	1880
Watson	2,271	1886
Hawkes and Bowman ..	2,402	1881
André	2,563	1881
Fluid electrode, Prosser ..	3,466	1875
Prosser and Moore ..	2,585	1877
Rapieff	4,432	1877
Fly and stop, Henley ..	120	1882
Four circular carbons, Heinrichs ..	4,608	1880
Friction arm and roller, Higgs	454	1879
Friction rollers, ratchet, pawl, and shunt solenoid, Siemens ..	4,614	1880
Friction rollers and toothed wheel, Houston and Thomson	4,400	1879
Fusible stops—		
Harding	4,590	1879
Gatehouse	4,796	1879
Gearing (*See also "Arc Lamps with Rack, &c."*)		
Whyte	5,152	1878
Chertemps	3,349	1881
Munro	1,626	1882
Gearing, bevel, Varley ..	5,665	1881
,, controlled by two flat coils, *Holcombe*	1,384	1881
Gearing and brake—		
Crompton	3,509	1879
Brockie	1,942	1881
,, ...	2,370	1882
Gearing and detent—		
Maxim	4,866	1880
Henley	5,137	1880
Gearing and escapement, Brockie	2,370	1882
Gearing and fly—		
Bright	377	1882
De Mersanne and Bertin	5,110	1878
Gearing, flyer and brake Siemens	4,208	1878
Gearing ratchet and pawl—		
Hatton	2,654	1882
Mathieson	2,734	1882
Moffatt	2,759	1882
Heinrichs	4,595	1878
Common and Joel ..	1,040	1881
Gearing and weight, Henley	130	1882
Globe charged with inert gas—		
Remington	192	1879
Varley	5,396	1881
Heating coil in expansible chamber, *Edison* ..	2,072	1882

ARC LAMPS with—

Entry	No.	Year
Helical carbons inclined to one another, *Pilleux* ..	636	1880
Helical carbons and stops, *Pilleux*	636	1880
Horizontal superposed carbons, Harding ..	4,590	1879
Incandescence lamp in its shunt circuit, Gatehouse and Kempe ..	2,569	1882
Inclined carbons—		
Wilson ..	4,347	1878
Rapieff ..	4,432	1877
Sellon, Ladd and Edmunds ..	4,646	1878
Von Alteneck ..	4,949	1878
Rapieff ..	211	1879
Brockie ..	3,771	1879
Gerard-Lescuyer ..	1,552	1879
Harding ..	4,191	1880
Gerard-Lescuyer ..	1,685	1881
Bouliguine ..	1,968	1881
Henley ..	130	1882
Spagnoletti ..	869	1882
Graham ..	1,171	1882
Jeffery ..	1,570	1882
Emmens ..	2,914	1882
Inclined carbons and refractory block—		
Wilde ..	5,197	1878
McCarty and Sellière ..	144	1879
Inner and outer tubes, André ..	5,206	1879
Intermittent feed, Brockie	1,713	1882
Jockey roller, Lea ..	1,919	1882
Lever adjustably fulcrumed		
Common ..	626	1882
Mackenzie ..	95	1882
Long solenoid—		
Krizik and Piette ..	1,397	1880
Thomas ..	578	1882
Long tapering solenoid core, *Krizik and Piette* ..	2,712	1882
Magnet core in two parts, Andrews ..	1,526	1881
Magnetic carbons, Hedges	925	1879
Magnetic clutch—		
Gülcher ..	2,038	1881
André ..	2,563	1881
Magnetism of solenoid automatically reversed, Brockie	4,504	1881
Main magnet automatically cut out, Crompton ..	346	1882
Main magnet normally overpowering shunt, *Waterhouse* ..	5,185	1881
Main and shunt magnets, *Brush* ..	947	1879
Main and shunt solenoids, *Von Alteneck* ..	2,652	1879
Make-and-break, Harding..	3,166	1881
Motor and relay, Mackenzie	95	1882
Motor, contact lever, and shunt solenoid, *Houston and Thomson* ..	315	1880
Motor, differential—		
André ..	2,764	1882
Tschikoleff and Kleiber ..	2,198	1881
Motor driven by heated air from arc, Varley ..	5,665	1881
Motor opposing weight of carbon, Andrews..	2,321	1879
Movable gearing frame—		
Bousfield ..	523	1879
Maxim ..	1,649	1880
Berjot ..	4,428	1880
Gumpel ..	253	1881
Common and Joel ..	1,040	1881
Wood ..	2,851	1881
Fyfe and Main ..	3,821	1841
Bürgin ..	4,820	1881
Varley ..	5,396	1881
Crompton ..	346	1882
Wood ..	2,503	1882
„ ..	2,570	1882
Moffatt ..	2,759	1882
Parallel carbons (*See also* " Candles, Electric.")		
Sabine ..	4,821	1878
Weston Co. ..	4,960	1878
Rapieff ..	211	1879
Mori, Hallewell, Milner, and Griffin ..	740	1879

ARC LAMPS with—

Entry	No.	Year
Parallel carbons,		
Jamin ..	863	1879
Berby ..	1,236	1881
Tubini ..	3,822	1881
Gerard-Lescuyer ..	2,144	1882
Parallel carbons and refractory stop, Hedges ..	81	1879
Parallel copper electrodes and stream of carbon, Wier	684	1879
Pawl and rack, *Roosevelt and Abdank* ..	3,070	1882
Peculiarly wound solenoid, Sperry ..	3,025	1882
Pendulum and escapement, *Von Alteneck* ..	4,949	1878
Pivotted arms operated by pinions, Dubos ..	427	1879
Pivotted guide tube, Hedges	925	1879
Pivotted + carbons inclined towards — carbon, André ..	830	1879
Plate electrodes, *Wallace*..	240	1878
Powdered substances fed to arc, Grant and Sennett..	2,267	1879
Pulleys and brake, Lande	834	1882
Pure calcium fed thereto, *McCarty and Sellière* ..	144	1879
Rack and gearing, Bright	3,305	1881
Rack and pinion, *Reynard* ..	5165	1878
Rack and spring pawl, Andrews ..	2,321	1879
Rack, pinion, and brake arrangement, *Union Co.*	392	1882
Rack, pinion, and pawl, Heinrichs ..	4,608	1880
Racks, pinion, and pendulum, *Million* ..	3,085	1879
Rack, pinion, and ratchet wheel, Gumpel ..	253	1881
Radial arms, pinions, and gearing, Heinrichs ..	4,589	1879
Radial electrode—		
Prosser ..	3,466	1875
Prosser and Moore ..	2,585	1877
Radius bars operated by chains and weights, *Puvilland and Raphaël* ..	2,111	1879
Radius rods, Dubos ..	427	1879
Ratchet feed, Gatehouse and Kempe ..	2,569	1882
Ratchet and pawl, *Von Alteneck* ..	4,949	1878
Refractory block—		
Bureau ..	1,704	1880
Wilde ..	5,197	1878
McCarty and Sellière ..	144	1879
Harrison ..	3,559	1881
Refractory electrodes—		
Heinke ..	1,910	1877
Hickley ..	4,132	1878
Sellon, Ladd, and Edmunds ..	4,646	1878
Delaye ..	1,536	1881
Daft ..	4,775	1881
Rogers ..	4,855	1881
Roche ..	5,477	1881
Mackenzie ..	95	1882
Refractory substance, gas or powder fed to arc, Rapieff ..	211	1879
Regulating cords, Rapieff..	831	1882
Regulating solenoids, Fyfe	774	1881
Relay and electric motor, Mackenzie ..	95	1882
Relay feed solenoid, André	4,948	1881
„ regulator, Bright ..	3,305	1881
Resistance, carbon, Greb	3,015	1881
Retarding apparatus, *Waterhouse* ..	5,185	1881
Ring series of carbons, Harrison ..	3,875	1879
Rotating arc—		
Jamin ..	863	1879
Andrews ..	2,321	1879
Rotating carbon helices, Freeman ..	5,307	1878
Rotating carbon ring and two discs, Harrison ..	3,875	1879
Rotating disc electrode—		
Whitehouse ..	1,820	1874
Weathers ..	610	1876

ARC LAMPS with—

Entry	No.	Year
Rotating disc electrode,		
Varley ..	4,905	1876
Heinke ..	1,910	1877
Reynier ..	2,982	1877
Heinke ..	4,275	1877
Ziffer ..	4,412	1877
Zanni ..	4,573	1878
Siemens ..	4,208	1878
Gye.. ..	4,473	1878
Forbes ..	4,116	1878
Whyte ..	5,152	1878
Heinrichs ..	4,595	1878
Rapieff ..	211	1879
Crompton and Willans ..	245	1879
Henley ..	51,37	1880
Etève ..	48	1881
Andrews ..	1,526	1881
Edison ..	2,495	1881
Gibbs ..	4,533	1881
Wier ..	2,686	1882
Rotating hollow carbons at right angles, Imray ..	382	1879
Screw feed—		
De Mersanne ..	3,315	1876
Ziffer ..	4,412	1877
De Mersanne and Bertin	5,110	1878
Reynard ..	5,165	1878
Mackenzie ..	1,635	1879
Roguier ..	75	1880
Muirhead and Hopkinson	153	1881
Cance ..	3,976	1881
Akester ..	2,419	1882
Bardon ..	3,079	1882
Semicircular carbons—		
Heinrichs ..	4,595	1878
Dubos ..	427	1879
Puvilland and Raphaël	2,111	1879
Heinrichs ..	4,589	1879
Gatehouse ..	4,796	1879
Mandon ..	2,147	1880
Heinrichs ..	4,608	1880
Mandon ..	4,914	1880
Separate feed current, André ..	4,948	1881
Shunt circuit automatically broken, *Scribner* ..	5,156	1879
Shunt feed magnet, *Moffatt*	2,759	1882
Shunt magnet, Krupp ..	3,837	1878
Shunt electro-magnet automatically cut out, Mathieson ..	2,734	1882
Shunt solenoid, *Gerard-Lescuyer* ..	1,685	1881
Six carbons fed to common centre, Brain ..	5,139	1878
Solenoid—		
Lontin ..	2,094	1877
Brush ..	2,003	1878
Dubos ..	2,401	1878
Von Alteneck ..	2,652	1879
Fyfe.. ..	774	1881
André ..	494	1881
Gerard-Lescuyer ..	1,685	1881
Solenoid automatically cut out—		
Andrews ..	2,321	1879
Million ..	2,788	1881
Solenoid, compound—		
Marcus and Egger ..	2,934	1877
Solenoid cores acting as dash-pots, Chadburn ..	2,755	1882
Solenoid, differential—		
Von Alteneck ..	4,949	1878
Kennedy ..	1,199	1882
Special core—		
Krizik and Piette ..	1,397	1880
Thomas ..	578	1882
Krizik and Piette ..	2,712	1882
Spiral carbon surrounding straight carbon, Pulvermacher ..	4,774	1878
Spring, coiled, Day.. ..	1,261	1874
Spring and dash-pot, Varley	5,396	1881
Step-by-step differentially wound solenoid, *Krizik and Piette*.. ..	2,712	1882
Stream of finely divided carbon, Varley ..	4,100	1878
Three carbon plates, Andrews ..	2,321	1879
Three insulated parallel plates, Andrews ..	2,321	1879
Tilting holders, *Weston Co.*	4,960	1878

ARC LAMPS with—	No.	Year
Train of wheels and fly, *De Mersanne and Bertin*	5,110	1878
Tubular electrodes for supply of carbonaceous matter, Tubini	3,635	1881
Two carbon rotating discs, Henley	5,137	1880
Two differential magnets and clutch, Levey and Lumley	1,249	1882
Two opposing motors, *Tschikoleff and Kleiber*	2,198	1881
Two pairs of carbons at right angles, Heinrichs	4,595	1878
Two pairs of inclined carbons, *Gerard-Lescuyer*	1,552	1880
Two pairs of semicircular carbons at right angles, Heinrichs	4,589	1879
Two parallel C carbons and rotating cylinder, Heinrichs	4,595	1878
Two parallel discs, Andrews	2,321	1879
Two + and one — electrode and abutment, Hedges	4,988	1880
Two semicircular and one straight electrode, Heinrichs	4,589	1879
Two upper inclined carbons, Jeffery	1,570	1882
Variable leverage, Mackenzie	95	1882
Verge escapement, Graham	1,171	1882
Vertical carbons and carbon disc, André	830	1879
Vertical carbon and horizontal carbons fed to fixed centre, Brain	5,139	1878
Vibrating armature, Hawkes and Bowman	2,402	1881
Vibrating electrodes, *De Mersanne*	2,787	1875
	3,315	1876
Vibrating electrodes, Wier	684	1879
Water-cased terminal, Siemens	2,110	1879
Water-jacketted carbons, Remington	192	1879
ARC INCANDESCENCE LAMPS (see also "Semi-Incandescence Lamps.")		
Werdermann	2,477	1878
Hickley	4,132	1878
Parallel, Werdermann	2,477	1878
ARMATURES—		
Tyer	1,845	1873
Siemens and Alteneck	2,006	1873
Gaumé	2,618	1873
Monckton	265	1874
"	3,509	1874
Chutaux	4,454	1874
Bürgin	3,243	1875
Fuller and Crandall	3,364	1875
Camacho	3,416	1875
Kimball	3,999	1875
Bastet	1,931	1876
Zanni	2,821	1876
Lontin	3,264	1876
Faure	3,670	1876
Varley	4,905	1876
"	270	1877
Jablochkoff	3,187	1877
Hickley	3,552	1877
Varley	4,435	1877
Schuckert	4,464	1877
Gramme and d'Ivernois	953	1878
Brush	2,003	1878
Société d'Electricité	4,066	1878
Higgs	4,206	1878
Verrue	4,287	1878
Formby	565	1879
Keith	1,387	1879
Chambrier	4,428	1879
Joel	5,157	1879
Slater	2,272	1880
Willatt	3,808	1880
Niaudet and Reynier	3,971	1880
Etéve	48	1881
Waller	803	1881
De Meritens	2,212	1881
Lachaussée	2,761	1881
Mignon and Rouart	3,400	1881

ARMATURES—	No.	Year
Arey	3,456	1881
Berthoud, Borel, and Co.	4,026	1881
Varley	5,665	1881
Williams	85	1881
Akester and Barnes	986	1882
Williams	1,174	1882
Matthews	1,201	1882
Williams	1,556	1882
Kennedy	1,640	1882
Chertemps and Dandeu	1,747	1882
Rogers	1,760	1882
Floyd and Kirkland	2,225	1882
Williams	2,558	1882
Schuyler and Waterhouse	3,002	1882
Annular, Werdermann	3,156	1874
Arrangement of, Richardson	5,681	1881
Automatically short-circuiting coils of, Chertemps and Dandeu	1,747	1882
Circuits, arrangement of (See also "Armatures, Winding.")		
Weston	*4,903	1877
Formby	565	1879
Bürgin	5,085	1879
Houston and Thomson	315	1880
Elphinstone and Vincent	2,893	1880
Elmore	3,832	1880
Heinrichs	4,608	1880
Lachaussée	2,761	1881
Edison	2,954	1881
Harling and Hartmann	3,472	1881
Fahrig	4,107	1881
Akester	5,525	1881
Williams	85	1882
Siemens	760	1882
Rogers	1,760	1882
Wood	2,623	1882
Fyfe and Main	2,636	1882
Willard	2,803	1882
Gordon	2,871	1882
Maquaire	2,885	1882
Ayrton and Perry	3,036	1882
Coils arranged in symmetrical sets, *Schuyler and Waterhouse*	3,002	1882
Coils, insulated metallic tape, Brougham	2,044	1882
Coils, insulating, *Williams*	85	1882
" manufacturing, Vincent and Elphinstone	2,340	1882
Coils, radially adjustable, *Maquaire*	2,885	1882
Compound pole—		
Akester and Barnes	986	1882
Piot	1,692	1882
Compound disc, *Edison*	2,052	1882
" ring—		
Bürgin	5,085	1879
Akester	5,525	1881
Chameroy	2,295	1882
Conductors, bar, *Edison*	1,139	1882
" silver wire, Thompson	4,988	1878
Cooling by water, Elmore	3,832	1880
Core of—		
Siemens and Alteneck	2,006	1873
Weston	*4,903	1877
Siemens and Alteneck	3,134	1878
Ward	3,976	1878
Heinrichs	4,589	1879
Perry	1,178	1880
Edison	1,385	1880
Zipernowsky	1,580	1880
Gumpel	253	1881
Jurgensen and Lorenz	2,416	1881
Gülcher	64	1882
Henley	130	1882
Sheridan	931	1882
Brockie	756	1882
Jarman	2,565	1882
Sperry	3,025	1882
Ayrton and Perry	3,036	1882
Core, brass, *De Meritens*	3,658	1878
" hollow ring containing the conductor, Farquharson	2,771	1882
Core, movable, *Schuyler and Waterhouse*	3,002	1882
Core, non-magnetic, *Siemens and Alteneck*	3,134	1878

ARMATURES—	No.	Year
Core, spiral, Chadburn	2,755	1882
Counter currents in, preventing, Wilde	5,008	1880
Cut-out, fusible, Edison	33	1880
Cutting out single coils of, Edison	3,964	1880
Cylindrical—		
Weston	*4,903	1877
Pulvermacher	4,844	1878
Weston Co.	4,960	1878
Rapieff	211	1879
Elphinstone and Vincent	332	1879
Bousfield	523	1879
Andrews	2,321	1879
Edison	2,402	1879
Houston and Thomson	4,400	1879
" "	315	1880
Zipernowsky "	1,580	1880
Gumpel	3,324	1880
Thompson	3,928	1880
Hussey and Dodd	4,265	1880
Gumpel	253	1881
Edison	1,240	1881
Hussey and Dodd	2,375	1881
Edison	2,954	1881
Newton	4,559	1881
Griscom	5,551	1881
Thomson	5,668	1881
Gravier	943	1882
Vincent and Elphinstone	2,340	1882
Ayrton and Perry	2,830	1882
Cylindrical cage, *Edison*	3,932	1881
" hollow, Lande	838	1882
" shell, *Siemens and Alteneck*	3,134	1878
Cylindrical, with diagonal coils, Elphinstone and Vincent	2,894	1880
Disc—		
Monckton	265	1874
Sprague	4,762	1878
Rapieff	211	1879
Seeley	1,998	1880
Edison	2,954	1881
Thomson	5,668	1881
Phillips and Johnson	1,347	1882
Rogers	1,760	1882
Gerard-Lescuyer	1,878	1882
Vincent and Elphinstone	2,340	1882
Jarman	2,565	1882
Fyfe and Main	2,636	1882
Driving—		
Zanni	4,232	1877
Thomson	5,668	1881
Drum, *Williams*	1,556	1882
Duplex, Brockie	756	1882
Ellipsoidal, Werdermann	3,156	1874
Grooved, *Paine and Paine*	2,049	1875
Helical—		
Werdermann	3,156	1874
Andrews	540	1882
Immersed in liquid, Millar	4,592	1881
Longitudinal coils, with—		
Solignac	740	1882
Brockie	756	1882
Spagnoletti	869	1882
Machine for winding annular, *Haase and Recker*	3,832	1881
Manufacture of, Dion	3,880	1881
Mounting—		
Thomson	5,668	1881
Brougham	2,044	1882
Wood	2,531	1882
Multiplex—		
Werdermann	3,156	1874
Kennedy	4,541	1881
Multiple pole—		
Heinrichs	4,595	1878
Bertin and De Mersanne	5,076	1878
Brain	5,139	1878
Rapieff	211	1879
Dubos	749	1879
Sellon and Edmunds	1,949	1879
Elmore	3,565	1879
Lamar	4,696	1870
Werdermann	79	1880
Cance	4,005	1880
Biloret and Mora	4,049	1880
Hopkinson and Muirhead	4,886	1889
Henley	5,137	1880
Gordon	78	1881
Wilde	497	1881

ARMATURES, Multiple pole—	No.	Year
Siemens and Halske	1,447	1881
Muller and Levett	1,787	1881
Lane-Fox	3,394	1881
Fahrig	4,107	1881
Aylesbury	4,304	1881
Piot	4,851	1881
Gordon	5,536	1881
Gerard-Lescuyer	5,593	1881
Cumine	2,318	1882
Hallett	2,573	1882
Gordon	2,871	1882
Oval, Raison	169	1882
Ring—		
Evans	1,970	1873
Monckton	4,597	1876
Wilde	1,228	1878
Siemens and Alteneck	3,134	1878
Ward	3,976	1878
Zanni	4,573	1878
Sprague	4,762	1878
Brain	5,139	1878
Rapieff	211	1879
Schuckert	960	1879
Heinrichs	4,589	1879
Glouchoff	478	1880
Perry	1,178	1880
Fitzgerald	872	1880
Gumpel	3,041	1880
Heinrichs	4,608	1880
Higgs	1,961	1881
Harling and Hartmann	3,472	1881
De Meritens	4,207	1881
Joel	4,607	1881
Henley	130	1882
Hussey and Dodd	234	1882
Sheridan	931	1882
Edison	1,139	1882
Gravier	1,211	1882
Levey and Lumley	1,249	1882
Rogers	1,760	1882
Emmens	2,349	1882
Willard	2,803	1882
Ayrton and Perry	2,830	1882
Maquaire	2,885	1882
Parker and Elwell	2,917	1882
La Cie. Electrique	2,990	1882
Ayrton and Perry	3,036	1882
Ring, elongated—		
Weston	*4,903	1877
Maxim	1,372	1880
Hussey and Dodd	4,265	1880
Moffatt and Chichester	3,441	1881
Little	497	1882
Brougham	2,044	1882
Werdermann	2,364	1882
Jarman	2,565	1882
Blyth and Peebles	2,661	1882
McTighe	2,744	1882
Schuyler and Waterhouse	3,002	1882
Sperry	3,025	1882
Rolling—		
Werdermann	1,829	1877
Smith	3,981	1877
Bear	3,283	1881
Spherical—		
Werdermann	3,156	1874
Bürgin	4,819	1881
Raison	169	1882
Spring mounted, Andrews	540	1882
Ventilating—		
Weston	*4,903	1877
Maxim	1,392	1880
Zipernowsky	1,580	1880
Hussey and Dodd	4,265	1880
Edison	3,932	1881
Gülcher	64	1882
Raison	169	1882
Lande	838	1882
Sheridan	931	1882
Vibrating, Pope	2,656	1873
Winding (See also "Armatures, Circuits, Arrangement of.")		
Weston	*4,903	1877
Weston Co...	4,960	1878
Thompson	3,928	1880
Cabanellas	200	1881
Siemens and Halske	1,447	1881
Muller and Levett	1,787	1881
Newton	4,559	1881
Joel	4,607	1881
Weston	1,614	1882

ARMATURE, Winding—	No.	Year
Werdermann	2,364	1882
Crompton	2,618	1882
Weston	2,694	1882
BATTERIES, SECONDARY—		
Planté	1,713	1873
Stearns	2,870	1873
Achard	1,387	1877
Lane-Fox	3,988	1878
Wilson	4,348	1878
Edison	5,306	1878
Faure	129	1881
	1,676	1881
La Société la Force et la Lumière, &c.	2,323	1881
Muirhead	2,606	1881
Tommasi	2,782	1881
Watt	4,255	1881
Laurie	4,310	1881
Volckmar	4,398	1881
Parod	4,508	1881
Rogers	4,854	1881
Fournier	5,104	1881
Volckmar	5,261	1881
Fitzgerald	5,481	1881
Kennedy	72	1882
Liardet and Donnithorne	120	1882
Tommasi	252	1882
Barrier and Vernède	538	1882
Williams	560	1882
"	700	1882
"	766	1882
De Meritens	1,173	1882
Molloy	1,455	1882
Faure	1,769	1882
Rogers	1,999	1882
Cuttriss	2,135	1882
Lory	2,409	1882
Bright	2,602	1882
Woodward	2,643	1882
Gumpel	2,688	1882
Epstein	2,807	1882
Nezeraux	3,039	1882
Alternating current from,		
Wilson	4,348	1878
Application of—		
Edison	4,226	1878
Rapieff	211	1879
Buell	5,656	1881
Duwelius, Goss, Higgs, Merrell, Peck, and Walter	5,751	1881
Arthur	2,128	1882
Rogers	1,390	1882
Charge of, indicating, Carus Wilson	5,687	1881
Charging—		
Houston and Thomson	4,400	1879
Edison	4,553	1882
Williams	85	1882
Woodward	2,643	1882
Charging and discharging—		
Thomson	3,032	1882
Deprez and Carpentier	4,128	1881
Parod	4,508	1881
Carus Wilson	5,681	1881
Perry	55	1882
Williams	2,558	1882
Emmens	2,912	1882
Combined with incandescence lamp, Gumpel	2,756	1882
Conical cells, with, La Société Anonyme la Force et la Lumière, &c.	1,097	1881
Condensers used with, Scott	4,671	1878
Coupling—		
Lane-Fox	4,043	1882
Brain	2,659	1882
Edison	5,306	1878
Williams	2,558	1882
Nezeraux	3,039	1882
Currents from lamps are switched through, Arthur	2,128	1882
Cut-out for, Munro	1,626	1882
Elastic separating pieces, Brain	1,548	1882
Electrodes for—		
Wilson	4,348	1878
Houston and Thomson	4,400	1879
Faure	129	1881
Faure	1,676	1881
Jablochkoff	1,745	1881

BATTERIES, Secondary—	No.	Year
Electrodes for,		
Swan	2,272	1881
Muirhead	2,606	1881
Tommasi	2,782	1881
Laurie	2,823	1881
Sellon	3,926	1881
"	3,987	1881
"	4,005	1881
De Kabath	4,037	1881
Watt	4,255	1881
Laurie	4,310	1881
Volckmar	4,398	1881
Swan	4,455	1881
Sellon	4,632	1881
Fournier	5,104	1881
Crompton and Fitzgerald	5,159	1881
Biggs and Beaumont	5,198	1881
Carpentier and Pezzer	5,322	1881
Fitzgerald, Biggs, and Beaumont	5,338	1881
Pitkin	5,451	1881
Fitzgerald	5,481	1881
Swan	5,494	1881
Grout and Jones	5,521	1881
Sellon	5,631	1881
Fitzgerald, Biggs, and Beaumont	29	1882
Kennedy	72	1882
Liardet and Donnithorne	120	1882
Boettcher	130	1882
Tommasi	252	1882
Humphreys	289	1882
Sellon	319	1882
Barrier and Vernède	538	1882
Williams	766	1882
Pulvermacher	837	1882
Swan	905	1882
Williams	1,174	1882
Lyte	1,363	1882
Cohne	1,437	1882
Lalande and Chaperon	1,464	1882
Brain	1,548	1882
Tribe	1,587	1882
Fitzgerald, &c.	1,875	1882
Cathcart	2,068	1882
Cuttriss	2,135	1882
Tribe	2,263	1882
Emmens	2,349	1882
Pitkin	2,391	1882
Lory	2,409	1882
Vincent and Elphinstone	2,491	1882
Williams	2,558	1882
Bright	2,602	1882
Woodward	2,643	1882
Brain	2,659	1882
Gerard-Lescuyer	2,676	1882
Fitzgerald	2,763	1882
Varley	2,776	1892
Epstein	2,807	1882
Sellon	2,818	1882
Eaton	2,826	1882
Emmens	2,913	1882
Parker and Elwell	2,917	1882
Aron	2,943	1882
Sorley	2,945	1882
Wilkinson	3,003	1882
Nezeraux	3,039	1882
Watt	3,097	1882
Electrodes, coiled—		
Varley	2,776	1882
Emmens	2,913	1882
Electrodes forming—		
Liardet and Donnithorne	120	1882
Boettcher	130	1882
Muirhead	2,654	1882
Gerard-Lescuyer	2,676	1882
Epstein	2,807	1882
Electrodes, protecting—		
Price	2,722	1882
Nezeraux	3,039	1882
Electrodes, separate charging and discharging, Williams	2,558	1882
Electrolytes for—		
Wilson	4,348	1878
Houston and Thomson	4,400	1879
Laurie	2,823	1881
Watt	4,255	1881
Laurie	4,310	1881
Sellon	4,005	1881
Fitzgerald, Biggs, and Beaumont	29	1882

	No.	Year
BATTERIES, Secondary—		
Electrolytes for,		
Cuttris	2,135	1882
Emmens	2,349	1882
Varley	2,776	1882
Gas—		
Biggs and Beaumont ..	5,198	1881
Morel	185	1882
Williams	856	1882
"	2,558	1882
Westphal	2,823	1882
Gülcher	2,875	1882
Lead fibres for, *Eaton* ..	2,826	1882
Portable—		
Sellon	4,632	1881
Gumpel	2,756	1882
Preparing lead for—		
Boys..	1,946	1882
Boggett	2,595	1882
Preventing reverse current		
entering, Munro	1,626	1882
Pyramidal - shaped cells,		
Faine	1,676	1881
Separate discharging elec-		
trodes for, Williams ..	2,558	1882
Small, fed by large, Rogers	1,999	1882
Regulating current from—		
Thomson	3,032	1881
De Kabath..	4,060	1881
Edison	4,553	1881
Williams	2,558	1882
Vessels for—		
Faure	1,769	1882
Parker and Elwell ..	2,917	1882
Battery, thermo-electric,		
Fauré	2,946	1875
BRAKE, Electric—		
Lontin	3,264	1876
Winter	3,146	1881
BRUSHES, Commutator—		
Schuckert	960	1879
Gumpel	3,041	1880
Hopkinson and Muir-		
head..	4,886	1880
André	5,268	1880
De Meritens	4,207	1881
Gordon	2,871	1882
Edison	3,804	1881
Gumpel	253	1881
Double, Schuyler and		
Waterhouse	3,002	1882
Embracing nearly the whole		
of the commutator, Rai-		
son	169	1882
Holders for—		
Hallett	2,573	1882
La Cie. Electrique ..	2,990	1882
Inclined, Edison	1,385	1880
Isolated, breaking contact,		
Edison	3,231	1881
Mounting—		
Heinrichs	4,595	1878
Weston and Co.	4,960	1878
Elphinstone and Vincent	332	1879
" "	2,893	1880
Edison	3,964	1880
De Meritens	4,207	1881
Joel	4,607	1881
Andrews	540	1882
Vincent and Elphinstone	2,340	1882
Werdermann	2,364	1882
Ayrton and Perry ..	2,830	1882
Gordon	2,871	1882
Sperry	3,025	1882
Mounting automatically ad-		
justed, *Houston and*		
Thomson	315	1880
Overlapping two plates,		
Weston	*4,903	1877
Placed at an angle, Willard	2,803	1882
Rotary, metallic wire, *Bi-*		
loret and Mora	4,049	1880
Rotating—		
Varley	5,667	1881
Thomson	5,668	1881
Subsidiary, Joel	4,607	1881
Two for each polar field,		
Andrews	540	1882
Electric hair, Ashworth ..	454	1882
CABLES, ELECTRIC (See also "Con-		
ductors, Compound.")		
Delany	333	1882

CABLES, Electric—	No.	Year
Clark	361	1882
Telegraph, Henley ..	386	1882
CANDLES, Electric—		
Jablochkoff	3,552	1876
Werdermann	4,805	1876
Jablochkoff	494	1877
Wilde	3,250	1878
De Meritens	178	1879
Jablochkoff	1,175	1879
De Meritens	2,339	1879
Gatehouse	4,796	1879
Berly	1,027	1881
"	1,236	1881
Williams	224	1882
Disc form, Andrews ..	416	1879
Holder, Denarouse ..	3,170	1879
In vacuum, Blamires ..	455	1880
Relighting, *Jablochkoff*	1,175	1879
With clockwork, Jabloch-		
koff	3,552	1876
CARBONS (See also " Electrodes ;" " In-		
candescence Lamps, Filaments		
for.")		
Monckton	4,597	1876
Prosser and Moore ..	2,585	1877
Reynier	2,982	1877
"	471	1878
Scott	861	1878
Brush	2,003	1878
Gray..	4,553	1878
Punshon	5,105	1878
Freeman	5,307	1878
Cohne	277	1879
Jamin	863	1879
Thompson	1,622	1879
De Hamel	2,543	1879
Werdermann	304	1881
Lorrain	2,848	1881
Bouteilloux and Laing ..	3,214	1881
James and Lee	4,396	1881
De Changy	4,405	1881
Allport and Punshon ..	4,850	1881
Liveing and Boys ..	69	1882
Munro	1,626	1882
Smith	1,465	1882
Maxim	1,619	1882
Hallet	2,560	1882
Varley	2,776	1882
Wilkinson	3,003	1882
Vogel	3,033	1882
Adjusting resistance of—		
Crookes	1,079	1882
Hallet	2,560	1882
Circular, Heinrichs ..	4,608	1880
Combined at ends with		
pyrotechnic composition,		
Holcombe	1,384	1881
Combined with liquid car-		
bonaceous substance,		
Brougham	630	1880
Concentric, Asten ..	2,020	1882
Covered with enamel or		
glass, *Mignon and Rouart*	3,402	1881
Endless, for incandescence		
lamp, Graham and Smith	1,392	1882
Flexible—		
Harrison	3,470	1878
Punshon	5,105	1878
Holders for—		
Brush	2,003	1878
Godfrey	4,718	1879
Bouteilloux and Laing ..	842	1880
Berjot	4,428	1880
Chertemps	3,349	1881
Bürgin	4,820	1881
Waterhouse	5,185	1881
Varley	5,396	1881
Hollow, De Hamel ..	2,543	1879
" for arc lamps, *King*	1,822	1882
" for incandescence		
lamps—		
Jameson	1,670	1882
Cruto	1,895	1882
Increasing conductivity of,		
King	1,822	1882
Insulated on three sides,		
Gibbs	4,533	1881
Joining, Siemens	3,315	1878
" to form con-		
tinuous, Siemens.. ..	4,614	1880
Manufacturing—		
Stuart-Wortley	3,656	1878

CABLES, Electric—	No.	Year
Manufacturing,		
Heinrichs	4,589	1879
Manufacturing hollow or		
cored, *Mignon and Rouart*	3,402	1881
Manufacturing, press for,		
Edison	2,402	1879
Measuring resistance of,		
Crookes	1,079	1882
Saturated with hydro-		
carbon, Varley	5,396	1881
Testing apparatus for,		
Rogers	1,288	1882
Thickening, at junction		
with conductors, Lane-		
Fox	1,647	1882
Spiral, Waller	803	1881
With wick, glass, *Siemens*	2,199	1879
" metallic, De		
Hamel	2,543	1879
Zig-zag, Waller	803	1881
CELL—		
Electro-plating, *Edison* ..	768	1881
Porous carbon for, Varley	2,776	1882
Cement, composition of,		
for mounting filaments,		
Nathomb	2,452	1882
Chandeliers for electric lamps,		
Edison	1,802	1881
Chinamen, hair of, Stanley ..	2,660	1882
CIRCUITS—		
Automatically opening,		
Edison	2,402	1879
Closer and breaker, Me-		
chanical, *Weston* ..	*4,903	1877
Preventing static charge in		
lighting, Crompton ..	346	1882
Coils, forming, Clarke and		
Leigh	1,483	1882
COILS, Induction (See " Gene-		
rator, Secondary ;" and		
" Induction Coil.")		
Insulating, Dion	3,880	1881
Intensity, Monckton ..	4,597	1876
Ruhmkorff, De Sussex ..	465	1879
COLLECTORS, Current (See also		
" Commutator.")		
Cochrane	4,313	1878
Liveing	4,833	1880
Parker and Elwell ..	2,917	1882
Disc running in mercury,		
Henley	5,137	1880
COMMUTATOR (See also " Collectors,		
Current," and " Switches.")		
Siemens and Alteneck ..	2,006	1873
Wallace	2,015	1873
Stone	94	1874
Monckton	265	1874
De Mersanne	1,446	1874
Werdermann	3,156	1874
Monckton	3,509	1874
Lartique	3,771	1874
Chutaux	4,454	1874
Hussey	2,043	1875
Clamond	2,205	1875
Kilner	2,996	1875
Fuller and Crandall ..	3,364	1875
Carmacho	3,416	1875
Edison	3,762	1875
Paine	4,118	1875
Bertin	4,311	1875
Lontin	3,264	1876
Faure	3,670	1876
Weston	4,280	1876
Monckton	4,597	1876
Varley	4,905	1876
Rapieff	4,432	1877
Weston	4,748	1877
"	*4,903	1877
Pulvermacher	1,587	1878
Brush	2,003	1878
Siemens and Alteneck ..	3,134	1878
De Meritens	3,658	1878
Ward	3,976	1878
Heinrichs	4,595	1878
Melhado	4,699	1878
Sprague	4,762	1878
Weston Co.	4,960	1878
Rapieff	211	1879
Formby	565	1879
Keith	1,387	1879
Edison	2,402	1879
Houston and Thomson ..	4,400	1879

COMMUTATORS—

	No.	Year
Lamar	4,696	1879
Bürgin	5,085	1879
Houston and Thomson	315	1880
Glouchoff	478	1880
De Meritens	1,136	1880
Perry	1,178	1880
Elphinstone and Vincent	2,893	1880
Gumpel	3,041	1880
Elmore	3,832	1880
Thompson	3,928	1880
Niaudet and Reynier	3,971	1880
Cance	4,005	1880
Johnson and Edison	4,621	1880
Henley	5,137	1880
Cabanellas	200	1881
Gumpel	253	1881
Siemens and Halske	1,447	1881
Muller and Levett	1,787	1881
Higgs	1,961	1881
De Meritens	2,212	1881
Hussey and Dodd	2,375	1881
Edison	2,954	1881
Lane-Fox	3,394	1881
Berthoud, Borel, and Co.	4,026	1881
Fabrig	4,107	1881
De Meritens	4,207	1881
Newton	4,559	1881
Joel	4,607	1881
Bürgin	4,819	1881
Piot	4,851	1881
Blake	5,096	1881
Varley	5,667	1881
Thomson	5,668	1881
Hussey and Dodd	234	1882
Siemens	760	1882
Matthews	1,201	1882
Kennedy	1,640	1882
Chertemps and Dandeu	1,747	1882
Gerard-Lescuyer	1,873	1882
Floyd and Kirkland	2,225	1882
Chameroy	2,295	1882
Vincent and Elphinstone	2,340	1882
Sperry	3,025	1882
Adjustable, Ayrton and Perry	2,830	1882
Brushes (See also "Brushes, Commutator.")		
Brushes, adjustable, Andrews	540	1882
Brushes, double, *Schuyler and Waterhouse*	3,002	1882
Brushes embrace nearly the whole of the, Raison	169	1882
Brushes, inclined, Edison	1,385	1880
" isolated, breaking contact, *Edison*	3,231	1881
Brushes, overlapping two plates, *Weston*	*4,903	1877
Brushes, subsidiary, Joel	4,607	1881
" two for each polar field, Andrews	540	1882
Cleaning—		
Monckton	265	1874
Raworth	27	1879
Connecting to armature coils, *Wood*	2,623	1882
Continuous or alternating current, *Johnson and Edison*	4621	1880
Disc form of—		
Werdermann	2,364	1882
Hallett	2,573	1882
Fyfe and Main	2,636	1882
McTighe	2,744	1882
Ayrton and Perry	2,830	1882
Gordon	2,871	1882
Schuyler and Waterhouse	3,002	1882
Elastic, *Siemens and Alteneck*	2,006	1873
Exciting alternate current generator, *Meritens*	1,136	1880
Heating, preventing in, *Laing*	3,169	1881
Lubricating, *Zipernowsky*	1,580	1880
Mounting, Joel	5,157	1879
Reducing resistance at, *Edison*	3,804	1881
Rotating collectors for—		
Varley	5,567	1881
Thomson	5,668	1881
Segments in two parts, connected by a resistance, Heinrichs	4,608	1880

COMMUTATORS—

	No.	Year
Segments, oblique, Heinrichs	4,608	1880
" spiral—		
Weston Co.	4,960	1878
Houston and Thomson	315	1880
Segments separated by oblique gaps, Thompson	3,928	1880
Segments, V shaped, Willard	2,803	1882
Sparking, preventing at—		
Sprague	4,762	1878
Edison	1,385	1880
Maxim	1,392	1880
Hopkinson and Muirhead	4,886	1880
Wilde	497	1881
Edison	3,231	1881
"	3,804	1881
Williams	1,174	1882
Edison	1,496	1882
Munro	1,626	1882
Chertemps and Dandeu	1,747	1882
Gordon	2,871	1882
Ventilating, *Houston and Thomson*	315	1880
Wear of, Insuring even, Edison	1,385	1880
Compensator, Pulvermacher	1,587	1878
CONDENSERS—		
Stearns	2,870	1873
Monckton	4,597	1876
Jablochkoff	3,839	1877
"	1,996	1877
Cadett	4,316	1878
Scott	4,671	1878
Parod	4,686	1878
Williams	2,558	1882
Applied to electric lighting, *Sawyer and Man*	4,705	1878
Used with secondary batteries, Scott	4,671	1878
CONDUCTORS—		
Edison	5,306	1878
Parod	4,686	1878
Garnier	4,952	1880
Perry and Ayrton	785	1881
Edison	3,483	1881
Crompton	5,080	1881
Delany	333	1882
Clark	361	1882
Arrangement of, for mine signalling, Cummings	2,888	1880
Arrangement of, for signalling on trains, Minchin and Despeissis	1,010	1882
Arrangement of, for signalling on railways, D'Auriac	1,683	1881
Attaching to insulators, Edwards	1,720	1880
Braiding, James	4,261	1874
" Ewen and James	1,719	1875
Chandelier arrangement, Common and Joel	1,040	1881
Coiling, Clarke and Leigh	1,483	1882
" for induction apparatus, Clarke and Leigh	3,652	1881
Collecting currents from, for railways, Putman	1,125	1880
Collecting currents from, for railways, Siemens	583	1880
Combined with secondary conductor, Perry and Ayrton	783	1881
Composition of—		
Higgs	1,961	1881
Weiller	1,821	1882
Composite, Zanni	1,855	1874
Compound—		
Lackersteen	3,666	1875
Rubery	2,759	1876
Rapieff	4,432	1877
Varley	4,435	1877
Berthoud and Borel	4,346	1878
Alberger and Pettit	4,601	1878
Linford	1,184	1880
Manly and Philips	1,475	1880
André	5,268	1880
Delany	1,809	1881
Henley	1,873	1881
Lugo	1,696	1881
Delany	2,217	1881
Mowbray	2,437	1881
Clark	2,592	1881
Brooks	3,254	1881
Stearns	5,468	1881

CONDUCTORS—

	No.	Year
Concentric, Van Rysselberge	2,804	1882
Conduits for (See also "Conductors, Laying Underground.")		
Hermann	2,962	1879
Prall and Obrick	3,831	1879
Linford	4,456	1879
Linford	1,184	1880
Watson	2,271	1880
Lamb	2,665	1880
Strohm	3,790	1881
Gore	4,797	1881
Connected to condensers, Edison	2,482	1881
Cooling, Higgs	1,961	1881
Coupling—		
Owen	534	1873
Radde	2,091	1873
Mackie	440	1874
Eustace	1,766	1874
Goldstone, Radcliffe, and Gray	3,381	1874
Preece, Goldstone, Radcliffe, and Gray	3,521	1874
Masin	792	1875
Richardson and Moffatt	3,085	1875
Tyer	557	1876
Pickersgill	1,693	1877
"	3,854	1877
Dibbin	4,036	1877
Nettlefold	2,962	1878
Winter	3,146	1881
Spagnoletti	869	1882
Coupling for automatic, Allen	2,215	1881
Coupling for, safety, Brewtnall	5,226	1881
Covering—		
Smith	3,622	1878
Glover and James	5,237	1880
Covering machine for, Truman	94	1874
Discharging surplus electricity from, *Richardson*	2,644	1882
Draw vices for erecting overhead, Fletcher	859	1881
Earth, Lane-Fox	3,988	1878
Electroliers, Brewtnall	2,934	1882
Electroplating, Rubery	4,193	1873
Fluid—		
Spalding	1,195	1878
Anderson	1,824	1880
For electric railways—		
Monckton	265	1874
Edison	3,894	1880
Perry and Ayrton	788	1881
Binko	3,073	1882
For railway signal and points, Denne	5,404	1880
Grapnels for submarine, &c., Anderson and Johnson	2,311	1882
Grooved, *Strickler*	3,099	1876
Hollow, for conveying gas, Wilkinson	3,472	1879
Hollow, forming also a pipe, Wilkinson	3,003	1882
Hollow, forming, out of wire, Wilkinson	3,003	1882
Induction in, preventing—		
Bell	4,341	1877
Chinnock and Harrison	995	1880
" "	699	1880
Gower	814	1880
Heaviside	1,407	1880
Lugo	1,119	1881
Barney	2,264	1881
Lugo	2,394	1881
Rysselberge	2,484	1881
Delany and Johnson	2,532	1881
Brooks	3,254	1881
Henck	4,058	1881
Stearns	5,468	1881
Delany	333	1882
Clark	361	1882
Smillie	1,850	1882
Briggs and Kinsman	1,774	1882
Smillie	1,850	1882
Inoxidisable, *Brownell*	2,302	1873
Insulating material fed to, Gray	5,056	1879
Insulating—		
Moseley	2,969	1873

CONDUCTORS—

	No.	Year
Insulating,		
Wilkinson	8,083	1873
Hooper	3,780	1873
Hooper and Dunlop	3,997	1873
Madsen	4,167	1873
Rubery	4,193	1873
Truman	94	1874
Monckton	265	1874
Walker	293	1874
Macintosh	447	1874
Zanni	1,855	1874
Conybeare and Naphegyi	2,106	1875
Ewen and James	1,719	1875
Field and Talling	1,938	1875
Greening	2,059	1875
Lackersteen	3,666	1875
Danckwerth	1,704	1876
Henley	1,944	1876
Deiss and Scaife	2,866	1876
Strickler	3,099	1876
Menier	4,705	1876
Beckingsale	555	1877
Henley	833	1877
Brooks	4,824	1877
Truman	375	1878
Wiebe	767	1878
Abbott	1,937	1878
Berthoud and Borel	4,346	1878
Truman	4,438	1878
Moseley	307	1879
Jack and Greening	718	1879
Haymen	959	1879
Blondot and Bourdin	2,629	1879
Spence	2,706	1879
Moseley	3,001	1879
Wilkinson	3,472	1879
Arbogast and McTighe	3,778	1879
Bell	4,549	1879
Heins	4,916	1879
Eaton	751	1880
Loeffler	905	1880
Quin	1,239	1880
Knudson and Kane	1,295	1880
Manly and Philips	1,475	1880
Truman	3,310	1880
Jacques	3,424	1880
Clark	229	1881
Solitdorf	*1,272	1881
Delany	1,809	1881
Sparling	*2,414	1881
Mackie	2,542	1881
Rinyard and Fleming	2,807	1881
Henak	4,058	1881
Millar	4,592	1881
Gore	4,797	1881
Johnson and Phillips	4,885	1881
Abbott and Field	5,407	1881
Smith	5,599	1881
Culbertson and Brown	5,615	1881
Labye and Locht-Labye	5,661	1881
Thomas and Requa	1,336	1882
Shippey and Punshon	2,293	1882
Wilkinson	3,003	1882
Insulating, of armature, *Edison*	3,932	1881
Insulating, of electric railways, Edison	1,862	1882
Insulating joints of, *Knudson and Kane*	1,295	1880
Insulating and laying in earthenware blocks, Meyer	232	1882
Insulating and laying for railways, Varley and Judd	441	1882
Insulating machine for (See also "Insulating Materials.")		
James	624	1880
"	1,333	1880
Clark	2,592	1888
Insulating material for—		
Mourlot	2,121	1880
Heyer	3,885	1879
Insulating and protecting for railways, Binko	3,073	1882
Insulating underground, Gray	3,863	1873
Insulating with glass, *Alberger and Pettit*	4,601	1878
Insulated metallic coated, *Menier*	755	1880
Insulation for non-inflammable, Phillips	2,571	1882

CONDUCTORS—

	No.	Year
Jointing,		
Truman	94	1874
Eustace	3,172	1874
Timmins	3,800	1874
Smith	4,311	1875
Hiboll	4,159	1876
Phillips	3,603	1878
Woods	3,971	1878
Alberger and Pettit	4,601	1878
Sheldon	2,397	1879
Blondot and Bourdin	2,629	1879
Young	3,368	1879
Arbogast and McTighe	3,778	1879
Crighton	4,189	1879
Allen	2,215	1881
Crompton	5,080	1881
Brewtnall	5,226	1881
Culbertson and Brown	5,615	1881
Bremner	1,345	1882
Laminated, Spalding	1,196	1878
Laying underground (See "Conductors, Conduits for.")		
Radde	2,091	1873
Gray	3,862	1873
Waters	862	1874
Prall	979	1874
Newton	1,569	1876
Henley	1,944	1876
O'Neill	4,380	1876
Baggeley	166	1877
Russell and Wilson	687	1877
Brooks	4,824	1877
Lane-Fox	3,988	1878
Lucas	5,270	1878
Pierson	5,321	1878
Edison	2,402	1879
"	602	1880
Espeut	1,257	1881
Bourdin and Maltzoff	1,474	1881
Delany	1,809	1881
Mackie	2,542	1881
Hussey and Dodd	2,573	1881
Detrick	3,388	1881
Edison	3,483	1881
Woodward	4,780	1881
Gore	4,797	1881
Johnson and Phillips	4,885	1881
Crompton	5,080	1881
Culbertson and Brown	5,615	1881
Labye and Locht-Labye	5,661	1881
Laying and insulating in earthenware blocks, Meyer	232	1882
Laying and insulating for railways, Varley and Judd	441	1882
Laying bare, *Richardson*	2,644	1882
" in conduits, *Thomas*	1,649	1882
" in earthenware blocks, Smith	1,132	1882
Laying in kerbways—		
White and Sears	347	1880
Townsend	39	1891
Smith	3,975	1881
Wheeler	*1,652	1882
Kincaid	2,911	1882
Banta	3,048	1882
Laying in troughs, *Brasseur and Dejaer*	4,296	1881
Laying in pipes, *Richardson*	2,644	1882
Multiple, Henley	336	1882
Ornamenting, Bossomaier and Schwegler	2,823	1879
Preserving, Saunders and Jamieson	1,416	1877
Preventing decomposition of caoutchouc, &c., Mowbray	*190	1881
Preventing excess of current in, Edison	5,306	1878
Preventing leakage from, of electric railways, *Edison*	1,862	1882
Preventing overcharging in, Williams	1,174	1882
Protecting—		
Zanni	1,855	1874
Phillips and Johnson	3,798	1875
Harrison	3,623	1876
Glover	855	1877
Lambert	759	1878
Lucas	3,464	1881
Henek	4,058	1881
Protecting, for railways, Binko	3,073	1882

CONDUCTORS—

	No.	Year
Protective composition for, Watson	390	1879
Restoring insulating material for, *Clark*	229	1881
Rails as—		
D'Arras	1,000	1873
Siemens	583	1880
Gassett and Fisher	3,768	1881
Rail for receiving composite, *Humebelle*	1,856	1881
Sheathing—		
Rowett	2,077	1873
Moseley	2,969	1873
Wilkinson	3,083	1873
Rowett	4,079	1873
Madsen	4,167	1873
Monckton	265	1874
Bullivant	1,159	1874
Highton	3,006	1875
Lucas	2,564	1875
Lackersteen	3,666	1875
Henley	4,115	1876
"	1,944	1876
Johnson and Phillips	3,533	1876
Gollner	131	1878
Siemens	251	1878
Clifford	312	1878
Truman	375	1878
Berthoud and Borel	4,346	1878
Alberger and Pettit	4,601	1878
Moseley	307	1879
Blondot and Bourdin	2,629	1879
Wilkinson	3,472	1879
Eaton	751	1880
Delany and Johnson	2,532	1881
Hooper	2,944	1881
Delany and Johnson	4,093	1881
Maxim	5,367	1881
Delany	333	1882
Henley	386	1882
Jeune	774	1882
Mewburn	866	1882
Sheathing, machinery for, Glover and James	913	1881
Sheathing with lead, Harrop	485	1874
Ships' logs, Massey	3,412	1876
Spiral, Chutaux	4,454	1874
Steel, Harvey	369	1877
Steel core and iron covering, *Wheeler*	3,023	1879
Strained, for suspending load, Jenkin	1,830	1882
Stretching, during laying, *Richardson*	2,644	1882
Submarine, protecting, Kendal	4,674	1880
Supports for, *Richard*	2,760	1882
Supporting aërial, *Allison*	2,256	1881
" centrally in insulating pipes, Dolbear	1,368	1882
Switch for connecting, *Blake*	5,096	1881
Swivel for, Bedwell	367	1876
Swivelling, Protheroe	2,725	1876
Tightening, Eustace	1,766	1874
Variable section, Edison	3,880	1880
Vulcanising, *Clark*	361	1882
Water employed as, Parod	4,686	1878
Winding or coiling, *Haase and Recker*	3,832	1881
Wound in helical coils, Gower	814	1880
CONTACT BREAKER (See "Commutator.")		
Automatic, Wilkins	4,306	1879
Contact maker for transforming continuous into alternating currents, Williams	700	1882
Contact, securing, in electric railways, *Edison*	1,862	1882
Contact surfaces, lubricating, Williams	856	1882
Current wheel, *Wallace*	2,015	1873
CURRENTS—		
Alternating, Siemens	3,315	1878
Collected in mercury, Courtenay	1,450	1873
Directing, Stern and Bylesby	2,336	1882
Distributing (See "Distributing Currents.")		

CURRENTS—	No.	Year
Dividing,		
Société l'Alliance ..	3,743	1877
Welch	4,114	1878
,, ...	4,278	1878
Shea	4,304	1878
Bell	4,403	1878
Thomson	4,462	1878
Stewart	4,466	1878
Bodmer	4,476	1878
Welch	4,635	1878
Sawyer and Man ..	4,705	1878
Coke	1,012	1879
Cook	2,769	1879
Rogers	3,809	1880
Dunstan and Pfankuche	3,655	1881
Dividing by mercurial re-		
sistances, Raworth ..	27	1879
Dynamo-electric, Spalding	1,467	1878
Induced, preventing (*See also*		
"Conductors, Induction		
Preventing.")		
Hussey	2,043	1875
Rapieff	4,432	1877
Intensifying, Rogers ..	3,809	1880
Leakage from, detecting,		
Raworth	27	1879
Measuring (*See* "*Meters*.")		
Regulating (*See* "*Regulators*.")		
Reversing, Varley ..	4,435	1877
CUT-OUTS (*See also* "*Switches*.")		
Reimenschneider and		
Christensen ..	4,693	1878
Automatic—		
Varley	4,435	1877
Edison	4,226	1878
,, ...	5,306	1878
André and Easton ..	2,236	1880
Gerard-Lescuyer ..	4,792	1881
Munro	1,626	1882
Williams	2,558	1882
Automatic, for incandescent		
lamps, Stokes ..	4,283	1878
DEVIATOR, Varley ..	4,435	1877
Distributing apparatus, De		
Mersanne	1,446	1874
DISTRIBUTING CURRENTS (*See*		
also "*Currents, Dividing*.")		
Jablochkoff	2,839	1877
Armand	4,074	1878
Edison	4,226	1878
Blandy	2,060	1879
Cook	2,769	1879
Edison	33	1880
,, ...	602	1880
Watson	2,271	1880
Avenarius	3,025	1880
Edison	3,880	1880
,, ...	3,964	1880
Cabanellas ..	200	1881
Gülcher	2,038	1881
Edison	2,482	1881
Hussey and Dodd ..	2,573	1881
Gravier	2,739	1881
Winter	3,146	1881
Edison	4,034	1881
Deprez and Carpentier	4,128	1881
Parod	4,508	1881
Edison	4,571	1881
Varley	5,667	1881
Carus-Wilson ..	5,687	1881
Rogers	621	1882
Williams	700	1882
,, ...	766	1882
,, ...	856	1882
,, ...	1,174	1882
Crossley, Harrison, and		
Emmott	1,327	1882
Williams	1,556	1882
Munro	1,626	1882
Williams	2,558	1882
By current breakers and con-		
densers, *Sawyer and Man*	4,705	1878
By combined electro-motor		
and generator, *Cabanellas*	200	1881
By feeding circuits, *Edison*	3,880	1880
Controlling by arrangement		
of generators, Perry ..	55	1882
Crossing mains connected,		
Edison	3,483	1881
Detecting leakage, *Edison*..	4,571	1881
Earth return, Vyle ..	5,002	1881

DISTRIBUTING CURRENTS—	No.	Year
For electric railways, Edison	3,894	1880
For telpherage system,		
Jenkin	1,830	1882
In railway trains, Stern and		
Byllesby	2,336	1882
Maintaining potential con-		
stant, *Maxim* ..	1,162	1882
Regulating energy of cur-		
rent, *Levy*	542	1882
To sidings and turntables of		
electrical railways, *Edison*	1,862	1882
Draw vices for erecting over-		
head conductors, Fletcher	859	1881
Dynamometer, Electric,		
Varley	4,905	1876
Dynamometer for indicating		
E.M.F. in circuit, Edison	33	1880
ELECTRICITY, COLLECTING,		
Cochrane	4,313	1878
ELECTRIC LAMP—		
Clark	3,991	1878
Kennedy	2,629	1882
Circuit to suspended, Rogers	4,855	1881
Lighting, Graham ..	3,274	1881
Mining, Graham ..	3,274	1881
ELECTRIC LIGHT (*See also* "*Arc*		
and Incandescence Lamps;"		
"*Light, Diffusing;*" "*Re-*		
flectors;" "*Lanterns;*"		
"*Optical Apparatus*.")		
Tyler	3,985	1878
Forbes	4,116	1878
Harrison	4,338	1878
Mackenzie	4,568	1878
Edwards and Normandy	4,611	1878
De Sussex	465	1879
Edison	33	1880
Werdermann	79	1880
Edison	602	1880
Gordon	1,826	1880
Ward	2,538	1881
Johnson and Phillips ..	2,635	1881
Winter	3,146	1881
De Changy	4,405	1881
Williams	560	1882
Pritchard	2,974	1882
ELECTRIC LIGHT applied to—		
Advertisements, Hickisson	2,421	1882
Balloons, Kinnear ..	4,684	1881
Buoys—		
Smith, Spruill, and Wood	4,312	1876
Roth..	*2,658	1881
Cities, Spalding ..	3,637	1880
Fishing, Allison and Hunter	2,772	1879
Igniting gas lamps, Bins-		
wanger	2,390	1882
Locomotives—		
Harding	4,192	1880
De Busscher ..	4,812	1881
Mariner's compasses, Cook	2,717	1874
Microscopes, *Molera and*		
Cabrian	1,217	1880
Mining, Monckton ..	3,509	1874
Photography—		
Winter	1,264	1877
Warlich	2,068	1880
Quarrying, Werdermann ..	1,438	1874
Railway trains—		
Tommasi	4,057	1881
De Changy	4,582	1881
Laybourne	5,316	1881
Signalling—		
Starr	819	1882
De Mersanne ..	2,787	1875
Moritz	4,636	1876
Haskins	3,016	1878
Shippey	879	1881
Jennings	1,699	1881
Tyer	2,879	1881
Roe	3,010	1881
Surgery, Monckton ..	265	1874
ELECTRIC LIGHT—		
Arranging, *Spalding* ..	65	1881
Combined with gas—		
Somzée	1,852	1881
Courtenay	4,659	1881
Discharge between balls, De		
Meritens	5,033	1880
Dispersing—		
Mangin, Lemonnier, and		
Co...	3,425	1879

ELECTRIC LIGHT—	No.	Year
Dispersing,		
Harrison	3,875	1879
Distributing through rail-		
way trains, Wier ..	2,557	1882
Dividing—		
Munro	4,006	1878
Clark	3,991	1878
Harrop	3,793	1879
Electrolier for, Berly & Hulett	4,755	1880
From Geissler tube, *Facio* .	3,462	1876
,, sparks, De Mersanne	1,446	1874
Globes for—		
Monckton	265	1874
Sabatou	469	1880
Globes for, double, Cadett..	4,022	1878
Motive power for—		
Birkhead	988	1879
Dillon	1,207	1879
Lambert and Iverneau ..	141	1880
Corbett and Lockhead ..	219	1880
Liardet and Donnithorne	5,216	1881
Preece and James ..	129	1882
Mines, *De Castro* ..	2,943	1878
Optical apparatus for,		
Monckton	3509	1874
Production of, Tambourin	1,235	1881
Regulated by expansion of		
metal, Siemens ..	2,281	1879
Regulating, Hopkinson ..	3,362	1881
System, *Sawyer and Man*	4,705	1878
,, *Avenarius* ..	3,025	1880
Supplying definite quantity		
of air to, André ..	1,507	1880
Switches for, automatic,		
Mackie	14	1882
Theatrical, Frome and Gibbs	3,416	1880
ELECTRIC LIGHTING—		
Jablochkoff	3,839	1877
By incandescence lamps,		
Edison	1,943	1881
Electric Motors (*See* "*Motors*.")		
,, Railway (*See* "*Rail-*		
way, Electric.")		
ELECTRODES (*See also* "*Carbons*.")		
Rapieff	4,432	1877
Harrison	3,470	1878
Stuart-Wortley ..	3,656	1878
Wilson	3,912	1878
Aronson and Farnie ..	4,163	1878
Siemens	4,208	1878
Sellon, Ladd, and Ed-		
munds	4,645	1878
Cohné	5,011	1878
Thompson and Earl ..	5,281	1878
McCarty and Sellière ..	144	1879
Harding	783	1879
Spence	876	1879
Thompson	1,622	1879
Desnos	2,340	1879
Portier	3,355	1879
Harrison	3,875	1879
De Changy	4,405	1881
Williams	766	1882
Gaseous, Wilson ..	3,912	1878
Hollow, *Jablochkoff* ..	3,552	1876
,, Grant and Sennett	2,267	1879
Inclined, *Siemens and Alte-*		
neck	3,134	1878
Inclined carbons, *Gerard-*		
Lescuyer	1,552	1880
Metal coated with carbon,		
Zanni	2,741	1882
Plate, Werdermann ..	4,805	1876
Powdered carbon in tubes,		
Thomson	927	1879
Protecting, Werdermann..	4,805	1876
Refractory material, *Sie-*		
mens and Alteneck ..	3,134	1878
Rotating—		
De Mersanne ..	1,446	1874
Whitehouse	1,820	1874
Marcus	4,006	1878
Vibrating—		
De Mersanne ..	1,446	1874
De Mersanne ..	2,787	1875
Siemens and Alteneck ..	3,134	1878
Water cased, Siemens ..	2,110	1879
Electro-dynamometer, Varley	4,905	1876
ELECTROLIERS—		
Berly and Hulett ..	4,755	1880
Brougham	1,302	1882
Brewtnall	2,934	1882

	No.	Year
Electro-Magnets (See " Magnets.")		
Electrometer, Dewar ..	2,886	1876
ELECTROMOTIVE FORCE—		
Keeping constant, Lane-		
Fox ..	3,988	1878
Standards of, Dewar	2,886	1876
ENGINES, Regulating (See also		
" Regulators.")		
Andrews ..	*2,321	1879
Richardson ..	288	1881
Levy ..	542	1882
Richardson ..	941	1882
FIELD MAGNETS (See "Magnets.")		
FILAMENTS (See also " Incandes-		
cence Lamps, Filaments		
for," and " Carbons.")		
Compound, for incande-		
scence lamps, Allport ..	2,192	1882
Fucusol, Smith ..	1,465	1882
Furfurol, Smith ..	1,465	1882
Fuze (See " Safety Fuze.")		
GALVANOMETERS—		
Sprague ..	1,558	1873
Ridout ..	3,003	1876
Pulvermacher ..	3,782	1876
Brasseur and De Sussex..	308	1878
Pulvermacher ..	1,587	1878
Obach ..	3,317	1878
Pulvermacher ..	4,094	1878
Scott ..	4,140	1878
Lane-Fox ..	4,626	1878
Sprague ..	4,762	1878
Raworth ..	27	1879
Elmore ..	3,565	1879
Glouchoff ..	478	1880
Sprague ..	4,454	1881
Carpentier ..	4,664	1881
André ..	4,948	1881
Thomson ..	5,668	1881
Faure ..	730	1882
Tangent, Sprague ..	1,558	1873
With movable coil, Sprague	1,558	1873
Geissler's tube applied to		
electric light, De Sussex	465	1879
GENERATORS (See also "Motors,"		
" Armatures," &c.)		
Higgs ..	4,206	1878
Cochrane ..	4,313	1878
Dessaigne ..	4,591	1881
Munro ..	1,626	1882
Millar ..	2,138	1882
Varley ..	2,148	1882
" ..	2,207	1882
Defty ..	2,628	1882
GENERATORS applied to—		
Air compressing machines,		
Maden ..	1,412	1876
Baths, *Barda* ..	1,107	1881
Curative purposes, Zanni ..	4,222	1876
Decomposing water, *Dering*	5,123	1878
Electrolytic bath, *Marchese*	1,884	1882
decomposition,		
Barlow ..	*1,897	1881
Electro-deposition of alu-		
miniuun and magnesium,		
Berthaut ..	4,087	1879
Galvanising iron, Elmore ..	922	1881
Gas engines—		
Otto.. ..	1,770	1878
Etéve and Lallement ..	*3,113	1881
Fielding ..	994	1882
Hair brush, Ashworth ..	454	1882
Heating metals, Williams..	*5,742	1881
Heating wire, Wier..	806	1875
Indicating presence of in-		
flammable gas, Liveing..	4,833	1880
Lighting trains, *Starr*	5,600	1881
Precipitating zinc, Létrange	3,211	1881
Protecting ships from cor-		
rosion, Lyte ..	5,358	1880
Purifying metals, *André* ..	4,053	1877
Railway brakes—		
Dibbin ..	4,036	1877
Bulkeley ..	4,464	1878
Stanford and Milligan ..	1,563	1880
Achard ..	2,453	1880
Sawiczeski ..	3,464	1881
Duwelius, Goss, Higgs,		
Merrell, Peck, and		
Walter ..	5,751	1881

	No.	Year
GENERATORS applied to—		
Railway signalling,		
Zanni ..	4,007	1880
Putnam ..	2,711	1881
Gary ..	4,069	1881
" ..	4,070	1881
Johnson ..	4,966	1881
Brear and Hudson ..	5,032	1881
Regulating motors, Cook ..	1,109	1880
Securing carriage doors,		
Meredith ..	4,707	1878
Signal buoys, Barr ..	455	1881
Signalling—		
Balnkiewicz ..	2,835	1880
Reese ..	51	1879
Speed indicators—		
Reynolds ..	515	1874
Cardew ..	5,354	1881
Stopping horses, Faucher..	2,750	1876
Telegraphy—		
Zanni ..	2,266	1873
" ..	4,277	1875
" ..	3,762	1875
Henley ..	833	1877
Lugo ..	5,352	1880
Hopkinson and Muirhead	*1,577	1881
Telephonic apparatus, *Herz*	93	1881
GENERATORS—		
Connecting terminals of, to		
a condenser, Weston ..	*4,903	1877
Controlling circuit from,		
Williams ..	85	1882
Cooling—		
Weston ..	4,748	1878
Werdermann ..	476	1873
Cutting out—		
Weston ..	*4,903	1877
Johnson and Edison ..	4,621	1880
Tommasi ..	4,057	1881
Driven from axles of railways—		
Wiles ..	65	1873
Laycock ..	478	1875
Davis ..	4,559	1878
Harding ..	4,192	1880
Liardet and Donnithorne	5,418	1881
Preece and James..	129	1882
Rogers ..	1,390	1882
Starr ..	819	1882
Driving—		
Courtenay ..	1,450	1873
Hedges ..	3,369	1881
Tambourin.. ..	1,235	1881
Driving from a water motor,		
Molera and Celrian ..	299	1879
Driving gear for, Swalwell	2,456	1882
Driving, motive power for—		
Dillon ..	1,207	1879
Dove ..	1,158	1879
Driving motors actuated by		
fall of sewage, Birk-		
head..	988	1879
Indicator, cessation of cur-		
rent from, Hedges ..	3,369	1881
Measuring speed of, Thom-		
son ..	5,668	1881
Mounted on same bedplate		
as driving engine, *Edison*	3,964	1880
Preventing heated bear-		
ings in, Williams ..	1,174	1882
Regulating power driving		
(See also " Engines.")		
Westinghouse ..	3,409	1881
Thomson ..	5,668	1881
Short-circuiting at variation		
in speed, *Weston*.. ..	*4,903	1877
Supported so as to oscillate,		
La Cie. Electrique ..	2,990	1882
With endless chain passing		
through helices, Harrison	886	1880
GENERATORS, Dynamo-electric—		
Wilde ..	618	1873
Evans ..	1,970	1873
Siemens and Alteneck ..	2,006	1873
Werdermann ..	3,156	1874
Lontin ..	473	1875
Burgin ..	3,243	1875
Fuller and Crandall ..	3,364	1875
Bertin ..	4,311	1875
Lontin ..	386	1876
Kilner ..	802	1876
Fuller and Crandall ..	1,557	1876
Zanni ..	2,821	1876
Lontin ..	3,099	1876

GENERATORS, Dynamo-electric—	No.	Year
Weston ..	4,280	1876
Monckton ..	4,597	1876
Varley ..	4,905	1876
"	270	1877
Jablochkoff ..	3,187	1877
Rapieff ..	4,432	1877
Varley ..	4,435	1877
Schuckert ..	4,164	1877
Weston ..	4,748	1877
Lontin and Co. ..	4,893	1877
Weston ..	*4,903	1877
Brush ..	2,003	1878
Siemens and Alteneck ..	3,134	1878
Siemens ..	3,315	1878
De Meritens ..	3,658	1878
Ward ..	3,976	1878
Société d' Electricité ..	4,066	1878
Zanni ..	4,573	1878
Heinrichs ..	4,595	1878
Sprague ..	4,762	1878
Pulvermacher ..	4,844	1878
Weston and Co. ..	4,960	1878
Bertin and de Mersanne	5,076	1878
Brain ..	5,139	1878
Rapieff ..	211	1879
Boustield ..	523	1879
Formby ..	565	1879
Dubos ..	749	1879
Schuckert ..	960	1879
Keith ..	1,387	1879
Andrews ..	2,321	1879
Edison ..	2,402	1879
Elmore ..	3,565	1879
Houston and Thomson ..	4,400	1879
Heinrichs ..	4,589	1879
Bürgin ..	5,085	1879
Joel ..	5,157	1879
Werdermann ..	79	1880
Houston and Thomson ..	315	1880
Fitzgerald ..	872	1880
Perry ..	1,178	1880
Griscom ..	1,259	1880
Edison ..	1,385	1880
Maxim ..	1,392	1880
Zipernowsky ..	1,580	1880
Slater ..	2,272	1880
André ..	2,764	1880
Gumpel ..	3,041	1880
" ..	3,324	1880
Willatt ..	3,808	1880
Elmore ..	3,832	1880
Thompson ..	3,928	1880
Edison ..	3,964	1880
Niaudet and Reynier ..	3,971	1880
Cance ..	4,005	1880
Biloret and Mora ..	4,049	1880
Hussey and Dodd ..	4,265	1880
Heinrichs ..	4,608	1880
Hopkinson and Muirhead	4,886	1880
Wilde ..	5,008	1880
Henley ..	5,137	1880
Gordon ..	78	1881
Gumpel ..	253	1881
Wilde ..	497	1881
Waller ..	803	1881
Edison ..	1,240	1881
Siemens and Halske ..	1,447	1881
Müller and Levett ..	1,787	1881
Gerard-Lescuyer ..	1,878	1881
Higgs ..	1,961	1881
Masson ..	2,013	1881
De Meritens ..	2,212	1881
Hussey and Dodd ..	2,375	1881
Lachaussée ..	2,761	1881
Edison ..	2,954	1881
Lane-Fox ..	3,394	1881
Mignon and Rouart ..	3,400	1881
Moffatt and Chichester ..	3,441	1881
Arey ..	3,456	1881
Harling and Hartmann ..	3,472	1881
Harborow ..	3,871	1881
Edison ..	3,932	1881
Fahrig ..	4,107	1881
De Meritens ..	4,207	1881
Aylesbury ..	4,304	1881
Kennedy ..	4,541	1881
Newton ..	4,559	1881
Joel ..	4,607	1881
Gordon ..	5,536	1881
Griscom ..	5,551	1881
Gerard-Lescuyer ..	5,593	1881
Varley ..	5,665	1881

GENERATORS, Dynamo-electric—

	No.	Year
Thomson	5,668	1881
Richardson	5,681	1881
Etève	48	1881
Gülcher	64	1882
Williams	85	1882
Raison	169	1882
Hussey and Dodd ..	234	1882
Andrews	540	1882
Chubb	761	1882
Lande	838	1882
Spagnoletti	869	1882
Sheridan	981	1882
Gravier	943	1882
Edison	1,139	1882
Gravier	1,211	1882
Matthews	1,201	1882
Levey and Lumley ..	1,249	1882
Phillips and Johnson	1,247	1882
Williams	1,556	1882
Weston	1,614	1882
Kennedy	1,640	1882
Piot	1,692	1882
Antill	1,787	1882
Voice	1,794	1882
Brougham	2,044	1882
Edison	2,052	1882
Floyd and Kirkland ..	2,225	1882
Chameroy	2,295	1882
Cumine	2,318	1882
Emmens	2,349	1882
Werdermann	2,364	1882
Williams	2,558	1882
Jarman	2,565	1882
Hallet	2,573	1882
Crompton	2,618	1882
Kennedy	2,629	1882
Fyfe and Main ..	2,636	1882
Blyth and Peebles ..	2,661	1882
Weston	2,694	1882
McTighe	2,744	1882
Chadburn	2,755	1882
Farquharson	2,771	1882
Gordon	2,871	1882
Maquaire	2,885	1882
Parker and Elwell ..	2,917	1882
La Cie. Electrique ..	2,990	1882
Schuyler and Waterhouse	3,002	1882
Ayrton and Perry ..	3,036	1882
Alloy for magnet cores of (See also "Magnets, Electro.")		
Elmore	4,821	1879
Alternate current—		
Siemens and Alteneck ..	8,134	1878
Ward	3,976	1878
Armature coils, serve also for induction of magnetism, *Siemens* ..	760	1882
Arrangement of, Edison ..	33	1880
Avoiding dead points of, *Trouvé*	4,009	1880
Combining two alternate current, *Deprez and Carpentier*	4128	1881
Compound, Akester ..	5,525	1881
Compound multiple circuit, *Weston Co.*	4,960	1878
Connecting armature coils to commutator, *Wood* ..	2,623	1882
Coupling-up, *Siemens and Alteneck*	3,134	1878
Dead points of, avoiding, *Trouvé*	4,009	1880
Duplex (See also "Generator, Dynamo Electric, Multiplex.")		
Heinrichs	4,589	1879
Henley	130	1882
Brockie	756	1882
Gravier	943	1882
Akester and Barnes ..	986	1882
Williams	1,174	1882
"	1,556	1882
Voice	1,794	1882
Jarman	2,565	1882
Hallet	2,573	1882
Maquaire ..	2,885	1882
Duplex unipolar—		
Ball ..	84	1882
Williams	85	1882
Williams	700	1882

GENERATORS, Dynamo-electric—

	No.	Year
Exciting by shunt current when starting, Lever ...	2,092	1882
Exciting alternate current, *Meritens*	1,136	1880
Indicating reversal of polarity, Elmore ..	3,565	1879
In series, Spalding	1,467	1881
Multiplex—		
Aylesbury	4,304	1881
Akester	5,525	1881
On railway axle, Evans	1,970	1873
Preventing reversal of—		
Edison	2,402	1879
Zanni	2,821	1879
Müller and Levett ..	1,787	1881
Fisher	1,727	1882
Stern and Byllesby ..	2,336	1882
Regulating current from (See also "Regulators.")		
Weston	*4,903	1877
Brush	849	1880
"	1,885	1881
Langley	4,168	1881
Edison	4,552	1881
"	1,496	1882
"	2,052	1882
Jarman	2,630	1882
Chameroy	2,295	1882
Securing armature to hub and armature coils, *Wood*	2,531	1882
Short-circuiting field coils of automatically, Raworth	27	1879
Shunt wound—		
Bertin	4,311	1875
Varley	4,905	1876
Brush	2,003	1878
Siemens	4,534	1879
Unipolar, *Siemens and Alteneck* ..	3,134	1878
Unipolar, duplex—		
Ball ..	84	1882
Williams	85	1882
GENERATORS, Dynamo-electric, with—		
Armature rolling in field magnets, *Bear* ..	3,283	1881
Duplicate armatures revolving in opposite directions, Stuart	2,232	1882
Field magnets, permanent and electro, Hallet ..	2,573	1882
Fixed armature and field magnets, Henley ..	130	1882
Internal field—		
Weston	*4,903	1877
Pulvermacher	4,844	1878
Thomson	5,668	1881
Blyth and Peebles ..	2,661	1882
Willard	2,803	1882
Internal and external field—		
Elphinstone and Vincent	332	1879
"	2,893	1880
Jurgensen and Lorenz ..	2,416	1881
Vincent and Elphinstone	2,340	1882
Sperry	3,025	1882
Obliquely fixed bobbins, *Jablochkoff*	2,769	1882
Oppositely rotating field and armature, Kelway ..	2,910	1882
Revolving field—		
Heinrichs	4,589	1879
Munro	1,626	1882
Gordon	2,871	1882
Revolving internal field, *Solignac*	740	1882
Revolving iron pieces, Henley	5,137	1880
Revolving iron shell, *Siemens*	760	1882
Rotating armature core—		
Brain	1,616	1882
Schuyler and Waterhouse	3,002	1882
Rotating field and armature—		
Vogel	4,812	1878
Wiles	644	1879
Sellon and Edmunds ..	1,949	1882
Raison	169	1882
Separate exciter on same shaft, Gordon ..	78	1881
Series of armature rings, Willard	2,803	1882

GENERATORS, Dynamo-electric, with—

	No.	Year
Silver wire conductors, Thompson	4,988	1878
Stationary armature—		
Weston	*4,903	1877
Chertemps and Dandeu..	1,747	1882
Rogers	1,760	1882
Unchanging polar field, Voice	1,794	1882
GENERATORS, Dynamo-electric without commutator—		
Courtenay	1,450	1873
Lontin	386	1876
Lontin and Co.	4,893	1877
GENERATORS, Electro-magnetic—		
Highton	4,277	1873
Wilde	618	1873
Monckton	265	1874
Gramme and d'Ivernois	953	1878
GENERATORS, Magneto-electric—		
Wiles	65	1873
Courtenay	1,450	1873
Tyer..	1,845	1873
Evans	1,970	1873
Pope	2,656	1873
Darlow and Fairfax	8,736	1873
Kilner	2,996	1875
Bertin	4,311	1875
Kilner	802	1876
Zanni	2,821	1876
Lontin	3,264	1876
Varley	270	1877
Werdermann	1,829	1877
Hickley	3,552	1877
Société l'Alliance.. ..	3,743	1877
Dibbin	4,036	1877
Zanni	4,232	1877
McTighe	162	1878
Wilde	1,228	1878
Otro	1,770	1878
De Meritens	3,658	1878
Verrue	4,287	1878
Gary	805	1879
Rosebrugh.. ..	1,476	1879
Sellon and Edmunds ..	1,949	1879
Joel	5,157	1879
Glouchoff	478	1880
Anders and Watson ..	1,958	1880
Seeley	1,998	1880
Williams, Harrington, and Lane	3,539	1881
Millar	5,566	1881
Schuyler and Waterhouse	3,002	1882
Alternating, De Meritens..	3,658	1878
For telegraphic purposes, *McTighe*	162	1878
Regulating, Kilner.. ..	2,996	1875
Tuning fork, Edison ..	4,226	1878
GENERATOR, Magneto-electric, with—		
Endless chain passing through solenoids, Harrison	3,496	1880
Movable soft iron rollers, Millar	4,592	1881
Revolving permanent magnets, Henley	5,137	1880
GENERATOR, Secondary (See also "Induction.")		
Rapieff	4,432	1877
Henley	5,137	1880
Gravier	1,211	1882
Static, *Smith*	380	1874
Static, electro, Varley ..	270	1877
GLOBES (See "Electric Light;" "Arc Lamps;" and "Incandescence Lamps, Globes for.")		
Of malleable glass, Prosser	3,466	1875
GOVERNORS (See also "Regulators," &c.)		
Spalding	1,197	1878
Jameson	2,618	1881
For electric motors, Jenkin	1,830	1882
Wheel for electric motors, Lamar	4,696	1879
Governing speed of trains on electric railways, *Edison*..	1,862	1882
Gutta-percha washing machine, Truman	94	1874
HEATING—		
Lane-Fox	4,043	1878

HEATING—

	No.	Year
Davis	4,407	1878

Holophotes—

	No.	Year
Wilde	618	1873
Mangin, Lemonnier, and Co.	3,425	1879

INCANDESCENCE LAMPS (See also "Incandescence and Electric Light;" and "Semi-Incandescence Lamp.")

	No.	Year
Lodighin	91	1873
Kosloff	2,767	1875
Lane-Fox	3,988	1878
Marcus	4,006	1878
Edwards and Normandy	4,611	1878
Scott	4,671	1878
Werdermann	79	1880
Apps	264	1881
Edison	792	1881
Lane-Fox	1,543	1881
Hallet	4,017	1881
Wright	4,778	1881
Gaulard and Gibbs	4,942	1881
Williams	5,229	1881
"	5,233	1881
Liveing and Boys	69	1882
Williams	224	1882
Aronson	359	1882
Wright and Mackie	1,029	1882
Crookes	1,079	1882
Wauthier	1,172	1882
Rogers	1,288	1882
Rapieff	2,136	1882
Stuart	2,233	1882
Williams	2,558	1882
Hallet	2,560	1882
Applied under water, Chauvin, Goizet, and Aubrey	2,410	1875
Arranged in shunt circuit of arc lamp, Gatehouse and Kempe	2,569	1882
Arranging in circuit, Cook	2,769	1879
Arranging, having different resistance in same circuit, Edison	3,765	1880
Brackets for, Maxim	3,189	1881
Charged with gas, Lodighin	91	1873

Combined with gas flame—

	No.	Year
Watson	2,271	1880
Courtenay	4,659	1881
Crastin	4,771	1881
Connecting to leads, Adie	5,211	1880
Rogers	1,618	1882

" " Cut-out for automatic (See also "Cut-Outs" and "Switches.")

	No.	Year
Edison	2,492	1881
Currents discharged between balls, De Meritens	5,033	1880
Electrodes fed by mercury float, Scott	4,671	1878

Exhausting (See also "Vacuum Pumps")

	No.	Year
Edison	2,402	1879
Maxim	1,649	1880
"	4,393	1880
Henley	5,137	1880
Edison	562	1881
Crookes	1,422	1881
André	4,654	1881
Schaeffer	4,294	1881

Exhausting during passage of current—

	No.	Year
Edison	2,402	1879
Swan	18	1880

Exhausting residual gas from—

	No.	Year
Edison	562	1881
Fitzgerald	3,890	1881

INCANDESCENCE LAMPS, Conductors of—

	No.	Year
Müller	3,711	1881
Crookes	3,799	1881
Edison	4,174	1881
Composition for glass surrounding, Nichols	5,495	1880
Covered with glass, Swan	250	1880
Hollow, Schaeffer	4,294	1881
" copper, André	2,432	1882
Iron, Ward	2,930	1881
Platinum, Debenham	3,010	1882
Preparing metallic, Swan	2,898	1882

INCANDESCENCE LAMPS (See also "Carbons" and "Electrodes.")

Filaments of—

	No.	Year
Lodighin	91	1873
Kosloff	441	1875
Konn	970	1875
Chauvin, Goizet, and Aubrey	2,410	1875
Lane-Fox	4,043	1878
Pulvermacher	4,180	1878
Van Choate	4,388	1878
Edison	4,502	1878
Sprague	4,662	1878
Scott	4,671	1878
Pulvermacher	4,774	1878
Edison	5,306	1878
Lane-Fox	1,122	1879
Edison	4,576	1879
"	5,127	1879
Clark	203	1880
Swan	250	1880
Edison	578	1880
Maxim	1,649	1880
Clingman	1,840	1880
Lane-Fox	3,494	1880
Edison	3,765	1880
Gordon	4,745	1880
Swan	4,933	1880
"	5,014	1880
Fitzgerald	5,275	1880
Muirhead and Hopkinson	153	1881
Gordon	218	1881
Lane-Fox	225	1881
Edison	562	1881
Maxim	639	1881
Scott and Akester	1,122	1881
Crookes	1,422	1881
Société la Force et la Lumière, &c.	1,653	1881
Edison	1,918	1881
"	2,492	1881
Crookes	2,612	1881
Courtenay	2,770	1881
André and Easton	2,883	1881
Lane-Fox	3,122	1881
Gatehouse	3,240	1881
Maxim	3,189	1881
Wright	3,437	1881
Pfannkuche	3,655	1881
Crookes	3,799	1881
Fitzgerald	3,890	1881
Hallet	4,017	1881
Swan	4,202	1881
Faure	4,311	1881
Lane-Fox	4,383	1881
Harrison	4,478	1881
André	4,654	1881
Allport and Punshon	4,850	1881
St. George	4,939	1881
Williams	5,229	1881
"	5,233	1881
Sennett	5,286	1881
Kennedy	5,524	1881
Williams	224	1882
Leipmann and Looker	1,036	1882
Crookes	1,079	1882
Wright and Mackie	1,274	1882
Gatehouse	1,400	1882
Woodhouse and Rawson	1,412	1882
Smith	1,465	1882
Salomans	1,580	1882
Maxim	1,619	1882
Jameson	1,670	1882
Cruto	1,895	1882
Jousselin	2,037	1882
Rapieff	2,136	1882
Allport	2,192	1882
Floyd and Probert	2,226	1882
Stuart	2,233	1882
Siemens	2,348	1882
Barrier and Lavernede	2,425	1882
André	2,432	1882
Nothomb	2,452	1882
Williams	2,558	1882
Bernstein	2,604	1882
Stanley	2,660	1882
Zanni	2,741	1882
Pfannkuche	2,845	1882
Debenham	3,010	1882
Willard	3,042	1882
Leask and Smith	3,099	1882

INCANDESCENCE LAMPS, Filaments of—

Automatically changing, Jameson

	No.	Year
Jameson	4,439	1881
Coating with phosphorous, &c., Barrier and Lavernede	2,425	1882
Compound, Gatehouse	1,400	1882
Salomons	1,580	1882
Allport	2,192	1882

Duplicate—

	No.	Year
Edison	3,765	1880
Sellon and Edmunds	1,791	1879
Harrison	4,478	1881
Enclosed in refractory casing, Tyler	4,575	1878
Endless, Graham and Smith	1,392	1882
Gauging, Evans	1,225	1882
Graphite, Konn	970	1875

Having extended surfaces—

	No.	Year
Williams	2,558	1882
Hollow, Jameson	1,670	1882

Iridium—

	No.	Year
Pfannkuche	2,845	1882
Gordon	4,745	1880
Liquid, Williams	5,229	1881

Mounting—

	No.	Year
Pulvermacher	4,180	1878
"	4,774	1878
Lane-Fox	1,122	1879
Edison	4,576	1879
"	5,127	1879
Swan	250	1880
Edison	578	1880
Guest	925	1880
Maxim	1,649	1880
Clingman	1,840	1880
Drew	2,037	1880
Lane-Fox	3,494	1880
Edison	3,765	1880
Swan	4,933	1880
"	5,014	1880
Fitzgerald	5,275	1880
Muirhead and Hopkinson	153	1881
Lane-Fox	225	1881
Edison	539	1881
"	562	1881
"	768	1881
Gimingham	2,079	1881
Crookes	2,304	1881
Nichols	3,187	1881
Wright	3,435	1881
Fitzgerald	3,890	1881
Hallet	4,017	1881
Fox	4,024	1881
Edison	4,174	1881
Gimingham	4,193	1881
Swan	4,202	1881
Faure	4,311	1881
Harrison	4,478	1881
André	4,654	1881
Sennett	5,286	1881
Kennedy	5,524	1881
Liveing and Boys	69	1882
Wright and Mackie	1,029	1882
Crookes	1,079	1882
Woodhouse and Lawson	1,412	1882
Akester	1,612	1882
Lane-Fox	1,647	1882
Brougham and Ormiston	1,697	1882
Leask	1,803	1882
Jousselin	2,037	1882
Emmens	2,348	1882
Barrier and Lavernede	2,425	1882
Nothomb	2,452	1882
Hallet	2,560	1882
Bernstein	2,604	1882
Pfannkuche	2,845	1882
Swan	2,898	1882
Debenham	3,010	1882
Mounting to allow for expansion, Willard	3,042	1882

Multiple—

	No.	Year
Kosloff	441	1875
Konn	970	1875
Sprague	4,662	1878
Hussey and Dodd	2,572	1881
Maxim	3,189	1881
Hallet	4,017	1881
Jameson	4,439	1881
Williams	5,233	1881
Aronson	359	1882
Andrews	1,324	1882
Floyd and Probert	2,226	1882

INCANDESCENCE LAMPS, Filaments of—   No.   Year

Multiple,
Emmens ... 2,348   1882
Barrier and Lavernede .. 2,425   1882
Nothomb .. 2,452   1882
Hussey and Dodd 2,572   1881
Phosphorescent, Bernstein 2,604   1882
Platinised, Muirhead and Hopkinson.. 153   1881
Platinum and carbon, Gatehouse 3,240   1881
Reducing resistance of, Scott 4,671   1878
Regulating resistance of—
Maxim .. 639   1881
Crookes .. 2,612   1881
Lane-Fox .. 4,383   1881
Regulating resistance of, automatically, Welch .. 4,689   1878
Replaceable—
Guest .. 925   1880
Edison .. 3,765   1880
Maxim .. 4,393   1880
Faure .. 4,311   1880
Liveing and Boys 69   1882
Lane .. 2,752   1882
Straight, Upton 1,232   1881
Thickening, Lane-Fox 225   1881

INCANDESCENCE LAMPS—
Fittings for, Brougham .. 1,302   1882
For heating, Lodighin 91   1873

INCANDESCENCE LAMPS, Globes of—
Drew .. 2,037   1880
Maxim .. 4,393   1880
Nichols .. 3,187   1881
De Changy.. 4,405   1881
Aronson .. 305   1882
Wright and Mackie 1,274   1882
Salomons .. 1,580   1882
Akester .. 1,642   1882
Stuart .. 2,233   1882
Swan .. 2,898   1882
Cleaning, Jameson .. 4,439   1882
Concentric, Beckinsale 2,512   1882
Containing inert vapour or gas—
Kosloff .. 441   1875
Bright .. 4,212   1878
Scott .. 4,671   1878
Clark .. 203   1880
Fitzgerald .. 5,275   1880
Gordon .. 218   1881
Rapieff .. 2,136   1882
Exhausting (See "Incandescence Lamp, Exhausting.")
Glass for—
Sellon .. 5,632   1881
Debenham .. 3,010   1882
Neck of—
Edison .. 4,174   1881
Lea .. 2,186   1882
Preventing blackening, Edison 2,492   1881
Preventing heating of, near conductors, Edison .. 539   1881
Reflecting—
Lorrain .. 5,738   1881
Woodhouse and Rawson 1,412   1882
Waters .. 1,462   1882
Floyd and Probert 2,226   1882
Sealing—
Kosloff .. 441   1875
Konn .. 970   1875
Chauvin, Goizet, and Aubrey 2,410   1875
Swan .. 250   1880
Edison .. 578   1880
Guest .. 925   1880
Maxim .. 1,649   1880
Drew .. 2,037   1880
Lane-Fox .. 3,494   1880
Edison .. 3,765   1880
Maxim .. 4,393   1880
Nichols .. 4,495   1880
Henley .. 5,137   1880
Edison .. 539   1881
Maxim .. 3,189   1881
Nichols .. 3,187   1881
Gatehouse .. 3,240   1881
Wright .. 3,437   1881
Fitzgerald .. 3,890   1881
Fox .. 4,024   1881
Edison .. 4,174   1881
Faure .. 4,311   1881

INCANDESCENCE LAMPS, Globes of—   No.   Year

Sealing,
Joussellin ... 2,037   1882

INCANDESCENCE LAMPS—
Holders for,
Edison .. 578   1880
Edison .. 1,802   1881
Graham .. 5,618   1881
Swan .. 5,702   1881
Johnson .. 1,094   1882
Rogers .. 1,288   1882
Woodhouse and Rawson 1,412   1882
Rogers .. 1,618   1882
Lea .. 2,186   1882
Defries .. 2,335   1882
Emmens .. 2,348   1882
André .. 2,432   1882
Hallet .. 2,560   1882
Debenham .. 3,018   1882
Lanterns for (See also "Lanterns.")
Graham .. 3,274   1881
Lighting by—
Edison .. 3,894   1880
" .. 1,943   1881
Manometer for, Fitzgerald 3,890   1881
Manufacture of—
Edison .. 2,492   1881
Crookes .. 3,799   1881
Brougham and Ormiston 1,697   1882
Leask .. 1,803   1882
Swan .. 2,898   1882
Medical, Nitze .. 153   1879
Mining, Edison .. 4,174   1881
Mounting, Hughes .. 3,190   1881
Portable—
Van Tenac .. 4,206   1875
Adie.. 5,211   1880
Regulating (See also "Regulators.")
Edison .. 539   1881
Hussey and Dodd .. 2,572   1881
Gatehouse .. 3,240   1881
Revolving, Werdermann 79   1880
Safety device for—
Fuller .. 5,183   1878
Edison .. 5,306   1878
Safety fuze for, Edison .. 1,802   1881
" with, André .. 2,432   1882
Sealing (See "Incandescence Lamp, Globes of, Sealing.")
Shunting portion of current by, Sellon and Edmunds 1,791   1879
Switch for (See also "Switches.")
Hughes .. 3,190   1881
Volk .. 2,962   1882
Switch for, automatic—
Sellon and Edmunds .. 1,791   1879
Swan .. 4,202   1881
Switch for multiple filament, Hussey and Dodd 2,572   1881
Supplying definite quantity of air to, André .. 1,507   1880
Terminals for, Leask .. 1,803   1882
Vacuum pumps for (See "Vacuum Pumps.")
Ventilating, André .. 1,507   1880
INCANDESCENCE LAMPS with—
Adjustable carbon shunt, Gatehouse .. 3,240   1881
Automatically fed carbon rod, Sawyer 3,587   1879
Enclosed gas rendered luminous, Werdermann 79   1880
Iridium filament and current of air, Gordon 4,745   1880
Sectional carbon plate, Andrews 1,324   1882
Small carbons, Graham and Smith 1,392   1882
Two sticks of carbon, Salomons 1,580   1882
Incandescence light, Jablochkoff 1,996   1877
India-rubber, curing, Henley 4,115   1875
Indicator current, Raworth.. 27   1879
" speed (See "Speed Indicator.")
INDUCED CURRENTS (See also Currents, Induced.")
Preventing, Rapieff .. 4,432   1877
INDUCTION (See also "Conductors, Induction in, Preventing.")

INDUCTION—   No.   Year

Apparatus,
Bright .. 4,219   1878
Fuller .. 5,183   1878
De Meritens .. 5,257   1878
Courtenay .. 3,543   1879
Apparatus for regulating current, Hopkinson 3,362   1881
Apparatus, coiling conductors for, Clark and Leigh 3,652   1881
Applied to electric lighting, Edwards and Normandy 4,611   1878
Coil—
Jablochkoff .. 1,996   1877
St. George.. 2,193   1878
Harrison .. 3,470   1878
Thompson .. 4,983   1878
Marx, Aklem, Kayser, and Tisdel 351   1880
Clarke and Leigh.. 245   1881
Jacobson .. 3,053   1881
Coil, winding, Grumpel .. 3,041   1880
Preventing, in conductors, Lugo 1,119   1881
Insulating cast iron, Shedlock 5,498   1880
conductors (See "Conductors, Insulating")
Insulating convolutions of coils, Dion 3,880   1881
INSULATING MATERIALS (See also "Conductors, Insulating.")
Truman .. 4,438   1878
Sanders and Danckwerth 2,016   1879
Wilkinson .. 3,472   1879
Heins .. 4,916   1879
Mourlot .. 4,846   1879
Quin .. 1,239   1880
Knudson and Kane .. 1,295   1880
Mourlot .. 2,121   1880
Truman .. 3,310   1880
Jacques .. 3,424   1880
Shedlock .. 5,498   1880
Fleming .. 1,762   1881
Masson .. 2,013   1881
Callender .. 4,409   1881
Woodward .. 4,780   1881
Fleming .. 5,309   1881
Allport .. 493   1882
Zingler .. 1,153   1882
Imray .. 1,516   1882
Fleming .. 2,414   1882
Field .. 2,480   1882
Rhodes and Bingswanger 2,501   1882
Page .. 2,516   1882
" .. 2,518   1882
Beckinsale .. 2,532   1882
Treating waste, Heyer 3,885   1879
INSULATORS—
Moseley .. 2,969   1873
Fottrell .. 3,086   1873
Capanema .. 4,171   1873
Sterne .. 3,974   1874
Johnson and Phillips .. 3,533   1876
Fuller and Fuller .. 76   1877
Cordeaux .. 522   1877
Boulton .. 805   1877
Kerr .. 2,527   1877
Oppenheimer .. 1,248   1878
Crighton .. 4,696   1878
Bloomfield .. 3,679   1879
Crighton .. 4,189   1879
Wells and Gilbert.. 1,510   1880
Edwards .. 1,720   1880
Wells and Gilbert .. 1,960   1880
Gilbert .. 3,438   1880
Fletcher .. 3,936   1890
Gilbert .. 4,507   1881
Lewis .. 1,017   1882
Curtoys .. 1,851   1882
Richards .. 2,760   1882
Attaching conductors to, Edwards 1,720   1880
Intensifying currents, Rogers 3,809   1880
Iodine, obtaining electrolytically, Reynoso .. 799   1873

JABLOCHKOFF CANDLE (See "Electric Candle.")

LAMPS, ARC (See "Arc Lamps.")
Lamps, arc incandescence (See "Arc Incandescence Lamps.")

	No.	Year
Lamps for candles, Mackenzie	4,568	1878
Lamps, extinguishing electric, Lane-Fox	225	1881
Lamps, incandescence (See "Incandescence Lamps.")		
Lamps, semi-incandescence (See "Semi-Incandescence Lamps.")		
Lamps, street, extinguishing alternate, Mackie	14	1882
LANTERNS (See also "Reflectors" and "Light Diffusing.")		
Tilleard	4,317	1878
Mangin, Lemonnier, and Co.	3,425	1879
Sabatou	469	1880
Berly and Hulett	4,755	1880
Graham	3,073	1881
„	3,274	1881
Air-tight, Brougham	832	1880
For supplying a definite quantity of air, André	1,507	1880
Submarine, Heinke and Lang	231	1880
Lead, automatically adjusting of brushes, Sperry	3,025	1882
LIGHT (See "Lamps," "Reflectors," "Lanterns," and "Holophotes.")		
Diffusing—		
Partz	3,455	1881
Wheeler	3,858	1881
Locomotive, electric (See also "Railways, Electric.")		
Monckton	265	1874
Locomotive, electro-magnetic, D'Arras	1,000	1873
Locomotive, electric, increasing grip of, Edison	3,894	1880
MAGNETS—		
Applied to separating metals, Leek and Edwards	2,890	1878
Insulating, Moore and Courtenay	3,078	1873
Preserving from oxidation, Byshe	4,961	1878
MAGNETS, Electro—		
Highton	1,178	1873
Camacho	3,461	1873
Stone	94	1874
Monckton	3,509	1874
Courtenay	1,487	1875
Faulkner	1,800	1875
Hequet	2,564	1875
Faure	2,946	1875
Smith	3,981	1877
Rapieff	4,432	1877
St. George	2,193	1878
Sprague	4,762	1878
Rapieff	211	1879
Dubos	749	1879
Chambrier	4,428	1879
Perry	1,178	1880
Gordon	78	1881
Apps	264	1881
Lockwood	2,398	1881
Markoff and de Kabath	4,271	1881
Little	497	1882
Matthews	1,201	1882
Weston	1,611	1882
Emmens	2,349	1882
Ayrton and Perry	2,613	1882
Gordon	2,871	1882
Conical, Chislett	87	1873
Cores of—		
Chislett	87	1873
Highton	1,178	1873
Camacho	3,461	1873
Slater	2,625	1874
Monckton	3,509	1874
Faulkner	1,800	1875
Paine and Paine	2,049	1875
Jablochkoff	836	1876
Varley	4,435	1877
Rapieff	4,432	1877
Cance	1,927	1878
De Meritens	3,658	1878
Elmore	4,821	1879
Henley	5,137	1880
„	130	1882
La Cie. Electrique	2,990	1882

	No.	Year
MAGNETS, Electro—		
Cores of,		
Werdermann	2,364	1882
Ayrton and Perry	3,036	1882
Employed as brakes, Groombridge	1,049	1880
For arc lamp, Ayrton and Perry	2,613	1882
Machine for winding, Cooke	1,903	1873
Shunting portion of current from, when armature attracted, André	5,206	1879
Tubular—		
Camacho	3,416	1875
Werdermann	1,829	1877
Bell	611	1878
Winding—		
Monckton	265	1874
„	3,509	1874
Jablochkoff	836	1876
Cance	1,927	1878
Bell and Scarlett	4,555	1879
Heinrichs	4,589	1879
Scarlett	1,585	1880
Henley	5,137	1880
MAGNETS, Field—		
Moore and Courtenay	3,078	1873
Evans	1,970	1873
Siemens and Alteneck	2,006	1873
Wallace	2,015	1873
Stone	94	1874
Werdermann	3,156	1874
Hussey	2,043	1875
Paine and Paine	2,049	1875
Fuller and Crandall	3,364	1875
Lontin	386	1876
Fuller and Crandall	1,557	1876
Zanni	2,821	1876
Monckton	4,597	1876
Varley	4,905	1876
Zanni	4,232	1876
Rapieff	4,432	1877
Varley	4,435	1877
Lontin and Co.	4,893	1877
Weston	*4,903	1877
Cane	1,927	1878
Brush	2,003	1878
Siemens and Alteneck	3,134	1878
De Meritens	3,658	1878
Société d'Electricité	4,066	1878
Higgs	4,206	1878
Verrue	4,287	1878
Zanni	4,573	1878
Heinrichs	4,595	1878
Sprague	4,762	1878
Pulvermacher	4,844	1878
Weston Co.	4,960	1878
Bertin and de Mersanne	5,076	1878
Brain	5,139	1878
Rapieff	211	1879
Elphinstone and Vincent	332	1879
Formby	565	1879
Dubos	749	1879
Schuckert	960	1879
Keith	1,387	1879
Sellon and Edmunds	1,949	1879
Andrews	2,321	1879
Edison	2,402	1879
Elmore	3,565	1879
Houston and Thomson	4,400	1879
Heinrichs	4,589	1879
Lamar	4,696	1879
Bürgin	5,085	1879
Joel	5,157	1879
Werdermann	79	1880
Houston and Thomson	315	1880
Glouchoff	478	1880
Fitzgerald	872	1880
Zipernowsky	1,580	1880
Seeley	1,998	1880
Slater	2,272	1880
Gumpel	3,041	1880
Thompson	3,928	1880
Niaudet and Reynier	3,970	1880
Cance	4,005	1880
Biloret and Mora	4,049	1880
Hussey and Dodd	4,265	1880
Heinrichs	4,608	1880
Hopkinson and Muirhead	4,886	1880
Henley	5,137	1880
Etéve	48	1881
Gordon	78	1881
Gumpel	253	1881

	No.	Year
MAGNETS, Field—		
Wilde	497	1881
Waller	803	1881
Siemens and Halske	1,447	1881
Müller and Levett	1,787	1881
Higgs	1,961	1881
De Meritens	2,212	1881
Hussey and Dodd	2,375	1881
Jurgensen and Lorenz	2,416	1881
Lachaussée	2,761	1881
Lane-Fox	3,394	1881
Mignon and Rouart	3,400	1881
Arey	3,456	1881
Harling and Hartmann	3,472	1881
Fahrig	4,107	1881
De Meritens	4,207	1881
Kennedy	4,541	1881
Joel	4,607	1881
Bürgin	4,819	1881
Piot	4,851	1881
Akester	5,525	1881
Gordon	5,536	1881
Gerard-Lescuyer	5,593	1881
Varley	5,665	1881
Thomson	5,668	1881
Richardson	5,681	1881
Hussey and Dodd	234	1882
Little	497	1882
Solignac	740	1882
Lande	838	1882
Spagnoletti	869	1882
Sheridan	931	1882
Gravier	943	1882
Matthews	1,201	1882
Gravier	1,211	1882
Williams	1,556	1882
Munro	1,626	1882
Kennedy	1,640	1882
Piot	1,692	1882
Rogers	1,760	1882
Gerard-Lescuyer	1,878	1882
Edison	2,052	1882
Floyd and Kirkland	2,225	1882
Cumine	2,318	1882
Vincent and Elphinstone	2,340	1882
Emmens	2,349	1882
Werdermann	2,364	1882
Williams	2,558	1882
Hallet	2,573	1882
Blyth and Peebles	2,661	1882
Chadburn	2,755	1882
Farquharson	2,771	1882
Willard	2,803	1882
La Cie. Electrique	2,990	1882
Connecting coils of, Gravier	943	1882
Connecting in several circuits, Maquaire	2,885	1882
Excited by thermopiles, Edison	2,402	1879
Exciting—		
Bürgin	5,085	1879
Elphinstone and Vincent	2,893	1880
Deprez and Carpentier	4,128	1881
Williams	1,174	1882
Maquaire	2,885	1882
Exciting by shunt circuit of variable, Brush	1,835	1881
Exciting, of alternate current generator, Meritens	1,136	1880
Preventing counter currents in, by slitting, Wilde	5,001	1880
Preventing reversal of polarity in, Weston	*4,903	1877
Rolling, Bear	3,283	1881
Shunt wound, Siemens	4,534	1879
Winding—		
Cabanellas	200	1881
Ayrton and Perry	3,636	1882
Mc'Ighe	2,744	1882
MAGNETS, Permanent—		
Monckton	3,509	1874
Sprague	4,762	1878
De Meritens	4,207	1881
Kotyra	3,047	1882
Compound—		
Fontaine	1,180	1873
Pritchard	2,816	1878
MAGNET POLES—		
Moore and Courtenay	3,078	1873
Monckton	265	1874
„	3,509	1874

MAGNET POLES—

	No.	Year
Fuller and Crandall ..	3,364	1875
Baxtet	1,931	1876
Lontin	3,264	1876
Faure	3,670	1876
Monckton	4,597	1876
Varley	4,905	1876
Rapieff	4,432	1877
Schuckert	4,464	1877
Siemens and Alteneck ..	3,134	1878
Ward	3,976	1878
Perry	1,178	1880
Griscom	1,259	1880
Moffatt and Chichester ..	3,441	1881
Edison	3,932	1881
Joel	4,607	1881
Gülcher	64	1882
Magnetic field, regulating intensity of (See "Regulator, Current.")		
Harling and Hartmann ..	3,472	1881
Magnetic field, unipolar, Werdermann	3,156	1874
Make and break, automatic, Thomson	3,032	1881
Manometer, Fitzgerald ..	3,890	1881
Metals, purifying, *André* ..	4,053	1877
Metals, purifying molten, Werdermann.. ..	476	1873
Metallic salts, decomposing, *Lontin*	473	1875

METERS, Current—

	No.	Year
Pulvermacher	3,782	1876
Lane-Fox	3,988	1878
Bright	4,212	1878
Watson	2,271	1880
Law	4,851	1880
Apps	264	1881
Sprague	4,454	1881
Lane-Fox	5,651	1881
Thomson	5,668	1881
Carus-Wilson	5,687	1881
Application of, *Edison*	4,571	1881
Automatically breaking circuit to, *Edison* ..	4,576	1881
Circulating electrolyte in, Sprague	2,902	1882
Clockwork governed by oscillating armature, Boys ..	4,472	1881
Clockwork; pendulum of, controlled by magnets, Ayrton and Perry..	2,642	1882
Clockwork automatically wound up, Boys ..	4,472	1881
Compensating for changes of temperature, *Edison*	1,783	1881
Sprague	4,454	1881
Edison	4,576	1881
Sprague	2,902	1882
Conductor rotating around a pole, Faure ..	730	1882
Electro-depositing—		
Edison	4,226	1878
Sprague	4,762	1878
Edison	4,391	1880
,, ...	1,016	1881
,, ...	1,783	1881
,, ...	4,576	1881
Carus-Wilson	5,623	1881
Sprague	2,902	1882
Electro-hydrometric, Sprague	4,454	1881
Endosmose, Carus-Wilson..	2,954	1882
Fluid passing through electrically regulated valve, Lane-Fox	4,626	1878
Freezing of, preventing—		
Sprague	4,454	1881
Edison	4,576	1881
Gearing constantly rotated drives through speed changers the registering mechanism—		
Lane-Fox	4,626	1878
Carus-Wilson	4,824	1881
Varley	2,248	1882
Indicating charge of secondary batteries, Carus-Wilson	5,687	1881
Lamp hour—		
Sawyer and Mann ..	4,705	1878
Swan	5,004	1880
Pel	3,380	1881

METERS, Current—

	No.	Year
Lamp hour,		
Swan	5,499	1881
Measuring relative intensity of two currents, *Carpentier*	4,664	1881
Motor controlled by electrical apparatus, Hopkinson	49	1882
Motor, copper disc, Munro	1,626	1882
,, works against a constant resistance—		
Edison	1,016	1881
Ayrton and Perry ..	2,642	1882
Oscillating balance wheel with length of torsional spring controlled, Boys ..	513	1882
Regulating current to, Swan	5,499	1881
Vibrating armature—		
Fuller	5,183	1878
Boys..	4,472	1881
Williams	1,174	1882
Ayrton and Perry ..	2,642	1882
Voltametric—		
Sprague	2,902	1881
Edison	1,016	1881
Swan	5,499	1881
Blythe and Peebles ..	2,661	1882

METERS, Electric—

	No.	Year
Metallic deposit, Sprague..	2,902	1882
Voltametric, Sprague ..	2,902	1882
Meters, electricity, motor controlled by electrical apparatus, Hopkinson ..	49	1882
Meters, electro, Dewar ..	2,886	1876
METERS, Power, Dewar ..	2,886	1876
Clockwork. pendulum of, electrically controlled, Ayrton and Perry ..	2,642	1882
Motor rotates against constant resistance, Ayrton and Perry	2,642	1882
Mines, exploding gases in, *Budenburg and Schäffer* ..	4,227	1880

MOTIVE POWER (See also "Generators.")

	No.	Year
Dove	1,158	1879
Whiteley	1,445	1879
Lambert and Invernau..	144	1880
Corbett and Lockhead ..	219	1880
Graddon	885	1880
Edison	3,964	1880

MOTORS, Electric—

	No.	Year
D'Arras	1,000	1873
Wallace	2,015	1873
Gaumé	2,618	1873
Moore and Courtenay ..	3,078	1873
Stone	94	1874
Monckton	265	1874
Wilde	1,554	1874
Monckton	3,509	1874
Chutaux	4,454	1874
Courtenay	1,487	1875
Hussey	2,043	1875
Paine and Paine ..	2,049	1875
Clamond	2,205	1875
Faure	2,946	1875
Camacho	3,416	1875
Edison	3,762	1875
Kimball	3,999	1875
Paine	4,118	1875
Bastet	1,931	1876
Lontin	3,264	1876
Faure	3,670	1876
Monckton	4,597	1876
Varley	4,905	1876
Lovel	732	1877
Werdermann	1,829	1877
Smith	3,981	1877
Spalding	*915	1878
Zanni	1,677	1878
Cance	1,927	1878
Harding and Bull ..	2,878	1878
Melhado	4,699	1878
Whyte	5,152	1878
Whiteley	1,445	1879
Heinrichs	2,317	1879
Edison	2,402	1879
Heinrichs	4,589	1879
Lamar	4,696	1879
Glouchoff	478	1880
Griscom	1,259	1880

MOTORS, Electric—

	No.	Year
Fonvielle	1,339	1880
Edison	1,385	1880
Edison	3,894	1880
,, ...	3,964	1880
Kuhlo	4,825	1880
Henley	5,137	1880
Richardson	2,703	1881
Hopkinson.. ..	2,969	1881
Thomson	3,032	1881
Berthoud, Borel, and Co.	4,026	1881
Millar	4,592	1881
Bürgin	4,819	1881
Piot	4,851	1881
Raison	169	1882
Hussey and Dodd ..	234	1882
Andrews	540	1882
Siemens	760	1882
Williams	856	1882
Spagnoletti	869	1882
Williams	1,174	1882
Varley	2,148	1882
Cunine	2,318	1882
Emmens	2,349	1882
Williams	2,558	1882
Parker and Elwell ..	2,917	1882
Ayrton and Perry ..	2,830	1882
La Cie. Electrique ..	2,990	1882

MOTORS, Electric, applied to—

	No.	Year
Brine evaporation, Biggs ..	2,106	1877
Coal cutting, Wilde ..	1,554	1874
Conveying small packages, *White and Hayden* ..	473	1881
Domestic tram, Harmant..	4,779	1880
Driving drills, Thomson ..	1,393	1881
Drills, Stanfield and Clark	*4,918	1881
Mining, *Lemaire-Douchy*..	2,107	1876
Musical box, Dudley ..	3,423	1881
Propellers, *Tissandier* ..	3,401	1881
Railways, *Siemens* ..	583	1880
Railway brakes, Siemens and Boothby ..	696	1881
Rock drills—		
Jones and Wild ..	5,614	1881
Finch	1,110	1879
Sewing machines, Journaux	4,518	1881
Tramways, Wigham ..	*2,414	1881
Teleradiophone, Mercadier	3,929	1881

MOTORS, Electric—

	No.	Year
Arranging in current with lamps, *Edison* ..	4,034	1881
Communicating power from, *Edison*	3,964	1880
Cooling, Little ..	497	1882
Dead points of, avoiding, *Trouvé*	4,009	1880
Driven by battery with adjustable electrodes, *Griscom*..	1,244	1880
Driving gearing for, Edison	1,385	1880
For obtaining power, Hosmer	4,220	1880
Locomotive, Jenkin ..	1,830	1882
Maintaining strength of field magnets of, *Edison*	1,862	1882
Mechanical starter for, *Lamar*	4,696	1879
Multiplex, *Chutaux* ..	4,454	1874
Obliquely fixed bobbin, *Jablochkoff*	2,769	1882
Regulating (See also "Regulators.")		
Monckton	3,509	1874
Paine and Paine ..	2,049	1875
Spalding	1,197	1878
Lamar	4,696	1879
Edison	33	1880
Deprez and Carpentier ..	4,128	1881
Langley	4,168	1881
Carus-Wilson	4,825	1881
Jarman	2,630	1882
Ayrton and Perry ..	2,830	1882
La Cie. Electrique ..	2,990	1882
Reversing—		
Wallace	2,015	1873
Hopkinson	2,481	1879
,, ...	4,653	1879
Lamar	4,696	1879
Gumpel	3,041	1880
Piot	4,851	1881
Ayrton and Perry ..	2,830	1882
Supplying current to, Binks	3,073	1882

	No.	Year
MOTOR, Electro-magnetic—		
Ludeke and Thornan ..	3,338	1878
North	4,041	1878
„	4,518	1878
Edison	33	1880
Little	497	1882
MOTOR, Magnetic—		
Andrews	292	1878
Hosmer	311	1878
La Cour	1,988	1878
Hosmer	2,930	1878
Sueur	3,857	1881
MOTOR, Magneto-electric—		
Ward and Ball	2,033	1878
Hosmer	3,676	1878
OPTICAL APPARATUS (*See also* "*Reflectors;*" "*Lanterns;*" "*Holophotes;*" *and* "*Photometers.*")		
Weyde	446	1878
For arc lamp, *De Mersanne*	3,315	1876
Oxyhydrogen light, Edwards and Normandy	4,611	1878
PERMUTATOR, *Wallace* ..	2,015	1878
Permanent magnets (*See* "*Magnets, Permanent.*")		
Permanent way for conductors, Meyer	232	1882
Phosphor-bronze, Highton ..	3,006	1874
Phosphorescent filament for incandescence lamps, Bernstein	2,604	1882
Photometer—		
Bolter and Webber ..	686	1873
Mucklow and Spurge ..	5,368	1881
Selenium, Limbeck	4,408	1880
Plating cell, electro, *Edison*	768	1881
Polarity. preventing reversal (*See also* "*Switches, Automatic.*")		
Weston	4,280	1876
„	4,748	1877
„	*4,903	1877
Stern and Byllesby ..	2,336	1882
Posts for conductors (*See also* "*Conductors.*")		
Richard	2,760	1882
Power. transmission of (*See* "*Transmission of Power.*")		
Pump. mercury (*See also* "*Vacuum Pumps.*")		
Akester	2,519	1882
RAILWAY CARRIAGE DOORS—		
Securing, Wiles	65	1873
RAILWAYS, Electric—		
Monckton	265	1874
Haddan	2,040	1879
Siemens	583	1880
Varley and Judd	441	1882
Edison	1,863	1882
Lighting carriages of—		
Preece and James ..	129	1882
Starr	819	1882
Ratchet, electric, *Cabanellas*..	200	1881
Reflecting refractory block for arc lamps, Harrison	3,559	1881
REFLECTORS (*See also* "*Optical Apparatus,*" "*Lanterns.*")		
Weyde	446	1878
Neale	902	1878
Alder and Clarke.. ..	1,442	1878
De Castro..	2,943	1878
Pulvermacher	4,079	1878
Sprague	4,662	1878
Molera and Celrian ..	299	1879
Young and Freeman ..	350	1879
Weyde	508	1879
Harding	783	1879
Holmes	920	1879
Werdermann	2,301	1879
Mangin, Lemonnier, and Co.	3,425	1879
Schumann..	2,179	1880
Heinrichs	4,608	1880
Benson	4,254	1880
Brush	*1,834	1881
Wheeler	3,911	1881
Varley	5,396	1881
Rowan	5,400	1881

	No.	Year
REFLECTORS—		
Williams	766	1882
Franke	849	1882
Wheeler	917	1882
Combined with globe, Aronson	305	1882
Fitted with coloured shade, Grieve	259	1879
For incandescence lamp, Lorrain	5,738	1881
Of silvered glass, *Watson* ..	389	1879
Regulating length of arc in lamps (*See also* "*Arc Lamps.*")		
Wood	2,563	1882
„	2,570	1882
REGULATORS, Current—		
Pulvermacher	1,900	1876
„	3,469	1877
Lane-Fox	4,043	1878
Harding	4,047	1878
Pulvermacher	4,094	1878
Scott	4,140	1878
Bright	4,212	1878
Edison	4,226	1878
Van Choate	4,388	1878
Fuller	5,183	1878
Blandy	2,060	1879
Edison	2,402	1879
Cabanellas..	200	1881
Lane-Fox	1,636	1881
Jameson	2,618	1881
Gravier	2,739	1881
Thompson	3,032	1881
Dunstan and Pfannkuche	3,655	1881
Edison	4,034	1881
Deprez and Carpentier..	4,128	1881
Carod	4,508	1881
Carus-Wilson	5,687	1881
Cruto	1,895	1882
Sperry	3,025	1882
Rigaud	3,054	1882
Automatic indicator for hand, *Edison*	1,023	1882
Acting on the driving motor—		
Edison	4,626	1878
Sawyer and Man.. ..	4,705	1878
Andrews	*2,321	1879
Richardson..	288	1881
Westinghouse.. ..	3,409	1881
Carus-Wilson	4,825	1881
Thomson	5,668	1881
Carus-Wilson	5,687	1881
Levy	542	1882
Richardson	941	1882
Adjusting the brushes—		
Edison	1,149	1882
Wood	2,526	1882
Adjusting field magnet poles—		
Chameroy	2,295	1882
Jarman	2,630	1882
Arrangement of generators, Perry	55	1882
Coils in circuit, Andrews ..	2,321	1879
Exciting current—		
Brush	849	1880
Maxim	1,392	1880
Edison	3,964	1880
Brush	1,835	1881
Edison	2,482	1881
Langley	4,168	1881
Edison	4,552	1881
Wright and Ormiston ..	5,006	1881
De Khotinsky	245	1882
Edison	1,191	1882
Edison	1,023	1882
„	2,052	1882
For use with gas engines, Carus-Wilson	5,687	1881
Hand, *Edison*	1,023	1882
Induction apparatus, Hopkinson	3,362	1881
Platino-carbon, Gatehouse	1,400	1882
Potential of—		
Cochrane	4,313	1878
Deprez and Carpentier ..	4,128	1881
Maxim	1,162	1882
Williams	1,556	1882
Potential of, by batteries—		
Williams	766	1882
„	700	1882
„	856	1882
Potential of, from batteries	2,558	1882

	No.	Year
REGULATORS, Current—		
Radial adjustment of armature coils, *Maquaire* ..	2,885	1882
Resistances—		
Lane-Fox	4,626	1878
De Mersanne and Bertin	5,053	1878
Sellon and Edmunds ..	1,692	1879
Cook	2,769	1879
Resistance of field magnet circuit of exciter—		
Edison	33	1880
„	602	1880
Secondary batteries—		
Thomson	3,032	1881
De Kabath	4,060	1881
Edison	4,553	1881
Varying relative speeds of motor and generator, *Rigaud*	3,054	1882
Varying length of arms of centrifugal governor, Varley	5,665	1881
REGULATORS, Current, with—		
Centrifugal governor driven by motor, Carus-Wilson	4,825	1881
Rapidly vibrating circuit controller, *Edison* ..	1,496	1882
Tilting cog-bar gearing with one of two pinions, *Sawyer and Man* ..	4,705	1878
REGULATORS—		
Expansion, *Lontin* ..	2,094	1877
Light—		
Sawyer and Man ..	4,704	1882
Edison	539	1881
Gatehouse	3,240	1881
Resistance, automatic, for incandescence lamps—		
Welch	4,689	1878
REGULATORS, Speed (*See also* "*Motors, Electric, Regulating.*")		
For electric motors—		
Edison	33	1880
„	602	1880
„	1,385	1880
Edison	3,964	1880
Deprez and Carpentier..	4,128	1881
Langley	4,168	1881
Carus-Wilson	4,825	1881
Jarman	2,630	1882
Ayrton and Perry ..	2,830	1882
Jenkin	3,007	1882
For electric motors, hand or automatic, *Le Cie. Electrique*	2,990	1882
For electric motors driven by batteries, *Griscom* ..	1,244	1880
For generator, *Cabanellas*	200	1881
For turbines, &c., *Williams*	1,174	1882
Relay, Brasseur and De Sussex	303	1878
Resistances—		
Cadett	4,022	1878
Thomson	3,032	1881
Wilkinson	3,003	1882
Binko	3,073	1882
Resistance balances—		
Abdank	339	1882
Roosevelt and Abdank ..	3,070	1882
Resistances, carbons for, Varley..	2,776	1882
Reversal of polarity, preventing (*See* "*Polarity, Preventing Reversal of.*")		
RHEOSTATS—		
Pulvermacher	1,900	1876
Coxeter	492	1878
Bright	596	1878
Cadett	4,022	1878
„	4,316	1878
Sellon and Edmunds ..	1,692	1879
Hadden	3,843	1879
Harrison	3,875	1879
Roguier	75	1880
Lugo	5,352	1880
Cuff	2,263	1881
Hussey and Dodd ..	2,572	1881
Gravier	2,739	1881
Thomson	3,032	1881
Pfannkuche and Dunstan	5,743	1881
Williams	856	1882
Munro	1,626	1882

RHEOSTATS—

	No.	Year
Metallic chain, *La Société la Force et la Lumière, &c.*	4,496	1881
Ruhmkorff's coil, application to electric light, De Sussex	465	1879

SAFETY FUZE—

	No.	Year
Lontin	2,094	1877
Inserted in main leads, *Edison*	1,023	1882
Secondary batteries (*See* "Batteries.")		

SEMI-INCANDESCENCE LAMPS (*See also* "Arc Incandescence Lamps.")

	No.	Year
Sawyer and Man	4,847	1878
De Mersanne and Bertin	5,044	1878
Rapieff	211	1879
Werdermann	2,301	1879
Joel	5,157	1879
André and Easton	2,236	1880
André	2,764	1880
Heinrichs	4,608	1880
Sachs	894	1881
Common and Joel	1,040	1881
Harrison and Blagburn	1,358	1881
Société La Force et la Lumière, &c.	1,653	1881
André	2,563	1881
Somzee	4,305	1881
De Changy	4,405	1881
Rogers	4,855	1881
Solignac	740	1882
Werdermann	1,444	1882
Salomons	1,580	1882
Zanni	2,740	1882
Automatically short-circuiting (*See* "Switches, Automatic.")		
Joel	5,157	1879
Automatic switch for, Werdermann	2,301	1879
Carbon coils for, Varley	2,776	1882
Feeding carbon to, André and Easton	2,236	1880
Preventing explosion in, Hedges	4,988	1880
With carbon descending against metallic disc, Hedges	4,988	1880
With electric motor, André	2,764	1880
With inclined carbons, Rogers	1,922	1881
With mercury float, Blamires	455	1880
With multiple carbons, *De Mersanne and Bertin*	5,044	1878
With Torricellian vacuum, Blamires	455	1880
Short-circuiting, avoiding in armature, *Weston*	1,614	1882
Shunt, automatic, Scott	4,140	1878

Signals, Electric, for railways—

	No.	Year
Putnam	1,125	1880
Perry and Ayrton	783	1881
Putnam	2,711	1881
D'Auriac	1,683	1881
Sparking, preventing at commutator (*See* "Commutator, Sparking at.")		

Speed Indicators—

	No.	Year
Reynolds	515	1874
Cardew	5,354	1881
Starr	5,600	1881
Rogers	1,390	1882

	No.	Year
Speed, measuring, of generators, Thomson	5,668	1881
Speed regulator (*See* "Regulators, Speed.")		
Standard of electromotive force, Dewar	2,886	1876
Starter, automatic, Edison	2,402	1879
Static charge, preventing, in lighting circuits, Crompton	346	1882

SWITCHES—

	No.	Year
Weston	4,280	1876
Lontin	2,094	1877
Rapieff	211	1879
Andrews	2,321	1879
Edison	2,402	1879
Hedges	3,369	1881
Faure	129	1881
Hussey and Dodd	2,572	1881
Gravier	2,739	1881
Williams	856	1882
Munro	1,626	1882
Brougham	2,030	1882
Defries	2,335	1882
Actuating electrically, Carus-Wilson	5,687	1881

SWITCHES for—

Arc lamp,

	No.	Year
Joel	5,157	1879
Gadot	2,344	1881
Wood	2,563	1882
Breaking circuit at several places, Edison	1,385	1880
Connecting circuits to earth, Howard	2,896	1882
Connecting up conductors, Andrews	2,321	1879

Incandescence lamp—

	No.	Year
Edison	1,802	1881
Hughes	3,190	1881
Berly	1,236	1881
Aronson	359	1882
Emmens	2,348	1882
Hallet	2,560	1882
Volk	2,962	1882
Regulating currents, Andrews	2,321	1879
Secondary batteries, *Tommasi*	252	1882
Short-circuiting lamp, Common and Joel	1,040	1881
Signalling apparatus, Lemair and Lebrun	4,081	1880
Successively completing number of circuits, Budenbury and Schäffer	4,227	1880
Turning current gradually on to lamps, *Sawyer and Man*	4,705	1878
Operated by compressed air, Berly	1,027	1881

SWITCHES, Automatic (*See also* "Batteries, Secondary, Charging, and Discharging;" "Cuts Outs;" and "Shunts.")

	No.	Year
Weston	4,748	1877
"	*4,903	1877
Scott	4,140	1878
Sawyer and Man	4,705	1878
Sellon and Edmunds	1,492	1879
Reynier	1,971	1879
Zanni	2,821	1879
Wilkins	4,306	1879
Godfrey	4,718	1879
Jablochkoff	725	1880

SWITCHES, Automatic—

	No.	Year
Berly	1,027	1881
Edison	2,492	1881
Hussey and Dodd	2,572	1881
Thomson	3,032	1881

SWITCHES, Automatic, for—

	No.	Year
Cutting changed battery out of circuit, *Houston and Thomson*	4,400	1879
Electric lamps, *Jablochkoff*	725	1880
Incandescence lamps, Sprague	4,662	1878
Incandescence lamps with multiple filaments, Aronson	359	1882
Multiplex lamps, Berly	1,236	1881
Preventing reversal of currents (*See also* "Polarity, Preventing Reversal of.")		
Zanni	2,821	1879
Fisher	1,727	1882
Replacing discharged batteries, *Lamar*	4,696	1879
Secondary batteries, Emmens	2,912	1882
Semi-incandescence lamp—		
Werdermann	2,301	1879
Joel	5,157	1879
Shunting excess of current, Mackie	14	1882
Shunting portion of current when armature attracted, André	5,206	1879

	No.	Year
TELERADIOPHONE, *Mercadier*	3,929	1881
Telpherage, Jenkin	1,830	1882
Tramcars, Richardson	2,703	1881

Transmission of power—

	No.	Year
Spalding	915	1878
Chretien and Felix	2,019	1879
Poulet and Commelin	1,046	1880
Edison	3,894	1880
Gumpel	253	1881
Hopkinson	2,969	1881

VACUUM PUMPS—

	No.	Year
Lane-Fox	3,494	1880
Akester	*4,458	1881
Stearn	*5,000	1881
Kennedy	5,524	1881
Goebel and Kulenkamp	*5,548	1881
Sennett	5,286	1881
Wright and Mackie	1,031	1882
Rogers	1,288	1882
Cruto	1,895	1882
Gimingham	2,375	1882
Akester	2,519	1882
Leask	2,701	1882
Vacuum tubes, Scantlebury	1,932	1878
" for theatrical effects, Aronson	2,008	1878

Voltameters—

	No.	Year
Pulvermacher	1,900	1876
Avenarius	3,025	1880

WATER—

	No.	Year
Utilising flow of, Williams	1,174	1882
Purifying, Atkins	556	1873

YARNS—

Rendering non-absorbent—

	No.	Year
Hooper	3,780	1873
Hooper and Dunlop	3,997	1873

	No.	Year
ZAPOTINE, Conybeare and Naphegyi	2,106	1874

APPENDIX A.

———:-:—--

MÜNICH EXHIBITION REPORTS:—DYNAMO AND LIGHT TESTS.

THE official reports of the tests made during the Electrical Exhibition at Münich, 1882, form a very valuable contribution to our literature on applied electricity. Münich could not vie with Paris in the brilliancy of the nightly display, and the variety and number of her exhibits. But the scientific importance of the Münich Exhibition, and the solid and thorough character of the investigations carried on during the short period that it was open—September 16th to October 15th, 1882—have never been questioned. We, therefore, think that the following summary will be interesting to our readers. We include in our abstract an able article from the pen of Dr. Hugo Kruess, of Hamburg, on the History of Electrical Photometry.

——————————————————— ..-

FROM the very commencement of the Münich Electrical Exhibition scheme, the idea of a rigorous scientific comparison of the various exhibits was prominently before the promoters, and as as soon as circumstances permitted, a general plan for the whole series of intended experiments was sketched out by Dr. Kittler, who was then on the staff of the Münich Technical College, although appointed to the electro-technical chair at the new college at Darmstadt, which was opened with the year 1883. This plan, having been approved of by the sectional committees, was published in pamphlet form and afterwards adhered to, few and slight alterations only suggesting themselves. Extensive and thorough tests of the instruments and methods to be employed, preceded the real work. It would lead us too far to dwell upon the theory and the execution of these calibrations. We may venture to state, however, to exemplify the character of the proceedings, that the graduation of the Wiedemann mirror galvanometer, which served for all determinations of current intensities, involved eight series of observations, each of them requiring numerous readings of various kinds.

It must be borne in mind that time and means were scanty, and circumstances not always so favourable as desired; yet it is remarkable how much was achieved. To arrive at absolutely comparable results, all the systems ought to have been tested under exactly the same conditions. This was possible only in the case of the electrical and optical measurements; as regards the power tests, it would have necessitated that all the dynamos should have been coupled to one and the same motor, which was practically impossible.

Power Tests.—The principle having once been adopted that none but simultaneous observations should be relied upon, the question as to the nature of the dynamometer to be employed had already been decided in favour of the exclusive use of transmission dynamometers, which are placed directly between the motor and the dynamo. In some other trials the power transmitted to the dynamo has been determined, afterwards and separately, by applying the Prony brake to rearrive at the conditions of the previous experiments. The practical difficulty is, of course, to reattain the same conditions. At Münich it was found necessary to carry the dynamometer from one dynamo to the other, as the dynamos could not be moved. The dynamometer of Mr. von Hefner Alteneck was the only one practicable under these conditions, and served for all the observations, except in one instance. In this case there was employed the registering dynamometer of Mr. Schuckert, the well-known electrician; it was constructed by Mr. Keck, of Nürnberg, but arrived too late. This instrument has on its main shaft two pulleys, one keyed to the shaft and revolving with the motor, the other loose and coupled with the dynamo; the connection between the two pulleys is effected by a strong spiral spring. When at work the loose pulley remains behind by a certain angle, varying with the pressure exerted upon the spring; and a special mechanism traces these variations on a drum. This dynamometer offers certain advantages and can be easily controlled, whilst the instrument devised by Mr. von Hefner Alteneck cannot quickly be verified. But Mr. Schuckert's dynamometer is not portable; to use it the dynamo had to be dismounted, mounted again for the tests, and removed to its original place after the tests. As these operations take up much time the Von Hefner Alteneck instrument was selected and bolted to a very heavy, yet portable, wooden block. The instrument was not supplied with a set of springs of various strengths. This defect was most particularly felt in connection with the most remarkable experiments conducted at Münich, the transmission of power by M. Deprez, where a comparatively small force at a very high speed had to be measured; we shall again refer to this point. Besides the dynamometer a tachymeter of Messrs. Buss and Sombart was used in the mechanical tests, which were mostly conducted by Professor Moritz Schroeder, of Münich.

Electrical Measurements.—Of the power given to the dynamo machine by the motor, and measured with the

help of the dynamometer, a certain portion is lost through friction, and the main part converted into electric energy. A part only of this electric energy, however, is rendered useful in the external circuit for the transmission of power or illumination, as currents in the iron and detrimental heating of the iron cores, and of the conductors, in consequence of the changes of polarity, cannot be avoided. The electrical tests are for the purpose of ascertaining what proportion of the electrical energy is really doing practical work. All tests were carried out, so far as possible, under the ordinary working conditions of the dynamos ; and whilst resistances and speed were varied, neither very high resistances nor abnormal speeds were tried. The respective dynamical and photometrical observations were always made at the same moment ; at first with the help of telephone signals, afterwards, when the telephone circuit had by carelessness been destroyed, at certain signals at every three or ten minutes.

The apparatus at the disposal of the investigators were very complete and of the best quality. Numerous copper and iron resistances were provided for them. Two thick copper wires of 7 millimetres diameter, supplied by Messrs. F. A. Hesse Söhne, of Heddernheim, could take the strongest current of 170 ampères without becoming hot. The increase of temperature of the bare thinner wires had previously been determined by a series of experiments undertaken by Professor Dorn, to whom, together with Professor Kittler, fell the main bulk of this work, and to whom we owe the very careful and extensive calculations. This increment was found to be a function of the square of the current intensity, and the hot resistances stated in the tables accompanying this article, are calculated on this assumption, which proved correct for the strongest currents available. All resistance coils, cables, &c., had been gauged and tested before use. The great rheostat consisted of a wire network on telegraph poles ; the whole of the resistances amounted to 18 kilometres (11 miles) length of wire. The resistance determinations were made with a great bridge of Messrs. Siemens and Halske, and the Universal Compensator, devised by Professor Von Beetz, the President of the test-committees, and one of the main promoters of the Exhibition. Current intensities were measured with a Wiedemann mirror galvanometer, placed in a shunt to a thick copper wire of 6.46 millimetres diameter, so that only a small portion passed through the galvanometer ; this arrangement was first suggested, for dynamo tests, by Professor Hagenbach, and the method yielded excellent results for currents between one half ampère and 170 ampères. The respective derivation points were soldered, and the readings taken with the aid of a telescope. There were, further, two electro-dynamometers of Siemens, and a Deprez galvanometer, and the results displayed a remarkable agreement between these four instruments. Potential differences were ascertained with the aid of a Siemens torsion galvanometer, and a box of resistances equal to 100,000 Siemens units. The determination of the units was left to a sub-committee, which agreed upon the ohm, ampère, and volt : the practical resistance unit to be the Siemens unit, equal to 0.95 ohm ; the ampère, that current which in a minute deposits 19.7 milligrammes of copper ; the volt to be derived as a product, 1 volt = 1 ohm × 1 ampère. It will be remembered that this plan differs from the method usually adopted which observes the volts and ohms, and calculates the ampères, as

$$\frac{\text{volts}}{\text{ohms}}$$

in this latter case, the determination of the differences of potential presupposes the most accurate knowledge of the electromotive force of the standard cells ; and this system is not so reliable as the calculation based upon the copper weight deposited by the current.

The Tests of Dynamo Machines were confined to continuous current machines, of which there were two classes, distinguished by the electro-magnets being placed either in the main circuit or in a branch circuit. Below we summarise shortly the formulæ employed to work out the experimental evidence, first, for dynamos whose electromagnets form part of the main circuit.

Let

R_1 be the resistance of the armature.
R_2 ,, ,, ,, electro-magnets.
R ,, total resistance of the dynamo.
r ,, external resistance.
J ,, total current intensity.
E ,, ,, electromotive force.
e ,, difference of potentials at the terminals.

Then R_1, R_2, R, J, and e can be measured directly, and r approximately, by introducing wire resistances ; this determination of r is, however, crude. As :

$$R = R_1 + R_2 \text{ and } J = \frac{e}{r} = \frac{E}{r + R},$$

we obtain

$$E = e + RJ, \text{ and } r = \frac{e}{J}.$$

This latter value represents the actual resistance of the external circuit, or in the case of lamps, the respective wire resistance to be substituted. The total electrical work per second will be :

$$L = E J \text{ (in volt-ampères)} = \frac{E J}{9.81} \text{ kilog. metres per sec.}$$

$$= \frac{E J}{736} \text{ HP. German or French} = \frac{E J}{746} \text{ HP. English.}$$

The external electrical work will be $l = e J$. If further A represents the power applied, as measured by the dynamometer, then the ratio $\frac{l}{L}$ will be the electrical efficiency, and $\frac{l}{A}$ the absolute efficiency. The latter quotient is the most important result, as it denotes the portion of the mechanical power, absorbed by the dynamo, which may be made useful for illumination, transmission of power, or other purposes. The other quotient, $\frac{l}{L}$ is equal $\frac{r}{r + R}$; as any machine may be worked with any external resistance, this latter quotient acquires signification only in connection with the other data.

The only machines of the second class, having their electro-magnets in a shunt circuit, experimented with were Edison dynamos. These dynamos are provided with a regulator, which, by means of resistances to be shunted in, varies the electromotive force. This regulator was removed during the tests. If we adhere to the above letters, and further introduce J_1 as the current in the armature, and J_2 as the current in the electro-magnets, then we shall again be able to measure R_1, R_2, J e, and to calculate $J_2 = \frac{e}{R_2}$ and $J_1 = J + J_2$; there results further $E = e + J_1 R_1$, and the external work $l = e J$. The electrical work in the machine itself comprises the sum of the heat evolved in the armature and the heat in the electro-magnets, so that the total electrical work is $L = e J + J_1^2 R_1 + J_2^2 R_2$.

When a complete light installation had to be tested, a first series of experiments was conducted with all the respective lamps in circuit, one of them being in the photometer room. These are the conditions of actual working. All resistances having first been determined cold, the intensity and electromotive force were measured, and simultaneously the photometrical observations were carried on. These latter we explain later. Then the hot resistance was ascertained. Fig. 1, page iii., illustrates the arrangements for these tests and gives also a diagrammatic sketch of the main switch devised by Professor Kittler. For a second series of experiments, all the lamps were replaced by wire resistances until the same intensity and electromotive force were reattained ; external resistance and speed were further varied to study the behaviour of the dynamos under different circumstances.

We have only a few remarks to add concerning the lamp tests, as we have simply to deal here with the power absorbed and not with the illuminative efficiency, which will be discussed later, as just mentioned. One arc lamp only, we said, was placed in circuit, and the others replaced by wire resistances to avoid current fluctuations. If current intensity J, and difference of potential at the

binding screws of the lamp ϵ, are known, the power absorbed by the lamp λ follows from $\lambda = J \epsilon$ volt ampères.

The tests of incandescence lamps proceeded in a similar way, but necessitated slightly different connections, which

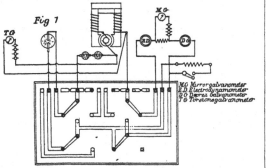

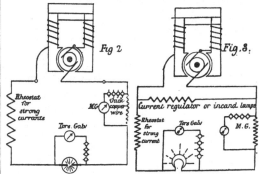

The arrangements for this examination are illustrated in Fig. 2.

are shown in Fig. 3. The circuit was double; the one part comprising all lamps but one or their equivalent in wire;

TABLE I.—DYNAMO TESTS.

MACHINES.	Arrangements.	Resistance in Ohms.								Current in Ampères.			
		Cold.			Warm.			External.		Total.				
		Armat. R_1.	El. M. R_2.	Total Machine. R.	R_1.	R_2.	R.	Calcul. $\frac{e}{J}$	Inserted r.	$r+R$.	M.G.	El. D. I.	El. D. II.	D.G.
I. Schuckert (for 7 arc lamps)	Wire ..	4.945	6.370	11.33	..	..	12.87	46.84	46.34	59.71	7.85	7.93	..	7.84
	7 lamps+ wire ..	..	..	..	..	..	..	50.94	..	..	8.03	8.08	8.07	
	Lamp+wire ..	..	..	..	..	..	13.25	47.65	..	..	7.68	7.77	7.83	
II. Schwerd (for 5 arc lamps)	4 lamps ..	..	..	..	..	..	3.57	13.45	..	..	14.80	14.75	14.73	
	Wire ..	1.14	1.85	2.97	..	..	3.51	12.59	12.69	16.10	16.05	16.17	16.11	16.19
	1lamp+wire ..	..	..	3.3	..	..	3.50	13.50	..	..	16.11	16.15	16.10	
III. Bürgin (Crompton)	3 Crompton lamps ..	..	..	..	..	..	4.19	6.99	..	11.18	22.63	22.76	..	22.58
	1lamp+wire ..	2.14	1.78	3.94	..	..	4.30	10.42	..	14.72	16.34	16.34		
IV. Schaeffer (Weston type)	Wire ..	0.785	0.456	1.216	..	..	1.218	7.88	8.01	9.10	12.20	12.0		
	Wire ..	0.459	1.051	1.498	..	..	1.55	10.72	10.93	12.27	13.59	13.88		
V. Schoenemann (Gramme type)	Wire ..	0.633	0.635	1.264	..	..	1.292	6.13	6.09	7.42	8.26	8.29		
VI. Edelmann ..	Wire ..	2.548	2.492	5.033	..	..	5.17	14.56	14.81	19.73	4.18	4.17		
VII. Edelmann (compound)	Wire ..	3.274	4.717	7.994	..	..	8.29	15.39	14.81	23.68	3.23	3.21	..	3.27
Einstein ..	Wire ..	1.516	2.150	3.620	..	..	3.73	12.79	12.75	16.52	4.81	4.82	4.84	

TABLE I.—*continued.*

MACHINES.	Electromotive Force. Volts.		Speed.	Electrical Effect.				Power Transmitted A.HP.	Ratios. Per Cent.		
				External. l.		Total. L.			$\frac{l}{L}$	$\frac{l}{A}$	$\frac{L}{A}$
	e.	E.		$e J$.	HP.	$E J$.	HP.				
I. {	367.6	468 6	799	2885	3.92	3678	5.00	6 09	78.5	64.4	82.0
	408.9	..	955	3284	4.46	..	..	6.66	..	67.4	
	366.0	467.8	800	2811	3.82	3593	4.88	5.30	78.2	72.07	92.1
II. {	199.0	251.8	992	2945	4.00	3728	5.06	5.95	79.0	67.3	85.1
	202 1	258.4	1025	3244	4.41	4148	5.64	6.14	78.2	71.8	91.8
	217.4	270.6	1099	3503	4.76	4359	5.92	7.17	80.3	66 4	82.6
III. {	158.1	252.9	1622	3578	4.86	5724	7.78	8.83	62.5	55.0	88.1
	170.0	240.2	1500	2778	3.78	3926	5.34	6.00	70.7	63.0	88.9
IV.	96.1	110.0	1397	1174	1.59	1356	1.84	..	86.6		
V.	145.8	166.9	802	1981	2.69	2269	3.08	..	87.3		
VI.	50.65	61.32	1195	418	0.568	507	0.688	1.50	82.5	37.9	45.9
VII.	60.89	82.49	1518	255	0.346	345	0.469	1.02	73.8	33.9	45.9
VIII.	49.74	76.53	1401	160.8	0.219	247.4	0.336	0.93	65.0	23.5	36.2
	61.6	79 6	802	296	0.403	383	0.520	2.00	77.4	20.1	26.1

and the other the great rheostat, the torsion galvanometer, the mirror galvanometer, and the one lamp to be tested. This arrangement permitted of taking easy measurements at different light intensities, and was considered by the Committee an improvement upon the method of the cognate tests at Paris.

The official reports, although filling a portly quarto volume of one hundred and fifty pages, give the *complete* data of one machine only; if the Committee had attempted to place before the reader all their original observations, their reports would have assumed quite abnormal dimensions. The one machine alluded to was a Schuckert dynamo; more than fifty readings were taken at all instruments, the greater number of them at given signals. Only those figures which represent the results of simultaneous tests in different portions of the system were made use of for the calculations. The two Tables which we subjoin are of a most condensed character. We greatly regret to be obliged to confine ourselves to so short a summary; a reprint *in extenso* is, however, out of the question.

The first Table, I., deals with machines whose electro-magnets form part of the main circuit. Most of the columns

whole, be acknowledged that the engineer who is chiefly interested to know what percentage of the power expended he gets back, will be less satisfied with these tests, than the electrician upon whose ampères and ohms every possible care was bestowed.

We cannot here enter into the theoretical analysis to which Professor Dorn further subjected the experimental results in order to test Dr. Froelich's well-known theory of the dynamo machine, which theory mainly rests on the experience gained with Siemens' generators. A few words will, however, suffice to elucidate the nature of this analysis. Dr. Froelich plots two curves for each machine; first, the curve of effective magnetism constructed with the current intensities as abscissæ; and the quotient of the total electromotive force divided by the number of revolutions, as ordinates. This effective magnetism soon attains its maximum with an increasing current, rises, therefore, slowly when near its maximum, and ought to decrease again slowly as the current still continues to augment. As most generators work at relatively high current intensities, the curve of effective maximum will soon become almost a horizontal line. The second curve is the current curve, whose ordinates are represented by

TABLE II.—DYNAMO TESTS.

MACHINES.	Resistance in Ohms. Cold. Armat. R_1	Cold. El. Mg. R_2	Hot. R_1	Hot. R_2	External. Calcul. eJ	External. Inserted r	Current in Ampères. External Circuit. M.G.	El. D. I.	El. D. II.	Machine. Armat. J_1	Machine. El. Mg. J_2	E.M.F. Volts. e	Volts. E
I. Edison K (250 A lamps) ..	0.0361	13.82	0.0416	13.84	1.298	..	100.6	..	..	110.0	9.4	130.6	135.2
II. Edison Z (60 A lamps) ..	0.142	40.1	0.161	40.5	4.55	..	27.67	27.9	27.7	30.77	3.10	125.8	130.7
III. Edison E (17 A lamps) ..	0.338	88.7	0.383	90.0	12.75	12.88	8.89	8.98	8.93	10.15	1.26	113.3	117.2

TABLE II.—continued.

MACHINES.	Electrical Effect. External. l. Volt. Amp.	External. l. HP.	In Armature. VA.	In Armature. HP.	Electro-Magnets. VA.	Electro-Magnets. HP.	Total. L. VA.	Total. L. HP.	Speed.	Power Transmitted A HP.	Ratio Percentages. $\dfrac{l}{L}$	$\dfrac{l}{A}$	$\dfrac{L}{A}$	$\dfrac{l}{A-2.19}$	$\dfrac{L}{A-2.19}$
I.	13,130	17.84	504	0.68	1231	1.67	14,870	20.20	914	..	88.4	53.5	61.9	71.2	82.3*
II.	3,480	4.73	152.2	0.207	390.2	0.530	4,022	5.47	1197	8.83	86.5	45.9	54.2		
III.	1,007	1.369	39.5	0.054	142.7	0.194	1,190	1.616	2409	2.98	84.7	45.9	54.2		

* This is the case above alluded to; the empty run absorbed as much as 2.19 HP., which extraordinarily high figure was probably due to a very stiff belt. It appears that if suitable deduction had been made on this account, the efficiencies $\dfrac{l}{A}$ and $\dfrac{L}{A}$ would have risen by about 20 per cent.

require no further explanation; the second column refers to the tests of complete light installations, specifying whether the current had to feed the actual lamps or an equivalent of wire resistance. The two sub-columns of the fourth vertical row demonstrate the great exactness with which this substitution could be effected. We read, for instance, in the top horizontal row, "external resistance calculated $r=46.84$; and inserted 46.34." The very close agreement of the four instruments indicating current intensities, is also very striking. These were the Wiedemann mirror galvanometer M.G., Fig. 1, two Siemens electro-dynamometers El. D. I. and El. D. II., and a galvanometer of M. Deprez D.G. The L, l, A have already been explained; also the e, difference of potential at the terminals, and E total electromotive force. We have to point out that no deductions were made for running the machine empty. This is all the more to be regretted as in the only case when the respective allowance for the empty run was determined, the efficiency percentages assumed considerably higher values. It must, on the

the current intensities, whilst the abscissæ depend upon the quotient of speed divided by the total resistance. Dr. Dorn found that all machines gave remarkably regular curves, the current curves being almost straight lines; the Bürgin, Schaeffer and Schoenemann machines yielded particularly fine curves. The theory of Dr. Froelich concerns only the generators, whose electro-magnets are included in their main circuit, such as the machines of Table I. A complete theory of the generators whose electro-magnets are placed in a shunt circuit, has not yet been written, and the Münich experiments are not complete enough to establish a law. It appears to be shown, however, that the total magnetism acting upon the armature is a linear function of the current in the electro-magnets; and the agreement between the current values derived from a formula based upon this assumption, and the experimental data of the two light Edison machines, which were subjected to a special series of trials, is so remarkable that we will add the two respective Tables.

Edison K, for 250A lamps.

J calculated : 100.5—107.0—123.7—142.8—157.9—168.4
observed : 100.6—107.2—123.2—142.5—157.5—169.3

Edison Z, for 60A lamps.

J calculated : 27.57—30.14—31.42—34.22—36.39—37.85
observed : 27.67—30.12—31.42—34.31—36.31—37.69

Transmission of Power.—The experiment of M. Deprez has been discussed so widely and by so many various pens in the scientific and technical press, that another notice in these columns might be dispensed with. The official reports issued by the Committee who watched over the experiment, deserve, however, due consideration, and our abstract would be incomplete in an important part if we omitted all mention of it. We shall, therefore, place the data, as they are given by Professor Dorn, once more before our readers.

The distance from Miessbach to Münich is 57 kilometres, or 35 miles. Both lines were ordinary telegraph wire of 950.2 ohms total line resistance, and proved at a preliminary test, conducted by Professor Kittler, to be well insulated. The primary machine at Miessbach had a resistance of 453.1 ohms., the secondary machine at Münich 453.4 ohms. A control test of the resistance, Münich + line, gave 1407 ohms. Further observations could not be taken on the first day, October 9, 1882, as the Münich machine refused to start. Nor did the experiments commence more successfully on the second day; and whilst various expedients were tried to ascertain the source of the disturbances, the Münich dynamo suddenly started, just when the galvanometer there was out of circuit, so that in the only useful series of tests no current measurement could be taken at Münich. When the speed at Miessbach was raised to 2000 revolutions, the secondary machine again failed ; nor would it work properly afterwards. Several other disturbing influences are to be mentioned. The Münich machine was not pro-

perly bolted down, and the Von Hefner Alteneck dynamometer, as we pointed out above, was not adapted for measurements of small power at high speeds. The latter defect may probably account for the altogether inadmissible result that the total electrical work (1.131) appeared to be greater than the power absorbed at Miessbach (1.065). An observation error of less than one division of the scale would have sufficed to make this anomaly disappear. It is of course manifest that under these circumstances the whole power tests are of a dubious value ; and our readers will be contented if we annex the means of the six observations taken during the half hour over which this only complete experiment extended.

The speed of the primary machine at Miessbach was 1608 ; of the secondary at Münich, 712 ; the power absorbed at Miessbach $A = 1.065$ horse-power ; that rendered at Münich $A_1 = 0.235$; therefore the ratio $\frac{A_1}{A} = 22.1$ per cent. The current J amounted to 0.527 ampère ; the difference of potential at Münich $E_2 = 843$ volts ; at Miessbach $E_1 = E_2 + 950.2 \, J = 1344$ volts ; the external electrical work of the primary machine $E_1 J = 708$ volt ampères = 0.961 horse-power; the total electrical work $E_1 J + J^2 \times 453.1 = 833$ volt ampères = 1.131 horse-power. There had to be deducted for the heating of the circuit $1856.7 \, J^2 = 515$ volt-ampères = 0.700 horse-power, leaving disposable for actual transmission of power 317 volt-ampères = 0.431 horse-power, or 38.1 per cent. of the total electrical work, of which available energy there could be only rendered useful at Münich 54.5 per cent. These 54.5 per cent. of the energy which did perform work at Münich constitute 20.8 per cent. of the total electrical work expended, the latter amounting to 833 volt-ampères. It will be in the recollection of our readers that a somewhat indiscriminate use was made immediately after the experiment of these various percentages, by some of our French contemporaries.

THE HISTORY OF PHOTOMETRY.

The second part of the report, viz., Photometrical Tests, is preceded by an Historical Sketch on Electrical Photometry, by Dr. H. Kruess.

PHOTOMETRY is a subject which has developed with the electric light. Formerly it did not appear to offer any great difficulties, and it has consequently been treated somewhat superficially in our text-books of science. For the electrician, photometry has a twofold importance ; the function of the photometer is, firstly, to show what quantity of light a certain lamp supplies, so that its economy may be judged ; and, secondly, to demonstrate the relations between the light produced and the other forces at work in a given system of machines and lamps. It was only when this latter point became better understood that photometers began to engage the attention of the electrician.

The first photometric tests generally referred to in text-books of science, are those of Fizeau and Foucault, of 1843. It must, however, be borne in mind, that what these eminent French scientists originally measured, is not that which really interests us at present in such experiments ; the decomposition of an iodine and silver combination, by means of light from various sources, indicates the chemical intensity of the rays, but not the optical intensity. The surprisingly low chemical in-

tensity of the lime light caused MM. Fizeau and Foucault to repeat their experiments, in order to determine the optical energy ; and the agreement of the new and old figures suggested to them that, for white light, the two determinations might practically be replaced by the one which is more convenient, that is the chemical test. It is evident, however, and needs no further support in these days, when we photograph the invisible ultra violet spectrum, that chemical tests cannot be relied upon ; and that photometers like Becquerel's electro-chemical actinometer, or Siemens' selenphotometer, cannot measure the illuminative power of a source of light, however perfect and ingenious they may be in other respects. We have still to depend upon the physiological action of the light rays upon the retina of our eye, untrustworthy as this evidently is, since different observers are not equally sensitive to the same degree, nor even the same observer so at all times.

The photometrical researches of Th. W. Casselmann, of Marburg, are of interest for the electrician, because they included the electric light, because they were prior to those of Fizeau and Foucault, and because in them

there was first introduced Bunsen's photometer, in favour of which Casselmann decided against Ritchie and Rumford.

No accurate tests appear to have taken place after that before 1855, when MM. Lacassagne and Thiers tried their electric lamps at Lyons. M. Edmond Becquerel reported on those trials to the Société d'Encouragement of Paris, and this report led to a contest between the interested parties, as Becquerel estimated the intensity at 350 candles, whilst the manufacturers claimed 600 and more. This contest was of importance, because the real point was the question of expense. Scarcely any tests of real scientific value were, however, undertaken before the Alliance magneto-electric machines in France, and those of Mr. Holmes and others in England, attracted attention. Then difficulties cropped up everywhere, and the main problem has not been solved up to the present day. Neither the French bec-carcel, nor the English standard candle, nor the German candle, can be considered as normal, as they are all variable; this is strikingly evidenced by the fact that the ratios between the different standards, as stated in text-books, do not agree. Many proposals have been made. MM. Rüdorff, and Methuen suggested that the middle part of a flame, as more steady than the flickering top and the lower zones, ought to be observed; Mr. Vernon Harcourt and others proposed to burn mixtures of air and normal gases, and the former gentleman exhibited a neat normal lamp of about three-candle power, at the Southport meeting of the British Association. But these devices were mostly too delicate and complicated; and Mr. Louis Schwendler, one of the many electricians whose deaths we have lately had to lament, perhaps made the most practical suggestion in once more drawing attention to Mr. J. W. Draper's idea of using a fine platinum wire, heated by a constant current. Schwendler's units are sheet platinum horseshoes, .017 millimetre thick; but these again are open to objections, as we shall find. If we return to our historical abstract, we find M. Tresca, in 1876, experimenting with a Foucault photometer, a modification of Rumford's instrument, which is largely employed in France, and comprises a milk glass disc, whose two halves are illuminated by the lamp to be examined and the standard candle respectively. Tresca experienced difficulty from the different colours of the lights, and interposed tinted glasses before them. Very instructive were the tests at the South Foreland lighthouses, conducted by Messrs. Tyndall and Douglass, and fully described in the Trinity House Report, 1876-77. The electric lights were compared to a powerful colza-oil lamp, kept as nearly as possible constant at 722 standard candles; this comparison was effected by a Bunsen photometer, the colza lamp being again controlled with the help of a Sugg photometer. As the arc itself emits very little light while the greater part comes from the negative carbon, and a smaller amount from the positive, the necessity arose of taking, even with the two carbons vertically above one another, observations in various horizontal planes; M. Allard has further pointed out that even the various points of the vertical plane of the normal candle do not emit equal quantities of light. The report on his very extensive tests at the French lighthouses to the French Ministry ("Mémoire sur les Phares Electriques," Paris, 1881), forms a very valuable contribution to the literature on photometry. M. Fontaine's observations on the Gramme machines and Serrin lamps, described in his "Eclairage à l'Electricité," 2nd edition, 1879, also deserve mention for their completeness. Both MM. Allard and Fontaine used Foucault's photometer, and suggested ways to arrive at mean values with very unsteady lamps; these proposals are, however, hardly of practical weight, nor could M. Allard's idea of verifying his figures, with the help of a Crookes radiometer, contribute much to their corroboration, as this comes scarcely within the functions of a radiometer. Foucault's photometer was likewise employed when MM. Sautter, Lemmonier, and Co., of Paris, were testing their photo-electric apparatus for military operations, the lamps being supplied with Colonel Mangin's aplanatic reflector; green glasses were also, in this case, interposed to equalise the colours.

A similar arrangement was adopted at Rouen in 1881, when, on behalf of the Société Industrielle, the systems of Jablochkoff, Gramme, and Siemens were subjected to a series of comparative tests. Here, again, a Foucault photometer was used, together with the ordinary bec-carcel, and the silvered-glass mirror to make the rays parallel. The loss from reflection was averaged at 30 per cent.; the observations were made at various distances, and a determination was made of the radius of that horizontal plane, which received the same quantity of light as a normal candle could supply at a distance of about 4 m. This was a step in the right direction; the report, "Rapport general sur l'Eclairage Electrique des Quais de Rouen," 1881, shows curves drawn to indicate by their ordinates the light falling upon the horizontal plane.

The municipality of Paris had for some years instituted annual tests of the Jablochkoff candles in the Avenue de l'Opera, which tests finally induced them to abandon those apparatus. The Jablochkoff candles shed their maximum light, of course, in the plane perpendicular to the line drawn through both candles; the minimum in this line was found to be 0.57 of the maximum; the mean intensity, however—not the mean between maximum and minimum—was equal to .9 of the maximum, as the intensity curve proved to be of the shape of a figure 8, and not an ellipsis. It also transpired that the air was less transparent to the red light of the Jablochkoff candles than to the gas rays.

The well-known experiments at Chatham of 1879 and 1880 ought to have been mentioned before this. The apparatus comprised a Rumford photometer and an Argand burner of 40 candle-power, with a Sugg's regulator. Photographs were taken at the same moment of the front and sides of the carbons, and the illuminated areas calculated from these photographs. The average illumination of a point was further derived from these calculations, under the questionable assumption that the light was evenly distributed over the whole plane.

We have already spoken of the difficulty which was met with in comparing lights of different colours. The method of action of the ether vibrations, which excite our optic nerves and create the sensation of sight, is unknown to us; but we know that this function depends upon the wave lengths of the rays. M. Purkinje has shown that two coloured planes which appear equally light at a certain distance, seem to lose their light in a different ratio if further removed from the eye. Two lights of different colour are therefore incommensurable. Mr. Dietrich has recently, with more perfect apparatus, repeated the tests by means of which Fraunhofer attempted to determine the illuminative power of the various parts of the spectrum, whose lines he so assiduously studied and noted, without in the least conceiving their character and importance. Fraunhofer had only an oil lamp at his disposal; and, just as we should expect, he was wrong by about 9 per cent. with reference to the rays from the line D in the yellow, as here the two lights were most homogeneous, but wrong by 60 per cent. when analysing the rays from lines B, G, and H. Spectrophotometric observations, such as first proposed by Vierordt, Glan, and others, may be perfected to a high degree of comparative accuracy; if we arrange the two spectra to be compared, above one another, and divide both by vertical lines into bands of one and the same tint, we may indeed achieve very exact measurements. But this can only be executed in the laboratory, and not in ordinary practice; while, after all, it yields only comparative values for the various colours.

Tinted glasses have often been employed to produce rays of equal colour; but such interposition means loss. Captain Abney conceived the interesting idea of watching lights through a photographically prepared glass plate with a scale of darker and darker bands, through the darkest of which even the sun was not visible. But in the dark the eye gets slowly capable of distinguishing details which were at first quite indiscernible. Further researches, moreover, convinced Captain Abney that the ratio between red and blue in the same electric lamp, varied very considerably as the speed of the generator in-

creased, so that the red rays, which originally were half as strong as the blue ones, finally possessed only one-fourth of the intensity of the blue, both, of course, increasing in intensity with the quicker revolutions. Professors Ayrton and Perry followed Captain Abney in making two series of tests, choosing, however, red and green lights, instead of red and blue. The difficulty remains, however, how really to compare and reduce to unit measurements those two sets of results. M. Crova ("Comptes Rend." xciii. p 512) went one step further in this direction. He watched the two half-discs of a Foucault photometer by means of two Nicol prisms, with their main sections vertical to one another, and between the pivots he put a quartz plate of 9 millimetres thickness. If the two lamps are placed in their proper positions both discs appear of a greenish-white tint, and may then easily be adjusted until equal illumination is attained. The theory of this apparatus is too complicated to be discussed here; the main point is that the quartz is designed to produce two broad interference bands towards the ends of the spectra; in the middle parts the intensity of the rays varies, but there must be one line at which the rays pass through the pivots without becoming weakened. This maximum of illumination is now, by adjustment of the second pivot, to be fixed at those rays whose comparison would yield the same result as that of the total intensities. The apparatus is ingenious, but in seeking for the districts of equal illumination in the two spectra, it presumes that the spectro-photometer received equal amounts of light from both sources, which anticipates the solution of the problem.

The newer photometers of both M. Cornu and Professors Ayrton and Perry, permit measurements of strong electric lamps being taken in small rooms without the awkward necessity of removing the lamps to great distances to bring them into comparison with the standard candle. M. Cornu intercalates between the rays of both lights an achromatic lens whose active aperture may be widened or lessened with a micrometer screw, and thus varies the quantity of light falling upon the photometer. Apparatus of this kind have often been thought of; the star photometers of Steinheil and Herschel are based upon the same principle. Messrs. Ayrton and Perry, in their dispersion photometer, use a concave lens to decrease the intensity of the rays. No loss of light was supposed to occur through absorption in this concave lens if it were only thin enough. Mr. Voller, of Hamburg, has, however, taken exception to this assumption, and pronounced the possibility of losses of 10 per cent.; and Messrs. Ayrton and Perry seem afterwards to have silently admitted this source of error by introducing a plane parallel glass plate between the screen and the standard candle, to weaken the intensity of the light standard. The losses through absorption in the air, Professors Ayrton and Perry observed to be strongest for green light. MM. Bouquer and Allard have further investigated the phenomena of absorption in air; the coefficients vary greatly with the conditions of the air, but they are sufficiently determined to show that in tests where the strong lamp is 50 metres distant from the screen, a loss of about 4 per cent. has to be taken into consideration.

We have finally to speak of the labours of the third section of the Congress at Paris in 1881. The candle found practically no advocate, although Dr. Werner Siemens declared that a good candle need not vary by more than 5 per cent. MM. Tschikoleff and Bède stood up for Schwendler's platinum unit, but M. Crova objected, because platinum had no constant molecular structure and consequently no constant emissive power; small differences of temperature would further lead to inexact figures. M. Violle recommended the use, as a unit, of the quantity of light radiated by one square centimetre of platinum at melting point. MM. Werner Siemens and Cornu assented, but preferred silver. Sir William Siemens proposed an iridium wire under the influence of the unit of current. MM. Neujean and Flamache caused slight surprise by praising the magnesium and lime lights. For want of anything better, the old beccarcel was finally left in office, although M. J. Dumas pronounced it too warm. The discussions on photometers were less easy. M. Bergè made the curious proposal to remove the lamp until a white screen would no longer be visible through a solution of sulphate of copper ammonia. The great problem of what to do with reference to the various colours also remained unsettled. M. Allard suggested the creation of a blinking effect, as then all colours would dissolve into one uniform grey, and Dr. Gladstone proposed the employment of long distances across which the differences of colours would disappear. The proposal of M. Rousseau that for each lamp the equation of the intensity curve $J = f(a)$ should be calculated was warmly supported and accepted.*

---

* Since the date of this report the Paris Congress has decided that a surface of one square centimetre of melted platinum shall be the standard of light.

PHOTOMETRICAL TESTS.

THE programme for Section IV., Photometry, had been arranged by Professor E. Voit, of Münich, Dr. Hugo Kruess, of Hamburg, and Dr. Bunte, of Münich. This Committee agreed upon the English spermaceti candle as the light standard, this being of all standards proposed by far the most largely used. As these candles are, however, not perfect as to constancy, the standard candle served only for the preliminary tests to regulate the flame of a one-hole soapstone gas burner of 1 millimetre orifice fed by a gas meter. Both these sources being thought too weak for powerful lamps, an Argand burner was employed in the incandescence lamp tests, and a Siemens regenerative gasburner of 100 candles for the arc lamps; the Argand burner was, however, afterwards discarded.

The question of the photometer to be selected was settled upon the principle of excluding all physical and chemical apparatus; of the physiological photometers left, Rumford's was rejected as not exact enough; Foucault's because of the difficulty experienced in equalising the two halves of the different tinges of the screen, and further, because it necessitates the shifting of either of the two lights; and a Bunsen photometer of Professor Ruedorf's modified type, having its screen in the line bisecting the angle formed by two mirrors inclined at 140 deg., was finally chosen. The incandescence lamps were subjected to two independent tests. The absolute measurements were taken in connection with the tests of the dynamos; the plane of the carbon filament was placed vertically

to the photometer scale and the intensity of the lamp was, by the interposition of resistances, reduced to certain values such as recommended by the manufacturer or by other considerations. The relative measurements had nothing to do with the electrical efficiency, but were only to decide the diffusion of the light in various directions; for this purpose, two lamps of the same construction were compared, in parallel circuit or in series, the one kept at zero position, that is, the plane of the filament perpendicular to the scale, whilst the other was turned round its vertical and also round its horizontal axis. The arc lamps were examined in the same way, the two tests here following one another immediately for practical reasons; the regulation of the arc lamp was left to the manufacturer to enable him to show his lamp to advantage; the two carbon pencils had always to be fixed most accurately vertical above one another. Determinations as to the intensity under various angles of inclination to the horizontal plane can be most conveniently made by raising and inclining the whole photometer; but as this was not feasible, a mirror, revolving round its horizontal axis, was fixed at the end of

the constancy of the one-hole burner II. was found sufficient as long as the height of the flame was maintained; the Argand burner, on the other hand, proved too troublesome, and the direct comparison between the gas-jet II. and the regenerative burner V perfectly reliable, so that the Argand burner was dispensed with. The actual test of the lamp VI. was effected according to M. Allard's method of arriving at a correct mean intensity of the always unsteady arc lamp; the observer moved the screen *b* backward and forward, adapting himself as much as possible to accord with the fluctuations; and took his observations, to the number of ten or fifteen, at signals given at intervals of ten or twenty seconds. This method promises a near average, and permits at the same time of a determination of the steadiness. Inquiries into the absorption of light in the atmosphere were at one time thought of, particularly with reference to the very powerful locomotive lamps exhibited by Mr. Schuckert; time, however, did not allow a full investigation, and a few observations would have been of no avail, since the absorptive capacity of the atmosphere depends upon many circumstances.

RESULTS OF TESTS.—INCANDESCENCE LAMPS.

I. *Relative Light Intensity.*—The nature and compass of the observations made to ascertain the relative light intensities, will be the best understood from the following complete series of tests of two Edison B lamps, in series: Lamp No. 1 at 0 metre, No. 2 at 2.10 metres.

TABLE III.—*Relative Light Intensity. Two Edison Lamps.*

1. Lamp No. 1 Revolving Round Vertical Axis, 0 deg.=Zero position (Carbon Filament Vertical to Photometer).				2. Lamp No. 1 Revolving Round Horizontal Axis, 90 deg.=Head of Lamp turned towards Photometer (Head Position).	
Angle.	Position of Screen.	Angle.	Position of Screen.	Angle.	Position of Screen.
deg.	m.	deg.	m.	deg.	m.
0	1.055	180	1.064	0	1.075
	4.057		65		72
	61		66		71
	58		63		72
22.5	1.115	202.5	1.093	22.5	1.036
	18		90		32
	21		95		35
	21		94		39
45.0	1.133	225	1.132	45	1.004
	37		36		01
	36		31		01
	37		35		03
67.5	1.147	247.5	1.142	67.5	0.967
	45		44		69
	49		41		69
	46		39		65
90	1.157	270	1.130	90	0.765
	58		29		64
	59		32		62
	56		30		65
112.5	1.145	292 5	1.143		
	47		45		
	40		44		
	45		40		
135.0	1.144	315	1.128		
	46		27		
	44		26		
	42		27		
157.5	1.092	327.5	1.066		
	95		71		
	92		64		
	90		64		
		360	1.036*		
			35		
			38		
			36		

* Not reliable, the current being interrupted.

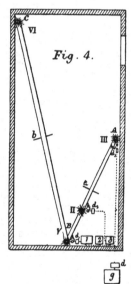

Fig. 4.

the scale B C (Fig. 4), and threw the rays falling from the arc lamp VI., which could be suspended at various heights, horizontally on the screen *b*. Fig. 4 illustrates the arrangement; A B and B C are the two scales, of 6 and 12 metres length; *a* and *b* the two photometers sliding on these scales. The gas main contained a meter *g*, and a pressure regulator *d*; in each of the three branches feeding the Siemens regenerative burner, the Argand, and the one-hole burner there was a gas-meter (1, 2, 3), a sensitive pressure regulator (d_1, d_2, d_3), and a regulating cock (h_1, h_2, h_3).

The observations for absolute intensity were at first taken in the following way. The English standard candle I. of 45 millimetres height, was compared to the one-hole gas burner II., by means of the photometer *a* on the scale A B; with the same apparatus, II. was compared to the Argand burner M, and III. to the regenerative burner V. Then followed on the scale B C and the photometer *b*, the comparison between the arc lamp VI., and the regenerative burner V., signals and time notations securing the synchronism of the observations. This whole series of tests was then checked by a repetition of all determinations in the reversed order, concluded by a second gauging of the gasburner II., by the standard candle I.; all gas flames were left burning throughout the experiment, and were screened off when not required. The results suggested a simplification of the arrangements;

From these figures, stating the screen positions at which equal illumination resulted, the following light intensities for lamp No. 1 in its various positions were calculated,

the intensity of lamp No. 2 being taken as unity. Table IV.

With the aid of the figures of this latter Table, there were then calculated the light intensities which the lamp emits in various directions ; the intensity which the lamp yielded when in zero position, that is, when the plane of its carbon filament was at right angles to the photometer, being taken as unity. Tables III. and IV. likewise show that there were three observations at zero positions, at

at the photometer, two others measured, simultaneously at given signals, the current passing through the lamp with the help of the mirror galvanometer and the difference of potential at the binding screws with the torsion galvanometer. The examination of an Edison A lamp, which is described in the report, involved fifty readings at each of the apparatus ; and from these observations were obtained the figures for the following abridged Table VI.

TABLE IV.—Relative Light Intensity. Two Edison Lamps.

Angle	0 deg.	22.5 deg.	45 deg.	67.5 deg.	90 deg.	112.5 deg.	135 deg.	157.5 deg.	
Revolving about vertical axis	1.019	1.281	1.373	1.449	1.505	1.438	1.432	1.174	
	1.027	296	94	38	11	49	43	87	
	43	311	89	60	17	10	32	74	
	31	311	94	43	00	38	21	65	
Mean	1.030	1.300	1.388	1.4775	1.508	1.434	1.432	1.175	
Angle	180 deg.	202.5 deg.	225 deg.	247.5 deg.	270 deg.	292.5 deg.	315 deg.	337.5 deg.	360 deg.
Vertical axis	1.055	1.178	1.368	1.421	1.357	1.427	1.347	1.063	0.948
	59	65	89	32	52	38	42	83	44
	63	87	62	16	68	32	36	55	55
	51	83	83	05	57	10	42	55	48
Mean	1.057	1.178	1.3775	1.4185	1.3585	1.427	1.342	1.064	0.949
Angle	0 deg.	22.5 deg.	45 deg.	67.5 deg.	90 deg.				
Revolving round horizontal axis	1.100	0.948	0.839	0.728	0.328				
	1.087	34	29	34	27				
	1.083	44	29	34	24				
	1.087	59	36	23	28				
Mean	1.089	0.946	0.833	0.730	0.327				

0 deg., 180 deg., 360 deg., and corresponding relations with regard to the other positions ; thus the means of three or two observations are noted in this table. The first column of Table V., page x., gives the name and distinctive number of the lamp in each case. The upper row indicates the values for the intensity of the lamp when revolving about its vertical axis, the lower line the same for the revolution about the horizontal axis.

This tabular matter is graphically arranged in the following diagrams, Figs. 5 to 22, illustrating the emission of light in the horizontal and vertical plane. The figures, it will be seen, are the same as in the preceding Table.

The report adds an interesting contribution from the able pen of Professor E. Hagenbach-Bischoff, of Basle, on some theoretical points, basing upon Lambert's law, according to which the quantity of light q reflected from one plane element df on to another df^1 is

$$q = \frac{i \, df \, df^1 \, \cos.e \, \cos.e^1}{r^2}$$

in which i signifies the intensity of the illuminating plane df, x its distance from df^1, and e and e^1, the angles which the central line forms with the normal. Professor Hagenbach found that a practically accurate mean horizontal intensity may be obtained by taking observations at 0 deg., 90 deg., and 45 deg, and calculating the average of those on the base of

$$\frac{0 \text{ deg.} + 2 \times 45 \text{ deg.} + 90 \text{ deg.}}{4},$$

that is, counting the intensity at 45 deg. twice when determining the mean horizontal intensity. This investigation was mainly confined to the Edison and Maxim lamps.

II. *Absolute Light Measurements.*—As above explained, the absolute light measurement formed part of the tests of the dynamos, and therefore partly engaged the members of Section I., who had to determine current intensity and potential difference. The lamp was in zero position, that is the plane of its carbon filament at right angles to the photometer screen, and whilst two observers were stationed

TABLE VI.—*Absolute Light Test.*
Edison A Lamp.—No. 1.

Time.	Intensity. Standard Candles.	Current.	Potential Difference.	Time.	Intensity. Standard Candles.	Current.	Potential Difference.
h. m.		ampères.	volts.	h. m.		ampères	volts
3 17	21.423	0.702	110.6	3 35	15.414		
	21.423	0.703	110.7	3 45	10.403	0.610	99.07
	21.037	0.706	110.6		10.519	0.610	99.07
	21.920	0.706	110.5		10.597	0.610	98.97
	21.620	0.702	110.3	3 47	10.636	..	99.07
3 19	22.122	0.704	110.7		10.327	0.608	98.99
3 24	18.985	0.699	107.8		10.176	0.609	98.99
	18.985	0.677	107.7		10.138	0.609	98.84
	18.413	0.678	107.7		10.289	0.614	98.84
3 26	18.818	0.678	107.5		10.289	0.608	
	18.985	0.678	107.7				
3 31	15.162	0.652	104.4				
	14.856	0.654	104.4				
	14.499	0.653	104.6				
3 33	15.351	0.651	104.6				
	14.152	0.648	104.2				
	14.676	0.652	104.4				
	18.737	..	107.5				
3 39	11.864	0.628	102.0				
	11.819	0.632	101.9				
	12.094	0.631	101.7				
3 41	12.094	0.632	101.9				
	12.140	0.632	101.7				
	12.048	0.632	101.7				
	14.856	0.652	104.4				

The report then proceeds to give, pages 118 to 120, the results of the tests with incandescence lamps, stating the time of observation, the light intensity in standard candles, the current intensity in ampères, the potential difference in volts as before, and further the electrical work in volt-ampères and in horse-power $\left(P = \dfrac{J \, e}{736} \right)$, and

TABLE V.—RELATIVE LIGHT INTENSITIES.—INCANDESCENCE LAMPS.

Angle	0 deg.	22.5 deg.	45 deg.	67.5 deg.	90 deg.	112.5 deg.	135 deg.	157.5 deg.	180 deg.	202.5 deg.	225 deg.	247.5 deg.	278 deg.	292.5 deg.	315 deg.	335.5 deg.	360 deg.	Revolving about Axis
Edison A, No. 1	0.958	1.014	1.199	1.195	1.201	1.209	1.175	1.009	0.965	1.053	1.165	1.176	1.163	1.166	1.128	1.015	1.047	Vertical.
	1.000	0.935	0.809	0.540	0.320												0.920	Horizontal.
Edison B, No. 1	·099	1.261	1.346	1.404	1.462	1.391	1.389	1.139	1.025	1.142	1.334	1.376	1.317	1.384	1.301	1.032	1.000	Vertical.
	1.056	0.917	0.808	0.708	0.137	1.030	1.028	1.011	1.031	1.031	1.020	0.999	0.963	0.965	0.959	0.982	0.571	Horizontal.
Müller, medium size, No. 1	1.043	1.038	1.043	1.015	1.016												0.992	Vertical.
	0.985	0.951	0.956	0.707	0.810		0.995		0.994		0.964		1.008		1.020		1.018	Horizontal.
Müller, Large No. 1	0.997	1.003	1.027	0.886	0.980												0.987	Vertical.
	0.992				0.947		0.935		0.946		0.961		0.929		0.988		0.951	Horizontal.
Cruto, No. 1	1.076	0.972	0.837	0.733	0.724	0.564	0.710	0.925	0.982	0.936	0.876	0.862	0.303	0.543	0.764	0.870	0.987	Vertical.
	1.002	0.936	0.807	0.591	0.403	0.423	0.733	0.935	1.004	0.924	0.856	0.471	0.524	0.522	0.862	0.935	0.951	Horizontal.
Maxim No. 1	1.007	0.985	0.536	0.578	0.370												0.998	Vertical.
	0.997	1.046	0.549	0.443	0.125												0.997	} Horizontal.
		601		400	201													
Maxim No. 2	0.997	0.895	0.741	0.421	0.330		0.709	0.944	1.011		0.915		0.337		0.750		1.009	Vertical.
	0.998	0.941	0.568	0.476	0.101													Horizontal.
	0.996	0.949	0.609	0.492	0.109													
Swan No. 1	1.010	0.905	0.890	0.980	1.064	1.077	1.093	1.101	1.025	0.951	0.927	0.950	1.023	1.027	1.068	1.038	0.959	Vertical.
	0.988	0.954	0.867	0.745	0.652	0.721	0.841	0.929	1.042						0.958	0.899		Horizontal.
Swan, No. 2	1.016	1.073	1.020	1.001	1.054	1.029	0.958	0.904	1.021	1.089	1.032	0.992	1.082	1.033	0.958		0.997	Vertical.
	0.997	0.957	0.887	0.783	0.615												0.990	Horizontal.

in the last column the resistance of the hot lamp derived from $R = \frac{c}{J}$. The figures show very considerable fluctuations, and the gradual and often very remarkable increase of intensity attained during the twenty or more minutes over which the observations extended. The relations between electrical work expended and light intensity realised are of particular interest. To render these relations clearer, curves were plotted out representing by their abscissæ, the electrical work in volt-ampères and horse-powers, and by their ordinates, the light intensities. Professor Voit thus followed a line of researches first entered upon by Mr. Andrew Jamieson. In analysing these curves it appears that of the various assumptions which have been made regarding the increase of light with increase of electrical work, the supposition of an increase of light proportional to the third power of work consumed, harmonises best with the observations. In a formula this law would be expressed thus : $L = a\,a^3$, L meaning the light intensity to be determined, a the intensity produced by the unit of work, and a the work expended. It is noteworthy that the Cruto lamp, which from its peculiar filament acquires a character of its own, does not agree to this law.

At the end of their exhaustive report, which in the abundance of its experimental facts and the circumspect character of the inferences based upon this material, is highly creditable to the members of the Committee, Table VII. is added permitting an easy comparison of the Münich tests with those of Paris. We must not forget, however, that if agreement seems to be missing even in the column which gives the light intensity per horse-power, that the respective lamps of the same name were not tested under equal conditions. The Edison B lamp, for instance, yielded at Münich more light per horse-power than the A lamp, contradictory to the experience gained at Paris, the explanation being that at Münich the lamp was excited to a higher degree.

The conclusions arrived at are :

1. The intensity of light emitted by an incandescence lamp in various directions of the horizontal plane depends upon the section of the filament; a circular section disperses the light equally.

2. The distribution of light in various directions depends upon the projection of the filament at each direction.

3. For the same lamp, the light intensity increases with the cube of the electrical work consumed.

4. The mean light intensities produced by the electrical unit of work are :

Edison	{ A	...	...	...	...	0.0000376
	{ B	...	...	...	...	0.0001106
Swan	{ small	...	...	..	...	0.0000848
	{ large	...	...	..	...	0.0000096
Maxim	...	...	...	...	..	0.0000148
Siemens	...	...	...	...	..	0.0000225
Müller	{ small	...	...	..	...	0.0000213
	{ medium	...	..	..	...	0.0000067
	{ large	...	...	..	...	0.0000021
Cruto	...	...	...	...	..	0.0000250

Arc Lamps.—The general arrangements have already been described. The plant comprised a reflector whose use involved a loss of light by absorption which had to be allowed for. The arc lamp tests are very complete in so far as photometry is concerned ; but they were limited to four lamps, two of which are hardly known in this country ; and they impart so little information about the power consumed that we may be brief. In glancing over the extensive tables, one is struck with the astonishing fluctuations in the light intensity occurring during the two or four minutes which one series of observations occupied. Although following one another quickly, as may be expected from the shortness of the period of observations, these changes are not abrupt, the intensity rising, in one instance, from 600 candles, to 700, 900, and 1100, and sinking again to 600 candles, during one minute or a little more. This case is perhaps the worst that could be quoted ; but fluctuations within one minute of 10 per cent., and more, are not rare, and they show the necessity of taking a very high number of readings.

DIAGRAMS OF PHOTOMETRIC VALUES OF INCANDESCENCE LAMPS.

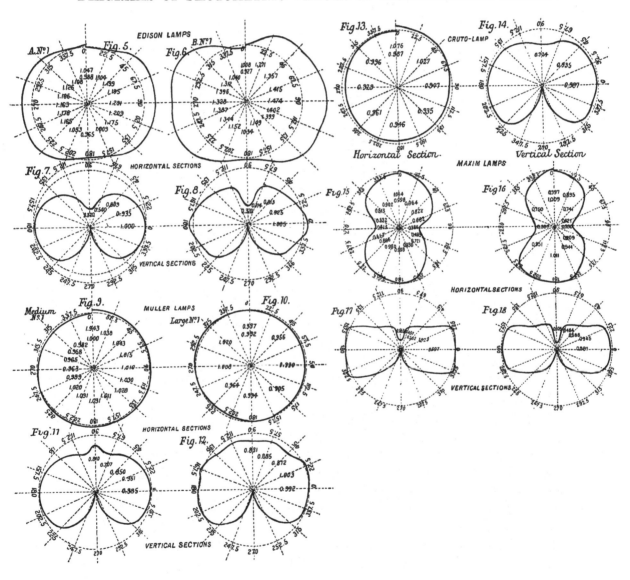

The Münich experimenters cannot be accused of having shirked this irksome duty; more than 400 observations, for instance, were made for the one Schuckert lamp. The means of a number of readings, on the other side, agree pretty well with one another, so that the variations must in general equalise one another within the space of a few minutes. Several causes may co-operate in effecting these changes; irregularities, even if slight, in the speed

English standard candles, and further, in its second half, the respective intensities for the same angles as multiples for the unit intensity in the horizontal plane itself.

The figures of the second half of this table were used in constructing the curves of the following diagrams (Fig. 23 to 26), and also in determining the mean illumination of a sphere surrounding the lamp. The maximum intensity seems here to have been emitted at an angle of about

TABLE VII.—RESULTS. MUNICH, 1882.

LAMPS.		Light Intensity.	Resistance.	Difference of Potential.	Current.	Electrical Work.		Light Intensity per Horse Power	Number of Lamps per Horse-Power.
		candles.	ohms.	volts.	ampères.	volt-amp.	HP.	candles.	
Edison	A	11.69	67.68	55.78	0.825	46.02	0.0625	186.90	26.361 (8 candles)
	B	15.32	139.60	103.05	0.755	77.80	0.1057	144.88	9.05 (16 „)
Maxim		13.34	47.01	65.07	1.384	90.06	0.1224	108.98	3.88 (28 „)
Swan	A	10.95	31.91	38.38	1.222	46.90	0.0637	171.78	17.18 (10 „)
	B	37.17	87.03	118.02	1.282	151.30	0.2056	180.75	4.52 (70 „)
Siemens		14.90	104.72	95.74	0.915	87.60	0.1191	125.14	7.82 (16 „)
Müller	A	18.43	58.62	74.04	1.263	93.51	0.1271	145.01	7.26 (20 „)
	B	43.08	59.52	105.22	1.779	187.19	0.2544	169.33	3.39 (50 „)
	C	102.35	65.41	155.15	2.367	367.24	0.4991	205.05	2.05 (100 „)
Cruto ..		8.47	8.16	22.15	2.715	60 14	0.0817	103.56	10.36 (10 „)

Results. Paris, 1881.

Edison	A	15.38	137.4	89.11	0.6510	57.98	0.0788	196.4	12.28 (16 candles)
	C	31.11	130.03	98.39	0.7585	74.62	0.0941	307.25	9.60 (32 „)
Swan	A	16.61	32.78	47.30	1.471	69.24	0.0945	177.92	11.12 (16 „)
	B	33.21	31.75	54.21	1.758	94.88	0.1059	262.49	8.20 (32 „)
Lane-Fox	A	16 36	27.40	43.63	1.593	69.53	0.1025	173.53	10.85 (16 „)
	B	32.71	26.59	48.22	1.815	87.65	0.1289	276.89	8.65 (32 „)
Maxim	A	15.96	41.11	56 49	1.380	78.05	0.1191	151.27	9.45 (16 „)
	B	31.93	39.60	62.27	1.578	98.41	0.1337	239.41	7.48 (32 „)

TABLE VIII.—ARC LAMPS; LIGHT INTENSITIES.

Lamp.	Intensity in Candles at Angles of								Intensity Ratios Compared to Intensity at 0 deg.							
	0 deg.	15 deg.	20 deg.	30 deg.	40 deg.	45 deg.	50 deg.	60 deg.	0 deg.	15 deg.	20 deg.	30 deg.	40 deg.	45 deg.	50 deg.	60 deg.
Schuckert	248	619	..	1037	1238	1464	1124	788	1	2.5	..	4.2	5.0	5 9	4.5	3.2
Schwerd	443	..	..	2859	..	3251	3250	1836	1	..	..	6.5	..	7.3	7.3	4.1
Crompton	452	1531	1116	2523	2116	3071	2155	1986	1	3.4	2.5	5.6	4.7	6.8	4.8	4.4
Schaeffer	745	..	..	875	..	..	1168	1227	1	..	..	1.2	..	..	1.6	1.6

of the motor, and hence in that of the dynamo, considerably influence the light intensity; and the behaviour of both the dynamo and the lamp regulator, and the quality of the carbons, are of course, most important factors.

For the incandescence lamps the dispersion of light in both the horizontal and vertical plane has to be studied. Arc lamps distribute their light evenly over the horizontal plane as long as their carbons are central. That great care was bestowed upon this point has already been remarked, and it appeared that the central disposition of the carbons, vertically above each other, secured uniform distribution. The suspended lamps were lowered and raised as usual, and the readings refer to inclinations of the light rays to the horizontal plane up to 60 deg. But the lamps were not lowered below the plane of the photometer, nor were any coloured glasses employed, since, as we have already said, conclusions as to the absorptive power of the atmosphere for rays of different refrangibility would demand the separation and elimination of several agents.

We quote from the report the following mean results:

The arc lamp of Messrs. Schwerd and Scharnweber contains two coils, whose common core moves in a tube partly filled with glycerine. The arc length is ordinarily kept constant by the play of this core, and occasionally, when necessary, is re-established by the iron core setting in action a clockwork. The report does not contain any information about the Schaeffer lamp, which did not always burn regularly, whilst the Crompton lamp, the Schuckert lamp (of the differential type), and the Schwerd lamp gave satisfaction. Table VIII. states the light intensities in

TABLE IX.—*Arc Lamp Tests.*

Emission of Light.	Intensity, English Standard Candles.	Electrical Work Delivered to the Lamp.	Intensity per Electrical Horse-Power at the Lamp.	Intensity per Horse-Power of the Dynamo.
	candles.	volt-ampères.	candles.	candles.
Schuckert Lamp.				
In horizontal direction ..	250	385	477	513
Maximum at 45 deg. ..	1464			
Mean intensity ..	470	394	878	579
Schwerd Lamp.				
In horizontal direction ..	456	728	461	278
Maximum at 45 deg.	3250	811	2950	1836
Mean intensity ..	1145	751	1121	788
Crompton Lamp.				
In horizontal direction ..	560	920	448	382
Maximum at 50 deg. ..	3071	1164	1942	..
Mean intensity ..	1221	958	939	814
Schaeffer Lamp.				
In horizontal direction ..	744	..	..	389
Maximum at 60 deg. ..	1227	..	..	634
Mean intensity ..	692	..	..	360

45 deg., whilst other electricians often recommend angles of 34 deg.

The diagrams Figs. 23 to 26 are liable to be misunderstood. The small space encircled by the Schaeffer curve might be interpreted as indicating a very low mean luminosity. By referring to Table IX., however, we perceive that the mean intensity of the Schaeffer lamp amounts to 692 candles, against 470 in the case of the Schukert lamp; nor have the respective values per horse-power, as also stated in Table IX., anything

With regard to the power tests we may also content ourselves with the summary of the last Table. Those who are practically interested in similar tests, we refer to the original reports. They form a lengthy document of 150 pages quarto. Theory may occasionally predominate and assume a speculative character, as is only natural in the case of so young a science. But there is no theory advanced without experimental foundation,

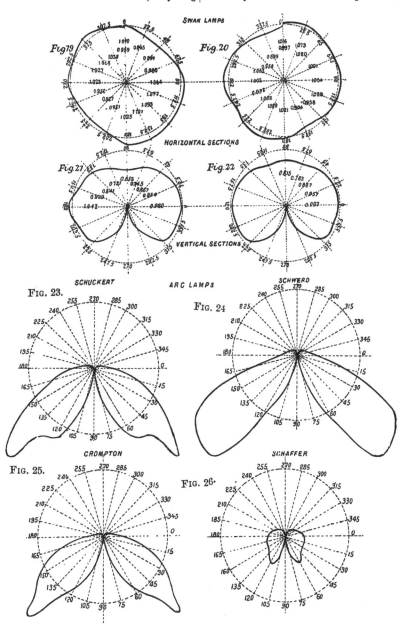

to do with these curves. They are simply based upon Table VIII.; the intensity in the horizontal line being taken as unity, the curves demonstrate how the Schwerd lamp emits seven times more light downward under an angle of 50 deg. than in the horizontal plane round it; whilst the Schaeffer lamp, at the same elevation, augments its intensity by one-half only, thus spreading its light much more uniformly.

and we are on the whole convinced that the Committee have done well in laying so complete a document before the public. This report, and the preliminary pamphlets, which minutely explained the distribution of the work between the various observers, really constitute a theoretical and practical treatise on electrical tests.

PRINTED AT
THE BEDFORD PRESS, 20 & 21, BEDFORDBURY,
LONDON, W.C.